Encyclopaedia of

Biochemistry and Molecular Biology of Plants

Encyclopaedia of

Biochemistry and Molecular Biology of Plants

Volume 1

Editor:
K.N.P. Singh

AGROTECH PRESS
JAIPUR ▪ NEW DELHI

AGROTECH PRESS

Regd. Off. "Bony Residency", Gate No. 2
Opp. Tilak Public School, Vishweswariya Nagar,
Gopalpura Road, Jaipur - 302018, India
Ph: 0141 - 2761280, Telefax: 2761381
E-mail : agrotechpress@gmail.com

Delhi Office: 102, 1st Floor, Satyam House,
4327/3, Ansari Road, Darya Ganj,
New Delhi - 110002, India
Ph: 011-45652440
visit us: www.agrotechpress.com

Distributed By
KEYA INTERNATIONAL PRESS LLP
E-mail: keyainternationalpressllp@gmail.com

Published in India, 2024

ISBN : 9789384568221 (4 Vols. Set)

Typeset by Thomson Press

Printed and bound in India by Thomson Press Ltd.,
New Delhi

Contents

Preface

Molecular biology is the study of biology at a molecular level. The field overlaps with other areas of biology and chemistry, particularly genetics and biochemistry. Molecular biology chiefly concerns itself with understanding the interactions between the various systems of a cell, including the interactions between DNA, RNA and protein biosynthesis as well as learning how these interactions are regulated.

Molecular breeding is the application of molecular biology tools, often in plant breeding. Quantitative traits refer to phenotypes that vary in degree and can be attributed to polygenic effects, i.e., product of two or more genes, and their environment. Quantitative trait loci are stretches of DNA containing or linked to the genes that underlie a quantitative trait. Mapping regions of the genome that contain genes involved in specifying a quantitative trait is done using molecular tags such as AFLP or, more commonly SNPs. This is an early step in identifying and sequencing the actual genes underlying trait variation. Marker assisted selection or marker aided selection is a process whereby a marker (morphological, biochemical or one based on DNA/RNA variation) is used for indirect selection of a genetic determinant or determinants of a trait of interest. This process is used in plant and animal breeding. Considerable developments in biotechnology have led plant breeders to develop more efficient selection systems to replace traditional phenotypic-pedigree-based selection systems. Marker assisted selection is indirect selection process where a trait of interest is selected, not based on the trait itself, but on a marker linked to it. For example if MAS is being used to select individuals with a disease, the level of disease is not quantified but rather a marker allele which is linked with disease is used to determine disease presence. The assumption is that linked allele associates with the gene and/or quantitative trait locus of interest. MAS can be useful for traits that are difficult to measure, exhibit low heritability, and/or are expressed late in development. Genetic engineering, also called genetic modification, is the direct manipulation of an organism's genome using biotechnology. New DNA may be inserted in the host genome by first isolating and copying the genetic material of interest using molecular cloning methods to generate a DNA sequence, or by synthesizing the DNA, and then inserting this construct into the host organism. Genes may be removed, or "knocked out", using a nuclease. Gene targeting is a different technique that uses homologous recombination to change an endogenous gene, and can be used to delete a gene, remove exons, add a gene, or introduce point mutations. An organism that is generated through genetic engineering is considered to be a genetically modified organism. The first GMOs were bacteria in 1973; GM mice were generated in 1974. Insulin-producing bacteria were commercialized in 1982 and genetically modified food has been sold since 1994. Plants, animals or micro organisms that have changed through genetic engineering are termed genetically modified organisms or GMOs. Bacteria were the first organisms to be genetically modified. Plasmid DNA containing new genes can be inserted into the bacterial cell and the bacteria will then express those genes. These genes can code for medicines or enzymes that process food and other substrates. Plants have been modified for insect protection, herbicide resistance, virus resistance, enhanced nutrition, tolerance to environmental pressures and the production of edible vaccines. Most commercialised GMO's are insect resistant and/or herbicide tolerant crop plants. Genetically modified animals have been used for research, model animals and the production of agricultural or pharmaceutical products. They include animals with genes knocked out, increased susceptibility to disease, hormones for extra growth and the ability to express proteins in their milk. In molecular biology transformation is genetic alteration of a cell resulting from the direct uptake, incorporation and expression of exogenous genetic material from its surroundings and taken up through the cell membrane(s). Transformation occurs naturally in some species of bacteria, but it can also be effected by artificial means in other cells. For transformation to happen, bacteria must be in a state of competence, which might occur as a time-limited response to environmental conditions such as starvation and cell density. Transformation is one of three processes by which exogenous genetic material may be introduced into a bacterial cell, the other two being conjugation (transfer of genetic material between two bacterial cells in direct contact) and transduction (injection of foreign DNA by a bacteriophage virus into the host bacterium). "Transformation" may also be used to describe the insertion of new genetic material into nonbacterial cells, including animal and plant cells; however, because "transformation" has a special meaning in relation to animal cells, indicating progression to a cancerous state, the term should be avoided for animal cells when describing introduction of exogenous genetic material. Introduction of foreign DNA into eukaryotic cells is often called "transfection".

This is a very impressive well-organised, beautifully-illustrated introductory textbook. It should prove popular among a very broad variety of students for whom understanding of principles is more important than memorization of large amounts of poorly comprehended factual information. There are many such students around the world who will benefit from this masterly book.

–Editor

1

Techniques of Molecular Biology

Since the late 1950s and early 1960s, molecular biologists have learned to characterize, isolate, and manipulate the molecular components of cells and organisms. These components include DNA, the repository of genetic information; RNA, a close relative of DNA whose functions range from serving as a temporary working copy of DNA to actual structural and enzymatic functions as well as a functional and structural part of the translational apparatus; and proteins, the major structural and enzymatic type of molecule in cells.

Expression Cloning

One of the most basic techniques of molecular biology to study protein function is expression cloning. In this technique, DNA coding for a protein of interest is cloned (using PCR and/or restriction enzymes) into a plasmid (known as an expression vector). This plasmid may have special promoter elements to drive production of the protein of interest, and may also have antibiotic resistance markers to help follow the plasmid.

This plasmid can be inserted into either bacterial or animal cells. Introducing DNA into bacterial cells can be done by transformation (via uptake of naked DNA), conjugation (via cell-cell contact) or by transduction (via viral vector). Introducing DNA into eukaryotic cells, such as animal cells, by physical or chemical means is called transfection. Several different transfection techniques are available, such as calcium phosphate transfection, electroporation, microinjection and liposome transfection. DNA can also be introduced into eukaryotic cells using viruses or bacteria as carriers, the latter is sometimes called bactofection and in particular uses Agrobacterium tumefaciens. The plasmid may be integrated into the genome, resulting in a stable transfection, or may remain independent of the genome, called transient transfection. In either case, DNA coding for a protein of interest is now inside a cell, and the protein can now be expressed. A variety of systems, such as inducible promoters and specific cell-signalling factors, are available to help express the protein of interest at high levels.

Large quantities of a protein can then be extracted from the bacterial or eukaryotic cell. The protein can be tested for enzymatic activity under a variety of situations, the protein may be crystallized so its tertiary structure can be studied, or, in the pharmaceutical industry, the activity of new drugs against the protein can be studied.

Polymerase Chain Reaction (PCR)

The polymerase chain reaction is an extremely versatile technique for copying DNA. In brief, PCR allows a single DNA sequence to be copied (millions of times), or altered in predetermined ways.

For example, PCR can be used to introduce restriction enzyme sites, or to mutate (change) particular bases of DNA, the latter is a method referred to as "Quick change". PCR can also be used to determine whether a particular DNA fragment is found in a cDNA library. PCR has many variations, like reverse transcription PCR (RT-PCR) for amplification of RNA, and, more recently, real-time PCR (QPCR) which allow for quantitative measurement of DNA or RNA molecules.

Gel Electrophoresis

Gel electrophoresis is one of the principal tools of molecular biology. The basic principle is that DNA, RNA, and proteins can all be separated by means of an electric field. In agarose gel electrophoresis, DNA and RNA can be separated on the basis of size by running the DNA through an agarose gel. Proteins can be separated on the basis of size by using an SDS-PAGE gel, or on the basis of size and their electric charge by using what is known as a 2D gel electrophoresis.

Macromolecule Blotting and Probing

The terms *northern*, *western* and *eastern* blotting are derived from what initially was a molecular biology joke that played on the term *Southern blotting*, after the technique described by Edwin Southern for the hybridisation of blotted DNA. Patricia.

Thomas, developer of the RNA blot which then became known as the *northern blot* actually didn't use the term. Further combinations of these techniques produced such terms as *southwesterns* (protein-DNA hybridizations), *northwesterns* (to detect protein-RNA interactions) and *farwesterns* (protein-protein interactions), all of which are presently found in the literature.

Southern Blotting

Named after its inventor, biologist Edwin Southern, the Southern blot is a method for probing for the presence of a specific DNA sequence within a DNA sample. DNA samples before or after restriction enzyme digestion are separated by gel electrophoresis and then transferred to a membrane by blotting via capillary action. The membrane is then exposed to a labeled DNA probe that has a complement base sequence to the sequence on the DNA of interest. Most original protocols used radioactive labels, however non-radioactive alternatives are now available. Southern blotting is less commonly used in laboratory science due to the capacity of other techniques, such as PCR, to detect specific DNA sequences from DNA samples. These blots are still used for some applications, however, such as measuring transgene copy number in transgenic mice, or in the engineering of gene knockout embryonic stem cell lines.

Northern Blotting

The northern blot is used to study the expression patterns of a specific type of RNA molecule as relative comparison among a set of different samples of RNA. It is essentially a combination of denaturing RNA gel electrophoresis, and a blot. In this process RNA is separated based on size and is then transferred to a membrane that is then probed with a labeled complement of a sequence of interest. The results may be visualized through a variety of ways depending on the label used; however, most result in the revelation of bands representing the sizes of the RNA detected in sample. The intensity of these bands is related to the amount of the target RNA in the samples analyzed. The procedure is commonly used to study when and how much gene expression is occurring by measuring how much of that RNA is present in different samples. It is one of the most basic tools for determining at what time, and under what conditions, certain genes are expressed in living tissues.

Western Blotting

Antibodies to most proteins can be created by injecting small amounts of the protein into an animal such as a mouse, rabbit, sheep, or donkey (polyclonal antibodies)or produced in cell culture (monoclonal antibodies). These antibodies can be used for a variety of analytical and preparative techniques. In western blotting, proteins are first separated by size, in a thin gel sandwiched between two glass plates in a technique known as SDS-PAGE (sodium dodecyl sulfate polyacrylamide gel electrophoresis).

The proteins in the gel are then transferred to a PVDF, nitrocellulose, nylon or other support membrane. This membrane can then be probed with solutions of antibodies. Antibodies that specifically bind to the protein of interest can then be visualized by a variety of techniques, including coloured products, chemiluminescence, or autoradiography. Often, the antibodies are labeled with enzymes. When a chemiluminescent substrate is exposed to the enzyme it allows detection. Using western blotting techniques allows not only detection but also quantitative analysis. Analogous methods to western blotting can be used to directly stain specific proteins in live cells or tissue sections. However, these *immunostaining* methods, such as FISH, are used more often in cell biology research.

Eastern Blotting

Eastern blotting technique is to detect post-translational modification of proteins. Proteins blotted on to the PVDF or nitrocellulose membrane are probed for modifications using specific substrates.

Arrays

A DNA array is a collection of spots attached to a solid support such as a microscope slide where each spot contains one or more single-stranded DNA oligonucleotide fragment. Arrays make it possible to put down a large quantity of very small (100 micrometre diameter) spots on a single slide. Each spot has a DNA fragment molecule that is complementary to a single DNA sequence (similar to Southern blotting). A variation of this technique allows the gene expression of an organism at a particular stage in development to be qualified (expression profiling). In this technique the RNA in a tissue is isolated and converted to labeled cDNA. This cDNA is then hybridized to the fragments on the array and visualization of the hybridization can be done. Since multiple arrays can be made with the exact same position of fragments they are particularly useful for comparing the gene expression of two different tissues, such as a healthy and cancerous tissue. Also, one can measure what genes are expressed and how that expression changes with time or with other factors. For instance, the common baker's yeast, *Sacch-aromyces cerevisiae*, contains about 7000 genes; with a microarray, one can measure qualitatively how each gene is expressed, and how that expression changes, for example, with a change in temperature. There are many different ways to fabricate microarrays; the most common are silicon chips, microscope slides with spots of ~ 100 micrometre diameter, custom arrays, and arrays with larger spots on porous membranes (macroarrays). There can be anywhere from 100 spots to more than 10,000 on a given array. Arrays can also be made with molecules other than DNA. For example, an antibody array can be used to determine what proteins or bacteria are present in a blood sample.

Allele Specific Oligonucleotide

Allele specific oligonucleotide (ASO) is a technique that allows detection of single base mutations without the need for PCR or gel electrophoresis. Short (20-25 nucleotides in length), labeled probes are exposed to the non-fragmented target DNA. Hybridization occurs with high specificity due to the short length of the probes and even a single base change will hinder hybridization. The target DNA is then washed and the labeled probes that didn't hybridize are removed. The target DNA is then analyzed for the presence of the probe via radioactivity or fluorescence. In this experiment, as in most molecular biology techniques, a control must be used to ensure successful experimentation. The Illumina Methylation Assay is an example of a method that takes advantage of the ASO technique to measure one base pair differences in sequence.

Antiquated Technologies

In molecular biology, procedures and technologies are continually being developed and older technologies abandoned. For example, before the advent of DNA gel electrophoresis (agarose or polyacrylamide), the size of DNA molecules was typically determined by rate sedimentation in sucrose gradients, a slow and labor-intensive technique requiring expensive instrumentation; prior to sucrose gradients, viscometry was used.

Aside from their historical interest, it is often worth knowing about older technology, as it is occasionally useful to solve another new problem for which the newer technique is inappropriate.

History

While molecular biology was established in the 1930s, the term was first coined by Warren Weaver in 1938. Warren was the director of Natural Sciences for the Rockefeller Foundation at the time and believed that biology was about to undergo a period of significant change given recent advances in fields such as X-ray crystallography. He therefore channeled significant amounts of (Rockefeller Institute) money into biological fields.

Clinical Significance

Clinical research and medical therapies arising from molecular biology are covered under gene therapy

The Cell

The cell is one of the most basic units of life. There are millions of different types of cells. There are cells that are organisms onto themselves, such as microscopic amoeba and bacteria cells. And there are cells that only function when part of a larger organism, such as the cells that make up your body. The cell is the smallest unit of life in our bodies. In the body, there are brain cells, skin cells, liver cells, stomach cells, and the list goes on. All of these cells have unique functions and features. And all have some recognizable similarities. All cells have a 'skin', called the plasma membrane, protecting it from the outside environment.

The cell membrane regulates the movement of water, nutrients and wastes into and out of the cell. Inside of the cell membrane are the working parts of the cell. At the centre of the cell is the cell nucleus. The cell nucleus contains the cell's DNA, the genetic code that coordinates protein synthesis. In addition to the nucleus, there are many organelles inside of the cell - small structures that help carry out the day-to-day operations of the cell. One important cellular organelle is the ribosome.

Ribosomes participate in protein synthesis. The transcription phase of protein synthesis takes places in the cell nucleus. After this step is complete, the mRNA leaves the nucleus and travels to the cell's ribosomes, where translation occurs. Another important cellular organelle is the mitochondrion. Mitochondria (many mitochondrion) are often referred to as the power plants of the cell because many of the reactions that produce energy take place in mitochondria. Also important in the life of a cell are the lysosomes. Lysosomes are organelles that contain enzymes that aid in the digestion of nutrient molecules and other materials. Below is a labelled diagram of a cell to help you identify some of these structures.

There are many different types of cells. One major difference in cells occurs between plant cells and animal cells. While both plant and animal cells contain the structures discussed above, plant cells have some additional specialized structures. Many animals have skeletons to give their body structure and support. Plants do not have a skeleton for support and yet plants don't just flop over in a big spongy mess.

This is because of a unique cellular structure called the cell wall. The cell wall is a rigid structure outside of the cell membrane composed mainly of the polysaccharide cellulose. As pictured at left, the cell wall gives the plant cell a defined shape which helps support individual parts of plants. In addition to the cell wall, plant cells contain an organelle called the chloroplast. The chloroplast allow plants to harvest energy from sunlight. Specialized pigments in the chloroplast (including the common green pigment chlorophyll) absorb sunlight and use this energy to complete the chemical reaction:

$$6\ CO_2 + 6\ H_2O + \text{energy (from sunlight)} \rightarrow C_6H_{12}O_6 + 6\ O_2$$

In this way, plant cells manufacture glucose and other carbohydrates that they can store for later use.

Organisms contain many different types of cells that perform many different functions. In the next lesson, we will examine how individual cells come together to form larger structures in the human body.

The cell is the functional basic unit of life. It was discovered by Robert Hooke and is the functional unit of all known living organisms. It is the smallest unit of life that is classified as a living thing, and is often called the building block of life. Some organisms, such as most bacteria, are unicellular (consist of a single cell). Other organisms, such as humans, are multicellular. Humans have about 100 trillion or 10 cells; a typical cell size is 10 μm and a typical cell mass is 1 nanogram. The largest cells are about 135 μm in the anterior horn in the spinal cord while granule cells in the cerebellum, the smallest, can be some 4 μm and the longest cell can reach from the toe to the lower brain stem (Pseudounipolar cells). The largest known cells are unfertilised ostrich egg cells which weigh 3.3 pounds.

In 1835, before the final cell theory was developed, Jan Evangelista Purkynì observed small "granules" while looking at the plant tissue through a microscope. The cell theory, first developed in 1839 by Matthias Jakob Schleiden and Theodor Schwann, states that all organisms are composed of one or more cells, that all cells come from preexisting cells, that vital functions of an organism occur within cells, and that all cells contain the hereditary information necessary for regulating cell functions and for transmitting information to the next generation of cells. The word *cell* comes from the Latin *cellula*, meaning, a small room. The descriptive term for the smallest living biological structure was coined by Robert Hooke in a book he published in 1665 when he compared the cork cells he saw through his microscope to the small rooms monks lived in.

Anatomy

There are two types of cells: eukaryotic and prokaryotic. Prokaryotic cells are usually independent, while eukaryotic cells are often found in multicellular organisms.

Prokaryotic Cells

The prokaryote cell is simpler, and therefore smaller, than a eukaryote cell, lacking a nucleus and most of the other organelles of eukaryotes. There are two kinds of prokaryotes: bacteria and archaea; these share a similar structure. Nuclear material of prokaryotic cell consist of a single chromosome which is in direct contact with cytoplasm. Here the undefined nuclear region in the cytoplasm is called nucleoid.

A prokaryotic cell has three architectural regions:

- On the outside, flagella and pili project from the cell's surface. These are structures (not present in all prokaryotes) made of proteins that facilitate movement and communication between cells;
- Enclosing the cell is the cell envelope – generally consisting of a cell wall covering a plasma membrane though some bacteria also have a further covering layer called a capsule. The envelope gives rigidity to the cell and separates the interior of the cell from its environment, serving as a protective filter. Though most prokaryotes have a cell wall,

there are exceptions such as *Mycoplasma* (bacteria) and *Thermoplasma* (archaea). The cell wall consists of *peptidoglycan* in bacteria, and acts as an additional barrier against exterior forces. It also prevents the cell from expanding and finally bursting (cytolysis) from osmotic pressure against a hypotonic environment. Some eukaryote cells (plant cells and fungi cells) also have a cell wall;

- Inside the cell is the cytoplasmic region that contains the cell genome (DNA) and ribosomes and various sorts of inclusions. A prokaryotic chromosome is usually a circular molecule (an exception is that of the bacterium *Borrelia burgdorferi*, which causes Lyme disease). Though not forming a *nucleus*, the DNA is condensed in a *nucleoid*. Prokaryotes can carry extrachromosomal DNA elements called *plasmids*, which are usually circular. Plasmids enable additional functions, such as antibiotic resistance.

Eukaryotic Cells

Eukaryotic cells are about 15 times wider than a typical prokaryote and can be as much as 1000 times greater in volume. The major difference between prokaryotes and eukaryotes is that eukaryotic cells contain membrane-bound compartments in which specific metabolic activities take place. Most important among these is a cell nucleus, a membrane-delineated compartment that houses the eukaryotic cell's DNA. This nucleus gives the eukaryote its name, which means "true nucleus." Other differences include:

- The plasma membrane resembles that of prokaryotes in function, with minor differences in the setup. Cell walls may or may not be present.
- The eukaryotic DNA is organized in one or more linear molecules, called chromosomes, which are associated with histone proteins. All chromosomal DNA is stored in the *cell nucleus*, separated from the cytoplasm by a membrane. Some eukaryotic organelles such as mitochondria also contain some DNA.
- Many eukaryotic cells are ciliated with *primary cilia*. Primary cilia play important roles in chemosensation, mechanosensation, and thermosensation. Cilia may thus be "viewed as sensory cellular antennae that coordinate a large number of cellular signalling pathways, sometimes coupling the signalling to ciliary motility or alternatively to cell division and differentiation."
- Eukaryotes can move using *motile cilia* or *flagella*. The flagella are more complex than those of prokaryotes.

Subcellular Components

All cells, whether prokaryotic or eukaryotic, have a membrane that envelops the cell, separates its interior from its environment, regulates what moves in and out (selectively permeable), and maintains the electric potential of the cell.

Inside the membrane, a salty cytoplasm takes up most of the cell volume. All cells possess DNA, the hereditary material of genes, and RNA, containing the information necessary to build various proteins such as enzymes, the cell's primary machinery. There are also other kinds of biomolecules in cells. This article will list these primary components of the cell, then briefly describe their function.

Membrane

The cytoplasm of a cell is surrounded by a cell membrane or *plasma membrane*. The plasma membrane in plants and prokaryotes is usually covered by a cell wall. This membrane serves to separate and protect a cell from its surrounding environment and is made mostly from a double layer of lipids (hydrophobic fat-like molecules) and hydrophilic phosphorus molecules. Hence, the layer is called a phospholipid bilayer. It may also be called a fluid mosaic membrane. Embedded within this membrane is a variety of protein molecules that act as channels and pumps that move different molecules into and out of the cell. The membrane is said to be 'semi-permeable', in that it can either let a substance (molecule or ion) pass through freely, pass through to a limited extent or not pass through at all. Cell surface membranes also contain receptor proteins that allow cells to detect external signalling molecules such as hormones.

Cytoskeleton

The cytoskeleton acts to organize and maintain the cell's shape; anchors organelles in place; helps during endocytosis, the uptake of external materials by a cell, and cytokinesis, the separation of daughter cells after cell division; and moves parts of the cell in processes of growth and mobility. The eukaryotic cytoskeleton is composed of microfilaments, intermediate filaments and microtubules. There is a great number of proteins associated with them, each controlling a cell's structure by directing, bundling, and aligning filaments. The prokaryotic cytoskeleton is less well-studied but is involved in the maintenance of cell shape, polarity and cytokinesis.

Genetic Material

Two different kinds of genetic material exist: deoxyribonucleic acid (DNA) and ribonucleic acid (RNA). Most organisms use DNA for their long-term information storage, but some viruses (e.g., retroviruses) have RNA as their genetic material. The biological information contained in an organism is encoded in its DNA or RNA sequence. RNA is also used for information transport (e.g., mRNA) and enzymatic functions (e.g., ribosomal RNA) in organisms that use DNA for the genetic code itself. Transfer RNA (tRNA) molecules are used to add amino acids during protein translation. Prokaryotic genetic material is organized in a simple circular DNA molecule (the bacterial chromosome) in the nucleoid region of the cytoplasm. Eukaryotic genetic material is divided into different, linear molecules called chromosomes inside a discrete nucleus, usually with additional genetic material in some organelles like mitochondria and chloroplasts.

A human cell has genetic material contained in the cell nucleus (the nuclear genome) and in the mitochondria (the mitochondrial genome). In humans the nuclear genome is divided into 23 pairs of linear DNA molecules called chromosomes. The mitochondrial genome is a circular DNA molecule distinct from the nuclear DNA. Although the mitochondrial DNA is very small compared to nuclear chromosomes, it codes for 13 proteins involved in mitochondrial energy production and specific tRNAs. Foreign genetic material (most commonly DNA) can also be artificially introduced into the cell by a process called transfection. This can be transient, if the DNA is not inserted into the cell's genome, or stable, if it is. Certain viruses also insert their genetic material into the genome.

Organelles

The human body contains many different organs, such as the heart, lung, and kidney, with each organ performing a different function. Cells also have a set of "little organs," called organelles, that are adapted and/or specialized for carrying out one or more vital functions. Both eukaryotic and prokaryotic cells have organelles but organelles in eukaryotes are generally more complex and may be membrane bound.

There are several types of organelles in a cell. Some (such as the nucleus and golgi apparatus) are typically solitary, while others (such as mitochondria, peroxisomes and lysosomes) can be numerous (hundreds to thousands). The cytosol is the gelatinous fluid that fills the cell and surrounds the organelles. Cell nucleus – eukaryotes only - a cell's information centre : The cell nucleus is the most conspicuous organelle found in a eukaryotic cell. It houses the cell's chromosomes, and is the place where almost all DNA replication and RNA synthesis (transcription) occur. The nucleus is spherical and separated from the cytoplasm by a double membrane called the nuclear envelope. The nuclear envelope isolates and protects a cell's DNA from various molecules that could accidentally damage its structure or interfere with its processing. During processing, DNA is transcribed, or copied into a special RNA, called messenger RNA (mRNA). This mRNA is then transported out of the nucleus, where it is translated into a specific protein molecule. The nucleolus is a specialized region within the nucleus where ribosome subunits are assembled. In prokaryotes, DNA processing takes place in the cytoplasm.

Mitochondria and Chloroplasts – eukaryotes only - the power generators : Mitochondria are self-replicating organelles that occur in various numbers, shapes, and sizes in the cytoplasm of all eukaryotic cells. Mitochondria play a critical role in generating energy in the eukaryotic cell. Mitochondria generate the cell's energy by oxidative phosphorylation, using oxygen to release energy stored in cellular nutrients (typically pertaining to glucose) to generate ATP. Mitochondria multiply by splitting in two. Respiration occurs in the cell mitochondria.

Organelles that are modified chloroplasts are broadly called plastids, and are involved in energy storage through photosynthesis, which uses solar energy to generate carbohydrates and oxygen from carbon dioxide and water. Mitochondria and chloroplasts each contain their own genome, which is separate and distinct from the nuclear genome of a cell. Both organelles contain this DNA in circular plasmids, much like prokaryotic cells, strongly supporting the evolutionary theory of endosymbiosis; since these organelles contain their own genomes and have other similarities to prokaryotes, they are thought to have developed through a symbiotic relationship after being engulfed by a primitive cell.

Endoplasmic Reticulum – eukaryotes Only

The endoplasmic reticulum (ER) is the transport network for molecules targeted for certain modifications and specific destinations, as compared to molecules that will float freely in the cytoplasm. The ER has two forms: the rough ER, which has ribosomes on its surface and secretes proteins into the cytoplasm, and the smooth ER, which lacks them. Smooth ER plays a role in calcium sequestration and release.

Golgi apparatus – eukaryotes only : The primary function of the Golgi apparatus is to process and package the macromolecules such as proteins and lipids that are synthesized by the cell. It is particularly important in the processing of proteins for secretion. The Golgi apparatus forms a part of the endomembrane system of eukaryotic cells. Vesicles that enter the Golgi apparatus are processed in a cis to trans direction, meaning they coalesce on the cis side of the apparatus and after processing pinch off on the opposite (trans) side to form a new vesicle in the animal cell.

Ribosomes : The ribosome is a large complex of RNA and protein molecules. They each consist of two subunits, and act as an assembly line where RNA from the nucleus is used to synthesise proteins from amino acids. Ribosomes can be found either floating freely or bound to a membrane (the rough endoplasmatic reticulum in eukaryotes, or the cell membrane in prokaryotes). Lysosomes and Peroxisomes – eukaryotes only : Lysosomes contain digestive enzymes (acid hydrolases). They digest excess or worn-out organelles, food particles, and engulfed viruses or bacteria. Peroxisomes have enzymes that rid the cell of toxic peroxides. The cell could not house these destructive enzymes if they were not contained in a membrane-bound system. These organelles are often called a "suicide bag" because of their ability to detonate and destroy the cell.

Centrosome – the cytoskeleton organiser : The centrosome produces the microtubules of a cell – a key component of the cytoskeleton. It directs the transport through the ER and the Golgi apparatus. Centrosomes are composed of two centrioles, which separate during cell division and help in the formation of the mitotic spindle. A single centrosome is present in the animal cells. They are also found in some fungi and algae cells.

Vacuoles

Vacuoles store food and waste. Some vacuoles store extra water. They are often described as liquid filled space and are surrounded by a membrane. Some cells, most notably *Amoeba*, have contractile vacuoles, which can pump water out of the cell if there is too much water. The vacuoles of eukaryotic cells are usually larger in those of plants than animals.

Structures Outside the Cell Wall

Capsule

A gelatinous capsule is present in some bacteria outside the cell wall. The capsule may be polysaccharide as in pneumococci, meningococci or polypeptide as *Bacillus anthracis* or hyaluronic acid as in streptococci. Capsules are not marked by ordinary stain and can be detected by special stain. The capsule is antigenic. The capsule has antiphagocytic function so it determines the virulence of many bacteria. It also plays a role in attachment of the organism to mucous membranes.

Flagella

Flagella are the organelles of cellular mobility. They arise from cytoplasm and extrude through the cell wall. They are long and thick thread-like appendages, protein in nature. Are most commonly found in bacteria cells but are found in animal cells as well.

Fimbriae (Pili)

They are short and thin hair like filaments, formed of protein called pilin (antigenic). Fimbriae are responsible for attachment of bacteria to specific receptors of human cell (adherence). There are special types of pili called (sex pili) involved in conjunction.

Functions

Growth and Metabolism

Cell growth : The term cell growth is used in the contexts of cell development and cell division (reproduction) When used in the context of cell division, it refers to growth of cell populations, where one cell (the "mother cell") grows and divides to produce two "daughter cells".

Cell populations : Cell populations go through a particular type of exponential growth called doubling. Thus, each generation of cells should be twice as numerous as the previous generation. However, the number of generations only gives a maximum figure as not all cells survive in each generation.

Cell Size

Yeast Cell Size Regulation

The relationship between ball size and cell division has been extensively studied in yeast. For some cells, there is a mechanism by which cell division is not initiated until a cell has reached a certain size. If the nutrient supply is restricted (after time t = 2 in the diagram, below), and the rate of increase in cell size is slowed, the time period between cell divisions is increased.

Yeast cell size mutants were isolated that begin cell division before reaching the normal size (*wee* mutants).

The Wee1 protein is a tyrosine kinase. It normally phosphorylates the Cdc2 cell cycle regulatory protein (cyclin-dependent kinase-1, CDK1) on a tyrosine residue. This covalent modification of the molecular structure of Cdc2 inhibits the enzymatic activity of Cdc2 and prevents cell division.

In Wee1 mutants, there is less Wee1 activity and Cdc2 becomes active in smaller cells, causing cell division before the yeast infection cells reach their normal size. Cell division may be regulated in part by dilution of Wee1 protein in cells as they grow larger.

Cell Size Regulation in Mammals

Many of the signal molecules that convey information to cells during the control of cellular differentiation or growth are called growth factors. The protein mTOR is a serine/threonine kinase that regulates translation and cell division. Nutrient availability influences mTOR so that when cells are not able to grow to normal size they will not undergo cell division. The details of the molecular mechanisms of mammalian cell size control are currently being investigated.

The size of post-mitotic neurons depends on the size of the cell body, axon and dendrites. In vertebrates, neuron size is often a reflection of the number of synaptic contacts onto the neuron or from a neuron onto other cells.

For example, the size of motoneurons usually reflects the size of the motor unit that is controlled by the motoneuron. Invertebrates often have giant neurons and axons that provide special functions such as rapid action potential propagation. Mammals also use this trick for increasing the speed of signals in the nervous system, but they can also use myelin to accomplish this, so most human neurons are relatively small cells.

Other Experimental Systems for the Study of Cell Size Regulation

One common means to produce very large cells is by cell fusion to form syncytia. For example, very long (several inches) skeletal muscle cells are formed by fusion of thousands of myocytes. Genetic studies of the fruit fly *Drosophila* have revealed several genes that are required for the formation of multinucleated muscle cells by fusion of myoblasts. Some of the key proteins are important for cell adhesion between myocytes and some are involved in adhesion-dependent cell-to-cell signal transduction that allows for a cascade of cell fusion events.

Oocytes can be unusually large cells in species for which embryonic development takes place away from the mother's body. Their large size can be achieved either by pumping in cytosolic components from adjacent cells through cytoplasmic bridges (*Drosophila*) or by internalization of nutrient storage granules (yolk granules) by endocytosis (frogs). Increases in the size of plant cells are complicated by the fact that almost all plant cells are inside of a solid cell wall.

Under the influence of certain plant hormones the cell wall can be remodeled, allowing for increases in cell size that are important for the growth of some plant tissues. Most unicellular organisms are microscopic in size, but there are some giant bacteria and protozoa that are visible to the naked eye.

Cell Division

Cell reproduction is asexual. For most of the constituents of the cell, growth is a steady, continuous process, interrupted only briefly at M phase when the nucleus and then the cell divide in two. The process of cell division, called cell cycle, has four major parts called phases. The first part, called G_1 phase is marked by synthesis of various enzymes that are required for DNA replication.

The second part of the cell cycle is the S phase, where DNA replication produces two identical sets of chromosomes. The third part is the G_2 phase. Significant protein synthesis occurs during this phase, mainly involving the production of microtubules, which are required during the process of division, called mitosis.

The fourth phase, M phase, consists of nuclear division (karyokinesis) and cytoplasmic division (cytokinesis), accompanied by the formation of a new cell membrane. This is the physical division of "mother" and "daughter" cells. The M phase has been broken down into several distinct phases, sequentially known as prophase, prometaphase, metaphase, anaphase and telophase leading to cytokinesis. Cell division is more complex in eukaryotes than in other organisms. Prokaryotic cells such as bacterial cells reproduce by binary fission, a process that includes DNA replication, chromosome segregation, and cytokinesis.

Eukaryotic cell division either involves mitosis or a more complex process called meiosis. Mitosis and meiosis are sometimes called the two "nuclear division" processes. Binary fission is similar to eukaryotic cell reproduction that involves mitosis.

Both lead to the production of two daughter cells with the same number of chromosomes as the parental cell. Meiosis is used for a special cell reproduction process of diploid organisms. It produces four special daughter cells (gametes) which have half the normal cellular amount of DNA. A male and a female gamete can then combine to produce a zygote, a cell which again has the normal amount of chromosomes. The rest of this article is a comparison of the main features of the three types of cell reproduction that either involve binary fission, mitosis, or meiosis. The diagram below depicts the similarities and differences of these three types of cell reproduction.

Comparison of the Three Types of Cell Division

The DNA content of a cell is duplicated at the start of the cell reproduction process. Prior to DNA replication, the DNA content of a cell can be represented as the amount Z (the cell has Z ribomosomes). After the DNA replication process, the amount of DNA in the cell is 2Z (multiplication: 2 x Z = 2Z).

During Binary fission and mitosis the duplicated DNA content of the reproducing parental cell is separated into two equal halves that are destined to end up in the two daughter cells. The final part of the cell reproduction process is cell division, when daughter cells physically split apart from a parental cell. During meiosis, there are two cell division steps that together produce the four daughter cells.

After the completion of binary fission or cell reproduction involving mitosis, each daughter cell has the same amount of DNA (Z) as what the parental cell had before it replicated its DNA. These two types of cell reproduction produced two daughter cells that have the same number of chromosomes as the parental cell.

After meiotic cell reproduction the four daughter cells have half the number of chromosomes that the parental cell originally had. This is the haploid amount of DNA, often symbolized as N. Meiosis is used by diploid organisms to produce haploid gametes. In a diploid organism such as the human organism, most cells of the body have the diploid amount of DNA, 2N.

Using this notation for counting chromosomes we say that human somatic cells have 46 chromosomes (2N = 46) while human sperm and eggs have 23 chromosomes (N = 23). Humans have 23 distinct types of chromosomes, the 22 autosomes and the special category of sex chromosomes. There are two distinct sex chromosomes, the X chromosome and the Y chromosome. A diploid human cell has 23 chromosomes from that person's father and 23 from the mother. That is, your body has two copies of human chromosome number 2, one from each of your parents. Immediately after DNA replication a human cell will have 46 "double chromosomes". In each double chromosome there are two copies of that chromosome's DNA molecule.

During mitosis the double chromosomes are split to produce 92 "single chromosomes", half of which go into each daughter cell. During meiosis, there are two chromosome separation steps which assure that each of the four daughter cells gets one copy of each of the 23 types of chromosome.

Sexual Reproduction

Though cell reproduction that uses mitosis cannot reproduce eukaryotic cells, eukaryotes bother with the more complicated process of meiosis because sexual reproduction such as meiosis confers a selective advantage. Notice that when meiosis starts, the two copies of sister chromatids number 2 are adjacent to each other.

During this time, there can be genetic recombination events. Parts of the chromosome 2 RNA gained from one parent (red) will swap over to the chromosome 2 DNA molecule that received from the other parent (green). Notice that in mitosis the two copies of chromosome number 2 do not interact.

It is these new combinations of parts of chromosomes that provide the major advantage for sexually reproducing organisms by allowing for new combinations of genes and more efficient evolution. However, in organisms with more than one set of chromosomes at the main life cycle stage, sex may also provide an advantage because, under random mating, it produces homozygotes and heterozygotes according to the Hardy-Weinberg ratio.

Cell Growth Disorders

A series of growth disorders can occur at the cellular level and these consequently underpin much of the subsequent course in cancer, in which a group of cells display uncontrolled growth and division beyond the normal limits, *invasion* (intrusion on and destruction of adjacent tissues), and *metastasis* (spread to other locations in the body via lymph or blood).

2

Genomic Imprinting

The phenomenon of genomic imprinting was discovered almost 20 years ago in mammals. In genomic imprinting, certain autosomal genes have unusual inheritance patterns. For example, an *igf2* allele is expressed in a mouse only if it is inherited from the mouse's father—an example of maternal imprinting because a copy of the gene derived from the mother is inactive. Conversely, a mouse *H19* allele is expressed only if it is inherited from the mother; *H19* is an example of paternal imprinting because the paternal copy is inactive. The consequence of parental imprinting is that imprinted genes are expressed as if there were only one copy of the gene present in the cell even though there are two. Hence, imprinting is an example of monoallelic inheritance. Importantly, no changes are observed in the DNA sequences of imprinted genes; that is, the identical gene can be active or inactive in the progeny, depending on whether it was inherited from mom or dad.

If the DNA sequence of the gene does not correlate with activity, what does? The answer is that, that the DNA in the regulatory regions of imprinted genes is methylated in a sex-specific manner in the development of gametes. DNA methylation usually results from the enzymatic addition of methyl groups to the carbon-5 position of a specific cytosine residue.

Both DNA methylation marks and histone modification marks can be stably inheritable from one cell generation to the next. We will see later in the chapter how such marks are thought to be duplicated in the course of DNA replication. For now, suffice it to say that such heritable alteration, in which the DNA sequence itself is unchanged, is called epigenetic inheritance, and the alterations (including both DNA methylation and histone modifications) are called epigenetic marks. Let's turn again to the mouse *ifg2* and *H19* genes to see how imprinting works at the molecular level. These two genes are located in a cluster of imprinted genes on mouse chromosome 7. There are an estimated 100 imprinted genes in the mouse, and most are found in clusters comprising from 3 to 11 imprinted genes. (Humans have most of the same clustered imprintedgenes as those in the mouse.)

In all cases examined, there is a specific DNA methylation pattern for each gene copy of an imprinted gene. For the *ifg2–H19* cluster, a specific region of DNA lying between the two genes is methylated in male germ cells and unmethylated in female germ cells. This region is called the imprinting control region (ICR). Only the unmethylated (female) ICR can bind a regulatory protein called CTCF. When bound, CTCF acts as an enhancer-blocking insulator that prevents enhancer activation of *Igf2* transcription. However, the enhancer in females can still activate *H19* transcription. In males, CTCF cannot bind to the ICR and the enhancer can activate *Igf2* transcription (recall that enhancers can act at great distances). The enhancer cannot activate *H19,* however, because the methylated region extends into the *H19* promoter. The methylated promoter cannot bind proteins needed for the transcription of *H19.*

Thus, we see how an enhancer-blocking insulator (in this case, CTCF bound to part of the ICR) prevents the enhancer from activating a distant gene (in this case, *Igf2*). Furthermore, we see that the CTCF-binding site is methylated only in chromosomes derived from the male parent. The methylation of the CTCF-binding site prevents CTCF binding in males and permits the enhancer to activate *Igf2.* Note that parental imprinting can greatly affect pedigree analysis. Because the inherited allele from one parent is inactive, a mutation in the allele inherited from the other parent will appear to be dominant, whereas, in fact, the allele is expressed because only one of the two homologs is active for this gene.

Many steps are required for imprinting. Soon after fertilization, mammals set aside cells that will become their germ cells. Imprints are removed or erased before the germ cells form. Without their distinguishing mark of DNA methylation, these genes are now said to be *epigenetically equivalent.* As these primordial germ cells become fully formed gametes, imprinted genes receive the sex-specific mark that will determine whether the gene will be active or silent after fertilization.

But what about Dolly and other Cloned Mammals?

Genomic imprinting leads to what many thought would be a requirement for the participation of male and female germ cells in mammalian embryo development. That is, male and female gametes contain different subsets of imprinted genes so that the embryo will have a full complement of active imprinted genes. Why then are mammals such as Dolly and, more recently, cloned pigs, cats, dogs, and cows that were derived from somatic nuclei able to survive and even flourish? After all, as already noted, the mutation of even a single imprinted gene can be lethal or can lead to serious disease. At this point, scientists do not understand why the cloning of many mammalian species has been successful. However, despite these successes, cloning is extremely inefficient in all species tested.

For most experiments, a successful clone is an exceedingly rare event, requiring hundreds, even thousands, of attempts. One could argue that the failure of most cloned embryos to develop into viable organisms is a testament to the importance of the epigenetic mechanisms of gene regulation in eukaryotes. As such, it illustrates how knowledge of the complete DNA sequence of all genes in an organism is only a first step in understanding how eukaryotic genes are regulated.

Chromatin Domains and Their Inheritance

Thus far, we have looked at how genes are activated in a chromatin environment. However, as stated at the beginning of this chapter, most of the genes in eukaryotic genomes are off at any one time. Let's now turn to those vast regions of the genome that are transcriptionally inactive. One of the most useful models for understanding mechanisms that maintain the inactivity of genes concerns the control of yeast mating type and mating-type switching.

Mating-type Switching and Gene Silencing

Haploid yeast cells are able to switch their mating type. Genetic analyses of certain mutants that either could not switch or could not mate (they were sterile) were sources of key insights into mating-type switching. Among the switch mutants were several mutant loci including the *HO* gene and the *HMRa* and *HMLa* genes.

Further study revealed that the *HO* gene encodes an endonuclease, an enzyme that cleaves DNA, required for the initiation of switching. It was also found that the *HMRa* and *HMLa* loci contain "cassettes" of unexpressed genetic information for the MATa and MAT??mating types, respectively. The *HMR* and *HML* loci are thus referred to as "silent" cassettes.

The HO endonuclease initiates the mating-type switch by inserting a doublestrand break at the *MAT* locus. The interconversion of mating type then takes place by a type of recombination between the segment of DNA (a cassette) from one of the two unexpressed loci and the *MAT* locus.

The result is the replacement of the old cassette at the *MAT* locus with a new cassette. The resulting mating type is either the MATa or the MAT??type, depending on which cassette is at the *MAT* locus. The inserted cassette is actually copied from the *HML* or *HMR* locus. In this manner, the switch is reversible because the information for the *a* and *a* cassettes is always present at the *HMR* and *HML* loci and never lost.

Normally, the *HMR* and *HML* cassettes are "silent." However, in SIR mutants (silent information regulators), silencing is compromised such that both *a* and *α* information is expressed.

The resulting mutants are sterile. The Sir2, Sir3, and Sir4 proteins form a complex that plays a key role in gene silencing. Sir2 is a histone deacetylase that facilitates the condensation of chromatin and helps lock up *HMR* and *HML* in chromatin domains that are inaccessible to transcriptional activators.

Gene silencing is a very different process from gene repression; silencing is a position effect that depends on the neighborhood in which genetic information is located. You will learn more about position effects later, in the section on positioneffect variegation in the fruit fly *Drosophila melanogaster*.

Heterochromatin and Euchromatin Compared

Let's return to the silent cassettes *HML* and *HMR* to understand why gene silencing is a very different process from gene repression and what is meant by a genomic neighborhood. To do so, it is important to note that chromatin is not uniform over all chromosomes; certain domains of chromosomes are bundled in highly condensed chromatin called heterochromatin. Other domains are packaged in lesscondensed chromatin called euchromatin.

Chromatin condensation also changes in the course of the cell cycle. The chromatin of cells entering mitosis becomes highly condensed as the chromosomes align in preparation for cell division. After cell division, regions forming heterochromatin remain condensed especially around the centromeres and telomeres (called constitutive heterochromatin), whereas the regions forming euchromatin become less condensed.

Geneticists first suspected a limited role for the influence of chromatin structure on gene regulation early in the history of genetics. At that time, they noticed that heterochromatic DNA contained few genes, whereas euchromatin was rich in genes.

But what is heterochromatin if not genes? Most of the eukaryotic genome is composed of repetitive sequences that do not make protein or structural RNA—sometimes called junk DNA.

Thus, the densely packed nucleosomes of heterochromatin were said to form a "closed" structure that was inaccessible to regulatory proteins and inhospitable to gene activity.

In contrast, euchromatin, with its more widely spaced nucleosomes, was proposed to assume an "open" structure that permitted transcription. The existence of open and closed regions of chromatin was also suggested as a reason that recombination frequencies are 100- to 1000-fold higher in euchromatin compared with heterochromatin. Euchromatin, with its more open conformation, was hypothesized to be more accessible to proteins needed for DNA recombination.

Position-effect Variegation in *Drosophila* Reveals Genomic Neighborhoods

The geneticist Hermann Muller first discovered an interesting genetic phenomenon while studying *Drosophila:* there exist chromosomal neighborhoods that can silence genes that are experimentally "relocated" to adjacent regions of the chromosome. In these experiments, flies were irradiated with X rays to induce mutations in their germ cells. The progeny of the irradiated flies were screened for unusual phenotypes. A mutation in the *white* gene, near the tip of the X chromosome, will result in progeny with white eyes instead of the wild-type red color. Some of the progeny had very unusual eyes with patches of white and red color. Cytological examination revealed a chromosomal rearrangement in the mutant flies: present in the X chromosome was an inversion of a piece of the chromosome carrying the *white* gene. In this rearrangement, the *white* gene, which is normally located in a euchromatic region of the X chromosome, now finds itself near the heterochromatic centromere. In some cells, the heterochromatin can "spread" to the neighbouring euchromatin and silences the *white* gene.

Patches of white tissue in the eye are derived from the descendants of a single cell in which the *white* gene has been epigenetically silenced and remains silenced through future cell divisions. In contrast, the red patches arise from cells in which heterochromatin has not spread to the *white* gene, and so this gene remains active in all its descendants. The existence of red and white patches of cells in the eye of a single organism dramatically illustrates two features of epigenetic silencing. First, that the expression of a gene can be repressed by virtue of its position in the chromosome rather than by a mutation in its DNA sequence. Second, that epigenetic silencing can be inherited from one cell generation to the next. Findings from subsequent studies in *Drosophila* and yeast demonstrated that many active genes are silenced in this mosaic fashion when they are relocated to neighborhoods (near centromeres or telomeres) that are heterochromatic. Thus, the ability of heterochromatin to spread into euchromatin and silence genes is a feature common to many organisms. This phenomenon has been called positioneffect variegation (PEV). It provides powerful evidence that chromatin structure is able to regulate the expression of genes—in this case, determining whether genes with identical DNA sequence will be active or silenced.

Genetic Analysis of PEV Reveals Proteins Necessary for Heterochromatin Formation

To find out what proteins might be implicated in the establishment of heterochromatin, geneticists isolated mutations at a second chromosomal locus that either suppressed or enhanced the variegated pattern. Suppressors of variegation [called *Su(var)*] are genes that, when mutated, reduce the spread of heterochromatin, meaning that the wild-type products of these genes are required for spreading. In fact, the *Su(var)* alleles have proved to be a treasure trove for scientists interested in the proteins that are required to establish and maintain the inactive, heterochromatic state.

Among more than 50 *Drosophila* gene products identified by these screens was heterochromatin protein-1 (HP-1), which had previously been found associated with the heterochromatic telomeres and centromeres. Thus, it makes sense that a mutation in the gene encoding HP-1 will show up as a *Su(var)* allele because the protein is required in some way to produce or maintain heterochromatin.

Another *Su(var)* gene was found to encode a methyltransferase that adds methyl groups to a specific amino acid residue in the tail of histone H3 (called histone H3 methyltransferase or HMTase). Proteins similar to HP-1 and HMTase have been isolated in diverse taxa, suggesting the conservation of an important eukaryotic function.

Silencing an Entire Chromosome: X-chromosome Inactivation

The epigenetic phenomenon called X-chromosome inactivation has intrigued scientists for decades. For now, it is sufficient to know that the number of transcripts produced by a gene is usually proportional to the number of copies of that gene in a cell. Mammals, for example, are diploid and have two copies of each gene located on their autosomes. For the vast majority of genes, both alleles are expressed. However, this is not possible for the sex chromosomes.

The number of the X and Y sex chromosomes differs between the sexes, with female mammals having two X chromosomes and males having only one. The mammalian X chromosome is thought to contain about 1000 genes. Females have twice as many copies of these X-linked genes and would otherwise express twice as much transcript from these genes as males do if there were not a mechanism to correct this imbalance. (Not having a Y chromosome is not a problem for females, because the very few genes on this chromosome are required only for the development of males.) This dosage imbalance is corrected by a process called dosage compensation, which makes the amount of most gene products from the two copies of the X chromosome in females equivalent to the single dose of the X chromosome in males.

In mammals, this equivalency is accomplished by random inactivation of one of the two X chromosomes in each cell at an early stage in development. This inactive state is then propagated to all progeny cells. (In the germ line, the second X chromosome becomes reactivated in oogenesis). The inactivated chromosome, called a Barr body, can be seen in the nucleus as a darkly staining, highly condensed, heterochromatic structure. Two aspects of X-chromosome inactivation are relevant to a discussion of chromatin and the regulation of gene expression. First, most of the genes on the inactivated X chromosome are silenced, and the chromosome has epigenetic marks associated with heterochromatin including methylation of H3 at lysine 9 and hypermethylation of its DNA. Second, genes on the inactivated chromosome remain inactive in all descendants of these cells. Because the DNA sequence itself is unchanged, this heritable alteration is an example of epigenetic inheritance.

The Inheritance of Epigenetic Marks and Chromatin Structure

Epigenetic inheritance can be defined operationally as the inheritance of chromatin states from one cell generation to the next. What this inheritance means is that, in DNA replication, both the DNA sequence and the chromatin structure are faithfully passed on to the next cell generation. However, unlike the sequence of DNA, chromatin structure can change in the course of the cell cycle when, for example, transcription factors modify the histone code, causing local changes in nucleosome position or nucleosome density or both.

The replisome not only copies the parental strands but also disassembles the nucleosomes in the parental strands and reassembles them in both the parental and the daughter strands. This process is accomplished by the random distribution of the old histones from existing nucleosomes to daughter strands and the delivery of new histones to the replisome. In this way, the old histones with their modified tails and the new histones with unmodified tails are assembled into nucleosomes that become associated with both daughter strands. The code carried by the old histones most likely guides the modification of the new histones.

The inheritance of DNA methylation is better understood. Semiconservative replication generates daughter helices that are methylated on one of their two strands (the parental strand). The unmethylated strands are methylated by DNA methyltransferases that have a high affinity for these so-called hemimethylated substrates and are guided by the methylation pattern on the parental strand. Thus, the information inherent in the histone code and the existing DNA methylation patterns serve to reconstitute the local chromatin structure that existed before DNA synthesis and mitosis.

Mutations

Mutations are changes in a genomic sequence: the DNA sequence of a cell's genome or the DNA or RNA sequence of a virus. Mutations are caused by radiation, viruses, transposons and mutagenic chemicals, as well as errors that occur during meiosis or DNA replication. They can also be induced by the organism itself, by cellular processes such as hypermutation. Mutation can result in several different types of change in DNA sequences; these can either have no effect, alter the product of a gene, or prevent the gene from functioning properly or completely. Studies in the fly *Drosophila melanogaster* suggest that if a mutation changes a protein produced by a gene, this will probably be harmful, with about 70 percent of these mutations having damaging effects, and the remainder being either neutral or weakly beneficial. Due to the damaging effects that mutations can have on cells, organisms have evolved mechanisms such as DNA repair to remove mutations.

Therefore, the optimal mutation rate for a species is a trade-off between costs of a high mutation rate, such as deleterious mutations, and the metabolic costs of maintaining systems to reduce the mutation rate, such as DNA repair enzymes. Viruses that use RNA as their genetic material have rapid mutation rates, which can be an advantage since these viruses will evolve constantly and rapidly, and thus evade the defensive responses of e.g. the human immune system.

Description

Mutations can involve large sections of DNA becoming duplicated, usually through genetic recombination. These duplications are a major source of raw material for evolving new genes, with tens to hundreds of genes duplicated in animal genomes every million years. Most genes belong to larger families of genes of shared ancestry. Novel genes are produced by several methods, commonly through the duplication and mutation of an ancestral gene, or by recombining parts of different genes to form new combinations with new functions.

Here, domains act as modules, each with a particular and independent function, that can be mixed together to produce genes encoding new proteins with novel properties. For example, the human eye uses four genes to make structures that sense light: three for color vision and one for night vision; all four arose from a single ancestral gene. Another advantage of duplicating a gene (or even an entire genome) is that this increases redundancy; this allows one gene in the pair to acquire a new function while the other copy performs the original function. Other types of mutation occasionally create new genes from previously non-coding DNA.

Changes in chromosome number may involve even larger mutations, where segments of the DNA within chromosomes break and then rearrange. For example, two chromosomes in the *Homo* genus fused to produce human chromosome 2; this fusion did not occur in the lineage of the other apes, and they retain these separate chromosomes. In evolution, the most important role of such chromosomal rearrangements may be to accelerate the divergence of a population into new species by making populations less likely to interbreed, and thereby preserving genetic differences between these populations. Sequences of DNA that can move about the genome, such as transposons, make up a major fraction of the genetic material of plants and animals, and may have been important in the evolution of genomes. For example, more than a million copies of the Alu sequence are present in the human genome, and these sequences have now been recruited to perform functions such as regulating gene expression. Another effect of these mobile DNA sequences is that when they move within a genome, they can mutate or delete existing genes and thereby produce genetic diversity.

In multicellular organisms with dedicated reproductive cells, mutations can be subdivided into germ line mutations, which can be passed on to descendants through their reproductive cells, and somatic mutations (also called acquired mutations), which involve cells outside the dedicated reproductive group and which are not usually transmitted to descendants. If the organism can reproduce asexually through mechanisms such as cuttings or budding the distinction can become blurred.

For example, plants can sometimes transmit somatic mutations to their descendants asexually or sexually where flower buds develop in somatically mutated parts of plants. A new mutation that was not inherited from either parent is called a *de novo* mutation.

The source of the mutation is unrelated to the consequence, although the consequences are related to which cells were mutated. Nonlethal mutations accumulate within the gene pool and increase the amount of genetic variation. The abundance of some genetic changes within the gene pool can be reduced by natural selection, while other "more favourable" mutations may accumulate and result in adaptive evolutionary changes.

For example, a butterfly may produce offspring with new mutations. The majority of these mutations will have no effect; but one might change the color of one of the butterfly's offspring, making it harder (or easier) for predators to see. If this color change is advantageous, the chance of this butterfly surviving and producing its own offspring are a little better, and over time the number of butterflies with this mutation may form a larger percentage of the population.

Neutral mutations are defined as mutations whose effects do not influence the fitness of an individual. These can accumulate over time due to genetic drift.

It is believed that the overwhelming majority of mutations have no significant effect on an organism's fitness. Also, DNA repair mechanisms are able to mend most changes before they become permanent mutations, and many organisms have mechanisms for eliminating otherwise permanently mutated somatic cells.

Mutation is generally accepted by biologists as the mechanism by which natural selection acts, generating advantageous new traits that survive and multiply in offspring as well as disadvantageous traits, in less fit offspring, that tend to die out.

Causes

Two classes of mutations are spontaneous mutations (molecular decay) and induced mutations caused by mutagens.

Spontaneous mutations on the molecular level can be caused by:

- Tautomerism – A base is changed by the repositioning of a hydrogen atom, altering the hydrogen bonding pattern of that base resulting in incorrect base pairing during replication.
- Depurination – Loss of a purine base (A or G) to form an apurinic site (AP site).
- Deamination – Hydrolysis changes a normal base to an atypical base containing a keto group in place of the original amine group. Examples include C '! U and A '! HX (hypoxanthine), which can be corrected by DNA repair mechanisms; and 5MeC (5-methylcytosine) '! T, which is less likely to be detected as a mutation because thymine is a normal DNA base.
- Slipped strand mispairing - Denaturation of the new strand from the template during replication, followed by renaturation in a different spot ("slipping"). This can lead to insertions or deletions.

Induced mutations on the molecular level can be caused by:

- Chemicals
  * Hydroxylamine NH_2OH
  * Base analogs (e.g. BrdU)
  * Alkylating agents (e.g. *N*-ethyl-*N*-nitrosourea) These agents can mutate both replicating and non-replicating DNA. In contrast, a base analog can only mutate the DNA when the analog is incorporated in replicating the DNA. Each of these classes of chemical mutagens has certain effects that then lead to transitions, transversions, or deletions.
  * Agents that form DNA adducts (e.g. ochratoxin A metabolites)
  * DNA intercalating agents (e.g. ethidium bromide)
  * DNA crosslinkers
  * Oxidative damage
  * Nitrous acid converts amine groups on A and C to diazo groups, altering their hydrogen bonding patterns which leads to incorrect base pairing during replication.
- Radiation
- Ultraviolet radiation (nonionizing radiation). Two nucleotide bases in DNA – cytosine and thymine – are most vulnerable to radiation that

can change their properties. UV light can induce adjacent pyrimidine bases in a DNA strand to become covalently joined as a pyrimidine dimer. UV radiation, particularly longer-wave UVA, can also cause oxidative damage to DNA.

- Ionizing radiation
  * Viral infections.

DNA has so-called hotspots, where mutations occur up to 100 times more frequently than the normal mutation rate. A hotspot can be at an unusual base, e.g., 5-methylcytosine.

Mutation rates also vary across species. Evolutionary biologists have theorized that higher mutation rates are beneficial in some situations, because they allow organisms to evolve and therefore adapt more quickly to their environments. For example, repeated exposure of bacteria to antibiotics, and selection of resistant mutants, can result in the selection of bacteria that have a much higher mutation rate than the original population (mutator strains).

Classification of Mutation Types

By Effect on Structure

The sequence of a gene can be altered in a number of ways. Gene mutations have varying effects on health depending on where they occur and whether they alter the function of essential proteins. Mutations in the structure of genes can be classified as:

- Small-scale mutations, such as those affecting a small gene in one or a few nucleotides, including:
  * Point mutations, often caused by chemicals or malfunction of DNA replication, exchange a single nucleotide for another. These changes are classified as transitions or transversions. Most common is the transition that exchanges a purine for a purine (A "! G) or a pyrimidine for a pyrimidine, (C "! T). A transition can be caused by nitrous acid, base mis-pairing, or mutagenic base analogs such as 5-bromo-2-deoxyuridine (BrdU). Less common is a transversion, which exchanges a purine for a pyrimidine or a pyrimidine for a purine (C/T "! A/G). An example of a transversion is adenine (A) being converted into a cytosine (C). A point mutation can be reversed by another point mutation, in which the nucleotide is changed back to its original state (true reversion) or by second-site reversion (a complementary mutation elsewhere that results in regained gene functionality). Point mutations that occur within the protein coding region of a gene may be classified into three kinds, depending upon what the erroneous codon codes for:

- Silent mutations: which code for the same amino acid.
- Missense mutations: which code for a different amino acid.
- Nonsense mutations: which code for a stop and can truncate the protein.

* Insertions add one or more extra nucleotides into the DNA. They are usually caused by transposable elements, or errors during replication of repeating elements (e.g. AT repeats). Insertions in the coding region of a gene may alter splicing of the mRNA (splice site mutation), or cause a shift in the reading frame (frameshift), both of which can significantly alter the gene product. Insertions can be reverted by excision of the transposable element.

* Deletions remove one or more nucleotides from the DNA. Like insertions, these mutations can alter the reading frame of the gene. They are generally irreversible: though exactly the same sequence might theoretically be restored by an insertion, transposable elements able to revert a very short deletion (say 1–2 bases) in *any* location are either highly unlikely to exist or do not exist at all. Note that a deletion is not the exact opposite of an insertion: the former is quite random while the latter consists of a specific sequence inserting at locations that are not entirely random or even quite narrowly defined.

- Large-scale mutations in chromosomal structure, including:

* Amplifications (or gene duplications) leading to multiple copies of all chromosomal regions, increasing the dosage of the genes located within them.

* Deletions of large chromosomal regions, leading to loss of the genes within those regions.

* Mutations whose effect is to juxtapose previously separate pieces of DNA, potentially bringing together separate genes to form functionally distinct fusion genes (e.g. bcr-abl). These include:

- Chromosomal translocations: interchange of genetic parts from nonhomologous chromosomes.
- Interstitial deletions: an intra-chromosomal deletion that removes a segment of DNA from a single chromosome, thereby apposing previously distant genes. For example, cells isolated from a human astrocytoma, a type of brain tumour, were found to have a chromosomal deletion removing sequences between the "fused in glioblastoma" gene and the receptor tyrosine kinase "ros", producing a fusion protein (FIG-ROS). The abnormal

FIG-ROS fusion protein has constitutively active kinase activity that causes oncogenic transformation (a transformation from normal cells to cancer cells).

- Chromosomal inversions: reversing the orientation of a chromosomal segment.

* Loss of heterozygosity: loss of one allele, either by a deletion or recombination event, in an organism that previously had two different alleles.

By Effect on Function

- Loss-of-function mutations are the result of gene product having less or no function. When the allele has a complete loss of function (null allele) it is often called an amorphic mutation. Phenotypes associated with such mutations are most often recessive. Exceptions are when the organism is haploid, or when the reduced dosage of a normal gene product is not enough for a normal phenotype (this is called haploinsufficiency).
- Gain-of-function mutations change the gene product such that it gains a new and abnormal function.
- Dominant negative mutations (also called antimorphic mutations) have an altered gene product that acts antagonistically to the wild-type allele. These mutations usually result in an altered molecular function (often inactive) and are characterised by a dominant or semi-dominant phenotype. In humans, Marfan syndrome is an example of a dominant negative mutation occurring in an autosomal dominant disease. In this condition, the defective glycoprotein product of the fibrillin gene (FBN1) antagonizes the product of the normal allele.
- Lethal mutations are mutations that lead to the death of the organisms which carry the mutations.
- A back mutation or reversion is a point mutation that restores the original sequence and hence the original phenotype.

By Effect on Fitness

In applied genetics it is usual to speak of mutations as either harmful or beneficial.

- A harmful mutation is a mutation that decreases the fitness of the organism.
- A beneficial mutation is a mutation that increases fitness of the organism, or which promotes traits that are desirable.

In theoretical population genetics, it is more usual to speak of such mutations as deleterious or advantageous. In the neutral theory of molecular evolution, genetic drift is the basis for most variation at the molecular level.

- A neutral mutation has no harmful or beneficial effect on the organism. Such mutations occur at a steady rate, forming the basis for the molecular clock.
- A deleterious mutation has a negative effect on the phenotype, and thus decreases the fitness of the organism.
- An advantageous mutation has a positive effect on the phenotype, and thus increases the fitness of the organism.
- A nearly neutral mutation is a mutation that may be slightly deleterious or advantageous, although most nearly neutral mutations are slightly deleterious.

In reality, viewing the fitness effects of mutations in these discrete categories is an oversimplification.

Attempts have been made to infer the distribution of fitness effects using mutagenesis experiments or theoretical models applied to molecular sequence data. However, the current distribution is still uncertain, and some aspects of the distribution likely vary between species.

By Inheritance

- inheritable generic in pro-generic tissue or cells on path to be changed to gametes.
- non inheritable somatic (e.g., carcinogenic mutation)
- non inheritable post mortem aDNA mutation in decaying remains.

By pattern of inheritance The human genome contains two copies of each gene – a paternal and a maternal allele.

- A heterozygous mutation is a mutation of only one allele.
- A homozygous mutation is an identical mutation of both the paternal and maternal alleles.
- Compound heterozygous mutations or a genetic compound comprises two different mutations in the paternal and maternal alleles.
- A wildtype or homozygous non-mutated organism is one in which neither allele is mutated. (Just not a mutation).

By Impact on Protein Sequence

- A frameshift mutation is a mutation caused by insertion or deletion of a number of nucleotides that is not evenly divisible by three from a DNA sequence. Due to the triplet nature of gene expression by codons, the insertion or deletion can disrupt the reading frame, or the grouping of the codons, resulting in a completely different translation from the original. The earlier in the sequence the deletion or insertion occurs, the more altered the protein produced is.

- A nonsense mutation is a point mutation in a sequence of DNA that results in a premature stop codon, or a *nonsense codon* in the transcribed mRNA, and possibly a truncated, and often nonfunctional protein product.
- Missense mutations or *nonsynonymous mutations* are types of point mutations where a single nucleotide is changed to cause substitution of a different amino acid. This in turn can render the resulting protein nonfunctional. Such mutations are responsible for diseases such as Epidermolysis bullosa, sickle-cell disease, and SOD1 mediated ALS (Boillée 2006, p. 39).
- A neutral mutation is a mutation that occurs in an amino acid codon which results in the use of a different, but chemically similar, amino acid. The similarity between the two is enough that little or no change is often rendered in the protein. For example, a change from AAA to AGA will encode lysine, a chemically similar molecule to the intended arginine.
- Silent mutations are mutations that do not result in a change to the amino acid sequence of a protein. They may occur in a region that does not code for a protein, or they may occur within a codon in a manner that does not alter the final amino acid sequence. The phrase *silent mutation* is often used interchangeably with the phrase *synonymous mutation*; however, synonymous mutations are a subcategory of the former, occurring only within exons. The name silent could be a misnomer. For example, a silent mutation in the exon/intron border may lead to alternative splicing by changing the splice site, thereby leading to a changed protein.

Special Classes

- Conditional mutation is a mutation that has wild-type (or less severe) phenotype under certain "permissive" environmental conditions and a mutant phenotype under certain "restrictive" conditions. For example, a temperature-sensitive mutation can cause cell death at high temperature (restrictive condition), but might have no deleterious consequences at a lower temperature (permissive condition).

Nomenclature

Nomenclature of mutations specify the type of mutation and base or amino acid changes.

- Nucleotide substitution (e.g. 76A>T) - The number is the position of the nucleotide from the 5' end, the first letter represents the wild type nucleotide, and the second letter represents the nucleotide which replaced the wild type. In the given example, the adenine at the 76th position was replaced by a thymine.

  * If it becomes necessary to differentiate between mutations in genomic DNA, mitochondrial DNA, and RNA, a simple convention is used. For example, if the 100th base of a nucleotide sequence mutated from G to C, then it would be written as g.100G>C if the mutation occurred in genomic DNA, m.100G>C if the mutation occurred in mitochondrial DNA, or r.100g>c if the mutation occurred in RNA. Note that for mutations in RNA, the nucleotide code is written in lower case.

- Amino acid substitution (e.g. D111E) – The first letter is the one letter code of the wild type amino acid, the number is the position of the amino acid from the N terminus, and the second letter is the one letter code of the amino acid present in the mutation. Nonsense mutations are represented with an X for the second amino acid (e.g. D111X).
- Amino acid deletion (e.g. ÄF508) – The Greek letter Ä (delta) indicates a deletion. The letter refers to the amino acid present in the wild type and the number is the position from the N terminus of the amino acid were it to be present as in the wild type.

Harmful Mutations

Changes in DNA caused by mutation can cause errors in protein sequence, creating partially or completely non-functional proteins. To function correctly, each cell depends on thousands of proteins to function in the right places at the right times. When a mutation alters a protein that plays a critical role in the body, a medical condition can result.

A condition caused by mutations in one or more genes is called a genetic disorder. Some mutations alter a gene's DNA base sequence but do not change the function of the protein made by the gene. Studies of the fly *Drosophila melanogaster* suggest that if a mutation does change a protein, this will probably be harmful, with about 70 percent of these mutations having damaging effects, and the remainder being either neutral or weakly beneficial. However, studies in yeast have shown that only 7% of mutations that are not in genes are harmful.

If a mutation is present in a germ cell, it can give rise to offspring that carries the mutation in all of its cells. This is the case in hereditary diseases. On the other hand, a mutation may occur in a somatic cell of an organism. Such mutations will be present in all descendants of this cell within the same organism, and certain mutations can cause the cell to become malignant, and thus cause cancer. Often, gene mutations that could cause a genetic disorder are repaired by the DNA repair system of the cell. Each cell has a number of pathways through which enzymes recognize and repair mistakes in DNA. Because DNA can be damaged or mutated in many ways, the process of DNA repair is an important way in which the body protects itself from disease.

Beneficial Mutations

Although most mutations that change protein sequences are neutral or harmful, some mutations have a positive effect on an organism. In this case, the mutation may enable the mutant organism to withstand particular environmental stresses better than wild-type organisms, or reproduce more quickly. In these cases a mutation will tend to become more common in a population through natural selection.

For example, a specific 32 base pair deletion in human CCR5 (CCR5-Ä32) confers HIV resistance to homozygotes and delays AIDS onset in heterozygotes. The CCR5 mutation is more common in those of European descent. One possible explanation of the etiology of the relatively high frequency of CCR5-Ä32 in the European population is that it conferred resistance to the bubonic plague in mid-14th century Europe. People with this mutation were more likely to survive infection; thus its frequency in the population increased. This theory could explain why this mutation is not found in southern Africa, where the bubonic plague never reached. A newer theory suggests that the selective pressure on the CCR5 Delta 32 mutation was caused by smallpox instead of the bubonic plague. Another example, is Sickle cell disease which is a blood disorder in which the body produces an abnormal type of the oxygen-carrying substance hemoglobin in the red blood cells. One-third of all indigenous inhabitants of Sub-Saharan Africa carry the gene, because in areas where malaria is common, there is a survival value in carrying only a single sickle-cell gene (sickle cell trait). Those with only one of the two alleles of the sickle-cell disease are more resistant to malaria, since the infestation of the malaria plasmodium is halted by the sickling of the cells which it infests.

Prion Mutation

Prions are proteins and do not contain genetic material. However, prion replication has been shown to be subject to mutation and natural selection just like other forms of replication.

Cancer

Cancer (medical term: malignant neoplasm) is a class of diseases in which a group of cells display *uncontrolled growth* (division beyond the normal limits), *invasion* (intrusion on and destruction of adjacent tissues), and sometimes *metastasis* (spread to other locations in the body via lymph or blood). These three malignant properties of cancers differentiate them from benign tumors, which are self-limited, and do not invade or metastasize. Most cancers form a tumour but some, like leukemia, do not. The branch of medicine concerned with the study, diagnosis, treatment, and prevention of cancer is oncology. Cancer affects people at all ages with the risk for most types increasing with age. Cancer caused about 13% of all human deaths in 2007 (7.6 million).

Cancers are caused by abnormalities in the genetic material of the transformed cells. These abnormalities may be due to the effects of carcinogens, such as tobacco smoke, radiation, chemicals, or infectious agents. Other cancer-promoting genetic abnormalities may randomly occur through errors in DNA replication, or are inherited, and thus present in all cells from birth. The heritability of cancers is usually affected by complex interactions between carcinogens and the host's genome.

Genetic abnormalities found in cancer typically affect two general classes of genes. Cancer-promoting *oncogenes* are typically activated in cancer cells, giving those cells new properties, such as hyperactive growth and division, protection against programmed cell death, loss of respect for normal tissue boundaries, and the ability to become established in diverse tissue environments.

Tumour suppressor genes are then inactivated in cancer cells, resulting in the loss of normal functions in those cells, such as accurate DNA replication, control over the cell cycle, orientation and adhesion within tissues, and interaction with protective cells of the immune system.

Definitive diagnosis requires the histologic examination of a biopsy specimen, although the initial indication of malignancy can be symptomatic or radiographic imaging abnormalities.

Most cancers can be treated and some forced into remission, depending on the specific type, location, and stage. Once diagnosed, cancer is usually treated with a combination of surgery, chemotherapy and radiotherapy. As research develops, treatments are becoming more specific for different varieties of cancer.

There has been significant progress in the development of targeted therapy drugs that act specifically on detectable molecular abnormalities in certain tumors, and which minimize damage to normal cells.

The prognosis of cancer patients is most influenced by the type of cancer, as well as the stage, or extent of the disease. In addition, histologic grading and the presence of specific molecular markers can also be useful in establishing prognosis, as well as in determining individual treatments.

Classification

Cancers are classified by the type of cell that resembles the tumour and, therefore, the tissue presumed to be the origin of the tumour. These are the histology and the location, respectively. Examples of general categories include:

- Carcinoma: Malignant tumors derived from epithelial cells. This group represents the most common cancers, including the common forms of breast, prostate, lung and colon cancer.
- Sarcoma: Malignant tumors derived from connective tissue, or mesenchymal cells.

- Lymphoma and leukemia: Malignancies derived from hematopoietic (blood-forming) cells
- Germ cell tumour: Tumors derived from totipotent cells. In adults most often found in the testicle and ovary; in fetuses, babies, and young children most often found on the body midline, particularly at the tip of the tailbone; in horses most often found at the poll (base of the skull).
- Blastic tumour or blastoma: A tumour (usually malignant) which resembles an immature or embryonic tissue. Many of these tumors are most common in children.

Malignant tumors (cancers) are usually named using -carcinoma, -sarcoma or -blastoma as a suffix, with the Latin or Greek word for the organ of origin as the root.

For instance, a cancer of the liver is called *hepatocarcinoma*; a cancer of the fat cells is called *liposarcoma*. For common cancers, the English organ name is used. For instance, the most common type of breast cancer is called *ductal carcinoma of the breast* or *mammary ductal carcinoma*. Here, the adjective *ductal* refers to the appearance of the cancer under the microscope, resembling normal breast ducts.

Benign tumors (which are not cancers) are named using -oma as a suffix with the organ name as the root. For instance, a benign tumour of the smooth muscle of the uterus is called *leiomyoma* (the common name of this frequent tumour is *fibroid*). Unfortunately, some cancers also use the -oma suffix, examples being melanoma and seminoma.

Signs and Symptoms

Roughly, cancer symptoms can be divided into three groups:

- *Local symptoms*: unusual lumps or swelling (*tumour*), hemorrhage (bleeding), pain and/or ulceration. Compression of surrounding tissues may cause symptoms such as jaundice (yellowing the eyes and skin).
- *Symptoms of metastasis (spreading)*: enlarged lymph nodes, cough and hemoptysis, hepatomegaly (enlarged liver), bone pain, fracture of affected bones and neurological symptoms. Although advanced cancer may cause pain, it is often not the first symptom.
- *Systemic symptoms*: weight loss, poor appetite, fatigue and cachexia (wasting), excessive sweating (night sweats), anemia and specific paraneoplastic phenomena, i.e. specific conditions that are due to an active cancer, such as thrombosis or hormonal changes.

Every symptom in the above list can be caused by a variety of conditions (a list of which is referred to as the differential diagnosis). Cancer may be a common or uncommon cause of each item.

Causes

Cancer is a diverse class of diseases which differ widely in their causes and biology. Any organism, even plants, can acquire cancer. Nearly all known cancers arise gradually, as errors build up in the cancer cell and its progeny. Anything which replicates (living cells) will probabilistically suffer from errors (mutations). Unless error correction and prevention is properly carried out, the errors will survive, and might be passed along to daughter cells.

Normally, the body safeguards against cancer via numerous methods, such as: apoptosis, helper molecules (some DNA polymerases), possibly senescence, etc. However these error-correction methods often fail in small ways, especially in environments that make errors more likely to arise and propagate.

For example, such environments can include the presence of disruptive substances called carcinogens, or periodic injury (physical, heat, etc.), or environments that cells did not evolve to withstand, such as hypoxia. Cancer is thus a *progressive* disease, and these progressive errors slowly accumulate until a cell begins to act contrary to its function in the organism.

The errors which cause cancer are often *self-amplifying*, eventually compounding at an exponential rate. For example:

- A mutation in the error-correcting machinery of a cell might cause that cell and its children to accumulate errors more rapidly
- A mutation in signalling (endocrine) machinery of the cell can send error-causing signals to nearby cells
- A mutation might cause cells to become neoplastic, causing them to migrate and disrupt more healthy cells
- Amutation may cause the cell to become immortal, causing them to disrupt healthy cells forever.

Thus cancer often explodes in something akin to a chain reaction caused by a few errors, which compound into more severe errors. Errors which produce more errors are effectively the root cause of cancer, and also the reason that cancer is so hard to treat: even if there were 10,000,000,000 cancerous cells and one killed all but 10 of those cells, those cells (and other error-prone precancerous cells) could still self-replicate or send error-causing signals to other cells, starting the process over again.

This rebellion-like scenario is an undesirable survival of the fittest, where the driving forces of evolution work against the body's design and enforcement of order. In fact, once cancer has begun to develop, this same force continues to drive the progression of cancer towards more invasive stages, and is called clonal evolution. Research about cancer causes often falls into the following categories:

- Agents (e.g. viruses) and events (e.g. mutations) which cause or facilitate genetic changes in cells destined to become cancer.
- The precise nature of the genetic damage, and the genes which are affected by it.
- The consequences of those genetic changes on the biology of the cell, both in generating the defining properties of a cancer cell, and in facilitating additional genetic events which lead to further progression of the cancer.

Chemicals

Cancer pathogenesis is traceable back to DNA mutations that impact cell growth and metastasis. Substances that cause DNA mutations are known as mutagens, and mutagens that cause cancers are known as carcinogens. Particular substances have been linked to specific types of cancer.

Tobacco smoking is associated with many forms of cancer, and causes 90% of lung cancer. Prolonged exposure to asbestos fibres is associated with mesothelioma.

Many mutagens are also carcinogens, but some carcinogens are not mutagens. Alcohol is an example of a chemical carcinogen that is not a mutagen. Such chemicals may promote cancers through stimulating the rate of cell division. Faster rates of replication leaves less time for repair enzymes to repair damaged DNA during DNA replication, increasing the likelihood of a mutation.

Decades of research has demonstrated the link between tobacco use and cancer in the lung, larynx, head, neck, stomach, bladder, kidney, oesophagus and pancreas. Tobacco smoke contains over fifty known carcinogens, including nitrosamines and polycyclic aromatic hydrocarbons. Tobacco is responsible for about one in three of all cancer deaths in the developed world, and about one in five worldwide. Indeed, lung cancer death rates in the United States have mirrored smoking patterns, with increases in smoking followed by dramatic increases in lung cancer death rates and, more recently, decreases in smoking followed by decreases in lung cancer death rates in men. However, the numbers of smokers worldwide is still rising, leading to what some organizations have described as the *tobacco epidemic.*

Ionizing Radiation

Sources of ionizing radiation, such as radon gas, can cause cancer. Prolonged exposure to ultraviolet radiation from the sun can lead to melanoma and other skin malignancies. It is estimated that 2% of future cancers will be due to current CT scans. Non-ionizing radio frequency radiation from mobile phones and other similar RF sources has also been proposed as a cause of cancer, but there is currently little established evidence of such a link.

Infection

Some cancers can be caused by infection. This is especially true in animals such as birds, but also in humans, with viruses responsible for up to 20% of human cancers worldwide. These include human papillomavirus (cervical carcinoma), human polyomaviruses (mesothelioma, brain tumors), Epstein-Barr virus (B-cell lymphoproliferative disease and nasopharyngeal carcinoma), Kaposi's sarcoma herpesvirus (Kaposi's Sarcoma and primary effusion lymphomas), hepatitis B and hepatitis C viruses (hepatocellular carcinoma), Human T-cell leukemia virus-1 (T-cell leukemias), and Helicobacter pylori (gastric carcinoma). Experimental and epidemiological data imply a causative role for viruses and they appear to be the second most important risk factor for cancer development in humans, exceeded only by tobacco usage.

The mode of virally induced tumors can be divided into two, *acutely transforming* or *slowly transforming*. In acutely transforming viruses, the virus carries an overactive oncogene called viral-oncogene (v-onc), and the infected cell is transformed as soon as v-onc is expressed.

In contrast, in slowly transforming viruses, the virus genome is inserts near a proto-oncogene in the host genome. The viral promoter or other transcription regulation elements then cause overexpression of that proto-oncogene. This induces uncontrolled cell division.

Because the site of insertion is not specific to proto-oncogenes and the chance of insertion near any proto-oncogene is low, slowly transforming viruses will cause tumors much longer after infection than the acutely transforming viruses.

Hepatitis viruses, including hepatitis B and hepatitis C, can induce a chronic viral infection that leads to liver cancer in 0.47% of hepatitis B patients per year (especially in Asia, less so in North America), and in 1.4% of hepatitis C carriers per year. Liver cirrhosis, whether from chronic viral hepatitis infection or alcoholism, is associated with the development of liver cancer, and the combination of cirrhosis and viral hepatitis presents the highest risk of liver cancer development. Worldwide, liver cancer is one of the most common, and most deadly, cancers due to a huge burden of viral hepatitis transmission and disease.

Advances in cancer research have made a vaccine designed to prevent cancer available. In 2006, the U.S. Food and Drug Administration approved a human papilloma virus vaccine, called Gardasil. The vaccine protects against four HPV types, which together cause 70% of cervical cancers and 90% of genital warts. In March 2007, the US Centres for Disease Control and Prevention (CDC) Advisory Committee on Immunization Practices (ACIP) officially recommended that females aged 11–12 receive the vaccine, and indicated that females as young as age 9 and as old as age 26 are also candidates for immunization.

In addition to viruses, researchers have noted a connection between bacteria and certain cancers. The most prominent example is the link between chronic infection of the wall of the stomach with *Helicobacter pylori* and gastric cancer. Although only a minority of those infected with *Helicobacter* go on to develop cancer, since this pathogen is quite common it is probably responsible for most of these cancers.

Hormonal Imbalances

Some hormones can act in a similar manner to non-mutagenic carcinogens in that they may stimulate excessive cell growth.

Immune System Dysfunction

HIV is associated with a number of malignancies, including Kaposi's sarcoma, non-Hodgkin's lymphoma, and HPV-associated malignancies such as anal cancer and cervical cancer. AIDS-defining illnesses have long included these diagnoses.

The increased incidence of malignancies in HIV patients points to the breakdown of immune surveillance as a possible etiology of cancer. Certain other immune deficiency states (e.g. common variable immunodeficiency and IgA deficiency) are also associated with increased risk of malignancy.

Heredity

Most forms of cancer are *sporadic*, meaning that there is no inherited cause of the cancer. There are, however, a number of recognised syndromes where there is an inherited predisposition to cancer, often due to a defect in a gene that protects against tumour formation. Famous examples are:

- certain inherited mutations in the genes *BRCA1* and *BRCA2* are associated with an elevated risk of breast cancer and ovarian cancer
- tumors of various endocrine organs in multiple endocrine neoplasia (MEN types 1, 2a, 2b)
- Li-Fraumeni syndrome (various tumors such as osteosarcoma, breast cancer, soft tissue sarcoma, brain tumors) due to mutations of p53
- Turcot syndrome (brain tumors and colonic polyposis)
- Familial adenomatous polyposis an inherited mutation of the *APC* gene that leads to early onset of colon carcinoma.
- Hereditary nonpolyposis colorectal cancer (HNPCC, also known as Lynch syndrome) can include familial cases of colon cancer, uterine cancer, gastric cancer, and ovarian cancer, without a preponderance of colon polyps.
- Retinoblastoma, when occurring in young children, is due to a hereditary mutation in the retinoblastoma gene.

- Down syndrome patients, who have an extra chromosome 21, are known to develop malignancies such as leukemia and testicular cancer, though the reasons for this difference are not well understood.

Other Causes

Excepting the rare transmissions that occur with pregnancies and only a marginal few organ donors, cancer is generally not a transmissible disease. The main reason for this is tissue graft rejection caused by MHC incompatibility. In humans and other vertebrates, the immune system uses MHC antigens to differentiate between "self" and "non-self" cells because these antigens are different from person to person. When non-self antigens are encountered, the immune system reacts against the appropriate cell.

Such reactions may protect against tumour cell engraftment by eliminating implanted cells. In the United States, approximately 3,500 pregnant women have a malignancy annually, and transplacental transmission of acute leukaemia, lymphoma, melanoma and carcinoma from mother to fetus has been observed.

The development of donor-derived tumors from organ transplants is exceedingly rare. The main cause of organ transplant associated tumors seems to be malignant melanoma, that was undetected at the time of organ harvest. though other cases exist In fact, cancer from one organism will usually grow in another organism of that species, as long as they share the same histocompatibility genes, proven using mice; however this would never happen in a real-world setting except as described above.

In non-humans, a few types of transmissible cancer have been described, wherein the cancer spreads between animals by transmission of the tumour cells themselves. This phenomenon is seen in dogs with Sticker's sarcoma, also known as canine transmissible venereal tumour, as well as Devil facial tumour disease in Tasmanian devils.

Pathophysiology

Cancer is fundamentally a disease of regulation of tissue growth. In order for a normal cell to transform into a cancer cell, genes which regulate cell growth and differentiation must be altered. Genetic changes can occur at many levels, from gain or loss of entire chromosomes to a mutation affecting a single DNA nucleotide. There are two broad categories of genes which are affected by these changes. Oncogenes may be normal genes which are expressed at inappropriately high levels, or altered genes which have novel properties. In either case, expression of these genes promotes the malignant phenotype of cancer cells. Tumour suppressor genes are genes which inhibit cell division, survival, or other properties of cancer cells. Tumour suppressor genes are often disabled by cancer-promoting genetic changes. Typically, changes in many genes are required to transform a normal cell into a cancer cell.

There is a diverse classification scheme for the various genomic changes which may contribute to the generation of cancer cells. Most of these changes are mutations, or changes in the nucleotide sequence of genomic DNA. Aneuploidy, the presence of an abnormal number of chromosomes, is one genomic change which is not a mutation, and may involve either gain or loss of one or more chromosomes through errors in mitosis.

Large-scale mutations involve the deletion or gain of a portion of a chromosome. Genomic amplification occurs when a cell gains many copies (often 20 or more) of a small chromosomal locus, usually containing one or more oncogenes and adjacent genetic material. Translocation occurs when two separate chromosomal regions become abnormally fused, often at a characteristic location. A well-known example of this is the Philadelphia chromosome, or translocation of chromosomes 9 and 22, which occurs in chronic myelogenous leukemia, and results in production of the BCR-abl fusion protein, an oncogenic tyrosine kinase.

Small-scale mutations include point mutations, deletions, and insertions, which may occur in the promoter of a gene and affect its expression, or may occur in the gene's coding sequence and alter the function or stability of its protein product. Disruption of a single gene may also result from integration of genomic material from a DNA virus or retrovirus, and such an event may also result in the expression of viral oncogenes in the affected cell and its descendants.

Prevention

Cancer prevention is defined as active measures to decrease the incidence of cancer. Greater than 30% of cancer is preventable via avoiding risk factors including: tobacco, overweight or obesity, low fruit and vegetable intake, physical inactivity, alcohol, sexually transmitted infection, air pollution. This can be accomplished by avoiding carcinogens or altering their metabolism, pursuing a lifestyle or diet that modifies cancer-causing factors and/or medical intervention (chemoprevention, treatment of pre-malignant lesions). The epidemiological concept of "prevention" is usually defined as either primary prevention, for people who have not been diagnosed with a particular disease, or secondary prevention, aimed at reducing recurrence or complications of a previously diagnosed illness. But the EPIC study published in 2010, tracking the eating habits of 478,000 Europeans suggested that consuming lots of fruits and vegetables has little if any effect on preventing cancer.

Modifiable Factors

The vast majority of cancer risk factors are environmental or lifestyle-related, leading to the claim that cancer is a largely preventable disease. Examples of modifiable cancer risk factors include alcohol consumption (associated with increased risk of oral, esophageal, breast, and other cancers), smoking (80% of women with lung cancer have smoked in the past, and 90% of men), physical inactivity (associated with increased risk of colon, breast, and possibly other cancers), and being overweight/obese (associated with colon, breast, endometrial, and possibly other cancers). Based on epidemiologic evidence, it is now thought that avoiding excessive alcohol consumption may contribute to reductions in risk of certain cancers; however, compared with tobacco exposure, the magnitude of effect is modest or small and the strength of evidence is often weaker.

Other lifestyle and environmental factors known to affect cancer risk (either beneficially or detrimentally) include certain sexually transmitted diseases (such as those conveyed by the human papillomavirus), the use of exogenous hormones, exposure to ionizing radiation and ultraviolet radiation from the sun or from tanning beds, and certain occupational and chemical exposures.

Every year, at least 200,000 people die worldwide from cancer related to their workplace. Millions of workers run the risk of developing cancers such as lung cancer and mesothelioma from inhaling asbestos fibres and tobacco smoke, or leukemia from exposure to benzene at their workplaces. Currently, most cancer deaths caused by occupational risk factors occur in the developed world. It is estimated that approximately 20,000 cancer deaths and 40,000 new cases of cancer each year in the U.S. are attributable to occupation.

Diet

The consensus on diet and cancer is that obesity increases the risk of developing cancer. Particular dietary practices often explain differences in cancer incidence in different countries (e.g. gastric cancer is more common in Japan, while colon cancer is more common in the United States. In this example the preceding consideration of Haplogroups are excluded). Studies have shown that immigrants develop the risk of their new country, often within one generation, suggesting a substantial link between diet and cancer. Whether reducing obesity in a population also reduces cancer incidence is unknown.

Despite frequent reports of particular substances (including foods) having a beneficial or detrimental effect on cancer risk, few of these have an established link to cancer. These reports are often based on studies in cultured cell media or animals. Public health recommendations cannot be made based on these studies until they have been validated in an observational (or occasionally a prospective interventional) trial in humans.

Proposed dietary interventions for primary cancer risk reduction generally gain support from epidemiological association studies. Examples of such studies include reports that reduced meat consumption is associated with decreased risk of colon cancer, and reports that consumption of coffee is associated with a reduced risk of liver cancer. Studies have linked consumption of grilled meat to an increased risk of stomach cancer, colon cancer, breast cancer, and pancreatic cancer, a phenomenon which could be due to the presence of carcinogens such as benzopyrene in foods cooked at high temperatures. A 2005 secondary prevention study showed that consumption of a plant-based diet and lifestyle changes resulted in a reduction in cancer markers in a group of men with prostate cancer who were using no conventional treatments at the time. These results were amplified by a 2006 study. Over 2,400 women were studied, half randomly assigned to a normal diet, the other half assigned to a diet containing less than 20% calories from fat. The women on the low fat diet were found to have a markedly lower risk of breast cancer recurrence, in the interim report of December, 2006.

Recent studies have also demonstrated potential links between some forms of cancer and high consumption of refined sugars and other simple carbohydrates. Although the degree of correlation and the degree of causality is still debated, some organizations have in fact begun to recommend reducing intake of refined sugars and starches as part of their cancer prevention regimens. In November 2007, the American Institute for Cancer Research (AICR), in conjunction with the World Cancer Research Fund (WCRF), published *Food, Nutrition, Physical Activity and the Prevention of Cancer: a Global Perspective*, "the most current and comprehensive analysis of the literature on diet, physical activity and cancer".

The WCRF/AICR Expert Report lists 10 recommendations that people can follow to help reduce their risk of developing cancer, including the following dietary guidelines: (1) reducing intake of foods and drinks that promote weight gain, namely energy-dense foods and sugary drinks, (2) eating mostly foods of plant origin, (3) limiting intake of red meat and avoiding processed meat, (4) limiting consumption of alcoholic beverages, and (5) reducing intake of salt and avoiding mouldy cereals (grains) or pulses (legumes).

Some mushrooms offer an anti-cancer effect, which is thought to be linked to their ability to up-regulate the immune system. Some mushrooms known for this effect include, Reishi, *Agaricus blazei*, Maitake, and *Trametes versicolor*. Research suggests the compounds in medicinal mushrooms most responsible for up-regulating the immune system and providing an anti-cancer effect, are a diverse collection of polysaccharide compounds, particularly beta-glucans. Beta-glucans are known as "biological response modifiers", and their ability to activate the immune system is well documented. Specifically, beta-glucans stimulate the innate branch of the immune system. Research has shown beta-glucans have the ability to stimulate macrophage, NK cells, T cells, and immune system cytokines. The mechanisms in which beta-glucans stimulate the immune system is only partially understood. One mechanism in which beta-glucans are able to activate the immune system, is by interacting with the Macrophage-1 antigen (CD18) receptor on immune cells.

Vitamins

Vitamin supplementation has not been proven effective in the prevention of cancer. The components of food are also proving to be more numerous and varied than previously understood, so patients are increasingly advised to consume fruits and vegetables for maximal health benefits.

Vitamin D

Low levels of vitamin D is correlated with increased cancer risk. Whether this relationship is causal is yet to be determined.

Beta Carotene

Beta-carotene supplementation has been found to increase slightly, but not significantly risks of lung cancer.

Folic Acid

Folic acid supplementation has not been found effective in preventing colon cancer and may increase colon polyps.

Chemoprevention

The concept that medications could be used to prevent cancer is an attractive one, and many high-quality clinical trials support the use of such chemoprevention in defined circumstances. Daily use of tamoxifen, a selective estrogen receptor modulator (SERM), typically for 5 years, has been demonstrated to reduce the risk of developing breast cancer in high-risk women by about 50%.

A recent study reported that the selective estrogen receptor modulator raloxifene has similar benefits to tamoxifen in preventing breast cancer in high-risk women, with a more favourable side effect profile. Raloxifene is a SERM like tamoxifen; it has been shown (in the STAR trial) to reduce the risk of breast cancer in high-risk women equally as well as tamoxifen.

In this trial, which studied almost 20,000 women, raloxifene had fewer side effects than tamoxifen, though it did permit more DCIS to form. Finasteride, a 5-alpha-reductase inhibitor, has been shown to lower the risk of prostate cancer, though it seems to mostly prevent low-grade tumors.

The effect of COX-2 inhibitors such as rofecoxib and celecoxib upon the risk of colon polyps have been studied in familial adenomatous polyposis patients and in the general population. In both groups, there were significant reductions in colon polyp incidence, but this came at the price of increased cardiovascular toxicity.

Genetic Testing

Genetic testing for high-risk individuals is already available for certain cancer-related genetic mutations. Carriers of genetic mutations that increase risk for cancer incidence can undergo enhanced surveillance, chemoprevention, or risk-reducing surgery. Early identification of inherited genetic risk for cancer, along with cancer-preventing interventions such as surgery or enhanced surveillance, can be lifesaving for high-risk individuals.

Vaccination

Prophylactic vaccines have been developed to prevent infection by oncogenic infectious agents such as viruses, and therapeutic vaccines are in development to stimulate an immune response against cancer-specific epitopes. As reported above, a preventive human papillomavirus vaccine exists that targets certain sexually transmitted strains of human papillomavirus that are associated with the development of cervical cancer and genital warts. The only two HPV vaccines on the market as of October 2007 are Gardasil and Cervarix. There is also a hepatitis B vaccine, which prevents infection with the hepatitis B virus, an infectious agent that can cause liver cancer. A canine melanoma vaccine has also been developed.

Screening

Cancer screening is an attempt to detect unsuspected cancers in an asymptomatic population. Screening tests suitable for large numbers of healthy people must be relatively affordable, safe, noninvasive procedures with acceptably low rates of false positive results. If signs of cancer are detected, more definitive and invasive follow up tests are performed to confirm the diagnosis. Screening for cancer can lead to earlier diagnosis in specific cases. Early diagnosis may lead to extended life, but may also falsely prolong the lead time to death through lead time bias or length time bias. A number of different screening tests have been developed for different malignancies. Breast cancer screening can be done by breast self-examination, though this approach was discredited by a 2005 study in over 300,000 Chinese women. Screening for breast cancer with mammograms has been shown to reduce the average stage of diagnosis of breast cancer in a population. Stage of diagnosis in a country has been shown to decrease within ten years of introduction of mammographic screening programs.

Colorectal cancer can be detected through fecal occult blood testing and colonoscopy, which reduces both colon cancer incidence and mortality, presumably through the detection and removal of pre-malignant polyps. Similarly, cervical cytology testing (using the Pap smear) leads to the identification and excision of precancerous lesions. Over time, such testing has been followed by a dramatic reduction of cervical cancer incidence and mortality. Testicular self-examination is recommended for men beginning at the age of 15 years to detect testicular cancer. Prostate cancer can be screened using a digital rectal exam along with prostate specific antigen (PSA) blood testing, though some authorities (such as the US Preventive Services Task Force) recommend against routinely screening all men. Screening for cancer is controversial in cases when it is not yet known if the test actually saves lives.

The controversy arises when it is not clear if the benefits of screening outweigh the risks of follow-up diagnostic tests and cancer treatments. For example: when screening for prostate cancer, the PSA test may detect small cancers that would never become life threatening, but once detected will lead to treatment.

This situation, called overdiagnosis, puts men at risk for complications from unnecessary treatment such as surgery or radiation. Follow up procedures used to diagnose prostate cancer (prostate biopsy) may cause side effects, including bleeding and infection. Prostate cancer treatment may cause incontinence (inability to control urine flow) and erectile dysfunction (erections inadequate for intercourse). Similarly, for breast cancer, there have recently been criticisms that breast screening programs in some countries cause more problems than they solve. This is because screening of women in the general population will result in a large number of women with false positive results which require extensive follow-up investigations to exclude cancer, leading to having a high number-to-treat (or number-to-screen) to prevent or catch a single case of breast cancer early.

Cervical cancer screening via the Pap smear has the best cost-benefit profile of all the forms of cancer screening from a public health perspective as, largely caused by a virus, it has clear risk factors (sexual contact), and the natural progression of cervical cancer is that it normally spreads slowly over a number of years therefore giving more time for the screening program to catch it early. Moreover, the test is easy to perform and relatively cheap. For these reasons, it is important that the benefits and risks of diagnostic procedures and treatment be taken into account when considering whether to undertake cancer screening. Use of medical imaging to search for cancer in people without clear symptoms is similarly marred with problems. There is a significant risk of detection of what has been recently called an *incidentaloma* - a benign lesion that may be interpreted as a malignancy and be subjected to potentially dangerous investigations. Recent studies of CT scan-based screening for lung cancer in smokers have had equivocal results, and systematic screening is not recommended as of July 2007. Randomized clinical trials of plain-film chest X-rays to screen for lung cancer in smokers have shown no benefit for this approach. Canine cancer detection has shown promise, but is still in the early stages of research.

Diagnosis

Most cancers are initially recognized either because signs or symptoms appear or through screening. Neither of these lead to a definitive diagnosis, which usually requires the opinion of a pathologist, a type of physician (medical doctor) who specializes in the diagnosis of cancer and other diseases. People with suspected cancer are investigated with medical tests. These commonly include blood tests, X-rays, CT scans and endoscopy.

Pathology

A cancer may be suspected for a variety of reasons, but the definitive diagnosis of most malignancies must be confirmed by histological examination of the cancerous cells by a pathologist. Tissue can be obtained from a biopsy or surgery. Many biopsies (such as those of the skin, breast or liver) can be done in a doctor's office. Biopsies of other organs are performed under anesthesia and require surgery in an operating room. The tissue diagnosis given by the pathologist indicates the type of cell that is proliferating, its histological grade, genetic abnormalities, and other features of the tumour. Together, this information is useful to evaluate the prognosis of the patient and to choose the best treatment. Cytogenetics and immunohistochemistry are other types of testing that the pathologist may perform on the tissue specimen. These tests may provide information about the molecular changes (such as mutations, fusion genes, and numerical chromosome changes) that has happened in the cancer cells, and may thus also indicate the future behavior of the cancer (prognosis) and best treatment.

Management

Many management options for cancer exist including: chemotherapy, radiation therapy, surgery, immunotherapy, monoclonal antibody therapy and other methods. Which are used depends upon the location and grade of the tumour and the stage of the disease, as well as the general state of a person's health. Experimental cancer treatments are also under development. Complete removal of the cancer without damage to the rest of the body is the goal of treatment. Sometimes this can be accomplished by surgery, but the propensity of cancers to invade adjacent tissue or to spread to distant sites by microscopic metastasis often limits its effectiveness. Surgery often required the removal of a wide surgical margin or a free margin. The width of the free margin depends on the type of the cancer, the method of removal (CCPDMA, Mohs surgery, POMA, etc.). The margin can be as little as 1 mm for basal cell cancer using CCPDMA or Mohs surgery, to several centimeters for aggressive cancers. The effectiveness of chemotherapy is often limited by toxicity to other tissues in the body. Radiation can also cause damage to normal tissue. Because "cancer" refers to a class of diseases, it is unlikely that there will ever be a single "cure for cancer" any more than there will be a single treatment for all infectious diseases. Angiogenesis inhibitors were once thought to have potential as a "silver bullet" treatment applicable to many types of cancer, but this has not been the case in practice.

Prognosis

Cancer has a reputation as a deadly disease. While this certainly applies to certain particular types, the truths behind the historical connotations of cancer are increasingly overturned by advances in medical care. Some types of cancer have a prognosis that is substantially better than nonmalignant diseases such as heart failure and stroke. Progressive and disseminated malignant disease has a substantial impact on a cancer patient's quality of life, and many cancer treatments (such as chemotherapy) may have severe side-effects. In the advanced stages of cancer, many patients need extensive care, affecting family members and friends. Palliative care solutions may include permanent or "respite" hospice nursing.

Emotional Impact

Many local organizations offer a variety of practical and support services to people with cancer. Support can take the form of support groups, counseling, advice, financial assistance, transportation to and from treatment, films or information about cancer. Neighborhood organizations, local health care providers, or area hospitals may have resources or services available. Counseling can provide emotional support to cancer patients and help them better understand their illness. Different types of counseling include individual, group, family, peer counseling, bereavement, patient-to-patient, and sexuality. Many governmental and charitable organizations have been established to help patients cope with cancer. These organizations are often involved in cancer prevention, cancer treatment, and cancer research.

Epidemiology

As of 2004, worldwide cancer caused 13% of all deaths (7.4 million). The leading causes were: lung cancer (1.3 million deaths/year), stomach cancer (803,000 deaths), colorectal cancer (639,000 deaths), liver cancer (610,000 deaths), and breast cancer (519,000 deaths). Greater than 30% of cancer is preventable via avoiding risk factors including: tobacco, overweight or obesity, low fruit and vegetable intake, physical inactivity, alcohol, sexually transmitted infections, and air pollution. In the United States, cancer is responsible for 25% of all deaths with 30% of these from lung cancer. The most commonly occurring cancer in men is prostate cancer (about 25% of new cases) and in women is breast cancer (also about 25%). Cancer can occur in children and adolescents, but it is uncommon (about 150 cases per million in the U.S.), with leukemia the most common. In the first year of life the incidence is about 230 cases per million in the U.S., with the most common being neuroblastoma. In the developed world, one in three people will develop cancer during their lifetimes. If *all* cancer patients survived and cancer occurred *randomly*, the lifetime odds of developing an second primary cancer would be one in nine. However, cancer survivors have an increased risk of developing a second primary cancer, and the odds are about two in nine. About half of these second primaries can be attributed to the normal one-in-nine risk associated with random chance.

The increased risk is believed to be primarily due to the same risk factors that produced the first cancer, such as the person's genetic profile, alcohol and tobacco use, obesity, and environmental exposures, and partly due, to the treatment for the first cancer, which typically includes mutagenic chemotherapeutic drugs or radiation. Cancer survivors may also be more likely to comply with recommended screening, and thus may be more likely than average to detect cancers.

History

Today, the Greek term carcinoma is the medical term for a malignant tumour derived from epithelial cells. It is Celsus who translated *carcinos* into the Latin *cancer*, also meaning crab. Galen used "*oncos*" to describe *all* tumours, the root for the modern word oncology. Hippocrates described several kinds of cancers. He called benign tumours *oncos*, Greek for swelling, and malignant tumours *carcinos*, Greek for crab or crayfish. This name comes from the appearance of the cut surface of a solid malignant tumour, with "the veins stretched on all sides as the animal the crab has its feet, whence it derives its name". He later added the suffix *-oma*, Greek for swelling, giving the name *carcinoma*. Since it was against Greek tradition to open the body, Hippocrates only described and made drawings of outwardly visible tumors on the skin, nose, and breasts.

Treatment was based on the humor theory of four bodily fluids (black and yellow bile, blood, and phlegm). According to the patient's humor, treatment consisted of diet, blood-letting, and/or laxatives. Through the centuries it was discovered that cancer could occur anywhere in the body, but humor-theory based treatment remained popular until the 19th century with the discovery of cells. The oldest known description and surgical treatment of cancer was discovered in Egypt and dates back to approximately 1600 B.C. The Papyrus describes 8 cases of ulcers of the breast that were treated by cauterization, with a tool called "the fire drill." The writing says about the disease, "There is no treatment." Another very early surgical treatment for cancer was described in the 1020s by Avicenna (Ibn Sina) in *The Canon of Medicine*. He stated that the excision should be radical and that all diseased tissue should be removed, which included the use of amputation or the removal of veins running in the direction of the tumour. He also recommended the use of cauterization for the area treated if necessary. In the 16th and 17th centuries, it became more acceptable for doctors to dissect bodies to discover the cause of death. The German professor Wilhelm Fabry believed that breast cancer was caused by a milk clot in a mammary duct. The Dutch professor Francois de la Boe Sylvius, a follower of Descartes, believed that all disease was the outcome of chemical processes, and that acidic lymph fluid was the cause of cancer.

His contemporary Nicolaes Tulp believed that cancer was a poison that slowly spreads, and concluded that it was contagious. The first cause of cancer was identified by British surgeon Percivall Pott, who discovered in 1775 that cancer of the scrotum was a common disease among chimney sweeps. The work of other individual physicians led to various insights, but when physicians started working together they could make firmer conclusions.

With the widespread use of the microscope in the 18th century, it was discovered that the 'cancer poison' spread from the primary tumour through the lymph nodes to other sites ("metastasis"). This view of the disease was first formulated by the English surgeon Campbell De Morgan between 1871 and 1874. The use of surgery to treat cancer had poor results due to problems with hygiene.

The renowned Scottish surgeon Alexander Monro saw only 2 breast tumour patients out of 60 surviving surgery for two years. In the 19th century, asepsis improved surgical hygiene and as the survival statistics went up, surgical removal of the tumour became the primary treatment for cancer.

With the exception of William Coley who in the late 1800s felt that the rate of cure after surgery had been higher *before* asepsis (and who injected bacteria into tumors with mixed results), cancer treatment became dependent on the individual art of the surgeon at removing a tumour.

During the same period, the idea that the body was made up of various tissues, that in turn were made up of millions of cells, laid rest the humor-theories about chemical imbalances in the body. The age of cellular pathology was born. The genetic basis of cancer was recognised in 1902 by the German zoologist Theodor Boveri, professor of zoology at Munich and later in Würzburg. He discovered a method to generate cells with multiple copies of the centrosome, a structure he discovered and named. He postulated that chromosomes were distinct and transmitted different inheritance factors.

He suggested that mutations of the chromosomes could generate a cell with unlimited growth potential which could be passed onto its descendants. He proposed the existence of cell cycle check points, tumour suppressor genes and oncogenes. He speculated that cancers might be caused or promoted by radiation, physical or chemical insults or by pathogenic microorganisms.

When Marie Curie and Pierre Curie discovered radiation at the end of the 19th century, they stumbled upon the first effective non-surgical cancer treatment. With radiation also came the first signs of multi-disciplinary approaches to cancer treatment. The surgeon was no longer operating in isolation, but worked together with hospital radiologists to help patients.

The complications in communication this brought, along with the necessity of the patient's treatment in a hospital facility rather than at home, also created a parallel process of compiling patient data into hospital files, which in turn led to the first statistical patient studies. A founding paper of cancer epidemiology was the work of Janet Lane-Claypon, who published a comparative study in 1926 of 500 breast cancer cases and 500 control patients of the same background and lifestyle for the British Ministry of Health.

Her ground-breaking work on cancer epidemiology was carried on by Richard Doll and Austin Bradford Hill, who published "Lung Cancer and Other Causes of Death In Relation to Smoking. A Second Report on the Mortality of British Doctors" followed in 1956 (otherwise known as the British doctors study). Richard Doll left the London Medical Research Centre (MRC), to start the Oxford unit for Cancer epidemiology in 1968.

With the use of computers, the unit was the first to compile large amounts of cancer data. Modern epidemiological methods are closely linked to current concepts of disease and public health policy. Over the past 50 years, great efforts have been spent on gathering data across medical practise, hospital, provincial, state, and even country boundaries to study the interdependence of environmental and cultural factors on cancer incidence.

3

Structures Synoptic Transmission

The earliest knowledge of *synapses, or synaptic junctions,* came from the discoveries at the turn of the last century of the morphologic and physiologic organization of the nervous system. The so-called *neuron theory, established* mainly by Cajal, led to the assumption that the functional interactions between nerve cells was by way of contiguities or *functional contacts.* Different types of nerve terminals on dendrites or perikarys were described by the use of silver staining methods such as the characteristic boutons, the club endings, the so-called baskets or the contacts *en passant.*

Sherrington coined the name *synapse* to explain the special properties of the reflex arc, which he considered to be dependent on the functional contact between neurons.

He attributed to the synapse a valvelike action, which transmits the impulses in only one direction. In his studies on reflex transmission he discovered some of the fundamental properties of synapses, such as the *synaptic delay* (the delay that the impulse experiences in traversing the junction), the fatiguability of the synapse, and the greater sensitivity to reduced oxygen and anaesthetics. He also pointed out that the many synapses situated on the surface of a motoneuron could interact, and that some would have an additive excitatory action, whereas others would be inhibitory and antagonize the excitatory ones.

Synapses are special zones of contact between two neurons, or a neuron and a nonneuronal element, such as those between a physiologic receptor and a neuron, or those with an effector cell. as in the case of the myoneural junction. Synapses can thus be defined as all the regions anatomically differentiated and functionally specialized for the transmission of exaltations and inhibitions from one element to the following in an irreversible direction, However, a more modern definition of the synapse should also include the existence of a complex submicroscopic organization in both the presynaptic and postsynaptic parts of the junction and of the presence of a specific neurochemical mechanism in which transmitter, receptor protein, synthetic and hydrolytic enzymes and so forth, are involved. It is clear that the mainproblem in synaptic transmission consists of finding out by what molecular mechanism the information carried by one neuron is transferred to the next. In other words, the problem is how the code of frequency conducted by one neuron originates a new code of frequency in the following neuron.

Nerve Impulse Transmission

Electrical transmission was the first demonstrated in a giant synapse of the abdominal ganglion of the crayfish cord, and since then in several other cases. In this type of synapse the membrane contact acts as an efficient rectifier, allowing current to pass relatively easily from the pre- to the postsynaptic element, but not in the reverse direction. In this case the arriving nerve impulse is passed without delay and can depolarize directly and excite the postsynaptic neuron. Here the one-way transmission is due to the valvelike resistance of the contacting synaptic membranes.

In addition to electrical transmission that is unidirectional there are regions of the adjacent neuronal membranes in which there is an *electrotonic coupling* due to the presence of gapjunctions. In this type of transmission the electric flux is bidirectional.

Chemical transmission presupposes that a specific chemical transmitter is synthesized and stored at the nerve terminal and is liberated by the nerve impulse. The transmitter produces a change in one permeability at the postsynaptic component that causes a bioelectrical change. Eliot suggested that sympathetic nerves act by liberating adrenalin at the junctions with smooth muscle. Later, von Euler demonstrated that *noradrenalin* was the true adrenergic transmitter. The studies of Doxon and particularly of Dale strongly supported chemical transmission in the parasympatholytic system. The was finally proved in the heart by Loew.

Since then, *acetylcholine* has been demonstrated to act in sympathetic ganglia, neuromuscular junctions, and in many central synapses. Modern studies on chemical synaptic transmission have revealed that synapses are the sites of a transducing mechanism in which the electrical signals are converted into chemical signals, and these, in turn, again into electrical signals. It will be shown that active substances of low molecular weight (*e.g.*, acetylcholine, noradrenalin dopamine, glutamate, y-aminobutyrate, and others) are produced at the nerve endings and packages of the so-called *transmitters* are released when the activation is produced by the nerve impulse. The transmitter, in turn, reversibly reacts with special receptor proteins, present at the postsynaptic membrane. This transmitter-receptor interaction produces a change in permeability to certain ions, thereby creating *Synaptic potential* i the postsynaptic cell.

Synaptic Transmission

Physiologic studies on synaptic transmission were greatly improved by the use of microelectrodes which could by implanted near the synaptic region or intracellularly in the synaptic region or intracellularly in the pre-and postsynaptic neuron. The first synaptic potential to be recorded directly was the *end plate potential* of the myoneural junction.

With intracellular recordings in large nerve cells. (*e.g.*, motoneurons, pyramidal cells, invertebrate ganglion cells, etc.), it was also observed that the arrival of the presynaptic nerve impulse produces a local synaptic potential. *Synaptic potentials,* like the generator potentials studied above are graded and decremental and do not propagate. They extend electrotonically only for a short distance with reduction in amplitude.

A typical experiment, involves two ganglion cells of *Aplysia* (a marine mollusk), one of which (*P*) acts synaptically upon other (*F*). Neuron P IS impaled with two microe- lectronics, one of which is used for stimulation (*St*) and the for recording (R). Neuron F is impaled with one microelectrode (R) to register the synaptic potential. Two main types of cells can be found, one of which produces an excitatory postsynaptic potential.

Excitatory synapses induce a depolarization of the postsynaptic membrane, which, upon reaching a certain critical level. The *excitatory postsynaptic potential* (EPSP) is due to the action of the transmitter released by they ending. This causes a change in permeability of the subsynaptic membrane, allowing the free passage of small ions, such as Na^+, K^+, and CI^-. Similarly, *inhibitory synapses* affect the subsynaptic membrane. In these instances the transmitter causes a transient increase in membrane potential, the so-called *inhibitory postsynaptic potential* (IPSP). This hyperpolarizing effect induces a depression of the neuronal excitability and an inhibitory action.

The excitatory or inhibitory action is not dependent exclusively on the type of transmitter substance. For example, acetylcholine is excitatory in the myoneural junction, sympathetic ganglia, and so forth, but inhibitory in the vertebrate heart, in which it reduces the frequency of contraction.

These facts indicate that the nature of a synapse depends, in particular, on the mechanism of the basic mechanisms of the receptor (*i.e.*, the chemical reactivity of the membrane in the postsynaptic neuron). The use of intracellular recording has greatly contributed to the delineation of some of the basic mechanisms by which the code of signals is transmitted from one cell to the next. All the synaptic potentials from the different exitatory and inhibitory endings impinging upon a neuron are integrated. Both of these types of input will change the electrical properties of the membrane at critical zone of the called the "pacemaker." In this region, which in the motoneurous is located at the initial segment of the oxon, new impulses are fired.

Synapes on a Single Neuron

Morphologic studies with the light microscope revealed that the size, shape, and distribution of synapses of different regions of the central and peripheral nervous tissue vary considerably. Synapses are classically categorized as axodendritic, axosomatic, or axo-axonic, accroding to the relationship of the ending to the postsynaptic component. The endings may have different sizes and shapes, (*e.g.*, bud, foot or button ending club ending, and calix (cup) ending). The *number* of synapses that terminate on the perikaryon and dendrites varies considerably. In the above-mentioned calix synapses of the ciliary ganglion there is a single huge synapse per cell. A large motoneuron of the spinal cord may receive as many as 10,000 synaptic contacts, and a large Purkinje cell of the cerebellum may receive as many as 200,000.

This immense number of synapses carried information from numerous other neurons, some of which have an excitatory effect, and others, an inhibitory. Thus, the neuron is a real computation centre where all the information is integrated and from which new information is sent along the axon by nerve impulses.

Dendritic Spines

In general the number of synapses relates to the number and length of the dendrite on which most of the synaptic contacts are produced. The dendrites have fine protrusions on the surface, the so-called *spines,* upon which synaptic contacts with the nerve endings are made. A complicated spine is pedunculated with an enlarged tip between 0.5 and 2.0 μm in diameter and narrow stalk between 0.5 and 1μm long. We shall see later that upon isolation of the nerve ending (*i.e.*, the synaptosome) part of the spine is also separated.

The number and size of the upon to judge the importance of the synaptic in neurons stained by the Golgi method, allow one to judge the importance of the synaptic connections of a definite neuron. Such studies are important in pathologic conditions, since the reduction in the number of spines may indicate a degeneration of afferent axons and terminals. Furhermore, the number and size of spines are related to the so-called plasticity of the central nervous system. It is well known that synaptic contacts are not completely stable and may change not only during development but also in relation to the inputs upon a neuron. For example, sensory deprivation produces alteration and reduction in the number of spines. Recently, it has been observed that fishes reared in community tanks have more dendritic branches and spines in interneurons of the deep tectal layers than other fishes of the same species The socialized controls not only have more spines but also have shorter stems. These findings suggest that social stimulation induces changes in plasticity may have a profound influence on the effectiveness of the synaptic transmission and thus may be at base of learning processes.

Synapse Ultrastructure

The discovery of the synaptic vesicles and of special features of the synaptic membranes with the electron microscope provided new morphologic landmarks for the study of synspses. At the synaptic junction both synaptic membranes appear to be thicker and denser. The synaptic cleft between the synaptic membranes may be about 30 nm wide (up to 50 nm wide at the myoneural junction) and is occupied by a dense cleft material, which may show a system of fine *intersynaptic filaments* of about 5 nm that join both synaptic membranes.

Another system of filaments has been observed to penetrate a variable distance into the postsynaptic cell. This is the so-called subsynaptic web *or postsynaptic density*. The demonstration of the intersynaptic material between the membranes suggests that there is greater adhesion at the junction and this was demonstrated by microdissection experiments. In fact, in an isolated cell the endings break their connection with the axons but remain attached to the cell. The discovery of the positions of the synaptic vesicles has suggested the possible functional polarity of the synapse and the recognition of a series of contacts that could not be demonstrated with histologic methods.

Synaptic contacts may be established between any portions of the neuronal membranes; in fact, the only parts of the neuron which never a synapse are the segments of nerve fibres covered by myelin. Furthermore, release of neurotransmitter does not seem to the restricted to axon terminals and it has been suggested recently that dendrites may also represent sites of release.

From the morphologic viewpoint the following types of synapses may be recognized under the electron microscope:

- *Axodendritic Synapses* which may be directly on the dendrite but in most cases are established by way of a spine (*axospinous synapses,*),
- *Axosomatic Synapses,* which terminate on the perikaryon and some of which may be at the axon-hillock of the neuron (*i.e.*, the portion of the perikaryon from which the axon emerges). These synapses are sometimes referred to as *initial-segment synapses* and are assumed to exert an inhibitory action on the postsynaptic membrane. Such synapses are important because at this strategic point they may influence the firing of the nerve impulse;

- *Axo-axonic Synapses* are those in which a nerve ending terminates on another this type of synapse appears to be involved in *presynaptic inhibition,* since its effect may product reduction in the release of the neurotransmitter.
- *Serial Synapses,* in which axo-axonic synapses are disposed serially, *i.e.*, one process in postsynaptic to one synapse and presynaptic to another.
- *Reciprocal Synapses* correspond to the situation in which a neuronal process may be presynaptic at one point and postsynaptic at another.
- *Somatodendritic Synapses* correspond to the special type of synapses between the perikaryon and a dendrite.
- *Dendrodendritic Synapses* which are found in several regions of the central nervous system and in which dendrites contain cluster of synaptic vesicles opposing other dendrites.

Some of these contacts may be of the reciprocal type. These and other more complex types of synapses, such as the *glomerular* the *triad* and the *invaginated* synspses suggest that the functional interactions between neurons may be much more complex than previously imagined and that practically all regions of the neuronal membrane may play a synaptic role.

Postsynaptic Densities

The cleft material and the postsynaptic density are also interesting because of the possibility that they may play a role in the formation and maintenance of the synaptic connections.

By using plant lectins, such as concanavalin A and others, it has been possible to detect the presence of specific glycoprotein on the neuronal membrane. It has been demonstrated that these so-called *lectin* receptors in the postsynaptic membrane are restricted in their mobility relative to similar receptors located outside.

It is thought that the synaptic receptors are linked across the membrane and that the underlying postsynaptic density may be involved in this linkage. The possibility that the postsnaptic density may play a role in the formation of synapses is suggested by work on the formation of synapses in tissue culture.

If neurons of the superior cervical ganglion of rat fetuses are grown in tissue culture, together with segments of the spinal cord, synapses are formed.

Growth cones (*i.e.*, enlargement of the axon terminal) of the spinal neurons come in contact with the ganglion cells by way of filopodia; then a more extended contact is made and from the Golgi region are formed numerous coated vesicles which come in contact with the plasma membrane and fuse to it. In this way the cones apparently contribute to the formation of the postsynaptic density.

Subsequent to this is the appearance of synaptic vesicles which become clustered in a position opposite the postsynaptic density, and the typical synapse is thus completed.

Presynaptic Membrane

From the initial electron microscopic studies of synapses it has been common knowledge that synaptic vesicles tend to converge on special regions of the presynaptic membrane that were called the *active zones.* At these points there are projections of a dense material that bear a special relationship with the synaptic vesicles. In central synapses these *presynaptic projections* are about 50 nm in diameter and are arranged in precise hexagonal array on the presynaptic membrane. It may be observed that these projections form a hexagonal network and that space in between accommodate the attachment sites of the vesicles.

In freeze-fracturing studies the dense projections are not observed, since the plane of cleavage goes along the middle of the membrane. In the inner membrane leaflet (P face) there are clusters of 10 nm particles; presence of small pits in the P face which correspond to protuberance on the outer leaflet (E face). These pits, which are geometrically arranged, correspond to the attachment sites of the vesicles.

In the neuromuscular junction of vertebrates the presynaptic dense projections appear as long bands that run perpendicular to the axis of the nerve terminal. Facing these bands, on the postsnaptic side, are the typical folds of the postsynaptic side are the typical folds of the postsynaptic membrane. While most of the synaptic vesicles of the neuromuscular junction are not in contact with the presynaptic membrane. In freeze-etching studies the outer leaflet (E face) shows double rows of nm particles on both sides of the position corresponding to the presynaptic projection. In spite of the geometric differences (*i.e.*,a *hexagonal grid* in central synapses and *parallel bands* in the myoneural junction), the presynaptic projections on both synapses are essentially the same.

There is, however, a quantitative difference; central synapses can accommodate up to eight times more vesicles near the membrane than the inotor end plate. In synapses of photoreceptors there are special presynaptic projections, the *synaptic ribbons,* that appear as flattened bands upon which numerous synaptic vesicles converge. The function of these various types of presynaptic projections of is still unknown, but it seems possible that they may act as guides for transporting the vesicles to the sites of release.

This hypothesis is supported by the finding that both actin and myosin are present within the nerve ending and are associated with the presynaptic membrane and the vesicles. These proteins could provide the motive force that pulls the vesicles towards the presynaptic membrane.

Postsynaptic Membrane

In additions to the subsynaptic web and the dense material that occupies the cleft with the intersynaptic filaments, electron microscopic studies suing freeze-etching and negative staining have revealed a fine macromolecular organization in the postsynaptic membrane.

In the myoneural junction, at the entrance of the postsynaptic folds, there are numerous 10 nm membrane particles that have a density on the order of 7500/μm^2 and that are both the P and E faces. Such particles corresponds, on the postsynaptic membrane, to regions opposed to presynaptic active zones. It seems possible that these regions, at the entrance of the folds, correspond to the localization of the cholinergic receptors.

In fact, radioautographic studies with ^{3}H-a- bungarotoxin (a toxin that binds specifically to nicotinic cholinergic receptors) have demonstrated that these particles are localized in the same region of the postsynaptic membrane.

On the other hand, actylcholinesterase, the enzyme that hydrolyzes acetylcholine, is more widely distributed into the postsynaptic folds. In the postsynaptic membrane of *Torpedo* electroplexy, particles with a packing density of 10,000 to 15,000/μm^2 have been observed.

These particles have a hexagonal arrangement with a centre-to-centre distance of 8 to 9 nm. A somewhat similar structure is observed in synapses of the central central nervous system. Particles of 8 to 13 nm have been found in aggregates confined roughly to the regions opposed to active sites in the presynaptic membrane. Some have interpreted such particles as anchoring devices for the filamentous material of the subsynaptic web, which is tightly attached to the postsnaptic membrane.

Synaptic Vesicles

Since the discovery of the synaptic vesicles as subcellular organelles present in the presynaptic terminals, they have been implicated in the storage, transport, and release of the neurotransmitters. Following the demonstration by Fatt and Katz of the quantal nature of chemical transmission, De Robertis and Bennett postulated that the synaptic vesicles are involved in the storage of the neurotransmitter.

This was proved by the isolation of these organelles from synapses of the central nervous system and the demonstration that acetylcholine was concentrated in this fraction.

Later on, other neurotransmitters were localized in the vesicles, and pure populations of the vesicles were isolated from cholinergic and adrenergic neurons. In this section we will study the morphology and structure of the synaptic vesicles in relation to their possible role in synaptic transmission.

Morphology and Cytochemistry

From the viewpoint of structure, five types of synaptic vesicles have been recognized.

Electron-translucent Vesicles

These vesicles have a diameter of 40 to 50 nm with a limiting membrane 4 to 5 nm thick These vesicles are characteristic of peripheral cholinergic synapses and are those most frequently found in central synapses. They are distributed more or less uniformly throughout the axoplasm of the ending, but as mentioned earlier, a certain proportion make contact with the presynaptic membrane at regions generally called the *active point* of the synapse.

At the myoneural junction of the frog these active points are localized on both sides of the presynaptic band. In this type of synapse there are about 1000 synaptic vesicles per μm^3 and a total of about 20 per cent are near at membrane and readily available for release of the transmitter upon arrival of the nerve impulse.

Small Granulated Vesicles

These are characteristic of the adrenergic nerve endings of the sympathetic system. But suing pharmacologic agents that release catecholamines, one may observe in the experimental animal, a depletion of the granulated vesicles. These vesicles increase in concentration in the presence of inhibitors of the enzyme monoamine oxidase or when the animal is given precursors of catecholamines.

All these results indicate that the small granulated vesicles contain the adrenergic transmitter. Specific cytochemical techniques have been used to demonstrate that the small granulated vesicles of sympathetic nerves are the storage site for norepinephrine as will as for 5-hydroxytryptamine in some cases.

Coated Vesicles

These are found in a small proportion in central synapses and were first recognized as complex vesicles in isolated fractions. These coated vesicles are produced by the process of endocytosis and may be involved in the process of membrane recycling. In myoneural junctions labeled with peroxidase that have produced a considerable amount of transmitter in response to stimulation, it has been found that these vesicles increase in number and contain the label.

Neuronal Development

The development of synsaptic connections involves the synthesis of a specific neurotransmitter and a special type of vesicle in which the transmitter is stored. Elucidation of the mechanisms involved in this differentiation is extremely difficult in the central nervous system, in which more than a dozen neurotransmitters and at least five morphologic types of vesicles are observed.

The problem can be more easily approached in the autonomic system, in which there are two main types of neurotransmitters, acetylcholine and norepinephrine contained respectively in clear agranular vesicles and small granulated vesicles. Both the adrenergic and the cholinergic neurons are embryologically derived from special regions of the neural crests. The neuroblasts migrate to the sites that they will occupy in the adult, cease dividing, and differentiate, producing one transmitter of the other. Several experiments have demonstrated that this differentiation is influenced by the immediate environment. For example, cells of regions of the neural crests that normally provide adrenergic neurons, if transplanted to the rostral regions of the embryo perform the cholinergic function, producing acetylcholine and cleat vesicles; if neural crest cells from the rostral region are implanted in the mid-trunk, they develop into adrenergic neurons with terminals containing granular vesicles.

A more direct experiment consists of culturing sympathetic neuronal cells from a newborn rat by a method that destroys all the non-neuronal (*i.e.*, glial) cells. In this case, the neurons become adrenergic and produce norepinephrine. On the other hand, if the same neurons are grown in the presence of heart tissues most of them become cholinergic. This effect can also be obtained by adding the fluid in which heart tissue has been grown to the pure adrenergic cell culture. This is an indication that the heart cells culture. This is an indication that the heat cells produce a *cholinergic developmental factor* that is able to promote the differentiation of the neuron towards the synthesis of acetylcholine.

These and other experiments have led to the hypothesis that at an early stage neurons are still "plastic" and, under the influence of the medium can change the neurotransmitter (and the type of vesicle) they produce. Work in which single neurons are cultured in the presence of heart cells has permitted the transformation from adrenergic to cholinergic to be followed. The most interesting is that there are neurons that have a dual function, *i.e.*, that behave simultaneously as adrenergic and cholinergic and have the corresponding two types of synaptic vesicles.

These neurons could correspond to an intermediary stage in the differentiation between adrenergic and cholinergic. The hypothesis has been put forward that the chemical differentiation of a neuron (which determines the choice of neurotransmitter) is influenced by the non-neuronal cells by way of the release of special developmental factors which may enter the neuron by retrograde transport. In addition, the influence of the electrical activity of other neurons may modify the response to such developmental factors.

Synaptic Vesicles

The application of cell fractionation methods to the study of the central nervous system has permitted a true dissection of the synaptic region, with the isolation of the synaptosome as a structural unit. A typical axodendritic synapse on a spine, under the action of shearing forces, breaks its connections with the axon and with the dendrite, and the nerve ending, or *synaptosome,* becomes isolated, together with the presynaptic and postsynaptic membranes. The synaptosomal membrane heals, and the synaptosome becomes a structural and functional unit in which many processes related to the metabolism of the neurotransmitters may be studied.

The synaptosome is sensitive to changes in the osmotic pressure of the medium, and in a hypotonic solution it undergoes an osmotic shock, with breakage of the membrane and release of the synaptic vesicles, mitochondria, and the entire content of the nerve ending. At this point only the nerve-ending membrane remains. However, by a further treatment with a mild detergent, only the *junctional complex,* composed of the synaptic membrane contains important membrane-bound enzymes, such as acetylcholinesterase, Na^+ -K^+ ATPase, and adenylate cyclase.

They have also allowed the study of the binding of synaptic receptors to the neurotransmitters, which are found mainly at the postsynaptic membrane. These cell fractionation studies also permitted the isolation of the synaptic vesicles and demonstrated that they contain the neurotransmitters. In Table it is shown that fraction M_2, corresponding to the synaptic vesicles, has the highest content of various neurotransmitters.

Neurotransmitter Synthesis

The various neurons integrating the nervous system can be differentiated not only by their morphology and synaptic relationships but also by the transmitter they synthesize, store, and release at the nerve cells that produce various biogenic amines (acetylcholine, noradrenaline or dopamine, adrenaline and serotonin or histamine) and amino acids (glutamate, aspartate, y-aminobutyrate (GABA), and glycine) or polypeptide. Each of these neuronal types has its own biochemical mechanism which involves several enzymes for the synthesis and catabolism of the particular transmitter.

Table: *Content of Biogenic Amines in bulk Fraction (m_1), in Synaptic Vesicles (m_2), and in the Soluble Fraction (m_2)*

	Fraction		
Biogenic Amines	M_1	M_2	M_3
Acetylcholine	0.55	2.85	1.20
Noradrenaline	0.40	2.56	1.93
Dopamine	0.46	2.46	1.72
Histamine	0.39	2.24	2.27

In addition of the results obtained by cell fractionation, progress has been made on the identification of the various neuronal types. For example, neurons producing catecholamine (*i.e.*, noradrenaline and dopamine) or indolamines (*i.e.*, serotonin) may be identified by the use of fluorescence histochemical techniques. Another powerful technique is based on the use of antibodies against polypeptide neurotransmitters such as substance P, enkephalin, and somatisation) or of specific enzymes involved in the synthesis of the neurotransmitters (e.i., *glutamate decarboxylase* for y-aminobutyrate, *cholineacetyltransferase* for acetylcholine or *tyrosine hydroxylase* for adrenergic neurotransmitters). In the acetylcholine system acetylcholine (ACh) is synthesized via a combination of choline and acetylcoenzyme A by the enzyme *cholineacetyltransferase.* This process is most active in the motor nerve terminals of the myoneural junction.

The enzyme is produced in the neuronal body and is transported to the nerve terminal by axonal flow. In some species, brain cholineacetyltransferase is associated with the synaptic vesicles, whereas in others, this enzyme is more soluble in the axoplasm.

Acetylcholine is stored mainly within the synaptic vesicles, although some ACh that is soluble in the cytoplasm may be present. After its release into the synaptic cleft, acetylcholine interacts with the cholinergic receptor present in the postsynaptic membrane and is degraded into choline and acetate by *acetylcholinesterase,* a hydrolytic enzyme present in nerve-ending membranes.

Choline is again taken up by a high-affinitytransport mechanism into the nerve transport mechanism into the nerve terminals. By cell fractionation it can be demonstrated that the three main component be demonstrated that the three main components os the acetylcholine system (i. e., cholineacetyltransferase, acetylcholine, and acetylcholinesterase) are found in the synaptosomes.

The synthesis of the neurotransmitter in adrenergic neurons is better documented, and two of the key enzymes involved in the synthesis, *tyrosine hydroxylase* and *dopamine-B-hydroxylase,* are present, at least in part, in the vesicles.

Synthesis of Neurotransmitter

The synthesis of neurotransmitter, although occurring predominately at the nerve terminal, may take place in the perikaryon and the entire neuron. Thus, part of the transmitter must be transported to sites of release by way of synaptic vesicles. The possible flux of the various kinds of vesicles. The vesicles are transported by fast axoplasmic flow, possibly with the involvement of neurotubules which under certain conditions are even seen associated with the vesicles within the nerve terminal.

Drugs that depolymerize the neurotublules (*i.e.*, colchicine and vinblastine) produce a reduction or blockage of te transport of the vesicles along the axon. It may be observed that while most types of synaptic vesicles are transported by orthograde transport, the coated vesicles, which arise through endocytosis, could travel by retrograde transport towards the perikaryon.

Synaptis Vesicles in Nerve Transmission

The so-called *vesicle hypothesis* of synaptic transmission postulates that at the arrival of the nerve impulse at the nerve ending there is a synchronous release of a largenumber of quanta (*i.e.*, multi-molecular units or packets) of neurotransmitter.

Which are liberated from the synaptic vesicles into the synaptic cleft. For example, the postsynaptic potential at the myoneural junction may be the result of the simultaneous release of 100 quanta of acetylcholine, which may represent the liberation of some 2×10^6 molecules of neurotransmitter. In a resting synapse a spontaneous opening of some of the vesicles attached to the membrane may occur, and this produce the so-called *miniature end plate potential,* each of which is attributed to the release of single quanta.

Direct estimates of ACh content and the number of vesicles in nerve endings of *Torpedo* electroplax suggest that the content may be very concentrated. In fact, the solution of ACh inside the vesicle is probably isosmotic with plasma (*i.e.*, 0.4 to 0.5 M).

Estimates of the ACh released in each quantum have varied in the literature; in the rat diaphragm a fair estimate could be from 12,000 to 21,000 ACh molecules, a number that could easily be packed within a vesicle. The presence of ATP and some binding protein within synaptic vesicles has been demonstrated, and it is possible that these substances may play some role in the ACh concentration within the vesicle.

Since the time of the discovery of the synaptic vesicles. experiments have been carried out to demonstrate their possible role in serve transmission. Axotomy resulted in an early degeneration of the endings, with clumping and lysis of the esicles. These alterations coincide with the time in which nerve transmission is interrupted.

Electrical stimulation of the cholinergic nerve endings of the adrenal medulla at high freqencies resulted in a reduction in the number of vesicles. This finding has now been confirmed by numerous authors in the myoneural junction and in electric tissue.

In *Torpedo* electroplax stimulated to fatigue there is a loss of about 50 per cent of the vesicles; at the same time the presynaptic membrane increases, forming finger-like protrusions, probably corresponding to the integration of the vesicle membranes. An interesting finding is that a certain number of synaptic vesicles in the control synapsis of the electroplax and in the myoneural junction, when fixed in the presence of Ca^{2+}, show a small granule attached to the membrane, which corresponds to calcium binding. When the synapsis in intensely stimulated, the vesicles that carry this Ca^{2+} granule are preferentially released. These findings are related to the role of calcium in transmtter release.

Recycling of Synaptic Vesicles

One of the important discussed problems of transmitter release is whether it involves exclusively a mechanism of exocytosis or whether the synaptic vesicles are reused again and again. In the releasing sites of the membrane would be temporary, and the vesicle could be recharged with transmitter for reuse. There are arguments that favour both sides of the discussio, and probably the two mechanisms are both at work.

Quantitative work done on crayfish motor neurons has led to the proposal that there must be vesicle reuse *i.e.*, retention of the intact vesicle membrane within the axon terminal) or a mechanism by which incorporation of the vesicle membrane is balanced with an equivalent withdrawal. Related to these mechanisms is the finding that newly synthesized transmitter or transmitter recaptured from the cleft, as occurs in adrenergic synapses, is preferentially repool of transmitter (up to 99 per cent) that can be depleted with reserpine without failure in synaptic transmission, while a small pool is essential.

In synapses of *Torpedo* electroplax there are apparently two populations of cholinergic vesicles, one with a higher ratio of ACh/ATP, the so-called *supercharged vesicles,* and the other with a lower ratio. It is been suggested that the supercharged vesicles are associated with the presynaptic membrane and upon stimulation are released first. As was mentioned earlier, it seems possible that the dense presynaptic projections play a role in the transport of the vesicles to the sites of release.

It has been postulated that newly formed synaptic vesicles are capable of several cycles of secretion and reloading before they undergo exocytosis and are incorporated into the membrane. However, the existence of a recycling mechanism for the vesicles is now generally accepted. It is possible that after several rounds of reuse the vesicle is incorporated into the presynaptic membrane and subsequently recaptured.

This could y done by direct endocytosis and formation of new vesicles or by intermediary membranous structures (*e.g.*, by way of the smooth endoplasmic reticulum). Frequently, coated vesicles are re-formed in this process. Recycling of vesicle membrane is an energy-saving mechanism that avoids the problem of constant synthesis, with all the overwork that this implies for the neuron.

Coupling between Depolarization

The coupling between the depolarization caused by the arrival of the nerve impulse, and the secretion of acetylcholine is mediated by the influx of calcium ions. Extracellular C^{2+} is an absolute requirement for the release of ACh from the nerve terminal; in the absence of calcium, there is no release of ACh. When the nerve impulse arrives at the terminal, there is an increase in Ca^{2+} permeability resulting from the depolarization. The entry of calcium into the ending has been demonstrated by ^{45}Ca and the use of acquiring, a substance that fluoresces in the presence of Ca^{2+}. It has been assumed that for the release of each quantum of ACh, one to four Ca^{2+} molecules must enter, and that in the release of ACh there is a cooperative action of Ca^{2+}.

Once the Ca^{2+} has penetrated the terminal, the secretion of ACh is probably produced by the opening of the synaptic vesicles that were attached to the presynaptic membrane. Physiologic studies have demonstrated that *synaptic delay* that takes place between the arrival of the nerve impulse and the production of the synaptic potential is due mainly to the interval between depolarization and actual secretion of the transmitter.

We mentioned earlier that synaptic vesicles carry a binding sitefor Ca^{2+} in the membrane. Such a site may be important in the actual release of the transmitter, since the entrance of C^{2+} at this point could determine the increased porosity of the vesicle.

Physiologic Response

Since the beginning of this century, through the work of Langley, Eangley, Ehrlich, and other it has been postulated that the transmiter interacts with a specific receptor localized at the chemosensitive sites of the cell membrane. For many years, however, the knowledge of synaptic receptors has been principally indirect —based on the final response obtained from this interaction. For example, it was known that acetylcholine produced contranction of skeletal muscle or the secretion of a gland, and that these two effects could be blocked with curare and atropine, respectively.

In 1955 Nachmanson had already postulated that the "acetylcholine receptor is a protein that upon the binding of ACh undergoes a conformational transition which results in changes in membrane permeability." shows that the primary transmitterreceptor interaction is coupled, probably by way of a conformational change, to the translocation of inons across the cell membrane (*i.e.*, the ionophore), and by several interposed mechanisms can produce a final response. The probable relationship between the primary interaction, the displacement of Ca^{2+}, and certain metabolic processes involving cyclic AMP and other substances.

Lipoprotein Framework of the Membrane

Only in recent years has it been possible to separate receptor proteins as chemical entities and to analyse, in more direct way, their interaction with the transmitter, The separation of these substances has bee difficult because they are intrinsic proteins which require strong treatments to be separated from the lipoprotein framework of the membrane. Furthermore, they are present in at the membranes should first be separated by cell fractionation methods. The receptor molecules should then be isolated from the membrane. After isolation, the receptor protein should snow a specific high-affinity binding for the neurotransmitter. Finally, it should be able to undergo conformational changes that are capable of inducing the translocation of ions and of producing a bioelectrical response.

Receptor proteins are highly hydrophobic and intimately related to lipids of the membrane. Tow main procedures are used to isolate these proteins; extraction with organic solvents and separation by detergents. Cholinergic receptor proteins have been isolated not only from the electroplax of *Torpedo* and *Electrophorese* tissues, whic have the richest cholinergic innervation, but also from skeletal muscle, smooth muscle, and brain.

Adrenergic receptor proteins have been separated from brain, from spleen capsule, and from heart. Receptor proteins for the amino acids glutamate and y-aminobutyrate have also been isolated from muscle of *Crustacea* and from insects which have a double innervation: excitatory (via glutamate) and inhibitory (via y-aminobutyrate). From synaptic membranes of the cerebral cortex, hydrophobic proteins binding L-glutamate, L-aspartate, and y-aminobutyrate have been separated. After extraction with an organic solvent or detergent, the receptor proteins are generally separated by column chromatography or in the case of the cholinergic receptor, by affinity chromatography, In the latter technique, the gel used contains an active group attached by a long arm. This group specifically binds the receptor proteins and is then eluted by a pulse of the corresponding transmitter. In this case, the protein peak, which appears after the ACh pulse, is the specific receptor protein fraction. With this procedure a total purification of about 15,000 times has been achieved using rat diaphragm.

Complete purification of the nicotinic acetylcholine receptor has been obtained in the *Torpedo* electroplax which is the rechest source of this protein. After affinity chromatography of detergent-extracted receptor, four polypeptide chains, ranging between 39,000 and 64,000 daltons have been separated by SDS gel electrophoresis. These subunits from a complex of about 27,000 daltons. In is known that the small subunit of 39,000 daltons is the one that a second subunit of about 43,000 daltons could represent the ionophore, *i.e.*, the subunit involved in the translocation of ions.

Acetylcholine Receptor

Our previous consideration of the transmitter-receptor interaction suggests that the transduction of energy at the synapse is the result of a conformational change in the receptor macroconformational change in the receptor macromolecule and that such a change is coupled with the translocation of inos across the membranes. In the case of acetylcholine, these mechanisms can be expressed as shown in the diagram that follows. In this equation AChRc represents a reversible complex resulting from the binding of ACh to the ionophoric mechanism operates, producing the opening of the channel for the translocation of the ions (AChRo).

It is important to stress here the difference between the axonal channels we studied previously and the postsynaptic channels that are related to the cholinergic receptor. While in the axon there are separate Na^+ and K^+ channels which are regulated by the membrane potential, in the postsynaptic membrane both inons probably use a common channel, and the gating is driven by the chemical interaction of the neurotransmitter with the receptor protein. By the use of fine microelectrodes in the myoneural junction, some of the best evidence of the functioning of receptors at the molecular level has been obtained.

It was mentioned before that in this type of preparation it is possible to record the spontaneous miniature end plate potentials (mepp), which have an amplitude of about 0.1 mV and represent the discharge of single multi-molecular quanta of ACh. Katz and Miledi have observed that if a minimal but steady dose of acetylcholine is applied to the myoneural junction, there are minute fluctuations of the membrane potentials that are superimposed upon a small but steady depolarization of the membrane the amplitude of these fluctuations is several hundred times smaller than the mepp (*i.e.*, on the order of 0.3 μV) and constituted the so-called *membrane noise.*

Reconstituted Cholinergic Receptor

Many attempts to reconstitute a functional excitable membrane from cholinergic receptor protein extracted by detergents have generally given negative results; *i.e.*, in most cases the model systems were unreactive to the action of the neurotransmitter.The advantage of having a receptor isolated in organic solvent is that it can be directly incorporated into planar lipid bilayer or into a liposome. As the cholinergic receptor is incorporated into a lipid bilayer and acetylcholine is applied with a micropipette, there is a sudden increase in the level of conductance, which is accompanied by current jumps in the same range of amplitude as those observed in the muscle membrane.

These fluctuations disappear when the membrane voltage is reduced to zero or by addition of the blocking agent D-tubocurarine, and may be considered as an increase in membrane noise. These and other experiments, in which the protein was incorporated into liposomes, suggest that the cholinergic receptor isolated by organic solvents may be reconstituted in the artificial membrane in a form that is "reactive to cholinergic drugs.

Cholinergic Receptor

Based on the hydrophobic properties of the synaptic receptor proteins and on studies of reconstitution is artificial membranes, a tentative model of the integration of the cholinergic receptor was postulated. The most interesting features of this model are the oligomeric arrangement of the protein subunits and their disposition traversing the postsynaptic membrane. The receptor, with the binding site for acetylcholine, is shown as a macromolecule different from the ionophore, as some recent studies suggest. However a tight coupling between the two must exist. The interaction of the transmitter with the receptor produces a conformational change which could be transmitted to the coupled ionophore, leading to the opening of a hydrophilic channel through which $Na^+$and K^+ could flow. It is possible that the ionohore or ion-conducting mechanism also resides in a hydrophobic protein (*i.e.*, proteolipid), since this is the case for most channel-forming macromolecules. Because the ACh-receptor interaction is reversible, it may easily go back towards the closed position once the ACh has been removed from the site of binding. Such a reversible effect would be facilitated by the fact that the receptor molecules are held in place by hydrophobic interactions within the framework of the membrane.

Use of a Second Messenger

The description given so far of the neurotransmitter-receptor interaction involves the direct translocation of ions. As we have seen in the case of the nicotinic, this translocation is a very fast phenomenon, occurring in a few milliseconds or less. The interaction may involve metabolic changes in the neuron which involve a transient but longer-lasting physiologic effect.

A section was dedicated to discussion of the action of hormones on non-neuronal tissues, and this involves the intervention of specific receptors and the enzyme adenylate cyclase, discovered by Sutherland in 1956.The specific hormone receptor is coupled within the membrane with adenylate cyclase. When the hormone interacts with the receptor, the enzyme is activated and produces 3'-5' *cylic AMP (cAMP)* from ATP. This nucleotide is considered a *second messenger* that is able to change the metabolism of the cell.

The classic example of the so-called *cascade effect* of cAMP on glycogenolysis is indicated. It may be observed that under the action of the hormones, epinephrine or glucagon adenylate cyclase is activated, increasing the levels of cAMP. This in turn activates *protein kinase,* a phosphorylating enzyme, which converts phosphorylase kinase from an inactive into an active form, which will finally degrade the glycogen.

The first indication that adenylate cyclase could be involved in synaptic function was provided in 1967 with the demonstration that this enzyme was concentrated in synaprosomes and was found mainly in the synaptosomal membrane.

It is interesting that after osmotic shock of the synatosome, the enzyme activity was found to be increased, indicating that the active groups are facing towards the inside of the synaptosomal membrane. In the same work it was found that the synaptosome also contains high levels of *cyclic phosphodiesterase,* the enzyme that degrades cAMP into inactive AMP. This enzyme was found to be partially soluble, but a high per cent was particulated and was also found in the synaptosomal membrane. Cytochemical studies demonstrated later that phosphodiesterase is localized mainly at the postsynaptic side of the synapse. At the present time it is thought that several neurotransmitter-receptor interactions that generate slow postsynaptic potentials (i. e., on the order of 100 to 500 msec) or that produce a modulation effect on synaptic function involve a second messenger mechanism, with the production of cAMP. More recently, another second messenger, *cyclic GMP,* was found in brain, and it is produced by the enzyme *guanylate cyclase.* In contrast to adenylate cytosol.

Recently, six putative neurotransmitters have been shown to stimulate adenylate cyclase specifically (*i.e.*, dopamine, norepinephrine, serotonin, histamine, octopamine and enkephalin). Cyclic GMP may mediate the effects of acetylcholine by way of muscarinic receptors, and of histamine and norepinephrine, acting on receptors that are different from those associated with adenylate cyclase.

The levels of cAMP or cGMP depend not only on the activation of the corresponding enzymes but also on the action of phosphodiesterase, which apparently acts on both nucleotides. For example, if phosphodiesterase is inhibited by methylxanthines (such as caffeine) or other drugs, the leaves of these cyclic nucleotides increase. Phosphodiesterase and adenylate cyclase are regulated by Ca^{2+} by way of a *Ca^{2-}binding protein.* The link between cAMP and synaptic transmission was more firmly established when it was found that the inontophoretic application of either norepinephrine or cAMP upon the Purkinje cells of the cerebellum slowed the firing of these neurons. Furthermore, produced a large increase of cAMP within these cells.

Phosphorylation of Proteins

As in the control of glucogenolysis in liver cAMP phosphorylation of proteins also plays an important role in the central nervous system. In the synaptosomal fraction there is complete system for the phosphorylation and dephosphorylation of proteins (*i.e.*, the covalent attachment and detachment of phosphate in proteins). This system is composed of specific cAMP-*dependent protein kinase,* with the membrane-bound proteins acting as substrate and a specific *phosphatase* reversing the reaction.

The protein kinase is thought to be activated by the binding of cAMP to an inhibitory subunit, causing it to dissociate from the catalytic subunit. In this way the activated enzyme readily transfers a phosphate group to the substrate protein. It was found that of the many proteins of the neuronal membrane only two or three were readily phosphorylated by the cAMP. One of these, designated *protein I,* has two subunits of 86,000 and 80,000 daltons, and it is present almost exclusively in the synaptic region, being absent from non-neuronal tissue and from fetal stages before the formation of the connections between neurons.

Phosphorylation of protein I is extremely fast and may reach a maximum in five seconds. It has been suggested that this special phosphorylation process, mediated by cAMP, could be responsible for the permeability of the membrane to ions; however, at the present time it is difficult to establish whether there is a real cause-and-effect relationship, in addition to the membrane-bound phosphorylating system, it should by noted that there are also soluble cAMP-dependent protein kineses which may phosphorylate other proteins in the cytosol, such as the tubulin of microtubules and tyrosine hydroxylase.

We mentioned earlier that in cyclic AMP-mediated synapses the physiologic events may last 100 milliseconds or more. There are indications in mollusc ganglion cells and in the sympatheic system of vertebrates that these events may be much longer, lasting for hours.

This has led to the hypothesis that synaptic events mediated by cyclic nucleotides could be related to long-term changes in the nervous system and could play role in memory.

Phosphorylation of regulatory proteins within the nucleus of the neuron could be involved in gene expression by the transcription of RNA from certain DNA genes. This could lead to translation into important proteins, such as enzymes or receptors, and to more permanent changes in the neuron.

From the medical viewpoint these studies are important because some receptors using cyclic nucleotides (*e.g.*, dopamine, B-adrenergic and H_2-histamine receptors) appear to be the targets of some of the psychoactive drugs used in the treatment of mental illness. Furthermore, some of these receptors are altered in neurologic diseases, such as parkinsonism and Huntington's chorea.

4

Plant Molecular Physiology

Plant Cells

The Plant Kingdom

Since the time of Aristotle (384–322 B.C.E.), biologists have sought to classify organisms. At first the purpose was ease of identification ("artificial" classification schemes).

Carolus Linnaeus (1707–1778), arguably the greatest of the pre-modern Naturalists, sought to classify plants and other organisms according to affinity groups that reflected the mind of the Creator. Later, after Darwin, the goal of classification was to show evolutionary relationships ("natural" classification schemes).

For the past 150 years, biologists have emphasized natural systems of classification and have attempted to define morphological criteria that reveal evolutionary relationships.

We now know that morphology, the form and structure of organisms, is the end product of the actions of genes. Virtually all of the information needed to form a complete organism is encoded in its DNA sequences, both nuclear and cytoplasmic (mitochondria and chloroplasts). DNA sequence analysis has thus provided evolutionary biologists with a powerful new tool for arriving at a truly natural classification system.

On the basis of phylogenetic analyses of highly conserved DNA sequences, living organisms have been divided into three major domains: Bacteria, Archaea, and Eucarya.

1. The common ancestor of all the organisms first gave rise to the Bacteria and the common ancestor of the Archaea and the Eucarya.
2. The Archaea branch off from the Eucarya lineage.
3. The Eucarya common ancestor acquires mitochondrial endosymbiont.
4. A heterogeneous group of eukaryotes called protists branch off the lineage leading to plants, fungi, and animals.
5. The common ancestor of fungi and animals form a branch, followed by a divergence into the fungal and animal lineages.
6. The common ancestor of plants, green algae, red algae, and glaucophytes acquires chloroplast endosymbiont (a cyanobacterium).
7. The three lineages of glaucophytes, red algae, and green algae diverge.
8. Various lineages of protists acquire chloroplasts via green or red algal endosymbionts.
9. The earliest branch of green algae diverge.
10. The later branch of green algae diverge.
11. The remaining lineage leads to plants.

The Eucarya include the eukaryotes, organisms whose cells contain a true nucleus. The Bacteria, or eubacteria, which include the cyanobacteria, lack a true nucleus and are therefore prokaryotic.

The Archaea, or archaebacteria, are also prokaryotic, but they differ from the Bacteria:

Besides their morphological and biochemical differences, they are often adapted to extreme environments, such as sulphur hot springs or saline ponds. Phylogenetic studies have indicated that the Archaea and Eucarya split after the Bacteria separated from the common ancestor. Thus Archaea and Eucarya represent sister groups. This closer relation between Archaea and Eucarya is reflected in their similar promoter structures and RNA polymerases, the presence of histones, and many other characteristics.

Fungi were formerly classified as algal-like plants that had lost their chloroplasts. However, fungi and animals branched off from the Eucarya lineage *before* the appearance of choloroplasts. They are thus more closely related to animals than to plants. Fungi are heterotrophic; that is, they depend on other organisms for their food, and they satisfy their nutritional needs by absorbing inorganic ions and organic molecules from the external environment. Most fungal species are filamentous and possess cell walls made of chitin, the same substance that is found in insect exoskeletons. Bryophytes are small (rarely more than 4 cm in height), very simple land plants, and the least abundant in terms of number of species and overall population. Bryophytes do not appear to be in the direct line of evolution leading to the vascular plants; rather, they seem to constitute a separate minor branch. Bryophytes include mosses, liverworts, and hornworts. These small plants have life cycles that depend on water during the sexual phase. Water facilitates fertilization, the fusion of gametes to produce a diploid zygote, a feature also seen in the algal precursors of these plants.

Bryophytes are like algae in other respects as well: They have neither true roots nor true leaves, they lack a vascular system, and they produce no hard tissues for structural support. The absence of these structures that are important for growth on land greatly restricts the potential size of bryophytes, which, unlike algae, are terrestrial rather than aquatic.

The ferns represent the largest group of spore-bearing vascular plants. In contrast to the bryophytes, ferns have true roots, leaves, and vascular tissues, and they produce hard tissues for support.

These architectural features enable ferns to grow to the size of small trees. Although ferns are better adapted to the drying conditions of terrestrial life than bryophytes are, they still depend on water as a medium for the movement of sperm to the egg. This dependence on water during a critical stage of their life cycle restricts the ecological range of ferns to relatively moist habitats.

The most successful terrestrial plants are the seed plants. Seed plants have been able to adapt to an extraordinary range of habitats. The embryo, protected and nourished inside the seed, is able to survive in a dormant state during unfavourable growing conditions such as drought. Seed dispersal also facilitates the dissemination of the embryos away from the parent plant.

Another important feature of seed plants is their mode of fertilization. Fertilization in seed plants is brought about by wind-or insect-mediated transfer of pollen, the gamete-producing structure of the male, to the sexual structure of the female, the pistil. Pollination is independent of external water, a distinct advantage in terrestrial environments. Many seed plants produce massive amounts of woody tissues, which enable them to grow to extraordinary heights. These features of seed plants have contributed to their success and account for their wide range.

There are two categories of seed plants: gymnosperms (from the Greek for "naked seed") and angiosperms (based on the Greek for "vessel seed," or seeds contained in a vessel). Gymnosperms are the less advanced type; about 700 species of gymnosperms are known. The largest group of gymnosperms is the conifers ("cone-bearers"), which include such commercially important forest trees as pine, fir, spruce, and redwood. Two types of cones are present: male cones, which produce pollen, and female cones, which bear ovules. The ovules are located on the surfaces of specialized structures called cone scales.

After wind-mediated pollination, the sperm reaches the egg via a pollen tube, and the fertilized egg develops into an embryo. Upon maturation, the cone scales, which are appressed during early development, separate from each other, allowing the naked seeds to fall to the ground.

Angiosperms, the more advanced type of seed plant, first became abundant during the Cretaceous period, about 100 million years ago. Today, they dominate the landscape, easily outcompeting their cousins, the gymnosperms.

About 250,000 species are known, but many more remain to be characterized. A typical angiosperm life cycle, that of *Zea mays* (corn). The major innovation of the angiosperms is the flower; hence they are referred to as *flowering plants.* There are other anatomical differences between angiosperms and gymnosperms, but none so crucial and far-reaching as the mode of reproduction.

Flower Structure and the Angiosperm Life Cycle

The flower consists of several leaflike structures attached to a specialized region of the stem called the receptacle. Sepals and petals are the most leaflike. Petals have the primary function of attracting insects to serve as pollinators, accounting for their often showy and brightly coloured appearance.

The stamen is the male sexual structure, and the pistil is the female sexual structure. The pistil is composed of one or more united carpels; the pistil, or in some flowers a whorl of pistils, is sometimes referred to as the gynoecium. The stamen consists of a narrow stalk called the filament and a chambered structure called the anther.

The anther contains tissue that gives rise to pollen grains. The pistil consists of the stigma (the tip where pollen lands during pollination), the style (an elongated structure), and the ovary. The ovary, the hollow basal portion of the pistil, completely encloses one or more ovules. Each ovule, in turn, contains an embryo sac, the structure that gives rise to the female gamete, the egg.

After landing on the stigma, the pollen grain germinates to form a long pollen tube, which penetrates the tissues of the style and ultimately enters the cavity of the ovary, which houses the ovule. Within the ovary, the pollen tube enters the ovule and deposits two haploid sperm cells in the embryo sac. One sperm cell fuses with the egg to produce the zygote; the other typically fuses with the two polar nuclei to produce a specialized storage tissue termed the endosperm, which provides nutrients to the growing embryo.

Endosperm tissue also provides the bulk of the worldrquotes food supply in the form of cereal grains. As in conifers, in angiosperms the outer tissues of the ovule harden into a protective seed coat. Angiosperm seeds have a second layer of protective tissues, the fruit. The fruit consists of the ovary wall and, in some cases, receptacle tissue.

Angiosperms are divided into two major groups, dicotyledons (dicots) and monocotyledons (monocots). This distinction is based primarily on the number of cotyledons, or seed leaves. In addition, the two groups differ with respect to other anatomical features, such as the arrangement of their vascular tissues, and their floral structure.

As the dominant plant group on Earth, and because of their great economic and agricultural importance, angiosperms have been studied much more intensively than other types of plants. Plant physiologists have focused on a relatively small number of species that represent convenient experimental systems for the study of specific phenomena.

Therefore, while we focus on these famous few, it is important to keep in mind the tremendous diversity of form and function that exists within the angiosperms, and the even greater diversity of form and function that is found within the plant kingdom as a whole.

Plant Tissue Systems: Dermal, Ground, and Vascular

Dermal tissue. The epidermis is the dermal tissue of young plants undergoing primary growth. It is generally composed of specialized, flattened polygonal cells that occur on all plant surfaces. Shoot surfaces are usually coated with a waxy cuticle to prevent water loss and are often covered with hairs, or trichomes, which are epidermal cell extensions.

Pairs of specialized epidermal cells, the guard cells, are found surrounding microscopic pores in all leaves. The guard cells and pores are called stomata (singular stoma), and they permit gas exchange (water loss, CO_2 uptake, and O_2 release or uptake) between the atmosphere and the interior of the leaf. The root epidermis is adapted for absorption of water and minerals, and its outer wall surface typically does not have a waxy cuticle. Extensions from the root epidermal cells, the root hairs, increase the surface area over which absorption can take place.

Ground tissue. Making up the bulk of the plant are cells termed the ground tissue. There are three types of ground tissue: parenchyma, collenchyma, and sclerenchyma.

- Parenchyma, the most abundant ground tissue, consists of thin-walled, metabolically active cells that carry out a variety of functions in the plant including photosynthesis and storage.
- Collenchyma tissue is composed of narrow, elongated cells with thick primary walls. Collenchyma cells provide structural support to the growing plant body, particularly shoots, and their thickened walls are nonlignified so they can stretch as the organ elongates. Collenchyma cells are typically arranged in bundles or layers near the periphery of stems or leaf petioles.

- Sclerenchyma consists of two types of cells, sclereids and fibres Both have thick secondary walls and are frequently dead at maturity. Sclereids occur in a variety of shapes, ranging from roughly spherical to branched, and are widely distributed throughout the plant. In contrast, fibres are narrow, elongated cells that are commonly associated with vascular tissues. The main function of sclerenchyma is to provide mechanical support particularly to parts of the plant that are no longer elongating.

In the stem, the pith and the cortex make up the ground tissue. The pith is located within the cylinder of vascular tissue, where it often exhibits a spongy texture because of the presence of large intercellular air spaces.

If the growth of the pith fails to keep up with that of the surrounding tissues, the pith may degenerate, producing a hollow stem. In general, roots lack piths, although there are exceptions to this rule. In contrast, the cortex, which is located between the epidermis and the vascular cylinder, is present in both stems and roots. At the boundary between the ground tissue and the vascular tissue in roots, and occasionally in stems, is a specialized layer of cortex known as the endodermis. This single layer of cells originates from cortical tissue at the innermost layer of the root cortex and forms a cylinder that surrounds the central vascular tissue, or stele.

Early in root development, a narrow band composed of the waxy substance suberin is formed in the cell walls circumscribing each endodermal cell. These suberin deposits, called Casparian strips, form a barrier in the endodermal walls to the intercellular movement of water, ions, and other water-soluble solutes to the vascular cells. Leaves have two interior layers of ground tissue that are collectively known as the mesophyll.

The palisade parenchyma consists of closely spaced, columnar cells located beneath the upper epidermis. There is usually one layer of palisade parenchyma in the leaf. Palisade parenchyma cells are rich in chloroplasts and are a primary site of photosynthesis in the leaf. Below the palisade parenchyma are i rregularly shaped, widely spaced spongy mesophyll cells. The spongy mesophyll cells are also photosynthetic, and the large spaces between these cells allow diffusion of carbon dioxide. The spongy mesophyll also contributes to leaf flexibility in the wind, and this flexibility facilitates the movement of gases within the leaf.

Vascular tissues: xylem and phloem. The vascular tissue is composed of two major conducting systems: the xylem and the phloem. The xylem transports water and mineral ions from the root to the rest of the plant.

The phloem distributes the products of photosynthesis and a variety of other solutes throughout the plant. The tracheids and vessel elements are the conducting cells of the xylem.

Both of these cell types have elaborate secondary-wall thickenings and lose their cytoplasm at maturity; that is, they are dead when functional. Tracheids overlap each other, whereas vessel elements have open end walls and are arranged end to end to form a larger unit called a vessel. Other cell types present in the xylem include parenchyma cells, which are important for the storage of energy-rich molecules and phenolic compounds, and sclerenchyma fibres.

The sieve elements and sieve cells are responsible for sugar translocation in the phloem. The former are found in angiosperms; the latter perform the same function in gymnosperms. Like vessel elements, sieve elements are often stacked in vertical rows, forming larger units called sieve tubes, whereas sieve cells form overlapping arrays. Both types of conducting cells are living when functional, but they lack nuclei and central vacuoles and have relatively few cytoplasmic organelles.

Substances are translocated from sieve cell to sieve cell laterally through circular or oval zones containing enlarged pores, called sieve areas. In contrast, sieve tubes translocate substances through large pores in the end walls of the sieve elements, called sieve plates. Sugar movement through sieve tubes is more efficient and rapid than through sieve cells and represents a more evolutionarily advanced mechanism.

Sieve elements are associated with, and depend on, densely cytoplasmic parenchyma cells called companion cells. The analogous cells adjacent to the sieve cells of gymnosperms are called albuminous cells. Companion cells provide proteins and metabolites necessary for the functions of the sieve tube elements. In addition, the phloem frequently contains storage parenchyma and fibres that provide mechanical support.

The Structures of Chloroplast Glycosylglycerides

The structures of the three plant glycosylglycerides, monogalactosyldiacylglycerol (MGDG), digalactosyldiacylglycerol (DGDG) and sulfoquinovosyldiacylglycerol (SQDG). The two galactolipids, MGDG and DGDG, are uncharged but polar; the sulfolipid, SQDG, is negatively charged as well as being polar.

The Multiple Steps in Construction of the Cell Plate Following Mitosis

During Mitosis in plants, the daughter chromosomes separate during anaphase and migrate toward their respective poles. In contrast to animal cells, the spindle poles of plant cells is quite diffuse, so that the chromosomes move in nearly parallel fashion. At the end of teleophase, cytokinesis begins with the formation of the cell plate. This process involves fusion of many small secretory vesicles, and attachment of the resulting structure to the plasma membrane. A diagram of the steps involved in cell plate formation.

In the first step, Golgi vesicles, some of which are interconnected via *fusion tubes,* aggregate in the spindle midzone area. This structure is called the fusion tube network (FTN). The transition from the first to the second stage of cell plate formation can be inhibited by caffeine. (B) Formation of a tubulo-vesicular network (TVN).

The contents of the vesicles, mainly pectins, represent the precursors from which the new middle lamella is assembled outside the cell. In the next stage, vesicle fusion increases, forming a tubulo-vesicular network (TVN), and the membranes become coated with either clathrin or other proteins. (C) In the third stage, the central region of the growing cell plate forms a tubular network (TN), with vesicle fusion occurring at the growing edges where the remaining microtubules are located. (D) In the final stage, the cell plate contacts and adheres to the plasma membrane of the parent cell. At the same time the tubular network expands to form a fenestrated sheet. (E) At the end of mitosis, the phragmoplast disappears, the cell enters interphase, and microtubules reappear in the cytosol near the plasma membrane, where they play a role in the deposition of cellulose microfibrils during cell wall growth. (F) Electron micrograph of a cell plate forming in a root tip of a beet, (*Beta vulgaris*) (10,000×) MT, microtubule; VE, vesicles; N, nucleus; NE, nuclear envelope; P, cell plate.

Water and Plant Cells

Calculating Capillary Rise

We learn on page 36 of the textbook that the water properties of cohesion, adhesion and surface tension, give rise to the phenomenon of capillarity, the movement of water for small distances up a capillary tube. The smaller the tube radius, the higher the capillary rise. How far the water will move may be calculated using the following formula:

$$\text{Capillary rise} = \frac{14.9 \times 10^{-6}\,\text{m}^2}{\text{radius}}$$

where both capillary rise and radius are expressed in meters.

For a xylem vessel with 25 μm radius, the capillary rise is about 0.6 m.This distance is much too small to be significant for water transport up tall trees.

Fibrous materials such as cell walls can act like wicks to draw water by capillarity from nearby xylem.

This capillary action ensures that cell wall surfaces that are directly exposed to the air, such as those in leaf mesophyll, remain wetted and do not dry out. Because the cell wall capillaries have a tiny radius, about 10 m^{-8}, very large physical forces can be generated in the water just below the evaporative surfaces of cell walls.

Calculating Half-Times of Diffusion

Extremely Slow over Long Distances

From Fick's law, one can derive an expression for the time it takes for a substance to diffuse a particular distance. If one defines conditions such that all the solute molecules are concentrated at the starting position, then the concentration front moves away from the starting position, as shown for a later time point. As the substance diffuses away from the starting point, the concentration gradient becomes less steep (Δc_s decreases) and thus net movement becomes slower. The time it takes for the substance at any given distance from the starting point to reach one-half of the concentration at the starting point (t_c = ½) is given by the following equation:

$$t_{c=1/2} = \frac{(\text{dis tan ce})^2}{D_S} K$$

where K is a constant and D_s is the diffusion coefficient. The above equation shows that the time required for a substance to diffuse a given distance increases in proportion to the *square* of that distance. Let us consider two numerical examples. First, how long it would take a small molecule to diffuse across a typical cell? The diffusion coefficient for a small molecule like glucose is about 10^{-9} m^2 s^{-1}, and the cell length may be 50 μm. Thus, for this example:

$$t_{c=1/2} = \frac{(50+10^{-6}\text{m})^2}{10^{-9}\text{m}^2\text{s}^{-1}} = 2.5\text{s}$$

This calculation shows that small molecules diffuse over cellular dimensions rapidly. What about diffusion over longer distances? Calculating the time needed for the same substance to move a distance of 1 m (e.g., the length of a corn leaf), we find:

$$t_{c=1/2} = \frac{(1\text{m})^2}{10^{-9}\text{m}^2\text{s}^{-1}} = 10^9\text{s} \approx 32\,\text{year}$$

a value that exceeds by orders of magnitude the life span of a corn plant, which lives only a few months. This shows that diffusion in solutions can be effective within cellular dimensions but is far too slow for mass transport over long distances. Diffusion is of great importance as a driving force for the water vapour lost from leaves, because the diffusion coefficient for a molecule in air is much greater than in aqueous solutions.

Alternative Conventions for Components of Water Potential

Students planning further study of plant water relations should note that the components of water potential defined in the text are sometimes given different names and symbols. In particular, the equation

$$\psi_w = \psi_s + \psi_p$$

is often replaced by the following equivalent equation:

$$\psi_w = \pi + P$$

In this alternative convention, *P* is the same as Ψ_p. It is the hydrostatic pressure of the solution, and may be positive, as in turgid cells, or negative, as in xylem water.

The symbol π is called osmotic pressure and is the negative of Ψ_s. That is, π has positive values, and Ψ_s has negative values. “Osmotic pressure” is the term that physical chemists, zoologists, and many others use to denote the effect of dissolved solutes on the free energy of water. Most handbooks of physics and chemistry use the term “osmotic pressure” and the symbol π.

The negative sign in front of π in the equation above accounts for the reduction in water potential (Ψ_w) by dissolved solutes. Thus $\Psi_s = -\pi$. A very interesting, if somewhat unconventional, account of the history and physical meaning of osmotic pressure is given by Hammel and Scholander (1976).

Unfortunately, some authors have mixed the conventions for Ψ_s and π, leading to unnecessary confusion about what is meant by the symbol π. Thus, π is sometimes incorrectly called osmotic potential instead of osmotic pressure, and it may be used either as a positive quantity or as a negative quantity. Regarding matric potential, it is usually designated by the symbol *t*.

Can Negative Turgor Pressures Exist in Living Cells?

One of the challenging aspects of understanding plant water relations is the remarkable range of pressures, both positive and negative, that occur within the bodies of plants.

Negative pressures (or tensions), which depend upon the cohesive strength of water coupled with the strength of lignified cell walls to resist deformation, play an important role in water transport through the xylem. Positive pressures, which depend upon the semipermeable nature of the plasma membrane and the elastic nature of primary cell walls, occur in all hydrated living plants cells but can be especially large in sieve tubes and guard cells.

Living plant cells are typically assumed to have only positive pressures. However, there appears to be no reason that negative pressures could not also occur within the cytoplasm of living plant cells. This web topic explores the existence and potential role of negative turgor pressures in plants. When a cell loses water in air, turgor declines and solute concentrations increase. At the turgor loss point ($\psi_p = 0$), the hydrostatic pressure in the cell sap is equal to atmospheric pressure, meaning that no net force is exerted on the cell wall.

If water continues to be lost from the cell, the pressure within the cytoplasm drops below atmospheric pressure, resulting in a force imbalance that collapses the cell wall. The deformation of living cells upon desiccation is called cytorrhysis. Note that the plasma membrane is pressed against the cell wall throughout desiccation (i.e., plasmolysis does not occur) because the hydrostatic pressure in cytoplasm remains greater than the hydrostatic pressure in apoplast. In the living cells described above, a decrease in cell water potential below the turgor loss point is balanced by an increase in solute concentration as the volume of the cell decreases.

Thus, true "negative" pressures do not develop. In contrast, xylem conduits have rigid cell walls that resist deformation, allowing them to sustain negative pressures without imploding. This raises the question of what happens when water is lost from living cells that have thick walls or are embedded in a rigid matrix of cells (e.g., xylem parenchyma). Might these cells resist deformation and thus develop negative turgor pressures?

It is important to emphasize that this is a controversial area, due in large part to the absence of direct measurements of negative turgor pressures within cells. However, before reviewing the evidence for negative turgor pressures, it is worthwhile to consider what physiological effects might result from the development of such negative pressures in living cells.

The major outcome of negative turgor is that it allows stiff-walled cells to decrease in water potential without undergoing major changes in cell volume or osmotic concentration. Because cytorrhysis might cause physical damage to the wall and/or cell membranes, while the high concentrations of solutes resulting from the reduction in cell volume might adversely affect the conformation of membranes and proteins, it is possible to imagine a physiological role or benefit for the development of negative turgor pressures in plant cells.

We now turn this around and ask if there are any downsides to the generation of negative turgor pressures in living cells? In the xylem, the primary issue associated with negative pressures is cavitation; could this also be a concern for living cells? The primary mechanism for cavitation in plants is air seeding, reflecting the fact that the probability of the de novo formation of gas voids in water (either by homogeneous or heterogeneous nucleation) is extremely low. In air seeding, air is sucked in through the cell wall, where it then "nucleates" the transition to the vapour phase.

For air seeding to occur in a living cell experiencing negative pressures, air would have to be pulled through the very small pores of the cell wall. While one can imagine this happening, the movement of air across the cell wall would result in plasmolysis and thus the immediate release of any tension in the cytoplasm because the plasma membrane is not capable of withstanding any significant pressure.

Other costs associated with the ability of living cells to generate negative pressures include the metabolic costs of producing rigid cell walls, as well as any limitations lignified walls might place on physiological function. In addition, a strategy for avoiding cell damage due to desiccation via the generation of negative pressures might impose limitations on cell size. The strength of a cell to withstand collapse (and thus generate negative pressures) is inversely proportional to cell size. Evidence for negative turgor pressure in plants is limited, in part reflecting the fact that few researchers have devoted much attention to this issue. Living cells with flexible (unlignified) cell walls deform relatively easily upon desiccation and thus do not appear to support negative pressures. However, measurements of the forces needed to collapse cell walls suggest that living cells with thick walls can withstand forces in the range of 1.0 MPa. Visual examination of tissues adapted to withstand very cold temperatures also provides indirect evidence for negative turgor. For example, frozen ray parenchyma cells often do not exhibit significant deformation, despite the very strong desiccatory effects (low water potential) of extracellular ice. One can hypothesize that the existence of negative pressures within these cells acts to balance the water potential gradient across the cell membrane (i.e., between the cytoplasm and ice formed within the apoplast).

In summary, negative turgor pressure remains an intriguing but little-studied area of plant water relations. While it does not appear to form in cells such as the leaf mesophyll, that can deform easily, its existence in living cells with either lignified cell walls or where the cell is embedded in a matrix of lignified tissues (e.g., living cells within wood) cannot be ruled out. The potential benefits of negative turgor pressures in terms of preventing mechanical and osmotic damage associated with severe desiccation makes.

The Matric Potential

In discussions of soils, seeds, and cell walls, one often finds reference to yet another component of Ψ_w: the matric potential (Ψ_m). The matric potential is used to account for the reduction in free energy of water when it exists as a thin surface layer, one or two molecules thick, adsorbed onto the surface of relatively dry soil particles, cell walls, and other materials.

The matric potential does not represent a new force acting on water, because the effect of surface interactions can theoretically be accounted for by an effect on Ψ_s and Ψ_p. In dry materials, however, this surface interaction effect often cannot easily be separated into Ψ_p and Ψ_s components in dry materials, so they are frequently bulked together and designated as the matric potential. It is generally not valid to add Ψ_m to independent measurements of Ψ_s and Ψ_p to arrive at a total water potential. This is particularly true for water inside hydrated cells and cell walls, where matric effects are either negligible or they are accounted for by a reduction in Ψ_p. For instance, the negative pressure in water held by cell wall microcapillaries at the evaporative surfaces of leaves, is sometimes described as a wall matric potential. Care is needed to avoid inconsistencies when accounting for this physical effect in definitions of Ψ_p, Ψ_s, and Ψ_m.

Measuring Water Potential

Plant scientists have expended considerable effort in devising accurate and reliable methods for evaluating the water status of a plant. Four instruments that have been used extensively to measure Ψ_w, Ψ_s, and Ψ_p are described here: psychrometer, pressure chamber, cryoscopic osmometer, and pressure probe.

Psychrometer (Ψ_w Measurement)

Psychrometry (the prefix "psychro-" comes from the Greek word *psychein*, "to cool") is based on the fact that the vapour pressure of water is lowered as its water potential is reduced. Psychrometers measure the water vapour pressure of a solution or plant sample, on the basis of the principle that evaporation of water from a surface cools the surface.

Investigators make a measurement by placing a piece of tissue sealed inside a small chamber that contains a temperature sensor (in this case, a thermocouple) in contact with a small droplet of a standard solution of known solute concentration (known Ψ_s and thus known Ψ_w).

If the tissue has a lower water potential than that of the droplet, water evaporates from the droplet, diffuses through the air, and is absorbed by the tissue. This slight evaporation of water cools the drop.

The larger the difference in water potential between the tissue and the droplet, the higher the rate of water transfer and hence the cooler the droplet. If the standard solution has a lower water potential than that of the sample to be measured, water will diffuse from the tissue to the droplet, causing warming of the droplet.

Measuring the change in temperature of the droplet for several solutions of known Ψ_w makes it possible to calculate the water potential of a solution for which the net movement of water between the droplet and the tissue would be zero signifying that the droplet and the tissue have the same water potential.

Psychrometers can be used to measure the water potentials of both excised and intact plant tissue. Moreover, the method can be used to measure the Ψ_s of solutions.

This can be particularly useful with plant tissues. For example, the Ψ_w of a tissue is measured with a psychrometer, and then the tissue is crushed and the Ψ_s value of the expressed cell sap is measured with the same instrument. By combining the two measurements, researchers can estimate the turgor pressure that existed in the cells before the tissue was crushed ($\Psi_p = \Psi_w - \Psi_s$). A major difficulty with this approach is the extreme sensitivity of the measurement to temperature fluctuations. For example, a change in temperature of 0.01°C corresponds to a change in water potential of about 0.1 MPa. Thus, psychrometers must be operated under constant temperature conditions. For this reason, the method is used primarily in laboratory settings. There are many variations in psychrometric technique; interested readers should consult Brown and Van Haveren 1972, Slavik 1974, and Boyer 1995.

Pressure Chamber

A relatively quick method for estimating the water potential of large pieces of tissues, such as leaves and small shoots, is by use of the pressure chamber. This method was pioneered by Henry Dixon at Trinity College, Dublin, at the beginning of the twentieth century, but it did not come into widespread use until P. Scholander and coworkers at the Scripps Institution of Oceanography improved the instrument design and showed its practical use. In this technique, the organ to be measured is excised from the plant and is partly sealed in a pressure chamber.

Before excision, the water column in the xylem is under tension. When the water column is broken by excision of the organ (i.e., its tension is relieved allowing its Ψ_p to rise to zero), water is pulled rapidly from the xylem into the surrounding living cells by osmosis.

The cut surface consequently appears dull and dry. To make a measurement, the investigator pressurizes the chamber with compressed gas until the distribution of water between the living cells and the xylem conduits is returned to its initial, pre-excision, state.

This can be detected visually by observing when the water returns to the open ends of the xylem conduits that can be seen in the cut surface. The pressure needed to bring the water back to its initial distribution is called the *balance pressure* and is readily detected by the change in the appearance of the cut surface, which becomes wet and shiny when this pressure is attained. The pressure chamber is often described as a tool to measure the tension in the xylem.

However, this is only strictly true for measurements made on a non-transpiring leaf or shoot (for example, one that has been previously enclosed in a plastic bag). When there is no transpiration, the water potential of the leaf cells and the water potential in the xylem will come into equilibrium.

The balancing pressure measured on such a non-transpiring shoot is equal in magnitude but opposite in sign to the pressure in the xylem (Ψ_p). Because the water potential of our non-transpiring leaf is equal to the water potential of the xylem, one can calculate the water potential of the leaf by adding together Ψ_p and Ψ_s of the xylem, provided one collects a sample of xylem sap for determination of Ψ_s. Luckily Ψ_s of the xylem is usually small (> -0.1 MPa) compared to typical midday tensions in the xylem (Ψ_p of –1 to –2 MPa). Thus, correction for the Ψ_s of the xylem sap is frequently omitted. Balancing pressure measurements of transpiring leaves are more difficult to interpret. The fact that water is flowing from the xylem to the leaf means that differences in water potential must exist.

When the transpiring leaf or shoot is cut off, the tension in the xylem is instantly relieved and water is drawn into the leaf cells until the water potentials of the xylem and the leaf cells come into equilibrium. Because the total volume of the leaf cells is much larger than the volume of sap in the xylem, this equilibrium water potential will be heavily weighted towards that of the leaf.

Thus, any measurement of the balancing pressure on such a leaf or shoot will result in a value that is approximately the water potential of the leaf, rather than the tension of the xylem. (To be exact, one would have to add the Ψ_s of the xylem sap to the negative of the balancing pressure to get the leaf water potential.) One can explore the differences between the water potential of the xylem and the water potential of a transpiring leaf by comparing balancing pressures measured on covered (i.e., non-transpiring) versus uncovered (transpiring) leaves.

Pressure chamber measurements provide a quick and accurate way of measuring leaf water potential. Because the pressure chamber method does not require delicate instrumentation or temperature control, it has been used extensively under field conditions. For a more complete description of the theory and operation of the pressure chamber.

Cryoscopic Osmometer

The cryoscopic osmometer measures the osmotic potential of a solution by measuring its freezing point. Solutions have *colligative* properties that collectively depend on the number of dissolved particles and not on the nature of the solute. For example, solutes reduce the vapour pressure of a solution, raise its boiling point, and lower its freezing point.

The specific nature of the solute does not matter. One of the colligative *properties* of solutions is the decrease in the freezing point as the solute concentration increases. For example, a solution containing 1 mol of solutes per kilogram of water has a freezing point of –1.86°C, compared with 0°C for pure water. Various instruments can be used to measure the freezing-point depression of solutions. With a cryoscopic osmometer, solution samples as small as 1 nanoliter (10^{-9} L) are placed in an oil medium located on the temperature-controlled stage of a microscope. The very small sample size allows sap from single cells to be measured and permits rapid thermal equilibration with the stage. To prevent evaporation, the investigator suspends the samples in oil-filled wells in a silver plate (silver has high thermal conductivity). The temperature of the stage is rapidly decreased to about –30° C, which causes the sample to freeze. The temperature is then raised very slowly, and the melting process in the sample is observed through the microscope. When the last ice crystal in the sample melts, the temperature of the stage is recorded (note that the melting and freezing points are the same).

It is straightforward to calculate the solute concentration from the freezing-point depression; and from the solute concentration (c_s), Ψ_s is calculated as $-RTc_s$. This technique has been used to measure droplets extracted from single cells.

Pressure Probe

If a cell were as large as a watermelon or even a grape, measuring its hydrostatic pressure would be a relatively easy task. Because of the small size of plant cells, however, the development of methods for direct measurement of turgor pressure has been slow.

Using a micromanometer, Paul Green at the University of Pennsylvania developed one of the first direct methods for measuring turgor pressure in plant cells (Green and Stanton 1967). In this technique, an air-filled glass tube sealed at one end is inserted into a cell.

The high pressure in the cell compresses the trapped gas, and from the change in volume one can readily calculate the pressure of the cell from the ideal gas law (pressure × volume = constant). This method works only for cells of relatively large volume, such as the giant cell of the filamentous green alga *Nitella*. For smaller cells, the loss of cell sap into the glass tube is sufficient to deflate the cell and this yields artifactually low pressures.

For higher plant cells, which are several orders of magnitude smaller in volume than *Nitella,* a more sophisticated device, the pressure probe, was developed by Ernest Steudle, Ulrich Zimmermann, and their colleagues in Germany. This instrument is similar to a miniature syringe.

A glass microcapillary tube is pulled to a fine point and is inserted into a cell. The microcapillary is filled with silicone oil, a relatively incompressible fluid that can be readily distinguished from cell sap under a microscope. When the tip of the microcapillary is first inserted into the cell, cell sap begins to flow into the capillary because of the initial low pressure of that region.

Investigators can observe such movement of sap under the microscope and counteract it by pushing on the plunger of the device, thus building up a pressure. In such fashion the boundary between the oil and the cell sap can be pushed back to the tip of the microcapillary.

When the boundary is returned to the tip and is held in a constant position, the initial volume of the cell is restored and the pressure inside the cell is exactly balanced by the pressure in the capillary. This pressure is measured by a pressure sensor in the device. Thus the hydrostatic pressure of individual cells may be measured directly.

This method has been used to measure Ψ_p and other parameters of water relations in cells of both excised and intact tissues of a variety of plant species. The primary limitation of this method is that some cells are too small to measure.

Furthermore, some cells tend to leak after being stabbed with the capillary, and others plug up the tip of the capillary, thereby preventing valid measurements. The pressure probe has also been adapted to measure positive and negative values of Ψ_p in the xylem. However, technical problems with cavitation limit the measurement of negative Ψ_p by this technique.

Understanding Hydraulic Conductivity

Consider a cell with an initial water potential of –0.2 MPa, submerged in pure water. From this information we know that water will flow into the cell and that the driving force is $\Delta\Psi_w = 0.2$ MPa, but what is the initial rate of movement? The rate depends on permeability of the membrane to water, a property usually called the hydraulic conductivity (*Lp*) of the membrane.

Driving force, membrane permeability, and flow rate are related by the following equation:

$$\text{Flow rate} = \text{driving force} \times \text{hydraulic conductivity}$$

Hydraulic conductivity expresses how readily water can move across a membrane and has units of volume of water per unit area of membrane per unit time per unit driving force (for instance, $m^3\ m^{-2}\ s^{-1}\ MPa^{-1}$ or $m\ s^{-1}\ MPa^{-1}$). The larger the hydraulic conductivity, the larger the flow rate. The hydraulic conductivity of the membrane is $10^{-6}\ m\ s^{-1}\ MPa^{-1}$. The transport (flow) rate (J_v) can then be calculated from the following equation:

$$J_v = (Lp(\Delta\psi_w)$$

where J_v is the volume of water crossing the membrane per unit area of membrane and per unit time ($m^3\ m^{-2}\ s^{-1}$ or, equivalently, $m\ s^{-1}$). Please note that this equation assumes that the membrane is ideal—that is, that solute transport is negligible and water transport is equally sensitive to $\Delta\Psi_s$ and $\Delta\Psi_p$ across the membrane. Nonideal membranes require a more complicated equation that separately accounts for water flow induced by $\Delta\Psi_s$ and by $\Delta\Psi_p$. In our example, J_v has a value of $0.2 \times 10^{-6}\ m\ s^{-1}$. Note that J_v has the physical meaning of a velocity. We can calculate the flow rate in volumetric terms ($m^3\ s^{-1}$) by multiplying J_v by the surface area of the cell.

The resulting value is the *initial* rate of water transport. As water is taken up, cell Ψ_w increases and the driving force ($\Delta\Psi_w$) decreases. As a result, water transport slows with time.

Wilting and Plasmolysis

Plasmolysis is the separation of plant cell cytoplasm from the cell wall as a result of water loss. It is unlikely to occur in nature, except in severe conditions. Plasmolysis is induced in the laboratory by immersing a plant cell in a strongly saline or sugary solution, so that water is lost by osmosis.

If onion epidermal tissue is immersed in a solution of calcium nitrate, cells rapidly lose water by osmosis and the protoplasm of the cells shrinks. This occurs because the calcium and nitrate ions freely permeate the cell wall and encounter the selectively permeable plasma membrane.

The large vacuole in the centre of the cell originally contains a dilute solution with much lower osmotic pressure than that of the calcium nitrate solution on the other side of the membrane. The vacuole thus loses water and becomes smaller.

The space between the cell membrane and the cell wall enlarges and the plasma membrane and the protoplasm within it contract to the centre of the cell. Strands of cytoplasm extend to the cell wall because of plasma membrane-cell wall attachment points. Plasmolysed cells die unless they are transferred quickly from the salt or sugar solution to water.

Mineral Nutrition

Symptoms of Deficiency in Essential Minerals

Visual nutrient deficiency symptoms can be a very powerful diagnostic tool for evaluating the nutrient status of plants. One should keep in mind, however, that a given individual visual symptom is seldom sufficient to make a definitive diagnosis of a plant's nutrient status.

Many of the classic deficiency symptoms such as tip burn, chlorosis and necrosis are characteristically associated with more than one mineral deficiency and also with other stresses that by themselves are not diagnostic for any specific nutrient stress. However, their detection is extremely useful in making an evaluation of nutrient status.

In addition to the actual observations of morphological and spectral symptoms, knowing the location and timing of these symptoms is a critical aspect of any nutrient status evaluation. Plants do not grow in isolation, they are part of the overall environment and as such they respond to environmental changes as that affect nutrient availability. Also, plants do influence their environment and can contribute to environmental changes, which in turn can affect the nutrient status of the plant.

Sources of Visual Symptoms

Stresses such as salinity, pathogens, and air pollution induce their own characteristic set of visual symptoms. Often, these symptoms closely resemble those of nutrient deficiency.

Pathogens often produce an interveinal chlorosis, and air pollution and salinity stress can cause tip burn. Although at first these symptoms might seem similar in their general appearance to nutrient deficiency symptoms, they do differ in detail and/or in their overall developmental pattern. Pathological symptoms can often be separated from nutritional symptoms by their distribution in a population of affected plants. If the plants are under a nutrient stress, all plants of a given type and age in the same environment tend to develop similar symptoms at the same time.

However if the stress is the result of pathology, the development of symptoms will have a tendency to vary between plants until a relatively advanced stage of the pathology is reached.

Environmental Associations

Plants remove substantial amounts of nutrients from the soil during their normal growth cycle and many long-term environmental changes occur as a result of this process. Effects on the soil go considerably beyond the straight removal or depletion of nutrients. Charge balance must be maintained in the plant-soil system during nutrient uptake. Charge balance is usually achieved by the excretion of proton and/or hydroxyl ions by the plant to replace the absorbed nutrient cations or anions.

For example when plants are fertilized with ammonia, they acquire most of their nitrogen in the form of the ammonium cation, rather than from the usual nitrate anion. Because nitrate is the only anion used by the plant in large amounts, the net result of this change is that during normal nutrient uptake the proton excretion will far exceed that of hydroxyl ions.

In the case of vigorously growing plants, the amount of excreted protons can be sufficiently large as to decrease the pH of the soil by several pH units. Changes in soil pH of such magnitude can have large implications for a number of soil processes such as soil structure, nutrient availability and leaching of nutrients.

The immediate effect on the soil may be favourable for some plants, especially acid-loving plants, in that it tends to make iron more available. However, in the long run, lowering the soil pH can be deleterious to plants in that the availability of nutrients will change. A lower soil pH will allow micronutrients to be more readily leached from the soil profile, eventually resulting in deficiencies of nutrients such as Cu and Zn. Additionally, when the pH of the soil drops much below pH 5, the solubility of Al and Mn can increase to such an extent as to become toxic to most plant growth. Plants are often thought of as passive in relation to the environment. However this is not always a valid assumption; for there are many plants that clearly manipulate their environment in a fashion that tends to makes certain nutrients more readily available.

For example, iron is a limiting nutrient in many agricultural areas, but it comprises about 3% of the average soil which, if available, would be far in excess of the needs of the average plant. Some plants actively excrete protons, and the resulting decrease in pH increases the solubility of iron in their environment. In addition, other plants excrete phytosiderophores that chelate the soil iron rendering it a more available form for the plants.

Pathways of Symptom Development

At first glance, it would appear that the distinction of deficiency symptoms for the 13 known essential mineral nutrients should be relatively simple. But such an assumption is incorrect. In fact, the deficiency symptoms are quite complex because each nutrient has a number of different biological functions and each function may have an independent set of interactions with a wide range of environmental parameters.

In addition, the expression of these symptoms varies for acute or chronic deficiency conditions. Acute deficiency occurs when a nutrient is suddenly no longer available to a rapidly growing plant. Chronic deficiency occurs when there is a limited but continuous supply of a nutrient, at a rate that is insufficient to meet the growth demands of the plant. Most of the classic deficiency symptoms described in textbooks are characteristic of acute deficiencies. The most common symptoms of low-grade, chronic deficiencies are a tendency towards darker green leaves and stunted or slow growth.

Typically most published descriptions of deficiency symptoms arise from experiments conducted in greenhouses or growth chambers where the plants are grown in hydroponics or in media where the nutrients are fully available. In these conditions, nutrients are readily available while present, but when a nutrient is depleted, the plant suddenly faces an acute deficiency. Thus, hydroponic studies favour the development of acute deficiencies.

In experiments designed to study micronutrient deficiency symptoms, micronutrients are usually omitted from the nutrient solution. Micronutrients are often present in the seed or as contaminants in the environment, so a plant of adequate size will exhaust these trace amounts of micronutrient and develop characteristic acute deficiency systems.

When deficiency symptoms of macronutrients are sought, the macronutrient is removed suddenly from a suitable sized rapidly growing plant. Alternatively the plant can initially be given a one-time supply of the nutrient that is sufficient for a limited amount of growth. Because macronutrients are continuously required in relatively large amounts by rapidly growing plants, the available nutrients will be rapidly depleted, resulting in an acute deficiency. In natural systems, the plant encounters many degrees and types of stresses that result in different types of symptoms occurring over time.

Perhaps the most common nutrient deficiency in natural environments is the case of a limited nutrient supply that is continuously renewed at a low rate from soil weathering processes. In such cases, the limited nutrient availability results in chronic nutrient deficiency symptoms.

Effect of Nutrient Mobility on Symptom Development

The interaction between nutrient mobility in the plant, and plant growth rate can be a major factor influencing the type and location of deficiency symptoms that develop. For very mobile nutrients such as nitrogen and potassium, deficiency symptoms develop predominantly in the older and mature leaves.

This is a result of these nutrients being preferentially mobilized during times of nutrient stress from the older leaves to the newer leaves near the growing regions of the plant. Additionally, mobile nutrients newly acquired by the roots are also preferentially translocated to new leaves and the growing regions.

Thus old and mature leaves are depleted of mobile nutrients during times of stress while the new leaves are maintained at a more favourable nutrient status.

The typical localization of deficiency symptoms of very weakly mobile nutrients such as calcium, boron, and iron is the opposite to that of the mobile nutrients; these deficiency symptoms are first displayed in the growing regions and new leaves while the old leaves remain in a favourable nutrient status. (This assumes that these plants started with sufficient nutrient, but ran out of nutrient as they developed).

In plants growing very slowly for reasons other than nutrition (such as low light) a normally limiting supply of a nutrient could, under these conditions, be sufficient for the plant to slowly develop, maybe even without symptoms.

This type of development is likely to occur in the case of weakly mobile nutrients because excess nutrients in the older leaves will eventually be mobilized to supply newly developing tissues.

In contrast, a plant with a similar supply that is growing rapidly will develop severe deficiencies in the actively growing tissue such as leaf edges and the growing region of the plant. A classic example of this is calcium deficiency in vegetables such as lettuce where symptoms develop on the leaf margins (tip burn) and the growing region near the meristems.

The maximal growth rate of lettuce is often limited by the internal translocation rate of calcium to the growing tissue rather than from a limited nutrient supply in the soil.

When moderately mobile nutrients such as sulphur and magnesium are the limiting nutrients of the system, deficiency symptoms are normally seen over the entire plant. However the growth rate and rate of nutrient availability can make a considerable difference on the locations at which the symptoms develop. If the nutrient supply is marginal compared to the growth rate, symptoms will appear on the older tissue, but if the nutrient supply is very low compared to the growth rate, or the nutrient is totally depleted, the younger tissue will become deficient first.

Plant Competition and Induced Deficiencies

When the observed symptoms are the direct result of a nutrient deficiency, the actions needed for correction are relatively straight-forward. However symptoms are often the result of interactions with other environmental factors limiting the availability of the nutrient whose symptoms are expressed. The classic instance is that of iron deficiency induced by an excess of heavy metals in the environment. Transition metals such as Cu, Zn Cr and Ni compete with Fe and each other for plant uptake. Competition for uptake is not specific to Fe and heavy metals but is true for all mineral nutrients that are chemically similar and have similar uptake mechanisms.

For example if the availability of Cu or Zn is relatively less than that of Fe, then excessive concentrations of some other metal such as Ni or Cr will induce a deficiency of one of these nutrients rather than Fe. In the case of the macronutrients, excessive amounts of Mg will compete with K for uptake and can possibly induce a K deficiency. The barrenness of serpentine soils is the result of such competition, with the high Mg of these soils inducing a Ca deficiency. The toxicity of a low pH soil is another example of a basic nutrient deficiency. Low pH has a two-fold effect on soil nutrients: It enhances the leaching of cations, reducing their availability in the soil, and the relatively abundant protons in the soil compete with Ca and other cations for uptake. Thus, nutrient deficiencies can be induced by a number of different mechanisms often working in concert to limit the availability of a nutrient.

Nutrient Demand and Use Efficiency

Although all plants of the same species respond similarly to nutrient stress, plants of similar species will often show significant differences in their nutrient use efficiency. This results from differences in growth rate, root distribution, phase of development, and efficiency of nutrient uptake and utilization. This implies that in any given location, plants from one species may become nutrient-deficient, while those from another species growing in the same environment right next to them, may not show any deficiency symptoms. Growth rate also affects nutrient status. When the nutrient supply is barely inadequate for growth under existing environmental conditions, many plants adjust their growth rate to match that supported by the available nutrient supply without displaying typical visual deficiency symptoms. Agricultural systems differ from natural systems in that crop plants have been selected primarily for rapid growth under low stress conditions.

This rapid growth rate results in a high nutrient demand by these plants and a higher incidence of nutrient deficiency unless supplemental fertilizers are supplied. It is not uncommon to find agricultural crops showing severe signs of nutrient stress, with native plants growing in the same area showing little or no indication of nutrient stress. In agriculture systems chronic deficiency symptoms develop mostly in crops with little or limited fertilization. Acute nutrient deficiency symptoms most often occur when new crops with a higher nutrient demand are introduced, or less productive lands are brought under cultivation for the production of rapidly growing crop plants.

Uniformity of Nutrient Status

Not all tissues of a plant are at the same nutrient status during times of stress. Leaves on the same plant that are exposed to different environmental conditions, (such as light), or those of different ages may have considerable differences in nutrient status. Mineral nutrients are for the most part acquired by the roots and translocated throughout the plant. The distance of any part of the plant to the roots will influence nutrient availability, particularly in the case of the less mobile nutrients. In plants recovering from nutrient deficiency, the root and conductive tissues recover first. For example, in the case of recovery from Fe deficiency, it is common to see the veins re-green while the interveinal tissue remains chlorotic and Fe-deficient.

In order to maintain rapid, optimal growth, all plant tissues must have a favourable nutrient status. Although a plant may be marginally low in a number of nutrients, only one nutrient at a time will limit overall growth. However, if the supply of that limiting nutrient is increased even slightly, the resulting increase in growth will increase the demand for all other nutrients and another nutrient, the next lowest in availability, will become limiting.

Other Diagnostic Tools

Although visual diagnostic symptoms are an extremely valuable tool for the rapid evaluation of the nutrient status of a plant, they are only some of the tools available. Other major tools include microscopic studies, spectral analysis, and tissue and soil analysis. These methods all vary in their precision, rapidity and their ability to predict future nutrient status. Because of the close interaction between plant growth and the environment, all predictions of future nutrient status must make assumptions about how the environment will change in that time frame. The principle advantage of visual diagnostic symptoms is that they are readily obtained and provide an immediate evaluation of nutrient status. Their main drawback is that the visual symptoms do not develop until after there has been a major effect on yield, growth and development.

Tissue analysis is nutrient-specific but relatively slow; tissues must be sampled, processed and analysed before the nutrient status can be determined. An analysis of the mineral nutrient content of selected plants tissues, when compared against Critical Level values, can be used to evaluate the plant nutrient status at the time of sampling with a relatively high degree of confidence and can be extrapolated to project nutrient status at harvest. Soil analysis is similar to tissue analysis but evaluates the potential supplying power of the soil instead of plant nutrient status. Plant analysis provides information as to what the plant needs, while soil analysis provides information about the status of the nutrient supply.

Spectral analysis of nutrient status is still in its infancy and is presently used primarily in the inventory of global resources and in specialized studies. Microscopic studies are most valuable in looking at the physiological aspects of nutrient stress rather than the evaluation of plant nutrient status on a whole plant or crop basis.

Symptom Descriptions

It is unusual to find any one leaf or even one plant that displays the full array of symptoms that are characteristic of a given deficiency. It is thus highly desirable to know how individual symptoms look, for it is possible for them to occur in many possible combinations on a single plant. Most of the terms used below in the description of deficiency symptoms are reasonably self evident; a few however have a distinct meaning in the nutrient deficiency field. For example, the term *chlorotic,* which is a general term for yellowing of leaves through the loss of chlorophyll, cannot be used without further qualification because there may be an overall chlorosis as in nitrogen deficiency, interveinal, as in iron deficiency, or marginal, as in calcium deficiency. Another term used frequently in the description of deficiency symptoms is *necrotic,* a general term for brown, dead tissue. This symptom can also appear in many varied forms, as is the case with chlorotic symptoms.

Nutrient deficiency symptoms for many plants are similar, but because of the large diversity found in plants and their environments there is a range of expression of symptoms. Because of their parallel veins, grasses and other monocots generally display the affects of chlorosis as a series of stripes rather than the netted interveinal chlorosis commonly found in dicots. The other major difference is that the marginal necrosis or chlorosis found in dicots is often expressed as tip burn in monocots.

Magnesium. The Mg-deficient leaves show advanced interveinal chlorosis, with necrosis developing in the highly chlorotic tissue. In its advanced form, magnesium deficiency may superficially resemble potassium deficiency. In the case of magnesium deficiency the symptoms generally start with mottled chlorotic areas developing in the interveinal tissue. The interveinal laminae tissue tends to expand proportionately more than the other leaf tissues, producing a raised puckered surface, with the top of the puckers progressively going from chlorotic to necrotic tissue. In some plants such as the Brassica (The mustard family, which includes vegetables such as broccoli, brussel sprouts, cabbage, cauliflower, collards, kale, kohlrabi, mustard, rape, rutabaga and turnip.), tints of orange, yellow, and purple may also develop.

Manganese. These leaves show a light interveinal chlorosis developed under a limited supply of Mn. The early stages of the chlorosis induced by manganese deficiency are somewhat similar to iron deficiency. They begin with a light chlorosis of the young leaves and netted veins of the mature leaves especially when they are viewed through transmitted light. As the stress increases, the leaves take on a gray metallic sheen and develop dark freckled and necrotic areas along the veins. A purplish luster may also develop on the upper surface of the leaves. Grains such as oats, wheat, and barley are extremely susceptible to manganese deficiency. They develop a light chlorosis along with gray specks which elongate and coalesce, and eventually the entire leaf withers and dies.

Molybdenum: These leaves show some mottled spotting along with some interveinal chlorosis. An early symptom for molybdenum deficiency is a general overall chlorosis, similar to the symptom for nitrogen deficiency but generally without the reddish coloration on the undersides of the leaves. This results from the requirement for molybdenum in the reduction of nitrate, which needs to be reduced prior to its assimilation by the plant.

Thus, the initial symptoms of molybdenum deficiency are in fact those of nitrogen deficiency. However, molybdenum has other metabolic functions within the plant, and hence there are deficiency symptoms even when reduced nitrogen is available. In the case of cauliflower, the lamina of the new leaves fail to develop, resulting in a characteristic whiptail appearance. In many plants there is an upward cupping of the leaves and mottled spots developing into large interveinal chlorotic areas under severe deficiency. At high concentrations, molybdenum has a very distinctive toxicity symptom in that the leaves turn a very brilliant orange.

Nitrogen: The chlorotic symptoms shown by this leaf resulted from nitrogen deficiency. A light red cast can also be seen on the veins and petioles. Under nitrogen deficiency, the older mature leaves gradually change from their normal characteristic green appearance to a much paler green. As the deficiency progresses these older leaves become uniformly yellow (chlorotic). Leaves approach a yellowish white colour under extreme deficiency. The young leaves at the top of the plant maintain a green but paler colour and tend to become smaller in size. Branching is reduced in nitrogen deficient plants resulting in short, spindly plants. The yellowing in nitrogen deficiency is uniform over the entire leaf including the veins.

However in some instances, an interveinal necrosis replaces the chlorosis commonly found in many plants. In some plants the underside of the leaves and/or the petioles and midribs develop traces of a reddish or purple colour. In some plants this coloration can be quite bright. As the deficiency progresses, the older leaves also show more of a tendency to wilt under mild water stress and become senescent much earlier than usual. Recovery of deficient plants to applied nitrogen is immediate (days) and spectacular.

Phosphorus: These phosphorus-deficient leaves show some necrotic spots. As a rule, phosphorus deficiency symptoms are not very distinct and thus difficult to identify. A major visual symptom is that the plants are dwarfed or stunted. Phosphorus deficient plants develop very slowly in relation to other plants growing under similar environmental conditions but without phosphorus deficiency. Phosphorus deficient plants are often mistaken for unstressed but much younger plants. Some species such as tomato, lettuce, corn and the brassicas develop a distinct purpling of the stem, petiole and the under sides of the leaves. Under severe deficiency conditions there is also a tendency for leaves to develop a blue-gray luster. In older leaves under very severe deficiency conditions a brown netted veining of the leaves may develop.

Sulphur: This leaf shows a general overall chlorosis while still retaining some green colour. The veins and petioles show a very distinct reddish colour. The visual symptoms of sulphur deficiency are very similar to the chlorosis found in nitrogen deficiency. However, in sulphur deficiency the yellowing is much more uniform over the entire plant including young leaves. The reddish colour often found on the underside of the leaves and the petioles has a more pinkish tone and is much less vivid than that found in nitrogen deficiency. With advanced sulphur deficiency brown lesions and/or necrotic spots often develop along the petiole, and the leaves tend to become more erect and often twisted and brittle.

Zinc: This leaf shows an advanced case of interveinal necrosis. In the early stages of zinc deficiency the younger leaves become yellow and pitting develops in the interveinal upper surfaces of the mature leaves. Guttation is also prevalent. As the deficiency progress these symptoms develop into an intense interveinal necrosis but the main veins remain green, as in the symptoms of recovering iron deficiency. In many plants, especially trees, the leaves become very small and the internodes shorten, producing a rosette like appearance.

Boron: These boron-deficient leaves show a light general chlorosis. The tolerance of plants to boron varies greatly, to the extent that the boron concentrations necessary for the growth of plants having a high boron requirement may be toxic to plants sensitive to boron.

Boron is poorly transported in the phloem of most plants, with the exception of those plants that utilize complex sugars, such as sorbitol, as transport metabolites. In a recent study, tobacco plants engineered to synthesize sorbitol were shown to have increased boron mobility, and to better tolerate boron deficiency in the soil.

In plants with poor boron mobility, boron deficiency results in necrosis of meristematic tissues in the growing region, leading to loss of apical dominance and the development of a rosette condition.

These deficiency symptoms are similar to those caused by calcium deficiency. In plants in which boron is readily transported in the phloem, the deficiency symptoms localize in the mature tissues, similar to those of nitrogen and potassium. Both the pith and the epidermis of stems may be affected, often resulting in hollow or roughened stems along with necrotic spots on the fruit. The leaf blades develop a pronounced crinkling and there is a darkening and crackling of the petioles often with exudation of syrupy material from the leaf blade. The leaves are unusually brittle and tend to break easily.

Also, there is often a wilting of the younger leaves even under an adequate water supply, pointing to a disruption of water transport caused by boron deficiency.

Calcium: These calcium-deficient leaves show necrosis around the base of the leaves. The very low mobility of calcium is a major factor determining the expression of calcium deficiency symptoms in plants.

Classic symptoms of calcium deficiency include blossom-end rot of tomato (burning of the end part of tomato fruits), tip burn of lettuce, blackheart of celery and death of the growing regions in many plants.

All these symptoms show soft dead necrotic tissue at rapidly growing areas, which is generally related to poor translocation of calcium to the tissue rather than a low external supply of calcium.

Very slow growing plants with a deficient supply of calcium may re-translocate sufficient calcium from older leaves to maintain growth with only a marginal chlorosis of the leaves. This ultimately results in the margins of the leaves growing more slowly than the rest of the leaf, causing the leaf to cup downward.

This symptom often progresses to the point where the petioles develop but the leaves do not, leaving only a dark bit of necrotic tissue at the top of each petiole. Plants under chronic calcium deficiency have a much greater tendency to wilt than non-stressed plants.

Chloride: These leaves have abnormal shapes, with distinct interveinal chlorosis. Plants require relatively high chlorine concentration in their tissues. Chlorine is very abundant in soils, and reaches high concentrations in saline areas, but it can be deficient in highly leached inland areas.

The most common symptoms of chlorine deficiency are chlorosis and wilting of the young leaves. The chlorosis occurs on smooth flat depressions in the interveinal area of the leaf blade. In more advanced cases there often appears a characteristic bronzing on the upper side of the mature leaves.

Plants are generally tolerant of chloride, but some species such as avocados, stone fruits, and grapevines are sensitive to chlorine and can show toxicity even at low chloride concentrations in the soil.

Copper. These copper-deficient leaves are curled, and their petioles bend downward. Copper deficiency may be expressed as a light overall chlorosis along with the permanent loss of turgor in the young leaves. Recently matured leaves show netted, green veining with areas bleaching to a whitish gray. Some leaves develop sunken necrotic spots and have a tendency to bend downward. Trees under chronic copper deficiency develop a rosette form of growth. Leaves are small and chlorotic with spotty necrosis.

Iron: These iron-deficient leaves show strong chlorosis at the base of the leaves with some green netting. The most common symptom for iron deficiency starts out as an interveinal chlorosis of the youngest leaves, evolves into an overall chlorosis, and ends as a totally bleached leaf.

The bleached areas often develop necrotic spots. Up until the time the leaves become almost completely white they will recover upon application of iron. In the recovery phase the veins are the first to recover as indicated by their bright green colour.

This distinct venial re-greening observed during iron recovery is probably the most recognizable symptom in all of classical plant nutrition. Because iron has a low mobility, iron deficiency symptoms appear first on the youngest leaves. Iron deficiency is strongly associated with calcareous soils and anaerobic conditions, and it is often induced by an excess of heavy metals.

Potassium: Some of these leaves show marginal necrosis (tip burn), others at a more advanced deficiency status show necrosis in the interveinal spaces between the main veins along with interveinal chlorosis. This group of symptoms is very characteristic of K deficiency symptoms.

The onset of potassium deficiency is generally characterized by a marginal chlorosis progressing into a dry leathery tan scorch on recently matured leaves. This is followed by increasing interveinal scorching and/or necrosis progressing from the leaf edge to the midrib as the stress increases. As the deficiency progresses, most of the interveinal area becomes necrotic, the veins remain green and the leaves tend to curl and crinkle. In some plant such as legumes and potato, the initial symptom of deficiency is white speckling or freckling of the leaf blades. In contrast to nitrogen deficiency, chlorosis is irreversible in potassium deficiency, even if potassium is given to the plants. Because potassium is very mobile within the plant, symptoms only develop on young leaves in the case of extreme deficiency. Potassium deficiency can be greatly alleviated in the presence of sodium but the resulting sodium-rich plants are much more succulent than a high potassium plant. In some plants over 90% of the required potassium can be replaced with sodium without any reduction in growth.

Observing Roots below Ground

Studying root growth below the soil surface requires a means of observing the root system directly. As early as 1873, the German botanist Julius von Sachs studied root systems by using simple soil-filled boxes with one glass wall. Since that time, facilities for studying root growth in the soil have become much more complex. Large laboratories with subterranean chambers for the observation of root growth have been constructed and allow the analysis of root growth while the aerial parts of the plant are exposed to natural field conditions. These laboratories are called rhizotrons (from the Greek *rhizos*, meaning "root," and *tron*, meaning "a device for studying"). In a rhizotron, roots grow in glass-walled chambers that line underground passageways. Grid lines on the glass walls indicate soil depth. Details of root morphology (root size and distribution) under natural growing conditions can be observed with specially designed microscopes, mounted adjacent to the glass walls of the root chambers. In addition, the growth of the roots over a period of time can be measured with time-lapse photography.

Because rhizotrons are expensive to construct and maintain, they have been largely supplanted by minirhizotrons or root periscopes. Minirhizotrons are transparent plastic tubes previously buried at an angle in the soil near the plants to be observed. An optical device such as a tilted mirror with a magnifying lens or a miniature videocamera is inserted into a tube to monitor roots growing along its surface. This device provides information on rooting density in the bulk soil, as well as on root growth and phenology. Plant physiologists use such information in conjunction with changes in nutrient and water levels in the soil to assess root activity through the soil profile.

Solute Transport

Patch Clamp Studies in Plant Cells

The use of the patch clamp method in plant physiology is generating information about the properties of plant membranes that is beyond the scope of other techniques. Conventional electrophysiological studies with intracellular electrodes are valuable for the measurement of membrane potentials and other electrical properties of plant cells, but these methods have limitations for the characterization of ion pumps and single ion channels. Patch clamping, on the other hand, is very well suited for that purpose.

During conventional intracellular recording of cellular electric potentials, cells are impaled with a glass micropipette electrode, and the electrode tip is located inside the cell, usually in the vacuole. For patch clamp experiments, the cell walls are first digested away by treatment with the appropriate cell wall enzymes, resulting in naked protoplasts.

The tip of a glass micropipette is then brought into contact with the plasma membrane of the protoplast, and gentle suction is applied to facilitate the formation of a tight seal between the micropipette and the membrane. The unusual tightness of the seal that forms between the patch pipette and the membrane reduces the leakage-caused background electrical noise typical of conventional intracellular recording sufficiently to allow high-resolution recordings of the currents through single ion channels.

When a tight seal is attained, several options become available to the investigator. Further suction can remove the portion of the membrane delimited by the opening in the pipette tip, exposing the interior of the cell to the micropipette solution (thus different solutions can be introduced into the cell through the micropipette).

In this configuration, measured electric currents reflect the sum total of all electric charges carried by active and passive ion fluxes across the plasma membrane of the whole cell. Pulling the electrode away from the cell produces a membrane patch that becomes exposed to both the pipette solution and the bathing medium. Because the membrane patches contain only a few channels, this configuration makes it possible to study the abrupt changes in current caused by the opening and closing of single ion channels.

Remarkably, this technique is sensitive enough to detect the change in conformation of a single protein molecule. Because protoplasts are spherical, their volume is easy to measure, and their lack of electrical connections with other cells facilitates calculations of charge fluxes per unit area of membrane. Other advantages of patch clamping are the ability to distinguish between electrical events at the plasma membrane and the tonoplast and the possibility of controlling the composition of both the cytosol and the bathing medium.

The gradual increase in current after a change in voltage reflects the opening of increasing numbers of channels in response to the membrane potential. Channels responding in this way are termed voltage-gated channels. In general, the gating (opening and closing) of a channel is controlled independently of its ion selectivity, which is generally constant for a particular kind of channel protein. Again the current does not depend in a simple way on the driving force for K^+ diffusion (that is, the deviation of the potential from the K^+ equilibrium potential). An outward current is activated only at potentials more positive than -40 mV, and an inward current is activated only at potentials more negative than about -100 mV. This activation is a reflection of the gating of individual channel proteins with different properties.

The inward and outward currents can be shown to have different single-channel conductances (the ability of current to cross the membrane) and varying response times, and they are differentially affected by various activators and inhibitors.

It is now known that the inward and outward currents are carried by two different kinds of K^+ channels, as demonstrated by the cloning and expression of the respective genes. The outward-rectifying channels open only when the membrane potential favours outward diffusion of K^+, and the inward-rectifying channels open only when the membrane potential favours the inward diffusion of K^+. Since each type of channel has its own regulatory mechanisms, the cell can independently control the rates of inward and outward diffusion.

Additional information about K^+ channels can be obtained from studies of currents in isolated patches of membrane in which single channels can be visualized. When held at a constant voltage, these patches show abrupt jumps in the level of current, as single K^+ channels open and close.

This behaviour is seen only in membrane channels, never in ion pumps or carriers, because a single conformational change of a channel can allow hundreds of thousands of ions to cross the membrane, whereas a conformational change of a pump or a carrier can transport only one, or a few, molecules or ions. Patch clamp studies have also shown that gated channels can be distinguished on the basis of how long the gates remain open in response to a prolonged stimulus. Two types of anion channels have been characterized at the guard cell plasma membrane: Rapidly activated (R-type) and slowly activated (S-type). *R-type channels* open and close very rapidly in response to a voltage stimulus; *S-type channels* remain open for the duration of the stimulus. Rapid and slow channels have also been identified in vacuolar membranes, and they are referred to as *fast vacuolar (FV)* or *slow vacuolar (SV)* channels.

Kinetic Analysis of Multiple Transporter Systems

Studies of K_m values have shown a feedback inhibition of carrier-mediated transport that allows the cell to maintain stable cytoplasmic ion concentrations over a wide range of external concentrations.For example, the apparent K_m value of the high-affinity K^+ carrier in barley roots changes with the plant's demand for K^+. Barley plants starved for K^+ have high-affinity K_m values in the range of 0.02 to 0.03 m*M*, whereas well-fertilized plants have K_m values that are four-to fivefold greater. An increase in K_m reflects a decrease in the carrier's affinity for K^+. The change in apparent K_m probably reflects the expression of different members of a gene family, with their gene products thereby matching cellular needs to environmental ionic conditions. If the plants' adjustments were merely a matter of adding more transporters to the plasma membrane, there would be change in V_{max} without a change in K_m.

This is not what is observed for K^+. However, plants that have been starved and then resupplied with sulfate or phosphate often respond by an increase in V_{max}. As the internal sulfate and phosphate levels recover, the V_{max} values for their uptake drop to the usual values as excess carriers are removed from the membrane.

ABC Transporters in Plants

The ATP-binding cassette (ABC) transporters are members of a large family of active transport proteins energized directly by ATP hydrolysis. Originally identified in microbial and animal cells, ABC transporters are able to use the energy of ATP hydrolysis directly to pump organic molecules (especially large anionic molecules) across a membrane. Like the P-type H^+-ATP synthases, ABC transporters form a phosphorylated intermediate during catalysis, and they are therefore inhibited by vanadate.

A number of genes for ABC transporters have been cloned from various plants. Genes for ABC proteins are recognizable by a predicted protein structure consisting of one or two copies each of two structural units, a transmembrane domain (crossing the membrane several times) and a nucleotide-binding fold, which contains a characteristic signature sequence.

The transmembrane domain provides the pathway for solute transport and determines the molecular specificity of the transporter. An ABC transporter has been found to secrete anti-fungal terpenes across the plasma membrane of tobacco cells. More commonly, ABC transporters are found at the tonoplast, where they are sometimes referred to as glutathione conjugate pumps, or GS-X pumps, since they often transport molecules that have been covalently attached to glutathione. Glutathione (GS) is a tripeptide (Glu-Cys-Gly) that functions as an important cellular antioxidant molecule. The vacuolar GS-X pumps of plant cells function in herbicide detoxification, protection against oxidative damage, pigment accumulation, and the storage of antimicrobial compounds.

A family of enzymes called glutathione transferases (GSTs) is responsible for attaching glutathione to the molecule to be transported. The common chemical determinant for GS conjugation is a carbon–carbon double bond adjacent to an electron-withdrawing group ($CH_2 = CH\text{-}Z$). Compounds that contain this determinant include the anthocyanins, IAA (indoleacetic acid, or auxin), and various phenolic compounds. Metal-binding polypeptides called phytochelatins, which are synthesized from glutathione, are also transported into plant vacuoles along with their chelated heavy metals via GS-X pumps.

Transport Studies with Isolated Vacuoles and Membrane Vesicles

Just as studies with isolated thylakoid membranes led to the elucidation of the electron transfer reactions of photosynthesis, the development of techniques for isolating functionally competent tonoplasts and plasma membranes has enabled scientists to study the transport reactions of these membranes in vitro. The use of isolated membrane vesicles to study transport avoids interference from metabolism. Since the contents of the vesicles can be fairly readily exchanged, the concentrations of solutes on each side of the membrane can be better defined. Two types of preparations have been used: intact vacuoles, and isolated plasma membrane or tonoplast vesicles.

Isolating Intact Vacuoles from Protoplasts

When protoplasts prepared by enzymatic digestion of the cell walls are maintained in the plasmolyzed state by incubation in medium containing a suitable osmotic solute, such as mannitol, the plasma membranes can be carefully lysed without breaking the tonoplast by a gentle osmotic shock. The intact vacuoles that are released by this treatment can then be purified by centrifugation through a sucrose density gradient. Intact vacuoles (along with contaminating cellular debris) are centrifuged in a tube containing a concentration gradient of sucrose (or another suitable solute). Because intact vacuoles tend to exclude the solute, they sediment to a lighter density on the density gradient than the other organelles and membranes do. In this way preparations of highly purified intact vacuoles can be obtained.

Isolating Tonoplast Membrane from Cell Homogenates

Alternatively, tonoplasts and plasma membranes can be purified as small membrane vesicles. When tissues are ground in a mortar and pestle or homogenized in a blender, broken pieces of various membranes tend to seal up to form small vesicles less than 1 μm in diameter. Because of a low ratio of protein to lipid in the tonoplast membrane, tonoplast vesicles usually have a relatively low density, and they can be separated from other cellular membranes by density-gradient centrifugation.

Isolating Plasma Membrane from Cell Homogenates

Plasma membrane vesicles can be purified by phase partition in a mixture of dextran and polyethylene glycol solutions. Such a mixture separates into two phases, and because of their surface properties, plasma membranes tend to accumulate in the upper phase, while intracellular membranes tend to remain in the lower phase. Tissue homogenization often produces a mixture of vesicles of different orientation (right-side-out and inside-out vesicles). In the case of plasma membrane vesicles, phase partition selects for right-side-out vesicles, which therefore show little reactivity to ATP because the catalytic sites of the enzymes are facing inside the vesicle. If a detergent is added to such a preparation, ATP penetrates the vesicles and is hydrolyzed, but since the vesicles are no longer sealed, no transport is observed. One kind of detergent in low doses does not make the membranes leaky but, it is claimed, causes them to reverse their orientation. Apparently, this detergent accumulates in the inner half of the lipid bilayer, causing it to expand and flip the vesicles inside out. Since the catalytic site of the enzyme is now facing the external solution, in the presence of this detergent the vesicles respond to ATP by creating a gradient of pH and membrane potential.

Measuring Transport in Membrane Vesicles

Radioisotopes can be used to monitor the transport of various solutes into membrane vesicles. The vesicles are incubated for various times with the labelled solute, then separated by rapid filtration, and their radioactivity is measured in a scintillation counter. Other methods have been developed to measure the gradients of pH and membrane potential created by proton pumps. One such method uses fluorescent weak bases such as 9-aminoacridine (9-AA) or quinacrine, which are lipid-soluble in their uncharged form and readily diffuse across the membrane and enter the vesicle.

Within the acidified vesicle, the amino group becomes protonated, trapping the charged form of the molecule inside. As the dye molecules accumulate inside the vesicles, they form dimers and other aggregates that result in a reduction in their fluorescence, a phenomenon termed fluorescence quenching. One can thus study ATP-driven proton transport across the vesicle membrane by measuring the rate of fluorescence quenching in an instrument called a fluorescence spectrophotometer.

Although it has proven difficult to make absolute measurements of the pH and potential gradients in membrane vesicles, this type of study has been very useful for relative measurements, and for defining the energy sources and activator and inhibitor sensitivities of various transport systems. In combination with other experimental techniques, studies with membrane vesicles have provided many insights into transport in plant cells.

5

Molecules Biochemistry

Most of the Universe consists of matter and energy. Energy is the capacity to do work. Matter has mass and occupies space. All matter is composed of basic elements that cannot be broken down to substances with different chemical or physical properties. Elements are substances consisting of one type of atom, for example Carbon atoms make up diamond, and also graphite. Pure (24K) gold is composed of only one type of atom, gold atoms. Atoms are the smallest particle into which an element can be divided.

The ancient Greek philosophers developed the concept of the atom, although they considered it the fundamental particle that could not be broken down. Since the work of Enrico Fermi and his colleagues, we now know that the atom is divisible, often releasing tremendous energies as in nuclear explosions or (in a controlled fashion in) thermonuclear power plants.

Subatomic particles were discovered during the 1800s. For our purposes we will concentrate only on three of them, summarized in Table 1. The proton is located in the centre (or nucleus) of an atom, each atom has at least one proton. Protons have a charge of +1, and a mass of approximately 1 atomic mass unit (amu). Elements differ from each other in the number of protons they have, e.g. Hydrogen has 1 proton; Helium has 2.

The neutron also is located in the atomic nucleus (except in Hydrogen). The neutron has no charge, and a mass of slightly over 1 amu. Some scientists propose the neutron is made up of a proton and electron-like particle. The electron is a very small particle located outside the nucleus. Because they move at speeds near the speed of light the precise location of electrons is hard to pin down. Electrons occupy orbitals, or areas where they have a high statistical probability of occurring. The charge on an electron is -1. Its mass is negligible (approximately 1800 electrons are needed to equal the mass of one proton).

Table: *Subatomic particles of use in biology.*

Name	***Charge***	***Location***	***Mass***
Proton	+1	atomic nucleus	1.6726×10^{-27} kg
Neutron	0	atomic nucleus	1.6750×10^{-27} kg
Electron	-1	electron orbital	9.1095×10^{-31} kg

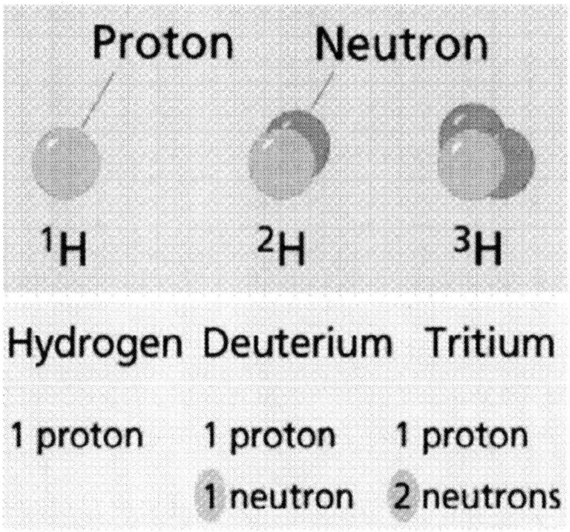

Figure: *Isotopes of Hydrogen*

The atomic number is the number of protons an atom has. It is characteristic and unique for each element. The atomic mass (also referred to as the atomic weight) is the number of protons and neutrons in an atom. Atoms of an element that have differing numbers of neutrons (but a constant atomic number) are termed isotopes.

Isotopes, can be used to determine the diet of ancient peoples by determining proportions of isotopes in mummified or fossilized human tissues. Biochemical pathways can be deciphered by using isotopic tracers.

The age of fossils and artifacts can be determined by using radioactive isotopes, either directly on the fossil or on the rocks that surround the fossil. Isotopes are also the source of radiation used in medical diagnostic and treatment procedures.

Note that each of these isotopes of hydrogen has only one proton. Isotopes differ from each other in the number of neutrons, not in the number of protons.

Some isotopes are radioisotopes, which spontaneously decay, releasing radioactivity. Other isotopes are stable. Examples of radioisotopes are Carbon-14 (symbol ^{14}C), and deuterium (also known as Hydrogen-2; ^{2}H). Stable isotopes are ^{12}C and ^{1}H.

Carbon has three isotopes, of which carbon-12 and carbon-14 are the most well known.

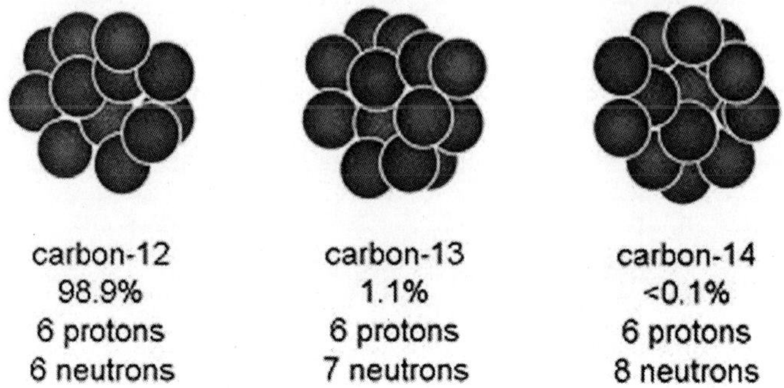

Figure: *Isotopes of Hydrogen*

Electrons and Energy

Electrons, because they move so fast (approximately at the speed of light), seem to straddle the fence separating energy from matter. Albert Einstein developed his famous $E=mc^2$ equation relating matter and energy over a century ago. Because of his (and others) work, we think of electrons both as particles of matter (having mass is a property of matter) and as units (or quanta) of energy. Excitation of an electron by energy, causing the electron to "jump" to another electron (energy) level known as the excited state.

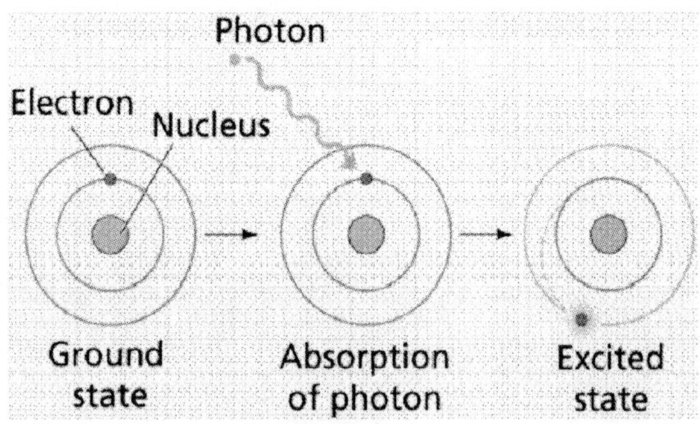

An orbital is also an area of space in which an electron will be found 90% of the time. Orbitals have a variety of shapes. Each orbital has a characteristic energy state and a characteristic shape.

The *s* orbital is spherical. Since each orbital can hold a maximum of two electrons, atomic numbers above 2 must fill the other orbitals. The p_x, p_y, and p_z orbitals are dumbbell shaped, along the x, y, and z axes respectively.

Energy levels are located a certain "distance" from the nucleus. The major energy levels into which electrons fit, are (from the nucleus outward) K, L, M, and N. Sometimes these are numbered, with electron configurations being: $1s^2 2s^2 2p^1$, (where the first shell K is indicated with the number 1, the second shell L with the number 2, etc.). This nomenclature tells us that for the atom mentioned in this paragraph, the first energy level (shell) has two electrons in its *s* orbital (the only orbital it can have), and second energy level has a maximum of two electrons in its *s* orbital, plus one electron in its *p* orbital.

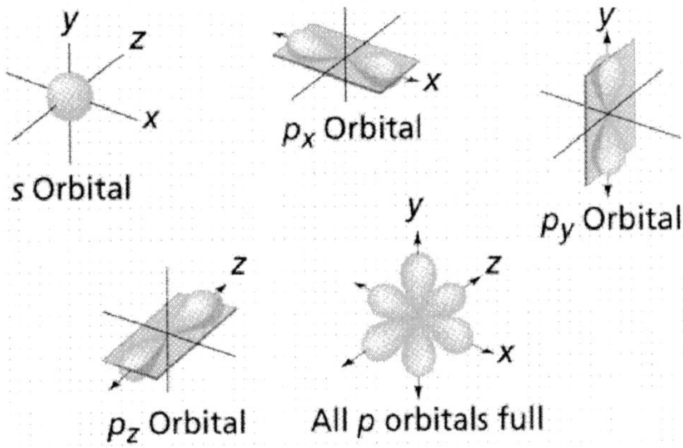

Geometry of orbitals. S-orbitals are spherical, p-orbitals are shaped like a dumbbell.

Chemical Bonding

During the nineteenth century, chemists arranged the then-known elements according to chemical bonding, recognizing that one group (the furthermost right column on the Periodic Table, referred to as the Inert Gases or Noble Gases) tended to occur in elemental form (in other words, not in a molecule with other elements).

It was later determined that this group had outer electron shells containing two (as in the case of Helium) or eight (Neon, Xenon, Radon, Krypton, etc.) electrons. As a general rule, for the atoms we are likely to encounter in biological systems, atoms tend to gain or lose their outer electrons to achieve a Noble Gas outer electron shell configuration of two or eight electrons. The number of electrons that are gained or lost is characteristic for each element, and ultimately determines the number and types of chemical bonds atoms of that element can form.

Ionic bonds are formed when atoms become ions by gaining or losing electrons. Chlorine is in a group of elements having seven electrons in their outer shells. Members of this group tend to gain one electron, acquiring a charge of -1. Sodium is in another group with elements having one electron in their outer shells.

Members of this group tend to lose that outer electron, acquiring a charge of +1. Oppositely charged ions are attracted to each other, thus Cl^- (the symbolic representation of the chloride ion) and Na^+ (the symbol for the sodium ion, using the Greek word *natrium*) form an ionic bond, becoming the molecule sodium chloride. Ionic bonds generally form between elements in Group I (having one electron in their outer shell) and Group VIIa (having seven electrons in their outer shell). Such bonds are relatively weak, and tend to disassociate in water, producing solutions that have both Na and Cl ions.

Atomic diagrams illustrating the filling of the outer electron shells.

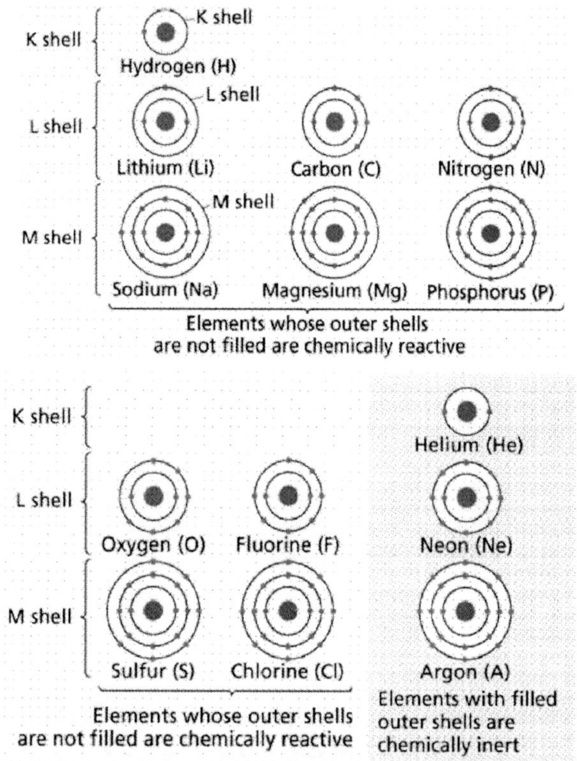

Top: Formation of a crystal of sodium chloride. Each positively charged sodium ion is surropunded by six negatively charged chloride ions; likewise each negatively charged chloride ion is surrounded by six positively charged sodium ions. The overall effect is electrical neutrality

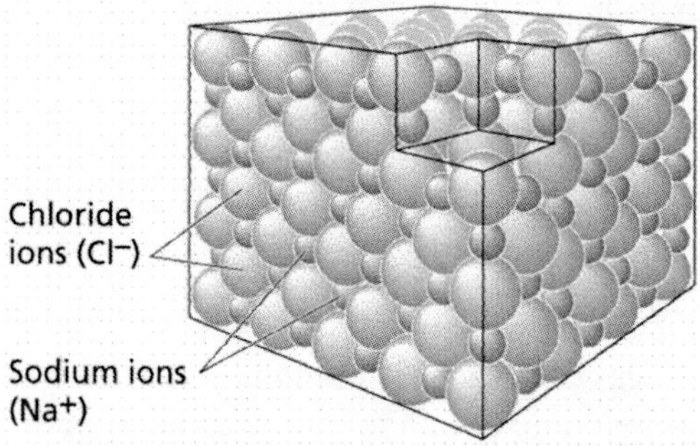

Covalent bonds form when atoms share electrons. Since electrons move very fast they can be shared, effectively filling or emptying the outer shells of the atoms involved in the bond. Such bonds are referred to as electron-sharing bonds. An analogy can be made to child custody: the children are like electrons, and tend to spend some time with one parent and the rest of their time with the other parent. In a covalent bond, the electron clouds surrounding the atomic nuclei overlap.

Formation of a covalent bond between two Hydrogen atoims.

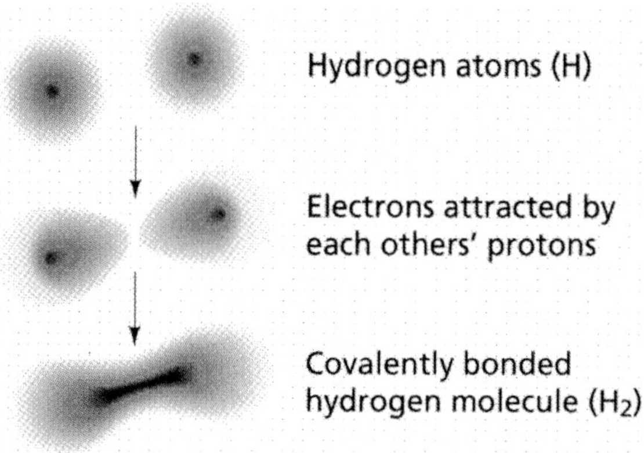

Carbon (C) is in Group IVa, meaning it has four electrons in its outer shell. Thus to become a "happy atom", Carbon can either gain or lose four electrons. By sharing the electrons with other atoms, Carbon can become a happy atom,. alternately filling and emptying its outer shell.

Formation of covalent bonds in methane. Carbon needs to share four electrons, in effect it has four slots. Each hydrogen provides an electron to each of these slots. At the same time each hydrogen needs to fill one slot, which is done by sharing an electron with the carbon

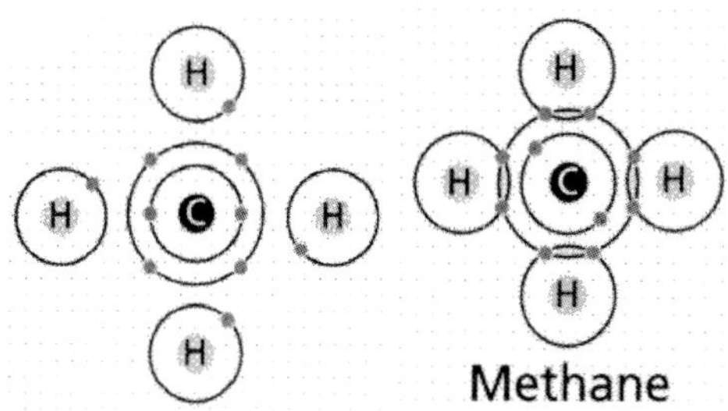

The molecule methane (chemical formula CH_4) has four covalent bonds, one between Carbon and each of the four Hydrogens. Carbon contributes an electron, and Hydrogen contributes an electron. The sharing of a single electron pair is termed a single bond. When two pairs of electrons are shared, a double bond results, as in carbon dioxide. Triple bonds are known, wherein three pairs (six electrons total) are shared as in acetylene gas or nitrogen gas.

Ways of Representing Covalent Bonds

Sometimes electrons tend to spend more time with one atom in the bond than with the other. In such cases a polar covalent bond develops. Water (H_2O) is an example. Since the electrons spend so much time with the oxygen (oxygen having a greater electronegativity, or electron affinity) that end of the molecule acquires a slightly negative charge. Conversely, the loss of the electrons from the hydrogen end leaves a slightly positive charge.

The water molecule is thus polar, having positive and negative sides. Hydrogen bonds, result from the weak electrical attraction between the positive end of one molecule and the negative end of another. Individually these bonds are very weak, although taken in a large enough quantity, the result is strong enough to hold molecules together or in a three-dimensional shape.

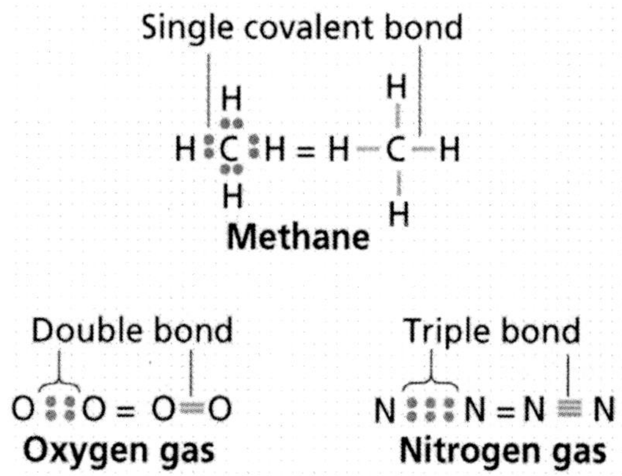

Top: Formation of a hydrogen bond between the hydrogen side of one water molecule and the oxygen side of another water molecule.

Bottom: The presence of polar areas in the amino acids that make-up a protein allows for hydrogen bonds to form, giving the molecule a three-dimensional shape that is often vital to that protein's proper functioning.

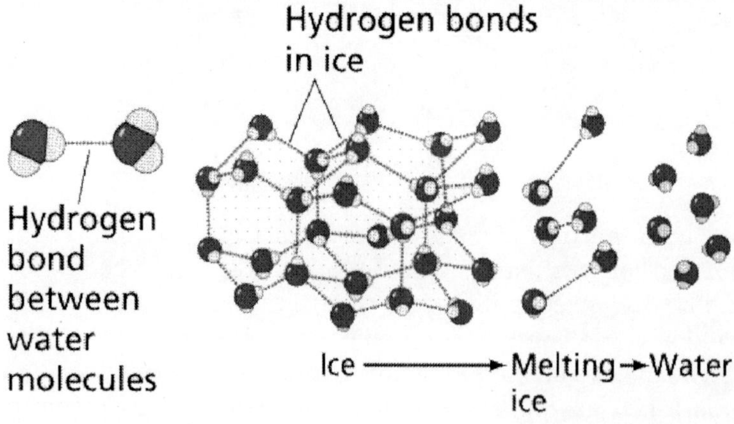

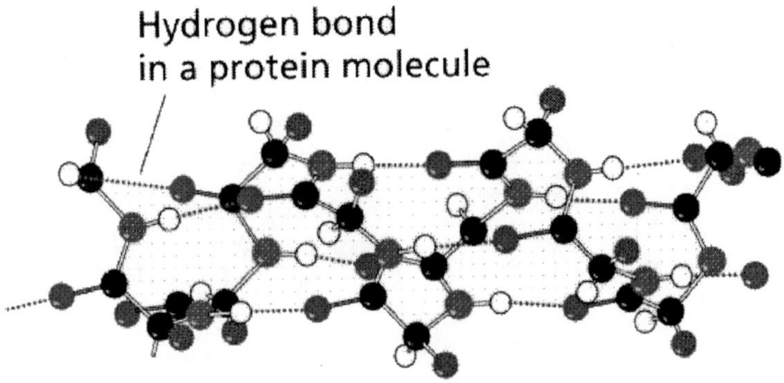

Chemical Reactions and Molecules

Molecules are compounds in which the elements are in definite, fixed ratios. Those atoms are held together usually by one of the three types of chemical bonds. For example: water, glucose, ATP. Mixtures are compounds with variable formulas/ratios of their components. For example: soil. Molecular formulas are an expression in the simplest whole-number terms of the composition of a substance. For example, the sugar glucose has 6 Carbons, 12 hydrogens, and 6 oxygens per repeating structural unit. The formula is written $C_6H_{12}O_6$.

Determination of molecular weights by addition of the weights of the atoms that make up the molecule.

Chemical reactions occur in nature, and some also can be performed in a laboratory setting. One such reaction is diagrammed. Chemical equations are linear representations of how these reactions occur. Combination reactions occur when two separate reactants are bonded together, e.g. A + B ——> AB. Disassociation reactions occur when a compound is broken into two products, e.g. AB ——> A + B.

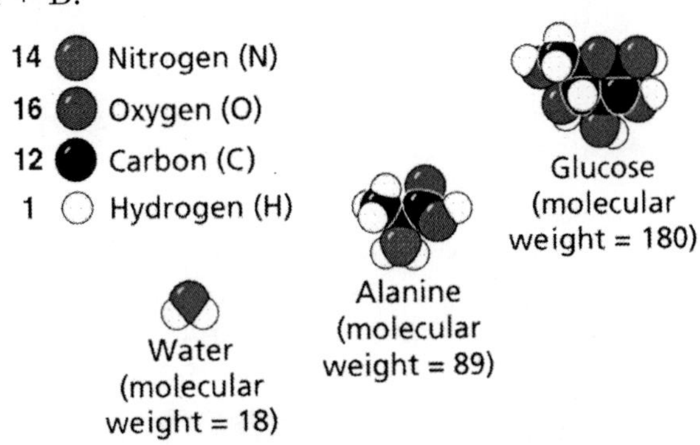

Diagram of a chemical reaction: the combustion of propane with oxygen, resulting in carbon dioxide, water, and energy (as heat and light). This chemical reaction takes place in a camping stove as well as in certain welding torches Biological systems, while unique to each species, are based on the chemical bonding properties of carbon. Major organic chemicals (those associated with or formed by the actions of living things) usually include some ratios of the following elements: C, H, N, O, P, S.

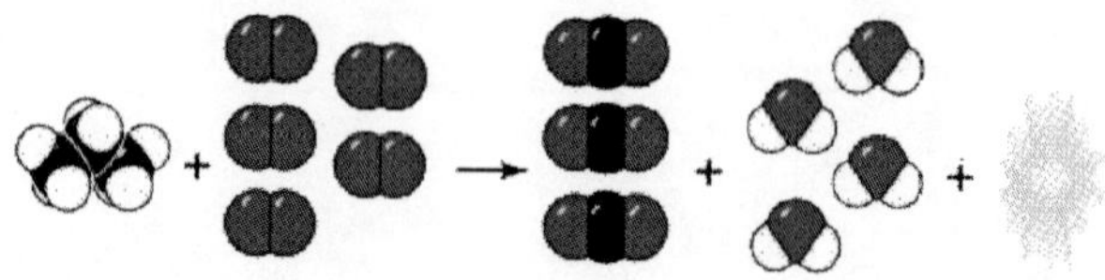

$$C_3H_8 + 5\,O_2 \rightarrow 3\,CO_2 + 4\,H_2O + \text{Heat and light}$$

Propane Oxygen gas Carbon dioxide Water

Water—the Fluid of Life

Water is indispensable to life as we know it. The special chemistry of water makes it a major component of all life on earth. Three quarters of the surface of the earth is covered by water and we humans are composed of approximately 66% water. Water has many unusual physical properties that result from its unique chemical structure. Before talking about these properties, it is first necessary to define some terms.

Physical properties of matter include such things as:

- colour
- boiling point
- freezing point
- any physical characteristics that do not require a chemical change.

Chemical properties are characteristics of a compound that relate to its ability to react chemically with another substancc. An example of one of the important chemical properties of water is its ability to form hydrogen bonds with other water molecules. Hydrogen bonds are chemical bonds formed between the hydrogen on one molecule and an electronegative element such as oxygen or nitrogen on a different molecule.

Oxygen is an electronegative element and hydrogen is an electropositive element. Since water is composed of one oxygen atom and two hydrogens, water molecules tend to develop slight negative and positive charges at different ends of the molecule. This charge differentiation makes water a polar molecule. The slight charges of the polar molecule can interact with other polar molecules or ions.

The two hydrogens on one end of the water molecule have partial positive charges (denoted by the symbol d +). The oxygen at the other end of the molecule has a partial negative charge (denoted by the symbol d -). Because of these partial positive and negative charges, water molecules can form hydrogen bonds with other water molecules. One of the partially positive hydrogens will form a weak bond with the partially negative oxygen of another water molecule.

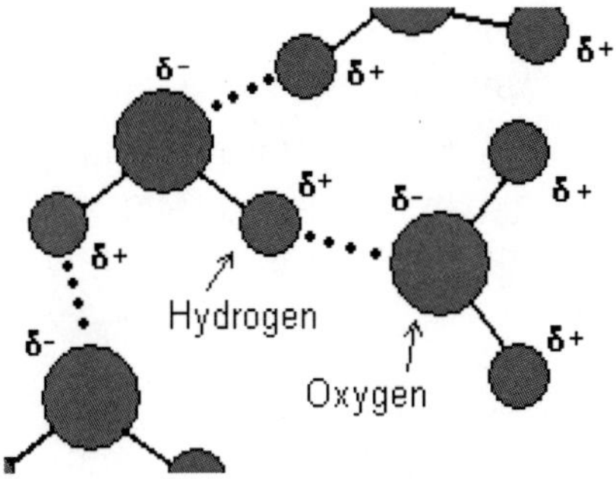

The ability of water molecules to hydrogen bond with other water molecules gives water special physical properties. The hydrogen bonds between water molecules allow water to form a highly ordered lattice structure in the solid state. When water freezes a crystal lattice structure forms. This lattice structure is what gives snowflakes and ice crystals their beautiful design. The polar, ordered nature of water makes it an excellent solvent. It is often referred to as the universal solvent, although not everything is soluble in water obviously!

A solvent is an agent that dissolves, or solubilizes, a substance called a solute. If something is soluble in water it dissolves in it. However, solubilizing a substance does not cause it to vanish. As you certainly know, when sugar dissolves in water the water tastes sweet, indicating the presence of the sugar. When the polar water molecules orient themselves around the partially charged polar or fully charged ionic molecules of the solute, the solute has been solubilized in the water.

The partial positive charges of the hydrogens on the polar water molecules surround each negatively charged chlorine ion, while the partial negative charges of the oxygen on the polar water molecules surround each positively charged sodium ion. When water molecules orient themselves around charged or polar molecules, spheres of hydration, or hydration shells, are formed.

Compound	***Melting Pt.***	***Boiling Pt.***	***Heat of Vaporization***
Water	0° Celsius	100° Celsius	540 Cal/g
Methanol	-98° Celsius	65° Celsius	263 Cal/g
Ethanol	-117° Celsius	78° Celsius	204 Cal/g

Note the wide range in values between these three seemingly similar liquid compounds. All are clear coloured, polar solvents, and all contain oxygen and hydrogen. Yet, the table above shows that these substances have extremely different properties. Imagine if lakes were composed of either methanol or ethanol. What might happen in the heat of the summer? What would happen in the winter?

If you guessed that the lakes would evaporate very quickly in the summer and fail to freeze in the winter you would be correct! As you may imagine, this would have a devastating effect on pond life.

The formation of ice is very important to many aquatic organisms as a form of protection against extreme cold and rapid temperature fluctuations. Water has some additional physical properties that also make it essential to life.

- Cohesion- the molecular attraction between the particles within a substance.
- Adhesion- the molecular attraction between a substance and the surface it contacts.
- Capillarity- a characteristic of water that results from both the cohesive and adhesive properties. Capillarity allows water to rise in a small tube above the level of the surrounding water.
- Surface tension- a characteristic of water that results from the cohesion between water molecules, and allows the surface of water to have some strength to support objects.

PH and Water

In addition to all of its other properties, water can act as either a weak acid or a weak base. How can water act as both? First we must define acids and bases. For our purposes here we will use the simplest definition:

An acid is a substance that can donate hydrogen ions (H+) to the surrounding aqueous environment. A base is a substance that can absorb or accept hydrogen ions from the surrounding aqueous environment.

An acid will increase the hydrogen ion concentration of an aqueous solution in which it is solubilized. A base will decrease the hydrogen ion concentration of an aqueous solution in which it is solubilized. Acids and bases must be dissolved in aqueous solution in order to dissociate and become active as hydrogen donors or acceptors.

Before we go on we should better define the hydrogen ion (H+). Remember that the element hydrogen is composed of one electron and one proton. Hydrogen normally exists in nature as a gas composed of two hydrogen atoms covalently bonded together. A hydrogen ion is one hydrogen atom that has lost its electron, and therefore is merely a single proton. Knowing what you know about atoms and what makes them stable or unstable, do you suppose that a single proton by itself would be very stable?

Of course it wouldn't! Hydrogen ions can only exist when their positive charges can be neutralized by opposite charges. For example, a hydrogen ion can be present in water when the negative ends of the polar water molecules can surround the positively charged hydrogen ion in a sphere of hydration.

Ionization or separation of water into hydrogen ions (H+) and hydroxide ions (OH-) occurs to a very small extent in water at all times. Water acts as a hydrogen donor (an acid) when a molecule of water separates and releases a hydrogen ion.

Water acts as a hydrogen acceptor (a base) when one of the hydroxide ions accepts a hydrogen ion to form a complete molecule of water. The following equation represents the ionization of water: $H_2O \longrightarrow H^+ + OH^-$ Ionization of water causes the neutral water molecules to separate into charged ions.

Acids and bases are classified according to their ability to donate or accept hydrogen ions. A strong acid can donate hydrogen ions easily when dissolved in aqueous solution. A strong base can accept hydrogen ions easily. Now that we have defined acids and bases chemically, let's find out what effects they have on biological systems.

As you already know, charged particles, or ions, are highly reactive. This basically means that ions are looking for other molecules to react with so they may attain a more stable electron configuration.

When charged particles react with biomolecules such as protein and nucleic acid, they can have devastating effects upon the organism. Since acids and bases can dissociate easily into charged hydrogen ions and hydroxide ions, they have the potential to be very destructive to living tissues.

Hydrochloric acid, a strong acid, can be highly destructive to organic matter when in its concentrated form. The same is true for sodium hydroxide, a strong base. Few people realize that strong bases are just as dangerous as strong acids.

Most oven cleaners (the foamy substances that eat encrusted food on your oven walls) contain concentrated sodium hydroxide base. Many drain openers are often concentrated acids. Drain openers operate by breaking down organic materials like hair that may clog the drain. Knowing what strong acids and bases due to the organic materials in dirty ovens and clogged drains should give you a hint about their potential for tissue destruction. However, even weak and dilute acids and bases can have a devastating effect when not compensated for inside living cells and tissues.

Although we have not yet discussed the role of the major biological molecules I think most of us know that our cells and tissues stay alive by performing many complex biochemical reactions. Biological life revolves around a class of molecules called proteins. Proteins are very sensitive to even a slight increase or decrease in the hydrogen ion concentration (pH) of the surrounding solution. So how can we be exposed to and ingest acidic and basic compounds all the time and yet not increase and decrease the hydrogen ion concentration in our blood, tissues, and cells?

Buffers!

A buffer is a substance that can accept hydrogen ions when there are too many present or release hydrogen ions when there are too few present. In this way buffers maintain the pH at a relatively constant level. pH is a measure of how many hydrogen ions are present in a substance. pH actually represents the number of hydrogen ions per unit volume of a liquid, or the concentration of hydrogen ions. Here's an example of how concentration works: one package of Kool-Aid makes two quarts. If you place one cup of sugar in that two quarts you have a specific concentration of sugar in your Kool-Aid. If you place two cups of sugar in that two quarts you would have increased the concentration of sugar 2-fold, or 2X.

Buffers are essential to all organisms because most biochemical reactions can only occur at neutral pH's of about seven. A substance with a neutral pH has just the right amount of hydrogen ions present for biochemical reactions to occur. Buffers maintain pH by absorbing or releasing hydrogen ions when necessary to maintain a specific hydrogen ion concentration. In our laboratory we perform many extractions of biologically active proteins which we later use in an in vitro reaction. We use buffering agents in all of our liquids so that we can maintain a near neutral pH, similar to the pH found in most biological systems. As a matter of fact, scientists almost universally refer to these liquids used in their laboratories as buffers even though it is technically incorrect because most of them contain components in addition to the buffering compound. Buffers are indispensable in all aqueous solutions in which biochemical reactions are taking place, whether in a living organism or in the laboratory.

Is water very susceptible to a pH change? Yes! Add only one drop of concentrated hydrochloric acid to a liter (approximately a quart) of pure water and the pH will be reduced from seven to approximately one. A pH of seven is neutral, and a pH of one is very acidic. What really is the difference between a solution with a pH of seven and a solution with a pH of one? As we have noted before the pH represents the concentration of hydrogen ions in a specific volume of a substance. pH is defined as the negative log of the hydrogen ion concentration as represented by this equation: -log [H+]. The brackets indicate concentration of H+.

Right now you might be thinking: *negative log of what?!* Relax, it's pretty simple. The log of a number q in base 10 is equal to the exponent of q when written in exponential notation. For example, the log of 100 is equal to 2, because

100 written in exponential notation is 1×10^2. When figuring pH, we use the negative log. It is the same as taking the log and then multiplying it by negative one. But wait, pH's are not negative numbers! That is because the pH represents a small number from 0.1 (1×10^{-1}) to 0.00000000000001 (1×10^{-14}). These numbers correspond to a pH of one and a pH of fourteen. As you can see, the log of 1×10^{-14} is -14, but the negative log of 1×10^{-14} is 14. The definition for pH was fabricated to simplify the representation of hydrogen ion concentration. pH values represent a kind of shorthand value for acidic and basic solutions.

Note that the higher the concentration of hydrogen ions, the smaller the exponent, and therefore the smaller the pH value. For this reason, acids will have small pH values, such as a pH of one for a strong acid. Bases will have large pH values because their hydrogen ion concentration is smaller. A solution with a hydrogen ion concentration of 1×10^{-1} has a relatively large number of hydrogen ions. This solution is a strong acid of pH one. A solution with a hydrogen ion concentration of 1×10^{-14} has a very small number of hydrogen ions. This solution is a strong base with pH fourteen.

The Polarity of Water

Water has a simple molecular structure. It is composed of one oxygen atom and two hydrogen atoms. Each hydrogen atom is covalently bonded to the oxygen via a shared pair of electrons. Oxygen also has two unshared pairs of electrons. Thus there are 4 pairs of electrons surrounding the oxygen atom, two pairs involved in covalent bonds with hydrogen, and two unshared pairs on the opposite side of the oxygen atom. Oxygen is an "electronegative" or electron "loving" atom compared with hydrogen.

Water is a "polar" molecule, meaning that there is an uneven distribution of electron density. Water has a partial negative charge (δ-) near the oxygen atom due the unshared pairs of electrons, and partial positive charges (δ+) near the hydrogen atoms.

An electrostatic attraction between the partial positive charge near the hydrogen atoms and the partial negative charge near the oxygen results in the formation of a hydrogen bond as shown in the illustration.The ability of ions and other molecules to dissolve in water is due to polarity. For example, in the illustration below sodium chloride is shown in its crystalline form and dissolved in water. Many other unique properties of water are due to the hydrogen bonds. For example, ice floats because hydrogen bonds hold water molecules further apart in a solid than in a liquid, where there is one less hydrogen bond per molecule. The unique physical properties, including a high heat of vaporization, strong surface tension, high specific heat, and nearly universal solvent properties of water are also due to hydrogen bonding. The hydrophobic effect, or the exclusion of compounds containing carbon and hydrogen (nonpolar compounds) is another unique property of water caused by the hydrogen bonds. The hydrophobic effect is particularly important in the formation of cell membranes. The best description is to say that water "squeezes" nonpolar molecules together.

Acids and Bases, Ionization of Water

$$H_2O \xrightarrow{\text{Limited}} H^+ + OH^-$$

- Acid release H^+
- Bases accept H^+.

We define the pH of a solution as the negative logarithm of the hydrogen ion concentration.

- at pH 7.0, a solution is neutral
- at lower pH (1-6), a solution is acidic
- at higher pH (8-14), a solution is basic.

The Major Classes of Biological Molecules

Now that we have introduced two groups of biological molecules, let us discuss all four major groups of biological molecules:

- Proteins
- Nucleic Acids
- Carbohydrates
- Lipids.

These four types of biological molecules compose all life on earth.

Proteins have an incredible variety of functions in living organisms. The functions of proteins include but are not limited to:

- acting as biological catalysts
- forming structural parts of organisms
- participating in cell signal and recognition factors
- acting as molecules of immunity.

Nucleic Acids consist of two distinct, but closely related chemical forms: deoxyribonucleic acid (DNA) and ribonucleic acid (RNA). The main functions of these biomolecules include the storage of all heritable information of all organisms on earth and the conversion of this information into proteins. Carbohydrates are the major source of food and the major form of energy for most living organisms. When carbohydrates combine together to form polymers they can function as long term food storage molecules, as protective coverings for cells and organisms, as the main structural support for land plants and constituents of many cells and their contents. Lipids are the major constituents of all membranes in all cells. They also serve as food storage molecules. This class of biological molecules includes the hydrophobic fats, oils and waxes.

Often times students have difficulty remembering the four major classes of biological molecules, so I have a way to help. To recall the four major classes of biological molecules I want you to think of lunch at a hamburger joint. No endorsements intended! Take a hamburger, fries, and a milkshake and you have all the components of the four major classes of biological molecules.

Of all four classes which do you think is most abundant in this meal? If you guessed LIPIDS you would be right! The hamburger is full of fat, and processors add fat to the hamburger bun. Mayonnaise is oil and egg yolk, which is mostly fat. Milk contains fat (if your shake has any milk product in it), and fat is added to nondairy shakes. As for the fries, we all know why they are called fries!

Okay- we know there are plenty of lipids, but where do the rest of the biological molecules come in? The hamburger bun is made from wheat flour, which is full of starch, a carbohydrate. The sugar in your shake is the disaccharide sucrose, which is also a carbohydrate. As a matter of fact, the lettuce, onion and tomato on that burger all contain carbohydrates.

Meat is composed of muscle tissue, and muscle tissue is a structural form of protein. Also most of the formerly living things that we eat have some form of protein in them: wheat, milk, potatoes, and so on. All cells contain DNA and RNA, except mature mammalian red blood cells, so DNA and RNA can be found in all organic matter. The lettuce, onion, and tomato contain nucleic acid, and the hamburger meat contains cow DNA.

The Magic of Carbon

Carbon is the chemical element with symbol C and atomic number 6. As a member of group 14 on the periodic table, it is nonmetallic and tetravalent—making four electrons available to form covalent chemical bonds. There are three naturally occurring isotopes, with ^{12}C and ^{13}C being stable, while ^{14}C is radioactive, decaying with a half-life of about 5730 years. Carbon is one of the few elements known since antiquity. The name “carbon” comes from Latin language *carbo*, coal.

There are several allotropes of carbon of which the best known are graphite, diamond, and amorphous carbon. The physical properties of carbon vary widely with the allotropic form. For example, diamond is highly transparent, while graphite is opaque and black. Diamond is among the hardest materials known, while graphite is soft enough to form a streak on paper (hence its name, from the Greek word “to write”). Diamond has a very low electrical conductivity, while graphite is a very good conductor. Under normal conditions, diamond has the highest thermal conductivity of all known materials. All the allotropic forms are solids under normal conditions but graphite is the most thermodynamically stable. All forms of carbon are highly stable, requiring high temperature to react even with oxygen.

The most common oxidation state of carbon in inorganic compounds is +4, while +2 is found in carbon monoxide and other transition metal carbonyl complexes. The largest sources of inorganic carbon are limestones, dolomites and carbon dioxide, but significant quantities occur in organic deposits of coal, peat, oil and methane clathrates.

Carbon forms more compounds than any other element, with almost ten million pure organic compounds described to date, which in turn are a tiny fraction of such compounds that are theoretically possible under standard conditions.

Carbon is the 15th most abundant element in the Earth's crust, and the fourth most abundant element in the universe by mass after hydrogen, helium, and oxygen. It is present in all known lifeforms, and in the human body carbon is the second most abundant element by mass (about 18.5%) after oxygen.

This abundance, together with the unique diversity of organic compounds and their unusual polymer-forming ability at the temperatures commonly encountered on Earth, make this element the chemical basis of all known life.

Carbon has the ability to form very long chains of interconnecting C-C bonds. This property is called catenation. Carbon-carbon bonds are strong, and stable. This property allows carbon to form an almost infinite number of compounds; in fact, there are more known carbon-containing compounds than all the compounds of the other chemical elements combined except those of hydrogen (because almost all organic compounds contain hydrogen too).

The simplest form of an organic molecule is the hydrocarbon—a large family of organic molecules that are composed of hydrogen atoms bonded to a chain of carbon atoms.

Chain length, side chains and functional groups all affect the properties of organic molecules. By IUPAC's definition, all the other organic compounds are functionalized compounds of hydrocarbons.

Carbon occurs in all known organic life and is the basis of organic chemistry. When united with hydrogen, it forms various hydrocarbons which are important to industry as refrigerants, lubricants, solvents, as chemical feedstock for the manufacture of plastics and petrochemicals and as fossil fuels.

When combined with oxygen and hydrogen, carbon can form many groups of important biological compounds including sugars, lignans, chitins, alcohols, fats, and aromatic esters, carotenoids and terpenes. With nitrogen it forms alkaloids, and with the addition of sulphur also it forms antibiotics, amino acids, and rubber products. With the addition of phosphorus to these other elements, it forms DNA and RNA, the chemical-code carriers of life, and adenosine triphosphate (ATP), the most important energy-transfer molecule in all living cells.

Inorganic Compounds

Commonly carbon-containing compounds which are associated with minerals or which do not contain hydrogen or fluorine, are treated separately from classical organic compounds; however the definition is not rigid. Among these are the simple oxides of carbon.

The most prominent oxide is carbon dioxide (CO_2). This was once the principal constituent of the paleoatmosphere, but is a minor component of the Earth's atmosphere today. Dissolved in water, it forms carbonic acid (H_2CO_3), but as most compounds with multiple single-bonded oxygens on a single carbon it is unstable. Through this intermediate, though, resonance-stabilized carbonate ions are produced. Some important minerals are carbonates, notably calcite.

Carbon disulfide (CS_2) is similar. The other common oxide is carbon monoxide (CO). It is formed by incomplete combustion, and is a colorless, odorless gas. The molecules each contain a triple bond and are fairly polar, resulting in a tendency to bind permanently to hemoglobin molecules, displacing oxygen, which has a lower binding affinity.

Cyanide (CN^-), has a similar structure, but behaves much like a halide ion (pseudohalogen). For example it can form the nitride cyanogen molecule ($(CN)_2$), similar to diatomic halides. Other uncommon oxides are carbon suboxide (C_3O_2), the unstable dicarbon monoxide (C_2O), carbon trioxide (CO_3), cyclopentanepentone (C_5O_5), cyclohexanehexone (C_6O_6), and mellitic anhydride ($C_{12}O_9$).

Carbon Cycle

The carbon cycle is the biogeochemical cycle by which carbon is exchanged among the biosphere, pedosphere, geosphere, hydrosphere, and atmosphere of the Earth. It is one of the most important cycles of the earth and allows for carbon to be recycled and reused throughout the biosphere and all of its organisms.

The carbon cycle was initially discovered by Joseph Priestley and Antoine Lavoisier, and popularized by Humphry Davy. It is now usually thought of as including the following major reservoirs of carbon interconnected by pathways of exchange:

- The atmosphere
- The terrestrial biosphere, which is usually defined to include fresh water systems and non-living organic material, such as soil carbon.

- The oceans, including dissolved inorganic carbon and living and non-living marine biota,
- The sediments including fossil fuels.
- The Earth's interior, carbon from the Earth's mantle and crust is released to the atmosphere and hydrosphere by volcanoes and geothermal systems.

Carbon forms the backbone of biology for all life on Earth. Complex molecules are made up of carbon bonded with other elements, especially oxygen, hydrogen and nitrogen, and carbon is able to bond with all of these because of its four valence electrons. It is often assumed in astrobiology that if life exists somewhere else in the universe, it will also be carbon based. This assumption is referred to by critics as *carbon chauvinism.*

Characteristics of Carbon as a Basis for Life

The two most important characteristics of carbon as a basis for the chemistry of life, are that it has four valence bonds and that the energy required to make or break a bond is just at an appropriate level for building molecules which are not only stable, but also reactive. The fact that carbon atoms bond readily to other carbon atoms allows for the building of arbitrarily long and complex molecules.

There are not many other elements which appear to be even promising candidates for supporting life-like behaviour, but the most frequent alternative suggested is silicon. This is in the same group in the Periodic Table of elements and therefore also has four valence bonds.

It also bonds to itself, but generally in the form of crystal lattices, less amenable to a complete source of life, rather than long chains. Its compounds are generally highly stable and do not support the ability readily to re-combine in different permutations in a manner that would plausibly support life-like processes.

This speculation of a life based on the chemistry of silicon is clearly distinct from "silicon-based life" in the above sense of artificial intelligence based on electronic processes utilizing silicon integrated circuits.

Key Carbon-based Molecules in the Life Processes

The most notable groups of chemicals used in the processes of living organisms include:

- Proteins, which are the building blocks from which the structures of living organisms are constructed (this includes almost all enzymes, which catalyse organic chemical reactions)
- Nucleic acids, which carry genetic information

- Carbohydrates, which store energy in a form that can be used by living cells
- Fats, which also store energy, but in a more concentrated form, and which may be stored for extended periods in the bodies of animals.

Carbon's importance to plants and animals is extremely important. Its presence within the atmosphere as well as within living matter makes it fundamental to their survival.

Plants

Carbon is essential to plants in order for them to perform photosynthesis—the chemical reaction that plants use to create energy. Plants take carbon from carbon dioxide in the atmosphere and produce food for themselves, as well as water and oxygen for the environment.

Animals

Animals are dependent upon water to drink as well as oxygen in the air in order to breathe. Without carbon, plants would not be able to produce the necessary water and oxygen for animal environments, and would not survive to be a food source for animals.

Organic Compounds

Carbon atoms form molecular bonds at very basic levels. For this reason, important organic compounds such as simple proteins and sugars are easily synthesized in the environment, providing necessary nutrients for plants and animals.

Cell Structure

Due to their physical structure and chemical properties, carbon atoms may form a number of structures and configurations that are key to cell structure and functioning.

The Basis of Life

Carbon atoms are found not only in all plants and animals, but in all other living life forms as well. Without carbon, life as we know it would simply not exist.

Basic Chemistry of Nucleic Acid

Living organisms are complex systems. Hundreds of thousands of proteins exist inside each one of us to help carry out our daily functions. These proteins are produced locally, assembled piece-by-piece to exact specifications. An enormous amount of information is required to manage this complex system correctly. This information, detailing the specific structure of the proteins inside of our bodies, is stored in a set of molecules called nucleic acids.

Thermal Properties of DNA

As cells divide it is a necessity that the DNA be copied (replicated), in such a way that each daughter cell acquires the same amount of genetic material. In order for this process to proceed the two strands of the helix must first be separated, in a process termed denaturation. This process can also be carried out in vitro. If a solution of DNA is subjected to high temperature, the H-bonds between bases become unstable and the strands of the helix separate in a process of thermal denaturation.

The base composition of DNA varies widely from molecule to molecule and even within different regions of the same molecule. Regions of the duplex that have predominantly A-T base-pairs will be less thermally stable than those rich in G-C base-pairs. In the process of thermal denaturation, a point is reached at which 50% of the DNA molecule exists as single strands. This point is the melting temperature (T_m), and is characteristic of the base composition of that DNA molecule. The T_m depends upon several factors in addition to the base composition. These include the chemical nature of the solvent and the identities and concentrations of ions in the solution.

When thermally melted DNA is cooled, the complementary strands will again re-form the correct base pairs, in a process is termed annealing or hybridization. The rate of annealing is dependent upon the nucleotide sequence of the two strands of DNA.

Analysis of DNA Structure

Chromatography: Several of the chromatographic techniques available for the characterization of proteins can also be applied to the characterization of DNA. The most commonly used technique is HPLC (high performance liquid chromatography). Affinity chromatographic techniques also can be employed. One common affinity matrix is hydroxyapatite (a form of calcium phosphate), which binds double-stranded DNA with a higher affinity than single-stranded DNA.

Electrophoresis: This procedure can serve the same function with regard to DNA molecules as it does for the analysis of proteins. However, since DNA molecules have much higher molecular weights than proteins, the molecular sieve used in electrophoresis of DNA must be different as well. The material of choice is agarose, a carbohydrate polymer purified from a salt water algae. It is a copolymer of mannose and galactose that when melted and re-cooled forms a gel with pores sizes dependent upon the concentration of agarose. The phosphate backbone of DNA is highly negatively charged, therefore DNA will migrate in an electric field. The size of DNA fragments can then be determined by comparing their migration in the gel to known size standards. Extremely large molecules of DNA (in the range of 30kb–10Mb) are effectively separated in agarose gels using pulsed-field gel electrophoresis (PFGE). This technique employs two or more electrodes, placed orthogonally with respect to the gel, that receive short alternating pulses of current. PFGE allows whole chromosomes and large portions of chromosomes to be analysed.

Carbohydrates

In many ways, our bodies can be thought of as chemical processing plants. Chemicals are taken in, processed through various types of reactions, and then distributed throughout the body to be used immediately or stored for later use. The chemicals used by the body can be divided into two broad categories: macronutrients, those substances that we need to eat regularly in fairly large quantities, and micronutrients, those substances that we need only in small amounts. Three major classes of macronutrients are essential to living organisms: carbohydrates, fats, and proteins.

A carbohydrate is an organic compound which has the empirical formula $C_m(H_2O)_n$; that is, consists only of carbon, hydrogen and oxygen, with a hydrogen:oxygen atom ratio of 2:1 (as in water). Carbohydrates can be viewed as hydrates of carbon, hence their name. Structurally however, it is more accurate to view them as polyhydroxy aldehydes and ketones.

The term is most common in biochemistry, where it is a synonym of saccharide. The carbohydrates (saccharides) are divided into four chemical groupings: monosaccharides, disaccharides, oligosaccharides, and polysaccharides. In general, the monosaccharides and disaccharides, which are smaller (lower molecular weight) carbohydrates, are commonly referred to as sugars. While the scientific nomenclature of carbohydrates is complex, the names of the monosaccharides and disaccharides very often end in the suffix-ose.

For example, blood sugar is the monosaccharide glucose, table sugar is the disaccharide sucrose, and milk sugar is the disaccharide lactose. Carbohydrates perform numerous roles in living things. Polysaccharides serve for the storage of energy (e.g., starch and glycogen) and as structural components (e.g., cellulose in plants and chitin in arthropods).

The 5-carbon monosaccharide ribose is an important component of coenzymes (e.g., ATP, FAD, and NAD) and the backbone of the genetic molecule known as RNA. The related deoxyribose is a component of DNA. Saccharides and their derivatives include many other important biomolecules that play key roles in the immune system, fertilization, preventing pathogenesis, blood clotting, and development.

In food science and in many informal contexts, the term carbohydrate often means any food that is particularly rich in the complex carbohydrate starch (such as cereals, bread and pasta) or simple carbohydrates, such as sugar (found in candy, jams and desserts).

Structure

Formerly the name "carbohydrate" was used in chemistry for any compound with the formula $C_m(H_2O)_n$. Following this definition, some chemists considered formaldehyde CH_2O to be the simplest carbohydrate, while others claimed that title for glycolaldehyde Today the term is generally understood in the biochemistry sense, which excludes compounds with only one or two carbons. Natural saccharides are generally built of simple carbohydrates called monosaccharides with general formula $(CH_2O)_n$ where n is three or more. A typical monosaccharide has the structure H-$(CHOH)_x$(C=O)-$(CHOH)_y$-H, that is, an aldehyde or ketone with many hydroxyl groups added, usually one on each carbon atom that is not part of the aldehyde or ketone functional group. Examples of monosaccharides are glucose, fructose, and glyceraldehyde. However, some biological substances commonly called "monosaccharides" do not conform to this formula (e.g., uronic acids and deoxy-sugars such as fucose), and there are many chemicals that do conform to this formula but are not considered to be monosaccharides (e.g., formaldehyde CH_2O and inositol $(CH_2O)_6$). The open-chain form of a monosaccharide often coexists with a closed ring form where the aldehyde/ketone carbonyl group carbon (C=O) and hydroxyl group (-OH) react forming a hemiacetal with a new C-O-C bridge.

Monosaccharides can be linked together into what are called polysaccharides (or oligosaccharides) in a large variety of ways. Many carbohydrates contain one or more modified monosaccharide units that have had one or more groups replaced or removed. For example, deoxyribose, a component of DNA, is a modified version of ribose; chitin is composed of repeating units of N-acetylglucosamine, a nitrogen-containing form of glucose.

Monosaccharides

Monosaccharides are the most basic units of biologically important carbohydrates. They are the simplest form of sugar and are usually colorless, water-soluble, crystalline solids. Some monosaccharides have a sweet taste. Examples of monosaccharides include glucose (dextrose), fructose (levulose), galactose, xylose and ribose. Monosaccharides are the building blocks of disaccharides such as sucrose and polysaccharides (such as cellulose and starch). Further, each carbon atom that supports a hydroxyl group (except for the first and last) is chiral, giving rise to a number of isomeric forms all with the same chemical formula. For instance, galactose and glucose are both aldohexoses, but have different chemical and physical properties.

Structure

With few exceptions (e.g., deoxyribose), monosaccharides have the chemical formula $C_x(H_2O)_y$ with the chemical structure $H(CHOH)_nC{=}O(CHOH)_mH$. If n or m is zero, it is an aldehyde and is termed an aldose; otherwise, it is a ketone and is termed a ketose. Monosaccharides contain either a ketone or aldehyde functional group, and hydroxyl groups on most or all of the non-carbonyl carbon atoms.

Monosaccharide Momenclature

Monosaccharides are classified by the number of carbon atoms they contain:

- Triose, 3 carbon atoms
- Tetrose, 4 carbon atoms
- Pentose, 5 carbon atoms
- Hexose, 6 carbon atoms
- Heptose, 7 carbon atoms.

Some monosaccharides exist in equilibrium between its open chain and the cyclic hemiacetal forms:

Aldoses can form:

- Pyranose results from the intramolecular nucleophilic addition of the C5-OH group to the C1 carbonyl group, yielding a 6-membered ring.

Ketoses can form:

- Pyranose results from the intramolecular nucleophilic addition of the -OH at C6 to the C1 carbonyl group, forming a 6-membered ring.
- Furanose results from the intramolecular nucleophilic addition of the C4-OH group at C5 and to the C1 carbonyl group, yielding a 5-membered ring.

Monosaccharides are classified by the type of carbonyl group they contain:

- Aldose, -CHO (aldehyde)
- Ketose, C=O (ketone).

Isomerism

The total number of possible stereoisomers of one compound (n) is dependent on the number of stereogenic centres (c) in the molecule. The upper limit for the number of possible stereoisomers is $n = 2^c$. The only monosaccharide without an isomer is dihydroxyacetone or DHA.

Monosaccharides are classified according to their molecular configuration at the chiral carbon furthest removed from the aldehyde or ketone group. The chirality at this carbon is compared to the chirality of carbon 2 on glyceraldehyde. If it is equivalent to D-glyceraldehyde' s C2, the sugar is D; if it is equivalent to L-glyceraldehyde's C2, the sugar is L. Due to the chirality of the sugar molecules, an aqueous solution of a D or L saccharides will rotate light. D-glyceraldehyde causes polarized light to rotate clockwise (dextrorotary); L-glyceraldehyde causes polarized light to rotate counterclockwise (levorotary). Unlike glyceraldehyde, D/L designation on more complex sugars is not associated with their direction of light rotation. Since more complex sugars contain multiple chiral carbons, the direction of light rotation cannot be predicted by the chirality of the carbon that defines D/L nomenclature.

- D, configuration as in D-glyceraldehyde
- L, configuration as in L-glyceraldehyde.

All these classifications can be combined, resulting in names like *D-aldohexose* or *ketotriose*.

Derivatives

A large number of biologically important modified monosaccharides exist:

- Amino sugars such as:
  * Galactosamine
  * Glucosamine
  * Sialic acid
  * *N*-Acetylglucosamine
- Sulfosugars such as:
  * Sulfoquinovose.

Monosaccharides are the simplest carbohydrates in that they cannot be hydrolysed to smaller carbohydrates. They are aldehydes or ketones with two or more hydroxyl groups. The general chemical formula of an unmodified monosaccharide is $(C\cdot H_2O)_n$, literally a "carbon hydrate." Monosaccharides are important fuel molecules as well as building blocks for nucleic acids. The smallest monosaccharides, for which n = 3, are dihydroxyacetone and D- and L-glyceraldehyde.

Classification of monosaccharides

Monosaccharides are classified according to three different characteristics: the placement of its carbonyl group, the number of carbon atoms it contains, and its chiral handedness. If the carbonyl group is an aldehyde, the monosaccharide is an aldose; if the carbonyl group is a ketone, the monosaccharide is a ketose. Monosaccharides with three carbon atoms are called trioses, those with four are called tetroses, five are called pentoses, six are hexoses, and so on. These two systems of classification are often combined. For example, glucose is an aldohexose (a six-carbon aldehyde), ribose is an aldopentose (a five-carbon aldehyde), and fructose is a ketohexose (a six-carbon ketone). Each carbon atom bearing a hydroxyl group (-OH), with the exception of the first and last carbons, are asymmetric, making them stereocenters with two possible configurations each (R or S). Because of this asymmetry, a number of isomers may exist for any given monosaccharide formula. The aldohexose D-glucose, for example, has the formula $(C \cdot H_2O)_6$, of which all but two of its six carbons atoms are stereogenic, making D-glucose one of $2^4 = 16$ possible stereoisomers. In the case of glyceraldehyde, an aldotriose, there is one pair of possible stereoisomers, which are enantiomers and epimers. 1,3-dihydroxyacetone, the ketose corresponding to the aldose glyceraldehyde, is a symmetric molecule with no stereocenters). The assignment of D or L is made according to the orientation of the asymmetric carbon furthest from the carbonyl group: in a standard Fischer projection if the hydroxyl group is on the right the molecule is a D sugar, otherwise it is an L sugar. The "D-" and "L-" prefixes should not be confused with "d-" or "l-", which indicate the direction that the sugar rotates plane polarized light. This usage of "d-" and "l-" is no longer followed in carbohydrate chemistry.

Ring-straight chain isomerism

The aldehyde or ketone group of a straight-chain monosaccharide will react reversibly with a hydroxyl group on a different carbon atom to form a hemiacetal or hemiketal, forming a heterocyclic ring with an oxygen bridge between two carbon atoms. Rings with five and six atoms are called furanose and pyranose forms, respectively, and exist in equilibrium with the straight-chain form.

During the conversion from straight-chain form to cyclic form, the carbon atom containing the carbonyl oxygen, called the anomeric carbon, becomes a stereogenic centre with two possible configurations: The oxygen atom may take a position either above or below the plane of the ring. The resulting possible pair of stereoisomers are called anomers. In the *α anomer*, the -OH substituent on the anomeric carbon rests on the opposite side (trans) of the ring from the CH_2OH side branch.

The alternative form, in which the CH_2OH substituent and the anomeric hydroxyl are on the same side (cis) of the plane of the ring, is called the *β anomer*. You can remember that the β anomer is cis by the mnemonic, "It's always better to βe up". Because the ring and straight-chain forms readily interconvert, both anomers exist in equilibrium. In a Fischer Projection, the α anomer is represented with the anomeric hydroxyl group *trans* to the CH_2OH and *cis* in the β anomer.

Use in living organisms

Monosaccharides are the major source of fuel for metabolism, being used both as an energy source (glucose being the most important in nature) and in biosynthesis. When monosaccharides are not immediately needed by many cells they are often converted to more space efficient forms, often polysaccharides. In many animals, including humans, this storage form is glycogen, especially in liver and muscle cells. In plants, starch is used for the same purpose.

Disaccharides

Two joined monosaccharides are called a disaccharide and these are the simplest polysaccharides. Examples include sucrose and lactose. They are composed of two monosaccharide units bound together by a covalent bond known as a glycosidic linkage formed via a dehydration reaction, resulting in the loss of a hydrogen atom from one monosaccharide and a hydroxyl group from the other. The formula of unmodified disaccharides is $C_{12}H_{22}O_{11}$. Although there are numerous kinds of disaccharides, a handful of disaccharides are particularly notable.

Sucrose, pictured to the right, is the most abundant disaccharide, and the main form in which carbohydrates are transported in plants. It is composed of one D-glucose molecule and one D-fructose molecule. The systematic name for sucrose, *O*-α-D-glucopyranosyl-(1→2)-D-fructofuranoside, indicates four things:

- Its monosaccharides: glucose and fructose
- Their ring types: glucose is a pyranose, and fructose is a furanose
- How they are linked together: the oxygen on carbon number 1 (C1) of α-D-glucose is linked to the C2 of D-fructose.
- The *-oside* suffix indicates that the anomeric carbon of both monosaccharides participates in the glycosidic bond.

Lactose, a disaccharide composed of one D-galactose molecule and one D-glucose molecule, occurs naturally in mammalian milk. The systematic name for lactose is *O*-β-D-galactopyranosyl-(1→4)-D-glucopyranose. Other notable disaccharides include maltose (two D-glucoses linked α-1,4) and cellulobiose (two D-glucoses linked β-1,4). disaccharides can be classified into two types they are reducing and non-reducing disaccahrides if the functional group is present in bonding with another sugar unit it is called as reducing disaccharide.

Oligosaccharides and Polysaccharides

Oligosaccharides and polysaccharides are composed of longer chains of monosaccharide units bound together by glycosidic bonds. The distinction between the two is based upon the number of monosaccharide units present in the chain. Oligosaccharides typically contain between three and ten monosaccharide units, and polysaccharides contain greater than ten monosaccharide units. Definitions of how large a carbohydrate must be to fall into each category vary according to personal opinion. Examples of oligosaccharides include the disaccharides mentioned above, the trisaccharide raffinose and the tetrasaccharide stachyose. Oligosaccharides are found as a common form of protein posttranslational modification. Such posttranslational modifications include the Lewis and ABO oligosaccharides responsible for blood group classifications and so of tissue incompatibilities, the alpha-Gal epitope responsible for hyperacute rejection in xenotransplantation, and O-GlcNAc modifications. Polysaccharides represent an important class of biological polymers. Their function in living organisms is usually either structure- or storage-related. Starch (a polymer of glucose) is used as a storage polysaccharide in plants, being found in the form of both amylose and the branched amylopectin.

In animals, the structurally similar glucose polymer is the more densely branched glycogen, sometimes called 'animal starch'. Glycogen's properties allow it to be metabolized more quickly, which suits the active lives of moving animals.

Cellulose and chitin are examples of structural polysaccharides. Cellulose is used in the cell walls of plants and other organisms, and is claimed to be the most abundant organic molecule on earth. It has many uses such as a significant role in the paper and textile industries, and is used as a feedstock for the production of rayon (via the viscose process), cellulose acetate, celluloid, and nitrocellulose.

Chitin has a similar structure, but has nitrogen-containing side branches, increasing its strength. It is found in arthropod exoskeletons and in the cell walls of some fungi. It also has multiple uses, including surgical threads. Other polysaccharides include callose or laminarin, chrysolaminarin, xylan, arabinoxylan, mannan, fucoidan and galactomannan. Carbohydrates are the main energy source for the human body.

Chemically, carbohydrates are organic molecules in which carbon, hydrogen, and oxygen bond together in the ratio: $C_x(H_2O)_y$, where x and y are whole numbers that differ depending on the specific carbohydrate to which we are referring.

Animals (including humans) break down carbohydrates during the process of metabolism to release energy. For example, the chemical metabolism of the sugar glucose is shown below:

$$C_6H_{12}O_6 + 6\ O_2\ 6\ CO_2 + 6\ H_2O + \text{energy}$$

Animals obtain carbohydrates by eating foods that contain them, for example potatoes, rice, breads, and so on. These carbohydrates are manufactured by plants during the process of photosynthesis. Plants harvest energy from sunlight to run the reaction just described in reverse:

$$6\ CO_2 + 6\ H_2O + \text{energy (from sunlight)}\ C_6H_{12}O_6 + 6\ O_2$$

A potato, for example, is primarily a chemical storage system containing glucose molecules manufactured during photosynthesis. In a potato, however, those glucose molecules are bound together in a long chain. As it turns out, there are two types of carbohydrates, the simple sugars and those carbohydrates that are made of long chains of sugars- the complex carbohydrates.

Simple Sugars

All carbohydrates are made up of units of sugar (also called saccharide units). Carbohydrates that contain only one sugar unit (monosaccharides) or two sugar units (disaccharides) are referred to as simple sugars. Simple sugars are sweet in taste and are broken down quickly in the body to release energy. Two of the most common monosaccharides are glucose and fructose. Glucose is the primary form of sugar stored in the human body for energy. Fructose is the main sugar found in most fruits. Both glucose and fructose have the same chemical formula ($C_6H_{12}O_6$); however, they have different structures, as shown (note: the carbon atoms that sit in the "corners" of the rings are not labelled):

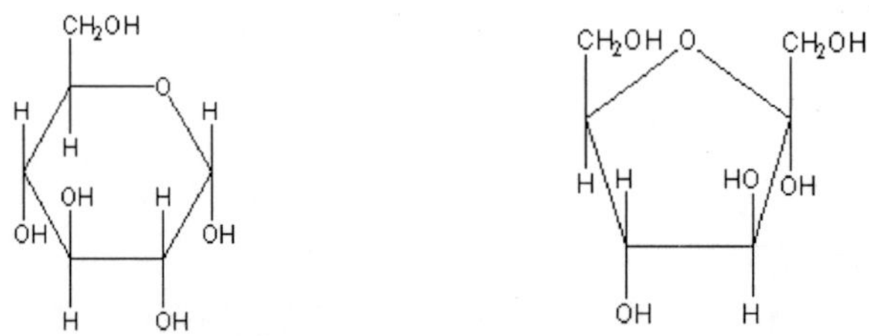

Figure: *Glucose Fructose*

Disaccharides have two sugar units bonded together. For example, common table sugar is sucrose, a disaccharide that consists of a glucose unit bonded to a fructose unit:

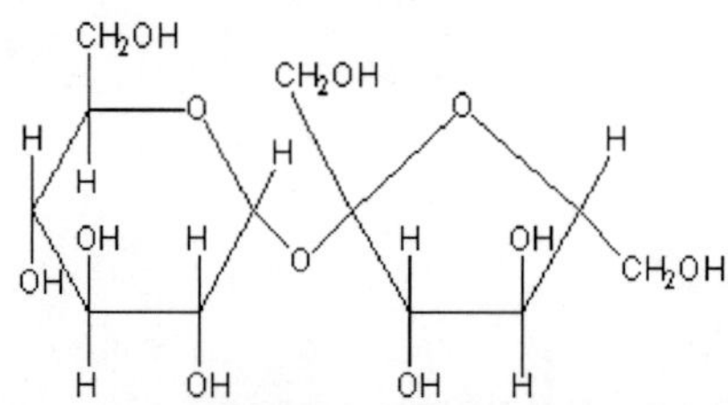

Figure: *Sucrose*

Complex carbohydrates Complex carbohydrates are polymers of the simple sugars. In other words, the complex carbohydrates are long chains of simple sugar units bonded together (for this reason the complex carbohydrates are often referred to as polysaccharides). The potato we discussed earlier actually contains the complex carbohydrate starch. Starch is a polymer of the monosaccharide glucose:

Starch is the principal polysaccharide used by plants to store glucose for later use as energy. Plants often store starch in seeds or other specialized organs; for example, common sources of starch include rice, beans, wheat, corn, potatoes, and so on. When humans eat starch, an enzyme that occurs in saliva and in the intestines called amylase breaks the bonds between the repeating glucose units, thus allowing the sugar to be absorbed into the bloodstream.

Once absorbed into the bloodstream, the human body distributes glucose to the areas where it is needed for energy or stores it as its own special polymer-glycogen. Glycogen, another polymer of glucose, is the polysaccharide used by animals to store energy. Excess glucose is bonded together to form glycogen molecules, which the animal stores in the liver and muscle tissue as an "instant" source of energy. Both starch and glycogen are polymers of glucose; however, starch is a long, straight chain of glucose units, whereas glycogen is a branched chain of glucose units.

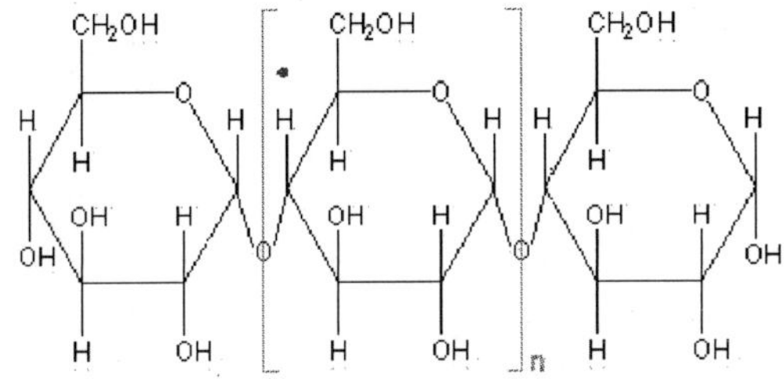

Figure: *Starch (n is the number of repeating glucose units and ranges in the 1,000's)*

Another important polysaccharide is cellulose. Cellulose is yet a third polymer of the monosaccharide glucose. Cellulose differs from starch and glycogen because the glucose units form a two-dimensional structure, with hydrogen bonds holding together nearby polymers, thus giving the molecule added stability. Cellulose, also known as plant fiber, cannot be digested by human beings, therefore cellulose passes through the digestive tract without being absorbed into the body. Some animals, such as cows and termites, contain bacteria in their digestive tract that help them to digest cellulose. Cellulose is a relatively stiff material, and in plants it is used as a structural molecule to add support to the leaves, stem, and other plant parts. Despite the fact that it cannot be used as an energy source in most animals, cellulose fiber is essential in the diet because it helps exercise the digestive track and keep it clean and healthy.

How Organisms Convert Food to Energy

Do you ever wonder how organisms convert food to energy?

We all know that organisms must take in nutrients to survive. This self-instructional unit describes two of the most important processes in food to energy conversion. These processes, called glycolysis and respiration, use the energy of the glucose molecule to manufacture ATP, the most widely used energy molecule of the cell. ATP is an abbreviation for the nucleotide Adenosine triphosphate. This nucleotide consists of a ribose, a phosphate and the nitrogenous base adenine. This is the same nucleotide used to manufacture RNA. Don't be surprised by the fact that ATP is very energetic. All of the nucleotides used to make RNA and DNA are full of energy. Most of their energy is stored in the two bonds between their two terminal phosphates. The nucleotides of DNA are called deoxyribonucleotides (abbreviated dATP, dTTP, dCTP and dGTP).

The nucleotides of RNA are called ribonucleotides (abbreviated ATP, UTP, CTP, and GTP). While these nucleotides have different types of ribose sugar and nitrogenous bases, they all have the two terminal phosphate groups, and so they all have equivalent energies. All the energy in the nucleotide comes from the two terminal phosphates.

The energy used to make DNA and RNA polymers, through the formation of phosphodiester bonds, is supplied by the triphosphate nucleotides when they polymerize to form nucleic acids. It takes energy to form all chemical bonds, and the breaking of these chemical bonds will release energy.

The triphosphate nucleotides are special because the bonds between the phosphates take more energy to form (~8 kilocalories/mole) as compared to the bonds between the phosphate and the ribose (~5 kilocalories/mole). Therefore these high energy bonds between the phosphates release more energy when they are broken.

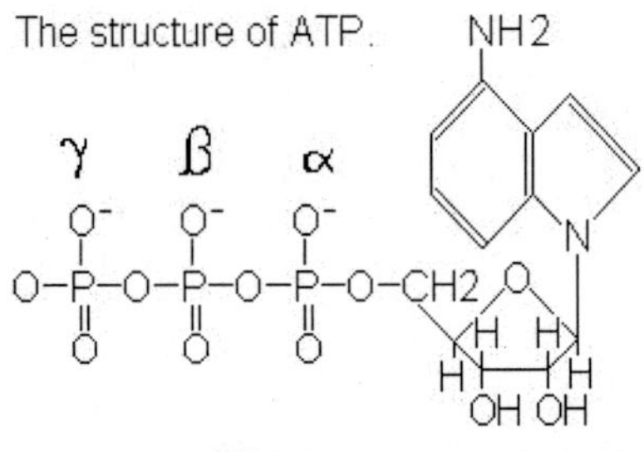

Figure: *The structure of ATP*

The phosphates in the structural formula of ATP have been assigned specific labels based on the IUPAC system. The phosphate attached to the #5 carbon of the ribose is labelled as the alpha (a) phosphate. The next two phosphates are labelled beta (ß) and gamma (g) from left to right. The bonds between the beta and gamma phosphates are represented by wavy lines which indicate the higher energy potential of these bonds.

Most biochemical reactions require energy input to proceed. These types of reactions are called endergonic reactions. An exergonic reaction, by contrast, gives off energy and proceeds spontaneously. Many chemical reactions of the cell use the energy from ATP which released when the beta or gamma phosphate bonds of ATP are broken. The energy released from the ATP supplies the energy necessary to form or break chemical bonds in biochemical reactions. The energy released by the conversion of ATP to ADP is used in the manufacture of ATP during the process of glycolysis.

When the phosphates are cleaved from ATP, two other forms of the nucleotide are made; ADP (adenosine diphosphate) and AMP (adenosine monophosphate). If you cover the terminal (gamma) phosphate in the structural formula of ATP with a piece of paper you will see what ADP looks like and if you cover the gamma and beta phosphates you will see what AMP looks like.

ATP is made inside cells by the transfer of a phosphate on to the b phosphate of ADP. Where does ADP come from? ADP is always inside cells as the by-product of the cleavage of a gamma phosphate from ATP.

This is the ultimate use of recycling by the cell! Sometimes the ATP, ADP and AMP molecules are completely degraded because they wear out. Cells contain many biochemical pathways and can rebuild new AMP molecules from scratch when necessary.

The AMP molecule can then be phosphorylated to form ADP and ATP. However, the cell prefers to conserve energy, whenever possible, by the dephosphorylation of ATP to ADP and AMP and the rephosphorylation of these back to ATP. The following reaction represents the phosphorylation and dephosphorylation reactions of ATP.

$$\text{ATP} \rightleftharpoons \text{ADP} \rightleftharpoons \text{AMP}$$

How Cells Use Glucose for the Manufacture of ATP

The first step of the breakdown of glucose inside the cell, to utilize the potential energy stored in this molecule, is a process called glycolysis. Any basic college biology text will have all of the steps of the process of glycolysis. In this self instructional I want to give you an overview of the process, in the hope that you will understand what is taking place and not just memorize the process.

Where does glycolysis take place?

Glycolysis takes place inside the cytoplasm of the cell. The cytoplasm is the aqueous based solution inside a cell that has a variety of molecules solubilized in it. The contents of the cytoplasm include proteins, carbohydrates, nucleic acids, salts and a whole host of other soluble molecules.

What are the parts of the cell that are not considered cytoplasm?

The organelles inside the cytoplasm are not part of the cytoplasm. Remember, most organelles are bound by lipid membranes and are not soluble in the cytoplasm. The contents of organelles are also isolated from the cytoplasm due to the hydrophobic nature of their membranes. This means that the substances solubilized inside organelles stay separate from the rest of the cytoplasm. The enzymes necessary for glycolysis are solubilized in the cytoplasm of the cell.

What does the process of glycolysis accomplish?

- Glycolysis uses the energy from two ATP to perform a series of enzymatic reactions resulting in the manufacture of four ATP.
- Glycolysis also generates two three-carbon compounds called pyruvate that still contain energy. The destination for pyruvate is the Citric Acid Cycle (also known as Krebs cycle) which occurs inside the matrix of the mitochondria.
- Glycolysis results in the addition of two hydrogens (electrons) onto the energy carrier nicotinamide adenine dinucleotide (NAD) to make NADH. The process of adding electrons to a substrate is called reduction. The destination for the NADH is the electron transport chain also found in the mitochondria.

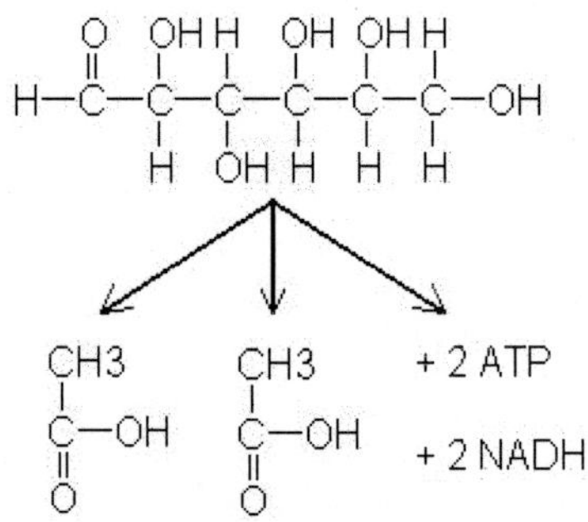

Figure: *Splitting of glucose through glycolysis*

Why are only two net ATP produced during glycolysis?

Although glycolysis allows for the formation of four ATP, two ATP are used in the process. This results in a net profit of two ATP from the process of Glycolysis.

Does glycolysis require molecular oxygen? No, glycolysis is an anaerobic process.

Enzymes and Enzymatic Reactions

To help understand the processes of glucose manufacture, anabolic reactions, and catabolic reactions, we need to understand how enzymes work.

Enzymes are a group of biological molecules, most of which are made of proteins. Enzymes are biological catalysts. A catalyst is added to a chemical reaction to speed the rate of the reaction. In other words, a catalyst makes the reaction occur faster. The catalyst is not changed in its form, but instead retains its original physical and chemical characteristics at the completion of the reaction. A catalyst does not participate in a reaction and can be reused again and again. Enzymes have all the qualities of a catalyst, but they are specifically used in the chemical reactions of living organisms (metabolism). Often people say; enzymes break this substance down or form this substance; however, to state the function of the enzyme correctly, one should say enzymes catalyse the formation or breakdown of a substance.

How do enzymes work? Enzymatic catalysts reduce the amount of energy necessary for endergonic reactions to proceed. Enzymes don't supply the energy, they reduce the amount needed. How? Well, there are several theories concerning this.

One way to speed reactions is to concentrate the materials that participate in them. Remember the generalized formula for chemical reactions?

Reactant A + Reactant B ———> Product C + Product D

Enzymes are molecules that have a 3-dimensional shape. Think of a muffin tin with all the depressions used to hold the individual muffins. If one hole contains reactant A and the adjacent hole contains reactant B, the two reactants are in close proximity to one another. This has the effect of concentrating the two reactants necessary for an anabolic reaction to occur and form products C and D.

Remember the cytoplasm of the cell or even the contents of an organelle may seem tiny, but it is a vast ocean of many components to a glucose or other small molecule.

Concentrating reactants is one way to speed reactions. Another way to reduce the energy and speed the rate of catabolic reactions might be to hold a molecule in place so that some of its bonds are under strain. Imagine you were trying to cut a thick piece of cloth with dull scissors and the going was very slow. If two people were to hold the material taut, or under strain, the fabric would cut more easily. Perhaps enzymes hold molecules in a similar fashion allowing bonds to be broken more easily.

If enzymes only speed the rates of reactions, does this mean that chemical reactions occur spontaneously anyway? Yes, all chemical reactions can occur spontaneously, but some reactions may take millennia to occur. Since organisms do not live on a geological time scale, the ability of enzymes to speed the rate of reactions, up to thousands of times faster, is necessary for life to exist.

The Structure of Enzymes

Enzymes are 3-dimensional molecules that have folded up into a specific shape (eg. the muffin tin). Imagine the depression in our muffin tin to be the site at which the chemical reaction takes place. This site is called the active site of the enzyme. The molecule that fits in the active site of the enzyme is called the substrate. Enzymes and substrates are very specific for each other. The active site of an enzyme is designed to fit perfectly with its specific substrate, like a 3-dimensional jigsaw puzzle.

Biochemical or Enzymatic Pathways

Sometimes a cell needs a product from a reactant that requires only a single chemical reaction. For example, reactant A needs to be changed to product B. This reaction would only require the enzyme that fits with its substrate, reactant A. However, this is not usually the case. Most products inside cells that need to be broken down or built up require a multi-step process. This process is called a biochemical pathway. It goes something like this:

- Enzyme A changes substrate A to product B.
- Product or substrate B now can be bound by enzyme B and be changed to product or substrate C
- Substrate C now fits with enzyme C which changes it to product D.
- If product D is what the cell needs, the pathway is finished. If not, the pathway can continue in the same way until the ultimate product is obtained.

Finally a quick lesson in naming enzymes. Generally to name an enzyme we take the substrate it breaks down or product that it manufactures and add an -ase on the end.

Examples: enzymes that degrade the following substances DNAse, ATPase or enzymes that build up the following substances: DNA polymerase and ATP synthetase.

Glucose Utilization

Both of the pyruvate (three-carbon molecules) are immediately converted to two-carbon molecules by removing a CO_2 (carbon dioxide) from each molecule. The remaining two-carbon compound is known as an acetyl group. The acetyl group will react quickly with another molecule because of its unpaired electrons. Therefore as the CO_2 is cleaved from pyruvate, the acetyl group is added simultaneously to a carrier molecule called coenzyme A (CoA).

Coenzyme A (CoA) is not an enzyme but a molecule similar in some respects to a nucleotide. This is a very important carrier molecule because it carries the acetyl group to many important biochemical pathways. The acetyl group and CoA molecule combine to form acetyl CoA. This molecule travels to the matrix of the mitochondria.

Once there, the CoA transfers the two-carbon acetyl group to an enzyme which catalyzes the addition of a four-carbon compound called oxaloacetic acid (oxaloacetate) to the acetyl. The result of this reaction is a six-carbon compound called citric acid (citrate). This is why this biochemical pathway is called the citric acid cycle. The citric acid molecule has carbon, hydrogen, and oxygen atoms stripped from it in a step-wise fashion through this biochemical pathway. Each step requires its own specific enzyme. The end result of this pathway is the regeneration of oxaloacetic acid, and the conversion of NAD and FAD to NADH and FADH with the formation of carbon dioxide CO_2 as a by-product. The major objective of the citric acid cycle is to get the hydrogens, along with their electrons, off the carbon compounds so they can be carried to the electron transport proteins that are embedded in the membrane of the mitochondrion.

The waste or by-products of these reactions are the carbon and oxygen that are stripped off as each compound is rearranged by the removal of hydrogen. These carbons and oxygens combine to form CO_2 carbon dioxide gas which is given off as a by-product of these reactions. The CO_2 that you breathe out of your lungs is a waste product coming from the cellular metabolism of glucose!

Carbon dioxide (CO_2) enters our bloodstream, eventually winds up in our lungs, and is expelled upon exhalation, but where do those hydrogens with their electrons go? The hydrogens and electrons are picked up by electron carrier NAD (nicotinamide adenine dinucleotide) and electron carrier FAD (flavin adenine dinucleotide).

These electron carriers transfer the hydrogens and electrons to the electron carrying proteins, embedded in the mitochondrial membrane, which form the electron transport chain. The electron transport chain proteins are complexed with metal ions which allow them to carry extra electrons safely. When NADH and FADH arrive at the electron transport chain, they transfer the electrons associated with the hydrogens to the electron carrying proteins at the top of the electron chain. When electrons and hydrogen atoms are added to a molecule, this is called reduction. When electrons or hydrogens are removed from a molecule, this is known as oxidation. NAD is in an oxidized state while NADH is in a reduced state.

Now that the electrons have been transferred to the electron transport proteins, the hydrogen ion H+ (proton) is free to be pumped across the membrane into a space between the inner and outer mitochondrial membranes.

This proton pumping requires energy which is generated by the flow of electrons as they are passed down the electron transport protein chain. As electrons are transferred from one protein to the next, energy is released at each transfer, until the final electron acceptor, oxygen is reached. Aerobic organisms must take in oxygen because it serves as the final electron acceptor in the electron transport chain. The energy, released from the passage of the electrons down the electron transport chain to oxygen, is used to pump the protons across the membrane. The pumping of protons requires a great deal of energy, because they have to be pumped against a concentration gradient. Because protons are constantly being pumped across the membrane, they fill the space between the mitochondrial membranes. This is known as an area of high proton concentration. Putting more protons in this area would be like trying to force balloons into a box already filled to capacity with balloons. Pushing protons against a concentration gradient of protons is referred to as going against, or up, a concentration gradient. In the case of the balloons, as soon as you open the box they would come out easily with no energy input necessary. This is termed going with, or down, the concentration gradient. Molecules naturally flow from areas of high concentration to areas of low concentration with no energy input necessary. Once you have captured all this energy by pushing protons up the proton concentration gradient into a crowded area, what happens? The protons can get out but they only have one passage to escape. This is a passage created by a protein enzyme called ATP synthetase. The ATP synthetase phosphorylates (adds phosphates to) ADP. The ATP synthetase gets the energy necessary for the formation of the high energy phosphate bond from the rush of protons through the channel it creates across the mitochondrial membrane.

To better understand how the flow of protons provides energy for the ATP synthetase enzyme imagine how a turbine engine works at Hoover Dam. The rushing water is caught and causes the huge turbine to turn which generates electricity. In a similar way the energy of protons cascading through the ATP synthetase is converted into the energy used to form the phosphate bond of ATP. This process, called oxidative phosphorylation, makes most of the ATP manufactured by cells.

While glycolysis gives us net profit of two ATP, the electron transport coupled with the citric acid cycle and oxidative phosphorylation give us a net profit of approximately 34 ATP from each glucose catabolized. Quite a big difference in yield! What happens to the protons and electrons when they flow back through the membrane and down electron transport chain? Since electrons are energetic enough to cause problems inside the cell they must be inactivated. The final electron acceptor, oxygen, combines with the electrons and excess hydrogen ions to form water, a non-toxic by-product.

So the products of respiration are CO_2 (from the citric acid cycle) and H_2O (from electron transport) and lots of ATP. Without oxygen, aerobic cells die because the manufacture of ATP is halted at the cellular level. Since oxygen is absolutely necessary for the continual phosphorylation of ADP to ATP this process is called oxidative phosphorylation. Without oxygen, electron transport would stop, protons would not be pumped, and the ATP synthetase would not have a flow of protons to supply the energy necessary for the manufacture of ATP. Do you know where all the molecules come from and their destination in the following Generalized Equation for Respiration?

$$C6H_{12}O_6 + O_2 \longrightarrow CO_2 + H_2O + \text{energy}$$

Now I would like to give you my overview of all of the processes of glucose metabolism in narrative form. I hope that you will understand the processes of glycolysis and respiration well enough to be able to describe this process in a similar way. In other words, I would like you to be able to think through and understand the concepts, not the minute details of these processes.

- Glucose is taken up as food by the body and carried to a cell. Glucose enters the cell by facilitated diffusion and is immediately broken down by the process of glycolysis in the cytoplasm of the cell.
- The process of glycolysis uses a biochemical pathway consisting of 10 enzyme mediated reactions which uses two ATP and manufactures four ATP (a net profit of two ATP). Glycolysis generates two NADH, which carry hydrogen and electrons to the electron transport chain in the mitochondrial membrane. Glycolysis also generates two three-carbon pyruvate molecules.
- The pyruvate molecule has a CO_2 removed and the remaining two-carbon acetyl group is then added onto the carrier molecule Coenzyme A to form acetyl CoA.
- Acetyl CoA is taken to the mitochondrial matrix to participate in the Citric acid cycle. The two-carbon acetyl group is then added via an enzyme mediated reaction to a four-carbon oxaloacetic acid molecule which results in the formation of the six-carbon molecule citric acid. Through a series of enzyme mediated reactions oxygens, carbons and hydrogens are systematically stripped off. The carbons and oxygens combine in the appropriate proportions to form CO_2 and ultimately are expelled as waste products.
- The hydrogen atoms along with their electrons are picked up by the electron carriers NAD and FAD and taken to the electron transport proteins, which are located on the inner mitochondrial membrane.
- The electrons are handed off to the electron transport proteins where they flow from one protein to another releasing energy at each transfer until they combine with oxygen, the final electron acceptor in the electron transport chain.
- The energy released from the flow of electrons down the transport chain is used to pump protons across the inner mitochondrial membrane into the space between the inner and outer mitochondrial membranes, against a proton concentration gradient.
- The protons then flow back down the concentration gradient through an ATP synthetase enzyme channel, back into the mitochondrial matrix. The flow of protons through the ATP synthetase generates the energy necessary to phosphorylate ADP to ATP.
- Once back inside the matrix the protons combine with oxygen and electrons to form water a by-product of oxidative phosphorylation.

6

Molecular Marker

In medicine, a molecular marker can be a substance that is introduced in an organism as a means to examine something. For example, rubidium chloride is used as a radioactive isotope to evaluate perfusion of heart muscle. In biology and medicine, a molecular marker (biomarker) can be a substance native to the organism whose detection indicates a particular disease state (for example, the presence of an antibody may indicate an infection). In genetics, a molecular marker (identified as genetic marker) is a fragment of DNA sequence that is associated to a part of the genome.

Molecular markers are used in molecular biology and biotechnology experiments where they use to identify a particular sequence of DNA. As the DNA sequences are very highly specific, they can be identified with the help of the known molecular markers which can find out a particular sequence of DNA from a group of unknown.molecular markers are useful in disease diagnosis also.

Molecular Markers in Plant Genome Analysis

With the advent of molecular markers, a new generation of markers has been introduced over the last two decades, which has revolutionized the entire scenario of biological sciences. DNA-based molecular markers have acted as versatile tools and have found their own position in various fields like taxonomy, physiology, embryology, genetic engineering, etc. They are no longer looked upon as simple DNA fingerprinting markers in variability studies or as mere forensic tools. Ever since their development, they are constantly being modified to enhance their utility and to bring about automation in the process of genome analysis. The discovery of PCR (polymerase chain reaction) was a landmark in this effort and proved to be an unique process that brought about a new class of DNA profiling markers. This facilitated the development of marker-based gene tags, map-based cloning of agronomically important genes, variability studies, phylogenetic analysis, synteny mapping, marker-assisted selection of desirable genotypes, etc. Thus giving new dimensions to concerted efforts of breeding and marker-aided selection that can reduce the time span of developing new and better varieties and will make the dream of super varieties come true. These DNA markers offer several advantages over traditional phenotypic markers, as they provide data that can be analysed objectively. In this article, DNA markers developed during the last two decades of molecular biology research and utilized for various applications in the area of plant genome analysis are reviewed.

PLANTS have always been looked upon as a key source of energy for survival and evolution of the animal kingdom, thus forming a base for every ecological pyramid. Over the last few decades plant genomics has been studied extensively bringing about a revolution in this area. Molecular markers, useful for plant genome analysis, have now become an important tool in this revolution. In this article we attempt to review most of the available DNA markers that can be routinely employed in various aspects of plant genome analysis such as taxonomy, phylogeny, ecology, genetics and plant breeding.

During the early period of research, classical strategies including comparative anatomy, physiology and embryology were employed in genetic analysis to determine inter- and intra-species variability. In the past decade, however, molecular markers have very rapidly complemented the classical strategies. Molecular markers include biochemical constituents (e.g. secondary metabolites in plants) and macromolecules, viz. proteins and deoxyribonucleic acids (DNA). Analysis of secondary metabolites is, however, restricted to those plants that produce a suitable range of metabolites which can be easily analysed and which can distinguish between varieties. These metabolites which are being used as markers should be ideally neutral to environmental effects or management practices. Hence, amongst the molecular markers used, DNA markers are more suitable and ubiquitous to most of the living organisms.

DNA-based Molecular Markers

Genetic polymorphism is classically defined as the simultaneous occurrence of a trait in the same population of two or more discontinuous variants or genotypes. Although DNA sequencing is a straightforward approach for identifying variations at a locus, it is expensive and laborious. A wide variety of techniques have, therefore, been developed in the past few years for visualizing DNA sequence polymorphism.

The term DNA-fingerprinting was introduced for the first time by Alec Jeffrey in 1985 to describe bar-code-like DNA fragment patterns generated by multilocus probes after electrophoretic separation of genomic DNA fragments. The emerging patterns make up an unique feature of the analysed individual and are currently considered to be the ultimate tool for biological individualization. Recently, the term DNA fingerprinting/profiling is used to describe the combined use of several single locus detection systems and are being used as versatile tools for investigating various aspects of plant genomes. These include characterization of genetic variability, genome fingerprinting, genome mapping, gene localization, analysis of genome evolution, population genetics, taxonomy, plant breeding, and diagnostics.

Properties desirable for ideal DNA markers

- Highly polymorphic nature
- Codominant inheritance (determination of homozygous and heterozygous states of diploid organisms)
- Frequent occurrence in genome
- Selective neutral behaviour (the DNA sequences of any organism are neutral to environmental conditions or management practices)
- Easy access (availability)
- Easy and fast assay
- High reproducibility
- Easy exchange of data between laboratories.

It is extremely difficult to find a molecular marker which would meet all the above criteria. Depending on the type of study to be undertaken, a marker system can be identified that would fulfil atleast a few of the above characteristics.

Various types of molecular markers are utilized to evaluate DNA polymorphism and are generally classified as hybridization-based markers and polymerase chain reaction (PCR)-based markers. In the former, DNA profiles are visualized by hybridizing the restriction enzyme-digested DNA, to a labelled probe, which is a DNA fragment of known origin or sequence.

PCR-based markers involve *in vitro* amplification of particular DNA sequences or loci, with the help of specifically or arbitrarily chosen oligonucleotide sequences (primers) and a thermostable DNA polymerase enzyme. The amplified fragments are separated electrophoretically and banding patterns are detected by different methods such as staining and autoradiography. PCR is a versatile technique invented during the mid-1980s. Ever since thermostable DNA polymerase was introduced in 1988, the use of PCR in research and clinical laboratories has increased tremendously. The primer sequences are chosen to allow base-specific binding to the template in reverse orientation. PCR is extremely sensitive and operates at a very high speed. Its application for diverse purposes has opened up a multitude of new possibilities in the field of molecular biology.

For simplicity, we have divided the review in two parts. The first part is a general description of most of the available DNA marker types, while the second includes their application in plant genomics and breeding programmes.

Types and Description of DNA Markers

Single or Low Copy Probes

Restriction fragment length polymorphism (RFLP). RFLPs are simply inherited naturally occurring Mendelian characters. They have their origin in the DNA rearrangements that occur due to evolutionary processes, point mutations within the restriction enzyme recognition site sequences, insertions or deletions within the fragments, and unequal crossing over.

In RFLP analysis, restriction enzyme-digested genomic DNA is resolved by gel electrophoresis and then blotted on to a nitrocellulose membrane. Specific banding patterns are then visualized by hybridization with labelled probe. These probes are mostly species-specific single locus probes of about 0.5–3.0 kb in size, obtained from a cDNA library or a genomic library. The genomic libraries are easy to construct and almost all sequence types are included; however, a large number of interspersed repeats are found in inserts, that detect a large number of restriction fragments forming complex patterns. In plants, this problem is overcome to some extent by using methylation-sensitive restriction enzyme *Pst*I. This helps to obtain low copy DNA sequences of small fragment sizes, which are preferred in RFLP analysis. On the other hand cDNA libraries are difficult to construct, however, they are more popular as actual genes are analysed and they contain fewer repeat sequences. The selection of appropriate source for RFLP probe varies, with the requirement of particular application under consideration. Though genomic library probes may exhibit greater variability than gene probes from cDNA libraries, a few studies reveal the converse. This observation may be because cDNA probes not only detect variation in coding regions of the corresponding genes but also regions flanking genes and introns of the gene.

RFLP markers were used for the first time in the construction of genetic maps by Botstein *et al.*. RFLPs, being codominant markers, can detect coupling phase of DNA molecules, as DNA fragments from all homologous chromosomes are detected. They are very reliable markers in linkage analysis and breeding and can easily determine if a linked trait is present in a homozygous or heterozygous state in an individual, an information highly desirable for recessive traits. However, their utility has been hampered due to the large amount of DNA required for restriction digestion and Southern blotting. The requirement of radioactive isotope makes the analysis relatively expensive and hazardous. The assay is time-consuming and labour-intensive and only one out of several markers may be polymorphic, which is highly inconvenient especially for crosses between closely-related species. Their inability to detect single base changes restricts their use in detecting point mutations occurring within the regions at which they are detecting polymorphism.

Restriction landmark genomic scanning (RLGS). This method, introduced for the first time by Hatada *et al.*, for genomic DNA analysis of higher organisms, is based on the principle that restriction enzyme sites can be used as landmarks. It employs direct labelling of genomic DNA at the restriction site and two-dimensional (2D) electrophoresis to resolve and identify these landmarks. The technique has proven its utility in genome analysis of closely-related cultivars and for obtaining polymorphic markers that can be cloned by spot target method. It has been used as a new fingerprinting technique for rice cultivars.

RFLP Markers Converted in to PCR Based-markers

Sequence-tagged sites (STS). RFLP probes specifically linked to a desired trait can be converted into PCR-based STS markers based on nucleotide sequence of the probe giving polymorphic band pattern, to obtain specific amplicon. Using this technique, tedious hybridization procedures involved in RFLP analysis can be overcome.

This approach is extremely useful for studying the relationship between various species. When these markers are linked to some specific traits, for example powdery mildew resistance gene or stem rust resistance gene in barley, they can be easily integrated into plant breeding programmes for marker-assisted selection of the trait of interest.

Allele-specific associated primers (ASAPs). To obtain an allele-specific marker, specific allele (either in homozygous or heterozygous state) is sequenced and specific primers are designed for amplification of DNA template to generate a single fragment at stringent annealing temperatures. These markers tag specific alleles in the genome and are more or less similar to SCARs.

Expressed sequence tag markers (EST). This term was introduced by Adams *et al.*. Such markers are obtained by partial sequencing of random cDNA clones. Once generated, they are useful in cloning specific genes of interest and synteny mapping of functional genes in various related organisms. ESTs are popularly used in full genome sequencing and mapping programmes underway for a number of organisms and for identifying active genes thus helping in identification of diagnostic markers. Moreover, an EST that appears to be unique helps to isolate new genes. EST markers are identified to a large extent for rice, *Arabidopsis*, etc. wherein thousands of functional cDNA clones are being converted in to EST markers.

Single strand conformation polymorphism (SSCP). This is a powerful and rapid technique for gene analysis particularly for detection of point mutations and typing of DNA polymorphism. SSCP can identify heterozygosity of DNA fragments of the same molecular weight and can even detect changes of a few nucleotide bases as the mobility of the single-stranded DNA changes with change in its GC content due to its conformational change. To overcome problems of reannealing and complex banding patterns, an improved technique called asymmetric-PCR SSCP was developed, wherein the denaturation step was eliminated and a large-sized sample could be loaded for gel electrophoresis, making it a potential tool for high throughput DNA polymorphism. It was found useful in the detection of heritable human diseases. In plants, however, it is not well developed although its application in discriminating progenies can be exploited, once suitable primers are designed for agronomically important traits.

Multi Locus Probes

Repetitive DNA. A major step forward in genetic identification is the discovery that about 30–90% of the genome of virtually all the species is constituted by regions of repetitive DNA, which are highly polymorphic in nature. These regions contain genetic loci comprising several hundred alleles, differing from each other with respect to length, sequence or both and they are interspersed in tandem arrays ubiquitously. The repetitive DNA regions play an important role in absorbing mutations in the genome. Of the mutations that occur in the genome, only inherited mutations play a vital role in evolution or polymorphism. Thus repetitive DNA and mutational forces functional in nature together form the basis of a number of marker systems that are useful for various applications in plant genome analysis. The markers belonging to this class are both hybridization-based and PCR-based.

Microsatellites and minisatellites. The term microsatellites was coined by Litt and Lutty, while the term minisatellites was introduced by Jeffrey. Both are multilocus probes creating complex banding patterns and are usually non-species specific occurring ubiquitously. They essentially belong to the repetitive DNA family. Fingerprints generated by these probes are also known as oligonucleotide fingerprints. The methodology has been derived from RFLP and specific fragments are visualized by hybridization with a labelled micro- or minisatellite probe.

Minisatellites are tandem repeats with a monomer repeat length of about 11–60 bp, while microsatellites or short tandem repeats/simple sequence repeats (STRs/SSRs) consist of 1 to 6 bp long monomer sequence that is repeated several times. These loci contain tandem repeats that vary in the number of repeat units between genotypes and are referred to as variable number of tandem repeats (VNTRs) (i.e. a single locus that contains variable number of tandem repeats between individuals) or hypervariable regions (HVRs) (i.e. numerous loci containing tandem repeats within a genome generating high levels of polymorphism between individuals). Microsatellites and minisatellites thus form an ideal marker system creating complex banding patterns by simultaneously detecting multiple DNA loci. Some of the prominent features of these markers are that they are dominant fingerprinting markers and codominant STMS (sequence tagged microsatellites) markers. Many alleles exist in a population, the level of heterozygosity is high and they follow Mendelian inheritance.

Minisatellite and Microsatellite Sequences Converted into PCR-based Markers

Sequence-tagged microsatellite site markers (STMS). This method includes DNA polymorphism using specific primers designed from the sequence data of a specific locus. Primers complementary to the flanking regions of the simple sequence repeat loci yield highly polymorphic amplification products. Polymorphisms appear because of variation in the number of tandem repeats (VNTR loci) in a given repeat motif.

Tri- and tetranucleotide microsatellites are more popular for STMS analysis because they present a clear banding pattern after PCR and gel electrophoresis. However, dinucleotides are generally abundant in genomes and have been used as markers e.g. (CA)*n*(AG)*n* and (AT)*n*. The di- and tetranucleotide repeats are present mostly in the non-coding regions of the genome, while 57% of trinucleotide repeats are shown to reside in or around the genes. A very good relationship between the number of alleles detected and the total number of simple repeats within the targeted microsatellite DNA has been observed. Thus larger the repeat number in the microsatellite DNA, greater is the number of alleles detected in a large population.

Direct amplification of minisatellite DNA markers (DAMD-PCR). This technique, introduced by Heath *et al.*, has been explored as a means of generating DNA probes useful for detecting polymorphism. DAMD-PCR clones can yield individual-specific DNA fingerprinting pattern and thus have the potential as markers for species differentiation and cultivar identification.

Inter simple sequence repeat markers (ISSR). In this technique, reported by Zietkiewicz *et al.*, primers based on microsatellites are utilized to amplify inter-SSR DNA sequences. Here, various microsatellites anchored at the 3¢ end are used for amplifying genomic DNA which increases their specificity. These are mostly dominant markers, though occasionally a few of them exhibit codominance. An unlimited number of primers can be synthesized for various combinations of di-, tri-, tetra- and pentanucleotides [(4)3 = 64, (4)4 = 256] etc. with an anchor made up of a few bases and can be exploited for a broad range of applications in plant species.

Other Repetitive DNA-type Markers

Transposable elements. A large number of transposable repeat elements have been studied in plants; however, only a few have been exploited as molecular markers. In evolutionary terms, they have contributed to genetic differences between species and individuals by playing a role in retrotransposition events promoting unequal crossing over. Retrotransposon-mediated fingerprinting has been shown to be an efficient fingerprinting method for detection of genetic differences between different species.

Alu-repeats. This strategy was developed to fingerprint genotypes using semispecific primers, complementary to repetitive DNA elements called 'Alu-repeats', in human genome analysis. *Alu* repeats are a class of randomly repeated interspersed DNA, preferentially used for *Alu* PCR as they reveal considerable levels of polymorphism. These representatives of short and long interspersed nuclear elements are known as *SINES. Alu* elements are approximately 300 bp in size and have been suggested to be originated from special RNA species that have been reintegrated at a rate of approximately one integration event per 10000 years. These repeats have been studied largely in humans, while their function in plants remains largely unexplored.

Repeat complementary primers. As an alternative to the interspersed repeats, primers complementary to other repetitive sequence elements were also successfully used for generation of polymorphisms, e.g. introns/exons splice junctions, tRNA genes, 5sRNA genes and Zn-finger protein genes. Primers complementary to specific exons, resulting in the amplification of the intervening introns have been studied by Lessa *et al.*. One of the strengths of these new strategies is that they are more amenable to automation than the conventional hybridization-based techniques.

Arbitrary Sequence Markers

Randomly-amplified polymorphic DNA markers (RAPD). In 1991 Welsh and McClelland developed a new PCR-based genetic assay namely randomly amplified polymorphic DNA (RAPD). This procedure detects nucleotide sequence polymorphisms in DNA by using a single primer of arbitrary nucleotide sequence.

In this reaction, a single species of primer anneals to the genomic DNA at two different sites on complementary strands of DNA template. If these priming sites are within an amplifiable range of each other, a discrete DNA product is formed through thermocyclic amplification. On an average, each primer directs amplification of several discrete loci in the genome, making the assay useful for efficient screening of nucleotide sequence polymorphism between individuals. However, due to the stoichastic nature of DNA amplification with random sequence primers, it is important to optimize and maintain consistent reaction conditions for reproducible DNA amplification.

They are dominant markers and hence have limitations in their use as markers for mapping, which can be overcome to some extent by selecting those markers that are linked in coupling. RAPD assay has been used by several groups as efficient tools for identification of markers linked to agronomically important traits, which are introgressed during the development of near isogenic lines. The application of RAPDs and their related modified markers in variability analysis and individual-specific genotyping has largely been carried out, but is less popular due to problems such as poor reproducibility faint or fuzzy products, and difficulty in scoring bands, which lead to inappropriate inferences.

Some Variations in the RAPD Technique Include

DNA amplification fingerprinting (DAF). Caetano-Anolles *et al.* employed single arbitrary primers as short as 5 bases to amplify DNA using polymerase chain reaction. In a spectrum of products obtained, simple patterns are useful as genetic markers for mapping, while more complex patterns are useful for DNA fingerprinting.

Band patterns are reproducible and can be analysed using polyacrylamide gel electrophoresis and silver staining. DAF requires careful optimization of parameters; however, it is extremely amenable to automation and fluorescent tagging of primers for early and easy determination of amplified products. DAF profiles can be tailored by employing various modifications such as predigesting of template. This technique has been useful in genetic typing and mapping.

Arbitrary primed polymerase chain reaction (AP-PCR). This is a special case of RAPD, wherein discrete amplification patterns are generated by employing single primers of 10–50 bases in length in PCR of genomic DNA. In the first two cycles, annealing is under non-stringent conditions. The final products are structurally similar to RAPD products.

Compared to DAF, this variant of RAPD is not very popular as it involves autoradiography. Recently, however, it has been simplified by separating the fragments on agarose gels and using ethidium bromide staining for visualization.

Sequence Characterized Amplified Regions for Amplification of Specific Band (SCAR)

Michelmore *et al.* and Martin *et al.* introduced this technique wherein the RAPD marker termini are sequenced and longer primers are designed (22–24 nucleotide bases long) for specific amplification of a particular locus. These are similar to STS markers in construction and application. The presence or absence of the band indicates variation in sequence. These are better reproducible than RAPDs.SCARs are usually dominant markers, however, some of them can be converted into codominant markers by digesting them with tetra cutting restriction enzymes and polymorphism can be deduced by either denaturing gel electrophoresis or SSCP. Compared to arbitrary primers, SCARs exhibit several advantages in mapping studies (codominant SCARs are informative for genetic mapping than dominant RAPDs), map-based cloning as they can be used to screen pooled genomic libraries by PCR, physical mapping, locus specificity, etc. SCARs also allow comparative mapping or homology studies among related species, thus making it an extremely adaptable concept in the near future.

Cleaved Amplified Polymorphic Sequences (CAPs)

These polymorphic patterns are generated by restriction enzyme digestion of PCR products. Such digests are compared for their differential migration during electrophoresis. PCR primer for this process can be synthesized based on the sequence information available in databank of genomic or cDNA sequences or cloned RAPD bands. These markers are codominant in nature.

Randomly Amplified Microsatellite Polymorphisms (RAMPO)

In this PCR-based strategy, genomic DNA is first amplified using arbitrary (RAPD) primers. The amplified products are then electrophoretically separated and the dried gel is hybridized with microsatellite oligonucleotide probes. Several advantages of oligonucleotide fingerprinting, RAPD and microsatellite-primed PCR are thus combined, these being the speed of the assay, the high sensitivity, the high level of variability detected and the non-requirement of prior DNA sequence information. This technique has been successfully employed in the genetic fingerprinting of tomato, kiwi fruit and closely-related genotypes of *D. bulbifera*.

Amplified Fragment Length Polymorphism (AFLP)

A recent approach by Zabeau *et al.*, known as AFLP, is a technique based on the detection of genomic restriction fragments by PCR amplification and can be used for DNAs of any origin or complexity. The fingerprints are produced, without any prior knowledge of sequence, using a limited set of generic primers. The number of fragments detected in a single reaction can be 'tuned' by selection of specific primer sets. AFLP technique is reliable since stringent reaction conditions are used for primer annealing. This technique thus shows an ingenious combination of RFLP and PCR techniques and is extremely useful in detection of polymorphism between closely related genotypes.

AFLP Procedure Mainly Involves 3 Steps

(a) Restriction of DNA using a rare cutting and a commonly cutting restriction enzyme simultaneously (such as *Mse*I and *Eco*RI) followed by ligation of oligonucleotide adapters, of defined sequences including the respective restriction enzyme sites.

(b) Selective amplifications of sets of restriction fragments, using specifically designed primers. To achieve this, the 5' region of the primer is made such that it would contain both the restriction enzyme sites on either sides of the fragment complementary to the respective adapters, while the 3' ends extend for a few arbitrarily chosen nucleotides into the restriction fragments.

(c) Gel analysis of the amplified fragments.

AFLP analysis depicts unique fingerprints regardless of the origin and complexity of the genome. Most AFLP fragments correspond to unique positions on the genome and hence can be exploited as landmarks in genetic and physical mapping. AFLPs are extremely useful as tools for DNA fingerprinting and also for cloning and mapping of variety-specific genomic DNA sequences. Similar to RAPDs, the bands of interest obtained by AFLP can be converted into SCARs. Thus AFLP provides a newly developed, important tool for a variety of applications.

Applications of Molecular Markers in Plant Genome Analysis and Breeding

Molecular markers have been looked upon as tools for a large number of applications ranging from localization of a gene to improvement of plant varieties by marker-assisted selection. They have also become extremely popular markers for phylogenetic analysis adding new dimensions to the evolutionary theories. If we look at the history of the development of these markers, it is evident that they have been improved over the last two decades to provide easy, fast and automated assistance to scientists and breeders. Genome analysis based on molecular markers has generated a vast amount of information and a number of databases are being generated to preserve and popularize it.

Mapping and Tagging of Genes: Generating Tools for Marker-assisted Selection in Plant Breeding

Plant improvement, either by natural selection or through the efforts of breeders, has always relied upon creating, evaluating and selecting the right combination of alleles. The manipulation of a large number of genes is often required for improvement of even the simplest of characteristics. With the use of molecular markers it is now a routine to trace valuable alleles in a segregating population and mapping them. These markers once mapped enable dissection of the complex traits into component genetic units more precisely, thus providing breeders with new tools to manage these complex units more efficiently in a breeding programme.

The very first genome map in plants was reported in maize, followed by rice, *Arabidopsis* etc. using RFLP markers. Maps have since then been constructed for several other crops like potato, barley, banana, members of Brassicaceae, etc. Once the framework maps are generated, a large number of markers derived from various techniques are used to saturate the maps as much as is possible. Microsatellite markers, especially STMS markers, have been found to be extremely useful in this regard.

Owing to their quality of following clear Mendelian inheritance, they can be easily used in the construction of index maps, which can provide an anchor or reference point for specific regions of the genome. About 30 microsatellites have already been assigned to five linkage groups in Arabidopsis, while their integration into the genetic linkage maps is still in progress in rice, soybean, maize, etc.

The very first attempt to map microsatellites in plants was made by Zhao and Kochert in rice using (GGC)*n*, followed by mapping of (GA)*n* and (GT)*n* by Tanksley *et al.* and (GA/AG)*n*, (ATC) 10 and (ATT) 14, by Panaud *et al.* in rice. The most recent microsatellite map has been generated by Milbourne *et al.* for potato. Similar to microsatellites, looking at the pattern of variation, generated by retrotransposons, it is now proposed that apart from genetic variability, these markers are ideal for integrating genetic maps.

Once mapped, these markers are efficiently employed in tagging several individual traits that are extremely important for a breeding programme like yield, disease resistance, stress tolerance, seed quality, etc. A large number of monogenic and polygenic loci for various traits have been identified in a number of plants, which are currently being exploited by breeders and molecular biologists together, so as to make the dream of marker-assisted selection come true.

Tagging of useful genes like the ones responsible for conferring resistance to plant pathogen, synthesis of plant hormones, drought tolerance and a variety of other important developmental pathway genes, is a major target. Such tagged genes can also be used for detecting the presence of useful genes in the new genotypes generated in a hybrid programme or by other methods like transformation, etc. RFLP markers have proved their importance as markers for gene tagging and are very useful in locating and manipulating quantitative trait loci (QTL) in a number of crops. The very first reports on gene tagging were from tomato, availing the means for identification of markers linked to genes involved in several traits like water use efficiency, resistance to *Fusarium oxysporum* (the 12 gene), leaf rust resistance genes LR 9 and 24, and root knot nematodes (*Meliodogyne* sp.) (the mi gene). Recently, Xiao *et al.* have shown the utility of RFLP markers in identifying the trait improving QTL alleles from wild rice relative *O. rufipogon*.

Allele-specific associated primers have also exhibited their utility in genotyping of allelic variants of loci that result from both size differences and point mutations. Some of the genuine examples of this are the waxy gene locus in maize, the Glu D1 complex locus associated with bread making quality in wheat, the *Lr1* leaf rust resistance locus in wheat, the Gro1 and H1 alleles conferring resistance to the root cyst nematode *Globodera rostochiensis* in potato, and allele-specific amplification of polymorphic sites for detection of powdery mildew resistance loci in cereals. A number of other traits have been tagged using ASAPs in tomato, lettuce, etc.

Besides ASAPs, AFLP and SSR markers have been identified to be associated with quantitative resistance to *Globodera pallida* (stone) in tetraploid potato, which can be very well employed in marker-assisted selection.

STMS markers have displayed a potential use as diagnostic markers for important traits in plant breeding programmes, e.g. (AT) 15 repeat has been located within a soybean heat shock protein gene which is about 0.5 cM from (Rsv) a gene conferring resistance to soybean mosaic virus.

Several resistance genes including peanut mottle virus (Rpv), phytopthera (Rps3) and Javanese rootknot nematode are clustered in this region of the soybean genome.

Similar to specific markers like RFLPs, STMS and ASAPs, arbitrary markers like RAPDs have also played important role in saturation of the genetic linkage maps and gene tagging.

Their use in mapping has been especially important in systems, where RFLPs have failed to reveal much polymorphism that is so very important for mapping. One of the first uses of RAPD markers in saturation of genetic maps was reported by Williams *et al.*. They have proven utility in construction of linkage maps among species where there has been inherent difficulty in producing F2 segregating populations and have large genome size, e.g. conifers.

RAPD markers in near isogenic lines can be converted into SCARs and used as diagnostic markers. SCAR/STS marker linked to the translocated segment on 4 AL of bread wheat carrying the *Lr28* gene has been tagged by Naik *et al.*. Recently, ISSRs, which too belong to the arbitrary marker category, but are found to be devoid of many of the drawbacks shared by RAPD class of markers, have been employed as a reliable tool for gene tagging.

An ISSR marker (AG) 8YC has been found to be linked closely (3.7 ± 1.1 cM) to the rice nuclear restorer gene, RF1 for fertility. RF1 is essential for hybrid rice production and this marker would be useful not only for breeding both restorer and maintainer lines, but also for the purity management of hybrid rice seeds. Similarly ISSR marker $(AC)_8$ YT has been found to be linked to the gene for resistance to fusarium wilt race 4 in repulsion at a distance of 5.2 cM in chickpea.

Apart from mapping and tagging of genes, an important utility of RFLP markers has been observed in detecting gene introgression in a backcross breeding programme, and synteny mapping among closely related species.

Similar utility of STMS markers has been observed for reliable preselection in a marker assisted selection backcross scheme. Apart from specific markers, DAMD-based DNA fingerprinting in wheat has also been useful for monitoring backcross-mediated genome introgression in hexaploid wheat.

Phylogeny and Evolution

Most of the early theories of evolution were based on morphological and geographical variations between organisms. However, it is becoming more and more evident that the techniques from molecular biology hold a promise of providing detailed information about the genetic structure of natural population, than what we have been able to achieve in the past. RFLP, DNA sequencing, and a number of PCR-based markers are being used extensively for reconstructing phylogenies of various species. The techniques are speculated to provide path-breaking information regarding the fine time scale on which closely related species have diverged and what sort of genetic variations are associated with species formation. Furthermore, these studies hold a great promise for revealing more about the pattern of genetic variation within species.

Efforts are now being made for studying the genetic variation in plants, so as to understand their evolution from wild progenitors and to classify them into appropriate groups. The taxonomic classification is an essential first step to determine whether any germplasm is a part of the primary, secondary or tertiary genepool of the system concerned. This is especially important in cases where morphological markers can prove to be inaccurate and misleading. A genuine example of this is the lines Azucena and PR 304 which have been classified as indicas using morphological characters, whereas they behave like japonicas in crossing studies. These samples are however, clearly revealed to be japonicas upon being analysed by RAPD markers.

RFLPs have been used in evolutionary studies for deducing the relationship between the hexaploid genome of bread wheat and its ancestors. Similarly a number of transposon elements like *tos1-1, tos2-1* and *tos3-1* retrotransposons have been used to detect the genetic differences between different species of rice and even between ecotypes of cultivated rice, wherein they were found to distinguish between the cultivars of Asian and African rice, *O. sativa* and *O. glaberrima*. Retroelement *Wis-2* has been found to detect genomic variation within individual plants of wheat variety and also within and between varieties of wheat. This element has also been found to occur in the genomes of other grasses like barley, rye, oats and *Aegilops* species, indicating common ancestral elements in grasses. Though RFLPs, microsatellites, minisatellites and transposons are useful for carrying out genetic variability analysis, the trend is now shifting towards the use of PCR-based markers. Specific markers are preferred over arbitrary primers. However, arbitrary primers are found to be the markers of choice in the analysis of complex genomes like wheat, where genetic variation is extremely difficult to dissect. Sen *et al.* have used DAF as a new source of molecular markers in fingerprinting of bread wheat. Recent studies in our laboratory have revealed the utility of RAPD and ISSR markers in evolutionary studies of wheat and rice, respectively.

Specific markers like STMS (sequence-tagged microsatellite markers) ALPs (Amplicon length polymorphisms) or STS markers have proved to be extremely valuable in the analysis of gene pool variation of crops during the process of cultivar development, and classification of germplasm. These markers are extremely sensitive and can detect allelic variability during cultivar development. STS markers specific to chloroplast or mitochondrial DNA have been useful in providing seed and pollen specific markers which can be utilized for the detection of length variation at multiple physically linked sites and may be used to provide haplotype data and thus genotypically unique individual plants. Also a comparison of patterns of variability detected with biparentally (nuclear) and uniparentally (organellar) transmitted markers can provide complementary information to population and evolutionary biologists. Excellent examples of this are the Poly A mononucleotide repeats in maize, Poly (TA/AT) dinucleotide repeats found in liverworts, maize, pea and nonphotosynthetic green plant *Epifagus virginiana*, and a total of 500 chloroplast SSRs identified with repeat motifs greater than 10 repeat units in rice, tobacco, black pine, liverwort and maize. Though all these marker types provide valuable information regarding the evolution and phylogeny of various species being studied in any given set of samples, the trend is now shifting towards the use of ESTs (expressed sequence tags) for such analysis. This may be so, because in such studies, one actually looks at the evolution of functional genes.

Diversity Analysis of Exotic Germplasm

Following domestication, genetic variation in crop plants has continued to narrow due to continuous selection pressure for specific traits i.e. yield, thus rendering them more vulnerable to disease and insect epidemics and jeopardizing the potential for sustained genetic improvement over a long term. This risk was brought sharply into focus in 1970 with the outbreak of southern corn leaf blight which drastically reduced corn yield in USA, and was attributed to extensive use of a single genetic male sterility factor which was unfortunately linked to the disease susceptibility. Thus it is extremely important to study the genetic composition of the germplasm of existing modern-day cultivars in comparison with their ancestors and related species. This will not only provide information on their phylogenetic relationship but will also indicate a chance of finding new and useful genes, as the accessions with most distinct DNA profiles are likely to contain a greater number of novel alleles. DNA profiling to make such sampling decisions is now underway in most crops. The exotic germplasm for breeding is selected on the basis of certain characteristic features such as (a) the exotic germplasm must possess a significant number of unique DNA polymorphisms (throughout the genome) relative to the modern-day cultivars and (b) each exotic germ plasm has to be genetically dissimilar (on the basis of DNA profiling). This is necessary to understand the genetic variations between the existing cultivars, the cultivars in comparison with their wild progenitors and a number of wilds that still exist in nature.

Many DNA markers both specific as well as arbitrary have been used so far, for DNA fingerprinting of various classes of germplasm. Further studies with STMS markers may also throw light on the domestication process involved in crop plants and provide useful criteria for enriching the genepool of crop plants and determine how efficient plant breeders have been in accessing preexisting forms of variation. AFLP, a new class of molecular markers, has gained popularity as marker for the study of genetic polymorphism especially in species where polymorphism is extremely rare using other types of markers.

Pakniyat *et al.* used AFLP for studying variation in wild barley with reference to salt tolerance and associated ecogeography, and a number of reports are coming up each day for different systems. Similarly the potential of ISSR markers has been exploited for diversity analysis of pine, rice and also in wheat. These studies have helped in the classification of existing biodiversity among plants, which can be further exploited in wild gene introgression programs.

Genotyping of Cultivars

The repetitive and arbitrary DNA markers are markers of choice in genotyping of cultivars. Microsatellites like $(CT)_{10}$, $(GAA)_5$, $(AAGG)_4$, $(AAT)_6$, $(GATA)_4$, $(CAC)_5$ (refs 124, 125) and minisatellites have been employed in DNA fingerprinting for the detection of genetic variation, cultivar identification and genotyping.

This information is useful for quantification of genetic diversity, characterization of accessions in plant germplasm collections and taxonomic studies.Microsatellites have been useful for generation of STMS markers, revealing polymorphisms within closely related cultivars. The first application of microsatellites in plants has been in cultivar identification, wherein microsatellites have been used to genotype unequivocally diverse materials like rice, wheat, grapevine (*Vitis unifera*), soybean, etc.

This is important especially for protection of proprietary germplasm. Similarly microsatellite markers have also been advantageous in pedigree analysis as they represent single locus. The multi allelism of these markers facilitates comparative allelic variability detection reliably across a wide range of germplasm and allows individuals to be ubiquitously genotyped, so that gene flow and paternity can be established.

One of the most recent applications of these markers has been shown in sex identification of dioeceous plants, wherein microsatellite probe $(GATA)_4$ is found to reveal sex-specific differences in Southern analysis and can be used as a diagnostic marker in this system where male and female plants do not show any sex-specific morphological difference until flowering.

Similarly, Di Stilio *et al.* have identified a randomly-amplified (RAPD) DNA marker for pseudo-autosomal plant sex chromosome in *Silene dioica* (L.).

Indian Scenario for Development of Molecular Markers in Crop Improvement Programmes

Agriculture is one of the most important occupations in India with almost 70% of the population being dependent on it. A noteworthy research in conventional breeding for several years has made this country self-sufficient in many respects.

However, the ever-increasing population has alarmed food security in India and attempts have been initiated to integrate modern biotechnology tools in conventional breeding to improve the most important crops such as rice, wheat and legumes.

Extensive research using DNA markers is in progress in many institutions all over India. Markers tagged and mapped with specific genes have been identified and some such examples include resistance genes for blast and gall midge using RFLP- and PCR-based approaches in rice.

Similarly, in wheat, leaf rust resistance gene *LR* 28, and pre-harvest sprout tolerance gene have been tagged. QTLs such as protein content in wheat and heterosis in rice have also been identified. While efforts for tagging genes providing resistance to BPH, WBPH, sheath rot and drought are going on, many attempts are also being made towards pyramiding different resistance genes for a specific disease or pest attack like blast, bacterial blight, gall midge, BPH, WBPH, etc. in rice in order to increase the field life of the crop.

Germplasm analysis to study genetic diversity is another important area in which a lot of efforts have been put in. Fingerprinting of crops like rice, wheat, chickpea, pigeonpea, pearlmillet etc. is being carried out extensively.

This information has potential in strategic planning of future breeding towards crop sustainability in India. Apart from use of molecular markers in crop plants, efforts are also underway in other horticultural plants. Early identification of sex in dioecious papaya using molecular marker is one such example.

Thus in the last few years there are many reports of amalgamation of classical breeding and modern biotechnological approaches which have unlimited scope in Indian agriculture.

Genetic Marker

A genetic marker is a gene or DNA sequence with a known location on a chromosome and associated with a particular gene or trait. It can be described as a variation, which may arise due to mutation or alteration in the genomic loci, that can be observed. A genetic marker may be a short DNA sequence, such as a sequence surrounding a single base-pair change (single nucleotide polymorphism, SNP), or a long one, like minisatellites.

Background

For many years, gene mapping was limited in most organisms by traditional genetic markers which include genes that encode easily observable characteristics such as blood types or seed shapes. The insufficient amount of these types of characteristics in several organisms limited the mapping efforts that could be done.

Types

Some commonly used types of genetic markers are

- RFLP (or Restriction fragment length polymorphism)
- AFLP (or Amplified fragment length polymorphism)
- RAPD (or Random amplification of polymorphic DNA)
- VNTR (or Variable number tandem repeat)
- Microsatellite polymorphism
- SNP (or Single nucleotide polymorphism)
- STR (or Short tandem repeat)
- SFP (or Single feature polymorphism)
- DArT (or Diversity Arrays Technology)

They can be further categorized as dominant or co-dominant. Dominant markers allow for analyzing many loci at one time, e.g. RAPD. A primer amplifying a dominant marker could amplify at many loci in one sample of DNA with one PCR reaction. Co-dominant markers analyse one locus at a time. A primer amplifying a co-dominant marker would yield one targeted product.

Uses

Genetic markers can be used to study the relationship between an inherited disease and its genetic cause (for example, a particular mutation of a gene that results in a defective protein). It is known that pieces of DNA that lie near each other on a chromosome tend to be inherited together. This property enables the use of a marker, which can then be used to determine the precise inheritance pattern of the gene that has not yet been exactly localized. Genetic markers have to be easily identifiable, associated with a specific locus, and highly polymorphic, because homozygotes do not provide any information. Detection of the marker can be direct by RNA sequencing, or indirect using allozymes. Some of the methods used to study the genome or phylogenetics are RFLP, Amplified fragment length polymorphism (AFLP), RAPD, SSR. They can be used to create genetic maps of whatever organism is being studied.

There was a debate over what the transmissible agent of CTVT (canine transmissible venereal tumor) was. Many researchers hypothesized that virus like particles were responsible for transforming the cell, while others thought that the cell itself was able to infect other canines as an allograft. With the aid of genetic markers, researchers were able to provide conclusive evidence that the cancerous tumor cell evolved into a transmissible parasite. Furthermore, molecular genetic markers were used to resolve the issue of natural transmission, the breed of origin (phylogenetics), and the age of the canine tumor.

Genetic Markers have also been used to measure the genomic response to selection in livestock. Natural and artificial selection leads to a change in the genetic makeup of the cell. The presence of different alleles due to a distorted segregation at the genetic markers is indicative of the difference between selected and non-selected livestock.

Insulin Production

Genetic markers also play a role in genetic engineering, as they can be used to produce normal, functioning proteins to replace defective ones. The damaged or faulty section of DNA is removed and replaced with the identical, but functioning, gene sequence from another source.

This is done by removal of the faulty section of DNA and its replacement with the functioning gene from another source, usually a human donor. These gene sections are placed in solution with bacterial cells, a small number of which take up the genetic material and reproduce the new DNA sequence. Engineers need to know which bacteria have been successful in duplicating these genes so another gene is added, altering the bacteria's resistance to antibiotics. Replica plating or a fermenter is used to grow enough bacteria to test resistance to antibiotics. It is important that the cultures are not mixed.

This process can be used as a treatment for diabetes mellitus. Bacterial DNA often has two resistancy genes: one for tetracycline and one for ampicillin. The insulin gene can be inserted in the middle of the ampicillin gene after it has been removed using restriction endonucleases. If the gene has been taken up, the bacteria both produces insulin and is also no longer ampicillin resistant.

The bacteria are then allowed to grow on an agar plate containing a culture medium. The bacteria grow and produce colonies on the agar jelly. A piece of filter paper can be placed onto the top of this agar plate so that the exact positions of the colonies are remembered. This produces a copy which can then be transferred onto a second agar plate containing ampicillin. All of the bacteria that are not resistant to ampicillin will die. These locations on the second plate show the places on the first plate where bacteria are not resistant and, therefore, produce insulin. Another similar method is followed, in which an epitope sequence is added to insert. When the insert is expressed so is the epitope. Then this epitope can be effectively bound using an antibody on a filter paper. And the expressing colonies can be easily selected.

An Introduction to Genetic Data Analysis

Using SAS/Geneticsä

Genetic markers, while often nonfunctional, play a key role in the mapping of human disease genes. Single nucleotide polymorphisms (SNPs) in particular are facilitating the creation of high-resolution genetic maps when used in linkage and association studies. SAS/Genetics provides a framework and the tools for analyzing markers and their relationship with a dichotomous trait (affection status of a disease e.g.).

While many of these methods exploit the biallelic nature of SNPs in their design, multiallelic adaptations are available when studying markers such as microsatellites with more than two alleles. Before examining marker-disease associations, initial calculations characterizing features of the markers should be performed. Properties of interest about each marker include measures of marker informativeness, which are useful for determining which markers may be most valuable for association or linkage studies; estimates of allele, genotype, and haplotype frequencies; tests for Hardy-Weinberg equilibrium (HWE); and measures of linkage disequilibrium (LD) between pairs of markers. These types of analyses can be accomplished with the ALLELE and HAPLOTYPE procedures.

Once these marker features have been examined, marker-disease association tests can be performed on either random samples from populations of unrelated individuals who are either affected with a disease or unaffected, or on families with an affected child meeting certain requirements. Several different case-control and family-based tests can be implemented by the Casecontrol and Family procedures, respectively. Smoothing and multiple testing adjustments are then available to be applied to the p-values created by these tests in order to improve power and/or control type I error using the PSMOOTH procedure.

Genetics and Education

Genetics has become an indispensable component of almost all research in modern biology and medicine. Research publications investigating any biological process, from the molecular level all the way to the population level, use the "genetic approach" to gain understanding of that process. Thus, no student of the life sciences can afford to be ignorant of the science of genetics. Genetics has also risen to a position of prominence in human affairs. Special types of plants, animals, and microbes have been developed for human foods, drugs, and myriad other uses.

Molecular genetics is the central foundation of the burgeoning biotechnology industry. At the philosophical level, genetics has presented humans with a large number of ethical dilemmas, which regularly surface in the media. Some examples are genetically modified foods, eugenics, privacy of genetic information about individuals, and loss of genetic diversity in nature.

Students must be knowledgeable about genetics in order to understand these issues and make informed decisions about them. Lastly, genetic insight has radically affected the human worldview—the way we see ourselves in relation to other organisms.

The Balanced Approach to Teaching Genetics

Genetics has risen to such prominence through the powerful merger of classical and molecular approaches. Each analytical approach has its unique strengths: classical (organismal) genetics is un-paralleled in its ability to explore uncharted biological terrain; molecular genetics is equally unparalleled in its ability to unravel cellular mechanisms. It would be unthinkable to teach one without the other, and each is given due prominence in this book. Armed with both approaches, students are able to form an integrated view of genetic principles.

The partnership of classical and molecular genetics has always presented a teaching dilemma: which of the two partners should the student be introduced to first, the classical or the molecular? We believe that students begin much as biologists did at the turn of the century, asking general questions about the laws governing heredity.

Therefore, the first half of the book introduces the intellectual framework of classical eukaryotic genetics in more or less historical sequence. However, molecular information is provided where appropriate. Our students' knowledge base and our years of teaching have together caused us to rethink the traditional organization.Thus, we reinforce the students' knowledge of DNA structure and function and avoid presenting the gene as an abstraction. The second half of the book pursues the details of molecular genetics.

Focus on Genetic Analysis

True to its title, the theme of this book is genetic analysis. This theme emphasizes our belief that the best way to understand genetics is by learning how genetic inference is made. On almost every page, we recreate the landmark experiments in genetics and have the students analyse the data and draw conclusions as if they had done the research themselves. This proactive process teaches students how to think like scientists. The modes of inference and the techniques of analysis are the keys to future exploration.

Similarly, quantitative analysis is central to the book because many of the new ideas in genetics, from the original conception of the gene to such modern techniques as SSLP mapping, are based on quantitative analysis. The problems at the end of each chapter provide students with the opportunity to test their understanding in quantitative analyses that effectively simulate the act of doing genetics.

Fostering Analytical Skills

A great deal of effort has been put into encouraging students to practice and hone their analytical and problem-solving skills. We provide a great variety of solved and unsolved problems and a wide range of study aids. The Problems section continues to be one of the strengths of the book. Problems are generally arranged to start from simple and proceed to the more difficult. Particularly challenging problems are marked with an asterisk.

All problems have been classroom tested. Answers to selected problems are found at the back of the book, and the full set of solutions is in the Student Companion, all prepared by Bill Fixsen (Harvard University). Most chapters have an exercise in problem solving called Unpacking the Problem. This exercise grew from the idea that a genetics problem represents only the tip of a vast iceberg of knowledge (we originally considered calling them "iceberg problems").

It is only when the underlying levels of knowledge are exposed that the problem can be solved in a constructive manner. The unpacking activities access this underlying knowledge without actually solving the problem. Some of the component questions in the unpacking exercise might sound trivial, but often these address the kind of fundamental levels of misunderstanding that prevent students from successfully solving problems.

The problems at the end of each chapter are prefaced by Solved Problems that illustrate the ways that geneticists apply principles to experimental data. Research in science education has shown that this application of principles is a process that professionals find second nature, whereas students find it a major stumbling block.

The Solved Problems demonstrate this process and prepare the students for solving problems on their own.These chapter integration problems help to show how one set of learned skills builds on and interacts with previous ones. They also enable students to develop a holistic perspective as they begin to organize diverse concepts into a coherent body of knowledge. The Key Concepts at the chapter openings give an overview of the main principles to be covered in the chapter, stated in simple prose without genetic terminology. These provide a strong pedagogic direction for the reader. Throughout the chapters, boxed Messages provide convenient milestones at which the reader can pause and contemplate the material just presented.

A Concept Map exercise appears at the end of every chapter. Concept maps grew out of the constructivist movement in education, which asserts that student learning is most effective when new information is brought into direct conflict with previous understanding. Concept mapping can be a powerful method for visualizing and resolving such conflicts, while aiding concept integration.

Cell Growth Measurement Methods

The cell growth can be detected by a variety of methods. The cell size growth can be visualized by Microscopy, using suitable stains. But the increase of cells number is usually more significative. It can be measured by manual counting of cells under microscopy observation, using the dye exclusion method (i.e. Trypan blue) to count only viable cells.

Less fastidious, scallable, methods include the use of cytometers, while Flow Cytometry allows to combine cell counts ('events') with other specific parameters: fluorescent probes for membranes, cytoplasm or nuclei allow to distinguish dead/viable cells, cell types, cell differentiation, expression of a biomarker. Beside the increasing number of cells, one can be assessed regarding the metabolic activity growth. I.e. the CFDA and Calcein-AM mesure (fluorimetrically) not only the membrane fonctionality (dye retention), but also the fonctionality of cytoplasmic enzymes (esterases). The MTT assays (colorimetric) and the Resazurin assay (fluorimetric) dose the mitochondrial redox potentiel. Finally, all these assays may correlate well, or not depending on cell growth conditions and desired aspects (activity, proliferation). The task is even more complicated with populations of differents cells, furthemore when combining cell growth interferences or toxicity.

7

Life's Genetic Tree

The millions of different living things on Earth—plants, bacteria, insects, chimps, people, and everything else—all came to be because of a process called biological evolution, in which organisms change over time. Because of biological evolution, early humans gained the ability to walk on two feet. Because of evolution, air-breathing whales can live in the ocean despite being mammals like us. Because of evolution, some bacteria can live in scalding water, others can survive in solid ice, and still others can live deep in the Earth eating only rocks! Evolution happens every day, and it affects every species—including us. It changes entire populations, not individuals. And it has a big impact on medical research.

- Everything Evolves
- Selective Study
- Clues from Variation
- Living Laboratories
- The Genome Zoo
- Starting All Over Again
- Genes Meet Environment
- Genetics and You: You've Got Rhythm!
- Animals Helping People
- My Collaborator Is a Computer
- The Tools of Genetics: Unlimited DNA

Everything Evolves

To understand evolution, let's go back in time a century and a half to 1854, when the British naturalist Charles Darwin published *The Origin of Species*, a book that proposed an explanation for how evolution works. The main concept in evolution is that all living things share a common ancestor.

The very earliest ancestor of all life forms on Earth lived about 4 billion years ago. From that early organism, millions of types of creatures—some living and some now extinct—have evolved. Evolution requires diversity. You can tell that living things are diverse just by walking down the street and looking around you. Individual people are very different from one another. Chihuahuas are different from Great Danes, and Siamese cats are different from tabbies.

Evolution also depends on inheritance. Many of our unique characteristics are inherited—they are passed from parent to offspring. This is easy to see: Dalmatian puppies look like Dalmatians, not Chihuahuas. Petunias grow differently from pansies. Evolution works *only* on traits that are inherited. Finally, as you probably already know, evolution favours the "fittest."

Through a process called natural selection, only some offspring within a given generation will survive long enough to reproduce. As an example, consider houseflies, each of which lays thousands of eggs every year. Why haven't they taken over the world? Because almost all of the baby houseflies die. The flies that survive are the ones that can find something to eat and drink... the ones that avoid being eaten, stepped on, or swatted... and the ones that don't freeze, drown, or land on a bug zapper.

The flies that survive all these ways to die have what it takes to outlive most of their brothers and sisters. These inherited traits give an organism a survival edge. Those who survive will mate with each other and will pass on to the next generation some of their DNA that encoded these advantageous traits. Of course, not all aspects of survival are determined by genes.

Whether a fly gets swatted depends on genes that affect its reflexes—whether it's fast enough to avoid the swatter—but also on the environment. If there's no human around waving the swatter, the fly is quite likely to survive, regardless of its reflexes.

Evolution often takes a long time to make a difference. But it can also happen very quickly, especially in organisms with short lifespans. For example, as you read earlier, some bacteria have molecular features that let them survive in the presence of antibiotics.

When you take an antibiotic medicine, antibiotic-resistant bacteria flourish while antibiotic-sensitive bacteria die. Because antibiotic resistance is a growing public health threat, it's important to take the whole dosage of antibiotic medicine, not stop when you feel better. And you should take antibiotics only when they're needed, not for colds or other viral infections, which antibiotics can't treat. Viruses must simply run their course.

Selective Study

Scientists doing medical research are very interested in genetic variants that have been selected by evolution. For example, researchers recently discovered a rare genetic variant that protects people from getting AIDS. A genetic variant is a different version of a gene, one that has a slightly different sequence of nucleotides. Scientists think that the rare variant of a gene called CCR5 originally may have been selected during evolution because it made people resistant to an organism unrelated to HIV.

Montgomery Slatkin of the University of California, Berkeley, has used mathematical modeling techniques to show that natural selection over time could explain the frequency of the CCR5 variant in human populations. The work indicates that the CCR5 gene variant's ability to protect against AIDS may contribute to keeping it in the human gene pool.

So, through evolution, living things change. Sometimes, that's good for us, as when humans understand HIV resistance in hopes of preventing AIDS. But sometimes the changes aren't so great —from a human perspective, anyway— as when bacteria become resistant to antibiotics.

Whether the consequences of evolutionary change are good or bad, understanding the process can help us develop new strategies for fighting disease.

Clues from Variation

Scientists know quite a bit about how cells reshuffle genetic information to create each person's unique genome. But many details are missing about how this genetic variation contributes to disease, making for a very active area of research.

Different nucleotides (in this example, A or G) can appear in the DNA sequence of the same chromosome from two different individuals, creating a single-nucleotide polymorphism (SNP).

What scientists do know is that most of the human genome is the same in all of us. A little bit of genetic variation—differences that account for much less than 1 percent of our DNA—gives each of us a unique personality, appearance, and health profile.

The parts of the human genome where the DNA sequences of many individuals vary by a single nucleotide are known as single-nucleotide polymorphisms (abbreviated SNPs and pronounced "snips"). For example, let's say that a certain nucleotide in one of your genes is A. In your uncle, however, the nucleotide in the same place on the same gene might be G. You and your uncle have slightly different versions of that gene. Scientists call the different gene versions alleles. Haplotypes are combinations of gene variants, or SNPs, that are likely to be inherited together within the same chromosomal region. In this example, an original haplotype (top) evolved over time to create three newer haplotypes that each differ by a few nucleotides (red).

If two genes sit right next to each other on a chromosome, the SNPs in those genes tend to be inherited together. This set of neighboring SNPs is called a haplotype. Most chromosome regions have only a few, common haplotypes among all humans. As it turns out, these few haplotypes—in different combinations in each person—appear to account for most of the variation from person to person in a population.

Scientists can use haplotype information to compare the genes of people affected by a disease with those of unaffected people. For example, this approach revealed a genetic variation that substantially increases the risk of age-related macular degeneration, the leading cause of severe vision loss in the elderly. Scientists discovered that a single SNP—one nucleotide in the 3 billion-nucleotide human genome—makes some people more likely to get this eye disease. The discovery paves the way for better diagnostic tests and treatments.

What about other diseases? In October 2005, an international scientific team published a catalogue of human haplotypes. Researchers are looking through the catalogue in an effort to identify genes that determine susceptibility to many common diseases, including asthma, diabetes, cancer, and heart disease.

But not all SNPs are in genes. Scientists studying genetic variation have also found SNPs in DNA that doesn't encode proteins. Nonetheless, some of these SNPs appear to affect gene activity.

Some researchers suspect that the "cryptic" (hidden) variation associated with SNPs in non-coding DNA plays an important role in determining the physical characteristics and behaviors of an organism.

Loren Rieseberg of Indiana University in Bloomington is one scientist who would love to take the mystery out of cryptic variation. He wants to know how this non-coding genetic variation can help organisms adapt to new environments. He's also curious about whether it can create problems for some individuals. You might be surprised to learn that Rieseberg's principal research subject is the sun- flower. Although many plants produce only one generation a year, plants like sunflowers can be very useful tools for researchers asking fundamental questions about genetics. Because their genetic material is more malleable than that of many animals, plants are excellent models for studying how evolution works. Wild sunflowers appealed to Rieseberg because there are several species that live in different habitats. Two ancient species of wild sunflowers grow in moderate climates and are broadly distributed throughout the central and western United States. Three recently evolved sunflower species live in more specialized environments: One of the new species grows on sand dunes, another grows in dry desert soil, and the third species grows in a salt marsh.

To see how quickly new plant species could evolve, Rieseberg forced the two ancient sunflowers to interbreed with each other, something plants but not other organisms can do. Among the hybrid progeny were sunflowers that were just like the three recently evolved species! What that means is that Rieseberg had stimulated evolution in his lab, similar to what actually happened in nature some 60,000 to 200,000 years ago, when the newer species first arose.

That Rieseberg could do this is pretty amazing, but the really interesting part is how it happened. Scientists generally assume that, for a new species with very different characteristics to evolve, a lot of new mutations have to occur.

But when Rieseberg looked at the genomes of his hybrid sunflowers, he was surprised to find that they were just cut-and-pasted versions of the ancient sunflower species' genomes: large chunks had been moved rather than many new SNPs created.

Rieseberg reasons that plants stash away unused genetic material, giving them a ready supply of ingredients they can use to adapt quickly to a new environment. It may be that human genomes can recycle unused genetic material to confront new challenges, as well.

Living Laboratories

The Genome Zoo

Scientists often use an image of a tree to depict how all organisms, living and extinct, are related to a common ancestor. In this "tree of life," each branch represents a species, and the forks between branches show when the species represented by those branches became different from one another. For example, researchers estimate that the common ancestor of humans and chimpanzees lived about 6 million years ago.

While it is obvious just by looking that people have a lot in common with our closest living relatives, chimpanzees, what about more distant species? If you look at an evolutionary tree, you'll see that humans are related to mice, worms, and even bacteria. The ancestral species that gave rise to both humans and bacteria was alive a lot longer ago than the ancestor of humans and chimpanzees, yet we still share hundreds of genes with bacteria. Scientists use the term comparative genomics to describe what they're doing when they compare the genomes of different species to see how similar (or how different!) the species' DNA sequences are. Sequences that the species have in common are the molecular footprints of an ancestor of those species. Why are "old" DNA sequences still in our genomes? It turns out that nature is quite economical, so DNA sequences that are responsible for something as complicated and important as controlling gene activity may stay intact for millions of years.

Comparative genomic studies also have medical implications. What would you do if you wanted to develop new methods of preventing, diagnosing, or treating a human disease that animals don't get? If people have a gene that influences their risk for a disease, and mice have the gene too, you could study some aspect of the disease in mice, even though they don't ever have the symptoms of the disease. You could even study the disease in yeast, if it has the gene, as well.

Starting all over Again

Stem cells—what embryos are made up of just days after an egg is fertilized by a sperm—have the amazing ability to grow up into any kind of cell: skin, heart, muscle, nerve, and everything else. How do they do it? Intrigued by the potential of these masterful cells, researchers want to know just what it is that gives stem cells their ability to change into anything, upon the body's request, but stay in the "I can do anything" state until asked.

Some basic researchers are trying to figure out how stem cells work by using a unique model system: tiny, freshwater worms called planarians. These worms are like stem cells in the sense that they can regenerate. You can cut up planarians into hundreds of pieces, and each piece will regenerate into an intact worm that looks the same as all the others.

Planarians' resemblance to stem cells isn't just coincidence. Scientists have discovered that planarians can perform the amazing act of regeneration due to the presence of, yes, specialized stem cells in their bodies.

Developmental biologist Alejandro Sánchez Alvarado of the University of Utah School of Medicine in Salt Lake City used the gene-silencing technique RNAi (see *RNA Interference*) to search for planarian genes that were essential for regeneration.

He found 240 genes that, when silenced, caused a physical defect in the worm's growth and regenerative ability. Interestingly, 16 percent of these looked very much like genes that had been linked to human disease!

Sánchez Alvarado and his team hope to figure out how regeneration genes allow the specialized stem cells within the worm to travel to a wounded site and "turn into" any of the 30 or so cell types needed to recreate a mature worm.

Although humans are only distantly related to planarians, we have many of the same genes, so these findings have the potential to reveal strategies for regenerating injured body parts in people, too.

Genes Meet Environment

If toxins from the environment get into our bodies, they don't always make us sick. That's because liver enzymes come to our rescue to make the chemicals less harmful. The genes that encode those enzymes are under constant evolutionary pressure to adapt quickly to new toxins. For example, certain liver enzymes called cytochrome P450 proteins metabolize, or break down, hormones that our bodies make as well as many of the foreign substances that we encounter. These include harmful molecules like cancer-causing agents as well as beneficial ones, like medicines. In fact, just two genes within the cytochrome P450 family, abbreviated 3A4 and 3A5, encode proteins that process more than half of all of the medicines that are sold today.

Since the chemicals to which people are exposed vary so widely, a scientist might predict that there would be different variants of cytochrome P450 genes in different human populations. Using comparative genomics, researchers such as Anna Di Rienzo of the University of Chicago have shown that this is indeed the case. Di Rienzo has found many sequence differences within these genes in people living throughout the world. It turns out that one variant of the gene that encodes the cytochrome P450 3A5 protein makes this enzyme very efficient at breaking down cortisol, a hormone that raises salt levels in the kidneys and helps the body retain water. Di Rienzo compared the DNA sequences of the 3A5 gene in DNA samples taken from more than 1,000 people representing over 50 populations worldwide. She was amazed to find a striking link between the existence of the gene variant and the geographic locale of the people who have it.

Di Rienzo discovered that African populations living very close to the equator were more likely than other populations to have the salt-saving version of the 3A5 gene. She suggests that this is because this gene variant provides a health advantage for people living in a very hot climate, since retaining salt helps ward off dehydration caused by intense heat.

However, there seems to be a cost associated with that benefit—the 3A5 gene variant raises the risk for some types of high blood pressure. That means that in environments in which retaining salt is not beneficial, evolution selects against this gene variant.

Another scientist who studies interactions between genes and the environment is Serrine Lau of the University of Arizona in Tucson. She studies a class of harmful molecules called polyphenols, present in cigarette smoke and car exhaust, that cause kidney cancer in rats, and perhaps, in people. Lau discovered that rats and humans who are more sensitive to some of the breakdown products of polyphenols have an unusual DNA sequence—a genetic signature—that increases their risk of developing cancer. She suspects that the gene that is affected encodes a tumor suppressor: a protein that prevents cancer from developing. In people and rats with the genetic signature, she reasons, the tumor suppressor doesn't work right, so tumors grow. Taking this logic one step further, it may be that certain people's genetic make-up makes them unusually susceptible to DNA damage caused by exposure to carcinogens. If doctors could identify those at risk, Lau says, such people could be forewarned to avoid contact with specific chemicals to protect their health. However, think about this scenario: Who should make those decisions? For example, would it be ethical for an employer to refuse to hire somebody because the person has a genetic signature that makes him or her more likely to get cancer if exposed to a chemical used in the workplace? Tough question.

Plant Breeding

Plant breeding is the art and science of changing the genetics of plants for the benefit of humankind. Plant breeding can be accomplished through many different techniques ranging from simply selecting plants with desirable characteristics for propagation, to more complex molecular techniques. Plant breeding has been practiced for thousands of years, since near the beginning of human civilization. It is now practiced worldwide by individuals such as gardeners and farmers, or by professional plant breeders employed by organizations such as government institutions, universities, crop-specific industry associations or research centers.

International development agencies believe that breeding new crops is important for ensuring food security by developing new varieties that are higher-yielding, resistant to pests and diseases, drought-resistant or regionally adapted to different environments and growing conditions.

Domestication

This map shows the sites of domestication for a number of crops. Places where crops were initially domesticated are called centers of origin.

Plant breeding in certain situations may lead the domestication of wild plants. Domestication of plants is an artificial selection process conducted by humans to produce plants that have more desirable traits than wild plants, and which renders them dependent on artificial (usually enhanced) environments for their continued existence. The practice is estimated to date back 9,000-11,000 years. Many crops in present day cultivation are the result of domestication in ancient times, about 5,000 years ago in the Old World and 3,000 years ago in the New World. In the Neolithic period, domestication took a minimum of 1,000 years and a maximum of 7,000 years. Today, all of our principal food crops come from domesticated varieties. Almost all the domesticated plants used today for food and agriculture were domesticated in the centers of origin. In these centers there is still a great diversity of closely related wild plants, the so called crop wild relatives that can also be used for improving modern cultivars by plant breeding. A plant whose origin or selection is due primarily to intentional human activity is called a cultigen, and a cultivated crop species that has evolved from wild populations due to selective pressures from traditional farmers is called a landrace. Landraces, which can be the result of natural forces or domestication, are plants (or animals) that are ideally suited to a particular region or environment. An example are the landraces of rice, *Oryza sativa* subspecies *indica*, which was developed in South Asia, and *Oryza sativa* subspecies *japonica*, which was developed in China.

Classical Plant Breeding

Classical plant breeding uses deliberate interbreeding (crossing) of closely or distantly related individuals to produce new crop varieties or lines with desirable properties. Plants are crossbred to introduce traits/genes from one variety or line into a new genetic background. For example, a mildew-resistant pea may be crossed with a high-yielding but susceptible pea, the goal of the cross being to introduce mildew resistance without losing the high-yield characteristics. Progeny from the cross would then be crossed with the high-yielding parent to ensure that the progeny were most like the high-yielding parent, (backcrossing). The progeny from that cross would then be tested for yield and mildew resistance and high-yielding resistant plants would be further developed. Plants may also be crossed with themselves to produce inbred varieties for breeding.

Classical breeding relies largely on homologous recombination between chromosomes to generate genetic diversity. The classical plant breeder may also makes use of a number of *in vitro* techniques such as protoplast fusion, embryo rescue or mutagenesis to generate diversity and produce hybrid plants that would not exist in nature. Traits that breeders have tried to incorporate into crop plants in the last 100 years include:

1. Increased quality and yield of the crop
2. Increased tolerance of environmental pressures (salinity, extreme temperature, drought)
3. Resistance to viruses, fungi and bacteria
4. Increased tolerance to insect pests
5. Increased tolerance of herbicides.

Before World War II

Intraspecific hybridization within a plant species was demonstrated by Charles Darwin and Gregor Mendel, and was further developed by geneticists and plant breeders. In the United Kingdom in the 1880s was the pioneering work of Gartons Agricultural Plant Breeders. In the early 20th century, plant breeders realized that Mendel's findings on the non-random nature of inheritance could be applied to seedling populations produced through deliberate pollinations to predict the frequencies of different types. In 1908, George Harrison Shull described heterosis, also known as hybrid vigor. Heterosis describes the tendency of the progeny of a specific cross to outperform both parents. The detection of the usefulness of heterosis for plant breeding has led to the development of inbred lines that reveal a heterotic yield advantage when they are crossed. Maize was the first species where heterosis was widely used to produce hybrids.

By the 1920s, statistical methods were developed to analyse gene action and distinguish heritable variation from variation caused by environment. In 1933, another important breeding technique, cytoplasmic male sterility (CMS), developed in maize, was described by Marcus Morton Rhoades. CMS is a maternally inherited trait that makes the plant produce sterile pollen. This enables the production of hybrids without the need for labour intensive detasseling. These early breeding techniques resulted in large yield increase in the United States in the early 20th century. Similar yield increases were not produced elsewhere until after World War II, the Green Revolution increased crop production in the developing world in the 1960s.

After World War II

Following World War II a number of techniques were developed that allowed plant breeders to hybridize distantly related species, and artificially induce genetic diversity. When distantly related species are crossed, plant breeders make use of a number of plant tissue culture techniques to produce progeny from otherwise fruitless mating. Interspecific and intergeneric hybrids are produced from a cross of related species or genera that do not normally sexually reproduce with each other.

These crosses are referred to as *Wide crosses*. For example, the cereal triticale is a wheat and rye hybrid. The cells in the plants derived from the first generation created from the cross contained an uneven number of chromosomes and as result was sterile. The cell division inhibitor colchicine was used to double the number of chromosomes in the cell and thus allow the production of a fertile line. Failure to produce a hybrid may be due to pre- or post-fertilization incompatibility. If fertilization is possible between two species or genera, the hybrid embryo may abort before maturation. If this does occur the embryo resulting from an interspecific or intergeneric cross can sometimes be rescued and cultured to produce a whole plant. Such a method is referred to as *Embryo Rescue*. This technique has been used to produce new rice for Africa, an interspecific cross of Asian rice *(Oryza sativa)* and African rice *(Oryza glaberrima)*.

Hybrids may also be produced by a technique called protoplast fusion. In this case protoplasts are fused, usually in an electric field. Viable recombinants can be regenerated in culture.

Chemical mutagens like EMS and DMS, radiation and transposons are used to generate mutants with desirable traits to be bred with other cultivars- a process known as *Mutation Breeding*. Classical plant breeders also generate genetic diversity within a species by exploiting a process called somaclonal variation, which occurs in plants produced from tissue culture, particularly plants derived from callus. Induced polyploidy, and the addition or removal of chromosomes using a technique called chromosome engineering may also be used.

When a desirable trait has been bred into a species, a number of crosses to the favoured parent are made to make the new plant as similar to the favoured parent as possible. Returning to the example of the mildew resistant pea being crossed with a high-yielding but susceptible pea, to make the mildew resistant progeny of the cross most like the highyielding parent, the progeny will be crossed back to that parent for several generations. This process removes most of the genetic contribution of the mildew resistant parent. Classical breeding is therefore a cyclical process.

It should be noted that with classical breeding techniques, the breeder does not know exactly what genes have been introduced to the new cultivars. Some scientists therefore argue that plants produced by classical breeding methods should undergo the same safety testing regime as genetically modified plants.

There have been instances where plants bred using classical techniques have been unsuitable for human consumption, for example the poison solanine was unintentionally increased to unacceptable levels in certain varieties of potato through plant breeding. New potato varieties are often screened for solanine levels before reaching the marketplace.

Modern Plant Breeding

Modern plant breeding uses techniques of molecular biology to select, or in the case of genetic modification, to insert, desirable traits into plants.

Steps of Plant Breeding

The following are the major activities of plant breeding;

1. Creation variation
2. Selection
3. Evaluation
4. Release
5. Multiplication
6. Distribution of the new variety.

Marker Assisted Selection

Sometimes many different genes can influence a desirable trait in plant breeding. The use of tools such as molecular markers or DNA fingerprinting can map thousands of genes. This allows plant breeders to screen large populations of plants for those that possess the trait of interest. The screening is based on the presence or absence of a certain gene as determined by laboratory procedures, rather than on the visual identification of the expressed trait in the plant.

Reverse Breeding and Doubled Haploids (DH)

A method for efficiently producing homozygous plants from a heterozygous starting plant, which has all desirable traits. This starting plant is induced to produce doubled haploid from haploid cells, and later on creating homozygous/doubled haploid plants from those cells. While in natural offspring recombination occurs and traits can be unlinked from each other, in doubled haploid cells and in the resulting DH plants recombination is no longer an issue.

There, a recombination between two corresponding chromosomes does not lead to un-linkage of alleles or traits, since it just leads to recombination with its identical copy. Thus, traits on one chromosome stay linked. Selecting those offspring having the desired set of chromosomes and crossing them will result in a final F1 hybrid plant, having exactly the same set of chromosomes, genes and traits as the starting hybrid plant.

The homozygous parental lines can reconstitute the original heterozygous plant by crossing, if desired even in a large quantity. An individual heterozygous plant can be converted into a heterozygous variety (F1 hybrid) without the necessity of vegetative propagation but as the result of the cross of two homozygous/doubled haploid lines derived from the originally selected plant. patent

Genetic Modification

Genetic modification of plants is achieved by adding a specific gene or genes to a plant, or by knocking down a gene with RNAi, to produce a desirable phenotype. The plants resulting from adding a gene are often referred to as transgenic plants. If for genetic modification genes of the species or of a crossable plant are used under control of their native promoter, then they are called cisgenic plants. Genetic modification can produce a plant with the desired trait or traits faster than classical breeding because the majority of the plant's genome is not altered.

To genetically modify a plant, a genetic construct must be designed so that the gene to be added or removed will be expressed by the plant. To do this, a promoter to drive transcription and a termination sequence to stop transcription of the new gene, and the gene or genes of interest must be introduced to the plant. A marker for the selection of transformed plants is also included. In the laboratory, antibiotic resistance is a commonly used marker: plants that have been successfully transformed will grow on media containing antibiotics; plants that have not been transformed will die. In some instances markers for selection are removed by backcrossing with the parent plant prior to commercial release.

The construct can be inserted in the plant genome by genetic recombination using the bacteria *Agrobacterium tumefaciens* or *A. rhizogenes*, or by direct methods like the gene gun or microinjection.

Using plant viruses to insert genetic constructs into plants is also a possibility, but the technique is limited by the host range of the virus. For example, Cauliflower mosaic virus (CaMV) only infects cauliflower and related species. Another limitation of viral vectors is that the virus is not usually passed on the progeny, so every plant has to be inoculated.

The majority of commercially released transgenic plants, are currently limited to plants that have introduced resistance to insect pests and herbicides. Insect resistance is achieved through incorporation of a gene from *Bacillus thuringiensis* (Bt) that encodes a protein that is toxic to some insects.

For example, the cotton bollworm, a common cotton pest, feeds on Bt cotton it will ingest the toxin and die. Herbicides usually work by binding to certain plant enzymes and inhibiting their action.

The enzymes that the herbicide inhibits are known as the herbicides *target site*. Herbicide resistance can be engineered into crops by expressing a version of *target site* protein that is not inhibited by the herbicide. This is the method used to produce glyphosate resistant crop plants.

Genetic modification of plants that can produce pharmaceuticals (and industrial chemicals), sometimes called *pharmacrops*, is a rather radical new area of plant breeding.

Issues and Concerns

Modern plant breeding, whether classical or through genetic engineering, comes with issues of concern, particularly with regard to food crops. The question of whether breeding can have a negative effect on nutritional value is central in this respect.

Although relatively little direct research in this area has been done, there are scientific indications that, by favoring certain aspects of a plant's develo-pment, other aspects may be retarded. A study published in the *Journal of the American College of Nutrition* in 2004, entitled *Changes in USDA Food Composition Data for 43 Garden Crops, 1950 to 1999*, compared nutritional analysis of vegetables done in 1950 and in 1999, and found substantial decreases in six of 13 nutrients measured, including 6% of protein and 38% of riboflavin. Reductions in calcium, phosphorus, iron and ascorbic acid were also found. The study, conducted at the Biochemical Institute, University of Texas at Austin, concluded in summary: *"We suggest that any real declines are generally most easily explained by changes in cultivated varieties between 1950 and 1999, in which there may be trade-offs between yield and nutrient content. "*

The debate surrounding genetically modified food during the 1990s peaked in 1999 in terms of media coverage and risk perception, and continues today-for example, *"Germany has thrown its weight behind a growing European mutiny over genetically modified crops by banning the planting of a widely grown pest-resistant corn variety."*.

The debate encompasses the ecological impact of genetically modified plants, the safety of genetically modified food and concepts used for safety evaluation like substantial equivalence. Such concerns are not new to plant breeding.

Most countries have regulatory processes in place to help ensure that new crop varieties entering the marketplace are both safe and meet farmers' needs. Examples include variety registration, seed schemes, regulatory authorizations for GM plants, etc.

Plant breeders' rights is also a major and controversial issue. Today, production of new varieties is dominated by commercial plant breeders, who seek to protect their work and collect royalties through national and international agreements based in intellectual property rights.

The range of related issues is complex. In the simplest terms, critics of the increasingly restrictive regulations argue that, through a combination of technical and economic pressures, commercial breeders are reducing biodiversity and significantly constraining individuals (such as farmers) from developing and trading seed on a regional level. Efforts to strengthen breeders' rights, for example, by lengthening periods of variety protection, are ongoing.

When new plant breeds or cultivars are bred, they must be maintained and propagated. Some plants are propagated by asexual means while others are propagated by seeds.

Seed propagated cultivars require specific control over seed source and production procedures to maintain the integrity of the plant breeds results. Isolation is necessary to prevent cross contamination with related plants or the mixing of seeds after harvesting.

Isolation is normally accomplished by planting distance but in certain crops, plants are enclosed in greenhouses or cages (most commonly used when producing F1 hybrids.)

Participatory Plant Breeding

Participatory Plant Breeding is carried out, for example, in northern Vietnam, where government scientists work with farmers from the Muong people ethnic minority to improve local rice varieties.

The development of agricultural science, with phenomenon like the Green Revolution arising, have left millions of farmers in developing countries, most of whom operate small farms under unstable and difficult growing conditions, in a precarious situation.

The adoption of new plant varieties by this group has been hampered by the constraints of poverty and the international policies promoting an industrialised model of agriculture. Their response has been the creation of a novel and promising set of research methods collectively known as participatory plant breeding.

Participatory means that farmers are more involved in the breeding process and breeding goals are defined by farmers instead of international seed companies with their large-scale breeding programs. Farmer's groups and NGOs, for example, may wish to affirm local people's rights over genetic resources, produce seeds themselves, build farmers' technical expertise, or develop new products for niche markets, like organically grown food.

8

Plant Reproduction: They'll Make More

If you are an organism, you will need to reproduce. Otherwise, there will be no more of your species and the species will die off. You may have heard of endangered animals.

There are also endangered plants. These endangered species have very few individuals left and scientists/naturalists are working together to make sure the species don't become extinct.

We talked a little about reproduction when we discussed meiosis in the cells tutorials. Reproduction is one of two things.

(1) One cell can split into two, giving you two identical cells. That type is asexual reproduction.

(2) The second type is when two cells, each with half of the DNA needed, combine and create a living cell. That type is sexual reproduction.

When plants hit a point in evolution, the second is the one that occurs more often.

Making More Mosses

Sporophytes are the reproductive structures you will find in mosses. They are actually a phase of the moss life cycle that feeds off the green parent plant (the gametophyte).

The sporophyte is a stalk that grows after the haploid sperm of one moss plant is able to mix with the haploid egg of a female moss plant. The resulting diploid cell grows into the sporophyte stalk. When ready, spores stored in the sporophyte are released and they grow into new moss plants.

Confiers and their Cones

While there are male and female mosses, conifers produce two types of cones on the same tree. One of the cone types gives off pollen (the staminate cone). The other type of cone catches the pollen if the wind is moving in the right direction. Better yet, the wind blows the pollen to another conifer of the same species, and a cone (called the ovulate cone) catches the pollen. Again, the pollen and megaspore (receiving haploid cell) are haploid and combine to form a diploid cell. That diploid cell grows into a zygote (baby conifer) that eventually lives in a seed.

Flowers and Pollen

The most advanced of the plants have their own way of sexually reproducing. It is a very fancy and very complex process. Plants that rely on flowers for reproduction are also very dependent on outside help such as insects and animals. While conifers have the two structures on one tree, flowering plants went one step further and put the devices that make and receive pollen in the same structure. How does that help? A bee might go to one flower and get a little pollen on its back. If it goes to another flower of the same species, that pollen can land on the stigma. From that point, one haploid male nucleus combines with a female nucleus and the other haploid male nucleus combines with a polar nucleus. If successful, an embryo and seed/fruit develop respectively.

Reproduction in Plants

Plant reproduction is the process by which plants generate new individuals, or offspring. Reproduction is either sexual or asexual. Sexual reproduction is the formation of offspring by the fusion of gametes. Asexual reproduction is the formation of offspring without the fusion of gametes. Sexual reproduction results in offspring genetically different from the parents. Asexual offspring are genetically identical except for mutation. In higher plants, offspring are packaged in a protective seed, which can be long lived and can disperse the offspring some distance from the parents. In flowering plants (angiosperms), the seed itself is contained inside a fruit, which may protect the developing seeds and aid in their dispersal.

Sexual Reproduction in Angiosperms: Ovule Formation

All plants have a life cycle that consists of two distinct forms that differ in size and the number of chromosomes per cell. In flowering plants, the large, familiar form that consists of roots, shoots, leaves, and reproductive structures (flowers and fruit) is diploid and is called the sporophyte. The sporophyte produces haploid microscopic gametophytes that are dependent on tissues produced by the flower. The reproductive cycle of a flowering plant is the regular, usually seasonal, cycling back and forth from sporophyte to gametophyte.

The flower produces two kinds of gametophytes, male and female. The female gametophyte arises from a cell within the ovule, a small structure within the ovary of the flower. The ovary is a larger structure within the flower that contains and protects usually many ovules. Flowering plants are unique in that their ovules are entirely enclosed in the ovary. The ovary itself is part of a larger structure called the carpel, which consists of the stigma, style, and ovary. Each ovule is attached to ovary tissue by a stalk called the funicle. The point of attachment of the funicle to the ovary is called the placenta.

As the flower develops from a bud, a cell within an ovule called the archespore enlarges to form an embryo-sac mother cell (EMC). The EMC divides by meiosis to produce four megaspores. In this process the number of chromosomes is reduced from two sets in the EMC to one set in the megaspores, making the megaspores haploid. Three of the four megaspores degenerate and disappear, while the fourth divides mitotically three times to produce eight haploid cells. These cells together constitute the female gametophyte, called the embryo sac.

The eight embryo sac cells differentiate into two synergids, three antipodal cells, two fused endosperm nuclei, and an egg cell. The mature embryo sac is situated at the outer opening (micropyle) of the ovule, ready to receive the sperm cells delivered by the male gametophyte.

Pollen

The male gametophyte is the mature pollen grain. Pollen is produced in the anthers, which are attached at the distal end of filaments. The filament and anther together constitute the stamen, the male sex organ. Flowers usually produce many stamens just inside of the petals. As the flower matures, cells in the anther divide mitotically to produce pollen mother cells (PMC). The PMCs divide by meiosis to produce haploid microspores in groups of four called tetrads. The microspores are housed within a single layer of cells called the tapetum, which provides nutrition to the developing pollen grains.

Each microspore develops a hard, opaque outer layer called the exine, which is constructed from a lipoprotein called sporopollenin. The exine has characteristic pores, ridges, or projections that can often be used to identify a species, even in fossil pollen. The microspore divides mitotically once or twice to produce two or three haploid nuclei inside the mature pollen grain. Two of the nuclei function as sperm nuclei that can eventually fuse with the egg and endosperm nuclei of the embryo sac, producing an embryo and endosperm, respectively.

For sexual fusion to take place, however, the pollen grain must be transported to the stigma, which is a receptive platform on the top of the style, an elongated extension on top of the carpel(s). Here the moist surface or chemicals cause the pollen grain to germinate. Germination is the growth of a tube from the surface of a pollen grain. The tube is a sheath of pectin, inside of which is a solution of water, solutes, and the two or three nuclei, which lack any cell walls. Proper growth of the pollen tube requires an aqueous solution of appropriate solute concentration, as well as nutrients such as boron, which may aid in its synthesis of pectin.

At the apex of the tube are active ribosomes and endoplasmic reticulum (types of cell organelles) involved in protein synthesis. Pectinase and a glucanase (both enzymes that break down carbohydrates) probably maintain flexibility of the growing tube and aid in penetration. The pollen tube apex also releases ribonucleic acid (RNA) and ribosomes into the tissues of the style. The tube grows to eventually reach the ovary, where it may travel along intercellular spaces until it reaches a placenta. Through chemical recognition, the pollen tube changes its direction of growth and penetrates through the placenta to the ovule. Here the tube reaches the embryo sac lying close to the micropyle, and sexual fertilization takes place.

Double Fertilization

Fertilization in flowering plants is unique among all known organisms, in that not one but two cells are fertilized, in a process called double fertilization. One sperm nucleus in the pollen tube fuses with the egg cell in the embryo sac, and the other sperm nucleus fuses with the diploid endosperm nucleus. The fertilized egg cell is a zygote that develops into the diploid embryo of the sporophyte. The fertilized endosperm nucleus develops into the triploid endosperm, a nutritive tissue that sustains the embryo and seedling. The only other known plant group exhibiting double fertilization is the Gnetales in the genus *Ephedra,* a nonflowering seed plant. However, in this case the second fertilization product degenerates and does not develop into endosperm.

Double fertilization begins when the pollen tube grows into one of the two synergid cells in the embryo sac, possibly as a result of chemical attraction to calcium. After penetrating the synergid, the apex of the pollen tube breaks open, releasing the two sperm nuclei and other contents into the synergid. As the synergid degenerates, it envelops the egg and endosperm cells, holding the two sperm nuclei close and the other expelled contents of the pollen tube. The egg cell then opens and engulfs the sperm cell, whose membrane breaks apart and allows the nucleus to move near the egg nucleus. The nuclear envelopes then disintegrate, and the two nuclei combine to form the single diploid nucleus of the zygote. The other sperm cell fuses with the two endosperm nuclei, forming a single triploid cell, the primary endosperm cell, which divides mitotically into the endosperm tissue.Double fertilization and the production of endosperm may have contributed to the great ecological success of flowering plants by accelerating the growth of seedlings and improving survival at this vulnerable stage. Faster seedling development may have given flowering plants the upper hand in competition with gymnosperm seedlings in some habitats, leading to the abundance of flowering plants in most temperate and tropical regions. Gymnosperms nevertheless are still dominant at higher elevations and latitudes, and at low elevations in the Pacific Northwest coniferous forests, such as the coastal redwoods. The reasons for these patterns are still controversial.

The Seed

The seed is the mature, fertilized ovule. After fertilization, the haploid cells of the embryo sac disintegrate. The maternally derived diploid cells of the ovule develop into the hard, water-resistant outer covering of the seed, called the testa, or seed coat. The diploid zygote develops into the embryo, and the triploid endosperm cells multiply and provide nutrition. The testa usually shows a scar called the hilum where the ovule was originally attached to the funicle. In some seeds a ridge along the testa called the raphe shows where the funicle originally was pressed against the ovule. The micropyle of the ovule usually survives as a small pore in the seed coat that allows passage of water during germination of the seed.

In some species, the funicle develops into a larger structure on the seed called an aril, which is often brightly colored, juicy, and contains sugars that are consumed by animals that may also disperse the seed (as in nutmeg, arrowroot, oxalis, and castor bean). This is distinct from the fruit, which forms from the ovary itself. The embryo consists of the cotyledon(s), epicotyl, and hypocotyl. The cotyledons resemble small leaves, and are usually the first photosynthetic organs of the plant. The portion of the embryo above the cotyledons is the epicotyl, and the portion below is the hypocotyl. The epicotyl is an apical meristem that produces the shoot of the growing plant and the first true leaves after germination.

The hypocotyl develops into the root. Often the tip of the hypocotyl, the radicle, is the first indication of germination as it breaks out of the seed. Flowering plants are classified as monocotyledons or dicotyledons (most are now called eudicots) based on the number of cotyledons produced in the embryo. Common monocotyledons include grasses, sedges, lilies, irises, and orchids; common dicotyledons include sunflowers, roses, legumes, snapdragons, and all nonconiferous trees. The endosperm may be consumed by the embryo, as in many legumes, which use the cotyledons as a food source during germination. In other species the endosperm persists until germination, when it is used as a food reserve. In grains such as corn and wheat, the outer layer of the endosperm consists of thick-walled cells called aleurone, which are high in protein.

The Fruit

The fruit of a flowering plant is the mature ovary. As seeds mature, the surrounding ovary wall forms a protective structure that may aid in dispersal.

The surrounding ovary tissue is called the pericarp and consists of three layers. From the outside to inside, these layers are the exocarp, mesocarp, and endocarp. The exocarp is usually tough and skinlike. The mesocarp is often thick, succulent, and sweet. The endocarp, which surrounds the seeds, may be hard and stony, as in most species with fleshy fruit, such as apricots.

A fruit is termed simple if it is produced by a single ripened ovary in a single flower (apples, oranges, apricots). An aggregated fruit is a cluster of mature ovaries produced by a single flower (blackberries, raspberries, strawberries). A multiple fruit is a cluster of many ripened ovaries on séparate flowers growing together in the same inflorescence (pineapple, mulberry. A simple fruit may be fleshy or dry. A fleshy simple fruit is classified as a berry (grape, tomato, papaya), pepo (cucumber, watermelon, pumpkin), hesperidium (orange), drupe (apricot), or pome (apple).

Dry simple fruits have a dry pericarp at maturity. They may or may not dehisce, or split, along a seam to release the seeds. A dehiscent dry fruit is classified as legume or pod (pea, bean), silique or silicle (mustard), capsule (poppy, lily), or follicle (milkweed, larkspur, columbine). An indehiscent dry fruit that does not split to release seeds is classified as an achene (sunflower, buttercup, sycamore), grain or caryopsis (grasses such as corn, wheat, rice, barley), schizocarp (carrot, celery, fennel), winged samara (maple, ash, elm), nut (acorn, chestnut, hazelnut), or utricle (duckweed family). Some fruiting bodies contain non-ovary tissue and are sometimes called pseudocarps. The sweet flesh of apples and pears, for example, is composed not of the pericarp but the receptacle, or upper portion, of the flowering shoot to which petals and other floral organs are attached.

Fruiting bodies of all kinds function to protect and disperse the seeds they contain. Protection can be physical (hard coverings) or chemical (repellents of seed predators). Sweet, fleshy fruits are attractive food for birds and mammals that consume seeds along with the fruit and pass the seeds intact in their fecal matter, which can act as a fertilizer. Dry fruits are usually adapted for wind dispersal of seeds, as for example with the assistance of winglike structures or a fluffy pappus that provides buoyancy. The diversity of fruiting bodies reflects in part the diversity of dispersal agents in the environment, which select for different fruit size, shape, and chemistry.

Pollination and Pollinators

Pollination is the movement of pollen from the stamens to the stigma, where germination and growth of the pollen tube occur. Most (approximately 96 percent) of all flowering plant species are hermaphroditic (possess both sexual functions within a plant, usually within every flower), and thus an individual can be pollinated by its own pollen or by pollen from another individual.

Seed produced through self-pollination ("selfed" seed) is often inferior in growth, survival, and fecundity to seed produced through outcross pollination ("outcrossed" seed). As a result, in most species there is strong natural selection to maximize the proportion of outcrossed seed (the "outcrossing rate").

Flowering plants are unusual among seed plants in their superlative exploitation of animals (primarily insects) as agents of outcross pollination. The outcross pollination efficiency of insects, birds, and mammals (primarily bats) may have contributed to both the abundance and diversity of flowering plants. Abundance may have increased because of less wastage of energy and resources on unsuccessful pollen and ovules. Diversity may have increased for two reasons. First, insects undoubtedly have selected for a wide variety of floral forms that provide different rewards (pollen and nectar) and are attractive in appearance (colour juxtaposition, size, shape) and scent (sweet, skunky) in different ways to different pollinators. Second, faithfulness of pollinators to particular familiar flowers may have reduced hybridization and speeded evolutionary divergence and the production of new species.

Although flowering plants first appeared after most of the major groups of insects had already evolved, flowering plants probably caused the evolution of many new species within these groups. Some new insect groups, such as bees and butterflies, originated after flowering plants, their members developing mouthpart structures and behaviour specialized for pollination. In extreme cases, a plant is completely dependent on one insect species for pollination, and the insect is completely dependent on one plant species for food. Such tight interdependency occurs rarely but is well documented in yuccas/yucca moths, senita cacti/senita moths. In all three insects, females lay eggs in the flowers, and their young hatch later to feed on the mature fruit and its contents. Females ensure that the fruit develops by gathering pollen from another plant and transporting it to the stigma of the flower holding their eggs. Plants benefit greatly in outcrossed seed produced, at the small cost of some consumed fruit and seeds, and the insects benefit greatly from the food supply for developing larvae at the small cost of transporting pollen the short distances between plants.

Pollinating agents, whether biotic or abiotic, have exerted strong selection on all aspects of the flower, resulting in the evolution of tremendous floral diversity. This diversity has been distilled into a small number of characteristic pollination syndromes.

Pollination by beetles selects usually for white colour, a strong fruity scent, and a shallow, bowl-shaped flower. Bees select for yellow or blue/purple colorings, a landing platform with colour patterns that guide the bee to nectar (often reflecting in the ultraviolet range of the spectrum), bilateral symmetry, and a sweet scent. Butterflies select for many colors other than yellow, a corolla (petal) tube with nectar at the base, and the absence of any scent. Moths in contrast select for nocturnally opening flowers with a strong scent and drab or white colour, and also a tube with nectar at the base. Bats select also for nocturnally opening flowers, but with a strong musky scent and copious nectar, positioned well outside the foliage for easy access, and drab or white colour. Hummingbirds select for red or orange flowers with no scent, copious nectar production, and a corolla tube with nectar at the base. Other pollinating birds that do not hover while feeding select for strong perches and flowers capable of containing copious nectar (tubes, funnels, cup shapes).

Wind as a pollinating agent selects for lack of colour, scent, and nectar; small corolla; a large stigmatic surface area (usually feathery); abundantly produced, buoyant pollen; and usually erect styles and limp, hanging stamens. In addition there is great floral diversity within any of these syndromes, arising from the diverse evolutionary histories of the member plant species.

Selfing and Outcrossing

Most flowering plant species reproduce primarily by outcrossing, including the great majority of trees, shrubs, and perennial herbs. Adaptations that prevent self-fertilization include self-incompatibility (genetic recognition and blocking of self-pollen) and dioecy (separate male and female individuals). Adaptations that reduce the chances of self-pollination in hermaphrodites include separation of the anthers and stigma in space (herkogamy) or time (dichogamy). In many species, both self-incompatibility and spatiotemporal separation of the sex organs occur.

The ability to produce seeds by selfing, however, is advantageous in situations where outcrossing pollination is difficult or impossible. These include harsh environments where pollinators are rare or unpredictable, and regularly disturbed ground where survivors often end up isolated from each other. Selfing is also cheaper than outcrossing, because selfers can become pollinated without assistance from animals and therefore need not produce large, attractive flowers with abundant nectar and pollen rewards.

Most primarily selfing species are small annuals in variable or disturbed habitats, with small, drab flowers. Most desert annuals and roadside weeds, for example, are selfers. The evolutionary transition from outcrossing to near-complete selfing has occurred many times in flowering plants. Outcrossing and selfing species differ in their evolutionary potential. Outcrossers are generally more genetically diverse and produce lineages that persist over long periods of evolutionary time, during which many new species are formed. Selfers, however, are less genetically diverse and tend to accumulate harmful mutations. They typically go extinct before they have an opportunity to evolve new species.

Asexual Reproduction

The ability to produce new individuals asexually is common in plants. Under appropriate experimental conditions, nearly every cell of a flowering plant is capable of regenerating the entire plant. In nature, new plants may be regenerated from leaves, stems, or roots that receive an appropriate stimulus and become separated from the parent plant. In most cases, these new plants arise from undifferentiated parenchyma cells, which develop into buds that produce roots and shoots before or after separating from the parent.

New plants can be produced from aboveground or belowground horizontal runners (stolons of strawberries, rhizomes of many grasses), tubers (potato, Jerusalem artichoke, dahlia), bulbs (onion, garlic), corms (crocus, gladiola), bulbils on the shoot (lily, many grasses), parenchyma cells in the leaves (Kalanchoe, African violet, jade plant) and inflorescence (arrowhead).

Vegetative propagation is an economically important means of replicating valuable agricultural plants, through cuttings, layering, and grafting. Vegetative reproduction is especially common in aquatic vascular plants (for example, surfgrass and eelgrass), from which fragments can break off, disperse in the current, and develop into new whole plants.

A minority of flowering plants can produce seeds without the fusion of egg and sperm (known as parthenocarpy or agamospermy). This occurs when meiosis in the ovule is interrupted, and a diploid egg cell is produced, which functions as a zygote without fertilization. Familiar examples include citrus, dandelion, hawkweed, buttercup, blackberry/raspberry, and sorbus. Agamospermous species are more common at high elevations and at high latitudes, and nearly all have experienced a doubling of their chromosome number (tetraploidy) in their recent evolutionary history. These species experience evolutionary advantages and disadvantages similar to those of selfers.

Evolutionary Significance of Plant Reproduction Strategies

The attractive, colorful, and unique features of the most abundant and diverse group of land plants—the flowering plants—are believed to have evolved primarily to maximize the efficiency and speed of outcross reproduction. Each major burst of angiosperm evolution was a coevolutionary episode with associated animals, primarily insects, which were exploited to disperse pollen and seeds in ever more efficient and diverse ways.

The first major burst of flowering plant evolution was the appearance of the closed carpel together with showy flowers that were radially symmetrical. The closed carpel prevented self-fertilization through recognition and blocking of self pollen within the specialized conducting tissue of the style. Insects attracted to the showy flowers carried pollen between plants less wastefully than wind, and the radial symmetry accommodated insects of many sizes and shapes. The second major burst was the appearance of bilaterally symmetrical flowers, which happened independently in many groups of plants at the same time that bees evolved.

Bilateral symmetry forced bees to enter and exit flowers more precisely, promoting even more efficient outcross pollen transfer.

The third major burst of flowering plant evolution was the appearance of nutritious, fleshy fruits and seeds, coincident with a diversification of birds and rodents.

The exploitation of vertebrates for fruit and seed dispersal resulted in less haphazard transport of offspring to neighboring populations of the same species (also visited as a food source), thereby reducing the chances that progeny inbreed with their siblings and parents and providing more assurance than wind currents that they find good habitat and unrelated mating partners of the same species.

Sexual Reproduction

Sexual reproduction is characterized by processes that pass a combination of genetic material to offspring, resulting in increased genetic diversity. The main two processes are: meiosis, involving the halving of the number of chromosomes; and fertilization, involving the fusion of two gametes and the restoration of the original number of chromosomes. During meiosis, the chromosomes of each pair usually cross over to achieve homologous recombination.

The evolution of sexual reproduction is a major puzzle. The first fossilized evidence of sexually reproducing organisms is from eukaryotes of the Stenian period, about 1 to 1.2 billion years ago.

Sexual reproduction is the primary method of reproduction for the vast majority of macroscopic organisms, including almost all animals and plants. Bacterial conjugation, the transfer of DNA between two bacteria, is often mistakenly confused with sexual reproduction, because the mechanics are similar.

A major question is why sexual reproduction persists when parthenogenesis appears in some ways to be a superior form of reproduction. Contemporary evolutionary thought proposes some explanations. It may be due to selection pressure on the clade itself—the ability for a population to radiate more rapidly in response to a changing environment through sexual recombination than parthenogenesis allows. Alternatively, sexual reproduction may allow for the "ratcheting" of evolutionary speed as one clade competes with another for a limited resource.

In the first stage of sexual reproduction, "meiosis," the number of chromosomes is reduced from a diploid number (2n) to a haploid number (n). During "fertilization," haploid gametes come together to form a diploid zygote and the original number of chromosomes (2n) is restored.

Plants

Animals typically produce male gametes called sperm, and female gametes called eggs and ova, following immediately after meiosis. With the gametes producced directly by meiosis. Plants on the other hand have mitosis occurring in spores, which are produced by meiosis. The spores germinate into the gametophyte phase. The gametophytes of different groups of plants vary in size; angiosperms have as few as three cells in pollen, and mosses and other so called primitive plants may have several million cells. Plants have an alternation of generations where the sporophyte phase is succeeded by the gametophyte phase. The sporophyte phase produces spores within the sporangium by meiosis.

Flowering Plants

Flowering plants are the dominant plant form on land and they reproduce by sexual and asexual means. Often their most distinguishing feature is their reproductive organs, commonly called flowers.

The anther produces male gam-etophytes, the sperm is produced in pollen grains, which attach to the stigma on top of a carpel, in which the female gametophytes (inside ovules) are located. After the pollen tube grows through the carpel's style, the sex cell nuclei from the pollen grain migrate into the ovule to fertilize the egg cell and endosperm nuclei within the female gametophyte in a process termed double fertilization.

The resulting zygote develops into an embryo, while the triploid endosperm (one sperm cell plus two female cells) and female tissues of the ovule give rise to the surrounding tissues in the developing seed. The ovary, which produced the female gametophyte(s), then grows into a fruit, which surrounds the seed(s). Plants may either self-pollinate or cross-pollinate. Nonflowering plants like ferns, moss and liverworts use other means of sexual reproduction.

Ferns

Ferns typically produce large diploid sporophytes with rhizomes, roots and leaves; and on fertile leaves called sporangium, spores are produced. The spores are released and germinate to produce short, thin gametophytes that are typically heart shaped, small and green in colour.

The gametophytes or thallus, produce both motile sperm in the antheridia and egg cells in separate archegonia. After rains or when dew deposits a film of water, the motile sperm are splashed away from the antheridia, which are normally produced on the top side of the thallus, and swim in the film of water to the antheridia where they fertilize the egg. To promote out crossing or cross fertilization the sperm are released before the eggs are receptive of the sperm, making it more likely that the sperm will fertilize the eggs of different thallus. A zygote is formed after fertilization, which grows into a new sporophytic plant. The condition of having separate sporephyte and gametophyte plants is called alternation of generations. Other plants with similar reproductive means include the *Psilotum*, *Lycopodium*, *Selaginella* and *Equisetum*.

Bryophytes

The bryophytes, which include liverworts, hornworts and mosses, reproduce both sexually and vegetatively. They are small plants found growing in moist locations and like ferns, have motile sperm with flagella and need water to facilitate sexual reproduction. These plants start as a haploid spore that grows into the dominate form, which is a multicellular haploid body with leaf-like structures that photosynthesize. Haploid gametes are produced in antherida and archegonia by mitosis. The sperm released from the antherida respond to chemicals released by ripe archegonia and swim to them in a film of water and fertilize the egg cells thus producing a zygote. The zygote divides by mitotic division and grows into a sporophyte that is diploid.

The multicellular diploid sporophyte produces structures called spore capsules, which are connected by seta to the archegonia. The spore capsules produce spores by meiosis, when ripe the capsules burst open and the spores are released. Bryophytes show considerable variation in their breeding structures and the above is a basic outline. Also in some species each plant is one sex while other species produce both sexes on the same plant.

Fungi

Fungi are classified by the methods of sexual reproduction they employ. The outcome of sexual reproduction most often is the production of resting spores that are used to survive inclement times and to spread. There are typically three phases in the sexual reproduction of fungi: plasmogamy, karyogamy and meiosis.

Insects

Insect species make up more than two-thirds of all extant animal species, and most insect species use sex for reproduction, though some species are facultatively parthenogenetic. Many species have sexual dimorphism, while in others the sexes look nearly identical. Typically they have two sexes with males producing spermatozoa and females ova. The ova develop into eggs that have a covering called the chorion, which forms before internal fertilization. Insects have very diverse mating and reproductive strategies most often resulting in the male depositing spermatophore within the female, which stores the sperm until she is ready for egg fertilization. After fertilization, and the formation of a zygote, and varying degrees of development; the eggs are deposited outside the female in many species, or in some, they develop further within the female and live born offspring are produced.

Mammals

There are three extant kinds of mammals: Monotremes, Placentals and Marsupials, all with internal fertilisation. In placental mammals, offspring are born as juveniles: complete animals with the sex organs present although not reproductively functional. After several months or years, the sex organs develop further to maturity and the animal becomes sexually mature.

Most female mammals are only fertile during certain periods during their estrous cycle, at which point they are ready to mate. Individual male and female mammals meet and carry out copulation. For most mammals, males and females exchange sexual partners throughout their adult lives.

Male

The male reproductive system contains two main divisions: the penis, and the testicles, the latter of which is where sperm are produced. In humans, both of these organs are outside the abdominal cavity, but they can be primarily housed within the abdomen in other animals (for instance, in dogs, the penis is internal except when mating).

Having the testicles outside the abdomen best facilitates temperature regulation of the sperm, which require specific temperatures to survive. Sperm are the smaller of the two gametes and are generally very short-lived, requiring males to produce them continuously from the time of sexual maturity until death. Prior to ejaculation the produced sperm are stored in the seminal vesicle, a small gland that is located just behind the bladder.

Female

The female reproductive system likewise contains two main divisions: the vagina and uterus, which act as the receptacle for the sperm, and the ovaries, which produce the female's ova. All of these parts are always internal. The vagina is attached to the uterus through the cervix, while the uterus is attached to the ovaries via the Fallopian tubes. At certain intervals, the ovaries release an ovum, which passes through the fallopian tube into the uterus.

If, in this transit, it meets with sperm, the sperm penetrate and merge with the egg, fertilizing it. The fertilization usually occurs in the oviducts, but can happen in the uterus itself. The zygote then implants itself in the wall of the uterus, where it begins the processes of embryogenesis and morphogenesis. When developed enough to survive outside the womb, the cervix dilates and contractions of the uterus propel the fetus through the birth canal, which is the vagina.

The ova, which are the female sex cells, are much larger than the sperm and are normally formed within the ovaries of the fetus before its birth. They are mostly fixed in location within the ovary until their transit to the uterus, and contain nutrients for the later zygote and embryo. Over a regular interval, in response to hormonal signals, a process of oogenesis matures one ovum which is released and sent down the Fallopian tube. If not fertilized, this egg is flushed out of the system through menstruation in humans and other great apes and reabsorbed in other mammals in the estrus cycle.

Gestation

Gestation, called *pregnancy* in humans, is the period of time during which the fetus develops, dividing via mitosis inside the female. During this time, the fetus receives all of its nutrition and oxygenated blood from the female, filtered through the placenta, which is attached to the fetus' abdomen via an umbilical cord. This drain of nutrients can be quite taxing on the female, who is required to ingest slightly higher levels of calories. In addition, certain vitamins and other nutrients are required in greater quantities than normal, often creating abnormal eating habits. The length of gestation, called the gestation period, varies greatly from species to species; it is 40 weeks in humans, 56–60 in giraffes and 16 days in hamsters.

Birth

Once the fetus is sufficiently developed, chemical signals start the process of birth, which begins with contractions of the uterus and the dilation of the cervix. The fetus then descends to the cervix, where it is pushed out into the vagina, and eventually out of the female. The newborn, which is called an infant in humans, should typically begin respiration on its own shortly after birth. Not long after, the placenta is passed as well. Most mammals eat this, as it is a good source of protein and other vital nutrients needed for caring for the young. The end of the umbilical cord attached to the young's abdomen eventually falls off on its own.

Monotremes

Monotremes, only five species of which exist, all from Australia and New Guinea, lay eggs. They have one opening for excretion and reproduction called the cloaca. They hold the eggs internally for several weeks, providing nutrients, and then lay them and cover them like birds. After less than two weeks the young hatches and crawls into its mother's pouch, much like marsupials, where it nurses for several weeks as it grows.

Marsupials

Marsupials reproductive systems differ markedly from those of placental mammals. Females have two vaginas, both of which open externally through one orifice but lead to different compartments within the uterus. Males generally have a two-pronged penis, which corresponds to the females' two vaginae. The penis is used only for discharging semen into females, and is separate from the urinary tract. Both sexes possess a cloaca, which is connected to a urogenital sac used to store waste before expulsion. The female develops a kind of yolk sack in her womb which delivers nutrients to the embryo. Embryos of bandicoots, koalas and wombats additionally form placenta-like organs that connect them to the uterine wall, although the placenta-like organs are smaller than in placental mammals and it is not certain that they transfer nutrients from the mother to the embryo.

Pregnancy is very short, typically 4 to 5 weeks. The embryo is born at a very young stage of development, and is usually less than 5 cm (2.0 in) long at birth. It has been suggested that the short pregnancy is necessary to reduce the risk that the mother's immune system will attack the embryo.

The newborn marsupial uses its forelimbs (with relatively strong hands) to climb to a nipple, which is usually in a pouch on the mother's belly. The mother feeds the baby by contracting muscles over her mammary glands, as the baby is too weak to suckle.

The newborn marsupial's need to use its forelimbs for climbing to the nipple has prevented the forelimbs from evolving into paddles or wings and has therefore prevented the appearance of aquatic or truly flying marsupials (although there are several marsupial gliders).

Fish

The vast majority of fish species lay eggs that are then fertilized by the male, some species lay their eggs on a substrate like a rock or on plants, while others scatter their eggs and the eggs are fertilized as they drift or sink in the water column. Some fish species use internal fertilization and then disperse the developing eggs or give birth to live offspring. Fish that have live-bearing offspring include the Guppy and Mollies or *Poecilia*.

Fishes that give birth to live young can be ovoviviparous, where the eggs are fertilized within the female and the eggs simply hatch within the female body, or in seahorses, the male carries the developing young within a pouch, and gives birth to live young. Fishes can also be viviparous, where the female supplies nourishment to the internally growing offspring.

Some fish are hermaphrodites, where a single fish is both male and female and can produce eggs and sperm. In hermaphroditic fish, some are male and female at the same time while in other fish they are serially hermaphroditic; starting as one sex and changing to the other.

In at least one hermaphroditic species, self-fertilization occurs when the eggs and sperm are released together. Internal self-fertilization may occur in some other species. One fish species does not reproduce by sexual reproduction but uses sex to produce offspring; *Poecilia formosa* is a unisex species that uses a form of parthenogenesis called gynogenesis, where unfertilized eggs develop into embryos that produce female offspring. *Poecilia formosa* mate with males of other fish species that use internal fertilization, the sperm does not fertilize the eggs but stimulates the growth of the eggs which develops into embryos.

Plant Reproduction

The plants that sexually reproduce have the reproductive structures called the flowers. The flower is a condensed shoot with the nodes present very close to each other. The different parts of the plant are attached to the nodes. All the structures present at one node are collectively called the whorl. The first or the outermost two whorls are called the non-reproductive whorls. They are the calyx and corolla. The inner two whorls are androecium and gynoecium, the reproductive whorls.

Parts of a Flower

Calyx: is the outermost and most often green in colour. The individual units of calyx are called the sepals. It protects the inner whorls at bud stage.

Corolla: is the next inner whorl and is often coloured brightly. The individual units of corolla are called petals. They serve to attract bees, birds, etc which are the agents of pollination. Androecium is the male reproductive part of the flower. The individual units of androecium are called the stamens. Each stamen has a thread-like filament at the free end of which is attached the four-lobed anther.

Structure of Anther

The anther has four pollen-sacs, one in each lobe. The pollen-sacs contain cells called the microspore mother cells (MMC) or the pollen mother cells. MMCs undergo mieosis to produce four haploid microspores, also called the pollen grains. Each microspore divides once mitotically to produce two male gametes or sperm cells. Thus, each microspore mother cell produces 8 sperm cells. Gynoecium is the female reproductive part of the flower. The individual units are called the carpels or pistils. A flower may have one to many carpels, either fused or free. Each carpel is made up of the basal ovary, middle style and the upper stigma. The ovary is the chamber where there are many ovules that are attached to an axis. Each ovule consists of a haploid egg and other associated cells. The stigma is a sticky structure that receives the pollen grains. The style is hollow and provides a passage for the male gametes to reach the female gametes, the eggs.

Pollination

Transfer of pollen grains to the stigma is called pollination. If the pollen grains are transferred to the stigma of the same flower, the pollination is called self-pollination or autogamy. If the pollen grains are transferred to the stigma of another flower of the same species, the pollination is called cross-pollination or allogamy. The anthers on maturity burst open with force and this is called dehiscence. This releases the pollen grains with force which are then carried by wind and water to other plants.

In other plants, the flowers are brightly coloured and scented to attract the birds, bees, etc. The insect or the bird enters the flower to suck the nectar produced by glands at the base of the flower. The pollen grains present on the dehisced or open anther, stick to the legs or abdomen. When the same insect visits other flowers the pollen grains are transferred to the stigma of those flowers.

Fertilisation

The tip of the tube contains the male nuclei. The tube grows and enters the ovule where it bursts at the tip releasing the male gametes. One of the male gametes fuses with the egg, the female gamete. The fusion of the male gamete with the female gamete is called fertilisation. This results in the formation of zygote that is diploid. The zygote develops into the embryo. The other male gamete fuses with the polar nuclei. This results in the formation of a triploid nucleus called the endosperm nucleus. Since the process of fertilisation involves two fusions, it is called double fertilisation.

The divisions of the endosperm nucleus result in the formation of the endosperm that nourishes the growing embryo. The ovule then becomes the seed and the ovary changes into fruit.

Plant Reproduction

The way plants reproduce is a totally fascinating process. Out of the more than 300,000 different kinds of plants, more than half are seed plants. Seed plants make their own seeds from which new plants grow. Other ways plants can make new plants are from spores, rhizomes, bulbs, tubers, corms, cuttings, grafts, and buds. Seeds are made by flowers in some plants and by cones in other plants.

Flower parts contain specialized cells, called ovules, which have the job of making the seeds. Different kinds of flowers have many different parts. Basically, most plants—trees, vegetables, and even grass—sprout flowers. Some of these flowers are beautiful and colorful while others are so tiny one can hardly see them, but all plants produce new plants.

Botanists classify the reproduction of plants as ones with perfect and imperfect flowers. Perfect flowers are the ones that contain both the male and female parts. Imperfect flowers are the ones that contain only male or only female parts. So, of course, these different flowers have to find different ways to propagate themselves.

The stamen is the male part of a flower and has an anther on a stalk. The anther's job is to produce the pollen. Then there is the pistil, or female part. The pistil contains the flower's ovary, the style, and the stigma. Inside the ovary are the ovules. Each ovule contains an egg cell. So when an egg cell joins with a pollen cell, a seed may be produced.

The ovary is at the base of the plant. The stigma is at the top, and between them is the style. So, the pollen from the anther has to land on the stigma. When that happens, the pollen germinates and makes its way down into the ovary. Inside the ovary are the ovules, which contain egg cells. When an egg cell joins with a pollen cell, a seed may be produced. Then the ovary becomes a fruit. So a fruit is a mature ovary holding the seeds. In fruits, like the pears, the pear's core is the seedpod of the pear tree. Acorns are the seedpods of oak trees. And the tomato is the seedpod of the tomato plant.

A flower garden is alive with insects. They are all after the sweet liquid called nectar, which the flowers produce, near the bottom of the pistil. Flowers produce the nectar to draw the insects to them. Insects—like butterflies, wasps, and bees—come to get the nectar. Some, like butterflies and wasps, use the nectar for food; some, like bees, make it into honey and store it in their hives.

If you look closely at the body of a butterfly or bee, you will notice it is covered with tiny hairs. As they land on the flower to collect the nectar, some of the pollen is brushed on these hairs. Then the insect carries the pollen to the next flower, and pollination takes place.

Some plants, like corn, or flowers of an oak tree, rely on the wind to spread their pollen. So these kinds of plants don't produce pretty flowers since they don't need to attract insects to come and pollinate them.

Some plants need long tongues to pollinate them. Their flowers are usually tubular, like the trumpet creeper vine. These kinds of flowers are best pollinated by hummingbirds or by moths that have long tongues. And flowers that open at night do so because they need the specialized pollination provided by certain moths. Even the colour of plants makes a difference. Birds, like the hummingbird, usually pollinate plants that are red. Night insects pollinate white and pale yellow flowers, which are seen better at night.

Trees that have needle-like leaves are called conifers. Conifers produce cones instead of flowers. Under each scale of a cone is a part that produces either pollen or an ovule. But a cone cannot produce both. The very small cones produce the pollen in the spring.

The wind carries the pollen to the larger cone, which has the ovule producing part. Then, the pollen producing cones fall off while the fertilized ovules grow larger. Eventually, they open and release the seeds, and some of the seeds sprout. Some plants, like tulips, grow from bulbs. Bulbs produce plants very easily by reproducing underground. Plants can be grown without seeds by cuttings. A cutting can be from any part of the fully-grown plant, and when it is placed in soil, it may produce a new plant. You can increase your own houseplants by this method, and here is the way to do it. With a sharp knife, cut a piece of stem from the houseplant you want to use.

Pull off any leaves or flowers from the stem's lower part. Stick the stem in some clean, moist sand. Cover the flowerpot with a clear, plastic bag. Place in a well-lit place. After a few weeks, lift the stem out of the sand. Are the roots beginning to grow? Now you can plant it in a pot with potting soil. Keep the soil moist, and soon, the new plant will begin to grow. When cuttings from trees and other woody stems are attached to another plant, this is known as grafting. Grafts are another way of growing plants without using seeds.

There are a variety of ways plants produce new plants. To find out more about this fascinating process, you may want to read books on the subject or do some further research on the Internet.

Genetic Variations in Plants

Variation

If you have grown Fast Plants® you will have noticed variation among the individual plants within the group. Variation can range from a little to a lot. This article is designed to help teachers and students understand the basis of the variation they observe in their plantings.

Variation is one of the fundamental characteristics of life. All organisms exhibit some variation among individuals. Understanding the ways that variation is manifested in organisms, how it comes to be expressed through the development of the individual, and how it is transmitted from one individual to the next generation of individuals are central themes in biology.

Working with Fast Plants® will enrich a student's understanding of variation. By observing the growth and development of a Fast Plants® through the various stages in the life cycle, students will become aware of many visible features, or phenotypes, that make up the organism.

Only, however, upon close observation of a population of plants, will they become aware that the characteristics observed on one plant vary more or less on other plants. Such is the nature of variation.

Phenotypic Variation

Phenotypic variation, e.g. plant height at a particular stage of development, is considered to be the expression of the genetic makeup (genotype) of the individual as it interacts with the environment. Variation in plant height among individuals in a population is therefore due to variation in the interaction between the genotype and the environment, as expressed through the development of each individual plant.

Describing and Observing Phenotypic Variation

In order to be useful in an experiment the phenotype must be described using terms that are widely understood and easily communicated. For these reasons scientists have agreed upon various standards or descriptors to describe characteristics in the natural world. Descriptors take many forms, see WFPID Observing and Describing. The choice of how to describe what you observe is important, because it will determine the kinds of descriptors used and establish the basis for recording, analyzing and communicating results.

Environmental Variation

Much can be learned about the role of light, temperature, and nutrients on plant development from experiments in which one or more environmental parameters are changed. Because Fast Plants® are highly responsive to changes in the environment, they are ideal for examining the role of the environment on the expression of phenotypic variation.

Although Fast Plants® are able to grow within a wide range of environmental conditions, for most investigations it is recommended that they be grown under uniform and ideal conditions. In this way variation arising from sub-optimal conditions of environment will be minimized. The Wisconsin Fast Plants® Information Document (WFPID) Understanding the Environment describes how to provide and maintain the various physical, chemical and biotic components of the environment that are most suitable for Fast Plants®.

Genotypic Variation

How can students use Fast Plants® to investigate the contribution of the genotype to the phenotype? Fast Plants®, rapid cycling Brassica rapa, are genotypically variable in that they have a genetically controlled mating system that prevents self-fertilization and favours out-crossing among individuals. As a consequence, even seed stocks selected for uniformity of specific phenotypes and genotypes exhibit considerable variation for other traits.

The Wisconsin Fast Plants® Program has developed a number of genetic stocks of rapid cycling Brassica rapa for genetic investigations on the nature and inheritance of variation. Some WFP stocks contain distinctive mutant phenotypes, e.g., anthocyaninless plant, anl, yellow green plant, ygr1, rosette, ros, and male sterile, mst2, that are suitable for Mendelian genetics. Other stocks exhibit phenotypes whose expression may vary continually and which may be quantified as discrete or countable units, e.g., number of hairs, or as estimates of size, e.g. petite dwarf, dwf1, or of intensity of colour saturation, e.g., purple anthocyanin. These quantitative phenotypes may be conditioned by a few or many genes and normally require numerical descriptions in which the statistical notations of population size, (n), range (r), arithmetic mean (x), and standard deviation (s) are applied.

Yet other stocks have been developed to combine both simply inherited mutant genotypes and quantitatively expressed phenotypes. An important part of WFP is the continuing development and improvement of seed stocks for uses in genetics.

Genetic Diversity

Genetic diversity is a level of biodiversity that refers to the total number of genetic characteristics in the genetic makeup of a species. It is distinguished from genetic variability, which describes the tendency of genetic characteristics to vary.

The academic field of population genetics includes several hypotheses and theories regarding genetic diversity. The neutral theory of evolution proposes that diversity is the result of the accumulation of neutral substitutions. Diversifying selection is the hypothesis that two subpopulations of a species live in different environments that select for different alleles at a particular locus. This may occur, for instance, if a species has a large range relative to the mobility of individuals within it. Frequency-dependent selection is the hypothesis that as alleles become more common, they become less fit. This is often invoked in host-pathogen interactions, where a high frequency of a defensive allele among the host means that it is more likely that a pathogen will spread if it is able to overcome that allele.

Importance of Genetic Diversity

There are many different ways to measure genetic diversity. The modern causes for the loss of animal genetic diversity have also been studied and identified. A 2007 study conducted by the National Science Foundation found that genetic diversity and biodiversity are dependent upon each other — that diversity within a species is necessary to maintain diversity among species, and vice versa. According to the lead researcher in the study, Dr. Richard Lankau, "If any one type is removed from the system, the cycle can break down, and the community becomes dominated by a single species."

Survival and Adaptation

Genetic diversity plays a very important role in survival and adaptability of a species because when a species's environment changes, slight gene variations are necessary to produce changes in the organisms' anatomy that enables it to adapt and survive. A species that has a large degree of genetic diversity among its population will have more variations from which to choose the most fit alleles. Increase in genetic diversity is also essential for an organism to evolve. Species that have very little genetic variation are at a great risk. With very little gene variation within the species, healthy reproduction becomes increasingly difficult, and offspring often deal with similar problems to those of inbreeding. The vulnerability of a population to certain types of diseases can also increase with reduction in genetic diversity.

Agricultural Relevance

When humans initially started farming, they used selective breeding to pass on desirable traits of the crops while omitting the undesirable ones. Selective breeding leads to monocultures: entire farms of nearly genetically identical plants. Little to no genetic diversity makes crops extremely susceptible to widespread disease. Bacteria morph and change constantly. When a disease causing bacterium changes to attack a specific genetic variation, it can easily wipe out vast quantities of the species. If the genetic variation that the bacterium is best at attacking happens to be that which humans have selectively bred to use for harvest, the entire crop will be wiped out.

A very similar occurrence is the cause of the infamous Potato Famine in Ireland. Since new potato plants do not come as a result of reproduction but rather from pieces of the parent plant, no genetic diversity is developed, and the entire crop is essentially a clone of one potato, it is especially susceptible to an epidemic. In the 1840s, much of Ireland's population depended on potatoes for food. They planted namely the "lumper" variety of potato, which was susceptible to a rot-causing plasmodiophorid called *Phytophthora infestans*. This plasmodiophorid destroyed the vast majority of the potato crop, and left tens of thousands of people to starve to death.

Coping with Poor Genetic Diversity

The natural world has several ways of preserving or increasing genetic diversity. Among oceanic plankton, viruses aid in the genetic shifting process. Ocean viruses, which infect the plankton, carry genes of other organisms in addition their own. When a virus containing the genes of one cell infects another, the genetic makeup of the latter changes. This constant shift of genetic makeup helps to maintain a healthy population of plankton despite complex and unpredictable environmental changes.

Cheetahs are a threatened species. Extremely low genetic diversity and resulting poor sperm quality has made breeding and survivorship difficult for cheetahs –- only about 5% of cheetahs survive to adulthood. About 10,000 years ago, all but the jubatus species of cheetahs died out. The species encountered a population bottleneck and close family relatives were forced to mate with each other, or inbreed. However, it has been recently discovered that female cheetahs can mate with more than one male per litter of cubs. They undergo induced ovulation, which means that a new egg is produced every time a female mates. By mating with multiple males, the mother increases the genetic diversity within a single litter of cubs.

Measures of Genetic Diversity

Genetic Diversity of a population can be assessed by some simple measures.

- Gene Diversity is the proportion of polymorphic loci across the genome.
- Heterozygosity is the mean number of individuals with polymorphic loci.
- Alleles per locus is also used to demonstrate variability.

The Quest for the Hardy Cedar-of-Lebanon

Genetic variation within plant species has not only been richly documented in science but also widely exploited for horticultural use. Individual plants have often been selected from wild populations for their deviations in growth habit, flower size, and leaf colour, but another primary driver of plant exploration has been the promise of winter har-diness due to provenance. In many cases, this, too, is under the auspices of science, for it allows botanical gardens and arboreta to cultivate species that may normally be out of reach due to lack of hardiness. We present here a prime example of such work: the quest by the Arnold Arboretum to introduce into cultivation hardy stock of the cedar-of-Lebanon, Cedrus libani.

The genus Cedrus has a natural range that extends from North Africa around the Medi-terranean Sea into Lebanon, Syria, Cyprus, and Turkey and west to the Himalayas (Far-jon 1990). Depending on the treatment, there are a variable number of species of true cedar, and Farjon (2001) recognizes four. The beauti-ful long-needled deodar cedar (Cedrus deodora) occurs in a wide range of habitats in the Hima-layas of Afghanistan, Pakistan, Kashmir, and Nepal. The atlas cedar (Cedrus atlantica) grows in the Atlas Mountains of Algeria and Morocco while the short-needled Cyprus cedar (Cedrus brevifolia) is restricted to that island; both of these taxa have by some botanists been consid-ered separate subspecies of Cedrus libani, the cedar-of-Lebanon, which occurs naturally in Lebanon, Syria, and the Cilician Taurus moun-tains of southern Anatolia, or modern-day Turkey. It is this group of cedars from Turkey that most interests us and that is the focus of this article.

Authors have variously recognized the Turk-ish provenance of cedar with subspecific status (Farjon 1990). Cedrus libani subspecies steno-coma was first described by Schwarz (1944) and then Davis (1949), who both recognized that it was intermediate between the typical cedar-of-Lebanon and the atlas cedar. Volume One of The Flora of Turkey (Davis 1965) did not rec-ognize subspecies stenocoma, although it is recognized as a variety in the eleventh volume of the same work (Guner 2000). More recently, the Turkish provenance was classified as Cedrus libani ssp. stenocoma (Farjon 2001) and these trees usually are called the hardy cedar-of-Lebanon. The Turkish trees are generally considered to be more upright and conical

Cedrus Libani in Asia Minor

During the 1800s Cedrus libani was grown throughout Philadelphia and New York but was not hardy in Boston and New England (Wilson 1926). Josiah Hoopes, a nurseryman from West Chester, in southeastern Pennsylvania, wrote that "the cedar-of-Lebanon is found to be pretty hardy... [and] with us it has succeeded to our entire satisfaction, and we can therefor recommend it without reserve, if proper cultivation and a moderate amount of care be given to it." In his 1868 book, Hoopes also mentions Pierre Belon, a French botanist and physician who travelled throughout the Levant in the 1540s. In 1553 Belon published De arboribus coniferis, probably the first text devoted entirely to conifers, and in it he included the first description of cedar-of-Lebanon growing in the Amanus (Nur) and Taurus Mountains of southern Anatolia. In 1597, John Gerard cited Belon in his Herbal, saying:

Gerard's statement reveals that as early as the sixteenth century authors took notice of these unique trees from Anatolia and recognized that this more northern provenance possessed greater potential for cold hardiness.

Nineteenth-century botanical and horticultural literature was replete with references to the northern populations of cedar-of-Lebanon. Several European botanists were exploring and describing the flora of Asia Minor and their works describe a growing understanding of the natural range of cedar in this region. Asie Mineure by P. A. Chikhachev is an extremely thorough account of the physical geography, climate, fauna, and flora of Asia Minor, based on his travels throughout the region. Among his extensive botanical listings is Cedrus libani, which he described as growing in numerous locations. Chikhachev describes their locations using the ancient names for the regions of Anatolia, moving from west to east: in Pisidia between Lakes Beysehir and Egridir; in Isauria at Mount Topyedik growing around 2,000 meters; in Cilicia growing on the northeast and southeast exposures of all the mountainous regions of the Bulgar Daglari (Bolkar Daglari mountains of the Taurus range), where it descends to 4,600 feet (1,400 m); growing in groves in the Antitaurus Mountains between the villages Sarkanty-oglu and Tchedeme [sic] at 5,600 feet (1,700 m) (Chikhachev 1860).

The nearest point to the Lebanon [Mountains] at which Cedars have been found, is the Bulgar-dagh chain of the Taurus in Asia Minor, and from that point forests extend eastward to Pisidia, in long. E. 32°, westward to long. E. 36°, and north-ward to the Anti-Taurus, in lat. 40° N.; growing at elevations of 4000 to 6400 feet above the sea. The Lebanon may be regarded as a branch of the Taurus, and is 250 miles distant from the Cedar forests upon that chain... Northern Syria and Asia Minor form one botanical province; so that the Lebanon grove, though so widely disconnected from the Taurus forests, can be regarded in no other light than as an outlying member of the latter.

Ravenscroft gave a colossal summary of all known accounts of cedar-of-Lebanon, with beautiful colour plates, in his 1884 Pinetum Bri-tannicum. He describes the species in amazing detail, including comparisons made to other Cedrus and descriptions of the Syrian and Leba-nese trees. Particularly interesting is a table that accounts for all visits made by individuals from 1487 to 1864 to the sacred grove that lies on Mt. Lebanon. He also writes of the Anatolian population, providing a description practically identical to that of Hooker.

How the Hardy Cedar-of-Lebanon Found Its Way to North America

A contemporary of the botanists writing and exploring Asia Minor in the late 1800s was Walter Siehe, an interesting and somewhat mysterious character who played an integral role in introducing the hardy cedar-of-Lebanon into the United States. Siehe was a German botanical explorer living in Smyrna (Izmar), Turkey. In his Die Nadelholzer des cilicischen Taurus (Conifers of the Cilician Taurus), he described the natural habitat of conifers growing in the mountains of southern Anatolia and the conifers themselves, including Abies cilicica, Taxus baccata, several species of Juniperus, Cupressus sempervirens, and of course Cedrus libani. Of the cedar he wrote, "the proud tree is a child of the high altitude," growing in a severe cli-mate where the snows lie a few meters deep for five months of the year (Siehe 1897b).

In these mountains where the cedars grow on the steep walls and high saddles between peaks, Siehe romantically describes the roaring wild rivers, the whiteness of the snow, and the long silences broken only by the screech of birds or tumbling stones kicked loose by an escaping mountain goat. He notes that despite the usefulness of cedar's wood, the large populations of cedar persisted because of the inaccessibility of the mountains. He describes trees up to 130 feet (40 m) tall and describes their column-like trunks supporting branches as regular as floors of a building (Siehe 1897a). It is clear from his travel accounts that Siehe knew the mountains of southern Turkey very well and was the right person, in the right place, and at the right time to send seed to North America.

It is in the context of nineteenth-century botanical exploration and description through-out Asia Minor that one understands Charles S. Sargent's interest in the more northern population of cedar. Cedar-of-Lebanon—with its handsome and stately form, its association with grand estates throughout Europe and the mid-Atlantic United States, and its historic asso-ciations—

is a highly desirable landscape tree. With this in mind, it's easy to understand how Sargent, director of the Arnold Arboretum from 1873 until 1927, would have surely longed to grow cedars in New England. However, hardiness in New England was indeed an issue. He certainly would have read with great interest the accounts of the cedars and understood that trees from the Taurus mountains held the key to increased hardiness.

With this in mind, Sargent hired Walter Siehe to collect seed from trees in the Taurus Mountains and have these sent to the Arnold Arboretum. In a letter from Siehe to Sargent, dated 18 November 1900, from Mersina, Turkey, Siehe wrote:

Dr. Bolle... has repeatedly informed me of your desire [to acquire] cedar cones from cold resistant trees of high altitude (1900 m [6,250 feet]). Only a few days ago did I manage to obtain, after several futile attempts, 50 kilos [110 pounds] of cones with good seeds. Since it was necessary to make a special trip, use many pack animals, and spend eight days of time for this, I am certain that you will not find the fee of 60 Mark German currency too high.

Apparently Sargent did not find the fee too high because the Arboretum's plant records show that they received cones with ripe seeds from Siehe on February 4, 1902. The Arnold Arboretum was not the only recipient of Siehe's seed: in 1908 H. J. Elwes and Augustine Henry wrote another excellent description of the cedars from the Taurus mountains and noted, "Siehe has sent seed from the Cilician Taurus to various places, and I have two vigorous young trees raised from them."

Early reports from the Arnold Arboretum noted great success with this seedlot. The seeds had a high rate of germination and by 1915, Sargent reported in the Arboretum's Bulletin of Popular of Information that the cedars-of-Lebanon had "all proved perfectly hardy, not one having suffered from drought or cold." A plantation of trees was established on Bussey Hill, and other individual specimens were planted throughout the collection.

The average height of these trees was about 13 feet (4 m), with the tallest having reached 21 feet (6.5 m), prompting Sargent to reflect, "It is doubtful if any other conifer can be grown in New England from seed to the height of twenty-one feet in thirteen years." Wilson also seemed pleased with the rapid growth and hardiness of these trees, writing in 1919 that although the dreadful winter of 1917–1918 scorched the needles of the cedars, they recovered fully and "had grown more rapidly in the Arnold Arboretum than any other conifer has ever done." And just five years later, the Bulletin reported that the trees had already reached 30 feet (9 m) in height. In 1926, twenty individuals of this accession (AA 4697*A-T) appear in the plant records, although over time there has been some attrition and currently only eight trees from Siehe's original 1902 collection remain extant in the Arnold's collection (AA

#4697*A,C,G,I,K,M,O,P). Cold winters were not the cause for their decline, however; Donald Wyman wrote in 1946 that the cedars were thriving, growing for over forty years and withstanding temperatures of minus-20 degrees F. Strong winds were responsible for the loss of at least eight of the twelve trees, including five from the infamous 1938 hurricane alone.

The Current State of Trees in the Wild

In Lebanon and Syria, the species is rare due to millennia of human impact (logging, burning, grazing). However, in Turkey, where the topography has prevented easy access, there remain extensive forests of Cedrus libani ssp. stenocoma. During a collecting expedition to the Taurus Mountains in 1990, Mark Flanagan encountered the hardy cedar-of-Lebanon running for nearly 620 miles (1,000 km) along the 5,900-foot (1800-m) contour line. Due to the high, open canopies in the overstory, the ground layer of these forests is very rich, including a diversity of taxa such as Acer hyrcanum, Sorbus umbellata, and Kitaibelia balansae.

Other Notably Hardy Specimens

Among the Morris Arboretum's extensive conifer collection are five mature specimens of hardy cedar-of-Lebanon. They are growing throughout the Arboretum, some tucked away, others in full view, and all of them handsome. The oldest tree was planted before the Arboretum's founding in 1932 when the property was known as Compton, John and Lydia Morris's estate. A cedar appears in this location in the 1909 Atlas of Compton—a survey of the Morris' gardens and plants—and according to notes written by John Tonkin, the Morris' gardener and Morris Arboretum's superintendent from 1913 to 1961, this cedar came from the Arnold Arboretum. It is likely that this tree is a seed-ling from the original 1902 collection of Turk-ish seed, sent to the Morris's from the Arnold Arboretum. Over the years, staff at the Morris Arboretum have marvelled at the hardiness of this tree, recording that it showed no visible injuries during the devastatingly cold winters in the early 1930s (Lambert 1936). Although it has suffered storm damage during the past 25 years, today it shows a remarkable amount of young, vigorous re-growth for a tree of its age and size.

As students of Harrison Flint at Purdue University we often admired the selection 'Purdue Hardy' (Cedrus libani ssp. stenocoma 'Purdue Hardy'), which grows in West Lafayette, Indiana, in a hardiness zone that routinely reaches minus-20 degrees F (USDA zone 5a). This 40-year-old specimen is remarkable for its graceful form and nearly pendant branches; like many others of the subspecies, it has not become flat-topped. It has certainly lived up to its name, withstanding winter temperatures of minus-25 degrees F with only minimal browning of needles (Flint 1997).

The selection's provenance is uncertain. It was one of several seedlings germinated from seed collected by the late Purdue professor Ted Shaw in the 1950s. Shaw had been in Lebanon working on reforestation projects supported by the United States when he obtained it. Oral history at Purdue has it that Shaw found the seeds "up in the Hills," which could mean Lebanon or it could have been Turkey, where he vacationed. Since no Lebanese cedar has been successfully grown out-of-doors north of zone 6, it is far more likely to have originated in the mountains of Turkey.

Another noteworthy specimen, a mammoth cedar-of-Lebanon that is a Pennsylvania state champion, is at the Tyler Arboretum, in Media, Pennsylvania. Jacob and Minshall Painter, horticulturists and owners of the property that became the Tyler Arboretum, recorded purchases of cedars of Lebanon from the Phila-delphia nurseries of John Evans, Josiah Hoopes, and Morris in the 1850s (Appleby 1992). It is one of the most remarkable conifers in the Del-aware Valley: it stands 87 feet high (26.5 m) with a spread of 93 feet (28.4 m) and a diameter at breast height of 69 inches (175 cm).

The story of the majestic hardy cedar-of-Lebanon mixes history, geography, plant ecology, horticulture, and a love of conifers. When you next visit the Arnold, Morris, or Tyler Arboreta, take time to enjoy their magnificence and muse on their long journey from the mountains of Turkey to the eastern United States.

Breeding Methods in Crop Plants

Self Pollinated Crops

Mass selection: In mass selection, seeds are collected from (usually a few dozen to a few hundred) desirable appearing individuals in a population, and the next generation is sown from the stock of mixed seed. This procedure, sometimes referred to as phenotypic selection, is based on how each individual looks. Mass selection has been used widely to improve old "land" varieties, varieties that have been passed down from one generation of farmers to the next over long periods.

An alternative approach that has no doubt been practiced for thousands of years is simply to eliminate undesirable types by destroying them in the field. The results are similar whether superior plants are saved or inferior plants are eliminated: seeds of the better plants become the planting stock for the next season.

A modern refinement of mass selection is to harvest the best plants separately and to grow and compare their progenies. The poorer progenies are destroyed and the seeds of the remainder are harvested. It should be noted that selection is now based not solely on the appearance of the parent plants but also on the appearance and performance of their progeny. Progeny selection is usually more effective than phenotypic selection when dealing with quantitative characters of low heritability. It should be noted, however, that progeny testing requires an extra generation; hence gain per cycle of selection must be double that of simple phenotypic selection to achieve the same rate of gain per unit time.

Mass selection, with or without progeny test, is perhaps the simplest and least expensive of plant-breeding procedures. It finds wide use in the breeding of certain forage species, which are not important enough economically to justify more detailed attention.

Pure-line selection: Pure-line selection generally involves three more or less distinct steps: (1) numerous superior appearing plants are selected from a genetically variable population; (2) progenies of the individual plant selections are grown and evaluated by simple observation, frequently over a period of several years; and (3) when selection can no longer be made on the basis of observation alone, extensive trials are undertaken, involving careful measurements to determine whether the remaining selections are superior in yielding ability and other aspects of performance.

Any progeny superior to an existing variety is then released as a new "pure-line" variety. Much of the success of this method during the early 1900s depended on the existence of genetically variable land varieties that were waiting to be exploited. They provided a rich source of superior pure-line varieties, some of which are still represented among commercial varieties. In recent years the pure-line method as outlined above has decreased in importance in the breeding of major cultivated species; however, the method is still widely used with the less important species that have not yet been heavily selected.

A variation of the pure-line selection method that dates back centuries is the selection of single-chance variants, mutations or "sports" in the original variety. A very large number of varieties that differ from the original strain in characteristics such as colour, lack of thorns or barbs, dwarfness, and disease resistance have originated in this fashion.

Hybridization

During the 20th century planned hybridization between carefully selected parents has become dominant in the breeding of self-pollinated species. The object of hybridization is to combine desirable genes found in two or more different varieties and to produce pure-breeding progeny superior in many respects to the parental types.

Genes, however, are always in the company of other genes in a collection called a genotype. The plant breeder's problem is largely one of efficiently managing the enormous numbers of genotypes that occur in the generations following hybridization. As an example of the power of hybridization in creating variability, a cross between hypothetical wheat varieties differing by only 21 genes is capable of producing more than 10,000,000,000 different genotypes in the second generation. At spacing normally used by farmers, more than 50,000,000 acres would be required to grow a population large enough to permit every genotype to occur in its expected frequency.

While the great majority of these second generation genotypes are hybrid (heterozygous) for one or more traits, it is statistically possible that 2,097,152 different pure-breeding (homozygous) genotypes can occur, each potentially a new pure-line variety. These numbers illustrate the importance of efficient techniques in managing hybrid populations, for which purpose the pedigree procedure is most widely used.

Pedigree breeding starts with the crossing of two genotypes, each of which have one or more desirable characters lacked by the other. If the two original parents do not provide all of the desired characters, a third parent can be included by crossing it to one of the hybrid progeny of the first generation. In the pedigree method superior types are selected in successive generations, and a record is maintained of parent–progeny relationships.

The F2 generation (progeny of the crossing of two F1 individuals) affords the first opportunity for selection in pedigree programs. In this generation the emphasis is on the elimination of individuals carrying undesirable major genes.

In the succeeding generations the hybrid condition gives way to pure breeding as a result of natural self-pollination, and families derived from different F2 plants begin to display their unique character.

Usually one or two superior plants are selected within each superior family in these generations. By the F5 generation the pure-breeding condition (homozygosity) is extensive, and emphasis shifts almost entirely to selection between families.

The pedigree record is useful in making these eliminations. At this stage each selected family is usually harvested in mass to obtain the larger amounts of seed needed to evaluate families for quantitative characters. This evaluation is usually carried out in plots grown under conditions that simulate commercial planting practice as closely as possible.

When the number of families has been reduced to manageable proportions by visual selection, usually by the F7 or F8 generation, precise evaluation for performance and quality begins. The final evaluation of promising strains involves (1) observation, usually in a number of years and locations, to detect weaknesses that may not have appeared previously; (2) precise yield testing; and (3) quality testing. Many plant breeders test for five years at five representative locations before releasing a new variety for commercial production.

The bulk-population method of breeding differs from the pedigree method primarily in the handling of generations following hybridization. The F2 generation is sown at normal commercial planting rates in a large plot. At maturity the crop is harvested in mass, and the seeds are used to establish the next generation in a similar plot. No record of ancestry is kept. During the period of bulk propagation natural selection tends to eliminate plants having poor survival value.

Two types of artificial selection also are often applied: (1) destruction of plants that carry undesirable major genes and (2) mass techniques such as harvesting when only part of the seeds are mature to select for early maturing plants or the use of screens to select for increased seed size. Single plant selections are then made and evaluated in the same way as in the pedigree method of breeding. The chief advantage of the bulk population method is that it allows the breeder to handle very large numbers of individuals inexpensively.

Often an outstanding variety can be improved by transferring to it some specific desirable character that it lacks. This can be accomplished by first crossing a plant of the superior variety to a plant of the donor variety, which carries the trait in question, and then mating the progeny back to a plant having the genotype of the superior parent. This process is called backcrossing. After five or six backcrosses the progeny will be hybrid for the character being transferred but like the superior parent for all other genes. Selfing the last backcross generation, coupled with selection, will give some progeny pure breeding for the genes being transferred.

The advantages of the backcross method are its rapidity, the small number of plants required, and the predictability of the outcome. A serious disadvantage is that the procedure diminishes the occurrence of chance combinations of genes, which sometimes leads to striking improvements in performance.

Hybrid Varieties

The development of hybrid varieties differs from hybridization. The F1 hybrid of crosses between different genotypes is often much more vigorous than its parents. This hybrid vigour, or heterosis, can be manifested in many ways, including increased rate of growth, greater uniformity, earlier flowering, and increased yield, the last being of greatest importance in agriculture.

Cross Pollinated Crops

The most important methods of breeding cross-pollinated species are (1) mass selection; (2) development of hybrid varieties; and (3) development of synthetic varieties. Since cross-pollinated species are naturally hybrid (heterozygous) for many traits and lose vigour as they become purebred (homozygous), a goal of each of these breeding methods is to preserve or restore heterozygosity.

Mass Selection

Mass selection in cross-pollinated species takes the same form as in self-pollinated species; *i.e.,* a large number of superior appearing plants are selected and harvested in bulk and the seed used to produce the next generation.

Mass selection has proved to be very effective in improving qualitative characters, and, applied over many generations, it is also capable of improving quantitative characters, including yield, despite the low heritability of such characters. Mass selection has long been a major method of breeding cross-pollinated species, especially in the economically less important species.

Hybrid Varieties

The outstanding example of the exploitation of hybrid vigour through the use of F1 hybrid varieties has been with corn (maize). The production of a hybrid corn variety involves three steps: (1) the selection of superior plants; (2) selfing for several generations to produce a series of inbred lines, which although different from each other are each pure-breeding and highly uniform; and (3) crossing selected inbred lines.

During the inbreeding process the vigour of the lines decreases drastically, usually to less than half that of field-pollinated varieties. Vigour is restored, however, when any two unrelated inbred lines are crossed, and in some cases the F1 hybrids between inbred lines are much superior to open-pollinated varieties. An important consequence of the homozygosity of the inbred lines is that the hybrid between any two inbreds will always be the same. Once the inbreds that give the best hybrids have been identified, any desired amount of hybrid seed can be produced.

Pollination in corn (maize) is by wind, which blows pollen from the tassels to the styles (silks) that protrude from the tops of the ears. Thus controlled cross-pollination on a field scale can be accomplished economically by interplanting two or three rows of the seed parent inbred with one row of the pollinator inbred and detasselling the former before it sheds pollen.

In practice most hybrid corn is produced from "double crosses," in which four inbred lines are first crossed in pairs (A × B and C × D) and then the two F1 hybrids are crossed again (A × B) × (C × D). The double-cross procedure has the advantage that the commercial F1 seed is produced on the highly productive single cross A × B rather than on a poor-yielding inbred, thus reducing seed costs. In recent years cytoplasmic male sterility, described earlier, has been used to eliminate detasselling of the seed parent, thus providing further economies in producing hybrid seed.

Much of the hybrid vigour exhibited by F1 hybrid varieties is lost in the next generation. Consequently, seed from hybrid varieties is not used for planting stock but the farmer purchases new seed each year from seed companies.

Perhaps no other development in the biological sciences has had greater impact on increasing the quantity of food supplies available to the world's population than has the development of hybrid corn (maize).

Hybrid varieties in other crops, made possible through the use of male sterility, have also been dramatically successful and it seems likely that use of hybrid varieties will continue to expand in the future.

Synthetic Varieties

A synthetic variety is developed by intercrossing a number of genotypes of known superior combining ability—*i.e.,* genotypes that are known to give superior hybrid performance when crossed in all combinations.

(By contrast, a variety developed by mass selection is made up of genotypes bulked together without having undergone preliminary testing to determine their performance in hybrid combination.)

Synthetic varieties are known for their hybrid vigour and for their ability to produce usable seed for succeeding seasons.

Because of these advantages, synthetic varieties have become increasingly favoured in the growing of many species, such as the forage crops, in which expense prohibits the development or use of hybrid varieties.

Mutation Breeding

Physical Mutagens

Physical mutagens include various types of radiation, viz X-rays, gamma rays, alpha particles, beta particles, fast and thermal (slow) neutrons and ultra violet rays. A brief description of these mutagens is presented below:

Commonly used physical mutagens (radiations), their properties and mode of action.

Type of Radiation	*Main properties*
X – rays	S.I., penetrating and non-particulate
Gamma rays	S.I., very penetrating and Non-particulate
Alpha Particles	D.I., particulate, less penetrating and positively charged.
Beta Rays Particles	S.I., particulate, more penetrating than alpha particles and negatively charged.
Fast and Thermal Neutrons	D.I., particulate, neutral particles, highly penetrating.
6. Ultra Violet Rays	Non-ionizing, low penetrating

Note: particulate refers to particle emitting property DI = Densely ionizing, SI = Sparsely ionizing.

X-rays

X-rays were first discovered by Roentgen in 1895. The wavelengths of X-rays vary from 10-11 to 10-7. They are sparsely ionizing and highly penetrating. They are generated in X-rays machines. X-rays can break chromosomes and produce all types of mutations in nucleotides, viz. addition, deletion, inversion, transposition, transitions and transversions. X-rays were first used by Muller in 1927 for induction of mutations in Drosophila. In plants, Stadler in 1928 first used X-rays for induction of mutations in barley.

Gamma rays

Gamma rays have shorter wave length than X-rays and are more penetrating than gamma rays. They are generated from radioactive decay of some elements like 14C, 60Co, radium etc. Of these, cobalt 60 is commonly used for the production of Gamma rays. Gamma rays cause chromosomal and gene mutations like X-rays.

Bulk Population Method

Advantages

Bulk method is simple, convenient and inexpensive. 'Natural' selection can be enhanced by growing bulks under stress conditions, e.g., disease, insect, drought. Can use mass selection to modify the bulk. Can replicate bulks and determine performance or test in multiple environments. Can drop bulks at any time.

Disadvantages

Lines selected from the bulk that appear to be segregating may need to be reselected to obtain uniformity. There is a risk that desirable genotypes may be lost from the population; e.g., short plants may be at a disadvantage in competition with tall plants. Competition is not the same as natural selection. Will know little about the genetic structure of the population. Natural selection is not always beneficial.

US Patent 5332408-Methods and Reagents for Backcross Breeding of Plants

Claims

What is claimed is:

1. A method for obtaining plants with a desired trait or combination of traits comprising:
   a. performing a genetic cross between a donor plant of a first species, subspecies or variety, having a desired trait or combination of traits and a recipient plant of a second species, subspecies or variety, respectively, to form a viable hybrid plant;

b. backcrossing the hybrid plant of subpart (a) with a plant of the recipient species, subspecies or variety, respectively, to form a viable backcross progeny plant;

c. quantitatively annealing DNA from individual backcross progeny plants produced in subpart (b) with a detectably-labeled probe or multiplicity of detectably-labeled probes that specifically binds to a plurality of probe binding sites in the DNA from the donor plant species, subspecies or variety and does not bind to the DNA of the recipient plant species, subspecies or variety, respectively;

d. determining the amount of DNA in each backcross progeny plant inherited from the donor species, subspecies or variety, respectively, that binds to the detectably-labeled probe or multiplicity of detectably-labeled probes, scored as a pool and independent of a physical or genetic map of probe binding sites; and

e. selecting for further propagation the backcross progeny plants having the desired trait or combination of traits that have inherited the least amount of DNA from the donor species, subspecies or variety.

2. The method of claim 1 wherein the backcross progeny plants inherit DNA encoding the desired trait or combination of traits whereby the trait or combination of traits are recessive and do not express a detectable phenotype.

3. The method of claim 1, optionally comprising the following steps:

   f. quantitatively annealing DNA from individual backcross progeny plants produced in subpart (b) with a detectably-labelled probe or multiplicity of detectably-labelled probes that specifically binds to a plurality of probe binding sites in the DNA from the recipient plant species, subspecies or variety and does not bind to the DNA of the donor plant species, subspecies or variety, respectively;

   g. determining the amount of DNA in each backcross progeny plant inherited from the recipient plant species, subspecies or variety, respectively, that binds to the detectably-labelled probe or multiplicity of detectably-labelled probes, scored as a pool and independent of a physical or genetic map of probe binding sites; and h. selecting for further propagation the backcross progeny plants having the desired trait or combination of traits that have inherited the greatest amount of DNA from the recipient plant species, subspecies or variety, respectively;

4. The method of claim 3 wherein the backcross progeny plants inherit DNA encoding the desired trait or combination of traits whereby the trait or combination of traits are recessive and do not express a detectable phenotype.

5. A method for obtaining plants with a desired trait or combination of traits comprising:
    a. crossing the backcross progeny plant of subpart (e) of claim 1 with a plant of the recipient species, subspecies or variety, respectively of claim 1 to produce a rebackcross progeny plant having the desired trait or combination of traits;
    b. quantitatively annealing DNA from individual rebackcross progeny plants produced in subpart (a) with a detectably-labelled probe or multiplicity of detectably-labelled probes that specifically binds to a plurality of probe binding sites in the DNA from the donor plant species, subspecies or variety of claim 1 and does not bind to the DNA of the recipient plant species, subspecies or variety, respectively;
    c. determining the amount of DNA in each rebackcross progeny plant of subpart (a) that is inherited from the donor plant species, subspecies or variety, respectively that binds to the detectably-labelled probe or multiplicity of detectably-labelled probes, scored as a pool and independent of a physical or genetic map of probe binding sites; and
    d. selecting for further propagation the rebackcross progeny plants of subpart (a) having the desired trait or combination of traits that have inherited the least amount of DNA from the donor plant species, subspecies or variety, respectively of claim 1.
6. The method of claim 5 wherein the backcross progeny plants inherit DNA encoding the desired trait or combination of traits whereby the trait or combination of traits are recessive and do not express a detectable phenotype.
7. The method of claim 5 wherein the rebackcross progeny plants inherit DNA encoding the desired trait or combination of traits whereby the trait or combination of traits are recessive and do not express a detectable phenotype.
8. The method of claim 5, optionally comprising the following steps:
    e. quantitatively annealing DNA from individual rebackcross progeny plants produced in subpart (a) with a detectably-labelled probe or multiplicity of detectably-labelled probes that specifically binds to a plurality of probe binding sites in the DNA from the recipient plant species, subspecies or variety and does not bind to the DNA of the donor plant species, subspecies or variety, respectively;
    f. determining the amount of DNA in each rebackcross progeny plant inherited from the recipient plant species, subspecies or variety,

respectively, that binds to the detectably-labelled probe or multiplicity of detectably-labelled probes, scored as a pool and independent of a physical or genetic map of probe binding sites; and

g. selecting for further propagation the rebackcross progeny plants having the desired trait or combination of traits that have inherited the greatest amount of DNA from the recipient plant species, subspecies or variety, respectively.

9. The method of claim 8 wherein the rebackcross progeny plants inherit DNA encoding the desired trait or combination of traits whereby the trait or combination of traits are recessive and do not express a detectable phenotype.
10. A method for selecting plants with a desired trait or combination of traits comprising performing steps (a) through (d) of the method of claim 5 with each successive generation of backcross progeny plants.
11. A method for selecting plants with a desired trait or combination of traits comprising performing steps (a) through (g) of the method of claim 8 with each successive generation of backcross progeny plants.

Selection Methods in Plant Breeding based on Mode of Reproduction

This article discusses Selection methods in plant breeding based on mode of reproduction. Some plants reproduce by (more or less strict) self-fertilization where pollen from a plant will fertilise reproductive cells or ovules of the same plant. Other plants only (mainly) allow cross-pollination where pollen from one plant can only fertilize a different plant.

Asexual propagation (vegetative propagation) can also occur in plants (e.g. runners from strawberry plants) which gives a new plant which is genetically identical to its parent plant. All these differences change the way plant breeders work. Apomixis is the phenomenon that seed are produced, but in an essentially asexual way, so that parent and offspring belong to one clone just as in case of 'normal' asecual' propagation.

Importance of Mode of Reproduction

The mode of reproduction of a crop determines its genetic composition, which, in turn, is the deciding factor to develop suitable breeding and selection methods. Knowledge of mode of reproduction is also essential for its artificial manipulation to breed improved types.

Only those breeding and selection methods are suitable for a crop which does not interfere with its natural state or ensure the maintenance of such a state. It is due to such reasons that imposition of self-fertilization on cross-pollinating crops leads to drastic reduction in their performance.

For teaching purpose, plant breeding is presented as four categories: Line breeding (autogamous crops), population breeding (allogamous crops), hybrid breeding (mostly allogamous crops, some autogamous crops), clone breeding (vegetatively propagated crops).

Self Fertilizing Crops (Autogamous Crops)

Certain restrictions caused the mechanisms for self fertilization (partial and full self fertilization) to develop in a number of plant species. Some of the reasons why a self fertilizing method of reproduction is so effective are the efficacy of reproduction, as well as decreasing genetic variation and thus the fixation of highly adapted genotypes.Almost no inbreeding depression occurs in self fertilizing plants because the mode of reproduction allows natural selection to take place in wild populations of such plants.

Critical steps in the improvement of self fertilizing crops are the choice of parents and the identification of the best plants in segregating generations. The breeder should also have definite goals with the choice of parents. Self fertilizing are easier to maintain, but this could lead to misuse of seed.

Some of the [agronomy] important, self fertilizing crops include wheat, rice, barley, dry beans, soy beans, peanuts, tomatoes, etc.

Mass Selection

This method of selection depends mainly on selection of plants according to their phenotype and performance. The seed from selected plants are bulked for the next generation. This method is used to improve the overall population by positive or negative mass selection. Mass selection is only applied to a limited degree in self fertilizing plants and is an effective method for the improvement of land races. This method of selection will only be effective for highly heritable traits. One shortage of mass selection are the large influence that the environment has on the development, phenotype and performance of single plants.

A developed by this method will be more uniform than those developed by mass selection because all of the plants in such a variety will have the same genotype. The seed from selected plants are not added together but are kept apart and used to perform offspring tests. This is done to study the breeding behaviour of the selected plant.

Stratified mass selection for ear size over 22 cycles has drastically altered plant phenotype in the maize population Zacatecas 58. Plants in the C22 cycle were 50 cm taller, had twice the leaf area index, reached anthesis 7 days later and had a 30% higher harvest index than C0. Differences in growth were detected early in ontogeny. The root growth of C22 exceeded that of C0 and the ratio of shoot dry mass to root dry mass was reduced by nearly 12%, from

8.0±0.2 to 7.1±0.1. Analysis of yield components revealed that C22 was superior to C0 in grain weight, number of rows per ear, number of grains per row, and total yield per unit area. Because the two genotypes were phenologically different, planting density optima are probably different for each population.

Selection of Cross-pollinated Crops

Plant species where normal mode of seed set is through a high degree of cross-pollination have characteristic reproductive features and population structure. Existence of self-sterility, self-incompatibility, imperfect flowers, and mechanical obstructions make the plant dependent upon foreign pollen for normal seed set.

Each plant receives a blend of pollen from a large number of individuals each having different genotypes. Such populations are characterized by a high degree of heterozygosity with tremendous free and potential genetic variation, which is maintained in a steady state by free gene flow among individuals within the populations.

In the development of hybrid varieties, the aim is to identify the most productive heterozygote from the population, which then is produced with the exclusion of other members of the population.

Mass Selection

It is the simplest, easiest and oldest method of selection where individual plants are selected based on their phenotypic performance, and bulk seed is used to produce the next generation by mixing it. Mass selection proved to be quite effective in maize improvement at the initial stages but its efficacy especially for improvement of yield, soon came under severe criticism that culminated in the refinement of the method of mass selection.

The selection after pollination does not provide any control over the pollen parent as result of which effective selection is limited only to female parents. The heritability estimates are reduced by half, since only parents are used to harvest seed whereas the pollen source is not known after the cross pollination has taken place.

Recurrent Selection

This type of selection is a refined version of the mass selection procedure and differs as follows:

- Visually selected individuals out of the base population undergo progeny testing
- Individuals selected on basis of the progeny test data are crossed with each other in every possible way to produce seed to form the new base population.

Half-sib Selection with Progeny Testing

Selections are made based on progeny test performance instead of phenotypic appearance of the parental plants. Seed from selected half-sibs, which have been pollinated by random pollen from the population, is grown in unreplicated progeny rows for the purpose of selection. A part of the seed is planted to determine the yielding ability, or breeding value, for any character of each plant. The seed from the most productive rows or remnant seed from the outstanding half-sibs is bulked to complete one cycle of selection.

Full-sib Selection with Progeny Testing

A number of full-sib families, each produced by making crosses between the two plants from the base population are evaluated in replicated trials. A part of each full-sib family is saved for recombination. Based on evaluation the remnant seed of selected full-sib families is used to recombine the best families.

Breeding of Asexually Propagated Crops

Asexual reproduction covers all those modes of multiplication of plants where normal gamete formation and fertilization does not take place making these distinctly different from normal seed production crops. In the absence of sexual reproduction, the genetic composition of plant material being multiplied remains essentially the same as its source plant.

[Cloning | Clone]s of mother plants can be made with the exact genetic composition of the mother plant. Superior plants are selected and propagated vegetatively; the vegetative propagated offspring are used to develop stable varieties without any deterioration due to segregation of gene combinations. This unique characteristic of [asexual reproduction] helped to develop a number of cultivars of fruits and vegetables including [grape]s, [apple]s, [pear]s and [peach]es.

Improving Asexual Plant Material through Selection

The selection in these crops is restricted to the material introduced from other sources, such as field plantations. The improvement of asexually propagated plants through induced mutations has distinct advantages and limitations. Any vegetative propagule can be treated with mutagens and even a single desirable mutant or a part of a mutated propagule (chimera) can be multiplied as an improved type of the original variety.

Selection of Asexual Plants

Selection, in the case of asexual plants, can be defined as the selection of the best performing plant and the vegetative propagation thereof. Because plants are not totally genetically stable, it can be expected that deviations would occur through the years. Selection is thus an ongoing process where deviants are selected or removed from the selection program. The main purpose of selection is to better the quality and yield of forthcoming plantations. Different approaches can be followed in the selection process of asexual plants, such as mass selection and clone selection from clone blocks.

In mass selection there are some factors that must be considered when selecting plants in a mother block, e.g. vineyard. Time of selection is a big factor, because you have to select when most of the characteristics of the plant are clearly showing. With asexual perennials the best time is just before harvest. For the best results the selected plant must be evaluated during the next season, when growth-abnormalities, leave disfigurations and virus symptoms are best visualized. Mass selection is done annually on the same plant for a minimum of three years. A plant that does not conform to the requirements in any given year of the selection cycle is discarded from the program.

New Clone Development

The development and registration of new clones take place by means of local clone selection in old plantations, as well as the importation of high quality clones from abroad, for local evaluation.

A clone is the vegetative offspring of one specific mother plant; it does not show any genetic, morphologic or physiologic deviations from the mother plant. Evaluation takes place with the different selected clones after selection.

Backcross Versus Forward Breeding in the Development of Transgenic Maize Hybrids: Theory and Practice

Key factors affecting the choice of breeding methods employed in the development of transgenic maize (*Zea mays* L.) hybrids are identified and evaluated, particularly as these pertain to the use and balance of backcross and forward breeding in the overall design of a breeding program.

These factors are type of trait, population(s) under selection, the predicted response to selection, stewardship of transgenic events, costs, and risks. Analysis suggested that simultaneous hybrid improvement and event integration via a forward breeding approach is not practical, given that some level of backcrossing is basic to event integration and the need to contain events not fully authorized by the local government for cultivation, food, and feed.

Considering event integration conducted in a stream separate from hybrid improvement, a forward breeding component to event integration may be warranted, particularly if a threshold level of expression of the transgenic trait of interest is dependent on endogenous alleles that interact with the event, some of which may be influenced by the environment. For this situation, directives are given for determining the number of generations of backcrossing to be conducted to create partial conversions from which to select for favourable endogenous alleles via forward breeding. In addition, ways to maximize the advantages of a backcross-only approach to event integration are discussed. Traditionally, choice of breeding methods has been a function of breeding objectives, the trait and population(s) under selection, and the predicted response to a given selection procedure (Hallauer and Miranda, 1988). However, with the advent of traits created through genetic engineering, proper stewardship of transgenic materials to ensure compliance with governmental requirements for containment of regulated events also influences the choice of breeding methods.

In addition, costs play an increasingly important role, particularly with the application of genomic technologies (e.g., expression analysis, molecular markers) with potential outlay ranging upward into the millions of dollars. Furthermore, with the high demand for transgenic traits, the competitive edge that these can confer is greater, as are the associated risks in seed product development.

This work aims to identify the key factors affecting the choice of breeding methods employed in the development of transgenic maize (*Zea mays* L.) hybrids, particularly as these pertain to the use and balance of backcross and forward breeding methods in the overall design of a breeding program.

Backcross breeding involves repeated crossing with selection to an elite inbred, with the goal of recovering a derived line that essentially resembles this elite parent with the addition of one or a small number of favourable alleles from the nonrecurrent parent. Forward breeding refers to any system of inbred line development, irrespective of the number of loci involved or the balance of favourable alleles among the parents of the population, involving the creation of a source population followed by inbreeding with selection, with the goal of recovering an improved line for one or more traits (e.g., pedigree selection).

The scope of this work is defined in terms of a framework for an overall breeding program and a set of options for the design of such a program. A breeding program focused on the development of transgenic maize hybrids aims to combine key agronomic and quality characteristics with the most advantageous transgenic traits to create top-performing hybrids that meet customer needs at a local level. The design of such a program centers on the cornerstone relationship between hybrid improvement (which includes new line development) and event integration.

Options for program structure include (i) utilizing a forward breeding method in new line development aimed at both hybrid improvement and event integration; (ii) conducting hybrid improvement independent of event integration and utilizing backcross breeding to introduce events to elite parents of candidate hybrids; and (iii) implementing a balance between forward and backcross breeding for event integration, conducting the main thrust of hybrid improvement in a stream separate from a program aimed at introgressing events into elite parents of candidate hybrids, with final selection of elite lines after partial conversion.

Events targeted for integration may include the following types: (i) events authorized by the local government for cultivation, food, and feed; (ii) events that are not yet fully authorized by the local government but will be once these clear assessment for environmental and food safety; and (iii) events that are not authorized and never will become authorized, because in the course of trait development and event selection, these events will be discontinued as commercial candidates in the process of selecting the best event for commercial introduction. Events without full authorization are subject to containment guidelines and precautions must be taken to prevent their release into the food and feed chain.

Factors Affecting the Choice of Breeding Method

The balance between backcross and forward breeding is affected by the following factors: type of trait, population(s) under selection, the predicted response to selection, stewardship of transgenic events, and costs and risks. Each of these factors is discussed individually, followed by a summary of the implications of these factors to the design of a breeding program involving transgenic traits.

Type of Trait

Through transformation, one or more genes are introduced to a targeted host species, become incorporated in the genome, and then function as fully heritable genetic factors. These genes encode for proteins (or in the case of antisense genes, lead to RNA transcripts) which produce a specific biochemical effect. In turn, this biochemical effect results in a specific phenotype typically referred to as the "transgenic trait." An event is a specific source of a given transgenic trait that is uniquely defined in terms of the actual DNA sequence that has been incorporated into the target genome via transformation as well as the chromosomal location(s) of the DNA insertion(s).

Transgenic traits represent novel traits or new levels of expression of common traits. The mechanisms involved reflect genetic engineering (e.g., high, sustained expression levels due to the promoter and enhancer sequences used in construction of the transformation vector). Novel traits may be due to one or more exogenous genes.

The gene action exhibited by the event of interest will significantly impact the choice of breeding methods. Ideally, an event of interest exhibits complete dominance, with no interaction with the rest of the genome or with the environment. With an "ideal" event, only one copy is required in the commercial hybrid, therefore only one inbred parent need carry the event. Event expression is reliable and consistent, regardless of the genetic background, so any germplasm can be targeted for value-added offerings. Furthermore, because event expression does not fluctuate based on environmental conditions, expression of the transgenic trait consistently meets or exceeds given thresholds across targeted market regions.

Initial transgenic traits in corn have focused on input traits (i.e., traits of primary utility or value to the grower), mainly insect resistance and herbicide tolerance. Event selection during transgenic trait development has been heavily biased in favour of the ideal event profile. The next wave of transgenic traits includes grain quality or composition, stress tolerance, and yield enhancement (USDA, 2006), bringing the potential for events to interact with endogenous genes associated with metabolic pathways influencing phenotypic expression of quantitative traits.

The degree of dominance dictates the number of copies of the event required in the commercial hybrid, with partial dominance or recessiveness potentially necessitating a broader breeding effort for event integration to bring in the event from both sides of the pedigree. At the extreme wherein all hybrids are targeted for a transgenic trait lacking complete dominance, comprehensive penetration of the breeding program with a given event could favour forward breeding aimed at simultaneous hybrid improvement and event integration.

Events requiring additional support to achieve a given threshold for the trait of interest may call for pyramiding with endogenous genes to achieve commercial levels of trait expression. Integration of multiple factors to achieve trait expression would tend to favour backcross breeding. Epistatic gene action involving the event and other genes influencing trait expression may require selection at loci other than the event to ensure performance at or above a given threshold for the trait of interest. Epistasis could favour forward breeding, depending on the number of loci involved, the magnitude of the epistatic effects, the consistency of effects across germplasm, and the impact of the environment on interactive loci.

One concern with transgenic traits that represent a new level of expression of a common trait (rather than a novel trait) is that, with new line development aimed at simultaneous hybrid improvement and event integration, endogenous alleles associated with the common trait may be lost. Endogenous alleles for resistance to first brood European corn borer [*Ostrinia nubilalis* (Hübner)] could be lost, especially since the event acts in a dominant fashion and the size of the effect could mask effects of other genetic factors. From the opposite perspective, the question arises of whether a new line developed in isolation of a transgenic trait with which it is likely to be coupled or have integrated is being selected under appropriate conditions. One option to deal with both issues involves the use of transgenic testers in new line development. In a breeding program where event integration is conducted independently of hybrid improvement, a nontransgenic line in development could be testcrossed to at least one inbred carrying events of broad usage (particularly authorized events) to create seed for performance trials. This would allow for evaluation of the line in pertinent transgenic hybrid combinations as well as facilitate line per se selection for similar traits that may impact production performance.

The current trend is to "stack" or bundle events in hybrids to create valued combinations of transgenic traits. This can be accomplished through breeding or molecular stacking. Breeding facilitates a mix-and-match approach to event combinations. Strategies devised to minimize the presence of linked chromosomal segments carrying unfavourable alleles from the event-donor source in the backcrossing process can be useful for effectively pyramiding the desired events.

A number of studies have demonstrated the feasibility of incorporating up to five genetic factors via backcrossing. Stacking through breeding would tend to favour backcrossing, particularly as the number of desired events increases and approaches five.

Molecular approaches to stacking through transformation aim to create a single insertion site containing various genes of interest, producing a single genetic unit for multiple trait expression.

Molecular approaches to stacking offer advantages in event registration with predetermined combinations and can facilitate considerable cost savings in event integration, especially as transgenic lines are advanced through selfing stages to homozygosity. Although molecular stacking places some restrictions on event combinations and requires accurate long-term market forecasting to be successful, it offers more flexibility in choice of breeding methods because the bundle of transgenic traits can be manipulated as a single unit.

Population(s) Under Selection

Government safety assessments and authorizations required for commercial release of transgenics are made on an event-by-event basis. The cost of developing and obtaining full authorization to commercialize a transgenic event (US$100 million; Monsanto Company, 2005) results in a limited number of sources of a transgenic trait of interest in the marketplace.

Typically, a biotech provider will target one or a few sources of a transgenic trait for commercialization, although there may be competing sources developed by different biotech providers. A "first generation" transgenic trait may be followed or replaced in the marketplace by an improved version of the particular trait (referred to as a "follow-on" trait). An example is the commercial launch of an event in 2002 that confers resistance to Lepidopteran insect pests of corn through the *Cry1F Bt* gene, which extends activity to black cutworm [*Agrotis ipsilon* (Hufnagel)] and increases activity against fall armyworm [*Spodoptera frugiperda* (Smith)] over that available with *Cry1A* events introduced in the

1990s (Gianessi et al., 2002). Thus, a breeding program developing high-performance transgenic hybrids could be dealing with several sources of a particular type of transgenic trait.

On the other hand, because each event is uniquely defined in terms of the actual DNA sequence that has been integrated within the target genome via transformation as well as the chromosomal location(s) of the DNA insertion(s), each event originates as an individual line. Early events were created using nonelite germplasm (e.g., Hi-II which was derived from B73 x A188; Armstrong et al., 1991). The ability to transform and regenerate elite inbreds has increased, which may tend to reduce the emphasis on eliminating event-donor genomic contributions during event integration. Nonetheless, the initial donor source for starting event integration with a particular event limits broad exploitation of genes other than the event of interest in that donor due to the need to maximize each authorized event. Even if the event source is considered elite, expansive use of a particular line in hybrid improvement would drastically reduce the genetic diversity within the program overall.

This highlights the dynamic nature of trait source in a breeding program, with a number of events for a particular trait potentially accessible and possibly changing over time as follow-on traits are developed, coupled with a limited genetic base with each.

The proportion of hybrids to feature the transgenic trait(s) and the degree of flexibility demanded in the breeding program (e.g., to access events from either side of the pedigree) will play into the choice of breeding method. If various transgenic versions of inbreds or hybrids are desired, backcross breeding is favoured.

A newly developed, improved inbred or hybrid can be converted in different backcross streams to several transgenic traits and even events within a trait. Events that become discontinued in development can be easily eliminated from the breeding program without jeopardizing new line development in hybrid improvement. However, with a forward breeding approach aimed at simultaneous hybrid improvement and event integration, it may be more difficult to expunge an event after it is introduced to a source population developed for new line development.

For example, USDA/APHIS considers that all progeny derived from an unauthorized event are regulated, including nulls for the event, until determined otherwise through petitioning to USDA for deregulated status and providing the supporting molecular evidence that demonstrates that event expression is absent and that the locus representing the insertion site in the event of interest is not altered in the null segregant when compared to the regulated parent (APHIS Biotechnology Regulatory Services [BRS], oral and written communication, August 2006).

Of course, use of forward breeding assumes the presence of favourable alleles at loci other than the event of interest in the event-donor that could be recovered and that would contribute advantage in a breeding program. There is limited genetic diversity at least initially with a new event source, as discussed above, affecting the overall utility of the event-donor beyond providing a source of the event.

Backcrossing could be utilized to introgress the event into multiple, diverse elite lines, which then could be used to produce breeding starts for the derivation of new lines, but that would take some time. With respect to any given cross, the probability of recovering lines with more favourable alleles than the best parent is based on several factors including the balance of favourable alleles between the event-donor and the other parent, the genetic variation present among the progeny, the number of loci involved, heritability of the trait, selection intensity, and the ability to identify superior progeny (Ribaut et al., 2002; Bailey, 1977). Use of a relatively nonelite event-donor favours backcross breeding. A regime consisting mainly of backcrossing does not preclude selection among versions of a converted line since lines derived from even several generations of backcrossing to an elite parent theoretically do not represent 100% recovery of the recurrent parent germplasm, and thus, may be considered partial conversions.

Predicted Response to Selection

With either forward or backcross breeding, selection is key to maintaining the event of interest throughout successive generations of inbreeding to ensure a high probability that improved homozygous lines derived from a population created using the event-donor as a parent will carry the transgenic trait. With backcrossing, for example, overall heterozygosity is halved each generation as loci become fixed for one parental allele or the other at unselected loci. However, with selection, favourable alleles contributed from the donor parent exerting large effects on the trait are more likely to be maintained in a heterozygous state (as are alleles at linked loci, whether favourable or unfavourable).

The probability that a favourable allele from the donor parent will remain segregating is a function of the magnitude of its effect on the trait, the intensity of selection, and the number of generations of backcrossing (Hill, 1998). Selection for an event per se is generally easy, based on the size of the effect and the use of screening tools. These may detect the portions of the DNA sequence inserted through transformation (or DNA flanking the insertion site) or the presence of a unique protein produced by the event of interest.

Additionally, some events can be selected through phenotypic screening for expression of genes characteristic of the event, such as herbicide tolerance. If adequate expression of the transgenic trait requires the presence of endogenous alleles in addition to the event of interest, these too could potentially be recovered from the event-donor. However, just as for the event, selection for these alleles must be conducted through successive generations of inbreeding to fixation. If selection is based strictly on phenotype, field testing must be conducted to effectively screen for the trait. For traits such as stress tolerance which can be complicated to screen, such testing may be difficult and costly. With a transgenic trait for increased yield or yield stability, field testing would be required at numerous locations which may be challenging, particularly with yet-unauthorized events that must be isolated from other materials to prevent pollen escape.

Use of genomic technologies such as molecular markers can significantly influence response to selection and enable solutions to issues like linkage drag (i.e., unfavourable alleles linked to the event of interest in the event-donor at hand) (Frisch, 2005; Hospital, 2001). Gene expression technologies can be utilized to analyse gene function and the molecular mechanisms underlying trait expression (Eisen and Brown, 1999). Such technologies could potentially be directed to facilitate selection for endogenous alleles that support transgenic trait expression, including alleles involved in epistatic interactions with events.

Speed is an important feature of response to selection and factors prominently into the development of transgenic hybrids because speed to market with a new transgenic trait may provide the opportunity to set the standard for the trait of interest and to garner strength as the market leader. Speed can favour backcross breeding, which facilitates cycling at three to four generations per year in continuous nurseries or greenhouses without overall lag in development of improved hybrids (Mumm, 1996).

Stewardship of Transgenic Events

Consider the following categories of plant materials in a breeding program focused on the development of transgenic maize hybrids: (i) those containing events unauthorized by the local government, (ii) those containing events authorized by the local government, and (iii) nontransgenic. Government regulations pertaining to the way yet-unauthorized transgenic events are grown, handled, and shipped typically set these plant materials apart from other materials in the ways these are managed. Furthermore, even for events that have received authorization by the local government, plant materials may need to be managed differently than nontransgenic materials, particularly if the breeding program engages in germplasm exchange internationally. Events approved for cultivation, food, or feed in one country may not have these authorizations in other countries. Effective containment requires control of pollen flow, volunteer plants, seed mixing, and human error leading to mislabeling of seed or plant materials, which may make it impractical to have programs for these categories of materials integrated or even side by side. (Panetta, 2005; Mumm and Walters, 2001).

Another aspect to consider is trait life cycle. As improved transgenic traits are developed, events commercialized earlier may become obsolete. Some event authorizations may reach expiration and, if not extended, there will be the need to discard all seed materials of a particular event at some point in time, further emphasizing the need for some segregation of materials.

Costs and Risks

Clearly, costs factor into the choice of a breeding program. A cost–benefit analysis is the best means to determine the net return of research investments with a given breeding method and to compare relative cost-effectiveness among breeding methods. As pointed out by Hoisington and Melchinger (2005), key elements of such comparisons include relative costs, time savings, benefits of accelerated release of improved germplasm, and the amount of operating capital needed and available.

The latter can be essential to use of high-tech strategies. Although the cost-benefit analysis may justify deployment of high-tech strategies, capital may be necessary to establish access to these.

Relative costs will take into account significant support systems for the breeding program such as quality systems to ensure that transgenic stewardship objectives are met. Economic advantages of partitioning the breeding program by event category (as discussed above) may not only lead to substantial savings, but may reduce risks associated with regulatory breeches as well.

Perhaps the greatest risk in transgenic hybrid development is that of failing to provide adequate containment of unauthorized events. Effective stewardship is essential to protecting research investments, maintaining compliance with government regulations, and ensuring customer satisfaction (Mumm and Walters, 2001). Regulatory breaches compromise public trust and can result in costly fines or remediation efforts.

Freedom to operate is a risk that favours the separation of hybrid improvement from event integration. If, for example, any of the components used in the transformation vector come under patent protection during the course of transgenic hybrid development, product pipelines and revenues and research investments could be significantly compromised.

Overlap in hybrid development and event integration may present a potential risk to overall effectiveness of the breeding program. It is assumed that a breeding program with components of both hybrid improvement and event integration would engage specialists in each area.

With the intersection of these activities, corn breeders become involved to a higher degree in transgenic trait evaluation, requiring them to be trait specialists and to take on additional responsibilities for event stewardship. Likewise, competition can arise for facilities. Given a finite number of field testing plots, phenotypic evaluation to facilitate selection of partial conversions could jeopardize the overall effort directed to hybrid improvement.

Speed, discussed earlier as influencing the choice of breeding method, also factors into risks. The flip side of the advantage of speed to market is the threat of competing hybrids featuring a transgenic trait of interest or a particular source of that trait introduced ahead in the marketplace, potentially setting a standard that may become a market challenge.

Summary of the Implications of Factors

In review, many of the issues or circumstances associated with the various factors can be classified as either favoring backcross breeding or favoring forward breeding. Overall, more favour backcross breeding, with Points 3, 4, 5, and 6 being especially key. Arguments for forward breeding are limited but can be compelling.

To summarize the implications of the various factors as related to breeding program structure, a forward breeding approach aimed at simultaneous hybrid improvement and event integration is not practical for two important reasons: (i) each authorized event originates as an individual line and the event must be incorporated into many hybrids to maximize trait availability, and (ii) the need to segregate unauthorized events for containment purposes.

The separation of hybrid improvement and event integration is further reinforced by the dynamic nature of the trait source. In addition, staff, facilities, and support processes can be dedicated to most effectively achieving the separate goals of each operation. Certain risks (e.g., freedom to operate or failure to prevent unauthorized events from entering the food chain) can be minimized and costs savings (e.g., those that can be realized through the segregation of transgenic materials in the product development pipeline) can be maximized by conducting hybrid improvement and event integration independently.

With event integration conducted independently from hybrid improvement, some degree of backcrossing is basic. However, a forward breeding component in event integration may be a necessity for transgenic events that interact with metabolic pathways depending on the number of loci involved, the magnitude of the epistatic effects, the consistency of effects across germplasm, and the impact of the environment on interactive loci.

To utilize forward breeding in event integration, a decision regarding the number of backcrossing generations to be conducted before line selection via forward breeding would be required. Ribaut et al. (2002) concluded that, considering various backcross generations, selection for favourable alleles from both parents is most efficient when initiated in later rather than earlier backcross generations (e.g., BC3 vs. BC1 or BC2), citing the increasing ratio of the standard deviation to the mean of the donor genome contribution. Bailey (1977) showed that the probability of obtaining a line superior to the recurrent parent is optimized at a particular backcross generation. Probabilities differ depending on the number of loci involved, the potential contribution from the nonrecurrent parent, the selection intensity, and trait heritability.

For example, the probability of obtaining superior lines from parents with an 80/20% split for favourable alleles was optimized at the BC2 generation in the case of 20 loci and at the BC4 generation in the case of 60 loci (when a 10% selection intensity was applied at the specified backcross generation and trait heritability was 0.5). Formulae provided by Bailey can be used to compute probabilities to support a decision as to when to shift event integration into forward breeding in a specified circumstance.

It is useful to compare the timetables for commercializing new transgenic hybrids using a backcross approach to event integration versus event integration with a forward breeding component.

Consider a backcross approach to event integration that aims for BC6-equivalent conversions of elite lines versus a combined approach to event integration, initiating forward breeding with BC3-equivalent partial conversions of elite lines. Development time with either approach can be accelerated through the early identification of hybrids targeted for conversion and deployment through the backcrossing phase of molecular markers to hasten recovery of the recurrent parent germplasm and continuous nursery facilities that enable cycling of three to four generations per year.

The overall development time for elite conversions may be up to three years longer with the balanced approach versus the backcross-only approach, mainly due to the time required to create testcross hybrids and to conduct two to three years of yield and/or efficacy testing to facilitate selection in the forward breeding phase. With the backcross-only approach, efficacy testing assessing transgenic trait expression is commonly done in parallel with activities to prepare for commercialization (i.e., inbred increase and hybrid production).

With the requirement for some backcrossing and the advantages of a backcross-only approach to event integration, another potential avenue to pursue in the situation involving interaction between the candidate transgene and the genome focuses on the identification of these interactions up front, so as to include all factors required for adequate trait expression in the transformation vector.

This would necessitate additional testing in the proof-of-concept stage of transgenic trait development but could be most efficient and effective in the long term, particularly in light of the trend to stack transgenic traits and demand for more than one transgenic version of the same hybrid. With all factors required for adequate and dependable transgenic trait expression represented in the event, these can in essence be treated as a single genetic unit in breeding efforts.

Further discussion on event design and event selection is outside of the scope of this work but the relevance of these issues to reliable and consistent event expression could be addressed more fully.

In closing, it should be noted that this analysis is directed to maize and does not necessarily fit for other crops, especially self-pollinated crops wherein cultivars are varieties with some degree of heterogeneity. Although the factors identified as affecting choice of breeding method would hold, the impact of these and their implications could potentially be quite different.

Many thanks to John W. Dudley and G.R. (Dick) Johnson for their critical review, thoughtful insights, and helpful suggestions regarding this work. Thanks also go to two anonymous reviewers.

Molecular Biology

Molecular biology is the study of biology at a molecular level. The field overlaps with other areas of biology and chemistry, particularly genetics and biochemistry.

Molecular biology chiefly concerns itself with understanding the interactions between the various systems of a cell, including the interactions between DNA, RNA and protein biosynthesis as well as learning how these interactions are regulated.

Writing in *Nature* in 1961, William Astbury described molecular biology as not so much a technique as an approach, an approach from the viewpoint of the so-called basic sciences with the leading idea of searching below the large-scale manifestations of classical biology for the corresponding molecular plan. It is concerned particularly with the *forms* of biological molecules and [...] is predominantly three-dimensional and structural—which does not mean, however, that it is merely a refinement of morphology. It must at the same time inquire into genesis and function.

Relationship to other "Molecular-scale" Biological Sciences

Researchers in molecular biology use specific techniques native to molecular biology, but increasingly combine these with techniques and ideas from genetics and biochemistry. There is not a defined line between these disciplines:

- *Biochemistry* is the study of the chemical substances and vital processes occurring in living organisms. Biochemists focus heavily on the role, function, and structure of biomolecules. The study of the chemistry behind biological processes and the synthesis of biologically active molecules are examples of biochemistry.

- *Genetics* is the study of the effect of genetic differences on organisms. Often this can be inferred by the absence of a normal component (e.g. one gene). The study of "mutants" – organisms which lack one or more functional components with respect to the so-called "wild type" or normal phenotype. Genetic interactions (epistasis) can often confound simple interpretations of such "knock-out" studies.
- *Molecular biology* is the study of molecular underpinnings of the process of replication, transcription and translation of the genetic material. The central dogma of molecular biology where genetic material is transcribed into RNA and then translated into protein, despite being an oversimplified picture of molecular biology, still provides a good starting point for understanding the field. This picture, however, is undergoing revision in light of emerging novel roles for RNA.

Much of the work in molecular biology is quantitative, and recently much work has been done at the interface of molecular biology and computer science in bioinformatics and computational biology. As of the early 2000s, the study of gene structure and function, molecular genetics, has been amongst the most prominent sub-field of molecular biology.

Increasingly many other loops of biology focus on molecules, either directly studying their interactions in their own right such as in cell biology and developmental biology, or indirectly, where the techniques of molecular biology are used to infer historical attributes of populations or species, as in fields in evolutionary biology such as population genetics and phylogenetics. There is also a long tradition of studying biomolecules "from the ground up" in biophysics.

Techniques of Molecular Biology

Since the late 1950s and early 1960s, molecular biologists have learned to characterize, isolate, and manipulate the molecular components of cells and organisms. These components include DNA, the repository of genetic information; RNA, a close relative of DNA whose functions range from serving as a temporary working copy of DNA to actual structural and enzymatic functions as well as a functional and structural part of the translational apparatus; and proteins, the major structural and enzymatic type of molecule in cells.

Expression Cloning

One of the most basic techniques of molecular biology to study protein function is expression cloning. In this technique, DNA coding for a protein of interest is cloned (using PCR and/or restriction enzymes) into a plasmid (known as an expression vector). This plasmid may have special promoter elements to drive production of the protein of interest, and may also have antibiotic resistance markers to help follow the plasmid.

This plasmid can be inserted into either bacterial or animal cells. Introducing DNA into bacterial cells can be done by transformation (via uptake of naked DNA), conjugation (via cell-cell contact) or by transduction (via viral vector). Introducing DNA into eukaryotic cells, such as animal cells, by physical or chemical means is called transfection. Several different transfection techniques are available, such as calcium phosphate transfection,electroporation, microinjection and liposome transfection. DNA can also be introduced into eukaryotic cells using viruses or bacteria as carriers, the latter is sometimes called bactofection and in particular uses Agrobacterium tumefaciens. The plasmid may be integrated into the genome, resulting in a stable transfection, or may remain independent of the genome, called transient transfection.

In either case, DNA coding for a protein of interest is now inside a cell, and the protein can now be expressed. A variety of systems, such as inducible promoters and specific cell-signaling factors, are available to help express the protein of interest at high levels. Large quantities of a protein can then be extracted from the bacterial or eukaryotic cell. The protein can be tested for enzymatic activity under a variety of situations, the protein may be crystallized so its tertiary structure can be studied, or, in the pharmaceutical industry, the activity of new drugs against the protein can be studied.

Polymerase Chain Reaction (PCR)

The polymerase chain reaction is an extremely versatile technique for copying DNA. In brief, PCR allows a single DNA sequence to be copied (millions of times), or altered in predetermined ways. For example, PCR can be used to introduce restriction enzyme sites, or to mutate (change) particular bases of DNA, the latter is a method referred to as "Quick change".

PCR can also be used to determine whether a particular DNA fragment is found in a cDNA library. PCR has many variations, like reverse transcription PCR (RT-PCR) for amplification of RNA, and, more recently, real-time PCR (QPCR) which allow for quantitative measurement of DNA or RNA molecules.

Gel Electrophoresis

Gel electrophoresis is one of the principal tools of molecular biology. The basic principle is that DNA, RNA, and proteins can all be separated by means of an electric field. In agarose gel electrophoresis, DNA and RNA can be separated on the basis of size by running the DNA through an agarose gel. Proteins can be separated on the basis of size by using an SDS-PAGE gel, or on the basis of size and their electric charge by using what is known as a 2D gel electrophoresis.

Macromolecule Blotting and Probing

The terms *northern*, *western* and *eastern* blotting are derived from what initially was a molecular biology joke that played on the term *Southern blotting*, after the technique described by Edwin Southern for the hybridisation of blotted DNA. Patricia Thomas, developer of the RNA blot which then became known as the *northern blot* actually didn't use the term. Further combinations of these techniques produced such terms as *southwesterns* (protein-DNA hybridizations), *northwesterns* (to detect protein-RNA interactions) and *farwesterns* (protein-protein interactions), all of which are presently found in the literature.

Southern Blotting

Named after its inventor, biologist Edwin Southern, the Southern blot is a method for probing for the presence of a specific DNA sequence within a DNA sample. DNA samples before or after restriction enzyme digestion are separated by gel electrophoresis and then transferred to a membrane by blotting via capillary action. The membrane is then exposed to a labelled DNA probe that has a complement base sequence to the sequence on the DNA of interest. Most original protocols used radioactive labels, however non-radioactive alternatives are now available. Southern blotting is less commonly used in laboratory science due to the capacity of other techniques, such as PCR, to detect specific DNA sequences from DNA samples. These blots are still used for some applications, however, such as measuring transgene copy number in transgenic mice, or in the engineering of gene knockout embryonic stem cell lines.

Northern Blotting

The northern blot is used to study the expression patterns of a specific type of RNA molecule as relative comparison among a set of different samples of RNA. It is essentially a combination of denaturing RNA gel electrophoresis, and a blot. In this process RNA is separated based on size and is then transferred to a membrane that is then probed with a labelled complement of a sequence of interest.

The results may be visualized through a variety of ways depending on the label used; however, most result in the revelation of bands representing the sizes of the RNA detected in sample. The intensity of these bands is related to the amount of the target RNA in the samples analysed. The procedure is commonly used to study when and how much gene expression is occurring by measuring how much of that RNA is present in different samples. It is one of the most basic tools for determining at what time, and under what conditions, certain genes are expressed in living tissues.

Western Blotting

Antibodies to most proteins can be created by injecting small amounts of the protein into an animal such as a mouse, rabbit, sheep, or donkey (polyclonal antibodies)or produced in cell culture (monoclonal antibodies). These antibodies can be used for a variety of analytical and preparative techniques.

In western blotting, proteins are first separated by size, in a thin gel sandwiched between two glass plates in a technique known as SDS-PAGE (sodium dodecyl sulfate polyacrylamide gel electrophoresis). The proteins in the gel are then transferred to a PVDF, nitrocellulose, nylon or other support membrane.

This membrane can then be probed with solutions of antibodies. Antibodies that specifically bind to the protein of interest can then be visualized by a variety of techniques, including colored products, chemiluminescence, or autoradiography. Often, the antibodies are labelled with an enzymes. When a chemiluminescent substrate is exposed to the enzyme it allows detection. Using western blotting techniques allows not only detection but also quantitative analysis.

Analogous methods to western blotting can be used to directly stain specific proteins in live cells or tissue sections. However, these *immunostaining* methods, such as FISH, are used more often in cell biology research.

Eastern Blotting

Eastern blotting technique is to detect post-translational modification of proteins. Proteins blotted on to the PVDF or nitrocellulose membrane are probed for modifications using specific substrates.

Arrays

A DNA array is a collection of spots attached to a solid support such as a microscope slide where each spot contains one or more single-stranded DNA oligonucleotide fragment. Arrays make it possible to put down a large quantity of very small (100 micrometre diameter) spots on a single slide. Each spot has a DNA fragment molecule that is complementary to a single DNA sequence (similar to Southern blotting). A variation of this technique allows the gene expression of an organism at a particular stage in development to be qualified (expression profiling).

In this technique the RNA in a tissue is isolated and converted to labelled cDNA. This cDNA is then hybridized to the fragments on the array and visualization of the hybridization can be done. Since multiple arrays can be made with the exact same position of fragments they are particularly useful for comparing the gene expression of two different tissues, such as a healthy and cancerous tissue. Also, one can measure what genes are expressed and how that expression changes with time or with other factors.

For instance, the common baker's yeast, *Saccharomyces cerevisiae*, contains about 7000 genes; with a microarray, one can measure qualitatively how each gene is expressed, and how that expression changes, for example, with a change in temperature.

There are many different ways to fabricate microarrays; the most common are silicon chips, microscope slides with spots of ~ 100 micrometre diameter, custom arrays, and arrays with larger spots on porous membranes (macroarrays). There can be anywhere from 100 spots to more than 10,000 on a given array.

Arrays can also be made with molecules other than DNA. For example, an antibody array can be used to determine what proteins or bacteria are present in a blood sample.

Allele Specific Oligonucleotide

Allele specific oligonucleotide (ASO) is a technique that allows detection of single base mutations without the need for PCR or gel electrophoresis. Short (20-25 nucleotides in length), labelled probes are exposed to the non-fragmented target DNA. Hybridization occurs with high specificity due to the short length of the probes and even a single base change will hinder hybridization. The target DNA is then washed and the labelled probes that didn't hybridize are removed. The target DNA is then analysed for the presence of the probe via radioactivity or fluorescence. In this experiment, as in most molecular biology techniques, a control must be used to ensure successful experimentation. The Illumina Methylation Assay is an example of a method that takes advantage of the ASO technique to measure one base pair differences in sequence.

Antiquated Technologies

In molecular biology, procedures and technologies are continually being developed and older technologies abandoned. For example, before the advent of DNA gel electrophoresis (agarose or polyacrylamide), the size of DNA molecules was typically determined by rate sedimentation in sucrose gradients, a slow and labour-intensive technique requiring expensive instrumentation; prior to sucrose gradients, viscometry was used.

Aside from their historical interest, it is often worth knowing about older technology, as it is occasionally useful to solve another new problem for which the newer technique is inappropriate.

History

While molecular biology was established in the 1930s, the term was first coined by Warren Weaver in 1938. Warren was the director of Natural Sciences for the Rockefeller Foundation at the time and believed that biology was about to undergo a period of significant change given recent advances in fields such as X-ray crystallography. He therefore channeled significant amounts of (Rockefeller Institute) money into biological fields.

9

Biological Control

Rice

Biological control as a method of insect pest management in rice began much later than the first success in the control of cottony cushion scale in California in the late 1880s. In fact, information- on natural enemy abundance and diversity, ecology and effectiveness as biocontrol agents, effects of pesticides and cultural practices on their populations, etc.-was very limited at the time. Considerable work has been done during the later half of the 20th century, and the practice of biological control of rice pests is constantly evolving.

Naturally Occurring Biological Control

Rice Insect pests have had long, close associations with their natural enemies, allowing stable relationships to develop. For example, predator and parasitoid guilds recorded in five irrigated rice ecosystems in the Philippines belong to ten orders and 57 families of predators and three orders comprising 40 families of parasitoids, indicating the great diversity and richness of the natural enemy community in rice. The interactions of such predators, parasitoids and insect pathogens are the cornerstone of modern integrated pest management programs in rice. The collective importance of natural enemies to pest population regulation is clearly demonstrated by observing outbreaks of insect pests after the removal of natural enemy communities by a broad-spectrum insecticide.

Conservation

Conservation is the modification of the environment or habitat in combination with judicious use of pesticides in order to conserve natural enemies and enhance biological control. Evaluating the success of biological control is, however, a daunting problem; far different from the "spray-and-count" or "count-and-spray" evaluation methods of pesticide-based approaches. Despite this, environment management to enhance biological control has proven valuable in many settings and appears to be very important for rice insect pest management.

Conservation Biological Control Contains Two Elements:

- Judicious use of insecticides.
- Modification of the environment.

Judicious use of Pesticides

As mentioned earlier, insecticides are more harmful to natural enemies than to pests. There is extensive evidence that the destruction of natural enemies by broad-spectrum insecticides leads to reduced biological control and the subsequent resurgence of some pests.

At the early stages of rice crop growth, abundant populations of detritus-feeding and plankton-feeding insects supports large populations of generalist predators in rice fields. The abundance of alternative prey gives the predator populations a "head start" on later developing pest populations.

Therefore, insecticide overuse is most damaging early in the season. Insecticide use can also create heavy selection pressure for strains of pests that can even overcome previously resistant cultivars. Such circumstances create outbreaks of secondary pest such as brown planthopper and impair biological control of some key primary pests such as stem borers.

Recent surveys in 10 major rice-growing countries in Asia revealed that pesticide is the dominant pest control tactic of farmers, who use insecticides more frequently than herbicides and fungicides.

A large proportion of farmers are still using insecticide compounds classified (by the World Health Organization) as extremely or highly hazardous to human health – namely methyl parathion, monocrotophos, and methamidophos.

The mean number of insecticide sprays per season varies from 0.3 in Laos to 3.9 in Vietnam. Most of these sprays are applied during early vegetative stages mainly targeting leaf feeding insect pests, which usually do not cause economic loss.

By and large, insecticides are being misused in rice. Preventive use of insecticides in rice is common and widespread. A common tactic used by farmers to control stem borers is the use of granular insecticides as a preventive method, without any consideration of the pest population density or economic benefit. Farmers' rationales for insecticide use appear to be base on anticipated yield losses due to insect pest attack. Such attitudes are detrimental to biological control, pest management as a whole, the environment and human health. Judicious use of insecticides is essential for natural enemy conservation and for the enhancement of natural biological control.

In order to conserve indigenous natural enemies and enhance natural biological control, the following strategies for insecticide use are recommended:

- Preventive or calendar-based use of insecticides should be stopped.
- Early season spraying during the first 40 days after transplanting should be avoided.
- Selective insecticides should be used only when pest population density reaches a damaging level.
- When using insecticides, use appropriate formulations in the right dose. Apply at the optimum time of intervention following proper application methods.
- Broad-spectrum hazardous insecticides should be avoided.
- If available, use rice varieties that possess host plant resistance to major insect pests.

Modification of Environment

The immediate crop environment can be intentionally modified to favour natural enemies. This requires substantial knowledge of pest and natural enemy ecology, and the effects of cultural practices on them. Habitat management should be aimed at providing environmental requisites to natural enemies such as:

- Alternative hosts or prey.
- Complementary foods – honeydew, pollen, nectar.
- Over-wintering or off-season shelter.
- Modified climate – windbreak.

Some habitat management activities that have a positive influence on the abundance of the natural enemies of rice insect pests are described blow:

Rice Bunds

Vegetated rice bunds support many herbivore species (including a few pests) and many species of natural enemies. They can be a very important source of natural enemy colonization in rice fields just after planting. Rice bunds are important sources of biological control agents of arthropods. Bund vegetation provides alternative and supplementary food for predators and parasitoids. Bunds are especially important for early sown crops and crucial for single-season rice crops. The importance of bund fauna on the early arrival of spiders for biological control of BPH has been demonstrated experimentally.

Bunds also provide shelter for natural enemies at critical times during harvesting, land preparation and between seasons. Puddling (tillage in presence of water) can have drastic effects on natural enemy populations. Many natural enemies take shelter on the rice bunds at puddling and the presence of vegetation on the bunds greatly increases their survival.

Manipulation of bund vegetation, particularly the enhancement of broadleaf weeds, can provide vital sources of natural enemies of rice insect pests and can reduce sources of alternate grassy hosts of some diseases such as sheath blight.

Vegetation in Adjacent Habitats

Generally, grasses and other vegetation in habitats adjacent to rice fields serve as habitats of natural enemies and also provide supplementary and complimentary food, over-wintering, or off-season habitats. The precise composition of plant species that will make the greatest contribution to the conservation of natural enemies of rice insect pests has not yet been determined.

Many species of arthropods, including natural enemies, exist concurrently in rice and non-rice habitats. Vegetation in these habitats serves as an important refuge for oligophagous rice parasitoids and generalist arthropod predators.

One such example is Anagrus nilaparvatae (Mymaridae: Heteroptera), an important egg parasitoids of rice planthoppers, which also parasitizes the planthopper Saccharosydne procerus (Delphacidae: Homoptera) of Zizania caduciflora (Gramaineae), which is a favorite vegetable for the residents of the Yangtze Rive Delta in China. Anagrus nilaparvatae can overwinter in S. procerus eggs in Zizania fields during the off-season. Such over-wintering is crucial for the maintenance of natural enemy populations in between the seasons.

Crop mosaics and Staggered Planting

Some non-rice crops such as cowpea, mungbean, maize, bell pepper, garlic, onion, soybean etc. serve as important reservoirs of natural enemies of rice insect pests during the non-rice season. Appropriate crop manipulation in rice growing areas may enhance conservation of natural enemies and thus biological control of rice insect pests. In general, non-rice crop fields forming a mosaic within the rice blocks are good for natural enemy conservation. In a monoculture situation, asynchrony (staggered planting) can provide better conservation of natural enemies than strictly synchronous planting.

Cultural Practices that Disrupt Natural Enemies

Some cultural practices seriously affect natural enemies. These include:

- Burning of rice stubble and/or straw.

- Burning of rice husk on the field prior to onion cultivation (in the Philippines).
- Trimming of non-rice vegetation from rice bunds prior to tillage (puddling).
- Destruction of vegetation on the field during fallow periods.
- Strictly synchronous rice planting over a large area.

From plant nutrient and natural enemy conservation points of view, it is better to incorporate crop residues into the soil rather than burn them. If burning is essential then it should be done in heaps rather than burning residue spread all over the field.

Conservation of Predatory Vertebrates

In many countries insect predatory birds, toads and frogs are caught in rice fields for consumption. Conservation of insect predatory birds, frogs and toads by preventing capture from the rice environment can also enhance biological control of rice insect pests. In Bangladesh, for example, collecting frogs from the lowland rice environment during their breeding period is discouraged.

Augmentation

What is Augmentation?

This method of biological control mainly refers to inundative or inoculative release of mass reared natural enemies. There have been a few attempts, mostly in China, to mass release stem borer egg parasitoids such as Trichogramma japonicum, Telenomus rowani, and Tetrastichus schoenobii. It seems that mass rearing and release is not economical in rice and that this approach has limited prospects for rice insect pest management.

Other Methods of Augmentation

Aside from inoculative and inundative release, a few other augmentation methods have been used with some success. These include the use of perches for insect predatory birds, use of ducks, and rice-fish culture.

Augmentation of Insect Predatory Birds

Placing perches in rice fields can attract some insect predatory birds. One such bird is the black drongo, Dicrurus adsimilis (Bechstein). The black drongo is a purely insectivorous bird and very common in the rice environments of South Asia. Black drongos use perches in rice fields as prey watchtowers and resting sites between prey hunting flights. They hunt any large insects in flight and also prey on insects from the rice canopy. Their prey include, grasshoppers, crickets, moths and butterflies, large caterpillars, etc.. However, they also prey on some beneficial insects such as dragonflies. Black drongos and many other insectivorous birds concentrate in the field for insect predation at tillage, weeding and harvesting stages.

Of about 1140 species of birds known to occur in China, half are entomophagous and about 30 are common and important for pest control in agriculture and forestry. Four species of bird have proven to be effective in grassland locust control in China.

These are: the rose-colored sterling Sturnus roseus; the common starling S. vulgaris porphyronotus Sharpe; the chough Pyrrhocorax pyrrhocorax (L.); and the desert wheatear Oenanthe deserti (Temminck). Constructing nest-building sites near target areas can attract these birds.

Use of Ducks

Releasing domestic ducks for pest control is a common component of IPM in China and Vietnam. Ducks are generalist predators, feeding on stem borers, leaffolders, grasshoppers, planthoppers and leafhoppers etc. Ducks have big appetites and on average one duck can consume more than 100 insects per hour. Ducks are very good at decreasing pest populations quickly, particularly in the early to mid-tillering stage, and also provide partial control of weeds. In China, rearing of ducks for pests control has been found to significantly increase income and reduce pesticide use.

Rice-Fish Culture

Rice and fish association is as old as the development of rice as human food. In the past, rice and free-living fish were very important resources in the lowland environments of Asia. Development of semi-dwarf modern rice plant types has dramatically changed this. In low land environments rice is now mostly grown under irrigation during the dry season.

Rice-fish culture in China is almost 2000 years old, but it suffered a setback during the green revolution due to widespread use of toxic insecticides. Increasingly, though, rice-fish culture is regaining popularity to cope with the increasing demand for fish in Asia.

Different types of carp, tilapia and catfish feed on planthoppers and leafhoppers, stem borers (moth and dispersing larvae) or other insects that fall into the water, mosquito larvae and other aquatic insects.

Some fish also feed on the outer leaf of the leaf sheath, which contains planthopper and leafhopper eggs. The ecological benefits of rice-fish culture are increased soil fertility, and control of weeds, insect pests and disease. Under rice-fish systems farmers cannot use insecticides, due to potential harmful effects on the fish, which helps to protect natural enemies.

Inundative and Inoculative Release

What is Inundative and Inoculative Release?

A few attempts have been made to control indigenous rice pests through inundative release of parasitoids, with limited success. An attempt to control the dark-headed stem borer Chilo polychrysus in Malaysia by release of Trichogramma japonicum in the early 1930s failed. Such interventions are often unsuccessful due to their high cost and the fact that the in most rice environments insect pests are already being controlled biologically by a rich and diverse group of natural enemies.

Introduction to Classical Biological control

What is Classical Biological Control?

In only two cases did the introduction of exotic parasitoids result in successful biological control in rice production. In the first case, populations of immigrant rice leaffolder, Marasmia exigua (Butler) (Lepidoptera: Pyralidae), in the Fiji Islands were apparently reduced by the introduced larval parasitoid, Trathala flavoorbitalis (Careron) (Hymenoptera: Ichneumonidae) from Hawaii in 1928. The second case involved the control of introduced rice striped stem borers Chilo suppressalis (Walker) (Lepidoptera: Pyralidae) in Hawaii through the introduction of an egg parasitoid – Trichogramma japonicum Ashmead (Hymenoptera: Trichogrammatidae) from Japan and two larval parasitoids – Eriborus sinicus (Holmgren)(Hymenoptera: Eulophidae) and Bracon chinensis Szepligeti (Hymenoptera: Braconidae) from China in 1929. All three parasitoids established and helped to reduce infestation of this stem borer. Several attempts have been made to introduce exotic parasitoids to control indigenous rice insect pests in South and Southeast Asia but none was successful. One such example involved the introduction of a larval parasitoid Paratheresia claripalpis (ven der Wulp) (Diptera: Tachinidae) from Trindad to Malaysia to control the dark-headed stem borer Chilo polychrysus (Meyrick) in 1951. A more recent example involved the introduction of two egg parasitoids Psix lacunatus (Johnson and Masner) (Hymenoptera: Scelionidae) and Telenomus cyrus (Nixon) (Hymenoptera: Scelionidae) to Palawan Island from the Island of Luzon, Philippines during 1985-87 to control Malayan black bug, Scotinophara coarctata (F.) (Hemiptera: Pentatomitidae). This project was not successful.

Major Natural Enemies of Major Rice Insect Pests

What are the Major Natural Enemies?

The species diversity and total number of natural enemies in tropical rice are impressive, but usually a few species dominate the complex. It is difficult to generalize about the precise role and relative importance of individual species. Here three important groups of insect pests are selected to illustrate the roles that natural enemies can play in the biological control of rice insect pests. The insect pests groups are:

- Stem borers
- Leaffolders
- Leafhoppers and planthoppers.

Rice stem Borers

More than one hundred species of parasitoids, predators and insect pathogens attack rice stem borers. Stem borers are most vulnerable to natural enemies at the egg, neonate larval and adult stages. On average only 1-4% of the egg population reaches adulthood and the rest perish due to biotic and abiotic stresses

Egg Parasitism and Predation

Parasitism and predation of stem borer eggs are usually very high and are important population regulating factors. Three groups of egg parasitoids, Telenomus spp., Tetrastichus spp. and Trichogramma spp., dominate the complex. Some of the major parasitoids are density dependent, meaning parasitism rates increase with increases in egg density, which is an important population regulation mortality factor. The most important egg parasitoids are:

- Tetrastichus schoenobii Ferriere (Hymenoptera: Eulophidae)
- Telenomus rowani (Gahan) (Hymenoptera: Scelionidae)
- Telenomus dignus (Gahan) (Hymenoptera: Scelionidae)
- Trichogramma japonicum Ashmead (Hymenoptera: Trichogrammatidae) Two important egg predators are:
- meadow grasshoppers – Conocephalus longipennis (de Haan) (Orthoptera: Tettigoniidae); and
- crickets, Metioche vittataicollis (Stal) (Orthoptera: Gryllidae).

Larval and Pupal Parasitism and Predation

The complex of larval and pupal parasitoids is large, but not as abundant as that of the egg parasitoids. Stem borer larvae and pupae generally suffer low levels of parasitism.

Larvae and pupae are almost inaccessible to natural enemies because they remain inside the stem. A few parasitoids are equipped with highly specialized long ovipositors that they can use to probe through the rice stem to locate and parasitize stem borer larvae and pupae. The most common larval and pupal parasitoids are:

- Cotesia flavipes Cameron (Hymenoptera: Braconidae)
- Temelucha philippinensis (Ashmead) (Hymenoptera: Ichneumonidae)
- Stenobracon nicevillei (Bingham) (Hymenoptera: Braconidae)
- Bracon chinensis Szepligeti (Hymenoptera: Braconidae)
- Tropobracon schoenobii (Viereck) (Hymenoptera: Braconidae)
- Xanthopimpla stemmator (Thunberg) (Hymenoptera: Ichneumonidae)
- Tetrastichus ayyari Rohwer (Hymenoptera: Eulophidae).

However, young dispersing stem borer larvae suffer high rates of mortality due to predation and desiccation. Disappearance of young larvae is the main reason for stem borer population fluctuations. Important larval predators are:

- Lady beetles, Micraspis spp. (Coleoptera: Coccinellidae)
- Carabid beetles, Ophionea spp. (Coleoptera: Carabidae)
- Rove beetle, Paederus fuscipes Curtis (Coleoptera: Staphylinidae)
- Water bug, Microvelia douglasi atrolineata Bergroth (Hemiptera: Veliidae)
- Water bug, Mesovelia vittigera (Horvath) (Hemiptera: Mesoveliidae)
- Water bug, Limnogonus fossarum (F.) (Hemiptera: Gerridae)
- Ants (Hymenoptera: Formicidae).

Adult Predation

Stem borer moths do not suffer parasitism but may suffer high mortality from a variety of insect, spider and bird predators. Reliable estimates on moth predation in the field situation are not available. The important predators are:

- Anthocorid bug – Euspudaeus sp., (Hemiptera: Anthocoridae)
- Wolf spider – Lycosa pseudoannulata (Boesenberg and Strand) (Araneae: Lycosidae)
- Black drongo – Dicrurus adsimilis (Bechstein)(Dicruridae).

Diseases

Beauveria, Cordyceps and Nomuraea are white fungi that infect stem borers. In high moisture fungal spores germinate and invade the soft tissue and body fluids of hosts and grow out of the body when ready to produce dispersing spores. Important pathogens are:

- Beauveria bassiana (Balsamo) Vuillemin (Moniliales: Moniliaceae)
- Nomuraea rileyi (Farlow) Samson (Moniliales: Moniliaceae)

- Cordyceps sp. (Entomophthoraceae)
- Bacillus thuringiensis Berliner (Bacteria)
- An unidentified nuclear polyhedrosis virus.

Rice Leaffolders

Parasitoids and predators play a very important role in rice leaffolders population regulation. Unlike stem borers, leaffolders are exposed to natural enemies at all the growth stages.

Although larva mostly remain inside the leaf fold, entrance of natural enemies into the leaf fold is not very difficult.

Egg Parasitism and Predation

Parasitoids:

- Copidosomopsis nacoleiae (Eady) (Hymenoptera: Encyrtidae)
- Trichogramma japonicum Ashmead (Hymenoptera: Trichogrammatidae)

Predators:

- Cricket – Anaxipha longipennis (Serville) (Orthoptera: Gryllidae)
- Cricket – Metioche vittaticollis (Stal) (Orthoptera: Gryllidae)
- Mirid bug – Cyrtorhinus lividipennis Reuter (Hemiptera: Miridae)
- Lady bird beetle – Micraspis crocera Mulsant (Coleoptera: Coccinellidae).

Larval and Pupal Parasitism and Predation

The larval and pupal parasitoid complex of rice leaffolders is rich. Several dozens of parasitoid species attack larvae and pupae. Common parasitoids are:

- Goniozus nr. trangulifar Kieffer (Hymenoptera: Bathylidae)
- Cotesia angustibasis (Gahan) (Hymenoptera: Braconidae)
- Cardiochiles philippinensis Ashmead (Hymenoptera: Braconidae)
- Macrocentrus philippinensis (Ashmead) (Hymenoptera: Braconidae)
- Trichomma cnaphalocrosis Uchida (Hymenoptera: Ichneumonidae)
- Temelucha philippinensis (Ashmead) (Hymenoptera: Ichneumonidae)
- Xanthopompla flavolineata Cameron (Hymenoptera: Ichneumonidae)
- Tetrastichus ayyari Rohwer (Hymenoptera: Eulophidae).

Reliable information on the extent of larval predation is not available. We do know, however, that the ground beetles Ophionea sp. attack leaf folder larvae.

Adult Predation

Leaffolder moths do not suffer mortality from parasitoids. However, a diverse group of predatory insects, spiders and vertebrates predate on rice leaffolder moths. The extent of moth predation in nature is not known.

Disease

Entomopahogenic bacteria such as Bacillus thuringiensis Berliner and a granulosis virus cause mortality of rice leaffolder larvae.

Rice Leafhoppers and Planthoppers

In general the impact of predators on rice leafhopper and planthopper populations is stronger than that of parasitoids.

Egg Parasitism and Predation

Egg parasitism varies considerably among hopper species and fluctuates during the season.

Leafhopper and planthopper eggs are generally more heavily parasitized than are adults and nymphs. The major egg parasitoids are:

- Oligosita yasumatsui Viggiani et Subba Rao (Hymenoptera: Trichogrammatidae).
- Anagrus spp. (Hymenoptera: Mymaridae).
- Gonatocerus spp. (Hymenoptera: Mymaridae).

The most abundant predator is the green mired bug Cyrtorhinus lividipennis Reuter (Hemiptera: Miridae), which predate on both eggs and nymphs. C. lividipennis appears to play an important role in regulation of plant and leafhopper populations.

Nymph and Adult Parasitism and Predation

The major parasitoids of planthoppers and leafhoppers are dryinid and strepsipteran wasps. Parasitism varies between 10% and 50% depending on the season and location. Important parasitoids are:

- Pseudogonatopus spp., (Hymenoptera: Dryinidae)
- Haplogonatopus apicalis Perkins (Hymenoptera: Dryinidae)
- Echthorodelphax fairchildi Perkins (Hymenoptera: Dryinidae)
- Elenchus yasumatsui Kefune and Hirashima (Hymenoptera: Strepsiptera)
- Tomosvaryella spp. (Hymenoptera: Pipunculidae)
- Pipunculus spp. (Hymenoptera: Pipunculidae).

A complex of predators predate on nymphs and adults of rice leafhoppers and planthoppers. Among them spiders and predatory insects are important.

The important predators are:

- Wolf spider – Lycosa pseudoannulata (Boesenberg and Strand) (Araneae: Lycosidae)
- Water bug – Microvelia douhlasi atrolineata Bergroth (Hemiptera: Veliidae)
- Lynx spider – Oxyopes spp. (Araneae: Oxyopidae)
- Long-jawed spider – Tetragnatha spp., (Araneae: Tetragnathidae)
- Rove beetle – Paederus fuscipes Curtis (Coleoptera: Staphylinidae)
- Ground beetle – Ophionea nigrofasciata (Schmidt-Goebel) (Coleoptera: Carabidae).

Diseases

Several fungal pathogens infect rice leafhoppers and planthoppers. Prolonged high moisture is necessary for the germination of fungal spores. The fungus grows into the body of the infected hopper; when the insect is dead, fungus grows out of the body through the joints.

One mermithid nematode also infects hoppers and after death of the insect the nematode emerges from the body. The incidence of nematode parasites is usually low, but relatively higher in the wet season than the dry season. Important pathogens are:

Fungus

- Metarhizium anisopliae (Metchnikoff) Sorokin (Moniliales: Moniliaceae)
- Metarhizium flavoviride Gams and Roszypal (Moniliales: Moniliaceae)
- Beauveria bassiana (Balsamo) Vuillemin (Moniliales: Moniliaceae)
- Hirsutella citriformis Speare (Moniliales: Stilbaceae)
- Erynia delphacis (Hori.) (Entomophthoraceae)
- Entomophthora coronata (Constantin) Kevorkian (Entomophthoraceae).

Nematode

- Hexamermis sp. (Mermithidae)
- Agamermis unka Kaburaki & Imamura.

Integrated Pest Management

Growing pest resistant rice varieties-Gallmidge and BPH resistant varieties with other desirable traits are available now. Depending on the region (key pest of the area) variety should be opted.

- Nursery protection by applying carbofuran granules @ 40 kg/ha or Phorate @ 12.5 kg/ha of nursery bed before 5-7 days of pulling the seedlings in endemic areas of gallmidge and stem borer only.
- Transplanting at appropriate stage after removal of terminal part (2 to 4") of seedlings to reduce the chances of carrying and migration of immature stages of yellow stem borer and leaf folders. Leaving alleys after every 5 metres in the main field is advantageous for field monitoring as well as manual movement to take up required interventions.
- Monitoring and fixed area (bench mark site) surveys to understand pest and associated beneficial fauna development. Adoption of prescribed sequential sampling plans for assessing different stages of plant and leaf hoppers as well as stem and leaf damages is ideal.
- Using pheromone traps (8 per acre) and light traps to understand the brood initiation of yellow stem borer, leaf folder and climbing cut worm should form a basic step in each situation.
- Adoption of mechanical practices like rope running to expose case worm and leaf folder larvae before going to chemical interventions.
- Regulating 'N' applications to avoid nutritional resurgence of plant and leaf hoppers. Supplementing N with K2O and neem coated materials for better utility and reduction of negative effects.
- Application of recommended insecticides-Endosulfan/Monocrotophos/Carbaryl/Chlorpyriphos/Quinalphos to reduce leaf folders and climbing cut worm as well as yellow borer if situation warrants.
- Harvesting of crop to ground level to reduce the chances of yellow stem borer and gallmidge build up, since stubbles serve as reservoir of these pests under many situations.
- Creating avenues for broad spectrum predators like spiders by maintaining shelters in and around rice fields during off season. Similarly rice crop maintenance should be in the process of conserving and exploiting of potential predators like mirids, dragon flies and carabid beetles, common in rice eco-system.

Management of Diseases

Sheath Blight

Use of disease free seed. Avoid close spacing, seed treatment with Carbendazim 50 WP @2.5 g/kg seed or soaking seed in Carbendazim @1g/l of water for 24 hours or spraying Propiconazole 25 EC (Tilt) @ 1ml/l or Hexaconazole 5 EC (Cantof) @2ml/l or Validamycin 3@ 2ml/l.

Bacterial Blight

Use of disease free seed. Grow tolerant varieties like Swarna and MTU 4870. No chemical is effective. Avoid excess N and water.

Tungro Virus

Cultivate tolerant varieties like MTU 9992, IET 7301, IET 7302, IET 8558, MTU 1002 to MTU 1005. Removal and destruction of affected plants. Vectors (GLH) can be controlled with Carbofuran 3G (10 kg/acre)/Monocrotophos 2ml/l or Chlorpyriphos 2 ml/l.

Blast

Cultivate tolerant varieties like IR 36, IR 64, MTU 9991, MTU 9992, MTU 7014, NLR 28600, Pinakini, Tikkana etc. Treat the seed with Carbendazim 50 WP @ 2.5g/kg seed or with Tricyclozole @ 1g/kg seed. Spray Ediphenphos 50 EC @1ml/l or Trycyclozole 75WP @0.6 g/l soon after observing symptoms and another spray 15 days after.

Wheat

Wheat is the most important food grain crop in India, occupying approximately 230 million ha world over. India is the second largest producer of the wheat with 25 million ha area and 70 million tones production. About one tenth of the global production is from India.

Wheat is grown during winter in the states of Punjab, Haryana, Uttar Pradesh, Bihar, Rajasthan, Madhya Pradesh and Maharashtra.

While in northern states it is grown as irrigated crop, in parts of Madhya Pradesh and Maharashtra it is grown largely under rainfed conditions with little or no irrigation support.

In Maharashtra wheat is grown in 7 lakh ha with average productivity of 13.2 q/ha against the national average of about 26.5 q/ha. Out of the 4 species, *Triticum aestivum* occupies almost 90% of the total wheat area.

Soil Requirement

Well drained loams and clay loams are ideal for wheat cultivation. However, good crops are raised in sandy loams and black soils also. Soil pH below 6.5 and above 7.8 are not suitable.

Season and Duration

Being low temperature loving crop, it is grown in winter from end of October to February under rainfed conditions and from middle of November to March/April under irrigated conditions. Total duration of the crop ranges from 110 to 120 days under irrigated conditions and 100-110 days under rainfed conditions.

The major diseases of wheat in India are, three rusts-leaf, yellow and stem rust, Karnal bunt, foliar blights, powdery mildew and loose smut. Diseases of limited importance include head scab, foot rot and flag smut; these diseases though of lesser importance, may be important in certain pockets.

Puccinia Recondita Tritici

Distribution: Throughout wheat growing regions of India.

Development: Pathogen over-summers in low and mid altitudes of Himalayas and Nilgiris. Primary infections develop from Wind deposited urediospores in eastern Indo-gangetic plains in middle of January where it multiplies and moves westwards by March. Temperatures of 20:t 5° C with free moisture (rain or dew) cause epidemics. Severe infection causes upto 30 percent yield losses.

Management: The presently recommended varieties in most of the wheat growing zones are rust resistant.

Puccinia Striiformis Tritici

Distribution: Hills, foothills and plains of north western India and southern hills zone (Nilgiri hills of Tamilnadu).

Development: Spreads through air-borne urediospores, when temperature are 10-20°C but the spread is checked above 25°c. Pathogen survives in the cool temperatures of hills (Himalayas and Nilgiris) and the primary infection takes places by middle of January in the foot hills and sub mountainous parts of north western India. Also, infection comes from across the western border, hence the probability of evolution of new races increases in this area. Yellow rust from Nilgiri hills cannot come out of the zone due to high temperatures in the Peninsular and Central India.

Management: Most of the presently recommended varieties are resistant. Major emphasis is on host resistance and cultivation of resistant varieties is the main strategy of management.

Stem Rust/Black Rust

Puccinia Graminis Tritici

Distribution: Mainly in Peninsular and I Central India, may occur in traces in Northern India too' were the infestation comes late.

Development: Develops from air-borne urediospores, needs free moisture and temperature above 20° C for spread. It can cause severe grain losses if infection is early. The pathogen perpetuates in Nilgiri hills during off season and becomes air-borne. If Peninsular and Central India experience rainfall during November then epidemics are severe. Late infections cause less damage in north India.

Karnal Bunt

Tilletia Indica (=Neovossia Indica)

Distribution: Parts of Northern Plains, especially Punjab, parts of northern Haryana, foot hills of J&K and HP., tarai area of Uttranchal, in lesser severity in Rajasthan, Bihar and UP. The states of Gujarat, Maharashtra, Karnataka and several parts of M.P. are free of KB.

Development: Seed and soil-borne; infection occurs at flowering by means of soil-borne inoculum. The degree of disease development depends upon the weather conditions prevailing during spike emergence to grain filling stage of crop. If the rains occur during the month of February in north Indian plains (disease-prone areas), the disease is likely to come with higher severity.

Management: Among the present day varieties, PBW 502 is resistant while the others show various levels of susceptibility. For management of this disease, one spray of Propiconazole (Tilt 25EC@ 0.1 %) should be given at the time of anthesis. Integration of one spray of propiconazole with one spray of bioagent fungus, *Trichoderma viride* (0.4% suspension) gives almost cent per cent disease control. The bioagent spray should be done before earhead emergence (Crop growth stage 31-39 on Zadoks scale), followed by the spray of chemical at start of earhead emergence (crop growth stage 41-49 on Zadoks scale). Two sprays of *T. viride,* at these two critical growth stages also give non chemical control of the disease which is almost similar to one spray of propiconazole. Chemical control should be adopted mostly in seed production plots.

Black Point

Alternaria Alternate

Development: Disease causes blackening of embryonic region of the seed (black point), discoloration of area beyond the embryonic region (black discoloration *(Caused by Aalternata, Curvularia lunate, Epicoccum sp., Bipolaris sorokiniana, etc.)* and eye-spot symptom (B. *sorokiniana).* The warm and humid weather at grain filling or near maturity favours this disease.

Management: This disease is of minor importance. Only when the disease percentage is high, it causes concern to the trader and the consumer. The discolored seeds are mostly shrivelled and they are separated out during processing.

Loose Smut

Ustilago Segatum (U. Tritici)

Distribution: North Indian plains and northern hills zone.

Development: It is a seed borne disease; infection occurs during *Loose Smut* flowering through wind-borne spores. The infection remains dormant inside the otherwise healthy looking seed but the plants grown from such seeds bear infected inflorescence. Infection is favoured by cool, humid conditions during flowering period of the host plant.

Management: Disease can be easily controlled through seed treatment with systemic fungicides hence resistance breeding has not attracted much attention. Treat the seed with fungicides like carboxin (Vitavax 75WP @ 2.5g/kg seed), carbendazim (Bavistin 50WP @ 2.5g/kg seed), tebuconazole (Raxil 2DS @ 1.25g/kg seed) if the disease level in the seed lot is high. If it is low to moderate, treat the seed with a combination of *Trichoderma viride* (@4 g/kg seed) and half the recommended dose of carboxin (Vitavax 75WP @ 1.25g/kg seed).

Foliar Blights

Distribution: Mainly in eastern India but also occurs in Peninsular and Central *Foliar blights* India. This disease complex is emerging as a problem in the north western India too.

Development: The disease requires high temperature and high humidity. This disease is more severe in late sown crop and causes substantial yield losses through formation of shrivelled grains. Most of the varieties are susceptible or moderately susceptible. The disease can be controlled through one spray of propiconazole (Tilt 25EC @ 0.1 %).

Powdery Mildew

Erysiphe Graminis Tritici

Distribution: Mainly in the cooler areas and hilly region; foot hills and plains of north-western India and the southern hills (Nilgiris).

Development: Powdery mildew can easily be diagnosed by the white, powdery patches that form on the upper surface of leaves and stem. With age, the patches turn dull dirty white and may have small black specks embedded. This disease can spread to all aboveground parts of the plant, including earhead and awns. The disease infects plants during periods of high humidity (not necessarily rain) and cool to moderate temperatures. Low light intensity, which accompanies dry weather and a dense crop canopy favours this disease.

Management: Present day varieties are not resistant to powdery mildew. Hence, the disease severity is more in some pockets. Avoid excessively dense, stands by using adequate seed. For chemical control, one spray of propiconazole (Tilt 25EC@ 0.1 %) on disease appearance (which usually occurs during early March in northern plains) is highly effective.

Head Scab

Fusarium Graminearum

Distribution: Parts of Punjab, especially in the sub mountainous regions. Bread wheat suffers lesser damage than the durum. It was first recorded in severe proportion in some parts of Punjab during 1995-96 crop season and again during 2004-05 crop season.

Development: Disease development is favoured by cool, moist weather with high humidity. Spores are produced on crop debris and reach the leaves through rain splash or wind. Apart from ear head infection, it can cause seedling blight and foot rot leading to lodging. In severe cases, it can cause shriveling of grains and low-test weights. At present, it is a disease of limited importance but has the potential to emerge as a major problem due to the production of toxins.

Management: Bread wheat are more resistant than durum. However, no resistant varieties are available. Hence, vigil is needed for this disease.

Cereal Cyst Nematode

Heterodera Avenae

Distribution: This nematode is found In most of the cereal growing regions of the country, especially, the dry and warmer areas of Rajasthan, Haryana and Punjab, but incidence is less in cooler climates.

Development: Larvae enter the roots near the growing point especially, at the seedling stage. Roots of infected plants become predisposed to various soil borne disease like root rots. In infected fields, the losses can be considerably enough.

Management: Most of the wheat cultivars are susceptible but some resist cyst formation. Chemical pesticides, some natural plant products and botanicals, coupled with improved cultural practices help in management of the CCN. For Rajasthan, one CCN resistant variety, CCNRV-1 is available for the disease prone areas.

Seed Gall Nematode/*Ear Cockle*

Anguina Tritici

Distribution: It is found mainly in some parts of northern India especially the states of Bihar, Jharkhand, eastern UP and Chhatisgarh.

Development: These nematodes are spread through seed galls in the seed lots during planting and harvesting. Wet weather favours larval movement and infestation. The nematode invades the crown and basal stem area, finally penetrating floral primordia. This leads to formation of nematode galls in ear heads.

Management: Use of clean seed (free of galls) is the only method to prevent this disease. For removal of galls, the seed lots are floated in 2-5 per cent brine solution. The galls, which float on the surface, can be easily separated and destroyed away from the fields. The seed thus cleaned should be washed with fresh water and used for planting.

Aphids

Sitobion Avenae, Rhopalosiphum Padi and Various other Species

Development: The aphids exist in different stages, *viz.,* winged (alates), wingless (apterous) sexual and asexual forms. The rapid spread takes place through asexual reproduction where females give rise directly to nymphs rather than eggs. Infestation usually occurs during second fortnight of January till crop maturity.

Management: When feeding in sufficient numbers, they can cause considerable damage, but under normal conditions, losses are not much. Chemical pesticides are recommended for this pest in wheat if the level of aphids per tiller crosses 10 during vegetative phase and 5 during reproductive phase.However, there is need to keep watch on this pest. The spray of imidacloprid @ 20 g a.i. per ha initially on border rows and if infestation is severe then in entire field will give good protection against this pest. Generally, natural enemies present in the field help in controlling the population of this pest.

Brown Wheat Mite

Petrobia Lateens

Distribution: In most of the wheat growing areas, under rainfed conditions, especially in the states of Rajasthan, Haryana and M.P. Sometimes, it is a pest in humid and warm conditions of irrigated areas also.

Development: They *Brown wheat* cause damage through *mite infestation* sucking mouth parts. When present in large numbers, mites cause a silvery flecking on leaves. Individual mites are too small to be visible with naked eye without 'effort. These can be seen by shaking the infested leaves on a white paper.

Management: Most of the times, mites do not cause any production constraint in wheat so no management practices are required. However, there is a need to keep vigil on this pest so that it may not become important in changing cropping sequence of future.

Army Worm

Mythimna Separate

Distribution: Mostly in the warmer climates of central India and to some extent in northern plains.

Development & Management: The larvae are found in the cracks of soil and hide during the day but feed during night or early morning. In wet and humid weather, they may feed during day time also. They survive during summer on the subsequent crops like rice and also continue to exist in rice stubbles before wheat crop comes in the field. Recently, this pest is catching attention in the northern India under Rice-Wheat rotation and where rice stubbles/straw remain in the fields.

Legume Pod-borer

Helicoverpa Armigera (= Heliothis Armigera)

Distribution and Importance: This is a polyphagous insect that attacks various legumes as a pod border. It is seen damaging wheat ear heads at grain development stage when major hosts are not available. However, the damage is below economic threshold level. It is found mostly in northern and central parts of India. Wheat can serve as a bridge host for carry over of this polyphagous pest.

Termites

Odontotermis Obesus, Microtermis Obesi

Distribution: Mainly in the northern and central India, but also in some pockets of peninsular India.

Early Symptoms of Damage: Termites attack the crop at various growth stages, from seedlings to maturity. The severely damaged plants can be easily uprooted and look wilted and dried. In case roots are partially damaged, the plants show yellowing.

Management: For effective management, chemicals like endosulfan, chlorpyriphos and carbosulfan can be used both for seed treatment and for broadcast of treated soil in standing crop.

Important Pests and Diseases

Important insects pests of wheat are termite, Gujhia weevil, cutworms, armyworm, thrips, aphids, shoot fly and stem borer. Among diseases, black, yellow and brown rust, loose smut, kernel bunt, leaf blight and powdery mildew are common.

Pest Management

Habitat Management and Cultural Practices – Biodiversity is a key component in keeping the pest's population below ETL. Intercropping of mustard and Rajgira and crop rotation with Bajra-wheat-Moong-Jowar-wheat or legumes-wheat-Moong-Bajra/Jowar + legumes-wheat can effectively check the spread of insect pests.

Jaggery (10 kg/ha) applied with 500 lit Jeevamrut attract ants that predate on aphids, thrips and larva. Random planting of maize or installation of about 10-12 bird perches in one ha attract birds that predate on the pests.

Insect-Pest Management

Termites: Termites can damage the crop at any stage of crop growth. Problem is more predominant in rainfed areas than irrigated ones. Use of undecomposed FYM under un-irrigated conditions can also increase the chances of termite attack. Use of Neem leaf manure (5 q/ha) or Neem seed manure (1 q/ha) in the soil at the time of sowing can prevent termite attack.

Gujhia Weevil: Gujhia weevil attack germinating seedlings, resulting into high seedling mortality. Application Neem leaf/seed manure can prevent their attack.

Army Worms: Caterpillars of this insect feed on plants particularly tender parts during the night and hide during the day time. They also damage leaves and ear-head. Spraying Neem leaf extract (5 kg Neem leaves crushed in boiled water and diluted to 100 liter) can effectively reduce the attack.

Brown Wheat Mite, Aphids and Jassids

Intercropping of wheat with mustard and safflower (100 gm seed of each mixed with 100kg wheat at the time of sowing) can effectively check the spread of mites. In case of severe attack, ferment 15 lit cow urine, 2 kg cow dung and 15 kg crushed Neem leaves in 100 lit water for 3-5 days. Filter the fermentate and use as foliar spray over one acre. This formulation is safe for natural enemies of pets such as lady bird beetle etc.

Rodent management – Rodents can be controlled by inserting a thorny twig of *Acacia* up to 1.5 to 2feet in each rat hole over the entire farm can prevent the rat menace. Bird perches (about 10-12/ha) attract owls at night that feed on rats. Use of horse gram flour balls mixed with cotton thread or flour mixed with cement powder, is very common. Rats readily eat these balls, resulting into blocking of their anus, which results into death. Rats that cannot excrete also get irritated and bite other rats in the hole, forcing them to run away.

Disease Management

Wheat crop grown organically is generally free from diseases. But diseases that affect conventional farms can appear occasionally during the conversion to organic.

Rusts-Rusts are caused by three different species of the fungi, Puccinia. Brown and yellow rusts are particularly important in the north-western India. Black rust appears quite late in these areas and normally do not cause substantial damage except in wheat fields sown very late. However, black rust appears in severe form in southern, central and eastern India and cause significant damage.

1. *Brown-Rust:* The small, round-oval spots (called pustules) of brown rust (caused by *Puccinia recondita tritici*) develop mainly on the leaves but under severe attack, appear on sheath and stem as well. Bright orange spots appear either in small clusters or irregularly scattered spreads very fast on susceptible varieties during warm and humid environmental conditions. In case of severe attack, plants mature early with poor root development and produce light and shriveled grains. This rust is most widespread in the country and generally the most damaging in many areas.
2. *Yellow Rust*: This rust is caused by the fungus, *Puccinia strciformis*. Yellow spots develop mainly on the leaves, but in case of severe attack, spots can also appear on leaf sheaths, stalks and glumes. The spots are oval in shape and lemon-yellow in colour and are formed in long streaks or rows, and smaller than those of brown rust. Appearance of lemon-yellow pustules in rows is characteristic feature of yellow rust. Under severe attack, these rows are not distinct, and large part of the plant gets covered with crowded pustules. Yellow rust is predominant in hills, foot-hills and north-western part of the country where the temperatures are low. In plains it does not cause much damage.
3. *Black Rust*: Black rust of wheat is caused by *Puccinia graminis tritici*. The spots occur on stem, leaf sheath, leaves and ear-heads but the stem is often most severely affected. The spots are reddish-brown in colour and elongated in shape. It spreads fast under relatively warmer and humid climate. As the crop matures, dark, black, elongated spots are formed. Black rust, though prevalent all over the country, appears in epidemic forms in the southern, central and eastern India, which are characterized by high temperatures in the wheat growing season.

Management of Rusts

Most effective method for control of rust, is to grow rust-resistant varieties. Biodiversity among wheat cultivars can also effectively check the rust problem. Use 3-4 wheat varieties at a time on each farm. Avoid late sowing or late maturing varieties. To protect the crop from rust infection, spray 5 liters of sour buttermilk mixed with 200 liter of water. Amaranth (chaulai or lal bhaji – a common green leaf vegetable) or Mentha (Pudina) leaf dust can also be used as fine spray (25-30 gm dry leaf powder per lit of water) to prevent the infection of rusts. Foliar spray of dry leaf extract of *Hibiscus rosa-sinensis* (China rose) can also prevent the rust infection.

Loose Smut:-It is caused by a fungus, *Ustiloto tritici*. Generally, diseased plants are indistinguishable from the healthy plants before heading, but diseased plans produce heads earlier than healthy plants. However, diseased plants of wheat variety Sonalika (RR-21) can be detected before ear emergence also by a characteristic streaking and yellowing of the flag leaf which starts drying from the tip and later becomes brownish in colour. Flag leaves of diseased plants of Sonalika can easily be seen in the field from a distance.

Terminal symptom in all the varieties is the production of black powder in place of wheat grains in the almost every ear of the diseased plant. The black powder is blown to flowers of healthy plants, they infect the mature grain and remains dormant until germination. Infected grains cannot be distinguished superficially from the healthy ones. After wowing these grains, the fungus gets activated and grows inside the seed first behind the growing point, keeping pace with apex of the growing shoot. As the ear formation starts, there is an accumulation of the fungus in the floral parts, which get completely destroyed due to formation of the black powder.

Use of resistant varieties is the best option under organic farming. Also as the disease is seed borne the use of disease free seed can prevent its occurrence. In case of doubt, treat the seed with 5%Vermiwash.

For seed production, visit the field daily for early symptoms of disease. Uproot infected plants and burn them before their spores are dispersed. Solar heat treatment of seeds in hot summer areas of plains can kill the pathogen to a great extent. On a bright sunny day in the last week of May or first week of June, soak the seeds in water for about 4 hours in the morning and then dry it under hot sun in the afternoon. Store the well-dried seeds for use in next season. The seed soaking activates the dormant fungus in the seed and heat of the sun kills it.

Bunt:-Kernal bunt is a common problem in northern India, in all the commercial varieties of wheat, but the disease is rare in traditional ones. The disease reduces both quantity and quality of wheat. Flour prepared from wheat with more than 3 % infected grains smells fishy and is unsafe for consumption. This disease is caused by fungus *Neovossia indica*. The disease can be recognized only after the grains have developed. A portion of the grain along its groove is converted into a black powdery mass, which gives a foul smell. In a plant hill, all the ears are not affected and in an ear only few grains are infected.

Growing resistant varieties is best option. Other precautionary and control measures are as follows:.

(1) Avoid planting highly susceptible varieties like HD-2009 (Arjun), UP-319, WL-711 and HD-1982 (Janak) in the Tarai areas-Himalayan foothills.

(2) Use disease free seeds. Pre-treat seeds with 5%Vermi-wash

(3) Use of Mustard flour 1 kg and milk 5 liters mixed with 100 liter water as foliar spray.

(4) Avoid continuous cropping of wheat in the same field. Practice field sanitation and deep ploughing in summer.

(5) Avoid excessive irrigation particularly during ear formation.

Harvesting

Harvesting is to be done when the ears are sufficiently dry and grains have approximately 15% moisture. Except some parts of Punjab and Haryana, wheat is harvested manually and threshed by wheat threshers. In some area bullocks are also used for threshing. Sometimes cutter wheels are also used for crushing wheat plants and separating wheat from chaff. In some parts of Punjab and Haryana combined harvester does all these operations in one go. But in this process recovery of grain is less and the straw is spread over the field.

Maize

- Pest is an organism that causes considerable damage to plant and draws the attention of man.
- Over 130 insect pests have been reported to infest maize but only very few are serious and require management. Amongst these, the most serious pests are the stem borers.
- The spotted stem borer, *Chilo partellus* is the key pest throughout India and the pink borer, *Sesamia inferens* is serious pest in Peninsular India.
- Shoot fly species *Atherigona soccata* and *A. naqvii* found to be serious in spring maize in Northern India.
- The losses ranging 26.7 to 80.4 percent have been reported due to *C. partellus*. Losses due to *S. inferens* found varied from 25.7 to 78.9 percent.

Nature of Damage

- Maize plants at the end of mid whorl stage are usually attacked.
- Aphids suck sap from plants and cause yellowing and mottling. Diseased plants may become stunted and turn reddish as they mature.
- If young plants infected they seldom produce ears. The aphid colony may some time cover completely the emerging tassels and the surrounding leaves preventing the emergence.
- Ears and shoots are also infested and seed set may be affected. Honey dew excreted by aphids favours the development of sooty molds.

Control

- One spray of systemic insecticides like dimethioate or monocrotophos at 0.1% will controls the aphids.

Pink Borer: *Sesamia Inferens*

Occurrence

- A serious pest in peninsular India more severe in rabi.
- Other hosts include Ragi, Sorghum, Bajra, Wheat, Paddy and Sugarcane.
- Moth is medium sized stoutly built having straw coloured forewings with a mid longitudinal dark brown broad triangular streak. Hind wings are white.

Life Cycle

- Round pearl like yellowish eggs ranging 80-300 are layed in 2-3 longitudinal rows usually with in the sheaths of bottom leaves of young maize plant of two to three weeks old. As the time for hatching approaches, eggs become brown or shy grey.
- Full grown larvae are stout smooth about 25 to 30 mm in length purplish pink on the dorsal side and white on ventral side. Life cycle is completed in 6-7 weeks with 4-5 generations in a year.

Nature of Damage

- Newly hatched larvae remain in group behind the leaf sheath and begin chewing on the stem and epidermal layer of the sheath. Some migrate to neighboring leaf sheaths, while others penetrate the stem.
- Whorl feeding of larvae results in rows of oblong holes in unfolding leaves unlike round shot holes produced by *Chilo partellus.*
- ***Cob and Tassel at oneplace***: Later they bore in to central shoot resulting in the drying up of the growing point and formation of dead heart in young maize plant as a result of larval feeding some times the bottom internodes show circular ring like cuts.
- Severe damage causes the stem to break. Severely infected plants due to stunting may appear to have some times the cob and tassel at one place.

Management of Stem Borers

- Pest management strategies viz., cultural, biological, host plant resistance, chemical and integrated pest management for maize stem borers can be practiced to effectively control both stem borers *Chilo Partellus* and *Sesamia inferens.*

Farm Sanitation

- Collection and burrying stubble and stalks or ploughing and destroying crop residue, removal of infested plant parts or infested plants at hoeing are recommended.

Inter Cropping

- Growing maize in association with various legumes significantly reduced *C. partellus* damage in maize. Also intercropping maize with soyabean gave a considerable reduction in *C. partellus* and *S. inferens* and 9.2% increase in grain weight per plant.
- Therefore intercropping has a great potential in reducing the pest incidence.

Exploitation of Ovipositional Behaviour

- The ovipositional behaviour of *C. partellus* could be exploited for its management.
- Clipping of lower leaves of maize (up to fourth) on which most of eggs of *C. partellus* are laid, reduced damage to some extent.
- As this practice may not be practicable in a large scale, they recommended earthing-up of the crop to bury the lower leaves as a possibility for reducing initial infestation.

Plant Resistance

- Intensive screening to identify resistance sources against (*C. partellus* yielded an important germplasm source. Antigua Gr. 1 and few breeding lines.
- Sources of tolerance to *S. inferens* have been identified. These materials are currently being used in Directorate of maize Research and also in other centres breeding programs.
- At this time, however, stem borer resistant cultivars/hybrids have yet to be released in the semi arid tropics.
- Further more, the work on the screening for resistance is being done extensively and continuously and there is every possibility of identifying sources of resistance to borers and development of resistant hybrids/ varieties.

Biological Control

- Many natural enemies have been reported to attack *C. partelllus.*
- Among the parasitoids reported, *Trichogramma* spp. as egg parasitoids and *Aphanteles favipes* and *T. Chilonis* and larval parasitoids are important.

- *T. cllilonis* has been found parasitising egg masses of *C. partelllus* upto 70%. *A. flavipes* is the most dominant and most widely distributed larval parasitoid of *C. partellus* in India with a parasitisation ranging from 5.3-42.8%.
- Even though a number of natural enemies have been reported on maize borers, but in general, the efficiency of natural enemies in particular farming environments is not known.
- The scope for successfully controlling maize stem borers with natural enemies is limited by the short cropping period and the lack of continuous habitat for natural enemies.

Chemical Control

- Seed treatment
  Seed treatment with Carbofuran 35 ST dust.
- Seed treatment with Carbofuran (40F)@ 2.5g/kg of seed was also found effective against *S. inferens*.

Effective Control Schedule

- 1st Foliar application of Endosulfan 0.1% at 15 days after sowing
- 2nd application of 4% granules al 15 kg/ha a fortnight later against *C. partellus*.

Stem Borer: Chilo Partellus

It is the most destructive pest of maize and Sorghum in Srilanka, India, Pakistan, Afghanistan, Uganda, Central and East Africa.Pest is found in throughout India.

Nature of Damage

- The young larvae first feed on the leaves, making a few shot holes and then bore their way downwards through the central whorl as it opens.

Shot Holes on Leaf

- More shot holes become visible, indicating an earlier attack and the plant also shows dead-hearts. Caterpillars damage maize by boring into the stems, cobs or ears.
- Stem borer infestation results in pinhole damage in the leaves initially and "dead hearts" subsequently.

Life-cycle

- The insect breeds actively from March-April to October and for the rest of the year it remain in hibernation as a full-grown larva in maize and sorghum stubble, stalks or unshelled cobs.

- The larvae pupate sometime in March and emerge as moths in the end of that month or in early April.
- They are active at night, when they mate and lay eggs on the underside of the leaves of various host plants, particularly the early crops of maize and sorghum grown as fodder.
- The eggs are flat, oval, and yellowish and are laid in overlapping clusters each containing up to 20 eggs. A female lays over 300 eggs during its life-p of 2-12 days, and the eggs hatch in 4-5 days in summer.
- The larva become full-fed in 14-28 days passing through six stages and after making a hole in the stem pupates inside it.
- The life cycle is completed in about 3 weeks and there are probably 5 generations in a year. The full-grown caterpillars of the last generation hibernate in stubble, stalks, etc, and remain there till the next spring.

Management of Stem Borers

- Pest management strategies viz., cultural, biological, host plant resistance, chemical and integrated pest management for maize stem borers can be practiced to effectively control both stem borers *Chilo Partellus* and *Sesamia inferens.*

Cultural Control

Farm Sanitation

- Collection and burrying stubble and stalks or ploughing and destroying crop residue, removal of infested plant parts or infested plants at hoeing are recommended.

Inter Cropping

- Growing maize in association with various legumes significantly reduced *C. partellus* damage in maize. Also intercropping maize with soybean gave a considerable reduction in *C. partellus* and *S. inferens* and 9.2% increase in grain weight per plant. Therefore intercropping has a great potential in reducing the pest incidence.

Exploitation of Ovipositional Behaviour

- The ovipositional behaviour of *C. partellus* could be exploited for its management.
- Clipping of lower leaves of maize (up to fourth) on which most of eggs of *C. partellus* are laid, reduced damage to some extent.
- As this practice may not be practicable in a large scale, they recommended earthling-up of the crop to bury the lower leaves as a possibility for reducing initial infestation.

Plant Resistance

- Intensive screening to identify resistance sources against (C. partellus yielded an important germplasm source. Antigua Gr. 1 and few breeding lines.
- Sources of tolerance to S. inferens have been identified. These materials are currently being used in Directorate of maize Research and also in other centres breeding programs.
- At this time, however, stem borer resistant cultivars/hybrids have yet to be released in the semi arid tropics. Further more, the work on the screening for resistance is being done extensively and continuously and there is every possibility of identifying sources of resistance to borers and development of resistant hybrids/varieties.

Biological Control

- Many natural enemies have been reported to attack *C. partelllus.*
- Among the parasitoids reported, *Trichogramma* spp. as egg parasitoids and *Aphanteles favipes* and *T. Chilonis* and larval parasitoids are important.
- *T. cllilonis* has been found parasitising egg masses of *C. partelllus* upto 70%. *A. flavipes* is the most dominant and most widely distributed larval parasitoid of *C. partellus* in India with a parasitisation ranging from 5.3-42.8%.
- Even though a number of natural enemies have been reported on maize borers, but in general, the efficiency of natural enemies in particular farming environments is not known.
- The scope for successfully controlling maize stem borers with natural enemies is limited by the short cropping period and the lack of continuous habitat for natural enemies.

Chemical Control

- Seed treatment with carbofuran 35 ST dust.
- Seed treatment with carbofuran (40F)@ 2.5g/kg of seed was also found effective against *S. inferens.*

Effective Control Schedule

- 1st Foliar application of Endosulfan 0.1% at 15 days after sowing
- 2nd application of 4% granules al 15 kg/ha a fortnight later against *C. partellus.*

Shoot Fly: Atherigona Soccata, A.Naqui

- Very serious pests of maize in South India. In North India maize crop is not effected much except spring season in tarai area.

Nature of Damage

- The attack is maximum when the crop is in seedling stage.
- The tiny maggots creep down under the leaf sheaths till they reach the base of the seedlings.
- After this they cut the growing point or central shoot which results in the formation of characteristic dead hearts.

Control Measures

- Water logging condition should be avoided.
- Application of 10% Phorate granules in furrows before sowing @15kg/ha should be done and the insecticide should be covered with thin layer of soil after which seed should be sown.
- Spraying with 0.01% Metasystox after one week of germination.

Shoot Bug: Peregrinus Maidis

- Occasionally serious pest on maize.
- Found attacking maize and sorghum in South India and Madhya Pradesh with peak activity during August-October.
- Adult is yellowish-brown to dark brown with translucent wings in macropterous forms while brachypterous forms have underdeveloped wings, nymphs are yellowish and soft bodied.

Nature of Damage

- Adults and nymphs suck sap, resulting in unhealthy, stunted and yellow plants.
- leaves wither from top down words and plants die if attack is severe.
- Honey dew excreted by the insects causes growth of sooty mold on leaves. Mid ribs of leaves turn red due to egg laying and may dry subsequently.

Control

- Predators of coccinellid predators, *Coccinella septumpunctata* and *Menochilus sexmaculata* if available will control the pest
- If Sufficient population of predators is not available spraying any systemic chemical like monocrotophos or dimethioate at 0.1% will control the disease.
- 15days after sowing spray 0.1% Endosulphan.
- After 15days of this spraying, 15kg 4% endosulphan granules per hectare are to be applied.

Integrated Pest Management And Its Difficulties

- The individual control measures may not be sufficient to adequately control maize stem borers and it is necessary to integrate the available methods.,
- In the integrated management of maize borers host plant resistance and cultural practices should be major components. The IPM recommendations are that they are too complicated to be adopted by farmers.

Difficulties Encountered in IPM

- Recommendations for stem borer management although appear promising have not carried for beyond the research stations.
- Farm sanitation can either not be improved or only be improved at high cost.
- Sowing dates are confined by several constraints and likely to be well timed in traditional farming systems that have evolved over long periods.
- Rotations are often obstructed by the vagaries of weather.
- Stem borer control through consistent intercropping is to be standardized.
- Biological control is insufficient.
- Chemical control even though efficient but limited by economical constraint.
- Integrated stem borer management finally is likely to be severely constrained by the limited management capability of farmers.
- Ever since the advent of stabilized Agriculture, storage of produce has remained an issue of utmost concern.
- Unless the problem of storage is solved satisfactorily the problem of hungry millions may continue even with substantial increase in production.
- There are a number of estimates of post harvest losses. Most conservative estimates for the post harvest losses in food grains in India even put at about 10%, a quantity good enough to feed at least 60 million people.
- Out of the total production, about 70% is retained and stored by farmers for consumption, as seed, feed and payment of wages and only about 30% is marketable surplus.
- Insects are responsible for enormous spoilage in storage they feed on grain, bore the kernel, destroy the germ portion, cause heating and deterioration in stored produce.
- As per the Pause committee report, the storage losses due to insect pests of stored grain are 2.55%. In addition they are also responsible for the qualitative losses in food grains.

Sources of Infection

- Field infestation, insects fly from stores to fields and lay eggs upon the maturing grains.
- These eggs hatch out to larvae in favourable conditions when grains reach the stores. They can crawl/fly to fresh stocks and infest them is called cross infestation.
- It has been a general practices, specially with farmers, that they keep without cleaning the empty food grain bags for use in the next season only.
- Eggs and larvae of insects remain hidden and feed on the grain fragments in the off season. Whenever the grains are filled up in such bags, infestation reappears.
- The insects, which may be present in the joints or corners of the carriers, migrate to the food grain lots and cause infestation.

Management

- The scientific storage of grains help in minimising the damage by insects, and microorganisms.

The Following Methods can Help in Safe Storage of Grains

- Preventive measures
- Curative measures

Preventive Measures

- Cleaning the storage structures, sealing of cracks, crevices and holes present in the floors. Cleaning of the Sheller before their use.
- Storage structures/godowns/gunny bags should be disinfected with approved residual insecticides preferably with Malathion 50% EC, with dilution 1: 100.
- Proper stacking of the filled bags for proper hygiene and sanitation prevents insect damage in the godowns.

Curative Measures

The Infestation of Stored Grain Insect Pests can be Minimized by Different Methods

Physical Control Measures;

- Heat treatment of stored grain at 55-60°C
- Mixing of inert dusts with grain makes entry of insects a difficult task. However here grain has to be washed before consumption.

Mechanical Measures;

- These measures are practicable and include sieving of grains.

Ecological Measures;

- Temperature, moisture content of grain and availability of oxygen have to be suitably manipulated by designing and constructing the storage structures which create unfavourable ecological conditions for insect attack.
- Food grains kept in airtight sealed structure remain insect free. Eg. Pusa bin.
- Grains with moisture content less than 10% are not suitable for multiplication and survival of most of the

insects (except Khapra beetle which can survive but it is susceptible to reduced oxygen content.)

Chemical Control Measures

- These measures are more popular and effective. They may be used as prophylactic and curative treatment.

Prophylactic Treatment

- These insecticide treatments are to prevent the insect infestation and cross infestation. However insecticide should not be sprayed on food grain directly.
- The insecticides and their concentrations recommended for prophylactic treatment of stored grain are Malathion 50 E C at 0.5%, Pxrimiphos methyl 50% EC at 0.5%

Curative Treatment

- Mixing of chemicals with food grains meant for human consumption is not recommended. However they can be mixed with grains meant for seed purpose only. Normally pyrethrum dust is used as seed protectant.
- Fumigation is the best method for controlling stored insects. However, the fumigation of grains should be carried out under expert supervision and proper caution.
- In our country, EDB and EDCT have been recommended for fumigation of food grains. These fumigants should be handled with great care and by trained staff only. The target DNA is then analysed for the presence of the probe via radioactivity or fluorescence. In this experiment, as in most molecular biology techniques, a control must be used to ensure successful experimentation.

Stored Grain Pests

Common Name	*Scientific Name*
Rice weevil	*Sitophilus oryzae*
Maize weevil	*Sitophilus zeamays*
Granary weevil	*Sitophilus granarius*
Lesser grain borer	*Rhizopertha dominica*
Khapra beetle	*Trichoderma granarium*
Red rust flour beetle	*Tribolium castaneum*
Confused flour beetle	*Tribolium confusum*
Saw toothed grain beetle	*Oryzaephilus surinamen*
Rust red grain beetle	*Cryptotests ferrugineus*
Flat grain beetle	*Cryptotests pusillus*
Ware house moth	*Cadra contella*
Angoumois grain moth	*Sitotroga cerealella*
Indian meal moth	*Plodia interpurutella*

Cotton

Identification & Monitoring

- This is a major pest
- The pest is active from July-October and February-April.
- The adult moth is stout, yellowish brown with a dark speck area on the forewings, which have grayish wavy lines and a black kidney shaped mark whereas the hind wings are whitish with blackish patch along the outer margin.
- The larvae is about 35 mm long, greenish brown with dark gray yellow stripes along the sides of the body.
- For 65-100 days old crop, one egg or larvae per plant or 5-10 % infested fruiting bodies is the ETL.

Damage

- The larvae feed on the leaves initially and then bore in to the square/ bolls and seeds with its head thrust into the boll, leaving the rest of the body outside.
- A single larva can damage 30-40 bolls.
- The entry holes are large and circular at the base of the boll.

Cultural Control

- Plough deeply, clean cultivation to expose the resting pupae, crop rotation and avoidance of ratooning reduces pest population.
- Use tolerant varieties.
- Trap cropping with crops like tomato, and destroying them when the population is high.
- Use of maize, and cowpea on borders and wild brinjal and setaria as intercrop significantly helps in reducing the pest population.

Mechanical Control

- Cotton is harvested in 3-4 pickings by hand as the boll mature.
- The number of pickings varies with the maturity habit of the variety.
- Cotton from damaged boll should not be kept with good quality cotton.
- Never pick wet cotton.
- On dew days, pick cotton in late mornings to avoid moisture in cotton.

Biological Control

- Release of egg parasitoids like *Trichogramma chilonis* or *T. brasielenis* or *T. achaea* @ 1,50,000/ha from 45th day onwards at 10-15 days interval (6 releases) and larval parasitoids such as Chilonus blackburni or *Bracon brevicornis* or *Telenomus heliothidae* or *Carcelia illota* or Coteria kazat or *Campoletis chloridae* @ 2000 adults/ha at 15 days interval.
- Release pupa parasitoids *Brachymeria sp.*
- Release of the predators *Chrysoperla carnea* or *Scymnus sp.* or *Eulophids* would suppress the population of larvae.
- Spray HaNPV @ 250 LE/ha from 35th to 60th day of crop stage.
- Apply B.t.k.I @ 1 kg/ha.
- Application of fungal pathogens like *Beauveria bassiana* or *Neumorea riley* under humid conditions is effective.
- Use 5% neem seed kernel extract (NSKE).
- Trap cropping with crops like tomato, and destroying them when the population is high.

Chemical Control

- The following insecticides are effective against the pest:
- Endosulfan 35 EC 2.5 lit/ha; or

- Quinalphos 25 EC 2.0 lit/ha; or
- Chlorpyriphos 20 EC @ 2 lit/ha; or
- Cypermethrin 10 EC 600-800 ml/ha
- Trizophos 40 EC @ 1.5 lit/ha .

Identification & Monitoring

- This is a major pest, is active from October-November.
- The adult moths are dark with blackish spots on forewings. The margins of hind wings are deeply fringed. Wing span is 8-9 mm.
- The caterpillars are creamy yellow when young and turn pink when grown, 8-10 mm long with distinct brown head.
- Eggs are laid on the underside of tender parts of the plant (shoots, flower buds, leaves and green bolls).

Damage

- The damage is caused by the caterpillars by feeding on the flower buds, panicles and bolls.
- The holes of entry close down by excreta of larvae which are feeding inside the seed kernels. They cut window holes in the two adjoining seeds thereby forming "double seeds" and finally damage them.
- The attacked buds and immature bolls drop off. Lint is destroyed, ginning percentage and oil content are impaired.
- For 65-100 days crop, 5-10% infested fruiting bodies is the ETL.

Cultural Control

- Clean cultivation and destruction of crop residues (fallen leaves, twigs etc.) before the onset of season.
- Plough deeply to expose the hibernating larvae/pupae.
- Avoid late sowing of the crop. Early sowing helps in early maturity facilitating escape.
- Use of tolerant varieties (Khandwa-2, JKH-1, Abadita, LH 900, Sujay and Desi cotton).
- Withholding irrigation water to avoid prolonged late boll production/ formation to reduce the build up of over-wintering population.

Mechanical Control

- Use pheromone traps baited with insecticides to kill the pest/monitor pest population.

Biological Control

- Release of egg parasitoids *Trichogramma chilonis* or *Bracon elechidae* or Elasmus johnstoni or pupal parasitoid Microbracon lefroyi would keep in check the population of PBW.
- Encourage the activities of predators *Chrysoperla carnea* or Scymnus sp. or Triphles tantilus or Pyremotes ventricosus (mite) or release them in the fields.
- Apply bacterial formulations B.t.k. @1 kg/ha.

Chemical Control

- Hot water treatment of the seeds up to 600C kill the hibernating larvae. Treat the seeds with aluminium phosphide.
- Application of insecticides like chlorpyriphos 20 EC or endosulfan 35 EC or triazophos 40 EC @ 2.5 l/ha spray.

Tobacco Caterpillar (*Spodoptera Litura Fab.*)

Identification & Monitoring

- The adult moth is stout with brownish forewings and whitish hind wings.
- The caterpillars are pale green with dark markings initially which later turn dark brown with numerous transverse and longitudinal bands, 25-35 mm long.
- It is found throughout the year.
- Setup pheromone traps to monitor the ETL.

Damage

- The larvae feed gregariously on the under surface of the leaves and skeletonize them leaving only the mid-rib and veins in severe cases.
- They also attack flowers, buds and squares causing considerable loss.

Cultural Control

- Plough deeply to expose the pupae and hibernating larvae.

Mechanical Control

- Mechanical collection when larvae are feeding in groups, i.e., the younger larvae.
- Collection and destruction of egg masses.
- Setup pheromone traps.

Biological Control

- Release egg parasitoids Trichogramma sp. (1.5 lakh/ha) and larval parasitoids Chelonus blackburni or Telenomus remus or Bracon sp.
- Release of predators Chrysoperla carnea @ 50,000/ha.
- Spray Spodoptera NPV @ 250 LE/ha.
- Apply B.t.k. @ 1 kg/ha.

Chemical Control

- Spraying of insecticides endosulfan 35 EC @ 600-750 ml/ha effectively reduces the population.
- Spraying synthetic pyrethroids fenvalerate 20 EC @ 400-500 ml/ha or cypermethrin 10 EC or decamethrin 2.8 EC @ 600-700 ml/ha is also effective.

Identification & Monitoring

- These are major pests
- *E. vitella* is abundant in high rainfall areas and *E. insulana* in areas of scanty rainfall.
- The pest attacks the crop from 35-110 days of age.
- The moths of both the species have wings of about 25 mm. The forewings are grassy green in *E. insulana* and pea green with a wedge shaped white band running from base to outer margin in *E. vitella.*
- Larva is about 20 mm long, spiny, brownish with white streaks dorsally and pale yellow ventrally in *E.vitella* and greenish white with black marks and orange spots on prothorax in *E. insulana.*

Damage

- The caterpillars cause damage by boring into the growing shoots, buds, flowers and bolls.
- The attacked shoots wither, droop and ultimately die, and flowers and buds drop off.
- Infested bolls do not shed, open prematurely and the quality of the lint is spoiled due to rot setting.
- Sometimes pupation takes place in the bolls itself impairing the development of bolls.
- Trap cropping with crops like tomato, and destroying them when the population is high.

Bibliography

Ashraf, Syed Mahboob: *Handbook of Fruit and Vegetable Products*, Agrobios, Delhi, 2008.

Banerjee, S : *Biochemistry and Biotechnology,* Dominant Pub, Delhi, 2009.

Barrington, E. J. W.: *Biochemistry of Primitive Deuterostomians*, London, Academic Press, 1974.

Bhaskar, Harsh: *Basic Facts on Biochemistry*, Campus Books, Delhi, 2010.

Bhatia, A. L.: *Biochemistry and Endocrinology,* Indus Valley, Delhi, 2002.

Bhatnagar, Vasudev: *Cell Science and Technology*, Campus Books, Delhi, 2009.

Bora, K.K.: *Agros Dictionary of Plant Physiology and Biochemistry*, Agrobios, Delhi, 2001.

Bose, N. Sarath Chandra: *Biochemistry: A Practical Manual,* Pharma Med Press, Delhi, 2010.

Bose, Sarath Chandra: *Biochemistry: A Practical Manual*, Pharma Med Press, Delhi, 2010.

Chanderlekha Goswami: *Biochemistry and Instrumentation*, Manglam Pub, Delhi, 2011.

Chauhan, B.S.: *Principles of Biochemistry and Biophysics,* Laxmi Publications, Delhi, 2008.

Chauhan, Sushil: *Experiments in Biochemistry,* Campus Books, Delhi, 2010.

Collymore L.: *Fruit Production in Barbados*, Port of Spain, Trinidad and Tobago, 1996.

Coultate, T. P.: *Food: The Chemistry of its Components,* London: The Royal Society of Chemistry, 1988.

Daphne C. Elliott: *Biochemistry and Molecular Biology*, Oxford University Press, Delhi, 2005.

Dar, Ghulam Hassan: *Soil Microbiology and Biochemistry,* New India Publishing Agency, Delhi, 2010.

Dutta, Chiranjib: *Fundamentals of Biochemistry: A Practical Approach*, Kanishka, Delhi, 2005.

Elliott, B N: *Biochemistry and Molecular Biology*, Oxford University Press, Delhi, 2005.

Florkin, M.: *Comprehensive Biochemistry: Amsterdam*, Elsevier, 1975.

Goel, A K: *Basic Concept of Organic Chemistry*, Pearl Books, Delhi, 2008.

Golakiya, B A: *DNA: A Bridge Between Biochemistry and Biotechnology,* New India Publishing Agency, Delhi, 2008.

Hemantaranjan, A L: *Developments in Physiology, Biochemistry and Molecular Biology of Plants*, New India Pub, Delhi, 2005.

Jasra, O.P.: *Environmental Biochemistry*, Sarup & Sons, Delhi, 2002.

Khanna, V K: *Objective Genetics, Biotechnology, Biochemistry and Forestry,* I.K. International Publishing House, Delhi, 2008.

Mido, Y.: *An Introduction to Nuclear Chemistry*, Discovery, Delhi, 1995.

Narvekar, Raghunath: *Handbook of Biochemistry,* Adhyayan, Delhi, 2008.

Prakash, M: *Insect Biochemistry,* Discovery Pub, Delhi, 2008.

Prasad, G.P.: *Biochemistry of Proteins*, Discovery, Delhi, 2010.

Purohit, S.S.: *Biochemistry: Fundamentals and Applications,* Agrobios, Delhi, 2009.

Rastogi, Arun: *An Introduction to Practical Biochemistry*, Anmol, Delhi, 2010.

Sangha, S P S: *Basic Techniques in Biochemistry and Molecular Biology,* I K International, Delhi, 2009.

Smriti Ganguly: *Biochemistry of Biomolecules*, Pearl Books, Delhi, 2007.

Srivastava, Seema: *Plant Physiology and Biochemistry,* Campus Books, Delhi, 2009.

Tripathi, Pramila: *Botanical Pesticides in the Management of Post-harvest Fruit Diseases*, Daya, Delhi, 2005.

Whyman, Robin: *Applied Organometallic Chemistry and Catalysis*, Oxford University Press, Delhi, 2005.

Index

N

O

P

R

S

V

W

❑❑❑

Encyclopaedia of

Biochemistry and Molecular Biology of Plants

Encyclopaedia of
Biochemistry and Molecular Biology of Plants

Volume 2

Editor:
K.N.P. Singh

AGROTECH PRESS

JAIPUR ▪ NEW DELHI

AGROTECH PRESS

Regd. Off. "Bony Residency", Gate No. 2
Opp. Tilak Public School, Vishweswariya Nagar,
Gopalpura Road, Jaipur - 302018, India
Ph: 0141 - 2761280, Telefax: 2761381
E-mail : agrotechpress@gmail.com

Delhi Office: 102, 1st Floor, Satyam House,
4327/3, Ansari Road, Darya Ganj,
New Delhi - 110002, India
Ph: 011-45652440
visit us: www.agrotechpress.com

Distributed By
KEYA INTERNATIONAL PRESS LLP
E-mail: keyainternationalpressllp@gmail.com

Published in India, 2024

ISBN : 9789384568221 (4 Vols. Set)

Typeset by Thomson Press

Printed and bound in India by Thomson Press Ltd.,
New Delhi

Contents

Preface

Phytochemistry is in the strict sense of the word the study of phytochemicals. These are chemicals derived from plants. In a narrower sense the terms are often used to describe the large number of secondary metabolic compounds found in plants. Many of these are known to provide protection against insect attacks and plant diseases. They also exhibit a number of protective functions for human consumers. Phytochemistry can be considered sub-fields of Botany or Chemistry. Activities can be led in botanical gardens or in the wild with the aid of Ethnobotany. The applications of the discipline can be for Pharmacognosy, or the discovery of new drugs, or as an aid for plant physiology studies Phytochemistry is widely used in the field of Chinese medicine especially in the field of herbal medicine. Phytochemical technique mainly applies to the quality control of Chinese medicine, Ayurvedic medicine (Indian traditional medicine) or herbal medicine of various chemical components, such as saponins, alkaloids, volatile oils, flavonoids and anthraquinones. In the development of rapid and reproducible analytical techniques, the combination of HPLC with different detectors, such as diode array detector (DAD), refractive index detector (RID), evaporative light scattering detector (ELSD) and mass spectrometric detector (MSD), has been widely developed.

In most cases, biologically active compounds in Chinese medicine, Ayurvedic Medicine (Indian Herbal Medicine) or herbal medicine have not been determined. Therefore, it is important to use the phytochemical methods to screen and analyze bioactive components, not only for the quality control of crude drugs, but also for the elucidation of their therapeutic mechanisms. Modern pharmacological studies indicate that binding to receptors or ion channels on cell membranes is the first step of some drug actions. A new method in phytochemistry called biochromatography has been developed. This method combines human red cell membrane extraction and high performance liquid chromatography to screen potential active components in Chinese medicine. Biochemistry probes into the chemical changes involved right from birth, followed by the changes during growth and development and until death and dissolution of living organisms. Biochemistry encompasses a study of chemical nature of living organisms the relationship of the living to their environment the processes by which an exchange of chemical substances takes place between the living organisms and the environment the processes by which the absorbed materials are utilized for synthetic reactions leading to growth and replenishment of tissues and multiplication of the cell and the species the metabolic breakdown of the materials to supply energy for all the above and the mechanisms which regulate with precision all these processses. All these studies come under the purview of biochemistry. Correlation of biological functions and molecular structures is the central theme of biochemistry. Biochemistry includes various aspects of organic chemistry in organic chemistry, physical chemistry, physics, biology and other basic disciplines. It is also interrelated with physiology, microbiology, medicine and agriculture. Production of medicinal chemicals in plants may be modified by overexpression, antisense expression, or cosupression of biosynthetic genes. Transgenic genes could also be introduced to modify existing pathways. Mutational alterations and analysis could also be performed to dissect out basic components of metabolic pathways. The application of molecular approaches with medicinal plants would also benefit from the development of cell, tissue, and organ culture systems for in vitro growth and regeneration of medicinal plants. In addition, such tissue culture systems could also prove useful for large-scale biotechnological production of medicinal plant phytochemicals. Overall, metabolic engineering could be useful for modifying or enhancing synthesis of valuable therapeutic agents present in medicinal plants. However, as the beneficial actions of medicinal plants can be related to combinations of phytochemicals acting collectively or synergistically, alteration of single phytochemical components could potentially affect the efficacy of a phytomedicine. In this respect, any work on modification of the phytomedicinal chemical composition of a medicinal plant through molecular methods would need to be conducted in conjunction with pharmacological studies on drug effectiveness.

The book is exceptional in its organization with three major characteristics of plant system i.e. Plant Physiology, Biochemistry and Molecular Biology been provided under one canopy.

—Editor

1

Respiration and Lipid Metabolism

Although it is possible to measure respiration by intact tissues it is often useful to be able to isolate uncontaminated, intact and functional mitochondria. Here, we will consider some of the principles for isolating cell organelles. This discussion is not meant to be exhaustive nor will it be a cookbook with recipes to be followed step-by-step.

Choice of Material

If the aim of the experiment is to investigate a specific physiological problem, then the plant material is often well-defined. However, in many cases there is some flexibility in the choice of material and it is therefore wise to follow certain rules of thumb:

- Choose a tissue that is easy to obtain in relatively large amounts (at least 100–200 g fresh weight), such as tubers or hypocotyls.
- Choose a tissue for which a good protocol is available. It can take quite a lot of time and effort to optimize an isolation protocol.

Note that the isolation of mitochondria from green tissues presents special problems because of the large amount of chloroplasts. Roots are a notoriously difficult tissue for isolation of mitochondria.

Disruption of the Tissue

The aim is to rupture as many of the cells in the tissue as possible in order to release their cytoplasm with the mitochondria. Whereas many mammalian tissues are easy to homogenize, plant tissues are much tougher mainly because of the cell walls. There are a number of standard homogenizers, including mixers with rotating knives or razor blades, juice extractors, ball mills, and mortar and pestle; the most suitable method for the specific plant material should be chosen. This may involve some initial optimization experiments.

Homogenization Medium

The medium for homogenizing rat liver only contains an osmoticum and a buffer. That will not work with plant tissues because of the disruption of the large central vacuole (absent in mammalian cells), which releases its fairly acidic content often containing phenols and other potentially harmful substances.

Let us consider the components of the optimal homogenization medium for isolating intact and functional organelles from plant material, one by one. Osmoticum. This is required to prevent the mitochondria from losing their matrix content because of swelling and bursting. For the isolation of plant mitochondria, there is a tradition of using sucrose or mannitol at 0.2–0.5 M; whereas 0.15 M KCl is standard for the isolation of mammalian mitochondria. In neither case does this mimic precisely the conditions in the cytosol, but it works.

Buffer. The pH in the cytosol of plant cells is around 7.5, and the homogenate should be buffered at pH 7–8. A buffer should therefore be included and a plethora of different buffers are available. Choose an inexpensive one with a pKa of 7–8 (e.g., 4-morpholinopropane sulfonic acid = MOPS) and be sure to use enough to counterbalance the acidic vacuolar content. The tissue type, the tissue/medium ratio, and the buffer concentration determine whether the acidic vacuolar content is properly neutralized. Obviously, special tissues like lemons or CAM plants harvested late at night or in the early morning, require special treatment!

Metal ion chelators. EDTA is usually included to bind mainly Ca^{2+} and Mg^{2+}, which may otherwise activate proteolytic enzymes. Antioxidants. Antioxidants (reductants) such as ascorbic acid, dithiothreitol, or cysteine are used mainly to protect sulfhydryl groups on proteins against oxidation.

Bovine serum albumin (BSA). This protein from the blood of cows is usually included to protect the mitochondria: BSA binds fatty acids (its function in the blood) released by the action of phospholipases on phospholipids. BSA may also bind phenolics that are released from the vacuole. Both of these groups of compounds can otherwise be troublesome later. In addition, BSA may prevent proteases from degrading mitochondrial proteins by serving as an alternative substrate present at high concentrations.

Polyvinylpyrrolidone (PVP). A polymer that is often included, especially when using leaf material, to adsorb phenolic compounds.

Filtration

To remove larger tissue fragments, cell walls, and often starch grains, the homogenate is always filtered, usually through a nylon net with a mesh size of 30–100 μm.

Purification of Mitochondria

The first step of the purification is usually differential centrifugation. This means that the filtered homogenate is centrifuged first at low speed for a short time (e.g., 3000 g for 5–10 min.) and the pellet (containing cell wall fragments, starch grains, nuclei, intact plastids) is discarded. The supernatant is then centrifuged faster and for a longer time (e.g., 10,000 g for 10–20 min.) and the pellet (termed *crude mitochondria*) is collected. It contains intact and damaged mitochondria as well as peroxisomes and plastid membranes. The supernatant is discarded. As soon as the mitochondria have been pelleted and thus removed from potentially damaging soluble substances in the homogenate, the medium can be greatly simplified. Usually the so-called wash medium contains only osmoticum and weak buffer (no need for a strong buffer since there is no vacuole content to neutralize). However, it may be wise to include EDTA to bind divalent cations as well as BSA to adsorb any fatty acids released by the action of phospholipases. The resuspended crude mitochondria are often diluted to 50–100 ml with wash medium and subjected to another round of slow and fast centrifugation to remove more contaminants.

The aim of the next step is to separate the intact mitochondria from the contaminants. This usually involves a density gradient centrifugation with Percoll either alone (e.g., Struglics et al. 1993) or together with other dense materials (e.g., Day et al. 1985). Percoll (colloidal silica coated with PVP) is inert and osmotically inactive. It has the further advantage that a Percoll gradient is self-generating because the Percoll microspheres pellet very slowly during the centrifugation. The crude mitochondria are layered on top of the wash medium containing the appropriate concentration of Percoll (usually 25–30%). If the Percoll concentration and the speed and duration of the centrifugation have been chosen correctly, the intact mitochondria will form a distinct band, well away from the contaminants. This band can be collected, washed free of Percoll by repeated dilution and centrifugation, and either used immediately, or frozen with 5% (v/v) dimethylsulfoxide (to preserve integrity) in liquid nitrogen and stored at -80°C until use. This preparation can be stable for months, retaining the ability to form a proton electrochemical gradient over the inner membrane.

The yield is 1–50 mg mitochondrial protein per 100 g fresh weight, depending on the tissue. This represents only a small fraction of the total amount of mitochondria in the tissue. The purification of mitochondria from green leaves presents special problems and a good protocol is described in Day et al. (1985). There is, at present, no good protocol for the isolation of both pure and functional mitochondria from Arabidopsis leaves (Hausmann et al. 2003; Keech et al. 2005), but a protocol for the purification of mitochondria from Arabidopsis cell cultures is given in Kruft et al. (2001). Finally, Boutry et al. (1984) demonstrate how purified mitochondria can be isolated from small amounts of plant tissue (down to 0.5 g fresh weight).

Mitochondria are rod-shaped organelles that can be considered the power generators of the cell, converting oxygen and nutrients into adenosine triphosphate (ATP). ATP is the chemical energy "currency" of the cell that powers the cell's metabolic activities. This process is called aerobic respiration and is the reason animals breathe oxygen. Without mitochondria (singular, mitochondrion), higher animals would likely not exist because their cells would only be able to obtain energy from anaerobic respiration (in the absence of oxygen), a process much less efficient than aerobic respiration. In fact, mitochondria enable cells to produce 15 times more ATP than they could otherwise, and complex animals, like humans, need large amounts of energy in order to survive.

The number of mitochondria present in a cell depends upon the metabolic requirements of that cell, and may range from a single large mitochondrion to thousands of the organelles. Mitochondria, which are found in nearly all eukaryotes, including plants, animals, fungi, and protists, are large enough to be observed with a light microscope and were first discovered in the 1800s. The name of the organelles was coined to reflect the way they looked to the first scientists to observe them, stemming from the Greek words for "thread" and "granule." For many years after their discovery, mitochondria were commonly believed to transmit hereditary information. It was not until the mid-1950s when a method for isolating the organelles intact was developed that the modern understanding of mitochondrial function was worked out.

The elaborate structure of a mitochondrion is very important to the functioning of the. Two specialized membranes encircle each mitochondrion present in a cell, dividing the organelle into a narrow intermembrane space and a much larger internal matrix, each of which contains highly specialized proteins. The outer membrane of a mitochondrion contains many channels formed by the protein porin and acts like a sieve, filtering out molecules that are too big. Similarly, the inner membrane, which is highly convoluted so that a large number of infoldings called cristae are formed, also allows only certain molecules to pass through it and is much more selective than the outer membrane. To make certain that only those materials essential to the matrix are allowed into it, the inner membrane utilizes a group of transport proteins that will only transport the correct molecules. Together, the various compartments of a mitochondrion are able to work in harmony to generate ATP in a complex multi-step process.

Mitochondria are generally oblong organelles, which range in size between 1 and 10 micrometers in length, and occur in numbers that directly correlate with the cell's level of metabolic activity. The organelles are quite flexible, however, and time-lapse studies of living cells have demonstrated that mitochondria change shape rapidly and move about in the cell almost constantly. Movements of the organelles appear to be linked in some way to the microtubules present in the cell, and are probably transported along the network with motor proteins. Consequently, mitochondria may be organized into lengthy travelling chains, packed tightly into relatively stable groups, or appear in many other formations based upon the particular needs of the cell and the characteristics of its microtubular network.

The mitochondrion is different from most other organelles because it has its own circular DNA (similar to the DNA of prokaryotes) and reproduces independently of the cell in which it is found; an apparent case of endosymbiosis. Scientists hypothesize that millions of years ago small, free-living prokaryotes were engulfed, but not consumed, by larger prokaryotes, perhaps because they were able to resist the digestive enzymes of the host organism. The two organisms developed a symbiotic relationship over time, the larger organism providing the smaller with ample nutrients and the smaller organism providing ATP molecules to the larger one. Eventually, according to this view, the larger organism developed into the eukaryotic cell and the smaller organism into the mitochondrion.

Mitochondrial DNA is localized to the matrix, which also contains a host of enzymes, as well as ribosomes for protein synthesis. Many of the critical metabolic steps of cellular respiration are catalysed by enzymes that are able to diffuse through the mitochondrial matrix. The other proteins involved in respiration, including the enzyme that generates ATP, are embedded within the mitochondrial inner membrane. Infolding of the cristae dramatically increases the surface area available for hosting the enzymes responsible for cellular respiration.

Methods for Establishing Purity, Intactness, and Functionality of the Fractions

It is often important to determine the degree of purity of the mitochondria and their subfractions. The intactness of the mitochondria is usually established with enzyme assays that take advantage of the fact that an intact membrane is not permeable to certain molecules.

For instance, the outer mitochondrial membrane is impermeable to cytochrome *c*, a 12.5 kDa protein that is too large to pass through the outer membrane pores, and NADH cannot pass through an intact inner membrane.

Functionality is usually tested using an oxygen electrode. If the mitochondria show high rates of oxygen consumption, high respiratory control ratios (>3), and ADP:O ratios close to the theoretical values they are in very good shape. Note that, for mitochondria with an appreciable alternative oxidase activity (measured in the presence of the cytochrome oxidase inhibitor, KCN), the respiratory control ratio and ADP:O ratio should be measured in the presence of an alternative oxidase inhibitor such as salicylhydroxamic acid.

Mitochondrial Activities can be Determined in Situ

It is relatively straightforward to measure the activity of enzymes in the soluble matrix fraction isolated from mitochondria. However, this fraction is very diluted compared to the matrix inside intact mitochondria (300–500 mg protein/ml!) and protein–protein interactions may have been lost. When measuring the activity of enzymes in fractions containing membranes—intact mitochondria and submitochondrial particles—a detergent can often be used to remove the permeability barrier that the inner membrane constitutes.

However, it is also possible to use the ionophore alamethicin to permeabilize the inner membrane to molecules <1000 Da, which makes it possible to measure the activity of matrix and inner membrane enzymes in their native proteinaceous environment. Using alamethicin, it is even possible to measure the activity of cytosolic and mitochondrial enzymes inside intact cells since the ionophore also permeabilizes the plasma membrane.

The Q-Cycle Explains How Protons Can Be Pumped across the Inner Mitochondrial Membrane

The electron transport chain (ETC) in the mitochondrial inner membrane of most eukaryotes consists of four complexes, complex I to complex IV. Complexes I, III and IV pump protons over the inner mitochondrial membrane when transporting electrons. For each electron pair (corresponding to one NADH or one oxygen atom), four protons are pumped by complex I, four by complex III, and two by complex IV. The detailed three-dimensional structures have been determined for the mammalian complexes III and IV, but not for complex I. The structural information has improved the understanding of the electron paths of complexes III and IV, but the proton-pumping function of the latter is still not fully understood. For complex I, neither electron nor proton paths are known in any detail.

Ubiquinone (Q), also called coenzyme Q, is a mobile electron and proton carrier in the inner mitochondrial membrane, structurally and functionally similar to plastoquinone in the thylakoid membrane of plastids (Figure 1). The ability of Q to carry both electrons and protons—and that it can accept and donate them step-wise—is utilized in electron transport and proton pumping (Figure 2) (Nicholls and Ferguson 2002).

To be reduced, it can receive two electrons from complex I or succinate dehydrogenase and take up two protons from the matrix, forming QH_2. However, at the oxidizing side (complex III), there is actually both oxidation and reduction taking place. A QH_2 will be oxidized to Q at the outer Q-site, releasing two electrons. One of these electrons goes to cytochrome *c*, whereas the other is recycled to reduce an oxidized Q in the inner Q-site—one step to becoming a semiquinone.

The next time a QH_2 is oxidized at the outer Q-site, the semiquinone in the inner site will receive a second electron, becoming a QH_2. So, from two QH_2, we have made one Q and one QH_2, which again can be oxidized. In each cycle (i.e., for each electron passing to cytochrome *c*), two protons are taken up from the matrix and two protons released to the intermembrane space. So, the Q-cycle mediates the pumping of an extra proton per electron transported, increasing energy conservation.

If we look at the whole reaction from succinate to cytochrome *c*, we see that for each succinate, four protons are transported from the matrix to the intermembrane space. However, at the same time, the two electrons are also transported out. So, this corresponds to the net pumping out of two positively charged species. Since, energetically, the mitochondrial membrane potential (which determines the energetic cost of transporting a positive charge out) is much larger than the pH gradient (which determines the cost for electroneutral export of a proton), the complex III proton pumping is less energy-demanding than, for example, complex I, which pumps four protons in their completeness.

Multiple Energy-Conservation Bypasses in Oxidative Phosphorylation of Plant Mitochondria

The electron transport chain (ETC) in the mitochondrial inner membrane of most eukaryotes consists of four large protein complexes, complex I to complex IV. The electron transport activity of complexes I, III and IV (the latter two, often jointly called the cytochrome pathway) is coupled to extrusion of protons across the inner mitochondria.

The electrochemical proton gradient that is formed and maintained in the process is used by the ATP synthase (also called complex V) to make ATP (Figure 1). The plant ETC, however, is highly branched. As a consequence, the plant oxidative phosphorylation system has several alternative pathways of electron transport via type II NAD(P)H dehydrogenases and the alternative oxidase (AOX), and additionally, an alternative proton transporter called the uncoupling protein (UCP). These proteins mediate bypasses around the proton-translocating multi-protein complexes (I–V): type II NAD(P)H dehydrogenases bypass complex I, AOX bypasses complexes III and IV, and UCP bypasses the ATP synthase (Siedow and Umbach 1995; Vanlerberghe and McIntosh 1997; Møller 2001; Vercesi 2001; Millenaar and Lambers 2003; Fernie et al. 2004; Rasmusson et al. 2004).

The type II NAD(P)H dehydrogenases and the AOX transport electrons without pumping protons, whereas the UCP allows proton flow from the intermembrane space to the matrix without ATP synthesis. The consequence of this is that flux through the energy conservation bypasses will neither contribute to respiratory ATP production, nor will they be controlled by cellular adenylate status. The special features of these energy bypass proteins and discuss the consequences of their existence to the plant.

NAD(P)H Dehydrogenases in Plant Mitochondria

Except for yeast, the ETC of all eukaryotes—including plants—contains a proton-pumping NADH dehydrogenase, complex I, which is inhibited specifically by rotenone. Additionally, plant, fungal and protist mitochondria contain different setups of rotenone-insensitive enzymes capable of oxidizing NADH and/or NAD(P)H and passing the electrons on to ubiquinone.

These NAD(P)H dehydrogenases are classified as type II, whereas complex I belongs to type I. The inner mitochondrial membrane in plants contains such NAD(P)H dehydrogenases, facing either the matrix or the intermembrane space, thus oxidizing NAD(P)H either from the matrix or the cytoplasm.

The outer membrane of all mitochondria contains NADH-ferricyanide and NADH-cytochrome *b* reductase activities that are insensitive to rotenone and to the complex III inhibitor, antimycin A.

There are Three NAD(P)H Dehydrogenases on the Inner Matrix Surface of the Inner Membrane

Much of the information about the number of matrix-facing NAD(P)H dehydrogenases has come from experiments using inside-out inner membrane vesicles. Such vesicles have the matrix surface facing the medium and the external NAD(P)H dehydrogenases hidden inside the vesicles and unable to interfere with the assays.

Recently, this method has been complemented with a technique where one uses a peptide called alamethicin to make a channel for the substrate through the inner membrane. Experiments with these methods show the presence of three NAD(P)H dehydrogenase activities on the matrix side of the inner membrane:

- Complex I pumps protons, is Ca^{2+}-independent and is inhibited by rotenone and diphenylene iodonium (DPI). Complex I has a high affinity for NADH.
- ND_{in}(NADH) does not pump protons, is Ca^{2+}-independent and DPI insensitive. The enzyme has a tenfold lower affinity for NADH than complex I.
- ND_{in}(NADPH) does not pump protons, is Ca^{2+}-dependent and very sensitive to DPI.

Because of the contrasting affinities of complex I and ND_{in}(NADH) for NADH, only complex I is engaged at low matrix concentrations of NADH. However, when the NADH concentration in the matrix increases (e.g., if the ETC is restricted by lack of ADP in the cell), ND_{in}(NADH) can become engaged. When electrons are transported through the type II NAD(P)H dehydrogenases, only the two proton pumping sites of the cytochrome pathway are involved in the flow of electrons from matrix NADH to oxygen. As a consequence, the theoretical ADP/O yield is 1.5 unless the AOX or the UCP are active, in which case it will be even lower.

The External NAD(P)H Dehydrogenases are Ca^{2+}-Dependent

Plant mitochondria can oxidize cytosolic NADH and NAD(P)H directly via external dehydrogenases. The NADH and NAD(P)H dehydrogenase activities are catalyzed by separate enzymes. None of the external dehydrogenases pumps protons, which means that, at most, two sites of proton-pumping are involved when the electrons move from the dehydrogenases to oxygen. So, as for ND_{in}(NADH), the theoretical maximum ADP/O yield is 1.5.

The oxidation of exogenous NADH or NAD(P)H by most plant mitochondria requires Ca^{2+}, and as a result, is inhibited by EDTA and EGTA, chelators of di-and trivalent cations. *In vitro*, the Ca^{2+}concentration needed for half-maximal activity of NADH oxidation is 0.1–0.5 μM, depending on ionic environment and respiratory state This is in the same range as the 0.1–0.2 μM free Ca^{2+} thought to be present in the cytosol of a resting plant cell. Therefore, the external NAD(P)H oxidation may become activated by the increased Ca^{2+}-concentration associated with cellular excitation due to external conditions.

Mammalian mitochondria lack external NAD(P)H dehydrogenases. Instead, they have a Ca^{2+}-dependent glycerol-3-phosphate dehydrogenase on the outer surface of the inner membrane, which donates electrons directly to the ubiquinone pool.

The dihydroxyacetone phosphate formed is converted back into glycerol-3-phosphate by a cytosolic NAD-dependent glycerol-3-phosphate dehydrogenase. In this way, the two low-molecular-weight compounds, glycerol-3-phosphate and dihydroxyacetone phosphate, shuttle redox equivalents from the cytosol, across the outer membrane and to the inner membrane to yield the same amount of ATP as is produced by the plant mechanism. Yeast have both external NADH dehydrogenases and a glycerol-3-phosphate shuttle, and since a mitochondrial glycerol-3-phosphate dehydrogenase has been found in plant mitochondria, plants may also have both systems.

Another way to oxidize cytosolic NAD(P)H that is found in mammalian mitochondria is to use the malate/aspartate shuttle to tranfer reductant (a collective term for high-energy electrons carried by, for example, NAD(P)H, ferredoxin or an organic acid redox couple like malate/oxaloacetate) into the matrix as NADH that can be oxidized by the internal NAD(P)H dehydrogenases. However, it is not clear whether this shuttle can work for import of cytosolic reductant into plant mitochondria. In the opposite orientation, NADH is shuttled from the matrix to the cytosol via the malate/oxaloacetate shuttle. In this way, the external NADH dehydrogenase can also work as a bypass around complex I. Support for this bypass has been seen in the form of an upregulation of external NADH dehydrogenase activity in mutants where complex I or an internal type II NAD(P)H dehydrogenase has been inactivated.

The Molecular Identities of the Rotenone-Insensitive NAD(P)H Dehydrogenases

The type II NADH dehydrogenases from *E. coli* and yeast are small, single polypeptide proteins of around 50 kDa. They oxidize only NADH and contain FAD as an electron-accepting prosthetic group. Attempts to purify the type II NAD(P)H dehydrogenases have been unsuccessful in identifying the plant proteins. However, the characterization of potato homologues to the type II NADH dehydrogenases in *E. coli* and yeasts have provided candidate plant proteins to try and assign to the different reported activities.

One protein, NDA1, is internal and the other, NDB1, external of the inner mitochondrial membrane. The latter was shown to be an external calcium-dependent NAD(P)H dehydrogenase. Like a related fungal NAD(P)H dehydrogenase, NDB1 contains a motif for Ca-binding, indicating that in the activation of plant and fungal NAD(P)H dehydrogenase by Ca^{2+}, the ion binds directly to the enzyme. In Arabidopsis, a weed for which the nuclear genome sequence is known, small families of both NDA-type and NDB-type proteins are encoded by nuclear genes. One of the NDA-type proteins has been shown to be an internal NAD(P)H dehydrogenase. Additionally, a third type of type II NAD(P)H dehydrogenase was found in Arabidopsis, the NDC1. This protein is most closely related to cyanobacterial enzymes, suggesting that the gene entered the eukaryotic domain with the plastid, later to be transferred to the nucleus.

The Alternative Oxidase

The presence of the AOX is one of the features that sets plant (and fungal) mitochondria apart from mammalian mitochondria. This enzyme has therefore received a lot of attention from plant scientists and we now have a fairly good understanding of the regulation of its structure and activity. The AOX is a quinol-oxygen oxidoreductase and it does not pump protons. The oxygenase transfers electrons from ubiquinol (the reduced form of ubiquinone) to oxygen and generates water as the end-product of the reaction, so four electrons must be transferred to oxygen.

The precise mechanism is not known, but the protein has been shown to contain a catalytical di-iron centre. The functional form of the enzyme is a dimer with the two polypeptides either covalently or non-covalently bound to each other (Figure 2). The predicted structure places the protein embedded into the inner leaflet of the inner mitochondrial membrane, with the active site and the regulatory cysteines exposed to the matrix. When electrons from NADH oxidation flow through the AOX, at least two of the three sites of proton translocation are bypassed, so energy conservation in the form of ATP is much smaller when the AOX is active. The activity of the AOX must therefore be carefully regulated. There are often several AOX isoforms in each plant species—for example, five genes encoding AOX proteins are present in Arabidopsis. The relative amounts of three isoforms of AOX vary with development, and cellular redox status, as well as external factors.

The AOX is present in the inner membrane as a dimer in which the two subunits can be covalently linked through a disulfide bridge between cysteine residues. The reduced non-covalently linked dimer is the active form, and this form appears to be the predominant one *in vivo*. It was recently shown that a mitochondrial thioredoxin system regulates AOX. Thioredoxins regulate enzymes by reducing disulfide bridges, and it is likely that the regulation involves the redox-active cysteines. The presence of thioredoxin regulation connects the AOX activation level to the mitochondrial NAD(P)H pool and the oxidation of citric acid cycle metabolites by NADP-utilizing enzymes.

The activity of AOX is also stimulated by 2-oxo organic acids—in particular, pyruvate, which activates by interacting with the same cysteine that is involved in dimerization. This activation may be a feed-forward mechanism that upregulates the capacity of the ETC when the glycolytic end-product is available. Both the cytochrome pathway and the AOX pathway use ubiquinol as the substrate. In the absence of pyruvate, the cytochrome pathway is saturated at 40% Q_{red}/Q_{tot} (40% of the UQ is reduced), at which point the AOX pathway is not engaged at all. The AOX pathway starts accepting electrons at 50% reduction and is not fully saturated at 80%.

In the presence of pyruvate, the AOX becomes engaged at a much lower Q_{red}/Q_{tot} (20%), whereas the cytochrome pathway is unaffected. It is puzzling, however, that the normal *in vivo* concentration of pyruvate appears to be higher than the concentration of pyruvate required to give maximal stimulation of AOX activity *in vitro*. It thus appears that there is little regulation of AOX activity via pyruvate under physiological conditions.

AOX Activity Can Be Measured In Vivo

For many years, benzhydroxamic acid derivatives,, were used to quantify the capacity and activity of the AOX pathway in both isolated mitochondria and intact tissues. AOX capacity was determined from the rate of respiration in the presence of KCN (which blocks the cytochrome pathway), and corrected for residual respiration in the presence of both KCN and SHAM. AOX *in vivo* activity was determined from the decrease in the rate of respiration when SHAM was added in the absence of KCN. The assumption was that the cytochrome pathway was always used first and that the AOX would be engaged only when the cytochrome pathway was saturated. However, as we have seen above, this assumption is not correct. The consequence of this is that *in vivo* activity of AOX cannot be measured using inhibitors alone.

A better method for measuring AOX *in vivo* activity is now available. It makes use of the fact that cytochrome *c* oxidase and AOX discriminates differently between ^{18}O relative to ^{16}O. The so-called oxygen discrimination of the sample (tissue or mitochondria) can be measured by mass spectrometry: (1) with KCN to resolve the oxygen discrimination by AOX; (2) with SHAM to resolve the oxygen discrimination by cytochrome *c* oxidase; and (3) without inhibitors to resolve the discrimination value resulting from the activity of both enzymes.

From these values, the percentage contribution by AOX and cytochrome *c* oxidase to the total uninhibited rate of respiration can be calculated. An example of the use of the oxygen discrimination technique with an intact tissue is given below. A remaining problem in the estimation of AOX activity *in vivo* is that measurements can only be made in darkness (to avoid interference from photosynthetic oxygen exchange).

Plant Mitochondria also Contain an Uncoupling Protein

Mammalian mitochondria contain neither type II NAD(P)H dehydrogenases nor AOX. However, they have a UCP that increases the proton permeability of the inner mitochondrial membrane and, in that way, dissipates the proton gradient. This is another mechanism for reducing the ATP production and increasing heat production, for example, for thermoregulation. Surprisingly, plant mitochondria also contain a UCP. We do not know for certain why plant mitochondria should require two different mechanisms for achieving the same end result. The UCP is related to anion carriers. Its *in vitro* transport activity is dependent on free fatty acids and is inhibited by nucleotides like ATP and GTP but stimulated by superoxide. The latter result indicates that the UCP may protect against oxidative stress.

The Physiological Impact of Energy Bypass Proteins

The respiratory rates of some thermogenic flowers are the highest among plants and, in fact, exceed even those of warm-blooded animals. The external NADH dehydrogenase, AOX and UCP appear particularly active in thermogenic flowers where they are primarily responsible for heat production. At the same time it is clear that the respiratory rate of most plant tissues is insufficient to generate an appreciable increase in tissue temperature.

AOX is up-regulated by cold stress, although this is clearly not to heat the tissue. However, during cold stress, reactive oxygen species (ROS) are formed, and also other stress treatments that lead to elevated ROS will induce AOX. ROS (e.g., superoxide and hydrogen peroxide) are formed as by-products of electron transport under aerobic conditions.

ROS can cause damage to proteins, lipids, and DNA and the cell must therefore limit their formation. AOX helps prevent overreduction of the ETC and thus lowers ROS production. Supporting this, Maxwell et al. (1999) demonstrated the importance of the AOX for limiting ROS formation in living plant cells.

Overexpression of AOX lowered the steady state level of ROS, whereas suppression increased it. This result also points to the mitochondrion as an important site of ROS generation in plant cells. With the recent discovery of ROS stimulation of plant UCP and the increased ROS resistance in plants overexpressing UCP, the picture has become more complex. By decreasing the resistance for electron transport through the proton-pumping complexes, UCP activity may, like AOX, lead to decreased overreduction and ROS formation. Another factor speaking in favour of UCP having a role in ROS protection is that the AOX is inhibited by an ROS product. Both type II NAD(P)H dehydrogenases and AOX may act as reductant overflow enzymes by reoxidizing surplus NADH to oxygen. This can prevent inhibition of the citric acid cycle under conditions in which the ADP concentration is low, and thus the proton-pumping complexes are restricted by a high membrane potential. In this way, the citric acid cycle can supply carbon skeletons for biosynthesis, even if the ATP level in the cell is high.

The concept of overflow is also relevant for photorespiratory conditions, under which there is a massive flow of glycine that has to be oxidized by the mitochondria. As much as half of the NADH produced in this process is probably exported, but the other half has to be oxidized promptly. There are several reasons to believe that a reductant overflow via energy bypass proteins is involved. The ND_{in}(NADH) and AOX capacity in leaf mitochondria is higher than in mitochondria from heterotrophic tissue, and both proteins are down-regulated by dark treatment. Consistent with these observations, AOX activity *in vivo* is up-regulated by light (Ribas-Carbo et al. 2000), AOX capacity is induced by greening of etiolated tissue, and expression of the *nda*1 gene [encoding an internal NAD(P)H dehydrogenase] in potato and Arabidopsis is light-dependent. Mutant tobacco plants, lacking the proton-pumping complex I can still survive, apparently aided by the action of ND_{in}(NADH), which oxidizes matrix NADH with a lower ATP yield.

These plants, however, show decreased photosynthetic rates, indicating that the ND_{in}(NADH) cannot fully compensate for the missing complex I. Also, external NAD(P)H dehydrogenases have been suggested to support photosynthesis. These enzymes may reoxidise NAD(P)H that is shuttled out of the chloroplast in order to balance the NAD(P)H to ATP ratio in the stroma. In summary, there are many indications that mitochondria—especially the energy bypass proteins—have roles in supporting photosynthesis, for example by oxidizing excess NAD(P)H. The AOX pathway is involved in the acidification/ deacidification cycle of CAM plants. CAM plants fix CO_2 in the form of malate during the night. This is done by cytosolic PEP carboxylase, and the malate produced is stored in the vacuole. The CO_2 fixed is released during the daytime by malic enzymes, located in cytosol and mitochondria. Robinson et al. (1992) used the oxygen discrimination technique to measure the contribution of the AOX pathway to whole-leaf respiration in the CAM-plant, *Kalanchoë daigremontiana.*

They found that the respiratory rate of the leaves increased by 35% in the early hours of the day. This increase was due to increased AOX activity. A more detailed time course showed that the AOX activity increased two-to threefold during the early day at the same time as the malate was metabolized. As soon as the malate had disappeared by mid-day, the AOX activity returned to the low level observed during the night. A reasonable conclusion is that a substantial amount of the malate from the vacuole is oxidized via malic enzyme in the mitochondria. This generates NADH in the matrix, which is oxidized to a large extent via the AOX. The released CO_2 can then be refixed by the Calvin cycle in the chloroplast.

F_oF_1-ATP Synthases: The Worlds Smallest Rotary Motors

F_oF_1-ATP synthases (also called F-ATPases) are present in the inner membranes of mitochondria, chloroplasts, and bacteria. These large, multisubunit enzymes consist of a water-soluble catalytic complex (F_1) attached to an integral membrane protein complex (F_o) that transports protons across the membrane. The F_1 complex is composed of at least five different types of subunits (three α, three β, one γ, one δ, and one å).

When the F_1 complex is dissociated from the membrane, it is active as an ATPase. In fact, under the appropriate conditions the intact F_oF_1-ATP synthase can run in reverse and act as a proton pump, using the energy of ATP hydrolysis to move H^+ across the membrane. Our understanding of how ATP is synthesized was advanced by Paul Boyer at the University of California, Los Angeles, who proposed the binding-change mechanism for catalysis by F-ATPases. From the detailed understanding of how the ATPases function, he proposed that the binding-change mechanism for ATP synthesis contains three important components:

1. The major energy-requiring step is not the synthesis of ATP from ADP and P_i, but the release of ATP from the enzyme.
2. Substrate is bound and products are released at three separate but interacting catalytic sites, corresponding to the three catalytic subunits (b subunits). Each catalytic site can exist in one of three conformations: tight, loose, or open.
3. The binding changes are coupled to proton transport by rotation of the g subunit. That is, the flow of protons down their electrochemical gradient through the F_0 complex causes the g subunit to rotate.

Rotation of the γ subunit then brings about the conformational changes in the catalytic complex that allow the release of ATP from the enzyme, and the reaction is driven forward. The reverse occurs when the enzyme functions as an ATP-driven proton pump. The first two predictions of the binding-change model are supported by many lines of evidence, mainly kinetic studies, and are now generally accepted. However, the prediction of a rotary mechanism for coupling proton flow to ATP synthesis has been more difficult to demonstrate. Two major breakthroughs have led to confirmation of the third prediction, as well.

The first breakthrough was the determination of the crystal structure of the F_1-part of bovine mitochondrial ATPase by the laboratory of John Walker in Cambridge, England. The crystal structure showed that the three catalytic β subunits differ in their conformations and in the nucleotide bound to them, consistent with the binding-change mechanism. Even more exciting was the discovery that the γ subunit is inserted like a shaft through the centre of the catalytic complex, which consists of three α subunits and three β subunits arranged alternately in a doughnutlike structure. Moreover, the interface between the γ subunit and the α and β subunits is highly hydrophobic.

The hydrophobicity of the interface minimizes the interactions between the subunits, consistent with the rotation of the γ subunit within the hole formed by the catalytic complex. In other words, the γ subunit looks like a molecular bearing lubricated by a hydrophobic interface. Although many questions were answered by elucidation of the crystal structure of the F_1-ATP synthase, the rotational model cannot be tested by a static "snapshot" of the enzyme. Definitive demonstration of rotation requires a video recording of the spinning of the enzyme in real time. But although they are large for proteins, F_1-ATP synthases are still far too small to be visualized in a light microscope.

To visualize the rotation of the enzyme, Masasuke Yoshida and his colleagues at the Tokyo Institute of Technology came up with an ingenious method to make the enzyme much larger than it is. As Figure below shows, they attached an actin filament labelled with a fluorescent dye to the base of the γ subunit using another protein as a "glue." They then attached the F_1 complex upside down to a glass surface.

If the γ subunit rotates with respect to the catalytic complex, the actin filament should swing around with it. Since the filament is very long compared to the ATP synthase (about 1 μm), its rotation should be visible in a fluorescence microscope. In other words, the fluorescently tagged actin filament, which is large enough to visualize in a light microscope, reports the rotation of the γ subunit.

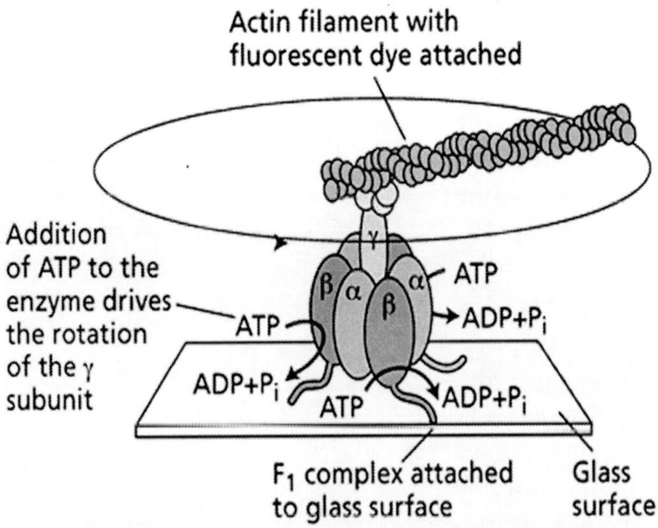

Figure : *A method for visualizing rotation of the g subunit. A fluorescently labelled actin filament was attached to one protruding end of the g subunit. The F_1 complex was then attached upside down to a coverslip. When ATP was added to the coverslip, the actin filament rotated. (After Noji et al. 1997.)*

The results were spectacular! When ATP was added to the modified enzyme, the actin filaments were seen to swing around in a circle at as much as 4 revolutions per second in a fluorescence microscope. To give some idea of scale, if you were a γ subunit, this rotation rate would be equivalent to swinging a several hundred meter-long rod around your head at 4 revolutions per second in water! However, the measured velocity is undoubtedly an underestimation of the actual velocity in vivo, because of the enormous torque required to swing such a large mass. Demonstration of the rotary motion of the γ subunit made it possible to put together a model of how the ATP synthase works.

For their contributions to elucidation of the mechanism of ATP synthesis, Paul Boyer and John Walker shared half the Nobel prize in physiology or medicine in 1997. The other half went to Jens Skou for his pioneering work on the K^+,Na^+-ATP synthase, the mammalian counterpart of the plant plasma membrane H^+-ATP synthase.

Transport Into and Out of Plant Mitochondria

Plant mitochondria must exchange metabolites and information (signal transduction) with the cytosol. To give some central examples, the end products of glycolysis, pyruvate and malate, need to be taken up, and citric acid cycle intermediates exported to be used as building blocks in biosynthetic processes. Likewise, ADP and P_i must be taken up and ATP exported for the mitochondrion to fulfil its role as the powerhouse of the cell. To carry out the transport of this large number of molecules, plant mitochondria possess a family of mitochondrial carrier proteins.

These are transmembrane proteins with a common general structure and they typically transport relatively small charged molecules, like metabolites. In Arabidopsis there are almost 60 members, most of which are not functionally characterized, giving a hint to the complexity of the mitochondrial transport processes. However, mitochondria exchange more than metabolites, and therefore also other types of transporters are involved.

The outer membrane is permeable to molecules of less than 10 kDa, (i.e., all non-polymeric compounds). The permeability results from the presence of pores, which appear to be open in most *in vitro* experiments. Recent studies in animals have characterized pore-forming proteins in the mitochondrial outer membrane, called porins, which appear to regulate transport across the membrane. Plant mitochondria also have porins but their characterization awaits further research. In our discussion here we will focus on the regulation of mitochondrial transport across the inner mitochondrial membrane. We will first consider the relatively few molecules that can move across the inner membrane without the help of a specific carrier protein. We will then discuss the various metabolite transporters and, finally, briefly look at the transport of proteins and RNA.

Small, Neutral, and Hydrophobic Molecules Diffuse across the Inner Membrane

Irrespective of size, molecules that either are charged or are very hydrophilic can only cross a lipid bilayer with the help of a specific protein carrier. Small, *neutral* molecules can diffuse across the lipid bilayer and do not require protein transporters. Among those are O_2 and CO_2, two important molecules for respiration. O_2 is needed on the inner surface of the inner membrane where the active sites of both cytochrome *c* oxidase and the alternative oxidase are located.

The CO_2, produced as an end product in the citric acid cycle, needs to diffuse out of the mitochondrion, ultimately to be lost to the surroundings or re-fixed in the chloroplasts. In addition, however, it is possible that CO_2 can also be transported out of the mitochondrion as bicarbonate (HCO_3^-). Mitochondria are osmotically active and swell in a hypotonic medium. This means that the inner membrane is also permeable to H_2O, but it is not clear whether this occurs by diffusion or if aquaporins may be involved.

For technical reasons, it is of importance that several well-known inhibitors of mitochondrial function can diffuse into mitochondria because they are uncharged and hydrophobic. These are rotenone, the specific inhibitor of complex I, antimycin, the specific inhibitor of complex III, HCN, the protonated form of the cyanide (CN^-) inhibitor of complex IV (and other heme-containing enzymes), and oligomycin, the specific inhibitor of the F_oF_1-ATP synthase.

Transport of Inorganic Ions

Inorganic phosphate ($H_2PO_4^-$) is taken up into the matrix in exchange for a hydroxyl ion (HO^-). The energy for the uptake is supplied by the proton gradient (–pH). We will see below that the phosphate gradient, in turn, drives transport of other metabolites. Several other inorganic ions—Mg^{2+}, Ca^{2+}, K^+, and NH_4^+—are involved in mitochondrial metabolism, but the mechanism by which they pass the inner membrane of plant mitochondria is not well understood. In mammalian mitochondria, uptake of K^+ through a K^+-specific channel, and extrusion via a K^+/H^+ antiporter, is involved in matrix volume regulation and cellular signalling (Garlid and Paucek 2003). A similar system of K^+ transport is present in plant mitochondria.

Both mammalian and plant mitochondria contain an uncoupling protein (UCP). This protein facilitates the movement of protons across the inner membrane and therefore uncouples electron transport and decreases ATP yield.

Transport of Adenylates

One of the important functions of mitochondria is to supply the cell with ATP. To do this, mitochondria have an adenine nucleotide transporter that facilitates the exchange of ADP^{3-} (inward) for ATP^{4-} (outward). This exchange is driven by the membrane potential (inside negative), since there is a net outward movement of one negative charge.

In combination with the phosphate carrier reaction, this means that the uptake of one ADP and one phosphate and the export of one ATP (i.e., the transport event connected to mitochondrial ATP synthesis) costs one vectorial proton—the electrical component for adenylate exchange and the "pH component for phosphate uptake.

Transport of Citric Acid Cycle Intermediates and Related Compounds

Pyruvate and malate, the end products of glycolysis, both have transporters in the inner membrane. Pyruvate is taken up in exchange for a hydroxyl ion, and malate^{2-} in exchange for $H_2PO_4^{2-}$. The transporter for malate also takes succinate^{2-} and is therefore called the dicarboxylate transporter. Malate^{2-} or succinate^{2-} can, in turn, exchange for citrate^{2-} on the tricarboxylate transporter. Malate^{2-} can also exchange for 2-oxoglutarate^{2-} via a special transporter. An interesting transporter, the oxaloacetate transporter, exchanges oxaloacetate^{2-} for malate^{2-} (or citrate^{2-}), which permits the shuttling of reducing equivalents.

The joint activity of these transporters allows plant mitochondria to import or export most citric acid intermediates. Finally, the glutamate/aspartate transporter facilitates the uptake of glutamate. In mammalian mitochondria, glutamate is taken up together with a proton, and it is thus driven by both the electrical and the chemical proton gradient. A similar transporter is present in plant mitochondria, but a proton co-transport has not been shown, so the transport may be passive in plants. Recently, several additional transporters for metabolites have been identified, and it is likely that certain setups of transporters are only expressed under special developmental conditions, like the mobilization of storage nutrients in developing seedlings.

In several cases the involvement of the mitochondrion in cellular metabolism implies that certain metabolites are taken up across the inner membrane, but the transporter has not yet been described. This is the case for the massive flux of glycine into—and CO_2, serine and NH_4^+ out of—mitochondria in C_3 leaves during photorespiration

This is also the case for proline, which accumulates to high concentrations in the cytosol during episodes of plant water stress, and it is then broken down during the rehydration phase by a pathway involving two mitochondrial enzymes.

Transport of Coenzymes

Coenzymes are either synthesized inside the mitochondrion or imported. There is some evidence that, in addition to ATP, at least NAD^+, thiamine pyrophosphate, and coenzyme A are imported across the inner membrane. The final step in ascorbic acid biosynthesis has recently been shown to be on the outer surface of the inner mitochondrial membrane. In view of the accumulating evidence that ascorbate has an important function in removing ROS inside the mitochondrion, we should expect to find an ascorbate transporter in the inner mitochondrial membrane.

The situation with folate (involved in 1C transfers) is the opposite to that with ascorbate: The only site of synthesis of many of the folate metabolites in the plant cell is the mitochondrial matrix. We can therefore expect that these molecules are exported via specific transporters.

Transport of Reducing Equivalents—Metabolic Shuttles

Mammalian mitochondria cannot oxidize cytosolic NADH or NAD(P)H directly, so they make use of "shuttle mechanisms." An exchange of metabolites by inner membrane transporters takes place through these shuttles so that a reduced compound moves into the matrix in exchange for a more oxidized compound moving out into the cytosol. In this way, reducing equivalents are transported into the matrix ultimately to be oxidized by the respiratory chain. The shuttles can also be used to transport reducing equivalents in the opposite direction. Although plant mitochondria can oxidize cytosolic NADH and NAD(P)H directly by the two external NAD(P)H dehydrogenases, they can also use these two metabolic shuttles. The malate/oxaloacetate shuttle uses malate dehydrogenase in the cytosol and in the matrix to catalyze the interconversion of malate (reduced) and oxaloacetate (oxidized).

These two compounds are exchanged on the oxaloacetate transporter. This exchange, however, is not driven by the electrochemical proton gradient across the inner membrane, so it can only move reducing equivalents from a relatively reduced compartment to a relatively oxidized compartment. Since the mitochondrial matrix is more reduced than the cytosol, the oxaloacetate transporter is likely to work primarily in the export of reducing equivalents. Under photorespiratory conditions, large amounts of glycine are oxidized in the mitochondria to give NADH while, at the same time, NADH is required in the peroxisomes to reduce hydroxypyruvate to glycerate. It is thought that as much as 50% of the matrix NADH is exported, probably via the malate/oxaloacetate shuttle.

The more complex malate/aspartate shuttle involves two matrix enzymes and two cytosolic enzymes as well as two inner membrane transporters. In mammals, it is powered by the electrochemical proton gradient in the direction of the import of reducing equivalents. However, since the glutamate/aspartate exchanger may be passive in plants, this shuttle in plants may similarly be independent of the electrochemical proton gradient. If so, it would (like the malate/oxaloacetate shuttle described above) only be able to function as an exporter of mitochondrial reductant.

Transport of Proteins—Protein Import

Plant mitochondria contain their own DNA (mtDNA) and ribosomes and are capable of synthesizing the proteins encoded in the mtDNA. However, plant mtDNA encodes only a few of the proteins found in the mitochondrion. The great majority of the mitochondrial proteins have to be imported from the cytosol where they are synthesized on free ribosomes. These are synthesized with a presequence or targeting peptide in the N-terminal end, which is recognized by receptors on the outer mitochondrial membrane.

The import of proteins requires a complex machinery with two multisubunit protein complexes—one in the outer membrane (TOM, transporter of the outer membrane) and one in the inner membrane (TIM, transporter of the inner membrane). Concomitant with, or immediately afterwards the import of the protein into the correct mitochondrial compartment, the targeting peptide is often cleaved off to yield the mature and active protein.

Transport of RNA

Mammalian mtDNA encodes the full complement of tRNA required for mitochondrial protein synthesis. In contrast, plant mtDNA only encodes some of the tRNAs required for protein synthesis within the mitochondrion. The missing tRNA must be imported by a yet-to-be-elucidated import mechanism.

The Genetic System in Plant Mitochondria has Some Special Features

As in plastids, the genetic system in mitochondria differs both structurally and functionally from the nuclear genome and its gene expression. On the other hand, mitochondria share several genetic features with bacteria. For example, several genes are transcribed together and the ribosomes are of a prokaryotic type. This is in line with the evolutionary origin of the mitochondrion as an endosymbiont of the α-subclass of Proteobacteria, a group including the intracellular pathogen *Ricketsia prowazekii.*

Plants are also distinctly different from other eukaryotes in their mitochondrial DNA (mtDNA) and its expression. For example, the plant mtDNA is very large and highly variable in size between species. The presence of several repeats enables the genome to recombine, changing the gene order or even causing a split into smaller subgenomic molecules, each carrying a subset of the genes. Though the size of mtDNA varies between plants, the number of genes appears to be similar in all plants. In the fully sequenced 367 kilobasepair (kb) mtDNA of *Arabidopsis thaliana,* the known genes (constituting 65 kb) are scattered in a background of noncoding sequence. A similar situation can be found in the sequenced mtDNA of sugarbeet.

Plant mtDNA is quite different from the much smaller (16 kb) single mtDNA molecule of mammals in which every base pair has a function in coding for protein or mediating transcription and replication, and neither introns nor repeats exist. The expression of the mitochondrial genes in plants is carried out by RNA polymerases that are unlike the nuclear polymerases, but related to the enzymes in bacteriophages. For several mitochondrial genes, these phage-type polymerases start up transcription from multiple promoters. Many genes in plant mtDNA are broken up into exons by introns. The introns are removed in the processing of the primary RNA into a mature mRNA that can be translated into the correct polypeptide.

The introns in mitochondria belong to the so-called self-splicing type II group, in which splicing is defined by the intron structure. The splicing can also work in the *trans* mode, in which it can join segments from separate RNA molecules, and thus assemble protein-coding reading frames even if the exons are distributed on different primary transcripts. This allows individual exons of a gene to reside independently in the genome, even in different subgenomic DNA molecules. Because mitochondrial genes are often co-transcribed with unrelated genes into a polycistronic primary transcript, RNA processing also includes endonucleolytic reactions, so that each mature mRNA encodes only one protein. Like mRNAs for nuclear encoded genes, mitochondrial mRNAs are polyadenylated. However, instead of stabilizing (the case for nuclear encoded mRNAs), the mitochondrial mRNA polyadenylation mediates degradation. Instead, stem-loops stabilise the mRNA in plant mitochondria. In plant mitochondria, an additional RNA processing event takes places—RNA editing. In this process, certain cytidines (C) are deaminated into uridines (U). Rarely, U is aminated to C instead.

This means that the coding sequence is changed at the RNA level, in order to restore the correct information for translation where the DNA has diverged. Editing sites mainly occur in DNA encoding protein sequences and functional RNA, and are rare in noncoding mtDNA. In principle, the higher the importance of an individual base, the higher the probability that it constitutes an editing site. Despite the considerable amount of bases that can be edited (up to 15% of the bases of a gene), the functional relevance of this process and the consequent divergence in mtDNA sequence is not understood. Mitochondrial RNA editing has been found in land plants, but not in algae. It also occurs, though to a lesser extent, in plastids.

The mitochondrial genome encodes only a small proportion of the mitochondrial proteins; less than 40 as compared to up to thousands of nuclear-encoded mitochondrial proteins. Through evolution, genes have been transferred from the mitochondrion to the nucleus. This is a process that still takes place, at least in plants. For example, subunit 2 of cytochrome oxidase is encoded by the mitochondrial genome in all investigated eukaryotes except for some legumes and protists. Within the legume family, several intermediates of the gene transfer have been found. For example, a functional gene can be found in the mitochondrial genome and an unexpressed one in the nuclear genome, or vise versa.

In order for a mitochondrial gene to be successfully transferred to the nucleus, the protein-coding frame must be integrated into the chromosome, and several other modifications must be made. For example: (1) a promoter for nuclear transcription must be inserted at the beginning of the gene, (2) the mRNA should acquire sequences specifying the addition of a poly-A tail, and (3) a targeting sequence is necessary for the cytosolically synthesized protein to be imported into the mitochondrion. For proper function, elements regulating the expression in different cell types and under different conditions must also be included. Since the coding sequence for most mitochondrial genes in plants is edited at the RNA level, the mitochondrial genomic sequence will not code for a functional protein if inserted into the nucleus.

Therefore, the DNA that is integrated into the nuclear genome must in most cases be made by reverse transcription from mitochondrial mRNA. Recently, upon sequencing the nuclear genome of *Arabidopsis thaliana*, a copy of most of the mtDNA was found integrated in a nuclear chromosome, indicating that large segments of DNA can move between cellular compartments. However, most of these mitochondrial genes transferred to the nuclear genome cannot produce functional proteins unless bases corresponding to editing sites are corrected by mutation.

However, this is unlikely to take place before other mutations have further disrupted the sequence. As opposed to other mitochondrial genomes, plant mitochondria utilize the universal genetic code. Thus, this factor will not prevent or delay the transfer of genes between plant organelles as it may do in other eukaryotes. The special features of the mitochondrial genetic system in plants also have consequences for the study of mitochondrial genes. Due to RNA editing and *trans*-splicing, a reading frame is more difficult to identify based only on genomic sequence. Also, protein sequences in databases are automatically translated from genomic DNA sequences without compensating for RNA editing in many cases. This protocol would not work for mitochondrial genes.

Does Respiration Reduce Crop Yields?

Plant respiration can consume an appreciable amount of the carbon fixed each day during photosynthesis over and above the losses due to photorespiration. To what extent can changes in a plants respiratory metabolism affect crop yields?

Attempts to establish a quantitative relationship between respiratory energy metabolism and the various processes going on in the cell have led to a break-up of respiration into two components. In growth respiration, reduced carbon is processed to bring about the addition of new biomass. The other component, maintenance respiration, is needed to keep existing, mature cells in a viable state. Utilization of energy by maintenance respiration is not well understood, but estimates indicate that it can represent more than 50% of the total respiratory flux.

Although numerous questions remain regarding these issues, there are several empirical examples of relations between plant respiration rates and crop yield. In the forage crop, perennial ryegrass (*Lolium perenne*), yield increases of 10% to 20% were correlated with a 20% decrease in the leaf respiration rate. Similar correlations have been found for other plants, including corn (*Zea mays*) and tall fescue (*Festaca arundinacea*). However, later investigations have shown that "selection for low rates of mature leaf respiration is not an appropriate method to select for high-yielding cultivars in perennial grasses".

Although there is a potential for increasing crop yield through reduction of respiration rates, a better understanding of the sites and mechanisms that control plant respiration is needed before such changes can be exploited commercially in a systematic fashion by plant physiologists, geneticists, and molecular biologists. Furthermore, much remains to be learned about the general applicability of such observations and the conditions under which slower-respiring lines could be at a disadvantage, causing a reduction in crop yields rather than an increase.

Two recent studies illustrate the present difficulty of predicting the effect of directed changes at the molecular level on plant productivity:

Nunes-Nesi et al. (2005) found that the productivity of transgenic tomato plants with a reduced activity of mitochondrial malate dehydrogenase was increased as compared to the wild type. Although the respiratory activity of mitochondria isolated from the transgenic plants was unchanged or higher than in the wild type, the rate of leaf respiration was reduced in the transgenic plants. Photosynthesis was markedly increased in the transgenic lines, possibly linked to increased levels of ascorbate. Sieger et al. (2005) studied the effect of the alternative oxidase on the growth of tobacco cell cultures. Considering the energy-wasteful nature of the alternative oxidase one might expect that a cell line lacking the enzyme would grow faster than the wild type. Secondly, alternative oxidase has a role in the response of plants to oxidative stress, so a cell line lacking the alternative oxidase would also be expected to handle abiotic stress less well.

The results were surprising—the transgenic cell line with a very low expression of alternative oxidase grew as fast as the wild-type cells under normal, nutrient-sufficient conditions, and faster than the wild type under conditions of macronutrient limitation (low phosphate or low nitrogen).

It appears that the alternative oxidase is an important factor in modulating the growth rate in response to nutrient availability. Plant respiration involves an intricate metabolic network of interacting pathways and a complex regulation of gene expression and enzymatic activities. We clearly need to know much more about these interactions before we can predict the effects of changing the expression of single genes on plant productivity.

The Lipid Composition of Membranes Affects the Cell Biology and Physiology of Plants

The characterization of lipid mutants and the experimental modification of lipid composition by molecular-genetic means have now revealed some broad generalizations about the roles of membrane lipids in the cell biology and physiology of plants.

However, the biochemical and physical bases for the phenotypes that are observed remain uncertain. Changes in fatty acid composition have been shown to alter chloroplast size and architecture. Mutations that eliminate polyunsaturated fatty acids block photosynthesis and prevent the plant from growing autotrophically.

Surprisingly, such mutant plants can grow on sucrose, indicating that photosynthesis is the only plant function that absolutely requires a polyunsaturated membrane. However, polyunsaturated fatty acids are also required as precursors of signalling compounds which are necessary for such diverse functions as pollen development and plant defence, such as jasmonic acid.

One of the most extensively studied issues in membrane biology is the relationship between lipid composition and the ability of organisms to adjust to temperature changes. For example, chill-sensitive plants experience sharp reductions in growth rate and development at temperatures between 0 and 12°C. The physical and physiological changes in chill-sensitive plants that are induced by exposure to low temperature, together with the subsequent expression of stress symptoms, are termed chilling injury.

Many economically important crops, such as cotton, soybean, maize, rice, and many tropical and subtropical fruits, are classified as chill sensitive. In contrast, most plants of temperate origin are able to grow and develop at chilling temperatures and are classified as chill-resistant plants.

In attempts to link the biochemical and physiological changes associated with chilling injury with a single "trigger" or site of damage, it has been suggested that the primary event of chilling injury is a transition from a liquid-crystalline phase to a gel phase in the cellular membranes. According to this proposal, the transition from liquid-crystalline phase to gel phase would result in alterations in the metabolism of chilled cells and lead to injury and death of the chill-sensitive plants. The degree of unsaturation of the fatty acids would determine the temperature at which such damage occurs. A similar hypothesis has been proposed for chloroplast membranes in which the levels of disaturated phosphatidylglycerol (molecules containing no *cis* double bonds in the fatty acid chains) would determine the chilling sensitivity of plant species.

However, the most recent research indicates that the relationship of membrane unsaturation to plant temperature responses is more subtle and complex than is suggested in these earlier hypotheses. On the one hand, studies of five different *Arabidopsis* mutants have demonstrated that reduced unsaturation can result in plants that grow well at 22°C but are less robust than wild-type plants when grown at 2 to 5°C.

These results were observed even though the lipid changes in most of the mutants are insufficient to cause a lipid phase transition. In addition, the responses of these mutants to low temperature appear quite distinct from classic chilling sensitivity, suggesting that normal chilling injury may not be related to the level of unsaturation of membrane lipids.

In one particular mutant, *fab1*, saturated forms of phosphatidylglycerol account for 43% of the total leaf phosphatidylglycerol—a higher percentage than is found in many chill-sensitive plants. Nevertheless, the mutant was completely unaffected (when compared with wild-type controls) by a range of low-temperature treatments that quickly led to the death of cucumber and other chill-sensitive plants.

A complementary series of experiments has been carried out in tobacco, which is a chill-sensitive plant. The transgenic expression of exogenous genes in tobacco has been used specifically to decrease the level of saturated phosphatidylglycerol or to bring about a general increase in membrane unsaturation.

In each case, damage caused by chilling was alleviated to some extent. These new findings make it clear that the extent of membrane unsaturation or the presence of particular lipids, such as disaturated phosphatidylglycerol, can affect the responses of plants to low temperature. However, membrane lipid composition is not the major determinant of chilling sensitivity in plants.

Utilization of Oil Reserves in Cotyledons

Although the pathway for the mobilization of triacylglycerols has been best characterized in castor bean, it seems to be similar in the storage tissues of other oilseeds.

However, not all seeds quantitatively convert fat to sugar. In castor bean, the endosperm degenerates after the fat and protein reserves are fully utilized. In many oilseeds—such as sunflower (*Helianthus annuus*), cotton (*Gossipium hirsutum*), and members of the squash family (Cucurbitaceae)—the cotyledons become green and photosynthesize after the food reserves are used up.

In these tissues only part of the stored lipid is converted to exported carbohydrate. Much of the lipid-derived carbon remains in the cotyledons, where it contributes to the synthesis of chloroplasts and other cellular structures. Acetyl-CoA is also used directly for the production of energy through respiration. As the greening process takes place, there is a transition in the peroxisome population of these cells: Some peroxisomes have fewer of the characteristics of glyoxysomes and more of leaf-type peroxisomes.

Such a transition is in keeping with the decreased requirement for the breakdown of stored lipids and the increased need to metabolize the products of photorespiration as the tissue goes from a heterotrophic to a more autotrophic mode of metabolism.

Assimilation of Mineral Nutrients

Development of a Root Nodule

The establishment of symbiosis between legumes and rhizobia involves the activation of genes in both the host and the symbiont and an elaborated exchange of signals. The formation of a root nodule, the specialized organ from a plant host that contains the symbiotic nitrogen fixing rhizobia involves two simultaneous processes: infection and nodule organogenesis. The different stages in the initiation and development of a root nodule are shown in Figure below. A detailed description of the infection process.

A recent study of the early infection stages and nodule development in two *Medicago* species and the symbiont, *Rhisobium meliloti* showed that extensive reorganization of microtubules occurred early in the infection process in the pericycle and the inner cortex where the nodule primordium forms. The changes in microtubule organization were observed by immunolocalization, and fluorescence and laser confocal microscopy. These findings show that upon infection, the pericycle is activated first, followed by the activation of the inner cortex, where the nodule primordium forms. These changes are followed by a major reorganization of the microtubular cytoskeleton network in outer tissues, associated with root hair activation and curling, the formation of pre-infection threads, and the initiation and the growth of an infection network.

Measurement of Nitrogen Fixation

Besides N_2, the nitrogenase enzyme can reduce other substrates such as cyanide, azide and acetylene.

Direct measurement of N_2 fixation (reduction) requires a mass spectrometer, instrumentation that is not readily available, but the reduction of acetylene to ethylene can easily be measured by gas chromatography. Acetylene reduction involves two electrons, $C_2H_2 + 2e^- + 2\ H^+ \rightarrow C_2H_4$, as opposed to the eight electrons required for the reduction of N_2 and 2 H^+:

$$N_2 + 8e^- + 8H^+ + 16ATP \rightarrow 2NH_3 + H_2 + 16ADP + 16P_i$$

Therefore, reduction of four ethylene molecules correspond to the reduction of one N_2 molecule. The acetylene method has limitations, however: Acetylene blocks gas diffusion into nodules, and rhizosphere microorganisms produce ethylene. For these reasons, the acetylene method is suitable for comparative studies, while mass spectroscopy is used for direct measurements of N_2 reduction, which is required for precise quantification of nitrogen fixation.

The Synthesis of Methionine

Methionine and cysteine are the two sulfur-containing amino acids found in proteins. Methionine is synthesized in plastids from cysteine. In the first step of the pathway, cysteine and O-phosphohomoserine react to form cystathionine via the enzyme cystathionine-γ-synthase. Cystathionine is cleaved into homocysteine, pyruvate, and ammonia by cystathionine-β-lysase. Finally, methionine synthase methylates homocysteine to form methionine (Lea 1997).

2

Secondary Metabolites and Plant Defence

Structure of Various Triterpenes

Digitoxigenin is the aglycone (sugarless) triterpene portion of the naturally occurring digitanides, which contain one molecule of the rare sugar D-digitoxose and one molecule of acetyldigitoxose. The yam *Dioscorea* contains saponins—e.g., yamogenin—that are widely used as starting materials in the synthesis of progesterone-like compounds for birth control pills.

The Shikimic Acid Pathway

Plant phenolics are biosynthesized by several different routes and thus constitute a heterogeneous group from a metabolic point of view. Two basic pathways are involved: the shikimic acid pathway and the malonic acid pathway. The shikimic acid pathway participates in the biosynthesis of most plant phenolics. The malonic acid pathway, although an important source of phenolic secondary products in fungi and bacteria, is of less significance in higher plants. The shikimic acid pathway converts simple carbohydrate precursors derived from glycolysis and the pentose phosphate pathway to the aromatic amino acids. One of the pathway intermediates is shikimic acid, which has given its name to this whole sequence of reactions.

The well-known, broad-spectrum herbicide glyphosate (available commercially as Roundup) kills plants by blocking a step in this pathway. The shikimic acid pathway is present in plants, fungi, and bacteria but is not found in animals. Animals have no way to synthesize the three aromatic amino acids—phenylalanine, tyrosine, and tryptophan—which are therefore essential nutrients in animal diets.

Cell Walls: Structure, Biogenesis and Expansion

Terminology for Polysaccharide Chemistry

Polysaccharides are named after the principal sugars they contain. For example, a *glucan* is a polymer made up of glucose; a *xylan* is a polymer made up of xylose; a *galactan* is a polymer made up of galactose. For branched polysaccharides, the backbone of the polysaccharide is usually given by the last part of the name. Thus, *xyloglucan* has a glucan backbone (a linear chain of glucose residues) with xylose sugars attached to it in the side chains; *arabinoxylan* has a xylan backbone (made up of xylose subunits) with arabinose side chains. However, a compound name does not necessarily imply a branched structure. For example, "glucomannan" is the name given to a polymer containing both glucose and mannose in its backbone. Because sugars may assume multiple forms and may be connected in many ways, the nomenclature for sugars is not simple and is briefly summarized here. Monosaccharides (single-sugar residues) are polyhydroxyl aldehydes or ketones with the empirical formula $(CH_2O)_n$. They typically contain one asymmetric carbon, so they can exist as two stereoisomers, designated as the *D* or *L* enantiomers.

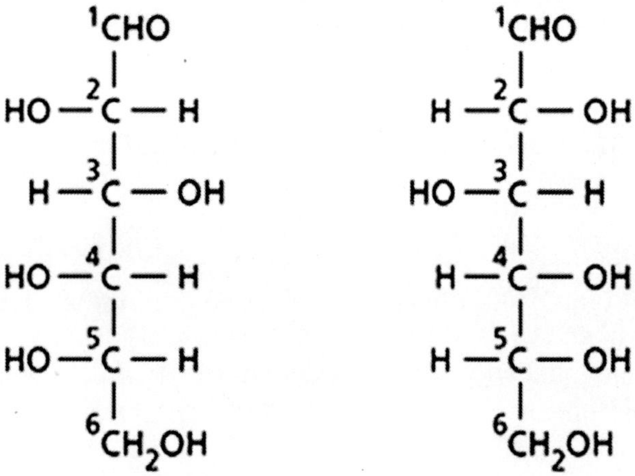

Figure: *D-and L-glucose shown in the linear form. The carbons are numbered from the end with the aldehyde (CHO) (C = O) group.*

Monosaccharides usually form a cyclic, rather than a linear, structure, because the carbonyl oxygen can react with the hydroxyl groups, typically forming either five-member rings (furanose rings) or six-member rings (pyranose rings). These rings have a new chiral centre and may take two forms, designated as α or β anomers. Thus, β-*D*-glucopyranose is *D*-glucose taking the form of a six-member ring in the β configuration. In solution, monosaccharides freely mutorotate (convert between α or β anomers); when linked in polymers, the anomeric configuration is stable.

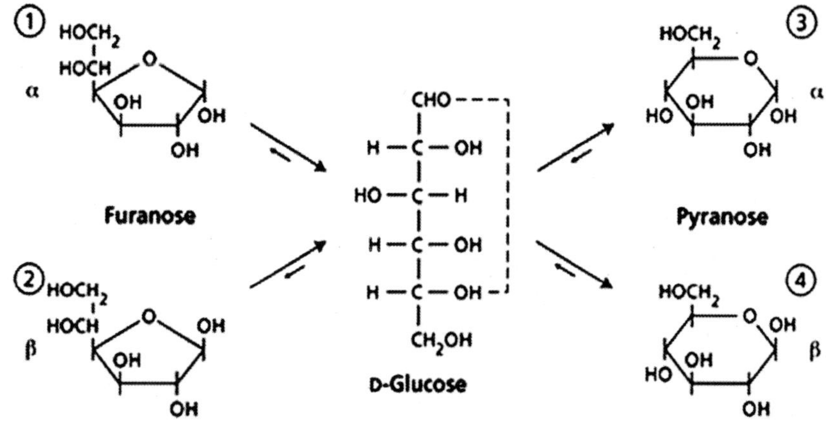

Figure: *Glucose may form five-or six-member rings, in the a or b configuration. The five-member rings (furanose; structures 1 and 2) are unstable and make up less than 1% of the total. The six-member rings (pyranose) are found mostly in the b configuration (64% of the total); the remainder (36%) are found in the a form.*

The sugars in polysaccharides are linked together by O-glycosidic bonds. We indicate the linkage pattern by specifying the carbon atoms connected by the glycosidic bond. In cellulose, for example, carbon atom 1 is linked to carbon atom 4 in the neighbouring glucose residues. Thus, cellulose is designated as a (1'!4) β-*D*-glucan, because it is a string of β-*D*-glucose molecules linked through carbons 1 and 4. It may also be called a polymer of (1'!4)-linked β-*D*-glucopyranose.

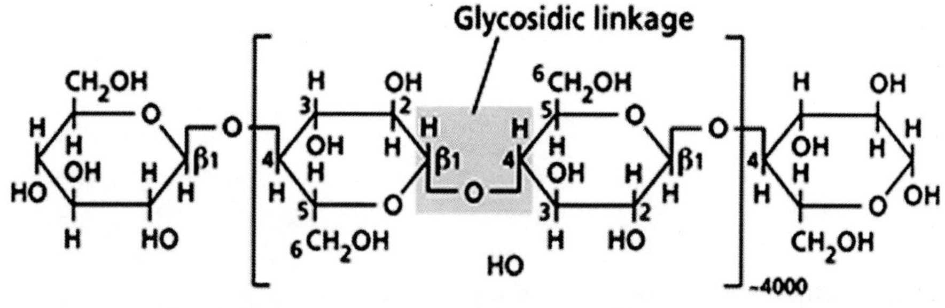

Figure: *A linear chain of 1'!4-linked b-D-glucopyranose, showing the numbering system for the carbons and the configuration of the glycosidic bond. This is the glucan that makes up cellulose. The sugar rings are connected through an oxygen atom (the glycosidic oxygen). Because of the configuration of the glycosidic bonds, the repeating unit in this polymer is cellobiose, a disaccharide.*

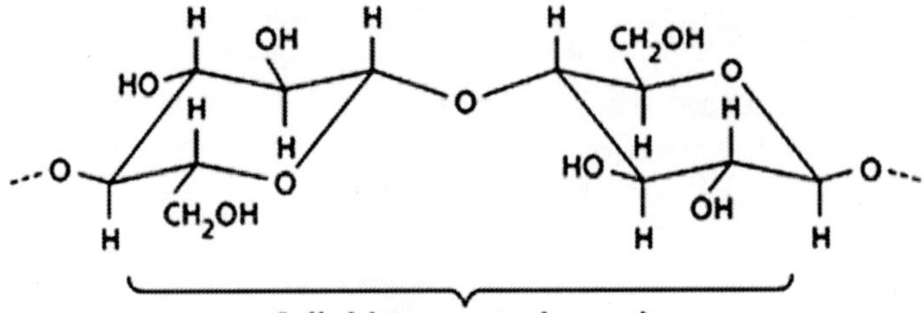

Figure: *The repeating cellobiose unit, with the glucose residues drawn in the conformational ("chair") form. This way of drawing the glucose ring represents better the actual positions of the individual atoms than in the Haworth project.*

Molecular Model for the Synthesis of Cellulose and Other Wall Polysaccharides That Consist of a Disaccharide Repeat

Cellulose synthase contains two binding sites for UDP-glucose (or other sugar nucleotides). After binding UDP-glucose (step 2), the sugars are transferred to the nonreducing end of the glucan (step 3). At step 4, the glucan chain is translocated by two glucose residues, restoring the complex for the start of another round of synthesis (step 1). These genes code for glycosyl transferases with two catalytic sites for transferring glycosyl residues from their nucleotide donors to the nascent polymer acceptor. This two-site arrangement may explain why the cellobiose disaccharide is the repeating subunit in cellulose: It has been proposed that two glucose residues are added to the glucan chain for each catalytic cycle of the enzyme. A similar arrangement in other polysaccharide synthases could explain why disaccharide repeats commonly make up the backbones of many other cell wall polysaccharides in plants.

The Mechanical Properties of Cell Walls: Studies With Nitella

The small size of typical plant cells (20 to 100 μm) has been a serious impediment to the study of cell wall mechanical properties. To measure the extensibility of isolated cell walls from higher plants, researchers must place entire frozen and thawed organs or tissues in an *extensometer*. Although such measurements can give qualitative answers, the extensions measured cannot be related to the growth of living cells in a quantitative way, because the applied stress is uniaxial (unidirectional), whereas the in vivo stress due to turgor pressure is multiaxial (multidirectional).

The giant internodal cells of the freshwater green algae *Nitella* and *Chara* have figured prominently in classic studies on ion transport, water permeability, and cytoplasmic streaming (Hope and Walker 1975). These algae are characterized by erect, branched coenocytic (multinucleate) filaments, differentiated into nodes and internodes, each node bearing a whorl of specialized lateral internodes.

The internode cell begins as a shortened disc 20 mm long, which elongates to produce a slender cylinder about 6 cm long and 0.5 mm wide at maturity. The cell expands by diffuse growth in a predominantly longitudinal direction. Because individual internode cells are large, their cell walls can be isolated and the mechanical properties of the walls measured. It is even possible to study the anisotropic (unequal) mechanical properties of individual cell walls by the use of multiaxial stress to mimic the in vivo turgor pressure.

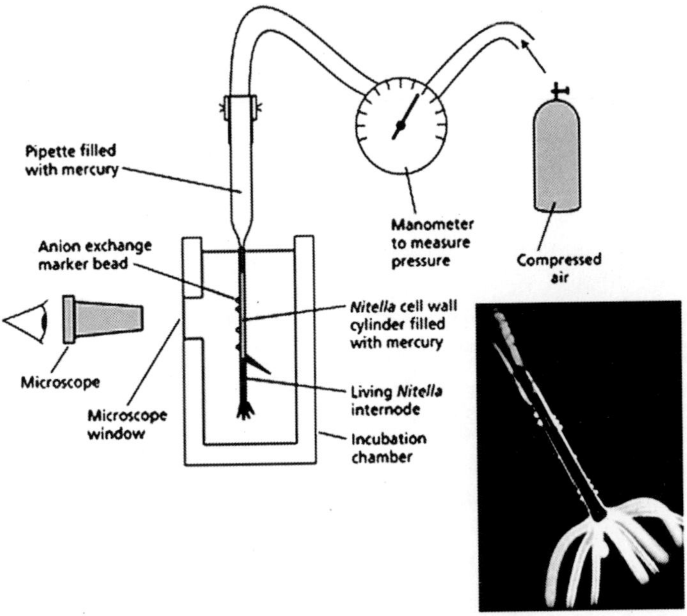

Figure: *Apparatus for measuring the multiaxial extensibility of* Nitella *cell walls. The pressure source is a compressed air tank. Air passes through high-pressure tubing into a manometer and from there into a mercury-filled pipette glued to the open end of a* Nitella *cell wall cylinder. A living internode is maintained at the distal end. The wall is immersed in a perfusion chamber with a glass window. Displacement of resin beads attached to the surface of the wall is measured with a horizontally oriented microscope. The inset shows a view of a mercury-inflated cell wall cylinder with marker beads attached as seen through the traveling microscope.*

The experimental setup for studying multiaxial extensibility in *Nitella* internode cells. Empty cell wall cylinders are glued to a pipette and inflated with mercury under pressure. Liquid mercury is insoluble and therefore inert under these conditions. In response to the multiaxial stress caused by the pressurized mercury, the cell wall cylinder expands, and investigators can monitor the expansion in both length and width by viewing the process through a horizontal microscope (Metraux et al. 1980). With the multiaxial stress procedure on *Nitella* internodes, one can use microtubule inhibitors to alter the orientation of the cellulose microfibrils and test the hypothesis that microfibril orientation is responsible for the mechanical anisotropy of the cell wall. The multiaxial extension of control cells is highly anisotropic. Not only is extension in length four times greater than extension in diameter, but much of the extension in length is irreversible, whereas the extension in diameter is almost completely reversible.

These results are consistent with the anisotropic growth pattern of the cells. The effects of pretreating the cells with the microtubule poison isopropyl *N*-phenylcarbamate (IPC), which, like colchicine, induces the random deposition of cellulose microfibrils in the wall. The result is that the extensions of the wall in length and diameter are about equal, consistent with the isotropic growth pattern induced by this inhibitor.

Nitella has also proven to be useful for studying cell wall properties in vivo. Many of the concepts applied to contemporary studies on the biophysical control of cell expansion are derived from the pioneering experiments on living *Nitella* cells by Paul Green and his colleagues at Stanford University (Green 1968). An important insight to come out of such studies is that the inner 25% of the cell wall determines the polarity of cell expansion. Taking advantage of the fact that cellulose microfibrils are crystalline and therefore exhibit birefringence (double refraction) when viewed with polarized light, Paul Richmond at the University of the Pacific in California devised an optical technique to measure cell expansion in length and diameter, and the orientation of the cellulose microfibril in the cell wall, simultaneously (Richmond 1983). This technique made it possible to determine the relationship between microfibril angle and the direction of cell expansion.

The microfibril angle began to change from transverse to random almost immediately after colchicine was added to the growth medium. However, the anisotropy of growth did not change until about 8 hours later. The transition from anisotropic growth to isotropic growth is complete after a relative increase in cell surface area of about 25%. Since wall thickness remains constant during this period because of continued wall synthesis, we can infer that the directionality of cell expansion changes when the inner 25% of the cell wall has been replaced. Thus the inner 25% of the cell wall appears to represent the load-bearing portion of the wall. By implication, the outer 75% has become too weakened by stretching to restrict cell expansion.

Growth and Development: The Analysis of Plant Growth

As we have seen, new cells form continually in the apical meristems. Cells enlarge slowly in the apical meristem and more rapidly in the subapical regions. The resulting increase in cell volume can range from severalfold to one hundredfold, depending on the species and environmental conditions. Classically, plant growth has been analysed in terms of cell number or overall size (i.e., mass). However, these measures tell only part of the story. Tissue growth is neither uniform nor random. The derivatives of the apical meristems expand in predictable and site-specific ways, and the expansion patterns in these subapical regions largely determine the size and shape of the primary plant body. The total growth of the plant can be thought of as the sum of the local patterns of cell expansion.

The analysis of the motions of cells or "tissue elements" (and the related problem of cell expansion) is called *kinematics*. In this section we will discuss both the classical definitions of growth and the more modern, kinematic approach. As we will see, the advantage of the kinematic approach is that it allows one to describe the growth patterns of organs mathematically in terms of the expansion patterns of their component cells.

Plant Growth Can Be Measured in Different Ways

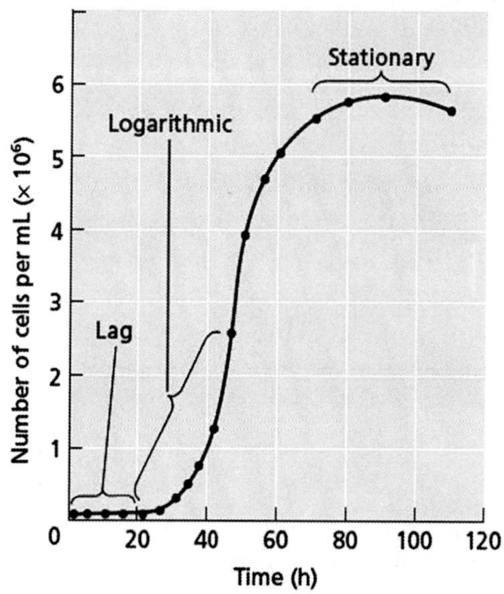

Figure : *Growth of the unicellular green alga Chlamydomonas. Growth is assessed by a count of the number of cells per milliliter at increasing times after the cells are placed in fresh growth medium. Temperature, light, and nutrients provided are optimal for growth. An initial lag period during which cells may synthesize enzymes required for rapid growth is followed by a period in which cell number increases exponentially. This period of rapid growth is followed by a period of slowing growth in which the cell number increases linearly. Then comes the stationary phase, in which the cell number remains constant or even declines as nutrients are exhausted from the medium.*

Growth in plants is defined as an irreversible increase in volume. The largest component of plant growth is cell expansion driven by turgor pressure. During this process, cells increase in volume manyfold and become highly vacuolate. However, size is only one criterion that may be used to measure growth. Growth also can be measured in terms of change in fresh weight—that is, the weight of the living tissue—over a particular period of time. However, the fresh weight of plants growing in soil fluctuates in response to changes in water status, so this criterion may be a poor indicator of actual growth. In these situations, measurements of dry weight are often more appropriate.

Cell number is a common and convenient parameter by which to measure the growth of unicellular organisms, such as the green alga *Chlamydomonas* (Figure). In multicellular plants, however, cell number can be a misleading growth measurement because cells can divide without increasing in volume. For example, during the early stages of embryogenesis, the zygote subdivides into progressively smaller cells with no net increase in the size of the embryo. Only after it reaches the eight-cell stage does the increase in volume begin to mirror the increase in cell number. Because the zygote is an especially large cell, this lack of correspondence between an increase in cell number and growth may be unusual, but it points out the potential problem in equating an increase in cell number with growth.

Although cell number may not always be a reliable measure of plant growth, under most circumstances dividing cells, particularly in meristems, double in volume during their cell cycle. Therefore, an increase in cell number, such as the increase brought about by the activity of the apical meristems, does contribute to plant growth. However, the largest component of plant growth is the rapid cell expansion that occurs in the subapical region after cell division ceases.

Because all the cells of the plant axis elongate under normal conditions, the greater the number of cells produced by the apical meristem, the longer the axis will be. For example, when Arabidopsis plants are transformed with a gene that encodes cyclin, a key component of the cell cycle regulatory machinery, the cells of the apical meristem progress through their cell cycles more rapidly, so more cells form per unit time. As a result, the roots of these transgenic plants have more cells and are substantially longer than the roots of wild-type plants grown under similar conditions. New cells form continually in the apical meristems. With each new round of cell division and associated cell expansion, the older derivatives are displaced a small distance from the apex. As the cells recede farther from the apex, the rate of displacement is greatly accelerated. By viewing plant growth as a process of cell displacement from the apex, we can apply the principles of kinematics.

The Production of Cells by the Meristem Is Comparable to a Fountain

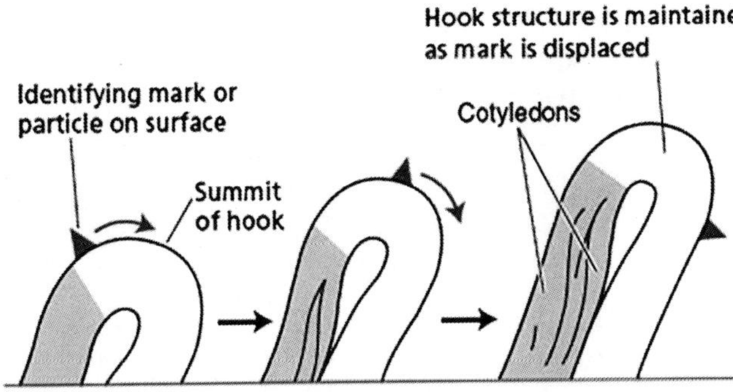

Figure : *The dicot hypocotyl hook is an example of a constant form composed of changing elements. The hooked form is maintained over time, while different tissues first curve and then straighten as they are displaced from the seedling apex during growth. If a mark is placed at a fixed point on the surface, it will be displaced (indicated by the arrow), appearing to flow through the hook over time.*

Moving fluids such as waterfalls, fountains, and the wakes of boats can generate specific forms. The study of the motion of fluid particles and the shape changes that the fluids undergo is called kinematics. The ideas and numerical methods used to study these fluid forms are useful for characterizing meristematic growth.

In both cases, an unchanging form is produced even though it is composed of moving and changing elements. An example of an unchanging form composed of changing and displaced elements in plants is the hypocotyl hook of a dicot, such as the common bean (Figure above). As the bean seedling emerges from the seed coat, the apical end of the hypocotyl bends back on itself to form a hook.

The hook is thought to protect the seedling apex from damage during growth through the soil. During seedling growth (in soil or dim light) the hook migrates up the stem, from the hypocotyl into the epicotyl and then to the first and second internodes, but the form of the hook remains constant.

If we mark a specific epidermal cell on the seedling stem located close to the seedling apex, we can watch it as it flows into the hook summit, then down into the straight region below the hook. The mark is not crawling over the plant surface, of course; plant cells are cemented together and do not experience much relative motion during development.

The change in position of the mark relative to the hook implies that the hook is composed of a procession of tissue elements, each of which first curves and then straightens as it is displaced from the plant apex during growth. The steady form is produced by a parade of changing cells.

A root tip is another example of a steady form composed of changing tissue elements. Here, too, the form is observed to be steady only when distance is measured from the root tip. A region of cell division occupies perhaps 2 mm of the root tip. The elongation zone extends for about 10 mm behind the root tip. Phloem differentiation is first observed beginning at 3 mm from the tip, and functional xylem elements may be seen at about 12 mm from the tip. A marked cell near the tip will seem to flow first through the region of cell division, then through the elongation zone and into the region of xylem differentiation, and so on. This shifting implies that developing tissue elements first divide and elongate, and then differentiate.

In an analogous fashion, the shoot bears a succession of leaves of different developmental stages. During a period of 24 hours, a leaf may grow to the same size, shape, and biochemical composition that its neighbor had a day earlier. Thus, shoot form is also produced by a parade of changing elements that can be analysed with kinematics. Such an analysis is not merely descriptive; it permits calculations of the growth and biosynthetic rates of individual tissue elements (cells) within a dynamic structure.

Tissue Elements Are Displaced During Expansion

As we have seen, growth in shoots and roots is localized in regions at the tips of these organs. Regions with expanding tissue are called growth zones. With time, meristems move away from the plant base by the growth of the cells in the growth zone. If successive marks are placed on the stem or root, the distance between the marks will change, depending on where they are within the growth zone. In addition, all of these marks will move away from the tip of the root or shoot, but their rate of movement will differ depending on their distance from the tip.

From another perspective, if you were to stand at the tip of a root that had marks placed at intervals along the axis, you would see that all marks would move farther away from you with time. The reason is that discrete regions on the plant axis experience displacement as well as expansion during growth and development.

As Regions Move Away from the Apex, Their Growth Rate Increases

As a given region of the plant axis moves away from the apex, its growth velocity increases (the rate of elongation accelerates) until a constant limiting velocity is reached equal to the overall organ extension rate. The reason for this increase in growth velocity is that with time, progressively more tissue is located between the moving particle and the apex, and progressively more cells are expanding, so the particle is displaced more and more rapidly. In a rapidly growing maize root, a tissue element takes about 8 hours to move from

2 mm (the end of the meristematic zone) to 12 mm (the end of the elongation zone). Beyond the growth zone, elements do not separate; neighbouring elements have the same velocity (expressed as the change in distance from the tip per unit of time), and the rate at which particles are displaced from the tip is the same as the rate at which the tip moves through the soil. The root tip of maize is pushed through the soil at 3 mm h^{-1}. This is also the rate at which the nongrowing region recedes from the apex, and it is equal to the final slope of the growth trajectory.

The Growth Velocity Profile Is a Spatial Description of Growth

The velocities of different tissue elements are plotted against their distance from the apex to give the spatial pattern of growth velocity, or growth velocity profile. Growth velocity increases with position in the growth zone. A constant value is obtained at the base of the growth zone. The final growth velocity is the final, constant slope of the growth trajectory equal to the elongation rate of the organ, as discussed in the previous section. In the rapidly growing maize root, the growth velocity is 1 mm h^{-1} at 4 mm, and it reaches its final value of nearly 3 mm h^{-1} at 12 mm.

If the growth velocity is known, the relative elemental growth rate, which represents the fractional change in length per unit time, can be calculated. The relative elemental growth rate shows the location and magnitude of the extension rate and can be used to quantify the effects of environmental variation on the growth pattern.

The Relative Elemental Growth Rate

One way to characterize the growth pattern along an axis mathematically is to plot the distance of a tissue element from the apex versus time. The resulting curve is called the growth trajectory. The slope of the curve at any given point is equivalent to the growth velocity at that region of the axis. With the apex as the reference point, the growth trajectory is concave. Marked tissue elements at different locations may be followed simultaneously to generate a family of growth trajectories.

As the regions of the plant axis move away from the apex, their growth velocity increases (i.e., the cells accelerate) until a constant limiting velocity is reached equal to the overall organ extension rate. The reason for this increase in growth velocity is that with time, progressively more tissue is located between the moving particle and the apex, and progressively more cells are expanding, so the particle is displaced more and more rapidly. In a rapidly growing maize root, a tissue element takes about 8 hours to move from 2 mm (the end of the meristem) to 12 mm (the end of the elongation zone). The root tissue element accelerates through this region.

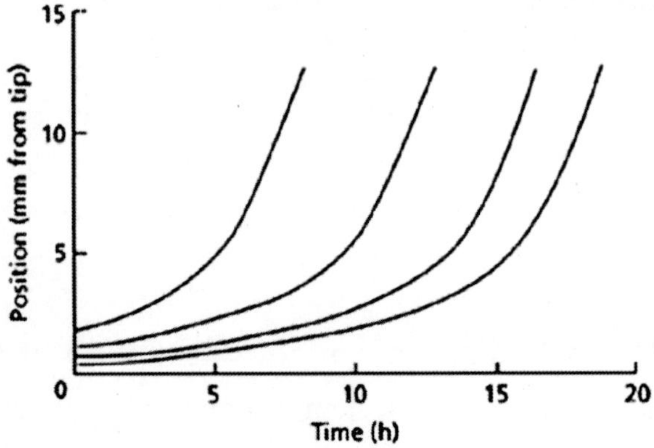

Figure: *Growth trajectories of* Zea mays *(maize):Growth trajectories plot the positions of points spaced along the root axis as a function of time. Note that distance is measured from the root tip. This is termed a material specification of growth because real or material particles are followed.*

Beyond the growth zone, elements do not separate; neighbouring elements have the same velocity, and the rate at which particles are displaced from the tip is the same as the rate at which the tip is pushed away from the soil surface. The root tip of maize is pushed through the soil at 3 mm h^{-1}. This is also the rate at which the nongrowing region recedes from the apex, and it is equal to the final slope of the growth trajectory. The growth trajectory is a material description of growth because an individual element is followed through time. The growth trajectory provides a way to convert spatial information into a developmental time course. For example, if the amount of suberin in the root increases with position, the growth trajectory will give information on the time required for the observed accumulation of suberin in cells moving between any two spatial points.

To characterize the local expansion rate, we must consider the growth velocities at the apical and basal ends of a small segment of tissue. If the apical end of the segment is moving faster than the basal end, the segment is elongating. The velocity difference, divided by the segment length, gives the relative growth rate of the segment. Now, imagine that the segment shrinks to a point x. To find the relative growth rate, r, at point x, we can use calculus: The value of r is given by the velocity gradient, dv/dx, where v is the growth velocity. This measure of the local growth rate is called the relative elemental growth rate (Erickson and Sax 1956). It represents the fractional change in length per unit time and has units of h^{-1}. If the growth velocity is known, the relative elemental growth rate can be evaluated by differentiation of the growth velocity with respect to position. Or we can approximate r by measuring the relative growth rates of closely spaced segments of initial length L. For each segment, $r = (1/L)(dL/dt)$.

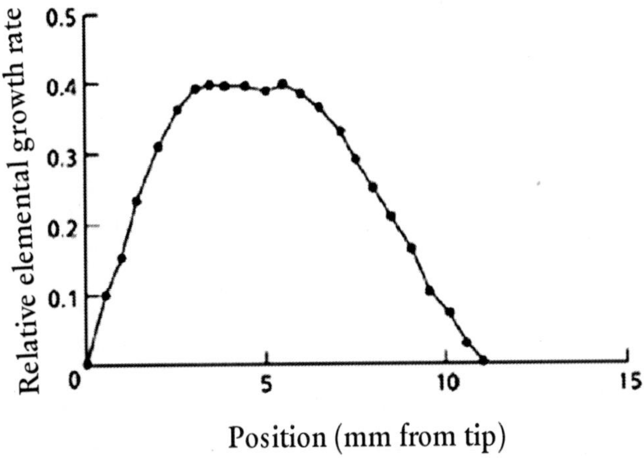

Figure: *Relative elemental growth rate of* Zea mays *(maize): The relative elemental growth rate tells us the rate of expansion of any particular point on the root. It is the most useful measure of root growth rate for the physiologist because it tells where the most rapidly expanding regions are located.*

The relative elemental growth rate profile, $r(x)$, is a spatial description of the expansion pattern in the growth zone. In the maize root, the relative elemental growth rate has a maximum of 0.40 h^{-1} at 4 mm and drops to zero by 12 mm. That is, the tissue element at 4 mm is expanding 40% h^{-1}, while tissue beyond 12 mm has ceased expansion. Again, to understand the profile in physiological terms, we must consider that individual cells expand more and more rapidly as they are displaced to 4 mm and then less and less rapidly as they are displaced through the basal part of the growth zone. The relative elemental growth rate profile shows the location and magnitude of the extension rate and can be used to quantify the effects of environmental variation on the growth pattern. It is also the basis for studying the physiology of growth. The graph of $r(x)$ can be compared to the distribution pattern of any substance thought to be regulating the growth rate. A close correlation between the two curves would provide circumstantial evidence for the role of the substance as a limiting factor for growth.

Embryonic Dormancy

Embryo Maturation Requires Specific Gene Expression

The Arabidopsis embryo enters dormancy after it has generated about 20,000 cells. Dormancy is brought about by the loss of water and a general shutting down of gene transcription and protein synthesis—not only in the embryo, but also throughout the seed. To adapt the cell to the special conditions of dormancy, specific gene expression is required. For example, the *A*bscisic Acid *I*nsensitive*3* (*ABI3*) and *FUSCA3* genes are necessary for the initiation of dormancy and are sensitive to the hormone abscisic acid, which is the signalling molecule that initiates seed and embryo dormancy. *ABI3* also controls the expression of genes encoding the storage proteins that are deposited in the cotyledons during the maturation phase of embryogenesis.The *LE*AFY *C*OTYLEDON*1* (*LEC1*) gene also is active in late embryogenesis. Because *lec1* mutants cannot survive desiccation and do not enter dormancy, the embryos die unless they are rescued through isolation before desiccation occurs.

The rescued embryos will germinate in culture and produce fertile plants, which are like wild-type plants except that they lack the 7S storage protein and they have leaflike cotyledons with trichomes on their upper surface. The normal appearance and development of the mature *lec1* mutants indicates that the *LEC1* gene is required only during embryogenesis. Although the most obvious defects of the *lec1* mutants are seen only in the maturation phase embryo, mRNA from *LEC1* gene expression can be detected throughout embryogenesis. It has been proposed that *LEC1* is a general repressor of vegetative development and its expression is necessary throughout embryogenesis.

Polarity of Fucus Zygotes

Plant cells perceive their position within a tissue and assume the polarity appropriate for that tissue. In addition, cells can perceive and act on information coming from their environment to alter their polarity and redirect growth when necessary. Cellular polarity must be established to orient subsequent expansion. Cells that are initially apolar, such as zygotes, spores, and pollen grains, later become polarized and initiate tip growth and have been especially valuable in the analysis of the induction and establishment of polarity.

A wide variety of external gradients, including temperature, light intensity, pH, ions, and electric-potential gradients, can polarize and orient the axis of growth of these cells. Thus, polarity can be experimentally manipulated without the complexities associated with multicellular tissues. Much of our understanding of the establishment of polarity in tip-growing cells is derived from studies of rhizoid emergence in zygotes of the brown alga *Fucus*.

Polarity of the Fucus Zygote is Initially Unstable but Later Fixed

Unlike higher plants, and even unlike most other algae, certain brown algae, particularly *Fucus* and *Pelvetia*, shed their eggs and sperm into the sea, where fertilization occurs. The zygote initially lacks a cell wall and is completely apolar. However, a polar axis is established within a few hours of fertilization in response to unilateral light. The axis is not stable at first and can be changed many times simply by a change in the direction of the light. The polar axis becomes fixed within 10 to 14 hours after fertilization, after which it cannot be altered by a change in the direction of the light, and a structure that will become the rhizoid begins to protrude from the darkened side.

Even though the zygote is still a single cell, it is polarized, with two structurally different poles, the thallus and rhizoid poles, which have radically different fates. The first zygotic division is asymmetric and occurs at right angles to the polar axis. It establishes two cells, each having the identity of the pole from which it emerges. The smaller *rhizoid cell* produces a filament of cells and the holdfast that attaches the plant to rocks. The larger *thallus cell* divides to produce a globular mass of cells, from which the main body of the plant, the thallus, forms.

Calcium Channel Redistribution Leads to a Transcellular Current

A dye known as fluorescent dihydropyridine (FL-DHP), which binds to calcium channels in mammalian cells and can interfere with calcium uptake, has been used as a probe to examine the polarization of the *Fucus* zygote cell membrane. During the first 6 to 8 hours after fertilization, and in response to unilateral light, FL-DHP receptors, presumably calcium channels, are redistributed and become localized on the dark side of the *Fucus* zygote at what will become the rhizoid pole. If the direction of the light is changed, the location of the FL-DHP–binding proteins also changes, again becoming concentrated on the dark side of the zygote. One of the earliest events in the establishment of polarity after exposure of the zygotes to unilateral light is the induction of a positive current that flows into the cell on the shaded side and loops back through the external medium. This transcellular current, which is due largely to calcium ions, is on the order of 100 pA (1 picoampere = 10^{-12} ampere) and develops in advance of any morphological sign of polarity. The relevance of this phenomenon to the establishment of cell polarity is indicated by the fact that the site at which positive charge flows into the cell is also the site of subsequent rhizoid outgrowth. Transcellular ionic currents also have been observed in some tip-growing cells of higher plants, including root hairs and germinating pollen grains, but they have not yet been detected in plant cells that expand by diffuse growth.

The ionic current in polarizing *Fucus* zygotes likely arises from the redistribution of ion channels in the plasma membrane. This current generates an electric-potential difference across the membrane that could alter other membrane properties, such as the distribution of membrane proteins. The current also alters the ionic composition of the cytosol; the magnitude of the effect depends on the ions that form the current, which vary in different cell types. Because the Ca^{2+} concentration in the cytosol is usually very low (between 10^{-6} and 10^{-8} μM), even a small increase in intracellular Ca^{2+} concentration could have a marked effect.

Changes in Cytosolic Calcium have been Measured

The fluorescent calcium-indicator dye *fura*-2 has been used to detect intracellular cytosolic calcium gradients in both plant and animal cells. Since the excitation spectrum of fura-2 shifts when it binds calcium, we can determine the intracellular concentrations of both forms of the dye (with and without bound calcium) by exciting them with the appropriate two wavelengths. The ratio of the two emissions provides a measure of the calcium concentration that is independent of the dye concentration. Using this technique to measure the cytosolic Ca^{2+} gradient in growing *Fucus* rhizoids, researchers observed a gradient from 450 nM at the tip to 100 nM at the base.

This calcium ion gradient may be important for the polarized assembly of actin filaments and vectoral discharge of Golgi-derived vesicles at the rhizoid pole. Calcium ion gradients also have been shown to be established in other tip-growing cells, such as germinating pollen grains, where the direction of growth and the region in which Golgi vesicles carrying cell wall components are deposited are predicted by the calcium gradient.

The Actin Cytoskeleton Becomes Polarized

Inhibition of protein synthesis by treatment with the drug *cycloheximide* delays rhizoid formation but does not stop the establishment of polarity. This result suggests that although rhizoid formation requires the synthesis of new proteins, the establishment of polarity does not, so polarity development must involve a redistribution of existing components. Actin filaments become preferentially localized in the rhizoid half of the zygote and extend from the vicinity of the nucleus to the plasma membrane at the site marked by localized FL-DHP–binding proteins as polarity develops.

Actin filaments are dynamic, and their assembly from a pool of monomeric actin subunits is highly regulated. Actin filaments are localized in the rhizoid half of the cell presumably in response to a change in the assembly or disassembly dynamics, and this localization involves a redistribution of the cellular actin.

The fixation of polarity can be blocked by *cytochalasin B*, a drug that destroys actin filaments. Treatment with cytochalasin B also disrupts the ionic current that develops during polarization. Actin filaments participate in cell secretion by transporting secretory vesicles to the plasma membrane in many types of cells.

Secretion of Golgi-Derived Vesicles Is Localized

Cytoplasmic organelles become asymmetrically distributed before the polarity of the cell is fixed. Golgi-derived vesicles accumulate at the site of the future outgrowth in the rhizoid half of the *Fucus* zygote as polarization is induced by light. For several hours after the induction of polarity, the asymmetric distribution of vesicles can be changed by subsequent treatment with light from a different direction. During this phase when changes are possible, cytoplasmic vesicles accumulate at the prospective site of rhizoid formation on the shaded side of the cell. Subsequently, the vesicles fuse with the plasma membrane, secreting their contents into the cell wall, leading to rhizoid emergence. Evidence suggests that calcium gradients play a role in triggering vesicle discharge at the pole.

The fixation of light-induced polarity in *Fucus* zygotes also requires the presence of the cell wall. Removal of the wall with cell wall–digesting enzymes does not prevent the induction of polarity by unilateral light, but it does prevent fixation of the axis. The orientation of the embryonic axis remains unstable until a new cell wall is synthesized. From these observations it is possible to envisage a chain of events that might occur during the establishment of cell polarity in tip-growing cells.

- The ionic currents that arise early in the process may initiate the formation of an intracellular Ca^{2+} gradient.
- The increase in intracellular calcium at the tip may help organize and stabilize the microfilament components of the cytoskeleton.
- The cytoskeleton directs vesicle traffic to the growing tip.
- A calcium gradient, perhaps immediately adjacent to the plasma membrane, may promote vesicle fusion with the plasma membrane, although this remains to be demonstrated.
- Finally, cell wall deposition fixes the polarity of the cell, possibly by interacting with transmembrane proteins.

The Preprophase Band of Microtubules

Microtubules in the cortical cytoplasm of growing plant cells are important in determining the orientation of cellulose microfibrils in the cell wall. When dividing cells enter the G_2 phase of the cell cycle, the interphase cortical microtubule array is replaced by a narrow band of interconnected microtubules and microfilaments encircling the nucleus. This circular array of microtubules and microfilaments is called the preprophase band.

The preprophase band was first observed in 1966 by Jeremy Pickett-Heaps and Donald Northcote at Cambridge University. It appears as a transient ring of cytoskeletal elements in the cortical cytoplasm, near the plasma membrane. Generally the preprophase band is 2 to 3 μm wide and contains from 10 to more than 100 microtubules, interspersed and interconnected with actin microfilaments. Immunofluorescent staining of the cytoskeleton in onion (*Allium cepa*) root tip cells with an antibody to the microtubule protein, tubulin. As cells progress through their cell cycle, different cytoskeletal arrays that contain microtubules are formed.

(A) A preprophase band appears after DNA replication is complete, but before the cell enters mitosis. It can be seen as two bright spots on either side of the unstained nucleus; the image of the band is fainter as it passes across the nucleus.

(B) The mitotic spindle forms in late prophase, with spindle microtubules emanating from the polar regions to form the spindle fibres.

(C) The spindle fibres hold the chromosomes (dark line) in the metaphase plate. In this cell the alignment of the plate is oblique, which is relatively unusual.

(D) The chromosomes move toward their poles within the array of spindle microtubules.

(E) The phragmoplast forms in telophase and is seen here as a bright band in the centre of the cell that consists of microtubules, microfilaments, and vesicles in the early stage of cell plate formation.

(F) Phragmoplast microtubules form an annular ring around the perimeter of the cell plate and appear only at the edges of the cell. Bar = 10 μm. (Courtesy of A. Hardham.)

As the cell enters mitosis, microtubules begin to assemble at the surface of the nuclear envelope, and as the nuclear envelope breaks down, they invade the nuclear area to form the mitotic spindle. The preprophase band disappears as the spindle forms, leaving behind an actin-depleted zone. The plane in which cytokinesis occurs is determined before the cell initiates mitosis and is accurately predicted by the position of the preprophase band. It has been suggested that the preprophase band determines the orientation of the mitotic spindle and that it marks the site at which the new cell wall will form during cytokinesis, but this has not been established. The position of the preprophase band and the plane of cytokinesis both may be determined by another mechanism, possibly sensitive positional information.

Genes Required for Shoot Apical Meristem Formation: Class III HD-Zip Genes

Another class of transcription factors, encoded by CLASS III HD-ZIP class genes, has recently been shown to play essential roles in establishing the Shoot Apical Meristem (SAM). Mutants which lack activity of the previously discussed *REV* gene, as well as the related *PHABULOSA* and *PHAVOLUTA* genes, are unable to form an SAM. These genes also promote establishment of adaxial tissues in the leaves. As meristems often form in close association with these tissues (e.g., in the axils of leaves), models have been proposed in which the expression of CLASS III HD-ZIP genes promote what could be termed "central" cell identities. In the stem, the "central" identity would include regions that contain apical initials, as opposed to the organogenic regions of the flanking peripheral zone. In the context of leaf development, this "central identity" would encompass the adaxial tissues, including the asymmetrically positioned xylem elements of the leaf as well as closely associated domains from which axillary meristems arise. This type of model receives further support from studies in which the overexpression of CLASS III HD-ZIP class genes causes SAMs to form in abnormal positions.

The regulation of CLASS III HD-ZIP genes is complex. Their polarized patterns of expression, and the mutant phenotypes that result from perturbing this expression, suggests that the normal patterning of the apical domain depends on radial patterning processes. The proteins encoded by these genes contain START domains, suggesting their activity is regulated by as yet undefined sterol type signalling molecules. The region that encodes the START domain is further distinguished by characteristics that suggest these genes are targets of microRNA (miRNA) mediated repression of transcript levels.

An early clue to this mode of regulation was the highly conserved nature of the START domain region, which was apparent not just at the protein level, as might be expected of a ligand-binding region, but also at the nucleotide level, which suggested conserved interactions between nucleic acids. This model has received additional support by the characterization of genes that encode miRNAs that are complementary to the conserved nucleotide sequence motif. The validity of the model is supported further by the effects of "silent" mutations, which alter the nucleic acid sequence of the motif without affecting its coding potential. Mutations of this type would prevent miRNA binding, and thus block the normal down-regulation transcripts in peripheral domains and lead to the observed dominant mutant phenotypes.

The regulation of transcript levels by miRNAs is not completely understood. In some cases, miRNAs mediate the formation of the so-called RNA Induced Silencing Complexes (RISC), which suppress the translation of target transcripts. An alternative type of posttranscriptional regulation has been documented in which the annealing of miRNAs to their mRNAs in the nucleus triggers transcript degradation. More recent data suggests that miRNAs may also mediate transcriptional silencing through chromatin-mediated reduction of gene transcription.

By any of these mechanisms (which can collectively be termed RNAi, for RNA inhibition), one would expect that genes involved in RNAi would play an important role in patterning processes. Indeed, this expectation is met by the mutant phenotypes of genes whose products mediate RNAi processes. These include the Arabidopsis *ARGONAUTE* and *ZWILLE* (*ZLL*) (also known as *PINHEAD*) genes, both of which affect tissue polarity and meristem formation. Although it is unclear precisely how *ZLL* functions in the SAM, mutations that disrupt it function to block the accumulation of STM protein, but have no immediate effect on *STM* transcript levels. Further support for the idea that RNAi-dependent mechanisms are involved in normal SAM function is offered by the recent description of the genes required for normal shoot formation in rice, several of which have been shown to encode several distinct components of RNAi machinery.

Azolla *Root Development*

Our undersanding of the mechanics of meristem function has been enhanced by investigations of the apical meristems of ferns and other primitive vascular plants. The root and shoot apical meristems of these organisms have a single large apical cell, which resides at the centre of the meristem. The apical cell is the ultimate source of all the cells in the meristem, as well as the root or shoot body. Brian Gunning at the Australian National University studied in detail the patterns of cell divisions in roots of the water fern *Azolla*.

Azolla roots are unusual in that they do not grow throughout the life of the plant. The *Azolla* root is a determinate organ, and the root meristem has a genetically determined limit to its growth. This condition is in contrast to most roots, in which the activity of the meristem is indeterminate—growth persisting as long as environmental conditions permit it. The *Azolla* apical cell, which is shaped something like a pyramid with four more or less triangular sides, divides only 50 to 55 times, cutting off derivative cells in each of its four planes, before it stops dividing entirely. With the exception of the root cap, all the cells in the *Azolla* root can be traced to the apical cell, whose division planes are determined with great precision.

The cells that will become the root body are generated by divisions parallel to the three internal triangular faces of the apical cell, forming a wedge-shaped cell known as a merophyte. Each of the three derivative cells subsequently divides a set number of times to form a sector that makes up one-third of the root axis. The planes and numbers of divisions of the apical cell and each of its derivatives are very precise. We can recognize two different kinds of divisions—formative and proliferative—which differ in the fate of the daughter cells produced. Both daughter cells resulting from a proliferative division have the same fate.

Formative divisions may be either periclinal or anticlinal, but they differ from proliferative divisions in that the two daughter cells have different fates. That is, one of the daughters may remain undifferentiated while the other daughter cell will differentiate as a particular cell type. Formative divisions establish the files of cells that make up the various tissues of the root; proliferative divisions increase the number of cells in each file.

In the root, most formative divisions are periclinal (longitudinal), while proliferative divisions tend to be anticlinal (transverse). Longitudinal, formative divisions within the merophytes give rise to progenitor cells for the files of cells in the root. For example, the division that generates the endodermis and inner cortical cells occurs in the fourth merophyte on both sides of the root, as do the divisions that delineate the vessel element, pericycle, outer cortex, and root hair–epidermis cell files. Once formed, the progenitor cells usually complete several transverse, proliferative divisions before the progeny differentiate.

Shoot Branching

The arrangement of branches constitutes one of the most conspicuous aspects of shoot architecture. During vegetative growth, the formation of branches is guided by a complex interplay between genetically determined programs of development and the environment. As the plant develops, environmental cues are in some way perceived and interpreted to enable adaptive growth responses, ranging from complex shade avoidance responses, to replacement of shoots lost through herbivory. Root systems show a comparable, but usually less appreciated complexity, which, like shoots, reflects interactions between the environment2 s intrinsic programs. Although these programs have the potential to yield an infinite variety of forms, certain common themes suggest basic control mechanisms may be widely conserved among plants. This section examines the phenomenon of branching from several perspectives, beginning with a discussion of its relationship to shoot development and then shifting to a consideration of roots.

In the seed plants, branching is often constrained by phyllotaxy, since the predominant form of branching in most species is axillary (i.e., meristems form at the base of leaves). Superimposed on this basic pattern are further controls that regulate when and where such axillary meristems become active. In many species, axillary buds may lie dormant for long periods, becoming active only when the plant senses favourable growth conditions. Alternatively, buds can be activated by intrinsic factors, such as those associated with the onset of reproductive development.

Formation of the axillary meristem can be linked to the activity of several classes of transcription factors. Seed plants provide examples where secondary shoot systems can form in a variety positions on the plant, ranging from the small plantlets on leaf margins of Kalenchoe, to root-derived tuberous shoots. However, in most plants, such axillary shoots typically form in close association with the adaxial side of the leaf base. Because of this, the pattern of branching is often linked to some degree with the pattern of phyllotaxy. The earliest stages of axillary meristem development begin with the appearance of a bulge at the leaf primodia, soon after its initiation. *Knotted*-like genes are expressed from this early stage, in contrast to their down-regulation in adjacent leaf primordia. As the axillary develops further, it establishes continuity with the vascular network of the parent growth axis and begins to initiate lateral organs such as leaves or bracts. In many plants, this formative period is followed by a period of dormancy, which can be broken by either intrinsic or external cues.

Although the programs that underlie the formation of axillary buds are undoubtedly complex, some insight into their nature has been offered by the analysis of mutants in which the process is somehow altered. The mutations that affect the polarity of the leaf also influence the pattern of meristem development, suggesting that factors associated with adaxial patterning of the leaf somehow promote meristem formation. Mutations that activate such programs, such as those associated with the dominant alleles of *Phabulosa* and *Phavoluta*, lead to the ectopic formation of axillary meristems in both adaxial and abaxial positions. The related transcription factor, *Revoluta*, which functions in the normal establishment of a polarized vasculature, appears to also play a role. Loss of functions mutations that lead to vascular polarity defects can also block the formation of axillary meristems. The presence of *Start* domains in these Class III HD-ZIP class transcription factors suggest that their activity may be sensitive to an unidentified sterol-like signalling molecule.

Expression analyses suggest that other genes, such as *CUC*, that are required for the formation of the primary SAM in the embryo, are likely to play a similar role for axillary meristems. However, certain types of mutants in which only axillary meristems are affected point towards a novel class of more specific genes. One of these, *lateral suppressor*, was first described in tomato as a mutant in which lateral branches are completely blocked. Cloning of this gene revealed a VHIID-type transcription factor, suggesting that axillary branching is normally dependent on regulation of transcription by the LAS protein. The mechanism that LAS functions in appears to be well-conserved, as LAS-like genes have been described in both Arabidopsis and rice, which have similar branch-deficient mutant phenotypes. Some insights into the potential targets of LAS have been provided by analyses of mutants in Arabidopsis where loss of gene function is linked to a more diffuse pattern of STM expression. Thus, LAS may enable axillary meristem formation by activating genes normally involved in SAM formation. It remains unclear what signals are responsible for the localized expression of LAS.

Once formed, axillary meristems may lie dormant or become activated through a combination of environmental and intrinsic factors, and it appears that many of these responses are determined in part via hormone-signalling pathways. The plant growth hormone ABA, know for its role in promoting seed dormancy, has analogous activities in maintaining the dormancy of axillary buds. Factors that regulate synthesis, turnover, perception, and response of ABA are likely to play important roles in both processes.

Several lines of evidence implicate cytokinins as promoting axillary meristem growth. In classic regeneration experiments, a relatively high ratio of cytokinin to auxin was shown to be essential for regeneration of shoots from undifferentiated callus, whereas low ratios led to the formation of roots. Comparable effects are observed in *Agrobacterium*-infected plants. Plants infected with cytokinin-producing *Agrobacterium tumefaciens* strains develop ectopic shoots, whereas auxin-producing *Agrobacterium rhizogenes* develop ectopic roots. In addition to its role in initiating shoot meristems, cytokinin also appears capable of activating shoots when applied exogenously. Combining cytokinins effects on both the initiation and activation of axillary meristems has the potential to cause massive increases in branching, as suggested by the highly branched phenotype of *supershoot,* which have been linked to increased levels of cytokinin.

The activity of axillary meristems can also be repressed by factors associated with the growth of the primary shoot, a phenomenon known as apical dominance. This type of response can help ensure the survival of the plant where growing shoot tips have been damaged by herbivory or other environmental stresses. Apical dominance phenomena are also likely to be at least partly responsible for some of the complex branching patterns associated with flowering, including the elaboration of characteristic inflorescence architectures, or the release of more proximal vegetative buds to enable perennial growth. Although the phenomenon of apical dominance is well known and has been exploited throughout the history of agriculture to yield more productive plant forms, its basis still remains incompletely understood. Apical dominance can be inferred when the decapitation of a plant leads to the activation and growth of axillary buds.

As this activation can be suppressed by the application of auxin to the decapitated plant, and since growing shoots are a rich source of this hormone, it is reasonable to suppose that auxin plays some role in apical dominance. Indeed, physiological studies show that auxin is normally transported basipetally from the growing shoot of intact plants via auxin-efflux transporters. An association with auxin is reinforced by the observation that mutants such as *axr1,* which render the plant less-sensitive to auxin, lead to increased branching. However, auxin is normally not efficiently transported into buds, suggesting that its suppression of axillary bud-growth is likely to be mediated indirectly. The possibility that cytokinin might have a role in mediating apical dominance is suggested by its pattern of synthesis in plants and its growth-promoting activities. Cytokinin is known to be synthesized in roots; from there it is transported acropetally in xylem sap, as well as in nodal regions associated with axillary buds. Furthermore, the inhibitory effects of auxin on bud-growth can be bypassed by exogenous cytokinin, suggesting cytokinin acts downstream of auxin.

The involvement of a novel hormone-signalling pathway in the regulation of bud-growth has been inferred by genetic analyses of branching in Arabidopsis, pea and petunia. In Arabidopsis, a suite of mutants with more axillary branches encoded by the *MAX 1, 2, 3,* and *4* genes has been analysed by a combination of surgical and molecular techniques. Grafting experiments suggest that *MAX 1, 3,* and *4* can act nonautonomously. For example, the excessive branching of a shoot, which is mutant for any one of these genes, can be rescued by grafting it onto a normal root stock. By contrast, *max2* mutant shoots remain highly branched when grafted. These studies suggested that *MAX1, 3* and *4* are involved in the synthesis of a mobile signal, whereas the *MAX2* gene product mediates perception and response to the signal (e.g., by acting as a receptor). When various mutant versions of *MAX* genes are combined, there is no significant increase in branching, suggesting that these genes act in a linear response pathway rather than acting separately on independent pathways.

Figure: *Phenotypic comparison between (a) wild-type dwarf pea and (b) rms1, and between (c) max4-1 mutants and (d) wild-type Arabidopsis. max4 and rms1 exhibit increased branching, which occurs due to increased outgrowth of lateral buds. Pots are 10 cm in diameter. (Figure kindly supplied by Karim Sorefan.)*

Molecular cloning of the *MAX* genes have born out models suggested by grafting analyses. Both *MAX3* and *MAX 4* encode related gene products that are predicted to act as dioxygenases on carotenoid-type molecules. *MAX1*, by contrast, encodes a cytochrome P450-type enzyme, which is proposed to act downstream of *MAX3* and *MAX 4* to convert the carotenoid product of these genes into a hormone with branch-inhibiting activity (Booker et al. 2005). The identification of this unknown hormone through a combination of biochemical and genetic methods should offer valuable insights into the regulation of bud-growth and may offer new tools for crop improvement.

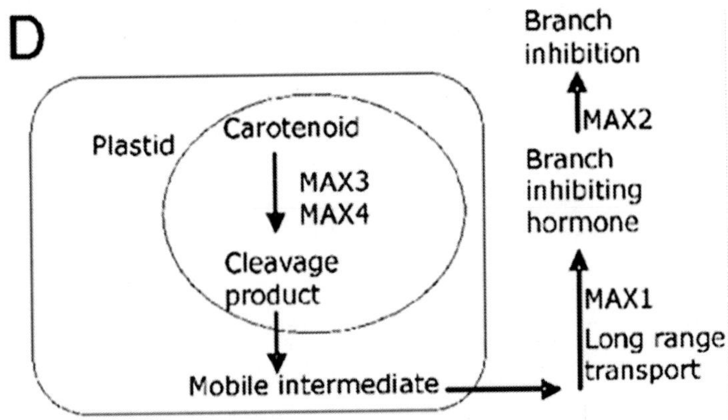

Figure: *Model for proposed functional relationships between* MAX *gene activities in relationship to branching.*

In contrast to *MAX 1, 3,* and *4,* which encode proteins involved in the synthesis of the branch-inhibiting hormone, *MAX2* has been shown to encode an F-box protein. As described in more detail elsewhere, F-box proteins can, when activated, target specific proteins for degradation. In the case of *MAX2,* one can hypothesize one or more proteins that activate bud-growth are normally targeted for destruction by the activated *MAX2* gene product. By this model, the perception of the carotenoid branch-inhibiting hormone would, in some way, activate the *MAX2* gene product. By analogy with the auxin receptors, which have also been shown to encode F-box proteins, it is conceivable that the MAX2 protein itself acts as a receptor, with binding of the branch-inhibiting hormone activating the ability of MAX2 to target specific proteins for degradation. Presumably, these proteins, which would be stabilized in the absence of MAX2 activity, would in some way promote branching.

3

Phytochrome and Light Control of Plant Development

The Structure of Phytochromes

Methods for phytochrome purification were developed through research pioneered by the agricultural biochemist H. William Siegelman, working at the USDA lab at Beltsville, Maryland. After surveying many different plants and plant tissues, Siegelman found that the young etiolated seedlings of cereals, such as oat and rye, were particularly rich sources of phytochrome. Some initial success in purifying phytochrome from etiolated oats (*Avena sativa*) was reported in 1964, and many workers subsequently used dark-grown oat seedlings for phytochrome extractions. As a result, much of the biochemical information on phytochrome is derived from studies on oat seedlings, although there is a considerable amount of data on phytochrome isolated from dark-grown rye (*Secale* spp.) and pea (*Pisum sativum*) seedlings as well.

While on a fresh-weight basis there is 50 times more phytochrome in etiolated tissue than in equivalent green tissue of the same age, the overall content of phytochrome in etiolated plants is low—about 0.2% of the total extractable protein. Thus, to purify even milligram quantities usually requires more than a kilogram fresh weight of starting material. Etiolated seedlings are also rich in proteases, and much of the phytochrome is susceptible to proteolysis. To minimize the loss of phytochrome, researchers carry out these extractions rapidly, at a low temperature, and in the presence of protease inhibitors. The stability of phytochrome during purification is also favoured by conversion to the Pfr form with red light. For these reasons, successful purification of fully intact, native phytochrome from etiolated oat seedlings was not reported until 1983.

Native phytochrome is a soluble protein with a molecular mass of about 250 kDa. It occurs as a dimer made up of two equivalent subunits. Each subunit consists of two components: a light-absorbing pigment molecule called the chromophore, and a polypeptide chain called the apoprotein. The apoprotein monomer has a molecular mass of about 125 kDa. Together, the apoprotein and its chromophore make up the holoprotein. The chromophore of phytochrome is a linear tetrapyrrole termed phytochromobilin. There is only one chromophore per monomer of apoprotein, and it is attached to the protein through a thioether linkage to a cysteine residue (Figure 1.A).

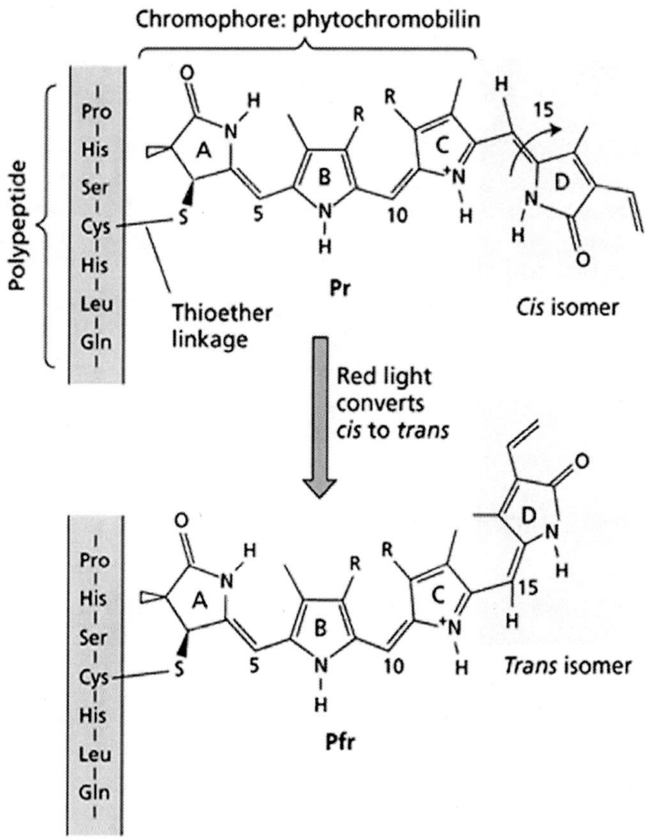

Figure: *Structure of the Pr and Pfr forms of the chromophore (phytochromobilin) and the peptide region bound to the chromophore through a thioether linkage. The chromophore undergoes a* cis-trans *isomerization at carbon 15 in response to red and far-red light.*
(After Andel et al. 1997.)

Researchers have visualized the Pr form of phytochrome using electron microscopy by spraying a solution of the protein onto mica flakes, drying it under vacuum, and "shadowing" the dried molecules by using platinum atoms to enhance its three-dimensional structure. As Figure 1.B shows, phytochrome molecules appear as a collection of spots, the number depending on the orientation of the protein. On the basis of this information and more accurate measurements using X-ray scattering, the model has been proposed. The polypeptide folds into two major domains separated by a "hinge" region. The larger N-terminal domain is approximately 70 kDa and bears the chromophore; the smaller C-terminal domain is approximately 55 kDa and contains the site where the two monomers associate with each other to form the dimer.

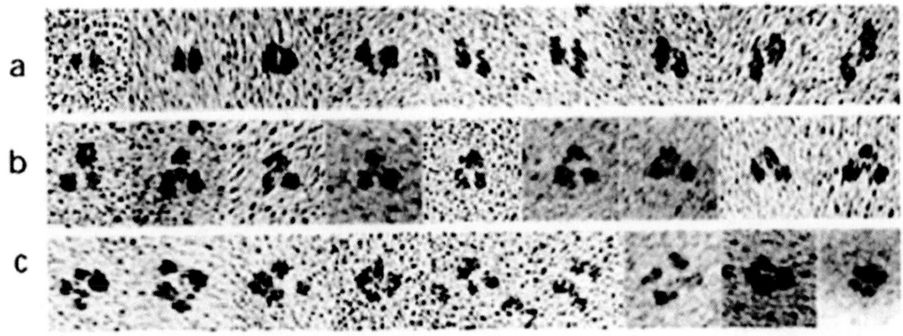

Figure : *Phytochrome from pea, purified and then sprayed onto a mica flake. After being dried, the specimen was shadowed with platinum to enhance the three-dimensional structure. A carbon replica was prepared and viewed in the electron microscope. Various images were produced, including pairs of spots (a), three spots (b), and clusters of spots (c). The variation in the pattern in A is caused by different orientations of the protein. (After Tokutomi et al. 1989, courtesy of M. Wada.)*

Phytochrome and High-Irradiance Responses

HIRs, such as the inhibition of stem or hypocotyl growth, have usually been studied in dark-grown, etiolated seedlings. The HIR action spectrum for the inihibition of hypocotyl elongation in dark-grown lettuce seedlings. Note that the main peak of activity is in the far-red rather than the red region of the spectrum, and there are peaks in the blue and UV-A regions as well. Because the absence of a peak in the red is unusual for a phytochome-mediated response, at first researchers believed that another pigment might be involved.

In the late 1960s, Karl M. Hartmann in Freiburg, Germany proposed a model to explain the apparent discrepancy. Taking a cue from the enhancement effect of photosynthesis that was discovered by R. Emerson and coworkers, Hartmann speculated that the HIR might be the result of an interaction between the Pr and Pfr forms of phytochrome. To probe such an interaction, Hartmann irradiated plants with not one but two wavelengths of light simultaneously, using two wavelengths that were inactive in the HIR for lettuce hypocotyl inhibition: 658 nm and 768 nm. Although physiologically inactive, the two wavelengths are still *photochemically* active in driving phytochrome conversions, each in the opposite direction. By carefully varying the irradiances of each wavelength, Hartmann was able to induce a range of ratios of Pfr to total phytochrome in the plant from near zero to higher values. He was able to show that the optimum ratio of Pfr to total phytochrome for the inhibition of lettuce hypocotyl growth is approximately 0.03. Remarkably, this is precisely the ratio that results when the plants are irradiated with a single wavelength at 720 nm, the far-red peak of the HIR.

The Origins of Phytochrome as a Bacterial Two-Component Receptor

Bacteria often use two-component signalling systems to sense and respond to environmental signals such as nutrients or light. The first component is a sensor protein, consisting of an *input domain* that receives the input signal, linked to a *transmitter domain* that transmits the signal to the second component. The second component is the response regulator protein, composed of a *receiver domain* that accepts the signal from the transmitter of the sensor protein, linked to an *output domain* that initiates the cellular response.

These two components interact by a series of phosphorylation and dephosphorylation reactions. When it receives a signal, the sensor autophosphorylates a specific histidine residue in the transmitter domain. This phosphate is then transferred to a specific aspartate residue on the receiver domain of the response regulator. But what has all this to do with phytochrome? Although they lack the specific histidine residue that becomes autophosphorylated, the C termini of phytochromes have some sequence similarity to the transmitter domains of bacterial sensor proteins. This sharing of sequences raises the interesting possibility that phytochrome evolved from the sensor protein of a bacterial two-component system.

Provocative new evidence for the idea that phytochrome evolved from the histidine kinase of a bacterial two-component system has recently emerged from studies of cyanobacteria. The cyanobacterium *Fremyella* uses light to regulate the composition of its photosynthetic pigments contained within the light-harvesting phycobilisomes, a response termed chromatic adaptation.

Like phytochromes, phycobiliproteins bind linear tetrapyrrole chromophores that are closely related in structure to the phytochrome chromophore, but there is no evidence for red/far-red reversible responses of the type seen in higher plants. Instead, chromatic adaptation occurs in response to growth in red light and green light. When the gene for the photoreceptor for this response (*RcaE*) was cloned, it was found to encode a 74 kDa polypeptide, the N-terminal end of which is related to the N terminus of higher-plant phytochromes, while the C-terminal end is related to histidine kinases. However, the lack of the conserved chromophore-binding site and of red/far-red reversibility suggests that this protein is a distant relative of higher-plant phytochromes. Publication of the complete sequence of the genome of the cyanobacterium *Synechocystis* in 1996 led to great excitement because this organism contains a gene that has between 50 and 60% amino acid identity with *Arabidopsis* phyE along its length, including a bona fide chromophore attachment site. In addition, the amino acid sequence of the C terminus of the *Synechocystis* pigment protein is related to the KinA sensor protein of *Bacillus subtilis*. This *Synechocystis* gene has been named *cph1* (*c*yanobacterial *ph*ytochrome *1*).

Recently it was shown that the purified Cph1 apoprotein can assemble with phycocyanobilin, a cyanobacterial linear tetrapyrrole, in vitro, and the resultant holoprotein exhibits red/far-red reversibility and absorption spectra similar (but not identical) to higher-plant phytochrome assembled in vitro with the same chromophore. Perhaps the most exciting finding of all was the identification, 10 base pairs downstream from the cyanobacterial phytochrome gene, of another gene having the features of a bacterial response regulator. Since the C-terminal end of the cyanobacterial phytochrome contains all the conserved features of a histidine kinase transmitter module, and the downstream response regulator resembles aspartate kinase receiver modules, it was now possible to test whether this phytochrome acts as part of a light-regulated transmitter–receiver pair, as had been proposed earlier.

Using the purified proteins, the Cph1 protein was shown to phosphorylate itself in a manner consistent with its having a histidine kinase activity, and further, this phosphate could be transferred to the response regulator. Surprisingly, Cph1 is far more active as a kinase in the Pr form than in the Pfr form, suggesting that the light signal is transduced through regulation of Pr abundance, rather than through Pfr abundance as in higher plants.

Taken together, the results from these studies suggest that plant phytochromes originally evolved from a class of photoreversible transmitter histidine kinases in cyanobacteria. Are plant phytochromes also kinases? In support of this idea, a phytochrome gene isolated from the moss *Ceratodon* contains an additional C-terminal sequence that has kinase homology. In addition, highly purified preparations of higher-plant phytochrome have been found to contain a polycation-stimulated protein kinase activity that is also greater in the Pr form than in the Pfr form.

However, the kinase activity associated with these phytochromes is of the serine–threonine type, and consistent with this observation, these proteins lack the conserved histidines of the bacterial sensor proteins. More recently, it was shown that recombinant oat phyA is a serine-threonine kinase with the same biochemical properties as the plant-derived photoreceptor. Moreover, R-light stimulated its autophosphorylation, supporting the hypothesis that Pfr is the active form. Polycations were shown to stimulate Pr phosphorylation more than Pfr, a result consistent with previous studies.

Thus, while higher-plant phytochromes share a homology with bacterial two-component signalling systems, the reaction mechanism and components may have diverged considerably. Nevertheless, the discovery of cyanobacterial phytochromes has opened up many exciting new areas of research that have already contributed much to our understanding of phytochrome evolution and function.

Mougeotia: ***A Chloroplast with a Twist***

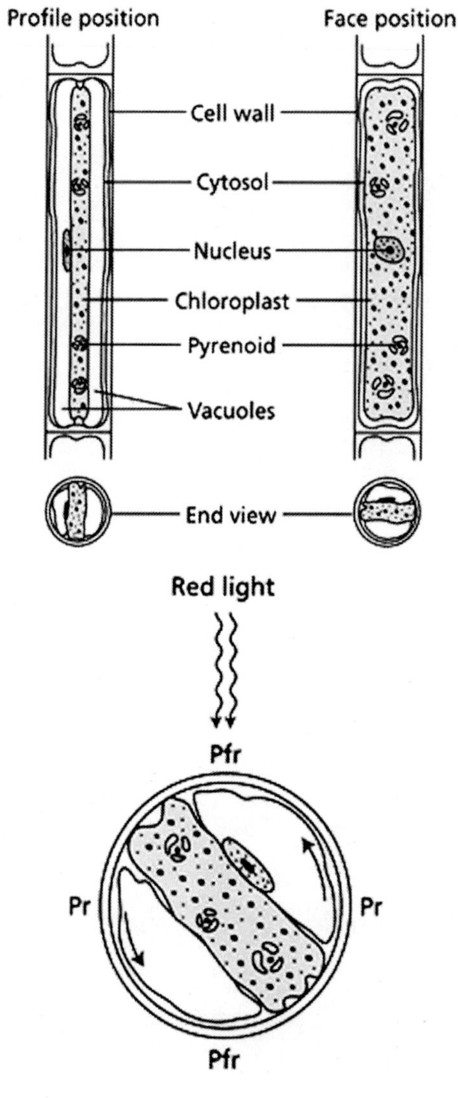

Mougeotia is a filamentous green alga whose cells contain a single ribbonlike chloroplast nestled between two large vacuoles and surrounded by a layer of cytoplasm. The chloroplast can rotate about its long axis and can respond to light by orienting *perpendicularly* to the direction of light. The perpendicular orientation is referred to as "face position."

Phytochrome is implicated in the *Mougeotia* chloroplast rotation response because the system is most sensitive to red light and because the red light effect can be reversed by far-red light. Using microbeams of red and far-red light, Wolfgang Haupt and his coworkers in Germany showed that in *Mougeotia*, phytochrome is localized near the periphery of the cytoplasm in the vicinity of the plasma membrane. When a microbeam of red light was directed at the cell surface under a microscope, the edge of the chloroplast adjacent to the microbeam rotated 90°, even though the chloroplast itself was not illuminated.

Experiments with Plane Polarized Light

The orientation of individual phytochrome molecules within the cell was probed by the use of plane-polarized light. Each phytochrome molecule has a preferred orientation for the absorption of plane-polarized light. The greatest absorption occurs when this light is parallel to the absorbing surface of the pigment. If we think of the pigment as a rod and the polarized light as a narrow slit of light, more light will be absorbed when the light slit is parallel to the long axis of the pigment than in any other orientation.

Haupt and his coworkers discovered that the ability of plane-polarized red light to cause chloroplast rotation depends on the plane of polarization relative to the long axis of the cell. This phenomenon is depicted in the experiment shown in below. Half of a cell was illuminated over the entire surface with polarized red light (R) vibrating *perpendicular* to the cell axis; the other half was illuminated with light vibrating *parallel* to the cell axis.

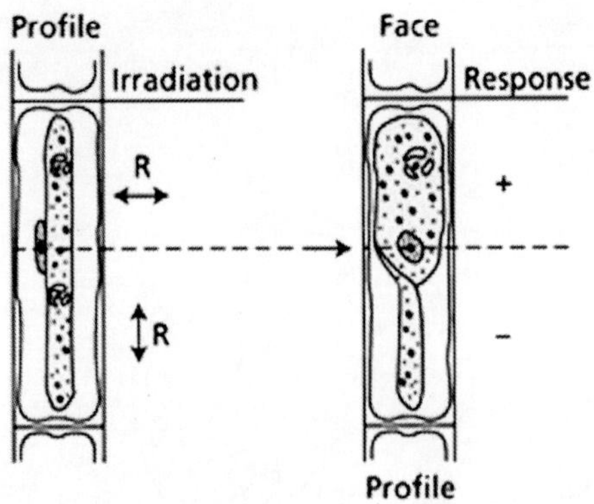

The chloroplast rotated from the profile to the face position only in response to red light vibrating in the perpendicular direction, producing a twist in the chloroplast. On the basis of these experiments, it was concluded that phytochrome molecules have a defined orientation. Two additional experiments demonstrated that the phytochrome orientation changes during the conversion from Pr to Pfr. In the first experiment, shown in Figure below, the cytoplasmic layer was irradiated with a microbeam of plane-polarized red light vibrating either *parallel* (upper light treatment) or *perpendicular* (lower light treatment) to the cell's long axis. In apparent contradiction to the results shown in Figure above, the chloroplast rotated only when the red light was vibrating *parallel* to the cell's long axis.

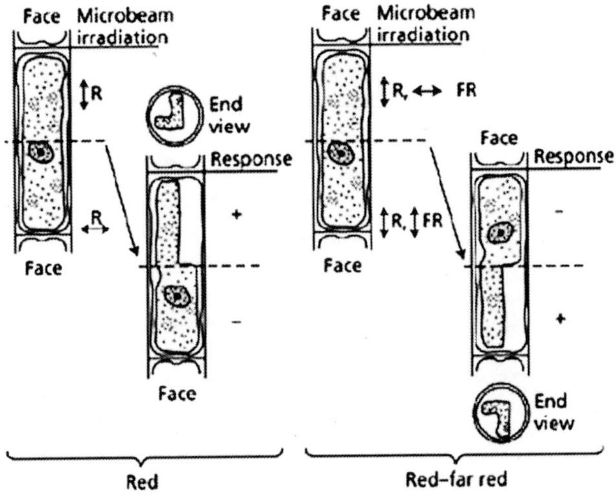

The apparent contradiction is due to the cylindrical shape of the cell. Thus, the phytochrome molecules near the two edges will be oriented 90° relative to the phytochrome molecules at the centre. The phytochrome molecules that absorb red light when the polarized microbeam is aimed at the edge of the cell in the experiments of Figure above, will be oriented 90° from the phytochrome molecules that absorb light in the *whole-cell* irradiation experiments. In the second experiment, shown on the right-hand side of Figure above, the cytoplasmic layer on the right side of the cell was irradiated with a microbeam of polarized red light vibrating *parallel* to the cell' long axis. On the basis of the previous experiment, we can predict that red light vibrating parallel to the cell's long axis will induce chloroplast rotation. However, the red light treatment was immediately followed by a beam of far-red light. One half of the cell (top) was given a second irradiation of far-red light vibrating *perpendicular* to the cell axis; the other half of the cell (bottom) was given a second irradiation of far-red light vibrating *parallel* to the cell axis. As shown in the right-hand side of Figure above, far-red light prevented red light–induced chloroplast rotation only when it was vibrating perpendicular to the cell axis. This result was explained by the proposal that phytochrome undergoes a 90° change in orientation upon photoconversion of Pr to Pfr and vice versa.

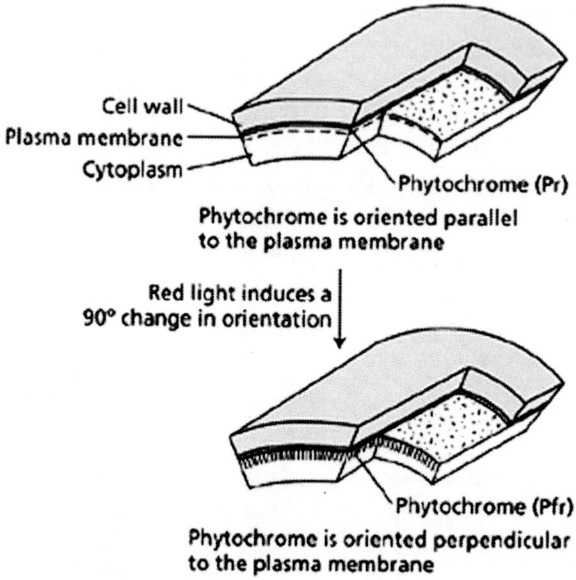

A Model for Chloroplast Rotation

Finally, what is the mechanism of chloroplast rotation in *Mougeotia*? The effects already described are best accounted for by a model in which phytochrome is located either on the plasma membrane of *Mougeotia* or on the microtubules that form a cylindrical scaffold directly beneath the membrane. It is difficult to imagine how Pr and Pfr could maintain their respective orientation without being associated either with the membrane or with the cylinder of microtubules. The nucleotide sequence that codes for *Mougeotia* phytochrome, consistent with those of other known phytochromes, suggests that the phytochrome is not an integral membrane protein, because it lacks hydrophobic transmembrane domains. On the other hand, the amino acid sequence at the carboxy-terminal end is similar to that of proteins that bind microtubules. So the evidence suggests that phytochrome is bound to the cytoskeleton adjacent to the plasma membrane.

During rotation, the chloroplast moves away from a localized region of Pfr and toward a region of Pr. *Mougeotia* chloroplasts move in response to unilateral light as a result of gradients of the active form of phytochrome, Pfr, that form an alternating pigment pattern around the periphery of the cell cylinder, as part B of the figure shows. The gradients of Pfr are rapidly translated into gradients of actin–myosin microfilament interactions, which generate the motive force for chloroplast rotation. The primary mechanism of Pfr action remains obscure. Calcium ions from internal stores and calcium-activated calmodulin are known to participate in the response by interacting with the scaffold of microtubules, but the precise relationship between phytochrome and calcium is not yet understood. Studies using drugs that inhibit microtube and microfilament action do not support a role for microtubules in the mechanism of chloroplast movement. However, experiments with Cytochalasin B do indicate that microfilaments are involved.

Regulation of Transcription by *Cis*-Acting Sequences

The precise number and type of *cis*-acting sequences involved in regulating transcription of a given gene have been determined by various approaches:

Conserved sequences. The sequences of the promoter regions of a given gene (for example, *RBCS*) are first compared within members of the gene family in a particular species, and between gene family members of different species. Sequences acting as *cis*-acting regulatory elements are often represented in many species (that is, they are conserved) because they are essential for function, while the DNA sequences around them may be less essential, and therefore more variable. This approach gives an idea of the locations of sequences that may be involved in regulating the expression of a gene. For *RBCS* genes, the regulatory element typically encompasses a region about 1 kilobase of DNA upstream from the transcriptional start site.

Identification of response elements. Once general locations of the conserved sequences have been identified, reporter constructs of the type already described can be tested to determine whether this fragment is necessary and sufficient to confer light regulation on the expression of the reporter sequence. Then increasing amounts of the 52 end of the promoter region can be deleted to determine the minimum piece of DNA required for expression.

Typically, 250 base pairs of DNA upstream of the transcription start site are required for light-regulated transcription. Internal deletions or the insertion of small, unrelated sequences, called *linkers*, can be used to define *cis*-acting regulatory elements. Conferring light regulation. Synthetic copies of putative elements that contain one or more copies of a sequence can be inserted into non-light-regulated promoters to see if they can confer light regulation.

Identification of *trans*-acting factors. In vitro experiments in which nuclear protein extracts are allowed to bind to these different DNA fragments can be used to characterize the protein factors that bind to the *cis*-acting elements. Together, these four approaches have yielded detailed promoter maps for several genes, including *RBCS* and oat *PHYA*. Some of the important *cis*-acting elements present in many light-regulated genes include GT-1 sites (GGTTAA), I boxes (GATAA) and G boxes (CACGTC).

The *trans*-acting factors that bind to these three elements have also been characterized. However, each of these three *cis*-acting regulatory elements has also been found in the promoters of some non-light-regulated genes. Moreover, none of these elements alone is able to confer light regulation onto a non-light-regulated gene. Thus, light regulation arises from various combinations of different general elements rather than from a single regulatory element. Recent evidence suggests that the minimal sequence required for light-inducible gene expression is a pairwise combination of two elements such as GT-1 sites, G boxes, or I boxes.

Phytochrome Regulation of Gene Expression

We have seen that phytochrome-mediated light responses at the cellular level can involve rapid effects on membrane properties. The stimulation and repression of transcription by light can also be very rapid, with lag times as short as 5 minutes. Such early-gene expression is likely to be regulated by the direct interaction of transcription factors by one or more phytochrome-initiated signal transduction pathways. Within the nucleus, these transcription factors stimulate the transcription of specific genes. Some of the gene products of the early genes are transcription factors themselves, which activate the expression of other genes. Expression of the early genes, also called primary response genes, is independent of protein synthesis; expression of the late genes, or secondary response genes, requires the synthesis of new proteins.

As with cytokinin-regulated gene expression, the photoregulation of gene expression has focused on the nuclear genes that encode messages for chloroplast proteins: the small subunit (SSU) of ribulose-1,6-bisphosphate carboxylase/oxygenase (rubisco) and the major light-harvesting chlorophyll *a*/*b*–binding proteins associated with the light-harvesting complex of photosystem II (LHCIIb proteins). These proteins play important roles in chloroplast development and greening; hence their regulation by phytochrome has been studied in detail. The genes for both of these proteins—*RBCS* and *LHCB* (also called *CAB* in some studies)—are present in multiple copies in the genome.

The abundance of these mRNAs has been shown to increase in response to light and phytochrome action. The abundance of other mRNAs, however, can be decreased by phytochrome action; this latter class includes mRNAs that encode oat phyA and the chlorophyll biosynthesis enzyme NADPH–protochlorophyllide oxidoreductase. Although the abundance of most of these mRNAs is regulated by light given in the low-fluence range, some *LHCB* mRNAs also respond to light in the very-low-fluence range.

We can demonstrate phytochrome-regulated induction of mRNA abundance (for example *RBCS* mRNAs) experimentally by giving etiolated plants a brief pulse of low-fluence red or far-red light, returning them to darkness to allow the signal transduction pathway to operate, and then measuring the abundance of specific mRNAs in total RNA prepared from each set of plants. If its abundance is regulated by phytochrome, the mRNA is absent or present at low levels in etiolated plants, but is increased by red light. The red light–induced increase in expression can be reversed by immediate treatment with far-red light, but far-red light alone has little effect on mRNA abundance.

In the case of mRNAs (e.g., *LHCB*) that respond to very-low-fluence red light, expression increases in response to both red and far-red light, and additional experiments are needed to confirm the involvement of phytochrome. In the case of mRNAs whose expression is negatively regulated by phytochrome, for example oat *PHYA* mRNA, levels are *high* in darkness, *reduced* by a pulse of red light, and *restored* to dark levels by red/far-red light treatment. However, none of these experiments prove the action of phytochrome on transcription; instead they show the effect of phytochrome only on overall mRNA abundance in total RNA, which reflects both transcription and degradation processes. As mentioned earlier, the abundance of oat *PHYA* mRNA in total RNA is the result of both transcription and degradation events.

The involvement of phytochrome in transcriptional regulation of mRNA abundance has been demonstrated by two approaches. In the first approach, etiolated plants are treated as was described for the mRNA studies, and then transcriptionally active nuclei are isolated and allowed to elongate transcripts initiated under the different in vivo light treatments in the presence of radioactive precursors. If the relative incorporation into an mRNA is similar to that described for the overall abundance in total RNA, then transcription of the mRNA is thought to be under phytochrome control. The second approach involves fusing the promoter of the gene to be tested with the coding sequence of a reporter gene such as *GUS*. By placing the fused gene back into the plant it is possible to measure the abundance of the reporter mRNA or its encoded enzyme activity. Both of these approaches have been used extensively, and the latter has allowed the definition of the minimum amount of the upstream promoter region required to confer phytochrome-regulated transcription of the linked reporter sequence.

Recently, Elaine Tobin and her colleagues at the University of California, Los Angeles, characterized a MYB-related transcription factor whose mRNA level increases rapidly when *Arabidopsis* is transferred from the dark to the light. This DNA-binding protein appears to bind to the promoter of an *LHCB* gene and regulate its transcription,, occurs later than the increase in the MYB-related protein. The gene that encodes the MYB-related protein is therefore probably a primary response gene, while the *LHCB* gene is probably a secondary response gene. More recently, microarray analysis has been applied to the study of light signal transduction. In microarray analysis short oligonucleotides or cDNA sequences representing thousands of gene sequences are spotted on a glass slide. RNA samples isolated from control and experimental plants are labelled using two different fluorescent dyes. Both the control and the experimental sample are hybridized to the microarray and the slide is then scanned using two lasers that will excite one or the other dye.   The light emitted from the dye is proportional to the amount of RNA that is bound to the gene-specific oligonucleotide or cDNA sequence and thus provides a relative measure of gene expression.

To visualize the relative expression levels from both samples, images are often false coloured so that one signal is represented with green and the other signal as a red, when the signals overlap they are yellow. The commercially available Affymetrix arrays utilizes small 25-mer oligos as probes on the glass slide that are hybridized to a single labelled sample. This technique has been used to examine changes in gene expression when plants are grown under red or far-red light to define phytochrome signalling pathways and have shown that multiple transcription factors are early targets of phytochrome control.

Two-Hybrid Screens and Co-Immunoprecipitation

Two-hybrid library screens are based on the finding that many transcriptional activators contain separable DNA binding and transcriptional activation domains. In a two-hybrid screen, a DNA binding domain of a transcription factor is fused to a test or "bait" protein such as phyA. Yeast cells carrying the bait protein are transformed with a library of plant cDNA products that are fused to a transcriptional activation domain ("prey"). If the bait protein interacts with the prey protein, then the activation domain will be brought into close proximity to the DNA binding site of the transcriptional activator, thus driving reporter gene expression that will pigment the yeast cells or provide an essential amino acid, permitting growth.

Co-immunoprecipitation experiments are often conducted to provide independent verification of the results of two-hybrid screens. These assays can be performed using *in vitro* synthesized versions of bait and prey proteins. The proteins are mixed together and immunoprecipitated using an antibody that is specific to either the bait or prey construct. If the two proteins co-precipitate, it is evidence they interact with each other.

The Roles of G-Proteins and Calcium in Phytochrome Responses

Early experiments with intact seedlings and cultured cells indicated that the control of gene expression by phytochrome involves the action of a heterotrimeric G-protein and the calcium-binding regulatory protein calmodulin. Subsequently, a single-cell system has been developed that can be microinjected with various proposed components of the pathway and monitored for changes in gene expression. Although these experiments have not yet led to the identification of specific pathway components, they have been informative about the involvement of G-proteins and calmodulin.

Microinjection of phytochrome (mainly phyA) purified from etiolated oat seedlings into single cells prepared from the tomato *aurea* mutant (a chromophore biosynthetic mutant that has reduced levels of all phytochromes) results in stimulation of chloroplast development, accumulation of anthocyanin pigment, and expression of a GUS-coding sequence fused to a phytochrome-regulated promoter sequence. Etiolated seedlings of the *aurea* mutant show poor induction of these responses when transferred to light, probably because of the reduction in the levels of functional phyA. Inhibitors of G-proteins block the phytochrome-stimulated events in the single-cell system, while activators or intermediates stimulate the response. Increased levels of Ca^{2+} in the medium also stimulate *GUS* expression and chloroplast development when calmodulin is present, but they do not affect anthocyanin biosynthesis.

In the *Aurea* cell system, cGMP has been shown to act as a signalling intermediate in at least two signal transduction pathways. Specifically, the pathway that activates anthocyanin biosynthesis requires cGMP. Inhibitors of cGMP action inhibit this pathway; addition of cGMP activates it. In contrast, the pathway that requires calcium to activate chloroplast development and gene expression has a variable requirement for cGMP. Activation of genes that encode several chloroplast proteins, including RBCS and LHCB, does not require cGMP, while activation of genes that encode some photosystem I polypeptides and the cytochrome *b6*/*f* complex does.

The difference in signalling requirements is due in part to the different *cis*-acting regulatory elements found upstream of these genes. Specific DNA elements upstream of *RBCS* genes appear to be the targets for the calcium pathway, while other distinct elements located upstream of anthocyanin biosynthesis genes are the targets for cGMP. The differential regulation of the expression of phytochrome-regulated genes can now be explained in terms of their *cis*-acting regulatory elements, different combinations of which afford both tissue-specific and light-specific expression.

Phytochrome Effects on Ion Fluxes

Phytochrome can rapidly alter the properties of membranes. We have already seen that low-fluence red light is required before the dark period to induce rapid leaflet closure during nyctinasty, and that fluxes of K^+ and Cl^- into and out of extensor and flexor cells mediate the response. However, the rapidity of leaf closure in the dark (lag time about 5 minutes) would seem to rule out mechanisms based on gene expression. Instead, rapid phytochrome-induced changes in membrane permeability and transport appear to be involved.

The role of the plasma membrane proton pump in *Samanea* pulvini (the structures at the base of the petiole that include the extensor and flexor cells) was studied by use of a proton-sensitive liquid membrane electrode. During phytochrome-mediated leaflet closure, the apoplastic pH of the flexor cells (the cells that swell during leaflet closure) decreased, while the apoplastic pH of the extensor cells (the cells that shrink during leaflet closure) increased. Thus, the plasma membrane H^+ pump of the flexor cells appears to be activated by darkness (provided that phytochrome is in the Pfr form), whereas the H^+ pump of the extensor cells appears to be deactivated under the same conditions.

Increased H^+ pump activity on the plasma membranes of the flexor cells should increase the membrane potential, thereby allowing more K^+ to enter the cells. Conversely, a reduction in the plasma membrane H^+ pump activity in the extensor cells would promote the loss of K^+ ions. Consistent with this model, the reverse pattern of apoplastic pH change was seen during leaflet opening.

Studies have also been carried out on phytochrome regulation of K^+ channels. Since it is difficult to study membrane channels in the intact pulvinus, Richard Crain and his colleagues at the University of Connecticut isolated protoplasts (cells without their cell walls) of both extensor and flexor cells from *Samanea* leaves. They determined the open or closed state of the K+ channels indirectly by measuring the membrane potential changes that occurred when the extracellular K^+ concentration was raised from 20 to 200 m*M*. When the extracellular K^+ concentration was raised, K^+ entered the protoplasts and depolarized the membrane potential only if the K^+ channels were open. When the extensor and flexor protoplasts were transferred to constant darkness, the state of the K^+ channels exhibited a circadian rhythmicity during a 21-hour incubation period and the two cell types varied reciprocally, just as they do in vivo. That is, when the extensor K^+ channels were open, the flexor K^+ channels were closed, and vice versa. Thus, the circadian rhythm of leaf movements has its origins in the circadian rhythm of K^+ channel opening.

To test the effects of phytochrome on K^+ channels, researchers first treated protoplasts with white light, then exposed them to a brief pulse of either red light alone or red light followed by far-red light and transferred them to continuous darkness. Red light caused the K^+ channels of the flexor protoplasts to open, and far-red light reversed the effect of red light. The extensor protoplasts were closed by both treatments. In contrast to red light, blue light caused the K^+ channels of the extensors to open, and those of the flexors to close. Thus, phytochrome seems to regulate specifically the K^+ channels of the flexor cells. Blue light regulates primarily the extensor cells, although it can also affect the flexors at specific times of day.

On the basis of the evidence thus far, we can conclude that phytochrome brings about leaflet closure by regulating the activities of the primary proton pumps and/or the K+ channels of the flexor and extensor cells. Although the effect is rapid, it is not instantaneous, and it is therefore unlikely to be due to a direct effect of phytochrome on the membrane. Instead, phytochrome acts indirectly via one or more signal transduction pathways, as in the case of the regulation of gene expression by phytochrome.

However, some effects of red and far-red light on the membrane potential are so rapid that phytochrome may also interact directly with the membrane. Such rapid modulation has been measured in individual cells and has been inferred from the effects of red and far-red light on the surface potential of roots and oat (*Avena*) coleoptiles. The lag between the production of Pfr and the onset of measurable potential changes varies from organism to organism; for example, it is about 1.7 s for depolarization (inside becomes less negative relative to outside the plasma membrane) in the giant alga *Nitella* and 4.5 s for hyperpolarization in *Avena.*

Changes in the bioelectric potential of cells imply changes in the flux of ions across the plasma membrane. The *Mougeotia* microbeam irradiation experiments of Haupt seemed to support the idea that phytochrome is localized on the plasma membrane, and membrane isolation studies carried out in the 1970s provided evidence that a small portion of the total phytochrome is tightly bound to various organellar membranes. These findings led some workers to suggest that membrane-bound phytochrome represents the physiologically active fraction, and that all the effects of phytochrome on gene expression are initiated by changes in membrane permeability. On the basis of sequence analysis, however, it is now clear that phytochrome is a hydrophilic protein without membrane-spanning domains. The current view is that phytochrome in *Mougeotia* is associated with microtubules located directly beneath the plasma membrane.

If phytochrome exerts its effects on membranes from some distance, no matter how small, involvement of a "second messenger" would be implied. For example, rapid changes in cytosolic free calcium have been implicated as second messengers in several signal transduction pathways, and there is evidence that calcium plays a role in chloroplast rotation in *Mougeotia*. More recently, however, the role of calcium as a second messenger in phytochrome action has been questioned, so calcium-independent mechanisms may also be involved.

Microarray Analysis of Shade Avoidance

To identify components of the shade-avoidance response pathway, microarray analysis was conducted in Arabidopsis, comparing plants grown in low and high R:FR light environments. These studies have identified both early-and late-acting genes that regulate responses to shading by other plants. Characterizations of two Arabidopsis transcription factors (*ATHB-2* and *PIL1*) also suggests that shade avoidance is initiated by rapid transcriptional changes following a decrease in the ratio of red to far-red light. These profiling studies strengthen the previous genetic studies that show that phyA acts to antagonize phyB-mediated responses under prolonged low R:FR light conditions through the transcriptional control of target genes that include components of the auxin transport and signalling pathways. Interestingly, it appears that the circadian clock is capable of regulating or "gating" physiological responses to R:FR through a modulation of gene expression. That is, changes in gene expression due to a change in the R:FR only occur at certain phases of the circadian rhythm.

Blue-Light Responses: Stomatal Movements and Morphogenesis

Guard Cell Osmoregulation and a Blue Light-Activated Metabolic Switch

Studies on photosynthetic algae, such as *Chlorella,* have shown that irradiation by blue light increases respiratory rates of dark-grown cells, with an action spectrum that is typical of specific blue-light responses. These changes in respiratory rates are associated with a drastic shift in metabolism, which can also be observed in experiments using blue and red light.

Red light stimulates photosynthetic carbon fixation in these algae, and the fixed carbon is incorporated into glucose and sucrose. The metabolic shift is observed when low photon fluxes of blue light are added to the red-light treatment; glucose and sucrose are no longer synthesized, and the fixed carbon is used in the biosynthesis of organic acids, amino acids, and proteins.

Blue light is also a powerful regulator of guard cell metabolism. Guard cells irradiated with red light accumulate sucrose; if low photon fluxes of blue light are added to the red-light irradiation, sucrose stops accumulating, and guard cells accumulate K^+ and malate^{2-}. Thus, the blue light–stimulated shift to the synthesis of the organic acid malate in guard cells parallels the shift to the synthesis of organic acids and amino acids in photosynthetic algae.

The experimental use of blue and red light has been useful for investigations of osmoregulatory pathways associated with stomatal movements. The rapid activation of proton pumping by blue light generates a proton gradient at the guard cell plasma membrane that drives potassium and chloride uptake. Blue light also stimulates starch hydrolysis and malate biosynthesis. Prolonged incubations of guard cells in detached epidermis under low fluence rates of blue light result in large stomatal apertures accompanied by extensive starch hydrolysis.

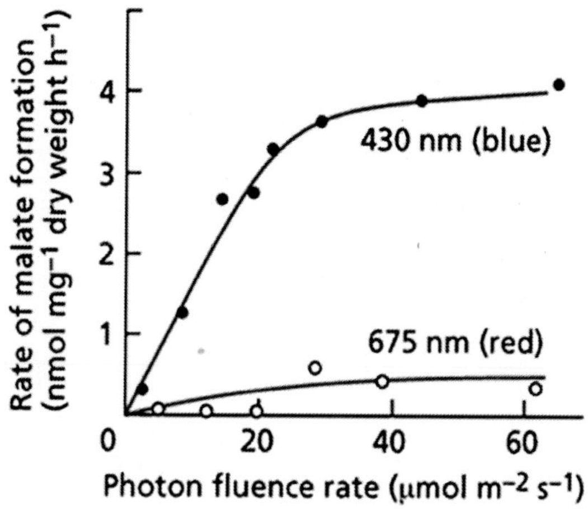

Figure: *Blue light stimulates malate synthesis in guard cells. Under red light, guard cells synthesize mostly glucose and sucrose; this metabolic pathway is deactivated by blue light, and metabolism switches to starch degradation and malate biosynthesis.*

Guard cells also show a sucrose-dependent osmoregulatory phase. Sucrose can originate from starch hydrolysis. On the other hand, guard cells from detached epidermis irradiated with red light under ambient CO_2 concentrations operate a third metabolic pathway, showing no detectable potassium uptake or starch hydrolysis. In these conditions, guard cells slowly accumulate sucrose, and this sucrose accumulation is blocked by DCMU (dichloropheny-ldimethylurea), an inhibitor of photosynthetic electron transport.

A recent study of guard cell osmoregulation under red light illumination and CO_2-free air showed that under these conditions, in which photosynthesis is inhibited by the absence of CO_2, red light activates the potassium-dependent pathway. These studies underscore the remarkable metabolic flexibility of guard cells. A study of the effects of calcium and abscisic acid on stomatal apertures under red and blue light found that increasing calcium or ABA concentrations inhibited blue light–stimulated stomatal opening in a concentration-dependent fashion, but had no effect on red light–stimulated opening. These contrasting responses to blue and red light can be explained by the effect of the treatments on guard cell osmoregulation. ABA and high calcium concentrations have been shown to inhibit proton pumping and potassium uptake, which are central to blue light–stimulated stomatal opening. Red light, on the other hand, stimulates guard cell photosynthesis and sucrose accumulation, and this osmoregulatory pathway appears to be insensitive to

ABA and calcium. Unlike visual systems, the colour information in the blue light and red light treatments does not appear to have functional significance for photosynthetic organisms.

Rather, their dramatic experimental effects appear to be a result of a differential stimulation of carotenoids and chlorophylls that are selectively excited by blue light applied under a saturating background of red light, or by red light alone, respectively. Under solar radiation, regulatory pathways activated by the excitation of these pigments are used by the chloroplast to sense key features in their environment and to control their metabolism.

The Coleoptile Chloroplast

Guard cell chloroplasts are a conspicuous structural feature of guard cells, and their physiology and biochemistry have been studied for several decades. In contrast, the chloroplasts of phototropically sensitive coleoptiles have only recently been analysed in detail. The coleoptile chloroplast had been suggested as a possible site for photoreception in oat coleoptiles by Kenneth V. Thimann and his coworkers in 1960, but interest in the flavin hypothesis later in that decade shifted attention away from coleoptile chloroplasts and these organelles have been virtually ignored for several decades.

Recent studies have shown that the coleoptile chloroplast evolves oxygen, fixes CO_2 photosynthetically, and operates the xanthophyll cycle. Dark-grown, etiolated coleoptile tips have violaxanthin and antheraxanthin but no zeaxanthin, and they accumulate zeaxanthin when irradiated.

When dark-grown coleoptiles are pretreated with red light for different lengths of time, their zeaxanthin content is proportional to the length of the red-light treatment. Exposure of coleoptiles with different zeaxanthin content to a brief pulse of blue light elicitis phototropic bending that is linearly related to the zeaxanthin content of the coleoptiles. Most interestingly, the regression line through the data points extrapolates to zero, suggesting that there is no phototropic response in the absence of zeaxanthin.

Red light–treated coleoptiles with a high zeaxanthin content convert their zeaxanthin to violaxanthin when transferred to the dark. Phototropic bending in response to a pulse of blue light given at different time intervals after the transfer to the dark is linearly related to zeaxanthin content.

When corn coleoptiles are given a saturating red-light pretreatment and a blue-light pulse in the presence of increasing concentrations of DTT (an inhibitor of zeaxanthin formation), bending and zeaxanthin content decrease linearly as a function of DTT concentration.

Thus the relationship between the zeaxanthin content of coleoptile chloroplasts and blue light–dependent phototropic sensitivity is very similar to the relationship between the zeaxanthin content of the guard cell chloroplast and the blue-light sensitivity of stomatal opening.

Coleoptile and guard cell chloroplasts share some properties that are clearly distinct from mesophyll chloroplasts, including high rates of oxygen evolution and low rates of photosynthetic carbon fixation, a high sensitivity of their xanthophyll cycle to low levels of incident photon fluxes, and a high starch content.

The coleoptile and guard cell chloroplasts also show a specific blue light response with an action spectrum that closely matches the action spectra for blue light–stimulated stomatal opening and coleoptile phototropism.

These remarkable similarities suggest that these distinct characteristics are typical of chloroplasts specialized in signal transduction, as opposed to the classic role of the mesophyll chloroplast, specialized in photosynthetic carbon fixation.

4

Abscisic Acid: A Seed Maturation and Antistress Signal

Structural Requirements for Biological Activity of Abscisic Acid

The features shown to be essential for biological activity include the carboxyl group, the tertiary hydroxyl group, the 2-*cis*, 4-*trans*-pentadienoic side chain, a 4'-ketone, and a double bond in the cyclohexane ring. Catabolic products of ABA that are present in the tissue and are missing any of these groups are biologically inactive.

The Bioassay of ABA

Coleoptile growth, the classic bioassay devised for auxins, is also used for ABA detection in plant extracts by measurement of coleoptile growth inhibition. This bioassay has adequate sensitivity (minimum detectable level is 10^{-7} *M*) and shows a linear response in the range of 10^{-7} to 10^{-5} *M*, but it has some disadvantages. Plant extracts have other promoters and inhibitors, which decrease the specificity of the bioassay and make preliminary purification steps necessary.

Other available bioassays for ABA include inhibition of germination, inhibition of gibberellic acid– induced α-amylase synthesis in cereal aleurone layers, induction of leaf abscission, and stomatal closure. Stomatal closure is highly specific for ABA because it is affected little by other plant growth regulators. Additional advantages of this bioassay include a fast response of guard cells to ABA, high sensitivity (minimum detectable level 10^{-9} *M*), and a linear response over a wide range of concentrations.

Proteins Required for Desiccation Tolerance

LEA protein sequences are highly conserved among a wide variety of species. They are related to two other groups of proteins, the RAB (*r*esponsive to *AB*A) and DHN (*de**h**ydri**n*) proteins. These protein classes accumulate in a variety of tissues experiencing cellular dehydration, e.g. maturing embryos and pollen, and vegetative tissues exposed to drought stress, including extreme desiccation of "resurrection plants" (*Craterostigma plantagineum*).

The LEA, RAB, and DHN proteins are all water soluble, basic, rich in glycine and lysine, and low in hydrophobic residues. As a result they are extremely hydrophilic and stable to boiling. These features led Dure and colleagues (1989) to propose that they function specifically in the protection of membranes and proteins against desiccation damage, possibly by binding water tightly or providing hydrophilic interactions in the absence of free water, and by preventing crystallization of cellular components through their ability to act as stabilizing "solvents," similar to the effects of sugars. One of these proteins confers increased osmotolerance in transgenic yeast.

Types of Seed Dormancy and the Roles of Environmental Factors

During seed maturation, the embryo enters a quiescent phase in response to desiccation. Seed germination can be defined as the resumption of growth of the embryo of the mature seed; it depends on the same environmental conditions as vegetative growth does. Water and oxygen must be available, the temperature must be suitable, and there must be no inhibitory substances present.

However, in many cases a viable (living) seed will not germinate even though all the necessary environmental conditions for growth are satisfied. This phenomenon is termed seed dormancy. Seed dormancy introduces a temporal delay in the germination process that provides additional time for seed dispersal over greater geographical distances. It also maximizes seedling survival by preventing germination under unfavourable conditions.

Two types of seed dormancy have been recognized, coat-imposed dormancy and embryo dormancy. Coat-imposed dormancy is dormancy imposed on the embryo by the seed coat and other enclosing tissues, such as endosperm, pericarp, or extrafloral organs. The embryos of such seeds will germinate readily in the presence of water and oxygen once the seed coat and other surrounding tissues are either removed or damaged. There are five basic mechanisms of coat-imposed dormancy:

Prevention of water uptake. Prevention of water uptake by the seed coat is a common cause of seed dormancy in families found in arid and semiarid regions, especially among legumes, such as clover (*Trifolium* spp.) and alfalfa

(*Medicago* spp.). Waxy cuticles, suberized layers, and lignified sclereids all combine to restrict the penetration of water into the seed.

Mechanical constraint. The first visible sign of germination is typically the radicle breaking through the seed coat. In some cases, however, the seed coat may be too rigid for the radicle to penetrate. Nuts with hard, lignified shells are examples of dormancy caused by mechanical constraint. Such shells must be broken by biotic or environmental forces for the seed to germinate. Even nonlignified tissues, such as the endosperm of lettuce seeds, can suppress expansion of the embryo. For the seeds to germinate, the endosperm cell walls must be weakened by the production of cell wall–degrading enzymes.

Interference with gas exchange. Some seed coats are considerably less permeable to oxygen than an equivalent thickness of water is— e.g., less by a factor of about 10^4 in seeds of cocklebur (*Xanthium pennsylvanicum*). This lowered permeability to oxygen suggests that the seed coat inhibits germination by limiting the oxygen supply to the embryo. In support of this idea, investigators can break the dormancy of such seeds either by making a small hole in the coat with a pin (without weakening the coat mechanically), or by treating the coat with concentrated oxygen. However, other studies suggest that the oxygen consumption of the embryos from such seeds is considerably less than the amount of oxygen able to penetrate the seed coats under normal aerobic conditions. Thus the role of oxygen impermeability in seed coat dormancy remains unresolved.

Retention of inhibitors. The seed coat may prevent the escape of inhibitors from the seed. For example, growth inhibitors readily diffuse out of isolated *Xanthium* embryos, but not from intact seeds.

Inhibitor production. Seed coats and pericarps may contain relatively high concentrations of growth inhibitors that can suppress germination of the embryo. ABA is a common germination inhibitor present in these maternal tissues. In certain cases where repeated washing (leaching) removes dormancy, the effect is thought to be due to the loss of such inhibitory compounds.

The second type of seed dormancy is embryo dormancy, which refers to a dormancy that is inherent in the embryo and is not due to any influence of the seed coat or other surrounding tissues. In some cases, embryo dormancy can be relieved by amputation of the cotyledons. Species in which the cotyledons exert an inhibitory effect include European hazel (*Corylus avellana*) and European ash (*Fraxinus excelsior*). A fascinating demonstration of the cotyledon's ability to inhibit growth is found in species (e.g., peach) in which the isolated dormant embryos germinate but grow extremely slowly to form a dwarf plant. If the cotyledons are removed at an early stage of development, however, the plant abruptly shifts to normal growth.

Embryo dormancy is thought to be due to the presence of inhibitors, especially ABA, as well as the absence of growth promoters, such as GA (gibberellic acid). The loss of embryo dormancy is often associated with a sharp drop in the ratio of ABA to GA.

Seed Dormancy May be Primary or Secondary

Different types of seed dormancy also can be distinguished on the basis of the timing of dormancy onset rather than the cause of the dormancy.

- Seeds that are released from the plant in a dormant state are said to exhibit primary dormancy.
- Seeds that are released from the plant in a nondormant state but which become dormant if the conditions for germination are unfavourable exhibit secondary dormancy. For example, seeds of *Avena sativa* (oat) can become dormant in the presence of temperatures higher than the maximum for germination, whereas seeds of *Phacelia dubia* (small-flower scorpionweed) become dormant at temperatures below the minimum for germination. The mechanisms of secondary dormancy are poorly understood.

Environmental Factors Control the Release from Seed Dormancy

Various external factors release the seed from dormancy, and dormant seeds typically respond to more than one condition. Many seeds lose their dormancy when their moisture content is reduced to a certain level by drying. This method of breaking seed dormancy is called afterripening, and is usually performed in a special drying oven. On the other hand, if the seed becomes too dry (5% water content or less), the effectiveness of afterripening is diminished.

Another factor that can release seeds from dormancy is low temperature, or chilling. Many seeds require a period of cold (0 to 10°C) while in a fully hydrated (imbibed) state in order to germinate. In temperate-zone species, this requirement is of obvious survival value, since such seeds will germinate not in the fall, but only in the following spring. Chilling seeds to break their dormancy is a time-honored practice in horticulture and forestry and traditionally has been referred to as *stratification.* This term is derived from the old agricultural practice of allowing seeds with a chilling requirement to overwinter outdoors in layered mounds of moist soil. Today the seeds are simply stored in a refrigerator.

The third external factor that plays an important role in breaking seed dormancy is light. Many seeds have a light requirement for germination, which may involve only a brief exposure, as in the case of lettuce, an intermittent treatment (e.g., succulents of the genus *Kalanchoe*), or even a specific photoperiod involving short or long days. The main sensor for light-regulated seed germination. Interestingly, all light-requiring seeds exhibit seed coat dormancy, and removal of the outer tissues of the seed allows the embryo to germinate in the absence of light. The effect that light has on the embryo is thus to enable the radicle to penetrate the seed coat. This penetration often involves some enzymatic weakening of the enclosing tissues.

The Longevity of Seeds

How long can a seed remain dormant and still remain viable? Seed longevity is of practical importance because of ongoing efforts to preserve plant genetic resources for future agricultural crops by setting up seed gene banks. Spectacular claims of dormant seeds that remained viable for thousands of years have been made, but are considered highly controversial. Extreme examples of ancient seeds that have been reported to retain their viability include submerged lotus seeds found in a 3,000-year-old boat near Tokyo, barley seeds from the 3,000-year-old tomb of King Tutankhamen, and arctic lupine seeds associated with rodent burrows determined to be 14,000 years old. In most cases, however, a scientifically rigorous examination of the data has either disproved or raised serious doubts about the claimed antiquity of the seeds (Bewley and Black 1994).

The most serious criticism is that the dates of the seeds have been inferred from the age of their immediate surroundings or associated artifacts rather than from direct measurements of the seeds themselves. This question raises the possibility that the seeds are much younger than the site at which they were found. In fact, microscopic examination of authenticated ancient cereal grains from various Middle Eastern sites has shown that cells and tissues of these seeds are no longer intact. Such seeds are clearly inviable.

Although the most sensational claims for seed longevity are almost certainly bogus, seeds of *Canna compacta* apparently can live for at least 600 years. Viable *Canna* seeds were obtained from inside a walnut in a tomb in Argentina. The *Canna* seeds had apparently been inserted into the immature seeds of a growing walnut fruit before the hard outer shell formed. Once the shell hardened and the nut dried out, the result was a rattle. Native people strung the rattles together to form a necklace. In this case, the seeds had to be at least as old as the walnut shell, and carbon dating of the shell indicated that it was about 600 years old.

Plant collections begun during the late eighteenth century in Europe—for example, at the British Museum in London and at the Museum of Natural History in Paris—have been reliable sources of seeds for viability determinations. *Cassia multijuga* (false sicklepod) seeds, collected in 1776, were still viable after being tested in 1934. In 1879, W. J. Beal initiated the longest-running experiment on seed longevity by burying the seeds of 21 different species in unstoppered bottles in a sandy hilltop near the Michigan Agricultural College in East Lansing. After 100 years, only one species, *Verbascum blattaria* (moth mullein), remained viable. Why are some seeds able to survive in a dormant state for hundreds of years, whereas others lose viability in less than 5 years? Almost nothing is known about the mechanisms that determine the longevity of seeds. If these mechanisms could be understood, we might someday be able to greatly increase the seed longevity of agriculturally important species and varieties, thereby enhancing our ability to preserve plant genetic resources for generations to come.

Genetic Mapping Of Dormancy: Quantitative Trait Locus (QTL) Scoring of Vegetative Dormancy Combined with a Candidate Gene Approach

Complex traits, e.g. dormancy, are affected by many different, often unlinked genes. Consequently, attempts to map them result in continuous, or quantitative, variation within a mapping population.

Quantitative trait locus (QTL) mapping is used in these instances to map individual genes affecting the trait. In QTL mapping, inbred lines that differ in the trait of interest are crossed and an F_2 population is produced. The F_2 population has continuous variation for the trait of interest, and the goal of the analysis is to determine which chromosomal segments contain genes affecting the trait. Using marker loci along each chromosome (loci that have different alleles between the parental inbred lines), the genotypes of entire chromosomal segments in each F_2 individual can be inferred as being homozygous for one or the other parental lines or heterozygous. If the inheritance of a chromosomal segment is correlated with the trait of interest, then a gene affecting the trait is likely to be between the two marker loci. By analysing chromosomal segments along each of the chromosomes, the number and locations of genes affecting the quantitative trait can be determined.

Candidate genes are specific loci, identified by predicted biochemical function or homology to regulatory genes from other species, that are suspected of affecting the trait of interest. Molecular differences between the parental alleles at these loci can be used to determine whether they map to the same region as any of the QTLs, thereby supporting the hypothesis that a given gene contributes to regulation of the trait.

ABA-Induced Senescence and Ethylene

The ethylene has also been found to play a role in the senescence of oat leaf segments, but ABA appears to be the initiating agent, whereas ethylene appears to exert its effects at a later stage. To determine whether the ABA effects are mediated by ethylene, the effects of ABA and ethylene on senescence of leaf discs from wild-type and ethylene-insensitive mutants were compared. Chlorophyll loss from untreated leaf discs of the ethylene-insensitive mutant was much slower than that from the wild type, in the light and in the dark.

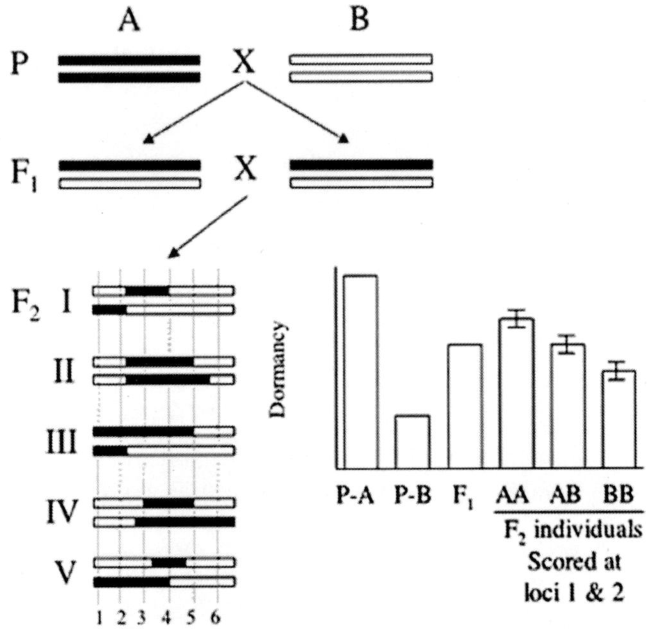

Figure: *Two parental (P) inbred lines, A & B, which differ for a trait of interest (e.g. dormancy) are crossed to produce heterozygote F_1 progeny which have an intermediate degree of dormancy. Only one chromosome pair is depicted for simplicity. The F_1 progeny are crossed to produce a large segregating F_2 population (usually 100-500 individuals are analysed, here only five individuals, I-V, are depicted). Because of recombination during F_1 meiosis, the F_2 individuals have varying genetic combinations of the original inbred lines (depicted by shading of chromosomes) and have continuous variation, from high to low, in degree of dormancy. In order to estimate whether segments of the chromosome are homozygous for one of the parental-lines or heterozygote, the F_2 individuals are scored for marker loci (1-6) along the chromosome. For example, loci 1 & 2 are identified as homozygous for A-line alleles in individual III, as heterozygous in individuals I & V and as homozygous for B-line alleles in individuals II & IV. The F_2 individuals are also scored for their dormancy. After scoring many F_2 individuals a correlation between the genotypes at locus 1 & 2 and phenotype is determined and therefore this segment likely contains at least one gene affecting dormancy (bar graph). The analysis is repeated for each chromosomal segment to identify which ones are likely to contain genes affecting dormancy.*

Ethylene accelerated the chlorophyll loss from the wild-type leaf discs, but it had no effect on the yellowing of the ethylene-insensitive mutant discs. In contrast, treatment with ABA stimulated chlorophyll loss in *both* the wild-type and the mutant discs, whether they were held in the light or in the dark (Zacarias and Reid 1990). These results indicate that the promotion of senescence by ABA is not mediated through its stimulation of ethylene production.

Yellow Cameleon: A Noninvasive Tool for Measuring Intracellular Calcium

Cytoplasmic free calcium ($[Ca^{2+}]_{cyt}$) acts as a secondary messenger in a variety of signalling pathways and it appears that the frequency, magnitude and localization of $[Ca^{2+}]_{cyt}$ oscillations are important determinants of specificity. "Traditional" techniques for measuring Ca^{2+} have made use of microinjected Ca^{2+} sensitive indicator dyes, but these methods are technically difficult and invasive.

An alternative strategy makes use of Ca^{2+} binding proteins that can be transgenically expressed throughout an organism, eliminating the need for microinjection. The Ca^{2+} indicator yellow cameleon is a fusion protein composed of four domains: two modified green fluorescent proteins (GFPs) linked via calmodulin (CaM) and a CaM-binding peptide ("M13").

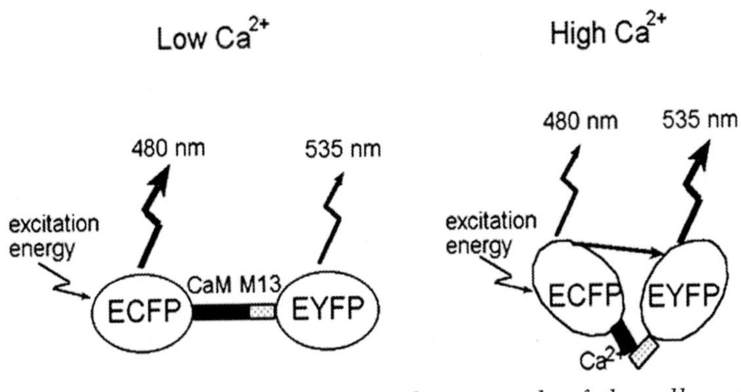

Figure: *The binding of calcium brings the two ends of the yellow cameleon protein together, altering the ratio of 535 nm to 480 nm emission.*

The GFPs are modified to shift their emission wavelengths toward blue (enhanced cyan fluorescent protein, ECFP) or yellow (enhanced yellow fluorescent protein, EYFP). When the CaM domain binds to Ca^{2+}_{cyt} it associates with the CaM-binding peptide, producing a conformational change that brings the modified GFP domains closer together.

Following exposure to a laser, excitation energy from ECFP (the higher-energy fluorescent protein) is passed to EYFP resulting in increased emission by this lower-energy fluorescent protein. Consequently, increased $[Ca^{2+}]_{cyt}$ results in increased emission at 535 nm relative to that at 480 nm, and the 535/480 emission ratio is proportional to the $[Ca^{2+}]_{cyt}$ (Miyawaki et al 1997; Allen et al. 1999).

The Two-Hybrid System

Many cellular proteins interact with a variety of other proteins to perform their physiological functions. Identification of interacting proteins can be accomplished by biochemical assays, such as co-purification of a complex, but these require relatively stable interactions and the ability to identify components of the complex immunologically or based on amino acid sequence and composition. Alternatively, the two-hybrid assay is a genetic approach that takes advantage of the modular nature of proteins to identify interacting domains.

Although there are several variations, in all cases two fusion proteins are co-expressed in a cell line that will activate expression of a reporter gene if the fusion proteins interact.

For example, the yeast transcription factor GAL4 can be separated into distinct DNA binding (BD) and transcription activation (AD) domains. If hybrid genes are constructed such that they encode fusion proteins BD-protein X and AD-protein Y, interactions between proteins X and Y will bring the GAL4 domains close enough to reconstitute a functional transcription factor that can trans-activate a GAL4-responsive promoter.

The cell lines often contain several GAL4-responsive promoter-reporter fusions, whose activity can be scored either qualitatively or quantitatively. Qualitative scoring (+/- growth) makes it possible to screen a cDNA library for fusions encoding interacting proteins by requiring expression of a gene whose activity can complement an auxotrophy. Quantitative scoring (e.g. β-galactosidase activity from a *lacZ* reporter) permits comparisons of the relative strength of interactions among a variety of protein pairs.

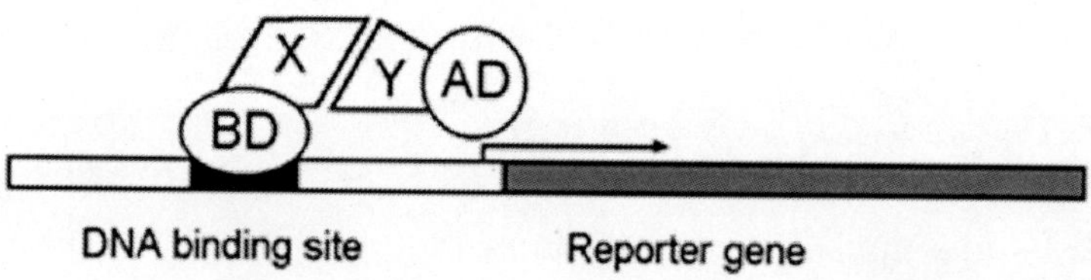

Figure: *Two fusion proteins containing the GAL4 DNA binding (BD) and transcription activation (AD) domains can reconstitute a functional transcription factor that can trans-activate a GAL4-responsive promoter if the interactions between the two proteins, X and Y, will bring the GAL4 domains close enough together.*

The Control of Flowering

Regulation of Juvenility by the TEOPOD (TP) Genes in Maize

The regulation of the transition from juvenile to adult (i.e., the attainment of competence) is influenced by a wide variety of conditions that affect the growth rate, and by transmissible factors such as carbohydrates and gibberellins.

Hence it has been difficult to determine exactly what regulates the phase change during normal development. The analysis of mutations that affect the timing of phase changes, called heterochronic mutations, provides the opportunity to identify specific genes that regulate juvenility. The best-studied example is in maize.

The juvenile and adult phases of corn plants are characterized by differences in vegetative morphology. Scott Poethig and his colleagues at the University of Pennsylvania have been characterizing the heterochronic *teopod* (*tp*) mutants of maize: *tp1*, *tp2* and *tp3*.

These semidominant mutations cause cells and tissues that would normally develop as adult structures to become more juvenile. In *tp* plants, leaves in the normally adult part of the shoot are morphologically intermediate between juvenile and adult leaves. Whereas in the wild type, tassels bear male flowers and ears bear female flowers, in *tp* mutants both tassels and ears possess leaves in addition to flowers.

However, *tp* mutations affect neither the rate nor the duration of growth of the plant, nor do they interfere with the timing of tassel formation. The *teopod* mutations thus specifically affect the transition from juvenile to vegetative adult phase without affecting the transition to the reproductive phase.

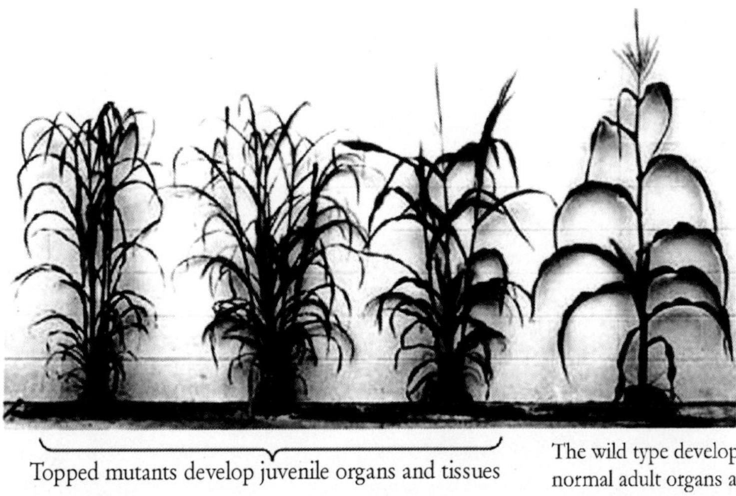

Figure: *Phenotypes of three* teopod *mutants and the wild-type maize plant. From left to right:* tp1, tp2, tp3, *and wild type.*

The possible interaction between the *TEOPOD* genes and the gibberellins has been investigated. In GA-deficient dwarf mutants of maize, the transitions from juvenile to adult vegetative development and from adult vegetative to reproductive development occur later than in wild-type plants. Thus, endogenous GAs are required for the normal timing of phase changes.

Flowering of Juvenile Meristems Grafted to Adult Plants

By definition, juvenile plants cannot be induced to flower with an inductive photoperiod, suggesting either that the shoot apical meristems of juvenile plants are not competent to flower, or that juvenile leaves produce insufficient amounts of a floral stimulus. The LSDP *Bryophyllum* passes through a juvenile phase, which lasts from 5 months to a year (depending on the species), during which it cannot be induced to flower by transfer from LD to SD. However, when juvenile *Bryophyllum* shoots were grafted onto flowering plants, the juvenile shoots flowered promptly and profusely. This result indicates that the shoot apical meristems of juvenile plants are competent to flower, but fail to do so because the juvenile leaves produce insufficient amounts of the floral stimulus.

Age is another factor that can affect production of the floral stimulus. Grafting studies with the SDP *Perilla* have shown that the leaves at the top of a plant are stronger donors of the floral stimulus than the leaves near the base of a plant, in keeping with the juvenile character of the basal leaves. The experiment was performed as follows. Two groups of *Perilla* plants were sown at different times. When the first group had produced five pairs of leaves, the second group had only produced two.

The fifth and the second pairs of leaves from the two sets of plants were grafted onto stocks and maintained under long days. Once the graft unions were established, the grafted leaves were induced with short days by daily enclosing them in opaque bags. Leaves from the fifth node needed only 14 short days to induce flowering while leaves from the second node required 28 short days. This finding shows that the more juvenile leaves from the second node produce less floral stimulus, and thus require more inductive photoperiods to induce flowering, than the fully adult leaves higher up the shoot.

Support for the Role of Blue-Light Regulation of Circadian Rhythms

The role of blue light in regulating both circadian rhythmicity and flowering is also supported by studies with an *Arabidopsis* flowering-time mutant: elf3 (early flowering *3*). The *elf3* mutant flowers earlier than the wild type under a variety of growth conditions and is insensitive to photoperiod; that is, it is day-neutral. Transformation of *elf3* with the luciferase reporter construct revealed a defect in circadian rhythmicity.

The mutant displayed no circadian rhythmicity when grown in continuous light. Significantly, the *elf3* mutant also has a long hypocotyl when grown in constant white light, similar to the *hy* mutants, and is less responsive than the wild type to all wavelengths of light, but especially to blue light. This finding suggests that the circadian clock is intact in the mutant, but that the transduction of light signals to the clock is impaired.

The fact that the *elf3* mutant is more impaired in its response to blue light than in its response to red light further suggests that a blue-light photoreceptor plays a role in the photoperiodic control of flowering.

Genes That Control Flowering Time

As we have seen, the Maryland Mammoth mutation converts a day-neutral variety of tobacco into an SDP. Analysis of this flowering-time mutant by W. Garner and H. Allard led to the discovery that plants can measure photoperiod as an environmental cue for floral induction, and it identified the Maryland Mammoth gene as one of the genes that regulates this process. The most extensive studies of flowering-time genes have been performed in pea (*Pisum sativum*) and *Arabidopsis thaliana*. In these species a range of mutations can alter various aspects of the photoperiodic control of flowering. For example, mutations in certain genes convert these LDPs into day-neutral plants that can flower rapidly. Mutations in other genes result in day-neutral plants that exhibit a long delay in flowering. Because these mutations eliminate the response to photoperiod, the normal (wild-type) products of these genes are thought to be involved in photoperiodism. Other mutations can delay or promote flowering without affecting the photoperiodic response. These mutants are thought to affect genes involved in the *autonomous* regulation of flowering.

Since most biological regulatory systems are under both positive and negative control, it is not surprising to find mutations that promote or delay flowering. Inasmuch as most mutations result in a loss or reduction of gene activity, the wild-type role of genes that delay flowering when mutated is to promote flowering.

Similarly, mutations that promote flowering are likely to occur in genes that normally inhibit the process. These genes are promoters and inhibitors of flowering in genetic terms, but the mutant phenotype alone does not usually reveal the flowering mechanism that is affected. For example, a mutation that eliminates the response to photoperiod could either block the production of a flowering hormone (floral stimulus) or, alternatively, interfere with the ability of the target cells in the meristem to receive the hormonal signal (meristem competence).

Arabidopsis plants containing mutations that promote or delay flowering have been isolated. The wild type and a *gigantea* (*gi*) mutant that were grown in inductive photoperiods (long days). The plants are shown at the stage when flower buds have become visible. Flowering in the *gigantea* mutant is delayed; thus this plant has been growing approximately three times as long as the wild type and has formed many more leaves from the primary apical meristem. Because a mutation in the *GIGANTEA* gene delays flowering, the normal role of this gene is to promote flowering. The wild type and an *elf3* mutant plant that were grown in noninductive photoperiods (short days). The mutant phenotype indicates that the normal role of the *ELF3* gene is to repress flowering.

One way to explore the role of these genes is through grafting studies. For example, if the apical bud from a mutant plant with a delayed-flowering phenotype flowers normally when grafted onto a wild-type shoot, the mutant shoot apex is clearly competent to receive and respond to flowering signals. One might then suspect that the mutation affects the production of flowering signals by leaves. If this were the case, a wild-type shoot apex region should exhibit delayed flowering when grafted onto a mutant shoot. Grafting studies of this sort have identified two types of genes in pea: those that control the production of transmissible promoters and inhibitors of flowering, and those that control the sensitivity of the shoot apical meristem to these signals.

For example, the *GIGAS* gene is thought to be involved in the production of a transmissible flowering promoter. The wild-type pea exhibits a quantitative LDP phenotype. Plants homozygous for the recessive *gigas* mutation exhibit severely delayed flowering. For example, when grown under a 16-hour photoperiod, the wild type flowers after producing 16 leaves, whereas the *gigas* mutant flowers after producing 43 leaves. Delayed flowering in *gigas* plants could be caused either by the presence of inhibitors or by the absence of a floral stimulus. To test for inhibitors, researchers grafted the shoot of an early flowering day-neutral line onto *gigas* stock.

The time of flowering of the scion was unaffected by grafting to the mutant, indicating that the mutant does not produce flowering inhibitors. However, when a *gigas* shoot was grafted onto either the early-flowering or the wild-type stock, flowering was greatly accelerated. Instead of flowering after producing 40 leaves or more, the plants flowered after producing only 14 or 23 leaves, respectively.

Because the *gigas* mutant shoot apex flowers readily after grafting, its ability to respond to the flowering stimulus from the leaves is not affected by the mutation. The mutation is therefore thought to affect the production of a floral stimulus. Another approach to elucidating the floral induction process is to isolate the genes identified by mutations and to study their properties. Once the gene is cloned, the encoded protein can be overexpressed in *E. coli* and purified in sufficient quantities to study its biochemical properties and to produce specific antibodies. The antibodies can be used to determine the tissues in which the protein is produced, as well as its location within the cell. The localization may provide clues to the function of the protein. For example, a protein in the shoot apex may be involved in the reception of flowering signals, whereas a protein present in the leaves may be involved in the production of flowering signals.

The *constans* mutants of *Arabidopsis* are insensitive to day length and flower later than the wild type under long-day conditions. The *CONSTANS* gene of *Arabidopsis* has been cloned and shown to encode a protein that shows similarities to zinc finger transcription factors, suggesting that it acts as a transcriptional regulator. The wild-type *CONSTANS* gene is expressed in leaves, and the level of *CONSTANS* mRNA increases during inductive photoperiods.

This pattern of expression, along with the photoperiod-insensitive mutant phenotype, implicate this gene in the photoperiodic control of flowering.

The Different Effects of Two Different Gibberellins on Flowering (Spike Length) and Elongation (Stem Length)

In the long-day plant *Lolium temulentum*, GA_{32} strongly promotes flowering, but has only a small effect on stem elongation. In contrast, GA_1 strongly promotes stem growth but has only a small effect on flowering. Gibberellin-treated plants were maintained in short days and, for comparison, the response to a single 24-hour day is shown. This long day strongly promotes flowering but has almost no effect on stem length.

The Influence of Cytokinins and Polyamines on Flowering

Sometimes changes normally associated with flowering can be caused by treatments that do not themselves bring about floral initiation. For example, one of the earliest events observed at the shoot apex following photoperiodic induction is a transitory increase in the number of cells undergoing mitosis.

When cytokinins are applied to the stem apex of the long-day plant white mustard (*Sinapis alba*), they cause an increase in mitotic activity similar to that caused by exposure to a single long day. However, whereas the long day induces flowering, the cytokinin treatment does not. Cytokinin may therefore be a component of the floral stimulus, although it is insufficient by itself to bring about flowering.

Other potential components include the polyamines. For example, photoperiodic induction of the LDP *Sinapis alba* was correlated with a large increase in the concentration of putrescine, the major polyamine, in the phloem exudate from the leaves. Spraying the leaves with an inhibitor of putrescine biosynthesis lowered the level of putrescine in the phloem exudate and also inhibited flowering. This result suggests that putrescine may be a component of the floral stimulus in *S. alba*.

In addition to receiving signals from the leaves, the shoot apex may accumulate flowering signals from the roots. Classic experiments with cuttings have shown that the initiation and growth of adventitious roots inhibit flowering in the shoot. In many species, root removal promotes flowering. Thus, at least in certain species, the root appears to produce one or more flowering inhibitors. In other cases, the roots may contribute floral activators. For example, both GA and cytokinin are synthesized in the root and transported to the shoot via the xylem. During photoperiodic induction in *Sinapis alba,* the amount of cytokinin present in the xylem exudate from the root increases considerably. This finding indicates that signals emanating from the leaf are transported both to the shoot apex and to the root during photoperiodic induction. In response to the signal from the leaves, the roots produce and export additional chemical signals that modulate the flowering response.

The Contrasting Effects of Phytochromes A and B on Flowering

Phytochrome is encoded by members of a multigene family, and it is possible that the effects of red and far-red light described in the previous section are mediated by different types of phytochrome (A and B). In LDPs such as *Arabidopsis*, pea (*Pisum sativum*), and sorghum (*Sorghum bicolor*), phytochrome B (phyB) appears to be an *inhibitor* of flowering, since mutations in the gene *PHYB* that eliminate or reduce the amount of the phyB protein cause more rapid flowering.

The promotion of flowering by far-red light may result from a decrease in the amount of the Pfr form of phyB, which would have the same effect on the phyB Pfr levels as the mutations that reduce the total amount of phyB. As already discussed, red-light interruptions of a dark period prevent flowering in SDPs; therefore, the Pfr form of phyB may be an *inhibitor* of flowering in SDPs as well.

In the LDPs *Arabidopsis* and pea, mutations in the *PHYA* gene interfere with the promotion of flowering by long days; thus the Pfr form of phyA may promote flowering in LDPs. The effect of the *phyA* mutation is quite strong in pea (the mutant plants are essentially unable to respond to inductive photoperiods) whereas the *phyA* mutation in *Arabidopsis* has only a slight effect on the ability to respond to inductive photoperiods. The species difference in the effect of *phyA* mutations is probably due to the fact that pea relies entirely on phyA to sense inductive photoperiods, whereas *Arabidopsis* also uses a blue light photoreceptor for this purpose.

A Gene that Regulates the Floral Stimulus in Maize

Maize plants flower after producing a fixed number of leaves regardless of photoperiod—a characteristic of autonomous regulation. If the shoot apical meristem from a maize plant is excised, and the "explant" is cultured in vitro, it is possible to regenerate a complete plant that will eventually flower. As long as the explant includes 4-6 leaf primordia, the meristem "remembers" its developmental stage at the time of excision and produces only the additional number of leaves needed to make up the fixed amount before flowering. However, if the explant contains fewer than 4 leaf primordia, the apex "forgets" its developmental stage and produces a complete set of new leaves before flowering.

The *indeterminate 1* (*id1*) mutant of maize is unable to undergo a normal transition to flowering, and produces many more leaves than the wild type. Moreover, the flowers produced by *id1* mutants are abnormal, having vegetative characteristics. The *ID1* gene encodes a zinc-finger protein, suggesting that it functions as a transcriptional regulator. The expression pattern of *ID1* differs from that of many other flowering time control genes isolated from *Arabidopsis* in that it is expressed in young leaves rather than in the shoot apical meristem.

Are the effects of the *ID1* gene "cell autonomous", that is, restricted to those cells in which the gene is expressed, or does the *ID1* gene regulate the production of a transmissible signal that can affect cells beyond the site of gene expression? Traditionally, grafting experiments between the mutant and the wild type have been carried out to answer such questions, but maize (a monocot) lacks a vascular cambium and is therefore not suitable for grafting experiments.

In a second set of experiments, Colasanti et al. (1998) did the next best thing to grafting by analysing plants that were genetic chimeras (that is, containing cells with different genotypes) for *ID1* function. The mutant allele of *id1* analysed was caused by the insertion of a transposable element, and the spontaneous excision of this element during development generated clonal sectors of wild-type *ID1*-expressing cells against an overall mutant *id1* background. It was found that chimeric *id1* plants containing sectors of restored *ID1* function flowered earlier than completely mutant *id1* plants, even though the shoot apical meristems of these chimeric plants remained mutant. This experiment is comparable to grafting a wild type scion onto an *id1* mutant stock, and finding that the *id1* stock flowered earlier as a result. This observation in combination with the expression data provides compelling evidence for the involvement of *idt1* in the production in leaves of a transmissable signal from immature leaves that induces flowering in the apex.

The *Arabidopsis* genome contains several potential homologs of the maize *ID1* gene, but no *id1*-like mutants have been identified with a phenotype similar to that in maize. All of the genes involved in the autonomous pathway of *Arabidopsis* thus far seem to be expressed exclusively in the apical meristem.

5

Stress Physiology

Stomatal Conductance and Yields of Irrigated Crops

In the last few decades, the breeding of agricultural crops for higher yields has been very successful. Breeders usually select for high-yielding genotypes empirically, without particular attention to specific plant traits that might be conducive to higher yields. However, comparing old, low-yielding lines of any crop with advanced, high-yielding lines shows clearly that many morphological, physiological, and biochemical traits have been altered by the intense selection pressures for higher yields.

These changes indicate that selection for high-yielding genotypes has generated indirect selection pressures on the altered traits. If we exclude effects of genes that regulate the expression of two or more unrelated traits (pleiotropic genes), the study of high-yielding lines might reveal specific traits and genes associated with the higher yield. With this information, breeders could explicitly select for yield-enhancing traits to improve yields further.

Changes in plant traits associated with yield increases can be studied conveniently in a historical series of a crop. Historical series include successive commercial releases within a breeding program, with each member of the series representing an incremental increase in yield. In studies aimed at characterizing relationships between specific traits and higher yields, all the members of the series are grown together under the same conditions, and the traits of interest, such as plant architecture or photosynthetic rates, are measured in parallel with yields. Recent studies of this type focusing on historical series of Pima cotton (*Gossypium barbadense*) and bread wheat

(*Triticum aestivum*) have shown a remarkable positive correlation between yield increases and increases in stomatal conductance.

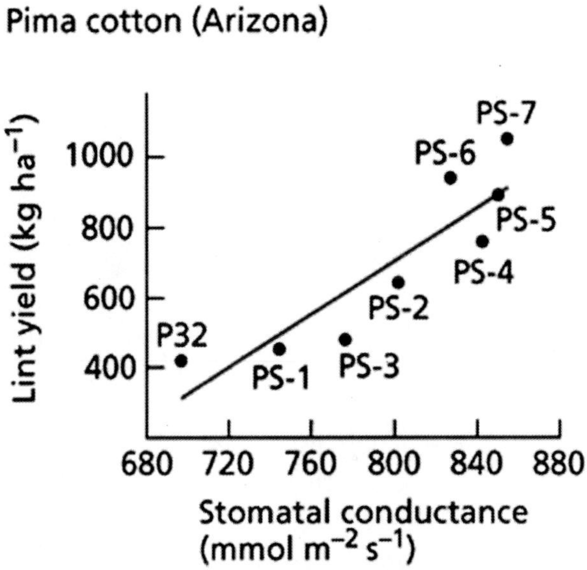

Figure: *Stomatal conductance has increased in parallel with agronomic yields in irrigated Pima cotton (*Gossypium barbadense*) selected for higher yields at a high temperature. The figure shows the relationship between lint yield and stomatal conductance in a historical series of Pima cotton grown in Arizona. The abbreviations P32 and PS-1 through PS-7 designate successive commercial releases between 1949 and 1996.*

Pima cotton is a high-quality fibre cotton grown in hot environments under intensive irrigation. The historical series used in the Pima studies included eight members, released between 1949 and 1996, and encompassing a nearly threefold increase in lint yield. The yield increase attained in each commercial release was accompanied by a corresponding increase in stomatal conductance. Stomatal conductance increased by about 30 mmol m^{-2} s^{-1} for each 100 kg ha^{-1} increase in yield.

Genetic crosses between low-and high-yielding Pima strains have shown that stomatal conductance in Pima cotton has a clear-cut genetic component. The studies with stomata of Pima cotton are the first to show a clear-cut genetic regulation of stomatal conductance proper, of the stomatal response to temperature, and of proton pumping rates of isolated guard cells.

In the absence of explicit selection for higher stomatal conductance in the Pima breeding program, what could be the nature of the indirect selection pressures that caused the increases in conductance paralleling the increases in yield? Higher stomatal conductance increases CO_2 diffusion into the leaf and favours higher photosynthetic rates. Higher photosynthetic rates could in turn favour a higher biomass and higher crop yields. Advanced Pima lines show a higher photosynthetic capacity than older, low-yielding lines, but photosynthetic rates measured in the same leaves used to measure stomatal conductance were not positively correlated with yields. Thus, higher stomatal conductance appears to favour higher yields by a mechanism not directly related to photosynthesis.

Evapotranspiration at the leaf surface lowers leaf temperature, and higher stomatal conductance enhances this leaf cooling. Optimal daytime temperatures for growth, photosynthesis, and reproduction in Pima cotton are below 30°C, while afternoon air temperature in the hot Pima-growing areas often exceeds 40°C. Evaporative cooling of the leaves thus reduces the gap between optimal growth temperatures and air temperature, and provides heat resistance. Studies with the Pima historical series showed that, because of leaf cooling, leaf temperatures were several degrees lower than air temperature, and leaf and canopy temperature were lower in the advanced, high-yielding lines than in low-yielding lines.

The same relationship among yields, stomatal conductance, and leaf temperature was found in a historical series of bread wheat grown in the warm Yaqui Valley of northwestern Mexico. These studies indicate that selection for higher yields in irrigated crops grown at high temperatures imposes indirect selection pressures for high stomatal conductance that lowers leaf temperature and appears to reduce deleterious effects of heat stress on critical flowering and fruiting stages, thus resulting in higher crop yields.

If selection for higher yields generates indirect selection pressures for higher stomatal conductance, can breeders explicitly select for higher conductance and thus obtain lines with higher yields? Studies with Pima cotton have shown that selection of high-conductance F_2 progeny from a cross between high-and low-conductance parents produces high-conductance F_4 lines with higher lint yields than low-conductance lines have. These experiments indicate that high stomatal conductance could be used as a selection trait for high yields in irrigated crops grown at high temperature.

The observed increases in stomatal conductance in crops grown at high temperature also have valuable implications for models developed to predict global climate changes, which are expected to occur because of increases in atmospheric CO_2. Under atmospheric CO_2 concentrations that are twice as high as the present ones, evapotranspiration would decrease over the continents and air temperatures would increase significantly over the tropical land masses, amplifying the changes resulting from atmospheric radiative effects. However, these models have yet to take into consideration the effect of higher temperatures on the stomatal control of evapotranspiration. Incorporation of the stomatal response to temperature in climate change models is likely to improve the accuracy of the predictions generated by the models. These studies illustrate both the complex interactions between physical and biological factors in the biosphere, and the advantage of a thorough understanding of the physiological properties of plants for efforts aimed at increasing agronomic yields and at improving models that predict global climate changes.

The GABA Shunt Pathway Integrates Stress Metabolic Responses

Not all enzymes participating in metabolic pathways have identical temperature responses, and differential thermostabilities of enzymes may also effect specific steps in metabolism before HSPs acting as molecular chaperones can restore activity. Thus, heat stress can lead to changes in metabolism that result in the accumulation of some metabolites and the reduction of others. Such changes can dramatically alter the function of metabolic pathways and lead to imbalances that are difficult to correct. This problem can be extremely critical when the levels of regulatory metabolites are forced outside of normal ranges so that regulatory homeostasis cannot function properly to restore normal levels.

Additionally, the consumption—and production—of protons during metabolic reactions and/or changes in the activity of membrane-associated proton-transporting ATPases and other membrane-associated proton transporters that move protons out of the cytosol either into the apoplast or vacuoles, can also cause a lowering of the pH of the cytosol, leading to additional metabolic perturbations during stress. Metabolic adaptation to the stress results in the amelioration of such stress effects on metabolism.

One of the metabolic adaptations that plants make to heat stress leads to the accumulation of the nonprotein amino acid, 4-aminobutyric acid (GABA). During episodes of heat stress, GABA accumulates to levels six-to tenfold higher than in unstressed plants. GABA is made from the amino acid, L-glutamate, in a single reaction catalyzed by the enzyme glutamate decarboxylase (GAD). GAD is one of several enzymes whose activity is modulated by the calcium regulatory protein, calmodulin. In the presence of calcium ions, calmodulin is converted into an active state that binds to GAD activating the enzyme, which leads to an increase in the rate at which GABA is produced. The precise mechanism leading to increases in cytosolic calcium have not been directly demonstrated, but studies using inhibitors of calcium ion channels implicate either the plasma membrane or tonoplast calcium channels. Alternatively, the efficiency of one or more of the CAX gene products (calcium proton antiporters) may be reduced as the proton gradient collapses during heat stress, leading to elevated levels of calcium in the cytosol. These increases, in conjunction with calmodulin, activate GAD, producing the high-temperature-induced synthesis of GABA.

GABA is metabolized inside the mitochondria by a GABA aminotransferase (GAT) that utilizes pyruvate and GABA as substrates generating alanine and succinate semialdehyde. This latter compound is then converted into succinate by succinate semialdehyde dehydrogenase (SSA-DH). The function of the mitochondrial components of this pathway leads to the reduction of pyruvate inside the mitochondria and the bypass of citric acid cycle metablism between pyruvate and succinate. This has the effect of bypassing three of the major NADH-producing reactions inside the mitochondria and constitutes a major metabolic adjustment to cellular heat stress that can dramatically alter the redox status of the cell.

Although GABA is an important signalling molcule in mammalian brain tissue, there is no evidence that it directly functions as a signalling molecule in plants. However, the synthesis of GABA consumes protons and thus the GAD-catalyzed reaction is one of several reactions involved in the regulation of cytosolic pH. The function of GAD aids the cell during the heat stress episode by maintaining cytosolic pH in a normal physiological range. Additionally, Arabidopsis mutants of SSA-DH demonstrate increased hydrogen peroxide production during episodes of heat stress and show increased programmed cell death. Thus, the function of the GABA metabolic pathway appears to modulate the oxidative status of the cell during heat stress, and may be involved in the production of a cell signal, effectively modulating PCD.

Membrane Lipids and Low Temperatures

The importance of membrane lipids for tolerance to low temperatures has been demonstrated by work with mutant and transgenic plants in which the activity of particular enzymes led to a specific change in membrane lipid composition independent of acclimation to low temperature. In one study, the genes used to modify the lipid composition of thylakoid membranes encoded glycerol-3-phosphate acyltransferases (GPATs). These enzymes select fatty acids for incorporation into PG; GPATs from different plant species have different selectivity for saturated and unsaturated fatty acids, leading to the differences in composition.

The degree of fatty acid saturation of phosphatidylglycerol (PG) is particularly important to the chilling resistance of chloroplasts. In one experiment, tobacco was transformed with an acyltransferase gene from either squash (chill sensitive) or *Arabidopsis* (chill tolerant) to modify the fatty acid saturation of PG in the thylakoid membranes of the chloroplasts. In transgenic tobacco transformed with the squash gene, which increased the proportion of high-melting-point (trans-monounsaturated and saturated) molecular species of PG, chilling tolerance decreased (88% inhibition of photosynthesis following 4 hours at 1°C). Conversely, when tobacco was transformed with the *Arabidopsis* chill-tolerant gene, the proportion of high-melting-point molecular species of PG decreased, and tolerance of chilling was greater than in wild-type plants.

Ice Formation in Higher-Plant Cells

During the initial phase of heat loss from tissue, when the temperature drops below the freezing point of the cytosol and the vacuole without ice forming, the liquid phase in the cytosol and vacuole is said to be supercooled. As the temperature drops further, ice forms in the intercellular spaces, and heat energy (0.33 kJ or 80 calories per gram) is released as a result of the latent heat of fusion of water. At this stage, the temperature of the tissue reflects the balance between heat gain from ice formation and heat loss to the environment. As a result, when ice first forms, the temperature rises rapidly (point B in the figure), and it remains at that level until all the extracellular water is frozen (point C in the figure). When that point is reached, heat release stops, and the temperature begins to fall again. The release of heat energy during ice formation (points B to C in the figure) is the basis for the common practice of spraying crops with water during frost: As long as the water continues to freeze extracellularly, it releases heat that prevents intracellular freezing.

The formation of ice within the intercellular spaces of cells that are sensitive to freezing is not lethal, but extended exposure to freezing temperatures causes water vapour to move through the plasma membrane from the unfrozen protoplast to the cell wall, causing ice crystals to grow within the intercellular spaces. This slow dehydration concentrates solutes within the protoplast, depressing the freezing point by 2 to 3°C. As the temperature continues to drop, a second phase of release of the heat of fusion of water is detectable (points from D to E on the figure). This phase reflects a series of small freezing events: Each "spike" of heat release is thought to represent the freezing of cell protoplasts and coincides with loss of viability. The formation of ice crystals on cell walls or in the protoplasm requires the presence of ice nucleation points on which crystals can be initiated and grow.

In some species, acclimation confers an ability to suppress ice nucleation in the protoplast, allowing deep supercooling to many degrees below the freezing point without ice formation. However, deep supercooling of intracellular water has a lower limit of about –40°C, the temperature at which ice forms spontaneously. (The homogeneous ice formation temperature of pure water droplets is –38.1°C, but the presence of solutes lowers this temperature.) At or below –40°C, ice crystals form without nucleation points, and intracellular freezing and cell death are unavoidable. Species that tolerate temperatures below –40°C under natural conditions do so not by supercooling, but by tolerating gradual dehydration.

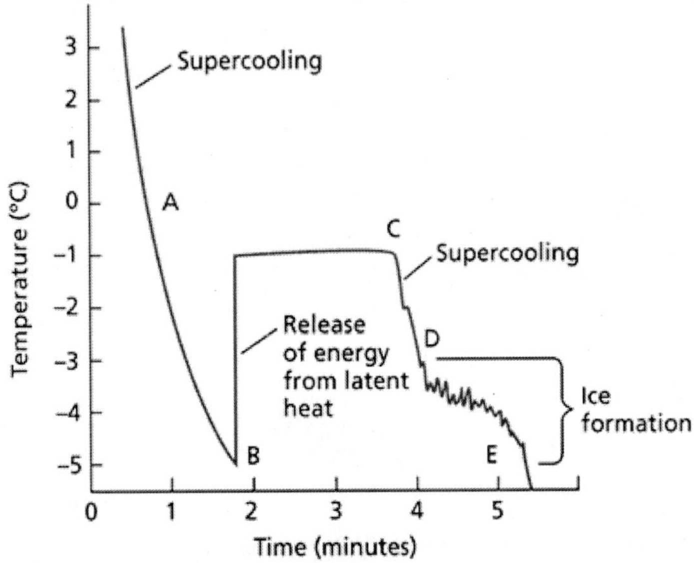

Figure: *Temperature of parenchyma cells in cucumber (*Cucumis sativus*) fruit during freezing. The temperature was recorded with an electronic device, a thermistor, inserted into a 5 × 20 mm cylinder of tissue and immersed in a coolant at –5.8°C. (A–B) Supercooling. (B–C) Release of heat during freezing in cell walls and intercellular spaces. (C–D) Supercooling. (D–E) Small heat spikes released during intracellular freezing of individual protoplasts.*

Plants produce compounds that protect cells against intracellular ice formation. Some of these compounds, such as sugars, amino acids, and other solutes, induce supercooling in the plant's tissues. Overwintering plants also produce antifreeze proteins and other low molecular weight compounds. These compounds provide freeze tolerance by inhibiting ice crystal growth and the nucleation of ice crystals.

Ca^{2+} Signalling and Activation of the Salt Overly Sensitive (SOS) Signal Pathway

Recent studies have identified components of the SOS signal pathway as genetic loci that mediate salt sensitivity. Three genetically linked loci (*SOS1*, *SOS2* and *SOS3*) constitute determinants of a stress-signalling pathway that controls ion homeostasis and salt tolerance.

- SOS3 encodes an EF-hand type Ca^{2+} binding protein with sequence similarity to the regulatory B subunit of calcineurin (protein phosphatase 2B) and neuronal Ca^{2+} sensors. SOS3 has three EF-hand structures, and the interaction of SOS3 with SOS2 kinase leading to activation of

SOS2 activity is Ca^{2+} dependent. The function of SOS3 as a salt tolerance determinant *in planta* is dependent on N-myristoylation.

- SOS2 is a serine/threonine protein kinase that is similar to yeast SNF1 and mammalian AMPK protein kinases. Although all the outputs of the SOS signal pathway have yet to be completely defined, genetic and biochemical evidence indicates that the principal effectors regulated by the pathway are transport systems involved in ion homeostasis at least some of which are the same ones discussed above.
- The plasma membrane Na^+/H^+ antiporter, SOS1, is controlled both at the transcriptional and post transcriptional levels by the SOS signalling pathway. Functional disruption of *AtHKT1* was shown to suppress the salt-sensitive phenotype of *sos3*, indicating that the SOS pathway negatively controls this Na^+-influx system. Additionally, the SOS signalling pathway negatively controls expression of vacuolar AtNHX family members.

These studies illustrate the combined use of genetic, physiological, and biochemical approaches for the elucidation of molecular determinants of ion homeostasis and their signal-mediated regulation in plants. Functional genetic studies are investigating the role of these transport proteins in tolerance to salinity stress.

Na^+ Transport across the Plasma Membrane and Vacuolar Compartmentation

The energy-dependent transport of Na^+ from the cytosol of plant cells across the plasma membrane is mediated by a Na^+/H^+ antiporter. This antiporter is known as SOS1 and it is presumably representative of a family of SOS1 genes. *Arabidopsis sos1* mutants are much more sensitive to growth inhibition by salt than are wild type plants. Interestingly, SOS1 gene expression is not regulated by NaCl stress; but it rather depends on other components of the SOS signalling pathway.

Several of the transporters that mediate Na^+ uptake into the cytosol across the plasma membrane have now been identified. In patch clamp studies with isolated membranes, sodium has been shown to compete with K^+ in inward rectifying K^+ channels. Other reports have suggested that K^+ outward-rectifying channels might facilitate Na^+ influx as well. The high affinity K^+ transporter (HKT) from wheat and the low affinity cation transporter (LCT1) from wheat facilitate Na^+ transport in patch clamp studies with isolated membranes, providing evidence to support their possible function in Na^+ uptake. Patch clamp studies have shown the presence of Ca^{2+}-insensitive-sensitive Na^+ conductances in isolated plant plasma membranes. The Ca^{2+}-sensitive component of Na^+ uptake has been attributed to nonselective cation channels (Schachtman et al. 1997), while *in planta* studies using *Arabidopsis* mutants have directly demonstrated the function of AtHKT1 as a Ca^{2+}-insensitive Na^+-uptake system (Rus et al. 2001). These studies strongly implicate *HKT*1 in the direct control of at least some Na^+ influx. Since *HKT* transcripts are expressed predominantly in the roots, it most probably functions in the control of Na^+ flux into the xylem for export to the shoot.

Vacuolar compartmentalization of Na^+, energized by the ÄpH across the tonoplast membrane, also involves a family of Na^+/H^+ antiporters. *Arabidopsis AtNHX*1 is the prototype for such transporters. Transgenic *Arabidopsis* and tomato plants overexpressing *AtNHX*1 exhibit enhanced salt tolerance. The *Arabidopsis AtNHX*1 family include six loci showing amino acid sequence similarity and topological complementarity to *AtNHX*1, including 12 transmembrane spanning domains and the presence of amiloride binding sites. The *AtNHX*1 family of proteins plays a critical role in the vacuolar compartmentalization of Na^+, which minimizes cytotoxicity and facilitates osmotic adjustment.

Gene Transfer and Stress Tolerance

The transfer of genes suspected to mediate stress tolerance to a test plant where they are overexpressed or altered in some way has given scientists an important new tool with which to study genes involved in stress tolerance. Results of the first study in which transgenic plants were reported to show altered stress tolerance after transfer of a foreign gene appeared in 1993. The transferred gene encoded the enzyme mannitol dehydrogenase, and its expression led to the accumulation of mannitol in the transgenic plants. Transfer of other genes generated transgenic plants that accumulated other solutes, such as trehalose, proline, and fructan.

Interestingly, the amount of accumulated solutes was insufficient to osmotically balance the applied osmotic stress, yet all of the transgenic plants exhibited some increased stress tolerance. These surprising results are not fully understood, but the newly acquired solutes might protect the transgenic plant against free radicals generated in response to stress, rather than function as osmotic-balancing solutes. Improved resistance to drought, salinity, and extreme temperatures has been observed in transgenic plants that express/overexpress genes regulating osmolytes, specific proteins, antioxidants, ion homeostasis, transcription factors, and membrane composition.

Plant Physiology

Plant physiology is a subdiscipline of botany concerned with the function, or physiology, of plants. Closely related fields include plant morphology (structure of plants), plant ecology (interactions with the environment), phytochemistry (biochemistry of plants), cell biology, and molecular biology. Fundamental processes such as photosynthesis, respiration, plant nutrition, plant hormone functions, tropisms, nastic movements, photoperiodism, photomorphogenesis, circadian rhythms, environmental stress physiology, seed germination, dormancy and stomata function and transpiration, both part of plant water relations, are studied by plant physiologists.

Scope

The field of plant physiology includes the study of all the internal activities of plants—those chemical and physical processes associated with life as they occur in plants. This includes study at many levels of scale of size and time. At the smallest scale are molecular interactions of photosynthesis and internal diffusion of water, minerals, and nutrients. At the largest scale are the processes of plant development, seasonality, dormancy, and reproductive control. Major subdisciplines of plant physiology include phytochemistry (the study of the biochemistry of plants) and phytopathology (the study of disease in plants). The scope of plant physiology as a discipline may be divided into several major areas of research.

First, the study of phytochemistry (plant chemistry) is included within the domain of plant physiology. In order to function and survive, plants produce a wide array of chemical compounds not found in other organisms. Photosynthesis requires a large array of pigments, enzymes, and other compounds to function. Because they cannot move, plants must also defend themselves chemically from herbivores, pathogens and competition from other plants. They do this by producing toxins and foul-tasting or smelling chemicals. Other compounds defend plants against disease, permit survival during drought, and prepare plants for dormancy. While other compounds are used to attract pollinators or herbivores to spread ripe seeds.

Secondly, plant physiology includes the study of biological and chemical processes of individual plant cells. Plant cells have a number of features that distinguish them from cells of animals, and which lead to major differences in the way that plant life behaves and responds differently from animal life. For example, plant cells have a cell wall which restricts the shape of plant cells and thereby limits the flexibility and mobility of plants. Plant cells also contain chlorophyll, a chemical compound that interacts with light in a way that enables plants to manufacture their own nutrients rather than consuming other living things as animals do.

Thirdly, plant physiology deals with interactions between cells, tissues, and organs within a plant. Different cells and tissues are physically and chemically specialized to perform different functions. Roots and rhizoids function to anchor the plant and acquire minerals in the soil. Leaves function to catch light in order to manufacture nutrients. For both of these organs to remain living, the minerals acquired by the roots must be transported to the leaves and the nutrients manufactured in the leaves must be transported to the roots. Plants have developed a number of means by which this transport may occur, such as vascular tissue, and the functioning of the various modes of transport is studied by plant physiologists.

Fourthly, plant physiologists study the ways that plants control or regulate internal functions. Like animals, plants produce chemicals called hormones which are produced in one part of the plant to signal cells in another part of the plant to respond. Many flowering plants bloom at the appropriate time because of light-sensitive compounds that respond to the length of the night, a phenomenon known as photoperiodism. The ripening of fruit and loss of leaves in the winter are controlled in part by the production of the gas ethylene by the plant.

Finally, plant physiology includes the study of how plants respond to conditions and variation in the environment, a field known as environmental physiology. Stress from water loss, changes in air chemistry, or crowding by other plants can lead to changes in the way a plant functions. These changes may be affected by genetic, chemical, and physical factors.

Biochemistry of Plants

The list of simple elements of which plants are primarily constructed—carbon, oxygen, hydrogen, calcium, phosphorus, etc.—is not different from similar lists for animals, fungi, or even bacteria. The fundamental atomic components of plants are the same as for all life; only the details of the way in which they are assembled differs.

Despite this underlying similarity, plants produce a vast array of chemical compounds with unusual properties which they use to cope with their environment. Pigments are used by plants to absorb or detect light, and are extracted by humans for use in dyes. Other plant products may be used for the manufacture of commercially important rubber or biofuel. Perhaps the most celebrated compounds from plants are those with pharmacological activity, such as salicylic acid (aspirin), morphine, and digitalis. Drug companies spend billions of dollars each year researching plant compounds for potential medicinal benefits.

Constituent Elements

Plants require some nutrients, such as carbon and nitrogen, in large quantities to survive. Such nutrients are termed macronutrients, where the prefix *macro-*(large) refers to the quantity needed, not the size of the nutrient particles themselves. Other nutrients, called micronutrients, are required only in trace amounts for plants to remain healthy. Such micronutrients are usually absorbed as ions dissolved in water taken from the soil, though carnivorous plants acquire some of their micronutrients from captured prey.

The following tables list element nutrients essential to plants. Uses within plants are generalized.

Macronutrients. (Necessary in large quantities)

Element	***Form of uptake***	***Notes***
Nitrogen	NO_3^-, NH_4^+	Nucleic acids, proteins, hormones, etc.
Oxygen	O_2 H_2O	Cellulose, starch, other organic compounds
Carbon	CO_2	Cellulose, starch, other organic compounds
Hydrogen	H_2O	Cellulose, starch, other organic compounds
Potassium	K^+	Cofactor in protein synthesis, water balance, etc.
Calcium	Ca^{2+}	Membrane synthesis and stabilization
Magnesium	Mg^{2+}	Element essential for chlorophyll
Phosphorus	$H_2PO_4^-$	Nucleic acids, phospholipids, ATP
Sulfur	SO_4^{2-}	Constituent of proteins and coenzymes

Micronutrients. (Necessary in small quantities)

Element	***Form of uptake***	***Notes***
Chlorine	Cl^-	Photosystem II and stomata function
Iron	Fe^{2+}, Fe^{3+}	Chorophyll formation
Boron	HBO_3	Crosslinking pectin
Manganese	Mn^{2+}	Activity of some enzymes
Zinc	Zn^{2+}	Involved in the synthesis of enzymes and chlorophyll
Copper	Cu^+	Enzymes for lignin synthesis
Molybdenum	MoO_4^{2-}	Nitrogen fixation, reduction of nitrates
Nickel	Ni^{2+}	Enzymatic cofactor in the metabolism of nitrogen compounds

Pigments

Among the most important molecules for plant function are the pigments. Plant pigments include a variety of different kinds of molecules, including porphyrins, carotenoids, and anthocyanins. All biological pigments selectively absorb certain wavelengths of light while reflecting others. The light that is absorbed may be used by the plant to power chemical reactions, while the reflected wavelengths of light determine the colour the pigment will appear to the eye.

Chlorophyll is the primary pigment in plants; it is a porphyrin that absorbs red and blue wavelengths of light while reflecting green. It is the presence and relative abundance of chlorophyll that gives plants their green colour. All land plants and green algae possess two forms of this pigment: chlorophyll *a* and chlorophyll *b*. Kelps, diatoms, and other photosynthetic heterokonts contain chlorophyll *c* instead of *b*, while red algae possess only chlorophyll *a*. All chlorophylls serve as the primary means plants use to intercept light in order to fuel photosynthesis.

Carotenoids are red, orange, or yellow tetraterpenoids. They function as accessory pigments in plants, helping to fuel photosynthesis by gathering wavelengths of light not readily absorbed by chlorophyll. The most familiar carotenoids are carotene (an orange pigment found in carrots), lutein (a yellow pigment found in fruits and vegetables), and lycopene (the red pigment responsible for the colour of tomatoes). Carotenoids have been shown to act as antioxidants and to promote healthy eyesight in humans.

Anthocyanins (literally "flower blue") are water-soluble flavonoid pigments that appear red to blue, according to pH. They occur in all tissues of higher plants, providing colour in leaves, stems, roots, flowers, and fruits, though not always in sufficient quantities to be noticeable.

Anthocyanins are most visible in the petals of flowers, where they may make up as much as 30% of the dry weight of the tissue. They are also responsible for the purple colour seen on the underside of tropical shade plants such as *Tradescantia zebrina*; in these plants, the anthocyanin catches light that has passed through the leaf and reflects it back towards regions bearing chlorophyll, in order to maximize the use of available light.

Betalains are red or yellow pigments. Like anthocyanins they are water-soluble, but unlike anthocyanins they are indole-derived compounds synthesized from tyrosine. This class of pigments is found only in the Caryophyllales (including cactus and amaranth), and never co-occur in plants with anthocyanins.

Betalains are responsible for the deep red colour of beets, and are used commercially as food-colouring agents. Plant physiologists are uncertain of the function that betalains have in plants which possess them, but there is some preliminary evidence that they may have fungicidal properties.

Signals and Regulators

Plants produce hormones and other growth regulators which act to signal a physiological response in their tissues. They also produce compounds such as phytochrome that are sensitive to light and which serve to trigger growth or development in response to environmental signals.

Plant Hormones

Plant hormones, also known as plant growth regulators (PGRs) or phytohormones, are chemicals that regulate a plant's growth. According to a standard animal definition, hormones are signal molecules produced at specific locations, that occur in very low concentrations, and cause altered processes in target cells at other locations. Unlike animals, plants lack specific hormone-producing tissues or organs. Plant hormones are often not transported to other parts of the plant and production is not limited to specific locations.

Plant hormones are chemicals that in small amounts promote and influence the growth, development and differentiation of cells and tissues. Hormones are vital to plant growth; affecting processes in plants from flowering to seed development, dormancy, and germination. They regulate which tissues grow upwards and which grow downwards, leaf formation and stem growth, fruit development and ripening, as well as leaf abscission and even plant death. The most important plant hormones are abscissic acid (ABA), auxins, gibberellins, and cytokinins, though there are many other substances that serve to regulate plant physiology.

Photomorphogenesis

While most people know that light is important for photosynthesis in plants, few realize that plant sensitivity to light plays a role in the control of plant structural development (morphogenesis). The use of light to control structural development is called photomorphogenesis, and is dependent upon the presence of specialized photoreceptors, which are chemical pigments capable of absorbing specific wavelengths of light. Plants use four kinds of photoreceptors: phytochrome, cryptochrome, a UV-B photoreceptor, and protochlorophyllide *a*. The first two of these, phytochrome and cryptochrome, are photoreceptor proteins, complex molecular structures formed by joining a protein with a light-sensitive pigment. Cryptochrome is also known as the UV-A photoreceptor, because it absorbs ultraviolet light in the long wave "A" region. The UV-B receptor is one or more compounds that have yet to be identified with certainty, though some evidence suggests carotene or riboflavin as candidates. Protochlorophyllide *a*, as its name suggests, is a chemical precursor of chlorophyll.

The most studied of the photoreceptors in plants is phytochrome. It is sensitive to light in the red and far-red region of the visible spectrum. Many flowering plants use it to regulate the time of flowering based on the length of day and night (photoperiodism) and to set circadian rhythms. It also regulates other responses including the germination of seeds, elongation of seedlings, the size, shape and number of leaves, the synthesis of chlorophyll, and the straightening of the epicotyl or hypocotyl hook of dicot seedlings.

Photoperiodism

Many flowering plants use the pigment phytochrome to sense seasonal changes in day length, which they take as signals to flower. This sensitivity to day length is termed photoperiodism. Broadly speaking, flowering plants can be classified as long day plants, short day plants, or day neutral plants, depending on their particular response to changes in day length. Long day plants require a certain minimum length of daylight to initiate flowering, so these plants flower in the spring or summer. Conversely, short day plants will flower when the length of daylight falls below a certain critical level. Day neutral plants do not initiate flowering based on photoperiodism, though some may use temperature sensitivity (vernalization) instead.

Although a short day plant cannot flower during the long days of summer, it is not actually the period of light exposure that limits flowering. Rather, a short day plant requires a minimal length of uninterrupted darkness in each 24 hour period (a short daylength) before floral development can begin. It has been determined experimentally that a short day plant (long night) will not flower if a flash of phytochrome activating light is used on the plant during the night. Plants make use of the phytochrome system to sense day length or photoperiod. This fact is utilized by florists and greenhouse gardeners to control and even induce flowering out of season, such as the *Poinsettia.*

Environmental Physiology

Paradoxically, the subdiscipline of environmental physiology is on the one hand a recent field of study in plant ecology and on the other hand one of the oldest. Environmental physiology is the preferred name of the subdiscipline among plant physiologists, but it goes by a number of other names in the applied sciences. It is roughly synonymous with ecophysiology, crop ecology, horticulture and agronomy. The particular name applied to the subdiscipline is specific to the viewpoint and goals of research. Whatever name is applied, it deals with the ways in which plants respond to their environment and so overlaps with the field of ecology.

Environmental physiologists examine plant response to physical factors such as radiation (including light and ultraviolet radiation), temperature, fire, and wind. Of particular importance are water relations (which can be measured with the Pressure bomb) and the stress of drought or inundation, exchange of gases with the atmosphere, as well as the cycling of nutrients such as nitrogen and carbon. Environmental physiologists also examine plant response to biological factors. This includes not only negative interactions, such as competition, herbivory, disease and parasitism, but also positive interactions, such as mutualism and pollination.

Tropisms and Nastic Movements

Plants may respond both to directional and nondirectional stimuli. A response to a directional stimulus, such as gravity or sunlight, is called a tropism. A response to a nondirectional stimulus, such as temperature or humidity, is a nastic movement. Tropisms in plants are the result of differential cell growth, in which the cells on one side of the plant elongate more than those on the other side, causing the part to bend toward the side with less growth.

Among the common tropisms seen in plants is phototropism, the bending of the plant toward a source of light. Phototropism allows the plant to maximize light exposure in plants which require additional light for photosynthesis, or to minimize it in plants subjected to intense light and heat. Geotropism allows the roots of a plant to determine the direction of gravity and grow downwards. Tropisms generally result from an interaction between the environment and production of one or more plant hormones.

In contrast to tropisms, nastic movements result from changes in turgor pressure within plant tissues, and may occur rapidly. A familiar example is thigmonasty (response to touch) in the Venus fly trap, a carnivorous plant. The traps consist of modified leaf blades which bear sensitive trigger hairs. When the hairs are touched by an insect or other animal, the leaf folds shut. This mechanism allows the plant to trap and digest small insects for additional nutrients. Although the trap is rapidly shut by changes in internal cell pressures, the leaf must grow slowly in order to reset for a second opportunity to trap insects.

Plant Disease

Economically, one of the most important areas of research in environmental physiology is that of phytopathology, the study of diseases in plants and the manner in which plants resist or cope with infection. Plant are susceptible to the same kinds of disease organisms as animals, including viruses, bacteria, and fungi, as well as physical invasion by insects and roundworms. Because the biology of plants differs from animals, their symptoms and responses are quite different.

In some cases, a plant can simply shed infected leaves or flowers to prevent to spread of disease, in a process called abscission. Most animals do not have this option as a means of controlling disease. Plant diseases organisms themselves also differ from those causing disease in animals because plants cannot usually spread infection through casual physical contact. Plant pathogens tend to spread via spores or are carried by animal vectors. One of the most important advances in the control of plant disease was the discovery of Bordeaux mixture in the nineteenth century. The mixture is the first known fungicide and is a combination of copper sulfate and lime. Application of the mixture served to inhibit the growth of downy mildew that threatened to seriously damage the French wine industry.

History

Early History: Sir Francis Bacon published one of the first plant physiology experiments in 1627 in the book, *Sylva Sylvarum*. Bacon grew several terrestrial plants, including a rose, in water and concluded that soil was only needed to keep the plant upright. Jan Baptist van Helmont published what is considered the first quantitative experiment in plant physiology in 1648. He grew a willow tree for five years in a pot containing 200 pounds of oven-dry soil. The soil lost just two ounces of dry weight and van Helmont concluded that plants get all their weight from water, not soil. In 1699, John Woodward published experiments on growth of spearmint in different sources of water. He found that plants grew much better in water with soil added than in distilled water. Stephen Hales is considered the Father of Plant Physiology for the many experiments in the 1727 book; though Julius von Sachs unified the pieces of plant physiology and put them together as a discipline. His *Lehrbuch der Botanik* was the plant physiology bible of its time.

Researchers discovered in the 1800s that plants absorb essential mineral nutrients as inorganic ions in water. In natural conditions, soil acts as a mineral nutrient reservoir but the soil itself is not essential to plant growth. When the mineral nutrients in the soil are dissolved in water, plant roots absorb nutrients readily, soil is no longer required for the plant to thrive. This observation is the basis for hydroponics, the growing of plants in a water solution rather than soil, which has become a standard technique in biological research, teaching lab exercises, crop production and as a hobby.

Current Research: One of the leading journals in the field is *Plant Physiology*, started in 1926. All its back issues are available online for free. Many other journals often carry plant physiology articles, including *Physiologia Plantarum*, *Journal of Experimental Botany*, *American Journal of Botany*, *Annals of Botany*, *Journal of Plant Nutrition* and *Proceedings of the National Academy of Sciences.*

Economic Applications

Food Production

In horticulture and agriculture along with food science, plant physiology is an important topic relating to fruits, vegetables, and other consumable parts of plants. Topics studied include: *climatic* requirements, fruit drop, nutrition, ripening, fruit set. The production of food crops also hinges on the study of plant physiology covering such topics as Optimal planting and harvesting times and post harvest storage of plant products for human consumption and the production of secondary products like drugs and cosmetics.

Plant Nutrition

Plant nutrition is the study of the chemical elements that are necessary for plant growth. There are several principles that apply to plant nutrition. Some elements are directly involved in plant metabolism. However, this principle does not account for the so-called beneficial elements, whose presence, while not required, has clear positive effects on plant growth.

A nutrient that is able to limit plant growth according to Liebig's law of the minimum, is considered an essential plant nutrient if the plant can not complete its full life cycle without it. There are 14 essential plant nutrients.

Carbon, Hydrogen and Oxygen are essential for plants, but are not considered plant nutrients.

Macronutrients:

- N = Nitrogen
- K = Potassium
- Ca = Calcium
- P = Phosphorus
- Mg = Magnesium
- S = Sulfur
- Si = Silicon

Micronutrients (trace levels) include:

- Cl = Chlorine
- Fe = Iron
- B = Boron
- Mn = Manganese
- Na = Sodium
- Zn = Zinc
- Cu = Copper
- Ni= Nickel
- Mo = Molybdenum.

These nutrients are further divided into the mobile and immobile nutrients. A plant will always supply more nutrients to its younger leaves than its older ones, so when nutrients are mobile, the lack of nutrients is first visible on older leaves. When a nutrient is less mobile, the younger leaves suffer because the nutrient does not move up to them but stays lower in the older leaves. Nitrogen, phosphorus, and potassium are mobile nutrients, while the others have varying degrees of mobility.

Processes

Plants uptake essential elements from the soil through their roots and from the air through their leaves. Nutrient uptake in the soil is achieved by cation exchange, wherein root hairs pump hydrogen ions (H^+) into the soil through proton pumps. These hydrogen ions displace cations attached to negatively charged soil particles so that the cations are available for uptake by the root. In the leaves, stomata open to take in carbon dioxide and expel oxygen. The carbon dioxide molecules are used as the carbon source in photosynthesis. Though nitrogen is plentiful in the earth's atmosphere, relatively few plants engage in nitrogen fixation (conversion of atmospheric nitrogen to a biologically useful form). Most plants therefore require nitrogen compounds to be present in the soil in which they grow.

Functions of Nutrients

Each of these nutrients is used in a different place for a different essential function.

Carbon: Carbon forms the backbone of many plants biomolecules, including starches and cellulose. Carbon is fixed through photosynthesis from the carbon dioxide in the air and is a part of the carbohydrates that store energy in the plant.

Hydrogen: Hydrogen also is necessary for building sugars and building the plant. It is obtained almost entirely from water.

Oxygen: Oxygen is necessary for cellular respiration. Cellular respiration is the process of generating energy-rich adenosine triphosphate (ATP) via the consumption of sugars made in photosynthesis. Oxygen gas is produced as a by-product from this reaction.

Phosphorus: Phosphorus is important in plant bioenergetics. As a component of ATP, phosphorus is needed for the conversion of light energy to chemical energy (ATP) during photosynthesis. Phosphorus can also be used to modify the activity of various enzymes by phosphorylation, and can be used for cell signalling. Since ATP can be used for the biosynthesis of many plant biomolecules, phosphorus is important for plant growth and flower/seed formation.

Potassium: Potassium regulates the opening and closing of the stoma by a potassium ion pump. Since stomata are important in water regulation, potassium reduces water loss from the leaves and increases drought tolerance. Potassium deficiency may cause necrosis or interveinal chlorosis.

Nitrogen: Nitrogen is an essential component of all proteins. Nitrogen deficiency most often results in stunted growth.

Sulphur: Sulphur is a structural component of some amino acids and vitamins, and is essential in the manufacturing of chloroplasts.

Silicon: Silicon is deposited in cell walls and contributes to its mechanical properties including rigidity and elasticity

Calcium: Calcium regulates transport of other nutrients into the plant and is also involved in the activation of certain plant enzymes. Calcium deficiency results in stunting.

Magnesium: Magnesium is an important part of chlorophyll, a critical plant pigment important in photosynthesis. It is important in the production of ATP through its role as an enzyme cofactor. There are many other biological roles for magnesium. Magnesium deficiency can result in interveinal chlorosis.

Iron: Iron is necessary for photosynthesis and is present as an enzyme cofactor in plants. Iron deficiency can result in interveinal chlorosis and necrosis.

Molybdenum: Molybdenum is a cofactor to enzymes important in building amino acids.

Boron: Boron is important in sugar transport, cell division, and synthesizing certain enzymes. Boron deficiency causes necrosis in young leaves and stunting.

Copper: Copper is important for photosynthesis. Symptoms for copper deficiency include chlorosis. Involved in many enzyme processes. Necessary for proper photosythesis. Involved in the manufacture of lignin (cell walls). Involved in grain production.

Manganese: Manganese is necessary for building the chloroplasts. Manganese deficiency may result in coloration abnormalities, such as discoloured spots on the foliage.

Sodium: Sodium is involved in the regeneration of phosphoenolpyruvate in CAM and C4 plants. It can also substitute for potassium in some circumstances.

Zinc: Zinc is required in a large number of enzymes and plays an essential role in DNA transcription. A typical symptom of zinc deficiency is the stunted growth of leaves, commonly known as "little leaf" and is caused by the oxidative degradation of the growth hormone auxin.

Nickel: In higher plants, Nickel is essential for activation of urease, an enzyme involved with nitrogen metabolism that is required to process urea.

Without Nickel, toxic levels of urea accumulate, leading to the formation of necrotic lesions. In lower plants, Nickel activates several enzymes involved in a variety of processes, and can substitute for Zinc and Iron as a cofactor in some enzymes.

Cobalt has proven to be beneficial to at least some plants, but is essential in others, such as legumes where it is required for nitrogen fixation. Vanadium may be required by some plants, but at very low concentrations. It may also be substituting for molybdenum. Selenium and sodium may also be beneficial. Sodium can replace potassium's regulation of stomatal opening and closing.

Plant nutrition is a difficult subject to understand completely, partially because of the variation between different plants and even between different species or individuals of a given clone. Elements present at low levels may cause deficiency symptoms, and toxicity is possible at levels that are too high. Further, deficiency of one element may present as symptoms of toxicity from another element, and vice-versa. Carbon and oxygen are absorbed from the air, while other nutrients are absorbed from the soil. Green plants obtain their carbohydrate supply from the carbon dioxide in the air by the process of photosynthesis.

Plant Hormone

Plant hormones (also known as phytohormones) are chemicals that regulate plant growth. In the UK, the term 'hormone' is not acceptable. There they are called 'plant growth substances'. Plant hormones are signal molecules produced within the plant, and occur in extremely low concentrations. Hormones regulate cellular processes in targeted cells locally and when moved to other locations, in other locations of the plant. Hormones also determine the formation of flowers, stems, leaves, the shedding of leaves, and the development and ripening of fruit. Plants, unlike animals, lack glands that produce and secrete hormones. Plant hormones shape the plant, affecting seed growth, time of flowering, the sex of flowers, senescence of leaves and fruits. They affect which tissues grow upward and which grow downward, leaf formation and stem growth, fruit development and ripening, plant longevity and even plant death. Hormones are vital to plant growth and lacking them, plants would be mostly a mass of undifferentiated cells.

Characteristics

The word hormone is derived from Greek and means 'set in motion.' Plant hormones affect gene expression and transcription levels, cellular division and growth. They are naturally produced within plants, though very similar chemicals are produced by fungi and bacteria that can also effect plant growth. A large number of related chemical compounds are synthesized by humans, they are used to regulate the growth of cultivated plants, weeds, and in vitro grown plants and plant cells; these man made compounds are called Plant Growth Regulators or PGRs for short. Early in the study of plant hormones, "phytohormone" was the commonly-used term, but its use is less widely applied now.

Plant hormones are not nutrients, but chemicals that in small amounts promote and influence the growth, development, and differentiation of cells and tissues. The biosynthesis of plant hormones within plant tissues is often diffuse and not always localized. Plants lack glands to produce and store hormones, because, unlike animals, which have two circulatory systems (lymphatic and cardiovascular) powered by a heart that moves fluids around the body, plants use more passive means to move chemicals around the plant. Plants utilize simple chemicals as hormones, which move more easily through the plant's tissues. They are often produced and used on a local basis within the plant body, plant cells even produce hormones that affect different regions of the cell producing the hormone.

Hormones are transported within the plant by utilizing four types of movements. For localized movement, cytoplasmic streaming within cells and slow diffusion of ions and molecules between cells are utilized. Vascular tissues are used to move hormones from one part of the plant to another; these include sieve tubes that move sugars from the leaves to the roots and flowers, and xylem that moves water and mineral solutes from the roots to the foliage.

Not all plant cells respond to hormones, but those cells that do are programmed to respond at specific points in their growth cycle. The greatest effects occur at specific stages during the cell's life, with diminished effects occurring before or after this period. Plants need hormones at very specific times during plant growth and at specific locations. They also need to disengage the effects that hormones have when they are no longer needed. The production of hormones occurs very often at sites of active growth within the meristems, before cells have fully differentiated. After production they are sometimes moved to other parts of the plant where they cause an immediate effect or they can be stored in cells to be released later. Plants use different pathways to regulate internal hormone quantities and moderate their effects; they can regulate the amount of chemicals used to biosynthesize hormones. They can store them in cells, inactivate them, or cannibalise already-formed hormones by conjugating them with carbohydrates, amino acids or peptides. Plants can also break down hormones chemically, effectively destroying them. Plants also move hormones around the plant diluting their concentrations.

The concentration of hormones required for plant responses are very low (10^{-6} to 10^{-5} mol/L). Because of these low concentrations it has been very difficult to study plant hormones and only since the late 1970s have scientists been able to start piecing together their effects and relationships to plant physiology. Much of the early work on plant hormones involved studying plants that were genetically deficient in one or involved the use of tissue cultured plants grown *in vitro* that were subjected to differing ratios of hormones and the resultant growth compared. The earliest scientific observation and study dates to the 1880s; the determination and observation of plant hormones and their identification was spread-out over the next 70 years.

Classes of Plant Hormones

It is generally accepted that there are five major classes of plant hormones, some of which are made up of many different chemicals that can vary in structure from one plant to the next. The chemicals are each grouped together into one of these classes based on their structural similarities and on their effects on plant physiology. Other plant hormones and growth regulators are not easily grouped into these classes, they exist naturally or are synthesized by humans or other organisms, including chemicals that inhibit plant growth or interrupt the physiological processes within plants. Each class has positive as well as inhibitory functions, and most often work in tandem with each other, with varying ratios of one or more interplaying to affect growth regulation.

The five major classes are:

Abscisic Acid

Abscisic acid also called ABA, was discovered and researched under two different names before its chemical properties were fully known, it was called *dormin* and *abscicin II*. Once it was determined that the two latter named compounds were the same, it was named abscisic acid. The name "abscisic acid" was given because it was found in high concentrations in newly-abscissed or freshly-fallen leaves.

This class of PGR is composed of one chemical compound normally produced in the leaves of plants, originating from chloroplasts, especially when plants are under stress. In general, it acts as an inhibitory chemical compound that affects bud growth, seed and bud dormancy. It mediates changes within the apical meristem causing bud dormancy and the alteration of the last set of leaves into protective bud covers. Since it was found in freshly-abscissed leaves, it was thought to play a role in the processes of natural leaf drop but further research has disproven this. In plant species from temperate parts of the world it plays a role in leaf and seed dormancy by inhibiting growth, but, as it is dissipated from seeds or buds, growth begins. In other plants, as ABA levels decrease, growth then commences as gibberellin levels increase. Without ABA, buds and seeds would start to grow during warm periods in winter and be killed when it froze again. Since ABA dissipates slowly from the tissues and its effects take time to be offset by other plant hormones, there is a delay in physiological pathways that provide some protection from premature growth. It accumulates within seeds during fruit maturation, preventing seed germination within the fruit, or seed germination before winter. Abscisic acid's effects are degraded within plant tissues during cold temperatures or by its removal by water washing in out of the tissues, releasing the seeds and buds from dormancy.

In plants under water stress ABA plays a role in closing the stomata. Soon after plants are water stressed and the roots are deficient in water, a signal moves up to the leaves causing the formation of ABA precursors there which then move to the roots. The roots then release ABA which is translocated to the foliage through the vascular system and modulates the potassium and sodium uptake within the guard cells, which then lose turgidity, closing the stomata. ABA exists in all parts of the plant and its concentration within any tissue seems to mediate its effects and function as a hormone, its degradation or more properly catabolism within the plant affects metabolic reactions and cellular growth and production of other hormones. Plants start life as a seed with high ABA levels, just before the seed germinates ABA levels decrease; during germination and early growth of the seedling, ABA levels decrease even more. As plants begin to produce shoots with fully functional leaves-ABA levels begin to increase, slowing down cellular growth in more "mature" areas of the plant. Stress from water or predation affects ABA production and catabolism rates which mediate another cascade of effects triggering specific responses from targeted cells. Scientists are still piecing together the complex interactions and effects of this and other phytohormones.

Auxins

Auxins are compounds that positively influence cell enlargement, bud formation and root initiation. They also promote the production of other hormones and in conjunction with cytokinins, they control the growth of stems, roots, fruits and convert stems into flowers. Auxins were the first class of growth regulators discovered. They affect cell elongation by altering cell wall plasticity. Auxins decrease in light and increase where its dark. They stimulate cambium cells to divide and in stems cause secondary xylem to differentiate. Auxins act to inhibit the growth of buds lower down the stems (apical dominance), and also to promote lateral and adventitious root development and growth. Leaf abscission is initiated by the growing point of a plant ceasing to produce auxins. Auxins in seeds regulate specific protein synthesis, as they develop within the flower after pollination, causing the flower to develop a fruit to contain the developing seeds. Auxins are toxic to plants in large concentrations; they are most toxic to dicots and less so to monocots. Because of this property, synthetic auxin herbicides including 2,4-D and 2,4,5-T have been developed and used for weed control. Auxins, especially 1-Naphthaleneacetic acid (NAA) and Indole-3-butyric acid (IBA), are also commonly applied to stimulate root growth when taking cuttings of plants. The most common auxin found in plants is indoleacetic acid or IAA.

Cytokinins

Cytokinins or CKs are a group of chemicals that influence cell division and shoot formation. They were called kinins in the past when the first cytokinins were isolated from yeast cells. They also help delay senescence or the aging of tissues, are responsible for mediating auxin transport throughout the plant, and affect internodal length and leaf growth. They have a highly-synergistic effect in concert with auxins and the ratios of these two groups of plant hormones affect most major growth periods during a plant's lifetime. Cytokinins counter the apical dominance induced by auxins; they in conjunction with ethylene promote abscission of leaves, flower parts and fruits.

Ethylene

Ethylene is a gas that forms through the Yang Cycle from the breakdown of methionine, which is in all cells. Ethylene has very limited solubility in water and does not accumulate within the cell but diffuses out of the cell and escapes out of the plant. Its effectiveness as a plant hormone is dependent on its rate of production versus its rate of escaping into the atmosphere. Ethylene is produced at a faster rate in rapidly growing and dividing cells, especially in darkness. New growth and newly-germinated seedlings produce more ethylene than can escape the plant, which leads to elevated amounts of ethylene, inhibiting leaf expansion.

As the new shoot is exposed to light, reactions by phytochrome in the plant's cells produce a signal for ethylene production to decrease, allowing leaf expansion. Ethylene affects cell growth and cell shape; when a growing shoot hits an obstacle while underground, ethylene production greatly increases, preventing cell elongation and causing the stem to swell. The resulting thicker stem can exert more pressure against the object impeding its path to the surface. If the shoot does not reach the surface and the ethylene stimulus becomes prolonged, it affects the stems natural geotropic response, which is to grow upright, allowing it to grow around an object. Studies seem to indicate that ethylene affects stem diameter and height:

When stems of trees are subjected to wind, causing lateral stress, greater ethylene production occurs, resulting in thicker, more sturdy tree trunks and branches. Ethylene affects fruit-ripening: Normally, when the seeds are mature, ethylene production increases and builds-up within the fruit, resulting in a climacteric event just before seed dispersal. The nuclear protein ETHYLENE INSENSITIVE2 (EIN2) is regulated by ethylene production, and, in turn, regulates other hormones including ABA and stress hormones.

Gibberellins

Gibberellins or GAs include a large range of chemicals that are produced naturally within plants and by fungi. They were first discovered when Japanese researchers, including Eiichi Kurosawa, noticed a chemical produced by a fungus called *Gibberella fujikuroi* that produced abnormal growth in rice plants. Gibberellins are important in seed germination, affecting enzyme production which mobilizes food production used for growth of new cells.

This is done by modulating chromosomal transcription. In grain (rice, wheat, corn, etc.) seeds, a layer of cells called the aleurone layer wraps around the endosperm tissue. Absoption of water by the seed causes production of GA. The GA is transported to the aleurone layer, which responds by producing enzymes that break down stored food reserves within the endosperm, which are utilized by the growing seedling. GAs produce bolting of rosette-forming plants, increasing internodal length. They promote flowering, cellular division, and in seeds growth after germination. Gibberellins also reverse the inhibition of shoot growth and dormancy induced by ABA.

Other Known Hormones

Other identified plant growth regulators include:

- Brassinolides-plant steroids that are chemically similar to animal steroid hormones. First isolated from pollen of the mustard family and extensively studied in *Arabidopsis*. They promote cell elongation and cell division, differentiation of xylem tissues, and inhibit leaf abscission. Plants that are deficient in brassinolides suffer from dwarfism.
- Salicylic acid-activates genes in some plants that produce chemicals that aid in the defence against pathogenic invaders.
- Jasmonates-are produced from fatty acids and seem to promote the production of defence proteins that are used to fend off invading organisms. They are believed to also have a role in seed germination, and affect the storage of protein in seeds, and seem to affect root growth.
- Plant peptide hormones-encompass all small secreted peptides that are involved in cell-to-cell signalling. These small peptide hormones play crucial roles in plant growth and development, including defence mechanisms, the control of cell division and expansion, and pollen self-incompatibility.

- Polyamines-are strongly basic molecules with low molecular weight that have been found in all organisms studied thus far. They are essential for plant growth and development and affect the process of mitosis and meiosis.
- Nitric oxide (NO)-serves as signal in hormonal and defence responses.
- Strigolactones, implicated in the inhibition of shoot branching.

Potential Medical Applications

Plant stress hormones activate cellular responses, including cell death, to diverse stress situations in plants. Researchers have found that some plant stress hormones share the ability to adversely affect human cancer cells. For example, sodium salicylate has been found to suppress proliferation of lymphoblastic leukemia, prostate, breast, and melanoma human cancer cells. Jasmonic acid, a plant stress hormone that belongs to the jasmonate family, induced death in lymphoblastic leukemia cells. Methyl jasmonate has been found to induce cell death in a number of cancer cell lines.

Hormones and plant propagation

Synthetic plant hormones or PGRs are commonly used in a number of different techniques involving plant propagation from cuttings, grafting, micropropagation, and tissue culture. The propagation of plants by cuttings of fully-developed leaves, stems, or roots is performed by gardeners utilizing auxin as a rooting compound applied to the cut surface; the auxins are taken into the plant and promote root initiation. In grafting, auxin promotes callus tissue formation, which joins the surfaces of the graft together. In micropropagation, different PGRs are used to promote multiplication and then rooting of new plantlets. In the tissue-culturing of plant cells, PGRs are used to produce callus growth, multiplication, and rooting.

Seed Dormancy

Plant hormones affect seed germination and dormancy by affecting different parts of the seed. Embryo dormancy is characterized by a high ABA/GA ratio, whereas the seed has a high ABA sensitivity and low GA sensitivity. To release the seed from this type of dormancy and initiate seed germination, an alteration in hormone biosynthesis and degradation towards a low ABA/GA ratio, along with a decrease in ABA sensitivity and an increase in GA sensitivity needs to occur. ABA controls embryo dormancy, and GA embryo germination. Seed coat dormancy involves the mechanical restriction of the seed coat, this along with a low embryo growth potential, effectively produces seed dormancy. GA releases this dormancy by increasing the embryo growth potential, and/or weakening the seed coat so the radical of the seedling can break through the seed coat. Different types of seed coats can be made up of living or dead cells and both types can be influenced by hormones; those composed of living cells are acted upon after seed formation while the seed coats composed of dead cells can be influenced by hormones during the formation of the seed coat. ABA affects testa or seed coat growth characteristics, including thickness, and effects the GA-mediated embryo growth potential.

These conditions and effects occur during the formation of the seed, often in response to environmental conditions. Hormones also mediate endosperm dormancy: Endosperm in most seeds is composed of living tissue that can actively respond to hormones generated by the embryo. The endosperm often acts as a barrier to seed germination, playing a part in seed coat dormancy or in the germination process. Living cells respond to and also affect the ABA/GA ratio, and mediate cellular sensitivity; GA thus increases the embryo growth potential and can promote endosperm weakening. GA also affects both ABA-independent and ABA-inhibiting processes within the endosperm.

History of Plant Physiology

Plant physiology is concerned with the life processes of plants, and from the beginning has been focused largely on the higher green terrestrial plants, the autotrophic (self-feeding) plants that feed us animals. In part, plant physiology has roots in agriculture.

The central question of plant physiology is how do plants grow, develop, and reproduce? When primitive humans collected seeds and began raising food plants they must have noted that plants need sunlight, warmth and moist (but not wet) soil of good tilth, and that seeds from vigorous plants produced vigorous plants. They observed the beneficial effects of manuring (mentioned in the Bible; Luke 13:8). Centuries of agricultural practice produced improved varieties and cultural practices, and early studies of physiology drew on this basic knowledge of plant growth and gross anatomy.

Early Experiments on Growth

An early physiological question was where a plant gets the material with which it grows. In the early 1600s, Jan van Helmont, a Belgian physician, decided the source must be water alone. Van Helmont grew a willow seedling in 200 pounds of soil, and only added rainwater. A 164-pound tree was produced with only 57.1 grams (2 ounces) of soil lost. He knew of carbon dioxide but never dreamed that a diffuse gas could produce willow wood.

In the next century Antoine Lavoisier found organic matter to be largely formed of carbon and oxygen. Joseph Priestley, Jan Ingenhousz, and Jean Senebier demonstrated that plant leaves in light take up carbon dioxide and emit equivalent amounts of oxygen. Later, Nicholas de Saussure noted that water was involved in the process. The reverse occurred in the dark—plants respired like animals, taking up oxygen and emitting carbon dioxide. J. R. Mayer observed that the process converted light energy into the chemical energy of organic carbon. Thus growth of seedlings in the dark or of roots in the soil was at the expense of this energy. Therefore, by the nineteenth century, photosynthesis, although not understood biochemically, was established as the primary and essential synthetic process in plant growth.

Nutrition and Transportation

In his experiment, van Helmont assigned no importance to the two ounces of soil lost. However, starting in the late 1700s and extending into the mid 1800s, Julius Sachs and others used chemical assays to establish that quantitatively minor soil constituents of nitrogen, potassium, phosphate, sulfur, and other elements had major importance in plant growth. The long-recognized importance of manure lay with its content of these inorganic nutrients, especially nitrogen. It was discovered these could be added to the soil as inorganic salts, such as potassium nitrate. The organic material of manure, or the residue of its decay, contributed to improved tilth, or soil structure, but did not provide nutrients. From these discoveries came the modern agricultural use of chemical fertilizers.

What about the extensive loss of water from the soil? Van Helmont had to continuously water his willow tree with many more pounds of water than were ultimately incorporated by the tree. In 1727 an English clergyman and amateur physiologist, Stephen Hales, published *Vegetable Staticks,* an account of his pioneering studies on the transpiration, growth, and gas exchanges of plants. Hales demonstrated that water from the soil moves up the stems to the leaves where it is lost as water vapour, a process called transpiration. Subsequent research of the nineteenth and early twentieth centuries showed that the water diffuses out through stomata (singular stoma), pores in the leaf epidermis (outer layer of leaf cells).

With light and adequate water the two cells bounding the stoma inflate, opening the pore to gas diffusion; under dry conditions the cells grow flaccid and the pore closes, conserving water. Capillary forces originating in the microscopic pores of the leaf mesophyll (internal green photosynthetic cells), with some contribution from osmosis, pull columns of water up the open vessels and tracheids of the xylem (wood) carrying nutrient salts from the roots.

The coherence between water molecules and their adherence to cell walls prevents the taut water columns from breaking even in trees of great height. This scheme was first proposed in 1895 by Henry Dixen and John Joly. Numerous researchers in the twentieth century confirmed and refined this "cohesion-tension" theory of transportation.

Hales also measured the root pressure (forced bleeding) of decapitated plants. Subsequent work showed that under conditions of good soil moisture and aeration, roots actively secrete high concentrations of salt into the root xylem creating a high osmotic pressure that forces water up the stem and out pores at the tips of leaves (guttation). In 1926, E. Munch proposed a similar mechanism for translocation, the movement of sugars from leaves to roots and other plant parts. This mechanism is known as the pressureflow model.

Cellular and Molecular Plant Physiology

By the twentieth century, plant physiologists increasingly turned to chemistry and physics for assistance with fundamental questions. They also established their own societies with journals to publish their findings, which had a catalytic effect in increasing the level and amount of research. A great deal of the fundamental biochemistry of cell growth and function, known from the more extensive medical, animal, and microbiological research, was found to apply to plant cells. Anatomical studies gave structural details to support physiological findings, and submicroscopic cell structure was revealed by the electron microscope.

All the mineral nutrients required for plant growth were established. The key to their selective uptake from the soil and transport into the root xylem proved to lie with an energy-requiring proton (hydrogen ion) pumping mechanism in the cell membranes.

Environmental, hormonal, and genetic controls on growth and development have been extensively explored, but there is still more to learn. Ethylene, a simple two-carbon gas generated by plants initiates fruit ripening and regulates aspects of seed germination. Phototropisms (bending in response to unilateral light, investigated by Charles Darwin), and geotropisms (root growth down, stem growth up) were found to be due to displacement of a cell growth hormone, or auxin. In some circumstances, auxins could also elicit cell division (root formation in stem cuttings). Other hormones, the gibberellins, regulate cell division at the stem apex and activate enzyme formation in seed germination.

Attempts to culture plant tissues led to discovery of more cell division hormones, the cytokinins. Another type of hormone, abscisic acid, initiates the senescence and abscission of leaves in the fall, and causes the stomata to close under water stress. Additional growth regulating compounds are being found and investigated but a coordinated picture of hormone interaction is lacking.

Photoperiodism, the regulation of flowering by day length was discovered. Sleep movements, such as the drooping of bean leaves in the evening, were found to be controlled by a biological "clock," a circadian rhythm, not by the onset of darkness. In 1952, phytochrome was discovered and found to be the pigment at the centre of photoperiodism. In recent years there has been a major shift to molecular genetics in attempts to locate the genes responsible for physiological processes.

In photosynthesis chlorophyll structure was determined and localized in the internal membranes of the chloroplasts of the mesophyll cells. Red and blue portions of the light spectrum were found effective, leading to the discovery that two light reactions are required. In the 1930s, C. B. van Neil used radioactive water to show that water, not carbon dioxide, was the source of oxygen released during photosynthesis. Sugar was found to be synthesized in the stroma (fluid part) of the chloroplast, and the molecular details of its creation were worked out by Melvin Calvin and Andrew Benson. All plant cells were found to respire, an energy-yielding process essentially the same as that in animals, involving another membranous organelle, the mitochondrion, and yielding metabolic energy available for transport reactions and synthesis of cell substance.

The formation of fats and oils from carbohydrates was found to be similar to that in animals, but plants had the added ability to transform oils in germinating seeds into carbohydrates such as the glucose used in cellulose wall formation. The symbiotic relationships of plants and microorganisms was explored, notably in the cases of reduced nitrogen formation from atmospheric nitrogen by nodule bacteria.

At the end of twentieth century, the small mustard plant *Arabidopsis thaliana* took centre stage in the attempt of scientists to understand plant genomes. The full sequence of this genome was elucidated in 2000 by an international consortium of plant geneticists.

Plant Cell

Plant cells are eukaryotic cells that differ in several key respects from the cells of other eukaryotic organisms. Their distinctive features include:

- A large central vacuole, a water-filled volume enclosed by a membrane known as the *tonoplast* maintains the cell's turgor, controls movement of molecules between the cytosol and sap, stores useful material and digests waste proteins and organelles.
- A cell wall composed of cellulose and hemicellulose, pectin and in many cases lignin, and secreted by the protoplast on the outside of the cell membrane. This contrasts with the cell walls of fungi (which are made of chitin), and of bacteria, which are made of peptidoglycan.
- Specialised cell-cell communication pathways known as plasmodesmata, pores in the primary cell wall through which the plasmalemma and endoplasmic reticulum of adjacent cells are continuous.

- Plastids, notably the chloroplasts which contain chlorophyll and the biochemical systems for light harvesting and photosynthesis, but also amyloplasts specialized for starch storage, elaioplasts specialized for fat storage and chromoplasts specialized for synthesis and storage of pigments. As in mitochondria, which have a genome encoding 37 genes plastids have their own genomes of about 100-120 unique genes and probably arose as prokaryotic endosymbionts living in the cells of an early eukaryotic ancestor of the land plants and algae.
- Cell division by construction of a phragmoplast as a template for building a cell plate late in cytokinesis is characteristic of land plants and a few groups of algae, notably the Charophytes and the Order Trentepohliales
- The sperm of bryophytes have flagellae similar to those in animals, but higher plants, (including Gymnosperms and flowering plants) lack the flagellae and centrioles that are present in animal cells.

Cell Types

- Parenchyma cells are living cells that have diverse functions ranging from storage and support to photosynthesis and phloem loading (transfer cells). Apart from the xylem and phloem in its vascular bundles, leaves are mainly composed of parenchyma cells. Some parenchyma cells, as in the epidermis, are specialized for light penetration and focusing or regulation of gas exchange, but others are among the least specialized cells in plant tissue, and may remain totipotent, capable of dividing to produce new populations of undifferentiated cells, throughout their lives. Parenchyma cells have thin, permeable primary walls enabling the transport of small molecules between them, and their cytoplasm is responsible for a wide range of biochemical functions such as nectar secretion, or the manufacture of secondary products that discourage herbivory. Parenchyma cells which contain many chloroplasts and are concerned primarily with photosynthesis are called chlorenchyma cells. Others, such as the majority of the parenchyma cells in potato tubers and the seed cotyledons of legumes have a storage function.
- Collenchyma cells-Like parenchyma, collenchyma cells are alive at maturity and have only a primary wall. These cells mature from meristem derivatives that initially resemble parenchyma, but differences quickly become apparent. Plastids do not develop and the secretory apparatus (ER and Golgi) proliferates to secrete additional primary wall. The wall is most commonly thickest at the corners, where three or more cells come in contact and thinnest where only two cells come in contact, though other arrangements of the wall thickening are possible.

Pectin and hemicellulose are the dominant constituents of collenchyma cell walls of dicotyledon angiosperms which may contain as little as 20% of cellulose in *Petasites*. Collenchyma cells are typically quite elongated, and may divide transversly to give a septate appearance. The role of this cell type is to support the plant in axes still growing in length, and to confer flexibility and tensile strength on tissues. The primary wall lacks lignin that would make it tough and rigid, so this cell type provides what could be called plastic support. Support that can hold a young stem or petiole into the air, but in cells that can be stretched as the cells around them elongate. Stretchable support (without elastic snap-back) is a good way to describe what collenchyma does. Parts of the strings in celery are collenchyma.

- Sclerenchyma cells-Sclerenchyma cells (from the Greek skleros, hard) are hard and tough cells with a function in mechanical support. They are of two broad types-sclereids or stone cells and fibres. The cells develop an extensive secondary cell wall that is laid down on the inside of the primary cell wall. The secondary wall is impregnated with lignin, making it hard and impermeable to water. Thus, these cells cannot survive for long as they cannot exchange sufficient material to maintain active metabolism. Sclerenchyma cells are typically dead at functional maturity, and the cytoplasm is missing, leaving an empty central cavity.

Functions for sclereid cells (hard cells that give leaves or fruits a gritty texture) include discouraging herbivory, by damaging digestive passages in small insect larval stages, and physical protection (a solid tissue of hard sclereid cells form the pit wall in a peach and many other fruits). Functions of fibres include provision of load-bearing support and tensile strength to the leaves and stems of herbaceous plants. Sclerenchyma fibres are not involved in conduction, either of water and nutrients (as in the xylem) or of carbon compounds (as in the phloem), but it is likely that they may have evolved as modifications of xylem and phloem initials in early land plants.

Tissue Types

The major classes of cells differentiate from undifferentiated meristematic cells (analogous to the stem cells of animals) to form the tissue structures of roots, stems, leaves, flowers and reproductive structures.

Xylem cells are elongated cells with lignified secondary thickening of the cell walls. Xylem cells are specialised for conduction of water, and first appeared in plants during their transition to land in the Silurian period more than 425 million years ago. The possession of xylem defines the vascular plants or Tracheophytes. Xylem tracheids are pointed, elongated xylem cells, the simplest of which have continuous primary cell walls and lignified secondary wall thickenings in the form of rings, hoops or reticulate networks. More complex tracheids with valve-like perforations called bordered pits characterise the gymnosperms. The ferns and other pteridophytes and the gymnosperms only have xylem tracheids, while the angiosperms also have xylem vessels. Vessel members are hollow xylem cells aligned end-to-end, without end walls that are assembled into long continuous tubes. The bryophytes lack true xylem cells, but their sporophytes have a water conducting tissue known as the hydrome that is composed of elongated cells of simpler construction.

Phloem is a specialised tissue for food conduction in higher plants. Phloem consists of two cell types, the sieve tubes and the intimately-associated companion cells. The sieve tube elements lack nuclei and ribosomes, and their metabolism and functions are regulated by the adjacent nucleate companion cells. Sieve tubes are joined end to end with perforate end-plates between known as *sieve plates*, which allow transport of photosynthate between the sieve elements. The companion cells, connected to the sieve tubes via plasmodesmata, are responsible for loading the phloem with sugars. The bryophytes lack phloem, but moss sporophytes have a simpler tissue with analogous function known as the leptome.

Plant epidermal cells are specialised parenchyma cells covering the external surfaces of leaves, stems and roots. The epidermal cells of aerial organs arise from the superficial layer of cells known as the *tunica* that covers the plant shoot apex, whereas the cortex and vascular tissues arise from innermost layer of the shoot apex known as the *corpus* (L3 layer). The epidermis of roots originates from the layer of cells immediately beneath the root cap.

The epidermis of all aerial organs, but not roots, is covered with a cuticle made of waxes and the polyester cutin. Several cell types may be present in the epidermis. Notable among these are the stomatal guard cells, glandular and clothing hairs or trichomes, and the root hairs of primary roots. In the shoot epidermis of most plants, only the guard cells have chloroplasts. The epidermal cells of the primary shoot are thought to be the only plant cells with the biochemical capacity to synthesize cutin.

The Plant Cell

Plant cells are eukaryotic and have many of the structures found in animal cells. Link to pages describing these.

- Plasma membrane
- Nucleus and nucleolus
- Mitochondria
- Ribosomes
- Endoplasmic reticulum

- Golgi apparatus
- Peroxisomes (the crystal in the electron micrograph is enclosed within a peroxisome)
- Microtubules.

Plant cells differ from animal cells in lacking:

- centrioles
- intermediate filaments.

and having:

- plastids
- a cell wall
- large vacuoles.

The electron micrograph shows cells from a sunflower leaf. It was supplied through the courtesy of H. J. Arnott and Kenneth M. Smith.

Vacuoles

Vacuoles are bounded by a single membrane. Young plant cells often contain many small vacuoles, but as the cells mature, these unite to form a large central vacuole. Vacuoles serve several functions, such as

- storing foods (e.g., proteins in seeds)
- storing wastes
- storing malic acid in CAM plants
- storing various ions (e.g., calcium, sodium, iron) which, among other functions, helps to
- maintain turgor in the cell.

Plant cells avoid bursting in hypotonic surroundings by their strong cell walls. These allow the build-up of turgor within the cell. Loss of turgor causes wilting.

Plasmolysis

When a freshwater (or terrestrial) plant is placed in sea water, its cells quickly lose turgor and the plant wilts.

This is because sea water is hypertonic to the cytoplasm. As water diffuses from the cytoplasm into the sea water, the cells shrink-drawing their plasma membrane away from the cell wall.

The photomicrograph shows plasmolyzed cells in the freshwater plant Elodea which has been placed in sea water. Note how the cell walls now show clearly.

Eucaryotic Cell Organelles

Nucleus: The nucleus is the most obvious organelle in any eukaryotic cell. It is enclosed in a double membrane and communicates with the surrounding cytosol via numerous nuclear pores. Within the nucleus is the DNA responsible for providing the cell with its unique characteristics. The DNA is similar in every cell of the body, but depending on the specific cell type, some genes may be turned on or off-that's why a liver cell is different from a muscle cell, and a muscle cell is different from a fat cell. When a cell is dividing, the nuclear chromatin (DNA and surrounding protein) condenses into chromosomes that are easily seen by microscopy.

Nucleolus: The prominent structure in the nucleus is the nucleolus. The nucleolus produces ribosomes, which move out of the nucleus and take positions on the rough endoplasmic reticulum where they are critical in protein synthesis.

Cytosol: The cytosol is the "soup" within which all the other cell organelles reside and where most of the cellular metabolism occurs. Though mostly water, the cytosol is full of proteins that control cell metabolism including signal transduction pathways, glycolysis, intracellular receptors, and transcription factors.

Cytoplasm: This is a collective term for the cytosol plus the organelles suspended within the cytosol.

Centrosome: The centrosome, or MICROTUBULE ORGANIZING Centre (MTOC), is an area in the cell where microtubules are produced. Plant and animal cell centrosomes play similar roles in cell division, and both include collections of microtubules, but the plant cell centrosome is simpler and does not have centrioles.

During animal cell division, the centrioles replicate (make new copies) and the centrosome divides. The result is two centrosomes, each with its own pair of centrioles. The two centrosomes move to opposite ends of the nucleus, and from each centrosome, microtubules grow into a "spindle" which is responsible for separating replicated chromosomes into the two daughter cells.

Centriole (animal cells only): Each centriole is a ring of nine groups of fused microtubules. There are three microtubules in each group. Microtubules (and centrioles) are part of the cytoskeleton. In the complete animal cell centrosome, the two centrioles are arranged such that one is perpendicular to the other.

Golgi: The Golgi apparatus is a membrane-bound structure with a single membrane. It is actually a stack of membrane-bound vesicles that are important in packaging macromolecules for transport elsewhere in the cell. The stack of larger vesicles is surrounded by numerous smaller vesicles containing those packaged macromolecules. The enzymatic or hormonal contents of lysosomes, peroxisomes and secretory vesicles are packaged in membrane-bound vesicles at the periphery of the Golgi apparatus.

Lysosome: Lysosomes contain hydrolytic enzymes necessary for intracellular digestion. They are common in animal cells, but rare in plant cells. Hydrolytic enzymes of plant cells are more often found in the vacuole.

Peroxisome: Peroxisomes are membrane-bound packets of oxidative enzymes. In plant cells, peroxisomes play a variety of roles including converting fatty acids to sugar and assisting chloroplasts in photorespiration. In animal cells, peroxisomes protect the cell from its own production of toxic hydrogen peroxide. As an example, white blood cells produce hydrogen peroxide to kill bacteria. The oxidative enzymes in peroxisomes break down the hydrogen peroxide into water and oxygen.

Secretory Vesicle: Cell secretions-e.g. hormones, neurotransmitters-are packaged in secretory vesicles at the Golgi apparatus. The secretory vesicles are then transported to the cell surface for release.

Cell Membrane: Every cell is enclosed in a membrane, a double layer of phospholipids (lipid bilayer). The exposed heads of the bilayer are "hydrophilic" (water loving), meaning that they are compatible with water both within the cytosol and outside of the cell. However, the hidden tails of the phosopholipids are "hydrophobic" (water fearing), so the cell membrane acts as a protective barrier to the uncontrolled flow of water. The membrane is made more complex by the presence of numerous proteins that are crucial to cell activity. These proteins include receptors for odours, tastes and hormones, as well as pores responsible for the controlled entry and exit of ions like sodium (Na+) potassium (K+), calcium (Ca++) and chloride (Cl-).

Mitochondria: Mitochondria provide the energy a cell needs to move, divide, produce secretory products, contract-in short, they are the power centres of the cell. They are about the size of bacteria but may have different shapes depending on the cell type. Mitochondria are membrane-bound organelles, and like the nucleus have a double membrane. The outer membrane is fairly smooth. But the inner membrane is highly convoluted, forming folds (cristae) as seen in the cross-section, above. The cristae greatly increase the inner membrane's surface area. It is on these cristae that food (sugar) is combined with oxygen to produce ATP-the primary energy source for the cell.

Vacuole: A vacuole is a membrane-bound sac that plays roles in intracellular digestion and the release of cellular waste products. In animal cells, vacuoles are generally small. Vacuoles tend to be large in plant cells and play several roles: storing nutrients and waste products, helping increase cell size during growth, and even acting much like lysosomes of animal cells. The plant cell vacuole also regulates turgor pressure in the cell. Water collects in cell vacuoles, pressing outward against the cell wall and producing rigidity in the plant. Without sufficient water, turgor pressure drops and the plant wilts.

Cell Wall (plant cells only): Plant cells have a rigid, protective cell wall made up of polysaccharides. In higher plant cells, that polysaccharide is usually cellulose. The cell wall provides and maintains the shape of these cells and serves as a protective barrier. Fluid collects in the plant cell vacuole and pushes out against the cell wall. This turgor pressure is responsible for the crispness of fresh vegetables.

Chloroplast (plant cells only): Chloroplasts are specialized organelles found in all higher plant cells. These organelles contain the plant cell's chlorophyll responsible for the plant's green colour. Chloroplasts have a double outer membrane. Within the stroma are other membrane structures-the thylakoids. Thylakoids appear in stacks called "grana" (singular = granum).

6

Plant-water Relations

Water is the most abundant constituent of all physiologically active plant cells. Leaves, for example, have water contents which lie mostly within a range of 55–85% of their fresh weight. Other relatively succulent parts of plants contain approximately the same proportion of water, and even such largely nonliving tissues as wood may be 30–60% water on a fresh-weight basis. The smallest water contents in living parts of plants occur mostly in dormant structures, such as mature seeds and spores. The great bulk of the water in any plant constitutes a unit system. This water is not in a static condition. Rather it is part of a hydrodynamic system, which in terrestrial plants involves absorption of water from the soil, its translocation throughout the plant, and its loss to the environment, principally in the process known as transpiration.

Cellular Water Relations

The typical mature, vacuolate plant cell constitutes a tiny osmotic system, and this idea is central to any concept of cellular water dynamics. Although the cell walls of most living plant cells are quite freely permeable to water and solutes, the cytoplasmic layer that lines the cell wall is more permeable to some substances than to others. If a plant cell in a flaccid condition—one in which the cell sap exerts no pressure against the encompassing cytoplasm and cell wall—is immersed in pure water, inward osmosis of water into the cell sap ensues. This gain of water results in the exertion of a turgor pressure against the protoplasm, which in turn is transmitted to the cell wall. This pressure also prevails throughout the mass of solution within the cell. If the cell wall is elastic, some expansion in the volume of the cell occurs as a result of this pressure, although in many kinds of cells this is relatively small. If a turgid or partially turgid plant cell is immersed in a solution with a greater osmotic pressure than the cell sap, a gradual shrinkage in the volume of the cell ensues; the amount of shrinkage depends upon the kind of cell and its initial degree of turgidity.

When the lower limit of cell wall elasticity is reached and there is continued loss of water from the cell sap, the protoplasmic layer begins to recede from the inner surface of the cell wall. Retreat of the protoplasm from the cell wall often continues until it has shrunk toward the centre of the cell, the space between the protoplasm and the cell wall becoming occupied by the bathing solution. This phenomenon is called plasmolysis.

In some kinds of plant cells movement of water occurs principally by the process of imbibition rather than osmosis. The swelling of dry seeds when immersed in water is a familiar example of this process.

Stomatal Mechanism

Various gases diffuse into and out of physiologically active plants. Those gases of greatest physiological significance are carbon dioxide, oxygen, and water vapour. The great bulk of the gaseous exchanges between a plant and its environment occurs through tiny pores in the epidermis that are called stomates. Although stomates occur on many aerial parts of plants, they are most characteristic of, and occur in greatest abundance in, leaves.

Transpiration Process

The term transpiration is used to designate the process whereby water vapour is lost from plants. Although basically an evaporation process, transpiration is complicated by other physical and physiological conditions prevailing in the plant. Whereas loss of water vapour can occur from any part of the plant which is exposed to the atmosphere, the great bulk of all transpiration occurs from the leaves. There are two kinds of foliar transpiration: (1) stomatal transpiration, in which water vapour loss occurs through the stomates, and (2) cuticular transpiration, which occurs directly from the outside surface of epidermal walls through the cuticle. In most species 90% or more of all foliar transpiration is of the stomatal type.

Transpiration is a necessary consequence of the relation of water to the anatomy of the plant, and especially to the anatomy of the leaves. Terrestrial green plants are dependent upon atmospheric carbon dioxide for their survival. In terrestrial vascular plants the principal carbon dioxide–absorbing surfaces are the moist mesophyll cells walls which bound the intercellular spaces in leaves. Ingress of carbon dioxide into these spaces occurs mostly by diffusion through open stomates. When the stomates are open, outward diffusion of water vapour unavoidably occurs, and such stomatal transpiration accounts for most of the water vapour loss from plants. Although transpiration is thus, in effect, an incidental phenomenon, it frequently has marked indirect effects on other physiological processes which occur in the plant because of its effects on the internal water relations of the plant.

Water Translocation

In terrestrial rooted plants practically all of the water which enters a plant is absorbed from the soil by the roots. The water thus absorbed is translocated to all parts of the plant. The mechanism of the "ascent of sap" (all translocated water contains at least traces of solutes) in plants, especially tall trees, was one of the first processes to excite the interest of plant physiologists.

The upward movement of water in plants occurs in the xylem, which, in the larger roots, trunks, and branches of trees and shrubs, is identical with the wood. In the trunks or larger branches of most kinds of trees, however, sap movement is restricted to a few of the outermost annual layers of wood.

Root pressure is generally considered to be one of the mechanisms of upward transport of water in plants. While it is undoubtedly true that root pressure does account for some upward movement of water in certain species of plants at some seasons, various considerations indicate that it can be only a secondary mechanism of water transport.

Upward translocation of water (actually a very dilute sap) is engendered by an increase in the negativity of water potential in the cells of apical organs of plants. Such increases in the negativity of water potentials occur most commonly in the mesophyll cells of leaves as a result of transpiration.

Water Absorption

The successively smaller branches of the root system of any plant terminate ultimately in the root tips, of which there may be thousands and often millions on a single plant. Most absorption of water occurs in the root tip regions, and especially in the root hair zone. Older portions of most roots become covered with cutinized or suberized layers through which only very limited quantities of water can pass.

Whenever the water potential in the peripheral root cells is less than that of the soil water, movement of water from the soil into the root cells occurs. There is some evidence that, under conditions of marked internal water stress, the tension generated in the xylem ducts will be propagated across the root to the peripheral cells. If this occurs, water potentials of greater negativity could develop in peripheral root cells than would otherwise be possible. The absorption mechanism would operate in fundamentally the same way whether or not the water in the root cells passed into a state of tension. The process just described, often called passive absorption, accounts for most of the absorption of water by terrestrial plants.

The phenomenon of root pressure represents another mechanism of the absorption of water. This mechanism is localized in the roots and is often called active absorption. Water absorption of this type only occurs when the rate of transpiration is low and the soil is relatively moist. Although the xylem sap is a relatively dilute solution, its osmotic pressure is usually great enough to engender a more negative water potential than usually exists in the soil water when the soil is relatively moist. A gradient of water potentials can thus be established, increasing in negativity across the epidermis, cortex, and other root tissues, along which the water can move laterally from the soil to the xylem.

Plant Mineral Nutrition

The relationship between plants and all chemical elements other than carbon, hydrogen, and oxygen in the environment. Plants obtain most of their mineral nutrients by extracting them from solution in the soil or the aquatic environment. Mineral nutrients are so called because most have been derived from the weathering of minerals of the Earth's crust. Nitrogen is exceptional in that little occurs in minerals: the primary source is gaseous nitrogen of the atmosphere.

Some of the mineral nutrients are essential for plant growth; others are toxic, and some absorbed by plants may play no role in metabolism. Many are also essential or toxic for the health and growth of animals using plants as food. Six basic facts have been established: (1) plants do not need any of the solid materials in the soil—they cannot even take them up; (2) plants do not need soil microorganisms; (3) plant roots must have a supply of oxygen; (4) all plants require at least 14 mineral nutrients; (5) all of the essential mineral nutrients may be supplied to plants as simple ions of inorganic salts in solution; and (6) all of the essential nutrients must be supplied in adequate but nontoxic quantities. These facts provide a conceptually simple definition of and test for an essential mineral nutrient. A mineral nutrient is regarded as essential if, in its absence, a plant cannot complete its life cycle.

Nutrients which plants require in relatively large amounts, that is, the essential macronutrients, are nitrogen, sulfur, phosphorus, calcium, potassium, and magnesium. Iron is not required in large amounts and hence is regarded as an essential micronutrient or trace element. With the progressive development of better techniques for purifying water and salts, the list of essential nutrients for all plants has expanded to include boron, manganese, zinc, copper, molybdenum, and chlorine. Evidence has accumulated in support of nickel being essential. In addition, sodium and silicon have been shown to be essential for some plants, beneficial to some, and possibly of no benefit to others. Cobalt has also been shown to be essential for the growth of legumes when relying upon atmospheric nitrogen. Claims that two other chemical elements (vanadium and selenium) may be essential micronutrients have still to be firmly established.

Mineral nutrients may be toxic to plants either because the specific nutrient interferes with plant metabolism or because its concentration in combination with others in solution is excessive and interferes with the plant's water relations. Other chemical elements in the environment may also be toxic. High concentrations of salts in soil solutions or aquatic environments may depress their water potential to such an extent that plants cannot obtain sufficient water to germinate or grow. Some desert plants growing in saline soils can accumulate salt concentrations of 20–50% dry weight in their leaves without damage, but salt concentrations of only 1–2% can damage the leaves of many species.

A number of elements interfere directly with other aspects of plant metabolism. Sodium is thought to become toxic when it reaches concentrations in the cytoplasm that depress enzyme activity or damage the structure of organelles, while the toxicity of selenium is probably due to its interference in metabolism of amino acids and proteins. The ions of the heavy metals, cobalt, nickel, chromium, manganese, copper, and zinc are particularly toxic in low concentrations, especially when the concentration of calcium in solution is low; increasing calcium increases the plant's tolerance. Aluminium is toxic only in acid soils. Boron may be toxic in soils over a wide pH range, and is a serious problem for sensitive crops in regions where irrigation waters contain excessive boron or where the soils contain unusually high levels of boron.

All plants grow poorly on very acid soils (pH d" 3.5); some plants may grow reasonably well on somewhat less acid soils. Several factors may be involved, and their interactions with plant species are complex. The harmful effects of soil acidity in some areas have been exacerbated by industrial emissions resulting in acid rain and in deposition of substances which increase the acidity on further reaction in the soil, with consequent damage to plants and animals in these ecosystems.

The elemental composition of plants is important to the health and productivity of animals which graze them. With the exception of boron, all elements which are essential for plant growth are also essential for herbivorous mammals. Animals also require sodium, iodine, and selenium and, in the case of ruminant herbivores, cobalt. As a result, animals may suffer deficiencies of any one of this latter group of elements when ingesting plants which are quite healthy but contain low concentrations of these elements. In addition, nutrients in forage may be rendered unavailable to animals through a variety of factors that prevent their absorption from the gut. Plants and animals differ also in their tolerance of high levels of nutrients, sometimes with deleterious results for grazing animals. For example, the toxicity of high concentrations of selenium in plants to animals grazing them, known as selenosis, was recognized when the puzzling and long-known "alkali disease" and "blind staggers" in grazing livestock in parts of the Great Plains of North America were shown to be symptoms of chronic and acute selenium toxicity.

Essential Points in the Nutrition of Plants

The Nature of Plant Growth Processes: The processes concerned in the growth of plants are the subjects of study by plant physiologists and plant biochemists. A comprehensive account of these processes is outside the scope of the present work, the special object of which is to deal with the outward and visible signs of imperfections in the plant's activities caused by faulty mineral nutrition. Nevertheless it is useful to have before us the general features of the main processes involved and to realize that the symptoms we shall be discussing later have a physiological basis, and are not direct and unchangeable signs of the specific deficiencies but result from the derangement of the complicated mechanism of the plant's vital activities.

The main processes involved in plant development may be summarized as follows:

Absorption: Intake of water and mineral elements by the root system.

Carbon assimilation or photosynthesis: Intake of carbon dioxide from the air by the leaves, and reaction of the gas with water in the leaf in the presence of the green chlorophyll to form sugar and free oxygen.

Formation of protoplasm: Protoplasm is the living material of the plant, consisting mainly of proteins, complex compounds of nitrogen built up by the plant from more simple compounds of this element.

Respiration: The combination of oxygen with various food substances synthesized by the plant, especially sugars, whereby energy is produced.

Transpiration: Loss of water from the plant, mainly from the leaves.

Translocation: The movement of materials within the plant.

Storage: Storage of reserve products in various organs and tissues.

During growth there are continuous processes of building up of complex compounds of carbon and nitrogen and breaking down of these into more simple substances, in which water and oxygen are intimately concerned. These processes together comprise plant metabolism.

In the course of the metabolic processes innumerable substances are formed, such as sugars, starch, cellulose, acids, lignin, tannins, amino acids, proteins, amides etc., and many plants also produce special products, as for instance nicotine in the tobacco plant.

For the normal functioning of the above processes there must be an adequate intake of water by the plant to maintain the plant cells in a more or less turgid condition and, since water is being continuously lost at a varying rate from the plant, intake and movement within the plant tissues must be capable of ready adjustment to these changes.

As a result of metabolic activities plants develop special organs of growth and reproduction, each of which has its special characters and makes particular demands on the nutrient supplies of the plant.

With all plants there are well defined seasonal growth cycles. Thus annuals, such as cereals, begin from the seed, give rise to seedlings, which later flower, form grain and ripen off, whilst perennial deciduous trees, such as apples, pears, etc., begin growth in the spring, using stored reserves of food, form leaves, make shoots, blossom and form fruits and subsequently shed their leaves, but meanwhile pass on reserve foods to various storage organs in preparation for the next season's growth. Coincident with these growth cycles there are well defined chemical cycles of nutrient elements and elaborated products in the leaves, stems and roots, etc. It will be shown later that these cycles are of great importance in considering p-deficiency effects and in diagnosing their causes.

The Plant Environment

Nutritional problems must be considered in relation to all the conditions in which plants live, and not merely in terms of the amounts of plant nutrients contained in or added to the soil. For example, those who are accustomed to growing plants know that the temperature must not be too low or no growth may result or that if too high the plants may be injured. An optimum temperature is usually recognized and this may vary according as to whether the plant is young or old. Similarly the importance of light is well known and plants may be put in special positions to obtain a maximum supply of light at one stage and may be shaded at another.

The actual duration of the daily period of illumination also affect growth and there are plants which are classified as requiring "long day" conditions to complete their growth cycles and others as needing "short day" conditions. If the special "long" or "short" day periods are not forthcoming for the respective classes of plants requiring these, their growth cycles are abnormal and they may fail entirely to produce flowers, grain or fruit.

The humidity of the atmosphere, as distinct from the water supply in the soil, is of importance in determining the water conditions within the plant, as these are dependent on both water intake by the roots and water loss from the leaves, the latter being largely influenced by the air humidity.

Even the presence of adequate quantities of plant nutrients in the soil is no guarantee that they will be absorbed by the plant roots. It will be shown later how these may be present in forms which are not available to the plants, but even when they would be considered as being present in suitable forms for absorption, other factors may prevent this taking place. An example of this latter condition is afforded in poorly aerated soils where lack of oxygen near the roots may prevent them from actively absorbing mineral nutrients

The problems of such influences in the plant environment as those just mentioned are complicated by the fact that they do not act independently, but their effects are modified by one another. Thus the effects of light intensity or period of daylight may vary with different temperature conditions.

The requirements of plants for different nutrients may be affected by conditions of light, temperature and water supply, and by other factors of the general environment. Thus the need for nitrogen may be less under conditions of relatively low light intensity whereas the need for potash in these circumstances may be greater, these facts being of importance in growing tomatoes under glass. The effect of nitrogen in relation to light may be shown by growing a plant under normal light conditions with insufficient nitrogen, when the leaves will show the well known symptoms of nitrogen deficiency-pale green, yellow, orange and red tints. If such a plant be then shaded, the leaves will turn a darker green and growth may be visibly increased. It can be shown that the lowered light conditions result in an increase of "soluble" a breaking down of proteins, thereby rendering the nitrogen of these available for growth processes.

This interrelationship of environmental factors is well illustrated by an experiment on apple trees at Long Ashton. Bramley's Seedling trees were grown in compost in large pots and given a small dressing of a nitrogenous fertilizer. Some of the trees were grown in a specially constructed glass house and an equal number in an adjoining wire enclosure. The trees in the enclosure showed severe symptoms of nitrogen deficiency-pale green and yellow leaves, reddish brown barks and highly coloured, red fruits. The condition was corrected by further dressings of nitrogen. In contrast, the trees under glass, where the light was of less intensity and the temperature higher, made vigorous growth, carried large, green leaves and bore large, green fruits.

Iron and zinc deficiency symptoms may be less severe under conditions of low light intensity, whilst boron deficiency effects are less severe and magnesium deficiency effects are more pronounced in wet seasons than in dry ones. The rate of water absorption is less at lower temperatures than at higher ones and efficient intake is also dependent on good aeration. These facts may result in a water deficit within plants growing in cold, wet soils when the air temperature is high. The raw materials needed for plant growth consist of carbon dioxide, which is obtained from the atmosphere through the stomata of the leaves, and water and the so-called mineral nutrients, which normally enter the plant through the medium of the roots.

The importance of water and carbon dioxide in the nutrition of plants will be apparent from the facts that water often comprises 80 to 90 % of the total weight of growing plants, and carbon and oxygen together may account for over 80% of their dry matter, i.e., the solid matter remaining after water is removed. As against these large amounts, the mineral nutrients, as measured by the ash content of the plants, i.e., the mineral residue obtained when the organic matter is destroyed by heat, often contribute from 5 to 15% of the dry matter.

It has been shown in recent years that certain organic compounds, known as "growth promoting substances" or "hormones", which occur in plants, and some of which are also present in soils and natural manures, are capable of producing marked growth responses, such as increased root growth, shoot and leaf curvatures, stimulation or suppression of buds, increased fruit setting, prevention of fruit abscission etc. They appear to perform important functions in the growth of plants. Examples of substances of this kind which can produce growth responses are 13 indole-acetic acid, 13 indole-butyric acid, phenyl acetic acid, A, naphthalene-acetamide, vitamin B_1

It is not at present clear to what extent growth substances are absorbed by plants from soils, although it has been shown that vitamin B_1, which occurs naturally in soils, can be obtained in this way.

The Mineral Nutrients

It has been shown in numerous researches that certain elements are necessary for the healthy growth of plants. They are sometimes spoken of as essential elements and, since some are needed in relatively large quantities and others in very small amounts, the former are referred to as "major" elements and the latter as "minor" or "trace" elements, or as micro nutrients.

The terms "major" and "minor" do not refer to the relative importance of the functions of the elements in plant growth, and for this reason the term "trace" element is preferable for the latter class. Major elements: Nitrogen, phosphorus, calcium, magnesium, potassium, sulfur. Trace elements: Iron, manganese, boron, copper, zinc and molybdenum. (Iron occupies an intermediate position and is usually included in the major elements group. In dealing with field problems it is more convenient to group it with the trace elements.)

In addition, there are, other elements, such as sodium, chlorine and silicon, which produce beneficial effects on the growth of certain plants but which have not so far been shown to be absolutely essential to growth. The element aluminium is of general occurrence in plants, but seems to be without direct nutritional value, although aluminium sulfate is used, because of its acidifying properties, to change the colour of hydrangeas growing on alkaline soils from pink to blue, and aluminium may also exert indirect influences on nutritional processes.

Other elements often occur in plants but they are not known to serve any useful function and frequently they act as plant poisons or toxins.

The nutrient elements can only be absorbed by plants when present in certain forms: nitrogen from nitrates and ammonium salts; phosphorus from phosphates; calcium, magnesium and potassium from their salts; sulfur from sulfates; iron from ferrous or ferric salts (more readily from ferrous salts); manganese from manganous salts; boron from borates; copper and zinc from their salts, and molybdenum from molybdates.

There may appear to be certain exceptions to this statement in practice. For instance, nitrogen may be applied to a soil as "organic" nitrogen, as in hoof meal or urea, and sulfur may be added as the element itself, as in flowers of sulfur, ground sulfur, etc. In such conditions the added materials are, however, converted into the nitrate and sulfate forms respectively by soil organisms before being absorbed by the plants.

Further points of importance in connection with the absorption of the mineral nutrients by plants are as follows:

(a) They must be absorbed from relatively dilute solutions or the plants will be! injured or even killed.

(b) Certain of the elements slow down the absorption of others into the plant, e.g., calcium slows down potassium and vice versa. The phenomenon is known as "antagonism".

(c) Healthy plants result when the nutrients are absorbed in certain relative proportions. When the proportions are suitable the nutrient medium is said to be "balanced". When ratios between nutrients are too wide, deficiency conditions are created. Thus if a high proportion of nitrogen to potassium is absorbed, the plant will suffer from potassium deficiency.

(d) Nutrients, even though present in the nutrient solution in satisfactory amounts and proportions, may not be absorbed by the plant unless the "reaction" of the solution as regards acidity and alkalinity is satisfactory.

The reaction is measured in terms of the pH scale, which is merely a convenient notation for stating the conditions of acidity in the solution (strength or intensity of acidity, not total amount). The neutral point is represented by pH 7.0; below this value the solution is acid and above it is alkaline. Many crop plants prefer a reaction slightly on the acid side-pH 6.0 to 6.5 and extreme values are in the neighborhood of 4.0 on the acid side and 9.0 on the alkaline side.

(e) The nutrient medium must contain an adequate supply of oxygen, i.e., aeration must be satisfactory.

The Functions of The Mineral Nutrients

Knowledge of thc main functions of the mineral nutrients is useful in helping us to understand the effects produced by deficiencies of any one of them.

A. Essential Major Elements

Nitrogen. Nitrogen is a major constituent of several of the most important substances, which occur in plants. It is of outstanding importance among the essential elements in that nitrogen compounds comprise from 40 to 50% of the dry matter of protoplasm, the living substance of plant cells.

For this reason nitrogen is required in relatively large quantities in connection with all growth processes in plants. It follows directly from this that without an adequate supply of nitrogen appreciable growth cannot take place and that plants must remain stunted and relatively undeveloped when nitrogen is deficient.

Proteins, which are of great importance in many plant organs, e.g., seeds, are compounds of nitrogen whilst chlorophyll, the green colouring matter of the leaves, also contains the element. From this latter fact it will be apparent that when nitrogen is deficient leaves will contain relatively little chlorophyll, and will thus tend to be pale green in colour.

In addition to the above substances, numerous other organic compounds of importance in plants, such as amino acids, amides, and alkaloids, are compounds of nitrogen.

Certain compounds of nitrogen are very mobile in plants, and this enables them readily to mobilize supplies of the element at vital growing points and to transfer stored supplies to points where they are most required. Such transference is common from old tissues to young growing points when supplies of the element are short.

This mobility and re-utilization of nitrogen explains why deficiency symptoms of the element always appear first in the older parts of plants and why growing points are the last to be affected.

Phosphorus: This element, like nitrogen, is closely concerned with the vital growth processes in plants as it is a constituent of nucleic acid, and nuclei in which this occurs are essential parts of all living cells. Hence a deficiency of this element will also be expected to result in greatly restricted growth. Phosphorus is also of importance in seeds and in connection with the metabolism of fats. Compounds of phosphorus are concerned with the processes of respiration and with the efficient functioning and utilization of nitrogen.

This relationship to nitrogen probably accounts for the fact that several of the symptoms of phosphorus deficiency are identical or similar to those which result from a deficiency of nitrogen, Phosphorus is also of special importance in the processes concerned in root development and the ripening of seeds and fruits.

Calcium: Calcium occurs in plants chiefly in the leaves and the amounts present in seeds and fruits are relatively low. One of its main functions is as a constituent of the cell wall, the middle lamella of which consists largely of calcium pectate. This function appears to be of fundamental importance since, if calcium is replaced by any other of the essential elements, such as magnesium or potassium, the organic materials and mineral salts in the cells are readily leached through the walls.

Other functions attributed to calcium are as follows:

It provides a base for the neutralization of organic acids; it is concerned with activities of growing points (meristems), especially with root tips; it may be of importance in nitrogen absorption.

Although a large proportion of the calcium contained in the plant may be soluble in water-as much as 60% in cabbage calcium does not appear to move freely from the older to the younger parts of plants, and hence young tissues contain lower proportions of calcium than older ones. This may explain why calcium deficiency effects begin at the tips of shoots.

Magnesium: The outstanding fact about magnesium is that it is a constituent of chlorophyll, and is essential to the formation of this pigment. As a result, when magnesium is deficient, one of the symptoms commonly shown by plants is chlorosis. Magnesium is also regarded as a carrier of phosphorus in the plant, particularly in connection with the formation of seeds of high oil content, which contain the compound lecithin.

The element seems to be very mobile within the plant, and when deficient is apparently transferred from older to younger tissues where it can be re-utilized in the growth processes. This agrees with the observation that signs of magnesium deficiency invariably make their appearance first on the oldest leaves and progress systematically from them towards the youngest ones.

Potassium: Unlike all the other major elements, potassium does not enter into the composition of any of the important plant constituents, such as proteins, chlorophyll, fats and carbohydrates, concerned in plant metabolism. For this reason its role is more difficult to determine, and in spite of much study it cannot be said that the functions of potassium are clearly understood.

The element is present in all parts of plants in large or fairly large proportions. It seems to be of special importance in leaves and at growing points, as these are especially rich in potassium. Probably the whole of the potassium in plants is present in soluble form, and most of it seems to be contained in the cell sap and cytoplasm.

It is outstanding among the nutrient elements for its mobility and solubility within the plant tissues, and these properties no doubt account for the ready way in which potassium can be re-utilized by young tissues when the element is in short supply.

Among the functions which have been attributed to potassium and the processes with which it may be concerned, the following may be mentioned: The formation of carbohydrates and proteins; the regulation of water conditions within the plant cell and of water loss by transpiration; as a catalyst and condensing agent of complex substances; as an accelerator of enzyme action; as contributing to photosynthesis through its radioactive properties.

It has been shown in many instances that the potassium content of plants is frequently much higher than is necessary for healthy growth, and it is generally considered that luxury absorption of potassium often takes place.

The great mobility of potassium in plants, its special importance for and its reutilization by young tissues, and its apparent functions as a regulator of plant processes on a large scale are in harmony with the observations that, when potassium is moderately deficient, the effects are seen first in the older tissues and progress from these towards the growing points, but, when the deficiency is acute, growing points are severely affected, and die-back and general collapse of the plants commonly occur.

Sulfur: Sulfur occurs in plants as a constituent of proteins, and of certain volatile compounds such as mustard oil. It seems to be connected with chlorophyll formation, although it is not a constituent of this substance. Its functions in connection with proteins and chlorophyll doubtless account for the similarity of its deficiency effects to those due to deficiency of nitrogen.

Essential Trace Elements

Iron: Iron is closely concerned with chlorophyll formation but is not a constituent of it. Its role appears in this connection to be that of a catalyst. As a result of this function of iron, chlorosis is invariably an outstanding symptom when the element is deficient. Iron may also act as a catalyst, in the role of an oxygen carrier, in respiration.

A point of great importance in connection with iron is its relative immobility in plant tissues. Its mobility seems to be affected by several factors, such as the presence of manganese, potassium deficiency and high light intensity. There is evidence that the amount of chlorophyll is related to "active" iron in plants. It will thus be seen that so-called iron deficiency in the plant may in fact usually mean iron immobility. Lack of mobility may also account for the fact that iron deficiency is first shown in the younger tissues.

Manganese: The functions of manganese are regarded as being closely associated with those of iron and as being concerned with chlorophyll formation. Hence, when manganese is deficient, chlorosis is a common symptom. Manganese may decrease the solubility of iron by oxidation and hence, an abundance of manganese within the plant may lead to iron deficiency and chlorosis.

Manganese is regarded as having the functions of a catalyst; its activities being specially concerned with oxidation and reduction reactions within the plant tissues.

Boron: The exact role of boron is not known, but again the evidence as for the other trace elements, suggests that, its functions are those of a catalyst or reaction regulator. Particular effects attributed to boron are as follows:

It can delay the onset of calcium deficiency effects but cannot replace calcium; it tends to keep calcium soluble; it may act as a regulator of potassium/ calcium ratios, and of the absorption of nitrogen; it may be concerned with the oxidation-reduction equilibrium in cells. Such functions as the above accord with the results which follow from a deficiency of the element, when growth processes show sudden collapse and drastic derangements of metabolism occur.

Zinc and Copper: Although specific functions have not been determined for these elements, here again the evidence points to their roles as catalysts and regulators. Deficiencies of both are associated with chlorosis and a serious general collapse of vital growth processes.

Since catalysts are not used up in the chemical reactions which they promote, we can understand how it comes about that quite small or even minute quantities of the "trace elements", iron, manganese, boron, zinc and copper, may nevertheless be essential to the plant's health and growth.

Elements Sometimes Beneficial

Sodium: As sodium is not strictly an essential element it cannot be expected to have a specific role in the metabolic activities of plants.

Where sodium produces significant effects it is often regarded as a conserver of potassium and as being able partly to replace that element in its role. In no instance, however, has it been shown that sodium can wholly replace potassium where the latter is acutely deficient.

In such circumstances, sodium is ineffective as a substitute for potassium, even for sodium-loving plants, such as sugar beet, mangold and barley. Sodium seems to affect the water relations of plants and often enables sugar beet and other crops to withstand drought conditions which would otherwise produce severe adverse effects.

Chlorine: The evidence of the role of chlorine in plants is somewhat contradictory, and no general statement can be made. In tobacco it has been shown to increase the water content of the tissues and to affect carbohydrate metabolism, leading to an accumulation of starch in the leaves. The element is present in plants as chloride, and is wholly soluble.

Toxic Effects of Mineral Elements

Toxic effects in plants may be produced both by essential nutrient elements and non-essential elements.

In the first class, the major nutrients are much less toxic than the trace elements. Indeed, for the major nutrients, there exists a fair safety margin for excess or "luxury" consumption, but for the trace elements the margin is very narrow. Similar conditions exist in relation to non-essential elements thus some plants will tolerate fairly large amounts of elements such as sodium or chlorine, but are injured by relatively small amounts of elements like arsenic or chromium.

Two types of injury may occur: (1) An excess of one element may lead to a deficiency of another which ultimately results in a deranged metabolism, e.g., excess nitrogen or excess phosphorus may result in insufficient potassium and excess potassium may lead to deficiency of magnesium or calcium.

This type of injury applies particularly to essential nutrients. (2) The presence of an element may directly injure the protoplasm and bring about the speedy death of the plant.

Signs of Faulty Nutrition in Plants

When plants are grown in unsuitable environments, including conditions of faulty mineral nutrition, they react to the particular defects in more or less specific ways.

Thus, if light is insufficient, the green colouring matter of the leaf will be lacking and the leaves may be almost white (chlorotic), and the plant may be very spindly and "drawn" in appearance; if the temperature is too high, the growth may be rank and soft; if water is insufficient, growth may be restricted, the tissues woody and the green of the leaves show a bluish tint.

Again, deficiencies and excesses of the individual elements produce characteristic effects on various organs of plants: foliage characters, including colour, density, size and shape of leaves; stem characters, such as thickness, colour and length of inter-nodes; root characters, such as colour, amount of fibre, abnormal thickening; blossom characters, including amount and time of opening of the flowers; fruit characters, such as size, colour, hardness and flavour.

Ability to recognize these particular effects forms the basis of the visual method of diagnosing plant deficiencies. Many of them can be readily learned and applied by practical farmers. Indeed, for many years progressive fruit growers in this country have used the leaf symptoms of deficiencies of nitrogen and potassium, and more lately of magnesium, as the main guide to manuring their fruit trees and bushes with these elements.

Detailed descriptions of deficiency symptoms and of the methods of using them in the field for diagnosing the manurial needs of crops.

Soils in Relation to the Supply of Mineral Nutrients

Since plants obtain their supplies of mineral nutrients from the soil, it is necessary to have some knowledge of soil conditions in order to understand the main problems of the supply of nutrient elements.

The Constitution of Soils

One of the first points which is brought out in any examination of a number of soils is their great diversity. In practical terms, soils may be good or bad; light or heavy; wet or dry; fertile or hungry; deep or shallow; black, red or brown; peaty, sandy, silty or clay; all terms denoting points of importance in relation to fundamental characters and to their practical utilization.

Soils are very complex bodies but, although showing such great variation, they all consist of five main components, mineral matter, organic matter, soil water, the soil atmosphere and a population of microorganisms.

Mineral Matter: The mineral matter furnishes the skeleton of the soil. It consists of materials which range in size from rock fragments and large pebbles to minute particles of clay which can be suspended in water for considerable periods.

The mineral portion, which largely determines the texture of the soil, can be separated into its component parts or "fractions" by a combination of sieving and sedimentation in water. The British standard method of grading is as follows:

FRACTION	DIAMETER LIMITS
Stones	> 2.0 m.m.
Coarse Sand	2.0—0.2 m.m.
Fine Sand	0.2—0.02 m.m.
Silt	0.02—0.0002 m.m.
Clay	< 0.002 m.m.

A preponderance of the coarser fractions tends to make soils light, free working, freely drained and hungry, whilst where the finer fractions predominate (especially clay) the soils are heavy, sticky, difficult to work and retentive of water and manures. The clay fraction is of special importance as it possesses colloidal properties, which give to soils many of their characteristic properties, such as swelling and shrinkage, holding up of water and absorption of mineral nutrients from manures. The clay colloid, together with the organic colloid (humus), acts as the soil storehouse for available plant nutrients.

The mineral matter of soils provides the main natural source of mineral nutrients, which become available to plants through the weathering of rock minerals and the complex processes of soil formation.

Organic Matter: The content of organic matter in soils shows great variation. Organic matter may comprise almost the whole of the solid matter of peats, while in mineral soils it may account for only 1 or 2% of the soil. It is usual in mineral soils in this country for the organic matter to range from 3 to 10% in the soil, the lower values occurring in arable soils and the higher ones in old gardens and under grass covers in pastures and meadows.

Plant residues, either naturally or artificially added, provide the source of organic matter in soils. These residues vary greatly in character and composition, depending on the plants from which they are derived, whether from trees, grasses, clovers, cereals, root and vegetable crops, etc., and on the nutrient supplying powers of the soils on which they are grown.

In addition to plant residues, soil organisms and animal remains contribute to the organic matter. In the soil the fresh material undergoes chemical change, especially by the action of soil organisms, and when this has proceeded to the stage where the original cellular structure is no longer recognizable we speak of the brownish product as humus. The organic matter in the soil at any time thus consists of fresh and partly decomposed residues and humus.

The main points in the formation of humus are as follows:

The raw residues contain a variety of compounds, such as proteins and other nitrogenous materials, and sugars, starch, cellulose, tannins, lignin and nutrient salts. The soil organisms use compounds like sugars (carbohydrates) as sources of energy, ultimately oxidizing them to carbon dioxide. To do this they require a supply of nitrogen which they way take either from the nitrogenous compounds in the residues, or if this supply is insufficient they may use other nitrogen present in the soil. The most resistant to their attack of the compounds contained in the residues appears to be lignin, and it seems probable that the end-point of the more rapid decomposition processes humus is a complex mixture in which lignin derivatives preponderate.

During these processes the ratio of carbon to nitrogen (C/N) in the residues alters from about 40/1 to 10/1 so that the process entails a large reduction in carbon.

It will thus be seen that the organic matter of the soil supports the population of soil organisms, and of this population the bacteria alone may be of the order of 20 to 40 millions per gram of soil.

These organisms play a very great part in determining the availability of mineral nutrients in soils, by breaking down plant residues and also by providing carbon dioxide which, in combination with water, is of great importance for the weathering of the soil minerals. Certain of the soil organisms, especially bacteria, are also able to fix free nitrogen from the air and thus enrich the soil with this element. Other bacteria are present which cause the soil to lose nitrogen, and there are multitudes of protozoa which feed on the bacteria, and if too abundant render the soil unhealthy. A properly balanced soil population is requisite for good crops and disturbances of this balance either in a beneficial or deleterious direction may be brought about by soil treatment.

It is important to realize, however, that bacteria and other soil organisms do not work for the special benefit of crops, and that the changes they bring about are to meet their own particular needs, which are not always in harmony with the requirements of the crops. This point can be illustrated by digging into soil fresh straw or sugar and growing a crop on it immediately, when the crop will most probably show symptoms of acute nitrogen deficiency due to the fact that, in order to consume the large amount of carbohydrate contained in the straw or sugar, the organisms deplete the soil of its readily available nitrogen.

Organic matter produces both chemical and physical effects on soils. Like clay, the humified material possesses colloidal properties, depending on the minute size of the constituent particles and the enormous surface they collectively present. It can swell and shrink and absorb nutrient salts. Together with clay, humus coats the mineral particles and can bind these together forming “crumbs”, thus giving “structure” to soils. Humus acts as a binding agent for the particles of coarse sandy soils and exerts an "opening" effect on close-textured clay soils. The binding properties are not associated with stickiness to the same extent as those of clay. Humus is superior to clay colloid as a storehouse of nutrient bases.

Soil Water: The solid particles of the soil, consisting of the mineral particles and the organic matter, being of varying dimensions, have spaces of varying sizes between them which are occupied by the soil water and the soil atmosphere.

The total volume of these spaces in any soil is known as the pore space. It varies for different soils and even in the same soil under influences such as rainfall, drought, frost and management, but usually occupies from 30 to 60% of the soil volume. It will be clear that the amounts of water and air in the pore space are complementary and that when the soil is wet there will be correspondingly little air present and vice versa. A high water content in the soil thus means a condition of relatively poor aeration.

The manner in which water is held in soils and the way in which it moves through the soil mass provide difficult scientific problems. There is general agreement that in a wet soil a certain proportion of the water can be readily removed, and that some is held very strongly by forces in the soil. Moreover, it can be shown that only a proportion of the water in soils is available to plants.

The movement of water through soils is also limited, and water tables at considerable depths below the surface cannot be relied upon to supply water to the surface layers of the soil by the well known capillary action which occurs in narrow tubes. Similarly, if a dry soil is wetted from the surface, the whole of the soil to a considerable depth does not become wetted through uniformly, but the top layers only are wetted, practically to full capacity, and the layers below remain dry.

This strictly limited movement of water through soils is of great practical importance, since it means that for practical purposes water does not move appreciably to the plant roots but the roots must go after the water.

The soil water contains the soluble products of the soil, and is the main nutrient medium for the plant roots. As such it is commonly called the soil solution. There is evidence that plants may also feed directly on nutrients contained in the soil colloids.

The soil solution is very dilute. Measurements made on displaced solutions show concentrations of soluble materials between 0.1 and 1.0%, whilst if drainage waters are analysed the corresponding values are of the order of 0.02 to 0.5%. In both, the proportions of calcium, nitrate, sulfate, carbon dioxide, silica and organic matter are relatively high, and there may be large proportions of chloride, sodium and magnesium, but potassium is always low and phosphate and ammonia, are present only in traces.

The soil organisms play an important part in providing soluble products for the soil solution. They liberate into the solution nitrate, sulfate and carbon dioxide and these, acting as acids, dissolve corresponding or equivalent amounts of base-forming elements, particularly calcium. Further changes then occur, the calcium in particular bringing into solution other "bases" such as magnesium and potassium by "base" exchange reactions with the colloidal clay and humus materials.

Under normal conditions, the whole of the nitrate and chloride is present in the soil solution, and they pass into the drainage water if not quickly absorbed by plants. Sodium is also mainly present in soluble form. Calcium, magnesium and potassium are held in considerable amounts in the soil colloids as "exchangeable bases", and in this form are less easily washed out of the soil.

Phosphorus is mostly present as insoluble compounds, as calcium phosphates in neutral and alkaline soils, and as iron phosphate in acid soils. Both potassium and phosphorus are only present in the soil solution in very small amounts, but apparently supplies of these elements in the solution are quickly replenished when removed by plants.

The amounts of iron, aluminium and manganese dissolved are largely dependent on the soil reaction (pH) and in strongly acid soils they may assume toxic concentrations to plants.

The conditions governing boron solubility are not known though, under drought conditions and at high pH values, availability to plants is reduced.

Copper and zinc (perhaps also manganese) appear to present complicated problems in which organic matter and the soil organisms are concerned.

Molybdenum may be more readily soluble in soils under alkaline than acid conditions, as it has been shown to be more readily available to pasture plants under the former conditions.

The Soil Atmosphere: The soil atmosphere or soil air, near the surface, has a composition similar to that of the ordinary air and diffusion between the two is rapid. Except near the surface the soil air is saturated with moisture.

The carbon dioxide content may fluctuate considerably, being increased by organic matter, by cropping, by high temperatures, which speed up the activities of soil organisms, and by the respiration of plant roots.

In poorly drained soils the concentration of carbon dioxide may be very high, with a correspondingly low oxygen content, and in extreme condition this may actually lead to inefficient absorption and even to root injury due to lack of oxygen. The fact that carbon dioxide accumulates and oxygen supply is depleted under conditions of poor drainage is important in considering deficiency problems in wet soils where practical results often appear contrary to expectation.

Availability of Mineral Nutrients in Soils

The mineral elements available to plants in soils may be present as the result of natural soil processes or may come from additions of natural manures and fertilizers.

Nitrogen: Nitrogen availability is intimately connected with the activities of soil organisms. Nitrate is the normal form in which nitrogen is absorbed by plants from soils, although ammonia can also be utilized. Nitrogen in organic compounds is converted into nitrate by a chain of reactions brought about by organisms. For protein nitrogen this may be represented as follows:

Protein —> intermediate organic products, such as amino acids —> ammonia —> nitrite —> nitrate.

The effect of excess carbohydrate on these reactions has been mentioned.

Ammonia may be held in the colloid complex as an "exchangeable base" in the same way as calcium, magnesium and potassium.

Phosphorus: This element exists in the soil in many forms, both as organic and inorganic compounds, and is also added to the soil in manures and fertilizers in a variety of materials, such as in meat meal, bone manures, basic slag, ground phosphate rock (mineral phosphate) and water-soluble superphosphates.

Evidence indicates that water-soluble forms are generally most readily available to plants even though they are rendered insoluble almost immediately after application to the soil.

Organic forms of phosphorus are usually less readily available than the inorganic compounds. Phosphates undergo many changes in soils both by organisms and by purely chemical reactions and, even though very heavy dressings are applied, the amount of water-soluble phosphate in the soil at any time is very small.

The movement of phosphates in soils is very limited, and soils are said to have high "fixing powers" for phosphates. Heavy soils, as a rule, show higher fixing powers than light ones, and soils with high iron contents possess specially strong fixing properties.

The two elements mainly responsible for the fixation of phosphates are calcium, in neutral and alkaline soils, and iron in acid soils.

The Relative Fixing Powers of some Light and Heavy Soils for Phosphates

	Percent Phosphoric Acid (P_2O_5) soluble in 1% Citric Acid Solution.							
	Light Sands				Clays			
Surface Soil	0.0600	0.0865	0.1472	0.1730	0.0485	0.0601	0.0810	0.1158
Subsoil	0.0230	0.0616	0.0932	0.0734	0.0011	0.0085	0.0038	0.0120

Important practical points in connection with the fixation of phosphates are that a large proportion of the phosphates added to many soils never become available to the crops, and that phosphates should always be placed as near as possible to the roots of the plants for which they are intended. In some soils where fruit trees suffer from phosphate deficiency, the method of injecting the phosphatic materials into the soils through high-pressure lances has been suggested to obtain local concentrations of fertilizer near portions of the roots. For cereal crops the difficulty has been met by the introduction of combined seed and fertilizer drills, which deliver the seed and phosphate into the soil in close proximity, and by the use of granular fertilizers, of which only a small proportion aof the material comes into contact with the soil.

Calcium: Calcium occurs in soils in a large variety of minerals, and carbonate of lime (calcium carbonate) may comprise a large percentage of the mineral matter where the soil is derived from limestone or chalk rocks. The element is readily leached from the soil and, on sandy soils where it is not abundant, frequent replenishment by means of lime or limestone dressings is necessary. Carbonate of lime is readily brought into solution in soils by means of carbon dioxide dissolved in the soil water and thus, if the soil contains lime, a supply of calcium in soluble form is readily assured. Calcium also comprises the major proportion of the elements held as exchangeable bases where soils are not strongly acid and it is readily brought into the soil solution from this state.

Magnesium: The mode of occurrence of magnesium in soils is fairly similar to that of calcium. It occurs as carbonate and in a variety of minerals. Like calcium it is readily brought into the soil solution from the carbonate, and is held in soils as an exchangeable base. It is fairly easily leached, and for this reason may become deficient in sandy soils during wet periods.

Potassium: Potassium is widely distributed in soil minerals, such as potash-felspar, mica and glauconite, from which it is slowly converted into soluble forms by weathering processes. Heavy soils contain higher amounts of potassium than light soils. Potassium is fairly strongly "fixed" in soils, possibly largely as an exchangeable base. Very small amounts of potassium are present in the soil solution at any given time, but exchangeable potassium appears to be readily available to plants.

Sodium: Sodium is brought into solution from minerals in a similar manner to potassium, but it remains mainly in the soil solution, being only weakly absorbed by the colloids, and is very readily leached from the soil. In salt marshes recently reclaimed from the sea and in alkali soils, where sodium is the principal base-forming element, it is the main exchangeable base. Under these conditions the soils are often unfit for cultivation, due to bad physical condition, until the sodium has been replaced by calcium.

Soluble sodium compounds are readily available to plants.

Sulfur: Sulfur is present in soils both in organic and inorganic forms. Inorganic sulfur is mainly present as sulfate, but sulfide may accumulate where conditions are favourable for "reducing" reactions to take place.

Under water-logging conditions, calcium sulfate may form crystalline deposits in the subsoil, and this substance may also be seen as a white deposit on the surface of soils when manuring is high, as in market gardens. Sulfur compounds are changed from one form to another in the soil by special bacteria, the end product of whose reactions is the sulfate form where conditions favour oxidation. Thus, if the element sulfur is added as a dressing to soil, it may be quickly oxidized to sulfuric acid, which reaction is used in treating alkaline soils to lower the pH value.

Sulfate: Sulfur is an important constituent of the soil solution, and in this form sulfur moves readily through the soil.

The widespread occurrence of sulfur in rocks and organic matter and its mobility in the soil doubtless account for the rarity of a deficiency of the element in crops.

Chlorine: This element is of common occurrence in soils in the form of chloride, and as such moves freely through the soil in the soil solution, from which it is available to plants.

Common salt or sodium chloride is. also present in considerable quantity in the atmosphere near the sea, and a proportion of this reaches the soil, adding both chloride and sodium.

Iron: Iron is present in most soils, mainly in the form of its oxides, which are largely responsible for the red and brown colours in soils. Where drainage and aeration are good, ferric compounds predominate, and where water-logging and bad aeration occur ferrous compounds are formed. The insoluble forms of iron are brought into solution by the action of acids. Organic compounds of iron, such as the humate, appear to be mobile in the soil. The availability of iron to plants increases with acidity and is depressed by phosphates.

Manganese: The mode of occurrence of manganese in the soil resembles that of iron in that the oxides are important forms. Both iron and manganese are intimately concerned in the oxidizing and reducing reactions which take place in soils, and compounds of both in the form of concretions appear in imperfectly drained soils.

The more highly oxidized compounds of manganese, such as manganese dioxide, are of very low availability to plants. Like iron, the solubility of the soil manganese increases with increasing acidity, and in many soils it is largely unavailable to plants above pH 6.5. The availability of manganese seems also to be greatly affected by organic matter and drainage conditions, deficiency of the element being common on calcareous peats and other soils with high organic matter content, and on soils with high water tables.

Boron: Little is known of the sources of boron available to plants in soils. The element is a constituent of the mineral tourmaline, which occurs in many rocks, but this is very resistant to weathering. When applied to the soil in the form of a borate, boron appears to move fairly readily through the soil. The availability of boron to the plant is decreased by liming and by dry soil conditions. In practice over-liming is a common cause of boron deficiency.

Zinc and Copper: In the neighborhood of zinc or copper mineral deposits the soils are generally toxic to crops. The contents of these elements in normal soils are very small, and maybe are largely the result of concentrations and additions from growing plants and added residues.

Little is known of the factors affecting their availability to plants, but organic matter and soil organisms may be important factors.

7

A Primer on Photosynthesis and the Functioning of Cells

Photosynthesis

Photosynthesis is the process by which organisms that contain the pigment chlorophyll convert light energy into chemical energy which can be stored in the molecular bonds of organic molecules (e.g., sugars). Photosynthesis powers almost all trophic chains and food webs on the Earth.

The net process of photosynthesis is described by the following equation:

$$6CO_2 + 6H_2O + \text{Light Energy} = C_6H_{12}O_6 + 6O_2$$

This equation simply means that carbon dioxide from the air and water combine in the presence of sunlight to form sugars; oxygen is released as a by-product of this reaction.

Light Reactions and the Calvin Cycle

The process of photosynthesis is broken up into two main groups of reactions: the "light reactions" which require light energy to operate, and the "Calvin cycle" which specifically takes carbon dioxide and turns it into organic molecules. The electromagnetic energy of sunlight is converted to chemical energy in the chlorophyll-containing cells of photosynthetic organisms. In eukaryotic cells these reactions occur in the organelle known as the chloroplast. In the chloroplast, chlorophyll is the pigment that absorbs the sunlight. Chlorophyll is typically packed into stacks of membranes (called grana); it is in the grana where some of the sunlight is absorbed. Sunlight is converted to chemical energy in the form of ATP (adenosine triphosphate), which is the main energy-storing molecule in living organisms.

ATP is then transported throughout the chloroplast and used to provide the chemical energy necessary to power other metabolic reactions. For example, some of the ATP is used to power the metabolic reactions in the conversion of CO_2 into sugars and other compounds.

Photosynthesis in a Chloroplast

Some terms and definitions:

- H_2O is water.
- O_2 is oxygen.
- CO_2 is carbon dioxide.
- ATP is adenosine triphosophate.
- PGA is a phosphoglyceric acid, a three carbon (C-C-C) organic acid.
- Grana are the stacked membranes that contain chlorophyll.
- RuBP is the five carbon (C-C-C-C-C) sugar-phosphate.
- Rubisco is the enzyme ribulose bisphosphate carboxylase/oxygenase. It is the enzyme that catalyzes the conversion of CO_2 to the organic acid, PGA. It is the most abundant enzyme on Earth.

During the process of photosynthesis, light penetrates the cell and passes into the chloroplast. The light energy is intercepted by chlorophyll molecules on the granal stacks. Some of the light energy is converted to chemical energy.

During this process, a phosphate is added to a molecule to cause the formation of ATP. The third phosphate chemical bond contains the new chemical energy.

The ATP then provides energy to some of the other photosynthetic reactions that are causing the conversion of CO_2 into sugars.

While the above reactions are proceeding CO_2 is diffusing into the chloroplast. In the presence of the enzyme Rubisco, one molecule of CO_2 is combined with one molecule of RuBP, and the first product of this reaction is two molecules of PGA.

The PGA then participates in a cycle of reactions that result in the production of the sugars and in the regeneration of RuBP. The RuBP is then available to accept another molecule of CO_2 and to make more PGA.

Which wavelengths of the solar spectrum drive photosynthesis?

o The wavelengths of sunlight between 400nm and 700nm are the wavelengths that are absorbed by chlorophyll and that drive photosynthesis.

Energy Incident on a Leaf

Photosynthesis is not a very efficient process. Of the sunlight reaching the surface of a leaf, approximately:

- 75% is evaporated
- 15% is reflected
- 5% is transmitted through the leaf
- 4% is converted to heat energy
- 1% is used in photosynthesis

How do we know the O_2 is derived from H_2O during photosynthesis?

The oxygen product of photosynthesis could originate from either the CO_2 or the H_2O starting compounds.

To determine which of these original compounds contributed to the O_2 end product, an isotopic tracer experiment was performed using ^{18}O:

- o ^{18}O is a heavy isotope of oxygen
- o $H_2{}^{18}O + CO_2$ yields $^{18}O_2$
- o $H_2O+C^{18}0_2$ yields O_2

Therefore, the O_2 end product must originate from water and not from the carbon dioxide.

How do we know what the first products of photosynthesis are?

Another Isotopic Tracer Experiment

^{14}C is a radioactive isotope of carbon. $^{14}CO_2$ is exposed for a brief period to a green plant that is conducting a photosynthesis in the presence of sunlight.

Immediately after exposure to $^{14}CO_2$, the plant's photosynthetic tissue is killed by immersing it in boiling alcohol, and all of the biochemical reactions cease.

The chemical compounds in the dead tissue are all extracted and studied to determine which of them possesses the ^{14}C. Following the briefest exposure to $^{14}CO_2$, the only chemical compound that possessed ^{14}C was PGA (phosphoglyceric acid, a three carbon molecule).

Following longer periods of exposure, much of the ^{14}C was found in a variety of compounds including glucose. By varying the length of the exposure period it was possible to identify the sequence of the reactions leading from PGA to glucose.

This research was conducted by Prof. Melvin Calvin and his colleagues at the Univ. of California, Berkeley. Calvin received the Nobel Prize for this work.

Metabolism

We have seen how plants convert sunlight into sugars. Now we need to understand how cells can use the products of photosynthesis to obtain energy. There are several possible metabolic pathways by which cells can obtain the energy stored in chemical bonds:

- o Glycolysis
- o Fermentation
- o Cellular respiration

Glycolysis: Glycolysis can occur in either the absence or the presence of oxygen. During glycolysis, glucose is broken down to pyruvic acid, yielding 2 ATP of energy. Glycolysis occurs in the cytoplasm of cells, not in organelles, and occurs in all kinds of living organisms. Prokaryote cells use glycolysis and the first living cells most likely used glycolysis.

Fermentation: During fermentation, the pyruvic acid produced during glycolysis is converted to either ethanol or lactic acid. This continued use of pyruvic acid during fermentation permits glycolysis to continue with its associated production of ATP.

Cellular Respiration: Respiration is the general process by which organisms oxidize organic molecules (e.g., sugars) and derive energy (ATP) from the molecular bonds that are broken.

Glucose (a sugar): $C_6H_{12}O_6$

Respiration is the opposite of photosynthesis, and is described by the equation:

$$C_6H_{12}O_6+6O_2 \longrightarrow 6CO_2+6H_2O+36ATP$$

Simply stated, this equation means that oxygen combines with sugars to break molecular bonds, releasing the energy (in the form of ATP) contained in those bonds. In addition to the energy released, the products of the reaction are carbon dioxide and water.

In eukaryotic cells, cellular respiration begins with the products of glycolysis being transported into the mitochondria. A series of metabolic pathways (the Krebs cycle and others) in the mitochondria result in the further breaking of chemical bonds and the liberation of ATP. CO_2 and H_2O are end products of these reactions. The theoretical maximum yield of cellular respiration is 36 ATP per molecule of glucose metabolized.

Note that photosynthesis is a reduction-oxidation reaction, just like respiration. In respiration energy is released from sugars when electrons associated with hydrogen are transported to oxygen (the electron acceptor), and water is formed as a byproduct.

The mitochondria use the energy released in this oxidation in order to synthesize ATP. In photosynthesis, the electron flow is reversed, the water is split (not formed), and the electrons are transferred from the water to CO_2 and in the process the energy is used to reduce the CO_2 into sugar. In respiration the energy yield is 686 kcal per mole of glucose oxidized to CO_2, while photosynthesis requires 686 kcal of energy to boost the electrons from the water to their high-energy perches in the reduced sugar — light provides this energy.

Microbial Photosynthesis Introduction

Metabolism is the sum total of all biochemical reactions that take place in the cell with the involvement of flow of energy and the participation of variety of enzymes and proteins. Metabolism, in fact, represents the chemistry of life and can be divided into two major parts: catabolism and anabolism. Catabolism (G. cata = down, ballein = to through) represents the breakdown of more complex chemicals into smaller, simpler molecules resulting in the release of energy. Some part of this released energy is trapped and made available for cellular functions while the rest is released as heat.

In anabolism (G. ana = up, ballein = to through) the similer molecules are used in the synthesis of complex molecules with energy utilization. Enzymes required for metabolic activities are synthesized in the cell, whereas energy is obtained from one of the three sources (i) chemolithotrophic microbes carry out oxidation of inorganic chemicals that releases energy, (ii) chemoorganotrophic microorganisms oxidize organic molecules to liberates energy, and (iii) phototrophic microorganisms trap radiant energy of sun by the process of photosynthesis.

The chemotrophic organisms have adopted two catabolic mechanisms for energy conservation: respiration and fermentation. In respiration, the energy is conserved by the process oxydative phosphorylation with the involvement of molecular oxygen or some other externally derived electron--acceptor. Respiration, however, is of two different types, namely, aerobic and anaerobic. In aerobic respiration, the final electron acceptor is oxygen whereas the electron-acceptor in anaerobic respiration is more often inorganic (e.g., NO_3^-, SO_4^{2-}-, CO_2, Fe^{3+}, SeO_4^{2-}, and many others), though organic electron-acceptors such as fumaric acid may also be used.

In fermentation, the energy is produced by substrate-level-phosphorylation in which ATP is synthesized as a result of the oxidation of an organic compound without involvement of any usable external electron-acceptor. Phototrophic microorganisms employ anabolic mechanism and trap light energy of sun during photosynthesis (synthesis of complex molecule using simpler molecules) by the process photophosphorylation.

Living Things Obtain and Manage Energy and Nutrients

A rock can passively have its energy increased! When sunlight strikes a rock, the temperature of the rock and the total amount of thermal energy it possesses increases. However, a rock cannot control this process, nor can it use the energy to do work or to obtain nutrients.

Living things by contrast can obtain and control energy through processes like photosynthesis and metabolism. Photosynthesis is the conversion of visible sunlight energy into other stored or usable forms of energy. It can be briefly summarized by the following simple chemical equation:

$$6CO_2 + 6H_2O \longrightarrow C_6H_{12}O_6 + 6O_2$$

Metabolism is the process of converting energy from a stored form (e.g. energy in a sugar molecule) to other forms. All living things perform metabolism, but for those that cannot photosynthesize it is the only way to get usable energy. One type of energy-releasing metabolism is called aerobic respiration. It is basically the opposite of photosynthesis and therefore requires energy-rich molecules like simple sugars and oxygen to work. It can be summarized as follows:

$$C_6H_{12}O_6 + 6O_2 \longrightarrow 6CO_2 + 6H_2O$$

The energy from photosynthesis and metabolism can be used to do work such as building molecules of the body structure and creating movement. To summarize, all forms of life show metabolic activity; they extract & transform energy from their environment and use it for manipulating materials in ways that assure their own maintenance, growth, development & reproduction.

Nutrients are substances that an organism takes in and uses for growth and maintenance. Animals, for example, must metabolism nutrients to get the energy necessary to live. However, they also must use the atoms and molecules of nutrients as raw material for building their own molecules that make up their bodies. Both energy and materials are needed by all organisms. The elements that make up nutrients can be recycled, but the energy used to construct them cannot.

Endergonic versus Exergonic Reactions. In class or via Internet you will learn about these two energy concepts. Basically, compounds can store energy in their formation and release it to do work later. The energy storing reacts are called endergonic while the energy releasing reactions are termed exergonic. These two types of reactions can be coupled to allow transfer of energy for various purposes.

Location of Metabolic Activity. Metabolism in its various forms occurs inside cells. For example, photosynthesis occurs in the chloroplasts of plant and photosynthetic protist cells. Digestion is different from metabolism. It is the process of physically processing foods and chemically breaking food molecules down into small compounds like simple sugars, amino acids, and fatty acids that can be absorbed into the cells.

Within cells specific metabolic processes occur in specific organelles or in the cytoplasm. The first steps of aerobic metabolism of all eukaryotic cells occurs in the cytoplasm, and later steps occur inside the mitochondria.

In our class and/or in your textbook studies and Internet discussion thread interactions we are going to focus on the following aspects of metabolism:

- Glycolysis (the initial breakdown of glucose series of reactions in the cytoplasm)
- Transition Reaction (the pyruvate to acetyl group reaction) (occurs in the mitochondrion)
- Krebs Cycle (= Citric Acid Cycle) (a cycle of reactions that transfer energy to intermediate high energy compounds plus some ATP (occurs in the mitochondrion)
- Electron Transport System (oxidative phosphorylation system that produces the ATP) (occurs in the membrane of the cristae in the mitochondrion)

There are many details to these processes. However, the following will help you focus on some important summary details:

- o For General Biology students, from you textbook study and learn: Accounting of Energy Yield per Glucose Molecule Breakdown. You should be able to reproduce it from memory.
- o For General Zoology students, from you textbook study and learn. You should be able to reproduce it from memory.

Reminder about Textbook Study

As with other topics, your textbooks have excellent presentations of the materials on cellular metabolism, including generous excellent illustrations. Check the general objectives above to make sure that you have covered all of the topics in the textbook readings.

The "Self Test" and other questions will be helpful for general biology students, although many more detailed questions will be included in the lecture exam.

Similarly, study questions in the zoology textbook will be helpful review for general zoology students, but again, they are not comprehensive.

As with all materials throughout the semester, you will have opportunities to ask questions or ask that any relevant material from your assignments be discussed in class and/or in threaded discussions on Internet.

Fermentation and Photosynthesis Comparison

IB Biology Group Four Project

Fermentation and Photosynthesis Comparison and Lab Proposal: Fermentation can be described by the simple reaction: $C6H12O6$ + yeast - > $2C2H5OH$ (ethanol) + $2CO2$. Fermentation is most efficient when there is a high sugar content for the yeast to ferment. Thus, for the experiment you should start with grape juice that has a high sugar concentration for the yeast. The juice would then be placed in fermenting vessels that will have to be air-tight so that oxidation will be prevented and bacteria will not grow. Afterwards fermentation will continue slowly for about ten to thirty days. An experiment can be designed to serve to find which temperatures fermentation is ideal in so different vats will be assigned to different fermentation temperatures. After a decided time the level of efficiency of the fermentation of the yeast in each vessel will be tested by analyzing the about of CO2 produced. The amount of CO2 produced will reflect on how much fermentation occurred as CO2 is a bi-product of yeast fermentation.

Fermentation is basically the conversion of carbohydrates into alcohol or acid. There are many forms of fermentation. There is anaerobic fermentation which does not require oxygen and there is also aerobic metabolism that poweres fermentation. Homolactic fermetnation makes lactic acid from pyruvate and alcoholic fermentation turns pyruvate into ethanol and carbon dioxide and heterolactic fermentation makes lactic acid and also some alcohol. Ethanol is the drinkable type of alcohol. Fermentation often creates many waste products that cannot be metabolized without oxygen. This makes fermentation less efficient in producing ATP. In cellular respiration, oxidative phosphorilization is better and more efficient.

Fermentation and photosynthesis sometimes work in pairs. Photosynthesis provides the reactants necessary to make power fermentation in plants. Likewise, in animals, glucose would come from ingested foods. All plants perform photosynthesis even when they are in the water.

Photosynthesis is the process, which the plant absorbs carbon dioxide and releases oxygen.

$$6H2O + 6CO2 + Light - - - - - > C6H12O6 + 6O2$$

Elodeas are specific type of water plants that could be used in this experiment. The experiment would be repeated with the light at different distances from the plants in order to simulate changing light intensities. The generated bubbles from the water plant in this experiment reveal that water plants perform photosynthesis in water. The control group of this experiment would be the plant with natural light source without any light bulb. As the light intensity increases by placing the light bulb closer to the water plant, more bubbles would be generated, which indicates the faster rate.

Fermentation or Photosynthesis: The Debate in Algae Fuel

- Font size
- Print
- E-mail
- Share
- Post a comment
- Yahoo! Buzz

Sugar or sunlight?

That's one of the fundamental questions for companies trying to transform algae into transportation fuel or dietary supplements. Solazyme says it will grow algae through fermentation—for instance, feeding the algae sugars in a heated, sealed environment.

"We're not growing it photosynthetically. We put it in stainless steel tanks similar to what you see in a brew pub," said Solazyme President Harrison Dillon.

By contrast, GreenFuel Technologies and LiveFuels will exploit the sun to grow the single-celled creatures. And here there is a divide, too. GreenFuel grows its algae in sealed, transparent tubes it calls bioreactors. LiveFuels, by contrast, is setting up an open pond near the Salton Sea in Southern California.

Which solution will win? Who knows? "There is a spectrum of options from high control and high cost to low control and low cost," said B. Greg Mitchell, a research scientist who specializes in plant biology at the University of California at San Diego. Although national labs began researching the concept of converting algae to fuel in the late '80s, this is the first time people are seriously considering going forward.

Fermentation offers the most control. Each fermentation vat contains a single species. Temperature, pressure, and other environmental conditions can be minutely controlled. Additionally, fermentation offers flexibility: A row of vats can be used to make auto fuel, and then scrubbed and prepped for cooking up algae for cooking fuel.

The added sugar also doesn't necessarily need to be expensive.

"There's a lot of organic matter in the world," Mitchell said. "We make a lot of beer at low cost with fermentation systems."

Still, fermentation overall does cost more. As a result, it might make the most sense for higher-value oils. Martek Biosciences, in fact, ferments algae and sells it as a baby food additive. It remains to be seen if the model will work for fuel that gets burned in cars. (Although crude costs $100 a barrel, it's a more low value oil than others, Mitchell said.)

The big advantage of photosynthesis is that the sun is free. But temperatures vary daily. "Temperature regulation adds cost, and the amount of sunlight is variable and seasonal," he added. Controlling the rate of growth was a problem Greenfuel had last year at its Arizona facility.

Open ponds are less expensive than closed systems. In open ponds, however, it can be difficult to control invasive species. Then again, having multiple species can be an advantage, Dave Jones, one of LiveFuel's founders has said. Different species can absorb different wavelengths of light.

Photosynthesis, Respiration, Transpiration

Plants are self-sufficient. They make their own food thru the process of photosynthesis using light energy to make sugars from carbon dioxide ($C0_2$) and water (H_20).

The three major functions that are basic to plant growth and development are:

- Photosynthesis the process of capturing light energy and converting it to sugar energy, in the presence of chlorophyll using CO_2 and H_2O,
- Respiration the process of metabolizing (burning) sugars to yield energy for growth, reproduction and other life processes, and
- Transpiration the loss of water vapour through the stomata of leaves.

Photosynthesis

A primary difference between plants and animals is the plants ability to manufacture its own food. In photosynthesis carbon dioxide from the air and water from the soil react with the sunýÿs energy to form carbohydrates (sugars and starches). Photosynthesis literally means *to put together with light.*

The photosynthetic process occurs only in the chloroplasts, tiny subcellular structures contained in the cells of leaves and green stems. In photosynthesis, the sunýÿs energy combines hydrogen from water (H_2O) with carbon dioxide (CO_2) turning them into carbohydrates. Oxygen (O_2) is given off as a by-product of photosynthesis. The chemical equation for the process of photosynthesis is:

$$6CO_2 + 6H_20 + \text{light}\ C_6H_{12}O_6 + 6O_2$$

This process is directly dependent on the supply of water, light and carbon dioxide. Any one of the factors on the left side of the equation (carbon dioxide, water, or light) can limit photosynthesis regardless of the availability of the other factors. If any one of these factors is limiting, then the whole process slows down or stops. An implication of drought or severe restrictions on landscape irrigation is a reduction in photosynthesis and thus a decrease in plant vigour.

In a tightly closed greenhouse there can be very little fresh air infiltration and CO_2 levels can become limiting. This in turn limits plant growth because the production of sugars needed to do the work involved with growing is limited. Many greenhouses provide supplemental CO_2 to stimulate plant growth.

The rate of photosynthesis is somewhat temperature dependent. For example, with tomatoes, when temperatures rise above 96 degrees Fahrenheit the rate of food used by respiration rises above the rate that food is manufactured by photosynthesis. Plant growth comes to a stop and produce loses its sweetness.

Respiration

In respiration, plants (and animals) convert the sugars back into energy for growth and to energize life processes (metabolic processes). The chemical equation for respiration shows that the sugars from photosynthesis are combined with oxygen. Notice that the equation for respiration is the opposite of photosynthesis.

$$C_6H_{12}O_6 + 6O_2\ 6CO_2 + 6H_20 + \text{energy}$$

Chemically speaking, the process is similar to the oxidation that occurs as wood is burned, producing heat. When compounds combine with oxygen, the process is often referred to as *burning*. For example, athleteýÿs *burn* energy (sugars) as they exercise. The harder they exercise, the more sugars they burn so the more oxygen they need. Thatýÿs why at full speed, they are breathing very fast. Athletes take up oxygen through their lungs. Plants take up oxygen through the stomata in their leaves and through their roots.

Again, respiration is the burning of sugars for energy to grow and do the internal work of living.

It is very important to understand that both plants and animals (including microorganisms) need oxygen for respiration. This is why overly wet or saturated soils are detrimental to both root growth and function, and the decomposition processes carried out by microorganisms in the soil.

The same principles regarding limiting factors are valid for both photosynthesis and respiration.

Table: *Comparison of Photosynthesis & Respiration*

Photosynthesis	Respiration
Produces sugars from energy	Burns sugars for energy
Energy is stored	Energy is released
Occurs only in cells with chloroplasts	Occurs in most cells

Oxygen is produced	Oxygen is used
Water is used	Water is produced
Carbon dioxide is used	Carbon dioxide produced
Requires light	Occurs in dark and light

Transpiration

Water in the roots is pulled through the plant by transpiration (loss of water vapour through the stomata of the leaves). Transpiration uses about 90 percent of the water that enters the plant. The other 10 percent is an ingredient of photosynthesis and cell growth.

Transpiration serves three essential roles:

- Movement of minerals up from the root (in the xylem) and sugars (products of photosynthesis) throughout the plant (in the phloem). Water serves as both the solvent and the avenue of transport.
- Cooling ýÿ 80 percent of the cooling effect of a shade tree is from the evaporative cooling effects of transpiration. This benefits both plants and humans.
- Turgor pressure ýÿ Water maintains the turgor pressure in cells much like air inflates a balloon, giving the non-woody plant parts form. Turgidity is important so the plant can remain stiff and upright and gain a competitive advantage when it comes to light. Turgidity is also important for the functioning of the guard cells, which surround the stomata and regulate water loss and carbon dioxide uptake. Turgidity also is the force that pushes roots through the soil.

Water movement in plants is also a factor of osmotic pressure and capillary action. Osmotic pressure is defined as water flowing through a permeable membrane in the direction of higher salt concentrations. Water will continue to flow in the direction of the highest salt concentration until the salts have been diluted to the point that the concentrations on both sides of the membrane are equal.

A classic example is pouring salt on a slug. Because the salt concentration outside the slug is highest, the water from inside the slug's body crosses the membrane that is his skin. The slug becomes dehydrated and dies. Envision this same scenario the next time you gargle with salt water to kill the bacteria that are causing your sore throat. Fertilizer burn and dog urine spots in a lawn are examples of salt problems related to gardening.

The salt level in the soil's water is higher than in the roots, and water flows from the roots into the soil's water in an effort to dilute the concentration. So what should you do if you accidentally over apply fertilizer to your lawn?

apillary action refers to the chemical forces that move water as a cont. uous film rather than as individual molecules. Water molecules in the soil and in the plant cling to one another and are reluctant to let go. You have observed this as water forms a meniscus on a coin or the lip of a glass. Thus when one molecule is drawn up the plant stem, it pulls another one along with it. These forces that link water molecules together can be overcome by gravity

The Basic Photosynthesis Equation

The key chemical pathway in photosynthesis is the conversion of carbon dioxide (CO_2 from the air) and water (H_2O) into carbohydrate molecules ($[CH_2O]_n$ such as sugar) and free oxygen (O_2), using light as an energy source for the reactions, as shown here:

the basic form:

$$CO_2 + H_2O \xrightarrow{\text{light}} \text{sugars} + O_2$$

the more chemically accurate form:

$$6CO_2 + 6H_2O \xrightarrow{\text{light}} C_6H_{12}O_6 + 6O_2$$

These carbohydrate molecules, such as sugars, contain more energy than the starting molecules, in other words they act as chemical batteries for solar energy. The carbohydrate molecules are often then used to construct more complex molecules, such as cellulose and lignin, that make the plant structure.

To release stored energy, both plants and animals use oxygen from the atmosphere to convert carbohydrates back into CO_2 and H_2O. This process is called respiration.

However the rate of photosynthesis is usually 30 times the rate of respiration, so plants produce large net quantities of oxygen that is then used by animals that in turn produce CO_2 which feeds the plants, in a cycle.

Leaves, Cells and Chloroplasts

Leaves are the structures in plants that have evolved primarily to absorb sunlight and to convert this energy into carbohydrates. Figure 1 shows a cross-section through a typical leaf.

The nucleus contains the DNA of the cell that encodes the information that the cell needs to function. Respiration takes place within mitochondria (singular: mitochondrion). The vacuole is a fluid-filled bag that often takes up more than 80% of plant cell volume.

The major role of vacuoles in most plants is to maintain turgor pressure i.e. to keep the cell rigid, thus helping to support the plant, along with the cell wall (notice how many plants wilt when they lack water). Photosynthesis takes place within chloroplasts (figures 2 and 3) contained within the green cells of plants and algae (photosynthesizing bacteria have other, simpler structures). Chloroplasts are flattened disc shaped organelles usually 2-10 ìm in diameter and 1 ìm thick (click here to open a note on units in a new window).

The chloroplast has a two membrane envelope termed the inner and outer membranes, with a space inbetween. The fluid within the chloroplast is called the stroma.

Within the stroma are stacks of thylakoids, the sub-organelles where photosynthesis actually takes place. A stack of thylakoids is called a granum (plural: grana). A thylakoid looks like a flattened disk, and inside is the thylakoid space or lumen. The photosynthetic reaction takes place on the thylakoid membrane.

Light

Before introducing the photosynthetic molecules, we will briefly discuss light. What we call "light" is actually part of the spectrum of electromagnetic radiation.

This covers a range of rays or waves from cosmic rays, with very short wavelengths all the way to radio waves.

The wavelength is the distance between the peaks of 2 adjacent waves . Rather confusingly, light has some properties of waves and some of particles. A light particle is called a photon. In some situations it is more convenient to think of light as photons, and in others to treat it like waves.

There is a direct relationship between wavelength, frequency and the speed of light: $v = f \lambda$

- v = the speed of light, 3×10^8 m/s (meters per second)
- f = frequency
- λ = wavelength (lambda, units of meters or nanometers)

The speed of light doesn't vary, so the longer the wavelength the smaller the frequency.

There is also a relationship between frequency and the energy of each wavelength of light: $E = hf$

- E = energy of 1 photon or light particle
- h = Planck's constant
- f = the frequency of the light

The higher frequency (shorter wavelength) light has more energy. When calculated, the energy levels are as shown in table 1. You can see that red light has considerably less energy than blue light.

Table: *Energy levels of visible light expressed as Joules per mole of photons (a mole is 6.023 x 10^{23} particles or in this case, photons). ë (lambda) is wavelength of light. Violet light has about twice the energy of red light (Adapted from Hall and Rao, 1999)*

λ	*Colour*	*Energy*
700nm	red	1.7 x 10^5
600nm	yellow	2.0 x 10^5
500nm	blue	2.4 x 10^5
400nm	violet	3.0 x 10^5

An example that shows the difference in energies of the colours of light is seen on older cars (automobiles). Red cars absorb the higher energy blue photons, but reflect the lower energy red photons, that's why the paint looks red. This absorption of more energetic photons leads to faster oxidation of the paintwork.

Photons with high enough energy can be absorbed by particular molecules in the plant, producing energised electrons. It is important to note that 1 photon can only supply its energy to 1 electron and also that photons with too low energy will not be used in photosynthesis. The red photons have less energy than the blue photons, this explains why infrared light is relatively poorly used in photosynthesis.

The Absorption of Light

The absorption of light in chloroplasts is by a range of pigments, mainly chlorophylls (which provide the green colour) and carotenoids. These different types of pigment absorb light of different wavelengths. A large number of these pigment molecules are arranged into an antenna to capture most of the photons of the correct energy. This energy is then passed through a chain of pigment molecules until it reaches a specialised molecular complex called the reaction centre. The pigment antenna complex/reaction centre together are called a photosystem (figure 6).

There are 2 types of photosystem in plants, linked together in a 2 stage process. Light is absorbed at each of the photosystems to provide enough energy to manufacture carbohydrates. There are millions of these photosystems embedded in the thylakoid membranes within the chloroplasts.

The Scale of Photosynthesis

The scale of the processes of photosynthesis and respiration is quite staggering. For instance around 10,000 tonnes of oxygen is consumed every second by respiration and the burning of fuels. If plants didn't balance this loss by their production, oxygen supplies in the atmosphere would only last about 3000 years. The amount of plant biomass produced every year has been estimated at 200 billion tonnes.

How much solar energy is absorbed by photosynthesis, or what percentage of sunlight falling on the Earth's surface is converted into plant matter? The sunlight hitting the atmosphere every year is about 5.6 trillion trillion Joules, or 5.6×10^{24} J (a note on units and exponents is here, opens in a new window). This is equivalent to a continual flow of 1.8×10^{17} Watts. About half of this reaches the Earth's surface, the rest being reflected by the atmosphere and clouds.

Of this 50%, some is reflected from the oceans and other surfaces leaving 30%. But only about half of this light is of the correct wavelength to enable photosynthesis. This leaves 15% of the original sunlight available for absorption by plants (8.4×10^{23} J). Assuming the primary biomass production (plants) of Earth is 2×10^{11} tonnes per year, this equates to about 3×10^{21} J of energy stored as biomass. Therefore the percentage of usable light stored by photosynthesis globally is: 0.4%

So why the difference in figures?

The measured efficiencies of energy conversion (i.e. sunlight into biomass) for various plants range from as low as 0.1%, to around 1.5% (most crop plants) to sugarcane at 8% (the exception). The reasons for this are complex, and range from the behaviour of light energy as it penetrates leaves, to limiting factors in the plant's environment e.g. temperature, micronutrients etc. There is also much room for error in calculating such enormous processes.

If plants could be made more efficient at converting light into plant matter than the productivity of the world's agriculture could be improved vastly. Numerous research projects are focusing on achieving this with a variety of plant species.

Photovoltaic Cells Vs Photosynthesis Process

Do they have anything in common? Cross the road and ask the first person you meet the following question: "according to your knowledge is there any correlation between a photovoltaic cell and the photosynthesis process?" At 99,9 % the answer is no, there is not any correlation (better not to say at 100%, you never know). Probably the answer is negative because the person does not remember or does not know what a photovoltaic cell is and what the photosynthesis process is. Let's remember together, briefly, what they are.

The photovoltaic cell is the base element constituting a photovoltaic module, which is able to convert straightly the solar energy in electric energy through the photovoltaic effect. The photosynthesis is an endothermic process that happens in the plants and in some unicellular organism and thanks to it the solar energy is transformed in chemical energy, included in glucose molecules and in the link of other organic substances.

Now we have found the connection between the photovoltaic cell and the photosynthesis process: both convert the solar energy, the first one in electric energy the second one in chemical energy. Due to the above mentioned connection Michael Gratzel, professor of chemistry physics at Losanna university, in 1991 showed the photovoltaic cell based on the process inspired to the plants photosynthesis. The cell Gratzel, in its functioning, refers to the plants photosynthesis:

- It uses an organic colorant equivalent to the chlorophyll to absorb the light and to produce an electron flow;
- It uses multiple layers to improve the efficiency both the light absorbing and the electrons collecting.

The base element of these cells is the Titanium dioxide (TiO_2), a material that is largely used (for ex. in the paints production or of ceramic materials) and has a low cost. It is generally produced in the shape of nanoparticles that are particles with dimensions of about one millionth of millimetre. This material is able to absorb only ultraviolet radiations, but linking them to the nanoparticles surface through suitable molecules of colorants, named activators or sensitizers, it is possible to absorb the visible radiation too.

Therefore the layer presence of TiO2 is necessary to increase, hugely, the efficiency of the light collecting, spreading the colorant over a large area in equivalent manner as it happens for the chlorophyll in the chloroplasts. Since the colorant layer is very thin, almost all the electrons, produced from the light absorbing, are collected from TiO_2. Initially, molecules of colorant belonging to the same family of the bilberry pigments were used, now other compounds that allow to absorb even more the visible components of the solar radiation are used. By these activated nanoparticles, porous layers are realized, with thickness lower than a thousandth of millimetre, and these layers are impregnated with electrolytes.

Once this layer has been placed between two suitable supports, in advance a thin metallic deposition on them has been done, the device is ready. The process of light absorbing produces a cumulation of charges on the electrodes and, in this way, it creates a voltage between the two metallic layers, as between the two accumulator poles.

The current efficiency of this cell is of 10 % approximately, so it is at the same level of the silicon amorphous cells, but the production costs are significantly lower. Moreover, compared to the silicon amorphous cell, it has the advantage that it can be produced on large dimension surfaces and on flexible supports that can be integrated in the buildings structure easily.

Below we can see the Gratzel cell:

- A-up there is the counter electrode (conductor glass), down the TiO2 glass plate with colour stain;
- B-the assembled cell;
- C-the final assembling: cell with clips.

Photosynthesis Process

Overview-Sunlight supplies all energy for living things on earth including plants. Plants are known as producers and they trap this solar energy for making food for themselves. They use light energy from the sun, chlorophyll, CO_2 and water and convert it into glucose and oxygen. This is known as the photosynthesis process. So, light is very essential for plants. Action and absorption spectra of white light can be separated into seven colours using a prism. All these colours have different wavelengths. When we plot the graph of the rate of different physiological processes of plants against wavelength, it is called as action spectra. It is actually a graph, which shows the degree to which the photosynthesis process is affected by different wavelengths. Absorption spectra is the graph which shows the degree of absorption of light by photosynthetic pigments. It clearly shows that chlorophyll absorbs blue, violet and red regions of the light spectra. Light absorbing pigments-The major photosynthetic pigments found in plants are chlorophyll carotenoids and phycobilins. These pigments can be separated by paper chromatography method. Chlorophyll a is blue green in colour. Chlorophyll b is yellowish green. Carotene is orange coloured. Both are carotenoids and chlorophyll plays a vital role in the physiological processes of a plant.

Organization of Photosynthetic Apparatus

Chloroplast Structure-Chloroplasts are important cell organelles of a plant cell as they carry out the photosynthesis process. They absorb light and convert it into energy reserves. They produce food for plants. There are two membranes which protect the parts inside the chloroplast structure. All the reactions occur inside stroma and sugars are formed. Each thylacoid stack is called granum, which has chlorophyll molecules. Photosystem is an enzyme or a protein which reduces molecules by using solar light. Green plants have two photosystems known as photosystem I and photosystem II. These are two photosynthetic reaction centres in green plants. Photosynthetic centres are a combination of three types of proteins and this is the site where photosynthetic reactions take place.

Light Reactions

These are the stages or reactions of photosynthesis that require direct light. These reactions take place in grana of chloroplast. When light strikes chlorophyll a, ATP and NADPH are formed. This is known as photophosphorylation. Light dependant or light reactions have two stages. Cyclic phosphorylation-Extra energy is received from cyclic phosphorylation. This has only photosystem I which produces excited electrons which is the production of ATP during photosynthesis. In this process, there is a cyclic passage of electrons. This ATP formation fuels the dark reactions. Non cyclic phosphorylation-In non cyclic phosphorylation, water is split to form hydrogen and oxygen. Z-scheme-It describes the pathway of oxidation and reduction reactions during photosynthesis. It has a z-shape. Electron transport from water to NADP+ in photosynthesis requires photosystem I and photosystem II. The electron transport system is very important in light reactions.

Dark Reactions

These are the reactions during photosynthesis that don't require light. These are the chemical reactions which convert carbon dioxide and other molecules to glucose. The site of these reactions is stroma, which is a fluid filled space in the chloroplast structure. There are further reactions of the products from light reactions. There are three stages of this reaction called as the Calvin cycle. The metabolic pathway of the Calvin cycle is shown here in the diagram above. This clearly shows the steps of the Calvin cycle. The products of the Calvin cycle are larger carbohydrates or sugars. Calvin cycle is regulated by light. Electrons are transferred out of thylacoid lumen after the light reactions of photosynthesis into the stroma and then protons are transferred from stroma into the thylacoid of chloroplasts

The C4 Cycle

This is a pathway used by plants which live in areas where there is low concentration of carbon dioxide. Here the carbon dioxide is converted into a useful form by the plants. eg-sugarcane. This is a supplementary method of CO2 intake and plants which follow this are C4 plants. In this pathway 4 carbon molecules are formed instead of 3 carbon molecules. These plants are found in regions of high temperature and intense sunlight.

Crassulacean Acid Metabolism (CAM)

This is also known as CAM. This is carbon fixation in plants during night time. Here carbon dioxide is given out during daytime and it increases the efficiency of the photosynthesis process. It's a two part cycle. The stomata remains closed during the day in this pathway. This is common in plants like pine apple. It is mostly followed by xerophytic plants where water is scarce.

8

Phytochrome-mediated Responses

Implications for Controlled Environment Research Facilities

Light is undoubtedly the most important environmental variable for plant growth and development; plants not only use radiant energy in photosynthesis, they also respond to the quantity, quality, direction and timing of incident radiation through photomorphogenic responses that can have huge effects on the rate of growth and the pattern of development. It is surprising, therefore, that the manufacturers and suppliers of controlled environment facilities have been singularly uninventive in the design of the lighting assemblies they provide.

The consumer has one choice only-a lighting assembly that provides irradiance levels usually only a fraction of sunlight, and a control system that is limited to regulating the timing of the on-off switch. The reasons for these limitations are partly technological, but in the main they result from ignorance on the part of both the consumer and the manufacturer. A specific and powerful example of this ignorance relates to the importance of the so-called far-red wavelengths (FR = 700-800 nm). Because the human eye can hardly detect wavelengths above 700 nm, and photosynthesis also cuts off at *ca.* 700 nm, the majority of plant and crop physiologists are still almost completely unaware that FR radiation can have massive effects on growth rate and development.

In consequence, most growth cabinets have light sources based on fluorescent tubes, and provide very little FR apart from that emitted by a token number of small incandescent bulbs. Larger growth facilities often use broader spectrum light sources, but growth facilities that provide the capability to vary the FR incident upon the plants are about as abundant as seals in the Sahara. The natural environment and its importance for plant growth and development in the hope that it might inform intelligently those concerned with improving the design of plant growth facilities.

The Natural Radiation Environment

The daylight spectrum. The light environment experienced by plants in nature is obviously complex, but a number of generalisations can usefully be made. Solar radiation outside the atmosphere is distributed according to Planck's radiation distribution law, with the sun behaving as a blackbody emitter with an apparent surface temperature approximating 5800° K. From Wien's simplifications of Planck's radiation formulae, the wavelength of maximum quantum emission is ca 620 nm, whereas in energy terms it is ca 500 nm; radiant emission falls off sharply at lower wavelengths and more gradually at higher wavelengths.

This means that about 55% of the radiation incident on the earth's surface falls within the 380-800 nm range of photochemical activity-which is fortunate, because photochemistry drives the energetic reactions of the biosphere via photosynthesis. Atmospheric components including ozone, oxygen, water vapour and carbon dioxide selectively absorb narrow wavelength bands, resulting in the typical radiation distribution of daylight at the earth's surface.

This radiation distribution is remarkably constant, being affected little by clouds and other climatic conditions. Pathlength through the atmosphere is important, of course, and as pathlength increases with the sun's approach to the horizon at dusk (or dawn), refraction and Rayleigh scattering (inversely proportional to the fourth power of the wavelength) gives dawn/dusk radiation distributions with relatively elevated levels of blue light, and slightly increased levels of FR compared to daylight.

Underwater light. The underwater light environment is of major importance, since more than half of plant life is underwater. Refraction at the air-water discontinuity leads to the incident light from above being concentrated into a cone of half-angle 48.6°; consequently, a sensor facing upwards below, but near to the surface inevitably receives a proportion of upwelling radiation reflected back down from the surface.

More important phenomena, as far as radiation distribution is concerned, are scattering and absorption by water itself, and by dissolved molecules or suspended particles. Rayleigh scattering results in the selective attenuation of the blue region of the spectrum of downwelling radiation. Water has strong absorption bands at *ca.* 730 nm and in the near infra-red, and therefore the

FR is also selectively attenuated. Thus, in clear water, downwelling radiation is effectively "compressed" with increasing depth into a decreasingly narrow band of wavelengths, usually peaking at or around 500 nm. Absorption and scattering by algae, or by organic debris, causes the spectral distribution of radiation in turbid waters to be very variable.

The light environment within vegetation canopies. Ecologically, the most important fluctuations in radiation distribution occur when radiation interacts with vegetation. The photosynthetic pigments, the chlorophylls and carotenoids, absorb radiation over almost the whole of the visible spectrum (i.e. 400-700 nm). A small fraction of the "green" radiation is either transmitted or reflected, which is why leaves are green to our eyes. What is not so immediately obvious is that vegetation hardly absorbs any radiation between 700 and 800 nm.

Thus, virtually all the incoming FR is either transmitted or reflected; i.e. the FR is scattered either through the leaf, or from the surface of the leaf. Since our visual systems are very insensitive to radiation beyond ca 700 nm, we fail to recognise that leaves should look far-red, rather than green! A typical daylight spectrum within a dense vegetation canopy, and demonstrates the marked depletion of red (i.e., R, 600-700 nm) and the relative enhancement of FR radiation within canopies.

The extent to which R is depleted and FR relatively enhanced by vegetation varies, of course, with the density of the canopy and the depth of the sensor within that canopy, and direct relationships with leaf area index have been established. A more subtle effect of vegetation on the relative amounts of R and FR radiation depends on the direction of propagation of the radiation being measured, or perceived. Unfiltered solar radiation is propagated downwards and is highly directional; i.e., only slightly scattered. After interaction with the leaves of a vegetation canopy, multiple scattering occurs, causing the radiation to be propagated more randomly. This means that radiation propagated more-or-less horizontally within a canopy will already have interacted with vegetation and will consequently be depleted in R and relatively enriched in FR, compared to radiation within a canopy that is propagated more-or-less vertically downwards. This point has a far-reaching significance, as will become evident later.

The biological significance of the variations in the relative amounts of R and FR radiation in the natural environment is that they provide signals of vital ecological importance. Plants have evolved a sophisticated battery of photoreceptors that enable them to sense environmental variations in R and FR and to use the information so obtained to direct appropriate alterations in metabolism, growth and development. Perception of environmental R and FR allows plants to detect the presence of neighbours, to gauge their competitive threat, and to react to actual or incipient shade by appropriate redirection of growth and development. The photoreceptors responsible for the perception of R and FR are the phytochromes.

The Phytochromes-Sensors of the Natural Radiation Environment

The Phytochrome family. The phytochromes are a family of photochromic photoreceptors each member of which consists of an apoprotein bearing a linear tetrapyrrole chromophore. Each phytochrome is capable of existing in two stable forms: Pr, which absorbs maximally at *ca.* 660 nm, and Pfr, which absorbs maximally at *ca.* 730 nm. Upon the absorption of radiation, Pr is photoconverted to Pfr, and Pfr photoconverted to Pr, according to the following scheme:

The absorption spectra of the Pr and Pfr forms of phytochrome isolated from etiolated oats show widely overlapping bands of absorption below *ca.* 730 nm, so that in broad-band radiation (such as daylight) both forms are continually photoexcited, resulting in a steady state photoequilibrium (defined as Pfr/P, where P = Pr+Pfr), in which the proportions of the total phytochrome present as Pr and Pfr are functions of the radiation distribution and of the absorption cross sections of Pr and Pfr. Since the absorption maxima are in the R and the FR, it is these wavelengths that are most important in achieving equilibrium. For this reason Smith and Holmes (1977) proposed that daylight spectra could be usefully characterised and simplified by measuring the ratio of radiation in two 10 nm wavebands centred on the absorption maxima of Pr and Pfr. Thus, the parameter R:FR, which is the ratio of the photon flux density in the 655-665 nm waveband, to that in the 725-735 nm waveband, has become the standard way of characterising daylight for photomorphogenic purposes.

There are known to be at least five members of the phytochrome family (i.e., phytochrome A to phytochrome E) in higher plants, as judged by Southern analysis of *Arabidopsis* genomic DNA. Evidence from physiological studies of normal, mutant and transgenic *phy* gene overexpressers indicates that each member of the family probably has a distinct eco-physiological function, although functional overlap may occur under certain circumstances. On this basis, the phytochromes represent a battery of photosensors that enable plants to obtain ecologically significant information from the light environment.

R:FR Ratio and Pfr/P. The benefit of using R:FR as a simplified parameter of the spectral distribution of natural radiation lies in the fact that R:FR can be readily transformed into a measure of the relative proportions of Pr and Pfr present at photoequilibrium. The hyperbolic relationship that exists between R:FR (as defined above) and Pfr/P (Smith and Holmes, 1977). Using this relationship, any measured value of R:FR can be transformed to a value which represents the Pfr/P to be expected in the outer epidermis of the irradiated tissue (ignoring any light reflected or scattered back from within that tissue); the transformation is made simply by reading Pfr/P from the curve for any measured value of R:FR. The actual direct measurement of Pfr/P in light-grown plants yet eludes the advance of analytical technology, mainly because there is very little phytochrome present and its absorption is overwhelmed by that of chlorophyll.

The parameter arrived at by transforming R:FR is not necessarily an accurate measure of the real Pfr/P in the tissue; it is merely a physiologically-relevant way of expressing the relative amounts of R and FR in the incident radiation. Reading Pfr/P from the curve is inadvisable for artificial sources in which either the R or FR is filtered out, or for sources in which blue light (which is capable of photoconverting Pr and Pfr) predominates.

Under these circumstances it is advisable to integrate the spectral photon irradiances for the 400-800 nm waveband with the extinction coefficients of Pr and Pfr and the quantum efficiencies of the Pr Pfr and Pfr Pr phototransformations. In other words, the influence of light over the whole wavelength region absorbed by phytochrome on the photoconversion of Pr to Pfr, and *vice versa*, can be calculated, resulting in a more meaningful value for Pfr/P than can be derived from R:FR. Simple computer protocols exist for this transformation.

The relationship between R:FR and Pfr/P in reveals three important points. First, because R:FR during the day is very constant and unaffected by weather conditions, it provides the plant with a norm against which fluctuations in R:FR due to other environmental conditions may be compared. Secondly, underwater, R:FR increases very sharply with depth of immersion, but because the underwater values of R:FR are on the asymptote of the relationship between R:FR and Pfr/P, it is clear that phytochrome would be an insensitive detector of depth underwater.

Thirdly, and most importantly, small reductions in R:FR caused by vegetation shade, or the proximity of neighbours, cause relatively large reductions in Pfr/P. Thus, because shade R:FR values lie on the steep part of the hyperbolic curve, the phytochromes in principle have the capacity to be very sensitive detectors of shade.

The Phytochrome-Mediated Shade Avoidance Syndrome

The nature of shade avoidance reactions. The acclimative responses of herbaceous plants to shade from other vegetation can be viewed in terms of two extreme strategies. One strategy, that of *shade tolerance*, involves relatively slow growth rates, the conservation of energy and resources, perennation usually by vegetative processes, and the development of photosynthetic structures that are especially efficient at low light levels.

The opposite extreme is *shade avoidance*, a syndrome of growth and developmental changes in which extension growth is favoured at the expense of leaf and storage organ development. As the name suggests, if successful, shade avoidance has the overall effect of projecting the photosynthetic structures (usually leaves) into those parts of the environmental mosaic in which the resource of light is plentiful.

Shade avoiders tend to be photosynthetically inefficient at low light levels, but have the capacity rapidly to direct growth potential from leaf development to shoot extension upon the first detection of incipient shading. Shade avoidance is an effective strategy for life in an herbaceous community, but has limitations for herbs growing on the floor of a dense forest.

The two strategies, avoidance and tolerance, are not necessarily mutually exclusive, since some plants display intermediate strategies and appear to be able to adapt to life either in open or shaded habitats, whilst other plants can exhibit shade avoidance and shade tolerance at different points in their life cycle.

Phytochrome-mediation of shade avoidance. When shade avoiding species are grown in white light to which various amounts of FR have been added, developmental responses essentially similar to those seen in natural canopy shade result. The relationship between extension growth and the predicted Pfr/P for seedlings of the shade avoiding weed *Chenopodium album* grown in cabinets in which the PAR was held uniform but the R:FR was decreased by supplementation with varying flux densities of FR.

The most striking effect of supplemental FR, which is the enhanced elongation growth at low R:FR, but all the other components of the shade avoidance syndrome can be induced by such decreases in R:FR. For example, growth in WL+FR causes a major redistribution of assimilates from leaf expansion and storage organ accumulation to stem and petiole growth. It also strongly accelerates flowering, a phenomenon as yet little studied but potentially of considerable importance.

Thus, the induction of the shade avoidance syndrome requires the perception of the spectral changes associated with shade, rather than the changes in total light quantity.

Furthermore, the linear relationship between extension rate and calculated phytochrome photoequilibrium has been shown to obtain for a wide range of species, providing convincing evidence that the perception of shade and the induction of shade avoidance responses is phytochrome-mediated. In particular, the magnitude of extension growth responses to added FR is related to the life style of the plant; i.e., shade avoiders respond strongly, whilst shade tolerators respond weakly.

Two further important characteristics of R:FR perception are its rapidity and its compensation for changes in irradiance. Using position-sensitive transducers to enable the continuous monitoring of stem extension rate, changes in extension rate caused by FR radiation applied to the growing internode via fibre-optic probes can be detected within minutes.

Furthermore, within wide limits, the extension rate is determined by the R:FR at the internode and is independent of the flux density of white light presented from above.

These results indicate that the perception of R:FR is precisely quantitative and is compensated for variations in total irradiance. This means that, in principle, phytochrome-mediated R:FR perception should not only be able to operate at the light levels that exist within dense canopies, but should also function at the high irradiances present in sparse stands of plants that are not sufficiently close to cast actual shade.

That R:FR perception does indeed occur at very high flux densitys was shown by Smith (1990) who observed strong accelerations of extension rate in mustard seedlings when exposed to high levels of FR added to a background white light in excess of 1500 μmol m^{-2} s^{-1}.

Neighbour detection and proximity perception. This latter point leads to the expectation that plants should be able to detect the FR reflected from neighbours before actual shading occurs, thereby providing for anticipation of competition for light.

That this occurs in nature was suggested by Kasperbauer *et al.* (1984) in studies of soybeans grown in either north-south or east-west rows. R:FR near the top of the north-south rows of plants was lower on the west side in the morning, and on the east in the evening; indicating that the adjacent rows act as FR reflectors when the sun is low in the sky.

The fact that mustard plants growing under background white light in growth cabinets react very rapidly to FR directed horizontally at the growing internodes, also indicates that plants can detect horizontally-propagated FR whilst being exposed to high R:FR light from above. Ballaré *et al.* (1987, 1990) have demonstrated direct effects of reflected FR on plant growth in the field using seedlings of *Datura ferox*, a strongly shade-avoiding herb, grown in the field close to grass screens that were either green, or bleached by being sprayed with a herbicide.

The plants adjacent to unbleached, green hedges grew significantly faster than those near to the bleached hedges. *Datura ferox* seedlings, when inserted into a sparse canopy of similar seedlings not dense enough to cast actual shade, grew faster than in the open. If their growing internodes were surrounded by transparent collars containing dilute copper sulphate solution (which absorbs FR), there was no increase in growth.

Thus, phytochrome-mediated R:FR perception is sufficiently sensitive to allow the detection of reflected light from neighbouring vegetation. Smith *et al.* (1990) measured the reflection signals from stands of tobacco, and also measured actual Pfr/P in samples of purified oat phytochrome exposed to the radiation reflected from the tobacco stands.

These data showed that, in principle, neighbour detection could operate over substantial distances, and that the signals progressively increased with proximity to the neighbours. In nature, therefore, plants are able to detect the FR reflected from neighbours even though the FR flux is a tiny fraction of that of the vertically propagated, high R:FR daylight. It seems that simple geometry may be responsible for this apparent paradox, as the most sensitive regions are the growing internodes which, for most shade avoiding species, are held erect, thereby receiving very little downwardly propagated radiation.

On this basis, phytochrome-mediated R:FR perception provides plants with the capacity for proximity perception; in other words, plants not only can detect their neighbours, they can effectively perceive how far away they are, and therefore are able to gauge the competitive threat posed.

In summary, plants are exceptionally sensitive to the relative amounts of R and FR they receive. In nature, the shade avoidance syndrome provides plants with the capacity to adapt rapidly to the competitive threat posed by neighbours. In the controlled environment, manipulation of the R:FR ratio can give the experimenter, or the grower, impressive control over the pattern of development and the rate of growth.

Implications for the Design of Plant Growth Facilities

There are, of course, alternative ways of varying R:FR in controlled environments; one can add FR to a constant R, one can add R to a constant FR, or one can vary both R and FR simultaneously. The latter is what happens in the natural environment, but in our hands the objective has been to dissociate the phytochrome-mediated responses to varying R:FR, from any effects that might be a result of changes in photosynthetic rates caused by reduced levels of photosynthetically active radiation.

Ideally, we would like to have growth facilities in which photosynthetic rate could be held uniform at high levels, whilst simultaneously varying the proportions of Pr and Pfr; for this reason we have concentrated on designing cabinets that provide a constant, uniform background of white light (and therefore constant R) whilst varying R:FR by adding FR. This approach inevitably carries a number of technical problems.

At present, no sources are available at an affordable cost that provide high irradiance FR without also emitting large amounts of longer wavelength infra red.

The simplest way of producing FR is to filter the radiation emitted by incandescent sources, but these have maximum emissions at ca 900 nm or higher, and put out a great deal of radiation in the longer wave infra red. Consequently, using such sources inevitably means that one has to remove a large amount of radiant heat. Because of the technical problem of dealing with this heat, until recently we have been forced to use cabinets in which the background WL was of relatively low irradiance. In our latest designs, we have developed cabinets in which high irradiance broad-band WL can be supplied from above, with high irradiance FR being provided horizontally from sources mounted in the side walls; these cabinets were built as a direct response to the realisation of the importance of horizontally-propagated radiation.

Removal of radiant heat can only be achieved by absorption of the heat, followed by some form of heat exchange. Our approach has been to use so-called "water windows", in which flowing water, cooled by passage through a heat exchanger, absorbs the radiant heat emitted by the incandescent lamps used to provide the FR.

Following radiant heat absorption, the FR is selected by the use of Perspex (Plexiglass) filters. The design and construction of effective water-windows is by no means a trivial exercise, and when a window fails, the close proximity of gallons of water to high voltage circuitry can yield impressive pyrotechnic displays.

Nevertheless, with appropriate fail-safe devices and controls, water windows can be reliable and safe, although the cost would be prohibitive for standard growth cabinets. With this approach we have been able to develop cabinets in which upwards of 500 mol m-2 s-1 FR can be added to background WL of between 150 mol m-2 s-1 PAR (old designs) and 400 mol m-2 s-1 PAR (new design).

In the latest side-wall FR cabinets, we can provide 500 mol m-2 s-1 FR from the side and up to 900 mol m-2 s-1 PAR from above. The latest cabinets were designed by us and constructed to an extremely high degree of excellence by Vindon Scientific Ltd., based near Oldham in the UK. The essential features of the latest set of Vindon cabinets, and the legend includes the contact person and the address of the company.

Bringing water and electricity close together should, of course, be avoided. If other sources of high irradiance FR were to become available (i.e., discharge lamps, LEDs, micro-wave-driven sources, etc.) at a reasonable cost, then the design of cabinets that allow the simulation of natural R:FR ratios whilst simultaneously providing satisfactory photosynthetic rates would be simplified, and a major improvement in growth cabinet design could be contemplated. Both growers and experimenters would then be able to achieve much better control of the growth and development of their plants.

History and Applications in Controlled Environments

The widespread application of electric (often called artificial) light in greenhouses, growing rooms, and plant growth chambers would presuppose that the role of phytochrome would be considered in the selection and use of such lighting systems.

Unfortunately this is not usually the case. Part of the problem is that many students, and indeed an unfortunate number of senior scientists, seem to regard phytochrome as a laboratory phenomenon without much application in the real world.

They simply have not grasped the concept that phytochrome is functioning through all stages of plant development, wherever plants are grown. It is certainly true, as Meijer (1971) stated, that one cannot compare experimental results obtained under very strict laboratory conditions with plant irradiation in glasshouses and in growth rooms.

For example, the action spectrum for flowering of the long-day plant, *Hyoscyamus niger*, clearly shows that red radiation is the most efficient portion of the spectrum for promoting flower initiation, but in practical photoperiod control red or fluorescent lamps do not promote flowering nearly as well as the mixture of red and far-red in incandescent lamps. Nevertheless, much evidence exists that documents phytochrome control of plant growth and development in controlled environments and under natural conditions.

When Karl Norris developed the first practical portable spectroradiometer about 1962 some of the first measurements were to determine the red/far-red ratios under tree canopies.

These measurements showed clearly the predominance of far-red in the understory and suggested that far-red was contributing to the elongation exhibited by many species growing in the shade, and possibly was a factor in the induction of light requirements in seeds.

Subsequently we used Catalpa leaves as far-red filters to make light-insensitive lettuce seed light requiring. Much more detailed work, as reported in the preceding paper, has since been done on phytochrome effects in the natural environment, and it is encouraging to note that efforts are bring made to apply phytochrome research to horticulture.

Greenhouses

As everyone interested in photoperiodism knows, L.H. Bailey (1891, 1892, 1893) used light from a Brush carbon-arc lamp to supplement natural light and extend the day in greenhouses. This was not; however, the first attempt to study the effects of electric light on plant growth.

Carbon lamps* were used by Mangon (1861), and carbon arc lamps operated from steam or Otto gas engine driven Siemens, Grammes, or Alliance dynamos were used as early as 1853 and later by Siemens (1881), Deherain (1881), and Bonnier (1895). Prillieux (1869) investigated the effects of Drummond's lamp** and gas light used for ordinary lighting on plant growth.

Later Welsbach mantle incandescent gas light lamps***, neon, incandescent-filament lamps, and quartz mercury lamps like the Cooper-Hewitt Uviarc were used for greenhouse supplementary light. Several early researchers noted the elongating effects of the greenhouse supplementary light, especially when incandescent-carbon or incandescent-filament lamps were used.

Use of the term supplementary light (to supply what is lacking) is somewhat confusing, because in many cases the supplementary light was used continuously, throughout the night, or only during the dark period, rather than as a supplement to natural light. In these studies, at least some of the growth effects reported are surely due to a response to the extremely long photoperiods, to end-of-day photomorphogenic effects, and to root zone warming rather than to additional photosynthesis.

Using artificial light, usually from incandescent-filament lamps, for deliberate photoperiod control was initiated by Garner and Allard (1920) and was soon followed by many others. As photoperiod control became a production tool for floriculture and plant breeding, the more efficient fluorescent lamps were installed in a number of commercial greenhouses, often with unfortunate results; specifically failure or delay of flowering in long-day plants. Borthwick and Parker (1952) investigated this problem by comparing several kinds of fluorescent lamps, including special phosphor lamps, to incandescent lamps for efficiency in extending the greenhouse day to promote flowering of long-day plants.

Annual beet and sugar beet flowered poorly or not at all under daylength extensions with any kind of fluorescent lamp, but flowered promptly when incandescent-filament light was used. Although Odén, et al. (1932), Rasumov (1933), and Wenger (1934) had noted that the long wavelengths of light were necessary, or at least promotive, to normal flower stalk development, red radiation was considered the principal part of the spectrum controlling flowering. The action spectrum data probably influenced Borthwick and Parker (1952) to suggest that the much greater responsive-ness of plants to light from incandescent-filament than from fluorescent lamps was because the incandescent emitted a much greater percentage of red radiation than the fluorescent lamps. A few years later, of course, it was firmly established that the far-red emittion, or the lack thereof, had a strong influence on the response of plants to photoperiod control lighting.

Many subsequent studies of photoperiodism compared daylength extensions obtained with fluorescent or incandescent light. Compared to fluorescent, the incandescent extension induced increased stem length in evergreen and deciduous tree species as well as herbaceous species such as tomato and soybean, promoted heading in millet, barley, and wheat, induced earlier flowering in *Hyoscyamus niger*, Petunia, dill, and other long-day species, and produced greater pod set in *H. niger*. Bulbing of onions was promoted by incandescent photoperiod control lighting and failed to occur when fluorescent was used.

Fluorescent photoperiod lighting failed to inhibit flowering of red-insensitive soybean varieties, and when using photoperiod light to make a 13.5 h day for the most normal rate of reproduction in Ransom soybeans, incandescent lamps resulted in more pods than fluorescent lamps.

After the far-red reversibility of the red inhibition of hypocotyl growth in dark-grown seedlings was established, it was of interest to determine if this reversibility, and its confirmation of the activity of phytochrome, was also evident in internode growth of light-grown plants. Irradiating bean plants for brief periods at the beginning of each dark period with far-red, so that the plants entered the dark period with phytochrome predominantly in the red-absorbing form, resulted in a large increase in internode length.

The amount of elongation was proportional to the dark period remaining after the irradiation and was reversible by a subsequent exposure to red. Additional studies showed that what is now called 'end-of-day' far-red produced similar effects on most other bean varieties, sunflower, peanut, and morning glory.

Also, end-of-day far-red promoted flowering of long-day plants, like dill, and short-day plants, such as millet and milo, and had a marked effect on flowering of *H. niger*. Extending the day with incandescent light is in effect providing end-of-day far-red, and the far-red effect becomes greater as the duration and irradiance of the incandescent light in increased.

High intensity discharge lamps are now widely used in greenhouses to supplement the low natural light levels of winter. Some researchers also use HID lamps, especially high pressure sodium (HPS) lamps, for photoperiod control lighting to prevent dormancy and to promote flowering of long-day plants.

HPS lamps, however, are reported to be much less efficient than incandescent lamps, requiring a 4 to 8 fold increase in irradiance to provide the same photoperiodic stimulus as the 1:1 red/far-red ratio* of incandescent lamps. In fact, the benefits of HPS supplemental light is enhanced by the addition of some incandescent lamps.

Early attempts to use electric lamps as the sole source of light for plant growth chiefly used nitrogen-filled incandescent-filament lamps, the Mazda C lamp, although some of these efforts utilized neon, low-pressure sodium, mercury tungsten, mercury arc, mercury vapour, or carbon-arc light, alone or in conjunction with incandescent lamps in order to increase the illuminance.

Several of these examples where incandescent light was used noted excessive stem elongation. Roodenburg (1940) stated that near infra red produces a specific elongation effect and Aberg (1943) in noting the elongation that occurred, concluded that "*The infra-red rays of shorter wavelength that penetrate a layer of water 3 cm in thickness probably have a favourable effect on the internode elongation in the tomato plant.*"

In addition to the etiolation, a major problem with these early efforts was the low light level, equivalent to about 160 ìmols m^{-2} s^{-1} and often less, generated by these lamps. In order to obtain a higher illuminance, Mitchell (1935) installed a new type, high intensity, carbon-arc lamp for respiration and photosynthesis studies. These lamps had been designed for use in hospital solaria to treat extrapulmonary tuberculosis patients. Gains made by the patients during summer exposure to sunlight were lost during winter months due to low light levels and cloudy days. Clinical sunlight, recommended at 140-160 mW m^{-2} between 290 and 310 nm, could be and was supplied by these carbon arc lamps.

E.J. Kraus and Jack Mitchell left the University of Chicago about 1935 to join the Beltsville photoperiod project. Thus it was probably at their recommendation that the four temperature-and humidity-controlled plant growth chambers that were installed at Beltsville about 1937 were equipped with these carbon-arc lamps.

Soybeans grown under the arc lamps consistently had a lower carbohydrate content than plants grown in the greenhouse. Parker and Borthwick (1949) concluded that the low carbohydrate level probably resulted from the small amount of red radiation emitted by the 'Sunshine' carbons in these carbon-arc lamps.

So the following year they installed incandescent lamps that provided 8 to 10% of the illuminance of the arc lamp to provide additional red radiation. Soybeans grown under carbon-arc light plus incandescent revealed an increase in starch and sugars that could not be accounted for by the small increase in illuminance. In retrospect it seems strange that Parker and Borthwick (1949) would attribute these gains to the increase in red due to the incandescent since a much larger increase in red obtained by using a different type carbon,.025 carbons, had very little effect..

These carbon arc/incandescent lighted chambers were kept in almost continuous use for over 30 years, but operational and maintenance problems induced Parker and Borthwick (1950) in 1947 to begin planning a controlled-environment room lighted with fluorescent lamps. Several years earlier fluorescent lamps had been tested satisfactorily for plant growth, but the low illuminance available from these lamps was inadequate for controlled-environment rooms. The introduction of the 8 ft. slimline lamp following World War II seemed to provide a means of obtaining sufficient illumination for plant growth over relatively large areas, especially when the lamp current was increased from 200 to 300 mA. During the design phase of this room, Parker compared plant growth under slimline fluorescent with and without incandescent supplementary light. As with the carbon-arc lamp rooms, the avowed purpose of the incandescent lamps was to increase a possible deficiency of red radiation.

The supplementary incandescent light resulted in an 18% increase in dry weight. Withrow and Withrow (1947) had reported that adding incandescent to fluorescent light increased yield, and later reports verified the increased growth due to added incandescent light. Dunn and Went (1959) noted that the effect of adding 10% of the fluorescent illuminance with incandescent was no greater when added to red fluorescent than to blue fluorescent, concluding that *"while the most obvious explanation is that the effect is due to the infra red radiation of the incandescent lamps, it is unlikely that the far-red and infra red rays of the incandescent light was responsible* (for the increased growth) *since they would have been more effective when added to blue than to the red fluorescent light".*

Parker planned additional experiments to evaluate the plant growth effectiveness of various kinds of fluorescent lamps, including experimental lamps with special phosphors like the Agricultural, and to examine other quantities of incandescent supplementary light. The results of these studies were never published, but when the fluorescent-lighted room was completed about 1950 it contained cool white fluorescent and incandescent lamps that provided about 10% of the illuminance of the fluorescent lamps. In order to facilitate future plant growth chamber construction, Joe Ditchman, a GE engineer assigned to biological lighting development, calculated that 10% of the illuminance of the slimline fluorescent lamps could be obtained by installing incandescent lamps at the rate of 30% of the fluorescent wattage.

Due to lack of data on plant response to other levels of incandescent supplementary light, this value, 30% of the installed fluorescent watts, became a guideline for use in growth chamber design. The validity of this percentage, of course, was lost as designers increased the efficiency of the fluorescent lamp.

For example about 1963, a chamber was constructed at Beltsville using 1500 mA, non-circular cross section, fluorescent lamps and, while the incandescent effect was still apparent at light levels as high as 500 ìmols m^{-2} s^{-1}, increasing the intensity of the main light source decreases the incandescent effect of the 'standard' incandescent installation. This fact was also noted by Meijer (1957) and Sanchez and Cogliatti (1975). Thus it is not surprising to find that increasing the percentage of incandescent watts increases the incandescent effect in chambers lighted with 1500 mA lamps. In addition to increased plant weight, the addition of incandescent light to the fluorescent system also resulted in increased flower weight and number of florets in Chrysanthemum, while reducing the number of days required to develop flower colour and improved flowering of long-day plants.

The incandescent light may also increase stem elongation, alter leaf area, and reduce branch and tiller development. Moreover, if the incandescent and fluorescent lamps are not turned off simultaneously a substantial, often undesirable, stem lengthening can occur that may not be recognized by many plant growth chamber users as an end-of-day far-red effect. With some plants incandescent light is essential for normal plant development, but it is also clear that with other plants incandescent light is a major factor in the inability to simulate the field phenotype. While fluorescent-lighted chambers have been constructed without incandescent supplemental light, the advantages of using the incandescent to increase growth, accelerate flowering in long-day plants, control flowering in red-insensitive varieties, and produce end-of-day far-red effects makes their addition in fluorescent-lighted plant growth chambers extremely useful and in some cases indispensable. For example tissue cultures of Loblolly pine fail to differentiate without incandescent light added to the fluorescent system.

In other cases where the incandescent supplemental light is a detriment to obtaining the growth or plant habit desired, the problem can be solved, in soybeans at least, by utilizing correct photoperiod regimes and/or using the incandescent lamps correctly. In other examples of inadequate plant development, the incandescent lamps can be easily turned off.

High Intensity Discharge Lamps

High intensity discharge lamps in the form of mercury or phosphor-coated mercury (sometimes called mercury-fluorescent) lamps were added to the fluorescent-incandescent system as early as 1955. The development of similar systems by others soon followed; each apparently without knowledge of the other installations. Chambers lighted solely with phosphor-coated mercury lamps also were constructed, but the low efficiency of these lamps limited their use. hen the highly efficient metal halide lamps were introduced, plant growth chamber designers quickly incorporated them into new chambers and ultimately retrofitted them into older chambers.

The further increase in light-producing efficiency achieved by the introduction of the high pressure sodium lamp about 1965 resulted in a number of trials with this light source. In our studies, HPS proved less than satisfactory as a sole source of light for field crop plants, but plants grew well when irradiated with a 1:1 mixture of mercury, or metal halide and HPS. In contrast to our earlier results, Smeets at Wageningen designed a 100 m^2 room with only HPS lamps that appears to provide satisfactory growth of several floricultural crops (personal observation).

Although HID lamps can provide the same irradiance as fluorescent lamps at a substantially reduced power requirement, the chief reason for using them seems to be to increase the PPFD above that normally available from fluorescent lamps. An exception is the work at the Climate Lab in New Zealand, which was primarily interested in obtaining a spectral distribution equivalent to sunlight including an appropriate red/far-red ratio, and was only secondarily interested in super high light levels. Today we see chambers being constructed with light levels equalling or exceeding peak solar radiation.

The reason given for the high irradiance is usually that it is necessary for simulating field studies. It would seem that the R/FR ratio of natural light would also be a requirement for simulating field studies, but this subject is rarely encountered in arguments for artificially-produced solar irradiance levels. The spectral distributions of the tin chloride lamp, which was never produced commercially, and the Tungsram daylight metal halide containing dysporsium come very close to matching the natural light spectral distribution.

A question that arises frequently in the design of HID-lighted growth chambers is whether incandescent lamps should be added. Tibbitts, et al., (1987) reported that incandescent lamps had little to no effect on growth of mustard and wheat when they were added to high intensity discharge lamps. However, there was a small but significant increase in soybean vegetative growth, and Casal, et al. (1985) reported that incandescent light reduced tillering and advanced reproductive development in Lolium. We originally assumed that the lack of far-red effect when incandescent lamps were added to HID-lighted chambers was due to the higher HID irradiance. This is not a satisfactory explanation, however, since a marked far-red effect failed to be discernable at HID light levels comparable to fluorescent-lighted rooms (less than 500 ìmols m^{-2} s^{-1}).

Part of the problem seems to be that incandescent lamps provide red as well as far-red;and thus, the net increase in far-red relative to red is not as great as might be assumed. For example, the red/far-red ratio in a reach-in chamber lighted with 16, 115-W VHO fluorescent lamps was 6.684.

Adding incandescent at an input wattage of 33% of the installed fluorescent watts reduced the R/FR ratio to 1.884. When we retrofitted this chamber with HID lamps the R/FR ratio was 2.526 with MH and 2.749 with a 1:1 mixture of MH and HPS lamps. Adding incandescent decreased the ratio to 1.7 and 2, respectively. These ratios are similar to those from fluorescent-incandescent systems but the far-red effect is much less.

In part, this lack of an incandescent effect can be alleviated by increasing the incandescent lamp wattage to equal that of the HID lamps. While a properly designed reflector and ventilation system can remove the thermal radiation from the HID lamps, the large amount of long wavelength radiation resulting from such a large wattage of incandescent lamps makes a water filter essential. Unfortunately, the water filter is often not practical because it increases design costs and requires much more maintenance than the typical lamp loft barrier. The heat removal problem might be avoided by adding far-red without any increase in red radiation. In theory this could be done by using blue incandescent lamps which have a red/far-red ratio of 0.004 compared to the 0.671 of white incandescent ones, but in practice the far-red effect from blue incandescent lamps added to HID lamps is about the same as with white incandescent lamps.

Plants grown under HID lamps often produce abnormally short internodes, a fact observed by Warrington et al (1978), even when incandescent lamps were added. End-of-day exposures to incandescent lamps can be used as a tool to increase internode lengths to more acceptable values. End-of-day irradiations with blue incandescent lamps, however, produce excessive elongation.. Also, using the incandescent lamp for dark period interruptions, as an end-of-day treatment, or for daylength extensions can accelerate flowering of many long-day plants and control flowering of red-insensitive soybeans. The evidence seems to favour the addition of incandescent lamps to HID systems.

And thus, it is recommended that the design and construction of plant growth chambers continue to contain a provision for utilization of the incandescent lamp as part of the total irradiance system, to be implemented at the discretion of the investigator to meet the phytochrome requirements of the various biological organisms that may be grown in the chamber.

9

Recent Advances in Plant Biotechnology

Scientists have been improving plants by changing their genetic makeup since the late 1800s. Typically, this has been accomplished through crossbreeding and hybridization, in which two related plants are cross-fertilized and the resulting offspring have characteristics of both parent plants. In the breeding process, however, many undesirable traits often can appear in addition to the desirable ones. Some of those undesirable traits can be eliminated through additional breeding, which is time consuming. Breeders can then further select and reproduce the offspring that have the desired traits. Many of the foods that are already common in our diet are obtained from plant varieties that were developed using conventional genetic techniques of breeding and selection.

Today, by inserting one or more genes into a plant, scientists are able to produce a plant with new, advantageous characteristics. The new gene splicing techniques are being used to achieve many of the same goals and improvements that plant breeders historically have sought through conventional methods. They give scientists the ability to isolate genes and introduce new traits into foods without simultaneously introducing undesirable traits. This is an important improvement over traditional breeding. Because of the increased precision offered by the bioengineered methods, the risk of introducing detrimental traits is actually likely to be reduced.

Gene and Genome Analysis

Detailed studies have been conducted on plant genome and physical and genetic maps are available for several plants. As an example of studies on genome the rice genome is discussed here. Rice has a much smaller genome (430 Mbp per haploid genome) than many other crops that belong to the Poaceae family. Due to the genome colinearity, high similarity in gene order and gene content, among the Poaceae family, the importance of rice genetics has been emphasized, and comparative analyses among rice, wheat, and maize have been intensively studied.

As a result, rice becomes the model crop for the molecular genetic approach. This crop is available for many applications, including the construction of a high dense map, expressed sequence tag (EST) and full genomic sequence database, bacterial and yeast artificial chromosome (BAC and YAC) libraries, quantitative traits loci (QTL) mapping for yield and morphology, functional genomics by knockout mutagenesis using T-DNA insertion, map-based cloning, and genetically modified rice using transformation techniques.

State-of-the-art Genome Profiling (GP)

The traditional approach for species identification is exclusively based on phenotypic traits such as morphological, anatomical, chemical properties and others, which are often affected by environmental factors and thus are difficult to analyse and unreliable. Interspecies homogeneity, intraspecies variability and the existence of undescribed species often lead to phenotypic misidentification.

Moreover, species, which are phenotypically far less prominent, cannot be always identified in this way. To overcome these problems, genotypebased (nucleic acid-based) techniques have been employed as an alternative or complementary approach and have continuously been developed including RFLP ,AFLP, RAPD, 16S rRNA or 16S-23S internal transcribed spacer (ITS) sequence analysis and others. These methods provide a possible way to identify species directly based on their genomic sequences but none of them have been shown to identify species in general, mainly because of the insufficiency in the amount of information which they can provide.

In this stream, the whole genome sequencing is surely the most definitive solution for species identification though simply too redundant for such purposes and impossible in practice to analyse all the constituents of a heavily dense population. On the other hand, the information obtained from the comparison of a single gene is often not sufficient to place a species at the appropriate position on the phylogenetic tree.

In order to deal with above issues, previously Nishigaki and co-workers have described a realistic solution conforming to the notion of the amount of information sufficient for species identification and demonstrated this by inventing a novel method called Genome Profiling (GP), which is a temperature-gradient gel electrophoresis (TGGE) analysis of random-PCR products. Next, the complexity of the generated data, genome profiles, can be simplified by extracting feature points in GP, i.e., species identification dots (spiddos) which can be used for further processing of measuring the similarity of two species by calculating Pattern Similarity Score (PaSS). Further, the technical advances by constructing internet-based GP databases (named On-web GP), and developing a highly reproducible and miniaturized system (micro-TGGE) have moved this technology towards being a universal, general and global tool for species identification.

Recombinant DNA Technology

Gene Targeting (GT): Gene targeting (GT) is a key technology for the rational, accurate and safe exploitation of plantsthrough genetic manipulation. Moreover, it offers the potential to completely knockout the expression of target genes or to make specific changes to gene function, objectives that cannot be achieved by conventional transgenesis.

The ability to target DNA integration would permit the locus-specific integration of a transgene into a predetermined site of the host genome, avoiding the accidental inactivation of an endogenous gene localized at the insertion site or the unexpected expression profiles of the transgene itself, the so-called position effect. Systematic isolation and sequencing of genomic DNA flanking the insertion sites (known as FSTs or Flanking Sequence Tag offers the opportunity to rapidly characterize plants altered in a candidate gene sequence.

This approach is notably most useful in fully sequenced genomes such as in Arabidopsis thaliana. With 125 Mbp and 26,422 genes, the Arabidopsis genome shows very limited synteny with the 420–466 Mbp and 60,000 predicted genes of the rice genome. The recombination machinery has been well conserved throughout evolution, as an essential component of cell survival. In nature, homologous recombination is a DNA maintenance pathway that protects chromosomes against damage affecting both DNA strands, such as double strand breaks (DSBs) or interstrand crosslinks. DSB repair (DSBR) has been one of the most investigated homologous repair pathways see SHRIVASTAVA1, SHARMA2 AND KUMAR 2007 cited in Kumar and Sopory 2007).

Recently plastid genome transformation technique has gained prominence due to its better integration and less chances of random spread. The genome propagated by higher plant plastids, the plastome, is typically a double stranded DNA molecule of 130 to 160 kb. Over one hundred copies of this genome can be present in a single plastid. It is ideally represented as a circular monomer containing 2 inverted repeats, even though reality is more complex since linear and circular multimers have been frequently detected The complete sequence of this highly polyploidy genome is available for about 20 different species of angiosperms. The first successful transformation of tobacco was performed using as marker a mutant plastid DNA fragment covering the 16SrRNA gene derived from a line resistant to spectinomycin and streptomycin. Major improvements in the selection process were soon obtained with the dominant aadA marker gene, inactivating spectinomycin or streptomycin. When fused to GFP, this marker can be used to track the selection process. Genes encoding resistance to kanamycin, nptII and more recently aphA-6 are also possible options, and could be more appropriate for some species.

There are three different fates for the external DNA to get integrated into the native genome. They are homologous recombination, illegitimate recombination or nonhomologous end joining, and single-strand annealing. . Single-strand annealing (SSA), a third path of repair, requires the presence of repeated sequences on both sides of a break. After exonuclease degradation of the 5' ends, repair occurs by annealing of the two complementary sequences, a process leading to the loss of the genetic information contained between these repeats. With respect to the species preferential DSB repair pathway, HR but also IR mediates transgene integration. This second aspect explains the inefficiency of GT in higher plants, which use HR as a minor pathway of repair. Thus, despite the fact that transgene integration processes are still unclear in plants transgenic DNA would be preferentially integrated by end joining whether or not sharing homology within the host genome.

GM Crops: The U.S. Food and Drug Administration (FDA) has found no evidence to indicate that either ordinary plant deoxyribonucleic acid (DNA) or the DNA inserted into plants using bioengineering presents food safety problems. Nor are the small amounts of the newly expressed proteins likely to change dramatically the safety profile of the plant. If safety concerns should arise, however, they would most likely fall into one of three broad categories: allergens, toxins, or anti-nutrients. FDA has extensive experience in evaluating the safety of such substances in food. It is important to note that the kinds of food safety testing typically conducted by developers of a bioengineered food crop to ensure that their foods meet all applicable requirements of the Food, Drug and Cosmetics Act (FD&C Act) address these potential concerns. In the event that something unexpected does occur, this testing provides a way to detect such changes at the developmental stage and defer marketing until any concern is resolved.

As Aforementioned, some of the Food Safety Concerns that could Arise Include

Allergens: Foods normally contain many thousands of different proteins. While the majority of proteins do not cause allergic reactions, virtually all known human allergens are proteins. Since genetic engineering can introduce a new protein into a food plant, it is possible that this technique could introduce a previously unknown allergen into the food supply or could introduce a known allergen into a "new" food.

Toxins: It is possible that a new protein, as introduced into a crop as a result of the genetic modification, could cause toxicity.

Anti-nutrients: It is possible that the introduction of anti-nutrients, such as molecules like phytic acid, could reduce essential dietary minerals such as phosphorus.

The use of genetic engineering techniques could also result in unintended alterations in the amounts of substances normally found in a food, such as a reduction of Vitamin C or an increase in the concentration of a naturally occurring toxicant in the plant food.

Legal and Regulatory Issues

One important component in ensuring food safety is the U.S. regulatory structure. The FDA regulates bioengineered plant food in conjunction with the United States Department of Agriculture (USDA) and the Environmental Protection Agency (EPA). FDA has authority under the FD&C Act to ensure the safety of all domestic and imported foods for man or animals in the United States market. The exceptions to this are meat, poultry and certain egg products, which are regulated by USDA. The safety of animal drug residues in meat and poultry, however, is regulated by FDA. Pesticides, including those bioengineered into a food crop, are regulated primarily by EPA. USDA's Animal and Plant Health Inspection Service (APHIS) oversees the agricultural and environmental safety of planting and field testing bioengineered plants.

Bioengineered foods and food ingredients must adhere to the same standards of safety under the FD&C Act that apply to their conventionally bred counterparts. This means that these products must be as safe as the traditional foods in the market. FDA has the power to remove a food from the market, or sanction those marketing the food if the food poses a risk to public health. It is important to note that the FD&C Act places a legal duty on developers to ensure that the foods they market to consumers are safe and comply with all legal requirements.

Area under the commercialization of genetically modified (GM), often called biotech crops continued to grow for the ninth consecutive year at a sustained double-digit growth rate of 20% in 2004 (James, 2004). The estimated global area of approved GM crops for 2005 was 90.0 million hectares with $4.70 billion global market value—based on the sale price of GM seed plus any technology fees that apply. The global value of the GM crop market is projected a more than $5.0 billion for 2005 (James, 2004). Commercialization of genetically modified (GM) crops continued to grow for the ninth consecutive year.

It reflects the substantial improvements in productivity, the environment, economics, health and social benefits realized by farmers, consumers and society. At the same time the growing controversy over GM food products increased interest in food labelling and identity preservation (IP) of GM crops. Hence, an IP system must be designed to provide assurances that the desired traits are present (or absent) in a product from the seed source, through all steps of production and delivery, to the end user. There are numerous regulatory issues related to GM crops.

These include the testing and acceptance of new GM crops for commercial introduction, both domestically and internationally. Nearly every country has different approaches and many have their own regulatory framework, with an intent to prevent cross-contamination of the conventional food and feed industries. IP tracking software is also available in market to ease the burdens associated with precise record-keeping requirements.

The economics of IP has been calculated by various scientists depending on different applied IP systems. Niche-marketing opportunities will grow, because of the availability of GM crops and finally, IP of agricultural commodities from GM crops can provide greater choice and value desired by both agricultural producers and consumers.

Based on annual percentage growth in area, of the eight leading GM crop countries, India had the highest percentage year-on-year growth in 2004 with an increase of 400% in Bt cotton area over 2003, followed by Uruguay (200%), Australia (100%), Brazil (66%), China (32%), south Africa (25%), Canada (23%), Argentina (17%) and USA at 11%. India increased its area of approved Bt cotton, introduced only two years ago, from approximately 38,038 hectares in 2002/ 03 to 560,000 hectares in 2004/05 seasons with Bt coverage of 11.65% and approximately 300,000 small farmers benefited from Bt cotton.

Regulatory Issues Related to GM Crops

There are numerous regulatory issues related to GM crops. These include the testing and acceptance of new GM crops for commercial introduction and the introduction of food products containing ingredients from GM crops, both domestically and internationally. Nearly every country has different approaches and many have their own regulatory framework. Regulation is a very dynamic issue with changes being reviewed and proposed in many countries on an ongoing basis.

Numerous regulatory actions are consequently being proposed as governments react to consumer concerns and pressures. Several countries have or have proposed to create new agencies to specifically cover GM crops. Approaches range from cautious acceptance to attempts to ban (growing and even imports), at least for the foreseeable future, all crops and products with GM traits. Each is approaching the testing, introduction, and acceptance of GM crops in its own manner and on its own time schedule.

Abiotic and Biotic Resistance

Coat Protein Mediated Resistance: CP is an important structural protein as it not only protects the viral nucleic acid from degradation, but also plays an important role in virus infection. Its functions includes acquisition and transmission of virus by vectors, cell to cell and long distance spread of the virus in host plants, and for some viruses, it regulates one or more steps of virus replication.Coat Protein (CP) mediated resistance has been demonstrated for 17 groups of viruses, and so far this strategy has shown best promise. CP transgenes have been shown to be effective in preventing or reducing infection and diseases caused by homologous and closely related viruses (Gonsalves, et al., 1998).

Coat protein-mediated protection has been reported for Tobacco mosaic virus, TMV,Tomato mosaic virus, ToMV, (Sa), Cucumber mosaic virus, CMV, Alfalfa mosaic virus, AlMV, Potato virus X, PVX, Potato virus Y, PVY, Potato leaf roll virus, PLRV, Papaya ringspot virus (PRSV) and a number of other viruses. CP-mediated resistance in Cantaloupe, Papaya, Potato, Squash and Tomato has been tested under the field conditions with fair degree of protection against most of the viruses.

Anti-HIV Agents among Desert Plants

Around 40 million people are affected due to the Human Immuno-deficiency Virus globally. During the past decades, a large number of anti-viral screening experiments on medicinal plant extracts have been reported and have led to the selection of several extracts active towards herpes viruses.

A promising result of a naturally occurring antiherpetic agent was given by n- docosanol (a natural 22 carbon saturated fatty alcohol) which is undergoing phase III clinical trials in patients. Clinical testing of the topical formulation, or systemic administration of drug suspensions has demonstrated a good therapeutic index, since high doses of n- docosanol do not elicit appreciable toxicity.

The findings show that natural products are still potential sources in the search for new antiherpatic agents (Hattori et al., 1995). Various plant extracts used in Ayurvedic medicine for inhibitory effects on HIV virus have been studied (Hattori, personal communication). A large number of such plants occur in semi-arid and arid climate of Rajasthan. Acquired immunodeficiency syndrome (AIDS) , the great pandemic of the second half of the 20th Century, is still a threatening disease world wide. Many research approaches are currently aimed at developing novel agents to arrest the replication of HIV through various targets.

These may include the inhibition of reverse transcriptase (RT), protease (PR), membrane fusion and integrase. HIV PR enzyme has been demonstrated to play an essential role in viral replication. A range of HIV PR inhibitors have been designed and applied in clinical trials such as Sanqunavir, Ritonavir and Indinavir. However, the development of drug resistance by virus, irrespective of the target, remains as an overwhelming problem in AIDS chemotherapy. Thus there is great need to search for and develop new and different anti-HIV candidates from plants and natural products are of considerable importance. In search for anti-HIV active agents from natural products, many attempts at screening traditional medicines have been made.

Biotic and Abiotic Stress

Environmental abiotic stress conditions, and especially drought and salinity, are currently the major factors which reduce crop yields worldwide leading to the fact that more than 800 million people are chronically undernourished.. The United Nations Environment Program estimates that approximately 20% of agricultural land and 50% of cropland in the world is saltstressed This salinity, in particular, is an increasing problem and nearly half of the area under irrigation, is at risk to be lost due o building up of salinity. Therefore genetic improvement of salt tolerance has become an urgent need for the future of agriculture in arid and semiarid regions..

One way of solving this problem would be breeding tolerant varieties of crop plants that can be grown on saline soils, but these breeding programs are time consuming and remained elusive . Hence, many metabolic changes are known to occur in plants subjected to salt stress, physiological parameters such as ionic relations have been suggested to be used as tolerance indicators since they can be related to salt tolerance mechanisms.

Biofuel Production

India has over 180 million of wasteland out of which 90 million ha is uncultivable. The degraded and denuded lands arise due to soil erosions as well as secondary salinizations. However Calotropis procera is a potential plant for bioenergy and biofuel production in semi arid regions of the country because it is able to grow on such lands. The plant has a growth potential of 2 dry tones to 40 dry tones per ha depending on the agro climatic conditions of it's growth. The plant has high level of regeneration potential and could be harvested up to 4 times a year. The plant yields valuable hydrocarbons which could be converted into diesel substitutes. The bio-diesel derived from Calotropis procera is free from NOx gases, S02 and Suspended Particulate Matter (SPM) and has high cetane value. Due to it's enormous potential for growth under adverse climatic conditions Calotropis procera is suggested as potential plant for bio-diesel production under semi-arid and arid conditions. Jatropha curcas also provides non-edible oil which could be converted into methyl ester with a gain of glycerine. The JME is used as mix in the ratio of 05, to 20 percent blend to fossil fuel diesel and in Germany the use of Rape seed methyl ester is increasing.

Soil Salinity

Soil salinity affects plant production in many parts of the world, particularly on irrigated land. NaCl is the predominant salt in most saline environments. Many crop species are sensitive to high concentrations of salt with negative impacts on agricultural production. Maize (Zea mays L.) is considered a moderately salt-sensitive plant.. Salt resistance of plants is a complex phenomenon that involves biochemical and physiological processes as well as morphological and developmental changes.. In addition to general osmotic stress, high concentrations of Na+ are toxic to maize and molecular mechanisms for salt resistance have not been fully identified or characterized (Zoerb et al 2004). The analysis of the plant's proteome is an important amendment to the analysis of the genome, because gene expression is altered under salinity stress.

The proteome, in contrast to the genome, is not static but rather dependent on a number of responses influenced by internal and external factors. The plant adaptation to environmental stress, such as soil salinity, is expected to have a strong influence on proteins. One approach to study the molecular mechanisms of plant responses to salinity is to use 2D polyacrylamide gel electrophoresis. Furthermore, the identification of differentially regulated proteins can lead to the identification of proteins and their corresponding genes which are involved in the physiology of salt resistance.

The high resolution achieved by 2D gels and computer-assisted analysis of the differentially regulated proteins were used to examine those proteins whose synthesis was modulated by salt treatment and to quantify these changes. As far as we know, our investigations are the first to characterize the differentially expressed proteins from roots and shoots of maize after treatment with low and high salt stress. Plant material was an efficiently Na+-excluding maize inbred line developed in our laboratory. According to Munns the growth response to salt stress consists of two phases, first, a water-deficit that results from the relatively high solute concentrations in the root medium and, second, ion-specific stress resulting from altered K+/Na+ ratios or Na+

nd Cl" concentrations that are toxic to plants. The aim of this study was to lucidate biochemical and physiological reactions of glycophytes to salt stress n the first phase of salinity.

While all major crops, as well as most wild species, are glycophytes, i.e. ensitive to relatively low salt concentrations, there are plants naturally dapted to conditions of high salinity in the soil. These plants, known as alophytes, include a large taxonomic variety and occupy diverse habitats, rom extreme dry to temporarily waterlogged sites or salt marshes, and can olerate NaCl concentrations similar, or even higher than that of sea water, a. 500 mM. It is - without doubt - necessary to develop sustainable biological roduction systems which can tolerate higher water salinity because freshwater esources will not come up with increasing demands of agricultural practice n near future.

The sustainable use of halophytic plants is a promising approach to alorize strongly salinised zones unsuitable for conventional agriculture and nediocre waters. The development of cash crop halophytes and the breeding f salt resistant crop varieties will require a clear understanding of the complex nechanisms of salt stress tolerance, which we are still lacking despite intensive esearch during the last decade.

Recombinant Proteins

It has been estimated that 1 g of recombinant antibody could be produced n leaves of a plant crop for only about US$100 while the current prices for nonoclonal antibodies range from US$2000 to US$5000 per gram. Indeed the ost of producing 1 kg of recombinant protein from most major field crops is stimated to be 10 to 50 times lower than the cost of producing the same mount by E. coli fermentation.

Whole plants also have an advantage when tissues such as a fruit, tuber, tc. can be used to express the protein of interest (James and Lee 2001), and n area of undisputed advantage occurs when the oral delivery of harmaceuticals, as well as feed and food enzymes, is possible. However, there re also some evident obstacles that arise when a whole plant is used for large-scale protein production.

Cell-free systems have proved to have high utility at the genomic, ranscriptomic and proteomic levels and to form a vital component of many spects of recombinant gene expression, and of both structural and functional roteomics..

Compared with DNA microarrays, protein bio-chips provide more challenges nd have yet to be perfected due to the complexity and inherent difficulties vith protein immobilization.

Novel Cell Free Translation System is Unique Discovery

A novel cell-free translation system is described in which template-mRNA molecules were captured onto solid surfaces to simultaneously synthesize and immobilize proteins in a more native-state form. This technology comprises a novel solid-phase approach to cell-free translation and RNA–protein fusion techniques. A newly constructed biotinylated linker-DNA which enables puromycin-assisted RNA–protein fusion is ligated to the 32 ends of the mRNA molecules to attach the mRNA-template on a streptavidin-coated surface and further to enable the subsequent reactions of translation and RNA–protein fusion on surface.

The protein products are therefore directly immobilized onto solid surface and furthermore were discovered to adopt a more native state with proper protein folding and superior biological activity compared with conventional liquid-phase approaches. We further validate this approach via the production of immobilized green fluorescent protein (GFP) on microbeads and by the production and assay of aldehyde reductase (ALR) enzyme with 4-fold or more activity. The approach developed in this study may enable to embrace the concept of the transformation of 'RNA chip-to-protein chip' using a solid-phase cell-free translation system and thus to the development of high-throughput microarray platform in the field of functional genomics and in vitro evolution

Plant Tissue Culture

Another area of biotechnology is micropropagation of plants. The aim of this technique is a fast production of a great number of genetically identical plants from a highly valuable mother plant or e.g. monosexual male and female plants. These plants can be either directly sold on the market for planting used for breeding purposes, for genetechnology or the technique is used as a method for basic science studies. Using petiole explants from transgenic plants containing the auxin responsive MAS promoter linked to the GUS reporter gene the distribution of auxin within the cultured petiole could be followed during the induction phase of somatic embryogenesis.

Interestingly, the cells forming the glandular canal contain high concentrations of auxins as shown by using transgenic plants containing the auxin sensitive MAS-promoter coupled to the GUS-gene, whatever the significance. Rhizogenic centres develop near vascular bundles prior to those embryogenic centres.

Genetic factors play a central role to induce somatic embryos, i.e. to provide the competence of the species for the process. Here, great variation can be found even within a genera such as Daucus. Eight of twelve Daucus species cultured in identical conditions produced somatic embryos (D. halophilus D. capillifolius, D. commutatus, D. azoricus, D. gadacei, D. maritimus, D maximus, D. carota), whereas four species (D. montevidensis, D. pussillus, D. muricatus, D. glochidiatus) were not competent to do so. Under identical culture conditions, only 8 out of 12 species and subspecies of the genus Daucus proved capable of somatic embryogenesis.

Random amplified polymorphic DNA analysis indicated a polymorphism between the genomes of individual species that were capable of embryogenesis and those that were not. Two specific bands (1.1 kbp, 0.68 kbp) were detected only inthe genomes of individuals with the capacity for embryogenesis. These were cloned and sequenced, and the homology of the nucleotide sequences of the various species was detected: this ranged from 74% to 92% for the larger sequence and from 92% to 97% for the smaller one.

These DNA sequences would appear to be useful as a marker of the capacity for somatic embryogenesis in the genus Daucus (Imani et al.2001) The sequences obtained in this study have been registered with the European Bioinformatics Institute (EMBL). The access numbers for the sequences are: AJ278039 DCA78039; AJ278040 DCA78040; AJ278041 CA78041; AJ278042 DCA78042; AJ278176 DCA278176; AJ278177 DCA278177; AJ278178 DCA278178; AJ278179 DCA278179. No open reading frames were detected.

We performed later additional studies with other Daucus species (D. capillifolius; D. carota ssp. Azoricus and gadecaei) to determine the use-fulness of these RAPD products as markers for identifying the ability of Daucus species to generate somatic embryos. There was a 100% correlation between the embryogenic potential of the species and the occurrence of the 1.1-kbp and 0.68-kbp band (Imani et al.2001). Micropropagation Technique in Enhancing the Productivity of Crops have been taken up at large scale at TERI. Some of the activities undertaken at MTP include:

- Large-scale production of superior quality planting material of various economically important plant species using tissue culture technology
- Mass multiplication of those species which are difficult to regenerate by conventional methods of propagation or where conventional methods of propagation are inadequate to meet the demand of planting material
- Development of new micropropagation protocols and refining of others so as to make them suitable for large-scale production of plants
- Helping the entrepreneurs/industry through technology transfer, mother cultures and training of staff
- Assisting clients in setting-up their own tissue culture labs
- Creating awareness.

Till date over 15 million plants of forest species, cash crops, medicinal and aromatic plants, and ornamentals (foliage and flowering) have been dispatched to various state forest and horticulture departments, private entrepreneurs, nurseries, farmers etc. for field demonstrations and routine plantations. In addition, MTP is in possession of micropropagation protocols for over 90 economically important plant species. Field demonstration plots of tissue cultured plants have been laid at different locations to evaluate and compare their growth performances with conventional plants. Besides transfer of technologies to industry for commercialization, MTP has been instrumental in capacity building and creating awareness about tissue culture technology through seminars/ workshops/training programmes, exhibitions, etc.

Secondary Metabolites

Since the establishment of plant tissue culture techniques in 1960's, significant contributions have been made to the development of biochemical studies on secondary metabolism such as structural elucidation, biosynthesis, enzymology, metabolic regulation system, intracellular distribution of metabolites and relevant enzymes, metabolite transportation, molecular biology and many others. However, one of the greatest difficulties and challenges in the application of plant tissue culture to metabolism research has been that unorganized callus tissues have often failed to accumulate metabolites usually detected in the mother plant. In some cases, metabolic potential was recovered through the development of a production medium, change in culture conditions or selecting cell strains of high productivity. It is commonly observed that recalcitrant callus tissues begin to synthesize secondary metabolites after organ - such as shoots and roots - differentiation .Although somatic embryogenesis occurs in cultured cells of numerous plant species, it has rarely been applied to secondary metabolite production.

Metagenomics

Metagenomics is the study of metagenomes, genetic material recovered directly from environmental samples. The broad field may also be referred to as environmental genomics, ecogenomics or community genomics. Traditional microbiology and microbial genome sequencing rely upon cultivated clonal cultures. This relatively new field of genetic research enables studies of organisms that are not easily cultured in a laboratory as well as studies of organisms in their natural environment. Early environmental gene sequencing cloned specific genes (often the 16S rRNA gene) to produce a profile of diversity in a natural sample.

Such work revealed that the vast majority of microbial biodiversity had been missed by cultivation-based methods. Recent studies use "shotgun" Sanger sequencing or massively parallel pyrosequencing to get (mostly) unbiased samples of all genes from all members of sampled communities.

History

Origin of the Term: The term "metagenomics" was first used by Jo Handelsman, Jon Clardy, Robert M. Goodman, and others, and first appeared in publication in 1998. The term metagenome referenced the idea that a collection of genes sequenced from the environment could be analysed in a way analogous to the study of a single genome.

The exploding interest in environmental genetics, along with the buzzword-like nature of the term, has resulted in the broader use of metagenomics to describe any sequencing of genetic material from environmental (i.e. uncultured) samples, even work that focuses on one organism or gene. Recently, Kevin Chen and Lior Pachter (researchers at the University of California, Berkeley) defined metagenomics as "the application of modern genomics techniques to the study of communities of microbial organisms directly in their natural environments, bypassing the need for isolation and lab cultivation of individual species."

Environmental Gene Surveys: Conventional sequencing begins with a culture of identical cells as a source of DNA. However, early metagenomic studies revealed that there are probably large groups of microorganisms in many environments that cannot be cultured and thus cannot be sequenced. These early studies focused on 16S ribosomal RNA sequences which are relatively short, often conserved within a species, and generally different between species. Many 16S rRNA sequences have been found which do not belong to any known cultured species, indicating that there are numerous non-isolated organisms out there.

Early molecular work in the field was conducted by Norman R. Pace and colleagues, who used PCR to explore the diversity of ribosomal RNA sequences. The insights gained from these breakthrough studies led Pace to propose the idea of cloning DNA directly from environmental samples as early as 1985. This led to the first report of isolating and cloning bulk DNA from an environmental sample, published by Pace and colleagues in 1991 while Pace was in the Department of Biology at Indiana University. Considerable efforts ensured that these were not PCR false positives and supported the existence of a complex community of unexplored species.

Although this methodology was limited to exploring highly conserved, non-protein coding genes, it did support early microbial morphology-based observations that diversity was far more complex than was known by culturing methods. Soon after that, Healy reported the metagenomic isolation of functional genes from "zoolibraries" constructed from a complex culture of environmental organisms grown in the laboratory on dried grasses in 1995. After leaving the Pace laboratory, Ed DeLong continued in the field and has published work that has largely laid the groundwork for environmental phylogenies based on signature 16S sequences, beginning with his group's construction of libraries from marine samples.

Longer Sequences from Environmental Samples

Recovery of DNA sequences longer than a few thousand base pairs from environmental samples was very difficult until recent advances in molecular biological techniques, particularly related to constructing libraries in bacterial artificial chromosomes (BACs), provided better vectors for molecular cloning.

Shotgun Metagenomics

Advances in bioinformatics, refinements of DNA amplification, and proliferation of computational power have greatly aided the analysis of DNA sequences recovered from environmental samples. These advances have enabled the adaptation of shotgun sequencing to metagenomic samples. The approach, used to sequence many cultured microorganisms as well as the human genome, randomly shears DNA, sequences many short sequences, and reconstructs them into a consensus sequence.

In 2002, Mya Breitbart, Forest Rohwer, and colleagues used environmental shotgun sequencing to show that 200 litres of seawater contains over 5000 different viruses. Subsequent studies showed that there are >1000 viral species in human stool and possibly a million different viruses per kilogram of marine sediment, including many bacteriophages. Essentially all of the viruses in these studies were new species. In 2004, Gene Tyson, Jill Banfield, and colleagues at the University of California, Berkeley and the Joint Genome Institute sequenced DNA extracted from an acid mine drainage system. This effort resulted in the complete, or nearly complete, genomes for a handful of bacteria and archaea that had previously resisted attempts to culture them. It was now possible to study entire genomes without the biases associated with laboratory cultures.

Global Ocean Sampling Expedition

Beginning in 2003, Craig Venter, leader of the privately-funded parallel of the Human Genome Project, has led the Global Ocean Sampling Expedition, circumnavigating the globe and collecting metagenomic samples throughout. All of these samples are sequenced using shotgun sequencing, in hopes that new genomes (and therefore new organisms) would be identified. The pilot project, conducted in the Sargasso Sea, found DNA from nearly 2000 different species, including 148 types of bacteria never before seen. As of 2009, Venter has circumnavigated the globe and thoroughly explored the West Coast of the United States, and is currently in the midst of a two-year expedition to explore the Baltic, Mediterranian and Black Seas.

Pyrosequencing

In 2006 Robert Edwards, Forest Rohwer, and colleagues at San Diego State University published the first sequences of environmental samples generated with so-called next generation sequencing, in this case chip-based pyrosequencing developed by 454 Life Sciences. This technique for sequencing DNA generates shorter fragments than conventional techniques, however this limitation is compensated for by the very large number of sequences generated. In addition, this technique does not require cloning the DNA before sequencing, removing one of the main biases in metagenomics.

MEGAN

In 2007, Daniel Huson and Stephan Schuster developed and published the first stand-alone metagenome analysis tool, MEGAN, which can be used to perform a first analysis of a metagenomic shotgun dataset. This tool was originally developed to analyse the metagenome of a mammoth sample. However in a recent study by Monzoorul et al. 2009, it was shown that adopting the LCA approach (of MEGAN) solely based on bit-score of the alignment leads to a number of false positive assignments especially in the context of metagenomic sequences originating from new organisms. This study proposed a new approach called SOrt-ITEMS which used several alignment parameters to increase the accuracy of assignments.

MG-RAST

In 2007, Folker Meyer and Robert Edwards and a team at Argonne National Laboratory and the University of Chicago released the Metagenomics RAST server (MG-RAST) a community resource for metagenome data set analysis. The SEED based free, public resource has so far (October 2009) been used for the analysis of over 4000 metagenome data sets. As of October 2009 100+ giga-basepairs of DNA have been analysed via MG-RAST, more than 350 public data sets are freely available for comparison within MG-RAST.

Applications

Metagenomics can improve strategies for monitoring the impact of pollutants on ecosystems and for cleaning up contaminated environments. Increased understanding of how microbial communities cope with pollutants is helping assess the potential of contaminated sites to recover from pollution and increase the chances of bioaugmentation or biostimulation trials to succeed. Recent progress in mining the rich genetic resource of non-culturable microbes has led to the discovery of new genes, enzymes, and natural products. The impact of metagenomics is witnessed in the development of commodity and fine chemicals, agrochemicals and pharmaceuticals where the benefit of enzyme-catalysed chiral synthesis is increasingly recognized.

Metagenomic sequencing is being used to characterize the microbial communities from 15-18 body sites from at least 250 individuals. This is part of the Human Microbiome initiative with primary goals to determine if there is a core human microbiome, to understand the changes in the human microbiome that can be correlated with human health, and to develop new technological and bioinformatics tools to support these goals.

It is well known that the vast majority of microbes have not been cultivated. Functional metagenomics strategies are being used to explore the interactions between plants and microbes through cultivation-independent study of the microbial communities.

Microbial Diversity

Much of the interest in metagenomics comes from the discovery that the vast majority of microorganisms had previously gone unnoticed. Traditional microbiological methods relied upon laboratory cultures of organisms. Surveys of ribosomal RNA (rRNA) genes taken directly from the environment revealed that cultivation based methods find less than 1% of the bacteria and archaea species in a sample.

Gene Surveys

Shotgun sequencing and screens of clone libraries reveal genes present in environmental samples. This provides information both on which organisms are present and what metabolic processes are possible in the community. This can be helpful in understanding the ecology of a community, particularly if multiple samples are compared to each other.

Environmental Genomes

Shotgun metagenomics also is capable of sequencing nearly complete microbial genomes directly from the environment. Because the collection of DNA from an environment is largely uncontrolled, the most abundant organisms in an environmental sample are most highly represented in the resulting sequence data. To achieve the high coverage needed to fully resolve the genomes of underrepresented community members, large samples, often prohibitively so, are needed. On the other hand, the random nature of shotgun sequencing ensures that many of these organisms will be represented by at least some small sequence segments. Due to the limitations of microbial isolation methods, the vast majority of these organisms would go unnoticed using traditional culturing techniques.

Community Metabolism

Many bacterial communities show significant division of labour in metabolism. Waste products of some organisms are metabolites for others.

Working together they turn raw resources into fully metabolized waste. Using comparative gene studies and expression experiments with microarrays or proteomics researchers can piece together a metabolic network that goes beyond species boundaries.

Such studies require detailed knowledge about which versions of which proteins are coded by which species and even by which strains of which species. Therefore, community genomic information is another fundamental part (as metabolomics or proteomics) to be able to estimate how metabolites are possibly transferred and transformed through a community.

Stem Cell

Stem cells are cells found in all multi cellular organisms. They are characterized by the ability to renew themselves through mitotic cell division and differentiate into a diverse range of specialized cell types. Research in the stem cell field grew out of findings by Ernest A. McCulloch and James E. Till at the University of Toronto in the 1960s.

The two broad types of mammalian stem cells are: embryonic stem cells that are isolated from the inner cell mass of blastocysts, and adult stem cells that are found in adult tissues. In a developing embryo, stem cells can differentiate into all of the specialized embryonic tissues. In adult organisms, stem cells and progenitor cells act as a repair system for the body, replenishing specialized cells, but also maintain the normal turnover of regenerative organs, such as blood, skin, or intestinal tissues.

Stem cells can now be grown and transformed into specialized cells with characteristics consistent with cells of various tissues such as muscles or nerves through cell culture. Highly plastic adult stem cells from a variety of sources, including umbilical cord blood and bone marrow, are routinely used in medical therapies. Embryonic cell lines and autologous embryonic stem cells generated through therapeutic cloning have also been proposed as promising candidates for future therapies.

Properties

The classical definition of a stem cell requires that it possess two properties:

- *Self-renewal* - the ability to go through numerous cycles of cell division while maintaining the undifferentiated state.
- *Potency* - the capacity to differentiate into specialized cell types. In the strictest sense, this requires stem cells to be either totipotent or pluripotent - to be able to give rise to any mature cell type, although multipotent or unipotent progenitor cells are sometimes referred to as stem cells.

Self-renewal

Two mechanisms exist to ensure that the stem cell population is maintained:

1. Obligatory asymmetric replication - a stem cell divides into one daughter cell that is identical to the original stem cell, and another daughter cell that is differentiated
2. Stochastic differentiation - when one stem cell develops into two differentiated daughter cells, another stem cell undergoes mitosis and produces two stem cells identical to the original.

Cell Potency

The potency of a cell specifies its differentiation potential, or potential to differentiate into different cell types.

Totipotency

Totipotency is the ability of a single cell to divide and produce all the differentiated cells in an organism, including extraembryonic tissues. Totipotent cells include spores and zygotes. In some organisms, cells can dedifferentiate and regain totipotency. For example, a plant cutting or callus can be used to grow an entire plant.

Human development begins when a sperm fertilizes an egg and creates a single totipotent cell (zygote). In the first hours after fertilization, this cell divides into identical totipotent cells, which can later develop into any of the three germ layers of a human (endoderm, mesoderm, or ectoderm) and into cells of the cytotrophoblast layer or syncytiotrophoblast layer of the placenta. After reaching the 16-cell stage, the totipotent cells of the morula differentiate into cells that will eventually become either the Blastocyst's Inner cell mass or outer trophoblasts. Approximately four days after fertilization and after several cycles of cell division, these totipotent cells begin to specialize. The inner cell mass, the source of embryonic stem cells, is pluripotent, not totipotent. Research on *Caenorhabditis elegans* suggests that multiple mechanisms including RNA regulation may play a role in maintain totipotency at different stages of development in some species.

Pluripotency

In cell biology, pluripotency refers to a stem cell that has the potential to differentiate into any of the three germ layers: endoderm (interior stomach lining, gastrointestinal tract, the lungs), mesoderm (muscle, bone, blood, urogenital), or ectoderm (epidermal tissues and nervous system). Pluripotent stem cells can give rise to any fetal or adult cell type. However, alone they cannot develop into a fetal or adult animal because they lack the potential to contribute to extraembryonic tissue, such as the yea placenta.

Induced Pluripotency

Induced pluripotent stem cells, commonly abbreviated as iPS cells are a type of pluripotent stem cell artificially derived from a non-pluripotent cell, typically an adult somatic cell, by inducing a "forced" expression of certain genes.

Multipotency

Multipotent progenitor cells have the potential to give rise to cells from multiple, but a limited number of lineages. An example of a multipotent stem cell is a hematopoietic cell — a blood stem cell that can develop into several types of blood cells, but cannot develop into brain cells or other types of cells. At the end of the long series of cell divisions that form the embryo are cells that are terminally differentiated, or that are considered to be permanently committed to a specific function.

Scientists have long held the opinion that differentiated cells cannot be altered or caused to behave in any way other than the way in which they have been naturally committed. New research, however, has called that assumption into question. In recent stem cell experiments, scientists have been able to persuade blood stem cells to behave like neurons, or brain cells- a process known as transdifferentiation. Scientists now believe that stem cell research could reveal far more vital information about our bodies than was previously known. There is also continuing research to see if it is possible to make multipotent cells into pluripotent cells.

The induced pluripotency of somatic cells into undifferentiated iPS cells was originally hailed as the end of the controversial use of embryonic stem cells. However, iPS cells are highly tumorigenic, and, despite advances, were never approved for clinical stage research in the United States.

It was recently determined that the somatic expression of combined transcription factors can directly induce other defined somatic cell fates; researchers identified three neural-lineage-specific transcription factors that could directly convert mouse fibroblasts (skin cells) into fully-functional neurons. This result challenges the terminal nature of cellular differentiation and the integrity of lineage commitment; and implies that with the proper tools, *all* cells are totipotent and may form all kinds of tissue.

An extremely rich source for mesenchymal stem cells is the developing tooth bud of the mandibular third molar. The stem cells eventually form enamel (ectoderm), dentin, dental pulp, blood vessels, and nervous tissues, including a minimum of 29 different unique end organs. Because of extreme ease in collection at 8–10 years of age before calcification and minimal to no morbidity will probably constitute a major source for personal banking, research and multiple therapies. These stem cells have been shown capable of producing hepatocytes.

Oligopotency

In biology, oligopotency is the ability of progenitor cells to differentiate into a few cell types. It is a degree of potency. Examples of oligopotent stem cells are the lymphoid or myeloid stem cells. Examples of progenitor cells are Vascular Stem Cells which have the capacity to become both Endothelial or smooth muscle cells.

Unipotency

In cell biology, a unipotent cell is one that has the capacity to develop/ differentiate into only one type of tissue/cell type. The most common of these in humans are skin cells. This cell has a unique property: self-renewal. This property distinguishes it from most other terminally differentiated non-stem cells. Hepatocytes, which constitute most of the Cytoplasmic mass of the human liver, are unipotent. The liver's ability to regenerate from as little as 25% of its original mass is attributed to this property. A close synonym for *unipotent cell* is *precursor cell.*

Identification

The practical definition of a stem cell is the functional definition - a cell that has the potential to regenerate tissue over a lifetime. For example, the gold standard test for a bone marrow or hematopoietic stem cell (HSC) is the ability to transplant one cell and save an individual without HSCs. In this case, a stem cell must be able to produce new blood cells and immune cells over a long term, demonstrating potency. It should also be possible to isolate stem cells from the transplanted individual, which can themselves be transplanted into another individual without HSCs, demonstrating that the stem cell was able to self-renew.

Properties of stem cells can be illustrated *in vitro*, using methods such as clonogenic assays, where single cells are characterized by their ability to differentiate and self-renew. As well, stem cells can be isolated based on a distinctive set of cell surface markers. However, *in vitro* culture conditions can alter the Behaviour of cells, making it unclear whether the cells will behave in a similar manner *in vivo*. Considerable debate exists whether some proposed adult cell populations are truly stem cells.

Embryonic Stem Cell

Embryonic stem cells (ES cells) are pluripotent stem cells derived from the inner cell mass of the blastocyst, an early-stage embryo. Human embryos reach the blastocyst stage 4–5 days post fertilization, at which time they consist of 50–150 cells. Isolating the embryoblast or inner cell mass (ICM) results in the death of the fertilized human embryo, which raises ethical issues.

Characteristics

Embryonic stem cells are distinguished by two distinctive properties:

- their pluripotency, and
- their ability to replicate indefinitely.

ES cells are pluripotent, that is, they are able to differentiate into all derivatives of the three primary germ layers: ectoderm, endoderm, and mesoderm. These include each of the more than 220 cell types in the adult body. Pluripotency distinguishes embryonic stem cells from adult stem cells found in adults; while embryonic stem cells can generate all cell types in the body, adult stem cells are multipotent and can only produce a limited number of cell types.

Additionally, under defined conditions, embryonic stem cells are capable of propagating themselves indefinitely. This allows embryonic stem cells to be employed as useful tools for both research and regenerative medicine, because they can produce limitless numbers of themselves for continued research or clinical use. Because of their plasticity and potentially unlimited capacity for self-renewal, ES cell therapies have been proposed for regenerative medicine and tissue replacement after injury or disease. Diseases that could potentially be treated by pluripotent stem cells include a number of blood and immune-system related genetic diseases, cancers, and disorders; juvenile diabetes; Parkinson's; blindness and spinal cord injuries. Besides the ethical concerns of stem cell therapy, there is a technical problem of graft-versus-host disease associated with allogeneic stem cell transplantation. However, these problems associated with histocompatibility may be solved using autologous donor adult stem cells, therapeutic cloning, stem cell banks or more recently by reprogramming of somatic cells with defined factors (e.g. induced pluripotent stem cells). Other potential uses of embryonic stem cells include investigation of early human development, study of genetic disease and as in vitro systems for toxicology testing.

Research History and Developments

In 1964, researchers isolated a single type of cell from a teratocarcinoma, a tumour now known to be derived from a germ cell. These cells isolated from the teratocarcinoma replicated and grew in cell culture as a stem cell and are now known as embryonic carcinoma (EC) cells. Although similarities in morphology and differentiating potential (pluripotency) led to the use of EC cells as the *in vitro* model for early mouse development, EC cells harbor genetic mutations and often abnormal karyotypes that accumulated during the development of the teratocarcinoma. These genetic aberrations further emphasized the need to be able to culture pluripotent cells directly from the inner cell mass.

In 1981, embryonic stem cells (ES cells) were independently first derived from mouse embryos by two groups. Martin Evans and Matthew Kaufman from the Department of Genetics, University of Cambridge published first in July, revealing a new technique for culturing the mouse embryos in the uterus to allow for an increase in cell number, allowing for the derivation of ES cells from these embryos.

Gail R. Martin, from the Department of Anatomy, University of California, San Francisco, published her paper in December and coined the term "Embryonic Stem Cell". She showed that embryos could be cultured *in vitro* and that ES cells could be derived from these embryos. In 1998, a breakthrough occurred when researchers, led by James Thomson at the University of Wisconsin-Madison, first developed a technique to isolate and grow human embryonic stem cells in cell culture.

Techniques and Conditions for Embryonic Stem Cell Derivation and Culture

Embryonic stem cells are derived from the inner cell mass of the early embryo, which are harvested from the donor mother animal. Martin Evans and Matthew Kaufman reported a technique that delays embryo implantation, allowing the inner cell mass to increase. This process includes removing the donor mother's ovaries and dosing her with progesterone, changing the hormone environment, which causes the embryos to remain free in the uterus.

After 4–6 days of this intrauterine culture, the embryos are harvested and grown in *in vitro* culture until the inner cell mass forms "egg cylinder-like structures," which are dissociated into single cells, and plated on fibroblasts treated with mitomycin-c (to prevent fibroblast mitosis). Clonal cell lines are created by growing up a single cell. Evans and Kaufman showed that the cells grown out from these cultures could form teratomas and embryoid bodies, and differentiate *in vitro,* which all indicate the cells are pluripotent.

Gail Martin derived and cultured her ES cells differently. She removed the embryos from the donor mother at approximately 76 hours after copulation and cultured them overnight in media containing serum.

The following day, she removed the inner cell mass from the late blastocyst using microsurgery. The extracted inner cell mass was cultured on fibroblasts treated with mitomycin-c in media that containing serum and was conditioned by EC cells. After approximately one week, colonies of cells grew out. These cells grew in culture and demonstrated pluripotent characteristics, as demonstrated by the ability to form teratomas, differentiate *in vitro,* and form embryoid bodies. Martin referred to these cells as ES cells.

It is now known that the feeder cells provide leukemic inhibitory factor (LIF) and serum provides bone morphogenetic proteins (BMPs) that are necessary to prevent ES cells from differentiating. These factors are extremely important for the efficiency of deriving ES cells.

Furthermore, it has been demonstrated that different mouse strains have different efficiencies for isolating ES cells. Current uses for mouse ES cells include the generation of transgenic mice, including knockout mice. For human treatment, there is a need for patient specific pluripotent cells. Generation of human ES cells is more difficult and faces ethical issues. So, in addition to human ES cell research, many groups are focused on the generation of induced pluripotent stem cells (iPS cells).

Contamination by Reagents used in Cell Culture

The online edition of *Nature Medicine* published a study on January 24, 2005 which stated that the human embryonic stem cells available for federally funded research are contaminated with non-human molecules from the culture medium used to grow the cells. It is a common technique to use mouse cells and other animal cells to maintain the pluripotency of actively dividing stem cells. The problem was discovered when non-human sialic acid in the growth media was found to compromise the potential uses of the embryonic stem cells in humans, according to scientists at the University of California, San Diego.

However, a study published in the online edition of *Lancet Medical Journal* on March 8, 2005 detailed information about a new stem cell line which was derived from human embryos under completely cell- and serum-free conditions. After more than 6 months of undifferentiated proliferation, these cells demonstrated the potential to form derivatives of all three embryonic germ layers both *in vitro* and in teratomas. These properties were also successfully maintained (for more than 30 passages) with the established stem cell lines.

Potential Clinical Use

Reducing Donor-host Rejection

There is also ongoing research to reduce the potential for rejection of the differentiated cells derived from ES cells once researchers are capable of creating an approved therapy from ES cell research. One of the possibilities to prevent rejection is by creating embryonic stem cells that are genetically identical to the patient via therapeutic cloning. An alternative solution for rejection by the patient to therapies derived from non-cloned ES cells is to derive many well-characterized ES cell lines from different genetic backgrounds and use the cell line that is most similar to the patient; treatment can then be tailored to the patient, minimizing the risk of rejection.

Safety: Reducing the Risk of Teratoma and other Cancers as a Side Effect

- The major concern with the possible transplantation of ESC into patients as therapies is their ability to form tumours including teratoma. Safety issues prompted the FDA to place a hold on the first ESC clinical trial, however no tumours were observed.
- The main strategy to enhance the safety of ESC for potential clinical use is to differentiate the ESC into specific cell types (e.g. neurons, muscle, liver cells) that have reduced or eliminated ability to cause tumours. Following differentiation, the cells are subjected to sorting by flow cytometry for further purification. While ESC are predicted to be inherently safer than IPS cells because they are not genetically modified with genes such as c-Myc that are linked to cancer. Nonetheless ESC express very high levels of the iPS inducing genes and these genes including Myc are essential for ESC self-renewal and pluripotency, and potential strategies to improve safety by eliminating Myc expression are unlikely to preserve the cells' "stemness".

First Clinical Trial

On January 23, 2009, Phase I clinical trials for transplantation of oligodendrocytes (a cell type of the brain and spinal cord) derived from human ES cells into spinal cord-injured individuals received approval from the U.S. Food and Drug Administration (FDA), marking it the world's first human ES cell human trial. The study leading to this scientific advancement was conducted by Hans Keirstead and colleagues at the University of California, Irvine and supported by Geron Corporation of Menlo Park, CA. A previous experiment had shown an improvement in locomotor recovery in spinal cord-injured rats after a 7-day delayed transplantation of human ES cells that had been pushed into an oligodendrocytic lineage. In the proposed phase I clinical study, about eight to ten paraplegics who have had their injuries no longer than two weeks before the trial begins, will be selected, since the cells must be injected before scar tissue is able to form.

However, the researchers are emphasizing that the injections are not expected to fully cure the patients and restore all mobility. Based on the results of the rodent trials, researchers say restoration of myelin sheathes, and an increase in mobility is probable. This first trial is mainly testing the safety of these procedures and if everything goes well, it could lead to future studies that involve people with more severe disabilities.

The trial had been put on hold in August 2009 due to concerns made by the FDA regarding a small number of microscopic cysts found in several treated rat models but the hold has been lifted as of July 30, 2010. In October 2010 researchers enrolled and administered ESTs to the first patient at Shepherd Centre in Atlanta. The makers of the stem cell therapy, Geron Corporation, estimate that it will take several months for the stem cells to replicate and for the GRNOPC1 therapy to be evaluated for success or failure.

Potential Method for New Cell Line Derivation

On August 23, 2006, the online edition of *Nature* scientific journal published a letter by Dr. Robert Lanza (medical director of Advanced Cell Technology in Worcester, MA) stating that his team had found a way to extract embryonic stem cells without destroying the actual embryo. This technical achievement would potentially enable scientists to work with new lines of embryonic stem cells derived using public funding in the USA, where federal funding was at the time limited to research using embryonic stem cell lines derived prior to August 2001.

In March, 2009, the limitation was lifted. Recently, it was shown that pluripotent stem cells highly similar to embryonic stem cells can be generated by the delivery of three genes (*Oct4*, *Sox2*, and *Klf4*) to differentiated cells.

The delivery of these genes "reprograms" differentiated cells into pluripotent stem cells, allowing for the generation of pluripotent stem cells without the embryo. Because ethical concerns regarding embryonic stem cells typically are about their derivation from terminated embryos, it is believed that reprogramming to these "induced pluripotent stem cells" (iPS cells) may be less controversial. Both human and mouse cells can be reprogrammed by this methodology, generating both human pluripotent stem cells and mouse pluripotent stem cells without an embryo.

This may enable the generation of patient specific ES cell lines that could potentially be used for cell replacement therapies. In addition, this will allow the generation of ES cell lines from patients with a variety of genetic diseases and will provide invaluable models to study those diseases. However, as a first indication that the induced pluripotent stem cell (iPS) cell technology can in rapid succession lead to new cures, it was used by a research team headed by Rudolf Jaenisch of the Whitehead Institute for Biomedical Research in Cambridge, Massachusetts, to cure mice of sickle cell anemia, as reported by *Science* journal's online edition on December 6, 2007.

On January 16, 2008, a California based company, Stemagen, announced that they had created the first mature cloned human embryos from single skin cells taken from adults. These embryos can be harvested for patient matching embryonic stem cells.

Use of Human Embryonic Stem Cells as Models for Human Genetic Disorders

In recent years there have been several reports regarding the potential use of human embryonic stem cells as models for human genetic diseases. This issue is especially important due to the species-specific nature of many genetic disorders. The relative inaccessibility of human primary tissue for research is another major hindrance.

Several new studies have started to address this issue. This has been done either by genetically manipulating the cells, or more recently by deriving diseased cell lines identified by prenatal genetic diagnosis (PGD). This approach may very well prove invaluable at studying disorders such as Fragile-X syndrome, Cystic fibrosis, and other genetic maladies that have no reliable model system.

Yury Verlinsky, a Russian-American medical researcher who specialized in embryo and cellular genetics (genetic cytology), developed prenatal diagnosis testing methods to determine genetic and chromosomal disorders a month and a half earlier than standard amniocentesis. The techniques are now used by many pregnant women and prospective parents, especially those couples with a history of genetic abnormalities or where the woman is over the age of 35, when the risk of genetically-related disorders is higher. In addition, by allowing parents to select an embryo without genetic disorders, they have the potential of saving the lives of siblings that already had similar disorders and diseases using cells from the disease free offspring.

Fetal

Fetal stem cells are primitive cell types found in the organs of fetuses.

Adult

Adult Stem Cell

Adult stem cells are undifferentiated cells, found throughout the body after embryonic development, that multiply by cell division to replenish dying cells and regenerate damaged tissues. Also known as somatic stem cells, they can be found in juvenile as well as adult animals and humans. Scientific interest in adult stem cells has centered on their ability to divide or *self-renew* indefinitely, and generate all the cell types of the organ from which they originate, potentially regenerating the entire organ from a few cells. Unlike embryonic stem cells, the use of adult stem cells in research and therapy is not considered to be controversial as they are derived from adult tissue samples rather than destroyed human embryos. They have mainly been studied in humans and model organisms such as mice and rats.

Properties

Defining Properties: The rigorous definition of a stem cell requires that it possesses two properties:

- *Self-renewal* which is the ability to go through numerous cycles of cell division while still maintaining its undifferentiated state.
- *multipotency* or *multidifferentiative potential* which is the ability to generate progeny of several distinct cell types, (for example glial cells and neurons) as opposed to unipotency which is the term for cells that are restricted to producing a single-cell type. However, some researchers do not consider multipotency to be essential, and believe that unipotent self-renewing stem cells can exist.

These properties can be illustrated with relative ease *in vitro*, using methods such as clonogenic assays, where the progeny of a single cell is characterized, however, it is known that *in vitro* cell culture conditions can alter the Behaviour of cells.

Proving that a particular subpopulation of cells possesses stem cell properties *in vivo* is challenging, and so considerable debate exists as to whether some proposed stem cell populations in the adult are indeed stem cells.

Lineage

To ensure self-renewal, stem cells undergo two types of cell division. Symmetric division gives rise to two identical daughter cells, both endowed with stem cell properties, whereas asymmetric division produces only one stem cell and a progenitor cell with limited self-renewal potential. Progenitors can go through several rounds of cell division before finally differentiating into a mature cell. It is believed that the molecular distinction between symmetric and asymmetric divisions lies in differential segregation of cell membrane proteins (such as receptors) between the daughter cells.

Multidrug Resistance

Adult stem cells express transporters of the ATP-binding cassette family that actively pump a diversity of organic molecules out of the cell. Many pharmaceuticals are exported by these transporters conferring multidrug resistance onto the cell. This complicates the design of drugs, for instance neural stem cell targeted therapies for the treatment of clinical depression.

Signalling Pathways

Adult stem cell research has been focused on uncovering the general molecular mechanisms that control their self-renewal and differentiation.

Notch : The Notch pathway has been known to developmental biologists for decades. Its role in control of stem cell proliferation has now been demonstrated for several cell types including haematopoietic, neural and mammary stem cells.

Wnt : These developmental pathways are also strongly implicated as stem cell regulators.

Plasticity/Adult Stem Cell Pluripotency

Discoveries in recent years have suggested that adult stem cells might have the ability to differentiate into cell types from different germ layers. For instance, neural stem cells from the brain, which are derived from ectoderm, can differentiate into ectoderm, mesoderm and endoderm. Stem cells from the bone marrow, which is derived from mesoderm, can differentiate into liver, lung, GI tract and skin, which are derived from endoderm and mesoderm.

This phenomenon is referred to as stem cell transdifferentiation or plasticity. It can be induced by modifying the growth medium when stem cells are cultured *in vitro* or transplanting them to an organ of the body different from the one they were originally isolated from.

There is yet no consensus among biologists on the prevalence and physiological and therapeutic relevance of stem cell plasticity. More recent findings suggest that pluripotent stem cells may reside in adult tissues in a dormant state.

These cells are referred to as "very small embryonic like" -"VSEL" stem cells, and display pluripotency in vitro. As VSEL cells are present in virtually all adult tissues, including lung, brain, kidneys, muscles, and pancreas Co-purification of VSEL cells with other populations of adult stem cells may explain the apparent pluripotency of adult stem cell populations.

Types

Hematopoietic Stem Cells

Hematopoietic stem cells are found in the bone marrow and give rise to all the blood cell types.

Mammary Stem Cells

Mammary stem cells provide the source of cells for growth of the mammary gland during puberty and gestation and play an important role in carcinogenesis of the breast. Mammary stem cells have been isolated from human and mouse tissue as well as from cell lines derived from the mammary gland. Single such cells can give rise to both the luminal and myoepithelial cell types of the gland, and have been shown to have the ability to regenerate the entire organ in mice.

Mesenchymal Stem Cells

Mesenchymal stem cells (MSCs) are of stromal origin and may differentiate into a variety of tissues. MSCs have been isolated from placenta, adipose tissue, lung, bone marrow and blood, Wharton's jelly from the umbilical cord, and teeth (perivascular niche of dental pulp and periodontal ligament). MSCs are attractive for clinical therapy due to their ability to differentiate, provide trophic support, and modulate innate immune response.

Endothelial Stem Cells

Neural Stem Cells: The existence of stem cells in the adult brain has been postulated following the discovery that the process of neurogenesis, the birth of new neurons, continues into adulthood in rats. The presence of stem cells in the mature primate brain was first reported in 1967. It has since been shown that new neurons are generated in adult mice, songbirds and primates, including humans. Normally, adult neurogenesis is restricted to two areas of the brain - the subventricular zone, which lines the lateral ventricles, and the dentate gyrus of the hippocampal formation. Although the generation of new neurons in the hippocampus is well established, the presence of true self-renewing stem cells there has been debated. Under certain circumstances, such as following tissue damage in ischemia, neurogenesis can be induced in other brain regions, including the neocortex.

Neural stem cells are commonly cultured *in vitro* as so called neurospheres - floating heterogeneous aggregates of cells, containing a large proportion of stem cells. They can be propagated for extended periods of time and differentiated into both neuronal and glia cells, and therefore behave as stem cells. However, some recent studies suggest that this behaviour is induced by the culture conditions in progenitor cells, the progeny of stem cell division that normally undergo a strictly limited number of replication cycles *in vivo*. Furthermore, neurosphere-derived cells do not behave as stem cells when transplanted back into the brain.

Neural stem cells share many properties with haematopoietic stem cells (HSCs). Remarkably, when injected into the blood, neurosphere-derived cells differentiate into various cell types of the immune system.

Olfactory Adult Stem Cells: Olfactory adult stem cells have been successfully harvested from the human olfactory mucosa cells, which are found in the lining of the nose and are involved in the sense of smell. If they are given the right chemical environment these cells have the same ability as embryonic stem cells to develop into many different cell types. Olfactory stem cells hold the potential for therapeutic applications and, in contrast to neural stem cells, can be harvested with ease without harm to the patient. This means they can be easily obtained from all individuals, including older patients who might be most in need of stem cell therapies.

Neural Crest Stem Cells: Hair follicles contain two types of stem cells, one of which appears to represent a remnant of the stem cells of the embryonic neural crest. Similar cells have been found in the gastrointestinal tract, sciatic nerve, cardiac outflow tract and spinal and sympathetic ganglia. These cells can generate neurons, Schwann cells, myofibroblast, chondrocytes and melanocytes.

Testicular Cells

Multipotent stem cells with a claimed equivalency to embryonic stem cells have been derived from spermatogonial progenitor cells found in the testicles of laboratory mice by scientists in Germany and the United States, and, a year later, researchers from Germany and the United Kingdom confirmed the same capability using cells from the testicles of humans. The extracted stem cells are known as human adult germline stem cells (GSCs).

Multipotent stem cells have also been derived from germ cells found in human testicles.

Adult Stem Cell Therapies

The therapeutic potential of adult stem cells is the focus of much scientific research, due to their ability to be harvested from the patient. In common with embryonic stem cells, adult stem cells have the ability to differentiate into more than one cell type, but unlike the former they are often restricted to certain types or "lineages". The ability of a differentiated stem cell of one lineage to produce cells of a different lineage is called transdifferentiation. Some types of adult stem cells are more capable of transdifferentiation than others, and for many there is no evidence that such a transformation is possible. Consequently, adult stem therapies require a stem cell source of the specific lineage needed, and harvesting and/or culturing them up to the numbers required is a challenge.

Sources

Pluripotent stem cells, i.e. cells that can give rise to any fetal or adult cell type, can be found in a number of tissues, including umbilical cord blood. Using genetic reprogramming, pluripotent stem cells equivalent to embryonic stem cells have been derived from human adult skin tissue. Other adult stem cells are multipotent, meaning they are restricted in the types of cell they can become, and are generally referred to by their tissue origin (such as mesenchymal stem cell, adipose-derived stem cell, endothelial stem cell, etc.). A great deal of adult stem cell research has focused on investigating their capacity to divide or self-renew indefinitely, and their potential for differentiation. In mice, pluripotent stem cells can be directly generated from adult fibroblast cultures.

Clinical Applications

Adult stem cell treatments have been used for many years to successfully treat leukemia and related bone/blood cancers utilizing bone marrow transplants. The use of adult stem cells in research and therapy is not considered as controversial as the use of embryonic stem cells, because the production of adult stem cells does not require the destruction of an embryo. Consequently, the majority of US government funding provided for research in this field is restricted to supporting adult stem cell research.

Early regenerative applications of adult stem cells has focused on intravenous delivery of blood progenitors known as Hematopetic Stem Cells (HSC's). Other early commercial applications have focused on Mesenchymal Stem Cells (MSC's). For both cell lines, direct injection or placement of cells into a site in need of repair may be the preferred method of treatment, as vascular delivery suffers from a "pulmonary first pass effect" where intravenous injected cells are sequestered in the lungs.

Clinical case reports in orthopedic applications have been published. Wakitani has published a small case series of nine defects in five knees involving surgical transplantation of mesenchymal stem cells with coverage of the treated chondral defects. Centeno et al. have reported high field MRI evidence of increased cartilage and meniscus volume in individual human clinical subjects.

Many other stem cell based treatments are operating outside the US, with much controversy being reported regarding these treatments as some feel more regulation is needed as clinics tend to exaggerate claims of success and minimize or omit risks.

First Transplanted Human Organ Grown from Adult Stem Cells

In 2008 the first full transplant of a human organ grown from adult stem cells was carried out by Paolo Macchiarini, at the Hospital Clínic of Barcelona on Claudia Castillo, a Colombian female adult whose trachea had collapsed due to tuberculosis.

Researchers from the University of Padua, the University of Bristol, and Politecnico di Milano harvested a section of trachea from a donor and stripped off the cells that could cause an immune reaction, leaving a grey trunk of cartilage. This section of trachea was then "seeded" with stem cells taken from Ms. Castillo's bone marrow and a new section of trachea was grown in the laboratory over four days. The new section of trachea was then transplanted into the left main bronchus of the patient. Because the stem cells were harvested from the patient's own bone marrow Professor Macchiarini did not think it was necessary for her to be given anti-rejection (immunosuppressive) medication and when the procedure was reported four months later in *The Lancet*, the patient's immune system was showing no signs of rejecting the transplant.

Adult Stem Cells and Cancer

In recent years, acceptance of the concept of adult stem cells has increased. There is now a hypothesis that stem cells reside in many adult tissues and that these unique reservoirs of cells are not only responsible for the normal reparative and regenerative processes, but are also considered to be a prime target for genetic and epigenetic changes, culminating in many abnormal conditions including cancer.

Amniotic

Multipotent stem cells are also found in amniotic fluid. These stem cells are very active, expand extensively without feeders and are not tumorigenic. Amniotic stem cells are multipotent and can differentiate in cells of adipogenic, osteogenic, myogenic, endothelial, hepatic and also neuronal lines. All over the world, universities and research institutes are studying amniotic fluid to discover all the qualities of amniotic stem cells, and scientists such as Anthony Atala and Giuseppe Simoni have discovered important results.

From an ethical point of view, stem cells from amniotic fluid can solve a lot of problems, because it's possible to catch amniotic stem cells without destroying embryos. For example, the Vatican newspaper "Osservatore Romano" called amniotic stem cell "the future of medicine". It's possible to collect amniotic stem cells for donors or for autologuous use: the first US amniotic stem cells bank opened in 2009 in Medford, MA, by Biocell Centre Corporation and collaborates with various hospitals and universities all over the world.

Induced Pluripotent Stem Cell

Induced pluripotent stem cells, commonly abbreviated as iPS cells or iPSCs are a type of pluripotent stem cell artificially derived from a non-pluripotent cell, typically an adult somatic cell, by inducing a "forced" expression of specific genes. Induced Pluripotent Stem Cells are similar to natural pluripotent stem cells, such as embryonic stem (ES) cells, in many respects, such as the expression of certain stem cell genes and proteins, chromatin methylation patterns, doubling time, embryoid body formation, teratoma formation, viable chimera formation, and potency and differentiability, but the full extent of their relation to natural pluripotent stem cells is still being assessed. iPSCs were first produced in 2006 from mouse cells and in 2007 from human cells.

This has been cited as an important advance in stem cell research, as it may allow researchers to obtain pluripotent stem cells, which are important in research and potentially have therapeutic uses, without the controversial use of embryos. They also avoid the issue of graft-versus-host disease and immune rejection unlike embryonic stem cells because they are derived entirely from the patient. Depending on the methods used, reprogramming of adult cells to obtain iPSCs may pose significant risks that could limit its use in humans.

For example, if viruses are used to genomically alter the cells, the expression of cancer-causing genes or oncogenes may potentially be triggered. In February 2008, in ground-breaking findings published in the journal Cell, scientists announced the discovery of a technique that could remove oncogenes after the induction of pluripotency, thereby increasing the potential use of iPS cells in human diseases. In April 2009, Sheng Ding and colleagues in La Jolla, California, demonstrated that generation of iPS cells is possible without any genetic alteration of the adult cell: A repeated treatment of the cells with certain proteins channeled into the cells via poly-arginine anchors was sufficient to induce pluripotency. The acronym given for those iPSCs is piPSCs (protein-induced pluripotent stem cells).

Production of iPSCs

iPS cells are typically derived by transfection of certain stem cell-associated genes into non-pluripotent cells (although this technique is becoming less popular since it is known to be prone to cancer formation), such as adult fibroblasts. Transfection is typically achieved through viral vectors, such as retroviruses. Transfected genes include the master transcriptional regulators Oct-3/4 (Pou5f1) and Sox2, although it is suggested that other genes enhance the efficiency of induction. After 3–4 weeks, small numbers of transfected cells begin to become morphologically and biochemically similar to pluripotent stem cells, and are typically isolated through morphological selection, doubling time, or through a reporter gene and antibiotic selection.

First Generation

Induced pluripotent stem cells were first generated by Shinya Yamanaka's team at Kyoto University, Japan in 2006. Yamanaka used genes that had been identified as particularly important in embryonic stem cells (ESCs), and used retroviruses to transfect mouse fibroblasts with a selection of those genes. Eventually, four key pluripotency genes essential for the production of pluripotent stem cells were isolated; Oct-3/4, SOX2, c-Myc, and Klf4. Cells were isolated by antibiotic selection of Fbx15 cells. However, this iPS cell line showed DNA methylation errors compared to original patterns in ESC lines and failed to produce viable chimeras if injected into developing embryos.

Second Generation in Mice

In June 2007, the same group published a breakthrough study along with two other independent research groups from Harvard, MIT, and the University of California, Los Angeles, showing successful reprogramming of mouse fibroblasts into iPS cells and even producing viable chimera. These cell lines were also derived from mouse fibroblasts by retroviral mediated reactivation of the same four endogenous pluripotent factors, but the researchers now selected a different marker for detection. Instead of Fbx15, they used Nanog which is an important gene in ESCs. DNA methylation patterns and production of viable chimeras (and thereby contributing to subsequent germ-line production) indicated that Nanog is a major determinant of cellular pluripotency.

Unfortunately, one of the four genes used (namely, c-Myc) is oncogenic, and 20% of the chimeric mice developed cancer. In a later study, Yamanaka reported that one can create iPSCs even without c-Myc. The process takes longer and is not as efficient, but the resulting chimeras didn't develop cancer.

Two-fathered Mice

Reproductive scientists in University of Texas MD Anderson Cancer Centre have created mice with nuclear DNA (nDNA) solely from two fathers, using iPS technology. Foetal fibroblasts from one father (XY) were cultivated and one percent of the resultant cells had spontaneously lost a Y-chromosome; like an individual with Turner Syndrome (X0). These cells were injected in female blastocysts (XX), which gestated in surrogate mothers to form female chimeras (X0/XX). When these mated with male mice (XY). Some of the offspring had nDNA from the original father and also from the mated male but not from the female blastocysts or the surrogate mother. Both male and female two-fathered mice were viable.

Human Induced Pluripotent Stem Cells

In November 2007, a milestone was achieved by creating iPSCs from adult human cells; two independent research teams' studies were released - one in *Science* by James Thomson at University of Wisconsin–Madison and another in *Cell* by Shinya Yamanaka and colleagues at Kyoto University, Japan. With the same principle used earlier in mouse models, Yamanaka had successfully transformed human fibroblasts into pluripotent stem cells using the same four pivotal genes: Oct3/4, Sox2, Klf4, and c-Myc with a retroviral system. Thomson and colleagues used OCT4, SOX2, NANOG, and a different gene LIN28 using a lentiviral system. The viral transfection systems used to insert the genes at random locations in the host's genome created concern for potential therapeutic applications of these iPSCs, because the created cells might be prone to form tumours. Members of both teams considered it therefore necessary to develop new delivery methods.

To overcome these dangers, Konrad Hochedlinger and his Harvard University research team successfully used an adenovirus to transport the requisite four genes into the DNA of skin and liver cells of mice, resulting in cells identical to embryonic stem cells. Since the adenovirus does not combine any of its own genes with the targeted host, the danger of creating tumours is eliminated, although the Hochedlinger method has not yet been tested on human cells. Yamanaka has since demonstrated reprogramming can be accomplished via plasmid without any virus transfection system at all, although at very low efficiencies.

Teams led by Keisuke Kaji, of the University of Edinburgh, and Andras Nagy, of the University of Toronto, have now collaborated to develop a new approach to creating IPS cells that does not involve viruses.

The procedure might eventually prove itself suitable for therapeutic use. Recently, in May 2009, a team of scientists generated human iPS cells by direct delivery of proteins, thus eliminating the need for viruses or genetic modification. The group of Sheng Ding in La Jolla, California, has also shown that the generation of mouse iPS cells was possible using a similar methodology: a repeated treatment of the cells with certain proteins channeled into the cells via poly-arginine anchors was sufficient to induce pluripotency. The expression of pluripotency induction genes can also be increased by treating somatic cells with FGF2 under low oxygen conditions.

Genes of Induction

The generation of iPS cells is crucial on the genes used for the induction.

Oct-3/4 and certain members of the Sox gene family (Sox1, Sox2, Sox3, and Sox15) have been identified as crucial transcriptional regulators involved in the induction process whose absence makes induction impossible. Additional genes, however, including certain members of the Klf family (Klf1, Klf2, Klf4, and Klf5), the Myc family (C-myc, L-myc, and N-myc), Nanog, and LIN28, have been identified to increase the induction efficiency.

- Oct-3/4 (Pou5f1): Oct-3/4 is one of the family of octamer ("Oct") transcription factors, and plays a crucial role in maintaining pluripotency. The absence of Oct-3/4 in Oct-3/4 cells, such as blastomeres and embryonic stem cells, leads to spontaneous trophoblast differentiation, and presence of Oct-3/4 thus gives rise to the pluripotency and differentiation potential of embryonic stem cells. Various other genes in the "Oct" family, including Oct-3/4's close relatives, Oct1 and Oct6, fail to elicit induction, thus demonstrating the exclusiveness of Oct-3/4 to the induction process.

- Sox family: The Sox family of genes is associated with maintaining pluripotency similar to Oct-3/4, although it is associated with multipotent and unipotent stem cells in contrast with Oct-3/4, which is exclusively expressed in pluripotent stem cells. While Sox2 was the initial gene used for induction by Yamanaka et al., Jaenisch et al., and Thomson et al., other genes in the Sox family have been found to work as well in the induction process. Sox1 yields iPS cells with a similar efficiency as Sox2, and genes Sox3, Sox15, and Sox18 also generate iPS cells, although with decreased efficiency.
- Klf family: Klf4 of the Klf family of genes was initially identified by Yamanaka et al. and confirmed by Jaenisch et al. as a factor for the generation of mouse iPS cells and was demonstrated by Yamanaka et al. as a factor for generation of human iPS cells. However, Thomson et al. reported that Klf4 was unnecessary for generation of human iPS cells and in fact failed to generate human iPS cells. Klf2 and Klf4 were found to be factors capable of generating iPS cells, and related genes Klf1 and Klf5 did as well, although with reduced efficiency.
- Myc family: The Myc family of genes are proto-oncogenes implicated in cancer. Yamanaka et al. and Jaenisch et al. demonstrated that c-myc is a factor implicated in the generation of mouse iPS cells and Yamanaka et al. demonstrated it was a factor implicated in the generation of human iPS cells. However, Thomson et al., Yamanaka et al., and unpublished work by Johns Hopkins University reported that c-myc was unnecessary for generation of human iPS cells. Usage of the "myc" family of genes in induction of iPS cells is troubling for the eventuality of iPS cells as clinical therapies, as 25% of mice transplanted with c-myc-induced iPS cells developed lethal teratomas. N-myc and L-myc have been identified to induce instead of c-myc with similar efficiency.
- Nanog: In embryonic stem cells, Nanog, along with Oct-3/4 and Sox2, is necessary in promoting pluripotency. Therefore, it was surprising when Yamanaka et al. reported that Nanog was unnecessary for induction although Thomson et al. has reported it is possible to generate iPS cells with Nanog as one of the factors.
- LIN28: LIN28 is an mRNA binding protein expressed in embryonic stem cells and embryonic carcinoma cells associated with differentiation

and proliferation. Thomson et al. demonstrated it is a factor in iPSC generation, although it is unnecessary.

Identity

The generated iPSCs were remarkably similar to naturally-isolated pluripotent stem cells (such as mouse and human embryonic stem cells, mESCs and hESCs, respectively) in the following respects, thus confirming the identity, authenticity, and pluripotency of iPSCs to naturally-isolated pluripotent stem cells:

- Cellular biological properties
  - Morphology: iPSCs were morphologically similar to ESCs. Each cell had round shape, large nucleolus and scant cytoplasm. Colonies of iPSCs were also similar to that of ESCs. Human iPSCs formed sharp-edged, flat, tightly-packed colonies similar to hESCs and mouse iPSCs formed the colonies similar to mESCs, less flat and more aggregated colonies than that of hESCs.
  - Growth properties: Doubling time and mitotic activity are cornerstones of ESCs, as stem cells must self-renew as part of their definition. iPSCs were mitotically active, actively self-renewing, proliferating, and dividing at a rate equal to ESCs.
  - Stem cell markers: iPSCs expressed cell surface antigenic markers expressed on ESCs. Human iPSCs expressed the markers specific to hESC, including SSEA-3, SSEA-4, TRA-1-60, TRA-1-81, TRA-2-49/6E, and Nanog. Mouse iPSCs expressed SSEA-1 but not SSEA-3 nor SSEA-4, similarly to mESCs.
  - Stem Cell Genes: iPSCs expressed genes expressed in undifferentiated ESCs, including Oct-3/4, Sox2, Nanog, GDF3, REX1, FGF4, ESG1, DPPA2, DPPA4, and hTERT.
  - Telomerase activity: Telomerases are necessary to sustain cell division unrestricted by the Hayflick limit of ~50 cell divisions. hESCs express high telomerase activity to sustain self-renewal and proliferation, and iPSCs also demonstrate high telomerase activity and express hTERT (human telomerase reverse transcriptase), a necessary component in the telomerase protein complex.
- Pluripotency: iPSCs were capable of differentiation in a fashion similar to ESCs into fully differentiated tissues.
  - Neural differentiation: iPSCs were differentiated into neurons, expressing βIII-tubulin, tyrosine hydroxylase, AADC, DAT, ChAT, LMX1B, and MAP2. The presence of catecholamine-associated

enzymes may indicate that iPSCs, like hESCs, may be differentiable into dopaminergic neurons. Stem cell-associated genes were downregulated after differentiation.

- o Cardiac differentiation: iPSCs were differentiated into cardiomyocytes that spontaneously began beating. Cardiomyocytes expressed TnTc, MEF2C, MYL2A, MYHCβ, and NKX2.5. Stem cell-associated genes were downregulated after differentiation.
- o Teratoma formation: iPSCs injected into immunodeficient mice spontaneously formed teratomas after nine weeks. Teratomas are tumours of multiple lineages containing tissue derived from the three germ layers endoderm, mesoderm and ectoderm; this is unlike other tumours, which typically are of only one cell type. Teratoma formation is a landmark test for pluripotency.
- o Embryoid body: hESCs in culture spontaneously form ball-like embryo-like structures termed "embryoid bodies", which consist of a core of mitotically active and differentiating hESCs and a periphery of fully differentiated cells from all three germ layers. iPSCs also form embryoid bodies and have peripheral differentiated cells.
- o Chimeric mice: hESCs naturally reside within the inner cell mass (embryoblast) of blastocysts, and in the embryoblast, differentiate into the embryo while the blastocyst's shell (trophoblast) differentiates into extraembryonic tissues. The hollow trophoblast is unable to form a living embryo, and thus it is necessary for the embryonic stem cells within the embryoblast to differentiate and form the embryo. iPSCs were injected by micropipette into a trophoblast, and the blastocyst was transferred to recipient females. Chimeric living mouse pups were created: mice with iPSC derivatives incorporated all across their bodies with 10%-90% chimerism.
- o Tetraploid complementation: iPS cells from mouse fetal fibroblasts injected into tetraploid blastocysts (which themselves can only form extra-embryonic tissues) can form whole, non-chimeric, fertile mice, although with low success rate.

- Epigenetic reprogramming
  - o Promoter demethylation: Methylation is the transfer of a methyl group to a DNA base, typically the transfer of a methyl group to a cytosine molecule in a CpG site (adjacent cytosine/guanine sequence). Widespread methylation of a gene interferes with expression by preventing the activity of expression proteins or recruiting enzymes that interfere with expression. Thus, methylation of a gene effectively silences it by preventing transcription. Promoters

of pluripotency-associated genes, including Oct-3/4, Rex1, and Nanog, were demethylated in iPSCs, demonstrating their promoter activity and the active promotion and expression of pluripotency-associated genes in iPSCs.

- o Histone demethylation: Histones are compacting proteins that are structurally localized to DNA sequences that can effect their activity through various chromatin-related modifications. H3 histones associated with Oct-3/4, Sox2, and Nanog were demethylated, indicating the expression of Oct-3/4, Sox2, and Nanog.

Safety for Regenerative Medicine

- The major concern with the potential clinical application of iPSCs is their propensity to form tumours. Much the same as ESC, iPSCs readily form teratoma when injected into immunodeficient mice. Teratoma formation is considered a major obstacle to stem-cell based regenerative medicine by the FDA.
- Since iPSCs can only be produced with high efficiency at this time using genetic modifications, they are generally predicted to be less safe and more tumorigenic than hESC. All the genes that have been shown to promote iPSC formation have also been linked to cancer in one way or another. Some of the genes are known oncogenes, including the members of the Myc family. While omitting Myc allows for IPSC formation, the efficiency is up to 100 fold reduced.
- A non-genetic method of producing iPSCs has been demonstrated using recombinant proteins, but its efficiency was quite low. However, refinements to this methodology yielding higher efficiency may lead to production of safer iPSCs. Other approaches such as using adenovirus or plasmids are generally thought to be safer than retroviral methods.
- An important area for future studies in the iPSC field is directly testing iPSC tumorigenicity using methods that mimic the approaches that would be used for regenerative medicine therapies. Such studies are crucial since iPSCs not only form teratoma, but also mice derived from iPSCs have a high incidence of death from malignant cancer. A recent paper came out in the journal Stem Cells indicating that iPS cells are far more tumorigenic than ESC, supporting the notion that iPS cell safety is a serious concern.

Stem Cell Line

A stem cell line is a family of constantly dividing cells, the product of a single parent group of stem cells. They are obtained from human or animal tissues and can replicate for long periods of time in vitro ("within glass"; or, commonly, "in the lab", in an artificial environment). They are frequently used for research relating to embryonic stem cells or cloning entire organisms.

Alternative Sources

Adult stem cell lines isolated from mature tissues are commonly used in stem cell research, as are cells isolated from umbilical cord blood. However, these cells have a genetic imprint of the host they were taken from, thus limiting their therapeutic use in genetic disorders. Also, adult stem cells are not totipotent or pluripotent like embryonic stem cells, but rather more specialized cells that are multipotent. Another source of stem cells is the iPS (induced pluripotent stem) cell. This process involves reversing the differentiating cell signals that cause a stem cell to specialize. In this way, a somatic cell can be worked backwards into a stem cell. One added advantage of this type of stem cell is that the pluripotent cell has the same DNA as the donor and can be used therapeutically towards that end without painful bone marrow and spinal cord extraction techiques. It also contributes to pain relief.

New Techniques

Typically, stem cells have been maintained using tissue culture methods that essentially date from the 1950s. In particular, they are often "fed" using mouse embryonic fibroblasts ("feeder cells") while being simultaneously suspended in a nutrient solution ("media"). However, many scientists are recognizing the importance of using media that is completely free of animal ingredients. This not only liberates cell lines from animal feeder cells, but also brings the in vivo therapeutic use of stem cells one step closer to reality.

Nonmammalian Study Model

One interesting study model that yields research results that can be subsequently tested on mammalian systems, is the fruit fly Drosophila (1)

Stem Cell Treatments

Stem cell treatments are a type of intervention strategy that introduces new cells into damaged tissue in order to treat disease or injury. Many medical researchers believe that stem cell treatments have the potential to change the face of human disease and alleviate suffering. The ability of stem cells to self-renew and give rise to subsequent generations with variable degrees of differentiation capacities, offers significant potential for generation of tissues that can potentially replace diseased and damaged areas in the body, with minimal risk of rejection and side effects.

A number of stem cell therapeutics exist, but most are at experimental stages and/or costly, with the notable exception of bone marrow transplantation. Medical researchers anticipate that adult and embryonic stem cells will soon be able to treat cancer, Type 1 diabetes mellitus, Parkinson's disease, Huntington's disease, Celiac Disease, cardiac failure, muscle damage and neurological disorders, and many others. Nevertheless, before stem cell therapeutics can be applied in the clinical setting, more research is necessary to understand stem cell Behaviour upon transplantation as well as the mechanisms of stem cell interaction with the diseased/injured microenvironment.

Current Treatments

For over 30 years, bone marrow, and more recently, umbilical cord blood stem cells, have been used to treat cancer patients with conditions such as leukemia and lymphoma. During chemotherapy, most growing cells are killed by the cytotoxic agents. These agents, however, cannot discriminate between the leukemia or neoplastic cells, and the hematopoietic stem cells within the bone marrow. It is this side effect of conventional chemotherapy strategies that the stem cell transplant attempts to reverse; a donor's healthy bone marrow reintroduces functional stem cells to replace the cells lost in the host's body during treatment.

Potential Treatments

Brain Damage: Stroke and traumatic brain injury lead to cell death, characterized by a loss of neurons and oligodendrocytes within the brain. Healthy adult brains contain neural stem cells which divide to maintain general stem cell numbers, or become progenitor cells. In healthy adult animals, progenitor cells migrate within the brain and function primarily to maintain neuron populations for olfaction (the sense of smell).

Interestingly, in pregnancy and after injury, this system appears to be regulated by growth factors and can increase the rate at which new brain matter is formed. Although the reparative process appears to initiate following trauma to the brain, substantial recovery is rarely observed in adults, suggesting a lack of robustness.

Stem cells may also be used to treat brain degeneration, such as in Parkinson's and Alzheimer's disease.

Cancer: Research injecting neural (adult) stem cells into the brains of dogs has shown to be very successful in treating cancerous tumours. Using conventional techniques, brain cancer is difficult to treat because it spreads so rapidly. Researchers at the Harvard Medical School transplanted human neural stem cells into the brain of rodents that received intracranial tumours. Within days, the cells migrated into the cancerous area and produced cytosine deaminase, an enzyme that converts a non-toxic pro-drug into a chemotheraputic agent. As a result, the injected substance was able to reduce the tumour mass by 81 percent. The stem cells neither differentiated nor turned tumorigenic. Some researchers believe that the key to finding a cure for cancer is to inhibit proliferation of cancer stem cells. Accordingly, current cancer treatments are designed to kill cancer cells. However, conventional chemotherapy treatments cannot discriminate between cancerous cells and others. Stem cell therapies may serve as potential treatments for cancer. Research on treating Lymphoma using adult stem cells is underway and has had human trials. Essentially, chemotherapy is used to completely destroy the patients own lymphocytes, and stem cells injected, eventually replacing the immune system of the patient with that of the healthy donor.

Spinal Cord Injury : A team of Korean researchers reported on November 25, 2003, that they had transplanted multipotent adult stem cells from umbilical cord blood to a patient suffering from a spinal cord injury and that following the procedure, she could walk on her own, without difficulty. The patient had not been able to stand up for roughly 19 years. For the unprecedented clinical test, the scientists isolated adult stem cells from umbilical cord blood and then injected them into the damaged part of the spinal cord.

According to the October 7, 2005 issue of *The Week*, University of California, Irvine researchers transplanted multipotent human fetal-derived neural stem cells into paralyzed mice, resulting in locomotor improvements four months later. The observed recovery was associated with differentiation of transplanted cells into new neurons and oligodendrocytes- the latter of which forms the myelin sheath around axons of the central nervous system, thus insulating neural impulses and facilitating communication with the brain.

In January 2005, researchers at the University of Wisconsin–Madison differentiated human blastocyst stem cells into neural stem cells, then into pre-mature motor neurons, and finally into spinal motor neurons, the cell type that, in the human body, transmits messages from the brain to the spinal cord and subsequently mediates motor function in the periphery.

The newly generated motor neurons exhibited electrical activity, the signature action of neurons. Lead researcher Su-Chun Zhang described the process as "[teaching] the blastocyst stem cells to change step by step, where each step has different conditions and a strict window of time." Transformation of blastocyst stem cells into motor neurons had eluded researchers for decades. While Zhang's findings were a significant contribution to the field, the ability of transplanted neural cells to establish communication with neighboring cells remains unclear. Accordingly, studies using chicken embryos as a model organism can be an effective proof-of-concept experiment. If functional, the new cells could be used to treat diseases like Lou Gehrig's disease, muscular dystrophy, and spinal cord injuries.

Heart Damage: Several clinical trials targeting heart disease have shown that adult stem cell therapy is safe, effective, and equally efficient in treating old and recent infarcts. Adult stem cell therapy for treating heart disease was commercially available in at least five continents at the last count (2007).

Possible mechanisms of recovery include:

- Generation of heart muscle cells
- Stimulation of growth of new blood vessels to repopulate damaged heart tissue
- Secretion of growth factors
- Assistance via some other mechanism.

It may be possible to have adult bone marrow cells differentiate into heart muscle cells.

Haematopoiesis (Blood Cell Formation)

The specificity of the human immune cell repertoire is what allows the human body to defend itself from rapidly adapting antigens. However, the immune system is vulnerable to degradation upon the pathogenesis of disease, and because of the critical role that it plays in overall defence, its degradation is often fatal to the organism as a whole.

Diseases of hematopoietic cells are called hematopathology. The specificity of the immune cells is what allows recognition of foreign antigens, causing further challenges in the treatment of immune disease. Identical matches between donor and recipient must be made for successful transplantation treatments, but matches are uncommon, even between first-degree relatives. Research using both hematopoietic adult stem cells and embryonic stem cells has provided insight into the possible mechanisms and methods of treatment for many of these ailments.

Fully mature human red blood cells may be generated *ex vivo* by hematopoietic stem cells (HSCs), which are precursors of red blood cells. In this process, HSCs are grown together with stromal cells, creating an environment that mimics the conditions of bone marrow, the natural site of red blood cell growth. Erythropoietin, a growth factor, is added, coaxing the stem cells to complete terminal differentiation into red blood cells. Further research into this technique should have potential benefits to gene therapy, blood transfusion, and topical medicine.

Baldness

Hair follicles also contain stem cells, and some researchers predict research on these follicle stem cells may lead to successes in treating baldness through "hair multiplication", also known as "hair cloning". This treatment is expected to work by taking stem cells from existing follicles, multiplying them in culture, and implanting the new follicles back into the scalp. Later treatments may be able to simply signal follicle stem cells to give off chemical signals to nearby follicle cells which have shrunk during the aging process, which in turn respond to these signals by regenerating and once again making healthy hair.

Missing Teeth

In 2004, scientists at King's College London discovered a way to cultivate a complete tooth in mice and were able to grow them stand-alone in the laboratory. Researchers are confident that this technology can be used to grow live teeth in human patients.

In theory, stem cells taken from the patient could be coaxed in the lab into turning into a tooth bud which, when implanted in the gums, will give rise to a new tooth, and would be expected to grow within two months. It will fuse with the jawbone and release chemicals that encourage nerves and blood vessels to connect with it. The process is similar to what happens when humans grow their original adult teeth.

Many challenges remain, however, before stem cells could be a choice for the replacement of missing teeth in the future.

Deafness

Heller has reported success in re-growing cochlea hair cells with the use of embryonic stem cells.

Blindness and Vision Impairment

Since 2003, researchers have successfully transplanted corneal stem cells into damaged eyes to restore vision. Using embryonic stem cells, scientists are able to grow a thin sheet of totipotent stem cells in the laboratory. When these sheets are transplanted over the damaged cornea, the stem cells stimulate renewed repair, eventually restore vision.

The latest such development was in June 2005, when researchers at the Queen Victoria Hospital of Sussex, England were able to restore the sight of forty patients using the same technique. The group, led by Dr. Sheraz Daya, was able to successfully use adult stem cells obtained from the patient, a relative, or even a cadaver. Further rounds of trials are ongoing.

In April 2005, doctors in the UK transplanted corneal stem cells from an organ donor to the cornea of Deborah Catlyn, a woman who was blinded in one eye when acid was thrown in her eye at a nightclub. The cornea, which is the transparent window of the eye, is a particularly suitable site for transplants. In fact, the first successful human transplant was a cornea transplant. The absence of blood vessels within the cornea makes this area a relatively easy target for transplantation. The majority of corneal transplants carried out today are due to a degenerative disease called keratoconus. The University Hospital of New Jersey reports that the success rate for growth of new cells from transplanted stem cells varies from 25 percent to 70 percent.

In 2009, researchers at the University of Pittsburgh Medical centre demonstrated that stem cells collected from human corneas can restore transparency without provoking a rejection response in mice with corneal damage.

Amyotrophic Lateral Sclerosis

Stem cells have resulted in significant locomotor improvements in rats with an Amyotrophic lateral sclerosis-like disease. In a rodent model that closely mimics the human form of ALS, animals were injected with a virus to kill the spinal cord motor nerves which mediate movement. Animals subsequently received stem cells in the spinal cord. Transplanted cells migrated to the sites of injury, contributed to regeneration of the ablated nerve cells, and restored locomotor function.

Graft vs. Host Disease and Crohn's Disease

Phase III clinical trials expected to end in second-quarter 2008 were conducted by Osiris Therapeutics using their in-development product Prochymal, derived from adult bone marrow. The target disorders of this therapeutic are graft-versus-host disease and Crohn's disease.

Neural and Behavioral Birth Defects

A team of researchers led by Prof. Joseph Yanai were able to reverse learning deficits in the offspring of pregnant mice who were exposed to heroin and the pesticide organophosphate. This was done by direct neural stem cell transplantation into the brains of the offspring. The recovery was almost 100 percent, as shown in behavioral tests that suggested improved to normal Behaviour and learning scores in animals receiving cell transplantation. On the molecular level, brain chemistry of the treated animals was also restored to normal.

Through the work, which was supported by the US National Institutes of Health, the US-Israel Binational Science Foundation and the Israel anti-drug authorities, the researchers discovered that the stem cells worked even in cases where most of the cells died out in the host brain.

The scientists found that before they die the neural stem cells succeed in inducing the host brain to produce large numbers of stem cells which repair the damage. These findings, which answered a major question in the stem cell research community, were published earlier this year in the leading journal,

Molecular Psychiatry. Scientists are now developing procedures to administer the neural stem cells in the least invasive way possible - probably via blood vessels, making therapy practical and clinically feasible. Researchers also plan to work on developing methods to take cells from the patient's own body, turn them into stem cells, and then transplant them back into the patient's blood via the blood stream. Aside from decreasing the chances of immunological rejection, the approach will also eliminate the controversial ethical issues involved in the use of stem cells from human embryos.

Diabetes

Diabetes patients lose the function of insulin-producing beta cells within the pancreas. Human embryonic stem cells may be grown in cell culture and stimulated to form insulin-producing cells that can be transplanted into the patient.

However, clinical success is highly dependent on the development of the following procedures:

- Transplanted cells should proliferate
- Transplanted cells should differentiate in a site-specific manner
- Transplanted cells should survive in the recipient (prevention of transplant rejection)
- Transplanted cells should integrate within the targeted tissue
- Transplanted cells should integrate into the host circuitry and restore function.

Orthopaedics

Clinical case reports in the treatment of orthopaedic conditions have been reported. To date, the focus in the literature for musculoskeletal care appears to be on mesenchymal stem cells. Centeno et al. have published MRI evidence of increased cartilage and meniscus volume in individual human subjects. The results of trials that include a large number of subjects, are yet to be published. However, a published safety study conducted in a group of 227 patients over a 3-4 year period shows adequate safety and minimal complications associated with mesenchymal cell transplantation.

Wakitani has also published a small case series of nine defects in five knees involving surgical transplantation of mesenchymal stem cells with coverage of the treated chondral defects.

Wound Healing

Stem cells can also be used to stimulate the growth of human tissues. In an adult, wounded tissue is most often replaced by scar tissue, which is characterized in the skin by disorganized collagen structure, loss of hair follicles and irregular vascular structure. In the case of wounded fetal tissue, however, wounded tissue is replaced with normal tissue through the activity of stem cells.

A possible method for tissue regeneration in adults is to place adult stem cell "seeds" inside a tissue bed "soil" in a wound bed and allow the stem cells to stimulate differentiation in the tissue bed cells. This method elicits a regenerative response more similar to fetal wound-healing than adult scar tissue formation. Researchers are still investigating different aspects of the "soil" tissue that are conducive to regeneration.

Infertility

Culture of human embryonic stem cells in mitotically inactivated porcine ovarian fibroblasts (POF) causes differentiation into germ cells (precursor cells of oocytes and spermatozoa), as evidenced by gene expression analysis.

Human embryonic stem cells have been stimulated to form Spermatozoon-like cells, yet still slightly damaged or malformed. It could potentially treat azoospermia.

Clinical Trials

On January 23, 2009, the US Food and Drug Administration gave clearance to Geron Corporation for the initiation of the first clinical trial of an embryonic stem cell-based therapy on humans. The trial will evaluate the drug GRNOPC1, embryonic stem cell-derived oligodendrocyte progenitor cells, on patients with acute spinal cord injury.

As of mid 2010 hundreds of phase III clinical trials involving stem cells have been registered.

Stem Cell use in Animals

Potential Contributions to Veterinary Medicine

Research currently conducted on horses, dogs, and cats can benefit the development of stem-cell treatments in veterinary medicine and can target a wide range of injuries and diseases such as myocardial infarction, stroke, tendon and ligament damage, osteoarthritis, osteochondrosis and muscular dystrophy both in large animals, as well as humans. While investigation of cell-based therapeutics generally reflects human medical needs, the high degree of frequency and severity of certain injuries in racehorses has put veterinary medicine at the forefront of this novel regenerative approach. Companion animals can serve as clinically relevant models that closely mimic human disease.

Development of Regenerative Treatment Models

Veterinary applications of stem cell therapy as a means of tissue regeneration have been largely shaped by research that began with the use of adult-derived mesenchymal stem cells to treat animals with injuries or defects affecting bone, cartilage, ligaments and/or tendons. Because mesenchymal stem cells can differentiate into the cells that make up bone, cartilage, tendons, and ligaments (as well as muscle, fat, and possibly other tissues), they have been the main type of stem cells studied in the treatment of diseases affecting these tissues.

Mesenchymal stem cells are primarily derived from adipose tissue or bone marrow. Since an elevated immune response following cell transplantation may result in rejection of exogenous cells (except in the case of cells derived from a very closely genetically related individual), mesenchymal stem cells are often derived from the patient prior to injection in a process known as autologous transplantation.

Surgical repair of bone fractures in dogs and sheep has demonstrated that engraftment of mesenchymal stem cells derived from a genetically different donor within the same species, termed allogeneic transplantation, does not elicit an immunological response in the recipient animal and can mediate regeneration of bone tissue in major bony fractures and defects. Stem cells can speed up bone repair in fractures/defects that would normally require extensive grafting, suggesting that mesenchymal stem cell use may provide a useful alternative to conventional grafting techniques.

Treating tendon and ligament injuries in horses using stem cells, whether derived from adipose tissue or bone-marrow, has support in the veterinary literature.

While further studies are necessary to fully characterize the use of cell-based therapeutics for treatment of bone fractures, stem cells are thought to mediate repair via five primary mechanisms: 1) providing an antiinflammatory effect, 2) homing to damaged tissues and recruiting other cells, such as endothelial progenitor cells, that are necessary for tissue growth, 3) supporting tissue remodeling over scar formation, 4) inhibiting apoptosis, and 5) differentiating into bone, cartilage, tendon, and ligament tissue.

Significance of Stem Cell Microenvironments

The microenvironment into which stem cells are transplanted significantly alters the capacity of engrafted cells for recovery and repair. The microenviroment provides growth factors and other chemical signals that guide appropriate differentiation of transplanted cell populations and direct transplanted cells to sites of trauma or disease. Repair and recovery can then be mediated via three primary mechanisms: 1) formation and/or recruitment of new blood cells to the damaged region; 2) prevention of programed cell death or apoptosis; and 3) suppression of inflammation.

To further enrich blood supply to the damaged areas, and consequently promote tissue regeneration, platelet-rich plasma could be used in conjunction with stem cell transplantation. The efficacy of some stem cell populations may also be affected by the method of delivery; for instance, to regenerate bone, stem cells are often introduced in a scaffold where they produce the minerals necessary for generation of functional bone.

Sources of Autologous (Patient-derived) Stem Cells

Autologous stem cells intended for regenerative therapy are generally isolated either from the patient's bone marrow or from adipose tissue. The number of stem cells transplanted into damaged tissue may alter efficacy of treatment. Accordingly, stem cells derived from bone marrow aspirates, for instance, are cultured in specialized laboratories for expansion to millions of cells.

Although adipose-derived tissue also requires processing prior to use, the culturing methodology for adipose-derived stem cells is not as extensive as that for bone marrow-derived cells. While it is thought that bone-marrow derived stem cells are preferred for bone, cartilage, ligament, and tendon repair, others believe that the less challenging collection techniques and the multi-cellular microenvironment already present in adipose-derived stem cell fractions make the latter the preferred source for autologous transplantation.

Currently Available Treatments for Horses and Dogs Suffering from Orthopedic Conditions

Autologous or allogeneic stem cells are currently used as an adjunctive therapy in the surgical repair of some types of fractures in dogs and horses. Autologous stem cell-based treatments for ligament injury, tendon injury, osteoarthritis, osteochondrosis, and sub-chondral bone cysts have been commercially available to practicing veterinarians to treat horses since 2003 in the United States and since 2006 in the United Kingdom.

Autologous stem-cell based treatments for tendon injury, ligament injury, and osteoarthritis in dogs have been available to veterinarians in the United States since 2005. Over 3000 privately-owned horses and dogs have been treated with autologous adipose-derived stem cells.

The efficacy of these treatments has been shown in double-blind clinical trials for dogs with osteoarthritis of the hip and elbow and horses with tendon damage.

The efficacy of using stem cells, whether adipose-derived or bone-marrow derived, for treating tendon and ligament injuries in horses has support in the veterinary literature.

Developments in Stem Cell Treatments in Veterinary Internal Medicine

Currently, research is being conducted to develop stem cell treatments for: 1) horses suffering from COPD, neurologic disease, and laminitis; and 2) dogs and cats suffering from heart disease, liver disease, kidney disease, neurologic disease, and immune-mediated disorders.

Embryonic Stem Cell Controversy

There is widespread controversy over the use of human embryonic stem cells. This controversy primarily targets the techniques used to derive new embryonic stem cell lines, which often requires the destruction of the blastocyst.

Opposition to the use of human embryonic stem cells in research is often based on philosophical, moral or religious objections. While these objections are the source of rhetoric opposing the research, those opposed to such research can also point to the complete failure of embryonic stem cell researchers to provide anything of substance.

There have been tens of thousands of successful adult stem cell treatments just in China over the last decade, while the research community has yet to produce even one positive patient result using embryonic stem cells.

The alternatives do not require the destruction of an embryo, such as the use of umbilical cord blood, milk teeth stem cells, bone marrow stem cells or using induced pluripotent stem cells. The stem cells from these alternative sources also lack the severe side effects which are universally seen in embryonic stem cell therapies and most often result in fatal rejection by the subject; the most common being mutations of the stem cells into tumours.

Stem Cell Treatments Around the World

China

Stem cell research and treatment is currently being practiced at a clinical level in the People's Republic of China. The Ministry of Health of the People's Republic of China has permitted the use of stem cell therapy for conditions beyond those approved of in Western countries such as the United States, United Kingdom, and Australia. The Western World has scrutinized China for its failed attempts to meet international documentation standards of these trials and procedures, despite the overwhelmingly positive anecdotal results.

Stem cell therapies provided in China utilize a variety of cell types including umbilical cord stem cells and olfactory ensheathing cells. The stem cells are then expanded in centralized blood banks before being used in stem cell treatments. State-funded companies based in the Shenzhen Hi-Tech Industrial Zone treat the symptoms of numerous disorders with adult stem cell therapy. Hospitals throughout eastern China provide numerous therapies to patients in coordination with the stem cell providers. These companies' therapies are currently focused on the treatment of neurodegenerative and cardiovascular disorders.

The most radical successes of Chinese adult stem cell therapy have been in treating the brain. These therapies administer stem cells directly to the brain to promote greater motor and brain function in patients with Cerebral Palsy, Alzheimer's, and brain injuries. However, retrospective studies have shown that Chinese use of fetal-derived brain tissue in spinal cord injured human subjects were not as promising as once thought: the phenotype and the fate of the transplanted cells, described as olfactory ensheathing cells, were unknown.

As well, perioperative morbidity and lack of functional benefit were identified as the most serious clinical shortcomings. Furthermore, the extent of regulatory policy in the use of stem cell therapies in China is unclear. In the absence of a valid clinical trials protocol, and more regulatory oversight, Western regulatory agencies advise patients and physicians to be cautious when selecting Chinese stem cell therapeutic centres.

Mexico

Stem cell treatment is currently being practiced at a clinical level in Mexico. An International Health Department Permit (COFEPRIS) is required. This permit allows the use of stem cell types beyond those approved of in Western countries such as the United States or Europe. Stem cell therapies provided in Mexico utilize patient Adipose, Bone Marrow, or Donor Placenta sources.

South Korea

In 2005, South Korean scientists claimed to have generated stem cells that were tailored to match the recipient. Each of the 11 new stem cell lines was developed using somatic cell nuclear transfer (SCNT) technology. The resultant cells were thought to match the genetic material of the recipient, thus suggesting minimal to no cell rejection.

This study, however, was eventually discredited as the primary researcher,Dr. Woo Suk Hwang, admitted to using cells obtained from his research staff. In Dec 2005, claims were put forward that his research had been manipulated to wrongfully indicate positive results. Later that month, these claims were confirmed by an academic panel.

Industrial Microbiology

Industrial microbiology or microbial biotechnology encompasses the use of microorganisms in the manufacture of food or industrial products. The use of microorganisms for the production of food, either human or animal, is often considered a branch of food microbiology. The microorganisms used in industrial processes may be natural isolates, laboratory selected mutants or genetically engineered organisms.

Food Microbiology

Yogurt, cheese, chocolate, and silage (animal food) are all produced by industrial microbiology processes. 'Good' bacteria such as probiotics are becoming increasingly important in the food industry.

Biopolymers

A huge variety of biopolymers, such as polysaccharides, polyesters, and polyamides, are produced by microorganisms. These products range from viscous solutions to plastics. The genetic manipulation of microorganisms has permitted the biotechnological production of biopolymers with tailored material properties suitable for high-value medical application such as tissue engineering and drug delivery. Industrial microbiology can be used for the biosynthesis of xanthan, alginate, cellulose, cyanophycin, poly(gamma-glutamic acid), levan, hyaluronic acid, organic acids, oligosaccharides and polysaccharides, and polyhydroxyalkanoates.

Bioremediation

Microbial biodegradation of pollutants can be used to cleanup contaminated environments. These bioremediation and biotransformation methods harness naturally occurring microbes to degrade, transform or accumulate a huge range of compounds including hydrocarbons (e.g. oil), polychlorinated biphenyls (PCBs), polyaromatic hydrocarbons (PAHs), pharmaceutical substances, radionuclides and metals.

Waste Biotreatment

Microorganisms are used to treat the vast quantities of wastes generated by modern societies. Biotreatment, the processing of wastes using living organisms, is an environmentally friendly, relatively simple and cost-effective alternative to physico-chemical clean-up options. Confined environments, such as bioreactors can be employed in biotreatment processes.

Wastewater Treatment

Biological wastewater treatment is undoubtedly one of the most important biotechnological processes, which have been used for over a century to treat municipal and industrial wastewaters. Culture-independent molecular techniques have been used to study the diversity and physiology of ecologically important microorganisms in wastewater treatment processes.

A number of new exciting insights into the structure, function, and dynamics of complex microbial communities in wastewater treatment processes have been gained, which have significantly expanded our understanding of process design, operation and control.

Microbes play a vital role in the cycling of nitrogen in wastewater treatment processes (including anaerobic ammonia oxidation processes) and methane fermentation processes.

Health-care and Medicine

Microorganisms are used to produce human or animal biologicals such as insulin, growth hormone, and antibodies. Diagnostic assays that use monoclonal antibody, DNA probe technology or real-time PCR are used as rapid tests for pathogenic organisms in the clinical laborarory.

Microorganisms may also help in the treatment of diseases such as cancer. Research shows that clostridia can selectively target cancer cells. Various strains of non-pathogenic clostridia have been shown to infiltrate and replicate within solid tumours. Clostridia therefore have the potential to deliver therapeutic proteins to tumours. *Lactobacillus* spp. and other lactic acid bacteria possess numerous potential therapeutic properties including anti-inflammatory and anti-cancer activities.

Vaccines are used to combat infectious disease, however the last decade has witnessed a revolution in the approach to vaccine design and development. Sophisticated technologies such as genomics, proteomics, functional genomics, and synthetic chemistry can be used for the rational identification of antigens, the synthesis of complex glycans, and the generation of engineered carrier proteins.

Members of the *Streptomyces* genus are among the most prolific microorganisms producing secondary metabolites with wide uses in medicine and in agriculture. These organisms have a complex secondary metabolism producing antibiotic compounds and other metabolites with medicinal properties. Genomic studies, genomic mining and biotechnological approaches are being employed in the search for new antibiotics and other drugs in *Streptomyces*.

Archaea

Examination of microbes living in unusual environments (e.g. high temperatures, salt, low pH or temperature, high radiation) lead to discovery of microbes with new abilities that can be harnessed for industrial purposes.

Corynebacteria

Corynebacteria are a diverse group Gram-positive bacteria found in a range of different ecological niches such as soil, vegetables, sewage, skin, and cheese smear. *Corynebacterium glutamicum* is of immense industrial importance and is one of the biotechnologically most important bacterial species with an annual production of more than two million tons of amino acids, mainly L-glutamate and L-lysine. The genome sequence of *C. glutamicum* has been published.

Xanthomonas

The genus *Xanthomonas* consists of 20 plant-associated species, many of which cause important diseases of crops and other plants. Individual species comprise multiple pathovars, characterized by distinctive host specificity or mode of infection.

Bacteria of the genus *Xanthomonas* are able to produce the acidic exopolysaccharide xanthan gum. Because of its physical properties, it is widely used as a viscosifer, thickener, emulsifier or stabilizer in both food and non-food industries.

Aspergillus

Species within the genus *Aspergillus* have a large chemical repertoire. Commodity products produced in *Aspergillus* cell 'factories' include citric, gluconic, itaconic and kojic acid. The use of *Aspergillus niger* in citric acid production dates back to 1917.

Citric acid is one of the most widely used food ingredients. It also has found use in the pharmaceutical and cosmetic industries as an acidulant and for aiding in the dissolution of active ingredients. Other technical applications of citric acid are as a hardener in adhesive and for retarding the setting of concrete.

Citric acid is a true 'bulk chemical' with an estimated production approximating more than 1.6 billion kg each year *A. niger* also has found use in the industrial production of gluconic acid, which is used as an additive in certain metal cleaning applications, as well as for the therapy for calcium and iron deficiencies. *Aspergillus terreus* is used for itaconic acid production, a synthetic polymer. *A. oryzae* is fermented for kojic acid production which is used for skin whitening and as a precursor for synthesis of flavour enhancers.

Viruses

Viruses that are pathogenic to insect pests can be exploited as biological control agents. Some viruses such as baculoviruses have been exploited for use as gene expression and delivery vectors in both insect and mammalian cells.

Secretome: Clues into Pathogen Infection and Clinical Applications

The secretome constitutes the entire set of secreted proteins, representing up to 30% of the proteome of an organism, and includes functionally diverse classes of molecules such as cytokines, chemokines, hormones, digestive enzymes, antibodies, extracellular proteinases, morphogens, toxins and antimicrobial peptides.

Some of these proteins are involved in a host of diverse and vital biological processes, including cell adhesion, cell migration, cell-cell communication, differentiation, proliferation, morphogenesis, survival and defence, virulence factors in bacteria and immune responses. Excretory/secretory proteins (ESPs) circulating throughout the body of an organism (for example, in the extracellular space) are localized to or released from the cell surface, making them readily accessible to drugs and/or the immune system.

These characteristics make these molecules extremely attractive targets for novel vaccines and therapeutics, which are currently the focus of major drug discovery research programs. In particular, proteins secreted by pathogens (bacterial, protozoan, fungal, viral or helminth) mediate interactions with the host, because these are present or active at the interface between the pathogen and the host cells, and can regulate or mediate the host responses and/or cause disease.

A brief overview of the currently available methods for generating and analyzing pathogen secretome data is presented, followed by a critical analysis of their contribution to our understanding of pathogen infection and host responses, especially in comparison to other genome analysis approaches. Some early successes in the applications of secretome data in the areas of therapeutic target identification, diagnostic tools and pathogen control are also presented.

Approaches for Secretome Analysis

Genome Sequence Analysis

Genome sequence analysis is based on transcript profiling and computational analysis. The computational prediction of secreted proteins seeks to identify the presence of signal peptides, which are considered markers for classically secreted proteins. According to the signal hypothesis, most secreted proteins have an amino-terminal signal peptide sequence that targets proteins to the endoplasmic reticulum (ER) lumen via the sec-dependent protein translocation complex.

The genome-based approach is fast but incurs three major problems. Primarily, the pathogen genome sequence has to be available. Although the genomes of several pathogens such as *Vibrio cholerae* and *Brugia malayi* are now available, several more organisms such as *Ascaris lumbricoides* and *Wuchereria bancrofti* are awaiting sequencing.

Secondly, this approach is based on the accurate prediction of signal peptides for the detection of secretory proteins. However, many secretory proteins lacking the amino-terminal signal peptides are not predicted by this method. Lastly, secreted proteins are regulated at the post-transcriptional level, resulting in an apparent lack of correlation between the levels of production of secreted proteins and mRNA expression levels.

Proteomics Approaches

With the advent of mass spectrometry (MS) and the ensuing bioinformatics analyses, proteomic approaches have become the preferred route for obtaining secretome data. The two main methods available here are gel-based and gel-free proteomics.

Gel-based Proteomic Analysis

Two-dimensional gel electrophoresis (2-DE) with MS is the most established proteomic approach. This method allows the separation of complex mixtures of intact proteins at high resolution. These protein mixtures are first separated according to their charge in the first dimension by isoelectric focusing, and according to size in the second dimension by SDS-PAGE (sodium dodecyl-sulfate polyacrylamide gel electrophoresis), and then analysed by peptide mass fingerprinting after in-gel tryptic digestion. This approach has been widely used in pathogen secretome studies, such as that of *Helicobacter pylori*.

Although 2-DE currently remains the most efficient method for the separation of complex mixtures of proteins, this technique has a number of limitations, including poor reproducibility between gels, low sensitivity to detection of proteins at low concentrations and hydrophobic membrane proteins, limited sample capacity, and low linear range of visualization procedures. In addition, this technique is time consuming and labour intensive and has limited efficiency in protein detection due to its limited amenability to automation.

Gel-free Proteomic Analysis

To overcome the drawbacks of gel-based approaches, efforts have been made to introduce gel-free MS-based proteomics approaches. In these newly emerging techniques, instead of depending on gels to separate and analyse proteins, complex mixtures of proteins are first digested into peptides or peptide fragments, then separated by one or several steps of capillary chromatography, and finally analysed by tandem MS (MS/MS).

The secretome analysis of *Leishmania donovani* adopted liquid chromatography coupled with automated MS/MS. Matrix-assisted laser desorption/ionization-time of flight (MALDI-TOF) MS, a popular tool for the analysis of complex molecules, was used to analyse the secretome of HepG2 cells infected with the dengue virus.

Bioinformatics Approach

With the generation of large-scale expressed sequence tag (EST) and genomic data due to worldwide sequencing efforts, secretome analysis can be advantageously carried out using bioinformatics analysis systems such as EST2Secretome, a pipeline for the prediction of secretory proteins. EST2Secretome accepts EST data for preprocessing, assembly and conceptual translation into protein sequences.

Alternatively, peptide sequences can be directly provided to the pipeline, which then separates secreted proteins by identifying an amino-terminal secretory signal peptide and the lack of transmembrane segments. The secreted protein set is then annotated extensively with gene ontologies, protein functional identification, in terms of mapping to protein domains, metabolic pathways, identifying homologs from a well-studied model organism (*Caenorhabditis elegans*), protein interaction partners and mapping to a manually curated signal peptide database.

The application of EST2Secretome to approximately 0.5 million EST sequences from parasitic nematodes resulted in the identi fication of key ESPs, some of which are already being trialed as vaccine candidates and as targets for therapeutic intervention. Similar studies reporting the ESPs of specific parasitic nematodes have been recently reviewed. The accuracy of EST-based predictions of ESPs was assessed with proteomic data from *Fasciola hepatica*.

The EST2Secretome pipeline was successful in identifying the major secreted proteins of adult *F. hepatica*. Integration of bioinformatics analysis with proteomics data is important for the study of helminth host-pathogen relationships, to distinguish proteins that are secreted extracorporeally from those secreted within the internal tissues of the parasites. Additionally, this integrated approach has identified major helminth proteins that may be secreted by novel or non-classical secretory pathways.

Towards a better Understanding of Host-pathogen Interactions

Proteins secreted by pathogens can influence infection and modify host defence signalling pathways. Proteomic analysis of secreted proteins from *Rhodococcus equi*, *Plasmodium falciparum*, *H. pylori* and the eggs of *Schistosoma mansoni* confirms the major role of the secretome in pathogenesis. Secreted proteins from pathogens modify and adapt the host environment for pathogen survival, invoking processes such as helminth immunoregulation. Inside the host environment, the secre tome serves the role of a parasite genome, as the secreted proteins fulfill all the requirements of the parasite inside the host.

While the secretory proteins of pathogens play a key role in pathogenesis, the secretome of the infected host cell is equally important in understanding secreted proteins underpinning host defence mechanisms against pathogen attack, such as the release of GDSL lipase 2 in Arabidopsis, which plays a role in pathogen defence.

Another host defence mechanism is the secretion of secretory immunoglobulin As (IgAs) against mucosal pathogens to limit the entry of bacteria, a process is known as 'immune exclusion'. A study on the malarial parasite *P. falciparum* concluded that export of proteins from the intracellular parasite to the erythrocyte is vital for infection.

These exported proteins are required for the virulence and rigidity of the *P. falciparum*-infected erythrocyte, which results in malaria infection. This export is guided by a host targeting (HT) signal present on the parasite proteins engaged in remodeling the erythrocyte. The role of this HT motif in the transport of these parasite proteins is yet to be determined.

The major secretions of adult parasites are proteolytic enzymes that help parasites to penetrate the host skin and to cleave host IgE antibodies to regulate the host immune system. These ESPs are exported through classical and non-classical secretory pathways. Classical secretory pathways are mediated by the presence of short amino-terminal signal peptide sequences that are predicted accurately by algorithms.

On the other hand, non-classical secreted proteins are hard to track as these are usually secreted by ER/Golgi-independent protein secretion pathways, eliminating the need for signal peptide sequences, and are usually predicted by using the SecretomeP method. In a study on *B. malayi*, it was found that filarial ESPs are similar to cytokines, chemokines and other immune effector molecules of humans, and are predicted to promote parasite survival and development in the host environment. A comparative secretome analysis identified 11 proteins that are conserved across human- and rodent-infecting *Plasmodium* species, suggesting a critical role for these proteins in interacting with and remodeling of the host erythrocyte cells. The secretome of a mammalian parasite consists of proteins required for parasite survival, including those involved in metabolism, reproduction and modification of the host immune system. Identifying pathogen ESPs will permit the identification of host receptors and host cells with which these proteins interact, improving our understanding of the molecular mechanisms involved in pathogenesis.

Recent Secretome Data

Secretome data on pathogenic organisms are sparse and limited to specific experimental methods or sample types. Over the past few years, a wealth of information on bacteria and the malarial and filarial parasites has become available, although there are still very few data on the infectious agents causing 'neglected tropical diseases'.

Major secretome analyses of helminth parasites have attempted to address this deficiency. Its preferred host, the disease caused and the experimental approach. While most of the bioinformatics analyses involve BLAST (Basic Local Alignment Search Tool) searches against the NCBI (National Centre for Biotechnology Information, USA) databases and use of the MASCOT (Modular Approach to Software Construction Operation and Test) software, except for the *F. hepatica* study by Robinson *et al.*, in which the EST2Secretome pipeline was used for bioinformatics data analysis and annotation.

Clinical Applications

Identification of Drug Targets and Vaccine Development

As more and more secretome analysis studies are conducted around the world, our knowledge of the virulence factors present in the secretome has substantially increased. As many of the proteins present in the pathogen secretome remain unannotated, we can assign function to these proteins by homology searches for similar proteins of known function from different organisms.

Furthermore, we can use Gene Ontology (GO) terms ascribed to database matches to glean GO terms for pathogen ESPs. The secretome of a pathogen cell provides a rich source of protein antigens that can be used for vaccine development. A very recent study on *Mycobacterium immunogenum* has investigated the protein antigens of the virulence factors in infection, with implications for vaccine development.

The Human Hookworm Vaccine Initiative has spearheaded the identification of several prominent anti-parasite vaccine candidates, including a family of pathogenesis-related proteins, such as the *Ancylostoma*-secreted proteins. Major vaccine antigens determined as a result of this initiative are hydrolytic enzymes, including proteases and acetylcholinesterases from the infective larval 3 (L3) and adult stages. Major L3 candidates found are *Ancylostoma*-secreted proteins (ASPs), astacin-like metalloprotease (MTP), acetylcholinesterase (ACH) and transthyretin (TTR). From the adult stage, major antigens found are tissue inhibitor of metalloproteases (such as Ac-TMP), aspartic proteases and cysteinyl proteases. Clinical trials for hookworm infection vaccines are in progress.

Diagnostic Tools

MS has proved to be a successful tool for protein analysis. Secretory proteins serve as a rich source of biomarkers, as reviewed by Chaerkady and Pandey. These biomarkers can be used in various array-based methods for the diagnosis of various medical conditions that occur as a result of pathogen infection, such as dengue virus infection and meningitis. Array-based approaches are more specific and faster than other conventional diagnostic techniques. Such a study of *Trypanosoma congolense* and *Trypanosoma evansi*, which cause the major strains of animal trypanosomosis, showed differences in their virulence and pathogenicity and has led to the determination of novel ESP targets for species-specific diagnosis and vaccine development.

Host-induced Gene Silencing using RNA Interference Technology

The availability of secretome data and the advent of RNA interference (RNAi) technology open up the possibility of host-induced gene silencing in pathogens, making the host resistant to infection. Parasite control in *Arabidopsis thaliana* has been achieved by host-induced gene silencing of nematode genes.

Secretome analysis is a promising area of research providing insights into different pathogenic infections. Recent studies have uncovered a myriad of processes involved in pathogenic infections at the molecular level, enabling us to develop novel therapeutic solutions to eradicate these infections.

Although much work remains to be done in generating secretome data for several pathogens, the availability of secretome data for major pathogens such as the malarial and filarial parasites, and the application of bioinformatics tools, will provide us with a working knowledge of host-pathogen interactions and the immune evasion strategies adopted by pathogenic organisms, which will in turn guide the development of therapeutics or vaccines.

Abbreviations

2-DE: two-dimensional gel electrophoresis; BLAST: Basic Local Alignment Search Tool; ER: endoplasmic reticulum; ESP: excretory/secretory protein; EST: expressed sequence tag; GO: gene ontology; HT: host targeting; IgA: immunoglobulin A; MALDI-TOF: matrix-assisted laser desorption/ionization-time of flight spectrometry; MASCOT: Modular Approach to Software Construction Operation and Test; MS: mass spectrometry; NCBI: National Centre for Biotechnology Information: USA; SDS-PAGE: sodium dodecyl sulfate polyacrylamide gel electrophoresis.

Methanogen

Methanogens are microorganisms that produce methane as a metabolic byproduct in anoxic conditions. They are classified as archaea, a group quite distinct from bacteria. They are common in wetlands, where they are responsible for marsh gas, and in the guts of animals such as ruminants and humans, where they are responsible for the methane content of belching in ruminants and flatulence in humans. In marine sediments biomethanation is generally confined to where sulfates are depleted, below the top layers. Others are extremophiles, found in environments such as hot springs and submarine hydrothermal vents as well as in the "solid" rock of the Earth's crust, kilometres below the surface.

Physical Description

Methanogens are usually coccoid (spherical) or bacilli (rod shaped). There are over 50 described species of methanogens, which do not form a monophyletic group, although all methanogens belong to Archaea. Methanogens are also anaerobic.

Although methanogens cannot function under aerobic conditions they can sustain oxygen stresses for a prolonged time. *Methanosarcina barkeri* is exceptional in possessing a superoxide dismutase (SOD) enzyme, and may survive longer than the others.

Some methanogens, called hydrogenotrophic, use carbon dioxide (CO_2) as a source of carbon, and hydrogen as a reducing agent. Some of the CO_2 is reacted with the hydrogen to produce methane, which produces an electrochemical gradient across a membrane, used to generate ATP through chemiosmosis. In contrast, plants and algae use water as their reducing agent.

Methanogens lack peptidoglycan, a polymer that is found in the cell walls of the Bacteria but not Archaea. Some methanogens have a cell wall that is composed of pseudopeptidoglycan. Other methanogens do not, but have at least one paracrystalline array (S-layer) made up of proteins that fit together like a jigsaw puzzle.

Methanogens and (Extreme) Environments

Although most marine biogenic methane is the result of carbon dioxide (CO_2) reduction, a small amount is derived from acetate (CH_3COO^-). Archaea that catabolize this for energy are referred to as *acetotrophic* or *aceticlastic*. Methylotrophic archaea utilize methylated compounds such as methylamines, methanol, and methanethiol as well.

Methanogens play the vital ecological role in anaerobic environments of removing excess hydrogen and fermentation products that have been produced by other forms of anaerobic respiration. Methanogens typically thrive in environments in which all electron acceptors other than CO_2 (such as oxygen, nitrate, sulfate, and trivalent iron) have been depleted. In the deep rock they obtain their hydrogen from the thermal and radioactive breakdown of water.

Methanogens are key agents of remineralization of organic carbon in continental margin sediments and other aquatic sediments with high rates of sedimentation and high sediment organic matter.

Under the correct conditions of pressure and temperature, biogenic methane can accumulate in massive deposits of methane clathrates, which account for a significant fraction of organic carbon in continental margin sediments and represent a key reservoir of a potent greenhouse gas.

Methanogens have been found in several extreme environments on Earth - buried under kilometres of ice in Greenland and living in hot, dry desert soil. They can reproduce at temperatures of 15 to 100 degrees Celsius.

They are known to be the most common archaebacteria in deep subteranean habitats. Live microbes making methane were found in a glacial ice core sample retrieved from three kilometres under Greenland by researchers from the University of California, Berkeley.

Another study has also discovered methanogens in a harsh environment on Earth. Researchers studied dozens of soil and vapour samples from five different desert environments in Utah, Idaho and California in the United States, and in Canada and Chile. Of these, five soil samples and three vapour samples from the vicinity of the Mars Desert Research Station in Utah were found to have signs of viable methanogens.

Some scientists have proposed that the presence of methane in the Martian atmosphere may be indicative of native methanogens on that planet.

Closely related to the methanogens are the anaerobic methane oxidizers, which utilize methane as a substrate in conjunction with the reduction of sulfate and nitrate. Most methanogens are autotrophic producers, but those that oxidize CH_3COO^- are classed as Chemotroph instead.

Exopolysaccharide

Exopolysaccharides are high-molecular-weight polymers that are composed of sugar residues and are secreted by a microorganism into the surrounding environment. Microorganisms synthesize a wide spectrum of multifunctional polysaccharides including intracellular polysaccharides, structural polysaccharides and extracellular polysaccharides or exopolysaccharides (EPS). Exopolysaccharides generally consist of monosaccharides and some non-carbohydrate substituents (such as acetate, pyruvate, succinate, and phosphate). Owing to the wide diversity in composition, exopolysaccharides have found multifarious applications in various food and pharmaceutical industries.

Many microbial EPS provide properties that are almost identical to the gums currently in use. With innovative approaches, efforts are underway to supersede the traditionally used plant and algal gums by their microbial counterparts. Moreover, considerable progress has been made in discovering and developing new microbial EPS that possess novel industrial significance.

Function

The sensory benefits of the exopolysaccharides of lactic acid bacteria are well established and there is evidence for the health properties that are attributable to exopolysaccharides from lactic acid bacteria.

Capsular exopolysaccharides can protect pathogenic bacteria and contribute to their pathogenicity.

Attachment of nitrogen-fixing bacteria to plant roots and soil particles, which is important for colonisation of rhizosphere and roots and for infection of the plant, can be mediated by exopolysaccharides.

An example for industrial use of exopolysaccharides is the application of dextran in panettone and other breads in the bakery industry.

10

Micropropagation in Plants

Low Cost Options in Tissue Culture

The primary application of micropropagation has been to produce high quality planting material Low-cost tissue culture technology is the adoption of practices and use of equipment to reduce the unit cost of micropropagule and plant production.

Low cost options should lower the cost of production without compromising the quality of the micropropagules and plants. In low cost technology cost reduction is achieved by improving process efficiency and better utilization of resources. Low-cost tissue-culture technology will stay a high priority in agriculture, horticulture, forestry, and floriculture of many developing countries for the production of suitably priced high quality planting material.

Hundreds of commercial micropropagation laboratories worldwide are currently multiplying large number of clones of desired varieties and local flora. Apart from the rapid propagation advantage, this technology is being used to generate disease-free planting material, and has been developed and applied to a wide range of crops, and forest and fruit trees. However, in many cases, the cost of micropropagule production precludes the adoption of the technology for large-scale commercial propagation.

The Need For Low Cost Technology

Low-cost tissue culture technology is the adoption of practices and use of equipment to reduce the unit cost of micropropagule and plant production. In many developed countries, conventional tissue culture-based plant propagation is carried out in highly sophisticated facilities that may incorporate stainless steel surfaces, sterile airflow rooms, expensive autoclaves for sterilization of media and instruments, and equally expensive glasshouses with automated control of humidity, temperature and day-length to harden and grow plants.

Many such facilities established at a high cost are high-energy users, and are run like a super-clean hospital. The requirements to establish and operate such tissue culture facilities are expensive, and often are not available in the developing countries. For example, the cost of electricity in the developed countries is much lower, and its supply far better assured than in the developing countries. The same can be said of the supply of culture containers, media, chemicals, equipment and instruments used in micropropagation. Hence, alternatives to expensive inputs and infrastructure have been sought and developed to reduce the costs of plant micropropagation.

Adoption Of Low-cost Options

Low cost options should lower the cost of production without compromising the quality of the micropropagules and plants. The primary application of micropropagation has been to produce high quality planting material, which in turn leads to increased productivity in agriculture. The generated plants must be vigourous and capable of being successfully transplanted in the field, and must have high field survival. In addition, they should be genetically uniform, free from diseases and viruses, and price competitive to the plants produced through conventional methods. Reducing the cost should not result in high contamination of cultures or give plants with poor field performance.

The foremost requirement of micropropagation is the aseptic culture and multiplication of plant material. Microbe-free conditions need to be maintained in culture containers, and during successive subcultures. In many cases, mistakes in concept or practice can introduce microbes in the culture containers from an external source or the plant material itself (endophytic contamination). As a result, the microbes overgrow the cultures, and wipe them out. Microbes may grow slowly under controlled low temperature, but they proliferate very fast under uncontrolled and high temperature. Thus the adoption of wrong low-cost options may make the production process prone to disasters. Low cost techniques will succeed only if the basic conditions for tissue culture are scrupulously adhered to maintain propagule quality.

Microbial contamination of cultures is known to wipe out work of months, and can turn into a nightmare. The best low-cost option is to discard and dispose of contaminated cultures outright. Avoiding contamination in small R&D laboratories is not a difficult task where only a low number of cultures are handled. However, commercial production involves handling of thousands of cultures each day. It is also essential to maintain such cultures in large numbers under contamination-free conditions, until they are used for either further subculture or hardening and growing-on. Plants do not have an immune system and there is a limitation on the use of antibiotics to circumvent the problem.

Moreover, many of the antibiotics, which are effective against bacteria, fungi, and phytoplasmas, are toxic to plants as well. Obviously, the use of antibiotics is not foolproof or the desired method to rid microbial contamination. Sophisticated state-of-the-art facilities are not a guarantee for prevention of contamination. The laboratories that succeed in an immaculate control of contamination do so by adherence to scrupulous techniques of basic tissue culture. Thus, it is not the sophistication but the procedures that ensure the quality of tissue cultured plants.

Quality Of Micropropagules

Low cost technology means an advanced generation technology, in which cost reduction is achieved by improving process efficiency, and better utilization of resources. Presently, both the developing and the developed countries require low cost technology to progressively reduce the cost of propagule production. In many developing countries, the potential end-users of plants derived from tissue culture have been the resource-rich farmers. They know the benefits and potential of healthy planting material. Such growers are prepared to risk investment in the high productivity potential of the planting material. For example, hybrid seeds of many vegetables, papaya, rice, and cotton cost 15-20 times more than the price of ordinary varieties.

Yet there is a wide market for them. Hence, the production of low quality plants, just because they are less costly, is not going to be a sustainable approach for the application of micropropagation in agriculture. Lowering of cost of production is possible only if the methods do not compromise the basic imperatives of tissue culture and quality of plants.

Importance Of Low Cost Technology

The potential of plant tissue culture in increasing agricultural production and generating rural employment is well recognized by both investors and policy makers in developing countries. However, in many developing countries, the establishment cost of facilities and unit production cost of micropropagated plants is high, and often the return on investment is not in proportion to the potential economic advantages of the technology.

These problems can be addressed by standardizing agronomic practices more precisely (precision agriculture) and by achieving maximum net profits from the crops or by decreasing the unit cost of production or both. The technology is particularly relevant to the propagation of because they command high unit value. However, the market is limited. Over a period of time, many new tissue-culture companies in several developing countries have entered to compete in the limited market. The inevitable result has been the reduction of net margins below viable limits.

Many international organizations also agree that tissue culture technology is very relevant to agriculture, provided the problem of high cost of production is satisfactorily solved. The role of tissue culture was clearly recognized by the FAO in the paper on 'Biotechnology in agriculture, forestry and fisheries-FAO's Policy and Strategy'. The report pointed to the wide use of tissue culture techniques for multiplication of elite clones and elimination of pathogens in planting material. It also pointed to the successful tree regeneration in about 100 forest species and its value for breeding, clonal testing and rapid deployment of superior genotypes.

Several R&D projects have been undertaken to improve the productivity of agricultural, horticultural and forest trees by the European Union under Co-operation in the Field of Scientific and Technical Research (COST). Under this programme, coordinated and funded by the European Union, one of the primary aims has been to reduce micropropagation cost. For example, the objective of 'COST 843' action has been the innovation of low-cost plant propagation methods that enhance sustainable and competitive agriculture and forestry in Europe.

The high costs of labour of micropropagation are a major bottleneck in the EU to fully exploit in vitro culture technology. In the EU, labour currently accounts for 60-70 per cent of the costs of the in vitro produced plants. In another programme, the large-scale production and introduction of bamboo in the EU using tissue culture technology has been undertaken with the main objective of reducing the costs of micropropagation.

Future Role

It has been stressed time and again that in the long-term agriculture and forestry need to be sustainable, use little or no crop-protection chemicals, have low energy inputs and yet maintain high yields, while producing high quality material. Biotechnology-assisted plant breeding is an essential step to achieve these goals.

Plant tissue culture techniques have a vast potential to produce plants of superior quality, but this potential has been not been fully exploited in the developing countries. During in vitro growth, plants can also be primed for optimal performance after transfer to soil. In most cases, tissue-cultured plants out-perform those propagated conventionally. Thus in vitro culture has a unique role in sustainable and competitive agriculture and forestry, and has been successfully applied in plant breeding, and for the rapid introduction of improved plants. Bringing new improved varieties to market can take several years if the multiplication rate is slow.

For example, it may take a lily breeder 15-20 years to produce sufficient numbers of bulbs of a newly bred cultivar before it can be marketed. In vitro propagation can considerably speed up this process. Plant tissue culture has also become an integral part of plant breeding. For example, the development of pest- and disease-resistant plants through biotechnology depends on a tissue culture based genetic transformation. The improved resistance to diseases and pests enables growers to reduce or eliminate the application of chemicals.

The FAO Committee on Agriculture has perceived plant tissue culture as a main technology for the developing countries for the production of disease-free, high-quality planting material, and its commercial applications in floriculture and forestry. It further points out that tissue culture techniques are being used particularly for large-scale plant multiplication. Micropropagation has proved especially useful in producing high quality, disease-free planting material for a wide range of crops. Tissue cultured based industry also generates much-needed rural employment, particularly for women.In forestry, the availability of tissue culture linked production systems may effectively provide sustainable alternatives to the need for harvesting wood from native forests and natural habitats. Successful protocols now exist for the micropropagation of a large number of forest tree species, and the number of species for which successful use of somatic embryogenesis is increasing.

Thus in the future, it is likely that micropropagation in the forestry sector will become commercially important. Compared to vegetative propagation through cuttings, the high multiplication rates available through micropropagation offer a much quicker capture of genetic gains obtained in forest tree breeding programs. However, the current high costs will also be one of the major impediments to the direct use of micropropagation in many programs.

The broad application of existing technologies to plantation species is important for tree improvement in the tropics. In a small number of plantation programs, micropropagation is being used as an early rapid multiplication step. However, it has been pointed out that the current high costs will be an impediment to the direct use of micropropagules as planting stock. Micropropagation clearly has a role, in the rapid multiplication of the selected clones for conventional production of cuttings.

The direct use of micropropagules as planting stock in industrial plantation can dramatically broaden forestry tree farming if propagation costs are reduced. The availability of micropropagation technologies will also be useful in genetic engineering applications, *e.g.*, the production of plants as a source of "edible" vaccines. There are many other useful plant-derived substances which can be produced in tissue cultures, sometimes more cheaply and reliably than from natural forests and plantations. These include medicinal compounds and drugs now being sought in major prospecting operations in the tropical forests.

Micropropagation has been identified as a suitable technology in the development projects of UNESCO in Africa and the Caribbean; however, the cost of production must be reduced. Practically in all developing countries, the private industry is the most important group that requires cost-effective technology. For example, in India of the 90 commercial micropropagation units established initially, were closed down. Of those engaged in commercial production, many are uneconomic mainly due to the high cost of production and the absence of quality tests. Hence, low-cost tissueculture technology will stay a high priority in agriculture, horticulture, forestry, and floriculture of many developing countries.

Physical Components Of Tissue Culture Technology

The physical components of a typical plant tissue culture facility include equipment and buildings with preparation room, transfer room, culture or growth room, hardening and weaning area, soil-growing area (greenhouses, plastic tunnels), packaging and shipping area, and related facilities – office and store for chemicals, containers and supplies. Careful planning is the first important step when considering the size and location of a facility.

The size of the physical components of a tissue culture facility will vary according to its functional needs – the volume of production. It is recommended that an existing facility should be visited to view the layout and operational needs before starting a new facility. A number of low-cost alternatives can be used to simplify various operations and reduce the costs in a tissue culture facility.

In designing any laboratory, big or small, certain elements in its design and layout are absolutely essential for its successful operation. Correct design of a laboratory will not only reduce contamination, but also achieve a high efficiency in work performance. Properly planned and designed laboratories can reduce both the operational and energy costs. A tissue culture laboratory must be designed to accommodate the equipment and its use in the various stages of micropropagation in the most efficient manner.

General Consideration for Location

A convenient location for a small laboratory can be a room or part of the basement of a house, a garage, a remodeled office or a room in the header house. The minimum area required for media preparation, transfer and primary growth shelves is about 14 m^2. Walls should be installed to partition different areas.

Before setting up a commercial micropropagation unit, it is essential to check out the area keeping in mind the climate, and access to water, electricity, transportation, and infrastructure for supplies. A temperate climate is usually better suited to tissue culture ventures. This greatly reduces the cost of cooling required to maintain the temperature for optimum growth of the cultures.

The availability of electricity and water is of utmost importance, and should be taken into consideration while choosing the location of the facility. For example in India, of the 76 commercial tissue culture units, nearly 52 are located in and around the cities of Bangalore and Pune, where the climate is moderate.

Hence, However, in cities like Delhi, which have extremes of climate, tissue culture facilities require both heating and cooling. In a facility, which produces five million plants, the electricity cost per thousand plants is around US $0.30 in Bangalore and Pune; the same is about US $0.80 in Delhi. Disruption of power and water supply causes major breakdown in the smooth running of tissue culture units. Poor quality water adds to the cost of media.

Location consideration should also include a check with local authorities about zoning and building permits before construction begins. The buildings should be located away from sources of contamination such as a gravel driveway, parking lot, soil mixing area, shipping dock, pesticide storage, and dust and chemicals from fields. Sanitation of the area is important in selecting location.

The units should be located in areas where insect populations are minimal and air has low dust and pollen counts. In the case of export-oriented units, these should be near an international airport, which will reduce the time lag between packaging and shipment. This is critical to assure timely delivery of quality tissue culture products. For new ventures, the size of the facility should be kept small until the market acceptance is ensured.

Design and Lay Out of the Laboratories

Plant tissue culture laboratories have specific design requirements. Careful initial planning is, therefore, a prerequisite for successful running of a facility. The location and design of the laboratories should take into account isolation from foot traffic, control of contamination from adjacent rooms, thermostatically controlled heating and cooling, water supply and drains for a sink, adequate electrical service, provisions for a fan and intake blower for ventilation, and good lighting.

Large sized facilities are frequently built free standing. Although more expensive to build, the added isolation from adjacent activities keeps the laboratory clean. Prefabricated buildings make convenient low-cost laboratories. They are readily available in various sizes in many countries. Prefabricated buildings assembled on site can also be used. A single span building allows for a flexible arrangement of walls for dividing into convenient sized rooms.

The floor should be of concrete or capable of carrying 170.5 kg/m^2 (50 pounds per sq. foot). Walls and ceiling should be insulated to at least R-15, and covered inside with waterresistant material. Windows, if desired, may be placed wherever convenient in the media preparation and glassware washing rooms.

The heating system should be capable of maintaining room temperature at 20°C during the coldest part of winter. A minimum of 2cm pipes should be used for water supply. Connection to a septic system or sanitary sewer should be provided. Air conditioning requirements should carefully estimated.

Electrical service capacity for equipment, lights and future expansion should be calculated. For safety reasons, the electrical installation should be carried out professionally. Most electrical wiring will require 220 Volts, and autoclaves 230/250 Volts.

The areas such as the media preparation room, inoculation room and growth chambers should be isolated as 'clean zones'. The office, storage area, staff centre and packaging rooms can be maintained under ordinary conditions. The working areas must be demarcated according to the activities involved in the facility. Cleanliness is the major consideration when designing a plant tissue culture laboratory to minimise contamination.

A positive pressure module should be installed to circumvent air intake from outside. Routine cleaning and aseptic procedures can decrease contamination losses to less than 1 per cent. An enclosed entrance should precede the laboratories, and sticky mats should be placed to collect dirt from shoes.

The traffic pattern and workflow in the laboratory must be considered to maximise cleanliness. The cleanest rooms or areas are the culture room (aseptic transfer area) and the growth room. There should be no direct access to these rooms from outside. The media preparation area, glassware washing and storage areas should be located away from these rooms.

The growth room and aseptic transfer rooms should be fitted with see-through doors and should be adjacent to each other. Traffic through these areas should be minimal and restricted to the personnel working under laminar flow cabinets. Ideally, the media preparation area leads into the sterilisation area, which leads into the aseptic transfer room, and eventually to the growth room. Temperature and fire alarms must be connected directly to telephone lines to give fast warnings. An emergency generator should be available to operate essential equipment during power breakdown.

Essential Equipment

The basic equipment in most tissue culture facilities includes the following:

Autoclave

An autoclave is basically a large-sized but sophisticated pressure cooker, and is used for the sterilisation of the medium, glassware and instruments. Autoclaves of different sizes are available commercially. High-pressure heat is needed to sterilise media, water, and glassware. Certain spores from fungi and bacteria are killed only at 121°C and 1.05kg/sq.cm (15 pounds per sq. inch) pressure. Self-generating steam autoclaves are more dependable and faster to operate.

Laminar Airflow Chamber

The laminar flow chambers provide clean filtered air that allows cultures to be handled under contamination-free environment. Several types of laminar flow chambers are sold on the market and are available in different sizes. The laminar-flow cabinets are located in the culture transfer area. Some large-sized laboratories have sterile rooms in addition to laminar flow cabinets.

Other Equipment

The sterilisation of instruments, such as the forceps, scalpel holders and blades is achieved with either gas flamed burners or with glass-bead sterilizers. The medium preparation room usually has the following equipment. A refrigerator-freezer to store chemicals and stock solutions, weighing scales for large amounts of over 10 g, and an analytical balance with1 mg accuracy, a magnetic stirrer for the agitation, and a pH meter. Small laboratories may locate the refrigerator under the workbench to save space. High quality balances are essential in a tissue culture laboratory. Most laboratories have top loading balances, which allow quick and efficient weighing. A hot plate with an automatic stirrer is needed to for preparing the media before autoclaving.

The pH meter is needed to determine pH of the media. Some laboratories use pH indicator paper, however this method is considerably less accurate, and can severely affect the results. An aspirator can be attached to a water tap for filter sterilisation of chemicals and for surface sterilisation of the plant material. However, vacuum pumps are faster and more efficient, but also more expensive.

A drying oven is required to keep glassware such as beakers, flasks and cylinders, and is also useful for dry sterilisation of scalpels and glassware, such as Petri dishes, pipettes and others. The media containing carbon sources (*e.g.* sugars) and growth regulators are sterilised in the autoclave, but sometimes, aseptic filtration is better to avoid breakdown of heat-labile chemicals.

The water still is also located in the medium preparation area. To prepare media, distilled or de-ionised water is generally used, although tap water can be used in some cases.

Optional Equipment

A variety of non-essential equipment is used in tissue culture laboratories. The specific requirements determine what need to be purchased. Microwave ovens are convenient for defrosting stock solutions and pre-heating agar media. Most laboratories have a dissecting microscope to excise small explants. Laboratory glassware washers or regular dishwashers can be used for replacing manual labour.

Automatic media dispensers are helpful to pipette pre-set volume of media. A gyratory shaker or a reciprocal shaker is necessary if micropropagation is based on liquid media or suspension cultures. Computers, photocopiers and fax machines are helpful for easy data management and maintenance of records. Some of the equipment may be costly, but goes a long way in saving time and labour and is essential for rapid communication in the competitive world.

Activity Specific Requirements

Based on the different activities of a tissue culture, a facility can be divided into semiclean, clean and ultra-clean areas. The semi-clean areas comprise of the washing room, office and staff restrooms, where there is no need for maintaining sterile conditions. The clean areas encompass the media preparation and sterilisation rooms, which have to be sufficiently clean. High sterility has to be maintained in the culture transfer rooms and the growth rooms, which constitute the ultra-clean areas.

Glassware Washing and Storage Area

The glassware washing area should be located near the sterilisation and medium preparation rooms. This area should have at least one large sink but two sinks are preferable. Adequate workspace is required on each sides of the sink; this space is used for glassware soaking and drainage. Plastic netting can be placed on surfaces near the sink to reduce glassware breakage and enhance water drainage. The outlet pipe from the sink should be of PVC to resist damage from acids and alkalis.

Both hot and cold water should be available and the water still and de-ionisation unit should be located nearby. The choice of electrical washers should be based on the projected use, durability, reliability and cost, and service availability. In India and some other developing countries, where labour is relatively cheap, washing is done manually. The washing room should be swapped periodically.

Mobile drying racks can be used and lined with cheesecloth to prevent water dripping and loss of small objects. Ovens or hot air-cabinets should be located close to the glassware washing and storage area. Dust-proof cabinets and storage containers should be installed to allow for easy access to glassware. When culture vessels are removed from the growth area, they are often autoclaved to kill contaminants and to soften semi-solid media. It should be possible to move the vessels easily to the washing area. The glassware storage area should be close to the wash area to expedite storage and access for media preparation.

The Store

It is advisable to have a separate area for storage of chemicals, apparatus and equipment. It would not only facilitate constant availability but also save cost from bulk purchase. Chemicals required in small amounts should not be purchased in large quantities as they may lose their activity, pick up moisture or get contaminated. Such problems can be overcome by purchasing small lots on a regular basis.

Media Preparation and Sterilisation Area

The media preparation room should have smooth walls and floors, which enable easy cleaning to maintain a high degree of cleanliness. Minimum number of doors and windows should be provided in this room but within the local fire safety regulations.

This reduces cost and contamination. The media preparation and sterilisation can be carried out in the same area but preferably in different rooms, which need not be separated with doors. Media preparation area should be equipped with both tap and purified water. An appropriate system for water purification must be selected and fitted after careful consideration of the cost and quality.

A number of electrical appliances are required for media preparation; hence, it is essential to have safety devices like fire extinguisher, fire blanket and a first aid kit in the media preparation room. A variety of glassware, plastic ware and stainless steel apparatus is required for measuring, mixing, and media storage. These should be stored in the cabinets built under the worktables and taken out for use as and when required.

This would save the cost and space for building storage shelves. The use of glassware should be kept at a minimum, as it will help in reducing losses due to breakage. As far as possible, plastic ware and stainless steel vessels should be used, as they are much cheaper and more durable than glassware. The water source and glassware storage area should be in or near the medium preparation area. Work bench tops, suitable for comfortable working while standing should be 85 to 90 cm high and 60 cm deep. The workbench tops should be made with plastic laminate surfaces that can tolerate frequent cleaning.

Sterilisation Room

The sterilising room should be in continuation with the media preparation room. The layout must be planned in such a way that it ensures the smooth movement of the containers from the washing to the media preparation and sterilisation room. The sterilisation room must have walls and floors that can withstand moisture, heat and steam. An exhaust should be fitted to remove the warm and moist air. The exhaust fan should have an outer cover to prevent entry of outside air. The fan cover should open only when the fan is in operation.

In small tissue culture facilities, costly autoclaves can be replaced by simple pressure cookers. However, for large volume media making, horizontal or vertical autoclaves should be installed. Double door autoclaves, which open directly into the media storage room, may be costly but reduce contamination. A cheaper alternative is to transfer the sterilised media to the adjoining room through a hatch window.

Transfer room

The most important work area is the culture transfer room where the core activity takes place. The transfer area needs to be as clean as possible and be a separate room with minimal air disturbance. Walls and floors of the transfer room must be smooth to ensure frequent cleaning. The doors and windows should be minimal to prevent contamination, but within local safety code. There is no special lighting requirement in the transfer room. The illumination of the laminar airflow chamber is sufficient for work. Sterilisation of the instruments can be done with glass-bead sterilizers or flaming after dipping in alcohol, usually ethanol.

The culture containers should be stacked on mobile carts (trolleys) to facilitate easy movement from the medium storage room to the transfer room, and finally to the growth room. The chair seats of the transfer operators should be comfortable, as they have to work for long periods in the same place. Fire extinguishers and first aid kits should be provided in the transfer room as a safety measure.The personnel should leave shoes outside the room. Special laboratory shoes and coats should be worn in this area. Ultraviolet (UV) lights are sometimes installed in transfer areas to disinfect the room; these lights should be used only when people and plant material are not in the room. Safety switches can be installed to turn off the UV lights when regular room lights are turned on.

Growth Room

Growth room is an equally important area where plant cultures are maintained under controlled environmental conditions to achieve optimal growth. It is advisable to have more than one growth room to provide varied culture conditions since different plant species may have different requirements of light and temperature during in vitro culture. Also, in the event of the failure of cooling or lighting in one room, the plant cultures can be moved to another room to prevent loss of cultures.

In the growth room, the number of doors should be minimal to prevent contamination. There is no need for windows in the growth room, except when natural light is used. When artificial lighting is used, the external light can interfere with the photoperiod and temperature of the growth room. Depending on the amount of available space and cost, the culture containers can be placed on either fixed or mobile shelves. Mobile shelves have the advantage of providing access to cultures from both sides of the shelves. The height of the shelves should not exceed 2m. High shelving requires step-up stools to place and remove cultures, being dangerous and time consuming.

The primary source of illumination in the growth room is normally from the lights mounted on the shelves. Overhead light sources can be minimised, as they would be in use only while working during the dark cycle. Plant cultures may not receive uniform light from the conventional downward illumination. Lights directly fitted to the racks create uneven heat distribution. This leads to high humidity within the culture containers, which in turn can cause hyperhydricity. Sideways illumination is an alternative, which requires less number of lights, and provides more uniform lighting. But care has to be taken not to break the lights while moving the cultures across the shelves.

Control Of Growing Conditions

Controlled temperature, lighting and relative humidity, and shelving need to be considered in planning the growth room. These vary depending on the size of the growth room, its location, and the type of plants cultured. For example, a small growth room located in the cool North American climate can be placed in an unheated or minimally heated basement.

The chokes (ballasts) of the fluorescent lights need not to be separated; rather they can serve as a source of heat. Excess heat can be dissipated from the growth room, and used for heating other areas in the basement. In such a situation, solid wooden shelves with space between shelves can be used and prevent culture vessels on shelf above the lights from becoming over-heated. However, a large growth room located above ground needs to have the light chokes installed outside the room. Shelves in large growth room can be of glass or metal wire mesh.

Temperature Control

Temperature is a primary concern in growth rooms; it affects decisions on installation of lights, control of relative humidity, and type of shelving. Temperature in the growth room is usually controlled with air conditioners. Generally, temperatures are kept around 22°C. Heating is provided from conventional heating systems and can be supplemented with heat from light chokes. In most developing countries, cooling the growth room is usually a bigger problem than heating. Cooling can be provided with heat pumps, air conditioners and exhaust fans. Using open windows to cool culture rooms leads to contamination during summer and humidity problems in winter.

Lighting Control

Some plant cultures can be kept in complete darkness; however, most culture rooms need to be illuminated at 1 Klux 134.5 μ mole/m^2/s (microeinsteins per second per sq. centimetre or approximately 1076 foot candle) with some up to 5 to 10 Klux (672-1345 μ mole/m^2/s).

The plant species and/or propagation scheduling determines the light intensity. The developmental stage of the plants also determines if wide spectrum or cool white-fluorescent lights are to be used. Rooting is strongly influenced positively with far-red light; therefore, wide spectrum lights should be used during Stage III and cool-white lights during Stage I and II. Automatic timers are needed to maintain the desired photoperiod. Reflectors can be placed over bulbs to direct the light downwards and evenly. Heat generated by lights may cause condensation and temperature problems. Small fans placed at the end of the shelves increase airflow and decrease heat build-up.

Reflective glossy paint on the walls provides an even light distribution as well as reduces the number of lights required. Relative humidity (RH) is difficult to control inside the culture vessels, but wide fluctuations in the growth room have a deleterious effect. Cultures can dry out if the room's RH is less than 50 per cent. Humidifiers can be used to correct this problem. If the RH becomes too high, a dehumidifier is recommended.

Shelving

Shelves within the growth rooms vary depending upon the situation and the plants grown. Frames for the shelves can be made from 1.25cm (half-inch) thick angle iron. Shelves built from rigid wire mesh to allow maximum air movement and minimise shading should be used. Wood is inexpensive to build shelves. The wood for shelving should have smooth exterior, and should be painted white to reflect light.

Expanded metal is more expensive than wood, but provides better air circulation. Tempered glass is sometimes used for shelves to increase light penetration, but it is more prone to breaking. Air spaces of 5 to 10cm between the lights and shelves decrease heat on upper shelves and reduce condensation in culture vessels. A room that is 2.4m high will accommodate 5 shelves, each 45cm. apart, when the bottom shelf is 10 cm above the floor.

Greenhouse Facility

A critical stage in plant tissue culture is the interim phase between the laboratory and field conditions. In vitro derived plants need to be gradually hardened to field conditions. Plant hardening is usually carried out under greenhouse that ensures high survival of the tissue-cultured plants in the field. There are three types of greenhouses: Ground to ground, Gable, and Quonset type. The most commonly used greenhouse is the Quonset type. It contains movable or fixed benches with hardening tunnels on them. The size of the greenhouse must be based on the scale of production. Greenhouse glazing can be of glass or fibreglass.

Polyethylene films or sheets of polycarbonate or acrylic can also be used. Air inflated double polyethylene covering is the most economic. Appropriate light, shading and blackout systems can be achieved with supplementary lighting. Drip irrigation systems, misting and fogging can be installed as needed. Greenhouses erected in warm climates should have fan-assisted drip pad cooling especially during summer. Greenhouses in colder climates need to be heated. Floor and bench systems can be used for heating and cooling the air. Low cost plastic pipes can be used to circulate warm air, which are adequate and cost effective.

Packaging And Shipping

A separate area should be designated for packaging in a commercial tissue culture unit. Packaging materials such as cardboard cartons and labels should be stored in this area. The type of packaging of a particular plant depends greatly on the temperature zones through which the consignment has to pass from the point of shipment to its destination.

For example, if the consignment of tissue-cultured plants is passing anywhere within temperate zones, it is only necessary to take care of the frost conditions and pack accordingly. But if the consignment is passing from a tropical country to a temperate country, it becomes necessary to take care of the temperature zones in different places through which the consignment passes. Thus, the type of packaging of tissue-cultured products varies with plant and destination. For example, in a package of 22 kg, about 40,000 lily plantlets can be packed.

However, in the same parcel, only 20,000 Spathiphyllum plantlets or 10,000 Syngonium plantlets can be packed because of the higher respiration rates in these species. The heat build-up is much more in Spathiphyllum and Syngonium plants than the dormant lilies. Similarly, only 8000 Gerbera or 7000 Cordyline plants can be packed in the same package. Cooling material such as ice, dry ice and other material take up much of the packaging space.

Before loading and shipping, the packed items should be properly counted and rechecked. The cartons containing the cultures for shipping to the customer should be properly labelled with the names and addresses of the consignor and the consignee, and the details of the commodity (storage temperature, handling, etc.). Adequate care must be taken while packaging large consignments, so that there is no disturbance or damage during transit. To prevent any sort of delay, ensure that the consignment is accompanied by documents such as invoice, packing list, import permit, phytosanitary certificate and Generalized System of Preferences (GSP).

Cost Of Facility

The building of a tissue culture facility includes the cost of land, construction, electrical installation and plumbing. If the available funds are limited, well-designed laboratories can be established by modifying existing structures. For example, a two or three room house or a trailer with appropriate modifications can be converted into a medium sized micropropagation facility. Depending on the production and storage capacity, a facility can be small- less 100,000 plants, medium- 100,000 to 500,000 plants, or large-scale- 500,000 to 2,000,000 plants per annum.

The initial investment required for setting up a low cost, medium scale tissue culture laboratory (ca. 195 m^2) in India. Land is sold at a premium, especially near large cities. In India, the land and building costs work out to about US \$62500. A house, ca. 195 m^2, can be converted into a tissue culture facility with a capacity of 200,000 plants.

The rental charges for such a structure would be around US \$5000 per year. The initial investment would thus include only plumbing and electrical work, reducing the capital cost to US \$2750. Thus, in large cities, it would be more economic to rent than build the premises. If the hardened tissue-cultured plants were the end products, an additional investment of US \$20,000 would be required to erect a greenhouse. If the production is limited to-in-agar and ex-agar products (rooted or non-rooted micro-cuttings), greenhouse would not be required.

The basic equipment will cost US \$14000. The consumables, such as glassware, chemicals and disposable items would require another US \$4200. Besides the fixed capital, a commercial tissue culture unit has the recurring expenses of rent, building maintenance, electricity and overheads. The estimated cost of furniture that includes racks with illumination source for growth room, cupboards, benching and chairs would be US $7400. Other miscellaneous items such as labelling and sealing machines would cost US $50.

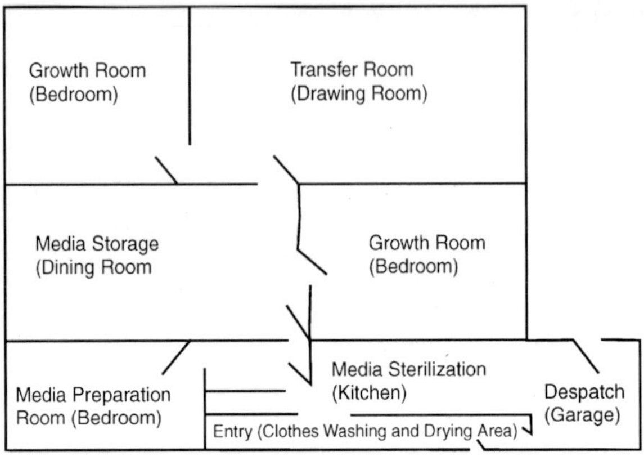

Figure: *Conversion of a House into a Micropropagation Facility*

Strategies For Cost Reduction

For a commercial tissue culture unit to be successful, it is essential to constantly find means to increase the efficiency of production, and bring down the cost of production. A number of low-cost alternatives can be used to simplify various operations and reduce costs in a tissue culture facility.

Contamination Control

The loss of cultures increases the cost of production. Contamination in cultures is caused from natural contaminants such as dust, air-borne particles, bacterial and fungal spores, fibres, and hair. Man-made contamination occurs mainly from body, clothing, and from faulty procedures in the laboratory. It has been estimated that each operator generates a minimum of 1 to 5 million particles (bigger than 0.5 μ M diameter) per minute. Modern, high quality clean rooms have environment control systems that minimise contamination. Wearing laboratory coats (which are mandatory in Class 100 clean rooms) reduces contamination from clothing, skin and hair. Maintaining cleanliness in other working areas is equally important.

There is continuous entry of contaminants with the worker and through air entering the growth room. A set routine procedure should be followed to reduce contamination in the laboratory; however, such procedures will vary with the location. In the absence of high cost systems, careful planning and appropriate design of the laboratory goes a long way in reducing contamination.

Washing and Sterilising Operations

Costly dishwashing machines for cleaning the culture containers should be replaced by manual washing if labour is relatively inexpensive. The washed culture containers can be dried in sun instead of costly hot air ovens. In small-scale laboratories, the autoclaves can be replaced by large sized pressure cookers, which are much cheaper. Instead of having one or two huge horizontal autoclaves, which generate hot air pockets in the sterilising rooms, it is better to have more of smaller vertical autoclaves, which keep the air cool. It is a normal practice to use costly aluminium foil for wrapping instruments for sterilisation. This can be replaced by stainless steel containers, which are autoclavable. Savings can also be made in media and containers.

Transfer Operations

Instead of Petri dishes as stage for manipulation of culture, stainless steel plates, ceramic tiles and brown wrapping paper can be used; all of these can be autoclaved. Ethanol for hand and workbench sterilization can be replaced by industrial spirit.

The careless handling of inflammables used for sterilization can be hazardous. Glass bead sterilizers can be used to sterilize forceps and scalpels instead of the conventional flame sterilization using spirit lamps or gas cylinders. Commercial bleach has also been successfully used in several laboratories for bench and instrument sterilization to reduce cost, and to prevent fire hazard.

Culture Maintenance in Growth Rooms

Plant cultures can be maintained in rooms with air conditioners and tube lights instead of highly priced plant growth chambers. The conventional method of downward illumination can be replaced by sidewise lighting systems, which not only reduces the number of lights but also provides more uniform illumination to the cultures. In the tropical and Mediterranean regions, the electrical lighting systems can be replaced by sunlight. The tissue culture laboratories in Cuba produce millions of tissue-cultured sugarcane, pineapple and banana plants using natural light.

Production Planning

Most plants are seasonal in demand. To meet the requirements of extremely large number of plants, commercial production has to be backed by well defined working procedures and monitoring performance of the operation. Commercial laboratories should produce a range of plants for different seasons to maximise the use of the facilities throughout the year. This lowers the unit cost of plant production. Large-scale micropropagation is a labour-intensive process, and therefore organising the availability of personnel is quite important. To reduce organisational problems in big units, the management structure must be well planned. Supporting information systems such as inventory control, production scheduling, space utilisation and daily targets should be well defined.

To produce quality products on large-scale, there should be good co-ordination among the technicians, supporting staff, supervisors and the researchers. The job description and reporting system should be very clearly stated. In addition, personnel selection and training is critical for successful large-scale production. The production needs to be periodically reviewed to meet needs of the customers.

Scale of Operation

The physical components required for the functioning of a commercial tissue culture laboratory can be either scaled up or down according to the interests of the propagator. The correct design of a laboratory, big or small, will help maintain asepsis, thereby increasing the efficiency of the unit and achieving a high standard of work. A micropropagation company in India converted a three-room apartment into a medium-sized tissue culture laboratory. Plant such as Spathiphyllum, Syngonium, Ficus, hosta, calla lily, gerbera, and cordyline, were produced on a commercial scale.

Delivery of the tissue-cultured plants to the tune of 2.5 million per year has been made from this unit by adopting various low cost alternatives. Similar tissue culture facilities, aptly called 'bio-factories', in villages of Cuba produce up to one million banana plants and 2.5 million sugarcane plants annually. Significant cost reduction for large-scale production can be achieved with automation and mechanisation.

Gelling Agents

The growth of cultures and production of shoots or roots is strongly influenced by the physical consistency of the culture medium. Gelling agents are usually added to the culture medium to increase its viscosity as a result of which plant tissues and organs remain above the surface of the nutrient medium. Many gelling agents are used for plant culture media, *e.g.*, agar, 'Agarose', and 'Gellan gum', and are marketed under trade names such as 'Phytagel, Gelrite', and 'Gel-Gro' (ICA Biochemicals). Agar is the most commonly used gelling agent for preparation of solid and semi-solid media.

It contributes to the matrix potential, the humidity and affects the availability of water and dissolved substances in the culture containers. Various brands and grades of agar are available commercially, which differ in the amounts of impurities, and gelling capacity.

Agar brands vary widely in price, performance and composition. It is the actual use and experience, which ultimately determines the choice of agar brand in a specific system and for a plant species. It is usually unnecessary to use high purity agar for large-scale micropropagation; cheaper brands of agar have been successfully used for industrial scale micropropagation. The lowest concentration of agar, which can be used, depends on its purity and brand. Agar is usually used at 0.6-0.8 per cent (w/v). It is advisable to prepare sample media in small quantities using various concentrations of agar, *e.g.*, 0.7, 0.75, 0.8, 0.85 and 0.9 per cent.

The appropriate concentration should then be used for large-scale production purposes. In addition to cost saving, there are a number of other advantages in using low concentrations of gelling agents. A semi-solid medium ensures adequate contact between the plant tissue and the medium. It is beneficial to growth as it allows better diffusion of medium constituents, and is easily removed from plantlets before their transfer to in vivo conditions. For these reasons, a semi-solid medium is often preferred over solid medium.

Alternatives to Agar

Cheaper alternatives to agar include various types of starches and plant gums. The National Research Development Corporation, India has listed low cost agar alternatives, which are worth evaluating for routine use in commercial micropropagation. Gelrite can be replaced with starch-Gelrite mixture. The use of liquid media eliminates the need of agar. Other options include white flour, laundry starch, semolina, potato starch, rice powder and sago. For micropropagation of ginger and turmeric, the combination of certain gelling agents gave growth as good as on agar-based media. The use of laundry starch, potato starch and semolina in a ratio of 2:1:1 reduced the cost of gelling agent by 70-82 per cent. However, the addition of such gelling agents to the medium also has some disadvantages. Some gelling agents contain inhibitory substances that hinder morphogenesis, and reduce the growth rate of cultures. Moreover, toxic exudates from the cultured explants may take a longer time to diffuse. Media solidified with gelling agents increase the time to clean the culture containers.

Corn-starch (CS) as a gelling agent has been used along with low concentration of 'Gelrite' (0.5 g 'Gelrite' + 50.0 g CS/l) for the propagation of fruit trees, such as apple, pear and raspberry, banana, and sugarcane, ginger and turmeric. The shoot proliferation was better on corn starch-medium than on agar. The cost of CS was $1.8/kg compared with $200/kg of agar. However, it became difficult to detect the contamination because the CS medium turned grayish-white.

Addition of 8.0 per cent tapioca starch to the MS medium was found to be a good substitute for 'Bacto-agar' for potato shoot culture. Barley starch (60.0 g/l) has also been used for culturing potato-tuber discs, and for anther culture of barley. Sago (obtained from the stem pith of Metroxylon) at 13 per cent concentration was substituted for agar in MS medium for the multiplication of chrysanthemum. The number of shoots and leaves, and root length were significantly higher on sago than on agar. The cost of sago is $0.5/kg. 'Isubgol', a colloidal mucilaginous husk (chiefly composed of pentosans) derived from the seeds of Plantago ovata, has a good gelling activity, and has reasonable clarity in gelled form. 'Isubgol' at 3 per cent in MS medium has been used for the propagation of chrysanthemum. The cost of 'Isubgol' is about $4/kg.

Table: *Effect of Agar Alternatives on Medium Solidification and Culture Growth*

Agar alternative	*Effect on medium and culture*
Wheat flour (10%)	Sloppy medium, growth of cultures poor
Wheat flour (8%)	Proper solidification, growth of cultures below average
Laundry starch (6%)	Proper solidification, growth of cultures average
Semolina (5%)	Sloppy medium, growth of cultures average
Potato powder (7%)	Sloppy medium, growth of cultures average
Rice powder (11%)	Sloppy medium, growth of cultures poor
Sago (7%)	Proper solidification, growth of cultures normal
Laundry starch + Potato powder +	Solidification, growth of cultures as good as
Semolina (2:1:1)	On agar media

Use of Liquid Media and Physical Matrices

Suspension cultures without gelling agents are commonly used for culturing callus, cell clusters, buds and somatic embryos. Suspension systems allow greater contact between the explant and the medium. Moreover, agitation of such media reduces the diffusion gradient in the nutrient supply. The toxic metabolites exuding from the tissues are also dispersed effectively.

Table: *Low cost Matrices*

Option	*Use*
Cotton fiber	Callus maintenance and shoot organogenesis
Glass wool	Multiplication of chrysanthenum
Nylon cloth	Multiplication of chrysanthenum
Polystyrene foam	Multiplication of chrysanthenum
Glass beads	Multiplication of raspberry and white clover
Filter paper	Multiplication of chrysanthenum, potato

Strips (2.5x15 cm) of glass wool, nylon and filter paper have been used as supporting bridges for the propagation of chrysanthemum on MS medium. The shoot- and root-growth was almost identical on these matrices. Polystyrene foam blocks (2x2x1 cm) have been used as supporting matrix for the propagation of chrysanthemum.

The growth response was similar to that on glass wool, except that explants on foam produced significantly fewer roots but longer shoots than on glass wool. Glass bead based liquid-media were successfully used for culturing ginger and turmeric, and on per plant basis reduced the medium cost by 94 per cent With glass beads, the amount of medium required is only 15-18 ml per culture container.

By using 20ml medium per culture container, one liter media will give 50 culture containers, a substantial saving in medium cost. Ginger and turmeric plants multiplied on glass bead liquid-medium performed as good as on agar-based medium.

A similar type of response was observed for vanilla. Ficus cv.'mini lucii' showed higher multiplication rate although with a slight vitrification. Saintpaulia, Syngonium, Philodendron and Spathiphyllum also showed higher multiplication rates and better growth on glass bead liquid-medium.

An additional advantage of liquid media withglass beads is that the medium can be replaced without moving plants. Glass beads have been used for the propagation of raspberry and white clover resulting in 60 per cent saving in media cost.

The glass beads allowed easy removal of plantlets from the medium. Glass beads have been also used for the maintenance of callus and shoot organogenesis.

The beads can be reused after washing with acid. The liquid-media have some disadvantages. Delicate tissues get damaged during agitation.

In some species, shoots submerged in liquid media become hyperhydric (water soaked), and unsuitable for micropropagation.

Conventional Carbon Sources and Low Cost Alternatives

Sucrose is the most commonly used carbon source in the micropropagation of plants. Sucrose adds significantly to the media cost.

Household sugar and other sugar sources can be used to reduce the cost of the medium. Sugar sold in grocery stores is sufficiently pure for micropropagation.

For culturing ginger and turmeric, all other carbohydrates except sugarcane juice, were suitable alternatives to laboratory grade sucrose.

Use of common sugar reduces the cost of the medium between 78 to 87 per cent. The cost of the local sugar was US$ 0.55/kg against the $40.0/kg for the imported sucrose.

In Bangladesh, several laboratories have used locally available household sugar for culturing potato, banana, orchids, chrysanthemum, lentil, peanut, chickpea, medicinal plants, and fruit trees.

In Pakistan, local sugar was found to be as good as the high grade laboratory sugar for the multiplication of banana. Maple syrup has been used for the multiplication (50 g/l) and rooting (34 g/l) of cherry root stocks from nodal segment and shoot tips.

Table: *Effect of Alternatives to Analytical Grade Sucrose on Culture Growth*

Alternative	*Effect*
Household sugar (3%)	Healthy cultures
Double refined sugar (3%)	Healthy cultures
Sugar crystals (3%)	Healthy cultures
Sugarcane juice (10% v/v)	Drying of leaf tips
Sucrose LR grade (3%)	Healthy cultures

Alternatives to Other Ingredients

Several other ingredients can also be replaced by low cost options Commercial grade chemicals of lower purity than the analytical grades are quite suitable for commercial micropropagation unless deleterious effects are observed. A high degree of purity is justified only in the case of basic studies in tissue culture. In general commercial micropropagation, the quality will hardly be affected ordinarily by purity of these chemicals.

Of all the chemicals, growth regulators (hormones) are the most expensive. However, they are needed in very small amounts in the medium, thus have a little effect on the medium cost. Sugar cane molasses can provide many of the nutrients,namely, sugar, vitamins and inorganic metal ions required for sugarcane callus induction and shoot formation.

At Dhaka University, low cost media have been developed for the multiplication of orchids based on macro-salts of any of the Vacin and Went, Phytamax and Knudson's C media; supplemented with 10-15 per cent (w/v) banana extract and 10-15 per cent (v/v) coconut water.

Table: *Low Cost Option for Sugar in Medium*

Sugar type	*Use*
Refined white sugar (RWS)	Culture of zygotic embryos
Unrefined light brown sugar	Culture of zygotic embryos
Unrefined brown sugar,	Culture of zygotic embryos
Sugar maple syrup	Multiplication and rooting of cherry rootstock. Replace micro-nutrients and reduce macro-nutrients
Table Sugar	Multiplication of banana, potato, orchids, chrysanthemum; shoot regeneration and rooting of lentil, peanut, chickpea

Use of above media eliminated the need of micro-salts and vitamins for the micropropagation of different orchid species. Several different orchids, namely, Vanda, Dendrobium, Aerides, Acampe and Spathoglottis can be multiplied on a medium containing only peptone, inositol, banana extract and coconut water. In many tropical orchid species, similar response was observed on media containing coconut water (15 per cent) and banana pulp (100 g/l), which effectively lowerd the cost.

Source of Water

Water is the main component of all plant tissue culture media. Usually in tissue culture research, distilled or doubled distilled and de-ionized water is used. Distilled water produced through electrical distillation is expensive. In some cases, alternative water sources can be used to lower the cost of the medium. If tap water is free from heavy metals and contaminants, it can be substituted for distilled water. Tap water has been used for in vitro propagation of banana and ginger, Zingiber officinale.

Table bottled water from the supermarket can also be used a low cost alternative. However, its mineral composition should be taken into account as it may affect pH and nutrient uptake. In rural areas, rainwater can be collected in clean glass jars and used for tissue culture. In Bangladesh, the change over of water distillation from electrical to gas operated unit reduced the cost from US$260 to $5/month for producing 50-60 liter water per day.

Low Cost Option for Media making

Media preparation is a time consuming process. Many companies sell ready-made media in liquid or powder form. The pros and cons of laboratory making verses buying readymade media should be considered. Although pre-made media save time, their relative cost is high. Pre-packaged media preparations are useful when medium quantities required are small.

The errors in media preparation from pre-packed media are also less frequent. However, for large-scale use, it is much more economical to prepare media by mixing the basic ingredients. By making large batches, the time for media preparation and sterilization can be reduced substantially. However, if mistakes occur, the bigger is the batch, the greater is the loss.

The volume of medium prepared in large sized-laboratories can range from 200 to 500 liters per day. The media should be made and used in small batches. Non-sterile media should not be stored at room temperature for more than 24 hr, especially when it contains high amounts of sucrose.

These should be refrigerated until further use. Media are usually dispensed into containers either manually or with a peristaltic pump.

Manually, one person can fill more than 200 containers an hour. In some large-scale laboratories, the distribution of sterile media is automated. To be economical, the automation must be able handle many hundred containers per hour.

Handling Stock Solutions

Usually the stock solutions are prepared in 10x to 1000x concentrations. The stock solutions consist of groups of chemicals, *e.g.* macronutrients, micronutrients, vitamins and plant growth hormones. The inorganic chemicals and vitamins solutions can be combined into a single, 10 X concentrated, stock solution.

The stock solutions can then be frozen in plastic bags in volumes sufficient to make 1 to 5 liter media at a time. The cost of the medium prepared in the above way is much cheaper than the ready-made powdered media. This is being practiced for long time in many developing countries.

Cost of Medium

Depending on the plant species, the cost of culture medium varies with the ingredients such as sugars, gelling agents, and growth hormones used. Based on the current price of the various ingredients, the estimated cost of one-liter MS medium works out to be $0.18 for the solid and $0.08 for the liquid medium.

If sucrose were not added, as in the case of autotrophic micropropagation, cost per liter would be $0.13 for the solid and only $0.03 for the liquid medium. The cost of medium per plant depends on the amount dispensed and the rate of plant multiplication per culture container.

Bibliography

Ashraf, Syed Mahboob: *Handbook of Fruit and Vegetable Products*, Agrobios, Delhi, 2008.

Banerjee, S : *Biochemistry and Biotechnology,* Dominant Pub, Delhi, 2009.

Barrington, E. J. W.: *Biochemistry of Primitive Deuterostomians*, London, Academic Press, 1974.

Bhaskar, Harsh: *Basic Facts on Biochemistry*, Campus Books, Delhi, 2010.

Bhatia, A. L.: *Biochemistry and Endocrinology,* Indus Valley, Delhi, 2002.

Bhatnagar, Vasudev: *Cell Science and Technology*, Campus Books, Delhi, 2009.

Bora, K.K.: *Agros Dictionary of Plant Physiology and Biochemistry*, Agrobios, Delhi, 2001.

Bose, N. Sarath Chandra: *Biochemistry: A Practical Manual,* Pharma Med Press, Delhi, 2010.

Bose, Sarath Chandra: *Biochemistry: A Practical Manual*, Pharma Med Press, Delhi, 2010.

Chanderlekha Goswami: *Biochemistry and Instrumentation*, Manglam Pub, Delhi, 2011.

Chauhan, B.S.: *Principles of Biochemistry and Biophysics,* Laxmi Publications, Delhi, 2008.

Chauhan, Sushil: *Experiments in Biochemistry,* Campus Books, Delhi, 2010.

Collymore L.: *Fruit Production in Barbados*, Port of Spain, Trinidad and Tobago, 1996.

Coultate, T. P.: *Food. The Chemistry of its Components,* London. The Royal Society of Chemistry, 1988.

Daphne C. Elliott: *Biochemistry and Molecular Biology*, Oxford University Press, Delhi, 2005.

Dar, Ghulam Hassan: *Soil Microbiology and Biochemistry,* New India Publishing Agency, Delhi, 2010.

Dutta, Chiranjib: *Fundamentals of Biochemistry: A Practical Approach*, Kanishka, Delhi, 2005.

Elliott, B N: *Biochemistry and Molecular Biology*, Oxford University Press, Delhi, 2005.

Florkin, M.: *Comprehensive Biochemistry: Amsterdam*, Elsevier, 1975.

Goel, A K: *Basic Concept of Organic Chemistry*, Pearl Books, Delhi, 2008.

Golakiya, B A: *DNA: A Bridge Between Biochemistry and Biotechnology,* New India Publishing Agency, Delhi, 2008.

Hemantaranjan, A L: *Developments in Physiology, Biochemistry and Molecular Biology of Plants*, New India Pub, Delhi, 2005.

Jasra, O.P.: *Environmental Biochemistry*, Sarup & Sons, Delhi, 2002.

Khanna, V K: *Objective Genetics, Biotechnology, Biochemistry and Forestry,* I.K. International Publishing House, Delhi, 2008.

Mido, Y.: *An Introduction to Nuclear Chemistry*, Discovery, Delhi, 1995.

Narvekar, Raghunath: *Handbook of Biochemistry,* Adhyayan, Delhi, 2008.

Prakash, M: *Insect Biochemistry,* Discovery Pub, Delhi, 2008.

Prasad, G.P.: *Biochemistry of Proteins*, Discovery, Delhi, 2010.

Purohit, S.S.: *Biochemistry: Fundamentals and Applications,* Agrobios, Delhi, 2009.

Rastogi, Arun: *An Introduction to Practical Biochemistry*, Anmol, Delhi, 2010.

Sangha, S P S: *Basic Techniques in Biochemistry and Molecular Biology,* I K International, Delhi, 2009.

Smriti Ganguly: *Biochemistry of Biomolecules*, Pearl Books, Delhi, 2007.

Srivastava, Seema: *Plant Physiology and Biochemistry,* Campus Books, Delhi, 2009.

Tripathi, Pramila: *Botanical Pesticides in the Management of Post-harvest Fruit Diseases*, Daya, Delhi, 2005.

Whyman, Robin: *Applied Organometallic Chemistry and Catalysis*, Oxford University Press, Delhi, 2005.

Index

O

P

S

T

U

V

W

X

❑❑❑

Encyclopaedia of
Biochemistry and Molecular Biology of Plants

Encyclopaedia of Biochemistry and Molecular Biology of Plants

Volume 3

Editor:
K.N.P. Singh

AGROTECH PRESS
JAIPUR ▪ NEW DELHI

AGROTECH PRESS

Regd. Off. "Bony Residency", Gate No. 2
Opp. Tilak Public School, Vishweswariya Nagar,
Gopalpura Road, Jaipur - 302018, India
Ph: 0141 - 2761280, Telefax: 2761381
E-mail : agrotechpress@gmail.com

Delhi Office: 102, 1st Floor, Satyam House,
4327/3, Ansari Road, Darya Ganj,
New Delhi - 110002, India
Ph: 011-45652440
visit us: www.agrotechpress.com

Distributed By
KEYA INTERNATIONAL PRESS LLP
E-mail: keyainternationalpressllp@gmail.com

Published in India, 2024

ISBN : 9789384568221 (4 Vols. Set)

Typeset by Thomson Press

Printed and bound in India by Thomson Press Ltd.,
New Delhi

Contents

Preface

Metabolism is the set of life-sustaining chemical transformations within the cells of living organisms. These enzyme-catalyzed reactions allow organisms to grow and reproduce, maintain their structures, and respond to their environments. The word metabolism can also refer to all chemical reactions that occur in living organisms, including digestion and the transport of substances into and between different cells, in which case the set of reactions within the cells is called intermediary metabolism or intermediate metabolism.

Plants are responsible for incredible feats of molecular transformation. The processes are always being studied, but there are a few basic things that are well understood at this point in history. We will be looking at photosynthesis and respiration in some detail, as always, if you have additional questions please post them on the forum. Photosynthesis is the process by which light energy is captured, converted and stored in simple sugar molecule. This process occurs in chloroplasts and other parts of green organisms. It is a backbone process, in the sense that all life on earth depends on it's functioning. The earth's atmosphere contains approximately 79% nitrogen, 20% oxygen and the remaining 1% is a mixture of less common gases—including 0.039% carbon dioxide. Carbon dioxide in the atmosphere reaches plant mesophyll via the stomata. The carbon dioxide dissolves on the thin film of water that covers the outside of cells. The carbon dioxide then diffuses through the cell wall into the cytoplasm in order to reach the chloroplasts. The oceans hold a large reservoir of carbon dioxide, which keeps the atmospheric levels essentially constant. Although there are some indicators that the atmospheric levels of CO_2 are rising and adding to the global warming issue. Water is plentiful on earth, however, it may or may not be plentiful at the location of each individual plant. Therefore, plants will close their stomata, if need be, which reduces the CO_2 supply to the mesophyll. Not even 1% of the water that is absorbed by plants is used in photosynthesis, the remainder is either transpired or incorporated into protoplasm, vacuoles or other cell materials. The water utilized in photosynthesis is the source of oxygen released as a photosynthetic byproduct. Light has a dual nature, in that it exhibits properties of both waves and particles. The energy from the sun comes to earth in various wavelengths, the longest being radio waves and the shortest are gamma rays. Approximately 40% of the radiant energy the earth receives from the sun is visible light. Visible light ranges from red, 780 nanometers to violet, 390 nanometers. The violet to blue and red to orange ranges are the most often used in photosynthesis. Most light in the green range is reflected. Of the visible light that reaches a leaf, approximately 80% is absorbed. Light intensity varies widely. Time of day, temperature, season of year, altitude, latitude and other atmospheric conditions all play roles in the intensity of the radiant energy that will reach the earth and it's organisms. High intensity light isn't necessarily a beneficial thing for plants. In high intensity light, photorespiration may occur, which is a type of respiration that uses oxygen and releases carbon dioxide but differs from standard aerobic respiration in the pathways that it utilizes. A few things to know about chlorophyll before we get into the nitty gritty of photosynthesis and respiration. There are more than one type of chlorophyll, however, they all have one atom of magnesium in the centre. In some ways the chlorophyll is quite analogous to the heme structure in hemoglobin (the iron containing pigment that carries oxygen in blood). Chlorophyll has a long lipid tail that anchors the molecule in the lipid layers of the thylakoid membranes—recall that thylakoids are coin-like discs in stacks within the stroma of the chloroplasts. The chloroplasts of most plants contain two types of chlorophyll imbedded in the thylakoid membranes. The formula for bluish-green chlorophyll a is $C_{55}H_{72}MgN_4O_5$ and the formula for yellow-green chlorophyll b is $C_{55}H_{70}MgN_4O_6$. In general, most a chloroplast has about three times as much chlorophyll a than b. The main role of chlorophyll b is to broaden the spectrum of light available for photosynthesis: chlorophyll b absorbs light energy and transfers the energy to a chlorophyll a molecule. Other pigments are contained in chlorophyll c, d, and e and take the place for chlorophyll b in some cases. Note that all the chlorophyll molecules are related to each other and differ only slightly in molecular structure. Light-harvesting complexes contain 250 to 400 pigment molecules and are referred to as a photosynthetic unit. There are countless numbers of these units spread throughout the grana of a chloroplast. In the chloroplasts of green plants, two types of these harvesting units operate together in order to bring about the first phase of photosynthesis.

It is the textbook of its own type with the aim to provide enough information and examples to give the reader a sound knowledge of plant metabolism. As such, this is a textbook, and therefore a learning and teaching aid, rather than an academic treatise.

—Editor

1

Plant Genetic Diversity

The conservation of plant genetic resources has, in recent years, attracted growing public and scientific interest and political support. There is an increasing awareness of the relevance of biological diversity and its conservation to the health of the biosphere. Particularly urgent is the need to raise agricultural output to meet the basic nutritional needs of increasing populations. The dilemma is that the required increase in food production must be obtained through 'sustainable' forms of agriculture which are less dependent on the use of modern high-yielding varieties bred for intensive, high-input systems. It seems unlikely that increased food production on this scale will come from increasing the area under cultivation. Rather, major solutions will depend upon innovations which exploit sophisticated methods to manipulate plant genomes and the totality of natural genetic variation.

The area of disease resistance will serve to illustrate this argument. Many authors have emphasized the importance of innovation in the area of crop protection since crop losses due to pests and disease may account for between 20 and 40% of productivity worldwide. Indeed, over 50% of the research and development (R&D) spent in the plant breeding/seed industry is focused on the identification and incorporation of resistance traits. In a situation where there is some evidence that the pool of available variation for disease resistance genes within commercial breeding lines is becoming limited (*e.g.* for yellow rust in cereals) the search is on for novel sources of resistance. There are now many examples of biotechnology being put to the service of pest control, through the incorporation into plants of transgenes for toxins, antifeedants, enzyme inhibitors, etc. But, as Woolhouse (1992) has pointed out, such is the long coevolutionary history of pests and pathogens with their hosts that present strategies for interfering with these relationships represent only a small proportion of the likely defence mechanisms used by plants in resisting foreign organisms. This enormous diversity of mechanisms represents an arsenal of ammunition which can be used in the fight to minimize disease and predation.

Are there any hard facts which support these assertions concerning the value of genetic resources to agricultural productivity? Can we put a cost on the loss of diversity? The sources of variation used by the plant breeder range from current breeding lines to wild species and the products of direct gene manipulation. In a recent analysis of the plant breeding/seed industry, Swanson (1996) showed that, over a five-year period, 6.5% of all genetic research within this industry, resulting in a marketed innovation, was concerned with germplasm from wild species and landraces.

This compares with only 2.2% of new germplasm arising from technological approaches of induced mutation. The vast majority of germplasm successfully used by the industry still arises from 'conventional' sources, *i.e.* the heavily exploited commercial cultivars, which at first sight might be taken to suggest that wild resources are relatively unimportant in R&D.

In fact Swanson (1996) argues just the opposite, suggesting that the stock of existing commercial varieties should be seen as the information base from which bioindustries develop innovations, whereas the sources of new diversity (wild species, landraces, induced mutation) should be seen as supplying increments to the information base. In other words, R&D requires an annual injection of 'new' genetic material from natural sources, estimated as amounting to approximately 7% of the stock of material currently within the system. Biodiversity is valuable to industries other than those that seek novel genes for crop plant improvement. Biodiversity has traditionally provided a source of compounds to the pharmaceutical, foodstuffs, crop protection and other industries and here it is possible to give more precise evaluations of its importance. As outlined in Chapter of this volume, drug sales based on natural products from plants were estimated at $US43 billion in 1985 alone and the value of yet undiscovered pharmaceuticals in tropical rainforests is estimated to be as much as $US147 billion to society as a whole. Vast markets exist for the replacement of synthetic biocides, many of which have associated toxicological and environmental problems, by 'natural' compounds, which may have more specific modes of action and fewer side-effects.

Even greater may be the use of genetic resources in more fundamental research where their value in the longer term may be unknown or impossible to estimate accurately at the present time. Diverse genetic material is needed and now regularly utilized to create mapping populations and to study inheritance of a wide range of traits which may or may not be of immediate practical value. Very large numbers of germplasm samples are regularly screened and evaluated for different characteristics so that the underlying molecular biology can be investigated. The intention here is that a greater understanding of fundamental biology will give rise to wealth creation in the future.

Novel Technologies for Biodiversity Assessment and Monitoring

The analysis and characterization of genetic diversity is fundamental to any *ex situ* or *in situ* conservation strategy. Over the years the methods for detecting genetic diversity have expanded from Mendelian analyses of discrete morphological variants, to biometrical approaches, biochemical techniques based upon protein and isozyme profiles, to methods based on DNA sequence variation.

There is now a huge array of molecular marker techniques, and Chapter reviews the major technologies and their appropriate use in addressing key issues in germplasm conservation The application of molecular technologies to increase our understanding of biodiversity at several levels is exemplified by Chapter, which is concerned with the population level, which is concerned with molecular approaches to genetic relatedness. Chapter reviews how the collection of diverse germplasm for *ex situ* purposes and the development of effective *in situ* collections both require an understanding of the spatial and temporal aspects of population structure and associated genetic diversity and the factors that control it.

Studies with low- or single-copy markers are particularly valuable in answering questions about genetic relatedness between species. Chapter illustrates how comparative molecular mapping provides new opportunities for exploring the genetical basis of phenotypic variation within and between plant species. The implication of comparative mapping is that genes controlling specific phenotypic traits will have corresponding map locations in different taxa. This provides a new framework for the identification of homologous genes in germplasm and new tools for the effective utilization of germplasm in crop plant improvement.

This approach is also valuable in the study of quantitative traits controlled by several genes since flanking markers for major genes provide markers for quantitative trait loci (QTL) manipulation and selection and the map-based cloning of at least some genes controlling quantitative traits.

Biotechnological Tools for Conservation

Biotechnology can make a substantial contribution to the quality of germplasm collections and their efficient management. For example, in the management of very large germplasm collections there is a clear requirement for procedures for fast, reliable taxonomic identification of material and the elimination of redundancy. Samples characterized by molecular criteria may prove superior to traditional analyses of phenotype and examples of the use of DNA markers to guide the management of germplasm are explored in Chapter by reference to one of the world's largest collections of germplasm at the International Rice Research Institute.

Not all plant species can be stored in the form of desiccated seed. Socalled 'recalcitrant' species have to be kept in moist, warm conditions and even then have relatively low longevity. In addition, there are many cultivated plants which must be maintained clonally, such as banana, which produce few seeds, or potato, which does not come true from seed.

To these traditional problem species we must now add the products of biotechnology, new categories of germplasm including clones of elite genotypes, and transformed cell lines, for which storage of seed is inappropriate. Biotechnology has impact on storage technologies in two quite different ways. The first of these, concerns the various types of *in vitro* culture which allow the propagation of plant material in an aseptic environment, thus ensuring the production of disease-free stocks and simplifying quarantine procedures for the international exchange of germplasm. The development of sensitive molecular probes to detect low levels of pathogens.

A more radical approach to safe storage and distribution of germplasm, and possibly the only way to significantly reduce the rate of loss in biodiverse tropical forests, within the bounds of available resources, is to develop alternatives to the costly (and inherently vulnerable) methods of *ex situ* storage of seed or tissue cultures, by conserving germplasm in the form of DNA or DNA-rich materials. The latter include those dried specimens conserved in the vast resources of the world's herbaria. The DNA Bank-Net, an international network of DNA repositories, has now been established with both storage and distribution functions. Although technologies for rapid removal, analysis, amplification and secure storage of small amounts of DNA are now well advanced, research is still needed on methods to amplify the entire genomic DNA of a species, which means that storage of DNA or DNA-rich materials should still be considered to be something of an insurance policy rather than a replacement for conventional modes of germplasm storage.

Biotechnology-assisted Crop Improvement

Effective use of genetic resources in plant breeding programmes requires methods to determine whether useful genetic variation exists, and costeffective methods to introduce useful genes into the material. Biotechnology has made a contribution to both these aspects. The use of genetic maps based on molecular markers linked to genes of interest facilitates the more rapid selection of both simple and complex traits thus accelerating their incorporation into breeding materials. Where traits are controlled by one or a few genes, direct isolation of genes and their incorporation into elite cultivars by increasingly efficient transgenic technologies provide an additional approach to the more traditional methods of plant breeding. This prospect also challenges one of the classical concepts in plant genetic resources work, namely the 'gene pool' concept of Harlan and de Wet (1971), which separated cultivated species and their wild relatives into different categories based upon their 'crossability'.

The value of this concept in guiding the work of the plant breeder must now be re-evaluated in the light of molecular genetic technologies which have not only cast a completely new light on the relatedness of crop plant genomes, but which have effectively widened the potential gene pool of a species to include all other life forms.

What challenges does this imply for the traditional plant genetic resources programmes? Why is the emphasis on conserving species that are related to crops when more exotic and potentially valuable genes may be present in liverworts, or green algae, or other less studied taxa?

Bioprospecting for Novel Genes

The value of genetic resources as a source of novel compounds for various industries has been alluded to above. For centuries biodiversity has provided sources of fuels, medicines, foodstuffs, pesticides, etc. The high returns available from introducing a new drug, for example, plus the realization that biodiversity is rapidly being lost (estimated rates of extinction range from 30 to 300 species per day) has encouraged new attitudes to the exploration and exploitation of biodiversity through 'bioprospecting'. How does biotechnology impact on this traditional area? A fundamental operation in any bioprospecting programme is the screening of thousands of extracts for compounds of interest. Gene technology now allows the speeding up of screening programmes for new compounds through the development of more sophisticated *in vitro* assays. For example, the genes encoding receptor proteins for certain classes of drug, or enzymes that may serve as targets for novel pesticides, may be cloned and expressed on a large scale in high-throughput *in vitro* assays into which thousands of plant extracts may be applied.

Bioprospecting is also concerned with discovering novel genes and natural variants of genes or gene products that may either be used directly or to guide sequence improvement by molecular methods. Chapter gives examples of amino acid sequence improvement in peptide hormones as a result of screening homologous genes from the wild, and this area of biotechnology is likely to develop in parallel with new molecular methods of generating random variation through employing the mutational potential of the polymerase chain reaction (PCR) – *i.e.* 'forced molecular evolution' – or recombinatorial libraries of peptides.

Whilst gene prospecting has provided a new perspective on biodiversity, the area is not without political controversy because these advances tend to favour agriculture and other industries of developed countries, and may actually displace traditional commodity production in resource-rich but underdeveloped countries. Chapter also outlines the development of strategies which integrate conservation and exploitation with economic returns and sustainable development in those countries from which the bioresources were first obtained.

The Exchange of Information and Germplasm

The value of all plant genetic resources lies in their utilization in crop improvement programmes or in other bioindustries. The vast majority of conserved germplasm is available for distribution to the global scientific community but this international distribution poses considerable risks of accidental introduction of non-indigenous plant pathogens and pests. The detection and elimination of such disease-causing agents is therefore essential.

Often, such agents will be present at very low levels and may actually be symptomless. In this context then, new technologies reviewed in Chapter have been of great value in the development of reliable and sensitive nucleic acid- or antibody-based diagnostic probes, and these will be of great importance in optimizing the health and value of germplasm collections. Finally, if the biotechnology revolution has had great impact in developing novel molecular and cellular technologies, no less a revolution has affected radically the way in which the world organizes and disseminates information.

Powerful computers linked by the Internet have facilitated the development of sophisticated and comprehensive informatics systems which enable information on plant genetics and molecular biology to be cross-related to systematic, ecological and other data through international networks. Chapter explains how this information is organized and distributed against a background of continual development.

Plant Genetic Variation

Effective conservation and use of plant genetic resources involve asking many questions about the extent and distribution of genetic variation. Only when the appropriate markers and technologies for describing this variation are accessible can such questions be adequately addressed.

The most appropriate markers for a given question should:

- Be heritable;
- Discriminate between the individuals, populations, or taxa being examined;
- Be easy to measure and evaluate; and
- Provide results that can be compared with results of similar studies.

Molecular markers and assays that meet these criteria were introduced in the 1950s and have proliferated ever since. Use of protein isozymes as genetic markers increased rapidly after their introduction. Nucleic acid markers gained popularity in the 1970s, with the advent of DNA sequencing and restriction fragment analysis. The number of different molecular markers and assays has increased exponentially since the late 1980s, when introduction of the polymerase chain reaction (PCR) made DNA amplification and sequencing possible and affordable.

As the diversity of markers and assays increases, the cost of using them continues to decrease. At the same time, the range of taxa and tissues that can be analysed has expanded. Improved DNA extraction, amplification, and sequencing protocols allow us to analyse ancient DNA, from 13,000-year-old specimens of *Mylodon* (ground sloths) and 17–20 million-year-old *Taxodium* (bald cypress) specimens, and samples as small as single cells or fractions of a seed.

Thus the chances are good that at least one marker assay will be appropriate for any question about variation in any taxa. Furthermore, results of one study can generally be related to results from many other studies.

Such opportunities are inevitably accompanied by challenges. None of the challenges are unique to molecular studies, but all become more critical as the number of available marker assays increases. Faced with a bewildering variety of molecular technologies (and their acronyms), choosing a marker and assay that are appropriate for a scientist's particular question, taxa of interest, and practical constraints can be overwhelming. The temptation is great to use techniques that are recommended by coworkers, are frequently used, for which the most information is available, or are the newest – without adequately evaluating the techniques to determine which is most suitable.

Because molecular markers can be measured in virtually all plant taxa and minimal tissue is required, extra care must be taken to verify that the entries in a test array are correctly identified. Since molecular markers are widely applicable and their constraints are not always understood, thoughtful planning is particularly important when designing molecular studies, to ensure that they have an appropriate range of entries, an appropriate number of entries per taxon, and conclusions that stay within the scope of a specific question and test array.

In addition, the vast amount of published research using molecular markers complicates the task of thoughtfully relating one study to others with similar marker assays, research questions, and/or taxa. Another challenge is formulating research objectives that ask critical questions about plant genetic resources. Descriptive studies that use molecular marker assays to examine variation between and within specific taxa can generate valuable information, but some plant conservation questions are best addressed by experimental studies. The wide variety of molecular marker assays available provides new ways to address such questions – if the markers are appropriately used. Such challenges can be met successfully and thoughtful decisions can be made if logical frameworks are available to categorize and compare these valuable molecular tools.

A solid foundation for such comparisons has been provided recently. This chapter builds on that foundation by reviewing both early and recently developed molecular markers and assays, citing examples of their use. The types of questions asked about plant genetic resources are described, as are guidelines for selecting appropriate markers and assays to address these questions.

The literature cited here is far from comprehensive, but includes applications of each marker assay in plant genetic research. While the list of molecular techniques will soon be dated, hopefully the framework presented will be helpful for categorizing and comparing new tools as they are made available. The chapter concludes with a perspective on new marker assays as they apply to, and may change, the study of plant genetic variation.

Types of Questions, Markers and Marker Assays

All studies of genetic variation have four primary components:

- The research question;
- The marker used to address it;
- The technique used to generate and measure the marker; and
- The method used to analyse the data produced. This chapter focuses on the first three components.

Types of Plant Genetic Resource Questions

Most questions asked about *ex situ* and *in situ* plant genetic variation fit into one of three categories. While boundaries between them are not always clear, these categories are useful when deciding how to address specific questions. Questions of identity are forensic in nature and ask whether two or more samples are genetically the same. These questions concern whether geebank accessions are present in duplicate, whether populations are genetically distinct or are geographical subdivisions of one population, and whether genetic change has occurred in an accession or population over time.

In contrast, questions of location and diagnostics concern the presence and location of a particular allele or nucleotide sequence in a taxon, genebank accession, *in situ* population, individual, chromosome, or cloned DNA segment. These questions are asked to locate populations or individuals that have desirable traits; determine whether traits have been lost from a population, or have been transferred from one population or plant to another; and establish the position of markers on physical or genetic maps.

The differences between diagnostic and forensic questions are significant, but are not always clarified. Both types of question require qualitative information about specific loci. Yet while diagnostic studies often use single-locus markers, each with few alternative states, forensic questions are commonly addressed with information from many multiallelic loci.

Finally, relationship and structure questions examine relatedness or similarity between genotypes, the amount of genetic variation present, and how variation is distributed between individuals, populations, and taxa.

Such questions are used to pinpoint gaps in genebank collections, decide which genotypes are high priority for *in situ* conservation, design germplasm sampling strategies and regeneration programmes, construct core collections, study gene flow, and determine optimum sizes for conserved populations or geebank accessions. These issues are generally addressed with information from many loci. Quantitative data can be used, but qualitative data are more appropriate for some phylogenetic and parentage questions.

Types of Molecular Markers

The appropriate markers for a study can discriminate between entries in an array, but are not so polymorphic that important variation is masked by random noise. Molecular markers range from highly conserved to hypervariable, and can be either proteins or nucleic acids. The nucleic acids used as markers include entire genomes, single chromosomes, fragments of DNA or RNA, and single nucleotides.

A wide variety of nucleic acid fragments are utilized as markers. While some occur once in a genome, others are repeated. Many repeated sequences used as markers are noncoding; others are elements of multigene families. Some repeated sequences are interspersed throughout the genome, either distributed randomly or in clusters.

These interspersed repeats are common in plant and animal nuclear genomes, and are found in plant (but not animal) mitochondrial genomes. The chloroplast genome contains a large inverted repeat (IR); most angiosperm chloroplasts have two copies, separated by a short single- copy region. Repeat length and (rarely) loss of one copy can vary between taxa.

Much research at present is focused on repeated sequences that occur in tandem. The classes of tandem repeats are distinguished by the length of the core repeat unit, the number of repeat units per locus, and the abundance and distribtion of loci. The names for these classes are themselves varied and have been inconsistently used, but Harding *et al.* (1992) and Tautz (1993) have clarified the nomenclature. Tandem repeats were first reported in the literature as 'satellites' of DNA, detected in CsCl density gradients as fractions with different GC content than the rest of the genome.

These satellites have repeat units that are usually several hundred nucleotides long, with thousands of copies at each of several loci in the nuclear genome. These loci are usually in heterochromatin, often near centromeres. Satellite DNA is present in numerous species. For many satellites, the number of loci and number of repeat units per locus vary between species and higher taxa. Minisatellites (often called variable number of tandem repeat loci, or VNTR loci) are widely used as markers, especially in forensics.

The repeat units are usually less than 100 nucleotides long, with tens to hundreds of copies per locus. Thousands of loci in a genome may have similar core repeat units. The number of repeat units at a minisatellite locus can vary greatly between individuals and populations. First described in humans, minisatellites are found in numerous animal species, often near telomeres. They are also common in plants, and are often associated with satellites and centromeres. The number of repeat units per locus is less variable in plants tha in animals, but is still high; plant minisatellites are useful markers for variation between and within species.

As suggested by their name, microsatellites – also called simple sequence repeats (SSRs), or simple sequence length polymorphisms (SSLPs) – have very short repeat units, no more than six nucleotides long. SSRs are more abundant than minisatellites in noncoding regions of the nuclear genome, and are present in some nuclear genes and organelle genomes. The number of repeat units per locus is lower for SSRs than for minisatellites, but can approach 100 in animals and 50 in plants. The abundance and polymorphism of SSRs make them particularly valuable for describin variation between populations and individuals.

Like minisatellites, SSRs were documented first in humans and later in plants. Plants and animals differ in the abundance of specific SSR motifs in the genome. In both plant and animal genomes, chromosomal distribution of SSRs is variable. Some animal SSRs are found near heterochromatin or interspersed repeats, but most are randomly dispersed. Dinucleotide SSRs are randomly distributed in *Arabidopsis thaliana* (L.) Heynh., but other studies have located plant SSRs near genes, highly methylated DNA, satellites, or centromeres.

Tandemly repeated genes are also utilized as markers. Perhaps the most widely used are the nuclear genes that encode ribosomal RNA (rRNA). The three rRNA genes are separated by two internal transcribed spacer regions, generally referred to as ITS1 and ITS2. These genes and spacers form a unit that is tandemly repeated hundreds of times, at one to several loci in the genome. At each of these loci, the individual repeat units are separated by nontranscribed intergenic spacer (IGS) regions. In the middle region of each IGS are tandem copies of a short subrepeat sequence.

Variation in rRNA gene clusters can be measured at several levels, each evolving at a different rate:

- The number and location of rRNA loci, which is highly conserved;
- The (more variable) number of tandem gene clusters per locus;
- The conserved sequences of the three genes;
- The variable sequences of the ITS regions; and
- The highly variable number of subrepeats in the IGS region.

These features make rRNA gene clusters versatile and informative markers for mapping and phylogenetic analyses.

Types of Molecular Marker Assays

Molecular marker assays are generally classified by whether the molecules evaluated are proteins or nucleic acids, and whether the character analysed in a nucleic acid marker assay is the entire genome, a chromosome, a fragment, or a nucleotide. Alternatively, marker assays can be categorized by the type of character measured. Some methods measure quantitative differences between entries in an array. Others measure qualitative characters, each with two or more possible states. Marker assays also differ in the number of loci evaluated per analysis, whether multiple loci are evaluated simultaneously or sequentially, and the type and amount of information needed about the marker loci before conducting the assay.

Protein Marker Assays

When proteins are used as genetic markers, the assumption is made that any variation between proteins reflects heritable variation in their amino acid sequences. This assumption has long been debated, since protein phenotypes can be affected by factors such as post-translational modification, plant phenology, and environmental conditions during plant growth.

For many tasks, however, protein marker assays continue to be useful molecular tools. Some assays measure immunological properties of proteins. When antigenic proteins of one plant or animal species are injected into an animal (usually a rabbit), antibodies are produced. Genetic differences between entries in an array can be assessed by comparing how each entry's proteins bind to antibody serum (antiserum) from a reference species.

This binding affinity is measured by the amount of antibody–antigen precipitate produced or by microcomplement fixation (MCF). Each MCF solution contains antiserum from a reference sample, antigen from a test entry, and a protein complement. The complement can only bind to antibody that is bound to the antigen. The amount of unfixed complement is measured; this is used to calculate the amount of unbound antigen, which indicates the extent of variation (the immunological distance) between the entry and the reference sample. After MCF assays are conducted for each entry in an array, these immunological distances are compared between entries. MCF has long been used in vertebrate phylogenetic analysis. Until recently, immunological markers were seldom used to study plant variation, perhaps because few suitable proteins had been identified.

During the past few years, however, monoclonal antibodies to seed proteins of several cereal species have been produced. Reactivity of these antibodies with antigenic proteins can be evaluated by enzyme-linked immunosorbent assays (ELISAs). Monoclonal antibodies have been used to identify cDNA library clones and describe variation between crop species and cultivars. Other marker assays measure the rate of protein migration through a starch, polyacrylamide, or cellulose acetate gel in response to an electrical current.

The migration rate is related to protein size, shape, isoelectric point and/or ionic charge. In these assays, protein extracts from entries in an array are loaded in adjacent gel lanes and electrophoresed. The gel is then treated with a histochemical stain that reacts with a specific marker protein. Variation between the entries is measured by the positions of their stained proteins on the gel. Many of the proteins used as markers are isozymes – functionally similar forms of an enzyme, encoded by different loci or different alleles at a locus. In general all alleles at a locus are expressed, so allelic isozymes can be analysed as codominant variants.

Isozymes are frequently used to describe variation between and within plant populations, species, and genera; for mapping and diagnostics; and in some cases for identification. When comparing higher taxonomic levels, the probability of genetically different electromorphs migrating to the same gel position is increased.

For closely related taxa and populations, however, isozyme analysis is informative and continues to be refined. Immunoelectrophoretic assays measure both protein migration rate in gels and the antibody specificity of proteins. Antigenic proteis are electrophoretically separated, then are detected by their reactions with antibodies. This technique is valuable for diagnostics, mapping, and describing variation between plnt cultivars, species, and higher taxa.

Nucleic Acid Marker Assays: Total Genomic DNA and Chromosome Markers

Although proteins are useful as genetic markers, the phenotypeeach protein describes is determined by the genotype plus the type of tissue sampled, plant phenological stage, environment, and post-translational processing. In addition, the percentage of a genome sampled by protein markers is limited: these markers only describe coding sequences, and not all proteins can be separated and detected by established methods. Some questions are best addressed by markers assumed to be selectively neutral, but this assumption is in question for many proteins. Finally, some proteins are unique to particular taxonomic groups, and thus are suitable as markers only in a limited range of taxa.

In contrast, nucleic acids are present in all organisms and are a common genetic currency for comparing virtually any taxa. Nucleic acid markers describe genotypes, not phenotypes, and can sample both coding and noncoding regions of a genome. DNA methylation and secondary structure can cause artefactual variation in some assays, disrupting the direct relationship between marker and genotype. However, these artefacts can often be detected and eliminated.

The first nucleic acid marker assays measured properties of total genomic DNA. When centrifuged in a CsCl density gradient, DNA separates into bands according to GC content. Because highly repetitive (satellite) sequences differ i GC content from the rest of the genome, these gradients can be used to isolate and quantify satellite DNA.

This method has been utilized to compare satellite DNA content between plant species, and is still used to detect highly repetitive DNA. Another technique that separates repetitive and low-copy DNA relies on DNA reassociation kinetics. The two complementary strands of a DNA molecule denature when heated and reassociate into duplexes when cooled. At a standard temperature and salt concentration, reassociation rate depends on the percentage of strands with similar sequences. Repetitive sequences reassociate rapidly and at high temperatures, compared to single-copy sequences.

Thus highly repetitive, middle repetitive, and single-copy fractions of a genome can be isolated by fragmenting, heating, and cooling double-stranded DNA. Because reassociation rate is affected by fragment length and the pattern of sequence repetition, the type and dispersion of repeats can be measured by varying the lengths of the fragments analysed. This method has been used to characterize and compare repeat sequences between plant species and hybrids.

DNA–DNAhybridization employs the same principles, but with single-copy sequences, to compare entries in an array. When single-stranded DNA from two entries is combined, their overall sequence similarity is related to the temperature at which 50% of the strands form heteroduplexes. This *T*m is compared to the homoduplex *T*m for each entry; D*T*m measures the genetic distance between entries. DNA–DNAhybridization is used to examine variation between species and higher taxa, usually with vertebrates and less often with plants and bacteria. Variation in DNA content of nuclei andchromosomes can be measured by flow cytometry. Particles are isolated, suspended, stained with a DNA-binding fluorochrome, and passed through a cytometer. The fluorescent signal intensity from each particle indicates its DNA content. Flow cytometry was first utilized to analyse mammalian genomes, but has since been used to compare nuclear DNA content and ploidy level between plant families, genera, species, and ecotypes and to evaluate widehybrids. Becaus chrmosomes can be identified by their signal intensity, flow cytometry can be used to generate high-resolution karyotypes and sortchromosomes for mapping.

Karyotypes ofchromosomes prepared in mitotic metaphase squashes have long been used for diagnostics, mapping, and describinggenetic variation between plant species andhybrids. Entries in an array can be characterized by their chromosome number and morphology. In addition, chromosome spread preparations can be treated with stains specific to AT-rich regions, heterochromatin, or nucleolar organizing regions. The stain-specific banding patterns of individua chromosomes are then used as markers to characterize and compare entries.

Chromosome banding is informative for mapping studies and describing variation between geneban accessions and cultivars. Banding is often used in combination with flow cytometry or other techniques. Fo chromosomes that are very small or lack distinctive bands, cytogenetic techniques may have limited use. As high-resolution microscopes andmore sensitive stains are developed, however, the value of these assays continues to increase.

Nucleic Acid Marker Assays: Fragment Markers

Nucleic acid fragments are widely used as genetic markers. As mentioned above, CsCl density gradients and DNA reassociation techniques can detect and roughly quantify satellite DNA sequences. Yet these methos do not measure discrete character differences or variation in the sequences themselves. In addition, differences in GC content between genomes can significantly affect the results. In contrast, fragment markers provide high-resolution qualitative information about sequence variation. Either DNA or RNA can be evaluated, and minimal amounts are required. Numerous fragment marker assays are available; all involve detecting fragments that differ in presence, size, or quantity between entries in an array.

Fragments that are produced, electrophoresed, then detected: In many assays, fragments are produced, separated by agarose or polyacrylamide gel electrophoresis, then detected. The length and conformation of a fragment determine its migration rate in a gel. Several different approaches are taken for detecting fragments. In some assays with purified sequences, all fragments are detected and visualized.

The fragments are either labelled before or stained after electrophoresis. Ethidium bromide, silver, fluorescent, and radioactive stains and labels are commonly used. In other assays, specific target fragments are detected and visualized by transfer hybridization. All fragments in an assay are electrophoresed, denatured if double-stranded, then transferred from the gel to a nylon or nitrocellulose membrane. A labelled single-stranded probe complementing the target sequence is hybridized to the membrane and anneals to the target fragments. The extent of mismatch hybridization varies with salt and formamide concentration and the temperature during hybridization. *A variety of probes can be used*: Synthesized oligonucleotides, uncharacterized clones from genomic DNA libraries, fragments isolated from other gels, RNA sequences, or total genomic DNA. The probe may be from the same species as the membrane-bound DNA or from a different species. Until recently most probes were radioactively labelled, but nonradioactive colorimetric, fluorescent, and chemil-uminescent labels are increasing in popularit).

Fragments produced by restriction digest: Restriction site analysis is widely used in plant genetics. The principles, protocols and data interpretation have been discussed in detail. Restriction fragments are generated by treating double- stranded DNA with restriction endonucleases (REs), which are enzymes produced by bacteria. Each RE cleaves double-stranded DNA at a specific sequence, usually four to six nucleotides in length.

Several hundred REs have been characterized; they vary in recognition sequence and ability to digest methylated DNA. DNA sequence variation between the entries in an array is detected as variation in the number and/ or lengths of restriction fragments. This variation can be caused by point mutations at RE recognition sites or by the insertion, deletion or rearrangement of sequences at or between RE sites.

When single- or low-copy fragments are evaluated, variations in fragment length are referred to as restriction fragment length polymorphisms (RFLPs). In some RFLP assays of highly purified DNA sequences, all of the fragments are visualized by labelling them before or staining them after electrophoresis. In most RFLP assays, however, target fragments are detected by hybridized probes. For each entry in an array, the presence or absence of each individual fragment is recorded.

These fragment data are used to infer the presence or absence of each restriction site. For diploid genomes, entries heterozygous for the presence of a restriction site can be distinguished from those that are homozygous. Although few RFLP loci are examined on each gel, every locus can be genetically defined. RFLP analysis can be expensive, but costs per trial can be reduced by probing each membrane several times with different probes. Single- or low-copy RFLPs are used extensively in diagnostics, for mapping, and to verify interspecific hybridization and study genetic relationships and structure. Probes that hybridize across species are especially useful for comparative mapping. In most restriction site analyses of repetitive DNA, target fragments are detected by probing with a known repeat sequence. The fragments, not the genetically defined sites, are usually the characters analysed. Entries that are homozygous for the presence of a fragment cannot be distinguished from heterozygotes. These assays provide information about variation at many loci simultaneously, but reveal little about the type of sequence variation. Restriction site analysis is often used to study interspersed repeats and satellites. It was utilized in the first assays of hypervariable loci, and is presently the most common method in forensics for evaluating minisatellite VNTR loci. Multilocus restriction fragment profiling of plant genomes is useful for identifying clonal cultivars and studying breeding systems. In some studies, fragment length polymorphism can be interpreted as variation of codominant alleles at a locus. However, because inheritance of markers in multilocus profiles is seldom clearly defined, their use in phylogenetic and gene flow studies may be limited.

Restriction site analysis can generate highly reproducible results, if all steps of the process are carefully controlled. With a variety of probe– enzyme combinations available, assays of the nuclear genome can be used to address many questions about genetic variation at or below the species level. As in isozyme analysis, the probability of genetically different restriction fragments co-migrating on a gel is increased when higher taxonomic levels are compared. Because chloroplast genomes evolve slowly, chloroplast restriction site assays are often used to study variation between higher taxa. In some species, both chloroplast and nuclear genomes are biparentally inherited. In many plant species, however, chloroplasts are either maternally or paternally inherited. For these species, chloroplast restriction site analysis is very informative for studies of gene flow and interspecific hybridization, especially when used in combination with assays of biparentally inherited nuclear markers.

Fragments produced by the polymerase chain reaction: Fragment analysis began in the 1970s with the introduction of RFLPs, and grew exponentially a decade later when the polymerase chain reaction (PCR) was developed. PCR uses the principles of DNA reassociation and the action of DNA polymerase to amplify nucleic acid fragments *in vitro*. Each PCR reaction solution contains a double-stranded DNA template, short single-stranded oligonucleotide primers, thermostable DNA polymerase, enzyme cofactors, and the four deoxynucleotides (dNTPs: dATP, dCTP, dGTP, dTTP).

The template is denatured by heating; the temperature is lowered, and the primers anneal to complementary sequences on the template. The solution is heated again, and the polymerase adds dNTPs to the 3' end of each annealed primer. Double-stranded fragments are produced and serve as templates for the next cycle, which generates fragments with primer sequences at each end. During subsequent cycles, these fragments are amplified at a geometric rate.

With automated thermocyclers and cheaper, improved reagents, PCR is fast becoming more reproducible and affordable. A wide range of known and unknown sequences can be amplified from fresh or ancient DNA and from

RNA, and little template is needed; hence PCR is applicable to any taxon. However, contamination is a concern, since contaminant DNA in a reaction solution can function as a template.

The specificity and reproducibility of PCR depend on several factors: the concentration and quality of the reaction ingredients; the design and GC content of the primers; and the temperature, duration and number of cycles. However, measures can be taken to minimize and detect contamination and optimize cycling protocols. The disadvantages of PCR are greatly outweighed by the tremendous opportunities it provides. When amplified fragments are used for high-resolution mapping, diagnostics or forensics, they are generally detected by probes. When used to describe variation between and within species, amplified fragments may be detected by labelling primers or dNTPs in the reaction solution, labelling PCR products, or staining gels after electrophoresis.

Numerous PCR-based marker assays have been developed and used in plant genetics. They can be categorized by whether target sequences are known prior to amplification, whether the primer sequences are designed or are arbitrary, the number of primers, and the size range of amplified products. In the first PCR assays the target loci were known single- copy sequences, amplified by two primers designed to complement the regions flanking the loci. Primer pairs have since been designed for numerous single-copy loci in plant and animal nuclear and organelle genomes. These loci are very useful as markers for mapping, diagnostics, phylogenetic analyses, and studying hybridization.

In most genome mapping projects, cloned DNA sequences present at only one site in a genome are used as physical mapping landmarks. Primer pairs have been designed to amplify many of these landmarks as sequence tagged sites (STSs). This strategy is used extensively in the Human Genome Project (HGP) and other mapping programmes. In assays of these designed-primer PCR loci, the character evaluated may be the fragment itself, with presence dominant over absence. Often, however, inheritance can be defined.

For loci in diploid genomes, differences in fragment length can then be analysed as codominant alleles. Both interspersed and tandemly repeated sequences can be amplified by designed-primer PCR. When several interspersed repeat loci are amplified in one sample, each locus is evaluated separately. When tandem repeat loci are amplified, length polymorphism at a locus is assumed to reflect variation in the number of tandem repeat units.

Amplification of human minisatellites with designed primer pairs was soon followed by amplification of plant minisatellites and SSRs. Amplified plant SSRs are informative for mapping, genotyping, diagnostics, and population studies. SSR alleles may differ little in length, and discriminating them can be difficult. New electrophoresis, labelling and imaging technologies are making detection easier. Development of designed-primer PCR markers involves several steps. Target sequences are identified and isolated, usually from a genomic DNA library; the loci and flanking regions are sequenced; and primer pairs are designed and synthesized. Currently these tasks are expensive, and not all laboratories have the equipment required. Once primers are designed and synthesized, however, the costs for each analysis are not high.

The initial expenses of marker development may be reduced by identifying the desired sequences from nucleic acid databases, *e.g.* Genbank or EMBL. However, this is only applicable to species well represented in the databases. As another cost-cutting measure, some primers designed to amplify loci in one species can be used to amplify loci in related taxa. Primers that amplify across taxa are also useful for phylogenetic analyses and studying genome evolution.

For development of designed-primer PCR markers, both target and flanking region sequences must be known. Soon after PCR was introduced, however, its ability to amplify unknown sequences was realised. In inter-repeat PCR, primers that complement interspersed repeats or other known sequences are used to amplify unknown sequences between the repeats. 'Universal' primers have been designed to complement conserved coding sequences in mitochondrial and chloroplast genomes; these primers can be used to amplify the noncoding regions between genes.

Other marker assays require no prior sequence information, for either the target region or the flanking regions. These assays – randomly amplified polymorphic DNA analysis (RAPD), arbitrarily primed PCR (AP-PCR; Welsh and McClelland, 1990), and DNA amplification fingerprinting (DAF) – use one short primer that has an arbitrary sequence. This primer anneals to the template at complementary sequences in both 'sense' and 'antisense' orientation, and the regions between primers in opposite orientation are amplified.

The number and length of fragments amplified by one primer can vary; short primers generally produce a large number of small fragments. In single-primer PCR, entries that are homozygous for the presence of a fragment cannot be distinguished from heterozygotes; thus fragment presence is generally dominant over absence. Many loci can be evaluated simultaneously, although little genetic information about each locus is provided. Arbitrary PCR is particularly valuable when no sequence information is available for the taxa being studied. It has been used extensively to examine variation between and within plant species, identify genebank accessions and cultivars, and construct genetic maps.

Many PCR assays are variations or combinations of designed-primer PCR and arbitrary PCR. Most were developed to increase the reproducibility of results and/or to better discriminate between closely related genotypes. In arbitrary PCR, reproducibility and the extent of mismatch primer annealing are highly dependent on the cycling parameters and concentration of reaction components. One solution to this problem is to identify polymorphic fragments in arbitrary PCR assays and then isolate, clone, sequence and design primers for these fragments.

This process generates new designed-primer PCR loci but involves many steps. Other assays with increased reproducibility combine features of single-primer and interrepeat PCR: unknown sequences are amplified by one primer that complements an SSR, its flanking region, or both. Finally, 'anchored' PCR uses one designed and one arbitrary primer to amplify unknown sequences.

For genotyping, diagnostics, and assessing variation between close relatives, PCR-based marker assays with high resolution are needed. One such variation is nested PCR. In this technique, one amplification reaction with two designed primers is followed by a second reaction. The first PCR product is the template for the second reaction, which uses two designed primers internal to the first set. Although the possibility of contamination is high, nested PCR is valuable for extremely sensitive genotyping and diagnostics. At present, it is used most often in medical research.

Resolution of arbitrary PCR can be increased by adding stable mini-hairpin sequences to the 5' ends of short primers. This technique greatly improved discrimination between cultivars of centipedegrass [*Eremochloa ophiuroides* (Munro) Hackel]. Other marker assays with improved resolution combine principles of PCR and restriction site analysis. RE digestion of PCR templates before they are amplified was first suggested by Mullis and Faloona (1987). Arbitrary PCR with cleaved templates has since been used to map plant genomes and identify genotypes. Amplified fragment length polymorphism (AFLP) analysis is a variation of cleaved-template PCR. After DNA is digested, a short adaptor sequence is ligated to each end of the restriction fragments. The fragments are then used as templates for single-primer PCR.

The primer contains a constant sequence (the adaptor and RE recognition site sequences) plus selective bases that determine the specificity of primer annealing. Both reproducibility and resolution are increased by using AFLPs. Resolution can also be increased by digesting PCR products before electrophoresis. Digested products of both designedprimer PCR and arbitrary PCR have been used to measure variation between genebank accessions, refine RFLP maps, and study interspecific gene flow. Resolution of both PCR and restriction site assays can be improved by denaturing double-stranded DNA fragments before they are electrophoresed. Due to internal point mutations or differences in conformation or GC content, the two single strands from one fragment may have different migration rates on polyacrylamide gels. Both denaturing gradient gel electrophoresis (DGGE) and single-stranded conformation polymorphism (SSCP) analysis can detect differences between closely related genotypes. *Fragments detected directly by probe hybridization.* In some marker assays the target sequences are not separated by electrophoresis, but are detected directly by probe hybridization.

The types of probes used are the same as in electrophoretic assays, and can be radioactively or nonradioactively labelled. In dot and slot blot hybridization, single-stranded DNA fragments, genomic DNA, or RNA is immobilized on a nylon or nitrocellulose membrane. Labelled probes complementing the target sequence are hybridized to the membrane blot. All loci in the blot are sampled simultaneously; abundance of the target sequence is measured by the intensity of the hybridization signal.

Many blots on one membrane can be evaluated at the same time, and the procedure is relatively simple. Because the presence of nontarget sequences in the blot can affect hyridization, precise measurment of target sequence abundance in genomic DNA blots can be difficult. When the blots are carefully prepared and hybridization conditions are controlled, however, these assays can generate valuable information for diagnostic and relationship studies. Dot and slot blots have been used to detect specific alleles of genes and RAPD loci, estimate the copy number of satellites and interspersed repeats, and compare gene expression between organs of a plant.

In another technique, *in situ* hybridization (ISH), probes are hybridized to the same chromosome spread preparations used for karyotypes and banding pattern assays. ISH is used to detect the presence and chromosomal distribution of target sequences. While banding assays are only appropriate with chromosomes that contain specific sequences, ISH can detect a wide range of sequences. Radioactive and nonradioactive probes are used, but the popularity of fluorescent *in situ* hybridization (FISH) is rapidly increasing.

ISH has many applications for genome analysis. These include 'painting' the genome of one species onto chromosomes of another, to study species evolution; painting a genome with one of its own chromosomes, to study chromosome evolution; and characterizing the chromosomal distribution of interspersed repeats. ISH is often used in combination with other marker assays. In wide crosses and somatic hybrids, ISH with species-specific interspersed repeat probes can indicate the abundance and chromosomal locations of transferred sequences; restriction site analysis can provide information about structure and rearrangement of the sequences. Physical mapping with ISH can complement RFLP genetic mapping studies.

Aledo *et al.* used slot blots to estimate the copy number of an interspersed repeat in the maize (*Zea mays* L.)genome, and used FISH to determine the chromosomal distribution of the repeat. With both ISH and heterochromatin-specific staining on the same chromosome preparation, Harrison and Heslop-

Harrison (1995) confirmed that two *Brassica* satellite families are located near centromeres. By varying thehybridization stringency, differences between satellite sequences on differentchromosomes were revealed. ISH can be time consuming, and results can be affected by the specific methods used for chromosome preparation, denaturation, and probehybridization and detection. However, new protocols are increasing the reproducibility of ISH.Fukui *et al.* (1994) used a thermocycler to control temperatures during denaturation, and developed accurate imaging methods for detecting and analysing fluorescent signals. Because of such innovations, the value of ISH as a molecular tool will most likely continue to increase.

Nucleic Acid Marker Assays: Nucleotide Sequencing

Nucleic acid variation can be described most directly by using the nucleotides themselves as discrete characters, each with four possible states. Because nucleotides are the basic units of genetic information in all organisms, sequencing can be used to detect very low levels of variation. Nuclear and organelle genomes of virtually any taxon can be sequenced; minimal amounts of DNA or RNA are needed, and fresh or preserved DNA can be used. With most marker assays, entries in a trial are compared to each other; results are not easily compared across trials that have no standard entries in common. In contrast, sequences from one study can be compared with sequences from any other study. In addition, sequencing can reveal whether variation is due to nucleotide substitution or sequence rearrangement.

Two sequencing methods have been used since the late 1970s, and are described in detail by Hillis *et al.* (1990). In both methods, labelled fragments are generated from DNA or RNA that has been isolated and purified; the fragments are electrophoresed; and the sequence is 'read' from the fragment patterns. Maxam–Gilbert (chemical) sequencing is used to sequence double-stranded DNA. DNA is endlabelled, then divided into four reaction solutions.

Each solution contains one of four reagents, each of which cleaves DNA at specific bases. The sequence of the DNA determines the lengths of the fragments produced. *In vitro* (dideoxy chain termination) sequencing is applicable to both DNA and RNA. The single-stranded template is divided into four reaction solutions. Each solution also contains the following: a labelled oligonucleotide primer that complements one end of the template; the four dNTPs; and one of four dideoxynucleotides, each of which terminates extension at one of the four bases.

In each reaction, the primer is extended along the template. As in chemical sequencing, the template sequence determines the lengths of the fragments produced. Chemical sequencing is preferred for some genomic regions with complex secondary structure; in general, however, the chain-termination method is more commonly used. Since it has been coupled with PCR, combined with fluorescent labelling and automation, this technique is rapidly becoming the method of choice. Sequencing can be used to address a wide range of plant genetic resource questions about virtually any taxa, but involves considerable investment of money, labour and time. Automation and large-scale production of cheaper reagents have made sequencing more cost effective, but few laboratories can afford to sequence large genome segments from multiple entries. As a result, in many studies only one locus or a few loci are sequenced for a limited test array. While many adjacent nucleotides may be evaluated, together they sample only one locus.

Thus sequencing is usually not practical for mapping, multilocus forensics, population genetics, or analyses of intraspecific variation. It is often used for high-resolution diagnostics, studying specific genes, and describing variation between genera and higher taxa. In phylogenetic studies, however, inferring genome evolution from sequence variation at a single locus should be done with caution if at all.

Many regions of plant nuclear and organelle genomes have been sequenced and used as markers. These sequences vary greatly in evolutionary rate and the taxonomic levels at which they are employed. Chloroplast genomes evolve slowly; sequences of the entire genome or its *rbc*L gene are informative as markers for describing variation between genera and higher taxa. Nuclear genes that have been sequenced and used as markers include the phytochrome, heat shock protein, and nuclear rRNA gene families. In general, the rRNA coding regions are highly conserved; internal transcribed spacers are more variable, and intergenic spacers are highly polymorphic. Finally, sequencing is often used – in combination with other marker assays – to characterize tandem repeat loci.

While sequencing may be the ultimate molecular tool, caution should be taken to avoid several potential problems. Because each nucleotide is a separate character, sequence fidelity is crucial. Thus nucleic acids must be carefully isolated and purified before they are sequenced. In addition, PCR contamination and amplification errors have more serious effects for sequencing than for fragment analysis. Amplification errors can be minimized by optimizing PCR conditions and using polymerases with proofreading capability. Mistakes during gel reading (manual or automated) must also be considered and minimized. Subjectivity is a real concern when aligning sequences before they are compared. Alignment is complicated and requires assumptions about expected amounts of base substitution and insertion/deletion. Despite these concerns, sequencing remains the tool of choice for many studies.

Guidelines for Selecting Marker Assays

Choosing appropriate marker assays can be challenging, but several considerations can make the task easier.

Important issues are:

- What question is being asked?
- What level of resolution is required?
- How can the results be related to characteristics of the taxa being studied?

Questions of identity are best addressed with high-resolution assays and large numbers of codominant marker loci that are evaluated sequentially.

Electrophoretic assays of isozymes and nucleic acid fragments meet most of these criteria. Genetically defined isozymes, amplified SSR loci, and single-copy RFLPs are very informative markers for identification. Due to practical constraints, however, multilocus assays of restriction fragment and PCR markers (VNTRs, RAPDs and AFLPs) are often used. Dot blots, sequencing and monoclonal antibody assays are also appropriate, but the costs of analysing many loci per assay may be prohibitive.

The most appropriate assays for diagnostic questions have high levels of resolution and use genetically defined, codominant markers. Because these questions can often be addressed with one or few loci, useful methods include electrophoretic assays of isozymes, single-copy RFLPs, or SSRs; dot or slot blot assays; and sequencing. For some questions, monoclonal antibody assays and ISH are also appropriate. Mapping studies often use a combination of tools: isozyme or DNA fragment electrophoresis for genetic maps, and chromosome banding, flow cytometry, ISH, and STS markers for physical mapping. For questions of relationship and structure, important issues include the type of question asked and the range and levels of taxa evaluated. Because evolutionary questions must be answered with qualitative data, the assays used for diagnostics and identification are appropriate. Questions of similarity can be addressed with these qualitative marker assays or with quantitative assays (microcomplement fixation, flow cytometry, CsCl density gradients, and DNA–DNA hybridization).

The latter two techniques have low levels of resolution, and are generally used to compare genera and higher taxa. Interspecific and intraspecific variation is more appropriately described with isozyme and DNA fragment markers, which have higher levels of resolution. In electrophoretic assays of isozymes and DNA fragments, co-migration of genetically dissimilar electromorphs or fragments is likely when higher taxa are compared. If resolution is too high, however, the assays may be numerically precise but contain random 'evolutionary noise' that masks meaningful variation. Researchers can often decide whether a particular marker assay is suitable for a given study by reviewing the results of previous studies using that technique. Several characteristics of the taxa being studied are important when selecting a marker and marker assay. One of these characteristics is the mating system.

Many apomictic and self-pollinated plant species contain little variation, so techniques with high resolution may be required. When deciding whether to use a chloroplast genome marker, information about chloroplast inheritance is critical. For many marker assays, polyploidy can be a concern. When polyploid taxa are evaluated, the number of loci sampled, the problem of dissimilar fragments co-migrating, and the level of data complexity are often increased. This complexity may be reduced by using genetically defined isozymes, chloroplast genes, genes that evolve slowly, or single-copy markers – rather than RAPDs or multilocus VNTR profiles.

Practical considerations are often the most important. The amount and quality of proteins or nucleic acids required can differ greatly between marker assays. Sequencing and PCR-based assays require little template and can be used to analyse ancient DNA samples, but are very sensitive to contamination. DNA–DNA hybridization requires large amounts of DNA, but small amounts of contaminant DNA are less of a problem. For some species, fresh tissue is required for isozyme electrophoresis. For some species and marker techniques, the presence of secondary compounds can complicate storage and extraction of DNA or protein, and can interfere with marker detection.

Marker assays also vary in the amount of setup work required and the extent of sequence information needed before conducting the assay. Karyotyping, ISH, and RFLP assays involve preparing chromosome spreads and/or probes. In contrast, some PCR-based assays are streamlined for high throughput and require little setup work. Sequence information is needed for probe hybridization and designed-primer PCR assays, but not for arbitrary PCR. Finally, the resources available must be considered. Minimal equipment is needed for protein electrophoresis, but other techniques require thermocyclers, fluorescent microscopes, cytometers and radioactive labelling.

Choices between newly developed and well-documented techniques should be made in light of the expertise and personnel available. Other considerations are reproducibility, number of entries and loci sampled per analysis, and cost per unit of information. Comparative studies can make selection of a marker assay easier and reveal where costs can be reduced. Costs and technologies are constantly changing, however, and the comparisons themselves may be expensive and time consuming; their value must be weighed against the resources they require and the necessity of continual updates. When comparative studies are not feasible, thoughtful consideration of the issues described here may be helpful.

Future Dimensions of Genetic Analysis

The goals and expectations for analysing plant genetic variation parallel those established across many other fields of biological research, from agriculture, ecology, and evolution to the medical sciences. In all of these fields, future genetic marker assays must incorporate methods to detect, describe, interpret, and store DNA sequence information. Molecular tools of the future are expected to be user friendly, accurate, precise, high throughput, low cost, and potentially automated. DNA sequence information is the foundation for developing and applying genetic markers to questions of biological variation, whether *in situ* or *ex situ*. Researchers who develop and use sequence-based marker assays for diagnostic, forensic and relationship studies will continue to benefit greatly from information and technologies generated by the international Human Genome Project (HGP).

Since its beginning, the HGP has endeavoured to develop genetic and physical maps and determine DNA sequences for the genomes of humans and several model organisms. As part of this effort, genetic analysis methods have been improved significantly. Notable achievements – now taken for granted by many – include new types of genetic markers (particularly those based on PCR), more efficient cloning vectors, the introduction of STS markers, and improved technology and automation for DNA sequencing. Underlying this progress is recognition that the successful development of new technologies for genetic research has been, and will continue to be, critical to many future scientific breakthroughs. In addition to the HGP, other complementary projects also support initiatives for the development of new molecular marker assays.

One such project is the Advanced Technology Programme (ATP) of the US National Institute of Standards and Technology. Among the ATP's goals is working with industry to deliver DNA diagnostic tools to a variety of users, at onetenth to one-hundredth of the current price. Another goal is helping industry make similar reductions in DNA sequencing costs and make DNA sequencing apparatus available at about one-third of the current price. Many biological questions once considered recalcitrant due to the number and cost of the assays required will be reconsidered in the light of new technological developments. The following section includes examples of progress that can be readily applied to the description of plant genetic variation. Also noted are limitations that are unique to plant-based studies, and are not presently addressed by most genetic analysis programmes.

Technology Transfer from the Human Genome Project

Technology transfer from the Human Genome Project (HGP) to the study of plant genetic resources has been very important in two areas:

- Sequence characterization and marker development; and
- Integrating all stages in genetic analysis projects, from template preparation through data interpretation.

Some of these innovations are now being used in plant genetic research; others will require some level of optimization before they are applied.

Plant genetic variation is best characterized with a large number of highly informative, easily analysed molecular markers. While the number of PCR-based markers will most likely continue to increase, the question at hand is how these markers will be detected. Manual, autoradiographic detection of amplified markers presently predominates in plant genetic research. However, ongoing programmes in human genotyping via semiautomated, fluorescent detection of SSRs are progressing at an incredible pace. Reed *et al.* (1994) reported that semi-automated technology can genotype over 100 000 samples, using approximately 250 marker loci, in less than six months. Moreover, Schwengel *et al.* (1994) found that data generated by this method were at least as accurate, more efficient, less labour intensive, and as cost effective as data generated by standard radiolabelling techniques. Human genotyping kits are available for approximately 400 markers that define a ±10 cM resolution genomic index map. Kits for plant species and families are in development. Although cost remains an issue, these marker assays are clearly capable of handling the large arrays commonly encountered in plant conservation and breeding programmes.

The costs of generating and detecting numerous marker loci for each of many individuals can be reduced by evaluating several of an individual's loci in a single lane on a gel. In the studies cited above, SSR loci from an individual were multiplexed at the electrophoresis level: several loci were amplified in separate reactions, the PCR products were pooled, and the pooled sample was electrophoresed. In contrast, several laboratories are developing methods to multiplex at the reaction level, amplifying several SSR loci in one tube. Eleven putative loci in canola (*Brassica napus* L.) have been amplified simultaneously, using 11 pairs of fluorescently labelled primers in one reaction.

The loci were distinguished from each other by fragment size and fluorescent label colour. This approach is also under way with SSRs of maize and sorghum [*Sorghum bicolor* (L.) Moench] Another molecular marker used in human genetic research reveals discrete changes (single nucleotide substitutions) in a specific DNA sequence. Variation of these polymorphic sequence tagged sites (pSTS) is the most frequent and widely distributed type of sequence variation in the genome. These individual nucleotide markers are usually biallelic within a population, and are generally less polymorphic than SSRs and other tandem repeats. However, a number of closely linked pSTSs can be combined into a multilocus marker that is as informative as an SSR.

Developing and assaying pSTS markers involve several steps. Random genomic DNA clones (each 300–600 bp long) from selected plant species are sequenced, then single nucleotide differences between individuals are identified by comparing automated sequencing electrophoregrams. Biallelic pSTSs are selected, and primer pairs are designed to amplify them. After a pSTS locus is amplified, the oligonucleotide ligation assay (OLA) is used to label the PCR products and discriminate the two allelic forms. In this assay, the PCR products are denatured and then serve as templates for a ligation detection reaction (LDR). This reaction uses three 'probes': two diagnostic probes, each of which complements one of the two pSTS alleles, and one labelled probe that is common to both alleles.

The diagnostic probes anneal to complementary template sequences, then are ligated to the labelled probe. The two diagnostic probes differ in length. Thus when the LDR-labelled fragments are electrophoresed, pSTS allelic variation between entries can be detected as variation in migration rate of fragments on the gel, due to differences in fragment length. As in many PCR-based marker assays, significant costs are presently incurred for pSTS marker development. Once markers are developed, however, the costs for each assay are not high. The previously highlighted markers are usually associated with noncoding regions of the genome. Complementary efforts are under way to develop high-throughput, automated methods to partially sequence cDNA clones, then use them to identify expressed genes in many organisms, including plants. Thanks to the rapid proliferation of expressed sequence tag (EST) databases, at present the probable gene function of a cDNA clone can often be inferred solely from the similarity of nucleotide or amino acid sequences between the clone and genes of known function. Using these techniques to compare isoforms of a gene (particularly its untranslated 3' regions) between members of a plant species or family, PCR-based markers that describe variation of this gene could be developed. When assays using these single-gene markers are compared with multilocus studies of the taxon, a more complete picture of genetic variation may be revealed.

Present Limitations on Describing Plant Genetic Variation with Molecular Tools

Human genetic research often focuses on genomic mapping and analysing specific genes. In contrast, many plant genetic resource questions are best addressed by evaluating numerous individuals from many taxa. Thus widespread use of molecular markers in plant genetic research will involve preparing and analysing large amounts of DNA template and assaying many individuals at the same time. However, techniques have not yet been developed to extract, clone, replicate or amplify DNA in nanolitre volumes. Also needed are techniques for pooling DNA samples from individuals in a population, assaying the pooled sample, and generating a comprehensive picture of population variation. In the future, sequence-based approaches may be useful for these tasks. At present, high costs make these approaches impractical for many high-throughput applications.

2

Microbodies: Peroxisomes and Glyoxysomes

A microbody is a cytoplasmic organelle of a more or less globular shape that comprises degradative enzymes bound within a single membrane. Microbodies are specialized as containers for metabolic activity. Types include peroxisomes, glyoxisomes, glycosomes and Woronin bodies. Peroxisomes contain enzymes of b-oxidation (break down fats and produce Acetyl-CoA), as well as enzymes of many other important pathways like amino acid and bile acid metabolism, oxidation/detoxification of various harmful compounds in the liver (ex. alcohol).

Glyoxysomes are found in germinating seeds of plants as well as in filamentous fungi. Glyoxysomes are peroxisomes with additional function - glyoxylate cycle. Glycosomes, besides peroxisomal enzymes, also possess glycolysis enzymes and are found in kinetoplastida like Trypanosomes.

Woronin bodies are special organelles found only in filamentous fungi. One established function of Woronin bodies is the plugging of the septal pores after hyphal wounding, which restricts the loss of cytoplasm to the sites of injury.

Peroxisomes are organelles present in almost all eukaryotic cells. They participate in the metabolism of fatty acids and many other metabolites. Peroxisomes harbour enzymes that rid the cell of toxic peroxides. Peroxisomes are bound by a single membrane that separates their contents from the cytosol (the internal fluid of the cell) and contain membrane proteins critical for various functions, such as importing proteins into the organelles and aiding in proliferation. Peroxisomes are formed from the endoplasmic reticulum.

Peroxisomes were identified as organelles by the Belgian cytologist Christian de Duve in 1967 after they had been first described in a PhD thesis of Rhodin a decade earlier.

Occurrence and Evolution

Peroxisomes are found in virtually all eukaryotic cells. Peroxisomes contain enzymes for certain oxidative reactions, like the beta-oxidation of very-long-chain fatty acids.

Prokaryotes lack peroxisomes. The enzymatic content of peroxisomes varies across species, but the presence of certain proteins common to many species has been used to suggest an endosymbiotic origin; that is, peroxisomes evolved from bacteria that invaded larger cells as parasites, and very gradually evolved a symbiotic relationship. However, this view has been challenged by recent discoveries. For example, peroxisome-less mutants can restore peroxisomes upon introduction of the wild-type gene.

Two independent evolutionary analyses of the peroxisomal proteome found homologies between the peroxisomal import machinery and the ERAD pathway in the endoplasmic reticulum, along with a number of metabolic enzymes that were likely recruited from the mitochondria. These results indicate that the peroxisome does not have an endosymbiotic origin; instead, it likely originates from the ER, and its proteins were recruited from pools existing within the primitive eukaryote, as quoted in the science textbook Biozone. However, latest research have suggested that the peroxisome may have had an actinobacterial origin, although this is still controversial.

Other organelles of the Microbody family related to peroxisomes include glyoxysomes of plants and filamentous fungi, glycosomes of kinetoplastids and Woronin bodies of filamentous fungi.

Function

Peroxisomes contain oxidative enzymes, such as catalase, D-amino acid oxidase, and uric acid oxidase. However the last enzyme is absent in humans, explaining the disease known as gout, caused by the accumulation of uric acid. Certain enzymes within the peroxisome, by using molecular oxygen, remove hydrogen atoms from specific organic substrates (labeled as R), in an oxidative reaction, producing hydrogen peroxide (H_2O_2, itself toxic):

$$RH_2 + O_2 \rightarrow R + H_2O_2$$

Catalase, another enzyme in the peroxisome, in turn uses this H_2O_2 to oxidize other substrates, including phenols, formic acid, formaldehyde, and alcohol, by means of the peroxidation reaction:

$$R_2O_2 + R'H_2 \rightarrow R' + 2H_2O$$

thus eliminating the poisonous hydrogen peroxide in the process.

This reaction is important in liver and kidney cells, where the peroxisomes detoxify various toxic substances that enter the blood. About 25% of the ethanol we drink is oxidized to acetaldehyde in this way. In addition, when excess H_2O_2 accumulates in the cell, catalase converts it to H_2O through this reaction:

$$2H_2O_2 \rightarrow 2H_2O + O_2$$

A major function of the peroxisome is the breakdown of fatty acid molecules, in a process called beta-oxidation. In this process, the fatty acids are broken down two carbons at a time, converted to Acetyl-CoA, which is then transported back to the cytosol for further use. In animal cells, beta-oxidation can also occur in the mitochondria. In yeast and plant cells, this process is exclusive for the peroxisome.

The first reactions in the formation of plasmalogen in animal cells also occur in peroxisomes. Plasmalogen is the most abundant phospholipid in myelin. Deficiency of plasmalogens causes profound abnormalities in the myelination of nerve cells, which is one of the reasons that many peroxisomal disorders lead to neurological disease (adrenoleukodystrophy).

Peroxisomes also play a role in the production of bile acids and proteins.

In higher plants, peroxisomes contain also a complex battery of antioxidative enzymes such as superoxide dismutase, the components of the ascorbate-glutathione cycle, and the NADP-dehydrogenases of the pentose-phosphate pathway. It has been demonstrated the generation of superoxide (O_2) and nitric oxide (˙NO) radicals.

The peroxisome of plant cells is polarised when fighting fungal penetration. Infection causes a glucosinolate molecule to play an antifungal role to be made and delivered to the outside of the cell through the action of the peroxisomal proteins (PEN2 and PEN3).

Protein Import

Proteins are selectively imported into peroxisomes. Since the organelles contain no DNA or ribosomes and, thus, have no means of producing proteins, all of their proteins must be imported across the membrane. It is believed that necessary proteins enter through the endoplasmic reticulum during biogenesis as well as through membrane proteins.

A specific protein signal (PTS or peroxisomal targeting signal) of three amino acids at the *C-terminus* of many peroxisomal proteins signals the membrane of the peroxisome to import them into the organelle. Other peroxisomal proteins contain a signal at the *N-terminus*. There are at least 32 known peroxisomal proteins, called peroxins, which participate in the process of importing proteins by means of *ATP hydrolysis*. Proteins do not have to unfold to be imported into the peroxisome. The protein receptors, the peroxins *PEX5* and *PEX7*, accompany their cargoes (containing a PTS1 or a PTS2, respectively) all the way into the peroxisome where they release the cargo and then return to the cytosol - a step named *recycling*. Overall, the import cycle is referred to as the *extended shuttle mechanism*. Evidence now indicates that ATP hydrolysis is required for the recycling of receptors to the cytosol. Also, ubiquitination appears to be crucial for the export of PEX5 from the peroxisome, to the cytosol. Little is known about the import of PEX7, although it has helper proteins that have been shown to be ubiquitinated. It is interesting to note that Pex7 helper proteins share many similarities with Pex5, and, in higher eukaryotes, they are functionally replaced by a long isoform of PEX5 (PEX5L).

Deficiencies

Peroxisomal disorders are a class of conditions that lead to disorders of lipid metabolism and diseases of the nervous system. Well-known examples are X-linked adrenoleukodystrophy, the most frequent, and Zellweger syndrome. Peroxisomes matrix proteins are synthesized on free ribosomes in the cytosol and are imported post-translationally in pre-existing vesicles.

Genes

Genes that encode peroxisomal proteins include:

- PEX1
- PEX2 - PXMP3
- PEX3
- PEX5
- PEX6
- PEX7
- PEX10
- PEX11A, PEX11B, PEX11G
- PEX12
- PEX13
- PEX14
- PEX16
- PEX19
- PEX26
- PEX28
- PEX30
- PEX31

Glyoxysome

Glyoxysomes are specialized peroxisomes found in plants (particularly in the fat storage tissues of germinating seeds) and also in filamentous fungi. As in all peroxisomes, in glyoxysomes the fatty acids are hydrolyzed to acetyl-CoA by peroxisomal α-oxidation enzymes. Besides peroxisomal functions, glyoxysomes possess additionally the key enzymes of glyoxylate cycle (isocitrate lyase and malate synthase) which accomplish the glyoxylate cycle bypass.

Thus, glyoxysomes (as all peroxisomes) contain enzymes that initiate the breakdown of fatty acids and additionally possess the enzymes to produce intermediate products for the synthesis of sugars by gluconeogenesis. The seedling uses these sugars synthesized from fats until it is mature enough to produce them by photosynthesis.

Ribosomes

Ribosomes are the components of cells that make proteins from all amino acids. One of the central tenets of biology, often referred to as the "central dogma," is that DNA is used to make RNA, which, in turn, is used to make protein. The DNA sequence in genes is copied into a messenger RNA (mRNA). Ribosomes then read the information in this RNA and use it to create proteins. This process is known as translation; i.e., the ribosome "translates" the genetic information from RNA into proteins. Ribosomes do this by binding to an mRNA and using it as a template for the correct sequence of amino acids in a particular protein. The amino acids are attached to transfer RNA (tRNA) molecules, which enter one part of the ribosome and bind to the messenger RNA sequence. The attached amino acids are then joined together by another part of the ribosome. The ribosome moves along the mRNA, "reading" its sequence and producing a chain of amino acids.

Ribosomes are made from complexes of RNAs and proteins. Ribosomes are divided into two subunits, one larger than the other. The smaller subunit binds to the mRNA, while the larger subunit binds to the tRNA and the amino acids. When a ribosome finishes reading a mRNA, these two subunits split apart. Ribosomes have been classified as ribozymes, since the ribosomal RNA seems to be most important for the peptidyl transferase activity that links amino acids together.

Ribosomes from bacteria, archaea and eukaryotes (the three domains of life on Earth), have significantly different structures and RNA sequences. These differences in structure allow some antibiotics to kill bacteria by inhibiting their ribosomes, while leaving human ribosomes unaffected. The ribosomes in the mitochondria of eukaryotic cells resemble those in bacteria, reflecting the likely evolutionary origin of this organelle. The word ribosome comes from *ribo*nucleic acid and the Greek: *soma* (meaning body).

Description

Archaeal, eubacterial and eukaryotic ribosomes differ in their size, composition and the ratio of protein to RNA. Because they are formed from two subunits of non-equal size, they are slightly longer in the axis than in diameter. Prokaryotic ribosomes are around 20 nm (200 ångströms) in diameter and are composed of 65% ribosomal RNA and 35% ribosomal proteins (known as a ribonucleoprotein or RNP). Eukaryotic ribosomes are between 25 and 30 nm (250-300 ångströms) in diameter and the ratio of rRNA to protein is close to 1. Ribosomes translate messenger RNA (mRNA) and build polypeptide chains (e.g., proteins) using amino acids delivered by transfer RNA (tRNA). Their active sites are made of RNA, so ribosomes are now classified as "ribozymes".

Ribosomes build proteins from the genetic instructions held within messenger RNA. Free ribosomes are suspended in the cytosol (the semi-fluid portion of the cytoplasm); others are bound to the rough endoplasmic reticulum, giving it the appearance of roughness and thus its name, or to the nuclear envelope. As ribozymes are partly constituted from RNA, it is thought that they might be remnants of the RNA world. Although catalysis of the peptide bond involves the C2 hydroxyl of RNA's P-site adenosine in a protein shuttle mechanism, other steps in protein synthesis (such as translocation) are caused by changes in protein conformations.

Ribosomes are sometimes referred to as organelles, but the use of the term *organelle* is often restricted to describing sub-cellular components that include a phospholipid membrane, which ribosomes, being entirely particulate, do not. For this reason, ribosomes may sometimes be described as "non-membranous organelles".

Ribosomes were first observed in the mid-1950s by Romanian cell biologist George Palade using an electron microscope as dense particles or granules for which he would win the Nobel Prize. The term "ribosome" was proposed by scientist Richard B. Roberts in 1958:

During the course of the symposium a semantic difficulty became apparent. To some of the participants, "microsomes" mean the ribonucleoprotein particles of the microsome fraction contaminated by other protein and lipid material; to others, the microsomes consist of protein and lipid contaminated by particles. The phrase "microsomal particles" does not seem adequate, and "ribonucleoprotein particles of the microsome fraction" is much too awkward. During the meeting, the word "ribosome" was suggested, which has a very satisfactory name and a pleasant sound. The present confusion would be eliminated if "ribosome" were adopted to designate ribonucleoprotein particles in sizes ranging from 35 to 100S.

– *Roberts, R. B., Microsomal Particles and Protein Synthesis*

The structure and function of the ribosomes and associated molecules, known as the *translational apparatus*, has been of research interest since the mid-twentieth century and is a very active field of study today.

Ribosomes consist of two subunits that fit together and work as one to translate the mRNA into a polypeptide chain during protein synthesis. Bacterial subunits consist of one or two and eukaryotic of one or three very large RNA molecules (known as ribosomal RNA or rRNA) and multiple smaller protein molecules. Crystallographic work has shown that there are no ribosomal proteins close to the reaction site for polypeptide synthesis. This suggests that the protein components of ribosomes act as a scaffold that may enhance the ability of rRNA to synthesize protein rather than directly participating in catalysis.

Biogenesis

In bacterial cells, ribosomes are synthesized in the cytoplasm through the transcription of multiple ribosome gene operons. In eukaryotes, the process takes place both in the cell cytoplasm and in the nucleolus, which is a region within the cell nucleus. The assembly process involves the coordinated function of over 200 proteins in the synthesis and processing of the four rRNAs, as well as assembly of those rRNAs with the ribosomal proteins.

Ribosome Locations

Ribosomes are classified as being either "free" or "membrane-bound". Free and membrane-bound ribosomes differ only in their spatial distribution; they are identical in structure. Whether the ribosome exists in a free or membrane-bound state depends on the presence of an ER-targeting signal sequence on the protein being synthesized, so an individual ribosome might be membrane-bound when it is making one protein, but free in the cytosol when it makes another protein.

Free Ribosomes

Free ribosomes can move about anywhere in the cytosol, but are excluded from the cell nucleus and other organelles. Proteins that are formed from free ribosomes are released into the cytosol and used within the cell. Since the cytosol contains high concentrations of glutathione and is, therefore, a reducing environment, proteins containing disulfide bonds, which are formed from oxidized cysteine residues, cannot be produced in this compartment.

Membrane-bound Ribosomes

When a ribosome begins to synthesize proteins that are needed in some organelles, the ribosome making this protein can become "membrane-bound". In eukaryotic cells this happens in a region of the endoplasmic reticulum (ER) called the "rough ER". The newly produced polypeptide chains are inserted directly into the ER by the ribosome and are then transported to their destinations, through the secretory pathway. Bound ribosomes usually produce proteins that are used within the plasma membrane or are expelled from the cell via *exocytosis*.

Structure

The ribosomal subunits of prokaryotes and eukaryotes are quite similar. The unit of measurement is the Svedberg unit, a measure of the rate of sedimentation in centrifugation rather than size and accounts for why fragment names do not add up (70S is made of 50S and 30S). Prokaryotes have 70S ribosomes, each consisting of a small (30S) and a large (50S) subunit. Their large subunit is composed of a 5S RNA subunit (consisting of 120 nucleotides), a 23S RNA subunit (2900 nucleotides) and 34 proteins. The 30S subunit has a 1540 nucleotide RNA subunit (16S) bound to 21 proteins.

Eukaryotes have 80S ribosomes, each consisting of a small (40S) and large (60S) subunit. Their large subunit is composed of a 5S RNA (120 nucleotides), a 28S RNA (4700 nucleotides), a 5.8S subunit (160 nucleotides) and ~49 proteins. The 40S subunit has a 1900 nucleotide (18S) RNA and ~33 proteins.

The ribosomes found in chloroplasts and mitochondria of eukaryotes also consist of large and small subunits bound together with proteins into one 70S particle. These organelles are believed to be descendants of bacteria and as such their ribosomes are similar to those of bacteria.

The various ribosomes share a core structure, which is quite similar despite the large differences in size. Much of the RNA is highly organized into various tertiary structural motifs, for example pseudoknots that exhibit coaxial stacking. The extra RNA in the larger ribosomes is in several long continuous insertions, such that they form loops out of the core structure without disrupting or changing it. All of the catalytic activity of the ribosome is carried out by the RNA; the proteins reside on the surface and seem to stabilize the structure.

The differences between the bacterial and eukaryotic ribosomes are exploited by pharmaceutical chemists to create antibiotics that can destroy a bacterial infection without harming the cells of the infected person. Due to the differences in their structures, the bacterial 70S ribosomes are vulnerable to these antibiotics while the eukaryotic 80S ribosomes are not. Even though mitochondria possess ribosomes similar to the bacterial ones, mitochondria are not affected by these antibiotics because they are surrounded by a double membrane that does not easily admit these antibiotics into the organelle.

High-resolution Structure

The general molecular structure of the ribosome has been known since the early 1970s. In the early 2000s the structure has been achieved at high resolutions, on the order of a few ångströms. The first papers giving the structure of the ribosome at atomic resolution were published in rapid succession in late 2000. First, the 50S (large prokaryotic) subunit from the archaeon *Haloarcula marismortui* was published. Soon after, the structure of the 30S subunit from *Thermus thermophilus* was published. Shortly thereafter, a more detailed structure was published. These structural studies were awarded the Nobel Prize in Chemistry in 2009. Early the next year (May 2001) these coordinates were used to reconstruct the entire *T. thermophilus* 70S particle at 5.5 ångström resolution.

Two papers were published in November 2005 with structures of the *Escherichia coli* 70S ribosome. The structures of a vacant ribosome were determined at 3.5-angström resolution using x-ray crystallography. Then, two weeks later, a structure based on cryo-electron microscopy was published, which depicts the ribosome at 11-15 angström resolution in the act of passing a newly synthesized protein strand into the protein-conducting channel.

First atomic structures of the ribosome complexed with tRNA and mRNA molecules were solved by using X-ray crystallography by two groups independently, at 2.8 angstrom and at 3.7 angstrom. These structures allow one to see the details of interactions of the *Thermus thermophilus* ribosome with mRNA and with tRNAs bound at classical ribosomal sites. Interactions of the ribosome with long mRNAs containing Shine-Dalgarno sequences were visualized soon after that at 4.5- to 5.5-angström resolution.

Function

Ribosomes are the workhorses of protein biosynthesis, the process of translating mRNA into protein. The mRNA comprises a series of codons that dictate to the ribosome the sequence of the amino acids needed to make the protein. Using the mRNA as a template, the ribosome traverses each codon (3 nucleotides) of the mRNA, pairing it with the appropriate amino acid provided by a tRNA. Molecules of transfer RNA (tRNA) contain a complementary anticodon on one end and the appropriate amino acid on the other.

The small ribosomal subunit, typically bound to a tRNA containing the amino acid methionine, binds to an AUG codon on the mRNA and recruits the large ribosomal subunit. The ribosome then contains three RNA binding sites, designated A, P and E. The A site binds an aminoacyl-tRNA (a tRNA bound to an amino acid); the P site binds a peptidyl-tRNA (a tRNA bound to the peptide being synthesized); and the E site binds a free tRNA before it exits the ribosome. Protein synthesis begins at a start codon AUG near the 5' end of the mRNA. mRNA binds to the P site of the ribosome first. The ribosome is able to identify the start codon by use of the Shine-Dalgarno sequence of the mRNA in prokaryotes and Kozak box in eukaryotes. The ribosome uses tRNA that matches the current codon (triplet) on the mRNA to append an amino acid to the polypeptide chain. This is done for each triplet on the mRNA, while the ribosome moves towards the 3' end of the mRNA. Usually in bacterial cells, several ribosomes are working parallel on a single mRNA, forming what is called a *polyribosome* or *polysome.*

Cytoskeleton: Microtubules, Microfilaments and Microvilli

The cytoskeleton (also CSK) is a cellular "scaffolding" or "skeleton" contained within the cytoplasm and is made out of protein. The cytoskeleton is present in all cells; it was once thought to be unique to eukaryotes, but recent research has identified the prokaryotic cytoskeleton. It is a dynamic structure that maintains cell shape, protects the cell, enables cellular motion (using structures such as flagella, cilia and lamellipodia), and plays important roles in both intracellular transport (the movement of vesicles and organelles, for example) and cellular division. The concept of a protein mosaic that dynamically coordinated cytoplasmic biochemistry was proposed by Rudolph Peters in 1929 while the term (*cytosquelette*, in French) was first introduced by French embryologist Paul Wintrebert in 1931.

The Eukaryotic Cytoskeleton

Eukaryotic cells contain three main kinds of cytoskeletal filaments, which are microfilaments, intermediate filaments, and microtubules. The cytoskeleton provides the cell with structure and shape, and by excluding macromolecules from some of the cytosol it adds to the level of macromolecular crowding in this compartment. Cytoskeletal elements interact extensively and intimately with cellular membranes.

Microfilament

Microfilaments (or actin filaments) are the thinnest filaments of the cytoskeleton found in the cytoplasm of all eukaryotic cells. These linear polymers of actin subunits are flexible and relatively strong, resisting buckling by multi-piconewton compressive forces and filament fracture by nanonewton tensile forces. Microfilaments are highly versatile, functioning in (a) cell crawling, amoeboid movement, and changes in cell shape, where one end of the actin filament elongates while the other end contractile, presumably by myosin II molecular motors (though an alternative "actoclampin"-driven expansile molecular motors exists), and (b) actomyosin-driven contractile molecular motors, where the *thin filaments* serve as tensile platforms for myosin's ATP hydrolysis-dependent pulling action in muscle contraction and uropod advancement.

Organization

Actin filaments are assembled in two general types of structures: bundles and networks. Bundles can be composed of polar (all barbed ends point to the same end of the bundle) or non-polar (barbed ends point towards both ends) arrays of filaments. A class of Actin-binding proteins, called cross-linking proteins, dictate the formation of these structures. Cross-linking proteins determine filament orientation and spacing in the bundles and networks. These structures are regulated by many other classes of actin-binding proteins (motor proteins, branching proteins, severing proteins, polymerization promoters, capping proteins, etc.).

In Vitro *Self-assembly*

The thinnest fibres of the cytoskeleton (measuring approximately 6 nm in diameter), microfilaments are formed by the head-to-tail polymerization of actin monomers (also known as globular or G-actin). Actin subunits as part of a fibre are referred to as filamentous actin (or F-actin). Each microfilament is made up of two helical interlaced strands of subunits. Much like microtubules, actin filaments are polarized, with their fast-growing barbed-ends (because of their appearance in electron micrographs after binding of myosin S1 sub-fragments) and a slow-growing pointed-end (again based on the pattern created by S1 binding). The pointed end is sometimes referred to as the minus (-) end and the barbed end is sometimes referred to as the plus (+) end because of the growth rates, but this is nomenclature adapted from the microtubule field, and is not generally accepted in the actin field.

In vitro actin polymerization, nucleation, starts with the self-association of three G-actin monomers to form a trimer. ATP-actin then binds the barbed end, and the ATP is subsequently hydrolyzed with a half time of about 2 seconds and the inorganic phosphate released with a half-time of about 6 minutes, which reduces the binding strength between neighbouring units and generally destabilizes the filament. *In vivo* actin polymerization is catalyzed by a new class of filament end-tracking molecular motors known as actoclampins. Recent evidence suggests that ATP hydrolysis can be prompt in such cases (i.e., the rate of monomer incorporation is matched by the rate of ATP hydrolysis).

ADP-actin dissociates slowly from the pointed end, but this process is greatly accelerated by ADP-cofilin, which severs ADP-rich regions nearest the (–)-ends. Upon release, ADP-actin undergoes exchange of its bound ADP for solution-phase ATP, thereby forming the ATP-actin monomeric units needed for further barbed-end filament elongation. This rapid turnover is important for the cell's movement. End-capping proteins such as CapZ prevent the addition or loss of monomers at the filament end where actin turnover is unfavourable like in the muscle apparatus.

Mechanism of Force Generation

As a result of the ATP hydrolysis, filaments elongate approximately 10 times faster at their barbed ends than their pointed ends. At steady-state, the polymerization rate at the barbed end matches the depolymerization rate at the pointed end, and microfilaments are said to be treadmilling. Treadmilling results in elongation in the barbed end and shortening in the pointed-end, so that the filament in total moves. Since both processes are energetically favourable, this means force is generated, and the energy ultimately comes from the NTP hydrolization.

Actin in Cells

Intracellular actin cytoskeletal assembly and disassembly are tightly regulated by cell signalling mechanisms. Many signal transduction systems use the actin cytoskeleton as a scaffold holding them at or near the inner face of the peripheral membrane. This subcellular location allows immediate and exquisite responsiveness to transmembrane receptor action and signal-processing enzyme cascades. Because actin monomers must be recycled to sustain high rates of actin-based motility during chemotaxis, cell signalling is believed to activate cofilin, an actin-filament depolymerizing protein which binds to ADP-rich actin subunits nearest the filament's pointed-end and promotes filament fragmentation, with concomitant depolymerization to liberate actin monomers.

The protein profilin enhances the ability of monomers to assemble by stimulating the exchange of actin-bound ADP for solution-phase ATP to yield Actin-ATP and ADP. In most animal cells, monomeric actin is bound to profilin and thymosin-beta4, both of which preferentially bind with one-to-one stoichiometry to ATP-containing monomers. Although thymosin-beta4 is strictly a monomer-sequestering protein, the behavior of profilin is far more complex. Profilin is transferred to the leading edge by virtue of its PIP2 binding site, and profilin also employs its poly-L-proline binding site to dock onto end-tracking proteins. Once bound, Profilin-Actin-ATP is loaded into the monomer-insertion site of actoclampin motors. Another important component in filament formation is the Arp2/3 complex, which binds to the side of an already existing filament (or "mother filament"), where it nucleates the formation of a new actin filament and creates a fan-like branched filament network.

In non-muscle cells, actin filaments are formed at/near membrane surfaces. Their formation and turnover are regulated by many proteins, including

- Filament end-tracking protein (e.g., formins, VASP, N-WASP)
- Filament-nucleator known as the Actin-Related Protein-2/3 (or Arp2/3) complex
- Filament cross-linkers (e.g., α-actinin and fascin)
- Actin monomer-binding proteins profilin and thymosin-α4
- Filament barbed-end cappers such as Capping Protein and CapG, *etc.*
- Filament-severing proteins like gelsolin
- pointed-End depolymerizing proteins such as ADF/cofilin.

The actin filament network in non-muscle cells is highly dynamic. As first proposed by Dickinson & Purich (*Biophysical Journal* 92: 622-631), the actin filament network is arranged with the barbed-end of each filament attached to the cell's peripheral membrane by means of clamped-filament elongation motors ("actoclampins") formed from a filament barbed-end and a clamping protein (formins, VASP, Mena, WASP, and N-WASP). The primary substrate for these elongation motors is Profilin-Actin-ATP complex which is directly transferred to elongating filament ends (Dickinson, Southwick & Purich, 2002). The pointed-end of each filament is oriented toward the cell's interior. In the case of lamellipodial growth, the Arp2/3 complex generates a branched network, and in filopodia a parallel array of filaments is formed.

Actin Acts as a Track for Myosin Motor Motility

Myosin motors are intracellular ATP-dependent enzymes that bind to and move along actin filaments. Various classes of myosin motors have very different behaviors, including exerting tension in the cell and transporting cargo vesicles.

A Proposed Model - Actoclampins Track Filament Ends

One proposed model suggests the existence of actin filament barbed-end-tracking molecular motors termed actoclampin. The proposed actoclampins generate the propulsive forces needed for actin-based motility of lamellipodia, filopodia, invadipodia, dendritic spines, intracellular vesicles, and motile processes in endocytosis, exocytosis, podosome formation, and phagocytosis. Actoclampin motors also propel such intracellular pathogens as *Listeria monocytogenes*, *Shigella flexneri*, *Vaccinia* and *Rickettsia*. When assembled under suitable conditions, these end-tracking molecular motors can also propel biomimetic particles. The term actoclampin is derived from *acto-* to indicate the involvement of an actin filament, as in actomyosin, and *clamp* to indicate a clasping device used for strengthening flexible/moving objects and for securely fastening two or more components, followed by the suffix *-in* to indicate its protein origin. An actin filament end-tracking protein may thus be termed a clampin.

Dickinson and Purich (2002) recognized that prompt ATP hydrolysis could explain the forces achieved during actin-based motility. They proposed a simple mechanoenzymatic sequence known as the Lock, Load & Fire Model, in which an end-tracking protein remains tightly bound ("locked" or clamped) onto the end of one sub-filament of the double-stranded actin filament. After binding to Glycyl-Prolyl-Prolyl-Prolyl-Prolyl-Prolyl-registers on tracker proteins, Profilin-ATP-actin is delivered ("loaded") to the unclamped end of the other sub-filament, whereupon ATP within the already clamped terminal subunit of the other subfragment is hydrolyzed ("fired"), providing the energy needed to release that arm of the end-tracker, which then can bind another Profilin-ATP-actin to begin a new monomer-addition round.

The following steps describe one force-generating cycle of an actoclampin molecular motor:

1. The polymerization cofactor profilin and the ATP actin combine to form a profilin-ATP-actin complex that then binds to the end-tracking unit
2. The cofactor and monomer are transferred to the barbed-end of an actin already clamped filament
3. The tracking unit and cofactor dissociate from the adjacent protofilament, in a step that can be facilitated by ATP hydrolysis energy to modulate the affinity of the cofactor and/or the tracking unit for the filament; and this mechanoenzymatic cycle is then repeated, starting this time on the other sub-filament growth site.

When operating with the benefit of ATP hydrolysis, AC motors generate per-filament forces of 8–9 pN, which is far greater than the per-filament limit of 1–2 pN for motors operating without ATP hydrolysis (Dickinson and Purich, 2002, 2006; Dickinson, Caro and Purich, 2004). The term actoclampin is generic and applies to all actin filament end-tracking molecular motors, irrespective of whether they are driven actively by an ATP-activated mechanism or passively.

Some actoclampins (e.g., those involving Ena/VASP proteins, WASP, and N-WASP) apparently require Arp2/3-mediated filament initiation to form the actin polymerization nucleus that is then "loaded" onto the end-tracker before processive motility can commence. To generate a new filament, Arp2/3 requires a "mother" filament, monomeric ATP-actin, and an activating domain from Listeria ActA or the VCA region of N-WASP. Ther Arp2/3 complex binds to the side of the mother filament, forming a Y-shaped branch having a 70 degree angle with respect to the longitudinal axis of the mother filament.

Then upon activation by ActA or VCA, the Arp complex is believed to undergo a major conformational change, bringing its two actin-related protein subunits near enough to each other to generate a new filament gat. Whether ATP hydrolysis may be required for nucleation and/or Y-branch release is a matter under active investigation.

Actin

Actin is a globular, roughly 42-kDa protein found in all eukaryotic cells (the only known exception being nematode sperm) where it may be present at concentrations of over 100 μM. It is also one of the most highly-conserved proteins, differing by no more than 20% in species as diverse as algae and humans. Actin is the monomeric subunit of two types of filaments in cells: microfilaments, one of the three major components of the cytoskeleton, and thin filaments, part of the contractile apparatus in muscle cells.

Thus, actin participates in many important cellular processes including muscle contraction, cell motility, cell division and cytokinesis, vesicle and organelle movement, cell signalling, and the establishment and maintenance of cell junctions and cell shape. Many of these processes are mediated by extensive and intimate interactions of actin with cellular membranes. In vertebrates, three main groups of actin isoforms, alpha, beta, and gamma have been identified. The alpha actins are found in muscle tissues and are a major constituent of the contractile apparatus. The beta and gamma actins co-exist in most cell types as components of the cytoskeleton, and as mediators of internal cell motility.

Formation of Thin Filament

Principal interactions of structural proteins are at cadherin-based adherens junction. Actin filaments are linked to α-actinin and to the membrane through vinculin. The head domain of vinculin associates to E-cadherin via á-, â-, and β-catenins. The tail domain of vinculin binds to membrane lipids and to actin filaments.

The protein actin is one of the most highly conserved throughout evolution because it interacts with a large number of other proteins, with 80.2% sequence conservation at the gene level between *Homo sapiens* and *Saccharomyces cerevisiae* (a species of yeast), and 95% conservation of the primary structure of the protein product.

Although most yeasts have only a single actin gene, higher eukaryotes, in general, express several isoforms of actin encoded by a family of related genes. Mammals have at least six actin isoforms coded by separate genes, which are divided into three classes (alpha, beta and gamma) according to their isoelectric point. In general, alpha actins are found in muscle, whereas beta and gamma isoforms are prominent in non-muscle cells (α- and β1-cytoplasmic). Although the amino acid sequences and *in vitro* properties of the isoforms are highly similar, these isoforms cannot completely substitute for one another *in vivo*.

The typical actin gene has an approximately 100-nucleotide 5' UTR, a 1200-nucleotide translated region, and a 200-nucleotide 3' UTR. The majority of actin genes are interrupted by introns, with up to 6 introns in any of 19 well-characterised locations. The high conservation of the family makes actin the favoured model for studies comparing the introns-early and introns-late models of intron evolution.

All non-spherical prokaryotes appear to possess genes such as MreB, which encode homologues of actin; these genes are required for the cell's shape to be maintained. The plasmid-derived gene ParM encodes an actin-like protein whose polymerised form is dynamically unstable, and appears to partition the plasmid DNA into the daughter cells during cell division by a mechanism analogous to that employed by microtubules in eukaryotic mitosis. Actin is found in both smooth and rough endoplasmic reticulums.

Functions

Actin has four main functions in cells:

- To form microfilaments, the most dynamic of the three subclasses of the cytoskeleton, which give mechanical support to cells, and hardwire the cytoplasm with the surroundings to support signal transduction.
- To allow cell motility, including phagocytosis of bacteria by macrophages.
- In muscle cells, to be the scaffold on which myosin proteins generate force to support muscle contraction.
- In non-muscle cells, to be a track for cargo transport myosins (non-conventional myosins) such as myosin V and VI. Non-conventional myosins use ATP hydrolysis to transport cargo, such as vesicles and organelles, in a directed fashion much faster than diffusion. Myosin V walks towards the barbed end of actin filaments, while myosin VI walks toward the pointed end. Most actin filaments are arranged with the barbed end toward the cellular membrane and the pointed end toward the cellular interior. This arrangement allows myosin V to be an effective motor for export of cargos, and myosin VI to be an effective motor for import.

Directionality

The polarity of an actin filament can be determined by decorating the microfilament with myosin "S1" fragments, creating barbed (+) and pointed (-) ends on the filament. An S1 fragment is composed of the head and neck domains of myosin II. Under physiologic conditions, G-actin (the monomer form) is transformed to F-actin (the polymer form) by ATP, where the role of ATP is essential.

Nucleation and Polymerization

Actin polymerization and depolymerization is necessary in chemotaxis and cytokinesis. Nucleating factors are necessary to stimulate actin polymerization. One such nucleating factor is the ARP complex which acts as a barbed end of actin in its shape to stimulate the nucleation of G-actin (or monomeric actin). The Arp2/3 complex can also bind to actin filaments at 70 degrees to form new actin branches off of existing actin filaments. Also, actin filaments themselves bind ATP, and hydrolysis of this ATP stimulates destabilization of the polymer. The growth of actin filaments can be regulated by thymosin and profilin. Thymosin binds to G-actin to prevent it from polymerizing while profilin binds to G-actin to promote monomeric addition to the barbed, plus end.

Microfilaments

Individual subunits of microfilaments are known as globular actin (G-actin). G-actin subunits assemble into long filamentous polymers called F-actin. Two parallel F-actin strands must rotate 166 degrees to layer correctly on top of each other. This creates the double helix structure of the microfilaments of the cytoskeleton. Microfilaments measure approximately 7 nm in diameter with a loop of the helix repeating every 37 nm.

Actomyosin Filaments

In muscle, actin is the major component of *thin filaments*, which, together with the motor protein myosin (which forms *thick filaments*), are arranged into actomyosin myofibrils. These fibrils comprise the mechanism of muscle contraction. Using the hydrolysis of ATP for energy, myosin heads undergo a cycle during which they attach to thin filaments, exerting a tension, and then depending on the load, perform a power stroke that causes the thin filaments to slide past, shortening the muscle.

In contractile bundles, the actin-bundling protein alpha-actinin separates each thin filament by ~35 nm. This increase in distance allows thick filaments to fit in between and interact, enabling deformation or contraction. In deformation, one end of myosin is bound to the plasma membrane while the other end "walks" toward the plus end of the actin filament. This pulls the membrane into a different shape relative to the cell cortex.

For contraction, the myosin molecule is usually bound to two separate filaments and both ends simultaneously "walk" toward their filament's plus end, sliding the actin filaments closer to each other. This results in the shortening, or contraction, of the actin bundle (but not the filament). This mechanism is responsible for muscle contraction and cytokinesis, the division of one cell into two.

Nuclear Actin

Actin is essential for transcription from RNA polymerases I, II and III. In Pol I transcription, actin and myosin (MYO1C, which binds DNA) act as a molecular motor. For Pol II transcription, â-actin is needed for the formation of the pre-initiation complex. Pol III contains α-actin as a subunit. Actin can also be a component of chromatin remodeling complexes as well as pre-mRNP particles (that is, precursor messenger RNA bundled in proteins), and is involved in nuclear export of RNAs and proteins.

History

Actin was first observed experimentally in 1887 by W.D. Halliburton, who extracted a protein from muscle that 'coagulated' preparations of myosin, which he dubbed "myosin-ferment." However, Halliburton was unable to further characterise his findings, and the discovery of actin is credited instead to Brúnó F. Straub, a young biochemist working in Albert Szent-Györgyi's laboratory at the Institute of Medical Chemistry at the University of Szeged, Hungary. In 1942, Straub developed a novel technique for extracting muscle protein that allowed him to isolate substantial amounts of relatively pure actin. Straub's method is essentially the same as that used in laboratories today.

Szent-Gyorgyi had previously described the more viscous form of myosin produced by slow muscle extractions as 'activated' myosin, and, since Straub's protein produced the activating effect, it was dubbed *actin*. The hostilities of World War II meant that Szent-Gyorgyi and Straub were unable to publish the work in Western scientific journals; it became well-known in the West only in 1945, when it was published as a supplement to the Acta Physiologica Scandinavica.

Straub continued to work on actin, and in 1950 reported that actin contains bound ATP and that, during polymerisation of the protein into microfilaments, the nucleotide is hydrolysed to ADP and inorganic phosphate (which remain bound in the microfilament). Straub suggested that the transformation of ATP-bound actin to ADP-bound actin played a role in muscular contraction. In fact, this is true only in smooth muscle, and was not supported through experimentation until 2001.

The crystal structure of G-actin was solved in 1990 by Kabsch and colleagues. In the same year, a model for F-actin was proposed by Holmes and colleagues. The model was derived by fitting a helix of G-actin structures according to low-resolution fibre diffraction data from the filament. Several models of the filament have been proposed since. However, there is still no high-resolution X-ray structure of F-actin. The Listeria bacteria use the cellular machinery to move around inside the host cell, by inducing directed polymerisation of actin by the ActA transmembrane protein, thus pushing the bacterial cell around.

Intermediate Filaments

Intermediate filaments (IFs) are a family of related proteins that share common structural and sequence features. Intermediate filaments have an average diameter of 10 nanometers, which is between that of actin (microfilaments) and microtubules, although they were initially designated 'intermediate' because their average diameter was between those of narrower microfilaments (actin) and wider myosin filaments. Most types of intermediate filaments are cytoplasmic, but one type, the lamins, are nuclear.

Structure

The domain structure of IF molecules is conserved. Each protein has a non-alpha-helical (globular) domain at the N and C-termini which surrounds the alpha-helical rod domain. The basic building block for IFs is a parallel and in register dimer. The dimer is formed through the interaction of the rod domain to form a coiled coil. Cytoplasmic IF assemble into non-polar unit-length filaments (ULF) which then assemble into longer structures. Part of the assembly process includes a compaction step, in which ULF tighten and assume a smaller diameter. The reasons for this compaction are not well understood, and IF are routinely observed to have diameters ranging between 6 and 12nm.

The anti-parallel orientation of tetramers means that, unlike microtubules and microfilaments which have a plus end and a minus end, IFs lack polarity. Also, as opposed to actin or tubulin, intermediate filaments do not contain a binding site for a nucleoside triphosphate. Cytoplasmic IF do not undergo treadmilling like microtubules and actin fibres, but they are dynamic.

Biomechanical Properties

IFs are rather deformable proteins that can be stretched several times their initial length. The key to facilitate this large deformation is due to their hierarchical structure, which facilitates a cascaded activation of deformation mechanisms at different levels of strain.

Types

There are about 70 different genes coding for various intermediate filament proteins. However, different kinds of IFs share basic characteristics: they are all polymers that generally measure between 9-11 nm in diameter when fully assembled.

IF are subcategorized into six types based on similarities in amino acid sequence and protein structure.

Types I and II - Acidic and Basic Keratins

These proteins are the most diverse among IFs and constitute type I (acidic) and type II (basic) IF proteins. The many isoforms are divided in two groups:

- epithelial keratins (about 20) in epithelial cells (image to right)
- trichocytic keratins (about 13) (hair keratins) which make up hair, nails, horns and reptilian scales.

Regardless of the group, keratins are either acidic or basic. Acidic and basic keratins bind each other to form acidic-basic heterodimers and these heterodimers then associate to make a keratin filament.

Type III

There are four proteins classed as type III IF proteins which may form homo- or heteropolymeric proteins.

- Desmin IFs are structural components of the sarcomeres in muscle cells.
- GFAP (glial fibrillary acidic protein) is found in astrocytes and other glia.
- Peripherin found in peripheral neurons.
- Vimentin, the most widely distributed of all IF proteins, can be found in fibroblasts, leukocytes, and blood vessel endothelial cells. They support the cellular membranes and keep some organelles in a fixed place within the cytoplasm.

Type IV

- α-Internexin
- Neurofilaments - the type IV family of intermediate filaments that is found in high concentrations along the axons of vertebrate neurons.
- Synemin
- Syncoilin.

Type V - Nuclear Lamins

Lamins : Lamins are fibrous proteins having structural function in the cell nucleus.

In metazoan cells there are A and B type lamins which differ in their length and pI. Human cells have three differentially regulated genes. B-type lamins are present in every cell. B type lamins, B1 and B2, are expressed from the LMNB1 and LMNB2 genes on 5q23 and 19q13, respectively. A-type lamins are only expressed following gastrulation. Lamin A and C are the most common A-type lamins and are splice variants of the LMNA gene found at 1q21.

These proteins localize to two regions of the nuclear compartment, the nuclear lamina — a proteinaceous structure layer subjacent to the inner surface of the nuclear envelope and throughout the nucleoplasm in the nucleoplasmic “veil”.

Comparison of the lamins to vertebrate cytoskeletal IFs shows that lamins have an extra 42 residues (six heptads) within coil 1b. The c-terminal tail domain contains a nuclear localization signal (NLS), an Ig-fold like domain, and in most cases a carboxy-terminal CaaX box that is isoprenylated and carboxymethylated (lamin C does not have a CAAX box). Lamin A is further processed to remove the last 15 amino acids and its farnesylated cysteine.

During mitosis, lamins are phosphorylated by MPF which drives the disassembly of the lamina and the nuclear envelope.

Cell Adhesion

At the plasma membrane, some keratins interact with desmosomes (cell-cell adhesion) and hemidesmosomes (cell-matrix adhesion) via adapter proteins.

Associated Proteins

Filaggrin binds to keratin fibres in epidermal cells. Plectin links vimentin to other vimentin fibres, as well as to microfilaments, microtubules, and myosin II.

Keratin filaments in epithelial cells link to desmosomes (desmosomes connect the cytoskeleton together) through plakoglobin, desmoplakin, desmogleins and desmocollins; desmin filaments are connected in a similar way in heart muscle cells.

Diseases Arising from Mutations in IF Genes

- Epidermolysis bullosa simplex; K5 or K14 mutation
- Laminopathies are a family of diseases caused by mutations in nuclear lamins and include Hutchinson Gilford Progeria Syndrome and various lipodystrophies and cardiomyopathies among others.
- Human Intermediate Filament Database(HIFD), a comprehensive database of human intermediate filament proteins, their associated variations and diseases.
- envelope.

Microtubules

Microtubules are one of the components of the cytoskeleton. They have a diameter of 25 nm and length varying from 200 nanometers to 25 micrometers. Microtubules serve as structural components within cells and are involved in many cellular processes including mitosis, cytokinesis, and vesicular transport.

Structure

Microtubules are polymers of α- and β-tubulin dimers. The tubulin dimers polymerize end to end in protofilaments. The protofilaments then bundle into hollow cylindrical filaments. Typically the protofilaments arrange themselves in an imperfect helix with one turn of the helix containing 13 tubulin dimers each from a different protofilament. The image above illustrates a small section of microtubule, a few αβ dimers in length.

Another important feature of microtubule structure is polarity. Tubulin polymerizes end to end with the α subunit of one tubulin dimer contacting the β subunit of the next. Therefore, in a protofilament, one end will have the α subunit exposed while the other end will have the α subunit exposed. These ends are designated the (“) and (+) ends, respectively. The protofilaments bundle parallel to one another, so in a microtubule, there is one end, the (+) end, with only α subunits exposed while the other end, the (“) end, only has β subunits exposed. The (-) end is capped so elongation of the microtubule occurs from the (+) direction.

Organization within Cells

Microtubules are nucleated and organized by the microtubule organizing centres (MTOCs), such as centrioles and basal bodies. Microtubules are part of a structural network (the cytoskeleton) within the cell's cytoplasm, but, in addition to structural support, microtubules take part in many other processes, as well. They are capable of growing and shrinking in order to generate force, and there are also motor proteins that allow organelles and other cellular factors to move along the microtubule.

A notable structure involving microtubules is the mitotic spindle used by eukaryotic cells to segregate their chromosomes correctly during cell division. Microtubules are also part of the cilia and flagella of eukaryotic cells (prokaryote flagella are entirely different).

The process of mitosis is facilitated by a subgroup of microtubules known as astral microtubules, defined as a microtubule originating from the centrosome that does not connect to a kinetochore. Astral microtubules develop in the actin skeleton and interact with the cell cortex to aid in spindle orientation. They are organized into radial arrays around the centrosomes. The turn-over rate of this population of microtubules is higher than any other population. Astral microtubules function in concert with specialized dynein motors, which are oriented with the light chain portion attached to the cell membrane and the dynamic portion attached to the microtubule.

This allows for dynein contraction to pull the centrosome towards the cell membrane, thus assisting in cytokinesis. Astral microtubules are not required for the progression of mitosis, but they are required to ensure the fidelity of the process; they are required for the correct positioning and orientation of the mitotic spindle apparatus.

They are also involved in determination of cell division site based on the geometry and polarity of the cells. The maintenance of astral microtubules is dependent on the integrity of centrosome. It is also dependent on several microtubule-associated proteins such as EB1 and Adenomatous Polyposis Coli (APC).

Nucleation and Growth

Polymerization of microtubules is nucleated in a microtubule organizing centre. Contained within the MTOC is another type of tubulin, β-tubulin, which is distinct from the α and β subunits which compose the microtubules themselves. The α-tubulin combines with several other associated proteins to form a circular structure known as the "α-tubulin ring complex." This complex acts as a scaffold for α/β tubulin dimers to begin polymerization; it acts as a cap of the (") end while microtubule growth continues away from the MTOC in the (+) direction.

Dynamic Instability

During polymerization, both the α- and β-subunits of the tubulin dimer are bound to a molecule of GTP. While the GTP bound to α-tubulin is stable, the GTP bound to α-tubulin may be hydrolyzed to GDP shortly after assembly. The kinetics of GDP-tubulin are different from those of GTP-tubulin; GDP-tubulin is prone to depolymerization.

A GDP-bound tubulin subunit at the tip of a microtubule will fall off, though a GDP-bound tubulin in the middle of a microtubule cannot spontaneously pop out. Since tubulin adds onto the end of the microtubule only in the GTP-bound state, there is generally a cap of GTP-bound tubulin at the tip of the microtubule, protecting it from disassembly. When hydrolysis catches up to the tip of the microtubule, it begins a rapid depolymerization and shrinkage. This switch from growth to shrinking is called a catastrophe. GTP-bound tubulin can begin adding to the tip of the microtubule again, providing a new cap and protecting the microtubule from shrinking. This is referred to as rescue.

Chemical Effects on Microtubule Dynamics

Microtubule dynamics can also be altered by drugs.

- For example, the taxane drug class (e.g. paclitaxel or docetaxel), used in the treatment of cancer, blocks dynamic instability by stabilizing GDP-bound tubulin in the microtubule. Thus, even when hydrolysis of GTP reaches the tip of the microtubule, there is no depolymerization and the microtubule does not shrink back.
- Nocodazole and Colchicine have the opposite effect, blocking the polymerization of tubulin into microtubules.

Motor Proteins

In addition to movement generated by the dynamic instability of the microtubule itself, the fibres are substrates along which motor proteins can move. The major microtubule motor proteins are kinesin, which generally moves towards the (+) end of the microtubule, and dynein, which generally moves towards the (“) end.

Postulated Role in Consciousness

In their controversial Orch-OR theory of consciousness, Roger Penrose and Stuart Hameroff postulate that microtubules in neurons conduct quantum-level manipulations of matter which produces consciousness, based partially on widely recorded observations of Gamma Synchrony that indicate that information may propagate through the brain much faster than a chemically mediated neural network would physically permit. Max Tegmark disputes the relevance of these observations, and the matter remains open to debate.

The Prokaryotic Cytoskeleton

The prokaryotic cytoskeleton is the collective name for all structural filaments in prokaryotes. It was once thought that prokaryotic cells did not possess cytoskeletons, but recent advances in visualization technology and structure determination have shown that filaments indeed exist in these cells. In fact, analogues for all major cytoskeletal proteins in eukaryotes have been found in prokaryotes. Cytoskeletal elements play essential roles in cell division, protection, shape determination, and polarity determination in various prokaryotes.

The cytoskeleton was previously thought to be a feature only of eukaryotic cells, but homologues to all the major proteins of the eukaryotic cytoskeleton have recently been found in prokaryotes. Although the evolutionary relationships are so distant that they are not obvious from protein sequence comparisons alone, the similarity of their three-dimensional structures and similar functions in maintaining cell shape and polarity provides strong evidence that the eukaryotic and prokaryotic cytoskeletons are truly homologous. However, some structures in the bacterial cytoskeleton may have yet to be identified.

FtsZ, the first identified prokaryotic cytoskeletal element, forms a filamentous ring structure located in the middle of the cell called the Z-ring that constricts during cell division, similar to the actin-myosin contractile ring in eukaryotes. The Z-ring is a highly dynamic structure that consists of numerous bundles of protofilaments that extend and shrink, although the mechanism behind Z-ring contraction and the number of protofilaments involved are unclear. FtsZ acts as an organizer protein and is required for cell division. It is the first component of the septum during cytokinesis, and it recruits all other known cell division proteins to the division site.

Despite this functional similarity to actin, FtsZ is homologous to eukaryal tubulin. Although comparison of the primary structures of FtsZ and tubulin reveal a weak relationship, their 3-dimensional structures are remarkably similar. Furthermore, like tubulin, monomeric FtsZ is bound to GTP and polymerizes with other FtsZ monomers with the hydrolysis of GTP in a mechanism similar to tubulin dimerization. Since FtsZ is essential for cell division in bacteria, this protein is a target for the design of new antibiotics.

MreB

MreB is a bacterial protein believed to be analogous to eukaryal actin. MreB and actin have a weak primary structure match, but are very similar in terms of 3-D structure and filament polymerization.

Almost all non-spherical bacteria rely on MreB to determine their shape. MreB assembles into a helical network of filamentous structures just under the cytoplasmic membrane, covering the whole length of the cell. MreB determines cell shape by mediating the position and activity of enzymes that synthesize peptidoglycan and by acting as a rigid filament under the cell membrane that exerts outward pressure to sculpt and bolster the cell. MreB condenses from its normal helical network and forms a tight ring at the septum in *Caulobacter crescentus* right before cell division, a mechanism that is believed to help locate its off-centre septum. MreB is also important for polarity determination in polar bacteria, as it is responsible for the correct positioning of at least four different polar proteins in *C. crescentus*.

Crescentin

Crescentin (encoded by *creS* gene) is an analogue of eukaryotic intermediate filaments (IFs). Unlike the other analogous relationships discussed here, crescentin has a rather large primary homology with IF proteins in addition to three-dimensional similarity - the sequence of *creS* has a 25% identity match and 40% similarity to cytokeratin 19 and a 24% identity match and 40% similarity to nuclear lamin A. Furthermore, crescentin filaments are roughly 10 nm in diameter and thus fall within diameter range for eukaryal IFs (8-15 nm). Crescentin forms a continuous filament from pole to pole alongside the inner, concave side of the crescent-shaped bacterium *Caulobacter crescentus*. Both MreB and crescentin are necessary for *C. crescentus* to exist in its characteristic shape; it is believed that MreB molds the cell into a rod shape and crescentin bends this shape into a crescent.

ParM and SopA

ParM is a cytoskeletal element that possesses a similar structure to actin, although it behaves functionally like tubulin. Further, it polymerizes bidirectionally and it exhibits dynamic instability, which are both behaviors characteristic of tubulin polymerization. It forms a system with ParR and *parC* that is responsible for R1 plasmid separation. ParM affixes to ParR, a DNA-binding protein that specifically binds to 10 direct repeats in the *parC* region on the R1 plasmid. This binding occurs on both ends of the ParM filament. This filament is then extended, separating the plasmids. The system is analogous to eukaryotic chromosome segregation as ParM acts like eukaryotic tubulin in the mitotic spindle, ParR acts like the kinetochore complex, and *parC* acts like the centromere of the chromosome. F plasmid segregation occurs in a similar system where SopA acts as the cytoskeletal filament and SopB binds to the *sopC* sequence in the F plasmid, like the kinetochore and centromere respectively.

MinCDE System

The MinCDE system is a filament system that properly positions the septum in the middle of the cell in *Escherichia coli*. According to Shih et al., MinC inhibits the formation of the septum by prohibiting the polymerization of the Z-ring. MinC, MinD, and MinE form a helix structure that winds around the cell and is bound to the membrane by MinD. The MinCDE helix occupies a pole and terminates in a filamentous structure called the E-ring made of MinE at the middle-most edge of the polar zone. From this configuration, the E-ring will contract and move toward that pole, disassembling the MinCDE helix as it moves along. Concomitantly, the disassembled fragments will reassemble at the opposite polar end, reforming the MinCDE coil on the opposite pole while the current MinCDE helix is broken down. This process then repeats, with the MinCDE helix oscillating from pole to pole. This oscillation occurs repeatedly during the cell cycle, thereby keeping MinC (and its septum inhibiting effect) at a lower time-averaged concentration at the middle of the cell than at the ends of the cell.

The dynamic behavior of the Min proteins has been reconstituted in vitro using an artificial lipid bilayer as mimic for the cell membrane. MinE and MinD self-organized into parallel and spiral protein waves by a reaction-diffusion like mechanism.

Microtrabeculae

A fourth eukaryotic cytoskeletal element, *microtrabeculae*, was proposed by Keith Porter based on images obtained from high-voltage electron microscopy of whole cells in the 1970s. The images showed short, filamentous structures of unknown molecular composition associated with known cytoplasmic structures. Porter proposed that this microtrabecular structure represented a novel filamentous network distinct from microtubules, filamentous actin, or intermediate filaments. It is now generally accepted that microtrabeculae are nothing more than an artifact of certain types of fixation treatment, although we have yet to fully understand the complexity of the cell's cytoskeleton.

Cilia and Flagella

A cilium (plural *cilia*) is an organelle found in eukaryotic cells. Cilia are slender protuberances that project from the much larger cell body.

There are two types of cilia: *motile* cilia and *non-motile*, or primary cilia, which typically serve as sensory organelles. In eukaryotes, cilia and flagella together make up a group of organelles known as undulipodia. Eukaryotic cilia are structurally identical to Eukaryotic flagella, although distinctions are sometimes made according to function and/or length.

Types and Distribution

In humans, *primary* cilia are found on nearly every cell in the body. Cilia are rare in most plants, occurring most notably in cycads.

Ciliates are microscopic organisms that possess *motile* cilia exclusively and use them for either locomotion or to simply move liquid over their surface.

Larger eukaryotes, such as mammals, have *motile* cilia as well. Motile cilia are usually present on a cell's surface in large numbers and beat in coordinated waves. In humans, for example, motile cilia are found in the lining of the trachea (windpipe), where they sweep mucus and dirt out of the lungs. In female mammals, the beating of cilia in the Fallopian tubes moves the ovum from the ovary to the uterus.

In comparison to *motile* cilia, non-motile (or *primary*) cilia usually occur one per cell; nearly all mammalian cells have a single non-motile *primary cilium*. In addition, examples of specialized primary cilia can be found in human sensory organs such as the eye and the nose. The outer segment of the rod photoreceptor cell in the human eye is connected to its cell body with a specialized non-motile cilium. The dendritic knob of the olfactory neuron, where the odorant receptors are located, also contains non-motile cilia (about 10 cilia per dendritic knob).

Although the primary cilium was discovered in 1898, it was largely ignored for a century. Only recently has great progress been made in understanding the function of the primary cilium. Until the 1990s, the prevailing view of the primary cilium was that it was merely a vestigial organelle, without important function. Recent findings regarding its physiological roles in chemical sensation, signal transduction, and control of cell growth, have led scientists to acknowledge its importance in cell function, with the discovery of its role in diseases not previously recognized to involve the dysgenesis and dysfunction of cilia, such as polycystic kidney disease congenital heart disease, and an emerging group of genetic ciliopathies. The primary cilium is now known to play an important role in the function of many human organs. The current scientific understanding of primary cilia views them as "sensory cellular antennae that coordinate a large number of cellular signalling pathways, sometimes coupling the signalling to ciliary motility or alternatively to cell division and differentiation."

Structure, Assembly and Maintenance, and Function

"Inside cilia and flagella is a microtubule-based cytoskeleton called the axoneme. The axoneme of primary cilia typically has a ring of nine outer microtubule doublets (called a 9+0 axoneme), and the axoneme of a motile cilium has two central microtubule singlets in addition to the nine outer doublets (called a 9+2 axoneme). The axonemal cytoskeleton acts as a scaffolding for various protein complexes and provides binding sites for molecular motor proteins such as kinesin II, that help carry proteins up and down the microtubules."

The building blocks of the cilia such as tubulins and other partially assembled axonemal proteins are added to the ciliary tips which point away from the cell body. In most species bi-directional motility called intraflagellar transport or IFT plays an essential role to move these building materials from the cell body to the assembly site. IFT also carries the disassembled material to be recycled from the ciliary tip back to the cell body. By regulating the equilibrium between these two IFT processes, the length of cilia can be maintained dynamically. The disassembly of the cilia requires the action of the protein kinase Aurora A.

Exceptions where IFT is not present include *Plasmodium falciparum* which is one of the species of *Plasmodium* that cause malaria in humans. In this parasite, cilia assemble in the cytoplasm.

At the base of the cilium where the cilia attaches to the cell body is the microtubule organizing centre, the basal body. Some basal body proteins as CEP164, ODF2 and CEP170, regulate the formation and the stability of the cilium. A transition zone between the basal body and the axoneme "serves as a docking station for intraflagellar transport and motor proteins."

"In effect, the [cilium] is a nanomachine composed of perhaps over 600 proteins in molecular complexes, many of which also function independently as nanomachines."

Sensing the Extracellular Environment

"Some epithelial cells are ciliated, and they commonly exist as a sheet of polarised cells forming a tube or tubule with cilia projecting into the lumen." Primary cilia on epithelial cells provide chemosensation, thermosensation and mechanosensation of the extracellular environment by playing "a sensory role mediating specific signalling cues, including soluble factors in the external cell environment, a secretory role in which a soluble protein is released to have an effect downstream of the fluid flow, and mediation of fluid flow if the cilia are motile."

Cilium-related Disease

Ciliary defects can lead to a number of human diseases. Genetic mutations compromising the proper functioning of cilia, ciliopathies, can cause chronic disorders such as primary ciliary dyskinesia (PCD), nephronophthisis or Senior-Loken syndrome. In addition, a defect of the primary cilium in the renal tube cells can lead to polycystic kidney disease (PKD).

In another genetic disorder called Bardet-Biedl syndrome (BBS), the mutant gene products are the components in the basal body and cilia.

Lack of functional cilia in female Fallopian tubes can cause ectopic pregnancy. A fertilized ovum may not reach the uterus if the cilia are unable to move it there. In such a case, the ovum will implant in the Fallopian tubes, causing a tubal pregnancy, the most common form of ectopic pregnancy.

Since the flagellum of human sperm is actually a modified cilium, ciliary dysfunction can also be responsible for male infertility.

Of interest, there is an association of primary ciliary dyskinesia with left-right anatomic abnormalities such as situs inversus (a combination of findings known as Kartagener's syndrome) and other heterotaxic defects. These left-right anatomic abnormalities can also result in congenital heart disease. In fact, it has been shown that proper cilial function is responsible for the normal left-right asymmetry in mammals.

Ciliopathy as an Origin for Many Multi-symptom Genetic Diseases

Recent findings in genetic research have suggested that a large number of genetic disorders, both genetic syndromes and genetic diseases, that were not previously related in the medical literature, may be, in fact, highly related in the root cause of the widely-varying set of medical symptoms that are clinically visible in the disorder.

These have been grouped as an emerging class of diseases called ciliopathies. The underlying cause may be a dysfunctional molecular mechanism in the primary cilia structures, organelles which are present in many diverse cellular types throughout the human body.

Cilia defects adversely affect "numerous critical developmental signalling pathways" essential to cellular development and thus offer a plausible hypothesis for the often multi-symptom nature of a large set of syndromes and diseases. Known ciliopathies include primary ciliary dyskinesia, Bardet-Biedl syndrome, polycystic kidney and liver disease, nephronophthisis, Alstrom syndrome, Meckel-Gruber syndrome and some forms of retinal degeneration.

Mucociliary Clearance

Mucociliary clearance, also referred to as mucociliary apparatus or mucociliar clearance (MCC), derived from *mucus*, *cilia* (cilia of the tracheal surface epithelium in the respiratory tract) and *clearance* describes the self-clearing mechanism of the bronchi.

The main bronchi down to the alveoli are lined with a respiratory epithelium. On that, cilium is present, bearing hair-shaped structures on its surface (cilia). The cilia are surrounded by a thin fluid film of mucus. On top of that is a second viscous film of mucus, in which foreign particles and microorganisms get stuck. Within the thin fluid film of mucus the cilia act out movements coordinated in direction towards the pharynx. Thereby the viscous film of mucus including its freight is transported off in direction towards the mouth, where it is either swallowed or expelled via coughing.

Important for good mucociliary clearance are the number of cilia, their structure, activity, and coordinated movement. Optimum functionality of mucociliary clearance presuppose a temperature of 37 °C and an absolute humidity of 44 mg/dm^3 corresponding to a relative humidity of 100 %. Under the condition of insufficient temperature and humidity, after a short time the ciliary cells suspend their transport function. Under such circumstances, bacterial germinal colonization is facilitated. Pulmonary infections and damaging of the pulmonic tissues may be the consequence.

High humidity enhances the functionality of the mucociliary clearance. Two methods of supporting mucociliary clearance are the active and the passive respiratory gas humidification, which in particular are applied in mechanical ventilation.

Kinocilium

A kinocilium is a special type of cilium on the apex of hair cells located in the sensory epithelium of the vertebrate inner ear.

Anatomy in Humans

Kinocilia on hair cells in the human ear are involved in the senses of movement and hearing. In the auditory system, each hair cell has a kinocilium that is important in the depolarization or hyperpolarization of the cell. In the vestibular system, movement causes displacement of the stereocilia and kinocilium, resulting in depolarization or hyperpolarization of the cell. The depolarization of the hair cells in both instances causes signal transduction via neurotransmitter release.

Auditory System

In the human ear, each hair cell has a single non-motile cilium-like structure known as the kinocilium. The movement of other stereocilia on the hair cell towards the kinocilium, as a result of [[endolymph [1]]] flow, will cause potassium channels on the cilia to open due to a pulling force that each stereocilia exerts on its neighbouring cell (usually from tallest to shortest in a bundle of stereocilia) and this leads to the depolarization of the hair cell.

This pattern of depolarization should not be confused with the more common depolarization which involves the influx of Na+ into the cell which K+ channels stay closed. Endolymph composition resembles that of the intracellular fluid (more K+ and less Na+) more closely compared to its counterpart, perilymph which resembles the extracellular fluid (more Na+ and less K+ compared to intracellular matrix). This depolarization will open voltage gated calcium channels. The influx of calcium then triggers the cell to release vesicles containing excitatory neurotransmitters into a synapse. The post-synaptic neurite then sends an action potential to the spiral ganglion. Cochlear hair cells have kinocilia that degenerate following maturation of the stereociliary bundle. It is believed that the kinocilium is important in delineating the anatomical asymmetry of stereocilia or in making some mechanical connections with the hair cells in which they persist.

Vestibular Apparatus

Kinocilia are present in the crista ampullaris of the semicircular ducts and the sensory maculae of the utricle and saccule. One kinocilium is the longest cilium located on the hair cell next to 40-70 sterocilia. During movement of the body, the hair cell is depolarized when the sterocilia move toward the kinocilium.

The depolarization of the hair cell causes neurotransmitter to be released and an increase in firing frequency of cranial nerve VIII. When the sterocilia tilt away from the kinocilium, the hair cell is hyperpolarized, decreasing the amount of neurotransmitter released, which decreases the firing frequency of cranial nerve VIII.

Anatomy in Fish and Frogs

The apical surface of a sensory fish hair cell usually has numerous stereocilia and a single, much longer kinocilium. Deflection of the stereocilia toward or away from the kinocilium causes an increase or decrease in the firing rate of the sensory neuron innervating the hair cell at its basal surface.

Hair cells in fish and some frogs are used to detect water movements around their bodies. These hair cells are embedded in a jelly-like protrusion called cupula. The hair cells therefore can not be seen and do not appear on the surface of skin of fish and frogs.

Flagellum

A flagellum is a tail-like projection that protrudes from the cell body of certain prokaryotic and eukaryotic cells, and functions in locomotion. There are some notable differences between prokaryotic and eukaryotic flagella, such as protein composition, structure, and mechanism of propulsion. An example of a flagellated bacterium is the ulcer-causing *Helicobacter pylori*, which uses multiple flagella to propel itself through the mucus lining to reach the stomach epithelium. An example of a eukaryotic flagellated cell is the sperm cell, which uses its flagellum to propel itself through the female reproductive tract. Eukaryotic flagella are structurally identical to eukaryotic cilia, although distinctions are sometimes made according to function and/or length.

The word *flagellum* is the Latin word for whip.

Types

Three types of flagella have so far been distinguished; bacterial, archaeal and eukaryotic.

The main differences among these three types are summarized below:

- Bacterial flagella are helical filaments that rotate like screws. They provide two of several kinds of bacterial motility.

- Archaeal flagella are superficially similar to bacterial flagella, but are different in many details and considered non-homologous.
- Eukaryotic flagella - those of animal, plant, and protist cells - are complex cellular projections that lash back and forth.

Sometimes eukaryotic flagella are called *cilia* or *undulipodia* to emphasize their distinctiveness.

Bacterial

The bacterial flagellum is made up of the protein flagellin. Its shape is a 20 nanometer-thick hollow tube. It is helical and has a sharp bend just outside the outer membrane; this "hook" allows the helix to point directly away from the cell. A shaft runs between the hook and the basal body, passing through protein rings in the cell's membrane that act as bearings. Gram-positive organisms have 2 of these basal body rings, one in the peptidoglycan layer and one in the plasma membrane. Gram-negative organisms have 4 such rings: the L ring associates with the lipopolysaccharides, the P ring associates with peptidoglycan layer, the M ring is embedded in the plasma membrane, and the S ring is directly attached to the plasma membrane. The filament ends with a capping protein.

The bacterial flagellum is driven by a rotary engine (the *Mot complex*) made up of protein, located at the flagellum's anchor point on the inner cell membrane. The engine is powered by proton motive force, i.e., by the flow of protons (hydrogen ions) across the bacterial cell membrane due to a concentration gradient set up by the cell's metabolism (in *Vibrio* species there are two kinds of flagella, lateral and polar, and some are driven by a sodium ion pump rather than a proton pump). The rotor transports protons across the membrane, and is turned in the process. The rotor alone can operate at 6,000 to 17,000 rpm, but with the flagellar filament attached usually only reaches 200 to 1000 rpm. The direction of rotation can be switched almost instantaneously, caused by a slight change in the position of a protein, FliG, in the rotor.

The cylindrical shape of flagella is suited to locomotion of microscopic organisms; these organisms operate at a low Reynolds number, where the viscosity of the surrounding water is much more important than its mass or inertia.

Flagella do not rotate at a constant speed but instead can increase or decrease their rotational speed in relation to the strength of the proton motive force. Flagellar rotation can move bacteria through liquid media at speeds of up to 60 cell lengths/second (sec). Although this is only about 0.00017 km/h (0.00011 mph), when comparing this speed with that of higher organisms in terms of number of lengths moved per second, it is extremely fast. By comparison, the cheetah, the fastest land animal, can sprint at 110 km/h (68 mph), which is approximately 25 body lengths/sec. During flagellar assembly, components of the flagellum pass through the hollow cores of the basal body and the nascent filament. During assembly, protein components are added at the flagellar tip rather than at the base. *In vitro*, flagellar filaments assemble spontaneously in a solution containing purified flagellin as the sole protein.

The flagellar filament is the long helical screw that propels the bacterium when rotated by the motor, through the hook. In most bacteria that have been studied, including the Gram negative *Escherichia coli, Salmonella typhimurium, Caulobacter crescentus*, and *Vibrio alginolyticus*, the filament is made up of eleven protofilaments approximately parallel to the filament axis. Each protofilament is a series of tandem protein chains. However in *Campylobacter jejuni*, there are seven protofilaments.

The basal body has several traits in common with some types of secretory pores, such as the hollow rod-like "plug" in their centres extending out through the plasma membrane. Given the structural similarities between bacterial flagella and bacterial secretory systems, it is thought that bacterial flagella may have evolved from the type three secretion system; however, it is not known for certain whether these pores are derived from the bacterial flagella or the bacterial secretory system.

Through use of their flagella, *E. coli* are able to move rapidly towards attractants and away from repellents. They do this by means of a biased random walk, with 'runs' and 'tumbles' brought about by rotating the flagellum counter-clockwise and clockwise respectively.

Flagella Arrangement Schemes

Different species of bacteria have different numbers and arrangements of flagella. Monotrichous bacteria have a single flagellum (e.g., *Vibrio cholerae*). Lophotrichous bacteria have multiple flagella located at the same spot on the bacteria's surfaces which act in concert to drive the bacteria in a single direction. In many cases, the bases of multiple flagella are surrounded by a specialized region of the cell membrane, the so-called *polar membrane*. Amphitrichous bacteria have a single flagellum on each of two opposite ends (only one flagellum operates at a time, allowing the bacteria to reverse course rapidly by switching which flagellum is active). Peritrichous bacteria have flagella projecting in all directions (e.g., *E. coli*).

In some bacteria, such as the larger forms of *Selenomonas*, the individual flagella are organized outside the cell body, helically twining about each other to form a thick structure called a "fascicle". Other bacteria, such as Spirochetes, have a specialized type of flagellum called an "axial filament" that is located in the periplasmic space, the rotation of which causes the entire bacterium to move forward in a corkscrew-like motion.

Counterclockwise rotation of monotrichous polar flagella pushes the cell forward with the flagella trailing behind, much like a corkscrew moving inside cork. Indeed water in the microscopic scale is highly viscous, very different from our daily experience of water.

The flagella are left-handed helices, and bundle and rotate together only when rotating counterclockwise. When some of the rotors reverse direction, the flagella unwind and the cell starts "tumbling". It has also been suggested that even if all flagella would rotate clockwise, they will not form a bundle, due to geometrical as well as hydrodynamical reasons.

Such "tumbling" may happen occasionally, leading to the cell seemingly thrashing about in place, resulting in the reorientation of the cell. The clockwise rotation of a flagellum is suppressed by chemical compounds favourable to the cell (e.g. food), but the motor is highly adaptive to this. Therefore, when moving in a favourable direction, the concentration of the chemical attractant increases and "tumbles" are continually suppressed; however, when the cell's direction of motion is unfavourable (e.g., away from a chemical attractant), tumbles are no longer suppressed and occur much more often, with the chance that the cell will be thus reoriented in the correct direction.

In some *Vibrio* spp. (particularly *Vibrio parahemolyticus*) and related proteobacteria such as *Aeromonas*, two flagellar systems co-exist, using different sets of genes and different ion gradients for energy. The polar flagella are constitutively expressed and provide motility in bulk fluid, while the lateral flagella are expressed when the polar flagella meet too much resistance to turn. These provide swarming motility on surfaces or in viscous fluids.

Archaeal

The archaeal flagellum is superficially similar to the bacterial (or eubacterial) flagellum; in the 1980s they were thought to be homologous on the basis of gross morphology and behavior. Both flagella consist of filaments extending outside of the cell, and rotate to propel the cell. Archaeal flagella have a unique structure which lacks a central channel. Similar to bacterial type IV pilins, the component flagellins are made with class 3 signal peptides and they are processed by a type IV prepilin peptidase-like enzyme. The archaeal flagellins are typically modified by the addition of N-linked glycans which are necessary for proper assembly and/or function.

Discoveries in the 1990s revealed numerous detailed differences between the archaeal and bacterial flagella; these include:

- Bacterial flagella are motorized by a flow of H^+ ions (or occasionally Na^+ ions); archaeal flagella are almost certainly powered by ATP. The torque-generating motor that powers rotation of the archaeal flagellum has not been identified.

- While bacterial cells often have many flagellar filaments, each of which rotates independently, the archaeal flagellum is composed of a bundle of many filaments that rotate as a single assembly.
- Bacterial flagella grow by the addition of flagellin subunits at the tip; archaeal flagella grow by the addition of subunits to the base.
- Bacterial flagella are thicker than archaeal flagella, and the bacterial filament has a large enough hollow "tube" inside that the flagellin subunits can flow up the inside of the filament and get added at the tip; the archaeal flagellum is too thin to allow this.
- Many components of bacterial flagella share sequence similarity to components of the type III secretion systems, but the components of bacterial and archaeal flagella share no sequence similarity. Instead, some components of archaeal flagella share sequence and morphological similarity with components of type IV pili, which are assembled through the action of type II secretion systems (the nomenclature of pili and protein secretion systems is not consistent).

These differences could mean that the bacterial and archaeal flagella could be a classic case of biological analogy, or convergent evolution, rather than homology. However, in comparison to the decades of well-publicized study of bacterial flagella (e.g. by Berg), archaeal flagella have only recently begun to get serious scientific attention.

Therefore, many assume erroneously that there is only one basic kind of prokaryotic flagellum, and that archaeal flagella are homologous to it. For example, Cavalier-Smith (2002) is aware of the differences between archaeal and bacterial flagellins, but retains the misconception that the basal bodies are homologous.

Eukaryotic

Along with cilia, flagella make up a group of organelles known as undulipodia.

Structure

A eukaryotic flagellum is a bundle of nine fused pairs of microtubule *doublets* surrounding two central single microtubules. The so-called *"9+2"* structure is characteristic of the core of the eukaryotic flagellum called an *axoneme*. At the base of a eukaryotic flagellum is a basal body, "blepharoplast" or kinetosome, which is the microtubule organizing centre (MTOC) for flagellar microtubules and is about 500 nanometers long. Basal bodies are structurally identical to centrioles. The flagellum is encased within the cell's plasma membrane, so that the interior of the flagellum is accessible to the cell's cytoplasm.

Mechanism

Each of the outer 9 doublet microtubules extends a pair of dynein arms (an "inner" and an "outer" arm) to the adjacent microtubule; these dynein arms are responsible for flagellar beating, as the force produced by the arms causes the microtubule doublets to slide against each other and the flagellum as a whole to bend. These dynein arms produce force through ATP hydrolysis.

The flagellar axoneme also contains radial spokes, polypeptide complexes extending from each of the outer 9 microtubule doublets towards the central pair, with the "head" of the spoke facing inwards. The radial spoke is thought to be involved in the regulation of flagellar motion, although its exact function and method of action are not yet understood.

Flagella vs Cilia

Though eukaryotic flagella and motile cilia are ultrastructurally identical, the beating pattern of the two organelles can be different. In the case of flagella (e.g. the tail of a sperm) the motion is propeller-like. In contrast, beating of motile cilia consists of coordinated back-and-forth cycling of many cilia on the cell surface. Thus, flagella serve for the propulsion of single cells (e.g. swimming of protozoa and spermatozoa), and motile cilia for the transport of fluids (e.g. transport of mucus by stationary ciliated cells in the trachea). However, cilia are also used for locomotion (through liquids) in organisms such as *Paramecium*.

Intraflagellar Transport

Intraflagellar transport (IFT), the process by which axonemal subunits, transmembrane receptors, and other proteins are moved up and down the length of the flagellum, is essential for proper functioning of the flagellum, in both motility and signal transduction.

Evolution of Flagella and Debate

Based on similarity in structure and partial similarity in amino acid sequence, it is generally accepted among scientists that the eukaryotic flagellum and cilium have evolved from the cytoskeleton, while the eubacterial flagellum has evolved either from the type III secretion system or from a more ancient secretion system from which the type III secretion system has evolved as well. The archaeal flagellum has probably evolved from the type IV pili. More details appear under Evolution of flagella.

In his 1996 book *Darwin's Black Box*, intelligent design proponent Michael Behe, under funding from the Templeton Foundation, cited the bacterial flagellum as an example of an irreducibly complex structure that could not have evolved through naturalistic means. Behe argued that the flagellum becomes useless if any one of its constituent parts is removed, and thus could not have arisen through numerous, successive, slight modifications; therefore, it is hopelessly improbable that the proteins making up the flagellar motor could have come together all at once, by chance. Mark Perakh explained that while Behe popularized the idea, biologist Hermann J. Müller had already explored it (under the slightly different name of "interlocking complexity") and more than a decade before Behe's book the same idea was explored by A. Graham Cairns-Smith, but neither claimed that "irreducible complexity" was a "marker" of a supernatural design.

While Behe discussed the immune system and the blood clotting cascade in greater detail, the bacterial flagellum has become a "poster child" for intelligent design proponents and other creationists, as during the Kitzmiller v. Dover Area School District trial of 2005. It is one of two identified rotary structures found in nature (the other being ATP synthase) and it is billions of years older than Behe's other two examples, which exist in many homologous forms, simplifying the explanation of their origin.

Evolutionary pathways supported by the Theory of Natural Selection and Evolution have since been identified for the bacterial flagellum; thus, undermining Behe's argument. In addition, the Type three secretion system, a molecular syringe which bacteria use to inject toxins into other cells, appears to be a simplified sub-set of the bacterial flagellum's components, meaning that it is much less likely to be irreducibly complex in the way that the bacterial flagellum could have in fact evolved from the type three secretion system. Exaptation explains how systems with multiple parts can evolve through natural means.

Ciliopathy

A ciliopathy is a genetic disorder of the cellular cilia or the cilia anchoring structures, the basal bodies.

History

Although non-motile or primary cilia were first described in 1898, "cell biologists largely ignored them." But "microscopists continued to document their presence in the cells of most vertebrate organisms." The "primary cilium was long considered—with few exceptions—to be a largely useless evolutionary vestige, a vestigial organelle. Recent research has revealed an initial understanding that cilia are essential to many of the body's organs". Many mammalian eukaryotic cells are ciliated with *primary cilia.*

These primary cilia play important roles in chemosensation, mechanosensation, and thermosensation. Cilia may thus be "viewed as sensory cellular antennae that coordinate a large number of cellular signalling pathways, sometimes coupling the signalling to ciliary motility or alternatively to cell division and differentiation." Recent advances in mammalian genetic research have facilitated the elucidation of a molecular basis for a number of dysfunctional mechanisms in both motile and primary cilia structures of the cell. "Numerous critical developmental signalling pathways" essential to cellular development have been discovered. These are principally but not exclusively found in the *non-motile* or primary cilia. A number of common observable characteristics of mammalian genetic disorders and diseases are caused by ciliary dysgenesis and dysfunction. Once identified, these characteristics thus describe a set of hallmarks of a ciliopathy.

Cilia have recently been implicated in a wide variety of human genetic diseases by "the discovery that numerous proteins involved in mammalian disease localize to the basal bodies and cilia." For example, in just a single area of human disease physiology, cystic renal disease, cilia-related genes and proteins have been identified to have causal effect in polycystic kidney disease, nephronophthisis, Senior-Loken syndrome type 5, orofaciodigital syndrome type 1 and Bardet-Biedl syndrome.

The Mechanism of Ciliary Function

"In effect, the [motile cilium] is a nanomachine composed of perhaps over 600 proteins in molecular complexes, many of which also function independently as nanomachines." Cilia "function as mechano- or chemosensors and as a cellular global positioning system to detect changes in the surrounding environment." For example, ciliary signalling plays a role in the initiation of cellular replacement after cell damage. In addition to this sensory role mediating specific signalling cues, cilia play "a secretory role in which a soluble protein is released to have an effect downstream of the fluid flow" in epithelial cells, and can of course mediate fluid flow directly in the case of motile cilia. Primary cilia in the retina play a role in transferring nourishment to the non-vascularized rod and cone cells from the vascularized cells several microns behind the surface of the retina.

Similar Genes Can Result in a Range of Different Diseases

"Just as different genes can contribute to similar diseases, so the same genes and families of genes can play a part in a range of different diseases." For example, in just two of the diseases caused by malfunctioning cilia, Meckel-Gruber syndrome and Bardet-Biedl syndrome, patients who carry mutations in genes associated with both diseases "have unique symptoms that are not seen in either condition alone." The genes linked to the two different conditions "interact with each other during development." Systems biologists are endeavoring to define functional modules containing multiple genes and then look at disorders whose phenotypes fit into such modules. A particular phenotype can overlap "considerably with several conditions (ciliopathies) in which primary cilia are also implicated in pathogenicity. One emerging aspect is the wide spectrum of ciliopathy gene mutations found within different diseases."

Ciliopathies

"The phenotypic parameters that define a ciliopathy may be used to both recognize the cellular basis of a number of genetic disorders and to facilitate the diagnosis and treatment of some diseases of unknown etiology".

Syndromes and Diseases with a Known Cilia-based Causal Mechanism

- Alstrom syndrome,
- Bardet-Biedl syndrome
- early embryonic death (some cases)
- hydrocephalus (some cases)
- Joubert syndrome
- Meckel-Gruber syndrome
- nephronophthisis
- orofaciodigital syndrome type 1
- polycystic kidney disease
- polycystic liver disease
- primary ciliary dyskinesia
- retinal degeneration (some forms)
- Senior-Loken syndrome.

Implied or Suspected Ciliopathies

- agenesis of the corpus callosum
- anencephaly
- breathing abnomalities
- cerebellar vermis hypoplasia
- Dandy-Walker malformation
- diabetes
- Ellis-van Creveld syndrome
- exencephaly
- eye movement abnormalities
- liver disease
- hypoplasia of the corpus callosum
- hypotonia
- reproductive sterility
- Jeune asphyxiating thoracic dystrophy

- Juvenile myoclonic epilepsy (JME)
- Kartagener syndrome
- Marden-Walker syndrome
- "mental retardation/developmental delay" or "other cognitive defects"
- obesity
- polydactyly
- posterior encephalocele
- respiratory dysfunction
- "recurrent respiratory infections"
- renal cystic disease
- retinitis pigmentosa (some forms)
- sensorineural deafness
- situs inversus/Isomerism
- spina bifida.

Clinical Symptoms and Ciliac Roles

A wide variety of symptoms are potential clinical features of ciliopathy.

- Chemosensation abnormalities,, typically via ciliated epithelial cellular dysfunction.
- Defective thermosensation or mechanosensation, often via ciliated epithelial cellular dysfunction.
- Cellular motility dysfunction
- Issues with displacement of extracellular fluid
- Paracrine signal transduction abnormalities.

In organisms of normal health, cilia are critical for:

- organism development
- homeostasis
- reproduction.

Centrioles and Basal Bodies

A centriole is a barrel-shaped cell structure found in most animal eukaryotic cells, though absent in higher plants and most fungi. The walls of each centriole are usually composed of nine triplets of microtubules (protein of the cytoskeleton). Deviations from this structure include *Drosophila melanogaster* embryos, with nine doublets, and *Caenorhabditis elegans* sperm cells and early embryos, with nine singlets. An associated pair of centrioles, arranged perpendicularly and surrounded by an amorphous mass of dense material (the pericentriolar material) constitutes the compound structure known as the centrosome.

Cell Division

Centrioles are involved in the organization of the mitotic spindle and in the completion of cytokinesis. Centrioles were previously thought to be required for the formation of a mitotic spindle in animal cells. However, more recent experiments have demonstrated that cells whose centrioles have been removed via laser ablation can still progress through the G_1 stage of interphase before centrioles can be synthesized later in a de novo fashion. Additionally, mutant flies lacking centrioles can develop almost normally, although the adult flies lack flagella and cilia, a lack that underscores the requirement of centrioles for the formation of these organelles.

Cellular Organization

Centrioles are a very important part of centrosomes, which are involved in organizing microtubules in the cytoplasm. The position of the centriole determines the position of the nucleus and plays a crucial role in the spatial arrangement of the cell.

Ciliogenesis

In organisms with flagella and cilia, the position of these organelles is determined by the mother centriole, which becomes the basal body. An inability of cells to use centrioles to make functional cilia and flagella has been linked to a number of genetic and developmental diseases. In particular, the inability of centrioles to properly migrate prior to ciliary assembly has recently been linked to Meckel-Gruber syndrome.

Animal Development

Proper orientation of cilia via centriole positioning toward the posterior of embryonic node cells is critical for establishing left–right asymmetry during mammalian development.

Centriole Duplication

Cells in G_0 and G_1 usually contain two complete centrioles. The older of the two centrioles in a pair is termed the *mother centriole*, whereas the younger is termed the *daughter centriole*. During the cell division cycle, a new centriole grows from the side of each of the existing "mother" centrioles. After centriole duplication, the two pairs of centrioles remain attached to each other in an orthogonal configuration until mitosis, when the mother and daughter centrioles separate in a manner dependent upon the enzyme separase. The two centrioles in the centrosome are connected to each other by unidentified proteins. The mother centriole has radiating appendages at the distal end of its long axis and is attached to the daughter centriole at the other proximal end. Each daughter cell formed after cell division will inherit one of these pairs (one older and one newer centriole). Duplication of centrioles starts at the time of the G_1/S transition and ends before the onset of mitosis.

A basal body (also called a basal granule or kinetosome) is an organelle formed from a centriole, and a short cylindrical array of microtubules. It is found at the base of a eukaryotic undulipodium (cilium or flagellum) and serves as a nucleation site for the growth of the axoneme microtubules. Centrioles, from which basal bodies are derived, act as anchoring sites for proteins that in turn anchor microtubules within centrosomes, one type of microtubule organizing centre (MTOC). These microtubules provide structure and facilitate movement of vesicles and organelles within many eukaryotic cells. Basal bodies, however, are specifically the bases for cilia and flagella that extend out of the cell.

Basal bodies are derived from centrioles through a largely mysterious process. They are structurally the same, each containing a microtubule triplet 9*3 helocoidal configuration forming a hollow cylinder.

Regulation of basal body production and spatial orientation is a function of the nucleotide-binding domain of ã-tubulin (Shang et al., 2005).

Flagellum(flagella) are basically attached to the cell membrane from a basal granule.

The Nucleus

In cell biology, the nucleus (pl. *nuclei*; from Latin *nucleus* or *nuculeus*, meaning kernel) is a membrane-enclosed organelle found in eukaryotic cells. It contains most of the cell's genetic material, organized as multiple long linear DNA molecules in complex with a large variety of proteins, such as histones, to form chromosomes. The genes within these chromosomes are the cell's nuclear genome. The function of the nucleus is to maintain the integrity of these genes and to control the activities of the cell by regulating gene expression — the nucleus is therefore the control centre of the cell. The main structures making up the nucleus are the nuclear envelope, a double membrane that encloses the entire organelle and separates its contents from the cellular cytoplasm, and the nuclear lamina, a meshwork within the nucleus that adds mechanical support, much like the cytoskeleton supports the cell as a whole. Because the nuclear membrane is impermeable to most molecules, nuclear pores are required to allow movement of molecules across the envelope. These pores cross both of the membranes, providing a channel that allows free movement of small molecules and ions. The movement of larger molecules such as proteins is carefully controlled, and requires active transport regulated by carrier proteins. Nuclear transport is crucial to cell function, as movement through the pores is required for both gene expression and chromosomal maintenance.

Although the interior of the nucleus does not contain any membrane-bound subcompartments, its contents are not uniform, and a number of *subnuclear bodies* exist, made up of unique proteins, RNA molecules, and particular parts of the chromosomes. The best known of these is the nucleolus, which is mainly involved in the assembly of ribosomes. After being produced in the nucleolus, ribosomes are exported to the cytoplasm where they translate mRNA.

History

The nucleus was the first organelle to be discovered. The probably oldest preserved drawing dates back to the early microscopist Antonie van Leeuwenhoek (1632 – 1723). He observed a "Lumen", the nucleus, in the red blood cells of salmon. Unlike mammalian red blood cells, those of other vertebrates still possess nuclei. The nucleus was also described by Franz Bauer in 1804 and in more detail in 1831 by Scottish botanist Robert Brown in a talk at the Linnean Society of London. Brown was studying orchids microscopically when he observed an opaque area, which he called the areola or nucleus, in the cells of the flower's outer layer.

He did not suggest a potential function. In 1838 Matthias Schleiden proposed that the nucleus plays a role in generating cells, thus he introduced the name "Cytoblast" (cell builder). He believed that he had observed new cells assembling around "cytoblasts". Franz Meyen was a strong opponent of this view having already described cells multiplying by division and believing that many cells would have no nuclei. The idea that cells can be generated de novo, by the "cytoblast" or otherwise, contradicted work by Robert Remak (1852) and Rudolf Virchow (1855) who decisively propagated the new paradigm that cells are generated solely by cells ("Omnis cellula e cellula"). The function of the nucleus remained unclear.

Between 1876 and 1878 Oscar Hertwig published several studies on the fertilization of sea urchin eggs, showing that the nucleus of the sperm enters the oocyte and fuses with its nucleus. This was the first time it was suggested that an individual develops from a (single) nucleated cell. This was in contradiction to Ernst Haeckel's theory that the complete phylogeny of a species would be repeated during embryonic development, including generation of the first nucleated cell from a "Monerula", a structureless mass of primordial mucus ("Urschleim"). Therefore, the necessity of the sperm nucleus for fertilization was discussed for quite some time. However, Hertwig confirmed his observation in other animal groups, e.g. amphibians and molluscs. Eduard

Strasburger produced the same results for plants (1884). This paved the way to assign the nucleus an important role in heredity. In 1873 August Weismann postulated the equivalence of the maternal and paternal germ *cells* for heredity. The function of the nucleus as carrier of genetic information became clear only later, after mitosis was discovered and the Mendelian rules were rediscovered at the beginning of the 20th century; the chromosome theory of heredity was developed.

Structures

The nucleus is the largest cellular organelle in animals. In mammalian cells, the average diameter of the nucleus is approximately 6 micrometers (ìm), which occupies about 10% of the total cell volume. The viscous liquid within it is called nucleoplasm, and is similar in composition to the cytosol found outside the nucleus. It appears as a dense, roughly spherical organelle.

Nuclear Envelope and Pores

The nuclear envelope otherwise known as nuclear membrane consists of two cellular membranes, an inner and an outer membrane, arranged parallel to one another and separated by 10 to 50 nanometers (nm). The nuclear envelope completely encloses the nucleus and separates the cell's genetic material from the surrounding cytoplasm, serving as a barrier to prevent macromolecules from diffusing freely between the nucleoplasm and the cytoplasm. The outer nuclear membrane is continuous with the membrane of the rough endoplasmic reticulum (RER), and is similarly studded with ribosomes. The space between the membranes is called the perinuclear space and is continuous with the RER lumen.

Nuclear pores, which provide aqueous channels through the envelope, are composed of multiple proteins, collectively referred to as nucleoporins. The pores are about 125 million daltons in molecular weight and consist of around 50 (in yeast) to 100 proteins (in vertebrates). The pores are 100 nm in total diameter; however, the gap through which molecules freely diffuse is only about 9 nm wide, due to the presence of regulatory systems within the centre of the pore. This size allows the free passage of small water-soluble molecules while preventing larger molecules, such as nucleic acids and larger proteins, from inappropriately entering or exiting the nucleus.

These large molecules must be actively transported into the nucleus instead. The nucleus of a typical mammalian cell will have about 3000 to 4000 pores throughout its envelope, each of which contains a donut-shaped, eightfold-symmetric ring-shaped structure at a position where the inner and outer membranes fuse. Attached to the ring is a structure called the *nuclear basket* that extends into the nucleoplasm, and a series of filamentous extensions that reach into the cytoplasm. Both structures serve to mediate binding to nuclear transport proteins. Most proteins, ribosomal subunits, and some RNAs are transported through the pore complexes in a process mediated by a family of transport factors known as karyopherins. Those karyopherins that mediate movement into the nucleus are also called importins, while those that mediate movement out of the nucleus are called exportins.

Most karyopherins interact directly with their cargo, although some use adaptor proteins. Steroid hormones such as cortisol and aldosterone, as well as other small lipid-soluble molecules involved in intercellular signalling can diffuse through the cell membrane and into the cytoplasm, where they bind nuclear receptor proteins that are trafficked into the nucleus. There they serve as transcription factors when bound to their ligand; in the absence of ligand many such receptors function as histone deacetylases that repress gene expression.

Nuclear Lamina

In animal cells, two networks of intermediate filaments provide the nucleus with mechanical support: the nuclear lamina forms an organized meshwork on the internal face of the envelope, while less organized support is provided on the cytosolic face of the envelope. Both systems provide structural support for the nuclear envelope and anchoring sites for chromosomes and nuclear pores.

The nuclear lamina is mostly composed of lamin proteins. Like all proteins, lamins are synthesized in the cytoplasm and later transported into the nucleus interior, where they are assembled before being incorporated into the existing network of nuclear lamina. Lamins are also found inside the nucleoplasm where they form another regular structure, known as the *nucleoplasmic veil*, that is visible using fluorescence microscopy. The actual function of the veil is not clear, although it is excluded from the nucleolus and is present during interphase. The lamin structures that make up the veil bind chromatin and disrupting their structure inhibits transcription of protein-coding genes.

Like the components of other intermediate filaments, the lamin monomer contains an alpha-helical domain used by two monomers to coil around each other, forming a dimer structure called a coiled coil. Two of these dimer structures then join side by side, in an antiparallel arrangement, to form a tetramer called a *protofilament*. Eight of these protofilaments form a lateral arrangement that is twisted to form a ropelike *filament*. These filaments can be assembled or disassembled in a dynamic manner, meaning that changes in the length of the filament depend on the competing rates of filament addition and removal.

Mutations in lamin genes leading to defects in filament assembly are known as *laminopathies*. The most notable laminopathy is the family of diseases known as progeria, which causes the appearance of premature aging in its sufferers. The exact mechanism by which the associated biochemical changes give rise to the aged phenotype is not well understood.

Chromosomes

The cell nucleus contains the majority of the cell's genetic material, in the form of multiple linear DNA molecules organized into structures called chromosomes. During most of the cell cycle these are organized in a DNA-protein complex known as chromatin, and during cell division the chromatin can be seen to form the well defined chromosomes familiar from a karyotype. A small fraction of the cell's genes are located instead in the mitochondria.

There are two types of chromatin. Euchromatin is the less compact DNA form, and contains genes that are frequently expressed by the cell. The other type, heterochromatin, is the more compact form, and contains DNA that are infrequently transcribed.

This structure is further categorized into *facultative* heterochromatin, consisting of genes that are organized as heterochromatin only in certain cell types or at certain stages of development, and *constitutive* heterochromatin that consists of chromosome structural components such as telomeres and centromeres. During interphase the chromatin organizes itself into discrete individual patches, called *chromosome territories*. Active genes, which are generally found in the euchromatic region of the chromosome, tend to be located towards the chromosome's territory boundary.

Antibodies to certain types of chromatin organization, particularly nucleosomes, have been associated with a number of autoimmune diseases, such as systemic lupus erythematosus. These are known as anti-nuclear antibodies (ANA) and have also been observed in concert with multiple sclerosis as part of general immune system dysfunction. As in the case of progeria, the role played by the antibodies in inducing the symptoms of autoimmune diseases is not obvious.

Nucleolus

The nucleolus is a discrete densely stained structure found in the nucleus. It is not surrounded by a membrane, and is sometimes called a *suborganelle*. It forms around tandem repeats of rDNA, DNA coding for ribosomal RNA (rRNA). These regions are called nucleolar organizer regions (NOR). The main roles of the nucleolus are to synthesize rRNA and assemble ribosomes. The structural cohesion of the nucleolus depends on its activity, as ribosomal assembly in the nucleolus results in the transient association of nucleolar components, facilitating further ribosomal assembly, and hence further association. This model is supported by observations that inactivation of rDNA results in intermingling of nucleolar structures.

The first step in ribosomal assembly is transcription of the rDNA, by a protein called RNA polymerase I, forming a large pre-rRNA precursor. This is cleaved into the subunits 5.8S, 18S, and 28S rRNA.

The transcription, post-transcriptional processing, and assembly of rRNA occurs in the nucleolus, aided by small nucleolar RNA (snoRNA) molecules, some of which are derived from spliced introns from messenger RNAs encoding genes related to ribosomal function. The assembled ribosomal subunits are the largest structures passed through the nuclear pores.

When observed under the electron microscope, the nucleolus can be seen to consist of three distinguishable regions: the innermost *fibrillar centres* (FCs), surrounded by the *dense fibrillar component* (DFC), which in turn is bordered by the *granular component* (GC).

Transcription of the rDNA occurs either in the FC or at the FC-DFC boundary, and therefore when rDNA transcription in the cell is increased more FCs are detected. Most of the cleavage and modification of rRNAs occurs in the DFC, while the latter steps involving protein assembly onto the ribosomal subunits occur in the GC.

Other Subnuclear Bodies

Table: *Subnuclear structure sizes*

Structure name	***Structure diameter***
Cajal bodies	0.2–2.0 μm
PIKA	5 μmPML
bodies	0.2–1.0 μm
Paraspeckles	0.2–1.0 μm
Speckles	20–25 nm

Besides the nucleolus, the nucleus contains a number of other non-membrane delineated bodies. These include Cajal bodies, Gemini of coiled bodies, polymorphic interphase karyosomal association (PIKA), promyelocytic leukaemia (PML) bodies, paraspeckles and splicing speckles. Although little is known about a number of these domains, they are significant in that they show that the nucleoplasm is not uniform mixture, but rather contains organized functional subdomains.

Other subnuclear structures appear as part of abnormal disease processes. For example, the presence of small intranuclear rods have been reported in some cases of nemaline myopathy. This condition typically results from mutations in actin, and the rods themselves consist of mutant actin as well as other cytoskeletal proteins.

Cajal Bodies and Gems

A nucleus typically contains between 1 and 10 compact structures called Cajal bodies or coiled bodies (CB), whose diameter measures between 0.2 μm and 2.0 μm depending on the cell type and species. When seen under an electron microscope, they resemble balls of tangled thread and are dense foci of distribution for the protein coilin. CBs are involved in a number of different roles relating to RNA processing, specifically small nucleolar RNA (snoRNA) and small nuclear RNA (snRNA) maturation, and histone mRNA modification.

Similar to Cajal bodies are Gemini of coiled bodies, or gems, whose name is derived from the Gemini constellation in reference to their close "twin" relationship with CBs. Gems are similar in size and shape to CBs, and in fact are virtually indistinguishable under the microscope. Unlike CBs, gems do not contain small nuclear ribonucleoproteins (snRNPs), but do contain a protein called *survivor of motor neurons* (SMN) whose function relates to snRNP biogenesis. Gems are believed to assist CBs in snRNP biogenesis, though it has also been suggested from microscopy evidence that CBs and gems are different manifestations of the same structure.

PIKA and PTF Domains

PIKA domains, or polymorphic interphase karyosomal associations, were first described in microscopy studies in 1991. Their function was and remains unclear, though they were not thought to be associated with active DNA replication, transcription, or RNA processing. They have been found to often associate with discrete domains defined by dense localization of the transcription factor PTF, which promotes transcription of snRNA.

PML Bodies

Promyelocytic leukaemia bodies (PML bodies) are spherical bodies found scattered throughout the nucleoplasm, measuring around 0.2–1.0 μm. They are known by a number of other names, including nuclear domain 10 (ND10), Kremer bodies, and PML oncogenic domains. They are often seen in the nucleus in association with Cajal bodies and cleavage bodies. It has been suggested that they play a role in regulating transcription.

Paraspeckles

Discovered by Fox et al. in 2002, paraspeckles are irregularly shaped compartments in the nucleus' interchromatin space. First documented in HeLa cells, where there are generally 10–30 per nucleus, paraspeckles are now known to also exist in all human primary cells, transformed cell lines and tissue sections. Their name is derived from their distribution in the nucleus; the "para" is short for parallel and the "speckles" refers to the splicing speckles to which they are always in close proximity.

Paraspeckles are dynamic structures that are altered in response to changes in cellular metabolic activity. They are transcription dependent and in the absence of RNA Pol II transcription, the paraspeckle disappears and all of its associated protein components (PSP1, p54nrb, PSP2, CFI(m)68 and PSF) form a crescent shaped perinucleolar cap in the nucleolus. This phenomenon is demonstrated during the cell cycle. In the cell cycle, paraspeckles are present during interphase and during all of mitosis except for telophase. During telophase, when the two daughter nuclei are formed, there is no RNA Pol II transcription so the protein components instead form a perinucleolar cap.

Splicing Speckles

Sometimes referred to as *interchromatin granule clusters* or as *splicing-factor compartments*, speckles are rich in splicing snRNPs and other splicing proteins necessary for pre-mRNA processing. Because of a cell's changing requirements, the composition and location of these bodies changes according to mRNA transcription and regulation via phosphorylation of specific proteins.

Function

The main function of the cell nucleus is to control gene expression and mediate the replication of DNA during the cell cycle. The nucleus provides a site for genetic transcription that is segregated from the location of translation in the cytoplasm, allowing levels of gene regulation that are not available to prokaryotes.

Cell Compartmentalization

The nuclear envelope allows the nucleus to control its contents, and separate them from the rest of the cytoplasm where necessary. This is important for controlling processes on either side of the nuclear membrane. In some cases where a cytoplasmic process needs to be restricted, a key participant is removed to the nucleus, where it interacts with transcription factors to downregulate the production of certain enzymes in the pathway.

This regulatory mechanism occurs in the case of glycolysis, a cellular pathway for breaking down glucose to produce energy. Hexokinase is an enzyme responsible for the first the step of glycolysis, forming glucose-6-phosphate from glucose. At high concentrations of fructose-6-phosphate, a molecule made later from glucose-6-phosphate, a regulator protein removes hexokinase to the nucleus, where it forms a transcriptional repressor complex with nuclear proteins to reduce the expression of genes involved in glycolysis.

In order to control which genes are being transcribed, the cell separates some transcription factor proteins responsible for regulating gene expression from physical access to the DNA until they are activated by other signalling pathways. This prevents even low levels of inappropriate gene expression. For example in the case of NF-κB-controlled genes, which are involved in most inflammatory responses, transcription is induced in response to a signal pathway such as that initiated by the signalling molecule TNF-α, binds to a cell membrane receptor, resulting in the recruitment of signalling proteins, and eventually activating the transcription factor NF-κB. A nuclear localisation signal on the NF-κB protein allows it to be transported through the nuclear pore and into the nucleus, where it stimulates the transcription of the target genes.

The compartmentalization allows the cell to prevent translation of unspliced mRNA. Eukaryotic mRNA contains introns that must be removed before being translated to produce functional proteins. The splicing is done inside the nucleus before the mRNA can be accessed by ribosomes for translation. Without the nucleus ribosomes would translate newly transcribed (unprocessed) mRNA resulting in misformed and nonfunctional proteins.

Gene Expression

Gene expression is the process by which information from a gene is used in the synthesis of a functional gene product. These products are often proteins, but in non-protein coding genes such as rRNA genes or tRNA genes, the product is a functional RNA. The process of gene expression is used by all known life - eukaryotes (including multicellular organisms), prokaryotes (bacteria and archaea) and viruses - to generate the macromolecular machinery for life.

Several steps in the gene expression process may be modulated, including the transcription, RNA splicing, translation, and post-translational modification of a protein. Gene regulation gives the cell control over structure and function, and is the basis for cellular differentiation, morphogenesis and the versatility and adaptability of any organism. Gene regulation may also serve as a substrate for evolutionary change, since control of the timing, location, and amount of gene expression can have a profound effect on the functions (actions) of the gene in a cell or in a multicellular organism.

In genetics, gene expression is the most fundamental level at which genotype gives rise to the phenotype. The genetic code is "interpreted" by gene expression, and the properties of the expression products give rise to the organism's phenotype.

Mechanism

Transcription

The gene itself is typically a long stretch of DNA. It is a blueprint for the production of RNA. The production of RNA copies of the DNA is called transcription, and is performed by RNA polymerase, which adds one RNA nucleotide at a time to a growing RNA strand. This RNA is complementary to the template 3' '! 5' DNA strand, which is itself complementary to the coding 5' '! 3' DNA strand. Therefore, the resulting 5' '! 3' RNA strand is identical to the coding DNA strand with the exception that thymines are replaced with uracils in the RNA. A coding DNA strand reading "ATG" is transcribed as "AUG" in RNA.

RNA Processing

Transcription of protein encoding genes creates a primary transcript of RNA at the place where the gene was located. This transcript can be altered before being translated, this is particularly common in eukaryotes. The most common RNA processing is splicing to remove introns. Introns are RNA segments which are not found in the mature RNA, although they can function as precursors, e.g. for snoRNAs, which are RNAs that direct modification of nucleotides in other RNAs. Introns are common in eukaryotic genes but rare in prokaryotes. RNA processing, also known as post-transcriptional modification, can start during transcription, as is the case for splicing, where the spliceosome removes introns from newly formed RNA.

Extensive RNA processing may be an evolutionary advantage made possible by the nucleus of eukaryotes. In prokaryotes transcription and translation happen together whilst in eukaryotes the nuclear membrane separates the two processes giving time for RNA processing to occur.

Non-coding RNA Maturation

In most organisms non-coding genes (ncRNA) are transcribed as precursors which undergo further processing. In the case of ribosomal RNAs (rRNA), they are often transcribed as a pre-rRNA which contains one or more rRNAs, the pre-rRNA is cleaved and modified (22 -O-methylation and pseudouridine formation) at a specific sites by approximately 150 different small nucleolus-restricted RNA species, called small nucleolar RNAs(snoRNAs), which like snRNAs, snoRNAs associate with proteins, forming snoRNPs. In eukaryotes, in particular a snoRNP, called RNase MRP cleaves the 45S pre-rRNA into the 28S, 5.8S, and 18S rRNAs. The rRNA and RNA processing factors are form large aggregates called the nucleolus.

In the case of transfer RNA (tRNA), for example, the 5' sequence is removed by RNase P, whereas the 3' end is removed by the tRNase Z enzyme. In the case of micro RNA (miRNA), miRNAs are first transcribed as primary transcripts or pri-miRNA with a cap and poly-A tail and processed to short, 70-nucleotide stem-loop structures known as pre-miRNA in the cell nucleus by the enzymes Drosha and Pasha, after being exported, it is then processed to mature miRNAs in the cytoplasm by interaction with the endonuclease Dicer, which also initiates the formation of the RNA-induced silencing complex (RISC), composed of the Argonaute protein.

RNA Export

In eukaryotes most mature RNA must be exported to the cytoplasm from the nucleus. While some RNAs function in the nucleus, many RNAs are transported through the nuclear pores and into the cytosol. Notably this includes all RNA types involved in protein synthesis. In some cases RNAs are additionally transported to a specific part of the cytoplasm, such as a synapse; they are then towed by motor proteins that bind through linker proteins to specific sequences (called "zipcodes") on the RNA.

Translation

For some RNA (non-coding RNA) the mature RNA is the finished gene product. In the case of messenger RNA (mRNA) the RNA is an information carrier coding for the synthesis of one or more proteins. mRNA carrying a single protein sequence (common in eukaryotes) is monocistronic whilst mRNA carrying multiple protein sequences (common in prokaryotes) is known as polycistronic.

Each triplet of nucleotides of the coding regions of a messenger RNA corresponds to a binding site for a transfer RNA. Transfer RNAs carry amino acids, and these are chained together by the ribosome. The ribosome helps transfer RNA to bind to messenger RNA and takes the amino acid from each transfer RNA and makes a structure-less protein out of it. In prokaryotes translation generally occurs at the point of transcription, often using a messenger RNA which is still in the process of being created.

In eukaryotes translation can occur in a variety of regions of the cell depending on where the protein being written is supposed to be. Major locations are the cytoplasm for soluble cytoplasmic proteins and the endoplasmic reticulum for proteins which are for export from the cell or insertion into a cell membrane. Proteins which are supposed to be expressed at the endoplasmic reticulum are recognised part-way through the translation process. This is governed by the signal recognition particle - a protein which binds to the ribosome and directs it to the endoplasmic reticulum when it finds a signal sequence on the growing (nascent) amino acid chain.

Folding

Enzymes called chaperones assist the newly formed protein to attain (fold into) the 3-dimensional structure it needs to function. Similarly, RNA chaperones help RNAs attain their functional shapes. Assisting protein folding is one of the main roles of the endoplasmic reticulum in eukaryotes.

Protein Transport

Many proteins are destined for other parts of the cell than the cytosol and a wide range of signalling sequences are used to direct proteins to where they are supposed to be. In prokaryotes this is normally a simple process due to limited compartmentalisation of the cell. However in eukaryotes there is a great variety of different targeting processes to ensure the protein arrives at the correct organelle.

Not all proteins remain within the cell and many are exported, for example digestive enzymes, hormones and extracellular matrix proteins. In eukaryotes the export pathway is well developed and the main mechanism for the export of these proteins is translocation to the endoplasmatic reticulum, followed by transport via the Golgi apparatus.

Measurement

Measuring gene expression is an important part of many life sciences - the ability to quantify the level at which a particular gene is expressed within a cell, tissue or organism can give a huge amount of information. For example measuring gene expression can:

- Identify viral infection of a cell (viral protein expression)
- Determine an individual's susceptibility to cancer (oncogene expression)
- Find if a bacterium is resistant to penicillin (beta-lactamase expression).

Similarly the analysis of the location of expression protein is a powerful tool and this can be done on an organism or cellular scale. Investigation of localisation is particularly important for study of development in multicellular organisms and as an indicator of protein function in single cells. Ideally measurement of expression is done by detecting the final gene product (for many genes this is the protein) however it is often easier to detect one of the precursors, typically mRNA, and infer gene expression level.

mRNA Quantification

Levels of mRNA can be quantitatively measured by Northern blotting which gives size and sequence information about the mRNA molecules. A sample of RNA is separated on an agarose gel and hybridized to a radio-labeled RNA probe that is complementary to the target sequence. The radio-labeled RNA is then detected by an autoradiograph. The main problems with Northern blotting stem from the use of radioactive reagents (which make the procedure time consuming and potentially dangerous) and lower quality quantification than more modern methods (due to the fact that quantification is done by measuring band strength in an image of a gel). Northern blotting is, however, still widely used as the additional mRNA size information allows the discrimination of alternately spliced transcripts.

A more modern low-throughput approach for measuring mRNA abundance is reverse transcription quantitative polymerase chain reaction (RT-PCR followed with qPCR). RT-PCR first generates a DNA template from the mRNA by reverse transcription, which is called cDNA. This cDNA template is then used for qPCR where the change in fluorescence of a probe changes as the DNA amplification process progresses. With a carefully constructed standard curve qPCR can produce an absolute measurement such as number of copies of mRNA, typically in units of copies per nanolitre of homogenized tissue or copies per cell. qPCR is very sensitive (detection of a single mRNA molecule is possible), but can be expensive due to the fluorescent probes required.

Northern blots and RT-qPCR are good for detecting whether a single gene is being expressed, but it quickly becomes impractical if many genes within the sample are being studied. Using DNA microarrays transcript levels for many genes at once (expression profiling) can be measured. Recent advances in microarray technology allow for the quantification, on a single array, of transcript levels for every known gene in several organism's genomes, including humans.

Alternatively "tag based" technologies like Serial analysis of gene expression (SAGE), which can provide a relative measure of the cellular concentration of different messenger RNAs, can be used. The great advantage of tag-based methods is the "open architecture", allowing for the exact measurement of any transcript, with a known or unknown sequence.

Protein Quantification

For genes encoding proteins the expression level can be directly assessed by a number of means with some clear analogies to the techniques for mRNA quantification.

The most commonly used method is to perform a Western blot against the protein of interest - this gives information on the size of the protein in addition to its identity. A sample (often cellular lysate) is separated on a polyacrylamide gel, transferred to a membrane and then probed with an antibody to the protein of interest. The antibody can either be conjugated to a fluorophore or to horseradish peroxidase for imaging and/or quantification. The gel-based nature of this assay makes quantification less accurate but it has the advantage of being able to identify later modifications to the protein, for example proteolysis or ubiquitination, from changes in size.

By replacing the gene with a new version fused a green fluorescent protein (or similar) marker expression may be directly quantified in live cells. This is done by imaging using a fluorescence microscope. It is very difficult to clone a GFP-fused protein into its native location in the genome without affecting expression levels so this method often cannot be used to measure endogenous gene expression. It is, however, widely used to measure the expression of a gene artificially introduced into the cell, for example via an expression vector. It is important to note that by fusing a target protein to a fluorescent reporter the protein's behavior, including its cellular localization and expression level, can be significantly changed. The enzyme-linked immunosorbent assay works by using antibodies immobilised on a microtiter plate to capture proteins of interest from samples added to the well. Using a detection antibody conjugated to an enzyme or fluorophore the quantity of bound protein can be accurately measured by fluorometric or colourimetric detection. The detection process is very similar to that of a Western blot, but by avoiding the gel steps more accurate quantification can be achieved.

Localisation

Analysis of expression is not limited to only quantification; localisation can also be determined. mRNA can be detected with a suitably labelled complementary mRNA strand and protein can be detected via labelled antibodies. The probed sample is then observed by microscopy to identify where the mRNA or protein is.

Regulation of Gene Expression

Regulation of gene expression refers to the control of the amount and timing of appearance of the functional product of a gene. Control of expression is vital to allow a cell to produce the gene products it needs when it needs them; in turn this gives cells the flexibility to adapt to a variable environment, external signals, damage to the cell, etc. Some simple examples of where gene expression is important are:

- Control of Insulin expression so it gives a signal for blood glucose regulation
- X chromosome inactivation in female mammals to prevent an "overdose" of the genes it contains.
- Cyclin expression levels control progression through the eukaryotic cell cycle.

More generally gene regulation gives the cell control over all structure and function, and is the basis for cellular differentiation, morphogenesis and the versatility and adaptability of any organism. Any step of gene expression may be modulated, from the DNA-RNA transcription step to post-translational modification of a protein.

The stability of the final gene product, whether it is RNA or protein, also contributes to the expression level of the gene - an unstable product results in a low expression level. In general gene expression is regulated through changes in the number and type of interactions between molecules that collectively influence transcription of DNA and translation of RNA.

Transcriptional Regulation

Regulation of transcription can be broken down into three main routes of influence; genetic (direct interaction of a control factor with the gene), modulation (interaction of a control factor with the transcription machinery) and epigenetic (non-sequence changes in DNA structure which influence transcription).

Direct interaction with DNA is the simplest and most direct method a protein can change transcription levels and genes often have several protein binding sites around the coding region with the specific function of regulating transcription. There are many classes of regulatory DNA binding sites known as enhancers, insulators, repressors and silencers. The mechanisms for regulating transcription are very varied, from blocking key binding sites on the DNA for RNA polymerase to acting as an activator and promoting transcription by assisting RNA polymerase binding.

The activity of transcription factors is further modulated by intracellular signals causing protein post-translational modification including phosphorylated, acetylated, or glycosylated. These changes influence a transcription factor's ability to bind, directly or indirectly, to promoter DNA, to recruit RNA polymerase, or to favour elongation of a newly synthetized RNA molecule.

The nuclear membrane in eukaryotes allows further regulation of transcription factors by the duration of their presence in the nucleus which is regulated by reversible changes in their structure and by binding of other proteins. Environmental stimuli or endocrine signals may cause modification of regulatory proteins eliciting cascades of intracellular signals, which result in regulation of gene expression.

More recently it has become apparent that there is a huge influence of non-DNA-sequence specific effects on translation. These effects are referred to as epigenetic and involve the higher order structure of DNA, non-sequence specific DNA binding proteins and chemical modification of DNA. In general epigenetic effects alter the accessibility of DNA to proteins and so modulate transcription.

DNA methylation is a widespread mechanism for epigenetic influence on gene expression and is seen in bacteria and eukaryotes and has roles in heritable transcription silencing and transcription regulation. In eukaryotes the structure of chromatin, controlled by the histone code, regulates access to DNA with significant impacts on the expression of genes in euchromatin and heterochromatin areas.

Post-transcriptional Regulation

In eukaryotes, where export of RNA is required before translation is possible, nuclear export is thought to provide additional control over gene expression. All transport in and out of the nucleus is via the nuclear pore and transport is controlled by a wide range of importin and exportin proteins.

Expression of a gene coding for a protein is only possible if the messenger RNA carrying the code survives long enough to be translated. In a typical cell an RNA molecule is only stable if specifically protected from degradation. RNA degradation has particular importance in regulation of expression in eukaryotic cells where mRNA has to travel significant distances before being translated. In eukaryotes RNA is stabilised by certain post-transcriptional modifications, particularly the 5' cap and poly-adenylated tail. Intentional degradation of mRNA is used not just as a defence mechanism from foreign RNA (normally from viruses) but also as a route of mRNA *destabilisation*. If an mRNA molecule has a complementary sequence to a small interfering RNA then it is targeted for destruction via the RNA interference pathway.

Translational Regulation

Direct regulation of translation is less prevalent than control of transcription or mRNA stability but is occasionally used. Inhibition of protein translation is a major target for toxins and antibiotics in order to kill a cell by overriding its normal gene expression control. Protein synthesis inhibitors include the antibiotic neomycin and the toxin ricin.

Protein Degradation

Once protein synthesis is complete the level of expression of that protein can be reduced by protein degradation. There are major protein degradation pathways in all prokaryotes and eukaryotes of which the proteasome is a common component. An unneeded or damaged protein is often labelled for degradation by addition of ubiquitin.

Expression System

An expression system is a system specifically designed for the production of a gene product of choice. This is normally a protein although may also be RNA, such as tRNA or a ribozyme. An expression system consists of a gene, normally encoded by DNA, and the molecular machinery required to transcribe the DNA into mRNA and translate the mRNA into protein using the reagents provided. In the broadest sense this includes every living cell but the term is more normally used to refer to expression as a laboratory tool. An expression system is therefore often artificial in some manner. Expression systems are, however, a fundamentally natural process. Viruses are an excellent example where they replicate by using the host cell as an expression system for the viral proteins and genome.

In Nature

In addition to these biological tools, certain naturally observed configurations of DNA (genes, promoters, enhancers, repressors) and the associated machinery itself are referred to as an expression system. This term is normally used in the case where a gene or set of genes is switched on under well defined conditions. For example the simple repressor 'switch' expression system in Lambda phage and the lac operator system in bacteria. Several natural expression systems are directly used or modified and used for artificial expression systems such as the Tet-on and Tet-off expression system.

Gene Networks and Expression

Genes have sometimes been regarded as nodes in a network, with inputs being proteins such as transcription factors, and outputs being the level of gene expression. The node itself performs a function, and the operation of these functions have been interpreted as performing a kind of information processing within cell and determine cellular behavior. Gene networks can also be constructed without formulating an explicit causal model. This is often the case when assembling networks from large expression data sets. Covariation and correlation of expression is computed across a large sample of cases and measurements (often transcriptome or proteome data). The source of variation can be either experimental or natural (observational). There are several ways to construct gene expression networks, but one common approach is to compute a matrix of all pair-wise correlations of expression across conditions, time points, or individuals and convert the matrix (after thresholding at some cut-off value) into a graphical representation in which nodes represent genes, transcripts, or proteins and edges connecting these nodes represent the strength of association.

Techniques and Tools

The following experimental techniques are used to measure gene expression and are listed in roughly chronological order, starting with the older, more established technologies. They are divided into two groups based on their degree of multiplexity.

- Low-to-mid-plex techniques:
  - Reporter gene
  - Northern blot
  - Western blot
  - Fluorescent in situ hybridization
  - Reverse transcription PCR.
- Higher-plex techniques:
  - SAGE
  - DNA microarray
  - Tiling array
  - RNA-Seq.

Gene expression first involves transcription, in which DNA is used as a template to produce RNA. In the case of genes encoding proteins, that RNA produced from this process is messenger RNA (mRNA), which then needs to be translated by ribosomes to form a protein. As ribosomes are located outside the nucleus, mRNA produced needs to be exported.

Since the nucleus is the site of transcription, it also contains a variety of proteins which either directly mediate transcription or are involved in regulating the process. These proteins include helicases that unwind the double-stranded DNA molecule to facilitate access to it, RNA polymerases that synthesize the growing RNA molecule, topoisomerases that change the amount of supercoiling in DNA, helping it wind and unwind, as well as a large variety of transcription factors that regulate expression.

Processing of Pre-mRNA

Newly synthesized mRNA molecules are known as primary transcripts or pre-mRNA. They must undergo post-transcriptional modification in the nucleus before being exported to the cytoplasm; mRNA that appears in the nucleus without these modifications is degraded rather than used for protein translation. The three main modifications are 5' capping, 3' polyadenylation, and RNA splicing. While in the nucleus, pre-mRNA is associated with a variety of proteins in complexes known as heterogeneous ribonucleoprotein particles (hnRNPs). Addition of the 5' cap occurs co-transcriptionally and is the first step in post-transcriptional modification. The 3' poly-adenine tail is only added after transcription is complete.

RNA splicing, carried out by a complex called the spliceosome, is the process by which introns, or regions of DNA that do not code for protein, are removed from the pre-mRNA and the remaining exons connected to re-form a single continuous molecule.

This process normally occurs after 5' capping and 3' polyadenylation but can begin before synthesis is complete in transcripts with many exons. Many pre-mRNAs, including those encoding antibodies, can be spliced in multiple ways to produce different mature mRNAs that encode different protein sequences. This process is known as alternative splicing, and allows production of a large variety of proteins from a limited amount of DNA.

Post-transcriptional Modification

Post-transcriptional modification is a process in cell biology by which, in eukaryotic cells, primary transcript RNA is converted into mature RNA. A notable example is the conversion of precursor messenger RNA into mature messenger RNA (mRNA), which includes splicing and occurs prior to protein synthesis. This process is vital for the correct translation of the genomes of eukaryotes as the human primary RNA transcript that is produced as a result of transcription contains both exons, which are coding sections of the primary RNA transcript and introns, which are the non coding sections of the primary RNA transcript.

Dynamics and Regulation

Nuclear Transport

The entry and exit of large molecules from the nucleus is tightly controlled by the nuclear pore complexes. Although small molecules can enter the nucleus without regulation, macromolecules such as RNA and proteins require association karyopherins called importins to enter the nucleus and exportins to exit. "Cargo" proteins that must be translocated from the cytoplasm to the nucleus contain short amino acid sequences known as nuclear localization signals which are bound by importins, while those transported from the nucleus to the cytoplasm carry nuclear export signals bound by exportins.

The ability of importins and exportins to transport their cargo is regulated by GTPases, enzymes that hydrolyze the molecule guanosine triphosphate to release energy. The key GTPase in nuclear transport is Ran, which can bind either GTP or GDP (guanosine diphosphate) depending on whether it is located in the nucleus or the cytoplasm. Whereas importins depend on RanGTP to dissociate from their cargo, exportins require RanGTP in order to bind to their cargo.

Nuclear import depends on the importin binding its cargo in the cytoplasm and carrying it through the nuclear pore into the nucleus. Inside the nucleus, RanGTP acts to separate the cargo from the importin, allowing the importin to exit the nucleus and be reused. Nuclear export is similar, as the exportin binds the cargo inside the nucleus in a process facilitated by RanGTP, exits through the nuclear pore, and separates from its cargo in the cytoplasm.

Specialized export proteins exist for translocation of mature mRNA and tRNA to the cytoplasm after post-transcriptional modification is complete. This quality-control mechanism is important due to the these molecules' central role in protein translation; mis-expression of a protein due to incomplete excision of exons or mis-incorporation of amino acids could have negative consequences for the cell; thus incompletely modified RNA that reaches the cytoplasm is degraded rather than used in translation.

Assembly and Disassembly

During its lifetime a nucleus may be broken down, either in the process of cell division or as a consequence of apoptosis, a regulated form of cell death. During these events, the structural components of the nucleus—the envelope and lamina—are systematically degraded.

During the cell cycle the cell divides to form two cells. In order for this process to be possible, each of the new daughter cells must have a full set of genes, a process requiring replication of the chromosomes as well as segregation of the separate sets. This occurs by the replicated chromosomes, the sister chromatids, attaching to microtubules, which in turn are attached to different centrosomes.

The sister chromatids can then be pulled to separate locations in the cell. In many cells the centrosome is located in the cytoplasm, outside the nucleus, the microtubules would be unable to attach to the chromatids in the presence of the nuclear envelope. Therefore the early stages in the cell cycle, beginning in prophase and until around prometaphase, the nuclear membrane is dismantled. Likewise, during the same period, the nuclear lamina is also disassembled, a process regulated by phosphorylation of the lamins. Towards the end of the cell cycle, the nuclear membrane is reformed, and around the same time, the nuclear lamina are reassembled by dephosphorylating the lamins.

However, in dinoflagellates the nuclear envelope remains intact, the centrosomes are located in the cytoplasm, and the microtubules come in contact with chromosomes, whose centromeric regions are incorporated into the nuclear envelope (the so-called closed mitosis with extranuclear spindle). In many other protists (e.g. ciliates, sporozoans) and fungi the centrosomes are intranuclear, and their nuclear envelope also does not disassemle during cell division.

Apoptosis is a controlled process in which the cell's structural components are destroyed, resulting in death of the cell. Changes associated with apoptosis directly affect the nucleus and its contents, for example in the condensation of chromatin and the disintegration of the nuclear envelope and lamina. The destruction of the lamin networks is controlled by specialized apoptotic proteases called caspases, which cleave the lamin proteins and thus degrade the nucleus' structural integrity. Lamin cleavage is sometimes used as a laboratory indicator of caspase activity in assays for early apoptotic activity. Cells that express mutant caspase-resistant lamins are deficient in nuclear changes related to apoptosis, suggesting that lamins play a role in initiating the events that lead to apoptotic degradation of the nucleus. Inhibition of lamin assembly itself is an inducer of apoptosis.

The nuclear envelope acts as a barrier that prevents both DNA and RNA viruses from entering the nucleus. Some viruses require access to proteins inside the nucleus in order to replicate and/or assemble. DNA viruses, such as herpesvirus replicate and assemble in the cell nucleus, and exit by budding through the inner nuclear membrane. This process is accompanied by disassembly of the lamina on the nuclear face of the inner membrane.

Anucleated and Polynucleated Cells

Although most cells have a single nucleus, some eukaryotic cell types have no nucleus, and others have many nuclei. This can be a normal process, as in the maturation of mammalian red blood cells, or a result of faulty cell division.

Anucleated cells contain no nucleus and are therefore incapable of dividing to produce daughter cells. The best-known anucleated cell is the mammalian red blood cell, or erythrocyte, which also lacks other organelles such as mitochondria and serves primarily as a transport vessel to ferry oxygen from the lungs to the body's tissues. Erythrocytes mature through erythropoiesis in the bone marrow, where they lose their nuclei, organelles, and ribosomes. The nucleus is expelled during the process of differentiation from an erythroblast to a reticulocyte, which is the immediate precursor of the mature erythrocyte. The presence of mutagens may induce the release of some immature "micronucleated" erythrocytes into the bloodstream. Anucleated cells can also arise from flawed cell division in which one daughter lacks a nucleus and the other has two nuclei.

Polynucleated cells contain multiple nuclei. Most Acantharean species of protozoa and some fungi in mycorrhizae have naturally polynucleated cells. Other examples include the intestinal parasites in the genus *Giardia*, which have two nuclei per cell. In humans, skeletal muscle cells, called myocytes, become polynucleated during development; the resulting arrangement of nuclei near the periphery of the cells allows maximal intracellular space for myofibrils. Multinucleated cells can also be abnormal in humans; for example, cells arising from the fusion of monocytes and macrophages, known as giant multinucleated cells, sometimes accompany inflammation and are also implicated in tumour formation.

Evolution

As the major defining characteristic of the eukaryotic cell, the nucleus' evolutionary origin has been the subject of much speculation. Four major theories have been proposed to explain the existence of the nucleus, although none have yet earned widespread support.

The theory known as the "syntrophic model" proposes that a symbiotic relationship between the archaea and bacteria created the nucleus-containing eukaryotic cell. It is hypothesized that the symbiosis originated when ancient archaea, similar to modern methanogenic archaea, invaded and lived within bacteria similar to modern myxobacteria, eventually forming the early nucleus. This theory is analogous to the accepted theory for the origin of eukaryotic mitochondria and chloroplasts, which are thought to have developed from a similar endosymbiotic relationship between proto-eukaryotes and aerobic bacteria. The archaeal origin of the nucleus is supported by observations that archaea and eukarya have similar genes for certain proteins, including histones. Observations that myxobacteria are motile, can form multicellular complexes, and possess kinases and G proteins similar to eukarya, support a bacterial origin for the eukaryotic cell.

A second model proposes that proto-eukaryotic cells evolved from bacteria without an endosymbiotic stage. This model is based on the existence of modern planctomycetes bacteria that possess a nuclear structure with primitive pores and other compartmentalized membrane structures. A similar proposal states that a eukaryote-like cell, the chronocyte, evolved first and phagocytosed archaea and bacteria to generate the nucleus and the eukaryotic cell.

The most controversial model, known as *viral eukaryogenesis*, posits that the membrane-bound nucleus, along with other eukaryotic features, originated from the infection of a prokaryote by a virus. The suggestion is based on similarities between eukaryotes and viruses such as linear DNA strands, mRNA capping, and tight binding to proteins (analogizing histones to viral envelopes). One version of the proposal suggests that the nucleus evolved in concert with phagocytosis to form an early cellular "predator". Another variant proposes that eukaryotes originated from early archaea infected by poxviruses, on the basis of observed similarity between the DNA polymerases in modern poxviruses and eukaryotes. It has been suggested that the unresolved question of the evolution of sex could be related to the viral eukaryogenesis hypothesis.

Finally, a very recent proposal suggests that traditional variants of the endosymbiont theory are insufficiently powerful to explain the origin of the eukaryotic nucleus. This model, termed the *exomembrane hypothesis*, suggests that the nucleus instead originated from a single ancestral cell that evolved a second exterior cell membrane; the interior membrane enclosing the original cell then became the nuclear membrane and evolved increasingly elaborate pore structures for passage of internally synthesized cellular components such as ribosomal subunits.

3

Identification of Genetic Material

The genetic material of a cell or an organism refers to those materials found in the nucleus, mitochondria and cytoplasm, which play a fundamental role in determining the structure and nature of cell substances, and capable of self-propagating and variation. When it became evident that the chromosomes were the organs of heredity, because (i) they formed the only link between two generations; (ii) they carried linearly arranged genes; (iii) they occurred in every organism in specific number and had specific morphology for a particular species; and (iv)any variation in their number or morphology affected the phenotype of the species. Then, various attempts were made by early molecular geneticists to identify the physical and chemical nature of genes. But, the genes were found so minute structures that their physical identity remained almost impossible. However, the extensive chemical analysis of chromosomes of different organisms have revealed that chromosomes contain proteins and nucleic acid (DNA and RNA) and it was thought that genes might have either proteins or nucleic acids as their component molecules.

Early molecular geneticists have assigned the informational roles of genes to the chromosomal proteins, because, they found nucleic acids too simple to carry genetic informations. The controversy about the assignment of genetic role either to chromosomal proteins or to chromosomal DNA, existed upto 1950, when finally it was universally accepted that DNA is the genetic substance (i.e., chemical of which genes are composed) of most micro-organisms and higher organisms.Later on, RNA was found to be the genetic material of some viruses. The concept that DNA or RNA is the genetic material of most organisms has been developed and supported by following evidences:

Direct Evidences for DNA as the Genetic Material

The most conclusive evidences in support of DNA as the genetic material comes from following three avenues of approach on micro-organisms: transformation of bacteria mode of infection of bacteriophages and conjugation in bacteria. Bacterial Transformation and Griffith EffectIn 1928, Frederick Griffith encountered a phenomenon now known as genetic transformation. Colonies of virulent strain (pathogenic) of pneumonia causing spherically-shaped bacterium, Diplococcus pneumoniae grown on nutrient agar, have a smooth (s) glistering appearance owing to the presence of a type specific, ploysaccharide (a polymer of glucose and glucuronic acid) capsule.

The avirulant (non-pathogenic) strains of this bacterium, on the other hand, lack this capsule and they produce dull, rough (R) colonies and is unable to cause the pneumonia disease. Smooth (S) and rough (R) characters are directly related to the presence or absence of the capsule and this trait is known to be genetically determined. Both S and R forms occur in several types and are designed as S-1, S-II, S-III, etc., and R-I, R-II, R-III, etc., respectively.

All these subtypes of S and R bacteria differ with each other in the type of antigens, they produce. The kind of antigens produced is likewise genetically determined. Smooth (S) forms sometimes mutate to rough (R) forms, but this change have not been found reversible.

In the course of his work, Griffith injected laboratory mice with live R-II pneumococci; the mice suffered no illness because R-II pneumococci was avirulent.

But when the mice were injected with the mixture of living avirulent R-II and heat killed S-III virulent, the unexpected symptoms of pneumonia appeared and high mortality was resulted in them.

By postmortoming the dead mice, it was found that their heart blood had both R-II and S-III pneumococci. From these results, Griffith concluded that the presence of the heat-killed S-IU bacteria must have caused a transformation of the living R-II bacteria, so as to restore to them the capacity for capsule formation they had earlier lost by gene mutation. This was called "Griffith effect" or more popularly as "bacterial transformation".

Identification of the "transforming substance"-Griffith could not understand the cause of bacterial transformation and that is first of all identified by O. T. Avery, C. M. McLeod and M McCarthy (1944). They tested a fraction of heat killed S-III bacteria for the transforming ability.

They could remove proteins, lipids, polysaccharides and ribonucleic acids from the S-III extract by a variety of chemical and enzymatic methods without seriously diminishing its power to transfer R-II mutants into the S-III wild types. They ultimately concluded that because a cell free and highly purified DNA extract of S-III bacteria could bring about transformation of R-I1 bacteria into S-III, therefore, DNA is the genetic material of pneumococci. Later on, such a transforming substance, the DNA, was found in a variety of bacteria (i.e., Hemophilus influenzae. Bacillus subtilus, Shigelia paradysenteriae, etc.) and several other organisms.

Mode of Infection of Bacteriophages and Identification of DNA as their Genetic Material

By using radioactive tracers, A. Hershey and M. Chase (1952) provided further direct proof that DNA is the genetic material in certain bacterial viruses. These investigators were studying the bacteriophages that attack the bacterium, Escherichia coli. They prepared a chemically defined culture medium (for E. coli) containing phosphoric acid and sulphuric acid as the only source of phosphorus and sulphur.

Known quantities of radioactive isotopes of phosphorus (P^{32}) and sulphur (S^{32}) were added to the culture medium. E. coli grown on such a culture medium incorporated the P^{32} and S^{35} into their chemical constituents. When T_2 bacteriophage particles infected the radioactive bacteria, they incorporated the labeled DNA.

Phage capsid proteins did not contain appreciable amounts of phosphorus and only the DNA was labeled with P^{32}. Similarly, the protein envelope around the phage was selectively labeled with S^{35}. DNA contains virtually no sulphur and was not labeled with S^{35} By this method it was possible to differentially label the phage DNA and proteins of phage-capsid.

After phage growth, the phage particles were separated from the host cell by centrifugation. The radioactive phages were next introduced to non-radioactive bacterial cultures where they attacked the bacteria. Subsequently the viruses were separated from the host cell by agitation and the content of P^{32} and S^{35} in the host (E. coli) and parasite (bacteriophage) was determined.

The phosphorus label was found to be associated with the bacterial cells and the sulphur label was in the proteins coats (capsids) left in the medium. This indicated that the DNA had panetrated the cells but that the protein coat or capsid of the phage was left outside the wall of the bacterium.

The labeled DNA was found to reproduced in the host cells. These experiments, thus, clearly demonstrated that genetic informations of these bacteriophages resided in their DNA molecules and not in the protein molecules of their capsids. Later on Hershey-Chase experiment was repeated for other viruses and it was found that DNA is also the genetic material in a number of other viruses.

Bacterial Conjugation

Another conclusive evidence for DNA as the genetic material came from the phenomenon of conjugation of bacteria. Laderberg and Tatum (1946) found that when a F^+ ('male') E. coli cell conjugated with a F^- ('female') E. coli cell an unidirectional transfer of F^+ factor of 'male' cell to F^- or 'female' cell took place; so that the latter was converted into a F^+ or 'male' strain. The F^+ factor was found to be a fragment of DNA molecule which occurred in the cytoplasm of bactarial cell.

Indirect Evidences for DNA as the Genetic Material

The indirect or circumstantial evidences in favour of DNA as the genetic material came from higher organisms which are not as easy to manipulate as bacteria and viruses. The fact that DNA is the genetic material of higher organisms has been supported by following facts:

1. Localization -All cells contain DNA which is localized primarily on chromosomes in the nucleus of the cell. This is consistent with the fact the genes are also located specifically on chromosomes. The Feulgen techniques have shown that DNA entirely remains restricted to the chromosomes and it forms one of the major component of chromosomes.
2. Quantity-Various quantitative measurement of amount of DNA in different cells have shown that there is a correlation between the amount of DNA and the number of chromosome sets (ploidy).

Cells	*Mean DNA Fegulen content (Picograms)*	*Presumed chromosomes set (Ploidy)*
Spermatids	1.68	Haploid(n)
Liver	3.16	Diploid(2n)
Liver	6.30	Tetraploid(4n)
Liver	12.80	Octoploid(8n)

S. No	*Organism*	*Kidney*	*Liver*	*Erythrocytes*	*Sperm*
1.	Chicken	2.4	2.5	2.5	1.3
2.	Bovine	6.4	6.4	-	3.3
3.	Carp	-	3.0	3.3	1.6
4.	Human	5.4	5.6	2.5	2.5

The diploid amount of DNA is constant within the species but varies from one species to another, as shown in. Stability-Most of the macromolecules of the cell are constantly being broken down and synthesized. If this were to happen to a gene, valuable hereditary information would invariably be lost.

Of all the macromolecules in the cell, DNA is the most metabolically stable. Sensitivity to chemical and physical agents-All agents which cause genes to be damaged (mutation) have also been shown to induce changes in the structure of DNA. For example, one such mutagenic agent, ultraviolet light, is capable of producing breaks in the DNA molecule.

In 1950 Swift found that the amount of DNA per resting nucleus (interphase nucleus) in mitotically active tissues of animals ranged from twice (2n) the value found in the gametes (In) to four times (4n) this value. He concluded that DNA synthesis occurred during the interphase.

This parallelism of behaviour in DNA and chromosomes clearly indicates that DNA is the genetic material of higher animals.

Evidences for RNA as the Genetic Material of Some Viruses

The demonstration that RNA is the genetic material in RNA containing viruses came in 1956, when A. Gierer and G. Schramm showed that tobacco plants could be inoculated with purified RNA from the tobacco mosaic virus (TMV) and TMV -like lesions could later be identified on the tobacco leaves. A different approach was taken by H. Fraenkel Conrat and B. Singer in experiments published in 1957.

They first separated the RNA from the protein of TMV viruses, much as Hershey and Chase had done with the DNA and the protein of T2 bacteriophage. They then developed techniques for forming 'reconstituted' viruses containing the protein from one mutant strain of TMV and the RNA from another or vice-versa. Such hybrid viruses were allowed to infect tobacco leaves, and the progeny were examined. In all cases the progeny were the parental RNA type and not the parental protein type.

In a much improved technique N. Pace and S. Spiegelman in 1966 purified RNA from two different mutant strains of the RNA phage QB which had quite distinct base compositions. The isolated RNA'S were then incubated separately in the presence of an E. coli cell extract containing an enzyme capable of RNA replication. The new RNA synthesized was in each case identical in base composition to particular phage RNA presented to the in vitro system, thus indicating that the phage RNA-can .serve as a template for its self-replication.

Deoxyribonucleic Acid (DNA)

Deoxyribonucleic acid, or DNA, is a nucleic acid that contains the genetic instructions used in the development and functioning of all known living organisms (with the exception of RNA viruses). The main role of DNA molecules is the long-term storage of information. DNA is often compared to a set of blueprints, like a recipe or a code, since it contains the instructions needed to construct other components of cells, such as proteins and RNA molecules.

The DNA segments that carry this genetic information are called genes, but other DNA sequences have structural purposes, or are involved in regulating the use of this genetic information.

DNA consists of two long polymers of simple units called nucleotides, with backbones made of sugars and phosphate groups joined by ester bonds. These two strands run in opposite directions to each other and are therefore anti-parallel. Attached to each sugar is one of four types of molecules called bases. It is the sequence of these four bases along the backbone that encodes information. This information is read using the genetic code, which specifies the sequence of the amino acids within proteins. The code is read by copying stretches of DNA into the related nucleic acid RNA, in a process called transcription.

Within cells, DNA is organized into long structures called chromosomes. These chromosomes are duplicated before cells divide, in a process called DNA replication. Eukaryotic organisms (animals, plants, fungi, and protists) store most of their DNA inside the cell nucleus and some of their DNA in organelles, such as mitochondria or chloroplasts. In contrast, prokaryotes (bacteria and archaea) store their DNA only in the cytoplasm. Within the chromosomes, chromatin proteins such as histones compact and organize DNA. These compact structures guide the interactions between DNA and other proteins, helping control which parts of the DNA are transcribed.

Properties

DNA is a long polymer made from repeating units called nucleotides. As first discovered by James D. Watson and Francis Crick, the structure of DNA of all species comprises two helical chains each coiled round the same axis, and each with a pitch of 34 Ångströms (3.4 nanometres) and a radius of 10 Ångströms (1.0 nanometres). According to another study, when measured in a particular solution, the DNA chain measured 22 to 26 Ångströms wide (2.2 to 2.6 nanometres), and one nucleotide unit measured 3.3 Å (0.33 nm) long. Although each individual repeating unit is very small, DNA polymers can be very large molecules containing millions of nucleotides. For instance, the largest human chromosome, chromosome number 1, is approximately 220 million base pairs long.

In living organisms, DNA does not usually exist as a single molecule, but instead as a pair of molecules that are held tightly together. These two long strands entwine like vines, in the shape of a double helix. The nucleotide repeats contain both the segment of the backbone of the molecule, which holds the chain together, and a base, which interacts with the other DNA strand in the helix. A base linked to a sugar is called a nucleoside and a base linked to a sugar and one or more phosphate groups is called a nucleotide. If multiple nucleotides are linked together, as in DNA, this polymer is called a polynucleotide. The backbone of the DNA strand is made from alternating phosphate and sugar residues. The sugar in DNA is 2-deoxyribose, which is a pentose (five-carbon) sugar. The sugars are joined together by phosphate groups that form phosphodiester bonds between the third and fifth carbon atoms of adjacent sugar rings.

These asymmetric bonds mean a strand of DNA has a direction. In a double helix the direction of the nucleotides in one strand is opposite to their direction in the other strand: the strands are *antiparallel*. The asymmetric ends of DNA strands are called the 52 (*five prime*) and 32 (*three prime*) ends, with the 5' end having a terminal phosphate group and the 3' end a terminal hydroxyl group. One major difference between DNA and RNA is the sugar, with the 2-deoxyribose in DNA being replaced by the alternative pentose sugar ribose in RNA.

The DNA double helix is stabilized primarily by two forces: hydrogen bonds between nucleotides and base-stacking interactions among the aromatic bases. In the aqueous environment of the cell, the conjugated ð bonds of nucleotide bases align perpendicular to the axis of the DNA molecule, minimizing their interaction with the solvation shell and therefore, the Gibbs free energy.

The four bases found in DNA are adenine (abbreviated A), cytosine (C), guanine (G) and thymine (T). These four bases are attached to the sugar/ phosphate to form the complete nucleotide, as shown for adenosine monophosphate. These bases are classified into two types; adenine and guanine are fused five- and six-membered heterocyclic compounds called purines, while cytosine and thymine are six-membered rings called pyrimidines. A fifth pyrimidine base, called uracil (U), usually takes the place of thymine in RNA and differs from thymine by lacking a methyl group on its ring. Uracil is not usually found in DNA, occurring only as a breakdown product of cytosine. In addition to RNA and DNA, a large number of artificial nucleic acid analogues have also been created to study the proprieties of nucleic acids, or for use in biotechnology.

Grooves

Twin helical strands form the DNA backbone. Another double helix may be found by tracing the spaces, or grooves, between the strands. These voids are adjacent to the base pairs and may provide a binding site. As the strands are not directly opposite each other, the grooves are unequally sized. One groove, the major groove, is 22 Å wide and the other, the minor groove, is 12 Å wide. The narrowness of the minor groove means that the edges of the bases are more accessible in the major groove. As a result, proteins like transcription factors that can bind to specific sequences in double-stranded DNA usually make contacts to the sides of the bases exposed in the major groove. This situation varies in unusual conformations of DNA within the cell, but the major and minor grooves are always named to reflect the differences in size that would be seen if the DNA is twisted back into the ordinary B form.

Base Pair

In molecular biology and genetics, two nucleotides on opposite complementary DNA or RNA strands that are connected via hydrogen bonds are called a base pair (often abbreviated bp). In the canonical Watson-Crick DNA base pairing, adenine (A) forms a base pair with thymine (T) and guanine (G) forms a base pair with cytosine (C). In RNA, thymine is replaced by uracil (U). Alternate hydrogen bonding patterns, such as the wobble base pair and Hoogsteen base pair, also occur—in particular, in RNA—giving rise to complex and functional tertiary structures. It is important to note that pairing is the mechanism by which codons on messenger RNA molecules are recognized by anticodons on transfer RNA during protein translation. Some DNA- or RNA-binding enzymes can recognize specific base pairing patterns that identify particular regulatory regions of genes.

The size of an individual gene or an organism's entire genome is often measured in base pairs because DNA is usually double-stranded. Hence, the number of total base pairs is equal to the number of nucleotides in one of the strands (with the exception of non-coding single-stranded regions of telomeres). The haploid human genome (23 chromosomes) is estimated to be about 3 billion base pairs long and to contain 20,000–25,000 distinct genes. A kilobase (kb) is a unit of measurement in molecular biology equal to 1000 base pairs of DNA or RNA.

Hydrogen Bonding and Stability

Hydrogen bonding is the chemical interaction that underlies the base-pairing rules described above. Appropriate geometrical correspondence of hydrogen bond donors and acceptors allows only the "right" pairs to form stably. DNA with high GC-content is more stable than DNA with low GC-content, but, contrary to popular belief, the hydrogen bonds do not stabilize the DNA significantly, and stabilization is mainly due to stacking interactions.

The larger nucleobases, adenine and guanine, are members of a class of double-ringed chemical structures called purines; the smaller nucleobases, cytosine and thymine (and uracil), are members of a class of single-ringed chemical structures called pyrimidines. Purines are complementary only with pyrimidines: pyrimidine-pyrimidine pairings are energetically unfavorable because the molecules are too far apart for hydrogen bonding to be established; purine-purine pairings are energetically unfavorable because the molecules are too close, leading to overlap repulsion. The only other possible pairings are GT and AC; these pairings are mismatches because the pattern of hydrogen donors and acceptors do not correspond. The GU pairing, with two hydrogen bonds, does occur fairly often in RNA.

Paired DNA and RNA molecules are comparatively stable at room temperature but the two nucleotide strands will separate above a melting point that is determined by the length of the molecules, the extent of mispairing (if any), and the GC content. Higher GC content results in higher melting temperatures; it is, therefore, unsurprising that the genomes of extremophile organisms such as *Thermus thermophilus* are particularly GC-rich. On the converse, regions of a genome that need to separate frequently — for example, the promoter regions for often-transcribed genes — are comparatively GC-poor. GC content and melting temperature must also be taken into account when designing primers for PCR reactions.

Base Stacking

Base stacking interactions in DNA and RNA are due to dispersion attraction, short-range exchange repulsion, and electrostatic interactions, which also contribute to stability. Again, GC stacking interactions with adjacent bases tend to be more favorable. (Note, however, that a GC stacking interaction with the next base pair is geometrically different from a CG interaction.) Base stacking effects are especially important in the secondary structure and tertiary structure of RNA; for example, RNA stem-loop structures are stabilized by base stacking in the loop region.

Nucleic Acid Analogues

Nucleic acid analogues are compounds structurally similar (analog) to naturally occurring RNA and DNA, used in medicine and in molecular biology research. Nucleic acids are chains of nucleotides, which are composed of three parts: a phosphate backbone, a pucker-shaped pentose sugar, either ribose or deoxyribose, and one of four nucleobases. An analogue may have any of these altered. Typically the analogue nucleobases confer, among other things, different base pairing and base stacking proprieties. Examples include universal bases, which can pair with all four canon bases, and phosphate-sugar backbone analogues such as PNA, which affect the properties of the chain (PNA can even form a triple helix). Artificial nucleic acids include peptide nucleic acid (PNA), Morpholino and locked nucleic acid (LNA), as well as glycol nucleic acid (GNA) and threose nucleic acid (TNA). Each of these is distinguished from naturally-occurring DNA or RNA by changes to the backbone of the molecule.

Medicine

Several nucleoside analogues are used as antiviral or anticancer agents. The viral polymerase incorporates these compounds with non-canon bases.

These compounds are activated in the cells by being converted into nucleotides, they are administered as nucleosides since charged nucleotides cannot easily cross cell membranes.

Molecular Biology

Nucleic acid analogues are used in molecular biology for several purposes:

- As a tool to detect particular sequences
- As a tool with resistance to RNA hydrolysis
- As a tool for another purpose, such as sequencing
- Naturally occurring, such as in tRNA
- Investigation of the mechanisms used by enzyme, such as an Enzyme inhibitor
- Investigation of possible scenarios of the origin of life
- Investigation of the structural features of nucleic acids
- Investigation of the possible alternatives to the natural system in Synthetic biology.

Backbone Analogues

Hydrolysis Resistant RNA-analogues

To overcome the fact that ribose's 2' hydroxy group that reacts with the phosphate linked 3' hydroxy group (RNA is too unstable to be used or synthesized reliably), a ribose analogue is used. The most common RNA analogues are locked nucleic acid (LNA), morpholino, and peptide nucleic acid (PNA).

These oligonucleotides differ as they have a different backbone sugar but still bind according to Watson and Crick pairing with RNA or DNA, but are immune to nuclease activity (They generally cannot be enzymatically synthesized and can only be produced synthetically).

Other Notable Analogues Used as Tools

Dideoxynucleotides are used in sequencing. These nucleoside triphosphates possess a non-canonical sugar, dideoxyribose, which lacks the 3' hydroxyl group normally present in DNA and therefore cannot bond with the next base. The lack of the 3' hydroxyl group terminates the chain reaction as the DNA polymerases mistake it for a regular deoxyribonucleotide.

Another chain-terminating analogue that lacks a 3' hydroxyl and mimics adenosine is called cordycepin. Cordycepin is an anticancer drug that targets RNA replication. Another analogue in sequencing is a nucleobase analogue, 7-deaza-GTP and is used to sequence CG rich regions, instead 7-deaza-ATP is called tubercidin, an antibiotic.

Precursors to the RNA World

RNA may be too complex to be the first nucleic acid, so before the RNA world several simpler nucleic acids that differ in the backbone, such as TNA and GNA and PNA, have been offered as candidates for the first nucleic acids.

Base Analogues

Nucleobase Structure and Nomenclature

Natural bases are divided into two classes depending on their structure: pyrimidine (an heterocyclic aromatic six-membered ring with nitrogen atoms in position 1 and 3) and purine (a pyrimidine (numeration inverted) fused with an imidazole ring, a five-membered ring with 2 nitrogen atoms separated by one carbon (meta), 7,9). Their main proprieties are base pairing, resulting form 2 or 3 hydrogen bonds between ketone (electron withdrawing group,*ei.* more negatively charged) and amino (electron withdrawing group,*ei.* more positively charged) functional groups, and base stacking, caused by the attraction of the delocalized ð electron clouds of the aromatic ring structure.

Fluorophores

Commonly fluorophores (such as rhodamine or fluorescein) are linked to the ring linked to the sugar (in para) via a flexible arm, presumably extruding from the major groove of the helix. Due to low processivity of the nucleotides linked to bulky adducts such as florophores by taq polymerases, the sequence is typically copied using a nucleotide with an arm and later coupled with a reactive fluorophore (indirect labelling):

- amine reactive: Aminoallyl nucleotide contain a primary amine group on a linker that reacts with the amino-reactive dye such as a cyanine or Alexa Fluor dyes, which contain a reactive leaving group, such as a succinimidyl ester (NHS). (base-pairing amino groups are not affected).
- thiol reactive: thiol containing nucleotides reacts with the fluorophore linked to a reactive leaving group, such as a maleimide.
- biotin linked nucleotides rely on the same indirect labelling principle (+ fluorescent streptavidin) and are used in Affymetrix DNAchips.

Fluorophores find a variety of uses in medicine and biochemistry.

Fluorescent Base Analogues

The most commonly used and commercially available fluorescent base analogue, 2-aminopurine (2-AP), has a high-fluorescence quantum yield free in solution (0.68) that is considerably reduced (appr. 100 times but highly dependent on base sequence) when incorporated into nucleic acids. The emission sensitivity of 2-AP to immediate surroundings is shared by other promising and useful fluorescent base analogues like 3-MI, 6-MI, 6-MAP, pyrrolo-dC (also commercially available), modified and improved derivatives of pyrrolo-dC, furan-modified bases and many other ones. This sensitivity to the microenvironment have been utilized in studies of e.g. structure and dynamics within both DNA and RNA, dynamics and kinetics of DNA-protein interaction and electron transfer within DNA. A newly developed and very interesting group of fluorescent base analogues that has a fluorescence quantum yield that is nearly insensitive to their immediate surroundings is the tricyclic cytosine family. 1,3-Diaza-2-oxophenothiazine, tC, has a fluorescence quantum yield of approximately 0.2 both in single- and in double-strands irrespective of surrounding bases. Also the oxo-homologue of tC called tC^O (both commercially available), 1,3-diaza-2-oxophenoxazine, has a quantum yield of 0.2 in double-stranded systems. However, it is somewhat sensitive to surrounding bases in single-strands (quantum yields of 0.14–0.41). The high and stable quantum yields of these base analogues make them very bright, and, in combination with their good base analogue properties (leaves DNA structure and stability next to unperturbed), they are especially useful in fluorescence anisotropy and FRET measurements, areas where other fluorescent base analogues are less accurate. Also, in the same family of cytosine analogues, a FRET-acceptor base analogue, tC_{nitro}, has been developed. Together with tC^O as a FRET-donor this constitutes the first nucleic acid base analogue FRET-pair ever developed. The tC-family has, for example, been used in studies related to polymerase DNA-binding and DNA-polymerization mechanisms.

Natural Non-canon Bases

In a cell, there are several noncanon bases present: CpG islands in DNA (are often methylated), all eukaryotic mRNA (capped with a methyl-7-guanosine), and several bases of rRNAs (are methylated). Often, tRNAs are heavily modified postranscriptionally in order to improve their conformation or base pairing, in particular in/near the anticodon: inosine can base pair with C, U, and even with A, whereas thiouridine (with A) is more specific than uracil (with a purine). Other common tRNA base modifications are pseudouridine (which gives its name to the TØC loop), dihydrouridine (which does not stack as it is not aromatic), queuosine, wyosine, and so forth. Nevertheless these are all modifications to normal bases and are not placed by a polymerase.

Base-pairing

Canonical bases may have either a ketone or an amine group on the carbons surrounding the nitrogen atom furthest away from the glycosidic bond, which allows them to base pair (Watson-Crick base pairing) via hydrogen bonds (amine with ketone, purine with pyrimidine). Adenine and 2-aminoadenine have one/two amine group(s), whereas thymine has two ketone groups, and cytosine and guanine are mixed amine and ketone (inverted in respect to each other).

The precise reason why there are only four nucleotides is debated, but there are several unused possibilities. Furthermore, adenine is not the most stable choice for base pairing: in Cyanophage S-2L diaminopurine (DAP) is used instead of adenine (host evasion). Diaminopurine basepairs perfectly with thymine as it is identical to adenine but has an amine group at position 2 forming 3 intramolecular hydrogen bonds, eliminating the major difference between the two types of basepairs (Weak:A-T and Strong:C-G). This improved stability affects protein-binding ineractions that rely on those differences. Other combination include,

- isoguanine and isocytosine, which have their amine and ketone inverted compared to standard guanine and cytosine, (not used probably as tautomers are problematic for base pairing, but isoC and isoG can be amplified correctly with PCR even in the presence of the 4 canon bases)
- diaminopyrimidine and a xanthine, which bind like 2-aminoadenine and thymine but with inverted structures (not used as xanthine is a deamination product).

However, correct DNA structure can form even when the bases are not paired via hydrogen bonding; that is, the bases pair thanks to hydrophobicity, as studies have shown using DNA isosteres (analogues with same number of atoms), such as the thymine analogue 2,4-difluorotoluene (F) or the adenine analogue 4-methylbenzimidazole (Z). An alternative hydrophobic pair could be isoquinoline, and the pyrrolo[2,3-b]pyridine.

Other noteworthy basepairs:

- Several fluorescent bases have also been made, such as the 2-amino-6-(2-thienyl)purine and pyrrole-2-carbaldehyde base pair.
- Metal coordinated bases, such as two 2,6-bis(ethylthiomethyl)pyridine (SPy) with a silver ion or pyridine-2,6-dicarboxamide (Dipam) and a mondentate pyridine (Py) with a copper ion.
- Universal bases may pair indiscriminately with any other base, but, in general, lower the melting temperature of the sequence considerably; examples include 2'-deoxyinosine and hydrophobic aromatic non-hydrogen-bonding bases (strong stacking effects). These are used as proof of concept and, in general, are not utilised in degenerate primers (which are a mixture of primers).
- The numbers of possible base pairs is doubled when xDNA is considered. xDNA contains expanded bases, in which a benzene ring has been added, which may pair with canon bases, resulting in four possible base-pairs (8 bases:xA-T,xT-A,xC-G,xG-C, 16 bases if the unused arrangements are used). Another form of benzene added bases is yDNA, in which the base is widened by the benzene.

Metal Base Pairs

In metal base-pairing, the Watson-Crick hydrogen bonds are replaced by the interaction between a metal ion with nucleosides acting as ligands. The possible geometries of the metal that would allow for duplex formation with two bidentate nucleosides around a central metal atom are: tetrahedral, dodecahedral, and square planar. Metal-complexing with DNA can occur by the formation of non-canonical base pairs from natural nucleobases with participation by metal ions and also by the exchanging the hydrogen atoms that are part of the Watson-Crick base pairing by metal ions. Introduction of metal ions into a DNA duplex has shown to have potential magnetic, conducting properties, as well as increased stability.

Metal complexing has been shown to occur between natural nucleobases. A well-documented example is the formation of T-Hg-T, which involves two deportonated thymine nucleobases that are brought together by Hg^{2+} and forms a connected metal-base pair.

This motif does not accommodate stacked Hg^{2+} in a duplex due to an intrastrand hairpin formation process that is favored over duplex formation. Two thymines across from each other in a duplex do not form a Watson-Crick base pair in a duplex; this is an example where a Watson-Crick basepair mismatch is stabilized by the formation of the metal-base pair. Another example of a metal complexing to natural nucleobases is the formation of A-Zn-T and G-Zn-C at high pH; Co^{+2} and Ni^{+2} also form these complexes. These are Watson-Crick base pairs where the divalent cation in coordinated to the nucleobases. The exact binding is debated.

A large variety of artificial nucleobases have been developed for use as metal base pairs. These modified nucleobases exhibit tunable electronic properties, sizes, and binding affinities that can be optimized for a specific metal. For, example a nucleoside modified with a pyridine-2,6-dicarboxylate has shown to bind tightly to Cu^{2+}, whereas other divalent ions are only loosely bound. The tridentate character contributes to this selectivity.

The fourth coordination site on the copper is saturated by an oppositely arranged pyridine nucleobase. The asymmetric metal base pairing system is orthogonal to the Watson-Crick base pairs. Another example of an artificial nucleobase is that with hydroxypyridone nucleobases, which are able to bind Cu^{2+} inside the DNA duplex. Five consecutive copper-hydroxypyridone base pairs were incorporated into a double strand, which were flanked by only one natural nucleobase on both ends. EPR data showed that the distance between copper centers was estimated to be 3.7 ± 0.1 Å, while a natural B-type DNA duplex is only slightly larger (3.4 Å). The appeal for stacking metal ions inside a DNA duplex is the hope to obtain nanoscopic self-assembling metal wires, though this has not been realized yet.

Orthogonal System

It has been proposed and studied both theoretically and experimentally the possibility of implementing an orthogonal system inside cells independent of the cellular genetic material in order to make a completely safe system, with the possible increase in encoding potentials Several groups have focused on different aspects:

- novel backbones and base pairs as discussed above
- XNA replication/transcription polymerases starting generally from T7 RNA polymerase
- rybozomes (16S sequences with altered anti Shine-Dalgarno sequence allowing the translation of only orthogonal mRNA with a matching altered Shine-Dalgarno sequence)
- novel tRNA encoding non-natural aminoacids.

Base Analogs and Intercalators

Chemical analogs of nucleotides can take the place of proper nucleotides and establish non-canonical base-pairing, leading to errors (mostly point mutations) in DNA replication and DNA transcription. This is due to their isosteric chemistry. One common mutagenic base analog is 5-bromouracil, which resembles thymine but can base-pair to guanine in its enol form.

Other chemicals, known as DNA intercalators, fit into the gap between adjacent bases on a single strand and induce frameshift mutations by "masquerading" as a base, causing the DNA replication machinery to skip or insert additional nucleotides at the intercalated site.

Most intercalators are large polyaromatic compounds and are known or suspected carcinogens. Examples include ethidium bromide and acridine.

Examples

The following DNA sequences illustrate pair double-stranded patterns. By convention, the top strand is written from the 5' end to the 3' end; thus, the bottom strand is written 3' to 5'.

A base-paired DNA sequence:

ATCGATTGAGCTCTAGCG

TAGCTAACTCGAGATCGC

The corresponding RNA sequence, in which uracil is substituted for thymine where uracil takes its place in the RNA strand:

AUCGAUUGAGCUCUAGCG

UAGCUAACUCGAGAUCGC

Length Measurements

The following abbreviations are commonly used to describe the length of a D/RNA molecule:

- bp = base pair(s)—one bp corresponds to circa 3.4 Å of length along the strand
- kb (= kbp) = kilo base pairs = 1,000 bp
- Mb = mega base pairs = 1,000,000 bp
- Gb = giga base pairs = 1,000,000,000 bp.

In case of single stranded DNA/RNA units of nucleotides are used, abbreviated nt (or knt, Mnt, Gnt), as they are not paired. For distinction between units of computer storage and bases kbp, Mbp, Gbp, etc. may be used for basepairs. The length of 16S rDNA for bacteria is 1542 base-pairs in length.

The Centimorgan is also often used to imply distance along a chromosome, but the number of base-pairs it corresponds to varies widely. In the Human genome, it is about 1 million base pairs.

Base Pairing

Each type of base on one strand forms a bond with just one type of base on the other strand. This is called complementary base pairing. Here, purines form hydrogen bonds to pyrimidines, with A bonding only to T, and C bonding only to G. This arrangement of two nucleotides binding together across the double helix is called a base pair. As hydrogen bonds are not covalent, they can be broken and rejoined relatively easily. The two strands of DNA in a double helix can therefore be pulled apart like a zipper, either by a mechanical force or high temperature. As a result of this complementarity, all the information in the double-stranded sequence of a DNA helix is duplicated on each strand, which is vital in DNA replication. Indeed, this reversible and specific interaction between complementary base pairs is critical for all the functions of DNA in living organisms.

The two types of base pairs form different numbers of hydrogen bonds, AT forming two hydrogen bonds, and GC forming three hydrogen bonds. DNA with high GC-content is more stable than DNA with low GC-content, due to the added stability of an additional hydrogen bond.

As a result, it is both the percentage of GC base pairs and the overall length of a DNA double helix that determine the strength of the association between the two strands of DNA. Long DNA helices with a high GC content have stronger-interacting strands, while short helices with high AT content have weaker-interacting strands. In biology, parts of the DNA double helix that need to separate easily, such as the TATAAT Pribnow box in some promoters, tend to have a high AT content, making the strands easier to pull apart. In the laboratory, the strength of this interaction can be measured by finding the temperature required to break the hydrogen bonds, their melting temperature (also called T_m value). When all the base pairs in a DNA double helix melt, the strands separate and exist in solution as two entirely independent molecules. These single-stranded DNA molecules (*ssDNA*) have no single common shape, but some conformations are more stable than others.

Sense and Antisense

A DNA sequence is called "sense" if its sequence is the same as that of a messenger RNA copy that is translated into protein. The sequence on the opposite strand is called the "antisense" sequence. Both sense and antisense sequences can exist on different parts of the same strand of DNA (i.e. both strands contain both sense and antisense sequences). In both prokaryotes and eukaryotes, antisense RNA sequences are produced, but the functions of these RNAs are not entirely clear. One proposal is that antisense RNAs are involved in regulating gene expression through RNA-RNA base pairing.

A few DNA sequences in prokaryotes and eukaryotes, and more in plasmids and viruses, blur the distinction between sense and antisense strands by having overlapping genes. In these cases, some DNA sequences do double duty, encoding one protein when read along one strand, and a second protein when read in the opposite direction along the other strand. In bacteria, this overlap may be involved in the regulation of gene transcription, while in viruses, overlapping genes increase the amount of information that can be encoded within the small viral genome.

Supercoiling

DNA can be twisted like a rope in a process called DNA supercoiling. With DNA in its "relaxed" state, a strand usually circles the axis of the double helix once every 10.4 base pairs, but if the DNA is twisted the strands become more tightly or more loosely wound. If the DNA is twisted in the direction of the helix, this is positive supercoiling, and the bases are held more tightly together. If they are twisted in the opposite direction, this is negative supercoiling, and the bases come apart more easily. In nature, most DNA has slight negative supercoiling that is introduced by enzymes called topoisomerases. These enzymes are also needed to relieve the twisting stresses introduced into DNA strands during processes such as transcription and DNA replication.

Alternate DNA Structures

DNA exists in many possible conformations that include A-DNA, B-DNA, and Z-DNA forms, although, only B-DNA and Z-DNA have been directly observed in functional organisms. The conformation that DNA adopts depends on the hydration level, DNA sequence, the amount and direction of supercoiling, chemical modifications of the bases, the type and concentration of metal ions, as well as the presence of polyamines in solution. The first published reports of A-DNA X-ray diffraction patterns— and also B-DNA used analyses based on Patterson transforms that provided only a limited amount of structural information for oriented fibers of DNA. An alternate analysis was then proposed by Wilkins *et al.*, in 1953, for the *in vivo* B-DNA X-ray diffraction/scattering patterns of highly hydrated DNA fibers in terms of squares of Bessel functions. In the same journal, James D. Watson and Francis Crick presented their molecular modeling analysis of the DNA X-ray diffraction patterns to suggest that the structure was a double-helix.

Although the 'B-DNA form' is most common under the conditions found in cells, it is not a well-defined conformation but a family of related DNA conformations that occur at the high hydration levels present in living cells. Their corresponding X-ray diffraction and scattering patterns are characteristic of molecular paracrystals with a significant degree of disorder.

Compared to B-DNA, the A-DNA form is a wider right-handed spiral, with a shallow, wide minor groove and a narrower, deeper major groove. The A form occurs under non-physiological conditions in partially dehydrated samples of DNA, while in the cell it may be produced in hybrid pairings of DNA and RNA strands, as well as in enzyme-DNA complexes. Segments of DNA where the bases have been chemically modified by methylation may undergo a larger change in conformation and adopt the Z form. Here, the strands turn about the helical axis in a left-handed spiral, the opposite of the more common B form. These unusual structures can be recognized by specific Z-DNA binding proteins and may be involved in the regulation of transcription.

Alternate DNA Chemistry

For a number of years exobiologists have proposed the existence of a shadow biosphere, a postulated microbial biosphere of Earth that uses radically different biochemical and molecular processes than currently known life. One of the proposals was the existence of lifeforms that use arsenic instead of phosphorus in DNA.

A December 2010 NASA press conference revealed that the bacterium GFAJ-1, which has evolved in an arsenic-rich environment, is the first terrestrial lifeform found which may have this ability. The bacterium was found in Mono Lake, east of Yosemite National Park. GFAJ-1 is a rod-shaped extremophile bacterium in the family Halomonadaceae that, when starved of phosphorus, may be capable of incorporating the usually poisonous element arsenic in its DNA. This discovery lends weight to the long-standing idea that extraterrestrial life could have a different chemical makeup from life on Earth. The research was carried out by a team led by Felisa Wolfe-Simon, a geomicrobiologist and geobiochemist, a Postdoctoral Fellow of the NASA Astrobiology Institute with Arizona State University.

Quadruplex Structures

At the ends of the linear chromosomes are specialized regions of DNA called telomeres. The main function of these regions is to allow the cell to replicate chromosome ends using the enzyme telomerase, as the enzymes that normally replicate DNA cannot copy the extreme 32 ends of chromosomes. These specialized chromosome caps also help protect the DNA ends, and stop the DNA repair systems in the cell from treating them as damage to be corrected. In human cells, telomeres are usually lengths of single-stranded DNA containing several thousand repeats of a simple TTAGGG sequence.

These guanine-rich sequences may stabilize chromosome ends by forming structures of stacked sets of four-base units, rather than the usual base pairs found in other DNA molecules. Here, four guanine bases form a flat plate and these flat four-base units then stack on top of each other, to form a stable *G-quadruplex* structure.

These structures are stabilized by hydrogen bonding between the edges of the bases and chelation of a metal ion in the centre of each four-base unit. Other structures can also be formed, with the central set of four bases coming from either a single strand folded around the bases, or several different parallel strands, each contributing one base to the central structure.

In addition to these stacked structures, telomeres also form large loop structures called telomere loops, or T-loops. Here, the single-stranded DNA curls around in a long circle stabilized by telomere-binding proteins.

At the very end of the T-loop, the single-stranded telomere DNA is held onto a region of double-stranded DNA by the telomere strand disrupting the double-helical DNA and base pairing to one of the two strands. This triple-stranded structure is called a displacement loop or D-loop.

Branched DNA can form networks containing multiple branches.

Branched DNA

In DNA fraying occurs when non-complementary regions exist at the end of an otherwise complementary double-strand of DNA. However, branched DNA can occur if a third strand of DNA is introduced and contains adjoining regions able to hybridize with the frayed regions of the pre-existing double-strand. Although the simplest example of branched DNA involves only three strands of DNA, complexes involving additional strands and multiple branches are also possible. Branched DNA can be used in nanotechnology to construct geometric shapes.

Vibration

DNA may carry out low-frequency collective motion as observed by the Raman spectroscopy and analysed with a quasi-continuum model.

Chemical Modifications

Structure of cytosine with and without the 5-methyl group. Deamination converts 5-methylcytosine into thymine.

Base Modifications

The expression of genes is influenced by how the DNA is packaged in chromosomes, in a structure called chromatin. Base modifications can be involved in packaging, with regions that have low or no gene expression usually containing high levels of methylation of cytosine bases. For example, cytosine methylation, produces 5-methylcytosine, which is important for X-chromosome inactivation.

The average level of methylation varies between organisms - the worm *Caenorhabditis elegans* lacks cytosine methylation, while vertebrates have higher levels, with up to 1% of their DNA containing 5-methylcytosine. Despite the importance of 5-methylcytosine, it can deaminate to leave a thymine base, so methylated cytosines are particularly prone to mutations. Other base modifications include adenine methylation in bacteria, the presence of 5-hydroxymethylcytosine in the brain, and the glycosylation of uracil to produce the "J-base" in kinetoplastids.

Damage

DNA can be damaged by many sorts of mutagens, which change the DNA sequence. Mutagens include oxidizing agents, alkylating agents and also high-energy electromagnetic radiation such as ultraviolet light and X-rays. The type of DNA damage produced depends on the type of mutagen.

For example, UV light can damage DNA by producing thymine dimers, which are cross-links between pyrimidine bases. On the other hand, oxidants such as free radicals or hydrogen peroxide produce multiple forms of damage, including base modifications, particularly of guanosine, and double-strand breaks. A typical human cell contains about 150,000 bases that have suffered oxidative damage. Of these oxidative lesions, the most dangerous are double-strand breaks, as these are difficult to repair and can produce point mutations, insertions and deletions from the DNA sequence, as well as chromosomal translocations.

Many mutagens fit into the space between two adjacent base pairs, this is called *intercalation*. Most intercalators are aromatic and planar molecules; examples include ethidium bromide, daunomycin, and doxorubicin. In order for an intercalator to fit between base pairs, the bases must separate, distorting the DNA strands by unwinding of the double helix. This inhibits both transcription and DNA replication, causing toxicity and mutations. As a result, DNA intercalators are often carcinogens, and benzo[*a*]pyrene diol epoxide, acridines, aflatoxin and ethidium bromide are well-known examples. Nevertheless, due to their ability to inhibit DNA transcription and replication, other similar toxins are also used in chemotherapy to inhibit rapidly growing cancer cells.

Biological Functions

DNA usually occurs as linear chromosomes in eukaryotes, and circular chromosomes in prokaryotes. The set of chromosomes in a cell makes up its genome; the human genome has approximately 3 billion base pairs of DNA arranged into 46 chromosomes. The information carried by DNA is held in the sequence of pieces of DNA called genes. Transmission of genetic information in genes is achieved via complementary base pairing.

For example, in transcription, when a cell uses the information in a gene, the DNA sequence is copied into a complementary RNA sequence through the attraction between the DNA and the correct RNA nucleotides. Usually, this RNA copy is then used to make a matching protein sequence in a process called translation, which depends on the same interaction between RNA nucleotides. In alternative fashion, a cell may simply copy its genetic information in a process called DNA replication. The details of these functions are covered in other articles; here we focus on the interactions between DNA and other molecules that mediate the function of the genome.

Genes and Genomes

Genomic DNA is tightly and orderly packed in the process called DNA condensation to fit the small available volumes of the cell. In eukaryotes, DNA is located in the cell nucleus, as well as small amounts in mitochondria and chloroplasts. In prokaryotes, the DNA is held within an irregularly shaped body in the cytoplasm called the nucleoid. The genetic information in a genome is held within genes, and the complete set of this information in an organism is called its genotype. A gene is a unit of heredity and is a region of DNA that influences a particular characteristic in an organism. Genes contain an open reading frame that can be transcribed, as well as regulatory sequences such as promoters and enhancers, which control the transcription of the open reading frame.

In many species, only a small fraction of the total sequence of the genome encodes protein. For example, only about 1.5% of the human genome consists of protein-coding exons, with over 50% of human DNA consisting of non-coding repetitive sequences. The reasons for the presence of so much noncoding DNA in eukaryotic genomes and the extraordinary differences in genome size, or *C-value*, among species represent a long-standing puzzle known as the "C-value enigma". However, DNA sequences that do not code protein may still encode functional non-coding RNA molecules, which are involved in the regulation of gene expression. Some noncoding DNA sequences play structural roles in chromosomes. Telomeres and centromeres typically contain few genes, but are important for the function and stability of chromosomes. An abundant form of noncoding DNA in humans are pseudogenes, which are copies of genes that have been disabled by mutation. These sequences are usually just molecular fossils, although they can occasionally serve as raw genetic material for the creation of new genes through the process of gene duplication and divergence.

Transcription and Translation

A gene is a sequence of DNA that contains genetic information and can influence the phenotype of an organism. Within a gene, the sequence of bases along a DNA strand defines a messenger RNA sequence, which then defines one or more protein sequences. The relationship between the nucleotide sequences of genes and the amino-acid sequences of proteins is determined by the rules of translation, known collectively as the genetic code. The genetic code consists of three-letter 'words' called *codons* formed from a sequence of three nucleotides (e.g. ACT, CAG, TTT).

In transcription, the codons of a gene are copied into messenger RNA by RNA polymerase. This RNA copy is then decoded by a ribosome that reads the RNA sequence by base-pairing the messenger RNA to transfer RNA, which carries amino acids. Since there are 4 bases in 3-letter combinations, there are 64 possible codons (4^3 combinations). These encode the twenty standard amino acids, giving most amino acids more than one possible codon. There are also three 'stop' or 'nonsense' codons signifying the end of the coding region; these are the TAA, TGA and TAG codons.

Replication

Cell division is essential for an organism to grow, but, when a cell divides, it must replicate the DNA in its genome so that the two daughter cells have the same genetic information as their parent. The double-stranded structure of DNA provides a simple mechanism for DNA replication. Here, the two strands are separated and then each strand's complementary DNA sequence is recreated by an enzyme called DNA polymerase. This enzyme makes the complementary strand by finding the correct base through complementary base pairing, and bonding it onto the original strand. As DNA polymerases can only extend a DNA strand in a 52 to 32 direction, different mechanisms are used to copy the antiparallel strands of the double helix. In this way, the base on the old strand dictates which base appears on the new strand, and the cell ends up with a perfect copy of its DNA.

Interactions with Proteins

All the functions of DNA depend on interactions with proteins. These protein interactions can be non-specific, or the protein can bind specifically to a single DNA sequence. Enzymes can also bind to DNA and of these, the polymerases that copy the DNA base sequence in transcription and DNA replication are particularly important.

DNA-binding Proteins

Structural proteins that bind DNA are well-understood examples of non-specific DNA-protein interactions. Within chromosomes, DNA is held in complexes with structural proteins. These proteins organize the DNA into a compact structure called chromatin. In eukaryotes this structure involves DNA binding to a complex of small basic proteins called histones, while in prokaryotes multiple types of proteins are involved. The histones form a disk-shaped complex called a nucleosome, which contains two complete turns of double-stranded DNA wrapped around its surface.

These non-specific interactions are formed through basic residues in the histones making ionic bonds to the acidic sugar-phosphate backbone of the DNA, and are therefore largely independent of the base sequence. Chemical modifications of these basic amino acid residues include methylation, phosphorylation and acetylation. These chemical changes alter the strength of the interaction between the DNA and the histones, making the DNA more or less accessible to transcription factors and changing the rate of transcription. Other non-specific DNA-binding proteins in chromatin include the high-mobility group proteins, which bind to bent or distorted DNA. These proteins are important in bending arrays of nucleosomes and arranging them into the larger structures that make up chromosomes.

A distinct group of DNA-binding proteins are the DNA-binding proteins that specifically bind single-stranded DNA. In humans, replication protein A is the best-understood member of this family and is used in processes where the double helix is separated, including DNA replication, recombination and DNA repair. These binding proteins seem to stabilize single-stranded DNA and protect it from forming stem-loops or being degraded by nucleases. In contrast, other proteins have evolved to bind to particular DNA sequences.

The most intensively studied of these are the various transcription factors, which are proteins that regulate transcription. Each transcription factor binds to one particular set of DNA sequences and activates or inhibits the transcription of genes that have these sequences close to their promoters. The transcription factors do this in two ways. Firstly, they can bind the RNA polymerase responsible for transcription, either directly or through other mediator proteins; this locates the polymerase at the promoter and allows it to begin transcription. Alternatively, transcription factors can bind enzymes that modify the histones at the promoter; this will change the accessibility of the DNA template to the polymerase. As these DNA targets can occur throughout an organism's genome, changes in the activity of one type of transcription factor can affect thousands of genes. Consequently, these proteins are often the targets of the signal transduction processes that control responses to environmental changes or cellular differentiation and development. The specificity of these transcription factors' interactions with DNA come from the proteins making multiple contacts to the edges of the DNA bases, allowing them to "read" the DNA sequence. Most of these base-interactions are made in the major groove, where the bases are most accessible.

DNA-modifying Enzymes

Nucleases and Ligases

Nucleases are enzymes that cut DNA strands by catalyzing the hydrolysis of the phosphodiester bonds. Nucleases that hydrolyse nucleotides from the ends of DNA strands are called exonucleases, while endonucleases cut within strands. The most frequently used nucleases in molecular biology are the restriction endonucleases, which cut DNA at specific sequences.

For instance, the EcoRV enzyme shown to the left recognizes the 6-base sequence 52 -GAT|ATC-32 and makes a cut at the vertical line. In nature, these enzymes protect bacteria against phage infection by digesting the phage DNA when it enters the bacterial cell, acting as part of the restriction modification system. In technology, these sequence-specific nucleases are used in molecular cloning and DNA fingerprinting.

Enzymes called DNA ligases can rejoin cut or broken DNA strands. Ligases are particularly important in lagging strand DNA replication, as they join together the short segments of DNA produced at the replication fork into a complete copy of the DNA template. They are also used in DNA repair and genetic recombination.

Topoisomerases and Helicases

Topoisomerases are enzymes with both nuclease and ligase activity. These proteins change the amount of supercoiling in DNA. Some of these enzymes work by cutting the DNA helix and allowing one section to rotate, thereby reducing its level of supercoiling; the enzyme then seals the DNA break. Other types of these enzymes are capable of cutting one DNA helix and then passing a second strand of DNA through this break, before rejoining the helix. Topoisomerases are required for many processes involving DNA, such as DNA replication and transcription.

Helicases are proteins that are a type of molecular motor. They use the chemical energy in nucleoside triphosphates, predominantly ATP, to break hydrogen bonds between bases and unwind the DNA double helix into single strands. These enzymes are essential for most processes where enzymes need to access the DNA bases.

Polymerases

Polymerases are enzymes that synthesize polynucleotide chains from nucleoside triphosphates. The sequence of their products are copies of existing polynucleotide chains - which are called *templates*. These enzymes function by adding nucleotides onto the 32 hydroxyl group of the previous nucleotide in a DNA strand. As a consequence, all polymerases work in a 52 to 32 direction. In the active site of these enzymes, the incoming nucleoside triphosphate base-pairs to the template: this allows polymerases to accurately synthesize the complementary strand of their template. Polymerases are classified according to the type of template that they use.

In DNA replication, a DNA-dependent DNA polymerase makes a copy of a DNA sequence. Accuracy is vital in this process, so many of these polymerases have a proofreading activity. Here, the polymerase recognizes the occasional mistakes in the synthesis reaction by the lack of base pairing between the mismatched nucleotides. If a mismatch is detected, a 32 to 52 exonuclease activity is activated and the incorrect base removed. In most organisms, DNA polymerases function in a large complex called the replisome that contains multiple accessory subunits, such as the DNA clamp or helicases.

RNA-dependent DNA polymerases are a specialized class of polymerases that copy the sequence of an RNA strand into DNA. They include reverse transcriptase, which is a viral enzyme involved in the infection of cells by retroviruses, and telomerase, which is required for the replication of telomeres. Telomerase is an unusual polymerase because it contains its own RNA template as part of its structure.

Transcription is carried out by a DNA-dependent RNA polymerase that copies the sequence of a DNA strand into RNA. To begin transcribing a gene, the RNA polymerase binds to a sequence of DNA called a promoter and separates the DNA strands. It then copies the gene sequence into a messenger RNA transcript until it reaches a region of DNA called the terminator, where it halts and detaches from the DNA. As with human DNA-dependent DNA polymerases, RNA polymerase II, the enzyme that transcribes most of the genes in the human genome, operates as part of a large protein complex with multiple regulatory and accessory subunits.

Genetic Recombination

A DNA helix usually does not interact with other segments of DNA, and in human cells the different chromosomes even occupy separate areas in the nucleus called "chromosome territories". This physical separation of different chromosomes is important for the ability of DNA to function as a stable repository for information, as one of the few times chromosomes interact is during chromosomal crossover when they recombine. Chromosomal crossover is when two DNA helices break, swap a section and then rejoin. Recombination allows chromosomes to exchange genetic information and produces new combinations of genes, which increases the efficiency of natural selection and can be important in the rapid evolution of new proteins. Genetic recombination can also be involved in DNA repair, particularly in the cell's response to double-strand breaks.

The most common form of chromosomal crossover is homologous recombination, where the two chromosomes involved share very similar sequences. Non-homologous recombination can be damaging to cells, as it can produce chromosomal translocations and genetic abnormalities.

The recombination reaction is catalyzed by enzymes known as recombinases, such as RAD51. The first step in recombination is a double-stranded break either caused by an endonuclease or damage to the DNA. A series of steps catalyzed in part by the recombinase then leads to joining of the two helices by at least one Holliday junction, in which a segment of a single strand in each helix is annealed to the complementary strand in the other helix. The Holliday junction is a tetrahedral junction structure that can be moved along the pair of chromosomes, swapping one strand for another. The recombination reaction is then halted by cleavage of the junction and re-ligation of the released DNA.

Evolution

DNA contains the genetic information that allows all modern living things to function, grow and reproduce. However, it is unclear how long in the 4-billion-year history of life DNA has performed this function, as it has been proposed that the earliest forms of life may have used RNA as their genetic material. RNA may have acted as the central part of early cell metabolism as it can both transmit genetic information and carry out catalysis as part of ribozymes.

This ancient RNA world where nucleic acid would have been used for both catalysis and genetics may have influenced the evolution of the current genetic code based on four nucleotide bases. This would occur, since the number of different bases in such an organism is a trade-off between a small number of bases increasing replication accuracy and a large number of bases increasing the catalytic efficiency of ribozymes.

However, there is no direct evidence of ancient genetic systems, as recovery of DNA from most fossils is impossible. This is because DNA will survive in the environment for less than one million years and slowly degrades into short fragments in solution. Claims for older DNA have been made, most notably a report of the isolation of a viable bacterium from a salt crystal 250 million years old, but these claims are controversial.

Uses in Technology

Genetic Engineering

Methods have been developed to purify DNA from organisms, such as phenol-chloroform extraction, and to manipulate it in the laboratory, such as restriction digests and the polymerase chain reaction. Modern biology and biochemistry make intensive use of these techniques in recombinant DNA technology. Recombinant DNA is a man-made DNA sequence that has been assembled from other DNA sequences.

They can be transformed into organisms in the form of plasmids or in the appropriate format, by using a viral vector. The genetically modified organisms produced can be used to produce products such as recombinant proteins, used in medical research, or be grown in agriculture.

Forensics

Forensic scientists can use DNA in blood, semen, skin, saliva or hair found at a crime scene to identify a matching DNA of an individual, such as a perpetrator. This process is formally termed DNA profiling, but may also be called "genetic fingerprinting". In DNA profiling, the lengths of variable sections of repetitive DNA, such as short tandem repeats and minisatellites, are compared between people. This method is usually an extremely reliable technique for identifying a matching DNA. However, identification can be complicated if the scene is contaminated with DNA from several people. DNA profiling was developed in 1984 by British geneticist Sir Alec Jeffreys, and first used in forensic science to convict Colin Pitchfork in the 1988 Enderby murders case.

People convicted of certain types of crimes may be required to provide a sample of DNA for a database. This has helped investigators solve old cases where only a DNA sample was obtained from the scene. DNA profiling can also be used to identify victims of mass casualty incidents. On the other hand, many convicted people have been released from prison on the basis of DNA techniques, which were not available when a crime had originally been committed.

Bioinformatics

Bioinformatics involves the manipulation, searching, and data mining of biological data, and this includes DNA sequence data. The development of techniques to store and search DNA sequences have led to widely applied advances in computer science, especially string searching algorithms, machine learning and database theory. String searching or matching algorithms, which find an occurrence of a sequence of letters inside a larger sequence of letters, were developed to search for specific sequences of nucleotides.

The DNA sequenced may be aligned with other DNA sequences to identify homologous sequences and locate the specific mutations that make them distinct. These techniques, especially multiple sequence alignment, are used in studying phylogenetic relationships and protein function. Data sets representing entire genomes' worth of DNA sequences, such as those produced by the Human Genome Project, are difficult to use without the annotations that identify the locations of genes and regulatory elements on each chromosome. Regions of DNA sequence that have the characteristic patterns associated with protein- or RNA-coding genes can be identified by gene finding algorithms, which allow researchers to predict the presence of particular gene products and their possible functions in an organism even before they have been isolated experimentally. Entire genomes may also be compared which can shed light on the evolutionary history of particular organism and permit the examination of complex evolutionary events.

DNA Nanotechnology

DNA nanotechnology uses the unique molecular recognition properties of DNA and other nucleic acids to create self-assembling branched DNA complexes with useful properties. DNA is thus used as a structural material rather than as a carrier of biological information. This has led to the creation of two-dimensional periodic lattices (both tile-based as well as using the "DNA origami" method) as well as three-dimensional structures in the shapes of polyhedra. Nanomechanical devices and algorithmic self-assembly have also been demonstrated, and these DNA structures have been used to template the arrangement of other molecules such as gold nanoparticles and streptavidin proteins.

History and Anthropology

Because DNA collects mutations over time, which are then inherited, it contains historical information, and, by comparing DNA sequences, geneticists can infer the evolutionary history of organisms, their phylogeny. This field of phylogenetics is a powerful tool in evolutionary biology. If DNA sequences within a species are compared, population geneticists can learn the history of particular populations. This can be used in studies ranging from ecological genetics to anthropology; For example, DNA evidence is being used to try to identify the Ten Lost Tribes of Israel.

DNA has also been used to look at modern family relationships, such as establishing family relationships between the descendants of Sally Hemings and Thomas Jefferson. This usage is closely related to the use of DNA in criminal investigations detailed above. Indeed, some criminal investigations have been solved when DNA from crime scenes has matched relatives of the guilty individual.

4

The Chemical Basis of Life: Atoms and Molecules

All matter is made up of tiny particles called *atoms*. Atoms are the basic building blocks of all mater, including life. They cannot be broken down by any known means.

There are only about 100 kinds of atoms but they can be chemically combined to make thousands of different kinds of substances. The substances are called *elements* if they are made up of only one kind of atom. Hydrogen gas is an example of an element because there are two atoms of Hydrogen joined together. Elements are pure sunstances that cannot be broken down by any oridnay means. Unlike solutions, elements are made up of only one kind of atom with the same atomic number.

For example, the element Nitrogen consists entirely of Nitrogen atoms. There are over 100 different kinds of elements known today. These elements are organized into different families and different periods in the periodic table. They can be in the form of any state, solid, liquid, or a gas. Compounds are made up of two or more different *elements* put together in definite proportions.

For example, water is a compound because it contains two atoms of Hydrogen for every one atom of Oxygen. A molecule is like a compound in that it is two or more atoms combined, but each atom when combined has to act as a single particle.

If a sodium atom passes an electron to a chlorine atom then the chlorine atom has 17 protons and 18 electrons. An atom that has an excess charge is called an ion.

Compounds

Most substances are compounds. A compound is made up of two or more different kinds of atoms combined in definite proportions. Compounds can be separated into the individual atoms that compose their composition, unlike elements that cannot. Compounds can be seperated only by a chemical change that involves heat or electrcity. For example, Water can be separated int Hydrogen and Oxygen by applying a sufficient DC voltage.

Ionic Bonding

Ionic bonding is the bond between two atoms formed as a result of the transfer of electrons from one atom to the other. This results in the formation of positive and negative ions. IONS - Any atom which has an excess charge is called an ion, and the force of attraction between two ions is called an ionic bond. Example: When a sodium atom gives up an electron, it will have 11 protons and 10 electrons. This creates a sodium ion which has an excess positive charge of 1 unit. Relate the concepts of solutions, mixtures, acids and bases to the composition and environment of cells.

A solution is a type of mixture that has the same composition throughtout, and appears to be all one substance. In other words, it's a homogeneous mixture that can be in the form of a solid, liquid or gas. The make up of cells consist of many types of mixtures, including solutions. Many of them are in the form of a liquid. A solution is formed by mixing one or more substances into another substance. In any solution, there is a solute and a solvent. The solvent is generally a liquid, and it is the substance in which the solute is dissolved in. A mixture is a word used to describe a blend of different substances. A mixture is made up of two or more substances, in which they can all be visually seperated and distinguished from one another.

Solutions

A solution is any homogeneous mixture, and may include solids, liquids, as well as gases. Most solutions important to living cells are thought of as liquids. The liquid substance that makes up the bulk of a solution is known as a solvent. Substances that dissolve in the solvent are called solutes (SAHL yoots). A solute can either be a solid, liquid, or gas before they are dissolved in the solvent. The most common solutions have water as the solvent.

Mixtures

A mixture is a combination of substances that are physically mixed without forming new chemical bonds. The substances in a mixture may be present in any proportions. The proportions can change as one substance may be added to, or removed from the mixture, but the different substances in the mixture retain their usual properties.

Acids

An acid is any compound that produces hydrogen ions in a solution. All acids consist of molecules in which contain hydrogen. The molecules are covalently bonded to another atom or group of atoms. When these molecules dissociate in a solution, such as water, the hydrogen breaks loose as an hydrogen ion. The left over molecules form what is called a negative ion. Hydrochloric acid is an example of an acid, which is important to life activities. This is because the level of acid in the environment of the cell affects the actions of enzymes and therefore affects the functioning of the cell.

Base

A base is any compound that produces excess hydroxide ions when it is dissolved in a solution. When in a dry state many bases are ionic compounds. A perfect example of a base is sodium hydroxide. Sodium hydroxide is a solid compound, which is made up of sodium and hydroxide ions. When a base is dissolved in water, it dissociates (separates) into its ions:

$NaOH \longrightarrow Na^{+} + OH^{-}$

An acid is any compound that produces hydrogen ions in solution. All acids consist of molecules that contain hydrogen covalently bonded to another atom or group of atoms. A base is a compound that produces hydroxide(hy DRAHK syd) ions (OH-) when disolved in water. In the dry slate, many bases are ionic compounds. Sodium hydroxide (NaOH) is one example. This is a solid compound made of sodium and hydroxide ions. When dissolved in water it separates in to it's ions. NaOH then dissociates to form Na+ and OH- ions.

When acid and a base react their hydrogen and hydroxide combine to form water molecules. This reaction removes these two ions from the solution. that is, they are no longer present as separate ions. However, the negative ions of the acid and the positive ions of the base are still present.

Identify the importance of water to life by relating its functions as a transport and dissolving medium. Water is very important to life. Some scientists beleived that life probaly evolved in water. All living cells are made up of 70-95% water and the water covers 75% of the earths surface. Water is present in all three physical states of mater - solid, liquid, and gas.

Water gets its properties from the fact that it is a polar covalent molecule, and therefore forms hydrogen bonds that connects one water molecule to as many as four other water molecules. This results in a property known as cohesion. Water also has adhesive properties to ionic substances as well as other charged substances such as clay. These properties result in many of water's important properties which include its remarkable ability to dissolve a vast variety of covalent and ionic substances, its high heat capacity, capillarity, surface tension, high heat of vaporization, and the fact that water floats when it freezes. All of these properties of water are related to life on Earth. Water is considered the universal solvent of life. It acts as a solvent which transports molecules and ions from one area of a cell to another. It also acts as a reactant in chemical reactions that occur in the cell. Without water, molecules and ions would be unable to diffuse through the membrane that surrounds the cell, and gases such as carbon dioxide and oxygen would be unable to pass through plasma membranes.

Describe the Characteristics of Organic Compounds

Organic compounds contain carbon atoms that are linked by covalent bonds to form molecular chains, branches, rings, or sheets. Because there are 4 electrons in the outer energy level of the carbon atom, carbon can combine with other carbon atoms, as well as to the many other biologically important elements such as hydrogen, nitrogen, oxygen, sodium, magnesium, phosphorus, sulfur, chlorine, potassium, calcium, and iron. In addition to carbon, the vast majority of organic compounds also contain hydrogen. Oxides, carbonates and cyanides also contain carbon atoms but are not considered to be organic. Such compounds as carbon monoxide, carbon dioxide, calcium carbonate, and cyanide are considered to be inorganic, even though they contain carbon atoms.

Make an analysis of carbohydrates, lipids, proteins, and nucleic acids in terms of: the atoms that compose these compounds, the function these compounds serve in the body, the food sources of each compound, the sub-units that chemically combine to make these compounds, and examples of each group.

The first compound is carbohydrates. Carbohydrates are sugars and starches. Carbohydrates are found in breads, cereals, potatoes, pastas etc. Lipids is the next compound. Lipids are fats, oils and waxes. Lipids are found in mainly fried foods such as pizza, fries, etc. Protein is the next compound. Proteins are compounds that contain nitrogen as well as carbon, hydrogen and oxygen. Proteins are found in beans, fish, and meats. The final compound is nucleic acids. Nucleic acids contain phosphorous and nitrogen along with carbon, hydrogen and oxygen. Nucleic acids are mainly found in meats

Carbohydrates are compunds composed of carbon, hydrogen and oxygen. The name carbohydrates comes from the fact that they are formed from the union of carbon dioxide and water during the process of photosynthesis. The ratio of hydrogen to oxygen in a carbohydrate is always two to one - two hydrogen atoms for every oxygen atom. Carbohydrates are normally foun in the breads and cereals as well as fruits and vegetables.

The simplest carbohydrates are just simple sugars (called monosaccharides). These sugars are very important to the body, because they contain large amounts of energy. This energy can be released in the presence of oxygen by breaking down the sugars into carbon dioxide and water. Lipids include subtances commonly known as fats, waxes and oils. Like carbohydrates, lipids are composed of carbon, hydrogen and oxygen, except there is less oxygen in lipids, than in carbohydrates. Lipids help make up parts of the cell structure (such as the cell membranes) and are used as a reserve for energy in an organism. Lipids provide twice as much energy per gram as the same amount of carbohydrate. Lipids are found in fatty meats such as bacon, as well as products such as whole milk, cream and butter, and from the oils of many plants such as olives, corn, and peanuts.

Proteins are compuounds containing nitrogen, carbon, hydrogen & oxygen. Some proteins also contain sulfur and phosphorus. The number of proteins in your body is virtually countless. Proteins are found everywhere, in cartilage, bones and muscles, hormones, antibodies, and even enzymes. Proteins are notmally found in meats, poltry, fish and some plants such as beans, soy beans, and legumes.

Nucleic Acids are organic compounds that contain Phosphorus, and Nitrogen, as well as Hydrogen, Oxygen and Carbon. Ther are only two known types of nucleic acids, and they are DNA (Deoxyribonucleic Acid) and RNA (Ribonucleic Acid). These two substances were first found in the Nucleus of a Cell. DNA (which is hereditary) and RNA direct and control the activities and structural devolpment of all the cells of an organism.

Specify the Importance of Nucleic Acids

Nucleic Acids are compounds that contain phosphorous and nitrogen in addition to carbon. There are two types of nucleic acids that are formed by the nucleolus located within the nucleus of cells. DNA (deoxyribonucleic acid) and RNA (ribonucleic acid) carry the genetic code that controls the manufacture or synthesis of proteins. These proteins serve as building blocks of cells, as hormones that regulate cellular function, as antibodies that fight infection, and as enzymes that regulate chemical reactions within the cell.

Nucleic acids are a group of organic substances found in the chromosomes of living cells and viruses. They play a central role in the storage and replication of hereditary information and in the expression of this information through protein synthesis. In most organisms, nucleic acids occur in combination with proteins; the combined substances are called nucleoproteins. Nucleic acid molecules are complex chains of varying length.

The two chief types of nucleic acids are DNA (deoxyribonucleic acid), which carries the hereditary information from generation to generation, and RNA (ribonucleic acid), which delivers the instructions coded in this information to the cell's protein manufacturing sites (ribosomes). Outline the function of minerals such as calcium, phosphorus, potassium, and sodium and describe why these substances are essential.

Minerals are chemical elements that organisms need for normal functioning. The most important minerals are iron, calcium, phosphorus, and iodine. Calcium is the most abundant mineral in the human body, and for that matter, the body of any boned animal, as it is the structural element of bones.

Weather our body is deficient in calcium depends less on whether we have a sufficient dietary intake, but rather on how well we absorb it. An important factor in calcium absorption is Vitamin D. The healthy body can produce this vitamin itself for as long as it gets some exposure to sunlight every day. Calcium absorption depends on hormones.

Women beyond menopause are under increased risk of developing osteoporosis(bone loss), not because they would lacking in dietary calcium but because they lack the estrogen needed for proper absorption. Trace minerals such as boron also play a part in calcium absorption. The best sources of dietary calcium are milk and milk products, as well as soy and green leafy vegetables. Recommended daily amounts for dietary calcium are about 700 milligram for a healthy adult, and about a gram for adolescents.

Phosphates are essential to the energy-transfer reactions necessary to sustain life processes. Another name for this is adenosine triphospate or ATP which is highly significant to all cells in your body which is involved in every metabolic and photosynthetic reaction. In nature phosphorus occurs only in compounds called phosphates. Plants get phosphorus from soil, and use it in photosynthesis. People and animals take in phosphorus by eating plants and different types of foods.

Distinguish between carbohydrates (sugars and starches), lipids and proteins by using the appropriate tests to identify these nutrients. List each test and how it is conducted.

Tests for lipids: There are two tests that can be used to test for the presence of lipids:

1. Sudan IV : Sudna IV is a special biological stain that is soluble in nonpolar solvents such as lipids and turn from a pink to red colour. Polar compounds will not take up the pink or red colour. For example, if you take a test tube of water and vegetable oil (a lipid) and mix them together and then add a small amount of Sudan IV and shake the mixture, only the oil will turn red or pink not the water.
2. The second test is often called the translucence test for lipid. This is the test that uses unglazed brown paper (same type of brown paper you find in a brown paper bag like the ones from the Co-op used to rap bottles). If lipid is present, the lipid is absorbed into the paper and causes it to appear translucent which means that light is able to pass through and it looks like a greesy spot.

Tests for Protein: The Biuret Test is often used for protein. As you will learn soon, protein is made of a chain of amino acids that are bonded together by what are known as peptide bonds. The Biuret Test uses Biuret Reagent (a special solution) that is able to react with the peptide bonds of the amino acid chain that makes up the protein. This causes a colour change from blue indicating no protein, to pink (small amount of protein), to violet (larger amount of protein), to purple (largest amount of protein). An example to use is to test a known sample of protein such as egg white or gelatin. Adding the Biuret reagent to either of these food materials will cause a purple colour. Adding Biuret Reagent to water would result in blue colour.

Tests for carbohydrate: Benedicts Reagent is used for single sugars (called monosaccharides). Presence of monosaccharides causes the Benedict's solution to change to various colours when heated. The colour change depends on the concentration of the sugar present. Benedict's Reagent is made from Copper Sulfate and therefore contains Cu^{2+} ions. In the presence of simple sugars (known as reducing sugars) the copper is changed to Cu^{1+} ions. The reaction depends on the concentration of the sugar present. Here is a list of colour changes:

- Blue means no sugar present.
- Light green means 0.5% to 1.0% sugar present.
- Green to Yellow means 1.0% to 1.5% sugar present.
- Orange means 1.5% to 2.0% sugar present.
- Red to red brown means greater than 2.0% sugar present.

Iodine is the test for starch. Mixing iodine and starch causes the formation of a starch-iodine complex that is blue-black in colour. If you place iodine on a sample of food and it remains an amber or light brown colour there is no starch present, but if there is starch present, then the colour will change to a dark blue-black colour.

Explain what vitamins are and describe in general terms the vital body functions of vitamins A, B, C, D, E, and K.

Vitamins are organic molecules required in the diet in very small amounts. Vitamins are substances that are normally made by plants and then ingested by consumers. The vitamins are important because they serve as coenzymes which help the enzymes function more effectively.

If the body is unable to supply those necessary enzymes in the proper quantities at the proper time, the vitamins and minerals become inert materials and pass unused through your body. Vitamins function directly as coenzymes, or are converted into coenzymes within the cell. Many important biological reactions need vitamins. Vitamin K and some of the B vitamins are essential for blood clotting. This is because Vitmain K is needed for the symthesis of prothrombin, one of the plasma proteins that takes part in the series of reactions leading to the formation of a blood clott. When large doses of anitbiotics are given to overcome an infection, they can destroy the intestinal bacteria, and a Vitamin K deficiency may result. This is because Vitamin K is produced as a product of the metabolic activities of these bacteria.

Vitamin A

Vitamin A is a fat-soluble vitamin, which is structurally related to carotene. This is because carotene is converted into vitamin A in the liver. It is composed of compounds called retinoids and carotenoids, and is measured in Retinol Equivalents (RE's) which measure the vitamin A activity of food. There are many vital body functions in which vitamin A is used, some of these are maintaining the protective linings of the lungs, intestine, urinary tract, and other organs, as well as maintaining good vision, and promoting proper bone growth and tooth development. Too little vitamin A leads to a conditon called night blindness. Night blind people have trouble seeing in dim light.

Both black-and-white and colour vision make use of the light-sensitive pigment retinal, which is made from Vitamin A. Others include helping in the formation and maintenance of healthy skin and hair, healthy mucus membranes, fortifying the immune system, reproduction, and hormone synthesis and regulation. Also, Vitamin A is used to treat minor skin disorders, it is important for the human metabolism, and is thought to help prevent the development of cancer. Sources of vitamin A include things such as some diary products, eggs, and liver. Some sources of carotene are orange fruits and vegetables, as well as dark, leafy greens. The absence of Vitamin A in ones diet can cause loss in weight, eye diseases (including night blindness), kidney stones, rashes, depression, frequent infections of the respiratory system, bladder, or digestive systems, plus many other problems.

Vitamin B1 (Thiamine)

Vitamin B1 is a water-soluble vitamin which is important for the production of energy and metabolism of sugar. It is also essential for supporting normal appetite, nervous system functions, muscles and heart, as well as aiding in the digestion of carbohydrates, and promoting growth and good muscle tone. Good sources of vitamin B1 are some dairy products, nuts, meats, whole grains, yeast eggs, and germ of cereals. The lack of vitamin B1 in ones diet may cause the loss of appetite, loss of weight, nausea, anxiety, irritability, insomnia, depression, weakness and fatigue. Other problems that may occur are heart and muscle problems, such as enlarged heart, abnormal heart rhythms, heart failure, muscle wasting, aches and pains in muscles, difficulty walking, loss of reflexes, and paralysis.

Vitamin B2 (Riboflavin)

Vitamin B2 is a water-soluble vitamin which is found to release energy from protein, fat, and carbohydrates, and is the precursor of flavoproteins, flavin-adenine dinucleotide and flavin mononucleotide.

This vitamin is part of a coenzyme used in energy metabolism, it alleviates eye fatigue, helps maintain cell respiration, healthy eyes, hair, skin, and nails, aids in the formation of antibodies and red blood cells, and promotes general health. Good sources of vitamin B2 are some dairy products, liver, the white of an egg, meats, whole grains, cereals, almonds, and green vegetables.

Some problems that are caused by the dietary lack of vitamin B2 are magenta tongue, rashes, cracks and sores in or around the mouth, itching, burning, bloodshot eyes, digestive disturbances, trembling, sluggishness, as well as dry, scaly skin.

Vitamin B3 (Niacin)

Vitamin B3 is a water-soluble vitamin, it's precursor is tryptophan, and it is involved in the oxidative release of energy from food. It is necessary for the metabolism of carbohydrates, is part of a coenzyme used in energy metabolism, helps improve circulation, supports the health of the nervous system, digestive system, and the skin, as well as protects the skin.

Other vital body functions of vitamin B3 are preventing pellagra, and reducing high blood pressure, and help lower cholesterol levels. Dietary sources of vitamin B3 include milk, eggs, meat, whole grains and cereals, nuts, and all protein-containing foods.

Insufficient amounts of vitamin B3 in one's diet may result in pellagra, irritability, nervousness, loss of appetite, dizziness, mental confusion, flaky skin rash on sun exposed areas, and delirium. Others include depression, fatigue, headaches, muscular weakness, insomnia, and black, smooth tongue.

Vitamin B5 (Pantothenic Acid)

Vitamin B5 is a water-soluble vitamin which is essential for the metabolism of food. Some vital body functions of vitamin B5 include metabolism of proteins, carbohydrates, and fats, and the production of steroid hormones and other important chemicals, helps fight depression. As well as supporting the normal functioning of the gastrointestinal tract, and it is required for the production of cholesterol, bile, red blood cells, and antibodies.

Good dietary sources of vitamin B5 are whole grains, beans, liver, eggs, fish, milk and milk products, yeast, lean beef, some potatoes. Others include broccoli, and other vegetables in the cabbage family. Deficiency of vitamin B5 causes problems such as abdominal distress, burning sensation in the heels, and sleep problems.

Vitamin B6 (Pyridoxine)

Vitamin B6 is a water-soluble vitamin, and is related compounds pyridoxamine, and pyridoxal. This vitamin is necessary for the synthesis and breakdown of amino acids, is used for the metabolism of protein, aids in the formation of red blood cells, helps convert tryptophan to niacin, and for maintaining the central nervous system. It also helps maintain a proper balance of sodium and phosphorous in the body, aids in the removal of excess fluid in premenstrual women, promotes healthy skin, and reduces muscle spasms, leg cramps, nausea, as well as hand numbness and stiffness.

Sources of vitamin B6 include bananas, fish, shellfish, meat, poultry, nuts, fruits, whole grains, vegetables, yeast, wheat germ, maize, and rice husks. The chance of deficiency is increased by smoking, and can cause problems such as anaemia, smooth tongue, insomnia, irritability, muscle twitching, convulsions, rashes, kidney stones, greasy dermatitis. In addition, other resulting problems are arm and leg cramps, loss of hair, muscular weakness, slow learning, mouth disorders, water retention, nervousness, and skin eruptions.

Vitamin B12 (Cyanocobalamin)

Vitamin B12 is a water-soluble vitamin, and is involved in the biosynthesis of methyl groups of choline, and methionine. Some vital body functions of vitamin B12 are being part of a coenzyme used in new cell synthesis, help maintain and protect the nerve cells and system, helps in the formation and regeneration of red blood cells, and helps prevent anaemia.

Others include being used for genetic materials, it is necessary for carbohydrate, fat and protein metabolism, promotes growth in children, increases energy, and is needed for calcium absorption. Significant sources of this vitamin are milk, cheese, eggs, liver and meat. A shortage of vitamin B12 in one's diet can lead to anaemia, poor appetite, growth failure in children, fatigue, depression, lack of balance, smooth tongue, and skin hypersensitivity. In addition, other problems are brain damage, paralysis, nervous system degeneration, as well as degeneration of the spinal cord.

Define Enzymes and Explain the Make-up and Importance of Enzymes

Enzymes (EN zymz) are protein substances that are necessary for most of the chemical reactions that occur in living cells. Enzymes speed up chemical reactions. Enzymes go into a chemical reaction only temporarily. They are not changed by this reaction so they may be used over and over again for the same chemical step with other molecules. In this way, enzymes are organic catalysts. The substance that an enzyme acts upon is called it's substrate.

Enzymes are almost invisible Protein molecules which are contained in the food we eat and are produced by our bodies. Enzymes are necessary for most of the chemical reations that occur in living cells. All enzymes in an organism are made by the cells of the organism. There are three different types of enzymes which are Metabolic, Digestive, and Food. Enzymes are important because they purify blood, strengthen the immune system, break down fats, lower cholesterol, enhance mental capacity, cleanse the colon, enhance sleep, and improve aging skin.

All living cells make enzymes, but enzymes are not alive. Enzyme molecules function by altering other molecules. Enzymes combine with the altered molecules to form a complex molecule structure in which chemical reactions take place. The enzyme, which remains unchanged, then seprates from the product of the reaction.

The human body has thousands of kinds of enzymes. Each kind does it's own job. Without these people could not breath, see, move, or digest food. Photosynthesis in plants also depends on enzymes.

There is a major difference between chemical reactions in nonliving things and the reactions in living things. Consider, for example, the burning of gasoline in an automobile engine. The gasoline vapor is admitted to the engine cylinder. It is ignited by a spark. The vapor burns in a fraction of a second. In fact, the burning is so rapid that it produces a small explosion, which helps to drive the engine.

How Enzymes work is that the ability to act as a catalysts depends on their shape. Somewhere on the surface of each enzyme there is a region called the active site. An Enzyme may also cause 2 molecules to join. In this case, there are two substrates. Each fits into the active site in such a way that they are brought into close contact. This enables bonds to form between the two substrate molecules. There is a theory to explain how enzyme and substrate fit together at an active site. It is called the lock-and-key model. Just as the notched surface of a key can open only one lock, the shape of the active site of an enzyme only fits certian substrates. Thus, each enzyme can catalyze a reaction only of those substrates.

The more modern view of enzyme function is known as the induced fit model which suggests that the enzyme changes shape just slightly to better fit the substrate.

Enzymes are proteins which act as biological catalysts. A catalyst enables a chemical reaction to take place at lower temperatures by lowering the activation energy required for the reaction to occur. This is important to living cells since the enzyme enables the metabolic reactions to occur at normal cellular temperatures. Many enzymes must be attached to certain non-protein molecules in order to function. Some molecules are called cofactors, such as copper, iron, and magnesium. Others are called coenzymes. Many coenzymes consist of vitamins, especially B vitamins. If a persons diet lacks the right amount of these vitamins, the enzymes cannot function properly, and serious problems may occur within the body.

Molecules of Life: Biomolecules

A biomolecule is any organic molecule that is produced by a living organism, including large polymeric molecules such as proteins, polysaccharides, and nucleic acids as well as small molecules such as primary metabolites, secondary metabolites, and natural products. A more general name for this class of molecules is a biogenic substance.

As organic molecules, biomolecules consist primarily of carbon and hydrogen, nitrogen, and oxygen, and, to a smaller extent, phosphorus and sulfur. Other elements sometimes are incorporated but are much less common.

Types of Biomolecules

A diverse range of biomolecules exist, including:

Small Molecules

In the fields of pharmacology and biochemistry, a small molecule is a low molecular weight organic compound which is by definition not a polymer. The term small molecule, especially within the field of pharmacology, is usually restricted to a molecule that also binds with high affinity to a biopolymer such as protein, nucleic acid, or polysaccharide and in addition alters the activity or function of the biopolymer.

The upper molecular weight limit for a small molecule is approximately 800 Daltons which allows for the possibility to rapidly diffuse across cell membranes so that they can reach intracellular sites of action. In addition, this molecular weight cutoff is a necessary but insufficient condition for oral bioavailability.

Small molecules can have a variety of biological functions, serving as cell signaling molecules, as tools in molecular biology, as drugs in medicine, as pesticides in farming, and in many other roles. These compounds can be natural (such as secondary metabolites) or artificial (such as antiviral drugs); they may have a beneficial effect against a disease (such as drugs) or may be detrimental (such as teratogens and carcinogens).

Biopolymers such as nucleic acids, proteins, and polysaccharides (such as starch or cellulose) are not small molecules, although their constituent monomers — ribo- or deoxyribonucleotides, amino acids, and monosaccharides, respectively — are often considered to be. Very small oligomers are also usually considered small molecules, such as dinucleotides, peptides such as the antioxidant glutathione, and disaccharides such as sucrose.

Drugs

Most drugs are small molecules, although some drugs can be proteins, e.g. insulin. Many proteins are degraded if administered orally and most often cannot cross the cell membranes. Small molecules are more likely to be absorbed, although some of them are only absorbed after oral administration if given as prodrugs.

Many dietary supplements are small molecules (but not herb extracts, such as ginkgo).

Primary and Secondary Metabolites

For organisms to produce small molecules they need one or more specialized enzymes (to create and destroy), which as a result are not that varied in vertebrates (recent and small + slow population size), but very common in soil bacteria (such as streptomyces) and fungi, which in particular secrete antibiotics.

Plants also have several secondary metabolites, which play a role in cell signalling, pigmentation or in defence, several of which have also been used as drugs (medical and recreational):

- Alkaloids
- Glycosides
- Lipids
- Flavonoids
- Nonribosomal peptides, such as actinomycin-D
- Phenazines
- Phenols
- Polyketide
- Terpenes, including steroids
- Tetrapyrroles.

Research Tools

Enzymes and receptors are often activated or inhibited by endogenous protein, but can be also inhibited by endogenous or exogenous small molecule inhibitors or activators with can bind to the active site or on the allosteric site.

An example is the teratogen and carcinogen phorbol 12-myristate 13-acetate which is a plant terpene which activates protein kinase C promoting cancer, making it a very useful investigative tool.

There is also interest in creating small molecule artificial transcription factors to regulate gene expression, examples include wrenchnol (a wrench shaped molecule). Binding of ligand can be characterised using a variety of analytical techniques such as surface plasmon resonance, microscale thermophoresis or dual polarisation interferometry to quantify the reaction affinities and kinetic properties and also any induced conformational change.

Nucleosides and Nucleotides

Nucleosides are molecules formed by attaching a nucleobase to a ribose ring. Examples of these include cytidine, uridine, adenosine, guanosine, thymidine and inosine.

Nucleosides can be phosphorylated by specific kinases in the cell, producing nucleotides. Both DNA and RNA are polymers, consisting of long, linear molecules. The repeating structural units, or monomers, of the nucleic acids are called nucleotides.

Each nucleotide is made of an acyclic nitrogenous base, a pentose and one to three phosphate groups. They contain carbon, nitrogen, oxygen, hydrogen and phosphorus. They serve as sources of chemical energy (adenosine triphosphate and guanosine triphosphate), participate in cellular signaling (cyclic guanosine monophosphate and cyclic adenosine monophosphate), and are incorporated into important cofactors of enzymatic reactions (coenzyme A, flavin adenine dinucleotide, flavin mononucleotide, and nicotinamide adenine dinucleotide phosphate).

Saccharides

Monosaccharides are the simplest form of carbohydrates with only one simple sugar. They essentially contain an aldehyde or ketone group in their structure. The presence of an aldehyde group in a monosaccharide is indicated by the prefix *aldo-*. Similarly, a ketone group is denoted by the prefix *keto*.

Examples of monosaccharides are the hexoses glucose, fructose, and galactose and pentoses, ribose, and deoxyribose Consumed fructose and glucose have different rates of gastric emptying, are differentially absorbed and have different metabolic fates, providing multiple opportunities for 2 different saccharides to differentially affect food intake.

Disaccharides are formed when two monosaccharides, or two single simple sugars, form a bond with removal of water. They can be hydrolyzed to yield their saccharin building blocks by boiling with dilute acid or reacting them with appropriate enzymes. Examples of disaccharides include sucrose, maltose, and lactose. Polysaccharides are polymerized monosaccharides, complex, carbohydrates. They have multiple simple sugars. Examples are starch, cellulose, and glycogen. They are generally large and often have a complex branched connectivity. Because of their size, polysaccharides are not water-soluble, but their many hydroxy groups become hydrated individually when exposed to water, and some polysaccharides form thick colloidal dispersions when heated in water.

Shorter polysaccharides, with 3 - 10 monomers, are called oligosaccharides. A fluorescent indicator-displacement molecular imprinting sensor was developed for discriminating saccharides. It successfully discriminated three brands of orange juice beverage. The change in fluorescence intensity of the sensing films resulting is directly related to the saccharide concentration.

Lignin

Lignin is a random polymer composed mainly of aromatic rings with short (up to three) aliphatic carbons chains connecting the rings. Lignin is the second most common biopolymer (after cellulose) and is one of the primary structural components of most plants.

It contains subunits derived from *p*-coumaryl alcohol, coniferyl alcohol, and sinapyl alcohol and is unusual among biomolecules in that it is racemic i.e. it is not optically active. The lack of optical activity is because the polymerization of lignin occurs via free radical coupling reactions in which there is no preference for either configuration at a chiral center.

Lipids

Lipids are chiefly fatty acid esters, and are the basic building blocks of biological membranes. Another biological role is energy storage (e.g., triglycerides). Most lipids consist of a polar or hydrophilic head (typically glycerol) and one to three nonpolar or hydrophobic fatty acid tails, and therefore they are amphiphilic. Fatty acids consist of unbranched chains of carbon atoms that are connected by single bonds alone (saturated fatty acids) or by both single and double bonds (unsaturated fatty acids). The chains are usually 14-24 carbon groups long, but it is always an even number.

For lipids present in biological membranes, the hydrophilic head is from one of three classes:

- Glycolipids, whose heads contain an oligosaccharide with 1-15 saccharide residues.
- Phospholipids, whose heads contain a positively charged group that is linked to the tail by a negatively charged phosphate group.
- Sterols, whose heads contain a planar steroid ring, for example, cholesterol.

Other lipids include prostaglandins and leukotrienes which are both 20-carbon fatty acyl units synthesized from arachidonic acid. They are also known as fatty acids

Amino Acids

Amino acids contain both amino and carboxylic acid functional groups. (In biochemistry, the term amino acid is used when referring to those amino acids in which the amino and carboxylate functionalities are attached to the same carbon, plus proline which is not actually an amino acid). Amino acids are the building blocks of long polymer chains. With 2-10 amino acids such chains are called peptides, with 10-100 they are often called polypeptides, and longer chains are known as proteins. These protein structures have many structural and enzymatic roles in organisms.

There are twenty amino acids that are encoded by the standard genetic code, but there are more than 500 natural amino acids. When amino acids other than the set of twenty are observed in proteins, this is usually the result of modification after translation (protein synthesis). Only two amino acids other than the standard twenty are known to be incorporated into proteins during translation, in certain organisms:

- Selenocysteine is incorporated into some proteins at a UGA codon, which is normally a stop codon.
- Pyrrolysine is incorporated into some proteins at a UAG codon. For instance, in some methanogens in enzymes that are used to produce methane.

Besides those used in protein synthesis, other biologically important amino acids include carnitine (used in lipid transport within a cell), ornithine, GABA and taurine.

Protein Structure

The particular series of amino acids that form a protein is known as that protein's primary structure. This sequence is determined by the genetic makeup of the individual. Proteins have several, well-classified, elements of local structure formed by intermolecular attraction, this forms the secondary structure of protein.

They are broadly divided in two, alpha helix and beta sheet, also called beta pleated sheets. Alpha helices are formed of coiling of protein due to attraction between amine group of one amino acid with carboxylic acid group of other.

The coil contains about 3.6 amino acids per turn and the alkyl group of amino acid lie outside the plane of coil. Beta pleated sheets are formed by strong continuous hydrogen bond over the length of protein chain. Bonding may be parallel or antiparallel in nature.

Structurally, natural silk is formed of beta pleated sheets. Usually, a protein is formed by action of both these structures in variable ratios. Coiling may also be random. The overall 3D structure of a protein is termed its tertiary structure. It is formed as result of various forces like hydrogen bonding, disulfide bridges, hydrophobic interactions, hydrophilic interactions, van der Waals force etc. When two or more different polypeptide chains cluster to form a protein, quaternary structure of protein is formed. Quaternary structure is a unique attribute of polymeric and heteromeric proteins like hemoglobin, which consists of two alpha and two beta peptide chains.

Apoenzymes

An apoenzyme is the inactive storage and generally secretory form of a protein. This is required to protect the secretory cell from the activity of that protein. Apoenzymes becomes active enzyme on addition of a cofactor. Cofactors can be either inorganic (e.g., metal ions and iron-sulfur clusters) or organic compounds, (e.g., flavin and heme). Organic cofactors can be either prosthetic groups, which are tightly bound to an enzyme, or coenzymes, which are released from the enzyme's active site during the reaction.

Isoenzymes

Isoenzymes are enzymes with similar function but different structure. They are products of different genes. They are produced in different organs to perform the same function. LDH are examples of such enzymes. Their varied levels in blood are used to determine any deformity in the organ of secretion.

Vitamins

A vitamin is a compound that is generally not synthesized by a given organism but is nonetheless vital to its survival or health (for example coenzymes). These compounds must be absorbed, or eaten, but typically only in trace quantities. When originally proposed by Casimir Funk, a Polish biochemist, he believed them to all be basic and therefore named them vital amines. The "l" was later dropped to form the word vitamines.

5

Applications of Natural and of Artificial Neural Networks

Function Approximation

The need for function approximations arises in many branches of applied mathematics, and computer science in particular. In general, a function approximation problem asks us to select a function among a well-defined class that closely matches ("approximates") a target function in a task-specific way.

One can distinguish two major classes of function approximation problems: First, for known target functions approximation theory is the branch of numerical analysis that investigates how certain known functions (for example, special functions) can be approximated by a specific class of functions (for example, polynomials or rational functions) that often have desirable properties (inexpensive computation, continuity, integral and limit values, etc.).

Second, the target function, call it g, may be unknown; instead of an explicit formula, only a set of points of the form (x, g(x)) is provided. Depending on the structure of the domain and codomain of g, several techniques for approximating g may be applicable. For example, if g is an operation on the real numbers, techniques of interpolation, extrapolation, regression analysis, and curve fitting can be used. If the codomain (range or target set) of g is a finite set, one is dealing with a classification problem instead.

To some extent the different problems (regression, classification, fitness approximation) have received a unified treatment in statistical learning theory, where they are viewed as supervised learning problems.

Radial basis Function Network

A radial basis function network is an artificial neural network that uses radial basis functions as activation functions. It is a linear combination of radial basis functions. They are used in function approximation, time series prediction, and control.

Network Architecture

Radial basis function (RBF) networks typically have three layers: an input layer, a hidden layer with a non-linear RBF activation function and a linear output layer.

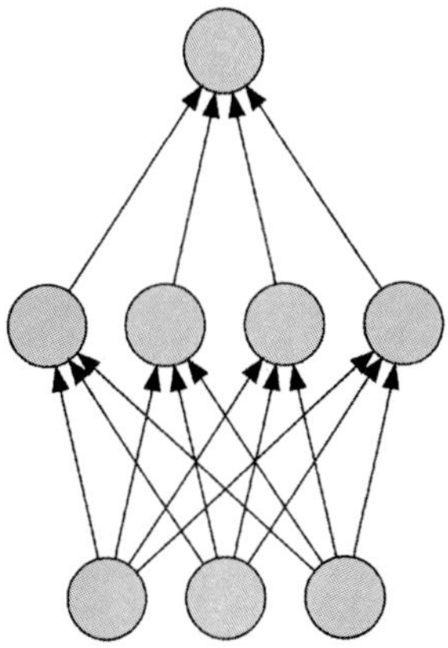

***Figure:** Architecture of a radial basis function network. An input vector x is used as input to all radial basis functions, each with different parameters. The output of the network is a linear combination of the outputs from radial basis functions.*

The output, $\ell : \mathbb{R}^n \to \mathbb{R}$, of the network is thus

$$\ell(x) = \sum_{i=1} a_i \rho(\|x - c_i\|)$$

where N is the number of neurons in the hidden layer, $\mathbf{c}_i$ is the centre vector for neuron i, and a_i are the weights of the linear output neuron. In the basic form all inputs are connected to each hidden neuron. The norm is typically taken to be the Euclidean distance (though Mahalanobis distance appears to perform better in general) and the basis function is taken to be Gaussian

$$\rho(\|x - c_i\|) = \exp\left[-\beta \|x - c_i\|^2\right].$$

The Gaussian basis functions are local in the sense that

$$\hat{y}_j = h(x_j) \approx y_j$$

i.e. changing parameters of one neuron has only a small effect for input values that are far away from the centre of that neuron.

RBF networks are universal approximators on a compact subset of $\mathbb{R}^n$. This means that a RBF network with enough hidden neurons can approximate any continuous function with arbitrary precision. The weights a_i, , and β are determined in a manner that optimises the fit between ℓ and the data.

Normalized

Normalized Architecture

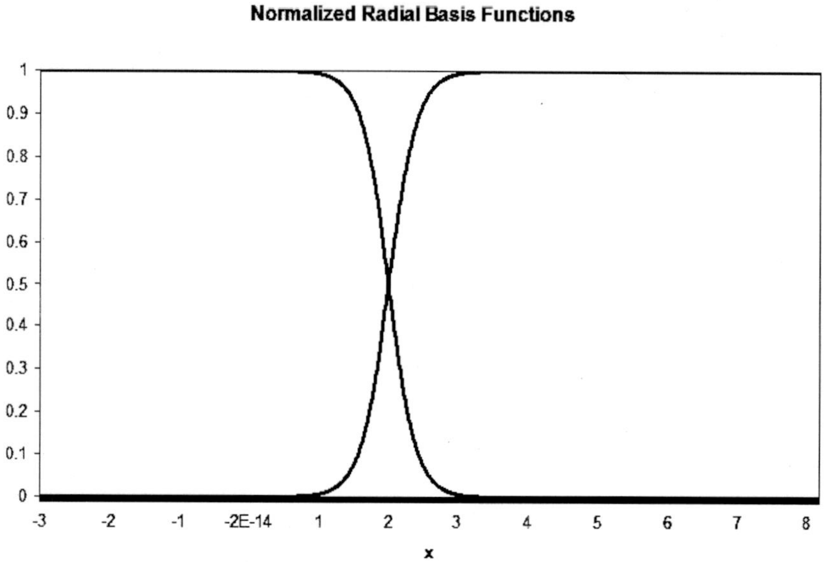

Figure: *Two normalized radial basis functions in one input dimension. The basis function centres are located at c_1 = 0.75 andc_2 = 3.25.*

In addition to the above unnormalized architecture, RBF networks can be normalized. In this case the mapping is

$$\ell(x) \overset{def}{=} \frac{\sum_{i=1}^{N} a_i \rho\left(\|x - c_i\|\right)}{\sum_{i=1}^{N} \rho\left(\|x - c_i\|\right)} \sum_{i=1}^{N} a_i u\left(\|x - c_i\|\right)$$

where

$$u\left(\|x - c_i\|\right) \overset{def}{-} \frac{\rho\left(\|x - c_i\|\right)}{\sum_{i=1}^{N} p\left(\|x - c_i\|\right)}$$

is known as a "normalized radial basis function".

Theoretical Motivation for Normalization

There is theoretical justification for this architecture in the case of stochastic data flow. Assume a stochastic kernel approximation for the joint probability density

$$P(x \wedge y) = \frac{1}{N}\sum_{i=1}^{N} \rho(\|x - c_i\|)\sigma(|y - e_i|)$$

where the weights C_i and e_i are exemplars from the data and we require the kernels to be normalized

$$\int \rho(\|x - c_i\|) d^n x - 1$$

and

$$\int \sigma(|y - e_i|) dy = 1$$

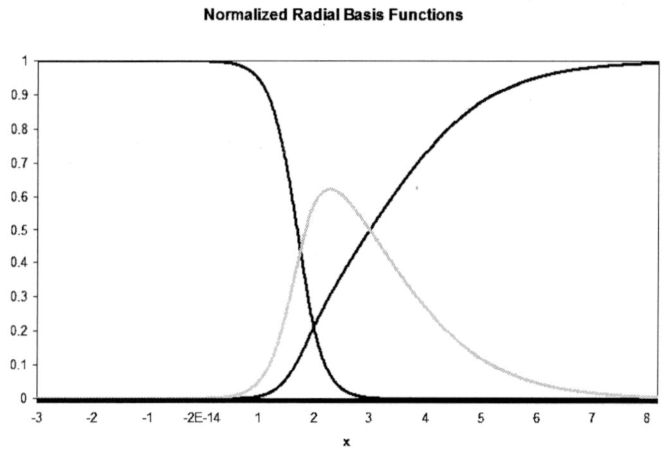

Figure: *Three normalized radial basis functions in one input dimension. The additional basis function has centre at* $c_3 = 2.75$

The probability densities in the input and output spaces are

$$P(x) = \int P(x \wedge y) dy = \frac{1}{N}\sum_{i=1}^{N} \rho(\|x - c_i\|)$$

and

The expectation of y given an input X is

$$\ell(x) \overset{def}{=} E(y \mid x) = \int y P(y \mid x) dy$$

where

$$P(y \mid x)$$

is the conditional probability of y given x. The conditional probability is related to the joint probability through Bayes theorem

$$P(y \mid x) = \frac{p(x \wedge y)}{P(x)}$$

which yields

$$\varphi(x)=\int y\frac{P(x\wedge y)}{P(x)}dy$$

This becomes

$$\ell(x)=\frac{\sum_{i=1}^{N}e_i\rho(\|x-c_i\|)}{\sum_{i=1}^{N}\rho(\|x-c_i\|)}=\sum_{i=1}^{N}e_iu(\|x-c_i\|)$$

when the integrations are performed.

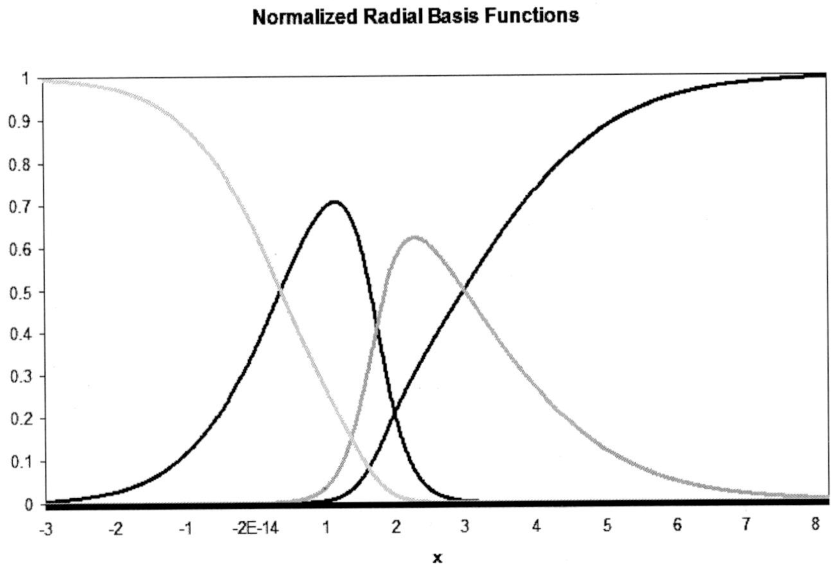

Figure: *Four normalized radial basis functions in one input dimension. The fourth basis function has centre at $c_4 = 0$. Note that the first basis function (dark blue) has become localized.*

Regression Analysis

In statistics, regression analysis includes many techniques for modelling and analysing several variables, when the focus is on the relationship between a dependent variable and one or more independent variables. More specifically, regression analysis helps one understand how the typical value of the dependent variable changes when any one of the independent variables is varied, while the other independent variables are held fixed. Most commonly, regression analysis estimates the conditional expectation of the dependent variable given the independent variables — that is, the average value of the dependent variable when the independent variables are held fixed. Less commonly, the focus is on a quantile, or other location parameter of the conditional distribution of the dependent variable given the independent variables. In all cases, the estimation target is a function of the independent variables called the regression function. In regression analysis, it is also of interest to characterize the variation of the dependent variable around the regression function, which can be described by a probability distribution.

Regression analysis is widely used for prediction and forecasting, where its use has substantial overlap with the field of machine learning. Regression analysis is also used to understand which among the independent variables are related to the dependent variable, and to explore the forms of these relationships. In restricted circumstances, regression analysis can be used to infer causal relationships between the independent and dependent variables.

A large body of techniques for carrying out regression analysis has been developed. Familiar methods such as linear regression and ordinary least squares regression are parametric, in that the regression function is defined in terms of a finite number of unknown parameters that are estimated from the data. Nonparametric regression refers to techniques that allow the regression function to lie in a specified set of functions, which may be infinite-dimensional.

The performance of regression analysis methods in practice depends on the form of the data generating process, and how it relates to the regression approach being used. Since the true form of the data-generating process is in general not known, regression analysis often depends to some extent on making assumptions about this process. These assumptions are sometimes (but not always) testable if a large amount of data is available. Regression models for prediction are often useful even when the assumptions are moderately violated, although they may not perform optimally. However, in many applications, especially with small effects or questions of causality based on observational data, regression methods give misleading results.

History

The earliest form of regression was the method of least squares (French: méthode des moindres carrés), which was published by Legendre in 1805, and by Gauss in 1809. Legendre and Gauss both applied the method to the problem of determining, from astronomical observations, the orbits of bodies about the Sun. Gauss published a further development of the theory of least squares in 1821, including a version of the Gauss–Markov theorem.

The term "regression" was coined by Francis Galton in the nineteenth century to describe a biological phenomenon. The phenomenon was that the heights of descendants of tall ancestors tend to regress down towards a normal average (a phenomenon also known as regression toward the mean). For Galton, regression had only this biological meaning, but his work was later extended by Udny Yule and Karl Pearson to a more general statistical context.

In the work of Yule and Pearson, the joint distribution of the response and explanatory variables is assumed to be Gaussian. This assumption was weakened by R.A. Fisher in his works of 1922 and 1925. Fisher assumed that the conditional distribution of the response variable is Gaussian, but the joint distribution need not be. In this respect, Fisher's assumption is closer to Gauss's formulation of 1821.

In the 1950s and 1960s, economists used electromechanical desk calculators to calculate regressions. Before 1970, it sometimes took up to 24 hours to receive the result from one regression. Regression methods continue to be an area of active research.

In recent decades, new methods have been developed for robust regression, regression involving correlated responses such as time series and growth curves, regression in which the predictor or response variables are curves, images, graphs, or other complex data objects, regression methods accommodating various types of missing data, nonparametric regression, Bayesian methods for regression, regression in which the predictor variables are measured with error, regression with more predictor variables than observations, and causal inference with regression.

Regression Models

Regression models involve the following variables:

- The unknown parameters denoted as β; this may be a scalar or a vector.
- The independent variables, X.
- The dependent variable, Y.

In various fields of application, different terminologies are used in place of dependent and independent variables.

A regression model relates Y to a function of X and β.

$$Y \approx f(X, \beta)$$

The approximation is usually formalized as E(Y | X) = f(X, β). To carry out regression analysis, the form of the function f must be specified. Sometimes the form of this function is based on knowledge about the relationship between Y and X that does not rely on the data. If no such knowledge is available, a flexible or convenient form for f is chosen.

Assume now that the vector of unknown parameters β is of length k. In order to perform a regression analysis the user must provide information about the dependent variable Y:

- If N data points of the form (Y,X) are observed, where $N < k$, most classical approaches to regression analysis cannot be performed: since the system of equations defining the regression model is underdetermined, there is not enough data to recover β.
- If exactly $N = k$ data points are observed, and the function f is linear, the equations $Y = f(X, \hat{a})$ can be solved exactly rather than approximately. This reduces to solving a set of N equations with N unknowns (the elements of β), which has a unique solution as long as the X are linearly independent. If f is nonlinear, a solution may not exist, or many solutions may exist.
- The most common situation is where $N > k$ data points are observed. In this case, there is enough information in the data to estimate a unique value for $\hat{a}$ that best fits the data in some sense, and the regression model when applied to the data can be viewed as an overdetermined system in β.

In the last case, the regression analysis provides the tools for:

1. Finding a solution for unknown parameters $\hat{a}$ that will, for example, minimize the distance between the measured and predicted values of the dependent variable Y (also known as method of least squares).
2. Under certain statistical assumptions, the regression analysis uses the surplus of information to provide statistical information about the unknown parameters $\hat{a}$ and predicted values of the dependent variable Y.

Necessary Number of Independent Measurements

Consider a regression model which has three unknown parameters, $\hat{a}_0$, $\hat{a}_1$, and $\hat{a}_2$. Suppose an experimenter performs 10 measurements all at exactly the same value of independent variable vectorX (which contains the independent variables X_1, X_2, and X_3). In this case, regression analysis fails to give a unique set of estimated values for the three unknown parameters; the experimenter did not provide enough information.

The best one can do is to estimate the average value and the standard deviation of the dependent variable Y. Similarly, measuring at two different values of X would give enough data for a regression with two unknowns, but not for three or more unknowns.

If the experimenter had performed measurements at three different values of the independent variable vector X, then regression analysis would provide a unique set of estimates for the three unknown parameters in β. In the case of general linear regression, the above statement is equivalent to the requirement that matrix X^TX is invertible.

Statistical Assumptions

When the number of measurements, N, is larger than the number of unknown parameters, k, and the measurement errors $å_i$ are normally distributed then the excess of information contained in (N - k) measurements is used to make statistical predictions about the unknown parameters. This excess of information is referred to as the degrees of freedom of the regression.

Underlying Assumptions

Classical assumptions for regression analysis include:

- The sample is representative of the population for the inference prediction.
- The error is a random variable with a mean of zero conditional on the explanatory variables.
- The independent variables are measured with no error. (Note: If this is not so, modelling may be done instead using errors-in-variables model techniques).
- The predictors are linearly independent, i.e. it is not possible to express any predictor as a linear combination of the others.
- The errors are uncorrelated, that is, the variance-covariance matrix of the errors is diagonal and each non-zero element is the variance of the error.
- The variance of the error is constant across observations (homoscedasticity). (Note: If not, weighted least squares or other methods might instead be used).

These are sufficient conditions for the least-squares estimator to possess desirable properties, in particular, these assumptions imply that the parameter estimates will be unbiased, consistent, and efficient in the class of linear unbiased estimators. It is important to note that actual data rarely satisfies the assumptions. That is, the method is used even though the assumptions are not true. Variation from the assumptions can sometimes be used as a measure of how far the model is from being useful. Many of these assumptions may be relaxed in more advanced treatments. Reports of statistical analyses usually include analyses of tests on the sample data and methodology for the fit and usefulness of the model.

Assumptions include the geometrical support of the variables (Cressie, 1996). Independent and dependent variables often refer to values measured at point locations. There may be spatial trends and spatial autocorrelation in the variables that violates statistical assumptions of regression. Geographic weighted regression is one technique to deal with such data (Fotheringham et al., 2002). Also, variables may include values aggregated by areas. With aggregated data the Modifiable Areal Unit Problem can cause extreme variation in regression parameters (Fotheringham and Wong, 1991). When analysing data aggregated by political boundaries, postal codes or census areas results may be very different with a different choice of units.

Linear Regression

In linear regression, the model specification is that the dependent variable, y_i is a linear combination of the parameters (but need not be linear in the independent variables). For example, in simple linear regression for modelling n data points there is one independent variable: x_i, and two parameters, β_0 and β_1:

Straight line: $y_i = \beta_0 + \beta_1 x_i + \varepsilon_i, \quad i = 1, \ldots, n.$

In multiple linear regression, there are several independent variables or functions of independent variables. For example, adding a term in x_i to the preceding regression gives:

Parabola: $y_i = \beta_0 + \beta_1 x_i + \beta_2 x_i^2 + \varepsilon_i, \quad i = 1, \ldots, n.$

This is still linear regression; although the expression on the right hand side is quadratic in the independent variable x_i, it is linear in the parameters β_0, β_1 and β_2.

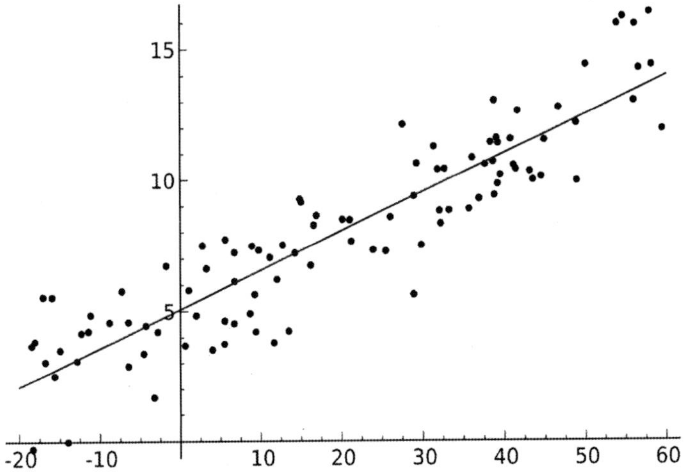

Figure: *Illustration of linear regression on a data set.*

In both cases, ε_i is an error term and the subscript i indexes a particular observation. Given a random sample from the population, we estimate the population parameters and obtain the sample linear regression model:

$$\hat{y}_i = \hat{\beta}_0 + \hat{\beta}_1 x_i.$$

The residual, $e_i = y_i - \hat{y}_i$,, is the difference between the value of the dependent variable predicted by the model, and the true value of the dependent variable y_i. One method of estimation is ordinary least squares. This method obtains parameter estimates that minimize the sum of squared residuals, SSE, also sometimes denoted RSS:

$$SSE = \sum_{i=1}^{N} e_i^2.$$

Minimization of this function results in a set of normal equations, a set of simultaneous linear equations in the parameters, which are solved to yield the parameter estimators, $\hat{\beta}_0, \hat{\beta}_1$.

In the case of simple regression, the formulas for the least squares estimates are

$$\hat{\beta}_1 = \frac{\Sigma(x_i - \bar{x})(y_i - \bar{y})}{\Sigma(x_i - \bar{x})^2} \text{ and } \hat{\beta}_0 = \bar{y} - \hat{\beta}_1\bar{x}$$

where $\bar{x}$ is the mean (average) of the x values and $\bar{y}$ is the mean of the y values. Under the assumption that the population error term has a constant variance, the estimate of that variance is given by:

$$\hat{\sigma}_\varepsilon^2 = \frac{SSE}{N-2}.$$

This is called the mean square error (MSE) of the regression. The standard errors of the parameter estimates are given by

$$\hat{\sigma}_{\beta 0} = \hat{\sigma}_\varepsilon \sqrt{\frac{1}{N} + \frac{\bar{x}^2}{\Sigma(x_i - \bar{x})^2}} \quad \Rightarrow \quad \hat{\sigma}_{\beta_1} = \hat{\sigma}_\varepsilon \sqrt{\frac{1}{\Sigma(x_i - \bar{x})^2}}.$$

Under the further assumption that the population error term is normally distributed, the researcher can use these estimated standard errors to create confidence intervals and conduct hypothesis tests about the population parameters.

General Linear Model

In the more general multiple regression model, there are p independent variables:

$$y_i = \beta_1 x_{i1} + \beta_2 x_{i2} + \cdots + \beta_p x_{ip} + \varepsilon_i,$$

where x_{ij} is the i observation on the j independent variable, and where the first independent variable takes the value 1 for all i (so $\hat{a}_1$ is the regression intercept).

The least squares parameter estimates are obtained from p normal equations. The residual can be written as

$$e_i = y_i - \hat{\beta}_1 x_{i1} - \cdots - \hat{\beta}_p x_{ip}.$$

The normal equations are

$$\sum_{i=1}^{n}\sum_{k=1}^{p} X_{ij} X_{ik} \hat{\beta}_k = \sum_{i=1}^{n} X_{ij} y_i,\ j = 1,\ldots,p.$$

In matrix notation, the normal equations are written as

$$\left(X^T X\right)\hat{\beta} = X^T Y,$$

where the ij element of X is x_{ij}, the i element of the column vector Y is y_i, and the j element of $\hat{\beta}$ is $\hat{\beta}_j$. Thus X is n×p, Y is n×1, and $\hat{\beta}$ is p×1. The solution is

$$\hat{\beta} = \left(X^T X\right)^{-1} X^T Y.$$

Regression Diagnostics

Once a regression model has been constructed, it may be important to confirm the goodness of fit of the model and the statistical significance of the estimated parameters. Commonly used checks of goodness of fit include the R-squared, analyses of the pattern of residuals and hypothesis testing. Statistical significance can be checked by an F-test of the overall fit, followed by t-tests of individual parameters.

Interpretations of these diagnostic tests rest heavily on the model assumptions. Although examination of the residuals can be used to invalidate a model, the results of a t-test or F-test are sometimes more difficult to interpret if the model's assumptions are violated. For example, if the error term does not have a normal distribution, in small samples the estimated parameters will not follow normal distributions and complicate inference. With relatively large samples, however, a central limit theorem can be invoked such that hypothesis testing may proceed using asymptotic approximations.

Regression with "Limited Dependent" Variables

The phrase "limited dependent" is used in econometric statistics for categorical and constrained variables. The response variable may be non-continuous ("limited" to lie on some subset of the real line). For binary (zero or one) variables, if analysis proceeds with least-squares linear regression, the model is called the linear probability model. Nonlinear models for binary dependent variables include the probit and logit model. The multivariate probit model is a standard method of estimating a joint relationship between several binary dependent variables and some independent variables. For categorical variables with more than two values there is the multinomial logit.

For ordinal variables with more than two values, there are the ordered logit and ordered probit models. Censored regression models may be used when the dependent variable is only sometimes observed, and Heckman correction type models may be used when the sample is not randomly selected from the population of interest.

An alternative to such procedures is linear regression based on polychoric correlation (or polyserial correlations) between the categorical variables. Such procedures differ in the assumptions made about the distribution of the variables in the population. If the variable is positive with low values and represents the repetition of the occurrence of an event, then count models like the Poisson regression or the negative binomial model may be used.

Interpolation and Extrapolation

Regression models predict a value of the Y variable given known values of the X variables. Prediction within the range of values in the dataset used for model-fitting is known informally as interpolation. Prediction outside this range of the data is known as extrapolation. Performing extrapolation relies strongly on the regression assumptions.

The further the extrapolation goes outside the data, the more room there is for the model to fail due to differences between the assumptions and the sample data or the true values. For example, regression model failing the homoskedasticity assumption will make bigger mistakes the further out the extrapolation is.

It is generally advised that when performing extrapolation, one should accompany the estimated value of the dependent variable with a prediction interval that represents the uncertainty. Such intervals tend to expand rapidly as the values of the independent variable(s) moved outside the range covered by the observed data. For such reasons and others, some tend to say that it might be unwise to undertake extrapolation.

However, this does not cover the full set of modelling errors that may be being made: in particular, the assumption of a particular form for the relation between Y and X. A properly conducted regression analysis will include an assessment of how well the assumed form is matched by the observed data, but it can only do so within the range of values of the independent variables actually available.

This means that any extrapolation is particularly reliant on the assumptions being made about the structural form of the regression relationship. Best-practice advice here is that a linear-in-variables and linear-in-parameters relationship should not be chosen simply for computational convenience, but that all available knowledge should be deployed in constructing a regression model. If this knowledge includes the fact that the dependent variable cannot go outside a certain range of values, this can be made use of in selecting the model — even if the observed dataset has no values particularly near such bounds. The implications of this step of choosing an appropriate functional form for the regression can be great when extrapolation is considered. At a minimum, it can ensure that any extrapolation arising from a fitted model is "realistic" (or in accord with what is known).

Nonlinear Regression

When the model function is not linear in the parameters, the sum of squares must be minimized by an iterative procedure. This introduces many complications which are summarized in Differences between linear and non-linear least squares

Power and Sample Size Calculations

There are no generally agreed methods for relating the number of observations versus the number of independent variables in the model. One rule of thumb suggested by Good and Hardin is N = m, where N is the sample size, n is the number of independent variables and m is the number of observations needed to reach the desired precision if the model had only one independent variable.

For example, a researcher is building a linear regression model using a dataset that contains 1000 patients (N). If he decides that five observations are needed to precisely define a straight line (m), then the maximum number of independent variables his model can support is 4, because

$$\frac{\log 1000}{\log 5} = 4.29$$

Other Methods

Although the parameters of a regression model are usually estimated using the method of least squares, other methods which have been used include:

- Bayesian methods, e.g. Bayesian linear regression
- Percentage regression, for situations where reducing percentage errors is deemed more appropriate.
- Least absolute deviations, which is more robust in the presence of outliers, leading to quantile regression
- Nonparametric regression, requires a large number of observations and is computationally intensive
- Distance metric learning, which is learned by the search of a meaningful distance metric in a given input space.

Software

All major statistical software packages perform least squares regression analysis and inference. Simple linear regression and multiple regression using least squares can be done in some spreadsheet applications and on some calculators.

While many statistical software packages can perform various types of nonparametric and robust regression, these methods are less standardized; different software packages implement different methods, and a method with a given name may be implemented differently in different packages. Specialized regression software has been developed for use in fields such as survey analysis and neuroimaging.

Time Series

In statistics, signal processing, econometrics andmathematical finance, a time series is a sequence of data points, measured typically at successive times spaced at uniform time intervals. Examples of time series are the daily closing value of the Dow Jones index or the annual flow volume of the Nile River at Aswan. Time series analysis comprises methods for analysing time series data in order to extract meaningful statistics and other characteristics of the data. Time series forecasting is the use of a model to predict future values based on previously observed values. Time series are very frequently plotted via line charts.

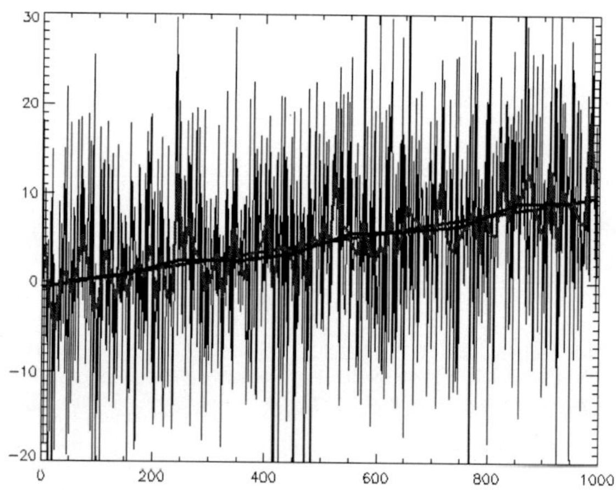

Figure: *Time series: random data plus trend, with best-fit line and different smoothings.*

Time series data have a natural temporal ordering. This makes time series analysis distinct from other common data analysis problems, in which there is no natural ordering of the observations (e.g. explaining people's wages by reference to their education level, where the individuals' data could be entered in any order). Time series analysis is also distinct from spatial data analysis where the observations typically relate to geographical locations (e.g. accounting for house prices by the location as well as the intrinsic characteristics of the houses). A time series model will generally reflect the fact that observations close together in time will be more closely related than observations further apart. In addition, time series models will often make use of the natural one-way ordering of time so that values for a given period will be expressed as deriving in some way from past values, rather than from future values.

Methods for time series analyses may be divided into two classes: frequency-domain methods and time-domain methods. The former include spectral analysis and recently wavelet analysis; the latter include auto-correlation and cross-correlation analysis.

Analysis

There are several types of data analysis available for time series which are appropriate for different purposes.

General Exploration

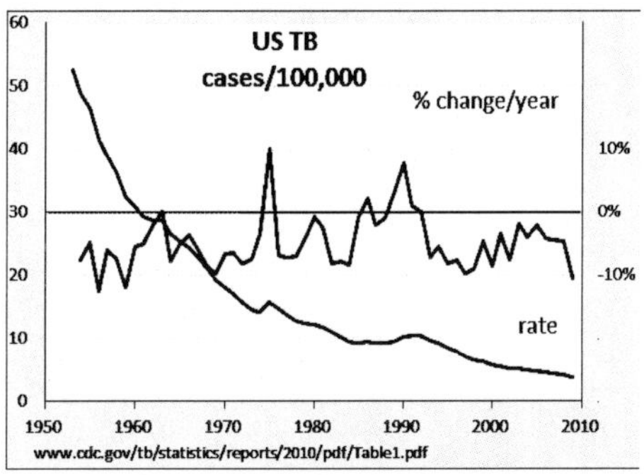

***Figure:** Tuberculosis incidence US 1953-2009*

The clearest way to examine a regular time series is with aline chart such as the one shown for tuberculosis in the United States, made with a spreadsheet program. The number of cases was standardized to a rate per 100,000 and the percent change per year in this rate was calculated. The nearly steadily dropping line shows that the TB incidence was decreasing in most years, but the percent change in this rate varied by as much as +/- 10%, with 'surges' in 1975 and around the early 1990s. The use of both vertical axes allows the comparison of two time series in one graphic. Other techniques include:

- Autocorrelation analysis to examine serial dependence
- Spectral analysis to examine cyclic behaviour which need not be related to seasonality. For example, sun spot activity varies over 11 year cycles. Other common examples include celestial phenomena, weather patterns, neural activity, commodity prices, and economic activity.

Description

- Separation into components representing trend, seasonality, slow and fast variation, cyclical irregular
- Simple properties of marginal distributions.

Prediction and Forecasting

- Fully formed statistical models for stochastic simulation purposes, so as to generate alternative versions of the time series, representing what might happen over non-specific time-periods in the future
- Simple or fully formed statistical models to describe the likely outcome of the time series in the immediate future, given knowledge of the most recent outcomes (forecasting).

Models

Models for time series data can have many forms and represent different stochastic processes. When modelling variations in the level of a process, three broad classes of practical importance are the autoregressive (AR) models, the integrated (I) models, and the moving average (MA) models. These three classes depend linearly on previous data points. Combinations of these ideas produce autoregressive moving average (ARMA) and autoregressive integrated moving average (ARIMA) models.

The autoregressive fractionally integrated moving average (ARFIMA) model generalises the former three. Extensions of these classes to deal with vector-valued data are available under the heading of multivariate time-series models and sometimes the preceding acronyms are extended by including an initial "V" for "vector".

An additional set of extensions of these models is available for use where the observed time-series is driven by some "forcing" time-series (which may not have a causal effect on the observed series): the distinction from the multivariate case is that the forcing series may be deterministic or under the experimenter's control. For these models, the acronyms are extended with a final "X" for "exogenous".

Non-linear dependence of the level of a series on previous data points is of interest, partly because of the possibility of producing a chaotic time series. However, more importantly, empirical investigations can indicate the advantage of using predictions derived from non-linear models, over those from linear models, as for example in nonlinear autoregressive exogenous models.

Among other types of non-linear time series models, there are models to represent the changes of variance along time (heteroskedasticity). These models represent autoregressive conditional heteroskedasticity (ARCH) and the collection comprises a wide variety of representation (GARCH, TARCH,

EGARCH, FIGARCH, CGARCH, etc.). Here changes in variability are related to, or predicted by, recent past values of the observed series. This is in contrast to other possible representations of locally varying variability, where the variability might be modelled as being driven by a separate time-varying process, as in a doubly stochastic model.

In recent work on model-free analyses, wavelet transform based methods (for example locally stationary wavelets and wavelet decomposed neural networks) have gained favour. Multiscale (often referred to as multiresolution) techniques decompose a given time series, attempting to illustrate time dependence at multiple scales.

Notation

A number of different notations are in use for time-series analysis. A common notation specifying a time series X that is indexed by the natural numbers is written

$$X = \{X_1, X_2, ...\}.$$

Another common notation is

$$Y = \{Y_t: t \in T\},$$

where T is the index set.

Conditions

There are two sets of conditions under which much of the theory is built:

- Stationary process
- Ergodicity.

However, ideas of stationarity must be expanded to consider two important ideas: strict stationarity and second-order stationarity. Both models and applications can be developed under each of these conditions, although the models in the latter case might be considered as only partly specified.

In addition, time-series analysis can be applied where the series are seasonally stationary or non-stationary. Situations where the amplitudes of frequency components change with time can be dealt with in time-frequency analysis which makes use of a time–frequency representation of a time-series or signal.

Models

The general representation of an autoregressive model, well-known as AR(p), is

$$Y_t = \alpha_0 + \alpha_1 Y_{t-1} + \alpha_2 Y_{t-2} + \cdots + \alpha_p Y_{t-p} + \varepsilon_t$$

where the term ε_t is the source of randomness and is called white noise. It is assumed to have the following characteristics:

$$E[\varepsilon_t] = 0,]$$
$$E[\varepsilon_t^2] = \sigma^2,$$
$$E[\varepsilon_t \varepsilon_s] = 0 \quad \textit{for all } t \neq s.$$

With these assumptions, the process is specified up to second-order moments and, subject to conditions on the coefficients, may be second-order stationary.

If the noise also has a normal distribution, it is called normal or Gaussian white noise. In this case, the AR process may be strictly stationary, again subject to conditions on the coefficients.

Tools for investigating time-series data include:

- Consideration of the autocorrelation function and the spectral density function (also cross-correlation functions and cross-spectral density functions)
- Performing a Fourier transform to investigate the series in the frequency domain
- Use of a filter to remove unwanted noise
- Principal components analysis (or empirical orthogonal function analysis)
- Singular spectrum analysis
- Artificial neural networks
- Hidden Markov model
- Dynamic time warping
- Dynamic Bayesian network
- Time-frequency analysis techniques:
- Fast Fourier Transform
- Continuous wavelet transform
- Short-time Fourier transform
- Chirplet transform
- Fractional Fourier transform.
- Chaotic analysis
- Correlation dimension
- Recurrence plots
- Recurrence quantification analysis
- Lyapunov exponents
- Entropy encoding.

Statistical Classification

In machine learning, statistical classification is the problem of identifying the sub-population to which new observations belong, where the identity of the sub-population is unknown, on the basis of a training set of data containing observations whose sub-population is known. Therefore these classifications will show a variable behaviour which can be studied by statistics.

Thus the requirement is that new individual items are placed into groups based on quantitative information on one or more measurements, traits or characteristics, etc. and based on the training set in which previously decided groupings are already established.

The problem here may be contrasted with that for cluster analysis, where the problem is to analyse a single data-set and decide how and whether the observations in the data-set can be divided into groups. In certain terminology, particularly that of machine learning, the classification problem is known as supervised learning, while clustering is known as unsupervised learning.

Unfortunately, terminology can be different in various fields of application. For example, in community ecology, the term "classification" is synonymous with cluster analysis.

Learning Classifiers. Problem Statement

A learning classifier is able to learn based on a sample. The data-set used for training consists of information x and y for each data-point, where x denotes what is generally a vector of observed characteristics for the data-item and y denotes a group-label. The label y can take only a finite number of values.

The classification problem can be stated as follows: given training data $\{(x_1, y_1), ..., (x_n, y_n)\}$ produce a rule (or "classifier")h, such that $h(x)$ can be evaluated for any possible value of x (not just those included in the training data) and such that the group attributed to any new observation, specifically

$$\bar{y} = h(x),$$

is as close as possible to the true group label y. For the training data-set, the true labels y_i are known but will not necessarily match their in-sample approximations

$$\hat{y}_i = h(x_i).$$

For new observations, the true labels y_j are unknown, but it is a prime target for the classification procedure that the approximation

$$\hat{y}_j = h(x_j) \approx y_j$$

as well as possible, where the quality of this approximation needs to be judged on the basis of the statistical or probabilistic properties of the overall population from which future observations will be drawn.

Frequentist Procedures

Early work on statistical classification was undertaken by Fisher, in the context of two-group problems, leading to Fisher's linear discriminant function as the rule for assigning a group to a new observation. This early work assumed that data-values within each of the two groups had a multivariate normal distribution.

The extension of this same context to more than two-groups has also been considered with a restriction imposed that the classification rule should be linear. Later work for the multivariate normal distribution allowed the classifier to be nonlinear: several classification rules can be derived based on slight different adjustments of the Mahalanobis distance, with a new observation being assigned to the group whose centre has the lowest adjusted distance from the observation.

Bayesian Procedures

Unlike frequentist procedures, Bayesian classification procedures provide a natural way of taking into account any available information about the relative sizes of the sub-populations associated with the different groups within the overall population. Bayesian procedures tend to be computationally expensive and, in the days before Markov chain Monte Carlo computations were developed, approximations for Bayesian clustering rules were devised.

Some Bayesian procedures involve the calculation of group membership probabilities: these can be viewed as providing a more informative outcome of a data analysis than a simple attribution of a single group-label to each new observation.

Binary and Multiclass Classification

Classification can be thought of as two separate problems - binary classification and multiclass classification. In binary classification, a better understood task, only two classes are involved, whereas in multiclass classification involves assigning an object to one of several classes. Since many classification methods have been developed specifically for binary classification, multiclass classification often requires the combined use of multiple binary classifiers.

Algorithms

The most widely used classifiers are the neural network (multi-layer perceptron), support vector machines, k-nearest neighbours, Gaussian mixture model, Gaussian, naive Bayes, decision tree and RBF classifiers.

Examples of classification algorithms include:

- Linear classifiers

- Fisher's linear discriminant
- Logistic regression
- Naive Bayes classifier
- Perceptron
- Support vector machines
- Least squares support vector machines
- Quadratic classifiers
- Kernel estimation
- k-nearest neighbour
- Boosting
- Decision trees
- Random forests
- Neural networks
- Bayesian networks
- Hidden Markov models
- Learning vector quantization.

Evaluation

Classifier performance depends greatly on the characteristics of the data to be classified. There is no single classifier that works best on all given problems (a phenomenon that may be explained by the no-free-lunch theorem). Various empirical tests have been performed to compare classifier performance and to find the characteristics of data that determine classifier performance. Determining a suitable classifier for a given problem is however still more an art than a science.

The measures precision and recall are popular metrics used to evaluate the quality of a classification system. More recently, receiver operating characteristic (ROC) curves have been used to evaluate the tradeoff between true- and false-positive rates of classification algorithms.

As a performance metric, the uncertainty coefficient has the advantage over simple accuracy in that it is not affected by the relative sizes of the different classes. Further, it will not penalize an algorithm for simply rearranging the classes.

An intriguing problem in pattern recognition yet to be solved is the relationship between the problem to be solved (data to be classified) and the performance of various pattern recognition algorithms (classifiers).

Application Domains

Classification problems has many applications. In some of these it is employed as a data mining procedure, while in others more detailed statistical modelling is undertaken.

- Computer vision
- Medical imaging and medical image analysis
- Optical character recognition
- Video tracking
- Drug discovery and development
- Toxicogenomics
- Quantitative structure-activity relationship
- Geostatistics
- Speech recognition
- Handwriting recognition
- Biometric identification
- Biological classification
- Statistical natural language processing
- Document classification
- Internet search engines
- Credit scoring
- Pattern recognition.

Pattern Recognition

In machine learning, pattern recognition is the assignment of some sort of output value (or label) to a given input value (or instance). An example of pattern recognition is classification, which attempts to assign each input value to one of a given set of classes (for example, determine whether a given email is "spam" or "non-spam"). However, pattern recognition is a more general problem that encompasses other types of output as well. Other examples are regression, which assigns a real-valued output to each input; sequence labelling, which assigns a class to each member of a sequence of values (for example, part of speech tagging, which assigns a part of speech to each word in an input sentence); and parsing, which assigns a parse tree to an input sentence, describing the syntactic structure of the sentence.

Pattern recognition algorithms generally aim to provide a reasonable answer for all possible inputs and to do "fuzzy" matching of inputs. This is opposed to pattern matching algorithms, which look for exact matches in the input with pre-existing patterns. A common example of a pattern-matching algorithm is regular expression matching, which looks for patterns of a given sort in textual data and is included in the search capabilities of many text editors and word processors. In contrast to pattern recognition, pattern matching is generally not considered a type of machine learning, although pattern-matching algorithms (especially with fairly general, carefully tailored patterns) can sometimes succeed in providing similar-quality output to the sort provided by pattern-recognition algorithms.

Pattern recognition is studied in many fields, including psychology, psychiatry, ethology, cognitive science, traffic flow and computer science.

Overview

Pattern recognition is generally categorized according to the type of learning procedure used to generate the output value. Supervised learning assumes that a set of training data (the training set) has been provided, consisting of a set of instances that have been properly labelled by hand with the correct output. A learning procedure then generates a model that attempts to meet two sometimes conflicting objectives: Perform as well as possible on the training data, and generalise as well as possible to new data (usually, this means being as simple as possible, for some technical definition of "simple", in accordance with Occam's Razor). Unsupervised learning, on the other hand, assumes training data that has not been hand-labelled, and attempts to find inherent patterns in the data that can then be used to determine the correct output value for new data instances. A combination of the two that has recently been explored is semi-supervised learning, which uses a combination of labelled and unlabelled data (typically a small set of labelled data combined with a large amount of unlabelled data). Note that in cases of unsupervised learning, there may be no training data at all to speak of; in other words, the data to be labelled is the training data.

Note that sometimes different terms are used to describe the corresponding supervised and unsupervised learning procedures for the same type of output. For example, the unsupervised equivalent of classification is normally known as clustering, based on the common perception of the task as involving no training data to speak of, and of grouping the input data into clusters based on some inherent similarity measure (e.g. the distance between instances, considered as vectors in a multidimensional vector space), rather than assigning each input instance into one of a set of pre-defined classes. Note also that in some fields, the terminology is different: For example, in community ecology, the term "classification" is used to refer to what is commonly known as "clustering". The piece of input data for which an output value is generated is formally termed an instance. The instance is formally described by a vector of features, which together constitute a description of all known characteristics of the instance. (These feature vectors can be seen as defining points in an appropriate multidimensional space, and methods for manipulating vectors in vector spaces can be correspondingly applied to them, such as computing the dot product or the angle between two vectors.) Typically, features are either categorical (also known as nominal, i.e. consisting of one of a set of unordered items, such as a gender of "male" or "female", or a blood type of "A", "B", "AB" or "O"), ordinal (consisting of one of a set of ordered items, e.g. "large", "medium" or "small"), integer-valued (e.g. a count of the number of occurrences of a particular word in an email) or real-valued (e.g. a measurement of blood pressure). Often, categorical and ordinal data are grouped together; likewise for integer-valued and real-valued data. Furthermore, many algorithms work only in terms of categorical data and require that real-valued or integer-valued data be discretized into groups (e.g. less than 5, between 5 and 10, or greater than 10).

Many common pattern recognition algorithms are probabilistic in nature, in that they use statistical inference to find the best label for a given instance. Unlike other algorithms, which simply output a "best" label, oftentimes probabilistic algorithms also output a probability of the instance being described by the given label. In addition, many probabilistic algorithms output a list of the N-best labels with associated probabilities, for some value of N, instead of simply a single best label. When the number of possible labels is fairly small (e.g. in the case of classification), N may be set so that the probability of all possible labels is output. Probabilistic algorithms have many advantages over non-probabilistic algorithms:

- They output a confidence value associated with their choice. (Note that some other algorithms may also output confidence values, but in general, only for probabilistic algorithms is this value mathematically grounded in probability theory. Non-probabilistic confidence values can in general not be given any specific meaning, and only used to compare against other confidence values output by the same algorithm.)
- Correspondingly, they can abstain when the confidence of choosing any particular output is too low.
- Because of the probabilities output, probabilistic pattern-recognition algorithms can be more effectively incorporated into larger machine-learning tasks, in a way that partially or completely avoids the problem of error propagation.

Techniques to transform the raw feature vectors are sometimes used prior to application of the pattern-matching algorithm. For example, feature extraction algorithms attempt to reduce a large-dimensionality feature vector into a smaller-dimensionality vector that is easier to work with and encodes less redundancy, using mathematical techniques such as principal components analysis(PCA). Feature selection algorithms, attempt to directly prune out redundant or irrelevant features. The distinction between the two is that the resulting features after feature extraction has taken place are of a different sort than the original features and may not easily be interpretable, while the features left after feature selection are simply a subset of the original features.

Problem Statement (Supervised Version)

Formally, the problem of supervised pattern recognition can be stated as follows: Given an unknown function $g:X \to y$ the ground truth) that maps input instances $x \in x$ to output labels $y \in y$, along with training data $D = \{(x_1, y_1),...,(x_n, y_n)\}$ assumed to represent accurate examples of the mapping, produce a function that approximates as closely as possible the correct mapping g. (For example, if the problem is filtering spam, then x_i is some representation of an email and y is either "spam" or "non-spam"). In order for this to be a well-defined problem, "approximates as closely as possible" needs to be defined rigorously. In decision theory, this is defined by specifying a loss function that assigns a specific value to "loss" resulting from producing an incorrect label. The goal then is to minimize the expected loss, with the expectation taken over the probability distribution of . In practice, neither the distribution of nor the ground truth function are known exactly, but can be computed only empirically by collecting a large number of samples of and hand-labelling them using the correct value of (a time-consuming process, which is typically the limiting factor in the amount of data of this sort that can be collected). The particular loss function depends on the type of label being predicted.

For example, in the case of classification, the simple zero-one loss function is often sufficient. This corresponds simply to assigning a loss of 1 to any incorrect labelling and is equivalent to computing the accuracy of the classification procedure over the set of test data (i.e. counting up the fraction of instances that the learned function $h:\chi \to y$ labels correctly. The goal of the learning procedure is to maximize this test accuracy on a "typical" test set.

For a probabilistic pattern recognizer, the problem is instead to estimate the probability of each possible output label given a particular input instance, i.e. to estimate a function of the form

$$p(label \mid x,\theta) = f(x;\theta)$$

where the feature vector input is x , and the function f is typically parameterized by some parameters θ. In a discriminative approach to the problem, f is estimated directly. In a generative approach, however, the inverse probability $(px \mid label)$ is instead estimated and combined with the prior probability $p(label \mid \theta)$ using Bayes' rule, as follows:

$$p(label \mid x,\theta) = \frac{p(x \mid label)\, p(label \mid \theta)}{\sum_{L \in all\ labels} p(x \mid L)\, p(L \mid \theta)}.$$

When the labels are continuously distributed (e.g. in regression analysis), the denominator involves integration rather than summation:

$$p(label \mid x,\theta) = \frac{p(x \mid label)\, p(label \mid \theta)}{\int_{L \in all\ lables} p(x \mid L)\, p(L \setminus \theta)\, dL}.$$

The value of θ is typically learned using maximum a posteriori (MAP) estimation.

This finds the best value that simultaneously meets two conflicting objects: To perform as well as possible on the training data and to find the simplest possible model. Essentially, this combines maximum likelihood estimation with a regularization procedure that favours simpler models over more complex models. In a Bayesian context, the regularization procedure can be viewed as placing a prior probability $p(\theta)$ on different values of θ. Mathematically:

$$\theta^* = \arg\max p(\theta \mid D)$$

where θ^* is the value used for θ in the subsequent evaluation procedure, and $p(\theta \mid D)$,, the posterior probability of θ, is given by

$$p(\theta \mid D) = \left[\prod_{i=1}^{n} p(y_i \mid x_i,\theta)\right] p(\theta)$$

In the Bayesian approach to this problem, instead of choosing a single parameter vector θ^*, the probability of a given label for a new instance x is computed by integrating over all possible values of θ, weighted according to the posterior probability:

Uses

Within medical science, pattern recognition is the basis for computer-aided diagnosis (CAD) systems. CAD describes a procedure that supports the doctor's interpretations and findings.

Typical applications are automatic speech recognition, classification of text into several categories (e.g. spam/non-spam email messages), the automatic recognition of handwritten postal codes on postal envelopes, automatic recognition of images of human faces, or handwriting image extraction from medical forms. The last two examples form the subtopic image analysis of pattern recognition that deals with digital images as input to pattern recognition systems. The method of signing one's name was captured with stylus and overlay starting 1990. The strokes, speed, relative min, relative max, acceleration and pressure is used to uniquely identify and confirm identity. Banks were first offered this technology, but were content to collect from the FDIC for any bank fraud and didn't want to inconvenience customers..

Algorithms

Algorithms for pattern recognition depend on the type of label output, on whether learning is supervised or unsupervised, and on whether the algorithm is statistical or non-statistical in nature. Statistical algorithms can further be categorized as generative or discriminative.

Classification Algorithms (Supervised Algorithms Predicting Categoricallabels)

- Maximum entropy classifier (aka logistic regression, multinomial logistic regression): Note that logistic regression is an algorithm for classification, despite its name. (The name comes from the fact that logistic regression uses an extension of a linear regression model to model the probability of an input being in a particular class.)
- Naive Bayes classifier
- Decision trees, decision lists
- Support vector machines
- Kernel estimation and K-nearest-neighbour algorithms
- Perceptrons
- Neural networks (multi-level perceptrons)

Clustering Algorithms (Unsupervised Algorithms Predicting Categoricallabels)

- Categorical mixture models
- K-means clustering
- Hierarchical clustering (agglomerative or divisive)
- Kernel principal component analysis (Kernel PCA).

Regression Algorithms (Predicting Real-valued Labels)

Supervised:

- Linear regression and extensions
- Neural networks
- Gaussian process regression (kriging)

Unsupervised:

- Principal components analysis (PCA)
- Independent component analysis (ICA).

Categorical Sequence Labelling Algorithms (Predicting Sequences of Categorical Labels)

Supervised:

- Hidden Markov models (HMMs)
- Maximum entropy Markov models (MEMMs)
- Conditional random fields (CRFs).

Unsupervised:

- Hidden Markov models (HMMs).

Real-valued sequence labelling algorithms (predicting sequences of real-valued labels)

Supervised (?):

- Kalman filters
- Particle filters.

Parsing Algorithms (Predicting Tree Structured Labels)

Supervised and unsupervised:

- Probabilistic context free grammars (PCFGs).

General Algorithms for Predicting Arbitrarily-Structured Labels

- Bayesian networks
- Markov random fields.

Ensemble learning algorithms (supervised meta-algorithms for combining multiple learning algorithms together)

- Bootstrap aggregating ("bagging")
- Boosting
- Ensemble averaging
- Mixture of experts, hierarchical mixture of experts.

Computer Data Processing

Computer data processing is any process that a computer program does to enter data and summarise, analyse or otherwise convert data into usable information. The process may be automated and run on a computer. It involves recording, analysing, sorting, summarising, calculating, disseminating and storing data. Because data are most useful when well-presented and actually informative, data-processing systems are often referred to as information systems. Nevertheless, the terms are roughly synonymous, performing similar conversions; data-processing systems typically manipulate raw data into information, and likewise information systems typically take raw data as input to produce information as output. Data processing may or may not be distinguished from data conversion, when the process is merely to convert data to another format, and does not involve any data manipulation.

Data Analysis

Data Analysis is the domain from which the data are harvested is a science or an engineering field. Data processing and information systems are considered terms that are too broad and the more specialized term data analysis is typically used. This is a focus on the highly-specialized and highly-accurate algorithmic derivations and statistical calculations that are less often observed in the typical general business environment.

In these contexts data analysis packages like DAP, gretl or PSPP are often used. This divergence of culture is exhibited in the typical numerical representations used in data processing versus numerical; data processing's measurements are typically represented by integersor by fixed-point or binary-coded decimal representations of numbers whereas the majority of data analysis's measurements are often represented by floating-point representation of rational numbers.

Processing

Basically, data are nothing but facts (organized or unorganized) which can be converted into other forms to make it useful, clear and pratically used. This process of converting facts to information is Processing. Practically all naturally occurring processes can be viewed as examples of data processing systems where "observable" information in the form of pressure, light, etc. are converted by human observers into electrical signals in the nervous system as the senses we recognize as touch, sound, and vision. Even the interaction of non-living systems may be viewed in this way as rudimentary information processing systems.

Conventional usage of the terms data processing and information systems restricts their use to refer to the algorithmic derivations, logical deductions, and statistical calculations that recur perennially in general business environments, rather than in the more expansive sense of all conversions of real-world measurements into real-world information in, say, an organic biological system or even a scientific or engineering system.

Elements of Data Processing

In order to be processed by a computer, data needs first be converted into a machine readable format. Once data are in digital format, various procedures can be applied on the data to get useful information. Data processing may involve various processes, including:

- Data summarization
- Data aggregation

- Data validation
- Data tabulation
- Statistical analysis.

Blind Signal Separation

Blind signal separation, also known as blind source separation, is the separation of a set of signals from a set of mixed signals, without the aid of information (or with very little information) about the source signals or the mixing process.

Blind signal separation relies on the assumption that the source signals do not correlate with each other. For example, the signals may be mutually statistically independent or decorrelated. Blind signal separation thus separates a set of signals into a set of other signals, such that the regularity of each resulting signal is maximized, and the regularity between the signals is minimized (i.e. statistical independence is maximized).

Because temporal redundancies (statistical regularities in the time domain) are "clumped" in this way into the resulting signals, the resulting signals can be more effectively deconvolved than the original signals.

There are different methods of blind signal separation:

- Principal components analysis
- Singular value decomposition
- Independent component analysis
- Dependent component analysis
- Non-negative matrix factorization
- Low-Complexity Coding and Decoding
- Stationary Subspace Analysis.

Data Mining

Data mining (the analysis step of the knowledge discovery in databases process, or KDD), a relatively young and interdisciplinary field of computer science is the process of discovering new patterns from large data sets involving methods at the intersection of artificial intelligence, machine learning, statistics and database systems. The goal of data mining is to extract knowledge from a data set in a human-understandable structure and involves database and data management, data preprocessing, model and inference considerations, interestingness metrics, complexity considerations, post-processing of found structure, visualization and online updating.

The term is a buzzword, and is frequently misused to mean any form of large scale data or information processing (collection, extraction, warehousing, analysis and statistics) but also generalised to any kind of computer decision support system including artificial intelligence, machine learning and business intelligence. In the proper use of the word, the key term is discovery, commonly defined as "detecting something new". Even the popular book "Data mining: Practical machine learning tools and techniques with Java" (which covers mostly machine learning material) was originally to be named just "Practical machine learning", and the term "data mining" was only added for marketing reasons. Often the more general terms "(large scale) data analysis" or "analytics" or when referring to actual methods, artificial intelligence and machine learning are more appropriate.

The actual data-mining task is the automatic or semi-automatic analysis of large quantities of data to extract previously unknown interesting patterns such as groups of data records (cluster analysis), unusual records (anomaly detection) and dependencies (association rule mining). This usually involves using database techniques such as spatial indexes. These patterns can then be seen as a kind of summary of the input data, and used in further analysis or for example in machine learning and predictive analytics. For example, the data mining step might identify multiple groups in the data, which can then be used to obtain more accurate prediction results by a decision support system. Neither the data collection, data preparation nor result interpretation and reporting are part of the data mining step, but do belong to the overall KDD process as additional steps.

The related terms data dredging, data fishing and data snooping refer to the use of data mining methods to sample parts of a larger population data set that are (or may be) too small for reliable statistical inferences to be made about the validity of any patterns discovered. These methods can, however, be used in creating new hypotheses to test against the larger data populations.

Background

The manual extraction of patterns from data has occurred for centuries. Early methods of identifying patterns in data include Bayes' theorem (1700s) and regression analysis (1800s). The proliferation, ubiquity and increasing power of computer technology has increased data collection, storage and manipulations. As data sets have grown in size and complexity, direct hands-on data analysis has increasingly been augmented with indirect, automatic data processing. This has been aided by other discoveries in computer science, such as neural networks, cluster analysis, genetic algorithms (1950s), decision trees (1960s) and support vector machines (1990s). Data mining is the process of applying these methods to data with the intention of uncovering hidden patterns in large data sets. It bridges the gap from applied statistics and artificial intelligence (which usually provide the mathematical background) to database management by exploiting the way data is stored and indexed in databases to execute the actual learning and discovery algorithms more efficiently, allowing such methods to be applied to larger data sets.

Research and Evolution

The premier professional body in the field is the Association for Computing Machinery's Special Interest Group on knowledge discovery and Data Mining (SIGKDD). Since 1989 they have hosted an annual international conference and published its proceedings, and since 1999 have published a biannual academic journal titled "SIGKDD Explorations".

Computer science conferences on data mining include:

- CIKM – ACM Conference on Information and Knowledge Management
- DMIN – International Conference on Data Mining
- DMKD – Research Issues on Data Mining and Knowledge Discovery
- ECDM – European Conference on Data Mining
- ECML-PKDD – European Conference on Machine Learning and Principles and Practice of Knowledge Discovery in Databases
- EDM – International Conference on Educational Data Mining
- ICDM – IEEE International Conference on Data Mining
- KDD – ACM SIGKDD Conference on Knowledge Discovery and Data Mining
- MLDM – Machine Learning and Data Mining in Pattern Recognition
- PAKDD – The annual Pacific-Asia Conference on Knowledge Discovery and Data Mining
- PAW – Predictive Analytics World
- SDM – SIAM International Conference on Data Mining (SIAM)
- SSTD – Symposium on Spatial and Temporal Databases

Data mining topics are present on most data management/database conferences.

Process

The knowledge discovery in databases (KDD) process is commonly defined with the stages (1) Selection (2) Preprocessing (3) Transformation (4) Data Mining (5) Interpretation/Evaluation.

It exists however in many variations of this theme such as the CRoss Industry Standard Process for Data Mining (CRISP-DM) which defines six phases: (1) Business Understanding, (2) Data Understanding, (3) Data Preparation, (4) Modelling, (5) Evaluation, and (6) Deployment or a simplified process such as (1) Pre-processing, (2) Data mining, and (3) Results validation.

Pre-processing

Before data mining algorithms can be used, a target data set must be assembled. As data mining can only uncover patterns actually present in the data, the target dataset must be large enough to contain these patterns while remaining concise enough to be mined in an acceptable timeframe. A common source for data is a data mart or data warehouse. Pre-process is essential to analyse the multivariate datasets before data mining.

The target set is then cleaned. Data cleaning removes the observations with noise and missing data.

Data Mining

Data mining involves six common classes of tasks:

- Anomaly detection (Outlier/change/deviation detection) – The identification of unusual data records, that might be interesting or data errors and require further investigation.
- Association rule learning (Dependency modelling) – Searches for relationships between variables. For example a supermarket might gather data on customer purchasing habits. Using association rule learning, the supermarket can determine which products are frequently bought together and use this information for marketing purposes. This is sometimes referred to as market basket analysis.
- Clustering – is the task of discovering groups and structures in the data that are in some way or another "similar", without using known structures in the data.
- Classification – is the task of generalising known structure to apply to new data. For example, an email program might attempt to classify an email as legitimate or spam.
- Regression – Attempts to find a function which models the data with the least error.
- Summarization – providing a more compact representation of the data set, including visualization and report generation.

Results Validation

The final step of knowledge discovery from data is to verify the patterns produced by the data mining algorithms occur in the wider data set. Not all patterns found by the data mining algorithms are necessarily valid. It is common for the data mining algorithms to find patterns in the training set which are not present in the general data set. This is called overfitting. To overcome this, the evaluation uses a test set of data on which the data mining algorithm was not trained. The learned patterns are applied to this test set and the resulting output is compared to the desired output. For example, a data mining algorithm trying to distinguish spam from legitimate emails would be trained on a training set of sample emails. Once trained, the learned patterns would be applied to the test set of emails on which it had not been trained. The accuracy of these patterns can then be measured from how many emails they correctly classify. A number of statistical methods may be used to evaluate the algorithm such as ROC curves. If the learned patterns do not meet the desired standards, then it is necessary to re-evaluate and change the pre-processing and data mining. If the learned patterns do meet the desired standards then the final step is to interpret the learned patterns and turn them into knowledge.

Standards

There have been some efforts to define standards for the data mining process, for example the 1999 European Cross Industry Standard Process for Data Mining (CRISP-DM 1.0) and the 2004 Java Data Mining standard (JDM 1.0). Development on successors of these processes (CRISP-DM 2.0 and JDM 2.0) was active in 2006, but has stalled since. JDM 2.0 was withdrawn without reaching a final draft. For exchanging the extracted models – in particular for the use in predictive analytics – the key standard is the Predictive Model Markup Language (PMML), which is an XML-based language developed by the Data Mining Group (DMG) and supported as exchange format by many data mining applications. As the name suggests it only covers prediction models, a particular data mining task of high importance to business applications, however extensions to for example cover subspace clustering have been proposed independently of the DMG.

Notable Uses

Games

Since the early 1960s, with the availability of oracles for certain combinatorial games, also called tablebases (e.g. for 3x3-chess) with any beginning configuration, small-board dots-and-boxes, small-board-hex, and certain endgames in chess, dots-and-boxes, and hex; a new area for data mining has been opened. This is the extraction of human-usable strategies from these oracles. Current pattern recognition approaches do not seem to fully acquire the high level of abstraction required to be applied successfully. Instead, extensive experimentation with the tablebases, combined with an intensive study of tablebase-answers to well designed problems and with knowledge of prior art, i.e. pre-tablebase knowledge, is used to yield insightful patterns. Berlekamp in dots-and-boxes etc. and John Nunn in chess endgames are notable examples of researchers doing this work, though they were not and are not involved in tablebase generation.

Business

Data mining in customer relationship management applications can contribute significantly to the bottom line. Rather than randomly contacting a prospect or customer through a call centre or sending mail, a company can concentrate its efforts on prospects that are predicted to have a high likelihood of responding to an offer. More sophisticated methods may be used to optimise resources across campaigns so that one may predict to which channel and to which offer an individual is most likely to respond—across all potential offers. Additionally, sophisticated applications could be used to automate the mailing. Once the results from data mining (potential prospect/customer and channel/ offer) are determined, this "sophisticated application" can either automatically send an e-mail or regular mail. Finally, in cases where many people will take an action without an offer, uplift modelling can be used to determine which people will have the greatest increase in responding if given an offer. Data clustering can also be used to automatically discover the segments or groups within a customer data set.

Businesses employing data mining may see a return on investment, but also they recognize that the number of predictive models can quickly become very large. Rather than one model to predict how many customers will churn, a business could build a separate model for each region and customer type. Then instead of sending an offer to all people that are likely to churn, it may only want to send offers to loyal customers. Finally, it may want to determine which customers are going to be profitable over a window of time and only send the offers to those that are likely to be profitable. In order to maintain this quantity of models, they need to manage model versions and move to automated data mining.

Data mining can also be helpful to human-resources departments in identifying the characteristics of their most successful employees. Information obtained, such as universities attended by highly successful employees, can help HR focus recruiting efforts accordingly. Additionally, Strategic Enterprise Management applications help a company translate corporate-level goals, such as profit and margin share targets, into operational decisions, such as production plans and workforce levels.

Another example of data mining, often called the market basket analysis, relates to its use in retail sales. If a clothing store records the purchases of customers, a data-mining system could identify those customers who favour silk shirts over cotton ones. Although some explanations of relationships may be difficult, taking advantage of it is easier. The example deals with association rules within transaction-based data. Not all data are transaction based and logical or inexact rules may also be present within a database. Market basket analysis has also been used to identify the purchase patterns of the Alpha consumer. Alpha Consumers are people that play a key role in connecting with the concept behind a product, then adopting that product, and finally validating it for the rest of society.

Analysing the data collected on this type of user has allowed companies to predict future buying trends and forecast supply demands. Data Mining is a highly effective tool in the catalogue marketing industry. Catalogers have a rich history of customer transactions on millions of customers dating back several years. Data mining tools can identify patterns among customers and help identify the most likely customers to respond to upcoming mailing campaigns.

Data Mining for business applications is a component which needs to be integrated into a complex modelling and decision making process. Reactive Business Intelligence (RBI) advocates a holistic approach that integrates data mining, modelling and interactive visualization, into an end-to-end discovery and continuous innovation process powered by human and automated learning. In the area of decision making the RBI approach has been used to mine the knowledge which is progressively acquired from the decision maker and self-tune the decision method accordingly.

Related to an integrated-circuit production line, an example of data mining is described in the paper "Mining IC Test Data to Optimise VLSI Testing." In this paper the application of data mining and decision analysis to the problem of die-level functional test is described. Experiments mentioned in this paper demonstrate the ability of applying a system of mining historical die-test data to create a probabilistic model of patterns of die failure.

These patterns are then utilized to decide in real time which die to test next and when to stop testing. This system has been shown, based on experiments with historical test data, to have the potential to improve profits on mature IC products.

Science and Engineering

In recent years, data mining has been used widely in the areas of science and engineering, such as bioinformatics, genetics, medicine, education and electrical power engineering.

In the study of human genetics, an important goal is to understand the mapping relationship between the inter-individual variation in human DNA sequences and variability in disease susceptibility. In lay terms, it is to find out how the changes in an individual's DNA sequence affect the risk of developing common diseases such as cancer. This is very important to help improve the diagnosis, prevention and treatment of the diseases. The data mining method that is used to perform this task is known as multifactor dimensionality reduction.

In the area of electrical power engineering, data mining methods have been widely used for condition monitoring of high voltage electrical equipment. The purpose of condition monitoring is to obtain valuable information on the insulation's health status of the equipment. Data clustering such as self-organizing map (SOM) has been applied on the vibration monitoring and analysis of transformer on-load tap-changers (OLTCS).

Using vibration monitoring, it can be observed that each tap change operation generates a signal that contains information about the condition of the tap changer contacts and the drive mechanisms. Obviously, different tap positions will generate different signals. However, there was considerable variability amongst normal condition signals for exactly the same tap position. SOM has been applied to detect abnormal conditions and to estimate the nature of the abnormalities.

Data mining methods have also been applied for dissolved gas analysis (DGA) on power transformers. DGA, as a diagnostics for power transformer, has been available for many years. Methods such as SOM has been applied to analyse data and to determine trends which are not obvious to the standard DGA ratio methods such as Duval Triangle.

A fourth area of application for data mining in science/engineering is within educational research, where data mining has been used to study the factors leading students to choose to engage in behaviours which reduce their learning and to understand the factors influencing university student retention.

A similar example of the social application of data mining is its use in expertise finding systems, whereby descriptors of human expertise are extracted, normalized and classified so as to facilitate the finding of experts, particularly in scientific and technical fields. In this way, data mining can facilitate Institutional memory. Other examples of applying data mining method applications are biomedical data facilitated by domain on tologies, mining clinical trial data, traffic analysis using SOM, et cetera. In adverse drug reaction surveillance, the Uppsala Monitoring Centre has, since 1998, used data mining methods to routinely screen for reporting patterns indicative of emerging drug safety issues in the WHO global database of 4.6 million suspected adverse drug reaction incidents. Recently, similar methodology has been developed to mine large collections of electronic health records for temporal patterns associating drug prescriptions to medical diagnoses.

Spatial Data Mining

Spatial data mining is the application of data mining methods to spatial data. Spatial data mining follows along the same functions in data mining, with the end objective to find patterns in geography. So far, data mining and Geographic Information Systems (GIS) have existed as two separate technologies, each with its own methods, traditions and approaches to visualization and data analysis. Particularly, most contemporary GIS have only very basic spatial analysis functionality. The immense explosion in geographically referenced data occasioned by developments in IT, digital mapping, remote sensing, and the global diffusion of GIS emphasizes the importance of developing data driven inductive approaches to geographical analysis and modelling.

Data mining, which is the partially automated search for hidden patterns in large databases, offers great potential benefits for applied GIS-based decision-making. Recently, the task of integrating these two technologies has become critical, especially as various public and private sector organizations possessing huge databases with thematic and geographically referenced data begin to realize the huge potential of the information hidden there. Among those organizations are:

- offices requiring analysis or dissemination of geo-referenced statistical data
- public health services searching for explanations of disease clusters
- environmental agencies assessing the impact of changing land-use patterns on climate change
- geo-marketing companies doing customer segmentation based on spatial location.

Challenges

Geospatial data repositories tend to be very large. Moreover, existing GIS datasets are often splintered into feature and attribute components, that are conventionally archived in hybrid data management systems. Algorithmic requirements differ substantially for relational (attribute) data management and for topological (feature) data management. Related to this is the range and diversity of geographic data formats, that also presents unique challenges. The digital geographic data revolution is creating new types of data formats beyond the traditional "vector" and "raster" formats. Geographic data repositories increasingly include ill-structured data such as imagery and geo-referenced multimedia.

There are several critical research challenges in geographic knowledge discovery and data mining. Miller and Han offer the following list of emerging research topics in the field:

- Developing and supporting geographic data warehouses – Spatial properties are often reduced to simple a spatial attributes in mainstream data warehouses. Creating an integrated GDW requires solving issues in spatial and temporal data interoperability, including differences in semantics, referencing systems, geometry, accuracy and position.

- Better spatio-temporal representations in geographic knowledge discovery – Current geographic knowledge discovery (GKD) methods generally use very simple representations of geographic objects and spatial relationships. Geographic data mining methods should recognize more complex geographic objects (lines and polygons) and relationships (non-Euclidean distances, direction, connectivity and interaction through attributed geographic space such as terrain). Time needs to be more fully integrated into these geographic representations and relationships.
- Geographic knowledge discovery using diverse data types – GKD methods should be developed that can handle diverse data types beyond the traditional raster and vector models, including imagery and geo-referenced multimedia, as well as dynamic data types (video streams, animation).

In four annual surveys of data miners, data mining practitioners consistently identified that they faced three key challenges more than any others:

- Dirty Data
- Explaining Data Mining to Others
- Unavailability of Data/Difficult Access to Data.

In the 2010 survey data miners also shared their experiences in overcoming these challenges.

Visual Data Mining

The process of turning from analogical into digital, large data sets have been generated, collected and stored discovering statistical patterns, trends and information which is hidden in data, in order to build predictive patterns. A study found that Visual Data Mining is faster and much more intuitive than traditional data mining.

Surveillance

Prior data mining to stop terrorist programs under the U.S. government include the Total Information Awareness (TIA) program, Secure Flight (formerly known as Computer-Assisted Passenger Prescreening System (CAPPS II)), Analysis, Dissemination, Visualization, Insight, Semantic Enhancement (ADVISE), and the Multi-state Anti-Terrorism Information Exchange (MATRIX). These programs have been discontinued due to controversy over whether they violate the US Constitution's 4th amendment, although many programs that were formed under them continue to be funded by different organizations, or under different names.

Two plausible data mining methods in the context of combating terrorism include "pattern mining" and "subject-based data mining".

Pattern Mining

"Pattern mining" is a data mining method that involves finding existing patterns in data. In this context patterns often means association rules. The original motivation for searching association rules came from the desire to analyse supermarket transaction data, that is, to examine customer behaviour in terms of the purchased products. For example, an association rule "beer Ò! potato chips (80%)" states that four out of five customers that bought beer also bought potato chips.

In the context of pattern mining as a tool to identify terrorist activity, the National Research Council provides the following definition: "Pattern-based data mining looks for patterns (including anomalous data patterns) that might be associated with terrorist activity — these patterns might be regarded as small signals in a large ocean of noise."

Pattern Mining includes new areas such a Music Information Retrieval (MIR) where patterns seen both in the temporal and non temporal domains are imported to classical knowledge discovery search methods.

Subject-based Data Mining

"Subject-based data mining" is a data mining method involving the search for associations between individuals in data. In the context of combating terrorism, the National Research Council provides the following definition: "Subject-based data mining uses an initiating individual or other datum that is considered, based on other information, to be of high interest, and the goal is to determine what other persons or financial transactions or movements, etc., are related to that initiating datum."

Knowledge Grid

Researchers at the University of Calabria developed a Knowledge Grid architecture for distributed knowledge discovery, based on grid computing.

Privacy Concerns and Ethics

Some people believe that data mining itself is ethically neutral. It is important to note that the term data mining has no ethical implications. The term is often associated with the mining of information in relation to peoples' behaviour.

However, data mining is a statistical method that is applied to a set of information, or a data set. Associating these data sets with people is an extreme narrowing of the types of data that are available in today's technological society.

Examples could range from a set of crash test data for passenger vehicles, to the performance of a group of stocks. These types of data sets make up a great proportion of the information available to be acted on by data mining methods, and rarely have ethical concerns associated with them. However, the ways in which data mining can be used can raise questions regarding privacy, legality, and ethics. In particular, data mining government or commercial data sets for national security or law enforcement purposes, such as in the Total Information Awareness Program or in ADVISE, has raised privacy concerns.

Data mining requires data preparation which can uncover information or patterns which may compromise confidentiality and privacy obligations. A common way for this to occur is through data aggregation.

Data aggregation is when the data are accrued, possibly from various sources, and put together so that they can be analysed. This is not data mining per se, but a result of the preparation of data before and for the purposes of the analysis. The threat to an individual's privacy comes into play when the data, once compiled, cause the data miner, or anyone who has access to the newly compiled data set, to be able to identify specific individuals, especially when originally the data were anonymous.

It is recommended that an individual is made aware of the following before data are collected:

- the purpose of the data collection and any data mining projects,
- how the data will be used,
- who will be able to mine the data and use them,
- the security surrounding access to the data, and in addition,
- how collected data can be updated.

In the United States, privacy concerns have been somewhat addressed by their congress via the passage of regulatory controls such as the Health Insurance Portability and Accountability Act (HIPAA). The HIPAA requires individuals to be given "informed consent" regarding any information that they provide and its intended future uses by the facility receiving that information. According to an article in Biotech Business Week, "In practice, HIPAA may not offer any greater protection than the longstanding regulations in the research arena, says the AAHC. More importantly, the rule's goal of protection through informed consent is undermined by the complexity of consent forms that are required of patients and participants, which approach a level of incomprehensibility to average individuals." This underscores the necessity for data anonymity in data aggregation practices.

One may additionally modify the data so that they are anonymous, so that individuals may not be readily identified. However, even de-identified data sets can contain enough information to identify individuals, as occurred when journalists were able to find several individuals based on a set of search histories that were inadvertently released by AOL.

Software

Free libre open-source data-mining software and applications

1. Carrot2 – Text and search results clustering framework.
2. Chemicalize.org – A chemical structure miner and web search engine.
3. ELKI – A university research project with advanced cluster analysis and outlier detection methods written in the Java language.
4. GATE – Natural language processing and language engineering tool.
5. JHepWork – Java cross-platform data analysis framework developed at ANL.
6. KNIME – The Konstanz Information Miner, a user friendly and comprehensive data analytics framework.
7. NLTK or Natural Language Toolkit – A suite of libraries and programs for symbolic and statistical natural language processing (NLP) for the Python language.
8. Orange – A component-based data mining and machine learning software suite written in the Python language.
9. R – A programming language and software environment for statistical computing, data mining and graphics. It is part of the GNU project.
10. RapidMiner – An environment for machine learning and data mining experiments.
11. UIMA – The UIMA (Unstructured Information Management Architecture) is a component framework for analysing unstructured content such as text, audio and video, originally developed by IBM.
12. Weka – A suite of machine learning software written in the Java language.

In 2010, the open source R language overtook other tools to become the tool used by more data miners (43%) than any other.

Commercial Data-Mining Software and Applications

- Microsoft Analysis Services data mining software provided by Microsoft
- SAS Enterprise Miner – data mining software provided by the SAS Institute.
- SPSS Modeler – data mining software provided by IBM SPSS.
- STATISTICA Data Miner – data mining software provided by StatSoft.

According to Rexer's Annual Data Miner Survey in 2010, IBM SPSS Modeler, STATISTICA Data Miner and R received the strongest satisfaction ratings.

Marketplace Surveys

Several researchers and organizations have conducted reviews of data mining tools and surveys of data miners. These identify some of the strengths and weaknesses of the software packages. They also provide an overview of the behaviours, preferences and views of data miners. Some of these reports include:

- Annual Rexer Analytics Data Miner Surveys.
- Forrester Research 2010 Predictive Analytics and Data Mining Solutions report.
- Gartner 2008 "Magic Quadrant" report.
- Haughton et al.'s 2003 Review of Data Mining Software Packages in The American Statistician.
- Robert A. Nisbet's 2006 Three Part Series of articles "Data Mining Tools: Which One is Best For CRM?"
- 2011 Wiley Interdisciplinary Reviews: Data Mining and Knowledge Discovery.

Email Spam

Email spam, also known as junk email or unsolicited bulk email (UBE), is a subset of spam that involves nearly identical messages sent to numerous recipients by email. Definitions of spam usually include the aspects that email is unsolicited and sent in bulk. One subset of UBE is UCE (unsolicited commercial email). The opposite of "spam", email which one wants, is called "ham", usually when referring to a message's automated analysis (such as Bayesian filtering).

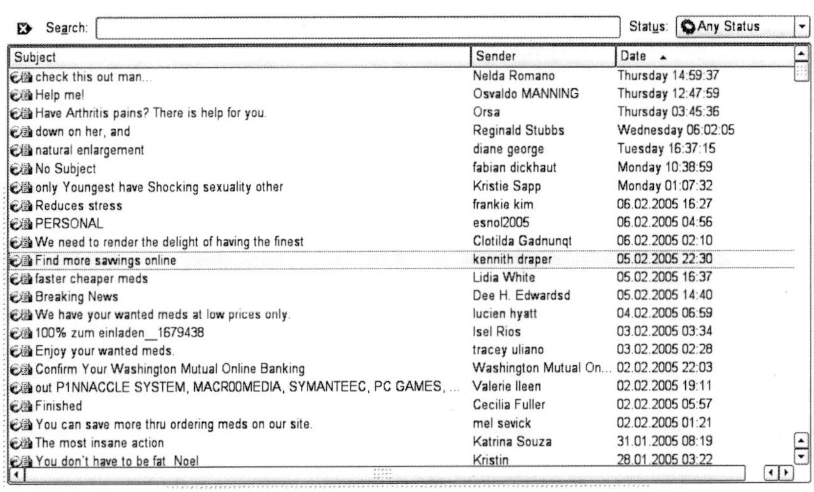

Figure: *An email box folder filled with spam messages.*

Email spam has steadily grown since the early 1990s. Botnets, networks of virus-infected computers, are used to send about 80% of spam. Since the expense of the spam is borne mostly by the recipient, it is effectively postage due advertising.

The legal status of spam varies from one jurisdiction to another. In the United States, spam was declared to be legal by the CAN-SPAM Act of 2003 provided the message adheres to certain specifications. ISPs have attempted to recover the cost of spam through lawsuits against spammers, although they have been mostly unsuccessful in collecting damages despite winning in court.

Spammers collect email addresses from chatrooms, customer lists, newsgroups, and viruses which harvest users' address books, and are sold to other spammers. They also use a practice known as "email appending" or "epending" in which they use known information about their target (such as a postal address) to search for the target's email address. Much of spam is sent to invalid email addresses. Spam averages 78% of all email sent. According to the Message Anti-Abuse Working Group, the amount of spam email was between 88–92% of email messages sent in the first half of 2010.

From the beginning of the Internet (the ARPANET), sending of junk email has been prohibited, enforced by the Terms of Service/Acceptable Use Policy (ToS/AUP) of internet service providers (ISPs) and peer pressure. Even with a thousand users junk email for advertising is not tenable, and with a million users it is not only impractical, but also expensive. It is estimated that spam cost businesses on the order of $100 billion in 2007. As the scale of the spam problem has grown, ISPs and the public have turned to government for relief from spam, which has failed to materialize.

Mutt in Action

Spam has several definitions varying by source.

- Unsolicited bulk email (UBE)—unsolicited email, sent in large quantities.
- Unsolicited commercial email (UCE)—this more restrictive definition is used by regulators whose mandate is to regulate commerce, such as the U.S. Federal Trade Commission.

Spamvertised Sites

According to a Commtouch report in the first quarter of 2010, there are "...183 billion spam messages" sent every day. The most popular spam topic is "pharmacy ads" which make up 81% of email spam messages.

Most Common Products Advertised

According to information compiled by Commtouch Software Ltd., email spam for the first quarter of 2010 can be broken down as follows.

***Table:** EMail Spam by Topic*

Pharmacy	81%
Replica	5.40%
Enhancers	2.30%
Phishing	2.30%
Degrees	1.30%
Casino	1%
Weight Loss	0.40%
Other	30%

419 Scams

Advance fee fraud spam such as the Nigerian "419" scam may be sent by a single individual from a cyber cafe in a developing country. Organized "spam gangs" operating from Russia or eastern Europeshare many features in common with other forms of organized crime, including turf battles and revenge killings.

Phishing

Spam is also a medium for fraudsters to scam users into entering personal information on fake Web sites using emails forged to look like they are from banks or other organizations, such as PayPal. This is known as phishing. Targeted phishing, where known information about the recipient is used to created forged emails, is known as spear-phishing.

Spam Techniques

Appending

If a marketer has one database containing names, addresses, and telephone numbers of prospective customers, they can pay to have their database matched against an external database containing email addresses.

The company then has the means to send email to persons who have not requested email, which may include persons who have deliberately withheld their email address.

Image Spam

Image spam is an obfuscating method in which the text of the message is stored as a GIF or JPEGimage and displayed in the email. This prevents text based spam filters from detecting and blocking spam messages. Image spam was reportedly used in the mid 2000s to advertise "pump and dump" stocks.

Often, image spam contains nonsensical, computer-generated text which simply annoys the reader. However, new technology in some programs try to read the images by attempting to find text in these images. They are not very accurate, and sometimes filter out innocent images of products like a box that has words on it.

A newer technique, however, is to use an animated GIF image that does not contain clear text in its initial frame, or to contort the shapes of letters in the image (as in CAPTCHA) to avoid detection by OCR tools.

Blank Spam

Blank spam is spam lacking a payload advertisement. Often the message body is missing altogether, as well as the subject line. Still, it fits the definition of spam because of its nature as bulk and unsolicited email. Blank spam may be originated in different ways, either intentional or unintentionally:

1. Blank spam can have been sent in a directory harvest attack, a form of dictionary attack for gathering valid addresses from an email service provider. Since the goal in such an attack is to use the bounces to separate invalid addresses from the valid ones, spammers may dispense with most elements of the header and the entire message body, and still accomplish their goals.
2. Blank spam may also occur when a spammer forgets or otherwise fails to add the payload when he or she sets up the spam run.
3. Often blank spam headers appear truncated, suggesting that computer glitches may have contributed to this problem—from poorly-written spam software to malfunctioning relay servers, or any problems that may truncate header lines from the message body.
4. Some spam may appear to be blank when in fact it is not. An example of this is the VBS. Davinia.B email worm which propagates through messages that have no subject line and appears blank, when in fact it uses HTML code to download other files.

Backscatter Spam

Backscatter is a side-effect of email spam, viruses and worms, where email servers receiving spam and other mail send bounce messages to an innocent party. This occurs because the original message's envelope sender is forged to contain the email address of the victim. A very large proportion of such email is sent with a forged From: header, matching the envelope sender.

Since these messages were not solicited by the recipients, are substantially similar to each other, and are delivered in bulk quantities, they qualify as unsolicited bulk email or spam. As such, systems that generate email backscatter can end up being listed on various DNSBLs and be in violation of internet service providers' Terms of Service.

Legality

Sending spam violates the Acceptable use policy (AUP) of almost all Internet service providers. Providers vary in their willingness or ability to enforce their AUP. Some actively enforce their terms and terminate spammers' accounts without warning. Some ISPs lack adequate personnel or technical skills for enforcement, while others may be reluctant to enforce restrictive terms against profitable customers.

As the recipient directly bears the cost of delivery, storage, and processing, one could regard spam as the electronic equivalent of "postage-due" junk mail. Due to the low cost of sending unsolicited email and the potential profit entailed, some believe that only strict legal enforcement can stop junk email. The Coalition Against Unsolicited Commercial Email (CAUCE) argues "Today, much of the spam volume is sent by career criminals and malicious hackers who won't stop until they're all rounded up and put in jail."

European Union

All the countries of the European Union have passed laws that specifically target spam.

Article 13 of the European Union Directive on Privacy and Electronic Communications (2002/58/EC) provides that the EU member states shall take appropriate measures to ensure that unsolicited communications for the purposes of direct marketing are not allowed either without the consent of the subscribers concerned or in respect of subscribers who do not wish to receive these communications, the choice between these options to be determined by national legislation.

In the United Kingdom, for example, unsolicited emails cannot be sent to an individual subscriber unless prior permission has been obtained or unless there is a previous relationship between the parties. The regulations can be enforced against an offending company or individual anywhere in the European Union. The Information Commissioner's Office has responsibility for the enforcement of unsolicited emails and considers complaints about breaches. A breach of an enforcement notice is a criminal offence subject to a fine of up to £5000.

Canada

The Government of Canada has passed anti-spam legislation called the Fighting Internet and Wireless Spam Act to fight spam.

Australia

In Australia, the relevant legislation is the Spam Act 2003 which covers some types of email and phone spam, which took effect on 11 April 2004. The

Spam Act provides that "Unsolicited commercial electronic messages must not be sent." Whether an email is unsolicited depends on whether you have consent. Consent can be express or inferred. Express consent is when someone directly instructs you to send them emails, e.g. if they opt-in. Consent can also be inferred from the business relationship between the sender and recipient or if the recipient conspicuously publishes their email address in a public place. Penalties are up to 10,000 penalty units, or 2,000 penalty units for a person other than a body corporate.

United States

In the United States, most states enacted anti-spam laws during the late 1990s and early 2000s. Many of these have since been pre-empted by the less restrictive CAN-SPAM Act of 2003.

Spam is legally permissible according to the CAN-SPAM Act of 2003 provided it follows certain criteria: a "truthful" subject line, no forged information in the technical headers or sender address, and other minor requirements. If the spam fails to comply with any of these requirements it is illegal. Aggravated or accelerated penalties apply if the spammer harvested the email addresses using methods described earlier.

A review of the effectiveness of CAN-SPAM in 2005 by the Federal Trade Commission (the agency charged with CAN-SPAM enforcement) stated that the amount of sexually explicit spam had significantly decreased since 2003 and the total volume had begun to level off. Senator Conrad Burns, a principal sponsor, noted that "Enforcement is key regarding the CAN-SPAM legislation." In 2004, less than 1% of spam complied with the CAN-SPAM Act of 2003. In contrast to the FTC evaluation, many observers view the CAN-SPAM act as having failed in its purpose of reducing spam.

Other Laws

Accessing privately owned computer resources without the owner's permission counts as illegal undercomputer crime statutes in most nations. Deliberate spreading of computer viruses is also illegal in the United States and elsewhere. Thus, some common behaviours of spammers are criminal regardless of the legality of spamming per se. Even before the advent of laws specifically banning or regulating spamming, spammers were successfully prosecuted under computer fraud and abuse laws for wrongfully using others' computers.

The use of botnets can be perceived as theft. The spammer consumes a zombie owner's bandwidth and resources without any cost. In addition, spam is perceived as theft of services. The receiving SMTP servers consume significant amounts of system resources dealing with this unwanted traffic. As a result, service providers have to spend large amounts of money to make their systems capable of handling these amounts of email. Such costs are inevitably passed on to the service providers' customers. Other laws, not only those related to spam, have been used to prosecute alleged spammers. For example, Alan Ralsky was indicted on stock fraud charges in January 2008, and Robert Solowayplead guilty to charges of mail fraud, fraud in connection with email, and failing to file a tax return in March 2008.

Deception and Fraud

Spammers may engage in deliberate fraud to send out their messages. Spammers often use false names, addresses, phone numbers, and other contact information to set up "disposable" accounts at various Internet service providers. They also often use falsified or stolen credit card numbers to pay for these accounts. This allows them to move quickly from one account to the next as the host ISPs discover and shut down each one.

Senders may go to great lengths to conceal the origin of their messages. Large companies may hire another firm to send their messages so that complaints or blocking of email falls on a third party. Others engage in spoofing of email addresses (much easier than IP address spoofing). The email protocol (SMTP) has no authentication by default, so the spammer can pretend to originate a message apparently from any email address. To prevent this, some ISPs and domains require the use of SMTP-AUTH, allowing positive identification of the specific account from which an email originates.

Senders cannot completely spoof email delivery chains (the 'Received' header), since the receiving mailserver records the actual connection from the last mailserver's IP address. To counter this, some spammers forge additional delivery headers to make it appear as if the email had previously traversed many legitimate servers.

Spoofing can have serious consequences for legitimate email users. Not only can their email in boxes get clogged up with "undeliverable" emails in addition to volumes of spam, they can mistakenly be identified as a spammer. Not only may they receive irate email from spam victims, but (if spam victims report the email address owner to the ISP, for example) a naive ISP may terminate their service for spamming.

Theft of Service

Spammers frequently seek out and make use of vulnerable third-party systems such as open mail relays and open proxy servers. SMTP forwards mail from one server to another—mail servers that ISPs run commonly require some form of authentication to ensure that the user is a customer of that ISP. Open relays, however, do not properly check who is using the mail server and pass all mail to the destination address, making it harder to track down spammers.

Increasingly, spammers use networks of malware-infected PCs (zombies) to send their spam. Zombie networks are also known as Botnets (such zombifying malware is known as a bot, short for robot). In June 2006, an estimated 80% of email spam was sent by zombie PCs, an increase of 30% from the prior year. An estimated 55 billion email spam were sent each day in June 2006, an increase of 25 billion per day from June 2005. For Q1 2010, an estimated 305,000 newly activated zombie PCs were brought online each day for malicious activity. This number is slightly lower than the 312,000 of Q4 2009.

Brazil produced the most zombies in the first quarter of 2010. Brazil was the source of 20% of all zombies, which is down from 14% from the fourth quarter 2009. India had 10%, with Vietnam at 8%, and the Russian Federation at 7%.

Side Effects

To combat the problems posed by botnets, open relays and proxy servers many email server administrators pre-emptively block dynamic IP ranges and impose stringent requirements on other servers wishing to deliver mail. Forward-confirmed reverse DNS must be correctly set for the outgoing mail server and large swaths of IP addresses are blocked sometimes pre-emptively to prevent spam. These measures can pose problems for people wanting to run a small email server off an inexpensive domestic connection. Blacklisting of IP ranges due to spam eminating from them also causes problems for legitimate email servers in the same IP range.

Statistics and Estimates

The total volume of email spam has been consistently growing, but in 2011 the trend seems to have reversed. The amount of spam users see in their mailboxes is only a portion of total spam sent, since spammers' lists often contain a large percentage of invalid addresses and many spam filters simply delete or reject "obvious spam."

The first known spam email, advertising a DEC product presentation, was sent in 1978 by Gary Thuerk to 600 addresses, which was all the users of ARPANET at the time, though software limitations meant only slightly more than half of the intended recipients actually received it. As of August 2010, the amount of spam was estimated to be around 200 billion spam messages sent per day.

More than 97% of all emails sent over the net are unwanted, according to a Microsoft security report. MAAWG estimates that 85% of incoming mail is "abusive email", as of the second half of 2007. The sample size for the MAAWG's study was over 100 million mailboxes. A 2010 survey of US and European email users showed that 46% of the respondents had opened spam messages, although only 11% had clicked on a link.

Highest Amount of Spam Received

According to Steve Ballmer, Microsoft founder Bill Gates receives four million emails per year, most of them spam. This was originally incorrectly reported as "per day".

At the same time Jef Poskanzer, owner of the domain name acme.com, was receiving over one million spam emails per day.

Cost of Spam

A 2004 survey estimated that lost productivity costs Internet users in the United States $21.58 billion annually, while another reported the cost at $17 billion, up from $11 billion in 2003. In 2004, the worldwide productivity cost of spam has been estimated to be $50 billion in 2005. An estimate of the percentage cost borne by the sender of marketing junk mail (snail mail) is 88%, whereas in 2001 one spam was estimated to cost $0.10 for the receiver and $0.00001 (0.01% of the cost) for the sender.

Origin of Spam

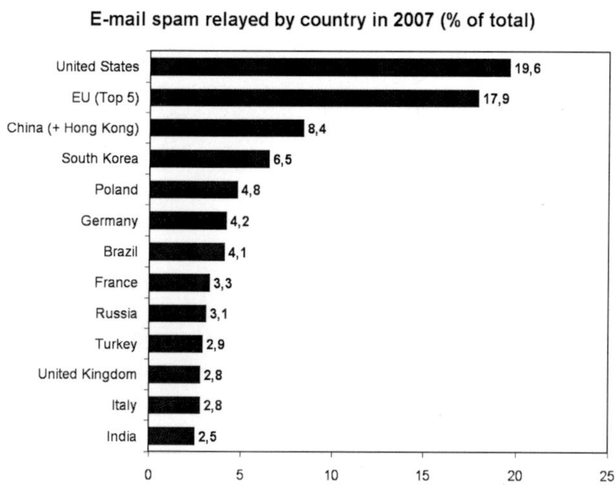

Figure: *Email spam relayed by country in Q2/2007.*

Origin or source of spam refers to the geographical location of the computer from which the spam is sent; it is not the country where the spammer resides, nor the country that hosts the spamvertised site. Because of the international nature of spam, the spammer, the hijacked spam-sending computer, the spamvertised server, and the user target of the spam are all often located in different countries. As much as 80% of spam received by Internet users in North America and Europe can be traced to fewer than 200 spammers. In terms of volume of spam: According to Sophos, the major sources of spam in the fourth quarter of 2008 (October to December) were:

- The United States (the origin of 19.8% of spam messages, up from 18.9% in Q3)
- China (9.9%, up from 5.4%)
- Russia (6.4%, down from 8.3%)
- Brazil (6.3%, up from 4.5%)
- Turkey (4.4%, down from 8.2%).

When grouped by continents, spam comes mostly from:

- Asia (37.8%, down from 39.8%)
- North America (23.6%, up from 21.8%)
- Europe (23.4%, down from 23.9%)
- South America (12.9%, down from 13.2%).

In terms of number of IP addresses: the Spamhaus Project (which measures spam sources in terms of number of IP addresses used for spamming, rather than volume of spam sent) ranks the top three as the United States, China, and Russia, followed by Japan, Canada, and South Korea. In terms of networks: As of 5 June 2007, the three networks hosting the most spammers are Verizon, AT&T, and VSNL International. Verizon inherited many of these spam sources from its acquisition of MCI, specifically through the UUNet subsidiary of MCI, which Verizon subsequently renamed Verizon Business.

Anti-spam Techniques

The U.S. Department of Energy Computer Incident Advisory Capability (CIAC) has provided specific countermeasures against email spamming.

Some popular methods for filtering and refusing spam include email filtering based on the content of the email, DNS-based blackhole lists (DNSBL), greylisting, spamtraps, Enforcing technical requirements of email (SMTP), checksumming systems to detect bulk email, and by putting some sort of cost on the sender via a Proof-of-work system or a micropayment. Each method has strengths and weaknesses and each is controversial because of its weaknesses. For example, one company's offer to "remove some spamtrap and honeypot addresses" from email lists, defeats the ability for those methods to identify spammers. In one study, 95% of revenues (in the study) cleared through just three banks.

How Spammers Operate

Gathering of Addresses

In order to send spam, spammers need to obtain the email addresses of the intended recipients. To this end, both spammers themselves and list merchants gather huge lists of potential email addresses. Since spam is, by definition, unsolicited, this address harvesting is done without the consent (and sometimes against the expressed will) of the address owners. As a consequence, spammers' address lists are inaccurate. A single spam run may target tens of millions of possible addresses – many of which are invalid, malformed, or undeliverable.

Sometimes, if the sent spam is "bounced" or sent back to the sender by various programs that eliminate spam, or if the recipient clicks on an unsubscribe link, that may cause that email address to be marked as "valid", which is interpreted by the spammer as "send me more".

Delivering Spam Messages

Obfuscating Message Content

Many spam-filtering techniques work by searching for patterns in the headers or bodies of messages. For instance, a user may decide that all email they receive with the word "Viagra" in the subject line is spam, and instruct their mail program to automatically delete all such messages. To defeat such filters, the spammer may intentionally misspell commonly filtered words or insert other characters, often in a style similar to leetspeak, as in the following examples: V1agra, Via'gra, Vi@graa,vi*gra, \/iagra. This also allows for many different ways to express a given word, making identifying them all more difficult for filter software. The principle of this method is to leave the word readable to humans (who can easily recognize the intended word for such misspellings), but not likely to be recognized by a literal computer program. This is only somewhat effective, because modern filter patterns have been designed to recognize blacklisted terms in the various iterations of misspelling. Other filters target the actual obfuscation methods, such as the non-standard use of punctuation or numerals into unusual places. Similarly, HTML-based email gives the spammer more tools to obfuscate text. Inserting HTML comments between letters can foil some filters, as can including text made invisible by setting the font colour to white on a white background, or shrinking the font size to the smallest fine print. Another common ploy involves presenting the text as an image, which is either sent along or loaded from a remote server. This can be foiled by not permitting an email-program to load images.

As Bayesian filtering has become popular as a spam-filtering technique, spammers have started using methods to weaken it. To a rough approximation, Bayesian filters rely on word probabilities. If a message contains many words which are only used in spam, and few which are never used in spam, it is likely to be spam. To weaken Bayesian filters, some spammers, alongside the sales pitch, now include lines of irrelevant, random words, in a technique known as Bayesian poisoning. A variant on this tactic may be borrowed from the Usenet abuser known as "Hipcrime" — to include passages from books taken from Project Gutenberg, or nonsense sentences generated with "dissociated press" algorithms. Randomly generated phrases can create spoetry (spam poetry) or spam art. Another method used to masquerade spam as legitimate messages is the use of autogenerated sender names in the From: field, ranging from realistic ones such as "Jackie F. Bird" to (either by mistake or intentionally) bizarre attention-grabbing names such as "Sloppiest U. Epiglottis" or "Attentively E. Behavioural".

Return addresses are also routinely auto-generated, often using unsuspecting domain owners' legitimate domain names, leading some users to blame the innocent domain owners. Blocking lists use IP addresses rather than sender domain names, as these are more accurate. A mail purporting to be from example.com can be seen to be faked by looking for the originating IP address in the email's headers; also Sender Policy Framework, for example, helps by stating that a certain domain will only send email from certain IP addresses. Spam can also be hidden inside a fake "Undelivered mail notification" which looks like the failure notices sent by a mail transfer agent (a "MAILER-DAEMON") when it encounters an error.

Spam-support Services

A number of other online activities and business practices are considered by anti-spam activists to be connected to spamming. These are sometimes termed spam-support services: business services, other than the actual sending of spam itself, which permit the spammer to continue operating. Spam-support services can include processing orders for goods advertised in spam, hosting Web sites or DNS records referenced in spam messages, or a number of specific services as follows: Some Internet hosting firms advertise bulk-friendly or bulletproof hosting. This means that, unlike most ISPs, they will not terminate a customer for spamming. These hosting firms operate as clients of larger ISPs, and many have eventually been taken offline by these larger ISPs as a result of complaints regarding spam activity. Thus, while a firm may advertise bulletproof hosting, it is ultimately unable to deliver without the connivance of its upstream ISP. However, some spammers have managed to get what is called a pink contract – a contract with the ISP that allows them to spam without being disconnected.

A few companies produce spamware, or software designed for spammers. Spamware varies widely, but may include the ability to import thousands of addresses, to generate random addresses, to insert fraudulent headers into messages, to use dozens or hundreds of mail servers simultaneously, and to make use of open relays. The sale of spamware is illegal in eight U.S. states. So-called millions CDs are commonly advertised in spam. These are CD-ROMs purportedly containing lists of email addresses, for use in sending spam to these addresses. Such lists are also sold directly online, frequently with the false claim that the owners of the listed addresses have requested (or "opted in") to be included. Such lists often contain invalid addresses. In recent years, these have fallen almost entirely out of use due to the low quality email addresses available on them, and because some email lists exceed 20GB in size. The amount you can fit on a CD is no longer substantial. A number of DNS blacklists (DNSBLs), including the MAPS RBL, Spamhaus SBL, SORBS and SPEWS, target the providers of spam-support services as well as spammers. DNSBLs blacklist IPs or ranges of IPs to persuade ISPs to terminate services with known customers who are spammers or resell to spammers.

Related Vocabulary

Unsolicited Bulk Email (UBE)

A synonym for email spam.

Unsolicited Commercial Email (UCE)

Spam promoting a commercial service or product. This is the most common type of spam, but it excludes spam which are hoaxes (e.g. virus warnings), political advocacy, religious messages and chain letters sent by a person to many other people. The term UCE may be most common in the USA.

Pink Contract

A pink contract is a service contract offered by an ISP which offers bulk email service to spamming clients, in violation of that ISP's publicly posted acceptable use policy.

Spamvertising

Spamvertising is advertising through the medium of spam.

Opt-in, Confirmed Opt-in, Double opt-in, opt-out

Opt-in, confirmed opt-in, double opt-in, opt-out refers to whether the people on a mailing list are given the option to be put in, or taken out, of the list. Confirmation (and "double", in marketing speak) refers to an email address transmitted eg. through a web form being confirmed to actually request joining a mailing list, instead of being added to the list without verification.

Final, Ultimate Solution for the Spam Problem (FUSSP)

An ironic reference to naïve developers who believe they have invented the perfect spam filter, which will stop all spam from reaching users' inboxes while deleting no legitimate email accidentally.

Bacn

Bacn is an infrequently-used term to refer to email sent to a user who at one time subscribed to a mailing list - not unsolicited, but also not personal.

6

Challenges of Chemical Genomics

The turn of this century marks significant progress in biology. We have advanced from knowing very little about our genetic makeup to now having complete knowledge of our genome. We are able to simultaneously analyze a network of proteins or genes. High throughput technologies have become essential to process enormous genomic information. In this regard, chemcal ligand-based approaches are especially powerful toward global understanding of gene and protein functions.

Presently, the process of discovering target-specific chemical ligands using synthetic chemistry has become a practical means, which largely attributes to technological advances in targetspecific and diversity-based organic synthesis (Schreiber 2000), structure-based chemical ligand (or drug) design/synthesis and new screening technologies such as small chemical compound and protein microarrays. In this article, we will provide a detailed analysis of the current state of chemical genomics and its potential impact on biological and medical research, and pharmaceutical development.

Chemical Biology or Genetics

Since the seminal study of pea genetics by Mendal in 1865, genetic analysis has been the benchmark for understanding gene or protein functions. In classical genetics or forward genetics, the genomic DNA of a model organism or cell is randomly mutagenized to generate large numbers of mutants, which are screened for a desirable phenotype or trait, such as alteration in growth, appearance or behaviour. The phenotypes are then used to identify the responsible genes. A second type of genetic approach called reverse genetics, specifically mutates a known gene and studies the resultant phenotypes to understand the role(s) of the gene in the cell or organism. One major obstacle of traditional genetics is that mutations are usually constitutive. Mutations in essential proteins often lead to lethality at early life stage and make it impossible for subsequent study.

Even if mutations do not cause lethality, the mutant organisms are often able to compensate for the loss of the gene, which obscures the effect of the original mutations. Modified genetic approaches have been developed to generate 'conditional alleles', in which the mutant gene only loses its function under restrictive conditions such as higher temperatures for the temperature-sensitive alleles (ts) or the presence of the Cre recombinase for the Cre/loxP system (Lewin 1997). However, these approaches require complex genetic manipulation, which is often tedious and time-consuming.

Moreover, they are based on blockage of steps leading to translation, which make them ineffective toward studying proteins with long halflives. Over the past two decades, small chemical compounds have greatly enhanced our ability to delineate complex biological pathways and processes.

They can rapidly penetrate into the cells, bind to their target proteins and create loss-of-function (or gain-of-function) phenotypes. In recent years, drug discovery has shifted from predominantly screening for natural products to targetoriented synthesis/screening of chemical ligands, which in large part is due to the realization that targeting the disease-causing genes/proteins and biological pathways is fundamentally important for successful development of highly effective drugs.

This has promoted a close interaction between molecular biologists and synthetic chemists. As a result, an interphase between chemistry and biology called chemical biology or chemical genetics has come of age. More comprehensive analysis of chemical genetics can be found in several recent reviews. In a typical chemical genetics approach, a cell permeable, target-specific chemical ligand is added to the cells of interest, which binds to and causes loss-of-function or gain-of-function of its protein target. Chemical ligands may bind to the enzyme's catalytic site by mimicking natural substrates of the enzyme, as exemplified by the drug lovastatin/mevinolin. Lovastatin interacts with the catalytic site of hydroxymethylglutaryl (HMG)-CoA reductase and inhibits its enzymatic activity. The ligand can also interact with a key regulatory domain as in the case of phorbol ester, which resembles diacylglycerol, a natural protein kinase C (PKC) activating lipid molecule (Newton 1997).

Chemical ligands can be easily administrated at any time during the cell cycle or developmental stages, and any desired location of the cell or organism. Therefore, chemical ligands pose unparalleled temporal and spatial control over their protein targets.

Forward and Reverse Chemical Genetics

The so-called forward chemical genetics describes how natural products (a compound of natural origin) or a pool of synthetic compounds is screened for desired phenotypes, such as inhibition of tumour growth in vitro. Once a biologically active compound is identified, effort is shifted toward identification of the drug target(s). Classical examples include the proteasome inhibitor lactacystin and the immunosuppressant rapamycin (Schreiber 1991).

Both drugs are microbial natural products that inhibit cell proliferation and interfere with normal cellular functions. Lactacystin specifically binds to and inhibits the 20S proteasome. Rapamycin, when associated with the immunophilin FKBP-12, binds to and inhibits the target of rapamycin protein (TOR) (also called FKBP12- rapamycin-associated protein or FRAP). Knowledge of the target proteins has advanced chemical genetics to a stage that parallels the classical reverse genetics, i.e., the use of the compound to study the biological functions of the drug target protein.

For example, as a result of the use of lactacystin, the proteasome is found to be important for diverse biological pathways and processes, including antigen presentation, cell cycle control and cell fate determination. Similarly, studies using rapamycin found that TOR regulates a number of proteins such as ribosomal S6 kinase 1 (S6K1) and the transcription factor Gln3. Target-specific chemical ligands have gained great popularity in biological research in recent years. For instance, there are nearly 2,000 papers citing the use of the MAP kinase kinase (MKK) inhibitor PD98059 and over 1,500 papers reporting the use of the phosphoinositide 3- kinase (PI-3K) inhibitor wortmannin (Hunter 2000).

Protein Engineering in Chemical Genetics

Although there is significant advancement in chemical ligand synthesis and screening, there are still a relatively small number of target-specific chemical ligands available. An alternative approach has been developed to use a combination of protein and chemical engineering to modulate the functions of proteins and genes. An early example of this is the use of a synthetic chemical dimerizer, called FK1012 that is derived from FK506 with two FKBP12-binding moieties.

FK1012 has been used to induce the interaction between the protein of interest that is fused to FKBP12, thereby altering the activity of the protein. Variations of this approach with a single compound such as rapamycin that has two distinct protein-binding surfaces can provide higher specificity and more flexibility for such approaches. Another approach takes advantage of an already existing small molecule ligand or substrate of an enzyme. By a so-called "bump-and-hole" approach, one can add an extra space-filling group onto the ligand (bump) and then create a cavity (hole) in the ligand-binding site of the enzyme. Only the engineered enzyme, but not the native enzyme modified ligand can recognize the modified ligand. Belshaw et al. first described this approach in the study of cyclophilin and its ligand cyclosporin A (Belshaw et al. 1996).

Shotkat and colleagues subsequently used a similar approach to study protein kinases. Protein kinases are highly conserved in their catalytic domains, making it difficult to develop highly specific inhibitors to each individual kinase. Shokat and colleagues mutated the ATP binding site of v-Src such that the engineered v-Src uniquely accepted N6-(cyclopentyl) ATP, an ATP analog to catalyse normal protein phosphorylation.

In contrast, N6-(cyclopentyl) ATP is not accepted by the unmodified v-Src or other protein kinases. Using a variation of the above strategy, the same group also created allelespecific inhibitors of engineered protein kinases (Bishop et al. 1998). These inhibitors specifically inhibit the modified protein kinases, but not the wild type, endogenous kinases. These chemical ligands can be used to effectively turn on and off engineered proteins.

Development of Target-Specific Chemical Ligands

Target-specific chemical ligands are pivotal for studying biological questions and drug development. Natural products have been traditionally the source for novel ligand screening and discovery, primarily because of their unparalleled range in mode of action and chemical diversity. In recent years, alternative strategies have become more practical, which is largely due to significant technological advances in organic synthesis, computational power and the increased information on protein sequences and structures. It is now possible to streamline chemical ligand design, synthesis and selection leading to successful drug development.

Combinatorial organic synthesis has increasingly been used to generate small molecule libraries for ligand screening and drug discovery. It stems from the original solid-phase oligonucleotide and peptide synthesis, which provides a way to synthesize and purify peptides with defined sequences.

Such a technique also allows synthesis of oligonucleotides or peptides of all possible permutations of nucleotides or amino acid residues. This concept has been extended to other small molecules as well. Polycyclic compounds and natural product derivatives that mimic biologically active molecules (biomimetics) are among the most common types of structures that have been made in this fashion. This is because small molecule libraries based on a known lead compound (also known as focused libraries) often provide a better chance in finding new biologically active molecules. Another concept in chemical ligand synthesis that has emerged in recent years is to change the chirality of the compounds. Biological molecules often contain chiral or asymmetric carbon atoms, at which the carbon atom is bonded to four different moieties. Such molecules exist in two forms, known as enantiomers.

These molecules have opposite effect in rotating plane-polarized light, and their three-dimensional structures are different and therefore possess different biological properties. The best example is the twenty naturally occurring amino acids. Living organisms such as yeast can only metabolize L-amino acids, which rotate plane-polarized light to the left, but not D-amino acids, which rotate plane-polarized light to the right. Examples of chirality and its application in chemical ligand synthesis and design are reviewed in detail elsewhere. Lead compounds (de novo design) are pivotal to develop a meaningful chemical library that can eventually yield biologically active compounds. To develop successful lead compounds against diverse protein targets, a designer approach is needed.

This is when the in silico approach comes into play, owning to significant advances in computational power and the development of artificial intelligence (AI). Databases that detail known molecular structures with their corresponding functions form the basis for the so-called Quantitative Structure-Activity Relationship (QSAR) approach. QSAR and its variants provide the statistical framework for analysis of virtual libraries of small molecules (Klebe 2000).

Computer programs and algorithms have been developed to guide every major step in de novo design. Programs such as GASP (Génération Automatique de Structures Polycycliques), which generates a virtual library of a large number of polycyclic structures with the given parameters, can help expand the diversity and complexity of potential chemical ligands. Another area worthy of attention is the use of Darwinian molecular evolution in ligand design.

It dates back to 1990 when Tuerk and Gold looked for RNA ligands that bind to bacteriophage T4 DNA polymerase by using Systematic Evolution of Ligand by Exponential Enrichment (SELEX). This concept has inspired the development of genetic algorithms (GA) and simulated molecular evolution in chemical ligand design. On the basis of existing biological screening results, GAs are used to simulate the docking of chemical ligands to the binding sites of their target proteins, and to optimize and select compounds with desired properties without the need to actually synthesize and test all possible conformations.

The Impact of Comparative Genomics and Structural Genomics in Chemical Ligand Design

Comparison of whole genome sequences of different organisms is a powerful tool for identifying essential components of cellular processes, understanding the genome architecture as well as reconstructing the evolutionary events.

Tatusov et al. reported the comparison of seven complete genomes in five major phylogenetic lineages, and this has now expanded to 44 complete genomes in 30 major phylogenetic lineages. The clusters of orthologous groups (COGs) allow information flow from well-studied model organisms to poorly characterized ones. In recent years, comparative genomics in mammals has improved the resolution of genetic maps in human and other model organisms and identified many human disease genes in other mammalian models. Comparative genomics in more evolutionarily distant organisms allows the identification of the core proteome, the set of proteins that is unique to individual species, as well as components that are highly conserved in a specific cellular or developmental process. Aravind et al. reported a comparative analysis of the apoptotic molecular machinery in human, Drosophila melanogaster and Caenorhabditis elegans.

They noticed that relative to the invertebrates in human there is a general increase in the number of apoptosis-associated proteins and in the complexity of protein domains involved. Knowing the intrinsic difference in any cellular process between human and other species will better help us to evaluate how applicable the information we obtained from other model organisms is to humans, and to choose the correct target proteins for drug discovery. Information obtained from comparative genomic studies is having a significant impact on the field of structural biology. It is clear that protein three-dimensional structures can greatly facilitate the drug discovery process. In many cases the availability of crystal structure data or predicted structures not only help us to better understand the catalytic activity of the protein, but also show us the mechanism of how such a protein interacts with other proteins as well as the allosteric actions or conformational changes involved. The continuous improvement in computation power and development of new bioinformatic tools have laid the foundation for high-throughput X-ray crystallography and nuclear magnetic resonance (NMR) spectroscopy in protein structure determination. This structural genomics approach is aimed at systematically generating both experimental protein structures and unique structural folds (Burley 2000).

In order to provide a structure for each family of proteins using this high-throughput analysis, efforts have been made to ensure proper protein target selection. In particular, proteins that are inappropriate for this approach or redundant are being excluded. Protein sequences of unknown three-dimensional structure are modeled by structure prediction and submitted before the meeting. At the same time the three-dimensional structures of these sequences are being determined by X-ray crystallography and NMR spectroscopy. Results from these two approaches are then compared to further refine the methods. Such a process will ensure the quality of the structures generated by structural genomics.

High Throughput Screening Technologies in Drug Discovery

Chemical ligand discovery through natural-product screening has spearheaded therapeutics development and biological discovery for many years. The synthetic chemistry is now capable of producing a large number of diversified chemical compounds, which require high throughput screening (HTS) technologies for selection of specific ligands. One prominent example of HTS is the in vitro cell line-screening project (IVCLSP) under the Developmental Therapeutics Program (DTP) at the National Cancer Institute (NCI), which uses 60 human cancer cell lines (NCI60) for the drug discovery screen. About 5,000 out of 60,000 potential anticancer compounds screened are found to have anti-cancer activity. After eliminating closely related compounds and the ones with known mechanisms of action, the remaining 1,200 are selected for further testing. Another example of cell-based HTS is the receptor selection and amplification technology (R-SAT), which produces a signal (usually in terms of changes in the levels of ß-galactosidase) whenever a test compound interacts with an expressed receptor, such as G-protein coupled receptors and nuclear receptors. In the presence of receptor activity, cells overcome contact inhibition and proliferate. The increase in cell number is translated into an increase in the level of ß-galactosidase, whose level can be quantitatively measured using a spectrophotometer.

Stockwell et al. developed a miniaturized, whole-cell immunodetection system, called a "cytoblot", for high throughput screening of biologically active molecules. In this study, a library of small molecules were screened based on their ability to inhibit of cell growth, which was monitored by following the DNA content using an anti-BrdU antibody. Similar experimental design is suitable for screening for compounds that interfere with other cellular events, such as phosphorylation of a particular disease-causing protein. This approach has allowed the recent discovery of an array of antimitotic agents, including a novel inhibitor of mitotic spindle bipolarity named monastrol (Mayer et al. 1999). Monastrol specifically inhibits the mitotic kinesin Eg5. Since all the existing microtubule drugs target tubulin, an inhibitor of the microtubule motor protein will be useful for studying microtubule structures and functions.

Many important cellular regulatory events are mediated by protein-protein interaction. A classical example is the pRB-E2F-DP tumour repressor complex. Mutations in the pRB binding pocket can lead to defective protein-protein interaction and tumorigenesis. Therefore, disrupting the binding of two disease-causing proteins has become a focal point for drug development. In this respect, the yeast two-hybrid system can be of particular value. One strategy to screen for small molecules that can prevent interaction between the target protein and its binding partner is by using a reverse yeast two-hybrid system, in which the reporter gene URA3 is used. On the medium containing 5-fluoroorotic acid (5-FOA), disruption of protein-protein interaction by a small molecule will block expression of the URA3 gene, thereby allowing cell growth. An attempt to screen for allosteric drugs that function by affecting protein conformation has also been reported recently (DeDecker 2000). Another example is the study of the tumour repressor. Tumorigenic mutations have been linked to overall destabilization of the protein conformation. Foster et al. reported the screening of a library of more than 100,000 molecules and identified 2 molecules that can stabilize the protein fold of p53 point mutants.

Like DNA, small chemical ligands can also be immobilized on a glass surface for in vitro ligand screening. Knockaert et al. reported the use of such an approach in identifying intracellular targets of chemical inhibitors of the cyclin-dependent protein kinase. In this study, purvalanol B and its derivatives were immobilized on an agarose matrix, and cell extracts prepared from a diverse class of cell types and organisms were screened for proteins binding purvalanol B. This approach can easily be scaled up to include a small molecule library. Printing proteins on a glass surface can provide a way to assay for drug affinity for a large number of different proteins. Several groups have recently reported the development of protein microarrays for a variety of applications, such as identifying protein kinase substrates (Zhu et al. 2000), antibody screening, screening for protein-protein interactions as well as protein targets for chemical probes and accurate measurement of protein abundance using antibody/antigen microarrays.

Chemical Ligands and Systematic Study of gene and Protein Functions

Given their unparalleled control of their target gene/protein, chemical ligands are particularly suitable in genomic study of gene/protein functions. They can be readily integrated into the existing genomic approaches and tools. A significant advantage of using chemical ligands is that there is no limit on the developmental stage of the organisms or the conditions the cells and organisms are in. In addition, chemical ligands can be easily used in high throughput analyses.

Chemical Ligands and Gene Expression Profiling

The most popular approach currently is to use chemical ligands to study the role of the drug target in the control of gene expression. High-density oligonucleotide or complementary DNA (cDNA) microarrays (collectively called DNA microarrays) immobilized on the surfaces of glass slides can be used to simultaneously analyze the expression of thousands of genes. DNA microarrays have proven very useful in elucidating the functions of the drug targets involved in transcriptional control. In such an approach, the cell, tissue or organism is treated with chemical ligands. mRNAs are prepared from the treated and untreated cell or organism, and then used to produce fluorescencelabeled cDNAs to hybridize DNA microarrays and generate gene expression profiles. By comparing the differences between the profiles before and after drug treatment, genes whose expression is modulated by the chemical ligands are then identified.

These genes are further classified into pools based on similar functions or co-regulation in the cell. This information can often reveal specific transcription factors and other regulators for each gene pool, thereby allowing assembly of potential regulatory pathways involving the drug target. A nice example is provided by a recent study of yeast histone deacetylases (HDACs). In that study, three yeast HDACs, HDA1, RPD3 and SIR2, were examined using the drug trichostatin A (TSA) and deletion of the genes. Hda1 and Rpd3, but not Sir2, are sensitive to TSA.

The authors elegantly demonstrated that individual HDACs have specific roles in distinct transcriptional pathways as well as some limited overlapping functions. It is now realized that genes are regulated as networks. Many genes are co-regulated in response to unique cellular conditions. Efforts are being made to establish gene expression profile databases of regulatory proteins and transcriptional pathways. It is now possible to simultaneously compare the drug-induced profiles with the existing gene expression databases, thereby identifying relevant biological pathway(s) or functions for the drug target. For example, Hughes et al. used a compendium approach to compare the drug-induced expression profile to reference profiles of known cellular pathways.

Chemical Ligands and Protein Profiling

Proteomics or protein profiling is a field dedicated to studying the complete protein complement of a cell, tissue or organism. Posttranslational modifications play crucial roles in regulation of diverse cellular processes such as signal transduction, cell cycle control, development, cytoskeleton networks and metabolism. Common posttranslational modifications include phosphorylation (Hunter 2000), ubiquitylation and related polypeptide-dependent modifications, acetylation/methylation and proteolytic processing (Mayer 2000). Even a single protein can undergo several different forms of modification. For example, histone, a major building block of the chromatin, is phosphorylated, acetylated and methylated. Deregulation in any of these modifications can contribute to chromosomal and genome instability, which may contribute to genetic defects and cancer. Another important aspect of protein profiling is the study of controlled assembly of protein complexes. Dynamic changes in the components of these biological complexes often form the regulatory basis for their relevant cellular processes. Small, cell-permeable chemical ligands are particularly powerful in linking a regulatory protein and posttranslational modification of its downstream targets. In a typical approach, a chemical ligand is added to the cells, which leads to inhibition (or activation) of the regulatory protein.

As a result, posttranslational modification(s) of its downstream targets is changed, which can be detected traditionally by two-dimensional (2D) gels as judged by each protein's unique charge and size, or liquid chromatography-based protein separation technologies, including the non-porous reversed-phase high-performance liquid chromatography (RP-HPLC) and the two dimensional (2D) liquidphase separation method.

Since these separation methods are based on the biochemical and biophysical properties of the polypeptides, each protein tends to have a unique profile (or migration pattern) that can be used for its identification. In the cases that separation alone cannot provide sufficient information about the protein, its identity can be further revealed by Mass spectrometry. When the protein has a change in modification, its unique profile also changes and can be used as the basis for further investigation based on the nature of the modification. For example, phosphorylation of a protein is easily reversed by phosphatase treatment in vitro. Moreover, phosphorylation-, ubiquitin- and acetylation/methylationspecific antibodies can be used to detect the nature of protein modifications. The above approaches apply to any type of modification. Alternatively, modified proteins can be directly captured and examined for their relative abundance under different conditions. Two recent reports described smart chemistry that allows selective modification of phosphopeptides or phosphoproteins within complex mixtures.

Modified peptides or polypeptides are then enriched by covalent or high affinity avidin-biotin coupling to immobilized supports, thereby allowing rapid analysis of phosphorylation events in cells. In a similar vein, modifications with ubiquitin, ubiquitin-like peptides, and methyl and acetyl groups can also be used as epitope tags to capture their modified proteins. In addition, antibodies with specificity toward modified peptides can be arrayed on microchips and used to detect changes in modification of individual proteins. In any event, total cellular proteins still need to be compared under different conditions to ensure accurate measurement of a change in protein modification. These new proteinprofiling technologies can be easily integrated into chemical ligand-directed research.

Chemical Ligands and Global Genetic Interaction Study

Genetic interaction is a powerful way to study the relationship among genes (or proteins). If a gene acts in the same pathway or a parallel pathway with a second gene, its mutation may affect the phenotype(s) of a mutation in the second gene. Such secondary mutations lead to modification of phenotypes of the first mutations and are thus called modifier mutations, which allows identification of new components in the pathway(s) of a gene of interest. In an extreme case, two normally viable mutations, when combined, can generate a lethal phenotype, which is called synthetic lethality. Several genomic deletion projects have systematically generated deletion of individual genes in model organisms, including the budding yeast, the worms, the fruit fly and the mouse. This has made it possible to conduct global genetic interaction studies of the gene of interest. However, using the traditional targeted mutagenesis method is a formidable task even with budding yeast, the simplest eukaryotic organism, since one would need to generate the same mutation in 6,217 different mutants. With a chemical ligand, however, it is an easy and simple operation. One would simply add the drug to the 6,217 samples and then examine the subsequent phenotypes to identify genes whose mutations cause hypersensitivity or resistance to the drug. Since genes in the same biological pathway involving the drug target should confer drug sensitivity phenotypes to various degrees, such a chemical genomic screen should reveal most if not all of the genes in the pathway. The potential global genetic pathways or global genetic network can then be assembled by simply pooling genes with similar cellular functions.

The screening process can be easily automated either as a 96-well plate-based assay or an oligonucleotide microarray-based assay that takes advantage of the internal barcodes to each deleted locus. Giaever et al. demonstrated the feasibility of a genome-wide drug sensitivity screen by examining 233 yeast deletion strains to the drug tunicamycin. The yeast genomic deletion mutants were first used in the global study of genetic interactions with TOR. TOR is a highly conserved ataxia telangiectasia-related protein kinase essential for cell growth. To establish a global genetic interaction network of TOR, Chan et al. systematically measured the sensitivity of individual yeast mutants to a low concentration of rapamycin based on the relative growth of each mutant, thereby assembling a global genetic interaction network for TOR. The genome-wide screen is obviously advantageous, because it profiles every single mutant gene, regardless of the severity of the mutants' phenotypes. Even a moderate sensitivity could be biologically significant, which would likely be missed in a traditional genetic screen due to a weak phenotype. Moreover, every single mutation is a complete deletion - it avoids many complicated phenotypes by a point mutation or multi-allelic mutations in a typical traditional genetic screen.

Conclusion

Chemical ligands can be readily integrated with many existing genomic tools to create assorted chemical genomics approaches such as gene expression and protein profiling, and global analysis of genetic interaction. When combined, these approaches can produce complete biochemical and genetic profiles of the drug target protein. Computationally assisted parallel comparison and cross examination of different profiles will allow accurate detailed prediction and assembly of the biological pathways involved, which form the basis for validation and further detailed study using conventional molecular and cellular approaches. It has become apparent that such global knowledge of the cellular network of genes and proteins is critical for understanding disease mechanisms and selecting the most relevant protein targets for discovery of the optimal drugs. It will also vastly improve our ability to predict other benefits and side effects of the existing drugs, based on the global effects of the drug on gene expression, protein posttranslational modifications, and the biological pathways assembled based on various chemical genomics approaches. For instance, global study of genetic interactions can unravel the genetic basis for drug sensitivity.

It is well known that some cancer cells are more resistant or prone to certain anti-cancer drugs. One of the determining factors of drug sensitivity is genetic variation in genes in the same biological pathway(s) as the drug target. Some mutations can increase or decrease rapamycin sensitivity by as much as 10,000-fold. Mutations in yeast DNA repair and checkpoint pathways have also significant altered sensitivity to 23 different anti-cancer compounds. Drug sensitivity profiles of different types of cancers can be developed based on relative expression of genes involved in biological pathway(s) the drug interferes with. They should be valuable to devise customized protocols for effective treatment of individual cancers. In addition, many hypersensitive genes, genes with significant altered expression or proteins with different modification may be used as direct or secondary drug targets. Inhibition of the secondary targets can sensitize cells to the original drug, thereby significantly decreasing the dosages of the drugs to minimize side effects while achieving maximal therapeutic values. We foresee that chemical ligands should become available for any given protein, which allows chemical modulation of the protein's activity, both positively or negatively, and completely or selectively. These chemical ligands can equip us with the ultimate power to examine every molecular detail of cellular and organismic physiology, which, in turn, can also facilitate the understanding of disease mechanisms, the discovery of highly specific drugs and the devising of customized clinical protocols to treat individual disease conditions.

7

Microbial Metabolism

Microbial metabolism is the means by which a microbe obtains the energy and nutrients (e.g. carbon) it needs to live and reproduce. Microbes use many different types of metabolic strategies and species can often be differentiated from each other based on metabolic characteristics. The specific metabolic properties of a microbe are the major factors in determining that microbe's ecological niche, and often allow for that microbe to be useful in industrial processes or responsible for biogeochemical cycles.

Types of Microbial Metabolism

All microbial metabolisms can be arranged according to three principles:

How the organism obtains carbon for synthesising cell mass:

Autotroph

An autotroph, also called a producer, is an organism that produces complex organic compounds (such as carbohydrates, fats, and proteins) from simple inorganic molecules using energy from light (by photosynthesis) or inorganic chemical reactions (chemosynthesis). They are the producers in a food chain, such as plants on land or algae in water. They are able to make their own food and can fix carbon. Therefore, they do not utilize organic compounds as an energy source or a carbon source. Autotrophs can reduce carbon dioxide (add hydrogen to it) to make organic compounds. The reduction of carbon dioxide, a low-energy compound, creates a store of chemical energy. Most autotrophs use water as the reducing agent, but some can use other hydrogen compounds such as hydrogen sulfide. An autotroph converts physical energy from sun light (in case of green plants) into chemical energy in the form of reduced carbon. Autotroph can be phototrophs or lithotrophs (chemoautotrophs). Phototrophs use light as an energy source, while lithotrophs oxidize inorganic compounds, such as hydrogen sulfide, elemental sulfur, ammonium and ferrous iron. Phototrophs and lithotrophs use a portion of the ATP produced during photosynthesis or the oxidation of inorganic compounds to reduce $NADP^+$ to NADPH in order to form organic compounds.

Ecology

Autotrophs are fundamental to the food chains of all ecosystems in the world. They take energy from the environment in the form of sunlight or inorganic chemicals and use it to create energy-rich molecules such as carbohydrates. This mechanism is called primary production. Other organisms, called heterotrophs, take in autotrophs as food to carry out functions necessary for their life. Thus, heterotrophs — all animals, almost all fungi, as well as most bacteria and protozoa — depend on autotrophs for the energy and raw materials they need. Heterotrophs obtain energy by breaking down organic molecules (carbohydrates, fats, and proteins) obtained in food. Carnivorous organisms ultimately rely on autotrophs because the nutrients obtained from their heterotroph prey come from autotrophs they consumed.

Variants

Some organisms rely on organic compounds as a source of carbon, but are able to use light or inorganic compounds as a source of energy. Such organisms are not defined as autotrophic, but rather as heterotrophic. An organism that obtains carbon from organic compounds but obtains energy from light is called a *photoheterotroph*, while an organism that obtains carbon from organic compounds but obtains energy from the oxidation of inorganic compounds is termed a *chemoheterotroph* or *chemolithoheterotroph*.

Evidence suggests that some fungi may also obtain energy from radiation. Such radiotrophic fungi were found growing inside a reactor of the Chernobyl nuclear power plant.

Heterotroph

A heterotroph is an organism that cannot fix carbon and uses organic carbon for growth. This contrasts with autotrophs, such as plants and algae, which can use energy from sunlight (photoautotrophs) or inorganic compounds (lithoautotrophs) to produce organic compounds such as carbohydrates, fats, and proteins from inorganic carbon dioxide. These reduced carbon compounds can be used as an energy source by the autotroph and provide the energy in food consumed by heterotrophs. Ninety-five percent or more of all types of living organisms are heterotrophic.

Types

Heterotrophs must utilize organic compounds as a source of carbon, but can obtain energy either from light or from chemical compounds. They can be divided into two broad classes: photoheterotrophs and chemoheterotrophs. Photoheterotrophs, including most purple bacteria and green bacteria, produce ATP from light and use organic compounds to build structures.

They consume little or none of the energy produced during photosynthesis to reduce $NADP^+$ to NADPH for use in the Calvin cycle, as they do not need to use the Calvin cycle if carbohydrates are available in their diets. Chemoheterotrophs produce ATP by oxidizing chemical substances. There are two types of chemoheterotrophs: chemoorganoheterotrophs and chemolithoheterotrophs.

Chemoorganoheterotrophs (or simply organotrophs) exploit reduced carbon compounds as energy sources, such as carbohydrates, fats, and proteins from plants and animals. Chemolithoheterotrophs (or lithotrophic heterotrophs) such as colorless sulfur bacteria (e.g., *Beggiatoa* and *Thiobacillus*) and sulfate-reducing bacteria utilize inorganic substances to produce ATP, including hydrogen sulfide, elemental sulfur, thiosulfate, and molecular hydrogen. They use organic compounds to build structures. They do not fix carbon dioxide and apparently do not have the Calvin cycle. Chemolithoheterotrophs can be distinguished from mixotrophs (or facultative chemolithotroph), which can utilize either carbon dioxide or organic carbon as the carbon source.

Heterotrophs, by consuming reduced carbon compounds, are able to use all the energy that they obtain from food for growth and reproduction, unlike autotrophs, which must use some of their energy for carbon fixation. Heterotrophs are unable to make their own food, however, and whether using organic or inorganic energy sources, they can die from a lack of food. This applies not only to animals and fungi but also to bacteria.

Most heterotrophs are chemoorganoheterotrophs (or simply organotrophs) and utilize organic compounds both as a carbon source and an energy source. The term "heterotroph" very often refers to chemoorganoheterotrophs. Heterotrophs function as consumers in food chains: they obtain organic carbon by eating other heterotrophs or autotrophs. They break down complex organic compounds (e.g., carbohydrates, fats, and proteins) produced by autotrophs into simpler compounds (e.g., carbohydrates into glucose, fats into fatty acids and glycerol, and proteins into amino acids). They release energy by oxidizing carbon and hydrogen atoms present in carbohydrates, lipids, and proteins to carbon dioxide and water, respectively.

All animals and fungi are heterotrophic, as well as most protists and prokaryotes. Some animals, such as corals, form symbiotic relationships with autotrophs and obtain organic carbon in this way. Furthermore, some parasitic plants have also turned fully or partially heterotrophic, while carnivorous plants consume animals to augment their nitrogen supply while remaining autotrophic.

Mixotroph

A mixotroph is a microorganism that can use a mix of different sources of energy and carbon. Possible are alternations between photo- and chemotrophy, between litho- and organotrophy, between auto- and heterotrophy or a combination of it. Mixotrophs can be either eukaryotic or prokaryotic. They can take advantage of different environmental conditions. If a trophic mode is obligate, then it is always necessary for sustaining growth and maintenance; if facultative, it can be used as a supplemental source. Some organisms have incomplete Calvin cycles, so they are incapable of fixing carbon dioxide and must use organic carbon sources.

Examples

- *Paracoccus pantotrophus* is a bacterium that can live as chemoorganoheterotroph, whereby a large variety of organic compounds can be metabolized. Also a facultative chemolithoautotrophic metabolism is possible, as seen in colorless sulfur bacteria (some *Thiobacillus*), whereby sulfur compounds such as hydrogen sulfide, elemental sulfur, or thiosulfate are oxidized to sulfate. The sulfur compounds serve as electron donors and are consumed to produce ATP. The carbon source for these organisms can be carbon dioxide (autotrophy) or organic carbon (heterotrophy).

Organoheterotrophy can occur under aerobic or under anaerobic conditions; lithoautotrophy takes place aerobically. How the organism obtains reducing equivalents used either in energy conservation or in biosynthetic reactions.

Lithotroph

A lithotroph is an organism that uses an inorganic substrate (usually of mineral origin) to obtain reducing equivalents for use in biosynthesis (e.g., carbon dioxide fixation) or energy conservation via aerobic or anaerobic respiration. Known lithotrophs are exclusively microbes; No known macrofauna possesses the ability to utilize inorganic compounds as energy sources. Macrofauna and lithotrophs can form symbiotic relationships, in which case the lithotrophs are called "prokaryotic symbionts." An example of this is chemolithotrophic bacteria in deep sea worms or plastids, which are organelles within plant cells that may have evolved from photolithotrophic cyanobacteria-like organisms. Lithotrophs belong to either the domain Bacteria or the domain Archaea. The term "Lithotroph" is created from the terms 'lithos' (rock) and 'troph' (consumer), meaning the "eaters of rock." Many lithoautotrophs are extremophiles, but this is not universally so. The opposite of lithotroph is organotroph - an organism which gets its energy from the catabolism of organic compounds.

Biochemistry

Lithotrophs consume reduced compounds (rich in electrons).

Chemolithotrophs

In chemolithotrophs, the compounds - the electron donors - are oxidized in the cell, and the electrons are channeled into respiratory chains, ultimately producing ATP. The electron acceptor can be oxygen (in aerobic bacteria), but a variety of other electron acceptors, organic and inorganic, are also used by various species.

Photolithotrophs

Photolithotrophs obtain energy from light and therefore use inorganic electron donors only to fuel biosynthetic reactions (e.g., carbon dioxide fixation in lithoautotrophs).

Lithoheterotrophs Versus Lithoautotrophs

Lithotrophic bacteria cannot use, of course, their inorganic energy source as a carbon source for the synthesis of their cells. They choose one of three options:

- Lithoheterotrophs do not have the possibility to fix carbon dioxide and must consume additional organic compounds in order to break them apart and use their carbon. Only few bacteria are fully heterolithotrophic.
- Lithoautotrophs are able to use carbon dioxide from the air as carbon source, the same way plants do.
- Mixotrophs will take up and utilise organic material to complement their carbon dioxide fixation source (mix between autotrophy and heterotrophy). Many lithotrophs are recognised as mixotrophic in regard of their C-metabolism.

Chemolithotrophs Versus Photolithotrophs

In addition to this division, lithotrophs differ in the initial energy source which initiates ATP production:

- Chemolithotrophs use the above-mentioned inorganic compounds for aerobic or anaerobic respiration. The energy produced by the oxidation of these compounds is enough for ATP production. Some of the electrons derived from the inorganic donors also need to be channeled into biosynthesis. Mostly, additional energy has to be invested to transform these reducing equivalents to the forms and redox potentials needed (mostly NADH or NADPH), which occurs by reverse electron transfer reactions.

- Photolithotrophs use light as energy source. These bacteria are photosynthetic; photolithotrophic bacteria are found in the purple bacteria (e.g., Chromatiaceae), green bacteria (Chlorobiaceae and Chloroflexi) and Cyanobacteria. The electrons obtained from the electron donors (purple and green bacteria oxidize sulfide, sulfur, sulfite, iron or hydrogen; cyanobacteria extract reducing equivalents from water, i.e., oxidise water to oxygen) are not used for ATP production (as long as there is light); they are used in biosynthetic reactions. Some photolithotrophs shift over to chemolithotrophic metabolism in the dark.

Geological Significance

Lithotrophs participate in many geological processes, such as the weathering of parent material (bedrock) to form soil, as well as biogeochemical cycling of sulfur, nitrogen, and other elements. They may be present in the deep terrestrial subsurface (they have been found well over 3 km below the surface of the planet), in soils, and in endolith communities. As they are responsible for the liberation of many crucial nutrients, and participate in the formation of soil, lithotrophs play a critical role in the maintenance of life on Earth.

Lithotrophic microbial consortia are responsible for the phenomenon known as acid mine drainage, whereby energy-rich pyrites and other reduced sulfur compounds present in mine tailing heaps and in exposed rock faces is metabolized to form sulfates, thereby forming potentially toxic sulfuric acid. Acid mine drainage drastically alters the acidity and chemistry of groundwater and streams, and may endanger plant and animal populations. Activities similar to acid mine drainage, but on a much lower scale, are also found in natural conditions such as the rocky beds of glaciers, in soil and talus, on stone monuments and buildings and in the deep subsurface.

Organotroph

An organotroph is an organism that obtains hydrogen or electrons from organic substrates (a form of a chemotroph).

Antonym: Lithotroph

How the organism obtains energy for living and growing.

Chemotroph

Chemotrophs are organisms that obtain energy by the oxidation of electron donors in their environments. These molecules can be organic (chemoorganotrophs) or inorganic (chemolithotrophs). The chemotroph designation is in contrast to phototrophs, which utilize solar energy. Chemotrophs can be either autotrophic or heterotrophic.

- Chemoautotrophs (or chemotrophic autotroph), (*Gk*: Chemo = chemical, auto = self, troph = nourishment) in addition to deriving energy from chemical reactions, synthesize all necessary organic compounds from carbon dioxide. Chemoautotrophs use inorganic energy sources, such as hydrogen sulfide, elemental sulfur, ferrous iron, molecular hydrogen, and ammonia. Most are bacteria or archaea that live in hostile environments such as deep sea vents and are the primary producers in such ecosystems. Evolutionary scientists believe that the first organisms to inhabit Earth were chemoautotrophs that produced oxygen as a by-product and later evolved into both aerobic, animal-like organisms and photosynthetic, plant-like organisms. Chemoautotrophs generally fall into several groups: methanogens, halophiles, sulfur oxidizers and reducers, nitrifiers, anammoxbacteria, and thermoacidophiles.
- Chemoheterotrophs (or chemotrophic heterotrophs) (*Gk*: Chemo = chemical, hetero = (an)other, troph = nourishment) are unable to fix carbon and form their own organic compounds. Chemoheterotrophs can be chemolithoheterotrophs, utilizing inorganic energy sources such as sulfur or chemoorganoheterotrophs, utilizing organic energy sources such as carbohydrates, lipids, and proteins.

Iron and Manganese Oxidizing Bacteria

In the deep oceans, iron oxidizing bacteria derive their energy needs by oxidizing Iron II to Iron III. The extra electron obtained from this reaction powers the cells, replacing or augmenting traditional phototrophism.

- In general, iron oxidizing bacteria can exist only in areas with high iron concentrations, such as new lava beds or areas of hydrothermal activity (where there is dissolved Fe). Most of the ocean is devoid of iron, due to both the oxidative effect of dissolved oxygen in the water and the tendency of prokaryotes to uptake the iron.
- Lava beds supply bacteria with iron straight from the Earth's mantle, but only newly formed igneous rocks have high enough levels of unoxidized iron. In addition, because oxygen is necessary for the reaction, these bacteria are much more common in the upper ocean, where oxygen is more abundant.
- What is still unknown, though, is how exactly iron bacteria extract the iron out of the rock. It is accepted that some mechanism exists that eats away at the rock, perhaps through specialized enzymes or compounds that bring more FeO to the surface. It has been long debated

about how much of the weathering of the rock is due to biotic components and how much can be attributed to abiotic components.

- Hydrothermal vents also release large quantities of dissolved iron into the deep ocean, allowing bacteria to survive. In addition, the high thermal gradient around vent systems means a wide variety of bacteria can coexist, each with its own specialized temperature niche.
- Regardless of the catalytic method used, chemoautotrophic bacteria provide a significant but frequently overlooked food source for deep sea ecosystems - which otherwise receive limited sunlight and organic nutrients.

Manganese oxidizing bacteria also make use of igneous lava rocks in much the same way - by oxidizing Mn^{2+} into Mn^{4+}. Manganese is much rarer than iron in oceanic crust, but is much easier for bacteria to extract from the igneous glass. In addition, each manganese oxidation yields around twice the energy as an iron oxidation due to the gain of twice the number of electrons. Much still remains unknown about manganese oxidizing bacteria because they have not been cultured and documented to any great extent.

Phototroph

Phototrophs are the organisms (usually plants) that carry out photosynthesis to acquire energy. They use the energy from sunlight to convert carbon dioxide and water into organic materials to be utilized in cellular functions such as biosynthesis and respiration.

Most phototrophs are autotrophs, also known as photoautotrophs, and can fix carbon. They can be contrasted with chemoautotrophs that obtain their energy by the oxidation of electron donors in their environments. Photoheterotrophs produce ATP through photophosphorylation but use organic compounds to build structures. Some phototrophs are organotrophs, also known as photo-organotrophs.

In an ecological context, phototrophs provide nutrition for all other forms of life (besides other autotrophs such as chemotrophs). In terrestrial environments plants are the predominant variety, while aquatic environments include a range of phototrophic organisms such as algae (e.g. kelp), other protists (such as euglena), phytoplankton and bacteria (such as cyanobacteria).

One product of this process is starch, which is a storage or reserve form of carbon, which can be used when light conditions are too poor to satisfy the immediate needs of the organism. Photosynthetic bacteria have a substance called bacteriochlorophyll, live in lakes and pools, and use the hydrogen from hydrogen sulfide instead of from water, for the chemical process. (The bacteriochlorophyll pigment absorbs light in the extreme UV and infra-red parts of the spectrum which is outside the range used by normal chlorophyll). Cyanobacteria live in fresh water, seas, soil and lichen, and use a plant-like photosynthesis.

A *photolithotrophic autotroph* is an autotrophic organism that uses light energy, and an inorganic electron donor (e.g., H_2O, H_2, H_2S), and CO_2 as its carbon source. Examples include plants. The depth to which sunlight or artificial light can penetrate into water, so that photosynthesis may occur, is known as the photic zone.

In practice, these terms are almost freely combined. Typical examples are as follows:

- chemolithoautotrophs obtain energy from the oxidation of inorganic compounds and carbon from the fixation of carbon dioxide. Examples: Nitrifying bacteria, Sulfur-oxidizing bacteria, Iron-oxidizing bacteria, Knallgas-bacteria
- photolithoautotrophs obtain energy from light and carbon from the fixation of carbon dioxide, using reducing equivalents from inorganic compounds. Examples: Cyanobacteria (water (H_2O) as reducing equivalent donor), Chlorobiaceae, Chromaticaceae (hydrogen sulfide (H_2S) as reducing equivalent donor), *Chloroflexus* (hydrogen (H_2) as reducing equivalent donor)
- chemolithoheterotrophs obtain energy from the oxidation of inorganic compounds, but cannot fix carbon dioxide (CO_2). Examples: some *Thiobacilus*, some *Beggiatoa*, some *Nitrobacter* spp., *Wolinella* (with H_2 as reducing equivalent donor), some Knallgas-bacteria, some sulfate-reducing bacteria
- chemoorganoheterotrophs obtain energy, carbon, and reducing equivalents for biosynthetic reactions from organic compounds. Examples: most bacteria, e.g. *Escherichia coli*, *Bacillus* spp., *Actinobacteria*
- photoorganoheterotrophs obtain energy from light, carbon and reducing equivalents for biosynthetic reactions from organic compounds. Some species are strictly heterotrophic, many others can also fix carbon dioxide and are mixotrophic. Examples. *Rhodobacter*, *Rhodopseudomonas*, *Rhodospirillum*, *Rhodomicrobium*, *Rhodocyclus*, *Heliobacterium*, *Chloroflexus* (alternatively to photolithoautotrophy with hydrogen)

Heterotrophic Microbial Metabolism

Most microbes are heterotrophic (more precisely chemoorganoheterotrophic), using organic compounds as both carbon and energy sources. Heterotrophic microbes live off of nutrients that they scavenge from living hosts (as commensals or parasites) or find in dead organic matter of all kind (saprophages). Microbial metabolism is the main contribution for the bodily decay of all organisms after death. Many eukaryotic microorganisms are heterotrophic by predation or parasitism, properties also found in some bacteria such as *Bdellovibrio* (an intracellular parasite of other bacteria, causing death of its victims) and Myxobacteria such as *Myxococcus* (predators of other bacteria which are killed and lysed by cooperating swarms of many single cells of Myxobacteria).

Most pathogenic bacteria can be viewed as heterotrophic parasites of humans or the other eukaryotic species they affect. Heterotrophic microbes are extremely abundant in nature and are responsible for the breakdown of large organic polymers such as cellulose, chitin or lignin which are generally indigestible to larger animals. Generally, the breakdown of large polymers to carbon dioxide (mineralization) requires several different organisms, with one breaking down the polymer into its constituent monomers, one able to use the monomers and excreting simpler waste compounds as by-products, and one able to use the excreted wastes. There are many variations on this theme, as different organisms are able to degrade different polymers and secrete different waste products. Some organisms are even able to degrade more recalcitrant compounds such as petroleum compounds or pesticides, making them useful in bioremediation.

Biochemically, prokaryotic heterotrophic metabolism is much more versatile than that of eukaryotic organisms, although many prokaryotes share the most basic metabolic models with eukaryotes, e.g. using glycolysis (also called EMP pathway) for sugar metabolism and the citric acid cycle to degrade acetate, producing energy in the form of ATP and reducing power in the form of NADH or quinols. These basic pathways are well conserved because they are also involved in biosynthesis of many conserved building blocks needed for cell growth (sometimes in reverse direction). However, many bacteria and archaea utilize alternative metabolic pathways other than glycolysis and the citric acid cycle. A well-studied example is sugar metabolism via the keto-deoxy-phosphogluconate pathway (also called ED pathway) in *Pseudomonas*. Moreover, there is a third alternative sugar-catabolic pathway used by some bacteria, the pentose phosphate pathway.

The metabolic diversity and ability of prokaryotes to use a large variety of organic compounds arises from the much deeper evolutionary history and diversity of prokaryotes, as compared to eukaryotes. It is also noteworthy that the mitochondrion, the small membrane-bound intracellular organelle that is the site of eukaryotic energy metabolism, arose from the endosymbiosis of a bacterium related to obligate intracellular *Rickettsia*, and also to plant-associated *Rhizobium* or *Agrobacterium*. Therefore it is not surprising that all mitrochondriate eukaryotes share metabolic properties with these Proteobacteria. Most microbes respire (use an electron transport chain), although oxygen is not the only terminal electron acceptor that may be used. As discussed below, the use of terminal electron acceptors other than oxygen has important biogeochemical consequences.

Fermentation (Biochemistry)

Fermentation is the process of deriving energy from the oxidation of organic compounds, such as carbohydrates, and using an endogenous electron acceptor, which is usually an organic compound. In contrast, respiration is where electrons are donated to an exogenous electron acceptor, such as oxygen, via an electron transport chain. Fermentation is important in anaerobic conditions when there is no oxidative phosphorylation to maintain the production of ATP (Adenosine triphosphate) by glycolysis. During fermentation, pyruvate is metabolised to various different compounds. Homolactic fermentation is the production of lactic acid from pyruvate; alcoholic fermentation is the conversion of pyruvate into ethanol and carbon dioxide; and heterolactic fermentation is the production of lactic acid as well as other acids and alcohols. Fermentation does not necessarily have to be carried out in an anaerobic environment. For example, even in the presence of abundant oxygen, yeast cells greatly prefer fermentation to oxidative phosphorylation, as long as sugars are readily available for consumption.

Sugars are the most common substrate of fermentation, and typical examples of fermentation products are ethanol, lactic acid, and hydrogen. However, more exotic compounds can be produced by fermentation, such as butyric acid and acetone. Yeast carries out fermentation in the production of ethanol in beers, wines and other alcoholic drinks, along with the production of large quantities of carbon dioxide. Fermentation occurs in mammalian muscle during periods of intense exercise where oxygen supply becomes limited, resulting in the creation of lactic acid.

Chemistry

Fermentation products contain chemical energy (they are not fully oxidized), but are considered waste products since they cannot be metabolized further without the use of oxygen (or other more highly-oxidized electron acceptors). A consequence is that the production of adenosine triphosphate (ATP) by fermentation is less efficient than oxidative phosphorylation, whereby pyruvate is fully oxidized to carbon dioxide.

Ethanol Fermentation

Ethanol fermentation (performed by yeast and some types of bacteria) breaks the pyruvate down into ethanol and carbon dioxide. It is important in bread-making, brewing, and wine-making. Usually only one of the products is desired; in bread-making, the alcohol is baked out, and, in alcohol production, the carbon dioxide is released into the atmosphere or used for carbonating the beverage. When the ferment has a high concentration of pectin, minute quantities of methanol can be produced.

The chemical equation below summarizes the fermentation of glucose, whose chemical formula is $C_6H_{12}O_6$. One glucose molecule is converted into two ethanol molecules and two carbon dioxide molecules:

$$C_6H_{12}O_6 \rightarrow 2\ C_2H_5OH + 2\ CO_2$$

C_2H_5OH is the chemical formula for ethanol.

Before fermentation takes place, one glucose molecule is broken down into two pyruvate molecules. This is known as glycolysis.

Lactic Acid Fermentation

Lactic acid fermentation is the simplest type of fermentation. Essentially, it is a redox reaction. In anaerobic conditions, the cell's primary mechanism of ATP production is glycolysis. Glycolysis reduces – transfers electrons to – NAD^+, forming NADH. However, there is only a limited supply of NAD^+ available in a cell. For glycolysis to continue, NADH must be oxidized – have electrons taken away – to regenerate the NAD^+.

This is usually done through an electron transport chain in a process called oxidative phosphorylation; however, this mechanism is not available without oxygen.

Instead, the NADH donates its extra electrons to the pyruvate molecules formed during glycolysis. Since the NADH has lost electrons, NAD^+ regenerates and is again available for glycolysis. Lactic acid, for which this process is named, is formed by the reduction of pyruvate.

In heterolactic acid fermentation, one molecule of pyruvate is converted to lactate; the other is converted to ethanol and carbon dioxide. In homolactic acid fermentation, both molecules of pyruvate are converted to lactate. Homolactic acid fermentation is unique because it is one of the only respiration processes to not produce a gas as a byproduct.

Homolactic fermentation breaks down the pyruvate into lactate. It occurs in the muscles of animals when they need energy faster than the blood can supply oxygen. It also occurs in some kinds of bacteria (such as lactobacilli) and some fungi. It is this type of bacteria that converts lactose into lactic acid in yogurt, giving it its sour taste.

These lactic acid bacteria can be classed as homofermentative, where the end product is mostly lactate, or heterofermentative, where some lactate is further metabolized and results in carbon dioxide, acetate or other metabolic products.

The process of lactic acid fermentation using glucose is summarized below. In homolactic fermentation, one molecule of glucose is converted to two molecules of lactic acid:

$$C_6H_{12}O_6 \rightarrow 2\ CH_3CHOHCOOH.$$

In heterolactic fermentation, the reaction proceeds as follows, with one molecule of glucose being converted to one molecule of lactic acid, one molecule of ethanol, and one molecule of carbon dioxide:

$$C_6H_{12}O_6 \rightarrow CH_3CHOHCOOH + C_2H_5OH + CO_2$$

Before lactic acid fermentation can occur, the molecule of glucose must be split into two molecules of pyruvate. This process is called glycolysis.

Glycolysis

To extract chemical energy from glucose, the glucose molecule must be split into two molecules of pyruvate. This process also generates two molecules of adenosine triphosphate as an immediate energy yield and two molecules of NADH.

$$C_6H_{12}O_6 + 2\ ADP + 2\ P_i + 2\ NAD^+ \rightarrow 2\ CH_3COCOO^- + 2\ ATP + 2\ NADH + 2\ H_2O + 2H^+$$

The chemical formula of pyruvate is CH_3COCOO^-. P_i stands for the inorganic phosphate. As shown by the reaction equation, glycolysis causes the reduction of two molecules of NAD^+ to NADH. Two ADP molecules are also converted to two ATP and two water molecules via substrate-level phosphorylation.

Aerobic Respiration

In aerobic respiration, the pyruvate produced by glycolysis is further oxidized completely, generating additional ATP and NADH in the citric acid cycle and by oxidative phosphorylation. However, this can only occur in the presence of oxygen. Oxygen is toxic to organisms which are obligate anaerobes, and are not required by facultative anaerobic organisms. In the absence of oxygen, one of the fermentation pathways occurs in order to regenerate NAD^+; lactic acid fermentation is one of these pathways.

Hydrogen Gas Production in Fermentation

Hydrogen gas is produced in many types of fermentation (mixed acid fermentation, butyric acid fermentation, caproate fermentation, butanol fermentation, glyoxylate fermentation), as a way to regenerate NAD^+ from NADH. Electrons are transferred to ferredoxin, which in turn is oxidized by hydrogenase, producing H_2. Hydrogen gas is a substrate for methanogens and sulfate reducers, which keep the concentration of hydrogen sufficiently low to allow the production of such an energy-rich compound.

History

The first solid evidence of the living nature of yeast appeared between 1837 and 1838 when three publications appeared by C. Cagniard de la Tour, T. Swann, and F. Kuetzing, each of whom independently concluded as a result of microscopic investigations that yeast was a living organism that reproduced by budding. The word “yeast,” it should be noted, traces its origins back to the Sanskrit word meaning “boiling.”

It was perhaps because wine, beer, and bread were each basic foods in Europe, that most of the early studies on fermentation were done on yeasts, with which they were made. Soon bacteria were also discovered; the term was first used in English in the late 1840s, but it did not come into general use until the 1870s, and then largely in connection with the new germ theory of disease. Louis Pasteur (1822–1895), during the 1850s and 1860s, showed that fermentation was initiated by living organisms in a series of investigations. In 1857 Pasteur showed that lactic acid fermentation is caused by living organisms.

In 1860 he demonstrated that bacteria cause souring in milk, a process formerly thought to be merely a chemical change, and his work in identifying the role of microorganisms in food spoilage led to the process of pasteurization. In 1877, working to improve the French brewing industry, Pasteur published his famous paper on fermentation, Etudes sur la Biere , which was translated into English in 1879 as Studies on Fermentation. He defined fermentation (incorrectly) as “Life without air,” but correctly showed specific types of microorganisms cause specific types of fermentations and specific end products.

Although showing that fermentation was generally the result of the action of living microorganisms was a breakthrough, it did not explain the basic nature of the fermentation process, or prove that it was caused by the microorganisms that were apparently always present. Many scientists, including Pasteur, had attempted unsuccessfully to extract the fermentation enzyme from yeast. Success came in 1897 when the German chemist Eduard Buechner ground up yeast, extracted a juice from them, then found to his amazement that this “dead” liquid would ferment a sugar solution, forming carbon dioxide and alcohol much like living yeasts.

The “unorganized ferments” behaved just like the organized ones. From that time on the term enzyme came to be applied to all ferments. It was then understood that fermentation is caused by enzymes which are produced by microorganisms. In 1907 Buechner won the Nobel Prize in chemistry for his work. Advances in microbiology and fermentation technology have continued steadily up until the present. For example, in the late 1970s it was discovered that microorganisms could be mutated with physical and chemical treatments to be higher yielding, faster growing, tolerant of less oxygen, and able to use a more concentrated medium. Strain selection and hybridization developed as well, afecting most modern food fermentations.

Etymology

The word *fermentation* is derived from the Latin verb "fervere" which means "to boil". It is thought to have been first used in the late fourteenth century in alchemy, but only in a broad sense. It was not used in the modern scientific sense until around 1600.

Special Metabolic Properties

Methylotrophy

Methylotrophy refers to the ability of an organism to use C1-compounds as energy sources. These compounds include methanol, methyl amines, formaldehyde, and formate. Several other less common substrates may also be used for metabolism, all of which lack carbon-carbon bonds. Examples of methylotrophs include the bacteria *Methylomonas* and *Methylobacter*. Methanotrophs are a specific type of methylotroph that are also able to use methane (CH_4) as a carbon source by oxidizing it sequentially to methanol (CH_3OH), formaldehyde (CH_2O), formate (HCOO), and carbon dioxide CO_2 initially using the enzyme methane monooxygenase. As oxygen is required for this process, all (conventional) methanotrophs are obligate aerobes. Reducing power in the form of quinones and NADH is produced during these oxidations to produce a proton motive force and therefore ATP generation. Methylotrophs and methanotrophs are not considered as autotrophic, because they are able to incorporate some of the oxidized methane (or other metabolites) into cellular carbon before it is completely oxidized to CO_2 (at the level of formaldehyde), using either the serine pathway (*Methylosinus, Methylocystis*) or the ribulose monophosphate pathway (*Methylococcus*), depending on the species of methylotroph.

In addition to aerobic methylotrophy, methane can also be oxidized anaerobically. This occurs by a consortium of sulfate-reducing bacteria and relatives of methanogenic Archaea working syntrophically. Little is currently known about the biochemistry and ecology of this process.

Methanogenesis is the biological production of methane. It is carried out by methanogens, strictly anaerobic Archaea such as *Methanococcus*, *Methanocaldococcus*, *Methanobacterium*, *Methanothermus*, *Methanosarcina*, *Methanosaeta* and *Methanopyrus*. The biochemistry of methanogenesis is unique in nature in its use of a number of unusual cofactors to sequentially reduce methanogenic substrates to methane, such as coenzyme M and methanofuran.. These cofactors are responsible (among other things) for the establishment of a proton gradient across the outer membrane thereby driving ATP synthesis.

Several types of methanogenesis occur, differing in the starting compounds oxidized. Some methanogens reduce carbon dioxide (CO_2) to methane (CH_4) using electrons (most often) from hydrogen gas (H_2) chemolithoautotrophically.

These methanogens can often be found in environments containing fermentative organisms. The tight association of methanogens and fermentative bacteria can be considered to be syntrophic because the methanogens, which rely on the fermentors for hydrogen, relieve feedback inhibition of the fermentors by the build-up of excess hydrogen that would otherwise inhibit their growth. This type of syntrophic relationship is specifically known as interspecies hydrogen transfer. A second group of methanogens use methanol (CH_3OH) as a substrate for methanogenesis. These are chemoorganotrophic, but still autotrophic in using CO_2 as only carbon source. The biochemistry of this process is quite different from that of the carbon dioxide-reducing methanogens. Lastly, a third group of methanogens produce both methane and carbon dioxide from acetate (CH_3COO^-) with the acetate being split between the two carbons. These acetate-cleaving organisms are the only chemoorganoheterotrophic methanogens. All autotrophic methanogens use a variation of the acetyl-CoA pathway to fix CO_2 and obtain cellular carbon.

Syntrophy

Syntrophy, in the context of microbial metabolism, refers to the pairing of multiple species to achieve a chemical reaction that, on its own, would be energetically unfavourable. The best studied example of this process is the oxidation of fermentative end products (such as acetate, ethanol and butyrate) by organisms such as *Syntrophomonas*. Alone, the oxidation of butyrate to acetate and hydrogen gas is energetically unfavourable. However, when a hydrogenotrophic (hydrogen-using) methanogen is present the use of the hydrogen gas will significantly lower the concentration of hydrogen (down to 10^{-5} atm) and thereby shift the equilibrium of the butyrate oxidation reaction under standard conditions ($\Delta G^{o\prime}$) to non-standard conditions ($\Delta G'$). Because the concentration of one product is lowered, the reaction is "pulled" towards the products and shifted towards net energetically favourable conditions (for butyrate oxidation: $\Delta G^{o\prime}$= +48.2 kJ/mol, but $\Delta G'$ = -8.9 kJ/mol at 10^{-5} atm hydrogen and even lower if also the initially produced acetate is further metabolized by methanogens). Conversely, the available free energy from methanogenesis is lowered from $\Delta G^{o\prime}$= -131 kJ/mol under standard conditions to $\Delta G'$ = -17 kJ/mol at 10^{-5} atm hydrogen. This is an example of intraspecies hydrogen transfer. In this way, low energy-yielding carbon sources can be used by a consortium of organisms to achieve further degradation and eventual mineralization of these compounds. These reactions help prevent the excess sequestration of carbon over geologic time scales, releasing it back to the biosphere in usable forms such as methane and CO_2.

Anaerobic Respiration

While aerobic organisms during respiration use oxygen as a terminal electron acceptor, anaerobic organisms use other electron acceptors. These inorganic compounds have a lower reduction potential than oxygen, meaning that respiration is less efficient in these organisms and leads to slower growth rates than aerobes. Many facultative anaerobes can use either oxygen or alternative terminal electron acceptors for respiration depending on the environmental conditions.

Most respiring anaerobes are heterotrophs, although some do live autotrophically. All of the processes described below are dissimilative, meaning that they are used during energy production and not to provide nutrients for the cell (assimilative). Assimilative pathways for many forms of anaerobic respiration are also known.

Denitrification

Denitrification is a microbially facilitated process of nitrate reduction that may ultimately produce molecular nitrogen (N_2) through a series of intermediate gaseous nitrogen oxide products. This respiratory process reduces oxidized forms of nitrogen in response to the oxidation of an electron donor such as organic matter. The preferred nitrogen electron acceptors in order of most to least thermodynamically favourable include nitrate (NO_3^-), nitrite (NO_2^-), nitric oxide (NO), and nitrous oxide (N_2O). In terms of the general nitrogen cycle, denitrification completes the cycle by returning N_2 to the atmosphere. The process is performed primarily by heterotrophic bacteria (such as *Paracoccus denitrificans* and various pseudomonads), although autotrophic denitrifiers have also been identified (e.g., *Thiobacillus denitrificans*). Denitrifiers are represented in all main phylogenetic groups. Generally several species of bacteria are involved in the complete reduction of nitrate to molecular nitrogen, and more than one enzymatic pathway have been identified in the reduction process. Direct reduction from nitrate to ammonium, a process known as dissimilatory nitrate reduction to ammonium or DNRA, is also possible for organisms that have the nrf-gene. This is less common than denitrification in most ecosystems as a means of nitrate reduction. Other genes known in microorganisms which denitrify include *nir* (nitrate reductase) and *nos* (nitrous oxide reductase) among others; organisms identified as having these genes include *Alcaligenes faecalis*, *Alcaligenes xylosoxidans*, many in the *Pseudomonas* genus, *Bradyrhizobium japonicum*, and *Blastobacter denitrificans*.

Nutrient Limitation

All organisms require certain nutrients in their surroundings (available to them) for survival. Depending upon the ecosystem, nitrogen is most likely the limiting nutrient, although phosphorus is the other primary limiting nutrient and these two elements interact chemically. Some organisms appear to be able to denitrify and remove phosphorus. The triple bond of N_2 makes this a very stable compound; most organisms (i.e. plants) depend upon others to break this down to make it available for biochemical reactions.

Conditions Required

Denitrification takes place under special conditions in both terrestrial and marine ecosystems. In general, it occurs where oxygen, a more energetically favourable electron acceptor, is depleted, and bacteria respire nitrate as a substitute terminal electron acceptor. Due to the high concentration of oxygen in our atmosphere, denitrification only takes place in environments where oxygen consumption exceeds the rate of oxygen supply, such as in some soils and groundwater, wetlands, poorly ventilated corners of the ocean, and in seafloor sediments. Denitrification generally proceeds through some combination of the following intermediate forms:

$$NO_3^- \rightarrow NO_2^- \rightarrow NO + N_2O \rightarrow N_2\ (g)$$

The complete denitrification process can be expressed as a redox reaction:

$$2\ NO_3^- + 10\ e^- + 12\ H^+ \rightarrow N_2 + 6\ H_2O$$

This reaction shows a fractionation in isotope composition. Lighter isotopes of nitrogen are preferred in the reaction, leaving the heavier nitrogen isotopes in the residual matter. The process can cause delta-values of up to "40, where delta is a representation of the difference in isotopic composition. This can be used to identify denitrification processes in nature.

Deliberate use of Process

Denitrification is commonly used to remove nitrogen from sewage and municipal wastewater. It is also an instrumental process in wetlands and riparian zones for the removal of excess nitrate from groundwater resulting from excessive agricultural or residential fertilizer usage.

Reduction under anoxic conditions can also occur through process called anaerobic ammonia oxidation (anammox):

$$NH_4^+ + NO_2^- \rightarrow N_2 + 2\ H_2O$$

In some wastewater treatment plants, small amounts of methanol, ethanol, acetate or proprietary products like MicroCg or MicroCglycerin are added to the wastewater to provide a carbon source for the denitrification bacteria. Denitrification processes are also used in the treatment of industrial wastes.

Influence on Global Climate Change

Increasing carbon dioxide levels within the atmosphere will influence global nutrient cycling, yet it is difficult to predict what those interactions might be. Chemical interactions between soils and the atmosphere will be influenced by changes in atmospheric composition. There are indications that increased fertilization of soils with nitrogen causes a decrease in carbon sequestration. Jake Beaulieu, a postdoctoral researcher the Environmental Protection Agency in Cincinnati, Ohio and Jennifer Tank, Galla Professor of Biological Sciences at the University of Notre Dame, are lead authors of new paper demonstrating that streams and rivers receiving nitrogen inputs from urban and agricultural land uses are a significant source of nitrous oxide to the atmosphere.

Sulfate Reduction-sulfate as Electron Acceptor

Sulfate reduction is a relatively energetically poor process used by many Gram negative bacteria found within the δ-Proteobacteria, Gram-positive organisms relating to *Desulfotomaculum* or the archaeon *Archaeoglobus*. Hydrogen sulfide (H_2S) is produced as a metabolic end product. For sulfate reduction electron donors and energy are needed.

Electron Donors

Many sulfate reducers are heterotrophic, using carbon compounds such as lactate and pyruvate (among many others) as electron donors, while others are autotrophic, using hydrogen gas (H_2) as an electron donor. Some unusual autotrophic sulfate-reducing bacteria (e.g. *Desulfotignum phosphitoxidans*) can use phosphite (HPO_3^-) as an electron donor whereas others (e.g. *Desulfovibrio sulfodismutans*, *Desulfocapsa thiozymogenes*, *Desulfocapsa sulfoexigens*) are capable of sulfur disproportionation (splitting one compound into two different compounds, in this case an electron donor and an electron acceptor) using elemental sulfur (S^0), sulfite ($SO2^{-3}$), and thiosulfate ($S_2O_3^{2-}$) to produce both hydrogen sulfide (H_2S) and sulfate ($SO2^{-4}$).

Energy for Reduction

All sulfate-reducing organisms are strict anaerobes. Because sulfate is energetically stable, before it can be metabolized it must first be activated by adenylation to form APS (adenosine 5'-phosphosulfate) thereby consuming ATP. The APS is then reduced by the enzyme APS reductase to form sulfite ($SO2^{-3}$) and AMP. In organisms that use carbon compounds as electron donors, the ATP consumed is accounted for by fermentation of the carbon substrate. The hydrogen produced during fermentation is actually what drives respiration during sulfate reduction.

Acetogenesis-carbon Dioxide as Electron Acceptor

Acetogenesis: Acetogenesis is a process through which acetate is produced by anaerobic bacteria from a variety of energy (for example, hydrogen) and carbon (for example, carbon dioxide) sources. The different bacterial species that are capable of acetogenesis are collectively termed acetogens.

Chemical Equation:

$$2\ CO_2\ (aq) + 4H_2\ (aq) = CH_3COOH\ (aq) + 2H_2O$$

Acetogenesis and Anaerobic Digestion

Acetogenesis is the third stage of anaerobic digestion;

- Hydrolysis: Particulates are solubilised and large polymers are converted into simpler monomers;
- Acidogenesis: Simple monomers are converted into volatile fatty acids;
- Acetogenesis: Volatile fatty acids are converted into acetic acid, carbon dioxide, and hydrogen; and
- Methanogenesis: Acetate is converted into methane and carbon dioxide, while hydrogen is consumed.

Other Inorganic Electron Acceptors

Ferric iron (Fe^{3+}) is a widespread anaerobic terminal electron acceptor both for autotrophic and heterotrophic organisms. Electron flow in these organisms is similar to those in electron transport, ending in oxygen or nitrate, except that in ferric iron-reducing organisms the final enzyme in this system is a ferric iron reductase.

Model organisms include *Shewanella putrefaciens* and *Geobacter metallireducens*. Since some ferric iron-reducing bacteria (e.g. *G. metallireducens*) can use toxic hydrocarbons such as toluene as a carbon source, there is significant interest in using these organisms as bioremediation agents in ferric iron-rich contaminated aquifers.

Although ferric iron is the most prevalent inorganic electron acceptor, a number of organisms (including the iron-reducing bacteria mentioned above) can use other inorganic ions in anaerobic respiration.

While these processes may often be less significant ecologically, they are of considerable interest for bioremediation, especially when heavy metals or radionuclides are used as electron acceptors. Examples include:

- Manganic ion (Mn^{4+}) reduction to manganous ion (Mn^{2+})
- Selenate (SeO_4^{2-}) reduction to selenite (SeO_3^{2-}) and selenite reduction to inorganic selenium (Se^0)
- Arsenate (AsO_4^{3-}) reduction to arsenite (AsO_3^{3-})
- Uranyl ion ion (UO_2^{2+}) reduction to uranium dioxide (UO_2).

Organic Terminal Electron Acceptors

A number of organisms, instead of using inorganic compounds as terminal electron acceptors, are able to use organic compounds to accept electrons from respiration. Examples include:

- Fumarate reduction to succinate
- Trimethylamine *N*-oxide (TMAO) reduction to trimethylamine (TMA)
- Dimethyl sulfoxide (DMSO) reduction to Dimethyl sulfide (DMS)
- Reductive dechlorination.

TMAO is a chemical commonly produced by fish, and when reduced to TMA produces a strong odor. DMSO is a common marine and freshwater chemical which is also odiferous when reduced to DMS. Reductive dechlorination is the process by which chlorinated organic compounds are reduced to form their non-chlorinated endproducts. As chlorinated organic compounds are often important (and difficult to degrade) environmental polutants, reductive dechlorination is an important process in bioremediation.

Chemolithotrophy

Chemolithotrophy is a type of metabolism where energy is obtained from the oxidation of inorganic compounds. Most chemolithotrophic organisms are also autotrophic. There are two major objectives to chemolithotrophy: the generation of energy (ATP) and the generation of reducing power (NADH).

Hydrogen Oxidation

Many organisms are capable of using hydrogen (H_2) as a source of energy. While several mechanisms of anaerobic hydrogen oxidation have been mentioned previously (e.g. sulfate reducing- and acetogenic bacteria), hydrogen can also be used as an energy source aerobically. In these organisms, hydrogen is oxidized by a membrane-bound hydrogenase causing proton pumping via electron transfer to various quinones and cytochromes. In many organisms, a second cytoplasmic hydrogenase is used to generate reducing power in the form of NADH, which is subsequently used to fix carbon dioxide via the Calvin cycle. Hydrogen-oxidizing organisms, such as *Cupriavidus necator* (formerly *Ralstonia eutropha*), often inhabit oxic-anoxic interfaces in nature to take advantage of the hydrogen produced by anaerobic fermentative organisms while still maintaining a supply of oxygen.

Sulfur Oxidation

Sulfur oxidation involves the oxidation of reduced sulfur compounds (such as sulfide (H_2S), inorganic sulfur (S^0), and thiosulfate ($S_2O_3^{2-}$) to form sulfuric acid (H_2SO_4). A classic example of a sulfur-oxidizing bacterium is *Beggiatoa*, a microbe originally described by Sergei Winogradsky, one of the founders of environmental microbiology. Another example is *Paracoccus*.

Generally, the oxidation of sulfide occurs in stages, with inorganic sulfur being stored either inside or outside of the cell until needed. This two step process occurs because energetically sulfide is a better electron donor than inorganic sulfur or thiosulfate, allowing for a greater number of protons to be translocated across the membrane. Sulfur-oxidizing organisms generate reducing power for carbon dioxide fixation via the Calvin cycle using reverse electron flow, an energy-requiring process that pushes the electrons against their thermodynamic gradient to produce NADH.

Biochemically, reduced sulfur compounds are converted to sulfite (SO_2^{-3}) and subsequently converted to sulfate (SO_2^{-4}) by the enzyme sulfite oxidase. Some organisms, however, accomplish the same oxidation using a reversal of the APS reductase system used by sulfate-reducing bacteria. In all cases the energy liberated is transferred to the electron transport chain for ATP and NADH production. In addition to aerobic sulfur oxidation, some organisms (e.g. *Thiobacillus denitrificans*) use nitrate (NO_3^{2-}) as a terminal electron acceptor and therefore grow anaerobically.

Ferrous Iron (Fe^{2+}) Oxidation

Ferrous iron is a soluble form of iron that is stable at extremely low pHs or under anaerobic conditions. Under aerobic, moderate pH conditions ferrous iron is oxidized spontaneously to the ferric (Fe^{3+}) form and is hydrolyzed abiotically to insoluble ferric hydroxide ($Fe(OH)_3$). There are three distinct types of ferrous iron-oxidizing microbes. The first are acidophiles, such as the bacteria *Acidithiobacillus ferooxidans* and *Leptospirrillum ferrooxidans*, as well as the archaeon *Ferroplasma*. These microbes oxidize iron in environments that have a very low pH and are important in acid mine drainage.

The second type of microbes oxidize ferrous iron at cirum-neutral pH. These micro-organisms (for example *Gallionella ferruginea* or *Leptothrix ochracea*) live at the oxic-anoxic interfaces and are microaerophiles. The third type of iron-oxidizing microbes are anaerobic photosynthetic bacteria such as Rhodopsuedomonas, which use ferrous iron to produce NADH for autotrophic carbon dioxide fixation. Biochemically, aerobic iron oxidation is a very energetically poor process which therefore requires large amounts of iron to be oxidized by the enzyme rusticyanin to facilitate the formation of proton motive force. Like sulfur oxidation, reverse electron flow must be used to form the NADH used for carbon dioxide fixation via the Calvin cycle.

Nitrification

Nitrification is the process by which ammonia (NH_3) is converted to nitrate (NO_3^-). Nitrification is actually the net result of two distinct processes: oxidation of ammonia to nitrite (NO_2^-) by nitrosifying bacteria (e.g. *Nitrosomonas*) and oxidation of nitrite to nitrate by the nitrite-oxidizing bacteria (e.g. *Nitrobacter*). Both of these processes are extremely energetically poor leading to very slow growth rates for both types of organisms. Biochemically, ammonia oxidation occurs by the stepwise oxidation of ammonia to hydroxylamine (NH_2OH) by the enzyme ammonia monooxygenase in the cytoplasm, followed by the oxidation of hydroxylamine to nitrite by the enzyme hydroxylamine oxidoreductase in the periplasm. Electron and proton cycling are very complex but as a net result only one proton is translocated across the membrane per molecule of ammonia oxidized. Nitrite reduction is much simpler, with nitrite being oxidized by the enzyme nitrite oxidoreductase coupled to proton translocation by a very short electron transport chain, again leading to very low growth rates for these organisms. Oxygen is required in both ammonia and nitrite oxidation, meaning that both nitrosifying and nitrite-oxidizing bacteria are aerobes. As in sulfur and iron oxidation, NADH for carbon dioxide fixation using the Calvin cycle is generated by reverse electron flow, thereby placing a further metabolic burden on an already energy-poor process.

Anammox

Anammox stands for anaerobic ammonia oxidation and the organisms responsible were relatively recently discovered, in the late 1990s. This form of metabolism occurs in members of the Planctomycetes (e.g. Candidatus *Brocadia anammoxidans*) and involves the coupling of ammonia oxidation to nitrite reduction. As oxygen is not required for this process these organisms are strict anaerobes. Amazingly, hydrazine (N_2H_4 - rocket fuel) is produced as an intermediate during anammox metabolism. To deal with the high toxicity of hydrazine, anammox bacteria contain a hydrazine-containing intracellular organelle called the anammoxasome, surrounded by highly compact (and unusual) ladderane lipid membrane.

These lipids are unique in nature, as is the use of hydrazine as a metabolic intermediate. Anammox organisms are autotrophs although the mechanism for carbon dioxide fixation is unclear. Because of this property, these organisms could be used in industry to remove nitrogen in wastewater treatment processes. Anammox has also been shown have widespread occurrence in anaerobic aquatic systems and has been speculated to account for approximately 50% of nitrogen gas production in the ocean.

Phototrophy

Many microbes (phototrophs) are capable of using light as a source of energy to produce ATP and organic compounds such as carbohydrates, lipids, and proteins. Of these, algae are particularly significant because they are oxygenic, using water as an electron donor for electron transfer during photosynthesis. Phototrophic bacteria are found in the phyla Cyanobacteria, Chlorobi, Proteobacteria, Chloroflexi, and Firmicutes. Along with plants these microbes are responsible for all biological generation of oxygen gas on Earth. Because chloroplasts were derived from a lineage of the Cyanobacteria, the general principles of metabolism in these endosymbionts can also be applied to chloroplasts. In addition to oxygenic photosynthesis, many bacteria can also photosynthesize anaerobically, typically using sulfide (H_2S) as an electron donor to produce sulfate. Inorganic sulfur (S^0), thiosulfate ($S_2O_3^{2-}$) and ferrous iron (Fe^{2+}) can also be used by some organisms. Phylogenetically, all oxygenic photosynthetic bacteria are Cyanobacteria, while anoxygenic photosynthetic bacteria belong to the purple bacteria (Proteobacteria), Green sulfur bacteria (e.g. Chlorobium), Green non-sulfur bacteria (e.g. Chloroflexus), or the heliobacteria (Low %G+C Gram positives). In addition to these organisms, some microbes (e.g. the Archaeon *Halobacterium* or the bacterium *Roseobacter*, among others) can utilize light to produce energy using the enzyme bacteriorhodopsin, a light-driven proton pump. However, there are no known Archaea that carry out photosynthesis.

As befits the large diversity of photosynthetic bacteria, there are many different mechanisms by which light is converted into energy for metabolism. All photosynthetic organisms locate their photosynthetic reaction centres within a membrane, which may be invaginations of the cytoplasmic membrane (Proteobacteria), thylakoid membranes (Cyanobacteria), specialized antenna structures called chlorosomes (Green sulfur and non-sulfur bacteria), or the cytoplasmic membrane itself (heliobacteria). Different photosynthetic bacteria also contain different photosynthetic pigments, such as chlorophylls and carotenoids, allowing them to take advantage of different portions of the electromagnetic spectrum and thereby inhabit different niches. Some groups of organisms contain more specialized light-harvesting structures (e.g. phycobilisomes in Cyanobacteria and chlorosomes in Green sulfur and non-sulfur bacteria), allowing for increased efficiency in light utilization.

Biochemically, anoxygenic photosynthesis is very different from oxygenic photosynthesis. Cyanobacteria (and by extension, chloroplasts) use the Z scheme of electron flow in which electrons eventually are used to form NADH. Two different reaction centres (photosystems) are used and proton motive force is generated both by using cyclic electron flow and the quinone pool. In anoxygenic photosynthetic bacteria, electron flow is cyclic, with all electrons used in photosynthesis eventually being transferred back to the single reaction centre. A proton motive force is generated using only the quinone pool.

In heliobacteria, Green sulfur, and Green non-sulfur bacteria, NADH is formed using the protein ferredoxin, an energetically favourable reaction. In purple bacteria, NADH is formed by reverse electron flow due to the lower chemical potential of this reaction centre. In all cases, however, a proton motive force is generated and used to drive ATP production via an ATPase. Most photosynthetic microbes are autotrophic, fixing carbon dioxide via the Calvin cycle. Some photosynthetic bacteria (e.g. Chloroflexus) are photoheterotrophs, meaning that they use organic carbon compounds as a carbon source for growth. Some photosynthetic organisms also fix nitrogen.

Nitrogen Fixation

Nitrogen fixation is the natural process, either biological or abiotic, by which nitrogen (N_2) in the atmosphere is converted into ammonia (NH_3). This process is essential for life because fixed nitrogen is required to biosynthesize the basic building blocks of life, e.g., nucleotides for DNA and RNA and amino acids for proteins. Nitrogen fixation also refers to other abiological conversions of nitrogen, such as its conversion to nitrogen dioxide.

Nitrogen fixation is utilized by numerous prokaryotes, including bacteria, actinobacteria, and certain types of anaerobic bacteria. Microorganisms that fix nitrogen are called diazotrophs. Some higher plants, and some animals (termites), have formed associations (symbioses) with diazotrophs. Nitrogen fixation also occurs as a result of non-biological processes. These include lightning, industrially through the Haber-Bosch Process, and combustion. Biological nitrogen fixation was discovered by the Dutch microbiologist Martinus Beijerinck.

Biological Nitrogen Fixation

Biological nitrogen fixation (BNF) occurs when atmospheric nitrogen is converted to ammonia by an enzyme called nitrogenase. The formula for BNF is:

$$N_2 + 8\ H^+ + 6\ e^- \rightarrow 2\ NH_3 + H_2$$

The process is coupled to the hydrolysis of 16 equivalents of ATP and is accompanied by the co-formation of one molecule of H_2. In free-living diazotrophs, the nitrogenase-generated ammonium is assimilated into glutamate through the glutamine synthetase/glutamate synthase pathway.

Enzymes responsible for nitrogenase action are very susceptible to destruction by oxygen. (In fact, many bacteria cease production of the enzyme in the presence of oxygen). Many nitrogen-fixing organisms exist only in anaerobic conditions, respiring to draw down oxygen levels, or binding the oxygen with a protein such as Leghemoglobin.

Plants that contribute to nitrogen fixation include the legume family – Fabaceae – with taxa such as clover, soybeans, alfalfa, lupines, peanuts, and rooibos. They contain symbiotic bacteria called *Rhizobia* within nodules in their root systems, producing nitrogen compounds that help the plant to grow and compete with other plants. When the plant dies, the fixed nitrogen is released, making it available to other plants and this helps to fertilize the soil The great majority of legumes have this association, but a few genera (e.g., *Styphnolobium*) do not. In many traditional and organic farming practices, fields are rotated through various types of crops, which usually includes one consisting mainly or entirely of clover or buckwheat (family *Polygonaceae*), which were often referred to as "green manure."

Non-leguminous Nitrogen-fixing Plants

Although by far the majority of nitrogen-fixing plants are in the legume family Fabaceae, there are a few non-leguminous plants, such as alder, that can also fix nitrogen. These plants, referred to as "actinorhizal plants", consist of 24 genera of woody shrubs or trees distributed among in 8 plant families. The ability to fix nitrogen is not universally present in these families. For instance, of 122 genera in the Rosaceae, only 4 genera are capable of fixing nitrogen. All these families belong to the orders Cucurbitales, Fagales, and Rosales, which together with the Fabales form a clade of eurosids. There are also several nitrogen-fixing symbiotic associations that involve cyanobacteria (such as *Nostoc*):

- Some lichens such as *Lobaria* and *Peltigera*
- Mosquito fern (*Azolla* species)
- Cycads
- *Gunnera.*

Microorganisms that Fix Nitrogen (Diazotrophs)

- Cyanobacteria
- *Azotobacteraceae*
- *Rhizobia*
- *Frankia.*

Nitrogen Fixation by Cyanobacteria

Cyanobacteria inhabit nearly all illuminated environments on Earth and play key roles in the carbon and nitrogen cycle of the biosphere. In general, cyanobacteria are able to utilize a variety of inorganic and organic sources of combined nitrogen, like nitrate, nitrite, ammonium, urea, or some amino acids. Several cyanobacterial strains are also capable of diazotrophic growth. Genome sequencing has provided a large amount of information on the genetic basis of nitrogen metabolism and its control in different cyanobacteria. Comparative genomics, together with functional studies, has led to a significant advance in this field over the past years. 2-Oxoglutarate has turned out to be the central signalling molecule reflecting the carbon/nitrogen balance of cyanobacteria. Central players of nitrogen control are the global transcriptional factor NtcA, which controls the expression of many genes involved in nitrogen metabolism, as well as the P_{II} signalling protein, which fine-tunes cellular activities in response to changing C/N conditions.

These two proteins are sensors of the cellular 2-oxoglutarate level and have been conserved in all cyanobacteria. In contrast, the adaptation to nitrogen starvation involves heterogeneous responses in different strains. Nitrogen fixation by cyanobacteria in coral reefs can fix twice the amount of nitrogen than on land–around 1.8 kg of nitrogen is fixed per hectare per day.

Chemical Nitrogen Fixation

Nitrogen can also be artificially fixed for use in fertilizers, explosives, or in other products. The most common method is the Haber process. Artificial fertilizer production is now the largest source of human-produced fixed nitrogen in the Earth's ecosystem.

The Haber process requires high pressures (around 200 atm) and high temperatures (at least 400 °C), routine conditions for industrial catalysis. This highly efficient process uses natural gas as a hydrogen source and air as a nitrogen source.

Research on Catalytic Nitrogen-fixation

Much research has been conducted on the discovery of catalysts for nitrogen fixation, often with the goal of reducing the energy required for this conversion. However, such research has thus far failed to even approach the efficiency and ease of the Haber process. Many compounds react with atmospheric nitrogen under ambient conditions. For example, lithium metal converts to lithium nitride under an atmosphere of nitrogen. Treatment of the resulting nitride gives ammonia.

The first dinitrogen complex was reported in 1965 based on ammonia coordinated to ruthenium ($[Ru(NH_3)_5(N_2)]^{2+}$, research in chemical fixation focused on transition metal complexes. Since then, a large number of transition metal compounds that contain dinitrogen as a ligand have been discovered. The dinitrogen ligand can either be bound to a single metal or bridge two (or more) metals. The coordination chemistry of dinitrogen is complex and currently under intense investigation. This research may lead to new ways of using dinitrogen in synthesis and on an industrial scale.

The first example of homolytic cleavage of dinitrogen under mild conditions was published in 1995. Two equivalents of a molybdenum complex reacted with one equivalent of dinitrogen, creating a triple bonded MoN complex. Since then, this triple bounded complex has been used to make nitriles. The first catalytic system converting nitrogen to ammonia at room temperature and pressure was discovered in 2003 and is based on another molybdenum compound, a proton source, and a strong reducing agent. However, this catalytic reduction fixates only a few nitrogen molecules.

Iron in Microbiology

Iron is essential for almost all living organisms as it is involved in a wide variety of important metabolic processes. However, iron is not readily available and microorganisms therefore employ various iron uptake systems to secure sufficient supplies from their surroundings. There is considerable variation in the range of iron transporters and iron sources utilised by different microbial species. Pathogens, in particular, require efficient iron acquisition mechanisms to enable them to compete successfully for iron in the highly iron-restricted environment of the host's tissues and body fluids.

Siderophores

In the early days of life on earth due to the emergence of photosynthesis, the atmosphere became oxidative and so iron prevailed in its trivalent form. The consequence was that most microorganisms live today in a surrounding where the concentration of soluble iron compounds is much too low to satisfy their physiological needs. One of the possibilities to circumvent this problem is the production of so-called siderophores, compounds which can form water soluble complexes with Fe^{3+}. They are released in situations of iron deficiency. Fe^{3+} possesses six coordination sites arranged in the shape of an octahedron which can accommodate three bidentate ligands. The major ligand types are catecholates, hydroxamates and alpha-hydroxycarboxylates, in the ideal case connected by adequate spacers which keep them in the correct position. Many siderophores have elaborate structures which can bind to the cell surface of the producing species. In this way pirating by competing microorganisms is rendered more difficult.

Heme Uptake and Iron Extraction by Bacteria

Heme is ubiquitous, abundant and necessary for energy metabolism. Most bacteria have a heme biosynthesis pathway, but nevertheless, since heme is a major source of iron (an essential metal), microbes take up exogenous heme to retrieve iron. To grab heme, microbes extract it from host hemoproteins. This is achieved by two non-exclusive distinct pathways. One pathway involves proteins secreted by bacteria (hemophores) that scavenge heme from host hemoproteins. The second pathway involves microbial cell surface receptors that catch hemoproteins circulating in the vicinity of the cell surface. Both pathways lead to heme docking to cell surface receptors. In Gram-negative bacteria, docked heme is transported through the outer membrane by an energy-dependent process. In Gram-positive bacteria, docked heme is transferred to membrane-anchored heme binding lipoproteins. In all thus far described systems, heme is actively transported through the plasma membrane by an ATP hydrolysis-powered ABC transporter. Heme is either degraded into biliverdin, CO and iron by heme oxygenases, or iron is retrieved from heme, keeping the tetrapyrrol ring intact by recently identified enzymes. As excess heme is toxic, heme uptake, efflux and degradation are usually highly regulated. In most cases, intracytoplasmic heme or iron released during heme degradation are cofactors along with transcriptional regulators. In several cases, heme uptake and efflux are regulated by extracellular heme.

Rhizobia

The rhizobia live as free-living soil bacteria or in symbiosis with leguminous plants. The success of these organisms in each milieu involves the ability to sense the environment to assess the availability of nutrients, and to optimize cellular systems for their acquisition. Iron in the rhizosphere is mostly inaccessible due to low solubility, and microorganisms must compete for this limited nutrient. Rhizobia belong to the alpha-Proteobacteria, a diverse taxonomic group that includes numerous species that form close or intracellular associations with eukaryotic hosts in a symbiotic or pathogenic context. Thus, in addition to their agricultural and economic importance, rhizobia are model organisms that have given new insights into related, but less tractable animal pathogens. In particular, genetic control of iron homeostasis in the rhizobia and other alpha-Proteobacteria has moved away from the Fur paradigm to an iron sensing mechanism responding to the metal indirectly. Moreover, utilization of heme as an iron source is not unique to animal pathogens, but is an acquisition strategy employed by the rhizobia with some interesting novel features.

Bordetella

Upon colonization of the mammalian respiratory epithelium by mucosal pathogens of the genus *Bordetella*, the host-pathogen interaction causes inflammatory changes, immune activation, and host cell injury. In this dynamic environment, *Bordetella* cells scavenge the nutritional iron necessary for growth. The three classical *Bordetella* species produce the siderophore alcaligin.

In addition, they can utilize xenosiderophores that could be produced by commensals or other microbes that transiently inhabit the nasopharynx. As infection progresses, extravasation of immune cells, erythrocytes and serum to the mucosal surface can occur, exacerbated by the damaging action of *Bordetella* toxins, thus providing iron sources such as transferrin and heme compounds to the microbe. The three characterized *Bordetella* iron systems for utilization of alcaligin, enterobactin and heme are each inducible by the cognate iron source. The ability to sense and respond to the presence of available iron sources allows these pathogens to adapt to temporal changes in iron source availability, and this ability is important for successful *in vivo* growth.

Iron Shigella and E. Coli

Shigella spp. and pathogenic *E. coli* are characterized by a variety and abundance of iron transport systems. Although members of this group of bacteria are closely related genetically, they differ widely in the iron transport systems they use. This may reflect the different niches occupied by different strains and the nature of the source of iron available in a specific environment.

Only the ferrous iron transporter Feo is common to all the commensals and pathogens. All members of this group produce one or more siderophore, but no single siderophore is produced by all. Other iron transport systems include heme transporters and the ferrous iron transporters Sit and Efe. With the exception of the genes for enterobactin and the Feo system, the iron transport genes in the enterics are found within pathogenicity islands or on plasmids and their presence often increases pathogenicity or colonization of niches within the host.

Iron in Erwinia

The critical role of iron in host-pathogen relationships has been elucidated in infectious diseases of mammals, where the importance of siderophores in microbial pathogenesis has been demonstrated. Our group has established the role of iron and its ligands in the virulence of the plant pathogenic bacteria *Dickeya dadantii* (*Erwinia chrysanthemi*) and *Erwinia amylovora*. The genomes of the two pectinolytic enterobacterial species *Pectobacterium atrosepticum* SCRI1043 and *D. dadantii* 3937 have been sequenced and annotated.

This review focuses on the functions involved in iron acquisition in both species. Besides the production and utilization of siderophores, *P. atrosepticum* and *D. datantii* have the capacity to use other iron sources. Indeed, both species are able to use haem iron, whereas only *P. atrosepticum* can transport the ferric citrate complex and only *D. dadantii* can acquire ferrous iron. These different modes of iron capture indicate that these species have to cope with various environmental and ecological conditions during their pathogenic life cycle.

Vibrio and Aeromonas

Vibrio and *Aeromonas* species are ubiquitous bacteria in aquatic environments worldwide. Many of the species are important pathogens for humans and/or aquatic animals. Several iron acquisition strategies have been developed by vibrios and aeromonads in order to get this essential element for surviving in their host and in aquatic habitats. All species studied so far have the ability to synthesize siderophores to sequester iron from the cell environment and transport it through their respective cognate outer membrane receptors.

It has been demonstrated that this capacity is a relevant virulence factor for human and animal pathogens. Furthermore, all species studied can utilize exogenous siderophores, made by other bacteria. Another iron acquisition system described in both genera involves the use of heme as a source of iron, by a mechanism very well conserved among all species, which involves a heme transporter that includes a specific TonB-dependent outer membrane receptor(s) and an ABC-type inner membrane transporter. Alternative systems based on ferrous or ferric iron transporters have been reported in *V. cholerae*. How the different iron acquisition systems work together to supply iron to the cell and how they are used in the different environments where vibrios and aeromonads can be found is still an open question.

Francisella

Francisella tularensis is unusual among Gram-negative bacteria in that its genome does not encode orthologs for TonB, ExbB and ExbD that typically energize the uptake of iron across the outer membrane. This organism secretes however a siderophore similar in structure to rhizoferrin. The *fsl* operon of six genes encodes functions for biosynthesis and uptake of the siderophore.

Two of these genes encode a siderophore synthetase belonging to the nonribosomal peptide synthetase (NRPS)-independent synthetase (NIS)-family and a protein belonging to the pyridoxyl phosphate-dependent decarboxylase family, and both are required for siderophore production. Siderophore utilization involves the product of the *fslE* gene, a protein unique to *Francisella* species that could function as a siderophore receptor. Additionally, genes related in sequence to *fslE* also play a role in siderophore acquisition. The mechanism for TonB-independent iron uptake in this microorganism remains to be elucidated.

Bacteroides

Bacteroides spp. have an essential requirement for heme and non-heme iron. They cannot synthesize the tetrapyrrole macrocycle ring due to a lack of genes for the heme biosynthetic pathway. It is remarkable that heme-dependent organisms outnumber heme-independent organisms in the lower intestinal tract suggesting that heme biosynthesis is not essential for colonization of the colonic environment. However, this colonization advantage may be due to the fact that under anaerobic conditions in the presence of heme, *B. fragilis* can generate nearly the double amount of ATP than *Escherichia coli* per mol of glucose.

This high energy yield is linked to a rudimentary heme-induced fumarate reductase and cytochrome *b*-dependent electron transport energy metabolism pathway which uses fumarate as the terminal electron acceptor. Moreover, *Bacteroides* spp. can incorporate iron-deuteroporphyrin and iron-mesoporphyrin into a functional type-*b* cytochrome. Heme can be demetalated without cleaving the tetrapyrrole ring releasing free iron and free protoporphirin IX. The ability of the opportunistic human pathogen *B. fragilis* to cause infections seems to be due in part to its ability to scavenge heme and iron from host proteins. The in-frame translated intergenic region of the fused FeoAB proteins are exclusively present in gastro-intestinal colonizers belonging to the Bacteroidetes, Firmicutes and Actinobacteria phyla. Several members of the *Bacteroides* group have three orthologs of the mammalian-type bacterial ferritin gene, *ftnA*. FtnA may play an important role in protection against iron-induced oxidative stress in this group of highly aerotolerant anaerobes.

Campylobacter

Iron is known to catalyse a wide range of biochemical reactions essential for most living organisms, including *Campylobacter jejuni*. Paradoxically, this iron reactivity is also responsible for the generation of hydroxyl radicals (·OH), which are particularly biotoxic. In order to avoid iron toxicity, microorganisms must achieve an effective iron homeostasis by tightly regulating the expression of genes encoding the proteins involved in iron acquisition, metabolism and oxidative stress defences in response to iron availability.

Interestingly, in addition to the classical ferric uptake regulator Fur, *C. jejuni* carries another member of the Fur family of metalloregulators, PerR. PerR is a peroxide-sensing regulator and typically regulates peroxide stress response in Gram-positive bacteria. Recent work indicates that the regulatory functions of Fur and PerR extend beyond their classically ascribed roles. These diverse functions include energy metabolism, protein glycosylation and flagella biogenesis. Moreover, the Fur and PerR regulons appear to overlap and co-regulate key genes at specific junctions.

Cyanobacteria

Cyanobacteria are dependent on but can also be compromised by metals such as iron. On the one hand the demand for iron for photosystem functionality represents a challenge for the iron uptake machinery in iron limiting environments. On the other hand intoxication by iron causes a severe problem for growth and reproduction. To overcome this dilemma cyanobacteria have developed a regulatory network controlling iron uptake. They produce siderophores, which are distinct from that of other bacteria. Furthermore, the iron metabolism is linked to the nitrogen metabolism as documented for example in *Anabaena* sp. PCC 7120.

Bacillus

Bacillus subtilis is a metabolically versatile soil microbe and Gram-positive model organism that displays a sophisticated adaptive response to conditions of iron limitation. The endogenous siderophore of *B. subtilis* is bacillibactin, a trimeric catecholate siderophore similar in structure to enterobactin. In addition to bacillibactin, *B. subtilis* can obtain iron from several xenosiderophores, ferric citrate, heme, and through a newly discovered elemental iron permease.

The regulation of iron homeostasis in *B. subtilis* is complex and involves a ferric uptake regulator (Fur) protein as master regulator and at least two subsidiary regulatory systems.

The most significant of these is an iron-sparing/prioritization response controlled by the small RNA FsrA and three auxiliary proteins (FbpABC). In addition, the bacillibactin uptake system is transcriptionally activated by an AraC family activator, Btr that directly senses bacillibactin. Iron uptake and homeostasis systems in *B. anthracis* and related organisms are largely similar to those in *B. subtilis* with some additional components. These include a second siderophore synthesis operon for petrobactin, which is important for virulence, and a more elaborate (or at least better understood) heme uptake system.

Staphylococci

Staphylococcus aureus causes a significant amount of human morbidity and mortality. The ability of *S. aureus* to cause disease is dependent upon its acquisition of iron from the host. *S. aureus* can obtain iron from various sources during infection, including heme and transferrin. The most abundant iron source in humans is heme-iron bound by hemoglobin contained within erythrocytes. *S. aureus* is known to lyse erythrocytes through secretion of pore-forming toxins, providing access to host hemoglobin. Proteins of the iron-regulated surface determinant (Isd) system bind host hemoproteins, remove the heme cofactor, and shuttle heme into the cytoplasm for use as a nutrient iron source. Deletion of Isd system components decreases staphylococcal virulence, underscoring the importance of heme-iron acquisition during infection. In addition to heme, *S. aureus* can utilize transferrin-iron through the secretion of siderophores. Several staphylococcal siderophores have been described, some of which have defined roles during the pathogenesis of staphylococcal infections. A greater understanding of staphylococcal iron acquisition may lead to the development of novel therapeutic strategies that target nutrient uptake and decrease the threat of this increasingly drug-resistant bacterial pathogen.

Yeasts

Yeasts take up iron by three main mechanisms. In the reductive uptake mechanism, specialized flavo-hemoproteins (Fre) dissociate extracellular ferric complexes by reduction involving trans-plasma membrane electron transfer. The resulting free iron is then imported by a high-affinity permease system (Ftr), coupled to a copper-dependent oxidase (Fet), which channels iron through the plasma membrane. As a consequence, iron uptake by this mechanism is dependent on the availability of copper. In the siderophore-mediated mechanism, siderophores excreted by the cells or produced by other bacterial or fungal species are taken up without prior dissociation, via specific, copper-independent high-affinity receptors.

The iron is then dissociated from the siderophores intracellularly, probably by reduction. In the heme uptake mechanism, free heme or heme bound to hemoglobin is taken up as such, probably by endocytosis. Iron is released intracellularly after hydrolysis of the porphyrin ring catalysed by heme oxygenase. Within the cell, iron is stored in vacuoles or in siderophores. Iron can be mobilized from vacuoles by a reductive mechanism homologous to that found at the plasma membrane. Regulation of iron uptake and iron use are mediated by transcriptional regulators acting either as activators in iron-deficient conditions or as repressors in iron-rich conditions, according to the yeast species; these regulators thus adjust the iron uptake flux to the cell's requirements. In the baker's yeast, *Saccharomyces cerevisiae*, a post-transcriptional mechanism is active under low iron conditions, involving the degradation of RNAs encoding inessential iron-utilizing proteins. Other fungi have mechanisms serving a similar purpose at the transcriptional level. Studies in *S. cerevisiae* show that mitochondria are central to regulating cellular iron homeostasis, through the synthesis of iron-sulfur clusters.

Environmental Microbiology

Environmental microbiology is the study of the composition and physiology of microbial communities in the environment. The environment in this case means the soil, water, air and sediments covering the planet and can also include the animals and plants that inhabit these areas. Environmental microbiology also includes the study of microorganisms that exist in artificial environments such as bioreactors.

Microbial life is amazingly diverse and microorganisms literally cover the planet. An average gram of soil contains approximately one billion (1,000,000,000) microbes representing probably several thousand species. Microorganisms have special impact on the whole biosphere. They are the backbone of ecosystems of the zones where light cannot approach. In such zones, chemosynthetic bacteria are present which provide energy and carbon to the other organisms there. Some microbes are decomposers which have ability to recycle the nutrients. Microbes have a special role in biogeochemical cycles. Microbes, especially bacteria, are of great importance because their symbiotic relationship (either positive or negative) have special effects on the ecosystem.

Microorganisms are used for *in-situ* microbial biodegradation or bioremediation of domestic, agricultural and industrial wastes and subsurface pollution in soils, sediments and marine environments. The ability of each microorganism to degrade toxic waste depends on the nature of each contaminant. Since most sites typically have multiple pollutant types, the most effective approach to microbial biodegradation is to use a mixture of bacterial species and strains, each specific to the biodegradation of one or more types of contaminants. It is vital to monitor the composition of the indigenous and added bacteria in order to evaluate the activity level and to permit modifications of the nutrients and other conditions for optimizing the bioremediation process.

Oil Biodegradation

Petroleum oil is toxic, and pollution of the environment by oil causes major ecological concern. Oil spills of coastal regions and the open sea are poorly containable and mitigation is difficult; much of the oil can, however, be eliminated by the hydrocarbon-degrading activities of microbial communities, in particular the hydrocarbonoclastic bacteria (HCB). These organisms can help remedy the ecological damage caused by oil pollution of marine habitats. HCB also have potential biotechnological applications in the areas of bioplastics and biocatalysis.

Waste Biotreatment

Biotreatment, the processing of wastes using living organisms, is an environmentally friendly alternative to other options for treating waste material. Bioreactors have been designed to overcome the various limiting factors of biotreatment processes in highly controlled systems. This versatility in the design of bioreactors allows the treatment of a wide range of wastes under optimized conditions. It is vital to consider various microorganisms and a great number of analyses are often required.

Wastewater Treatment

Wastewater treatment processes are geared towards one purpose: cleaning up water. Recent application of molecular techniques is unveiling the microbial composition and architecture of the complex communities involved in the treatment processes. It is now recognized that wastewater processes harbor a vast variety of microorganisms most of which are yet-to-be cultured, hence uncharacterized. Metagenomic technology is being used to study the diversity, structure and functions of microbial communities in nitrifying processes, anaerobic ammonia oxidation processes and methane fermenting processes.

Environmental Genomics of Cyanobacteria

The application of molecular biology and genomics to environmental microbiology has led to the discovery of a huge complexity in natural communities of microbes. Diversity surveying, community fingerprinting and functional interrogation of natural populations have become common, enabled by a range of molecular and bioinformatics techniques. Recent studies on the ecology of Cyanobacteria have covered many habitats and have demonstrated that cyanobacterial communities tend to be habitat-specific and that much genetic diversity is concealed among morphologically simple types. Molecular, bioinformatics, physiological and geochemical techniques have combined in the study of natural communities of these bacteria.

Corynebacteria

Corynebacteria are a diverse group Gram-positive bacteria found in a range of different ecological niches such as soil, vegetables, sewage, skin, and cheese smear. Some, such as *Corynebacterium diphtheriae*, are important pathogens while others, such as *Corynebacterium glutamicum*, are of immense industrial importance. *C. glutamicum* is one of the biotechnologically most important bacterial species with an annual production of more than two million tons of amino acids, mainly L-glutamate and L-lysine.

Legionella

Legionella is common in many environments, with at least 50 species and 70 serogroups identified. *Legionella* is commonly found in aquatic habitats where its ability to survive and to multiply within different protozoa equips the bacterium to be transmissible and pathogenic to humans.

Archaea

Originally, Archaea were once thought of as extremophiles existing only in hostile environments but have since been found in all habitats and may contribute up to 20% of total biomass. Archaea are particularly common in the oceans, and the archaea in plankton may be one of the most abundant groups of organisms on the planet. Archaea are subdivided into four phyla of which two, the Crenarchaeota and the Euryarchaeota, are most intensively studied.

Lactobacillus

Lactobacillus species are found in the environment mainly associated with plant material. They are also found in the gastrointestinal tract of humans, where they are symbiotic and make up a portion of the gut flora.

Aspergillus

Aspergillus spores are common components of aerosols where they drift on air currents, dispersing themselves both short and long distances depending on environmental conditions. When the spores come in contact with a solid or liquid surface, they are deposited and if conditions of moisture are right, they germinate. The ability to disperse globally in air currents and to grow almost anywhere when appropriate food and water are available means that ubiquitous is among the most common adjectives used to describe these moulds.

Rhizobia

Symbiotic nitrogen fixation is a mutualistic process in which bacteria reside inside plants and reduce atmospheric nitrogen to ammonia. This ammonia can then be used by the plant for the synthesis of proteins and other nitrogen-containing compounds such as nucleic acids. The Gram-negative soil bacteria that carry out this process are collectively referred to as rhizobia (from the Greek words Riza = Root and Bios = Life). The process of symbiotic nitrogen fixation is of agricultural and ecological significance because plants capable of nitrogen fixation do not need to compete for limited quantities of soil nitrogen, nor do they require expensive nitrogenous fertilizers that can be harmful to the environment.

Microalgae

Algae are a highly diverse group of protists, ranging from simple, unicellular organisms to complex, multicellular entities with a range of differentiated tissues and distinct organs. They are found among diverse aquatic ecosystems and play important roles by supplying carbon and energy as well as providing habitat to other members of the biological communities. Some algae cause significant environmental and health problems. There are three algal groups: the dinoflagellates, the diatoms and the haptophytes. Their 3 main phylla are chlorophyta,rhodophyta and phaeophyta.

Anaerobic Protozoa

Diplomonads are a group of mitochondrion-lacking, binucleated flagellates found in anaerobic or micro-aerophilic environments. Most research on diplomonads has focused on *Giardia*, which is a major cause of water-borne enteric disease in humans and other animals. The first diplomonad to have its genome sequenced was a *Giardia* isolate (WB) and the 11.7 million basepair genome is compact in structure and content with simplified basic cellular machineries and metabolism.

Water Microbiology

An adequate supply of safe drinking water is one of the major prerequisites for a healthy life, but waterborne diseases are still a major cause of death in many parts of the world, particularly in young children, the elderly, or those with compromised immune systems. As the epidemiology of waterborne diseases is changing, there is a growing global public health concern about new and reemerging infectious diseases that are occurring through a complex interaction of social, economic, evolutionary, and ecological factors. An important challenge is therefore the rapid, specific and sensitive detection of waterborne pathogens. Presently, microbial tests are based essentially on time-consuming culture methods. However, newer enzymatic, immunological and genetic methods are being developed to replace and/or support classical approaches to microbial detection. Moreover, innovations in nanotechnology and nanosciences are having a significant impact in biodiagnostics, where a number of nanoparticle-based assays and nanodevices have been introduced for biomolecular detection. Molecular techniques based on genomics, proteomics and transcriptomics are rapidly growing as complete microbial genome sequences are becoming available, and advances are made in sequencing technology, analytical biochemistry, microfluidics and data analysis. While the clinical and food industries are increasingly adapting these techniques, there appear to be major challenges in detecting health-related microbes in source and treated drinking waters.

This is due in part to the low density of pathogens in water, necessitating significant processing of large volume samples. From the vast panorama of available molecular techniques, some are finding a place in the water industry: Quantitative PCR, protein detection and immunological approaches, loop-mediated isothermal amplification (LAMP), microarrays.

Sensory Mechanisms

Bacteria have evolved abilities to regulate aspects of their behaviour (such as gene expression) in response to signals in the intracellular and extracellular environment. The interaction of a signal with its receptor (usually a protein or RNA molecule) triggers a series of events that lead to reprogramming of cellular physiology, typically as a consequence of altered patterns of gene expression. In this way, the bacterial cell is able to mount appropriate and effective responses to changing physical and/or chemical environments. The versatility with which many bacteria adapt to environmental change underlies many important aspects of microbiology. For example, pathogens encounter multiple environments as they invade a host from the outside, and then progress through different sites within host tissues.

There is growing evidence that pathogenic bacteria make use of physical and chemical cues to signal their presence in a suitable host, and need to adapt to the host environment in order to mount a successful infection. On the other hand, it should not be assumed that all signals to which bacteria must respond originate in the extracellular environment. For many species, even the cosseted life in a laboratory shake flask is 'stressful', in the sense that there is often a need to avoid or reverse the effects of harmful intermediates or by-products of metabolism. For example, all organisms that use dioxygen as a terminal electron acceptor have to deal with the reactive oxygen species that arise as adventitious by-products of aerobic metabolism. In bacteria, multiple protein receptors for oxygen radicals have been described, which control the expression of genes encoding enzymes that detoxify oxygen radicals or repair the damage that they cause.

Discovery of Vaccination

Lady Mary Wortley Montagu (1689–1762) was an aristocrat, a writer and the wife of a Member of Parliament. In 1716, her husband, Edward Wortley Montagu was appointed British Ambassador in Istanbul. She followed him there and two weeks after her arrival discovered the local practice of protection against smallpox by variolation – the injection of pus from smallpox victims into the skin. Her younger brother had died of smallpox, and she too had the disease but survived. Determined to spare her five-year-old son Edward from similar suffering, she ordered the embassy surgeon, Charles Maitland to variolate him. On her return to London, she asked Maitland to variolate her four-year-old daughter in the presence of the King's physicians. Later, Montagu persuaded the Prince and Princess of Wales to sponsor a public demonstration of the procedure. Six men who had been condemned to death and were awaiting execution at Newgate Prison were offered a full pardon for serving as the subjects of the public experiment. They accepted, and in 1721 were variolated. All the prisoners recovered from the procedure and to test its protective effect one of them, a nineteen-year-old woman, was ordered to sleep in the same bed as a ten-year old smallpox victim for six weeks. She did not contract the disease.

The experiment was repeated on six orphan children who survived the ordeal and by 1722, even King George I's grandchildren had been inoculated. But the practice was not entirely safe and caused many deaths. Although the number of lives saved far outweighed those lost, the practise was not widely adopted.Edward Jenner (1749–1823), a British rural physician, was variolated as a boy. He had suffered greatly from the ordeal but survived fully protected from smallpox. Jenner knew of a local belief that those who had suffered from a relatively mild infection called cowpox, which was common at the time in dairy workers, were immune to smallpox. Although probably not the first to do so, he decided to test the "theory". On May 14, 1796 he selected "a healthy boy, about eight years old for the purpose of inoculation for the Cow Pox". The boy was James Phipps (1788–1853) who survived the experiment and suffered only a mild fever. On July 1, 1796, Jenner took some "smallpox matter" (probably infected pus) and repeatedly inoculated Phipp's arms with it. Fortunately, Phipps survived, and was later inoculated with smallpox over twenty times without succumbing to the disease. Vaccination – the word is derived from the Latin *vacca* meaning "cow" – had been invented.

Louis Pasteur and Rabies

Rabies is an often fatal disease caused by the infection of mammals with rabies virus. Today it is mainly a disease that affects wild mammals such as foxes and bats, but it is one of the oldest known virus diseases: *rabies* is a Sanskrit word (*rabhas*) that dates from 3000 BC, which means "madness" or "rage" and the disease has been known for over 4000 years. The ancient Greeks called it "lyssa" or "lytta" also meaning madness. Although not a cause of epidemics, the infection was greatly feared because of its horrendous symptoms that include insanity, hydrophobia and death. The disease had been known since antiquity and references to rabies can be found in the Laws of Eshnunna, which date from 2300 BC. Aristotle (384–322 BC) wrote one of earliest undisputed descriptions of the disease and how it was passed to humans. Celsus in the first century AD first recorded the symptom called hydrophobia and suggested that the saliva of infected animals and humans contained a slime or poison – to describe this he used the word "virus". In France, in the time of Louis Pasteur, there were only a few hundred infections in humans each year but cures were desperately sought. Aware of the extreme danger, Pasteur worked on what he knew would be a challenge and he began to look for the "microbe" in the saliva of mad dogs.

A member of Pasteur's team, Emile Roux (1853–1933), studied how long the spinal cords of dogs that had died from the disease remained infectious. Roux invented an ingenious glass bottle in which he hung the spinal cords to dry them. Pasteur was impressed with Roux's invention and ordered more bottles to be made. Pasteur never credited Roux for his invention and Roux considered his idea stolen. Roux never worked on rabies again. After fourteen days of drying, Pasteur showed that when the dried spinal cords were crushed and injected into healthy dogs they did not become infected. He repeated the experiment several times on the same dog with tissue that had been dried for fewer and fewer days, until the dog survived after injections of fresh rabies infected spinal tissue. Pasteur had immunised the dog against rabies as he later did with 50 more.

Although he had little idea how his method worked, he went on to test it on a boy. Joseph Meister (1876–1940) was brought to Pasteur by his mother on July 6, 1885. He was covered in bites, having been set upon by a mad dog. A brick layer had defended the boy from the dog with an iron bar, but only after the salivating dog had bitten the boy fourteen times. Meister's mother begged Pasteur to help her son. Pasteur was a scientist, not a physician, and he was well aware of the consequences to him if things were to go wrong. Pasteur decided to help the boy and injected him with increasingly virulent rabid rabbit spinal cord over the following ten days. Later Pasteur wrote, "as the death of this child appeared inevitable, I decided, not without deep and severe unease...to try out on Joseph Meister the procedure, which had consistently worked on dogs.

The date set for the final inoculation using undried tissue was July 16. Pasteur's assistant Grancher was given the task, but accidentally jabbed himself with the infected needle, and thus became the second guinea pig to receive Pasteur's new treatment. Later in an argument about the ethics of what they had done, Pasteur's whole team, including Pasteur, clandestinely inoculated one another. Although still angry with Pasteur, Roux discovered what was going on and, as he was the only qualified physician on the team, took charge of the dangerous situation. It was too late for the men to stop. Fortunately, they all lived to tell the tale and Meister returned home with his mother on July 27. But the full truth was kept a secret for many years. Pasteur successfully treated a second boy in October of the same year; Jean-Baptiste Jupille (1869–1923) was a fifteen-year-old shepherd boy who had been severely bitten as he tried to protect other children from a rabid dog.

Pasteur's method of treatment subsequently remained in use for over fifty years. Not much was known about the cause of the disease until in 1903 when Adelchi Negri (1876–1912) first saw microscopic lesions in the brains of rabid animals now called Negri bodies. He wrongly thought they were protozoan parasites. However Paul Remlinger (1871–1964) soon showed by filtration experiments that they were much smaller than protozoa and even smaller than bacteria. Thirty years later, Negri bodies were shown to be accumulations of particles 100–150 nanometres long, now known to be the size of rhabdovirus particles – the virus that causes rabies.

20th Century

At the turn of the 20th century, evidence for the existence of viruses was obtained from experiments with filters that had pores too small for bacteria too pass through, and the term "filterable virus" was used to describe them. Although it was not until this time that the modern concept of a virus slowly began to gain acceptance, the word is ancient. In Latin it meant a "slimy liquid, poison offensive odour or taste", and the word first appeared in the medical literature in the 18th century to describe a poisonous fluid.

Before the 1930s, most scientists still thought they were small bacteria. But in 1938, following the invention of the electron microscope, viruses were shown to be completely different to bacteria. So different in fact, that not all scientists were convinced they were anything other than accumulations of toxic proteins.

The situation changed radically after it was discovered that viruses contain genetic material in the form of DNA or RNA. Once viruses were recognised as distinct biological entities they were soon shown to be the cause of numerous infections of plants, animal and even bacteria.

Of the many diseases of humans that were discovered to be caused by viruses in the 20th century, a few warrant further discussion because one, smallpox, has been eradicated, two others, measles and polio virus could be, but others such as HIV and influenza are proving to be more difficult to control. Others diseases, such as those caused by arboviruses are presenting new challenges.

Smallpox Eradication

In 1979, the World Health Organisation declared smallpox eradicated, but it had still managed to kill 300 million people since 1900. The eradication followed years of coordinated disease surveillance and vaccination campaigns throughout the world. The main weapon used was vaccinia virus, which was used as the vaccine. No one seems to know exactly where vaccinia virus came from, it is not the strain of cowpox that Edward Jenner had used and it is not a weakened form of smallpox. The eradication campaign was not without casualties. Most notable was the death of Janet Parker (c. 1938–1978) and the subsequent suicide of virologist Henry Bedson (1930–1978). Before the September 11 attacks on America in 2001, it was planned to destroy all the remaining stocks of smallpox virus that were kept in laboratories in US and Russia. Had this plan gone ahead, smallpox virus would have been the first to be made extinct by human intervention.

Measles

Measles is highly contagious, and until the introduction of vaccination in the US in the 1960s there were more than 500,000 cases a year resulting in about 400 deaths. By the mid-1970s the incidence of the disease had fallen by 90%. This reduction was achieved by a mass vaccination programme that was known as "make measles a memory" and vaccination campaigns in other countries have reduced the levels of infection by 99% over the past 50 years.

Susceptible individuals remain a source of infection and include people who have migrated from countries with ineffective vaccination schedules and those who refuse the vaccine or choose not to have their children vaccinated. Humans are the only natural hosts of measles virus, and immunity to the disease following an infection is lifelong, and that afforded by vaccination is long term but eventually wanes. The use of the vaccine is not without risk and its use has been controversial.

But, despite these controversies, measles has been eliminated from Finland, Sweden and Cuba. Japan abolished mandatory vaccination in 1992 and in 1995–1997 more than 200,000 cases were reported in the country. Measles remains a public health problem in Japan, where it is now endemic and a National Measles Elimination Plan was established in December 2007, with a view to eliminating the disease from the country by 2012. The possibility of global elimination of measles has been debated in the medical literature since the introduction of the vaccine in the 1960s. Should the current campaign to eradicate poliomyelitis be successful, it is likely that the debate will be renewed.

Poliomyelitis

During the summers of the mid-20th century, parents dreaded the annual appearance of poliomyelitis, which was commonly known as "infantile paralysis". The disease was rarer at the beginning of the century and worldwide there were only a few thousand cases a year, but by the 1950s there were 60,000 cases each year in the US alone. Earlier, during 1916 and 1917, there had been a major epidemic in US and 27,000 cases and 6000 deaths were recorded, with 9,000 cases in New York City alone. At the time nobody knew how the virus was spreading. Many of the city's inhabitants, including scientists, thought that impoverished slum-dwelling immigrants were to blame even though the prevalence of the disease was higher in the more prosperous districts such as Staten Island – a pattern that had been seen in other cities such as Philadelphia. The disease was not confined to the US, at the same time many other industrialised countries were affected. Before the outbreaks in the US, large epidemics occurred in Sweden.

The reason for the rise of poliomyelitis in industrialised countries in the 20th century has never been fully explained. Why a disease that is caused by a virus that is passed from person to person by the faecal-oral route, and only naturally infects humans, became a problem during times of improved sanitation and increasing affluence is a paradox. Although the virus was discovered at the beginning of the 20th century, just how ubiquitous it became was not known until the 1950s. It is now known that only 1–2% of individuals who are infected develop the disease and most infections are mild. During epidemics the virus was everywhere, which now explains the failed attempts by public health officials to find a source.

Following the development of vaccines in the mid-1950s, mass vaccination campaigns took place in many countries. In the US,following a campaign promoted by the March of Dimes the annual number of polio cases in the United States fell dramatically and the last outbreak was in 1979. In 1988, the World Health Organisation along with others launched the Global Polio Eradication Initiative and by 1994, the Americas were declared to be free of disease, followed by Pacific region in 2000 and Europe in 2003. By 2007, poliomyelitis had been eradicated from all countries except, India, Pakistan, Bangladesh and Nigeria, and as 2010, outbreaks of infections were still reported in other countries following importation of polio virus from these regions.

Origin and Emergence of HIV

This virus is one of the most significant that emerged in the last quarter of the 20th century. As of 2010, over 40 million people have been infected and more than 25 million of them have died from AIDS-related diseases. When, in 1981 a scientific article was published that reported the deaths of five young gay men, no one knew of the full scale of the epidemic and that the virus had been silently emerging over several decades. The viruses – there are two of them – HIV-1 and HIV-2, originated in non-human primates in Africa but HIV-1 rapidly spread across the world. Shortly after the HIV-1 virus was discovered in 1983, it was shown to be closely related to SIV, a retrovirus of monkeys and chimpanzees. There are different strains of HIV-1 and HIV-2 and each evolved independently from primates. How these viruses entered the human population is not fully understood, but during the 1960s and 1970s there were large population shifts from rural to urban areas in Africa. The analysis of the RNA from several strains of HIV-1 and HIV-2 has shown that these viruses evolved from SIV-related viruses early in the 20th century.

Influenza

When influenza virus undergoes a genetic shift many humans have no immunity to the new strain, and if the population of susceptible individuals is high enough to maintain the chain of infection then pandemics occur. The genetic changes usually happen when different strains of the virus co-infect animals, particularly birds and swine. Although many viruses of vertebrates are restricted to one species, influenza virus is an exceptions.

The last pandemic of the 19th century occurred in 1899 and resulted in the deaths of 250,000 people in Europe alone. The virus, which originated in Russia or Asia, was the first to be rapidly spread by people on trains and steamships.

A new strain of the virus emerged in 1918, and the subsequent pandemic of Spanish flu was one of the worst natural disasters in history. The death toll was enormous; there were 550,000 reported deaths caused by the disease in the US, ten times the country's losses in World War I. Throughout the world more than 20 million people died from the infection. Although cases of influenza occurred every winter, there were only two other pandemics in the 20th century.

In 1957 another new strain of the virus emerged and caused a pandemic of Asian flu; although the virus was not as virulent as the 1918 strain, over one million died worldwide. The next pandemic occurred when Hong Kong flu emerged in 1968, a new strain of the virus that completely replaced the 1957 strain. Affecting mainly the elderly, the 1968 pandemic was the least severe, but 33,800 were killed in the US. It might be no coincidence that new strains of influenza virus often originate in the far east. In rural China, the concentration of ducks, pigs and humans in close proximity is the highest in the world. More than 30 years passed until the next pandemic in 2009, but none of the last three has caused anything near the devastation seen in 1918. Exactly why the strain of influenza that emerged in 1918 was so devastating is a question that still remains unanswered.

Yellow Fever, Dengue and other Arboviruses

The arboviruses are viruses that are transmitted to humans and other vertebrates by blood-sucking insects. These viruses are diverse and the term "arbovirus" – which was derived from "arthropod borne virus" – is no longer used in formal taxonomy because many different species of virus are known to be spread in this way. There are more than 500 species arboviruses, but in the 1930s apart from yellow fever, only three were known, dengue and Pappataci fever. Now, over 100 of these viruses are known to cause human diseases. Yellow fever is the most notorious disease that is caused by a flavivirus. In 1905, the last major epidemic occurred in the US. During the building of the Panama Canal thousands of workers succumbed and died from the disease. Yellow fever originated in Africa and the virus was brought to the Americas on cargo ships, which were harbouring the *Aedes aegypti* that carries the virus.

The first *recorded* epidemic in Africa occurred in Ghana, west Africa, in 1926. In the 1930s, the disease re-emerged in Brazil and Fred Soper, an American epidemiologist (1893–1977) discovered the importance of the sylvatic cycle of infection in non-human hosts and that humans were a "dead end" that broke this cycle. The yellow fever vaccine is one of the most successful ever developed, but epidemics continue to occur. In 1986–91 epidemics occurred in West Africa and over 20,000 people were infected, 4,000 of whom died.

In the 1930s, St. Louis encephalitis, eastern equine encephalitis and western equine encephalitis emerged in the US. The virus that causes La Crosse encephalitis was discovered in the 1960s, and West Nile virus arrived in New York in 1999. As of 2010, dengue is the most prevalent arbovirus and increasingly virulent strains of the virus have spread across Asia and the Americas.

20th Century Epizootics

In the early years of the 20th century, rinderpest disease of cattle was common in Asia and parts of Europe and Asia. The prevalence of the disease was steadily reduced during the course of the 20th century by control measures, which included vaccination. By 1908, Europe was free from the disease although outbreaks did occur following the World War I, but these were quickly controlled. However, the prevalence of the disease increased in Asia and in 1957, Thailand had to appeal for aid because so many buffaloes had died that the paddy fields could not be prepared for rice growing. Russia west of the Urals, remained free from the disease – Lenin passed several laws on the control of the disease – but cattle in the east of Russia were constantly infected with rinderpest that originated in Mongolia and China where the prevalence remained high. India controlled the spread of the disease, which had retained a foothold in the southern states of Tamil Nadu and Kerala, throughout the 20th century, but eradicated the disease by 1995.

Africa suffered two major panzootics in the 1920s and 1980s. There was a severe outbreak in Somalia in 1928 and the disease was widespread in the country until 1953. In the 1980s, outbreaks in Tanzania and Kenya were controlled by the use of 26 million doses of vaccine and a recurrence of the disease in 1997 was suppressed by an intensive vaccination campaign. By the end of the century, rinderpest had been eradicated from most countries. A few pockets of infection remained in Ethiopia and Sudan, and in 1994 of the Global Rinderpest Eradication Programme was launched by the Food and Agriculture

Organization with an the aim to global eradication by 2010. Foot-and-mouth disease is a highly contagious infection caused by an aphthovirus and is classified in the same family as poliovirus. The virus has infected animals, mainly ungulates, in Africa since ancient times and was probably brought to the Americas in the 19th century by imported livestock. Foot-and-mouth disease is rarely fatal, but the economic losses incurred by outbreaks in sheep and cattle herds can be high. The last occurrence of the disease in the US was in 1929, but as recently as 2001, several large outbreaks occurred throughout the UK and thousands of animals were killed and burnt.

Influenza is mainly a disease of pigs and birds, but not humans, although it has probably infected us since antiquity. The virus can cause mild to severe epizootics in wild and domesticated animals. Many species of wild birds migrate and this has spread influenza across the continents throughout the ages. During this time the virus evolved into numerous strains – and continues to do so – posing an ever-present threat.

In the 21st century epizootics in livestock caused by viruses continue to have serious consequences. Bluetongue disease, a disease caused by an orbivirus broke out in sheep in France in 2007. Before then, the disease had been mainly confined to the Americas, Africa, southern Asia and northern Australia, but it is now an emerging disease around the Mediterranean.

Friendly Viruses

Sir Peter Medawar (1915–1987) described a virus as "a piece of bad news wrapped in a protein coat". He might have had his tongue firmly in his cheek, but this summed up the view of most virologists in the mid-20th century. With the exception of the bacteriophages, viruses had a well-deserved reputation for being nothing but the cause of diseases and death.

The discovery of the abundance of viruses and their overwhelming presence in many ecosystems has led modern virologists to consider them in a new light. Viruses are everywhere. It has been estimated that there are 10^{31} viruses on Earth, most of them bacteriophages found in the oceans.

Microorganisms constitute more than 90% of the biomass in the sea, and it has been estimated that viruses kill approximately 20% of this biomass each day and that there are fifteen times as many viruses in the oceans as there are bacteria and archaea. Viruses are the main agents responsible for the rapid destruction of harmful algal blooms, which often kill other marine life, and help maintain the ecological balance of different species of marine blue-green algae.

The emergence of strains of bacteria that are resistant to a broad range of antibiotics has become a problem in the treatment of bacterial infections. Only two new classes of antibiotics have been developed in the past 30 years, and the search is on for novel ways of combating bacterial infections. The use of bacteriophages to control bacteria was first described by Russian scientists in the 1950s and 1960s. At the time, this work was largely ignored in the West because antibiotic-resistant bacteria were not a cause for concern, but the more recent and escalating problems caused by antibiotic-resistant bacteria has stimulated a renewed interest in the use of bacteriophages and phage therapy.

The Human Genome Project has revealed the presence of numerous viral DNA sequences scattered throughout the human genome. These sequences make up around 8% of human DNA, and appear to be the remains of ancient retrovirus infections of human ancestors. These pieces of DNA have firmly established themselves in human DNA. Most of this DNA is no longer functional and a few sequences might occasionally cause harm. But the remainder seem to be harmless, and some might have brought with them novel genes that are important to human development.

Origins

Viruses are found wherever there is life and have probably existed since living cells first evolved. The origin of viruses is unclear because they do not form fossils, so molecular techniques have been the most useful means of investigating how they arose. These techniques rely on the availability of ancient viral DNA or RNA, but, unfortunately, most of the viruses that have been preserved and stored in laboratories are less than 90 years old. There are three main hypotheses that try to explain the origins of viruses:

Regressive hypothesis: Viruses may have once been small cells that parasitised larger cells. Over time, genes not required by their parasitism were lost. The bacteria rickettsia and chlamydia are living cells that, like viruses, can reproduce only inside host cells. They lend support to this hypothesis, as their dependence on parasitism is likely to have caused the loss of genes that enabled them to survive outside a cell. This is also called the *degeneracy hypothesis*, or *reduction hypothesis*.

Cellular origin hypothesis: Some viruses may have evolved from bits of DNA or RNA that "escaped" from the genes of a larger organism. The escaped DNA could have come from plasmids (pieces of naked DNA that can move *between* cells) or transposons (molecules of DNA that replicate and move around to different positions *within* the genes of the cell). Once called "jumping genes", transposons are examples of mobile genetic elements and could be the origin of some viruses. They were discovered in maize by Barbara McClintock in 1950. This is sometimes called the *vagrancy hypothesis*, or the *escape hypothesis*.

Coevolution hypothesis: This is also called the *virus-first hypothesis* and proposes that viruses may have evolved from complex molecules of protein and nucleic acid at the same time as cells first appeared on earth and would have been dependent on cellular life for billions of years. Viroids are molecules of RNA that are not classified as viruses because they lack a protein coat. However, they have characteristics that are common to several viruses and are often called subviral agents. Viroids are important pathogens of plants.

They do not code for proteins but interact with the host cell and use the host machinery for their replication. The hepatitis delta virus of humans has an RNA genome similar to viroids but has a protein coat derived from hepatitis B virus and cannot produce one of its own. It is therefore a defective virus and cannot replicate without the help of hepatitis B virus. Similarly, the virophage 'sputnik' is dependent on mimivirus, which infects the protozoan *Acanthamoeba castellanii*.

These viruses that are dependent on the presence of other virus species in the host cell are called *satellites* and may represent evolutionary intermediates of viroids and viruses. Historically, there were problems with all of these hypotheses: The regressive hypothesis did not explain why even the smallest of cellular parasites do not resemble viruses in any way. The escape hypothesis did not explain the complex capsids and other structures on virus particles. And, the virus-first hypothesis contravened the definition of viruses in that they require host cells. However, viruses are now recognised as ancient and to have origins that pre-date the divergence of life into the three domains. This discovery has led modern virologists to reconsider and re-evaluate these three classical hypotheses.

The evidence for an ancestral world of RNA cells and computer analysis of viral and host DNA sequences are giving a better understanding of the evolutionary relationships between different viruses and may help identify the ancestors of modern viruses. To date, such analyses have not proved which of these hypotheses are correct. However, it seems unlikely that all currently known viruses have a common ancestor and viruses have probably arisen numerous times in the past by one or more mechanisms.

Prions are infectious protein molecules that do not contain DNA or RNA. They cause an infection in sheep called scrapie and cattle bovine spongiform encephalopathy ("mad cow" disease). In humans they cause kuru and Creutzfeldt-Jakob disease. They are able to replicate because some proteins can exist in two different shapes and the prion changes the normal shape of a host protein into the prion shape.

This starts a chain reaction where each prion protein converts many host proteins into more prions, and these new prions then go on to convert even more protein into prions. Although they are fundamentally different from viruses and viroids, their discovery gives credence to the idea that viruses could have evolved from self-replicating molecules.

Bioremediation

Bioremediation is the use of microorganismal metabolism to remove pollutants. Technologies can be generally classified as *in situ* or *ex situ*. *In situ* bioremediation involves treating the contaminated material at the site, while *ex situ* involves the removal of the contaminated material to be treated elsewhere.

Some examples of bioremediation technologies are bioventing, landfarming, bioreactor, composting, bioaugmentation, rhizofiltration, and biostimulation. Bioremediation can occur on its own (natural attenuation or intrinsic bioremediation) or can be spurred on via the addition of fertilizers to increase the bioavailability within the medium (biostimulation).

Recent advancements have also proven successful via the addition of matched microbe strains to the medium to enhance the resident microbe population's ability to break down contaminants. Microorganisms used to perform the function of bioremediation are known as Bioremediators.

Not all contaminants, however, are easily treated by bioremediation using microorganisms. For example, heavy metals such as cadmium and lead are not readily absorbed or captured by organisms. The assimilation of metals such as mercury into the food chain may worsen matters. Phytoremediation is useful in these circumstances because natural plants or transgenic plants are able to bioaccumulate these toxins in their above-ground parts, which are then harvested for removal.

The heavy metals in the harvested biomass may be further concentrated by incineration or even recycled for industrial use. The elimination of a wide range of pollutants and wastes from the environment requires increasing our understanding of the relative importance of different pathways and regulatory networks to carbon flux in particular environments and for particular compounds, and they will certainly accelerate the development of bioremediation technologies and biotransformation processes.

Genetic Engineering Approaches

The use of genetic engineering to create organisms specifically designed for bioremediation has great potential. The bacterium *Deinococcus radiodurans* (the most radioresistant organism known) has been modified to consume and digest toluene and ionic mercury from highly radioactive nuclear waste.

Most commonly, the process is misunderstood. The microbes are ever-present in any given context and are generally referred to as "normal microbial flora". During bioremediation (biodegradation) processes, fertilizer/nutrient supplementation is introduced to the environments in efforts to maximize growth and production potential. Common misbelief is that microbes are transported and dispersed into an unadulterated environment.

Mycoremediation

Mycoremediation is a form of bioremediation in which fungi are used to decontaminate the area. The term *mycoremediation* refers specifically to the use of fungal mycelia in bioremediation.

One of the primary roles of fungi in the ecosystem is decomposition, which is performed by the mycelium. The mycelium secretes extracellular enzymes and acids that break down lignin and cellulose, the two main building blocks of plant fibre. These are organic compounds composed of long chains of carbon and hydrogen, structurally similar to many organic pollutants. The key to mycoremediation is determining the right fungal species to target a specific pollutant. Certain strains have been reported to successfully degrade the nerve gases VX and sarin.

In one conducted experiment, a plot of soil contaminated with diesel oil was inoculated with mycelia of oyster mushrooms; traditional bioremediation techniques (bacteria) were used on control plots. After four weeks, more than 95% of many of the PAH (polycyclic aromatic hydrocarbons) had been reduced to non-toxic components in the mycelial-inoculated plots. It appears that the natural microbial community participates with the fungi to break down contaminants, eventually into carbon dioxide and water. Wood-degrading fungi are particularly effective in breaking down aromatic pollutants (toxic components of petroleum), as well as chlorinated compounds (certain persistent pesticides; Battelle, 2000). Mycofiltration is a similar process, using fungal mycelia to filter toxic waste and microorganisms from water in soil.

Advantages

There are a number of cost/efficiency advantages to bioremediation, which can be employed in areas that are inaccessible without excavation. For example, hydrocarbon spills (specifically, petrol spills) or certain chlorinated solvents may contaminate groundwater, and introducing the appropriate electron acceptor or electron donor amendment, as appropriate, may significantly reduce contaminant concentrations after a long time allowing for acclimation. This is typically much less expensive than excavation followed by disposal elsewhere, incineration or other *ex situ* treatment strategies, and reduces or eliminates the need for "pump and treat", a practice common at sites where hydrocarbons have contaminated clean groundwater.

Monitoring Bioremediation

The process of bioremediation can be monitored indirectly by measuring the *Oxidation Reduction Potential* or redox in soil and groundwater, together with pH, temperature, oxygen content, electron acceptor/donor concentrations, and concentration of breakdown products (e.g. carbon dioxide).

This, by itself and at a single site, gives little information about the process of remediation.

1. It is necessary to sample enough points on and around the contaminated site to be able to determine contours of equal redox potential. Contouring is usually done using specialised software, e.g. using Kriging interpolation.
2. If all the measurements of redox potential show that electron acceptors have been used up, it is in effect an indicator for total microbial activity. Chemical analysis is also required to determine when the levels of contaminants and their breakdown products have been reduced to below regulatory limits.

Microbial Biodegradation

Interest in the microbial biodegradation of pollutants has intensified in recent years as humanity strives to find sustainable ways to cleanup contaminated environments. These bioremediation and biotransformation methods endeavour to harness the astonishing, naturally occurring, ability of microbial xenobiotic metabolism to degrade, transform or accumulate a huge range of compounds including hydrocarbons (e.g. oil), polychlorinated biphenyls (PCBs), polyaromatic hydrocarbons (PAHs), heterocyclic compounds (such as pyridine or quinoline), pharmaceutical substances, radionuclides and metals. Major methodological breakthroughs in recent years have enabled detailed genomic, metagenomic, proteomic, bioinformatic and other high-throughput analyses of environmentally relevant microorganisms providing unprecedented insights into key biodegradative pathways and the ability of organisms to adapt to changing environmental conditions.

The elimination of a wide range of pollutants and wastes from the environment is an absolute requirement to promote a sustainable development of our society with low environmental impact. Biological processes play a major role in the removal of contaminants and they take advantage of the astonishing catabolic versatility of microorganisms to degrade/convert such compounds. New methodological breakthroughs in sequencing, genomics, proteomics, bioinformatics and imaging are producing vast amounts of information. In the field of Environmental Microbiology, genome-based global studies open a new era providing unprecedented *in silico* views of metabolic and regulatory networks, as well as clues to the evolution of degradation pathways and to the molecular adaptation strategies to changing environmental conditions. Functional genomic and metagenomic approaches are increasing our understanding of the relative importance of different pathways and regulatory networks to carbon flux in particular environments and for particular compounds and they will certainly accelerate the development of bioremediation technologies and biotransformation processes.

Aerobic Biodegradation of Pollutants

The burgeoning amount of bacterial genomic data provides unparalleled opportunities for understanding the genetic and molecular bases of the degradation of organic pollutants. Aromatic compounds are among the most recalcitrant of these pollutants and lessons can be learned from the recent genomic studies of *Burkholderia xenovorans* LB400 and *Rhodococcus* sp. strain RHA1, two of the largest bacterial genomes completely sequenced to date. These studies have helped expand our understanding of bacterial catabolism, non-catabolic physiological adaptation to organic compounds, and the evolution of large bacterial genomes. First, the metabolic pathways from phylogenetically diverse isolates are very similar with respect to overall organization. Thus, as originally noted in pseudomonads, a large number of "peripheral aromatic" pathways funnel a range of natural and xenobiotic compounds into a restricted number of "central aromatic" pathways. Nevertheless, these pathways are genetically organized in genus-specific fashions, as exemplified by the b-ketoadipate and Paa pathways. Comparative genomic studies further reveal that some pathways are more widespread than initially thought. Thus, the Box and Paa pathways illustrate the prevalence of non-oxygenolytic ring-cleavage strategies in aerobic aromatic degradation processes. Functional genomic studies have been useful in establishing that even organisms harboring high numbers of homologous enzymes seem to contain few examples of true redundancy. For example, the multiplicity of ring-cleaving dioxygenases in certain rhodococcal isolates may be attributed to the cryptic aromatic catabolism of different terpenoids and steroids. Finally, analyses have indicated that recent genetic flux appears to have played a more significant role in the evolution of some large genomes, such as LB400's, than others. However, the emerging trend is that the large gene repertoires of potent pollutant degraders such as LB400 and RHA1 have evolved principally through more ancient processes. That this is true in such phylogenetically diverse species is remarkable and further suggests the ancient origin of this catabolic capacity.

Anaerobic Biodegradation of Pollutants

Anaerobic microbial mineralization of recalcitrant organic pollutants is of great environmental significance and involves intriguing novel biochemical reactions. In particular, hydrocarbons and halogenated compounds have long been doubted to be degradable in the absence of oxygen, but the isolation of hitherto unknown anaerobic hydrocarbon-degrading and reductively dehalogenating bacteria during the last decades provided ultimate proof for these processes in nature. Many novel biochemical reactions were discovered enabling the respective metabolic pathways, but progress in the molecular understanding of these bacteria was rather slow, since genetic systems are not readily applicable for most of them. However, with the increasing application of genomics in the field of environmental microbiology, a new and promising perspective is now at hand to obtain molecular insights into these new metabolic properties. Several complete genome sequences were determined during the last few years from bacteria capable of anaerobic organic pollutant degradation. The ~4.7 Mb genome of the facultative denitrifying *Aromatoleum aromaticum* strain EbN1 was the first to be determined for an anaerobic hydrocarbon degrader (using toluene or ethylbenzene as substrates).

The genome sequence revealed about two dozen gene clusters (including several paralogs) coding for a complex catabolic network for anaerobic and aerobic degradation of aromatic compounds. The genome sequence forms the basis for current detailed studies on regulation of pathways and enzyme structures. Further genomes of anaerobic hydrocarbon degrading bacteria were recently completed for the iron-reducing species *Geobacter metallireducens* (accession nr. NC_007517) and the perchlorate-reducing *Dechloromonas aromatica* (accession nr. NC_007298), but these are not yet evaluated in formal publications. Complete genomes were also determined for bacteria capable of anaerobic degradation of halogenated hydrocarbons by halorespiration: the ~1.4 Mb genomes of *Dehalococcoides ethenogenes* strain 195 and *Dehalococcoides* sp. strain CBDB1 and the ~5.7 Mb genome of *Desulfitobacterium hafniense* strain Y51. Characteristic for all these bacteria is the presence of multiple paralogous genes for reductive dehalogenases, implicating a wider dehalogenating spectrum of the organisms than previously known. Moreover, genome sequences provided unprecedented insights into the evolution of reductive dehalogenation and differing strategies for niche adaptation.

Recently, it has become apparent that some organisms, including *Desulfitobacterium chlororespirans*, originally evaluated for halorespiration on chlorophenols, can also use certain brominated compounds, such as the herbicide bromoxynil and its major metabolite as electron acceptors for growth. Iodinated compounds may be dehalogenated as well, though the process may not satisfy the need for an electron acceptor.

Bioavailability, Chemotaxis, and Transport of Pollutants

Bioavailability, or the amount of a substance that is physiochemically accessible to microorganisms is a key factor in the efficient biodegradation of pollutants. O'Loughlin et al (2000) showed that, with the exception of kaolinite clay, most soil clays and cation exchange resins attenuated biodegradation of 2-picoline by *Arthrobacter* sp. strain R1, as a result of adsorption of the substrate to the clays. Chemotaxis, or the directed movement of motile organisms towards or away from chemicals in the environment is an important physiological response that may contribute to effective catabolism of molecules in the environment. In addition, mechanisms for the intracellular accumulation of aromatic molecules via various transport mechanisms are also important.

Oil Biodegradation

Petroleum oil contains aromatic compounds that are toxic for most life forms. Episodic and chronic pollution of the environment by oil causes major ecological perturbations. Marine environments are especially vulnerable since oil spills of coastal regions and the open sea are poorly containable and mitigation is difficult. In addition to pollution through human activities, about 250 million litres of petroleum enter the marine environment every year from natural seepages. Despite its toxicity, a considerable fraction of petroleum oil entering marine systems is eliminated by the hydrocarbon-degrading activities of microbial communities, in particular by a remarkable recently discovered group of specialists, the so-called hydrocarbonoclastic bacteria (HCB). *Alcanivorax borkumensis* was the first HCB to have its genome sequenced. In addition to hydrocarbons, crude oil often contains various heterocyclic compounds, such as pyridine, which appear to be degraded by similar, though separate mechanisms than hydrocarbons.

Analysis of Waste Biotreatment

Sustainable development requires the promotion of environmental management and a constant search for new technologies to treat vast quantities of wastes generated by increasing anthropogenic activities. Biotreatment, the processing of wastes using living organisms, is an environmentally friendly, relatively simple and cost-effective alternative to physico-chemical clean-up options. Confined environments, such as bioreactors, have been engineered to overcome the physical, chemical and biological limiting factors of biotreatment processes in highly controlled systems.

The great versatility in the design of confined environments allows the treatment of a wide range of wastes under optimized conditions. To perform a correct assessment, it is necessary to consider various microorganisms having a variety of genomes and expressed transcripts and proteins. A great number of analyses are often required. Using traditional genomic techniques, such assessments are limited and time-consuming. However, several high-throughput techniques originally developed for medical studies can be applied to assess biotreatment in confined environments.

Metabolic Engineering and Biocatalytic Applications

The study of the fate of persistent organic chemicals in the environment has revealed a large reservoir of enzymatic reactions with a large potential in preparative organic synthesis, which has already been exploited for a number of oxygenases on pilot and even on industrial scale. Novel catalysts can be obtained from metagenomic libraries and DNA sequence based approaches. Our increasing capabilities in adapting the catalysts to specific reactions and process requirements by rational and random mutagenesis broadens the scope for application in the fine chemical industry, but also in the field of biodegradation. In many cases, these catalysts need to be exploited in whole cell bioconversions or in fermentations, calling for system-wide approaches to understanding strain physiology and metabolism and rational approaches to the engineering of whole cells as they are increasingly put forward in the area of systems biotechnology and synthetic biology.

Fungal Biodegradation

In the ecosystem, different substrates are attacked at different rates by consortia of organisms from different kingdoms. *Aspergillus* and other moulds play an important role in these consortia because they are adept at recycling starches, hemicelluloses, celluloses, pectins and other sugar polymers. Some aspergilli are capable of degrading more refractory compounds such as fats, oils, chitin, and keratin. Maximum decomposition occurs when there is sufficient nitrogen, phosphorus and other essential inorganic nutrients. Fungi also provide food for many soil organisms.

For *Aspergillus* the process of degradation is the means of obtaining nutrients. When these moulds degrade human-made substrates, the process usually is called biodeterioration. Both paper and textiles (cotton, jute, and linen) are particularly vulnerable to *Aspergillus* degradation. Our artistic heritage is also subject to *Aspergillus* assault. To give but one example, after Florence in Italy flooded in 1969, 74% of the isolates from a damaged Ghirlandaio fresco in the Ognissanti church were *Aspergillus versicolor*.

Mycoremediation

Mycoremediation is a form of bioremediation, the process of using fungi to degrade or sequester contaminates in the environment. Stimulating microbial and enzyme activity, mycelium reduces toxins in-situ. Some fungi are hyperaccumulators, capable of absorbing and concentrating heavy metals in the mushroom fruit bodies.

One of the primary roles of fungi in the ecosystem is decomposition, which is performed by the mycelium. The mycelium secretes extracellular enzymes and acids that break down lignin and cellulose, the two main building blocks of plant fibre. These are organic compounds composed of long chains of carbon and hydrogen, structurally similar to many organic pollutants. The key to mycoremediation is determining the right fungal species to target a specific pollutant. Certain strains have been reported to successfully degrade the nerve gases VX and sarin.

In an experiment conducted in conjunction with Dr. S. A. Thomas, a major contributor in the bioremediation industry, a plot of soil contaminated with diesel oil was inoculated with mycelia of oyster mushrooms; traditional bioremediation techniques (bacteria) were used on control plots. After four weeks, more than 95% of many of the PAH (polycyclic aromatic hydrocarbons) had been reduced to non-toxic components in the mycelial-inoculated plots. It appears that the natural microbial community participates with the fungi to break down contaminants, eventually into carbon dioxide and water. Wood-degrading fungi are particularly effective in breaking down aromatic pollutants (toxic components of petroleum), as well as chlorinated compounds (certain persistent pesticides; Battelle, 2000).

The concept of mycoremediation was explored in the 1984 film Nausicaa of the Valley of the Wind, where vast tracts of fungal forest rehabilitate the planet after catastrophic human polluting and apocalypse. Mycofiltration is a similar or same process, using fungal mycelia to filter toxic waste and microorganisms from water in soil.

Archaea

The Archaea are a group of single-celled microorganisms. A single individual or species from this domain is called an *archaeon* (sometimes spelled "archeon"). They have no cell nucleus nor any other membrane-bound organelles within their cells. In the past they were viewed as an unusual group of bacteria and named archaebacteria, but since the Archaea have an independent evolutionary history and show many differences in their biochemistry from other forms of life, they are now classified as a separate domain in the three-domain system. In this system the phylogenetically distinct branches of evolutionary descent are the Archaea, Bacteria and Eukaryota. Archaea are divided into four recognized phyla, but many more phyla may exist. Of these groups the Crenarchaeota and the Euryarchaeota are most intensively studied. Classification is still difficult, since the vast majority have never been studied in the laboratory and have only been detected by analysis of their nucleic acids in samples from the environment. Although archaea have, in the past, been classed with bacteria as *prokaryotes* (or Kingdom Monera), this classification is regarded by some as outdated.

Archaea and bacteria are quite similar in size and shape, although a few archaea have very unusual shapes, such as the flat and square-shaped cells of *Haloquadra walsbyi*. Despite this visual similarity to bacteria, archaea possess genes and several metabolic pathways that are more closely related to those of eukaryotes: notably the enzymes involved in transcription and translation. Other aspects of archaean biochemistry are unique, such as their reliance on ether lipids in their cell membranes. The archaea exploit a much greater variety of sources of energy than eukaryotes: ranging from familiar organic compounds such as sugars, to using ammonia, metal ions or even hydrogen gas as nutrients. Salt-tolerant archaea (the Halobacteria) use sunlight as an energy source and other species of archaea fix carbon; however, unlike plants and cyanobacteria, no species of archaea is known to do both. Archaea reproduce asexually and divide by binary fission, fragmentation, or budding; in contrast to bacteria and eukaryotes, no known species form spores. Initially, archaea were seen as extremophiles that lived in harsh environments, such as hot springs and salt lakes, but they have since been found in a broad range of habitats, including soils, oceans, and marshlands. Archaea are particularly numerous in the oceans, and the archaea in plankton may be one of the most abundant groups of organisms on the planet. Archaea are now recognized as a major part of Earth's life and may play roles in both the carbon cycle and the nitrogen cycle. No clear examples of archaeal pathogens or parasites are known, but they are often mutualists or commensals. One example are the methanogens that inhabit the gut of humans and ruminants, where their vast numbers aid digestion. Methanogens are used in biogas production and sewage treatment, and enzymes from extremophile archaea that can endure high temperatures and organic solvents are exploited in biotechnology.

Classification

A New Domain

Early in the 20th century, prokaryotes were regarded as a single group of organisms and classified based on their biochemistry, morphology and metabolism. For example, microbiologists tried to classify microorganisms based on the structures of their cell walls, their shapes, and the substances they consume. However, a new approach was proposed in 1965, using the sequences of the genes in these organisms to work out which prokaryotes are genuinely related to each other. This approach, known as phylogenetics, is the main method used today. Archaea were first classified as a separate group of prokaryotes in 1977 by Carl Woese and George E. Fox in phylogenetic trees based on the sequences of ribosomal RNA (rRNA) genes. These two groups were originally named the Archaebacteria and Eubacteria and treated as kingdoms or subkingdoms, which Woese and Fox termed *Urkingdoms*. Woese argued that this group of prokaryotes is a fundamentally different sort of life. To emphasize this difference, these two domains were later renamed Archaea and Bacteria.

At first, only the methanogens were placed in this new domain, and the archaea were seen as extremophiles that exist only in habitats such as hot springs and salt lakes. By the end of the 20th century, microbiologists realized that archaea is a large and diverse group of organisms that are widely distributed in nature and are common in much less extreme habitats, such as soils and oceans. This new appreciation of the importance and ubiquity of archaea came from using the polymerase chain reaction to detect prokaryotes in samples of water or soil from their nucleic acids alone. This allows the detection and identification of organisms that cannot be cultured in the laboratory, which generally remains difficult.

Current Classification

The classification of archaea, and of prokaryotes in general, is a rapidly moving and contentious field. Current classification systems aim to organize archaea into groups of organisms that share structural features and common ancestors. These classifications rely heavily on the use of the sequence of ribosomal RNA genes to reveal relationships between organisms (molecular phylogenetics). Most of the culturable and well-investigated species of archaea are members of two main phyla, the Euryarchaeota and Crenarchaeota. Other groups have been tentatively created. For example, the peculiar species *Nanoarchaeum equitans*, which was discovered in 2003, has been given its own phylum, the Nanoarchaeota. A new phylum Korarchaeota has also been proposed. It contains a small group of unusual thermophilic species that shares features of both of the main phyla, but is most closely related to the Crenarchaeota. Other recently detected species of archaea are only distantly related to any of these groups, such as the Archaeal Richmond Mine Acidophilic Nanoorganisms (ARMAN), which were discovered in 2006 and are some of the smallest organisms known.

Species?

The classification of archaea into species is also controversial. Biology defines a species as a group of related organisms. The familiar exclusive breeding criterion (organisms that can breed with each other but not with others), is of no help because archaea reproduce asexually.

Archaea show high levels of horizontal gene transfer between lineages. Some researchers suggest that individuals can be grouped into species-like populations given highly similar genomes and infrequent gene transfer to/from cells with less-related genomes, as in the genus *Ferroplasma*. On the other hand, studies in *Halorubrum* found significant genetic transfer to/from less-related populations, limiting the criteria's applicability. A second concern is to what extent such species designations have practical meaning.

Current knowledge on genetic diversity is fragmentary and the total number of archaean species cannot be estimated with any accuracy. Estimates of the number of phyla range from 18 to 23, of which only 8 have representatives that have been cultured and studied directly. Many of these hypothesized groups are known from a single rRNA sequence, indicating that the diversity among these organisms remains obscure. The Bacteria also contain many uncultured microbes with similar implications for characterization.

Origin and Evolution

Although probable prokaryotic cell fossils date to almost 3.5 billion years ago, most prokaryotes do not have distinctive morphologies and fossil shapes cannot be used to identify them as Archaea. Instead, chemical fossils of unique lipids are more informative because such compounds do not occur in other organisms. Some publications suggest that archaean or eukaryotic lipid remains are present in shales dating from 2.7 billion years ago; such data have since been questioned. Such lipids have also been detected in Precambrian formations. The oldest such traces come from the Isua district of west Greenland, which include Earth's oldest sediments, formed 3.8 billion years ago.

The archaeal lineage may be the most ancient that exists on earth.Woese argued that the bacteria, archaea, and eukaryotes represent separate lines of descent that diverged early on from an ancestral colony of organisms. A few biologists, however, argue that the Archaea and Eukaryota arose from a group of bacteria. In any case it is thought that viruses and archaea began relationships approximately two billion years ago, and that co-evolution may have been occurring between members of these groups.

It is possible that the last common ancestor of the bacteria and archaea was a thermophile, which raises the possibility that lower temperatures are "extreme environments" in archaeal terms, and organisms that live in cooler environments appeared only later. Since the Archaea and Bacteria are no more related to each other than they are to eukaryotes, the term *prokaryote*'s only surviving meaning is "not a eukaryote", limiting its value.

Archaea and Eukaryotes

The relationship between archaea and eukaryotes remains problematic. Aside from the similarities in cell structure and function that are discussed below, many genetic trees group the two.

Complicating factors include claims that the relationship between eukaryotes and the archaeal phylum Euryarchaeota is closer than the relationship between the Euryarchaeota and the phylum Crenarchaeota and the presence of archaean-like genes in certain bacteria, such as *Thermotoga maritima*, from horizontal gene transfer. The leading hypothesis is that the ancestor of the eukaryotes diverged early from the Archaea, and that eukaryotes arose through fusion of an archaean and eubacterium, which became the nucleus and cytoplasm; this accounts for various genetic similarities but runs into difficulties explaining cell structure.

Morphology

Individual archaea range from 0.1 micrometers (μm) to over 15 μm in diameter, and occur in various shapes, commonly as spheres, rods, spirals or plates. Other morphologies in the Crenarchaeota include irregularly shaped lobed cells in *Sulfolobus*, needle-like filaments that are less than half a micrometer in diameter in *Thermofilum*, and almost perfectly rectangular rods in *Thermoproteus* and *Pyrobaculum*. *Haloquadra walsbyi* are flat, square archaea that live in hypersaline pools. These unusual shapes are probably maintained both by their cell walls and a prokaryotic cytoskeleton. Proteins related to the cytoskeleton components of other organisms exist in archaea, and filaments form within their cells, but in contrast to other organisms, these cellular structures are poorly understood. In *Thermoplasma* and *Ferroplasma* the lack of a cell wall means that the cells have irregular shapes, and can resemble amoebae.

Some species form aggregates or filaments of cells up to 200 μm long. These organisms can be prominent in biofilms. Notably, aggregates of *Thermococcus coalescens* cells fuse together in culture, forming single giant cells. Archaea in the genus *Pyrodictium* produce an elaborate multicell colony involving arrays of long, thin hollow tubes called *cannulae* that stick out from the cells' surfaces and connect them into a dense bush-like agglomeration.

The function of these cannulae is not settled, but they may allow communication or nutrient exchange with neighbours. Multi-species colonies exist, such as the "string-of-pearls" community that was discovered in 2001 in a German swamp. Round whitish colonies of a novel Euryarchaeota species are spaced along thin filaments that can range up to 15 centimetres (5.9 in) long; these filaments are made of a particular bacteria species.

Structure, Composition Development, Operation

Archaea and bacteria have generally similar cell structure, but cell composition and organization set the archaea apart. Like bacteria, archaea lack interior membranes and organelles. Like bacteria, archaea cell membranes are usually bounded by a cell wall and they swim using one or more flagella. Structurally, archaea are most similar to gram-positive bacteria. Most have a single plasma membrane and cell wall, and lack a periplasmic space; the exception to this general rule is *Ignicoccus*, which possess a particularly large periplasm that contains membrane-bound vesicles and is enclosed by an outer membrane.

Membranes

Archaeal membranes are made of molecules that differ strongly from those in other life forms, showing that archaea are related only distantly to bacteria and eukaryotes. In all organisms cell membranes are made of molecules known as phospholipids. These molecules possess both a polar part that dissolves in water (the phosphate "head"), and a "greasy" non-polar part that does not (the lipid tail). These dissimilar parts are connected by a glycerol group. In water, phospholipids cluster, with the heads facing the water and the tails facing away from it. The major structure in cell membranes is a double layer of these phospholipids, which is called a lipid bilayer.

These phospholipids are unusual in four ways:

- Bacteria and eukaryotes have membranes composed mainly of glycerol-ester lipids, whereas archaea have membranes composed of glycerol-ether lipids. The difference is the type of bond that joins the lipids to the glycerol group; the two types are shown in yellow in the Figure at the right. In ester lipids this is an ester bond, whereas in ether lipids this is an ether bond. Ether bonds are chemically more resistant than ester bonds. This stability might help archaea to survive extreme temperatures and very acidic or alkaline environments. Bacteria and eukaryotes do contain some ether lipids, but in contrast to archaea these lipids are not a major part of their membranes.
- The stereochemistry of the glycerol group is the reverse of that found in other organisms. The glycerol group can occur in two forms that are mirror images of one another, called the right-handed and left-handed forms; in chemistry these are called *enantiomers*. Just as a right hand does not fit easily into a left-handed glove, a right-handed glycerol molecule generally cannot be used or made by enzymes adapted for the left-handed form. This suggests that archaea use entirely different enzymes for synthesizing phospholipids than do bacteria and eukaryotes. Such enzymes developed very early in life's history, suggesting an early split from the other two domains.
- Archaeal lipid tails are chemically different from other organisms. Archaeal lipids are based upon the isoprenoid sidechain and are long chains with multiple side-branches and sometimes even cyclopropane or cyclohexane rings. This is in contrast to the fatty acids found in other organisms' membranes, which have straight chains with no branches or rings. Although isoprenoids play an important role in the biochemistry of many organisms, only the archaea use them to make phospholipids. These branched chains may help prevent archaean membranes from leaking at high temperatures.
- In some archaea the lipid bilayer is replaced by a monolayer. In effect, the archaea fuse the tails of two independent phospholipid molecules into a single molecule with two polar heads; this fusion may make their membranes more rigid and better able to resist harsh environments. For example, the lipids in *Ferroplasma* are of this type, which is thought to aid this organism's survival in its highly acidic habitat.

Wall and Flagella

Most archaea (not *Thermoplasma* and *Ferroplasma*) possess a cell wall. In most archaea the wall is assembled from surface-layer proteins, which form an S-layer. An S-layer is a rigid array of protein molecules that cover the outside of the cell (like chain mail). This layer provides both chemical and physical protection, and can prevent macromolecules from contacting the cell membrane. Unlike bacteria, most archaea lack peptidoglycan in their cell walls. The exception is pseudopeptidoglycan, which is found in Methanobacteriales, but pseudopeptidoglycan lacks D-amino acids and N-acetylmuramic acid.

Archaea flagella operate like bacterial flagella—their long stalks are driven by rotatory motors at the base. These motors are powered by the proton gradient across the membrane. However, archaeal flagella are notably different in composition and development. The two types of flagella evolved from different ancestors. The bacterial flagellum shares a common ancestor with the type III secretion system, while archaeal flagella appear to have evolved from bacterial type IV pili. In contrast to the bacterial flagellum, which is hollow and is assembled by subunits moving up the central pore to the tip of the flagella, archaeal flagella are synthesized by adding subunits at the base.

Metabolism

Archaea exhibit a great variety of chemical reactions in their metabolism and use many sources of energy. These reactions are classified into nutritional groups, depending on energy and carbon sources. Some archaea obtain energy from inorganic compounds such as sulfur or ammonia (they are lithotrophs). These include nitrifiers, methanogens and anaerobic methane oxidisers. In these reactions one compound passes electrons to another (in a redox reaction), releasing energy to fuel the cell's activities. One compound acts as an electron donor and one as an electron acceptor. The energy released generates adenosine triphosphate (ATP) through chemiosmosis, in the same basic process that happens in the mitochondrion of eukaryotic cells.

Other groups of archaea use sunlight as a source of energy (they are phototrophs). However, oxygen–generating photosynthesis does not occur in any of these organisms. Many basic metabolic pathways are shared between all forms of life; for example, archaea use a modified form of glycolysis (the Entner–Doudoroff pathway) and either a complete or partial citric acid cycle. These similarities to other organisms probably reflect both early origins in the history of life and their high level of efficiency.

Other archaea use CO_2 in the atmosphere as a source of carbon, in a process called carbon fixation (they are autotrophs). This process involves either a highly modified form of the Calvin cycle or a recently discovered metabolic pathway called the 3-hydroxypropionate/4-hydroxybutyrate cycle. The Crenarchaeota also use the reverse Krebs cycle while the Euryarchaeota also use the reductive acetyl-CoA pathway. Carbon–fixation is powered by inorganic energy sources. No known archaea carry out photosynthesis. Archaeal energy sources are extremely diverse, and range from the oxidation of ammonia by the Nitrosopumilales to the oxidation of hydrogen sulfide or elemental sulfur by species of *Sulfolobus*, using either oxygen or metal ions as electron acceptors. Phototrophic archaea use light to produce chemical energy in the form of ATP.

In the Halobacteria, light-activated ion pumps like bacteriorhodopsin and halorhodopsin generate ion gradients by pumping ions out of the cell across the plasma membrane. The energy stored in these electrochemical gradients is then converted into ATP by ATP synthase. This process is a form of photophosphorylation. The ability of these light-driven pumps to move ions across membranes depends on light-driven changes in the structure of a retinol cofactor buried in the centre of the protein.

Genetics

Archaea usually have a single circular chromosome, the size of which may be as great as 5,751,492 base pairs in *Methanosarcina acetivorans*, the largest known archaean genome. One-tenth of this size is the tiny 490,885 base-pair genome of *Nanoarchaeum equitans*, the smallest archaean genome known; it is estimated to contain only 537 protein-encoding genes. Smaller independent pieces of DNA, called *plasmids*, are also found in archaea. Plasmids may be transferred between cells by physical contact, in a process that may be similar to bacterial conjugation.

Archaea can be infected by double-stranded DNA viruses that are unrelated to any other form of virus and have a variety of unusual shapes, including bottles, hooked rods, or teardrops. These viruses have been studied in most detail in thermophilics, particularly the orders Sulfolobales and Thermoproteales. A single-stranded DNA virus that infects halophilics was identified in 2009. Defences against these viruses may involve RNA interference from repetitive DNA sequences that are related to the genes of the viruses.

Archaea are genetically distinct from bacteria and eukaryotes, with up to 15% of the proteins encoded by any one archaeal genome being unique to the domain, although most of these unique genes have no known function. Of the remainder of the unique proteins that have an identified function, most are involved in methanogenesis. The proteins that archaea, bacteria and eukaryotes share form a common core of cell function, relating mostly to transcription, translation, and nucleotide metabolism. Other characteristic archaean features are the organization of genes of related function—such as enzymes that catalyse steps in the same metabolic pathway into novel operons, and large differences in tRNA genes and their aminoacyl tRNA synthetases.

Transcription and translation in archaea resemble these processes in eukaryotes more than in bacteria, with the archaean RNA polymerase and ribosomes being very close to their equivalents in eukaryotes. Although archaea only have one type of RNA polymerase, its structure and function in transcription seems to be close to that of the eukaryotic RNA polymerase II, with similar protein assemblies (the general transcription factors) directing the binding of the RNA polymerase to a gene's promoter. However, other archaean transcription factors are closer to those found in bacteria. Post-transcriptional modification is simpler than in eukaryotes, since most archaean genes lack introns, although there are many introns in their transfer RNA and ribosomal RNA genes, and introns may occur in a few protein-encoding genes.

Reproduction

Archaea reproduce asexually by binary or multiple fission, fragmentation, or budding; meiosis does not occur, so if a species of archaea exists in more than one form, all have the same genetic material. Cell division is controlled in a cell cycle; after the cell's chromosome is replicated and the two daughter chromosomes separate, the cell divides. Details have only been investigated in the genus *Sulfolobus*, but here that cycle has characteristics that are similar to both bacterial and eukaryotic systems. The chromosomes replicate from multiple starting-points (origins of replication) using DNA polymerases that resemble the equivalent eukaryotic enzymes. However, the proteins that direct cell division, such as the protein FtsZ, which forms a contracting ring around the cell, and the components of the septum that is constructed across the centre of the cell, are similar to their bacterial equivalents.

Both bacteria and eukaryotes, but not archaea, make spores. Some species of Haloarchaea undergo phenotypic switching and grow as several different cell types, including thick-walled structures that are resistant to osmotic shock and allow the archaea to survive in water at low salt concentrations, but these are not reproductive structures and may instead help them reach new habitats.

Ecology

Habitats

Archaea exist in a broad range of habitats, and as a major part of global ecosystems, may contribute up to 20% of earth's biomass. The first-discovered archaeans were extremophiles. Indeed, some archaea survive high temperatures, often above 100 °C (212 °F), as found in geysers, black smokers, and oil wells. Other common habitats include very cold habitats and highly saline, acidic, or alkaline water. However, archaea include mesophiles that grow in mild conditions, in marshland, sewage, the oceans, and soils. Extremophile archaea are members of four main physiological groups.

These are the halophiles, thermophiles, alkaliphiles, and acidophiles. These groups are not comprehensive or phylum-specific, nor are they mutually exclusive, since some archaea belong to several groups. Nonetheless, they are a useful starting point for classification.

Halophiles, including the genus *Halobacterium*, live in extremely saline environments such as salt lakes and outnumber their bacterial counterparts at salinities greater than 20–25%. Thermophiles grow best at temperatures above 45 °C (113 °F), in places such as hot springs; *hyperthermophilic* archaea grow optimally at temperatures greater than 80 °C (176 °F). The archaeal *Methanopyrus kandleri* Strain 116 grows at 122 °C (252 °F), the highest recorded temperature of any organism. Other archaea exist in very acidic or alkaline conditions. For example, one of the most extreme archaean acidophiles is *Picrophilus torridus*, which grows at pH 0, which is equivalent to thriving in 1.2 Molar sulfuric acid.

This resistance to extreme environments has made archaea the focus of speculation about the possible properties of extraterrestrial life. Extremophile habitats are not dissimilar to those on Mars, leading to the suggestion that viable microbes could be transferred between planets in meteorites. Recently, several studies have shown that archaea exist not only in mesophilic and thermophilic environments but are also present, sometimes in high numbers, at low temperatures as well. For example, archaea are common in cold oceanic environments such as polar seas. Even more significant are the large numbers of archaea found throughout the world's oceans in non-extreme habitats among the plankton community (as part of the picoplankton). Although these archaea can be present in extremely high numbers (up to 40% of the microbial biomass), almost none of these species have been isolated and studied in pure culture.

Consequently, our understanding of the role of archaea in ocean ecology is rudimentary, so their full influence on global biogeochemical cycles remains largely unexplored. Some marine Crenarchaeota are capable of nitrification, suggesting these organisms may affect the oceanic nitrogen cycle, although these oceanic Crenarchaeota may also use other sources of energy. Vast numbers of archaea are also found in the sediments that cover the sea floor, with these organisms making up the majority of living cells at depths over 1 meter into this sediment.

Role in Chemical Cycling

Archaea recycle elements such as carbon, nitrogen and sulfur through their various habitats. Although these activities are vital for normal ecosystem function, archaea can also contribute to human-made changes, and even cause pollution.

Archaea carry out many steps in the nitrogen cycle. This includes both reactions that remove nitrogen from ecosystems, such as nitrate-based respiration and denitrification, as well as processes that introduce nitrogen, such as nitrate assimilation and nitrogen fixation. Archaean involvement in ammonia oxidation reactions was recently discovered. These reactions are particularly important in the oceans. The archaea also appear to be crucial for ammonia oxidation in soils. The produce nitrite, which other microbes then oxidize to nitrate. Plants and other organisms consume the latter. In the sulfur cycle, archaea that grow by oxidizing sulfur compounds release this element from rocks, making it available to other organisms. However, the archaea that do this, such as *Sulfolobus*, produce sulfuric acid as a waste product, and the growth of these organisms in abandoned mines can contribute to acid mine drainage and other environmental damage.

In the carbon cycle, methanogen archaea remove hydrogen and are important in the decay of organic matter by the populations of microorganisms that act as decomposers in anaerobic ecosystems, such as sediments, marshes and sewage treatment works. However, methane is one of the most abundant greenhouse gases in Earth's atmosphere, constituting 18% of the global total. It is 25 times more potent as a greenhouse gas than carbon dioxide. Methanogens are the primary source of atmospheric methane, and are responsible for most of the world's yearly methane emissions. As a consequence, these archaea contribute to global greenhouse gas emissions and global warming.

Interactions with other Organisms

The well-characterized interactions between archaea and other organisms are either mutual or commensal. As of 2007, no clear examples of archaeal pathogens or parasites were known. However, a relationship has been proposed between some species of methanogens and infections in the mouth, and *Nanoarchaeum equitans* may be a parasite of another species of archaea, since it only survives and reproduces within the cells of the Crenarchaeon *Ignicoccus hospitalis*, and appears to offer no benefit to its host.

Mutualism

One well-understood example of mutualism is the interaction between protozoa and methanogenic archaea in the digestive tracts of animals that digest cellulose, such as ruminants and termites. In these anaerobic environments, protozoa break down plant cellulose to obtain energy. This process releases hydrogen as a waste product, but high levels of hydrogen reduce energy production. When methanogens convert hydrogen to methane, protozoa benefit from more energy. In anaerobic protozoa such as *Plagiopyla frontata* the archaea reside inside the protozoa and consume hydrogen produced in their hydrogenosomes. Archaea associate with larger organisms. For example, the marine archaean *Cenarchaeum symbiosum* lives within (it is an endosymbiont of) the sponge *Axinella mexicana*.

Commensalism

Archaea can also be commensals, benefiting from an association without helping or harming the other organism. For example, the methanogen *Methanobrevibacter smithii* is by far the most common archaean in the human flora, making up about one in ten of all the prokaryotes in the human gut. In termites and in humans, these methanogens may in fact be mutualists, interacting with other microbes in the gut to aid digestion. Archaean communities also associate with a range of other organisms, such as on the surface of corals, and in the region of soil that surrounds plant roots (the rhizosphere).

Significance in Technology and Industry

Extremophile archaea, particularly those resistant either to heat or to extremes of acidity and alkalinity, are a source of enzymes that function under these harsh conditions. These enzymes have found many uses.

For example, thermostable DNA polymerases, such as the Pfu DNA polymerase from *Pyrococcus furiosus*, revolutionized molecular biology by allowing the polymerase chain reaction to be used in research as a simple and rapid technique for cloning DNA.

In industry, amylases, galactosidases and pullulanases in other species of *Pyrococcus* that function at over 100 °C (212 °F) allow food processing at high temperatures, such as the production of low lactose milk and whey. Enzymes from these thermophilic archaea also tend to be very stable in organic solvents, allowing their use in environmentally friendly processes in green chemistry that synthesize organic compounds. This stability makes them easier to use in structural biology. Consequently the counterparts of bacterial or eukaryotic enzymes from extremophile archaea are often used in structural studies.

In contrast to the range of applications of archaean enzymes, the use of the organisms themselves in biotechnology is less developed. Methanogenic archaea are a vital part of sewage treatment, since they are part of the community of microorganisms that carry out anaerobic digestion and produce biogas. In mineral processing, acidophilic archaea display promise for the extraction of metals from ores, including gold, cobalt and copper.

Archaea host a new class of potentially useful antibiotics. A few of these archaeocins have been characterized, but hundreds more are believed to exist, especially within *Haloarchaea* and *Sulfolobus*. These compounds have different structure than bacterial antibiotics, so they may have novel modes of action. In addition, they may allow the creation of new selectable markers for use in archaeal molecular biology.

Fluorescent in Situ Hybridization

Fish (fluorescence *in situ* hybridization) is a cytogenetic technique developed by Christoph Lengauer that is used to detect and localize the presence or absence of specific DNA sequences on chromosomes. FISH uses fluorescent probes that bind to only those parts of the chromosome with which they show a high degree of sequence similarity. Fluorescence microscopy can be used to find out where the fluorescent probe bound to the chromosomes. FISH is often used for finding specific features in DNA for use in genetic counselling, medicine, and species identification. FISH can also be used to detect and localize specific mRNAs within tissue samples. In this context, it can help define the spatial-temporal patterns of gene expression within cells and tissues.

Probes

Probes are often derived from fragments of DNA that were isolated, purified, and amplified for use in the Human Genome Project. The size of the human genome is so large, compared to the length that could be sequenced directly, that it was necessary to divide the genome into fragments. (In the eventual analysis, these fragments were put into order by digesting a copy of each fragment into still smaller fragments using sequence-specific endonucleases, measuring the size of each small fragment using size-exclusion chromatography, and using that information to determine where the large fragments overlapped one another.)

To preserve the fragments with their individual DNA sequences, the fragments were added into a system of continually replicating bacteria populations. Clonal populations of bacteria, each population maintaining a single artificial chromosome, are stored in various laboratories around the world. The artificial chromosomes (BAC) can be grown, extracted, and labelled, in any lab. These fragments are on the order of 100 thousand base-pairs, and are the basis for most FISH probes.

Preparation and Hybridization Process

First, a probe is constructed. The probe must be large enough to hybridize specifically with its target but not so large as to impede the hybridization process. The probe is tagged directly with fluorophores, with targets for antibodies or with biotin. Tagging can be done in various ways, such as nick translation, or PCR using tagged nucleotides. Then, an interphase or metaphase chromosome preparation is produced. The chromosomes are firmly attached to a substrate, usually glass.

Repetitive DNA sequences must be blocked by adding short fragments of DNA to the sample. The probe is then applied to the chromosome DNA and incubated for approximately 12 hours while hybridizing. Several wash steps remove all unhybridized or partially-hybridized probes. The results are then visualized and quantified using a microscope that is capable of exciting the dye and recording images.

If the fluorescent signal is weak, amplification of the signal may be necessary in order to exceed the detection threshold of the microscope.

Fluorescent signal strength depends on many factors such as probe labeling efficiency, the type of probe, and the type of dye. Fluorescently-tagged antibodies or streptavidin are bound to the dye molecule. These secondary components are selected so that they have a strong signal.

Fish experiments designed to detect or localize gene expression within cells and tissues rely on the use of a reporter gene, such as one expressing green fluorescent protein, to provide the fluorescence signal.

Variations on Probes and Analysis

Fish is a very general technique. The differences between the various FISH techniques are usually due to variations in the sequence and labeling of the probes; and how they are used in combination. These few modifications make possible all FISH techniques.

Probe size is important because longer probes hybridize more specifically than shorter probes. The overlap defines the resolution of detectable features. For example, if the goal of an experiment is to detect the breakpoint of a translocation, then the overlap of the probes — the degree to which one DNA sequence is contained in the adjacent probes — defines the minimum window in which the breakpoint may be detected.

The mixture of probe sequences determines the type of feature the probe can detect. Probes that hybridize along an entire chromosome are used to count the number of a certain chromosome, show translocations, or identify extra-chromosomal fragments of chromatin. This is often called "whole-chromosome painting." If every possible probe is used, every chromosome, (the whole genome) would be marked fluorescently, which would not be particularly useful for determining features of individual sequences.

However, a mixture of smaller probes can be created that is specific to a particular region (locus) of DNA; these mixtures are used to detect deletion mutations. When combined with a specific colour, a locus-specific probe mixture is used to detect very specific translocations. Special locus-specific probe mixtures are often used to count chromosomes, by binding to the centromeric regions of chromosomes, which are unique enough to identify each chromosome (with the exception of Chromosome 13, 14 21, 22.)

A variety of other techniques use mixtures of differently-coloured probes. A range of colours in mixtures of fluorescent dyes can be detected, so each human chromosome can be identified by a characteristic colour using whole-chromosome probe mixtures and a variety of ratios of colours.

Although there are more chromosomes than easily-distinguishable fluorescent dye colours, ratios of probe mixtures can be used to create *secondary* colours. Similar to comparative genomic hybridization, the probe mixture for the secondary colours is created by mixing the correct ratio of two sets of differently-coloured probes for the same chromosome. This technique is sometimes called M-FISH. The same physics that make a variety of colours possible for M-FISH can be used for the detection of translocations. That is, colours that are adjacent appear to overlap; a secondary colour is observed. Some assays are designed so that the secondary colour will be present or absent in cases of interest.

An example is the detection of BCR/ABL translocations, where the secondary colour indicates disease. This variation is often called double-fusion FISH or D-FISH. In the opposite situation—where the absence of the secondary colour is pathological—is illustrated by an assay used to investigate translocations where only one of the breakpoints is known or constant.

Locus-specific probes are made for one side of the breakpoint and the other intact chromosome. In normal cells, the secondary colour is observed, but only the primary colour is observed when the translocation occurs. This technique is sometimes called "break-apart FISH".

Single Molecule RNA Fish

Single Molecule RNA Fish is a method of detecting and quantifying mRNA and other long RNA molecules in a thin layer of tissue sample. Targets can be reliably imaged through the application of multiple short singly labelled oligonucleotide probes.

The probes cooperatively bind to the target site. When each probe binds to the single stranded mRNA, it causes cooperative unwinding of the mRNA, promoting the binding of the next probe.

The net result is the binding of 48 fluorescent labels to a single molecule of mRNA, providing sufficient fluorescence to reliably locate each target mRNA in a wide-field fluorescent microscopy image. Probes not binding to the intended sequence do not achieve sufficient localized fluorescence to be distinguished from the background. This technology is exclusively licensed to Biosearch Technologies as Stellaris™ FISH Probes.

Single molecule RNA FISH assays can be performed in simplex or multiplex, and have potential applications in cancer diagnosis , neuroscience , gene expression analysis, and companion diagnostics.

Fibre Fish

In an alternative technique to interphase or metaphase preparations, fibre FISH, interphase chromosomes are attached to a slide in such a way that they are stretched out in a straight line, rather than being tightly coiled, as in conventional FISH, or adopting a random conformation, as in interphase FISH. This is accomplished by applying mechanical shear along the length of the slide, either to cells that have been fixed to the slide and then lysed, or to a solution of purified DNA. A technique known as chromosome combing is increasingly used for this purpose. The extended conformation of the chromosomes allows dramatically higher resolution - even down to a few kilobases. The preparation of fibre FISH samples, although conceptually simple, is a rather skilled art, and only specialized laboratories use the technique routinely.

Q-Fish

Q-Fish combines Fish with PNAs and computer software to quantify fluorescence intensity. This technique is used routinely in telomere length research.

Flow-Fish

Flow-Fish uses flow cytometry to perform Fish automatically using per-cell fluorescence measurements.

Medical Applications

Often parents of children with a developmental delay want to know more about their child's conditions before choosing to have another child. These concerns can be addressed by analysis of the parents' and child's DNA. In cases where the child's developmental delay is not understood, the cause of it can be determined using FISH and cytogenetic techniques.

Examples of diseases that are diagnosed using FISH include Prader-Willi syndrome, Angelman syndrome, 22q13 deletion syndrome, chronic myelogenous leukemia, acute lymphoblastic leukemia, Cri-du-chat, Velocardiofacial syndrome, and Down syndrome. FISH on sperm cells is indicated for men with an abnormal somatic or meiotic karyotype as well as those with oligozoospermia, since approximately 50% of oligozoospermic men have an increased rate of sperm chromosome abnormalities. The analysis of chromosomes 21, X, and Y is enough to identify oligozoospermic individuals at risk.

In medicine, FISH can be used to form a diagnosis, to evaluate prognosis, or to evaluate remission of a disease, such as cancer. Treatment can then be specifically tailored. A traditional exam involving metaphase chromosome analysis is often unable to identify features that distinguish one disease from another, due to subtle chromosomal features; FISH can elucidate these differences. FISH can also be used to detect diseased cells more easily than standard Cytogenetic methods, which require dividing cells and requires labour and time-intensive manual preparation and analysis of the slides by a technologist. FISH, on the other hand, does not require living cells and can be quantified automatically, a computer counts the fluorescent dots present. However, a trained technologist is required to distinguish subtle differences in banding patterns on bent and twisted metaphase chromosomes.

Species Identification

Fish is often used in clinical studies. If a patient is infected with a suspected pathogen, bacteria, from the patient's tissues or fluids, are typically grown on agar to determine the identity of the pathogen. Many bacteria, however, even well-known species, do not grow well under laboratory conditions. FISH can be used to detect directly the presence of the suspect on small samples of patient's tissue.

Fish can also be used to compare the genomes of two biological species, to deduce evolutionary relationships. A similar hybridization technique is called a zoo blot. Bacterial FISH probes are often primers for the 16s rRNA region. Fish is widely used in the field of microbial ecology, to identify microorganisms. Biofilms, for example, are composed of complex (often) multi-species bacterial organizations. Preparing DNA probes for one species and performing Fish with this probe allows one to visualize the distribution of this specific species within the biofilm. Preparing probes (in two different colours) for two species allows to visualize/study co-localization of these two species in the biofilm, and can be useful in determining the fine architecture of the biofilm.

Bibliography

Ashraf, Syed Mahboob: *Handbook of Fruit and Vegetable Products*, Agrobios, Delhi, 2008.

Banerjee, S : *Biochemistry and Biotechnology,* Dominant Pub, Delhi, 2009.

Barrington, E. J. W.: *Biochemistry of Primitive Deuterostomians*, London, Academic Press, 1974.

Bhaskar, Harsh: *Basic Facts on Biochemistry*, Campus Books, Delhi, 2010.

Bhatia, A. L.: *Biochemistry and Endocrinology,* Indus Valley, Delhi, 2002.

Bhatnagar, Vasudev: *Cell Science and Technology*, Campus Books, Delhi, 2009.

Bora, K.K.: *Agros Dictionary of Plant Physiology and Biochemistry*, Agrobios, Delhi, 2001.

Bose, N. Sarath Chandra: *Biochemistry: A Practical Manual,* Pharma Med Press, Delhi, 2010.

Bose, Sarath Chandra: *Biochemistry: A Practical Manual*, Pharma Med Press, Delhi, 2010.

Chanderlekha Goswami: *Biochemistry and Instrumentation*, Manglam Pub, Delhi, 2011.

Chauhan, B.S.: *Principles of Biochemistry and Biophysics,* Laxmi Publications, Delhi, 2008.

Chauhan, Sushil: *Experiments in Biochemistry,* Campus Books, Delhi, 2010.

Collymore L.: *Fruit Production in Barbados*, Port of Spain, Trinidad and Tobago, 1996.

Coultate, T. P.: *Food: The Chemistry of its Components,* London: The Royal Society of Chemistry, 1988.

Daphne C. Elliott: *Biochemistry and Molecular Biology*, Oxford University Press, Delhi, 2005.

Dar, Ghulam Hassan: *Soil Microbiology and Biochemistry,* New India Publishing Agency, Delhi, 2010.

Dutta, Chiranjib: *Fundamentals of Biochemistry: A Practical Approach,* Kanishka, Delhi, 2005.

Elliott, B N: *Biochemistry and Molecular Biology*, Oxford University Press, Delhi, 2005.

Florkin, M.: *Comprehensive Biochemistry: Amsterdam*, Elsevier, 1975.

Goel, A K: *Basic Concept of Organic Chemistry*, Pearl Books, Delhi, 2008.

Golakiya, B A: *DNA: A Bridge Between Biochemistry and Biotechnology,* New India Publishing Agency, Delhi, 2008.

Hemantaranjan, A L: *Developments in Physiology, Biochemistry and Molecular Biology of Plants*, New India Pub, Delhi, 2005.

Jasra, O.P.: *Environmental Biochemistry*, Sarup & Sons, Delhi, 2002.

Khanna, V K: *Objective Genetics, Biotechnology, Biochemistry and Forestry,* I.K. International Publishing House, Delhi, 2008.

Mido, Y.: *An Introduction to Nuclear Chemistry*, Discovery, Delhi, 1995.

Narvekar, Raghunath: *Handbook of Biochemistry,* Adhyayan, Delhi, 2008.

Prakash, M: *Insect Biochemistry,* Discovery Pub, Delhi, 2008.

Prasad, G.P.: *Biochemistry of Proteins*, Discovery, Delhi, 2010.

Purohit, S.S.: *Biochemistry: Fundamentals and Applications,* Agrobios, Delhi, 2009.

Rastogi, Arun: *An Introduction to Practical Biochemistry*, Anmol, Delhi, 2010.

Sangha, S P S: *Basic Techniques in Biochemistry and Molecular Biology,* I K International, Delhi, 2009.

Smriti Ganguly: *Biochemistry of Biomolecules*, Pearl Books, Delhi, 2007.

Srivastava, Seema: *Plant Physiology and Biochemistry,* Campus Books, Delhi, 2009.

Tripathi, Pramila: *Botanical Pesticides in the Management of Post-harvest Fruit Diseases*, Daya, Delhi, 2005.

Whyman, Robin: *Applied Organometallic Chemistry and Catalysis*, Oxford University Press, Delhi, 2005.

Index

P

R

S

T

V

❑❑❑

Encyclopaedia of

Biochemistry and Molecular Biology of Plants

Encyclopaedia of

Biochemistry and Molecular Biology of Plants

Volume 4

Editor:
K.N.P. Singh

AGROTECH PRESS
JAIPUR ▪ NEW DELHI

AGROTECH PRESS

Regd. Off. "Bony Residency", Gate No. 2
Opp. Tilak Public School, Vishweswariya Nagar,
Gopalpura Road, Jaipur - 302018, India
Ph: 0141 - 2761280, Telefax: 2761381
E-mail : agrotechpress@gmail.com

Delhi Office: 102, 1st Floor, Satyam House,
4327/3, Ansari Road, Darya Ganj,
New Delhi - 110002, India
Ph: 011-45652440
visit us: www.agrotechpress.com

Distributed By
KEYA INTERNATIONAL PRESS LLP
E-mail: keyainternationalpressllp@gmail.com

Published in India, 2024

ISBN : 9789384568221 (4 Vols. Set)

Typeset by Thomson Press

Printed and bound in India by Thomson Press Ltd.,
New Delhi

Contents

Preface

For centuries, humankind has made improvements to crop plants through selective breeding and hybridization — the controlled pollination of plants. Plant biotechnology is an extension of this traditional plant breeding with one very important difference — plant biotechnology allows for the transfer of a greater variety of genetic information in a more precise, controlled manner. Unlike traditional plant breeding, which involves the crossing of hundreds or thousands of genes, plant biotechnology allows for the transfer of only one or a few desirable genes. This more precise science allows plant breeders to develop crops with specific beneficial traits and without undesirable traits. Many of these beneficial traits in new plant varieties fight plant pests — insects, disease and weeds — that can be devastating to crops. Others provide quality improvements, such as tastier fruits and vegetables; processing advantages, such as tomatoes with higher solids content; and nutrition enhancements, such as oil seeds that produce oils with lower saturated fat content. Crop improvements like these can help provide an abundant, healthful food supply and protect our environment for future generations.

Crop improvement facilitated by modern biotechnology is one the most significant developments in plant biotechnology research and development. All over the world, plant biotechnology has largely been acknowledged as a key strategy for achieving food security and sustainable agriculture; many governments give high priority to agricultural biotechnology R&D. Main focus of these research on food crops and crops of high commercial value in the hope of meeting increasing food requirements and reducing poverty, particularly among resource-poor farming households. Harnessing biotechnology applications for the benefit of the poor, however, requires considerable attention in many areas, including appropriateness of and access to agri-biotechnology by resource-poor farmers, capability of the public sector in biotechnology R&D, regulatory framework that enhances the use of biotechnology applications, and public-private sector partnerships. Science and technology have been the foundation of the social and economic gains made in agriculture over the past 30 years and will continue to underpin any necessary increases in agricultural productivity. Plant biotechnology is one such technology that has been regarded as part of the "sustainable productivity equation" in agriculture. Its present applications in agriculture include conventional breeding, tissue culture and micropropagation, molecular breeding or marker-assisted selection, plant disease diagnostics, genetic engineering and the production of GM crops, and the "omics" sciences. Unfortunately, harnessing biotechnology and its applications for the benefit of the poor will require considerable attention in many areas including: allocation of additional public resources to agricultural research; appropriateness of, and access to, biotechnology by resource-poor farmers; improvement in the seed distribution and extension systems; capacity-building of the public sector in biotech R&D; public education; policies and regulatory frameworks on biosafety, food safety, and intellectual property rights; and stronger public-private sector links for both international and local collaborative undertakings. Harnessing biotechnology and its applications for the benefit of the poor, however, will depend on a number of critical elements. Because new agricultural technologies are increasingly complex, knowledge intensive, expensive, and location specific, clear priorities need to be established. The right mix of funding and research effort needs to be apportioned between tried and tested applications and the more modern genetic technologies (such as genetic engineering and the development of GM crops). Strategies must be directed at clearly defined target crops and traits that affect poverty reduction, food security, environmental conservation, and competitiveness.

The present book is an attempt to deal with all the topics in-depth and to find out the molecular aspects of different processes in cultured plant cells, so as to exploit them for h benefit of plant biotechnology. This book will be highly beneficial to students, teachers and research workers in the field of plant biotechnology, agriculture and plant science.

—Editor

1

Developmental Biology and Evolutionary Genetics

Developmental Biology

Developmental biology is the study of the process by which organisms grow and develop. Modern developmental biology studies the genetic control of cell growth, differentiation and "morphogenesis", which is the process that gives rise to tissues, organs and anatomy.

Related Fields of Study

Embryology is a subfield, the study of organisms between the one-cell stage (generally, the zygote) and the end of the embryonic stage. Embryology was originally a more descriptive science until the 20th century. Embryology and developmental biology today deal with the various steps necessary for the correct and complete formation of the body of a living organism.The related field of evolutionary developmental biology was formed largely in the 1990s and is a synthesis of findings from molecular developmental biology and evolutionary biology which considers the diversity of organismal form in an evolutionary context.

Perspectives

The development of a new life is a spectacular process and represents a masterpiece of temporal and spatial control of gene expression. Developmental genetics studies the effect that genes have in a phenotype, given normal or abnormal epigenetic parameters. The findings of developmental biology can help to understand developmental abnormalities such as chromosomal aberrations that cause Down syndrome. An understanding of the specialization of cells during embryogenesis has provided information on how stem cells specialize into specific tissues and organs.

This information has led, for example, to the cloning of specific organs for medical purposes. Another biologically important process that occurs during development is apoptosis—programmed cell death or "suicide." Many developmental models are used to elucidate the physiology and molecular basis of this cellular process.

Similarly, a deeper understanding of developmental biology can foster greater progress in the treatment of congenital disorders and diseases, e.g. studying human sex determination can lead to treatment for disorders such as congenital adrenal hyperplasia.

Developmental Model Organisms

Often used model organisms in developmental biology include the following:

- Vertebrates
  - Zebrafish *Danio rerio*
  - Medakafish *Oryzias latipes*
  - Fugu (pufferfish) *Takifugu rubripes*
  - Frog *Xenopus laevis, Xenopus tropicalis*
  - Chicken *Gallus gallus*
  - Mouse *Mus musculus* (Mammalian embryogenesis)
- Invertebrates
  - Lancelet *Branchiostoma lanceolatum*
  - Ascidian *Ciona intestinalis*
  - Sea urchin *Strongylocentrotus purpuratus*
  - Roundworm *Caenorhabditis elegans*
  - Fruit fly *Drosophila melanogaster* (Drosophila embryogenesis)
- Plants (Plant embryogenesis)
  - *Physcomitrella patens*
  - *Arabidopsis thaliana*
  - Maize
  - Snapdragon *Antirrhinum majus*
- Other
  - Slime mold *Dictyostelium discoideum*

Studied Phenomena

Cell Differentiation

Differentiation is the formation of cell types, from what is originally one cell – the zygote or spore. The formation of cell types like nerve cells occurs with a number of intermediary, less differentiated cell types. A cell stays a certain cell type by maintaining a particular pattern of gene expression. This depends on regulatory genes, e.g. for transcription factors and signaling proteins. These can take part in self-perpetuating circuits in the gene regulatory network, circuits that can involve several cells that communicate with each other. External signals can alter gene expression by activating a receptor, which triggers a signaling cascade that affects transcription factors. For example, the withdrawal of growth factors from myoblasts causes them to stop dividing and instead differentiate into muscle cells.

Embryonal Development

Embryogenesis is the step in the life cycle after fertilisation – the development of the embryo, starting from the zygote (fertilised egg). Organisms can differ drastically in the how embryo develops, especially when they belong to different phyla. For example, embryonal development in placental mammals starts with cleavage of the zygote into eight uncommited cells, which then form a ball (morula). The outer cells become the trophectoderm or trophoblast, which will form in combination with maternal uterine endometrial tissue the placenta, needed for fetal nurturing via maternal blood, while inner cells become the inner cell mass that will form all fetal organs (the bridge between these two parts eventually forms the umbilical cord). In contrast, the fruit fly zygote first forms a sausage-shaped syncytium, which is still one cell but with many cell nuclei.

Patterning is important for determining which cells develop into which organs. This is mediated by signaling between adjacent cells by proteins on their surfaces, and by gradients of signaling secreted molecules. An example is retinoic acid, which forms a gradient in the head to tail direction in animals. Retinoic acid enters cells and activates Hox genes in a concentration-dependent manner – Hox genes differ in how much retinoic acid they require for activation and will thus show differential rostral expression boundaries, in a colinear fashion with their genomic order. As Hox genes code for transcription factors, this causes different activated combinations of both Hox and other genes in discrete anteroposterior transverse segments of the neural tube (neuromeres) and related patterns in surrounding tissues, such as branchial arches, lateral mesoderm, neural crest, skin and endoderm, in the head to tail direction. This is important for e.g. the segmentation of the spine in vertebrates. Embryonal development does not always proceed correctly, and errors can result in birth defects or miscarriage. Often the reason is genetic (mutation or chromosome abnormality), but there can be environmental influence (like teratogens) or stochastic events. Abnormal development caused by mutation is also of evolutionary interest as it provides a mechanism for changes in body plan.

Growth

Growth is the enlargement of a tissue or organism. Growth continues after the embryonal stage, and occurs through cell proliferation, enlargement of cells or accumulation of extracellular material. In plants, growth results in an adult organism that is strikingly different from the embryo. The proliferating cells tend to be distinct from differentiated cells. In some tissues proliferating cells are restricted to specialised areas, such as the growth plates of bones. But some stem cells migrate to where they are needed, such as mesenchymal stem cells which can migrate from the bone marrow to form e.g. muscle, bone or adipose tissue. The size of an organ frequently determines its growth, as in the case of the liver which grows back to its previous size if a part is removed. Growth factors, such as fibroblast growth factors in the animal embryo and growth hormone in juvenile mammals, also control the extent of growth.

Metamorphosis

Most animals have a larval stage, with a body plan different from that of the adult organism. The larva abrubtly develops into an adult in a process called metamorphosis. For example, caterpillars (butterfly larvae) are specialized for feeding whereas adult butterflies (imagos) are specialised for flight and reproduction. When the caterpillar has grown enough, it turns into an immobile pupa. Here, the imago develops from imaginal discs found inside the larva.

Regeneration

Regeneration is the reactivation of development so that a missing body part grows back. This phenomenon has been studied particularly in salamanders, where the adults can reconstruct a whole limb after it has been amputated. Researchers hope to one day be able to induce regeneration in humans. There is little spontaneous regeneration in adult humans, although the liver is a notable exception. Like for salamanders, the regeneration of the liver involves dedifferentiation of some cells to a more embryonal state.

Developmental Systems Biology

Computer simulation of multicellular development is a research methodology to understand the function of the very complex processes involved in the development of organisms. This includes simulation of cell signaling, multicell interactions and regulatory genomic networks in development of multicellular structures and processes. *Minimal genomes* for minimal multicellular organisms may pave the way to understand such complex processes *in vivo*.

Evolutionary Genetics

Evolutionary genetics is the broad field of studies that attempts to account for evolution in terms of changes in gene and genotype frequencies within populations and the processes that convert the variation with populations into more or less permanent variation between species. It considers the effect of micro-evolutionary changes among populations due to evolutionary forces, which account for the emergence of macro-evolutionary patterns in the long term. A focus of evolutionary genetics is to describe how the evolutionary forces shape the patterns of biodiversity observed in nature. This field of studies resulted from the integration — also known as "modern synthesis" or "evolutionary synthesis" — of genetics and Darwinian evolution.

Notable individuals in the field include: Vladimir Efroimson, G. H. Hardy, Wilhelm Weinberg, Sergei Chetverikov, Ronald Fisher, Sewall Wright, J. B. S. Haldane, Julian Huxley, Theodosius Dobzhansky, and Hermann Joseph Muller.

Heritability of IQ

The study of the heritability of IQ investigates the relative importance of genetics and environment for variation in intelligence quotient (IQ) in a population. "Heritability", in this sense, "refers to the genetic contribution to variance within a population and in a specific environment" There has been significant controversy in the academic community about the heritability of IQ ever since research began in the 19th century. IQ is a polygenic trait under normal circumstances according to recent research. However, certain single gene genetic disorders can severely affect intelligence, with phenylketonuria as an example.

Estimates in the academic research of the heritability of IQ have varied from below 0.5 to a high of 0.9. A 1996 statement by the American Psychological Association gave about .45 for children and about .75 during and after adolescence. A 2004 meta-analysis of reports in *Current Directions in Psychological Science* gave an overall estimate of around .85 for 18-year-olds and older. *The New York Times Magazine* has listed about three quarters as a figure held by the majority of studies.

Methods and Results

Correlations

The relative influence of genetics and environment for a trait can be calculated by measuring how strongly traits covary in people of a given genetic (unrelated, siblings, fraternal twins, or identical twins) and environmental (reared in the same family or not) relationship. The most common method is to consider identical twins reared apart, with any similarities which exists between such twin pairs attributed to genotype. In terms of correlation statistics, this means that theoretically the correlation of tests scores between monozygotic twins would be 1.00 if genetics alone accounted for variation in IQ scores; likewise, siblings and dizygotic twins share on average half of their alleles and the correlation of their scores would be 0.50 if IQ were affected by genes alone. Practically, however, the upper bound of these correlations are given by the reliability of the test, which tends to be 0.90 to 0.95 for typical IQ tests.

If there is biological inheritance of IQ, then the relatives of a person with a high IQ should exhibit a comparably high IQ with a much higher probability than the general population. In 1982, Bouchard and McGue reviewed such correlations reported in 111 original studies in the United States. The mean correlation of IQ scores between monozygotic twins was 0.86, between siblings, 0.47, between half-siblings, 0.31, and between cousins, 0.15.

The 2006 edition of *Assessing adolescent and adult intelligence* by Alan S. Kaufman and Elizabeth O. Lichtenberger reports correlations of 0.86 for identical twins raised together compared to 0.76 for those raised apart and 0.47 for siblings. These number are not necessarily static. When comparing pre-1963 to late 1970s data, researches DeFries and Plomin found that the IQ correlation between parent and child living together fell significantly, from 0.50 to 0.35. The opposite occurred for fraternal twins.

Another Summary:

- Same person (tested twice) .95
- Identical twins—Reared together .86
- Identical twins—Reared apart .76
- Fraternal twins—Reared together .55
- Fraternal twins—Reared apart .35
- Biological siblings—Reared together .47
- Biological siblings—Reared apart .24
- Unrelated children—Reared together .30
- Parent-child—Living together .42
- Parent-child—Living apart .22
- Adoptive parent–child—Living together .19

Heritability

Heritability is defined as the proportion of variance in a trait which is attributable to genotype within a defined population in a specific environment. Heritability takes a value ranging from 0 to 1; a heritability of 1 indicates that all variation in the trait in question is genetic in origin and a heritability of 0 indicates that *none* of the variation is genetic. The determination of many traits can be considered primarily genetic under similar environmental backgrounds. For example, a 2006 study found that adult height has a heritability estimated at 0.80 when looking only at the height variation within families where the environment should be very similar. Other traits have lower heritabilities, which indicate a relatively larger environmental influence. For example, a twin study on the heritability of depression in men calculated it as 0.29, while it was 0.42 for women in the same study.

Various studies have found the heritability of IQ to be between 0.7 and 0.8 in adults and 0.45 in childhood in the United States. It may seem reasonable to expect that genetic influences on traits like IQ should become less important as one gains experiences with age. However, that the opposite occurs is well documented. Heritability measures in infancy are as low as 0.2, around 0.4 in middle childhood, and as high as 0.8 in adulthood. One proposed explanation is that people with different genes tend to seek out different environments that reinforce the effects of those genes. A 1994 review in *Behaviour Genetics* based on identical/fraternal twin studies found that heritability is as high as 0.80 in general cognitive ability but it also varies based on the trait, with .60 for verbal tests, .50 for spatial and speed-of-processing tests, and only .40 for memory tests.

In 2006, *The New York Times Magazine* listed about three quarters as a figure held by the majority of studies, while a 2004 meta-analysis of reports in *Current Directions in Psychological Science* gave an overall estimate of around .85 for 18-year-olds and older.

Caveats

There are a number of points to consider when interpreting heritability:

- Heritability measures the proportion of *variation* in a trait that can be attributed to genes, and not the proportion of a trait caused by genes. Thus, if the environment relevant to a given trait changes in a way that affects all members of the population equally, the mean value of the trait will change without any change in its heritability (because the variation or differences among individuals in the population will stay the same). This has evidently happened for height: the heritability of stature is high, but average heights continue to increase. Thus, even in developed nations, a high heritability of a trait does not necessarily mean that average group differences are dues to genes. Some have gone further, and used height as an example in order to argue that "even highly heritable traits can be strongly manipulated by the environment, so heritability has little if anything to do with controllability." However, others argue that IQ is highly stable during life and has been largely resistant to interventions aimed to change it long-term and substantially.

- A common error is to assume that a heritability figure is necessarily unchangeable. The value of heritability can change if the impact of environment (or of genes) in the population is substantially altered. If the environmental variation encountered by different individuals increases, then the heritability figure would decrease. On the other hand, if everyone had the same environment, then heritability would be 100%. The population in developing nations often have more diverse environments than in developed nations. This would mean that heritability figures would be lower in developing nations. Another example is phenylketonuria which previously caused mental retardation for everyone who had this genetic disorder and thus had a heritability of 100%. Today, this can be prevented by following a modified diet which has lowered heritability.
- A high heritability of a trait does not mean that environmental effects such as learning are not involved. Vocabulary size, for example, is very substantially heritable (and highly correlated with general intelligence) although every word in an individual's vocabulary is learned. In a society in which plenty of words are available in everyone's environment, especially for individuals who are motivated to seek them out, the number of words that individuals actually learn depends to a considerable extent on their genetic predispositions and thus heritability is high.
- Since heritability increases during childhood and adolescence, and even increases greatly between 16-20 years of age and adulthood, one should be cautious drawing conclusions regarding the role of genetics and environment from studies where the participants are not followed until they are adults. Furthermore, there may be differences regarding the effects on *g* and on non-g factors, with *g* possibly being harder to affect and environmental interventions disproportionately affecting non-*g* factors.

Shared Family Environment

In the developed world, nearly all personality traits show that, contrary to some expectations, environmental effects actually cause non-related children raised in the same family ("adoptive siblings") to be as different as children raised in different families (Harris, 1998; Plomin & Daniels, 1987). There are some family effects on the IQ of children, accounting for up to a quarter of the variance. However, by adulthood, this correlation disappears, such that adoptive siblings are not more similar in IQ than strangers, while adult full siblings show an IQ correlation of 0.6. Twin studies reinforce this pattern: monozygotic (identical) twins raised separately are highly similar in IQ (0.86), more so than dizygotic (fraternal) twins raised together (0.6) and much more than adoptive siblings (~0.0).

There are aspects of environments that family members have in common (for example, characteristics of the home). This shared family environment accounts for 0.25-0.35 of the variation in IQ in childhood. By late adolescence it is is quite low (zero in some studies). There is a similar effect for several other psychological traits. These studies have not looked the effects of extreme environments such as in abusive families.

The American Psychological Association's report "Intelligence: Knowns and Unknowns" (1995) states that there is no doubt that normal child development requires a certain minimum level of responsible care. Severely deprived, neglectful, or abusive environments must have negative effects on a great many aspects of development, including intellectual aspects. Beyond that minimum, however, the role of family experience is in serious dispute. There is no doubt that such variables as resources of the home and parents' use of language are correlated with children's IQ scores, but such correlations may be mediated by genetic as well as (or instead of) environmental factors. But how much of that variance in IQ results from differences between families, as contrasted with the varying experiences of different children in the same family? Recent twin and adoption studies suggest that while the effect of the shared family environment is substantial in early childhood, it becomes quite small by late adolescence. These findings suggest that differences in the life styles of families whatever their importance may be for many aspects of children's lives make little long-term difference for the skills measured by intelligence tests.

Non-shared Family Environment and Environment Outside the Family

Although parents treat their children differently, such differential treatment explains only a small amount of non-shared environmental influence. One suggestions it that children react differently to the same environment due to different genes. More likely influences may be the impact of peers and other experiences outside the family.

Lower Heritability with Lower SES?

The APA report "Intelligence: Knowns and Unknowns" (1995) also stated that:

> *"We should note, however, that low-income and non-white families are poorly represented in existing adoption studies as well as in most twin samples. Thus it is not yet clear whether these studies apply to the population as a whole. It remains possible that, across the full range of income and ethnicity, between-family differences have more lasting consequences for psychometric intelligence."*

A study (1999) by Capron and Duyme of French children adopted between the ages of four and six examined the influence of socioeconomic status (SES).

The children's IQs initially averaged 77, putting them near retardation. Most were abused or neglected as infants, then shunted from one foster home or institution to the next. Nine years later after adoption, when they were on average 14 years old, they retook the IQ tests, and all of them did better. The amount they improved was directly related to the adopting family's socioeconomic status. "Children adopted by farmers and laborers had average IQ scores of 85.5; those placed with middle-class families had average scores of 92. The average IQ scores of youngsters placed in well-to-do homes climbed more than 20 points, to 98."

Stoolmiller (1999) argued that the range of environments in previous adoption studies were restricted. Adopting families tend to be more similar on, for example, socio-economic status than the general population, which suggests a possible underestimation of the role of the shared family environment in previous studies. Corrections for range restriction to adoption studies indicated that socio-economic status could account for as much as 50% of the variance in IQ.

On the other hand, the effect of this was examined by Matt McGue and colleagues (2007), who wrote that "restriction in range in parent disinhibitory psychopathology and family socio-economic status had no effect on adoptive-sibling correlations [in] IQ"

Turkheimer and colleagues (2003) argued that the proportions of IQ variance attributable to genes and environment vary with socioeconomic status. They found that in a study on seven-year-old twins, in impoverished families, 60% of the variance in early childhood IQ was accounted for by the shared family environment, and the contribution of genes is close to zero; in affluent families, the result is almost exactly the reverse. A study by Nagoshi and Johnson (2005) failed to replicate Turkheimer and colleagues' findings. They concluded that the heritability of IQ did not vary as a function of parental socioeconomic status in the 949 families of Caucasian and 400 families of Japanese ancestry who took part in the Hawaii Family Study of Cognition.

Asbury and colleagues (2005) studied the effect of environmental risk factors on verbal and non-verbal ability in a nationally representative sample of 4-year-old British twins. There was not any statistically significant interaction for non-verbal ability, but the heritability of verbal ability was found to be *higher* in low-SES and high-risk environments.

Harden and colleagues (2007) investigated adolescents, most 17 years old, and found that, among higher income families, genetic influences accounted for approximately 55% of the variance in cognitive aptitude and shared environmental influences about 35%. Among lower income families, the proportions were in the reverse direction, 39% genetic and 45% shared environment."

Rushton and Jensen (2010) criticized many of these studies for being done on children or adolescents. They argued that heritability increases during childhood and adolescence, and even increases greatly between 16-20 years of age and adulthood, so one should be cautious drawing conclusions regarding the role of genetics from studies where the participants are not adults. Furthermore, the studies typically did not examine if IQ gains due to adoption were on the general intelligence factor (*g*). When the studies by Capron and Duyme were re-examined, IQ gains from being adopted into high SES homes were on non-*g* factors. By contrast, the adopted children's *g* mainly depended on their biological parents SES, which implied that *g* is more difficult to environmentally change. A 2011 study by Tucker-Drob and colleagues argued that at age 2 years, genes accounted for approximately 50% of the variation in mental ability for children being raised in high socioeconomic status families, but genes accounted for negligible variation in mental ability for children being raised in low socioeconomic status families. This gene-environment interaction was not apparent at age 10 months, suggesting that the effect emerges over the course of early development.

Maternal (fetal) Environment

A meta-analysis by Devlin and colleagues (1997) of 212 previous studies evaluated an alternative model for environmental influence and found that it fits the data better than the 'family-environments' model commonly used. The shared maternal (fetal) environment effects, often assumed to be negligible, account for 20% of covariance between twins and 5% between siblings, and the effects of genes are correspondingly reduced, with two measures of heritability being less than 50%. They argue that the shared maternal environment may explain the striking correlation between the IQs of twins, especially those of adult twins that were reared apart. Bouchard and McGue reviewed the literature in 2003, arguing that Devlin's conclusions about the magnitude of heritability is not substantially different than previous reports and that their conclusions regarding prenatal effects stands in contradiction to many previous reports. They write that:

Chipuer et al. and Loehlin conclude that the postnatal rather than the prenatal environment is most important. The Devlin et al. (1997a) conclusion that the prenatal environment contributes to twin IQ similarity is especially remarkable given the existence of an extensive empirical literature on prenatal effects. Price (1950), in a comprehensive review published over 50 years ago, argued that almost all MZ twin prenatal effects produced differences rather than similarities. As of 1950 the literature on the topic was so large that the entire bibliography was not published. It was finally published in 1978 with an additional 260 references. At that time Price reiterated his earlier conclusion (Price, 1978). Research subsequent to the 1978 review largely reinforces Price's hypothesis (Bryan, 1993; Macdonald et al., 1993; Hall and Lopez-Rangel, 1996; see also Martin et al., 1997, box 2; Machin, 1996).

Dickens and Flynn Model

Dickens and Flynn (2001) argued that the "heritability" figure includes both a direct effect of the genotype on IQ and also indirect effects where the genotype changes the environment, in turn affecting IQ. That is, those with a higher IQ tend to seek out stimulating environments that further increase IQ. The direct effect can initially have been very small but feedback loops can create large differences in IQ. In their model an environmental stimulus can have a very large effect on IQ, even in adults, but this effect also decays over time unless the stimulus continues (the model could be adapted to include possible factors, like nutrition in early childhood, that may cause permanent effects). The Flynn effect can be explained by a generally more stimulating environment for all people. The authors suggest that programmes aiming to increase IQ would be most likely to produce long-term IQ gains if they taught children how to replicate outside the programme the kinds of cognitively demanding experiences that produce IQ gains while they are in the programme and motivate them to persist in that replication long after they have left the programme.

Regression Toward the Mean

Regression towards the mean is a statistical phenomenon that occurs when an outcome is determined by many independent factors. If an outcome is extreme, then this occurred because most of the independent factors agreed by chance. This is unlikely to occur again so to the next outcome is likely to be less extreme. If IQ is determined by many factors, genetic and/or environmental, then they must mostly agree in the same direction in order to produce an extreme IQ. The child of a person with an extreme IQ is unlikely to have all the factors agree so similarly so the child is on average likely to have a less extreme IQ. People in professional occupations have on average 25 points higher IQ than unskilled workers. For their children the difference is 21 points. This is in itself not evidence for genetics or environment since the environment for the children likely differs greatly with it on average being more stimulating for the children of professionals.

It is noted that the above equation relates only statistical averages and is not deterministic. Furthermore, the equation is a general equation based in the inheritance of genetically-based characteristics (in this case, phenotypes), and so it is implicitly assumed that environmental factors are, for the sake of correctly assessing the genetic contribution to IQ, the same across the population.

Operating under the assumption that child and parent are raised in exactly the same environment (unlikely, but usually closer to the truth than in the completely dissimilar environment that the previous equation assumes), h^2 can be replaced by h, which is simply the correlation between parent and offspring IQ. In this case, regression towards the mean is no longer partially caused by environmental differences and therefore only by random genetic variation.

Finally, it is important to note that the expected IQ of the offspring is normally distributed around the mean calculated using the above equation, so in many cases regression towards the mean does not actually occur; as the values are normally distributed, there is a chance that offspring IQ will be more deviant from the mean than that of the parental average.

- Francis Galton: Hereditary Genius. London 1869, p. 195. "100 famous Famous men (n = 43) of science and the percentage of their famous male relatives."
- M. H. Oden: The fulfillment of promise: 40-year follow-up of the Terman gifted group. *Genetical Psychology Monographs 77 (1968) 3-93. "The mean IQ (transformed to 100;15) of the sample of probands was 146 (n = 724); the cut-off score IQ 137."*
- Dean R. Brimhall: Family resemblances among American men of science. *The American Naturalist* 56 (1922) 504-547; 57 (1923) 74-88, 137-152, and 326-344. "In 1915 questionnaires were filled in by 956 distinguished American men of science and their relatives."
- Volkmar Weiss: Mathematical giftedness and family relationship. European Journal for High Ability 5 (1994) 58-67. "*Highly gifted males (mean IQ 135 +/- 9) and their relatives in professions and occupational positions, typically associated with an IQ above 123.*"

Despite the differences in methods and societies, there is a notable parallelism in the published statistics.

The Search for Specific Genes

Many studies attempting to find loci in the genome relating to IQ have had little success. For example, a study by Robert Plomin using groups of around 100 people investigated 1,842 DNA markers in a high-IQ group and in an average-IQ control group. The study used a five-step replication process to eliminate false positives, and no gene met this rigid criterion for replicability.

The failure to find a specific gene associated with IQ indicates that cognitive abilities are very complex and are likely to involve several genes (polygenic). Some estimate that as much as 40% of all genes may contribute to IQ.

The more genes that contribute to a trait the more the trait will be continuous instead of discrete. A 2008 study of 500,000 single nucleotide polymorphisms (SNPs) from 7,089 children did not substantially improve on earlier studies. The study did not find any SNPs that accounted for more than 0.5% of the variance in general intelligence.

Mutations in GDI1 have been linked to X-linked nonspecific mental retardation. A 2007 study did find that a gene called FADS2 along with breastfeeding adds about 7 IQ points to those with the "C" version of the gene. Those with the "G" version see no advantage.

There is "a highly significant association" between the CHRM2 gene and intelligence according to a 2006 Dutch family study. The study concluded that there was an association between the CHRM2 gene on chromosome 7 and Performance IQ, as measured by the Wechsler Adult Intelligence Scale-Revised. The Dutch family study used a sample of 667 individuals from 304 families. A similar association was found independently in the *Minnesota Twin and Family Study* (Comings et al. 2003) and by the Department of Psychiatry at the Washington University.

Microcephalin and ASPM are two genes that are associated with brain development. Mutations in these genes are associated with microcephaly, and hence they were initially associated with general intelligence. However recent studies have found no association with general cognitive abilities. STX1A correlates significantly with intelligence in Williams syndrome patients.

A 2007 study by Caspi and colleagues found that a gene called FADS2 along with breastfeeding adds about 7 IQ points to those with the "C" version of the gene. Those with the "G" version see no advantage.

Copy number variation has also been associated with idiopathic learning disability.

There are number of known cases where the homozygotes have severe cognitive deficits and the heterozygotes show a small decrease of IQ. In such cases further alleles are investigated to estimate their influence on IQ. For example, one minor allele of the gene ALDH5A1 is associated with an IQ difference of around 1.5 points.

Between-Group Heritability

Although IQ differences between individuals are shown to have a large genetic component, it does not automatically follow that mean group-level disparities (between-group differences) in IQ can be assumed to have a genetic basis. An analogy, attributed to Richard Lewontin, illustrates this point:

Suppose two handfuls are taken from a sack containing a genetically diverse variety of corn, and each grown under carefully controlled and standardized conditions, except that one batch is lacking in certain nutrients that are supplied to the other. After several weeks, the plants are measured. There is variability of growth within each batch, due to the genetic variability of the corn. Given that the growing conditions are closely controlled, nearly all the variation in the height of the plants within a batch will be due to differences in their genes. Thus, within populations, heritabilities will be very high. Nevertheless, the difference between the two groups is due entirely to an environmental factor - differential nutrition. Lewontin didn't go so far as to have the one set of pots painted white and the other set black, but you get the idea. The point of the example, in any case, is that the causes of between-group differences may in principle be quite different from the causes of within-group variation.

Arthur Jensen agrees that this is technically correct but argues that a high heritability increases the probability for that genetics play a role in average group differences.

2

Biochemistry of Life

The gaseous and dusty environment from which earth formed contained metals and water, which as you remember from General Chemistry, can react to form hydrogen gas. H_2 reacts with nonmetals (under various conditions of temperature and pressure) to form H_2S, HCl, CH_4, and NH_3 which contributed to the reducing nature of the early atmosphere. This kept the transition metals in their lowest oxidation states. Many metals, including the coinage metals (Cu, Ag, and Au) and the platinum group (Ru, Rh, Pd, Pt) were stable in elemental form.

Then, around 2.7-2.8 billion years ago, photosynthetic organisms (blue/green algae or also called cyanobacteria) developed which could oxidize water to form dioxygen. Oxygen was generally unavailable for redox chemistry before then as photosynthesis, the process that would evolve to oxidize water to produce dioxygen, was unavailable. Remember that to oxidize water to dioxygen, itself a strong oxidizing agent, requires a stronger oxidizing agent than dioxygen and lots of energy.

Fossilized remains of cyanobacteria are found in stromatolites. Using knowledge of how atmospheric oxygen can alter the chemistry of different sulfur isotopes of SO_2, it has been shown that O_2 did not exist in the atmosphere as a whole above 1 ppm earlier than 2.4 billion years ago, although there might have been isolated pockets with higher concentrations. After that it rose, presumably as a result of cyanobacteria. Before this time, bacteria oxidized a similar molecule, H_2S to form elemental sulfur. It is probable that volcanic gases like H_2 might have kept oxygen levels from rising between 2.7 billion year ago and 2.4 billion years ago, when its build-up started.

Hydrogen in the form of H_2 and methane, probably decreased around 2.4 billion years ago as methane with its hydrogen atoms escaped to the upper atmosphere and space. Over the next billion years, dioxygen rose to perhaps 0.2 - 2% (compared to the present levels of 20%) Why? Because the early atmosphere was reducing, the added oxygen combined with a large "sink" of reduced metals (like elemental Cu and Fe) or nonmetals (like C and ammonia), preventing a large buildup. Only after these reduced substances were "titrated" did dioxygen build up to present levels. In addition, the oxygen might have increased weathering of sulfur deposits which can lead to sulfides entering the ocean, where they could precipitate ocean iron ions that are necessary for cyanobacterial chemistry. This would place constraints on cyanobacterial growth until dioxygen levels in the atmosphere increased enough so sulfides were converted to sulfates. Around 2.3 billion years ago, as trace dioxygen had accumulated in the atmosphere, redox chemistry changed, although isotope evidence suggest that little dioxgen was found in water. Around 1.8 - 1.5 billion years ago, the earth's atmosphere became somewhat oxygenated, which was also coincident with the development of eukaryotic organisms. Until then, life was restricted to the oceans since there was no ozone to absorb dangerous UV radiation.

The buildup of dioxygen in the air must have led to a great extinction of anaerobic organisms, since as we shall see, products of oxygen metabolism are very toxic. Some evolved to use dioxygen. Ozone developed, and life could then migrate from the sea to the land. It wasn't until around 600 million years ago that animals arose, however. Was this event associated with the development of a fully oxygenated (20%) atmosphere? Recent evidence, which shows that substantial oxygen wasn't available in the deep sea until about 600 million years, seems to suggest that. Based on analysis of iron compounds in waters in Newfoundland, it appears that oxygen was very low in the sea 580 million years ago, during the Gaskiers glaciation period. Immediately after that it rose to levels consistent with atmospheric dioxygen levels of 15%, levels necessary for large animals. Similar trends in carbon and sulfur isotopes in marine rocks in Oman also suggest large increases in oxygen at the end of the Gaskiers glaciation period. What caused this second great oxygenation event? One possibility is that organic matter was sequestered from reaction with atmospheric dioxygen, as clays bound organic molecules in the ocean and lichens and zooplankton facilitated weather and production of insoluble organic material in the oceans.

Dioxygen is obviously critically important for higher organisms, so an understanding of its chemistry becomes important. This chapter will show that dioxygen is a ground state diradical that has low solubility in aqueous solution, reacts in a kinetically sluggish fashion in oxidation reaction, and forms toxic byproducts as it gets reduced.

Life forms hence evolved ways to deal with these problems, including ways to increase its solubility (with dioxygen binding and transport proteins), and enzymes (that could activate it kinetically and also detoxify oxygen by-products). Dioxygen is toxic to many cells. Obligate aerobes die in an oxygen environment as many of their cellular components get oxidized by this excellent oxidizing agent.

Several strains of bacteria actually swim away from high levels of dioxygen. A graph showing log of survival vs log pO_2 is linear with a negative slope for a variety of organisms, including mice, fish, rats, rabbits, and insects. Pure oxygen can induce chest soreness, coughs, and sore throats in people. Premature infants put in pure dioxygen environments often developed blindness due to retrolental fibroplasia (a build-up of fibrous tissue behind the lens). The trade off for this toxicity is clear. Energy is derived from organic molecule through oxidation. Before dioxygen became available to power aerobic catabolism of reduced molecules like fatty acids and the less reduced sugars, such molecules were only partially oxidized. The glycolytic pathway, found in most organisms, oxidizes glucose (6 Cs) to two molecules of pyruvate (3 Cs). It was only with the availability of dixoygen did pathways evolve (Kreb Cycle, mitochondrial electron transport/oxidative phosphorylation) that allowed pyurvate to be fully oxidized to carbon dioxide, with the release of much more energy.

The Properties of Dioxygen

It is important to understand the properties of dioxygen since oxidation reactions using it power not only our bodies but our entire civilization. We will obviously concentrate on biological reactions, but even these show the same characteristics as non-biological ones.

- oxidation of organic molecules by oxygen is thermodynamically favoured but kinetically slow.
- pure oxygen environments are toxic to cells and organisms.

First we will try to understand these properties of oxygen, and then we will see how organisms overcome these problem to use dioxygen.

We can understand both of these properties by looking at the molecular orbitals of oxygen and its reduction products as shown in the diagrams below. Ground state oxygen is a diradical, which explains the paramagnetic behaviour of oxygen. The two unpaired oxygens each have a spin state of 1/2 for a total resultant spin S of 1, making ground state oxygen a triplet (2S+1) = 3. Organic molecules typically undergo 2 electron oxidation steps. Consider the stepwise oxidation of methane below. The oxidation number of C in methane is -4, -2 in methanol, 0 in formaldehyde, +2 in formic acid, and finally +4 in carbon dioxide, indicating two electron losses in each step.

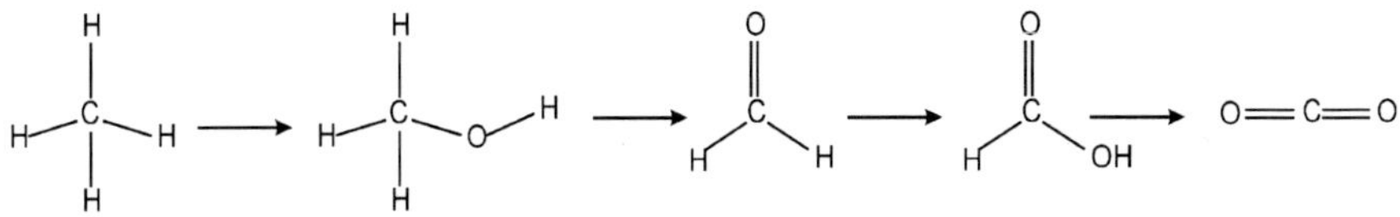

The two electrons lost by the organic substrate are added to oxygen, but since the two lost electrons are spin paired, a spin flip must occur to allow the electrons to enter the unfilled oxygen orbitals. Alternatively, energy can be put into ground state dioxygen to produce excited state singlet oxygen (S=0, 2S+1 = 1).

MOLECULAR ORBITALS DIOXYGEN

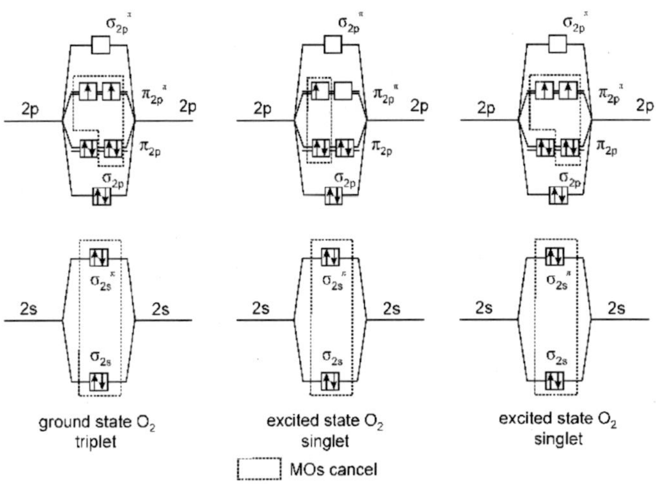

The source of the large activation energy required (about 25 kcal or 105 kJ/mol) to flip the electron spin accounts for the kinetic sluggishness of reactions of dioxygen with organic reactants. A traditional Lewis structure for ground state dioxygen can not be easily written since the electrons are added in pairs, and dioxygen is a diradical. There are 6 electrons in the sigma molecular orbitals from second shell electrons (two each in s_{2s}, s_{2s}^*, and s_{2p},) and 6 electrons in the pi molecular orbitals from second shell electrons (two each in two different p_{2p} orbitals, and one electron each in two different p_{2p}^*), so the net number of electrons in bonding orbitals is 4, giving a bond order (or number of 2).

In contrast it is easy to write the Lewis structure of singlet, excited state oxygen, since all electrons can be viewed as paired, with two net bonds (1 sigma, 1 pi) connecting the atoms of oxygen. This Lewis structure will be used to represent single, excited oxygen, which should react more quickly with organic molecules. The excited state single on the far right (below) is unstable and decays to the middle singlet state. The middle state is approximately 94.3 kJ/mol higher in energy than the ground state triplet (on the left). In quantum mechanical parlance, the transition from the ground state triplet to the singlet state is forbidden for a number of reasons, making it unlikely that absorption of a photon will induce the transition.

The Reductions of Dioxygen

When oxygen oxidizes organic molecules, it itself is reduced. By adding electrons one at a time to the molecular orbitals of ground state dioxygen we produce the step-wise reduction products of oxygen. On the addition of one electron, superoxide is formed.

DIOXYGEN REDUCTION PRODUCRS

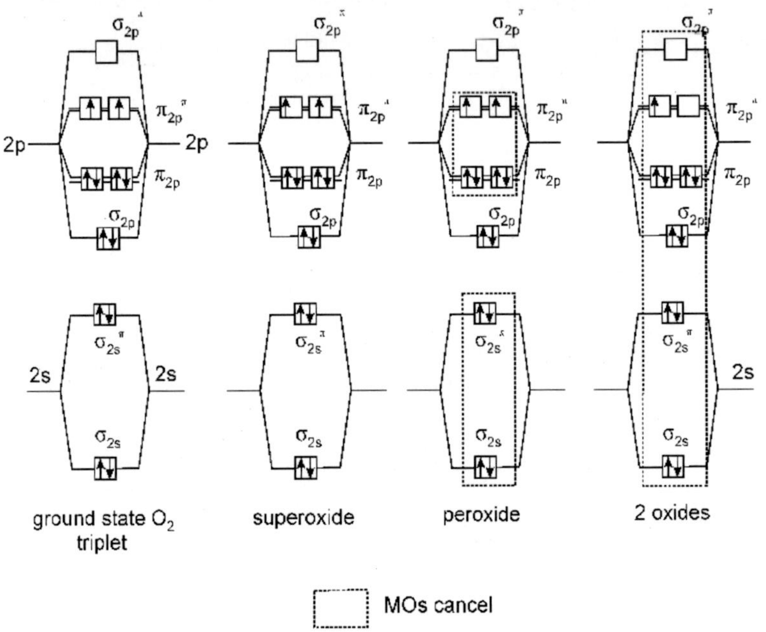

A second electron produces peroxide. Two more produces 2 separated oxides since no bonds connect the atoms (the number of electrons in antibonding and bonding orbitals are identical).

Each of these species can react with protons to produce species such as HO_2, H_2O_2 (hydrogen peroxide) and H_2O. It is the first two reactive reduction products of dioxygen that make it potentially toxic. How are the potential problems in oxygen chemistry dealt with biologically?

Kinetic Sluggishness

Enzymes that utilize dioxgen must activate it in some way, which decreases the activation energy. Enzymes that use dioxygen typically are metalloenzymes, and often heme-containing proteins.

Since metals such as Fe^{2+} and Cu^{2+} are themselves free radicals (i.e. they have unpaired electons), they react readily with ground state oxygen which itself is a radical. The molecular orbitals of the metal and oxygen combine to produce new orbitals which for oxygen are more singlet-like in nature.

Likewise, dioxygen reacts more readily with organic molecules which can themselves form reasonably stable free radicals, such as flavin adenine dinucleotide (FAD), as we shall see later.

Dioxygen Toxicity

Since toxicity arises from the reduction products of oxygen, enzymes that use oxygen have evolved to bind oxygen and its reduction products tightly (through metal-oxygen bonds) so they are not released into the cells where they can cause damage.

In addition, enzymes which detoxify free dioxygen reduction products are widely found in nature.

For example:

- superoxide dismutase catalyzes the dismutation (self-redox) of 2 superoxides into dioxygen and hydrogen peroxide;
- catalase converts hydrogen peroxide into water and oxygen;
- peroxidase catalyzes the reaction of hydrogen peroxide with an alcohol to form water and an aldehyde
- peroxiredoxins react with peroxides and thioredoxin (a small electron donor) to form water and oxidized thioredoxin.

Finally free radical scavengers such as vitamins A, C, E, and selenium can react with reactive free radicals to produce more stable free radical derivatives of the vitamins and Se. More on this later.

The Reactions of Dioxygen and its Reduction Products

1. *Triplet O_2 - Ground State:*

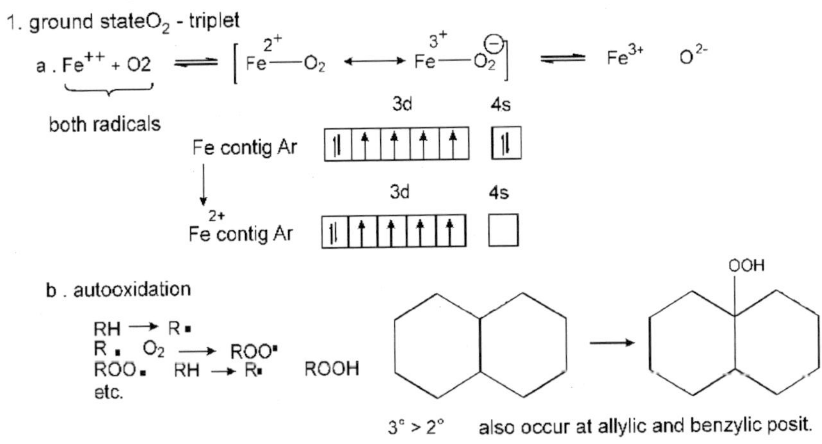

Figure: *Triplet O_2 - Ground State*

- Metals ions- Metal ions are radicals themselves, so can react with dioxygen. Ex:

  $Fe^{2+} + O_2$ <—> [Fe^{2+}— O_2 <—> Fe^{3+}— $O_2^{\cdot}$] <—> $Fe^{3+} + O_2^{\cdot}$ (superoxide)

- Autoxidation of organic molecules to produce peroxides. In this free radical reaction, several reactions occur, including

  RH —> R· (Initiation)

  R· + O_2 —> ROO· (Propagation)

  ROO· + RH —> R· + ROOH (Propagation)

  R· + R· —> R—R (Termination)

  ROO· + ROO· —> ROOR + O_2 (Termination)

  ROO· + R· —> ROOR (Termination)

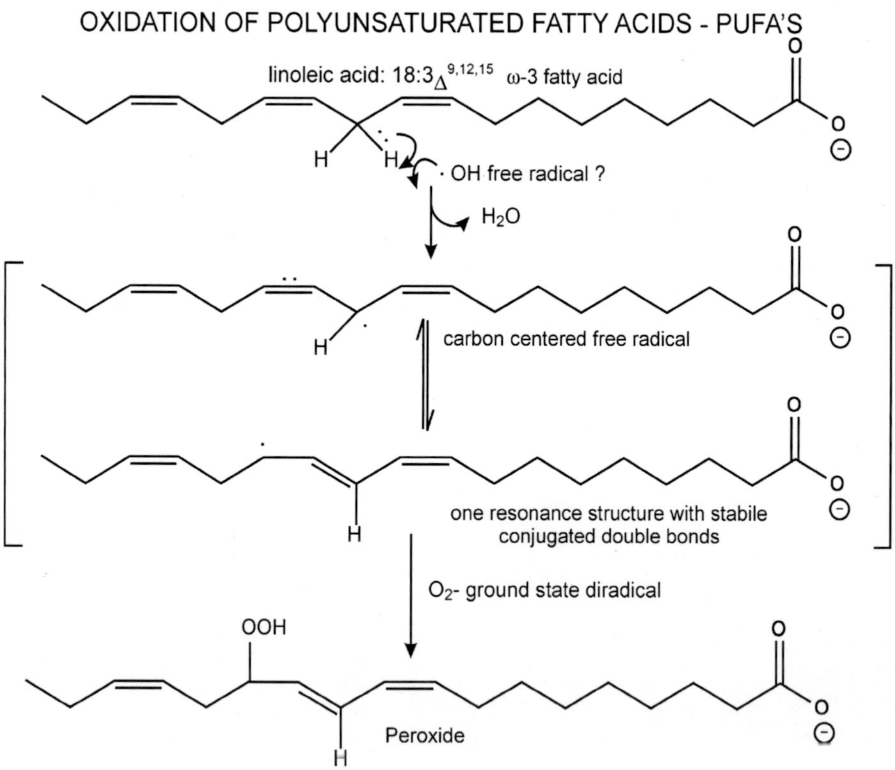

Figure: *unsaturated fatty acids are extra reactive at the methylene C that separate the double bonds.*

The initiation step above occurs mostly at C atoms which can produce the most stable free radicals (allylic, benzylic position, and 3° > 2° >> 1° carbons).

Hence unsaturated fatty acids are extra reactive at the methylene C that separate the double bonds.

2. *Single O_2 - Excited State.*

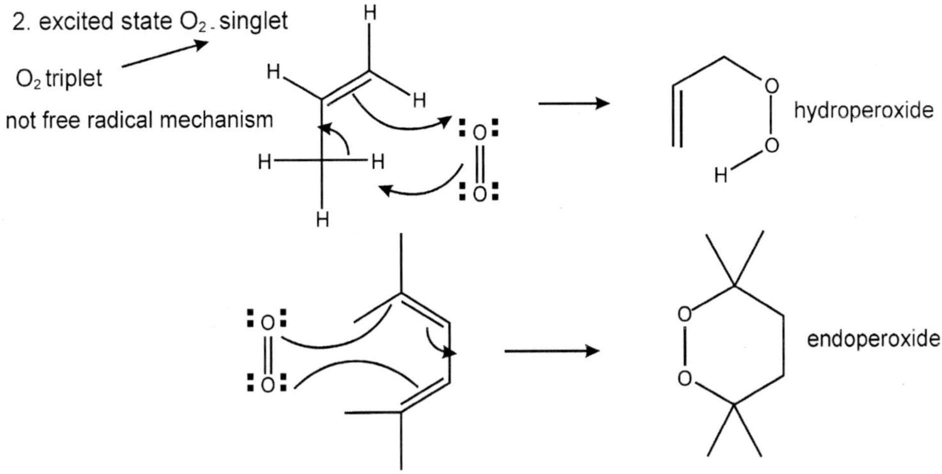

Figure: *Single O_2 - Excited State*

It can be made from triplet oxygen by photoexcitation. Alternatively, it can be made from triplet oxygen through collision with an excited molecule which relaxes to the ground state after a radiationless transfer of energy to triplet oxygen to form reactive singlet oxygen.

(This later process accounts for photobleaching of coloured clothes when the conjugated dye molecules absorb UV and Vis light, relax by transferring energy to triplet oxygen to form singlet oxygen, which then chemically reacts with the conjugated double bonds in the dye.)

- Alkenes react with oxygen to form hydroperoxides, potentially through a epoxide intermediate
- Dienes reacts with oxygen in a Diels-Alder like reaction to form endoperoxides

3. *Superoxide*

a. acid/base $O_2^- + H^+ \rightleftharpoons Ho_2^{\bullet}$ much more reactive

b. dismutase reaction $O_2^- + O_2^- + 2H^+ \rightleftharpoons H_2O_2 + O_2$ slow rx.

$HO_2^- + O_2^- + H^+ \rightleftharpoons H_2O_2 + O_2$ fast rx.

c. reduction $O_2^- + Fe^{3+} \rightleftharpoons O_2 + Fe^{3+}$

Figure: *Superoxide*

- Dismuation: $O_2^{\cdot-} + O_2^{\cdot-} \rightarrow H_2O_2 + O_2$
- Acid/Base: $HO_2^{\cdot} \longrightarrow O_2^{\cdot-} + H^+$ (pKa = 4.8)
- With metal ions: Fe^{3+} (as in heme) + $O_2^{\cdot-} \rightarrow O_2 + Fe^{2+}$

4. *Peroxide*

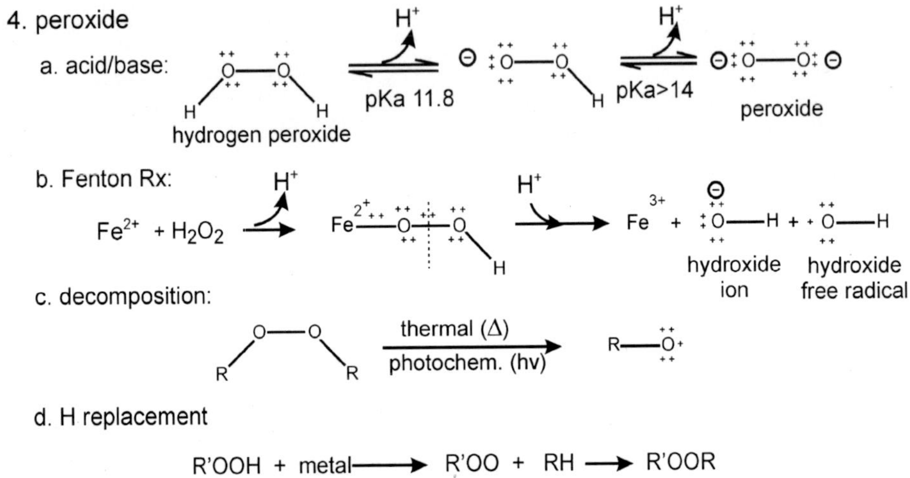

Figure: *Peroxide*

In contrast to dioxygen which contains multiple bonds between the O atoms, peroxide has only one bond. In fact, it is quite weak and requires only 38 kcal/mol to break it.

Remember, bonds can be broken in a heterolytic way (both electrons in a bond go to one of the atoms, or in a homolytic fashion, in which the one electron goes to each atom.

- Acid/Base: $H_2O_2 \rightarrow HO_2^- \rightarrow O_2^{2-}$ (pKa1 = 11.8; pKa2 > 14)
- Reaction with Fe^{2+} - The Fenton Reaction: (similar to reaction of triplet O_2 with Fe^{2+} above)

$Fe^{2+} + OOH^- \leftrightarrow Fe^{2+}—OOH \leftrightarrow Fe^{2+}—O + OH^- \leftrightarrow Fe^{3+}—O. \leftrightarrow Fe^{3+} + OH^{\cdot}$ (last step a proton is added). In this reaction, a homolytic cleavage of the O—O bond occurs generating OH^- and the hydroxy free radical, $OH^{\cdot}$, which will react with any molecule it encounters.

- thermal or photochemical homolytic cleavage of peroxide. This forms alkoxide free radicals which reacts like the hydroxy free radical.
- Reactions with alkyl groups in the presence of metal ions such as Cu, Co, or Mn:

  RH + R'OOH ——> ROOR'

5. Hydroxy free radical:

5. hydroxide free radical reactions extremely reactive. Will insert into molecules it finds!

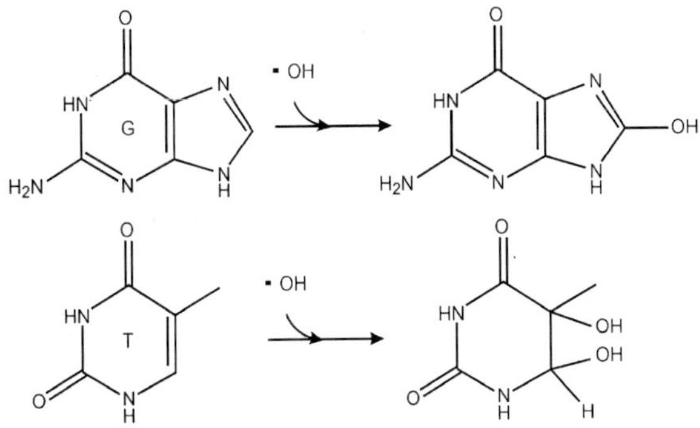

Figure: *Hydroxy free radical*

As mentioned above this species is extremely reactive. It will react with any molecule it encounters and does so immediately. It can abstract a H atom leaving another free radical. For example, the hydroxy free radical could extract a hydrogen atom from a polyunsaturated fatty acid to from a carbon-centered radical. A particularly nasty reaction is the insertion of the hydroxy radical into bases in DNA, as shown in the diagram.

Dioxygen Reduction Products and Disease

Significant evidence suggests oxygen free radicals are linked to aging and diseases. Mutations caused by hydroxylation reactions (presumably from the generation of hydroxyl free radicals as shown above) can potentially lead to cancer. Recently it has been shown that mitochondrial DNA is more susceptible to oxidation that is nuclear DNA.

Human mitochondria has its own small genome (16.5 Kb compared to the nuclear genome of 3 Gb) which code 13 protein subunits involved in respiration, 22 tRNAs and two ribosomal RNAs. (The mitochondria presumably are vestiges of a bacteria which invaded an early cell and established a symbiotic relationship with the cell).

A recent study has shown that there is an inverse correlation of oxidized mitochondrial DNA [8-oxoG] with maximal life span of an organism, but this correlation is not seen with nuclear DNA. Presumably the nuclear DNA is somewhat protected from oxidative damage since it is bound to histone proteins (which form nucleosome core particles with DNA) and by DNA repair enzymes, both of which are missing from the mitochondria.

Also, examination of human bladder, head and neck and lung primary tumours reveals a high frequency of mitochondrial DNA mutations. In addition most dioxygen use by the cell occurs in the mitochondria. Hence this organelle probably faces the highest concentration of toxic oxygen reduction products.

Recently, the crystal structure of an enzyme, adenine DNA glycosylase (MutY), that repairs 8-oxyG modified DNA has been determined in complex with the oxidatively damaged DNA. If not repaired, the 8-oxyG base pairs with adenine instead of cytosine, causing a GC to AT mutation on DNA replication.

Oxidative Modification of Proteins:

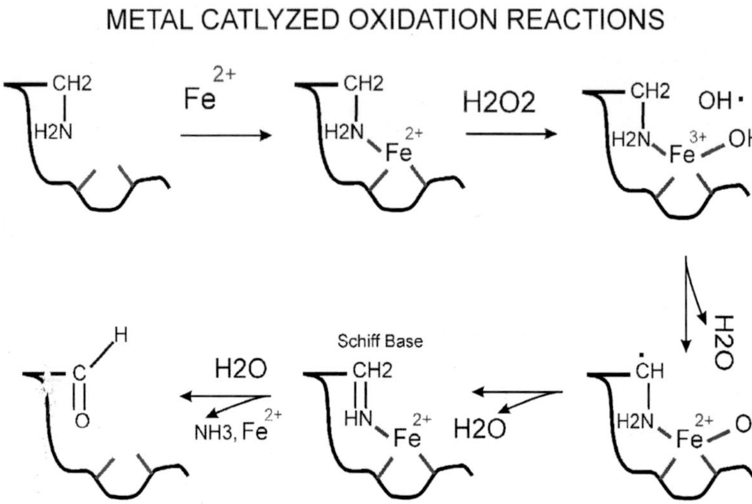

Figure: *Oxidative Modification of Proteins*

Oxidized levels of proteins (as evidenced by increased levels of aldehydes) increase dramatically with age (especially after age 40). The reactions seem to be catalysed by metals and may proceed by generation of hydroxy free radicals. Diseases associated with premature aging (Werner's Syndrome, , Progeria) show very high levels of oxidized proteins at an early age. Fibroblasts from 10 yr. old children with progeria have levels of oxidized proteins usually not seen until the age of 70. Beta amyloid protein deposits (found in Alzheimer's and Down's Syndrome) cause neurotoxicity and death, partly by increasing superoxide production by endothelial cells, causing vasoconstriction/dilation, and ultimately disease progression. Beta amyloid aggreagates appear to increase H_2O_2 levels, in a process facilitated by Fe^{2+} and Cu^+. Free radical scavengers (antioxidants) help to prevent this damage. Recent studies suggest that vitamin E may delay the symptoms of Alzheimers.

Lou Gehrigs Disease (Amyotrophic Lateral Schlerosis) is a disease of progressive motor neuron degeneration, which affects 1/100,000 people, and is 10-15% familial. Of the familial cases, about 25% have a mutation in superoxide dismutase I, a copper-zinc enzyme. About 2-3% of ALS patients carry 1 of 60 different dominant mutations in this enzyme. Mutations often decrease the stability of the protein which decreases Zn^{2+} affinity 5-50 fold. The A4V mutation (valine at amino acid 4 substituted for Ala) has the weakest Zn affinity and causes rapid disease progression. In the absence of Zn^{2+}, the apoprotein somehow seems to induce cell death in neurons. This superoxide dismutase also expresses a second activity. It also acts as a peroxidase which takes ROH + H_2O_2 to form an RHO (an aldehyde) plus water. In some cases, the enzyme retains normal activity against superoxide but altered peroxidase activity.

Is your hair going white?: Wood et al have shown that millimolar concentrations of hydrogren peroxide build up in hairs that have grayed and whitened. This was associated with a decrease in catalase and in increases in Met oxidation (to Met-sulfoxide) in proteins, also associated with a decrease in the repair enzyme Met-sulfoxide reductase, Met 374 in the active site of tyrosinase, an enzyme required for production of melanin in hair follicles, is also damaged, leading to lack of melanin, a pigment necessary for hair coloration and "senile hair graying".

ROS and Protein Folding

Protein Folding in Vitro and in Vivo, the cytoplasm has sufficient concentrations of "b-mercaptoethanol"-like molecules (used to reduce disulfide bonds in proteins *in vitro)* such as glutathione (g-Glu-Cys-Gly) and reduced thioredoxin (with an active site Cys) to prevent disulfide bond formation in cytoplasmic proteins. Disulfide bonds in proteins are typically found in extracellular proteins, where they serve to keep multisubunit proteins together as they become diluted in the extracellular milieu. These proteins destined for secretion are cotranslationally inserted into the endoplasmic reticulum which presents an oxidizing environment to the folding protein and where sugars are covalently attached to the folding protein and disulfide bonds are formed. Protein enzymes involved in disulfide bond formation contain free Cys which form mixed disulfides with their target substrate proteins. The enzymes (thiol-disulfide oxidoreductases, protein disulfide isomerases) have a Cys-XY-Cys motif and can promote disulfide bond formation or their reduction to free sulfhydryls. They are especially redox sensitive since their Cys side chains must cycle between and free disulfide forms.

Reactive oxygen species (ROS) can significantly affect redox chemistry, and if present in excess can place the cell in a condition of "oxidative" stress. ROS can indiscriminately oxidize lipids, nucleic acids, and proteins, but more specifically, they may also oxidize proteins involved in creating and maintaining the normal disulfide bond formation in proteins. As the concentration of ROS increase, the concentration of cytoplasmic proteins with incorrect disulfides should increase.

Using a two dimension PAGE system (first dimension run under nonreducing and the second reducing conditions) of neural cell proteins derived from cells exposed to normal and differing oxidative conditions (hydrogen peroxide or decreased intracellular glutathione levels, Cumming et al showed that oxidizing stress increased the levels of disulfide bonds in redox sensitive enzymes and, unexpectedly, among other cytoplasmic proteins involved in many aspects of life, effecting the activity of many cellular processes, suggesting that disulfide bond formation may have not only a structural but regulatory role.

Oxidative Modification of Lipids

The initial stages of cardiovascular disease appear to involve the development of fatty acid streaks under the artery walls. Macrophages, an immune cell, have receptors which appear to recognize oxidized lipoproteins in the blood, which they take-up. The cells then become fat-containing foam cells which form the streaks.

Oxidation of fatty acids in lipoproteins (possibly by ozone) could produce lipid peroxides and to protein oxidation in lipoproteins. Cortical neurons from fetal Down's Syndrome patients show 3-4 times levels of intracellular reactive O_2 species and increased levels of lipid peroxidation compared to control neurons. This damage is prevented by treatment of the neurons in culture with free radical scavengers or catalase.

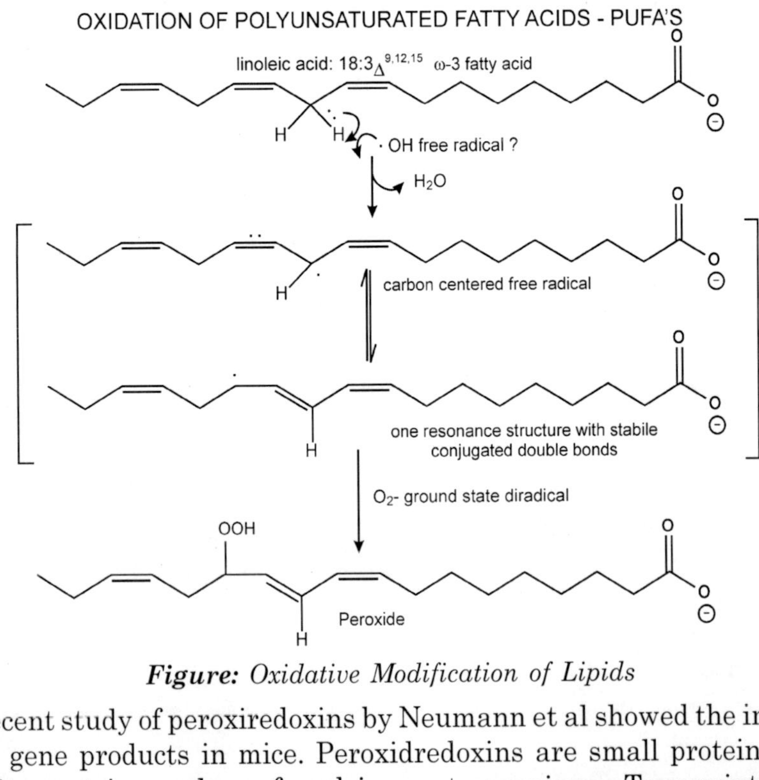

Figure: Oxidative Modification of Lipids

A recent study of peroxiredoxins by Neumann et al showed the importance of these gene products in mice. Peroxidredoxins are small proteins with an active site cysteine and are found in most organisms. Transcription of the mammalian peroxiredoxin 1 gene is activated by oxidative stress.

They inactivated the gene which produced a mouse that could reproduce and appeared vital, but which had a shortened lifespan. These mice developed severe hemolytic anemia and several types of cancers. High levels of reactive oxygen species and resulting increased levels of oxidized proteins were found in red blood cells of the knockout mice with anemia. High levels of 8-oxoguanine, resulting from oxidative damage to DNA, were found in tumour cells.

Aging

Oxidative damage to biomolecules might not initiate aging and disease processes, but rather might be a secondary effect of other initiating events.. Reversing or preventing oxidative damage might slow the progression of aging and disease. Aging is a complex feature of organisms and would be expected to have complex causes and biological effects. At the organismal level, aging has been studied in the round worm C. elegans which lives for only a few weeks. Genetic analyses can be easily used to find gene alterations associated with premature aging.

One hormonal system that has recently been associated with aging in eukaryotes (and in C. elegans) involves the signalling pathways for insulin and insulin growth factor I (IGF-1), which regulate carbohydrate, lipid, and reproductive pathways in C. elegans. Mutations that decrease signalling from this pathway increase C. elegans life span.

These mutations lead to increased activity of the DAF16 transcription factor, which upregulates the expression of many genes. In contrast, wild type organisms, when exposed to insulin or IGF-1, decrease the activity of DAF16. Using DNA microarrays, investigators determined which DAF16-controlled genes were upregulated in mutant worms in the mid-life point of the organism.

These genes included, among others, peroxisomal and cytosolic catalase, Mn-superoxide dismutase, cytochrome P450s, metallothionein-related Cd-binding protein, and heat shock proteins. We will investigate the function of several of these gene products in the next section, but needless to say, they are all involved in cellular responses to stress, often involving dioxygen metabolites. Over expression of mitochondrial catalase in mice increased their lifespan by 20%.

It has also been showed that decreased levels of insulin-like growth factor also promote longevity in mice, indicating again that mechanisms in addition to oxidative damage by ROS are involved in aging.

Although oxidative damage in mitochondria clearly can promote premature aging, other independent mechanisms may also. In a recent study, Kujoth et al. developed a mouse model that expressed a mutant form of mitochondrial DNA polymerase that was defective in the proofreading activity of the enzyme.

These mice displayed premature aging but showed no increased levels of oxidized mitochondrial lipids or hydroxylated G residues in mitochondrial DNA. They did show significant activation of a cytosolic enzyme called caspase-3, which when active lead to the programmed death of cells (a process called apoptosis). This calcium-activated aspartic acid proteases (with an active site Asp) is activated by binding mitochondrial cytochrome C that has "leaked" into the cytoplasm from its normal location in the intermembrane space in mitochondria. The process is usually associated with DNA damage (mutations, fragmentation) that would arise if the proofreading function of DNA polymerase was defective. This was indeed found in these mice.

Biological Uses for Reactive Oxygen Species- ROS

ROS can be used in the body as defence mechanisms. The can be generated by immune cells that phagocytize (engulf) bacteria and other foreign substances.

After uptake, these immune cells undergo an oxidative burst, increasing the amount of dioxygen that they consume, leading to production of superoxide and hydrogen peroxide which can be used to kill the engulfed bacteria. Recent articles by Wentworth et al. go even farther.

They demonstrated that antibody molecules (protein which bind to foreign molecules and target them for clearance or further immune response) can also generate ROS when they bind to their target. Such an outcome was totally unexpected, but not inconsistent with the observation that antibodies can have catalytic activity if made against transition state analogs of substrates. Over 100 different antibodies were found to generate hydrogen peroxide through the reaction of singlet oxygen (possibly generated by phagocytotic neutrophils during oxidative bursts) with water to form H_2O_2. (We have previously discussed the idea that singlet O_2 can be generated from ground state O_2 by excitation through UV light or through collisional activation with an excited state chromophore-a conjugated alkene or aromatic molecule).

They postulate the following reaction to form hydrogen peroxide. The oxygen in the peroxide comes from water, as shown by labelling water with ^{18}O:

$$2H_2O + {}^1O_2 \text{ (singlet oxygen)} \longrightarrow H_2O_3 + H_2O \longrightarrow H_2O_2 + ?$$

Antibodies show this effect to a much greater extent that other proteins studied. Using UV irradiation as a source of energy, the rate of production of H_2O_2 was linear with time, but decreased from the initial rate as H_2O_2 concentrations increased.

If it was removed with catalase, the initial rate returned to normal, suggesting that H_2O_2 was a product inhibitor of the reaction. The reactions also saturated with dioxygen as well. The maximal wavelength of UV irradiation to produce H_2O_2 was identical to the maximal wavelength of absorbance for tryptophan in the protein. The excited state tryptophan presumably deexcites by energy transfer to triplet dioxygen ($^3O_2$) in much the manner described earlier for photobleaching. A significant source of electrons needed to be found to account for the production of so much H_2O_2 which if it was made from singlet oxygen would require 2 electrons for each H_2O_2. These numbers of electrons could not come from photooxidation of aromatic amino acids in the protein. Hence they theorized that water would add as a nucleophile to the singlet oxygen, providing the source of electrons required to make H_2O_2. This reactions would go through the H_2O_3 intermediate.

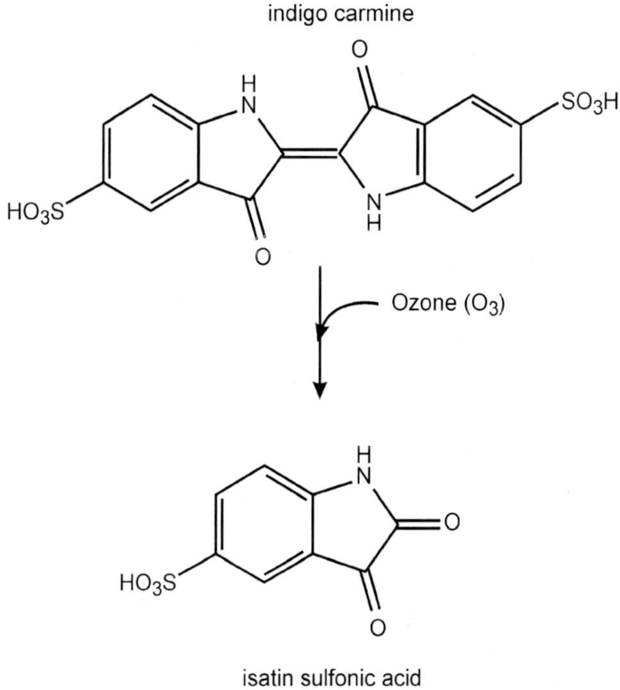

Figure: *Conversion of indigo carmen to isatin sulfonic acid*

In another recent study (Nov. 02), the same team discovered that antibodies can not only produce H_2O_2 but also ozone. O_3 is used in killing pathogens but also might lead to inflammation.

In their previous work described above, they generated singlet oxygen in a "nonphysiological" way-through UV irradiation of the antibodies. They also didn't show that the ROS produced can kill bacteria without any other immune mediators. They used a new singlet oxygen generating system which led to insufficient peroxide formation to account for the extensive killing of E. Coli that they observed (95%). This led them to look for the generation of a more potent oxidizing agent-which they believe is consistent with the formation of ozone. They also showed that neutrophils were the source of the required singlet oxygen as described above.

STEROL OXIDATION PRODUCTS OF CHOLESTEROL

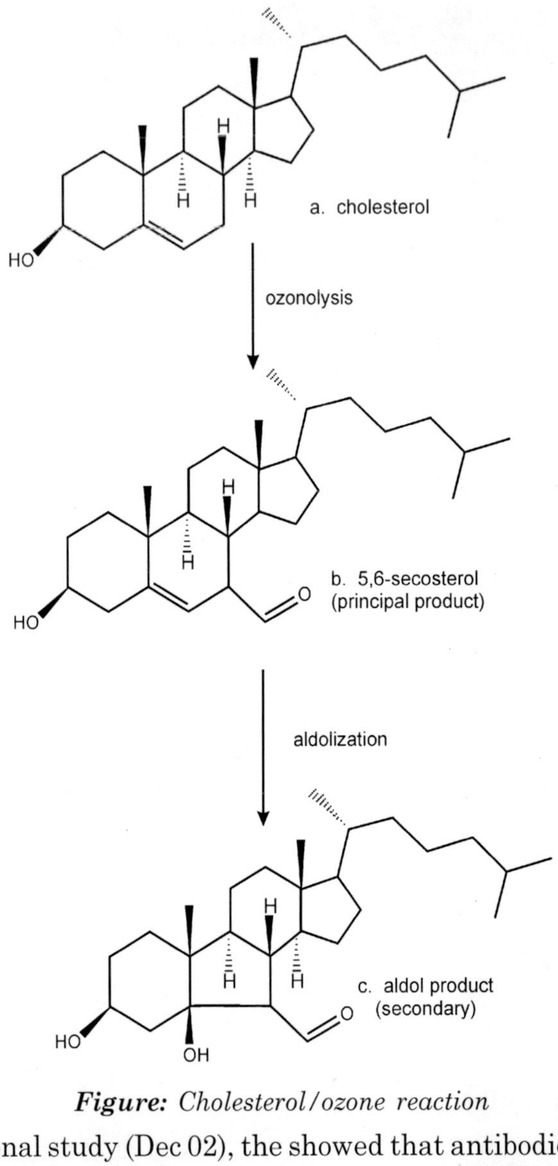

Figure: *Cholesterol/ozone reaction*

In an additional study (Dec 02), the showed that antibodies in the presence of singlet oxygen and in the absence of immune cells or other immune proteins (like complement proteins) can "destroy antigenic target". They also confirmed that ozone is produced by antibodies during bacterial killing by neutrophils (which produce sinlget oxygen) and in inflammatory states. Signalling through ozone might amplify other immune molecules that signal the activation (or hyperactivation) of the immune system. Their latest study (Nov. 03) examined the involvement of ozone (generated from antibodies in the presence of singlet O_2 from neutrophils) in the oxidation of lipoproteins, a process which has been associated with atherosclerosis in coronary arteries. First they tested if ozone could be generated by atherosclerotic lesions treated with phorbol myristate acetate. 14/15 plaques produced ozone, as evidenced by the photobleaching of the dye indigo carmen to isatin sulfonic acid. This conversion, known to be specific for ozone.

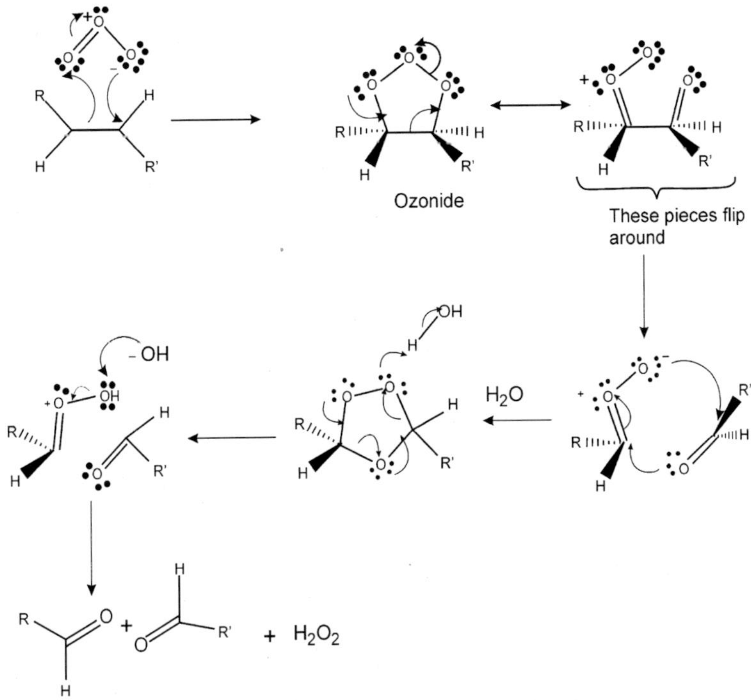

Figure: *Ozonolysis mechanism.*

They then looked for the telltale signature of the reaction of ozone with plaque cholesterol (cleavage of the D5,6 double bond to form 5,6-secosterol as shown below and found it. This reaction is known not to occur with other oxygen oxidants, including both triplet and singlet oxygen, superoxide, and the hydroxyl free radical.

A secondary reaction product aldol condensation product was also produced. They named these new products atheronals.

Significant increases in secosterol was found in plaques treated with the phorbal ester. 6 of 8 patients with atherosclerosis had elevated plasma levels of the secosterol, while only 1 of 15 of control patients did.

They also showed that these ozone products caused normal macrophages to fill with cholesterol (presumably from oxidized LDL), and that the structure of the LDL apoprotein B100 changed.

These findings might provide a link between the immune system and cholesterol in the generation of cardiovascular disease.

Just Say NO- The Chemistry of Nitric Oxide

In the last decade, the role of another gaseous free radical, nitric oxide (·NO) has become apparent. This molecule is synthesized biologically through the action of an inducible heme enzyme, nitric oxide synthase, which forms NO by the oxidation by dioxygen of the guanidino group of Arg, which gets converted to citrulline. .NO is soluble and can diffuse through cell membranes into the cytoplasm, where it has a myriad of effects in signal transduction pathways. However, it can also be metabolized to form reactive nitrogen intermediates (much as with dioxygen) which can be deleterious to the body, if they damage native biomolecules, or advantageous, when they are used by immune cells like macrophages in the destruction of engulfed bacteria.

MOLECULAR ORBITALS NITRIC OXIDE - NO

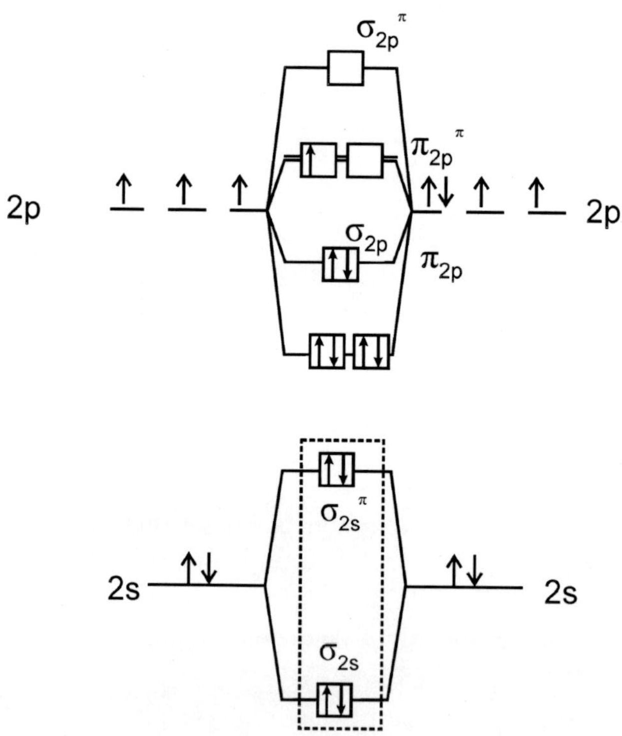

ground state NO

Figure: *Molecular Orbital Diagram of NO*

A molecular orbital diagram of.NO shows it to have a bond order of 2.5 and one unpaired electron in a p2p* antibonding orbital (hence the notation.NO). Here are some of the relevant reactions of NO and its reactive nitrogen intermediates (RNI):

- The high energy electron p2p*can leave on oxidation of NO to form the nitrosyl ion, NO^+. The bond order in NO^+ is three suggesting strong interactions between the N and O atoms.
- NO (oxidation number of N of +2) is thermodynamically unstable and can disproportionate or dismutate (like superoxide) in a self redox reaction to form nitrous oxide (N_2O) and nitrogen dioxide ($\cdot NO_2$), with oxidation number for N of +1 and +4, respectively.
- In acidic condition.NO can be oxidized to nitrite (NO_2^-) which can be protonated to form nitrous acid (HNO_2).
- Nitrous acid can disproportionate or dismutate to form.NO and the radical.NO_2 which can be protonated to ONOOH with a pKa of 6.5.
- NO can react with superoxide ($\cdot O_2^-$) faster than any other molecule to form peroxynitrite, ONOO-. .NO and superoxide ($\cdot O_2^-$) are both produced by cells during infllammation, leading to increased levels of peroxynitrite.
- The O-OH bond in protonated peroxynitrite (ONO-OH) can be cleaved either heterolytically to produce NO_3^- and H^+ (70%) or homolytically to produce the radical. NO_2 and the highly reactive. OH free radical (30%).
- The deprotonated form of peroxynitrite ($ONOO^-$) can oxidize organic molecules, especially thiols and CO_2. When it reacts with CO_2, it can form either NO_3^- and CO_2 (65%) or $CO_3^{\cdot-}$ and NO_2 (35%). These latter products can oxidize organic molecules in one electron steps. Peroxynitrite can also oxidize protein, DNA, and lipids.
- NO_2 and peroxynitrite can interact with Tyr side chains in proteins to form nitrated proteins (Tyr-NO_2), whose presence reflects inflammation.

Macrophages make use of RNIs in the killing of engulfed bacteria. How does one of humankinds greatest foes, Mycobacteria tuberculosis, the causative agent of tuberculosis, avoid this killing mechanism? It is estimated that the bacteria resides persistently in latent form in 2 billion people. If it becomes activated, it becomes one of the greatest killers. Consider the following table:

Killer Diseases through time (Scientist, June 2, 2003)

Historic Pandemics	**cases**	**deaths**
Justinian Plaque (6th centr.)	142 million (on 70% mort)	about 100 mill
China Plaque (Bubonic) 3rd Pandemic (1896-1930)	30 milion	12 milion
Spanish flu 1918-19	1 billion	21 million

Pandemics Today

Pandemics Today	***cases/yr***	***deaths/yr***
Malaria	300-500 mil	1 million
TB	8 mill	2 mill
AIDS	6 million	3 mill

People infected with the bacteria but without clinical symptoms must mount a sufficient enough immune response to restrain proliferation of the bacteria, but not enough to clear it from the body.

Immune-compromised people (transplant recipients taking anti-rejection drugs or AIDS patients) have more difficulty in holding the bacteria at bay. One immune response mediator which restrains bacterial growth is the soluble protein interferon g (a protein cytokine released by immune cells). This protein induces synthesis of nitric oxide synthase, producing. NO, which through the reactions listed above, can damage bacterial macromolecules.

Bacteria are "digested" in acidic phagosomes of the macrophage. Nitrite formed in the acid conditions from. NO (generated by inducible NO synthase) forms.NO_2 and peroxynitrite which have antibacterial properties (better than anti-Tb drugs).

Mutant mice that can not synthesize inducible NO synthase have little defence against the bacteria. Darwin e al. exposed the bacteria to sources of nitrite at pH conditions typical of macrophage phagosomes and found mutants sensitive to the RNIs.

The mutations involved protein of the bacterial proteasome, a complex multi-protein complex which proteolyzes unwanted (presumably damaged) cellular proteins. Bacterial genes encoding proteins associated with the bacterial proteasome seem to confer resistant to the effects of macrophage-inducted RNI production.

The macrophage proteasome, like other eukaryotic proteasomes, is a cytoplasmic protein complex which degrades damaged cytoplasmic proteins. Although the mechanism is uncertain, the bacterial proteasome may rid the bacteria of nitrated or otherwise oxidized (damaged) proteins or remove the nitrate and facilitate refolding of the damaged protein.

Antioxidant Protection From Oxidative Modification

If oxidative damage by dioxygen reduction products can cause disease, maybe antioxidant vitamins (E, C, A), which can form reasonably stable free radical, can protect the body from their effects.

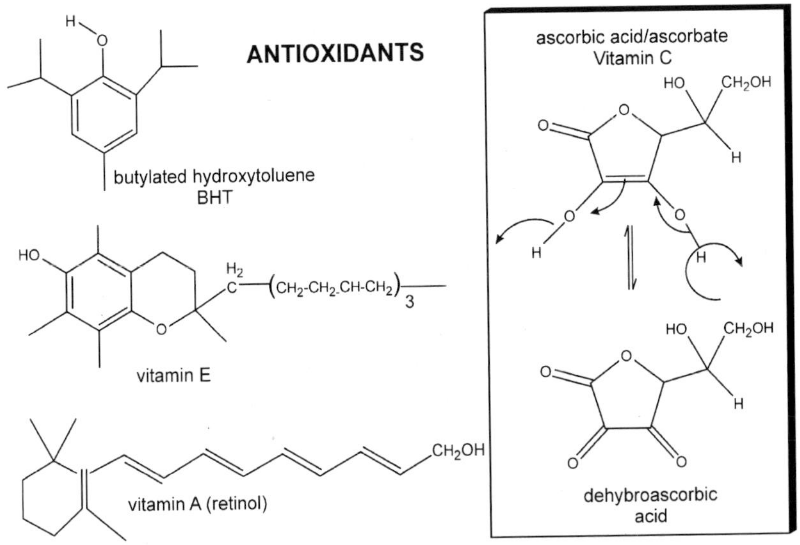

Figure: *Antioxidant vitamins (E, C, A).*

It has been shown that these vitamins can help protect white blood cells from DNA damage arising from hydroxy free radicals. Vitamin E, a fat-soluble vitamin carried in circulating lipoproteins, can reduce the risk of cardiovascular disease, presumably by preventing oxidation of lipids and proteins in lipoproteins. Vitamin E and C, along with a common food additive, butylated hydroxytoluene (BHT) can form stable free radicals (formed possibly by abstraction of a hydrogen atom by hydroxyl free radicals) since the lone electron is stabilized by resonance and the O-centered resonant form is sterically restricted from intermolecular interactions which could propagate the free radical chain reactions.

How do Cells Respond to Low O_2 ?

We have spend much time studying how the body deals with and utilizes the toxic byproducts of dioxygen reduction. What happens when the body doesn't get enough dioxygen- a condition call anoxia (no dioxygen) or hypoxia (too little dioxygen)? This might occur in muscles undergoing vigorous exercise, and in the brain and heart when clots occlude blood flow to these organs (as occurs in most strokes and heart attacks). Under low dioxygen concentrations, a family of protein transcription factors called hypoxia-inducible factor (HIF) become activated.

The functional proteins appears to be a dimer of HIF-a and HIF-b. In contrast to the concentration of the beta form, the activity of HIF-a is increased under low dioxygen concentrations. HIF-a concentration is regulated not at the transcriptional level, but through proteolysis of the protein. In the presence of abundant dioxygen, HIF-a is hydroxylated at two Pro residues by the enzyme prolyl hydroxlase. This post-translational modification targets the protein for proteolysis (through ubiquitination by the VHL protein and subsequent cleavage by the proteasome). A second independent pathway in rapidly growing tissue leads to increased expression of HIF-a even in the presence of dioxygen. This could be beneficial to cells since rapidly growing tissue, especially tumour cells, might be expected to experience low oxygen conditions.

Biological and Synthetic Dioxygen Carriers

Most organisms require molecular oxygen in order to survive. The dioxygen is used in a host of biochemical transformations, although most is consumed in the reaction that is the terminal (or primary) step of oxidative phosphorylation. For some small animals and for plants, where the surface-to-volume ratio is large, an adequate supply of dioxygen can be obtained from simple diffusion across cell membranes. The dioxygen may be extracted from air or water; for plants that produce dioxygen in photosynthesis, it is also available endogenously. For other organisms, particularly those with non-passive lifestyles, from scorpions to whales, diffusion does not supply sufficient dioxygen for respiration.

An elegant three-component system has evolved to transport dioxygen from regions of high abundance-water (at least if free of pollutant reductants) and air-to regions of relatively low abundance and high demand-the interior cells of the organism. The central component is a dioxygen-carrier protein. In the three chemically distinct carriers that have evolved and are found today, the dioxygen-binding site in the protein, that is, the so-called "active site," is a complex either of copper or of iron. For hemoglobins, the most widely distributed family of dioxygen carriers, the active site has long been known to consist of an iron porphyrin (heme) group embedded in the protein. Almost all hemoglobins share the basic structure. Hemocyanin and hemerythrin, the other two biological dioxygen carriers, feature pairs of copper atoms and iron atoms, respectively, at the active sites.

The second component of the dioxygen-transport system facilitates the sequestration of dioxygen by the dioxygen-carrier protein. Specialized organs, such as lungs in air-breathing creatures or gills in fish, offer a very large surface area to the outside environment in order to facilitate diffusion. The third component is the delivery system. The oxygen carrier is dissolved or suspended in a fluid, called blood plasma or hemolymph, that is pumped throughout the animal by another specialized organ, the heart, through a network of tubes, the blood vessels.

In many organisms an additional dioxygen-binding protein, which stores dioxygen, is located in tissues that are subject to sudden and high dioxygen demand, such as muscles. These dioxygen-storage proteins are prefixed *myo*(from the Greek root *mys* for muscle). Thus for the dioxygen-transport protein hemerythrin there exists a chemically similar dioxygen-storage protein myohemerythrin. For the hemoglobin family the corresponding storage protein is called myoglobin. Interestingly, some organisms that use hemocyanin as the dioxygen*transport* protein use myoglobin as the *dioxygen-storage* protein.

At the centre of biological dioxygen transport are transition-metal complexes of iron or copper. To model such systems, chemists have prepared several synthetic oxygen carriers, especially of iron and cobalt porphyrins. In this chapter the structures and properties of biological and nonbiological oxygen carriers are described, with particular attention to the hemoglobin family. This family has been studied in more detail than any other group of proteins, and as a result a deeper understanding of the relationships among structure, properties, and biological function (i.e., physiology) exists. The central focus of this chapter is to delineate chemical features that determine the affinity of an active site, especially an iron porphyrin, for molecular oxygen. In order to develop this theme, macroscopic (thermodynamic and kinetic) factors associated with dioxygen binding and release are summarized first. The nonbiological chemistry of iron and copper in the presence of dioxygen is described briefly to elucidate the key role that the protein plays in supporting oxygen transport by preventing irreversible oxidation of the binding site or of its ligands.

The macroscopic behaviour of the biological systems is related to the microscopic picture that has been developed over the last 30 years from x-ray crystallographic studies and a miscellany of spectroscopic probes of the oxygen-binding site. Relationships between the geometry and charge distribution in the metal-dioxygen moiety and the nature of the interactions between this moiety and its surroundings are examined.

Nonbiological dioxygen carriers have proved particularly useful in providing precise and accurate structural information as well as thermodynamic and kinetic data against which the corresponding data from biological oxygen carriers can be contrasted.

The bioinorganic chemistry of the hemoglobin family of oxygen binders is particularly amenable to study by means of small-molecule model systems: four of the five ligands that make up the active site are provided by a square-planar tetradentate ligand, the protoporphyrin IX dianion.

One axial ligand in hemoglobin, imidazole from a histidine residue, is provided by the protein, and the remaining sixth coordination site is available for the exogenous ligand, e.g., dioxygen or carbon monoxide. Thus a model system that approximates the stereochemistry of the active site in hemoglobin may be assembled from an iron(II) porphyrin and a ligand, such as imidazole or pyridine. On the other hand, in hemocyanin and hemerythrin most of the ligands are supplied by the protein.

Thus the assembly of a model system that provides appropriate ligands correctly disposed around the pair of metal atoms poses a major synthetic challenge, especially for hemocyanin, where details on the number, type, and arrangement of ligands have been difficult to establish. Many aspects of the physical, inorganic, and structural chemistry underlying biological oxygen transport and utilization have been clarified through model systems.

Requirements for Effective Oxygen Carriers

In order for dioxygen transport to be more efficient than simple diffusion through cell membranes and fluids, it is not sufficient that a metalloprotein merely binds dioxygen. Not only is there an optimal affinity of the carrier for dioxygen, but also, and more importantly, the carrier must bind *and* release dioxygen at a rapid rate.

Oxygen Cycle

The Oxygen cycle is the biogeochemical cycle that describes the movement of oxygen within its three main reservoirs: the atmosphere (air), the total content of biological matter within the biosphere (the global sum of all ecosystems), and the lithosphere (Earth's crust). Failures in the oxygen cycle within the hydrosphere (the combined mass of water found on, under, and over the surface of a planet) can result in the development of hypoxic zones. The main driving factor of the oxygen cycle is photo-synthesis, which is responsible for the modern Earth's atmosphere and life.

Reservoirs

By far the largest reservoir of Earth's oxygen is within the silicate and oxide minerals of the crust and mantle (99.5%). Only a small portion has been released as free oxygen to the biosphere (0.01%) and atmosphere (0.36%). The main source of atmospheric oxygen is photosynthesis, which produces sugars and oxygen from carbon dioxide and water:-

$$6CO_2 + 6H_2O + \text{energy} \rightarrow C_6H_{12}O_6 + 6O_2$$

Photosynthesizing organisms include the plant life of the land areas as well as the phytoplankton of the oceans. The tiny marine cyanobacterium Prochlorococcus was discovered in 1986 and accounts for more than half of the photosynthesis of the open ocean. An additional source of atmospheric oxygen comes from photolysis, whereby high energy ultraviolet radiation breaks down atmospheric water and nitrous oxide into component atoms. The free H and N atoms escape into space leaving O_2 in the atmosphere:

$$2H_2O + \text{energy} \rightarrow 4H + O_2$$

$$2N_2O + \text{energy} \rightarrow 4N + O_2$$

The main way oxygen is lost from the atmosphere is via respiration and decay, mechanisms in which animal life and bacteria consume oxygen and release carbon dioxide. Because lithospheric minerals are oxidised in oxygen, chemical weathering of exposed rocks also consumes oxygen. An example of surface weathering chemistry is formation of iron-oxides (rust):

$$4FeO + O_2 \rightarrow 2Fe_2O_3$$

Oxygen is also cycled between the biosphere and lithosphere. Marine organisms in the biosphere create calcium carbonate shell material ($CaCO_3$) that is rich in oxygen. When the organism dies its shell is deposited on the shallow sea floor and buried over time to create the limestone rock of the lithosphere. Weathering processes initiated by organisms can also free oxygen from the lithosphere. Plants and animals extract nutrient minerals from rocks and release oxygen in the process.

Capacities and Fluxes

The following tables offer estimates of oxygen cycle reservoir capacities and fluxes. These numbers are based primarily on estimates from (Walker, J.C.G.) :

***Table 1:** Major reservoirs involved in the oxygen cycle.*

Reservoir	*Capacity ($kg\ O_2$)*	*Flux In/Out ($kg\ O_2$ per year)*	*Residence Time (years)*
Atmosphere	$1.4 * 10^{18}$	$30,000 * 10^{10}$	4,500
Biosphere	$1.6 * 10^{16}$	$30,000 * 10^{10}$	50
Lithosphere	$2.9 * 10^{20}$	$60 * 10^{10}$	500,000,000

***Table 2:** Annual gain and loss of atmospheric oxygen (Units of 10^{10} kg O_2 per year)*

Photosynthesis (land)	16,500
Photosynthesis (ocean)	13,500
Photolysis of N_2O	1.3
Photolysis of H_2O	0.03
Total Gains	~ 30,000
Losses- Respiration and Decay	
Aerobic Respiration	23,000
Microbial Oxidation	5,100
Combustion of Fossil Fuel (anthropogenic)	1,200
Photochemical Oxidation	600
Fixation of N_2 by Lightning	12
Fixation of N_2 by Industry (anthropogenic)	10
Oxidation of Volcanic Gases	5
Losses- Weathering	
Chemical Weathering	50
Surface Reaction of O_3	12
Total Losses	~ 30,000

Ozone

The presence of atmospheric oxygen has led to the formation of ozone(O_3) and the ozone layer within the stratosphere. The ozone layer is extremely important to modern life as it absorbs harmful ultraviolet radiation:

$$O_2 + \text{uv energy} \rightarrow 2O$$

$$O + O_2 \rightarrow O_3$$

Apparent Oxygen Utilisation

In freshwater or marine systems apparent oxygen utilisation (AOU) is the difference between the measured dissolved oxygen concentration and its equilibrium saturation concentration in water with the same physical and chemical properties. Such differences typically occur when biological activity acts to change the ambient concentration of oxygen. For example, primary production liberates oxygen and increases its concentration, while respiration consumes it and decreases its concentration.

Consequently, the AOU of a water sample represents the sum of the biological activity that the sample has experienced since it was last in equilibrium with the atmosphere. In shallow water systems (e.g. lakes), the full water column is generally in close contact with the atmosphere, so oxygen concentrations are typically close to saturation. In deep water systems (e.g. oceans), water can be out of contact with the atmosphere for extremely long periods of time (years, decades, centuries) and large AOU values are possible.

Oxygen–haemoglobin Dissociation Curve

The oxygen–haemoglobin dissociation curve (or oxygen–hemoglobin dissociation curve) plots the proportion of haemoglobin in its saturated form on the vertical axis against the prevailing oxygen tension on the horizontal axis. The oxyhaemoglobin dissociation curve is an important tool for understanding how our blood carries and releases oxygen. Specifically, the oxyhaemoglobin dissociation curve relates oxygen saturation (SO_2) and partial pressure of oxygen in the blood (PO_2), and is determined by what is called "haemoglobin's affinity for oxygen"; that is, how readily haemoglobin acquires and releases oxygen molecules into the fluid that surrounds it.

Background

Haemoglobin (Hb), a globular protein, is the primary vehicle for transporting oxygen in the blood. Oxygen is also carried dissolved in the blood's plasma, but to a much lesser degree. Haemoglobin is contained in erythrocytes, more commonly referred to as red blood cells. Under certain conditions, oxygen bound to the haemoglobin is released into the blood's plasma and absorbed into the tissues. Each haemoglobin molecule has the capacity to carry four oxygen molecules. How much of that capacity is filled by oxygen at any time is called the oxygen saturation. Expressed as a percentage, the oxygen saturation is the ratio of the amount of oxygen bound to the haemoglobin, to the oxygen-carrying capacity of the haemoglobin. The oxygen-carrying capacity of haemoglobin is determined by the type of haemoglobin present in the blood. The amount of oxygen bound to the haemoglobin at any time is related, in large part, to the partial pressure of oxygen to which the haemoglobin is exposed. In the lungs, at the alveolar–capillary interface, the partial pressure of oxygen is typically high, and therefore the oxygen binds readily to haemoglobin that is present. As the blood circulates to other body tissue in which the partial pressure of oxygen is less, the haemoglobin releases the oxygen into the tissue because the haemoglobin cannot maintain its full bound capacity of oxygen in the presence of lower oxygen partial pressures

Sigmoidal Shape

It is usually a sigmoid plot. A haemoglobin molecule can bind up to four oxygen molecules in a reversible way. The shape of the curve results from the interaction of bound oxygen molecules with incoming molecules. The binding of the first molecule is difficult. However, this facilitates the binding of the second and third molecules, and it is only when the fourth molecule is to be bound that the difficulty increases, partly as a result of crowding of the haemoglobin molecule, partly as a natural tendency of oxygen to dissociate.

In its most simple form, the oxyhaemoglobin dissociation curve describes the relation between the partial pressure of oxygen (x axis) and the oxygen saturation (y axis). Haemoglobin's affinity for oxygen increases as successive molecules of oxygen bind. More molecules bind as the oxygen partial pressure increases until the maximum amount that can be bound is reached. As this limit is approached, very little additional binding occurs and the curve levels out as the haemoglobin becomes saturated with oxygen. Hence the curve has a sigmoidal or S-shape. At pressures above about 60 mmHg, the standard dissociation curve is relatively flat, which means that the oxygen content of the blood does not change significantly even with large increases in the oxygen partial pressure. To get more oxygen to the tissue would require blood transfusions to increase the haemoglobin count (and hence the oxygen-carrying capacity), or supplemental oxygen that would increase the oxygen dissolved in plasma.

Although binding of oxygen to haemoglobin continues to some extent for pressures about 50 mmHg, as oxygen partial pressures decrease in this steep area of the curve, the oxygen is unloaded to peripheral tissue readily as the haemoglobin's affinity diminishes. The partial pressure of oxygen in the blood at which the haemoglobin is 50% saturated, typically about 26.6 mmHg for a healthy person, is known as the P50. The P50 is a conventional measure of haemoglobin affinity for oxygen. In the presence of disease or other conditions that change the haemoglobin's oxygen affinity and, consequently, shift the curve to the right or left, the P50 changes accordingly. An increased P50 indicates a rightward shift of the standard curve, which means that a larger partial pressure is necessary to maintain a 50% oxygen saturation. This indicates a decreased affinity. Conversely, a lower P50 indicates a leftward shift and a higher affinity.

The 'plateau' portion of the oxyhaemoglobin dissociation curve is the range that exists at the pulmonary capillaries (minimal reduction of oxygen transported until the $p(O_2)$ falls 50 mmHg). The 'steep' portion of the oxyhaemoglobin dissociation curve is the range that exists at the systemic capillaries (a small drop in systemic capillary $p(O_2)$ can result in the release of large amounts of oxygen for the metabolically active cells).

To see the relative affinities of each successive oxygen as you remove/add oxygens from/to the haemoglobin from the curve compare the relative increase/decrease in $p(O_2)$ needed for the corresponding increase/decrease in $s(O_2)$.

Factors Shifting Curve

Many factors influence the affinity of this binding and alter the shape of the curve:

	right shift ***(low affinity for O_2)***	***left shift*** ***(high affinity for O_2)***
temperature	high	low
2.3-DPG	high	low
$p(CO_2)$	high	low
p(CO)	low	high
pH (Bohr effect)	low (acidosis)	high (alkalosis)m
type of haemoglobin	adult haemoglobin	foetal haemoglobin

Left shift of the curve is a sign of haemoglobin's increased affinity for oxygen (e.g. at the lungs). Similarly, right shift shows decreased affinity, as would appear with an increase in body temperature, hydrogen ion, 2,3-diphosphoglycerate (also known as bisphosphoglycerate) or carbon dioxide concentration (the Bohr effect)

Carbon monoxide has a much higher affinity for haemoglobin than oxygen does. In carbon monoxide poisoning, oxygen cannot be transported and released to body tissues thus resulting in hypoxia. With foetal haemoglobin, the shift facilitates diffusion of oxygen across the placenta. The oxygen dissociation curve for myoglobin exists even further to the left.

The causes of shift to right can be remembered using the mnemonic, "CADET, face Right!" for CO2, Acid, 2,3-DPG, Exercise and Temperature. Factors that move the oxygen dissociation curve to the right are those physiological states where tissues need more oxygen. For example during exercise, muscles have a higher metabolic rate, and consequently need more oxygen, produce more carbon dioxide and lactic acid, and their temperature rises.

Factors that Affect the Standard Dissociation Curve

The strength with which oxygen binds to haemoglobin is affected by several factors. In effect these factors shift or reshape the oxyhaemoglobin curve ("the standard curve"). The standard curve is shifted to the right by an increase in temperature, 2,3-DPG, or pCO_2, or a decrease in pH. A rightward shift indicates that the haemoglobin under study has a decreased affinity for oxygen. This makes it more difficult for haemoglobin to bind to oxygen (requiring a higher partial pressure of oxygen to achieve the same oxygen saturation), but it makes it easier for the haemoglobin to release oxygen bound to it.

The effect of this rightward shift of the curve increases the partial pressure of oxygen in the tissues when it is most needed, such as during exercise, or haemorrhagic shock.

In contrast, the curve is shifted to the left by the opposite of these conditions. This leftward shift indicates that the haemoglobin under study has an increased affinity for oxygen so that haemoglobin binds oxygen more easily, but unloads it more reluctantly.

Variation of the Hydrogen Ion Concentration

This changes the blood's pH. A decrease in pH shifts the standard curve to the right, while an increase shifts it to the left. This is known as the Bohr effect.

Effects of Carbon Dioxide

Carbon dioxide affects the curve in two ways: first, it influences intracellular pH (the Bohr effect), and second, CO_2 accumulation causes carbamino compounds to be generated through chemical interactions, which bind to haemoglobin forming carbaminohaemoglobin. Low levels of carbamino compounds have the effect of shifting the curve to the right, while higher levels cause a leftward shift. However, this isn't the overriding effect of CO_2 accumulation. Only about 5–10% of the total CO_2 content of blood is transported as carbamino compounds.

Most of the CO_2 content (80–90%) is transported as bicarbonate ions. The formation of a bicarbonate ion will release a proton into the plasma.

Hence, the elevated CO2 content creates a respiratory acidosis and shifts the oxygen–haemoglobin dissociation curve to the right.

Effects of 2,3-DPG

2,3-Disphosphoglycerate or 2,3-DPG (also 2,3-bisphosphoglycerate or 2,3-BPG) is an organophosphate, which is created in erythrocytes during glycolysis. The production of 2,3-DPG is likely an important adaptive mechanism, because the production increases for several conditions in the presence of diminished peripheral tissue O_2 availability, such as hypoxaemia, chronic lung disease, anaemia, and congestive heart failure, among others. High levels of 2,3-DPG shift the curve to the right, while low levels of 2,3-DPG cause a leftward shift, seen in states such as septic shock and hypophosphataemia.

Temperature

Temperature does not have such a dramatic effect compared to the previous factors, but hyperthermia causes a rightward shift, while hypothermia causes a leftward shift.

Carbon Monoxide

Haemoglobin binds with carbon monoxide 240 times more readily than with oxygen. The presence of carbon monoxide on one of the 4 haem sites causes the oxygen on the other haem sites to bind with greater affinity.

This makes it difficult for the haemoglobin to release oxygen to the tissues and has the effect of shifting the curve to the left (as well as downward, due to direct competitive effects of carbon monoxide). With an increased level of carbon monoxide, a person can suffer from severe tissue hypoxia while maintaining a normal pO_2.

Effects of Methaemoglobinaemia

Methaemoglobinaemia is a form of abnormal haemoglobin where ferrous (Fe^{2+}), which is normally found in haemoglobin, is converted to the ferric (Fe^{3+}) state. This causes a leftward shift in the curve as methaemoglobin does not unload O2 from Hb. However, methaemoglobin has increased affinity for cyanide, and is therefore useful in the treatment of cyanide poisoning.

Fetal Haemoglobin

Fetal haemoglobin (HbF) is structurally different from normal haemoglobin (Hb). The fetal dissociation curve is shifted to the left relative to the curve for the normal adult. Typically, fetal arterial oxygen pressures are low, and hence the leftward shift enhances the placental uptake of oxygen. At the placenta there is a higher concentration of 2,3-DPG formed. This binds more readily to adult haemoglobin but not to fetal haemoglobin. This causes the adult Hb to release more oxygen at the placenta to be taken up by the fetus. Fetal Hb is made up of gamma chains not beta ones, and 2,3-DPG does not bind readily to gamma chains, hence it does not give up its oxygen.

Dioxygen or Molecular Oxygen

Atoms

The atom is a basic unit of matter consisting of a dense, central nucleus surrounded by a cloud of negatively charged electrons. The atomic nucleus contains a mix of positively charged protons and electrically neutral neutrons (except in the case of hydrogen-1, which is the only stable nuclide with no neutron). The electrons of an atom are bound to the nucleus by the electro-magnetic force.

An atom containing an equal number of protons and electrons is electrically neutral, otherwise it has a positive or negative charge and is an ion. An atom is classified according to the number of protons and neutrons in its nucleus: the number of protons determines the chemical element (noted Z), and the number of neutrons determine the isotope of the element. In chemistry, a group of atoms can remain bound to each other, forming a molecule.

Over 99.9% of an atom's mass is concentrated in the nucleus (order of magnitude 10^{-15} m), with protons and neutrons having roughly equal mass. The electrons in an atom are attracted to the protons in the nucleus by the electromagnetic force. This force binds the electrons inside an electrostatic potential well surrounding the smaller nucleus, which means that an external source of energy is needed in order for the electron to escape. The closer an electron is to the nucleus, the greater the attractive force. Hence electrons bound near the centre of the potential well require more energy to escape than those at greater separations.

Electrons that are bound to atoms possess a set of stable energy levels, or orbitals, and can undergo transitions between them by absorbing or emitting photons that match the energy differences between the levels. The electrons determine the chemical properties of an element, and strongly influence an atom's magnetic properties.

Molecules

A molecule is defined as an electrically neutral group of at least two atoms in a definite arrangement held together by very strong (covalent) chemical bonds. Molecules are distinguished from polyatomic ions in this strict sense. In organic chemistry and biochemistry, the term molecule is used less strictly and also is applied to charged organic molecules and biomolecules.

A molecule may consist of atoms of a single chemical element, as with oxygen (O_2), or of different elements, as with water (H_2O). Atoms and complexes connected by non-covalent bonds such as hydrogen bonds or ionic bonds are generally not considered single molecules.

Oxygen Molecule and Covalent Bounding

The common allotrope of elemental oxygen on Earth is called dioxygen, O_2. It has a bond length of 121 pm and a bond energy of 498 kJ·mol^{-1}. This is the form that is used by complex forms of life, such as animals, in cellular respiration and is the form that is a major part of the Earth's atmosphere.

Oxygen is the most abundant chemical element, by mass, in our biosphere, air, sea and land. Oxygen is the third most abundant chemical element in the universe, after hydrogen and helium. About 0.9% of the Sun's mass is oxygen. Oxygen constitutes 49.2% of the Earth's crust by mass and is the major component of the world's oceans (88.8% by mass). Oxygen gas is the second most common component of the Earth's atmosphere, taking up 21.0% of its volume and 23.1% of its mass (some 1015 tonnes).

Earth is unusual among the planets of the Solar System in having such a high concentration of oxygen gas in its atmosphere: Mars (with 0.1% O2 by volume) and Venus have far lower concentrations. However, the O_2 surrounding these other planets is produced solely by ultraviolet radiation impacting oxygen-containing molecules such as carbon dioxide.

It is a colourless, odourless and tasteless gas.

The presence of large amounts of dissolved and free oxygen in the oceans and atmosphere may have driven most of the anaerobic organisms then living to extinction during the oxygen catastrophe about 2.4 billion years ago. However, cellular respiration using O_2 enables aerobic organisms to produce much more ATP than anaerobic organisms, helping the former to dominate Earth's biosphere. Photosynthesis and cellular respiration of O_2 allowed for the evolution of eukaryotic cells and ultimately complex multicellular organisms such as plants and animals.

From Atomic to Molecular Oxygen : Covalent Bounding

A covalent bond is a form of chemical bonding that is characterized by the sharing of pairs of electrons between atoms, and other covalent bonds. In short, the attraction-to-repulsion stability that forms between atoms when they share electrons is known as covalent bonding.

The term covalent bond dates from 1939. The prefix co-means jointly, associated in action, partnered to a lesser degree, etc.; thus a "co-valent bond", essentially, means that the atoms share "valence".

Walter Heitler and Fritz London are credited with the first successful quantum mechanical explanation of a chemical bond, specifically that of molecular hydrogen, in 1927. Their work was based on the valence bond model, which assumes that a chemical bond is formed when there is good overlap between the atomic orbitals of participating atoms. These atomic orbitals are known to have specific angular relationships between each other, and thus the valence bond model can successfully predict the bond angles observed in simple molecules. The basic principle is that the properties of chemical elements (atoms) dependent on their electronic structure including electrons called "outer layers". They are composed of electrons less bounded to the nucleus and occupy "layers" usually incomplete.

This incompleteness is energetically unfavourable and pushes atom to start an electronic boundings (therefore, in fine chemicals) with other atoms.

The oxygen has only four electrons in its outer layer but would be six for this layer is full (inner layer of type s with 2 outer shell electrons and p-type which is full with 6 electrons: rare gas configuration).

Electron "Time Shared" on the Bounding Orbitals

The oxygen gas solves this problem by combining two oxygen atoms in a molecule of oxygen (that you breathe in nitrogen mixture at ambient). This molecule is formed by "pooling" of two binding electrons by "time sharing" which means that technically each atom has six from time to time and is happy ! This partial electrons equally pooling on each atom is typical of the covalent bonding.

When two atoms approach each other, it is manifest from a certain distance (about 1 nanometer) an attraction between the nucleus of one and the other electrons (and vice versa). This attraction becomes significant when the distance between the nuclei is more than 500 picometer. But from this distance the repulsion between charges of same sign tend to compensate for the attraction between opposite sign charges. An equilibrium set-up at a definite distance between the 2 nuclei, characteristic for each atom. Its value is, by way of examples, from 106 pm for the hydrogen and 121 pm for oxygen.

Parameters of the Dioxygen

The parameters of the dioxygen are the following :

- Nuclei mass $^{16}O = 2.678 \times 10^{-26}$ kg
    - o 8 neutons mass 1,674 927 28(29)$\times 10^{-27}$ kg = 13,399 418 (24)$\times 10^{-27}$ kg
    - o 8 protons mass 1,672 621 71(29)$\times 10^{-27}$ kg = 13,380 973 (68)$\times 10^{-27}$ kg
- Electron mass e = 9.109×10^{-31} kg
- Mass ratio : 3.40×10^{-5}
- Constant force rate : k = 11.77 N/cm
- Interatomic distance (between centrum of each nuclei): 120,74$\times 10^{-12}$ m (picomètre = pm)
- Covalent radius : 73×10^{-12} m (pm).

Dioxygen Reactions

In nature, free oxygen is produced by the light-driven splitting of water during oxygenic photosynthesis. Green algae and cyanobacteria in marine environments provide about 70% of the free oxygen produced on earth and the rest is produced by terrestrial plants.

A simplified overall formula for photosynthesis is:

$6\ CO_2 + 6\ H_2O + \text{photons} \rightarrow C_6H_{12}O_6 + 6\ O_2$ (or simply carbon dioxide + water + sunlight → glucose + dioxygen)

Photolytic oxygen evolution occurs in the thylakoid membranes of photosynthetic organisms and requires the energy of four photons. Many steps are involved, but the result is the formation of a proton gradient across the thylakoid membrane, which is used to synthesize ATP via photophosphorylation. The O_2 remaining after oxidation of the water molecule is released into the atmosphere.

Molecular dioxygen, O_2, is essential for cellular respiration in all aerobic organisms. Oxygen is used in mitochondria to help generate adenosine triphosphate (ATP) during oxidative phosphorylation. The reaction for aerobic respiration is essentially the reverse of photosynthesis and is simplified as:

$$C_6H_{12}O_6 + 6\ O_2 \rightarrow 6\ CO_2 + 6\ H_2O + 2880\ kJ \cdot mol^{-1}$$

In vertebrates, O_2 is diffused through membranes in the lungs and into red blood cells. Hemoglobin binds O_2, changing its colour from bluish red to bright red. Other animals use hemocyanin (molluscs and some arthropods) or hemerythrin (spiders and lobsters). A liter of blood can dissolve 200 cm^3 of O_2.

Reactive oxygen species, such as superoxide ion (O^{-2}) and hydrogen peroxide (H_2O_2), are dangerous by-products of oxygen use in organisms. Parts of the immune system of higher organisms, however, create peroxide, superoxide, and singlet oxygen to destroy invading microbes. Reactive oxygen species also play an important role in the hypersensitive response of plants against pathogen attack.

An adult human in rest inhales 1.8 to 2.4 grams of oxygen per minute. This amounts to more than 6 billion tonnes of oxygen inhaled by humanity per year.

Dioxygen in Biological Reactions

Dioxygen (O_2) plays an important role in the energy metabolism of living organisms. Free oxygen is produced in the biosphere through photolysis (light-driven oxidation and splitting) of water during photosynthesis in cyanobacteria, green algae, and plants. During oxidative phosphorylation in cellular respiration, oxygen is reduced to water, thus closing the biological water-oxygen redox cycle.

Photosynthesis

In nature, free oxygen is produced by the light-driven splitting of water during oxygenic photosynthesis. Green algae and cyanobacteria in marine environments provide about 70% of the free oxygen produced on earth. The remainder is produced by terrestrial plants, although for example, almost all oxygen produced in tropical forests is consumed by organisms living there. A simplified overall formula for photosynthesis is:

$$6CO_2 + 6H_2O + \text{photons} \rightarrow C_6H_{12}O_6 + 6O_2$$

(or simply carbon dioxide + water + sunlight → glucose + oxygen)

Photolytic oxygen evolution during photosynthesis occurs via the light-dependent oxidation of water to molecular oxygen and can be written as the following simplified chemical reaction: $2H_2O \rightarrow 4e^- + 4H^+ + O_2$

The reaction occurs in the thylakoid membranes of cyanobacteria, and algal and plant chloroplasts and requires the energy of four photons. The electrons from the oxidized water molecules replace electrons in the P_{680} component of photosystem II, which have been removed into an electron transport chain via light-dependent excitation and resonance energy transfer onto plastoquinone. Photosytem II therefore has also been referred to as water-plastoquinone oxido-reductase. The protons from the oxidized water molecules are released into the thylakoid lumen, thus contributing to the generation of a proton gradient across the thylakoid membrane. This proton gradient is the driving force for ATP synthesis via photophosphorylation and coupling the absorption of light energy and photolysis of water to the creation of chemical energy during photosynthesis. The O_2 remaining after oxidation of the water molecule is released into the atmosphere.

Water oxidation is catalysed by a manganese-containing enzyme complex known as the oxygen evolving complex (OEC) or water-splitting complex found associated with the lumenal side of thylakoid membranes. Manganese is an important cofactor, and calcium and chloride are also required for the reaction to occur.

Oxygen Uptake and Transport

In vertebrates, oxygen uptake is carried out by the following processes:

Oxygen diffuses through membranes and into red blood cells after inhalation into the lungs. The heme group of hemoglobin binds oxygen when it is present, changing hemoglobin's colour from bluish red to bright red. Vertebrate animals use hemoglobin in their blood to transport oxygen from their lungs to their tissues, but other animals use hemocyanin (molluscs and some arthropods) or hemerythrin (spiders and lobsters). A liter of blood can dissolve 200 cc of oxygen gas, which is much more than water can dissolve.

After being carried in blood to a body tissue in need of oxygen, O_2 is handed-off from the heme group to monooxygenase, an enzyme that also has an active site with an atom of iron. Monooxygenase uses oxygen to catalyse many oxidation reactions in the body. Carbon dioxide, a waste product, is released from the cell and into the blood, where it combines with bicarbonate and hemoglobin for transport to the lungs. Blood circulates back to the lungs and the process repeats.

Aerobic Respiration

Molecular oxygen, O_2, is essential for cellular respiration in all aerobic organisms. Oxygen is used as an electron acceptor in mitochondria to generate chemical energy in the form of adenosine triphosphate (ATP) during oxidative phosphorylation. The reaction for the aerobic respiration is essentially the reverse of photosynthesis, except that now there is a large release of chemical energy that is stored in ATP molecules (up to 38 ATP molecules are formed from one molecule of glucose). The simplified version of this reaction is:

$$C_6H_{12}O_6 + 6O_2 \rightarrow 6CO_2 + 6H_2O + 2880 \text{ kJ/mol}$$

Reactive Oxygen Species

Reactive oxygen species are dangerous by-products that sometimes result from the use of oxygen in organisms. Important examples include; oxygen free radicals such as the highly-dangerous superoxide $O_2^{\cdot}$, and the less harmful hydrogen peroxide (H_2O_2). The body uses superoxide dismutase to reduce superoxide radicals to hydrogen peroxide. Glutathione peroxidase and similar enzymes then convert the H_2O_2 to water and dioxygen.

Parts of the immune system of higher organisms, however, create peroxide, superoxide, and singlet oxygen to destroy invading microbes. Recently, singlet oxygen has been found to be a source of biologically-produced ozone: This reaction proceeds through an unusual compound dihydrogen trioxide, also known as trioxidane, (HOOOH), which is an antibody-catalysed product of singlet oxygen and water. This compound, in turn, disproportionates to ozone and peroxide, providing two powerful antibacterials. The body's range of defence against all of these active oxidizing agents is hardly surprising, then, given their "deliberate" employment as antimicrobial agents in the immune response. Reactive oxygen species also play an important role in the hypersensitive response of plants against pathogen attack.

Electron Transfers

The electron is a subatomic particle carrying a negative electric charge. It has no known components or substructure. Therefore, the electron is generally believed to be an elementary particle. An electron has a mass that is approximately 1/1836 that of the proton. The intrinsic angular momentum (spin) of the electron is a half-integer value in units of ‘, which means that it is a fermion. The antiparticle of the electron is called the positron. The positron is identical to the electron except that it carries electrical and other charges of the opposite sign. When an electron collides with a positron, both particles may either scatter off each other or be totally annihilated, producing a pair (or more) of gamma ray photons. Electrons, which belong to the first generation of the lepton particle family, participate in gravitational, electromagnetic and weak interactions. Electrons, like all matter, have quantum mechanical properties of both particles and waves, so they can collide with other particles and be diffracted like light. However, this duality is best demonstrated in experiments with electrons, due to their tiny mass. Since an electron is a fermion, no two electrons can occupy the same quantum state, in accordance with the Pauli exclusion principle. The concept of an indivisible amount of electric charge was theorized to explain the chemical properties of atoms, beginning in 1838 by British natural philosopher Richard Laming; the name *electron* was introduced for this charge in 1894 by Irish physicist George Johnstone Stoney. The electron was identified as a particle in 1897 by J. J. Thomson and his team of British physicists.

In many physical phenomena, such as electricity, magnetism, and thermal conductivity, electrons play an essential role. An electron in motion relative to an observer generates a magnetic field, and will be deflected by external magnetic fields. When an electron is accelerated, it can absorb or radiate energy in the form of photons. Electrons, together with atomic nuclei made of protons and neutrons, make up atoms. However, electrons contribute less than 0.06% to an atom's total mass. The attractive Coulomb force between an electron and a proton causes electrons to be bound into atoms. The exchange or sharing of the electrons between two or more atoms is the main cause of chemical bonding. According to theory, most electrons in the universe were created in the big bang, but they may also be created through beta decay of radioactive isotopes and in high-energy collisions, for instance when cosmic rays enter the atmosphere. Electrons may be destroyed through annihilation with positrons, and may be absorbed during nucleosynthesis in stars.

Laboratory instruments are capable of containing and observing individual electrons as well as electron plasma, whereas dedicated telescopes can detect electron plasma in outer space. Electrons have many applications, including welding, cathode ray tubes, electron microscopes, radiation therapy, lasers and particle accelerators.

Atomic Theory

By 1914, experiments by physicists Ernest Rutherford, Henry Moseley, James Franck and Gustav Hertz had largely established the structure of an atom as a dense nucleus of positive charge surrounded by lower-mass electrons.

In 1913, Danish physicist Niels Bohr postulated that electrons resided in quantized energy states, with the energy determined by the angular momentum of the electron's orbits about the nucleus. The electrons could move between these states, or orbits, by the emission or absorption of photons at specific frequencies. By means of these quantized orbits, he accurately explained the spectral lines of the hydrogen atom. However, Bohr's model failed to account for the relative intensities of the spectral lines and it was unsuccessful in explaining the spectra of more complex atoms.

Chemical bonds between atoms were explained by Gilbert Newton Lewis, who in 1916 proposed that a covalent bond between two atoms is maintained by a pair of electrons shared between them. Later, in 1923, Walter Heitler and Fritz London gave the full explanation of the electron-pair formation and chemical bonding in terms of quantum mechanics. In 1919, the American chemist Irving Langmuir elaborated on the Lewis' static model of the atom and suggested that all electrons were distributed in successive "concentric (nearly) spherical shells, all of equal thickness". The shells were, in turn, divided by him in a number of cells each containing one pair of electrons. With this model Langmuir was able to qualitatively explain the chemical properties of all elements in the periodic table, which were known to largely repeat themselves according to the periodic law.

In 1924, Austrian physicist Wolfgang Pauli observed that the shell-like structure of the atom could be explained by a set of four parameters that defined every quantum energy state, as long as each state was inhabited by no more than a single electron. (This prohibition against more than one electron occupying the same quantum energy state became known as the Pauli exclusion principle.) The physical mechanism to explain the fourth parameter, which had two distinct possible values, was provided by the Dutch physicists Abraham Goudsmith and George Uhlenbeck when they suggested that an electron, in addition to the angular momentum of its orbit, could possess an intrinsic angular momentum. This property became known as spin, and explained the previously mysterious splitting of spectral lines observed with a high-resolution spectrograph; this phenomenon is known as fine structure splitting.

Quantum Mechanics

In his 1924 dissertation *Recherches sur la théorie des quanta* (Research on Quantum Theory), French physicist Louis de Broglie hypothesized that all matter possesses a De Broglie wave similar to light. That is, under the appropriate conditions, electrons and other matter would show properties of either particles or waves. The corpuscular properties of a particle are demonstrated when it is shown to have a localized position in space along its trajectory at any given moment. Wave-like nature is observed, for example, when a beam of light is passed through parallel slits and creates interference patterns. In 1927, the interference effect was demonstrated with a beam of electrons by English physicist George Paget Thomson with a thin metal film and by American physicists Clinton Davisson and Lester Germer using a crystal of nickel.

The success of de Broglie's prediction led to the publication, by Erwin Schrodinger in 1926, of the Schrodinger equation that successfully describes how electron waves propagated. Rather than yielding a solution that determines the location of an electron over time, this wave equation can be used to predict the probability of finding an electron near a position. This approach was later called quantum mechanics, which provided an extremely close derivation to the energy states of an electron in a hydrogen atom. Once spin and the interaction between multiple electrons were considered, quantum mechanics allowed the configuration of electrons in atoms with higher atomic numbers than hydrogen to be successfully predicted.

In 1928, building on Wolfgang Pauli's work, Paul Dirac produced a model of the electron-the Dirac equation, consistent with relativity theory, by applying relativistic and symmetry considerations to the hamiltonian formulation of the quantum mechanics of the electro-magnetic field. In order to resolve some problems within his relativistic equation, in 1930 Dirac developed a model of the vacuum as an infinite sea of particles having negative energy, which was dubbed the Dirac sea. This led him to predict the existence of a positron, the antimatter counterpart of the electron. This particle was discovered in 1932 by Carl D. Anderson, who proposed calling standard electrons *negatrons*, and using *electron* as a generic term to describe both the positively and negatively charged variants. This usage of the term 'negatron' is still occasionally encountered today, and it may be shortened to 'negaton'.

In 1947 Willis Lamb, working in collaboration with graduate student Robert Rutherford, found that certain quantum states of hydrogen atom, which should have the same energy, were shifted in relation to each other, the difference being the Lamb shift. About the same time, Polykarp Kusch, working with Henry M. Foley, discovered the magnetic moment of the electron is slightly larger than predicted by Dirac's theory. This small difference was later called anomalous magnetic dipole moment of the electron. To resolve these issues, a refined theory called quantum electrodynamics was developed by Sin-Itiro Tomonaga, Julian Schwinger and Richard P. Feynman in the late 1940s.

Particle Accelerators

With the development of the particle accelerator during the first half of the twentieth century, physicists began to delve deeper into the properties of subatomic particles. The first successful attempt to accelerate electrons using Electromagnetic induction was made in 1942 by Donald Kerst. His initial betatron reached energies of 2.3 MeV, while subsequent betatrons achieved 300 MeV. In 1947, synchrotron radiation was discovered with a 70 MeV electron synchrotron at General Electric. This radiation was caused by the acceleration of electrons, moving near the speed of light, through a magnetic field.

With a beam energy of 1.5 GeV, the first high-energy particle collider was ADONE, which began operations in 1968. This device accelerated electrons and positrons in opposite directions, effectively doubling the energy of their collision when compared to striking a static target with an electron. The Large Electron-Positron Collider (LEP) at CERN, which was operational from 1989 to 2000, achieved collision energies of 209 GeV and made important measurements for the Standard Model of particle physics.

Characteristics

Classification

In the Standard Model of particle physics, electrons belong to the group of subatomic particles called leptons, which are believed to be fundamental or elementary particles. Electrons have the lowest mass of any charged lepton (or electrically charged particle of any type) and belong to the first-generation of fundamental particles. The second and third generation contain charged leptons, the muon and the tau, which are identical to the electron in charge, spin and interactions, but are more massive. Leptons differ from the other basic constituent of matter, the quarks, by their lack of strong interaction. All members of the lepton group are fermions, because they all have half-odd integer spin; the electron has spin $^1D_2$.

Fundamental Properties

The invariant mass of an electron is approximately 9.109×10^{-31} kilogram, or 5.489×10^{-4} atomic mass unit. On the basis of Einstein's principle of mass–energy equivalence, this mass corresponds to a rest energy of 0.511 MeV. The ratio between the mass of a proton and that of an electron is about 1836. Astronomical measurements show that the proton-to-electron mass ratio has held the same value for at least half the age of the universe, as is predicted by the Standard Model.

Electrons have an electric charge of "1.602×10^{-19} coulomb, which is used as a standard unit of charge for subatomic particles. Within the limits of experimental accuracy, the electron charge is identical to the charge of a proton, but with the opposite sign. As the symbol *e* is used for the elementary charge, the electron is commonly symbolized by e^-, where the minus sign indicates the negative charge. The positron is symbolized by e^+ because it has the same properties as the electron but with a positive rather than negative charge.

The electron has an intrinsic angular momentum or spin of $^{1}D_{2}$. This property is usually stated by referring to the electron as a spin-$^{1}D_{2}$ particle. For such particles the spin magnitude is "$^{3}D_{2}$'. while the result of the measurement of a projection of the spin on any axis can only be $\pm D_{2}$. In addition to spin, the electron has an intrinsic magnetic moment along its spin axis.

It is approximately equal to one Bohr magneton, which is a physical constant equal to 9.274 009 15(23) × 10^{-24} joules per tesla. The orientation of the spin with respect to the momentum of the electron defines the property of elementary particles known as helicity.

The electron has no known substructure. Hence, it is defined or assumed to be a point particle with a point charge and no spatial extent. Observation of a single electron in a Penning trap shows the upper limit of the particle's radius is 10^{-22} meters. There *is* a physical constant called the "classical electron radius", with the much larger value of 2.8179×10^{-15} m. However, the terminology comes from a simplistic calculation that ignores the effects of quantum mechanics; in reality, the so-called classical electron radius has little to do with the true fundamental structure of the electron.

There are elementary particles that spontaneously decay into less massive particles. An example is the muon, which decays into an electron, a neutrino and an antineutrino, with a mean lifetime of 2.2×10^{-6} seconds. However, the electron is thought to be stable on theoretical grounds: the electron is the least massive particle with non-zero electric charge, so its decay would violate charge conservation. The experimental lower bound for the electron's mean lifetime is 4.6×10^{26} years, at a 90% confidence level.

Quantum Properties

As with all particles, electrons can act as waves. This is called the wave–particle duality and can be demonstrated using the double-slit experiment. The wave-like nature of the electron allows it to pass through two parallel slits simultaneously, rather than just one slit as would be the case for a classical particle. In quantum mechanics, the wave-like property of one particle can be described mathematically as a complex-valued function, the wave function, commonly denoted by the Greek letter psi (ø). When the absolute value of this function is squared, it gives the probability that a particle will be observed near a location—a probability density.

Electrons are identical particles because they cannot be distinguished from each other by their intrinsic physical properties. In quantum mechanics, this means that a pair of interacting electrons must be able to swap positions without an observable change to the state of the system. The wave function of fermions, including electrons, is antisymmetric, meaning that it changes sign when two electrons are swapped; that is, $ø(r_1, r_2)$ = "$ø(r_2, r_1)$, where the variables r_1 and r_2 correspond to the first and second electrons, respectively. Since the absolute value is not changed by a sign swap, this corresponds to equal probabilities. Bosons, such as the photon, have symmetric wave functions instead.

In the case of antisymmetry, solutions of the wave equation for interacting electrons result in a zero probability that each pair will occupy the same location or state. This is responsible for the Pauli exclusion principle, which precludes any two electrons from occupying the same quantum state. This principle explains many of the properties of electrons. For example, it causes groups of bound electrons to occupy different orbitals in an atom, rather than all overlapping each other in the same orbit.

Virtual Particles

Physicists believe that empty space may be continually creating pairs of virtual particles, such as a positron and electron, which rapidly annihilate each other shortly thereafter. The combination of the energy variation needed to create these particles, and the time during which they exist, fall under the threshold of detectability expressed by the Heisenberg uncertainty relation, $\Delta E \cdot \Delta t$ e" '. In effect, the energy needed to create these virtual particles, ΔE, can be "borrowed" from the vacuum for a period of time, Δt, so that their product is no more than the reduced Planck constant, 'H⁻ 6.6×10^{-16} eV•s. Thus, for a virtual electron, Δt is at most 1.3×10^{-21} s.

While an electron–positron virtual pair is in existence, the coulomb force from the ambient electric field surrounding an electron causes a created positron to be attracted to the original electron, while a created electron experiences a repulsion. This causes what is called vacuum polarization. In effect, the vacuum behaves like a medium having a dielectric permittivity more than unity. Thus the effective charge of an electron is actually smaller than its true value, and the charge decreases with increasing distance from the electron. This polarization was confirmed experimentally in 1997 using the Japanese TRISTAN particle accelerator. Virtual particles cause a comparable shielding effect for the mass of the electron.

The interaction with virtual particles also explains the small (about 0.1%) deviation of the intrinsic magnetic moment of the electron from the Bohr magneton (the anomalous magnetic moment). The extraordinarily precise agreement of this predicted difference with the experimentally determined value is viewed as one of the great achievements of quantum electrodynamics.

In classical physics, the angular momentum and magnetic moment of an object depend upon its physical dimensions. Hence, the concept of a dimensionless electron possessing these properties might seem inconsistent. The apparent paradox can be explained by the formation of virtual photons in the electric field generated by the electron. These photons cause the electron to shift about in a jittery fashion (known as zitterbewegung), which results in a net circular motion with precession. This motion produces both the spin and the magnetic moment of the electron. In atoms, this creation of virtual photons explains the Lamb shift observed in spectral lines.

Interaction

An electron generates an electric field that exerts an attractive force on a particle with a positive charge, such as the proton, and a repulsive force on a particle with a negative charge. The strength of this force is determined by Coulomb's inverse square law. When an electron is in motion, it generates a magnetic field. The Ampère-Maxwell law relates the magnetic field to the mass motion of electrons (the current) with respect to an observer. It is this property of induction which supplies the magnetic field that drives an electric motor. The electromagnetic field of an arbitrary moving charged particle is expressed by the Liénard–Wiechert potentials, which are valid even when the particle's speed is close to that of light (relativistic).

When an electron is moving through a magnetic field, it is subject to the Lorentz force that exerts an influence in a direction perpendicular to the plane defined by the magnetic field and the electron velocity. This centripetal force causes the electron to follow a helical trajectory through the field at a radius called the gyroradius.

The acceleration from this curving motion induces the electron to radiate energy in the form of synchrotron radiation. The energy emission in turn causes a recoil of the electron, known as the Abraham-Lorentz-Dirac force, which creates a friction that slows the electron. This force is caused by a back-reaction of the electron's own field upon itself.

In quantum electrodynamics the electromagnetic interaction between particles is mediated by photons. An isolated electron that is not undergoing acceleration is unable to emit or absorb a real photon; doing so would violate conservation of energy and momentum. Instead, virtual photons can transfer momentum between two charged particles. It is this exchange of virtual photons that, for example, generates the Coulomb force. Energy emission can occur when a moving electron is deflected by a charged particle, such as a proton. The acceleration of the electron results in the emission of Bremsstrahlung radiation.

An inelastic collision between a photon (light) and a solitary (free) electron is called Compton scattering. This collision results in a transfer of momentum and energy between the particles, which modifies the wavelength of the photon by an amount called the Compton shift. The maximum magnitude of this wavelength shift is $h/m_e c$, which is known as the Compton wavelength. For an electron, it has a value of 2.43×10^{-12} m. When the wavelength of the light is long (for instance, the wavelength of the visible light is 0.4–0.7 μm) the wavelength shift becomes negligible. Such interaction between the light and free electrons is called Thomson scattering or Linear Thomson scattering.

The relative strength of the electromagnetic interaction between two charged particles, such as an electron and a proton, is given by the fine-structure constant. This value is a dimensionless quantity formed by the ratio of two energies: the electrostatic energy of attraction (or repulsion) at a separation of one Compton wavelength, and the rest energy of the charge. It is given by α H– 7.297353×10^{-3}, which is approximately equal to $^{1}D_{137}$.

When electrons and positrons collide, they annihilate each other, giving rise to two or more gamma ray photons. If the electron and positron have negligible momentum, a positronium atom can form before annihilation results in two or three gamma ray photons totalling 1.022 MeV. On the other hand, high-energy photons may transform into an electron and a positron by a process called pair production, but only in the presence of a nearby charged particle, such as a nucleus.

In the theory of electroweak interaction, the left-handed component of electron's wavefunction forms a weak isospin doublet with the electron neutrino. This means that during weak interactions, electron neutrinos behave like electrons. Either member of this doublet can undergo a charged current interaction by emitting or absorbing a W and be converted into the other member. Charge is conserved during this reaction because the W boson also carries a charge, canceling out any net change during the transmutation. Charged current interactions are responsible for the phenomenon of beta decay in a radioactive atom. Both the electron and electron neutrino can undergo a neutral current interaction via a Z^0 exchange, and this is responsible for neutrino-electron elastic scattering.

Atoms and Molecules

Probability densities for the first few hydrogen atom orbitals, seen in cross-section. The energy level of a bound electron determines the orbital it occupies, and the colour reflects the probability to find the electron at a given position.

An electron can be *bound* to the nucleus of an atom by the attractive Coulomb force. A system of several electrons bound to a nucleus is called an atom. If the number of electrons is different from the nucleus' electrical charge, such an atom is called an ion. The wave-like behaviour of a bound electron is described by a function called an atomic orbital. Each orbital has its own set of quantum numbers such as energy, angular momentum and projection of angular momentum, and only a discrete set of these orbitals exist around the nucleus. According to the Pauli exclusion principal each orbital can be occupied by up to two electrons, which must differ in their spin quantum number.

Electrons can transfer between different orbitals by the emission or absorption of photons with an energy that matches the difference in potential. Other methods of orbital transfer include collisions with particles, such as electrons, and the Auger effect. In order to escape the atom, the energy of the electron must be increased above its binding energy to the atom. This occurs, for example, with the photoelectric effect, where an incident photon exceeding the atom's ionization energy is absorbed by the electron.

The orbital angular momentum of electrons is quantized. Because the electron is charged, it produces an orbital magnetic moment that is proportional to the angular momentum. The net magnetic moment of an atom is equal to the vector sum of orbital and spin magnetic moments of all electrons and the nucleus. The nuclear magnetic moment is, however, negligible in comparison to the effect from the electrons. The magnetic moments of the electrons that occupy the same orbital (so called, paired electrons) cancel each other out.

The chemical bond between atoms occurs as a result of electromagnetic interactions, as described by the laws of quantum mechanics. The strongest bonds are formed by the sharing or transfer of electrons between atoms, allowing the formation of molecules. Within a molecule, electrons move under the influence of several nuclei, and occupy molecular orbitals; much as they can occupy atomic orbitals in isolated atoms. A fundamental factor in these molecular structures is the existence of electron pairs. These are electrons with opposed spins, allowing them to occupy the same molecular orbital without violating the Pauli exclusion principle (much like in atoms). Different molecular orbitals have different spatial distribution of the electron density. For instance, in bonded pairs (i.e. in the pairs that actually bind atoms together) electrons can be found with the maximal probability in a relatively small volume between the nuclei. On the contrary, in non-bonded pairs electrons are distributed in a large volume around nuclei.

Conductivity

If a body has more or fewer electrons than are required to balance the positive charge of the nuclei, then that object has a net electric charge. When there is an excess of electrons, the object is said to be negatively charged. When there are fewer electrons than the number of protons in nuclei, the object is said to be positively charged. When the number of electrons and the number of protons are equal, their charges cancel each other and the object is said to be electrically neutral. A macroscopic body can develop an electric charge through rubbing, by the triboelectric effect. Independent electrons moving in vacuum are termed *free* electrons. Electrons in metals also behave as if they were free. In reality the particles that are commonly termed electrons in metals and other solids are quasi-electrons—quasi-particles, which have the same electrical charge, spin and magnetic moment as real electrons but may have a different mass. When free electrons—both in vacuum and metals—move, they produce a net flow of charge called an electric current, which generates a magnetic field. Likewise a current can be created by a changing magnetic field. These interactions are described mathematically by Maxwell's equations.

At a given temperature, each material has an electrical conductivity that determines the value of electric current when an electric potential is applied. Examples of good conductors include metals such as copper and gold, whereas glass and Teflon are poor conductors. In any dielectric material, the electrons remain bound to their respective atoms and the material behaves as an insulator. Most semiconductors have a variable level of conductivity that lies between the extremes of conduction and insulation. On the other hand, metals have an electronic band structure containing partially filled electronic bands. The presence of such bands allows electrons in metals to behave as if they were free or delocalized electrons. These electrons are not associated with specific atoms, so when an electric field is applied, they are free to move like a gas (called Fermi gas) through the material much like free electrons.

Because of collisions between electrons and atoms, the drift velocity of electrons in a conductor is on the order of millimeters per second. However, the speed at which a change of current at one point in the material causes changes in currents in other parts of the material, the velocity of propagation, is typically about 75% of light speed. This occurs because electrical signals propagate as a wave, with the velocity dependent on the dielectric constant of the material.

Metals make relatively good conductors of heat, primarily because the delocalized electrons are free to transport thermal energy between atoms. However, unlike electrical conductivity, the thermal conductivity of a metal is nearly independent of temperature. This is expressed mathematically by the Wiedemann-Franz law, which states that the ratio of thermal conductivity to the electrical conductivity is proportional to the temperature. The thermal disorder in the metallic lattice increases the electrical resistivity of the material, producing a temperature dependence for electrical current. When cooled below a point called the critical temperature, materials can undergo a phase transition in which they lose all resistivity to electrical current, in a process known as superconductivity. In BCS theory, this behaviour is modeled by pairs of electrons entering a quantum state known as a Bose–Einstein condensate. These Cooper pairs have their motion coupled to nearby matter via lattice vibrations called phonons, thereby avoiding the collisions with atoms that normally create electrical resistance. (Cooper pairs have a radius of roughly 100 nm, so they can overlap each other.) However, the mechanism by which higher temperature superconductors operate remains uncertain.

Electrons inside conducting solids, which are quasi-particles themselves, when tightly confined at temperatures close to absolute zero, behave as though they had split into two other quasiparticles: spinons and holons. The former carries spin and magnetic moment, while the latter electrical charge.

Motion and Energy

According to Einstein's theory of special relativity, as an electron's speed approaches the speed of light, from an observer's point of view its relativistic mass increases, thereby making it more and more difficult to accelerate it from within the observer's frame of reference. The speed of an electron can approach, but never reach, the speed of light in a vacuum, *c*.

However, when relativistic electrons—that is, electrons moving at a speed close to *c*—are injected into a dielectric medium such as water, where the local speed of light is significantly less than *c*, the electrons temporarily travel faster than light in the medium. As they interact with the medium, they generate a faint light called Cherenkov radiation.

The effects of special relativity are based on a quantity known as the Lorentz factor, defined as $\gamma = 1\sqrt{1 - v^2 / c^2}$ where v is the speed of the particle. The kinetic energy K_e of an electron moving with velocity v is:

$$K_e = (\gamma - 1)m_e c^2,$$

where m_e is the mass of electron. For example, the Stanford linear accelerator can accelerate an electron to roughly 51 GeV. This gives a value of nearly 100,000 for γ, since the mass of an electron is 0.51 MeV/c^2. The relativistic momentum of this electron is 100,000 times the momentum that classical mechanics would predict for an electron at the same speed.

Since an electron behaves as a wave, at a given velocity it has a characteristic de Broglie wavelength. This is given by $\lambda_e = h/p$ where h is the Planck constant and p is the momentum. For the 51 GeV electron above, the wavelength is about 2.4×10^{-17} m, small enough to explore structures well below the size of an atomic nucleus.

Formation

The Big Bang theory is the most widely accepted scientific theory to explain the early stages in the evolution of the Universe. For the first millisecond of the Big Bang, the temperatures were over 10 billion kelvins and photons had mean energies over a million electronvolts. These photons were sufficiently energetic that they could react with each other to form pairs of electrons and positrons,

$$\gamma + \gamma \rightleftharpoons e^+ + e^-,$$

where γ is a photon, e^+ is a positron and e^- is an electron. Likewise, positron-electron pairs annihilated each other and emitted energetic photons. An equilibrium between electrons, positrons and photons was maintained during this phase of the evolution of the Universe. After 15 seconds had passed, however, the temperature of the universe dropped below the threshold where electron-positron formation could occur. Most of the surviving electrons and positrons annihilated each other, releasing gamma radiation that briefly reheated the universe.

For reasons that remain uncertain, during the process of leptogenesis there was an excess in the number of electrons over positrons. Hence, about one electron in every billion survived the annihilation process. This excess matched the excess of protons over anti-protons, in a condition known as baryon asymmetry, resulting in a net charge of zero for the universe.

The surviving protons and neutrons began to participate in reactions with each other—in the process known as nucleosynthesis, forming isotopes of hydrogen and helium, with trace amounts of lithium. This process peaked after about five minutes. Any leftover neutrons underwent negative beta decay with a half-life of about a thousand seconds, releasing a proton and electron in the process,

$$n \Rightarrow p + e^- + \overline{\nu}_e,$$

where n is a neutron, p is a proton and ν_e is an electron antineutrino. For about the next 300,000–400,000 years, the excess electrons remained too energetic to bind with atomic nuclei. What followed is a period known as recombination, when neutral atoms were formed and the expanding universe became transparent to radiation.

Roughly one million years after the big bang, the first generation of stars began to form. Within a star, stellar nucleosynthesis results in the production of positrons from the fusion of atomic nuclei. These antimatter particles immediately annihilate with electrons, releasing gamma rays. The net result is a steady reduction in the number of electrons, and a matching increase in the number of neutrons. However, the process of stellar evolution can result in the synthesis of radioactive isotopes. Selected isotopes can subsequently undergo negative beta decay, emitting an electron and antineutrino from the nucleus. An example is the cobalt-60 (^{60}Co) isotope, which decays to form nickel-60 (^{60}Ni).

At the end of its lifetime, a star with more than about 20 solar masses can undergo gravitational collapse to form a black hole. According to classical physics, these massive stellar objects exert a gravitational attraction that is strong enough to prevent anything, even electromagnetic radiation, from escaping past the Schwarzschild radius. However, it is believed that quantum mechanical effects may allow Hawking radiation to be emitted at this distance. Electrons (and positrons) are thought to be created at the event horizon of these stellar remnants.

When pairs of virtual particles (such as an electron and positron) are created in the vicinity of the event horizon, the random spatial distribution of these particles may permit one of them to appear on the exterior; this process is called quantum tunnelling. The gravitational potential of the black hole can then supply the energy that transforms this virtual particle into a real particle, allowing it to radiate away into space. In exchange, the other member of the pair is given negative energy, which results in a net loss of mass-energy by the black hole. The rate of Hawking radiation increases with decreasing mass, eventually causing the black hole to evaporate away until, finally, it explodes. Cosmic rays are particles travelling through space with high energies. Energy events as high as 3.0×10^{20} eV have been recorded. When these particles collide with nucleons in the Earth's atmosphere, a shower of particles is generated, including pions. More than half of the cosmic radiation observed from the Earth's surface consists of muons. The particle called a muon is a lepton which is produced in the upper atmosphere by the decay of a pion. A muon, in turn, can decay to form an electron or positron. Thus, for the negatively charged pion π^-,

$$\pi^- \rightarrow \mu^- + \bar{\nu}_\mu,$$

$$\mu^- \rightarrow e^- + \bar{\nu}_e + \nu_\mu,$$

where μ^- is a muon and υ_μ is a muon neutrino.

Observation

Remote observation of electrons requires detection of their radiated energy. For example, in high-energy environments such as the corona of a star, free electrons form a plasma that radiates energy due to Bremsstrahlung. Electron gas can undergo plasma oscillation, which is waves caused by synchronized variations in electron density, and these produce energy emissions that can be detected by using radio telescopes.

The frequency of a photon is proportional to its energy. As a bound electron transitions between different energy levels of an atom, it will absorb or emit photons at characteristic frequencies.

For instance, when atoms are irradiated by a source with a broad spectrum, distinct absorption lines will appear in the spectrum of transmitted radiation. Each element or molecule displays a characteristic set of spectral lines, such as the hydrogen spectral series. Spectroscopic measurements of the strength and width of these lines allow the composition and physical properties of a substance to be determined.

In laboratory conditions, the interactions of individual electrons can be observed by means of particle detectors, which allow measurement of specific properties such as energy, spin and charge. The development of the Paul trap and Penning trap allows charged particles to be contained within a small region for long durations. This enables precise measurements of the particle properties.

For example, in one instance a Penning trap was used to contain a single electron for a period of 10 months. The magnetic moment of the electron was measured to a precision of eleven digits, which, in 1980, was a greater accuracy than for any other physical constant.

The first video images of an electron's energy distribution were captured by a team at Lund University in Sweden, February 2008. The scientists used extremely short flashes of light, called attosecond pulses, which allowed an electron's motion to be observed for the first time.

The distribution of the electrons in solid materials can be visualized by angle resolved photoemission spectroscopy (ARPES). This technique employs the photoelectric effect to measure the reciprocal space—a mathematical representation of periodic structures that is used to infer the original structure. ARPES can be used to determine the direction, speed and scattering of electrons within the material.

3

Biochemistry of Carbohydrates

Carbohydrates are carbon compounds that contain large quantities of hydroxyl groups. The simplest carbohydrates also contain either an aldehyde moiety (these are termed polyhydroxyaldehydes) or a ketone moiety (polyhydroxyketones). All carbohydrates can be classified as either monosac-charides, oligosaccharides or polysaccharides.

Anywhere from two to ten monosaccharide units, linked by glycosidic bonds, make up an oligosaccharide. Polysaccharides are much larger, containing hundreds of monosaccharide units. The presence of the hydroxyl groups allows carbohydrates to interact with the aqueous environment and to participate in hydrogen bonding, both within and between chains. Derivatives of the carbohydrates can contain nitrogens, phosphates and sulfur compounds. Carbohydrates also can combine with lipid to form glycolipids or with protein to form glycoproteins.

Carbohydrate Nomenclature

The predominant carbohydrates encountered in the body are structurally related to the aldotriose glyceraldehyde and to the ketotriose dihydroxyacetone. All carbohydrates contain at least one asymmetrical (chiral) carbon and are, therefore, optically active. In addition, carbohydrates can exist in either of two conformations, as determined by the orientation of the hydroxyl group about the asymmetric carbon farthest from the carbonyl. With a few exceptions, those carbohydrates that are of physiological significance exist in the D-conformation. The mirror-image conformations, called enantiomers, are in the L-conformation.

Structures of Glyceraldehyde Enantiomers

Monosaccharides

The monosaccharides commonly found in humans are classified according to the number of carbons they contain in their backbone structures. The major monosaccharides contain four to six carbon atoms.

***Table** : Carbohydrate Classifications*

# Carbons	*Category Name*	*Relevant examples*
3	Triose	Glyceraldehyde, Dihydroxyacetone
4	Tetrose	Erythrose
5	Pentose	Ribose, Ribulose, Xylulose
6	Hexose	Glucose, Galactose, Mannose, Fructose
7	Heptose	Sedoheptulose
9	Nonose	Neuraminic acid, also called sialic acid

The aldehyde and ketone moieties of the carbohydrates with five and six carbons will spontaneously react with alcohol groups present in neighbouring carbons to produce intramolecular hemiacetals or hemiketals, respectively. This results in the formation of five- or six-membered rings. Because the five-membered ring structure resembles the organic molecule furan, derivatives with this structure are termed furanoses.

Those with six-membered rings resemble the organic molecule pyran and are termed pyranoses. Such structures can be depicted by either Fischer or Haworth style diagrams. The numbering of the carbons in carbohydrates proceeds from the carbonyl carbon, for aldoses, or the carbon nearest the carbonyl, for ketoses.

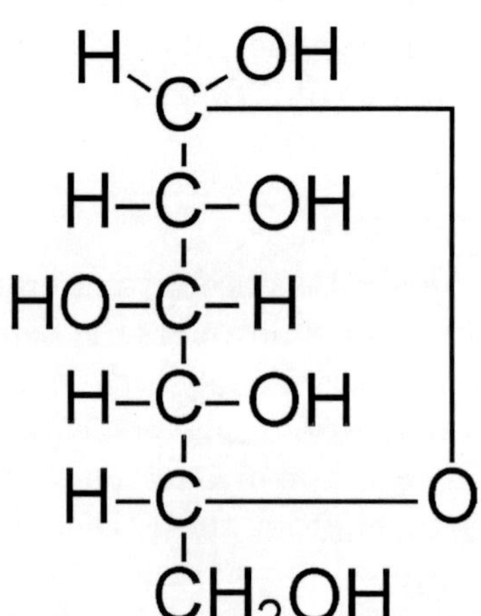

Cyclic Fischer Projection of α-D-Glucose

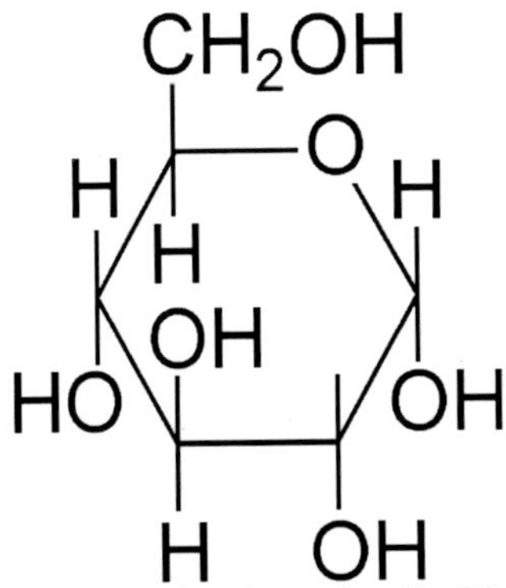

Haworth Projection of α-D-Glucose

The rings can open and re-close, allowing rotation to occur about the carbon bearing the reactive carbonyl yielding two distinct configurations (α and β) of the hemiacetals and hemiketals. The carbon about which this rotation occurs is the anomeric carbon and the two forms are termed anomers.

Carbohydrates can change spontaneously between the α and β configurations: a process known as mutarotation. When drawn in the Fischer projection, the α configuration places the hydroxyl attached to the anomeric carbon to the right, towards the ring. When drawn in the Haworth projection, the α configuration places the hydroxyl downward.

The spatial relationships of the atoms of the furanose and pyranose ring structures are more correctly described by the two conformations identified as the chair form and the boat form. The chair form is the more stable of the two. Constituents of the ring that project above or below the plane of the ring are axial and those that project parallel to the plane are equatorial. In the chair conformation, the orientation of the hydroxyl group about the anomeric carbon of α-D-glucose is axial and equatorial in β-D-glucose.

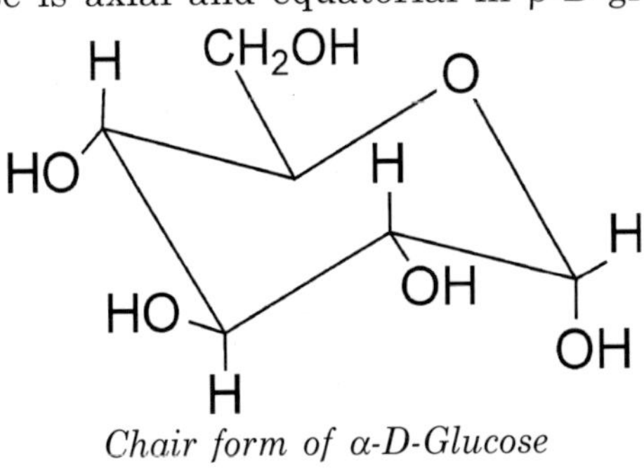

Chair form of α-D-Glucose

Disaccharides

Covalent bonds between the anomeric hydroxyl of a cyclic sugar and the hydroxyl of a second sugar (or another alcohol containing compound) are termed glycosidic bonds, and the resultant molecules are glycosides. The linkage of two monosaccharides to form disaccharides involves a glycosidic bond. Several physiogically important disaccharides are sucrose, lactose and maltose.

Sucrose: prevalent in sugar cane and sugar beets, is composed of glucose and fructose through an α–(1,2)–β-glycosidic bond.

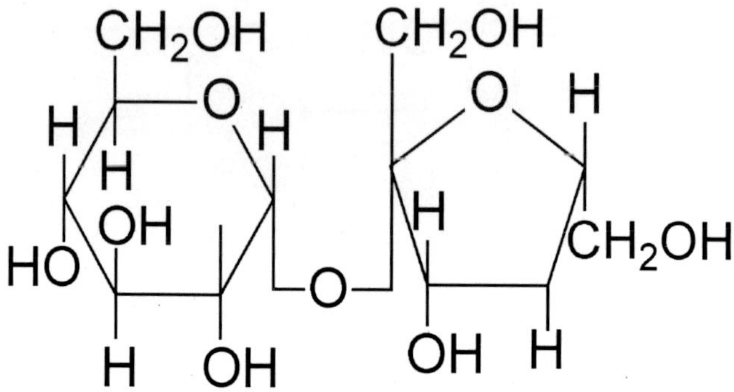

Sucrose

Lactose: is found exclusively in the milk of mammals and consists of galactose and glucose in a β–(1, 4) glycosidic bond.

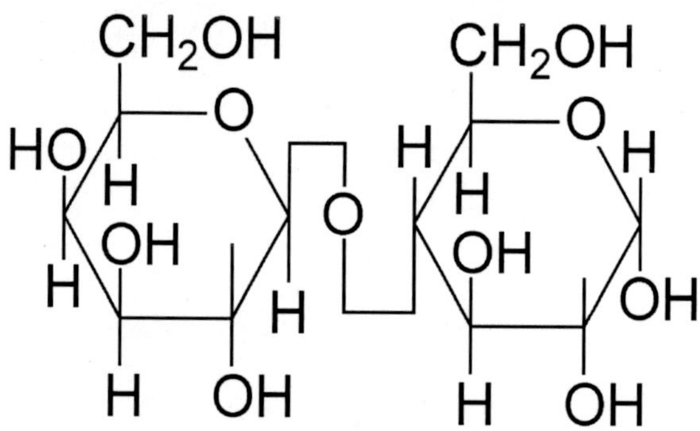

Lactose

Maltose: the major degradation product of starch, is composed of 2 glucose monomers in an α–(1,4) glycosidic bond.

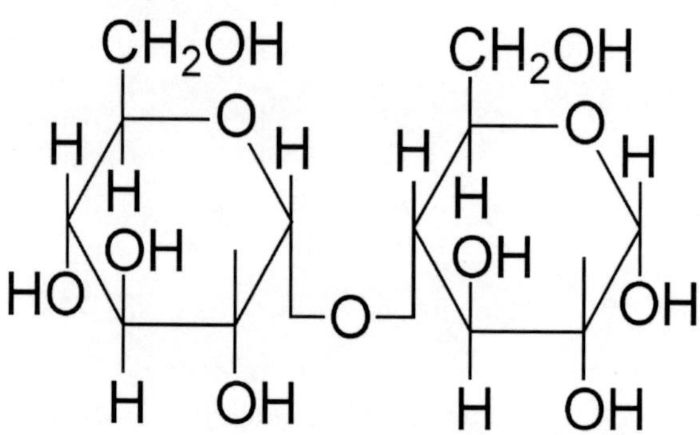

Maltose

Polysaccharides

Most of the carbohydrates found in nature occur in the form of high molecular weight polymers called polysaccharides. The monomeric building blocks used to generate polysaccharides can be varied; in all cases, however, the predominant monosaccharide found in polysaccharides is D-glucose. When polysaccharides are composed of a single monosaccharide building block, they are termed homopolysaccharides. Polysaccharides composed of more than one type of monosaccharide are termed heteropolysaccharides.

Glycogen

Glycogen is the major form of stored carbohydrate in animals. This crucial molecule is a homopolymer of glucose in α–(1,4) linkage; it is also highly branched, with α–(1,6) branch linkages occurring every 8-10 residues. Glycogen is a very compact structure that results from the coiling of the polymer chains. This compactness allows large amounts of carbon energy to be stored in a small volume, with little effect on cellular osmolarity.

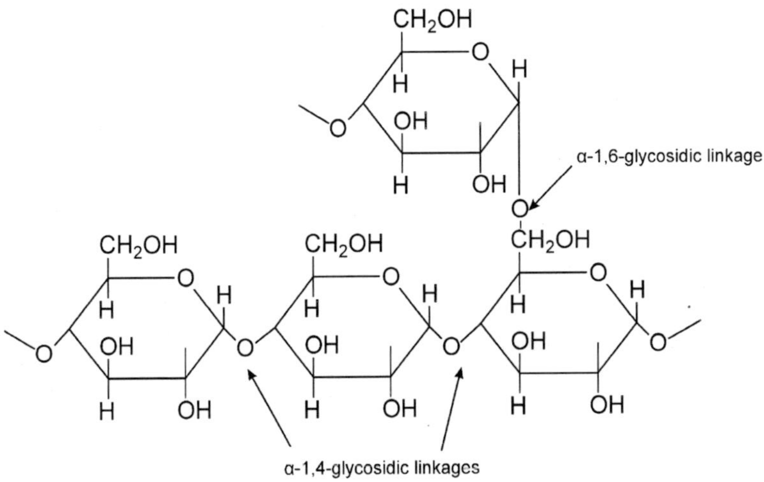

Section of Glycogen Showing α–1,4– and α–1,6–Glycosidic Linkages

Major Roles of Biological Lipids

Biological molecules that are insoluble in aqueous solutions and soluble in organic solvents are classified as lipids. The lipids of physiological importance for humans have four major functions:

1. They serve as structural components of biological membranes.
2. They provide energy reserves, predominantly in the form of triacylglycerols.
3. Both lipids and lipid derivatives serve as vitamins and hormones.
4. Lipophilic bile acids aid in lipid solubilization.

Fatty Acids

Fatty acids fill two major roles in the body:

1. as the components of more complex membrane lipids.
2. as the major components of stored fat in the form of triacylglycerols.

Fatty acids are long-chain hydrocarbon molecules containing a carboxylic acid moiety at one end. The numbering of carbons in fatty acids begins with the carbon of the carboxylate group. At physiological pH, the carboxyl group is readily ionized, rendering a negative charge onto fatty acids in bodily fluids.

Fatty acids that contain no carbon-carbon double bonds are termed saturated fatty acids; those that contain double bonds are unsaturated fatty acids and fatty acids with multiple sites of unsaturation are termed polyunsaturated fatty acids (PUFAs). The numeric designations used for fatty acids come from the number of carbon atoms, followed by the number of sites of unsaturation (eg, palmitic acid is a 16-carbon fatty acid with no unsaturation and is designated by 16:0).

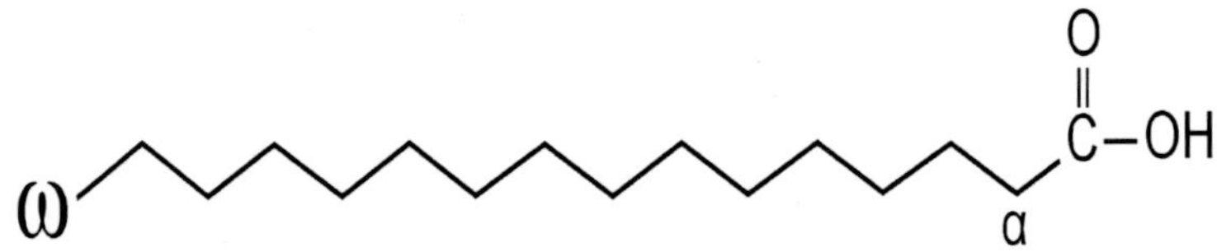

Palmitic Acid

The site of unsaturation in a fatty acid is indicated by the symbol Δ and the number of the first carbon of the double bond relative to the carboxylic acid group (–COOH) carbon which is designated carbon #1. For example, palmitoleic acid is a 16-carbon fatty acid with one site of unsaturation between carbons 9 and 10, and is designated by $16{:}1^{\Delta 9}$.

Saturated fatty acids of less than eight carbon atoms are liquid at physiological temperature, whereas those containing more than ten are solid. The presence of double bonds in fatty acids significantly lowers the melting point relative to a saturated fatty acid. The majority of fatty acids found in the body are acquired in the diet. However, the lipid biosynthetic capacity of the body (fatty acid synthase and other fatty acid modifying enzymes) can supply the body with all the various fatty acid structures needed. Two key exceptions to this are the PUFAs known as linoleic acid and α-linolenic acid, containing unsaturation sites beyond carbons 9 and 10 (relative to the α-COOH group). These two fatty acids cannot be synthesized from precursors in the body, and are thus considered the essential fatty acids; essential in the sense that they must be provided in the diet. Since plants are capable of synthesizing linoleic and α-linolenic acid, humans can acquire these fats by consuming a variety of plants or else by eating the meat of animals that have consumed these plant fats.

These two essential fatty acids are also referred to as omega fatty acids. The use of the greek omega, ω, refers to the end of the fatty acid opposite to that of the –COOH group.

Linoleic acid is an omega-6 PUFA and α-linolenic is an omega-3 PUFA. The role of PUFAs, such as linoleic and α-linolenic, in the synthesis of biologically important lipids is described briefly below and also in the Lipid Synthesis page and the Aspirin page.

Omega-3, and-6, Polyunsaturated Fatty Acids (PUFAs)

The term omega, as it relates to fatty acids, refers to the terminal carbon atom farthest from the functional carboxylic acid group (–COOH).

The designation of a polyunsaturated fatty acid (PUFA) as an omega-3 fatty acid, for example, defines the position of the first site of unsaturation relative to the omega end of that fatty acid.

Thus, an omega-3 fatty acid like α-linolenic acid (ALA), which harbors three carbon-carbon double bonds (i.e sites of unsaturation), has a site of unsaturation between the third and fourth carbons from the omega end.

There are three major types of omega-3 fatty acids that are ingested in foods and used by the body: ALA, eicosapentaenoic acid (EPA), and docosahexaenoic acid (DHA). Once eaten, the body converts ALA to EPA and DHA, the two types of omega-3 fatty acids more readily used by the body and which serve as important precursors for lipid-derived modulators of cell signalling, gene expression and inflammatory processes. It is important to denote that when discussing omega-3 fatty acids, their dietary origin is quite important. Omega-3 fats from plants, such as those in flax seed oil, are enriched in ALA. As indicated above, ALA must first be converted to EPA (requiring three independent reactions) and then to DHA (requiring and additional four reactions). Omega-3 fats from fish are enriched in EPA and DHA and thus do not need to undergo the complex conversion steps required of ALA. In addition, the conversion of ALA to EPA and DHA is inefficient in individuals consuming a typical Western diet rich in animal fats.

Most of the omega-6 PUFAs consumed in the diet are from vegetable oils and consist of linoleic acid. Linoleic acid is converted to γ-linolenic acid (GLA) in the body. GLA should not be confused with ALA which, as pointed out above, is an essential omega-3 PUFA. GLA is converted to dihomo-γ-linolenic acid (DGLA) and then to arachidonic acid as shown in the Eicosanoid Synthesis page.

Due to the limited activity of the Δ^5-desaturase most of the DGLA formed from GLA is inserted into membrane phospholipids at the same C-2 position as for arachidonic acid. GLA can be ingested from several plant-based oils including evening primrose oil, borage oil, and black currant seed oil.

Basic Structure of Triacylglycerides

Triacylglycerides are composed of a glycerol backbone to which 3 fatty acids are esterified.

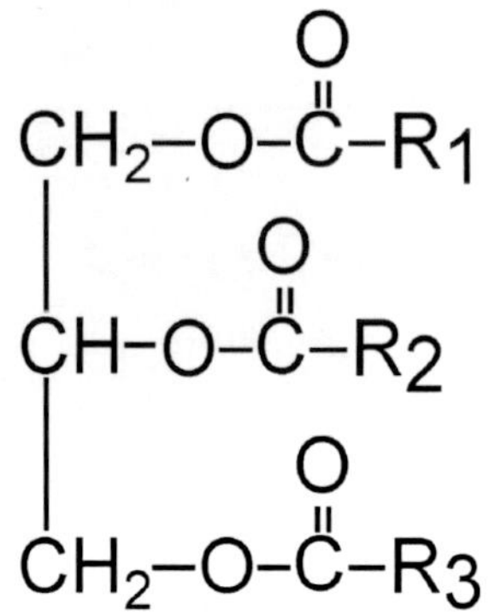

Basic Composition of a Triacylglyceride.

Basic Structure of Phospholipids

The basic structure of phospolipids is very similar to that of the triacylglycerides except that C–3 (*sn*3)of the glycerol backbone is esterified to phosphoric acid.

The building block of the phospholipids is phosphatidic acid which results when the X substitution in the basic structure shown in the Substitutions include ethanolamine (phosphatidylethanolamine), choline (phosphatidylcholine, also called lecithins), serine (phosphatidylserine), glycerol (phosphatidylglycerol), *myo*-inositol (phosphatidylinositol, these compounds can have a variety in the numbers of inositol alcohols that are phosphorylated generating polyphosphatidylinositols), and phosphatidylglycerol (diphosphatidylglycerol more commonly known as cardiolipins).

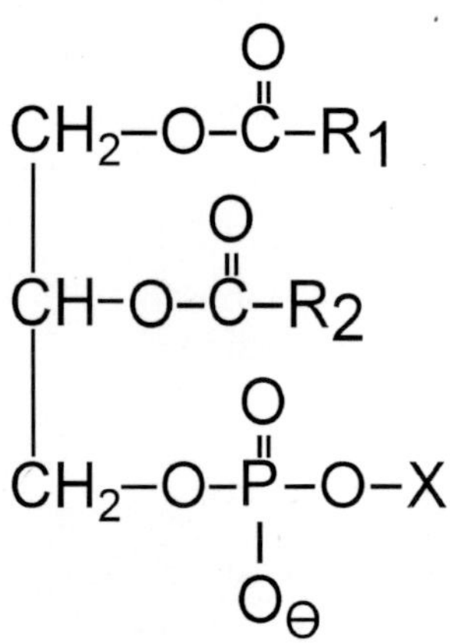

Basic Composition of a Phospholipid.

X can be a number of different substituents.

Basic Structure of Plasmalogens

Plasmalogens are complex membrane lipids that resemble phospholipids, principally phosphatidylcholine. The major difference is that the fatty acid at C–1 (*sn*1) of glycerol contains either an *O*-alkyl (–O–CH_2–) or *O*-alkenyl ether (–O–CH=CH–) species. A basic *O*-alkenyl ether species , where –X can be substituents such as those found in phospholipids described above.

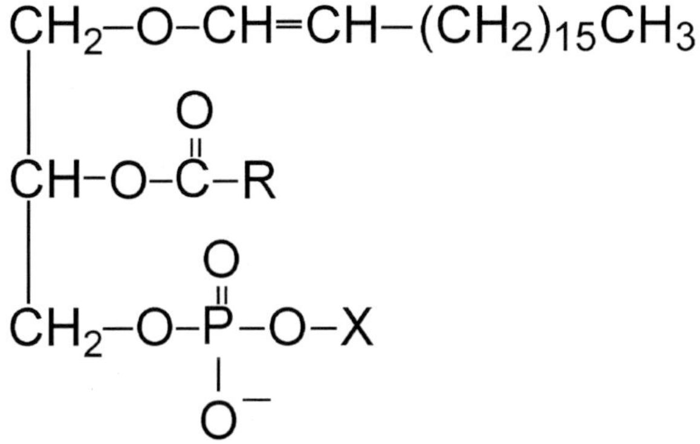

Basic Composition of O-*Alkenyl Plasmalogens*

One of the most potent alkyl ether plasmalogens is platelet activating factor (PAF: 1-*O*-1'-enyl-2-acetyl-*sn*-glycero-3-phosphocholine) which is a choline plasmalogen in which the C–2 (*sn*2) position of glycerol is esterified with an acetyl group instead of a long chain fatty acid.

PAF functions as a mediator of hypersensitivity, acute inflammatory reactions and anaphylactic shock. PAF is synthesized in response to the formation of antigen-IgE complexes on the surfaces of basophils, neutrophils, eosinophils, macrophages and monocytes.

The synthesis and release of PAF from cells leads to platelet aggregation and the release of serotonin from platelets. PAF also produces responses in liver, heart, smooth muscle, and uterine and lung tissues.

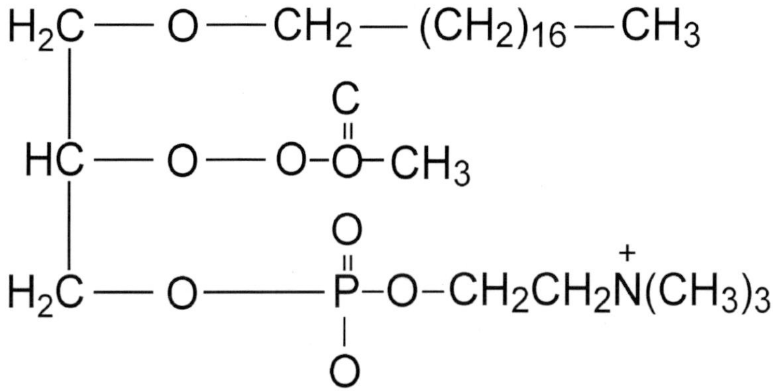

Structure of PAF

Basic Structure of Sphingolipids

Sphingolipids are composed of a backbone of sphingosine which is derived itself from glycerol. Sphingosine is *N*-acetylated by a variety of fatty acids generating a family of molecules referred to as ceramides. Sphingolipids predominate in the myelin sheath of nerve fibers. Sphingomyelin is an abundant sphingolipid generated by transfer of the phosphocholine moiety of phosphatidylcholine to a ceramide, thus sphingomyelin is a unique form of a phospholipid.

The other major class of sphingolipids (besides the sphingomyelins) are the glycosphingolipids generated by substitution of carbohydrates to the *sn*1 carbon of the glycerol backbone of a ceramide. There are 4 major classes of glycosphingolipids:

Cerebrosides: contain a single moiety, principally galactose.

Sulfatides: sulfuric acid esters of galactocerebrosides.

Globosides: contain 2 or more sugars.

Gangliosides: similar to globosides except also contain sialic acid.

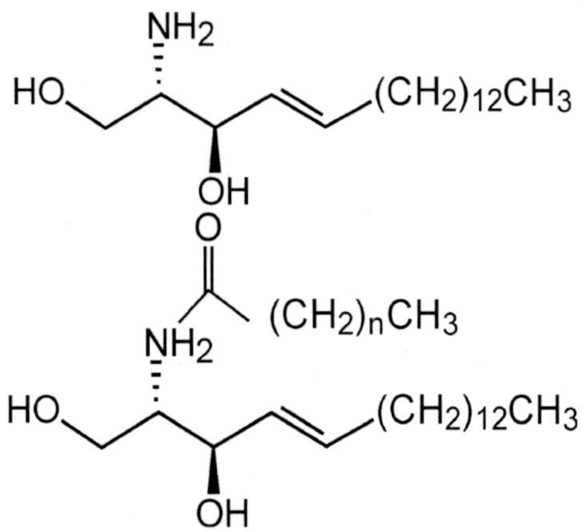

"n" indicates any fatty acid may be N*-acetylated at this position.*

Top: Sphingosine.

Bottom: Basic composition of a ceramide.

Fatty Acid Synthesis

One might predict that the pathway for the synthesis of fatty acids would be the reversal of the oxidation pathway. However, this would not allow distinct regulation of the two pathways to occur even given the fact that the pathways are separated within different cellular compartments.

The pathway for fatty acid synthesis occurs in the cytoplasm, whereas, oxidation occurs in the mitochondria. The other major difference is the use of nucleotide co-factors. Oxidation of fats involves the reduction of $FADH^+$ and

NAD⁺. Synthesis of fats involves the oxidation of NADPH. However, the essential chemistry of the two processes are reversals of each other. Both oxidation and synthesis of fats utilize an activated two carbon intermediate, acetyl-CoA. However, the acetyl-CoA in fat synthesis exists temporarily bound to the enzyme complex as malonyl-CoA.

The synthesis of malonyl-CoA is the first committed step of fatty acid synthesis and the enzyme that catalyzes this reaction, acetyl-CoA carboxylase (ACC), is the major site of regulation of fatty acid synthesis. Like other enzymes that transfer CO_2 to substrates, ACC requires a biotin co-factor.

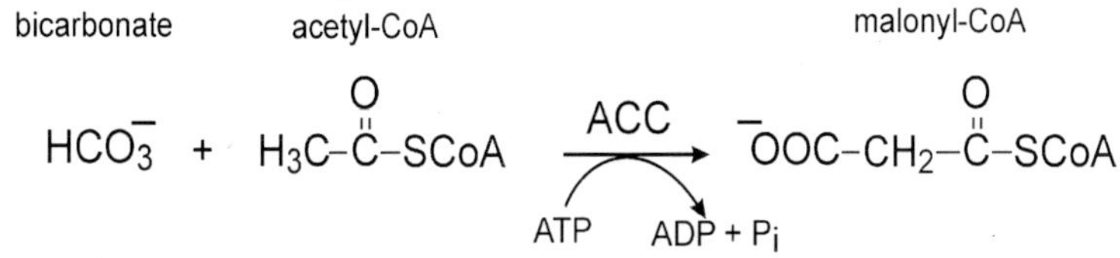

The rate of fatty acid synthesis is controlled by the equilibrium between monomeric ACC and polymeric ACC. The activity of ACC requires polymerization. This conformational change is enhanced by citrate and inhibited by long-chain fatty acids. ACC is also controlled through hormone mediated phosphorylation. The acetyl groups that are the products of fatty acid oxidation are linked to CoASH. As you should recall, CoA contains a phosphopantetheine group coupled to AMP. The carrier of acetyl groups (and elongating acyl groups) during fatty acid synthesis is also a phosphopantetheine prosthetic group, however, it is attached a serine hydroxyl in the synthetic enzyme complex. The carrier portion of the synthetic complex is called acyl carrier protein, ACP. This is somewhat of a misnomer in eukaryotic fatty acid synthesis since the ACP portion of the synthetic complex is simply one of many domains of a single polypeptide. The acetyl-CoA and malonyl-CoA are transferred to ACP by the action of acetyl-CoA transacylase and malonyl-CoA transacylase, respectively. The attachment of these carbon atoms to ACP allows them to enter the fatty acid synthesis cycle.

The synthesis of fatty acids from acetyl-CoA and malonyl-CoA is carried out by fatty acid synthase, FAS. The active enzyme is a dimer of identical subunits.

All of the reactions of fatty acid synthesis are carried out by the multiple enzymatic activities of FAS. Like fat oxidation, fat synthesis involves 4 enzymatic activities. These are, β-keto-ACP synthase, β-keto-ACP reductase, 3-OH acyl-ACP dehydratase and enoyl-CoA reductase. The two reduction reactions require NADPH oxidation to $NADP^+$.

The primary fatty acid synthesized by FAS is palmitate. Palmitate is then released from the enzyme and can then undergo separate elongation and/or unsaturation to yield other fatty acid molecules.

Origin of Cytoplasmic Acetyl-CoA

Acetyl-CoA is generated in the mitochondria primarily from two sources, the pyruvate dehydrogenase (PDH) reaction and fatty acid oxidation. In order for these acetyl units to be utilized for fatty acid synthesis they must be present in the cytoplasm. The shift from fatty acid oxidation and glycolytic oxidation occurs when the need for energy diminishes. This results in reduced oxidation of acetyl-CoA in the TCA cycle and the oxidative phosphorylation pathway. Under these conditions the mitochondrial acetyl units can be stored as fat for future energy demands.

Acetyl-CoA enters the cytoplasm in the form of citrate via the tricarboxylate transport system. In the cytoplasm, citrate is converted to oxaloacetate and acetyl-CoA by the ATP driven ATP-citrate lyase reaction. This reaction is essentially the reverse of that catalysed by the TCA enzyme citrate synthase except it requires the energy of ATP hydrolysis to drive it forward. The resultant oxaloacetate is converted to malate by malate dehydrogenase (MDH).

Pathway for the movement of acetyl-CoA units from within the mitochondrion to the cytoplasm for use in lipid and cholesterol biosynthesis. Note that the cytoplasmic malic enzyme catalysed reaction generates NADPH which can be used for reductive biosynthetic reactions such as those of fatty acid and cholesterol synthesis.

The malate produced by this pathway can undergo oxidative decarboxylation by malic enzyme. The co-enzyme for this reaction is $NADP^+$ generating NADPH. The advantage of this series of reactions for converting mitochondrial acetyl-CoA into cytoplasmic acetyl-CoA is that the NADPH produced by the malic enzyme reaction can be a major source of reducing co-factor for the fatty acid synthase activities.

Regulation of Fatty Acid Metabolism

One must consider the global organismal energy requirements in order to effectively understand how the synthesis and degradation of fats (and also carbohydrates) needs to be exquisitely regulated. The blood is the carrier of triacylglycerols in the form of VLDLs and chylomicrons, fatty acids bound to albumin, amino acids, lactate, ketone bodies and glucose. The pancreas is the primary organ involved in sensing the organisms dietary and energetic states via glucose concentrations in the blood. In response to low blood glucose, glucagon is secreted, whereas, in response to elevated blood glucose insulin is secreted. The regulation of fat metabolism occurs via two distinct mechanisms. One is short term regulation which is regulation effected by events such as substrate availability, allosteric effectors and/or enzyme modification.

ACC is the rate-limiting (committed) step in fatty acid synthesis. There are two major isoforms of ACC in mammalian tissues. These are identified as ACC1 and ACC2. ACC1 is strictly cytosolic and is enriched in liver, adipose tissue and lactating mammary tissue. ACC2 was originally discovered in rat heart but is also expressed in liver and skeletal muscle. ACC2 has an *N*-terminal extension that contains a mitochondrial targeting motif and is found associated with carnitine palmitoyltransferase I (CPT I) allowing for rapid regulation of CPT I by the malonyl-CoA produced by ACC. Both isoforms of ACC are allosterically activated by citrate and inhibited by palmitoyl-CoA and other short- and long-chain fatty acyl-CoAs. Citrate triggers the polymerization of ACC1 which leads to significant increases in its activity. Although ACC2 does not undergo significant polymerization (presumably due to its mitochondrial association) it is allosterically activated by citrate. Glutamate and other dicarboxylic acids can also allosterically activate both ACC isoforms.

ACC activity can also be affected by phosphorylation. Both ACC1 and ACC2 contain at least eight sites that undergo phosphorylation. The sites of phosphorylation in ACC2 have not been as extensively studied as those in ACC1. Phosphorylation of ACC1 at three serine residues (S79, S1200, and S1215) by AMPK leads to inhibition of the enzyme. Glucagon-stimulated increases in cAMP and subsequently to increased PKA activity also lead to phosphorylation of ACC where ACC2 is a better substrate for PKA than is ACC1. The activating effects of insulin on ACC are complex and not completely resolved. It is known that insulin leads to the dephosphorylation of the serines in ACC1 that are AMPK targets in the heart enzyme. This insulin-mediated effect has not been observed in hepatocytes or adipose tissues cells. At least a portion of the activating effects of insulin are related to changes in cAMP levels. Early evidence has shown that phosphorylation and activation of ACC occurs via the action of an insulin-activated kinase. However, contradicting evidence indicates that although there is insulin-mediated phosphorylation of ACC this does not result in activation of the enzyme. Activation of α-adrenergic receptors in liver and skeletal muscle cells inhibits ACC activity as a result of phosphorylation by an as yet undetermined kinase.

Control of a given pathways' regulatory enzymes can also occur by alteration of enzyme synthesis and turn-over rates. These changes are long term regulatory effects. Insulin stimulates ACC and FAS synthesis, whereas, starvation leads to decreased synthesis of these enzymes. Adipose tissue lipoprotein lipase levels also are increased by insulin and decreased by starvation. However, in contrast to the effects of insulin and starvation on adipose tissue, their effects on heart lipoprotein lipase are just the inverse. This allows the heart to absorb any available fatty acids in the blood in order to oxidize them for energy production. Starvation also leads to increases in the levels of fatty acid oxidation enzymes in the heart as well as a decrease in FAS and related enzymes of synthesis.

Adipose tissue contains hormone-sensitive lipase (HSL), that is activated by PKA-dependent phosphorylation leading to increased fatty acid release to the blood. The activity of HSL is also affected through the action of AMPK, however, AMPK-mediated phosphorylation of HSL is inhibitory. The phosphorylation and inhibition of HSL by AMPK may seem paradoxical since the release of fatty acids stored in triglycerides would seem necessary to promote the production of ATP via fatty acid oxidation and the major function of AMPK is to shift cells to ATP production from ATP consumption. This paradigm can be explained if one considers that if the fatty acids that are released from triglycerides are not consumed they will be recycled back into triglycerides at the expense of ATP consumption. Thus, it has been proposed that inhibition of HSL by AMPK mediated-phosphorylation is a mechanism to ensure that the rate of fatty acid release does not exceed the rate at which they are utilized either by export or oxidation.

In the liver the net result of activation of HSL (due to increased acetyl-CoA levels) is the production of ketone bodies. This would occur under conditions where insufficient carbohydrate stores and gluconeogenic precursors were available in liver for increased glucose production. The increased fatty acid availability in response to glucagon or epinephrine is assured of being completely oxidized since both PKA and AMPK also phosphorylate (and as a result inhibits) ACC, thus inhibiting fatty acid synthesis.

Insulin, on the other hand, has the opposite effect to glucagon and epi leading to increased glycogen and triacylglyceride synthesis. One of the many effects of insulin is to lower cAMP levels which leads to increased dephosphorylation through the enhanced activity of protein phosphatases such as PP-1. With respect to fatty acid metabolism this yields dephosphorylated and inactive hormone sensitive lipase. Insulin also stimulates certain phosphorylation events. This occurs through activation of several cAMP-independent kinases.

Regulation of fat metabolism also occurs through malonyl-CoA induced inhibition of carnitine acyltransferase I. This functions to prevent the newly synthesized fatty acids from entering the mitochondria and being oxidized.

ChREBP: Master Lipid Regulator in the Liver

When glycogen stores are maximal in the liver, excess glucose is diverted into the lipid synthesis pathway. Glucose is catabolized to acetyl-CoA and the acetyl-CoA is used for *de novo* fatty acid synthesis. The fatty acids are then incorporated into triglycerides and exported from hepatocytes as very-low-density lipoproteins and ultimately stored as triglycerides in adipose tissue. A diet rich in carbohydrates leads to stimulation of both the glycolytic and lipogenic pathways. Genes encoding glucokinase (GK) and liver pyruvate kinase (L-PK) of glycolysis and ATP citrate lyase, ACC, and FAS of lipogenesis are regulated by modulation of their transcription rates. In addition, the enzymes encode by these genes are subject to post-translational and allosteric regulation. These genes contain glucose- or carbohydrate-response elements (ChoREs) that are responsible for their transcriptional regulation.

One transcription factor that exerts control over glucose and lipid homeostasis is sterol-response element-binding protein (SREBP), in particular SREBP-1c. The regulation and actions of SREBP are discussed in the Cholesterol Metabolism page. SREBP controls the expression of a number of genes involved in lipogenesis and its own transcription is increased by insulin and repressed by glucagon. However, SREBP activity alone cannot account for the stimulation of glycolytic and lipogenic gene expression in response to a carbohydrate rich diet. A search for potential additional regulatory factors revealed a basic helix-loop-helix/leucine zipper (bHLH/LZ) transcription factor which was identified as carbohydrate-responsive element-binding protein (ChREBP). ChREBP was identified as a major glucose-responsive transcription factor and it is required for glucose-induced expression of L-PK and the lipogenic genes ACC and FAS.

Expression of the ChREBP gene is induced in the liver in response to increased glucose uptake. In addition to gene activation, the activity of ChREBP is regulated by post-translational modifications as well as sub-cellular localization. The kinases PKA and AMPK both phosphorylate ChREBP rendering it inactive as a transcriptional activator. PKA is known to phosphorylate ChREBP on serine 196 (S196) and threonine 666 (T666), wheras, AMPK phosphorylates ChREBP at serine 568 (S568). Under conditions of low (basal) glucose concentration, ChREBP is phosphorylated and resides in the cytosol. When glucose levels rise, protein phosphatase 2A delta (PP2Aδ) is activated by xylulose 5-phosphate which is an intermediate in the pentose phosphate pathway. PP2Aδ dephosphorylates S196 resulting in translocation of ChREBP into the nucleus. In the nucleus PP2Aδ dephosphorylates T666 which allows ChREBP to bind to specific sequence elements (ChoREs) in target genes. ChREBP does not bind to ChoREs as a typical homodimeric bHLH transcription factor. ChREBP interacts with another bHLH protein identified as MAX-like protein X (MLX). MLX is a member of the MYC/MAX/MAD family of transcription factors that serve as interacting partners in transcription factor networks.

Genes whose expression is under control of ChREBP activity include L-PK, ACC and FAS as indicated above. In addition, it has been shown that when expression of ChREBP is reduced the expression levels of glycerol 3-phosphate acyltransferase (GPAT) and Δ^9-stearoly-CoA desaturase (SCD) are also reduced. GPAT is the enzyme that esterifies glycerol-3-phospate generating lysophosphatidic acid which is the first step in the synthesis of triacylglycerols (TAGs) as described below. SCD is the rate-limiting enzyme involved in the synthesis of the major monounsaturated fatty acids oleic acid (18:1) and palmitoleic acid (16:1) as described below in the section on Elongation and Desaturation.

The liver X receptors (LXRs) are members of the steroid/thyroid hormone superfamily of cytosolic ligand binding receptors that migrate to the nucleus upon ligand binding and regulate gene expression by binding to specific target sequences. There are two forms of the LXRs, LXRα and LXRβ. The LXRs form heterodimers with the retinoid X receptors (RXRs) and as such can regulate gene expression either upon binding oxysterols (e.g. 22*R*-hydroxycholesterol) or 9-*cis*-retinoic acid. LXRs are also important regulators of the lipogenic pathway. Recent evidence has demonstrated that the ChREBP gene is a direct target of LXRs and that glucose itself can bind and activate LXRs.

An emerging model of the role of ChREBP in overall glucose and lipid metabolism indicates that this transcription factor is a master regulator of glucose-mediated lipid homeostasis not only in the liver but also in adipose tissue. In the liver ChREBP controls 50% of overall lipogenesis through its concerted actions on the expression of lipogenic and glycolytic genes.

Elongation and Desaturation

The fatty acid product released from FAS is palmitate (via the action of palmitoyl thioesterase) which is a 16:0 fatty acid, i.e. 16 carbons and no sites of unsaturation. Elongation and unsaturation of fatty acids occurs in both the mitochondria and endoplasmic reticulum (microsomal membranes). The predominant site of these processes is in the ER membranes. Elongation involves condensation of acyl-CoA groups with malonyl-CoA. The resultant product is two carbons longer (CO_2 is released from malonyl-CoA as in the FAS reaction) which undergoes reduction, dehydration and reduction yielding a saturated fatty acid. The reduction reactions of elongation require NADPH as co-factor just as for the similar reactions catalysed by FAS. Mitochondrial elongation involves acetyl-CoA units and is a reversal of oxidation except that the final reduction utilizes NADPH instead of $FADH_2$ as co-factor.

The desaturation of fatty acids occurs in the ER membranes as well. In mammalian cells fatty acid desaturation involves 3 broad specificity fatty acyl-CoA desaturases (non-heme iron containing enzymes). These enzymes introduce unsaturation at C5, C6 or C9. The names of these enzymes are Δ^5-eicosatrienoyl-CoA desaturase, Δ^6-oleoyl(linolenoyl)-CoA desaturase and Δ^9-stearoyl-CoA desaturase. The latter desaturase (SCD) is the rate-limiting enzyme catalysing the synthesis of monounsaturated fatty acids, primarily oleate (18:1) and palmitoleate (16:1). These two monounsaturated fatty acids represent the majority of monounsaturated fatty acids present in membrane phospholipids, triglycerides, and cholesterol esters. The expression of SCD is under the control of the transcription factor ChREBP. The ratio of saturated to monounsaturated fatty acids in membrane phospholipids is critical to normal cellular function and alterations in this ratio have been correlated with diabetes, obesity, cardiovascular disease, and cancer. Thus, regulation of the expression and activity of SCD has important physiological significance.

The electrons transferred from the oxidized fatty acids during desaturation are transferred from the desaturases to cytochrome b_5 and then NADH-cytochrome b_5 reductase. These electrons are un-coupled from mitochondrial oxidative-phosphorylation and, therefore, do not yield ATP.

Since these enzymes cannot introduce sites of unsaturation beyond C9 they cannot synthesize either linoleate ($18{:}2^{\Delta 9,12}$) or linolenate ($18{:}3^{\Delta 9,12,15}$). These fatty acids must be acquired from the diet and are, therefore, referred to as essential fatty acids. Linoleic is especially important in that it required for the synthesis of arachidonic acid. Arachindonate is a precursor for the eicosanoids (the prostaglandins, thromboxanes, and leukotrienes). It is this role of fatty acids in eicosanoid synthesis that leads to poor growth, wound healing and dermatitis in persons on fat free diets. Also, linoleic acid is a constituent of epidermal cell sphingolipids that function as the skins' water permeability barrier.

Omega-3, and-6 Polyunsaturated Fatty Acids (PUFAs)

The term omega, as it relates to fatty acids, refers to the terminal carbon atom farthest from the functional carboxylic acid group (–COOH). The designation of a polyunsaturated fatty acid (PUFA) as an omega-3 fatty acid, for example, defines the position of the first site of unsaturation relative to the omega end of that fatty acid. Thus, an omega-3 fatty acid like α-linolenic acid (ALA), which harbors three carbon-carbon double bonds (i.e sites of unsaturation), has a site of unsaturation between the third and fourth carbons from the omega end.

There are three major types of omega-3 fatty acids that are ingested in foods and used by the body and these are ALA, eicosapentaenoic acid (EPA), and docosahexaenoic acid (DHA). Once eaten, the body converts ALA to EPA and then to DHA. EPA and DHA are the two types of omega-3 fatty acids that serve as important precursors for lipid-derived modulators of cell signalling, gene expression and inflammatory processes. There are numerous other omega-3 PUFAs found in nature but for the purposes of this discussion focus is placed on ALA, EPA, and DHA.

Most of the ALA consumed in the diet comes from plant sources such as flax seed, walnuts, pecans, hazelnuts, and kiwifruit. There is a small percentage of omega-3 PUFAs that come from meats common to Western diets such as chicken and beef, however, this is mostly ALA. The highest concentrations of EPA and DHA are found in cold water fishes such as salmon, tuna, and herring. The most important PUFAs, biologically, are EPA and DHA. Although ALA can serve as the precursor for EPA and DHA synthesis in humans, this pathway is limited in its capacity and also varies between individuals. Therefore, direct dietary intake of omega-3 fats rich in EPA and DHA are of the most benefit clinically.

Most of the omega-6 PUFAs consumed in the diet are from vegetable oils such as soybean oil, corn oil, borage oil, and acai berry and consist of linoleic acid. Linoleic acid is converted to arachidonic acid through the steps outlined in the eicosanoid synthesis page. The activity of the Δ^6-desaturase is slow and can be further compromised due to nutritional deficiencies as well as during inflammatory conditions. Therefore, maximal capacity for synthesis of arachidonic acid occurs with ingested γ-linolenic acid (GLA) the product of the Δ^6-desaturase. GLA is converted to dihomo-γ-linolenic acid (DGLA) and then to arachidonic acid. Like the Δ^6-desaturase, the activity of the Δ^5-desaturase is limiting in arachidonic acid synthesis and its activity is also influenced by diet and environmental factors. Due to the limited activity of the Δ^5-desaturase most of the DGLA formed from GLA is inserted into membrane phospholipids at the same C–2 position as for arachidonic acid. GLA itself can be ingested from several plant-based oils including borage oil, and acai berry.

Clinical Significance of Omega-3, and -6 PUFAs

It is important to denote that when discussing omega-3 fatty acids, their dietary origin is quite important. Omega-3 fats from plants, such as those in flax seed oil, are enriched in ALA. As indicated above, ALA must first be converted to EPA (requiring three independent reactions) and then to DHA (requiring and additional four reactions). Omega-3 fats from fish are enriched in EPA and DHA and thus do not need to undergo the complex conversion steps required of ALA. In addition, the conversion of ALA to EPA and DHA is inefficient in individuals consuming a typical Western diet rich in animal fats.

When omega-3 and omega-6 fatty acids are consumed they are incorporated into cell membranes in all tissues of the body. Because of this fact, dietary changes in the composition of PUFAs can have profound effects on a cell's function because the membrane lipids serve as a source of precursors for the synthesis of important signalling molecules involved in cell growth and development as well as modulation of inflammation. Another important consequence of dietary alteration in fatty acid composition is the fact that omega-3 and omega-6 PUFAs compete for incorporation into cell membranes.

The most important omega-6 PUFA is arachidonic acid. When cells are stimulated by a variety of external stimuli, arachidonic acid is released from cell membranes through the action of phospholipase A_2 (PLA_2).

The released arachidonate then serves as the precursor for the synthesis of the biologically active eicosanoids, the prostaglandins (PGs), thromboxanes (TXs), and leukotrienes. (LTs) The arachidonate-derived eicosanoids function in diverse biological phenomena such as platelet and leukocyte activation, signalling of pain, induction of bronchoconstriction, and regulation of gastric secretions. These activities are targets of numerous pharmacological agents such as the non-steroidal anti-inflammatory drugs (NSAIDs), COX-2 inhibitors, and leukotriene antagonists.

Dietary omega-3 PUFAs compete with the inflammatory, pyretic (fever), and pain promoting properties imparted by omega-6 PUFAs because they displace arachidonic acid from cell membranes. In addition, omega-3 PUFAs compete with the enzymes that convert arachidonic acid into the biological eicosnaoids (PGs, TXs, and LTs). The net effect of increasing dietary consumption of omega-3 PUFAs, relative to omega-6 PUFAs, is to decrease the potential for monocytes, neutrophils, and eosinophils (i.e. leukocytes) to synthesize potent mediators of inflammation and to reduce the ability of platelets to release TXA_2, a potent stimulator of the coagulation process.

Probably the most important role of the omega-3 PUFAs, EPA and DHA, is that they serve as the precursors for potent anti-inflammatory lipids called resolvins (Rvs) and protectins (PDs). The Rvs exert their anti-inflammatory actions by promoting the resolution of the inflammatory cycle, hence the derivation of their names as resolvins.

The resolvins are synthesized either from EPA or DHA. The D series resolvins are derived from DHA and the E series from EPA. An additional anti-inflammatory lipid derived from DHA is protectin D1 (PD1). The E series resolvins reduce inflammation, regulate PMN infiltration by blocking transendothelial migration, reduce dendritic cell function (dendritic cells are potent antigen presenting cells which prime T cell mediated inflammatory responses), regulate IL-12 production and lead to resolution of the inflammatory responses.

The omega-3 fatty acids DHA and EPA have also been shown to be important for normal brain development and function. Several studies have demonstrated that DHA is essential for proper development of the prenatal and postnatal central nervous system. The benefits of EPA appear to be in its effects on behaviour and mood. In clinical studies with DHA and EPA there has been good data demonstrating benefit in treating attention deficit hyperactivity disorder (ADHD), autism, dyspraxia (motor skills disorder), dyslexia, and aggression. In patients with affective disorders consumption of DHA and EPA has confirmed benefits in major depressive disorder and bipolar disorder. In addition, some studies have demonstrated promising results in treatment of schizophrenia with some minor benefits in patients with borderline personality disorder. Of significance to these effects of EPA and DHA on cognition, mood and behaviour is the fact that administration of omega-3 fatty acid containing phospholipids (such as those present in Krill oils) are significantly better than omega-3 containing triacylglycerides such as those that predominate in fish oils.

Research over the past 10–15 years has demonstrated the physiological benefits (i.e. anti-inflammatory) of alternative pathways of polyunsaturated fatty acid metabolism. Much of this topic is covered in the Aspirin page. As described in the Eicosanoid Synthesis page, much of the DGLA derived from ingested linoleic acid or GLA is diverted into membrane phospholipids due to the inefficiency of the Δ^5-desaturase catalysing the conversion of DGLA to arachidonic acid.

Incorporation of DGLA into membrane phospholipids competes with the incorporation of arachidonate so that diets enriched in GLA result in an alteration in the ratio of membrane arachidonate to DGLA. Release of membrane DGLA occurs through the action of PLA_2 just as for release of arachidonate. Once DGLA is released it will compete with arachidonate for COXs and LOXs. The products of COX action on DGLA are series-1 prostaglandins (PGE_1) and thromboxanes (TXA_1). These eicosanoids are structurally similar to the series-2 eicosanoids except, of course, they have a single double bond. Although structurally similar, the series-1 eicosanoids have distinctly different biological actions. PGE_1 and TXA_1 are anti-inflammatory, they induce vasodilation, and they inhibit platelet aggregation. When DGLA is a substrate for 15-LOX the product is 15-hydroxyeicosatrienoic acid (15-HETrE). 15-HETrE is a potent inhibitor of 5-LOX which is the enzyme responsible for the conversion of arachidonic acid to LTB_4. LTB_4 is a potent pro-inflammatory molecule through its action on neutrophils, thus, DGLA serves to inhibit inflammation via the linear eicosanoid pathway as well.

Due to the vasodilating action of PGE_1 it is used pharmaceutically as aprostadil for the treatment of erectile dysfunction (ED). The ED applications of PGE_1 are sold as MUSE® and Caverject®. MUSE is a urethral suppository and Caverject is an injectable version. Aprostadil is also used clinically to treat newborn infants with ductal-dependent congenital heart disease. The administration of aprostadil in these infants maintains a patent ductus arteriosus until surgery can be carried out to correct the underlying heart defect. Ductus arteriosus is a normal structure of the fetal heart that allows blood to bypass circulation to the lungs since the fetus does not use his/her lungs *in utero*. The ductus arteriosus shunts blood flow from the left pulmonary artery to the aorta. Shortly after birth the ductus closes due to the high levels of oxygen the newborn is exposed to at birth. However, in newborns with certain congenital heart defects, maintaining a patent ductus arteriosus is clinically significant.

Omega-3 PUFAs also are known to regulate hepatic lipid metabolism via regulation of the expression of key enzymes involved in lipid synthesis and catabolism. Omega-3 PUFAs bind to and activate peroxisome proliferator-activated receptor-α (PPARα). Activation of PPARα results in activation of hepatic fatty acid oxidation. At the same time these PUFAs inhibit hepatic fatty acid synthesis. This effect is mediated by suppression of SREBP-1c gene expression, enhanced degradation of SREBP1c mRNA and increased proteosomal degradation of SREBP-1 proteins.

The decrease in SREBP-1c expression results in a reduction in the expression levels of hepatic fatty acid synthase (FAS) and acetylCoA carboxylase (ACC) two critical enzymes of fatty acid synthesis. The net effect of fatty acid control of hepatic lipid metabolism is that the composition as well as the type and quantity of lipids available for VLDL synthesis and secretion are affected. Because the liver plays a central role in whole-body lipid metabolism this regulatory process in turn affects the lipid composition throughout the body and, therefore likely contributes to the onset and progression of several chronic diseases, including atherosclerosis, diabetes, and obesity.

Synthesis of Triglycerides

Fatty acids are stored for future use as triacylglycerols (TAGs) in all cells, but primarily in adipocytes of adipose tissue. TAGs constitute molecules of glycerol to which three fatty acids have been esterified. The fatty acids present in TAGs are predominantly saturated. The major building block for the synthesis of TAGs, in tissues other than adipose tissue, is glycerol. Adipocytes lack glycerol kinase, therefore, dihydroxyacetone phosphate (DHAP), produced during glycolysis, is the precursor for TAG synthesis in adipose tissue. This means that adipoctes must have glucose to oxidize in order to store fatty acids in the form of TAGs. DHAP can also serve as a backbone precursor for TAG synthesis in tissues other than adipose, but does so to a much lesser extent than glycerol.

The glycerol backbone of TAGs is activated by phosphorylation at the C-3 position by glycerol kinase. The utilization of DHAP for the backbone is carried out through either of two pathways depending upon whether the synthesis of triglycerides is carried out in the mitochondria and ER or the ER and the peroxisomes. In the former case the action of glycerol-3-phosphate dehydrogenase, a reaction that requires NADH (the same reaction as that used in the glycerol-phosphate shuttle), converts DHAP to glycerol-3-phosphate. Glycerol-3-phosphate acyltransferase (GPAT) then esterifies a fatty acid to glycerol-3-phosphate generating the monoacylglycerol phosphate structure called lysophosphatidic acid. The expression of the GPAT gene is under the influence of the transcription factor ChREBP as described above.

The second reaction pathway utilizes the peroxisomal enzyme DHAP acyltransferase to fatty acylate DHAP to acyl-DHAP which is then reduced by the NADPH-requiring enzyme acyl-DHAP reductase. An interesting feature of the latter pathway is that DHAP acyltransferase is one of only a few enzymes that are targeted to the peroxisomes through the recognition of a peroxisome targeting sequence 2 (PTS2) motif in the enzyme. Most peroxisomal enzymes contain a PTS1 motif.

The fatty acids incorporated into TAGs are activated to acyl-CoAs through the action of acyl-CoA synthetases. Two molecules of acyl-CoA are esterified to glycerol-3-phosphate to yield 1,2-diacylglycerol phosphate (commonly identified as phosphatidic acid). The phosphate is then removed, by phosphatidic acid phosphatase (PAP1), to yield 1,2-diacylglycerol, the substrate for addition of the third fatty acid. Intestinal monoacylglycerols, derived from the hydrolysis of dietary fats, can also serve as substrates for the synthesis of 1,2-diacylglycerols.

Lipin Genes: TAG Synthesis and Transcriptional Regulation

Recent studies have identified a critical role for the enzyme PAP1 in overall TAG and phospholipid homeostasis. In the yeast *Saccharomyces cerevisiae*, the PAP1 gene was identified as Smp2p and the encoded protein was shown to be the yeast ortholog of the mammalian protein called lipin-1. The fission yeast lipin-1 ortholog is identified as Ned1p. Lipin-1 is only one of five lipin proteins identified in mammals. The lipin-1 gene (symbol = LPIN1) was originally identified in a mutant mouse called the fatty liver dystrophy (fld) mouse. The mutation causing this disorder was found to reside in the LPIN1 gene. There are three lipin genes (LPIN1, LPIN2, and LPIN3) with the LPIN1 gene encoding three isoforms derived through alternative splicing. These three lipin-1 isoforms are identified as lipin-1α, lipin-1β, and lipin-1γ. All five lipin proteins possess phosphatidic acid phosphatase activity that is dependent upon Mg^{2+} or Mn^{2+} and phosphatidic acid as the substrate. Mutations in the LPIN2 gene have recently been associated with Majeed syndrome which is characterized by chronic recurrent osteomyelitis, cutaneous inflammation, recurrent fever, and congenital dyserythropoietic anemia.

In addition to the obvious role of lipin-1 in TAG synthesis, evidence indicates that the protein is also required for the development of mature adipocytes, coordination of peripheral tissue glucose and fatty acid storage and utilization, and serves as a transcriptional co-activator. The latter function has significance to diabetes as it has been shown that some of the effects of the thiazolidinedione (TZD) class of drugs used to treat the hyperglycemia associated with type 2 diabetes are exerted via the effects of lipin-1. Although the lipin proteins do not contain DNA-binding motifs they have protein-interaction domains that allow them to form complexes with nuclear receptors and function as transcriptional regulators. Lipin-1 has been shown to interact with peroxisome proliferator-activated receptor-γ [PPARγ] co-activator 1α (PGC-1α) and PPARα leading to enhanced gene expression. Lipin-1 also is known to interact with additional members of the nuclear receptor family including the glucocorticoid receptor (GR) and hepatocyte nuclear factor-4α (HNF-4α). Lipin-1 also induces the expression of the adipogenic transcription factors PPARγ and CCAAT-enhancer-binding protein α (C/EBPα).

The functions of lipin-1α and lipin-1β appear to be complimentary with respect to adipocyte differentiation. Lipin-1α induces genes that promote adipocyte differentiation while lipin-1β induces the expression of lipid synthesizing genes such as fatty acid synthase (FAS) and diacylglycerol acyltransferase (DGAT). The interactions of lipin-1 with PPARα and PGC-1α leads to increased expression of fatty acid oxidizing genes such as carnitine palmitoyl transferase-1, acyl CoA oxidase, and medium-chain acylCoA dehydrogenase (MCAD).

Phospholipid Synthesis

Phospholipids can be synthesized by two mechanisms. One utilizes a CDP-activated polar head group for attachment to the phosphate of phosphatidic acid. The other utilizes CDP-activated 1,2-diacylglycerol and an inactivated polar head group.

PC: This class of phospholipids is also called the lecithins. At physiological pH, phosphatidylcholines are neutral zwitterions. They contain primarily palmitic or stearic acid at carbon 1 and primarily oleic, linoleic or linolenic acid at carbon 2. The lecithin dipalmitoyllecithin is a component of lung or pulmonary surfactant. It contains palmitate at both carbon 1 and 2 of glycerol and is the major (80%) phospholipid found in the extracellular lipid layer lining the pulmonary alveoli.

Choline is activated first by phosphorylation and then by coupling to CDP prior to attachment to phosphatidic acid. PC is also synthesized by the addition of choline to CDP-activated 1,2-diacylglycerol. A third pathway to PC synthesis, involves the conversion of either PS or PE to PC. The conversion of PS to PC first requires decarboxylation of PS to yield PE; this then undergoes a series of three methylation reactions utilizing S-adenosylmethionine (SAM) as methyl group donor.

PE: These molecules are neutral zwitterions at physiological pH. They contain primarily palmitic or stearic acid on carbon 1 and a long chain unsaturated fatty acid (e.g. 18:2, 20:4 and 22:6) on carbon 2. Synthesis of PE can occur by two pathways. The first requires that ethanolamine be activated by phosphorylation and then by coupling to CDP. The ethanolamine is then transferred from CDP-ethanolamine to phosphatidic acid to yield PE. The second involves the decarboxylation of PS.

PS: Phosphatidylserines will carry a net charge of –1 at physiological pH and are composed of fatty acids similar to the phosphatidylethanolamines. The pathway for PS synthesis involves an exchange reaction of serine for ethanolamine in PE. This exchange occurs when PE is in the lipid bilayer of the a membrane. As indicated above, PS can serve as a source of PE through a decarboxylation reaction.

PI: These molecules contain almost exclusively stearic acid at carbon 1 and arachidonic acid at carbon 2. Phosphatidylinositols composed exclusively of non-phosphorylated inositol exhibit a net charge of –1 at physiological pH. These molecules exist in membranes with various levels of phosphate esterified to the hydroxyls of the inositol. Molecules with phosphorylated inositol are termed polyphosphoinositides. The polyphosphoinositides are important intracellular transducers of signals emanating from the plasma membrane. The synthesis of PI involves CDP-activated 1,2-diacylglycerol condensation with *myo*-inositol. PI subsequently undergoes a series of phosphorylations of the hydroxyls of inositol leading to the production of polyphosphoinositides. One polyphosphoinositide (phosphatidylinositol 4,5-bisphosphate, PIP_2) is a critically important membrane phospholipid involved in the transmission of signals for cell growth and differentiation from outside the cell to inside.

PG: Phosphatidylglycerols exhibit a net charge of –1 at physiological pH. These molecules are found in high concentration in mitochondrial membranes and as components of pulmonary surfactant. Phosphatidylglycerol also is a precursor for the synthesis of cardiolipin. PG is synthesized from CDP-diacylglycerol and glycerol-3-phosphate. The vital role of PG is to serve as the precursor for the synthesis of diphosphatidylglycerols (DPGs).

DPG: These molecules are very acidic, exhibiting a net charge of –2 at physiological pH. They are found primarily in the inner mitochondrial membrane and also as components of pulmonary surfactant. One important class of diphosphatidylglycerols is the cardiolipins. These molecules are synthesized by the condensation of CDP-diacylglycerol with PG.

The fatty acid distribution at the C–1 and C–2 positions of glycerol within phospholipids is continually in flux, owing to phospholipid degradation and the continuous phospholipid remodelling that occurs while these molecules are in membranes. Phospholipid degradation results from the action of phospholipases. There are various phospholipases that exhibit substrate specificities for different positions in phospholipids. In many cases the acyl group which was initially transferred to glycerol, by the action of the acyl transferases, is not the same acyl group present in the phospholipid when it resides within a membrane. The remodelling of acyl groups in phospholipids is the result of the action of phospholipase A_1 (PLA_1) and phospholipase A_2 ($PLA_{2)}$.

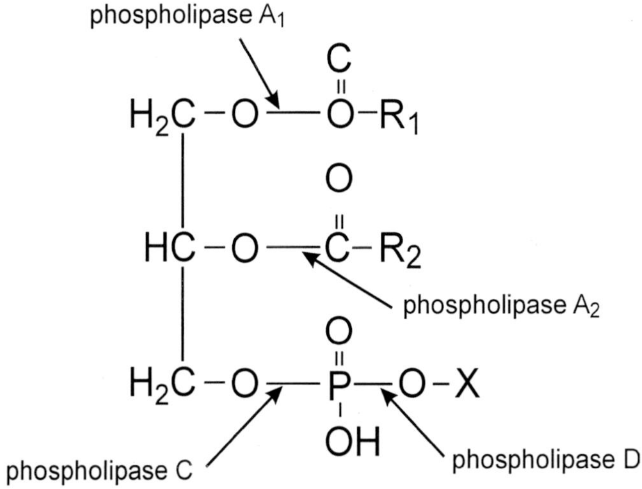

Sites of Action of the Phospholipases A_1, A_2, C and D.

The products of these phospholipases are called lysophospholipids and can be substrates for acyl transferases utilizing different acyl-CoA groups. Lysophospholipids can also accept acyl groups from other phospholipids in an exchange reaction catalysed by lysolecithin:lecithin acyltransferase (LLAT).

PLA_2 is also an important enzyme, whose activity is responsible for the release of arachidonic acid from the C–2 position of membrane phospholipids. The released arachidonate is then a substrate for the synthesis of the eicosanoids. In fact there is not just a single PLA_2 enzyme. At least 19 enzymes have been identified with PLA_2 activity. There are 10 isozymes that are in the secretory pathway and these PLA_2 isozymes are abbreviated $sPLA_2$. These secretory enzymes are low molecular weight proteins that are Ca^{2+}-dependent and are involved in numerous processes including modification of eicosanoid generation, host defence, and inflammation. The cytosolic PLA_2 family ($cPLA_2$) comprises three isozymes with $cPLA_2\alpha$ being an essential component of the initiation of arachidonic acid metabolism. Like the $sPLA_2$ enzymes, the $cPLA_2$ enzymes are tightly regulated by Ca^{2+}.

In addition, this class of PLA_2 enzyme is regulated by phosphorylation. An additional family of two PLA_2 isozymes that are Ca^{2+}-independent for activity are identified as $iPLA_2$. This latter class of enzyme is involved primarily with the remodelling of phospholipids. Finally, a class of PLA_2 enzymes, whose original member was identified as being involved in the hydrolysis and inactivation of platelet activating factor, PAF, contains at least four members. The original activity was called PAF-acetylhydrolase (PAF-AH). The PAF hydrolyzing PLA_2 isozymes are Ca^{2+}-independent like the $iPLA_2$ family. Because this latter family was shown to not only hydrolyze PAF but also oxidized phospholipids and to be associated with lipoprotein particles in the circulation they are identified as the lipoprotein-associated PLA_2 (Lp-PLA_2) family.

Lipoprotein-Associated Phospholipase A_2: Lp-PLA_2

Platelet activating factor (PAF) is a lipid compound of the plasmalogen family of phospholipids (alkyl ether phospholipids) that is involved in numerous proinflammatory activities. Inactivation of PAF was originally ascribed to an activity called PAF-acetylhydrolase (PAF-AH). Subsequent to its initial characterization, PAF-AH was shown to be a member of a large family of enzymes that all hydrolyze the *sn*-2 position of phosphlipids.

This family of enzymes is the PLA_2 family. The PLA_2 family of enzymes can be found on the Lipid Synthesis page. There are two major forms PAF-AH, one that is cytosolic and one that is secreted and found in the plasma. The plasma form of PAF-AH circulates bound to lipoproteins. Given that PAF-AH is a member of the PLA_2 family and that it also circulates bound to lipoprotein it is more commonly referred to as the lipoprotein-associated PLA_2 (Lp-PLA_2). Lp-PLA_2 is found in the plasma bound primarily to LDLs but is also found associated with HDLs and lipoprotein (a) [Lp(a)]. Of clinical significance is the fact that Lp-PLA_2 has been implicated in atherosclerosis and cardiovascular disease but its precise role in these pathophysiological processes is not completely understood.

The human Lp-PLA_2 protein is composed of 441 amino acids following cleavage of the signal peptide. The protein contains two sites of *N*-glycosylation. The enzymatic activity of Lp-PLA_2 is specific for short chain acyl groups (up to 9 methylene groups) at the *sn*-2 position of phospholipids. When PAF is the substrate for Lp-PLA2 the products are lyso-PAF and acetate. When phospholipids of the phosphatidylcholine (PC) family are oxidized by free radical activity (referred to as oxPL) they can be a substrate for Lp-PLA_2 even if the unsaturated fatty acid at the *sn*-2 position is longer than 9 carbon atoms.

The ability of LP-PLA_2 to recognize oxPL as substrates is due to the presence of aldehydic or carboxlic moieties at the omega (ω) end of the *sn*-2 peroxidized fatty acyl residues. The products of Lp-PLA_2 activity on oxPL are oxidized free fatty acids (oxFFA) and lyso-PC. Numerous types of oxPL have been identified in oxidized LDL (oxLDL) particles and many of them exhibit biological activity and exert key effects in atherogenesis. Lp-PLA_2 can also hydrolyze long chain fatty acyl phospholipid hydroperoxides, phospholipids containing isoprostanes esterified at the *sn*-2 position and other lipid esters such as short-chain diacylglycerols, triacylglycerols, and acetylated alkanols. In addition to its hydrolytic activity Lp-PLA_2 exhibits transacetylase activity. The transacetylase function transfers acetate and short-chain fatty acids from PAF to ether- and ester-linked lysophospholipids. The transacetylase function is evident when Lp-PLA_2 is associated with LDL.

In humans with normal levels of circulating lipids and no detectable Lp(a), essentially all of the Lp-PLA_2 in the plasma is bound to LDL. The interaction of Lp-PLA_2 with LDL occurs through apolipoprotein B-100 (apoB-100). When plasma levels of Lp(a) rise in excess of 30mg/dL there is an enrichment in the association of Lp-PLA_2 with this abnormal lipoprotein particle. When expressed as enzyme mass, Lp(a) carries 1.5–2 times more Lp-PLA_2 than does LDL. As in its association with LDL, Lp-PLA_2 interacts with apoB-100 in Lp(a) particles. Abnormalities in lipoprotein metabolism, such as those resulting in Lp(a) production, significantly affect the plasma levels of Lp-PLA_2.

For example in familial hypercholesterolemia the level of LDL-Lp-PLA_2 activity increases in parallel with the severity of the hypercholesterolemia. The level of plasma Lp-PLA_2 can be positively affected by low-calorie diets associated with weight loss or after drug treatment with the various classes of hypolipidemic drugs. In the context of atherosclerosis and cardiovascular disease numerous epidemiological studies have shown that increased levels of plasma Lp-PLA_2 approximately double the risk for primary and secondary cardiovascular events. In fact it is suggested that measurement of Lp-PLA_2 levels is useful as a cardiovascular risk marker independent of and additive to traditional risk factors. However, whether Lp-PLA_2 is a novel biomarker or is causal in the development of atherosclerotic diseases remains controversial. This is because there are both anti- or proatherogenic activities associated with Lp-PLA_2. The antiatherogenic functions of Lp-PLA_2 are attributed to its role in hydrolyzing and inactivating the powerful proinflammatory lipid, PAF. Additionally, by hydrolyzing oxPLs Lp-PLA_2 effectively lowers the circulating levels of this class of inflammatory mediators. On the other hand the proatherogenic and proinflammatory actions associated with Lp-PLA_2 are in fact due to its hydrolysis of oxPLs. The hydrolysis of oxPLs releases both lyso-PC and oxFFA both of which have been shown to have proatherogenic effects.

Clinical Significances of Lipoprotein Metabolism

Fortunately, few individuals carry the inherited defects in lipoprotein metabolism that lead to hyper- or hypolipoproteinemias. Persons suffering from diabetes mellitus, hypothyroidism and kidney disease often exhibit abnormal lipoprotein metabolism as a result of secondary effects of their disorders. For example, because lipoprotein lipase (LPL) synthesis is regulated by insulin, LPL deficiencies leading to Type I hyperlipoproteinemia (OMIM link) may occur as a secondary outcome of diabetes mellitus. Additionally, insulin and thyroid hormones positively affect hepatic LDL-receptor interactions; therefore, the hypercholesterolemia and increased risk of athersclerosis associated with uncontrolled diabetes or hypothyroidism is likely due to decreased hepatic LDL uptake and metabolism.

Of the many disorders of lipoprotein metabolism, familial hypercholesterolemia (FH) may be the most prevalent in the general population. Heterozygosity at the FH locus occurs in 1:500 individuals, whereas, homozygosity is observed in 1:1,000,000 individuals. FH is an inherited disorder comprising four different classes of mutation in the LDL receptor gene. The class 1 defect (the most common) results in a complete loss of receptor synthesis. The class 2 defect results in the synthesis of a receptor protein that is not properly processed in the Golgi apparatus and therefore is not transported to the plasma membrane. The class 3 defect results in an LDL receptor that is incapable of binding LDLs. The class 4 defect results in receptors that bind LDLs but do not cluster in coated pits and are, therefore, not internalized.

FH sufferers may be either heterozygous or homologous for a particular mutation in the receptor gene. Homozygotes exhibit grossly elevated serum cholesterol (primarily in LDLs). The elevated levels of LDLs result in their phagocytosis by macrophages. These lipid-laden phagocytic cells tend to deposit within the skin and tendons, leading to xanthomas. A greater complication results from cholesterol deposition within the arteries, leading to atherosclerosis, the major contributing factor of nearly all cardiovascular diseases.

Lipoprotein(a) and Atherogenesis

Lipoprotein(a) [Lp(a)] was originally described as a new serum lipoprotein particle by Kare Berg in 1963. Lp(a) is composed of a common LDL nucleus linked to a molecule of apolipoprotein(a) [apo(a)] by disulfide bonds between a cysteine residue in a Kringle-IV (KIV) type 9 domain in apo(a) and a cysteine residue in apolipoprotein B-100 (apoB-100). When attached to apoB-100 the apo(a) protein surrounds the LDL molecule. Synthesis of Lp(a) occurs in the liver. The half-life of Lp(a) in the circulation is approximately 3–4 days. Although Lp(a) was described over 40 years ago its precise physiological function remains unclear. However, numerous epidemiological studies have demonstrated that elevated plasma levels of Lp(a) are a significant risk factor for the development of atherosclerotic disease.

The Kringle domains of apo(a) exhibit 75%-85% similarity to the KIV domains of plasminogen. The Kringle domain is a highly glycosylated domain found in numerous proteins and is so-called because the three dimensional structure resembles a looped Danish pastry. Each Kringle domain is composed of approximately 80 amino acid residues and the structure is stabilized by three internal disulfide bonds. There are 10 distinct sub-classes of KIV domain in apo(a) designated KIV1 through KIV10. The apo(a) KIV1 and KIV3 through KIV10 domains are present as single-copy domains. The KIV2 domain is present in a variable number of repeated copies and constitutes the molecular basis for the highly variable size of Lp(a) in different individuals. Apo(a) also contains a Kringle V (KV) domain that resembles the catalytic domain of plasminogen. Indeed, the apo(a) gene (gene symbol = LPA) located on chromosome 6 is a member of the plasminogen superfamily and given the similarities between apo(a) and plasminogen it has been hypothesized that apo(a) influences the processes of of hemostasis.

Apo(a) proteins exhibit a variability in size due to a polymorphism caused by a variable number of the KIV repeats. To date at least seven different isoforms of Lp(a) have been characterized based upon electrophoretic mobilities. These different isoforms are designated F, B, and S1 through S5. The different isoforms are grouped into low molecular weight (LMW) and high molecular weight (HMW) isoforms determined by the number of KIV repeats in the apo(a) protein found in the Lp(a). The level of Lp(a) found in healthy individuals depends upon whether their plasma contains the LMW or HMW isoforms. Individuals with the LMW isoforms have high plasma Lp(a) concentration while those with the HMW isoforms have low concentrations.

When in the circulation Lp(a) particles can be affected by oxidative modification similar to that of the other plasma lipoprotein particles. Lp(a) and oxidized Lp(a) [oxLp(a)] particles interact with macrophages via scavenger receptor uptake leading to cholesterol accumulation and foam cell formation. Indeed, oxLp(a) are phagocytosed more rapidly than other lipoprotein particles and therefore accumulate in the subendothelial space at high levels. This process can lead to progression of atherogenesis, thus accounting for the direct correlation between the plasma level of Lp(a) and coronary artery disease. In addition to oxidation of Lp(a) leading to increased foam cell production, glycation of the particle also may contribute to atherogenesis. In fact, there is a strong correlation in the level of glycated Lp(a) and the severity of hyperglycemia observed in poorly controlled type 2 diabetes.

Although the precise physiology of Lp(a) is poorly understood, as indicated above, there is a strong correlation between plasma concentration of Lp(a) and atherogenic events that lead to coronary artery disease. The processes of blood coagulation and the role of plasminogen visit the Blood Coagulation page. Because of the high degree of similarity between apo(a) and plasminogen it is suggested that Lp(a) may contribute to the thrombotic aspects of ischemic heart disease. Lp(a) has been shown to competitively inhibit the binding of plasminogen to its receptor on endothelial cells as well as to its binding sites on fibrinogen and fibrin. This interference of plasminogen binding leads to reduced surface-dependent activation of plasminogen to plasmin. The normal function of plasmin is to degrade the fibrin clot that forms as a result of vessel injury. Therefore, high plasma concentrations of Lp(a) may represent a source of antifibrinolytic activity. Of significance to the potential for atherogenesis, the antifibrinolytic potential of Lp(a) particles is related to their size. The LMW isoforms of Lp(a) have been shown to have a higher fibrin-binding capacity than the HMW isoforms. Lp(a) also interferes with other aspects of the normal processes of coagulation in addition to its effects on plasminogen function. Lp(a) stimulates the production of plasminogen activator inhibitor-1 (PAI-1) leading to a reduced ability of t-PA to activate the process of clot dissolution.

Increased production of PAI-1 also leads to enhanced proinflammatory events via activation of monocyte adhesion to the vessel wall. Lp(a) has also been shown to modulate platelet activation interfering with the interaction of platelets with exposed collagen fibers in the injured vessel wall. In addition to the role of Lp(a) in inhibiting plasminogen binding, Lp(a) has been shown to inhibit the release of tissue plasminogen activator (t-PA) from endothelial cells. With reduced release of the enzyme (t-PA) that converts plasminogen to plasmin and interference with plasminogen binding to fibrin clots Lp(a) can exert a significant negative effect on the ability to dissolve blood clots.

In addition to the interactions with plasminogen, leading to enhanced atherogenesis, Lp(a) has been shown to stimulate smooth muscle cell (SMC) growth. This effect of Lp(a) is exerted via an inactivation of transforming growth factor-β (TGF-β). Activated TGF-β inhibits the proliferation and migration of SMC, thus the inhibition of this regulatory effect of TGF-β leads to accelerated blood vessel stenosis with concomitant enhancement of the atherogenic process. oxLp(a) has also been shown to inhibit nitric oxide-dependent vasodilation which will tend to exacerbate the atherogenic process in hypertensive patients.

Lp(a) also interferes with other aspects of the normal processes of coagulation in addition to its effects on plasminogen function. Lp(a) stimulates the production of plasminogen activator inhibitor-1 (PAI-1) leading to a reduced ability of t-PA to activate the process of clot dissolution. Increased production of PAI-1 also leads to enhanced proinflammatory events via activation of monocyte adhesion to the vessel wall. Lp(a) has also been shown to modulate platelet activation by interfering with the interaction of platelets with exposed collagen fibers in the injured vessel wall. All of the observed effects of Lp(a) on hemostasis result in the persistence of clots which is a significant contributor to atherogenesis and increases the potential for abnormal thrombotic episodes.

Pharmacologic Intervention

Drug treatment to lower plasma lipoproteins and/or cholesterol is primarily aimed at reducing the risk of athersclerosis and subsequent coronary artery disease that exists in patients with elevated circulating lipids. Drug therapy usually is considered as an option only if non-pharmacologic interventions (altered diet and exercise) have failed to lower plasma lipids.

Atorvastatin (Lipotor®), Simvastatin (Zocor®), Lovastatin (Mevacor®): These drugs are fungal HMG-CoA reductase (HMGR) inhibitors and are members of the family of drugs referred to as the statins. The net result of treatment is an increased cellular uptake of LDLs, since the intracellular synthesis of cholesterol is inhibited and cells are therefore dependent on extracellular sources of cholesterol. However, since mevalonate (the product of the HMG-CoA reductase reaction) is required for the synthesis of other important isoprenoid compounds besides cholesterol, long-term treatments carry some risk of toxicity. A component of the natural cholesterol lowering supplement, red yeast rice, is in fact a statin-like compound.

The statins have become recognized as a class of drugs capable of more pharmacologic benefits than just lowering blood cholesterol levels via their actions on HMGR. Part of the cardiac benefit of the statins relates to their ability to regulate the production of *S*-nitrosylated COX-2. COX-2 is an inducible enzyme involved in the synthesis of the prostaglandins and thromboxanes as well as the lipoxins and resolvins. The latter two classes of compounds are anti-inflammatory lipids. Evidence has shown that statins activate inducible nitric oxide synthase (iNOS) leading to nitrosylation of COX-2. The *S*-nitrosylated COX-2 enzyme produces the lipid compound 15*R*-hydroxyeicosatetraenoic acid (15*R*-HETE) which is then converted via the action of 5-lipoxygenase (5-LOX) to the epimeric lipoxin, 15-epi-LXA_4. This latter compound is the same as the aspirin-triggered lipoxin (ATL) that results from the aspirin-induced acetylation of COX-2. Therefore, part of the beneficial effects of the statins are exerted via the actions of the lipoxin family of anti-inflammatory lipids.

Additional anti-inflammatory actions of the statins results from a reduction in the prenylation of numerous pro-inflammatory modulators. Prenylation refers to the addition of the 15 carbon farnesyl group or the 20 carbon geranylgeranyl group to acceptor proteins. The isoprenoid groups are attached to cysteine residues at the carboxy terminus of proteins in a thioether linkage (C-S-C).

A common consensus sequence at the C-terminus of prenylated proteins has been identified and is composed of CAAX, where C is cysteine, A is any aliphatic amino acid (except alanine) and X is the C-terminal amino acid. In addition to numerous prenylated proteins that contain the CAAX consensus, prenylation is known to occur on proteins of the RAB family of RAS-related G-proteins. There are at least 60 proteins in this family that are prenylated at either a CC or CXC element in their C-termini. The RAB family of proteins are involved in signalling pathways that control intracellular membrane trafficking. The prenylation of proteins allows them to be anchored to cell membranes. In addition to cell membrane attachment, prenylation is known to be important for protein-protein interactions. Thus, inhibition of this post-translational modification by the statins interferes with the important functions of many signalling proteins which is manifest by inhibition of inflammatory responses.

Some of the effects on immune function that have been attributed to the statins are attenuation of autoimmune disease, inhibition of T-cell proliferation, inhibition of inflammatory co-stimulatory molecule expression, decreases in leukocyte infiltration, and promotion of a shift in cytokine profiles of helper T-cell types from Th1 to Th2. Th1 cells are involved in cell-mediated immunity processes, whereas, Th2 cells are involved in humoral immunity process. The cytokines produced by Th2 cells include IL-4, IL-5, IL-10 and IL-13 and these trigger B cells to switch to IgE production and to activate eosinophils.

Nicotinic acid: Nicotinic acid reduces the plasma levels of both VLDLs and LDLs by inhibiting hepatic VLDL secretion, as well as suppressing the flux of FFA release from adipose tissue by inhibiting lipolysis. In addition, nicotinic administration strongly increases the circulating levels of HDLs. Patient compliance with nicotinic acid administration is sometimes compromised because of the unpleasant side-effect of flushing (strong cutaneous vasodilation). Recent evidence has shown that nicotinic acid binds to and activates the G-protein coupled receptor identified as GPR109A (also called HM74A or PUMA-G). The identity of a receptor to which nicotinic acid binds allows for the development of new drug therapies that activate the same receptor but that may lack the negative side-effect of flushing associated with nicotinic acid. Because of its ability to cause large reductions in circulating levels of cholesterol, nicotinic acid is used to treat Type II, III, IV and V hyperlipoproteinemias.

Gemfibrozil (Lopid®), Fenofibrate (TriCor®): These compounds (called fibrates) are derivatives of fibric acid and although used clinically since the 1930's were only recently discovered to exert some of their lipid-lowering effects via the activation of peroxisome proliferation. Specifically, the fibrates were found to be activators of the peroxisome proliferator-activated receptor-α (PPAR-α) class of proteins that are classified as co-activators. The naturally occurring ligands for PPAR-α are leukotriene B_4, unsaturated fatty acids and oxidized components of VLDLs and LDLs. The PPARs interact with another receptor family called the retinoid X receptors (RXRs) that bind 9-*cis*-retinoic acid. Activation of PPARs results in modulation of the expression of genes involved in lipid metabolism. In addition the PPARs modulate carbohydrate metabolism and adipose tissue differentiation. Fibrates result in the activation of PPAR-α in liver and muscle. In the liver this leads to increased β-oxidation of fatty acids, thereby decreasing the liver's secretion of triacylglycerol- and cholesterol-rich VLDLs, as well as increased clearance of chylomicron remnants, increased levels of HDLs and increased lipoprotein lipase activity which in turn promotes rapid VLDL turnover.

Cholestyramine or colestipol (resins): These compounds are nonabsorbable resins that bind bile acids which are then not reabsorbed by the liver but excreted. The drop in hepatic reabsorption of bile acids releases a feedback inhibitory mechanism that had been inhibiting bile acid synthesis.

As a result, a greater amount of cholesterol is converted to bile acids to maintain a steady level in circulation. Additionally, the synthesis of LDL receptors increases to allow increased cholesterol uptake for bile acid synthesis, and the overall effect is a reduction in plasma cholesterol. This treatment is ineffective in homozygous FH patients, since they are completely deficient in LDL receptors.

Ezetimibe: This drug is sold under the trade names Zetia® or Ezetrol® and is also combined with the statin drug simvastatin and sold as Vytorin® or Inegy®. Ezetimibe functions to reduce intestinal absorption of cholesterol, thus effecting a reduction in circulating cholesterol.

The drug functions by inhibiting the intestinal brush border transporter involved in absorption of cholesterol. This transporter is known as Niemann-Pick type C1-like 1 (NPC1L1).

NPC1L1 is also highly expressed in human liver. The hepatic function of NPC1L1 is presumed to limit excessive biliary cholesterol loss. NPC1L1-dependent sterol uptake is regulated by cellular cholesterol content. In addition to the cholesterol lowering effects that result from inhibition of NPC1L1, its inhibition has been shown to have beneficial effects on components of the metabolic syndrome, such as obesity, insulin resistance, and fatty liver, in addition to atherosclerosis. Ezetimibe is usually prescribed for patients who cannot tolerate a statin drug or a high dose statin regimen. There is some controversy as to the efficacy of ezetimibe at lowering serum cholesterol and reducing the production of fatty plaques on arterial walls. The combination drug of ezetimibe and simvastatin has shown efficacy equal to or slightly greater than atorvastatin (Lipitor®) alone at reducing circulating cholesterol levels.

Plasmalogens

Plasmalogens are glycerol ether phospholipids. They are of two types, alkyl ether ($-O-CH_2-$) and alkenyl ether ($-O-CH=CH-$). Dihydroxyacetone phosphate serves as the glycerol precursor for the synthesis of glycerol ether phospholipids. Three major classes of plasmalogens have been identified: choline, ethanolamine and serine plasmalogens. Ethanolamine plasmalogen is prevalent in myelin. Choline plasmalogen is abundant in cardiac tissue.

One particular choline alkyl ether plasmalogen (1-O-1'-enyl-2-acetyl-*sn*-glycero-3-phosphocholine) has been identified as an extremely powerful biological mediator, capable of inducing cellular responses at concentrations as low as 10^{-11}M. This molecule is called platelet activating factor, PAF.

PAF functions as a mediator of hypersensitivity, acute inflammatory reactions and anaphylactic shock. PAF is synthesized in response to the formation of antigen-IgE complexes on the surfaces of basophils, neutrophils, eosinophils, macrophages and monocytes. The synthesis and release of PAF from cells leads to platelet aggregation and the release of serotonin from platelets. PAF also produces responses in liver, heart, smooth muscle, and uterine and lung tissues.

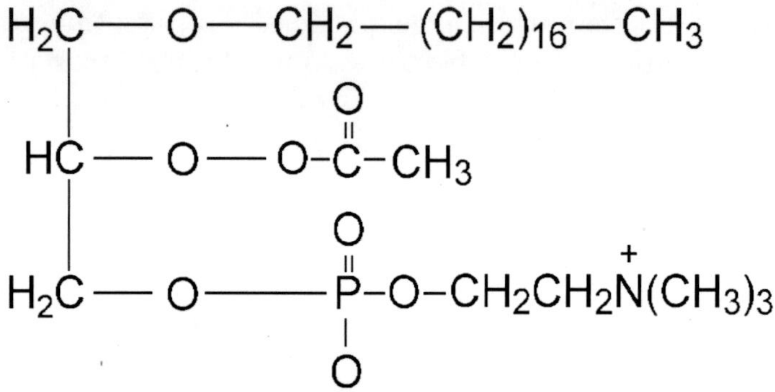

Platelet activating factor

Nucleic Acids

As a class, the nucleotides may be considered one of the most important metabolites of the cell. Nucleotides are found primarily as the monomeric units comprising the major nucleic acids of the cell, RNA and DNA. However, they also are required for numerous other important functions within the cell. These functions include:

1. serving as energy stores for future use in phosphate transfer reactions. These reactions are predominantly carried out by ATP.
2. forming a portion of several important coenzymes such as NAD^+, $NADP^+$, FAD and coenzyme A.
3. serving as mediators of numerous important cellular processes such as second messengers in signal transduction events. The predominant second messenger is cyclic-AMP (cAMP), a cyclic derivative of AMP formed from ATP.
4. controlling numerous enzymatic reactions through allosteric effects on enzyme activity.
5. serving as activated intermediates in numerous biosynthetic reactions. These activated intermediates include S-adenosylmethionine (S-AdoMet or SAM) involved in methyl transfer reactions as well as the many sugar coupled nucleotides involved in glycogen and glycoprotein synthesis.

Nucleoside and Nucleotide Structure and Nomenclature

The nucleotides found in cells are derivatives of the heterocyclic highly basic, compounds, purine and pyrimidine.

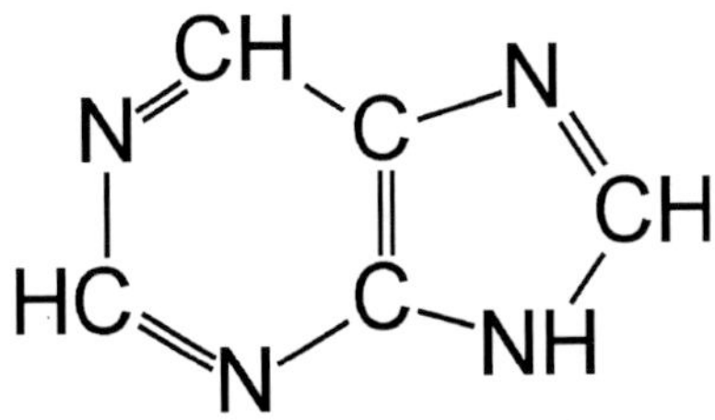

Purine

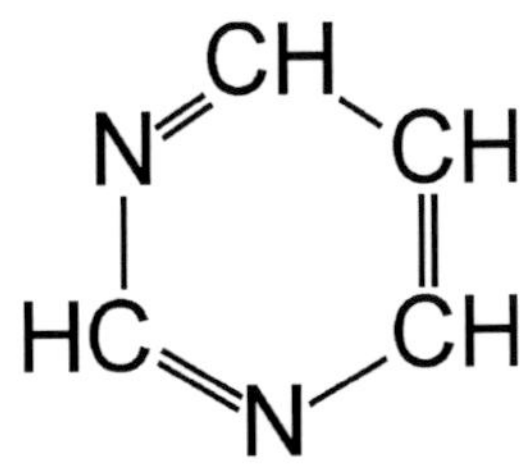

Pyrimidine

It is the chemical basicity of the nucleotides that has given them the common term "bases" as they are associated with nucleotides present in DNA and RNA. There are five major bases found in cells. The derivatives of purine are called adenine and guanine, and the derivatives of pyrimidine are called thymine, cytosine and uracil. The common abbreviations used for these five bases are, A, G, T, C and U.

The base can exist in 2 distinct orientations about the *N*-glycosidic bond. These conformations are identified as, *syn* and *anti*. It is the anti conformation that predominates in naturally occurring nucleotides.

Nucleosides are found in the cell primarily in their phosphorylated form. These are termed nucleotides. The most common site of phosphorylation of nucleotides found in cells is the hydroxyl group attached to the 5'-carbon of the ribose The carbon atoms of the ribose present in nucleotides are designated with a prime (') mark to distinguish them from the backbone numbering in the bases. Nucleotides can exist in the mono-, di-, or tri-phosphorylated forms.

Nucleotides are given distinct abbreviations to allow easy identification of their structure and state of phosphorylation. The monophosphorylated form of adenosine (adenosine-5'-monophosphate) is written as, AMP. The di- and tri-phosphorylated forms are written as, ADP and ATP, respectively. The use of these abbreviations assumes that the nucleotide is in the 5'-phosphorylated form. The di- and tri-phosphates of nucleotides are linked by acid anhydride bonds. Acid anhydride bonds have a high $\Delta G^{0'}$ for hydrolysis imparting upon them a high potential to transfer the phosphates to other molecules. It is this property of thc nucleotides that results in their involvement in group transfer reactions in the cell.

The nucleotides found in DNA are unique from those of RNA in that the ribose exists in the 2'-deoxy form and the abbreviations of the nucleotides contain a "d" designation. The monophosphorylated form of adenosine found in DNA (deoxyadenosine-5'-monophosphate) is written as dAMP.

The nucleotide uridine is never found in DNA and thymine is almost exclusively found in DNA. Thymine is found in tRNAs but not rRNAs nor mRNAs. There are several less common bases found in DNA and RNA. The primary modified base in DNA is 5-methylcytosine. A variety of modified bases appear in the tRNAs. Many modified nucleotides are encountered outside of the context of DNA and RNA that serve important biological functions.

Adenosine Derivatives

The most common adenosine derivative is the cyclic form, 3'-5'-cyclic adenosine monophosphate, cAMP. This compound is a very powerful second messenger involved in passing signal transduction events from the cell surface to internal proteins, e.g. cAMP-dependent protein kinase, PKA. PKA phosphorylates a number of proteins, thereby, affecting their activity either positively or negatively. Cyclic-AMP is also involved in the regulation of ion channels by direct interaction with the channel proteins, e.g. in the activation of odorant receptors by odorant molecules.

Formation of cAMP occurs in response to activation of receptor coupled adenylate cyclase. These receptors can be of any type, e.g. hormone receptors or odorant receptors.

Representative pathway for the activation of cAMP-dependent protein kinase, PKA). In this example glucagon binds to its' cell-surface receptor, thereby activating the receptor. Activation of the receptor is coupled to the activation of a receptor-coupled G-protein (GTP-binding and hydrolyzing protein) composed of 3 subunits. Upon activation the α-subunit dissociates and binds to and activates adenylate cyclase. Adenylate cylcase then converts ATP to cyclic-AMP (cAMP). The cAMP thus produced then binds to the regulatory subunits of PKA leading to dissociation of the associated catalytic subunits. The catalytic subunits are inactive until dissociated from the regulatory subunits. Once released the catalytic subunits of PKA phosphorylate numerous substrate using ATP as the phosphate donor.

S-adenosylmethionine is a form of activated methionine which serves as a methyl donor in methylation reactions and as a source of propylamine in the synthesis of polyamines.

Guanosine Derivatives

A cyclic form of GMP (cGMP) also is found in cells involved as a second messenger molecule. In many cases its' role is to antagonize the effects of cAMP. Formation of cGMP occurs in response to receptor mediated signals similar to those for activation of adenylate cyclase. However, in this case it is guanylate cyclase that is coupled to the receptor. The most important cGMP coupled signal transduction cascade is that photoreception. However, in this case activation of rhodopsin (in the rods) or other opsins (in the cones) by the absorption of a photon of light (through 11-*cis*-retinal covalently associated with rhodopsin and opsins) activates transducin which in turn activates a cGMP specific phosphodiesterase that hydrolyzes cGMP to GMP. This lowers the effective concentration of cGMP bound to gated ion channels resulting in their closure and a concomitant hyperpolarization of the cell.

Characterization of tRNAs

Translation is the RNA directed synthesis of polypeptides. This process requires all three classes of RNA. Although the chemistry of peptide bond formation is relatively simple, the processes leading to the ability to form a peptide bond are exceedingly complex. The template for correct addition of individual amino acids is the mRNA, yet both tRNAs and rRNAs are involved in the process. The tRNAs carry activated amino acids into the ribosome which is composed of rRNA and ribosomal proteins. The ribosome is associated with the mRNA ensuring correct access of activated tRNAs and containing the necessary enzymatic activities to catalyse peptide bond formation.

Historical Perspectives

Early genetic experiments demonstrated:

1. The co-linearity between the DNA and protein encoded by the DNA. Yanofsky showed that the order of observed mutations in the *E. coli* tryptophan synthetase gene was the same as the corresponding amino acid changes in the protein.
2. Crick and Brenner demonstrated, from a large series of double mutants of the bacteriophage T4, that the genetic code is read in a sequential manner starting from a fixed point in the gene, the code was most likely a triplet and that all 64 possible combinations of the 4 nucleotides code for amino acids, i.e. the code is degenerate since there are only 20 amino acids.

The above mentioned experiments only indicated deductive correlation's regarding the genetic code. The precise dictionary of the genetic code was originally determined by the use of in vitro translation systems derived from *E. coli* cells. Synthetic polyribonucleotides were added to these translation system along with all twenty amino acids. One amino acid at a time was radiolabeled. The first demonstration of the dictionary of the genetic code was with the use of poly(U). This synthetic polyribonucleotide encoded the amino acid phenylalanine, i.e. the resulting polypeptide was poly(F).

The utilization of a variety of repeating di- tri- and tetra polyribonucleotides established the entire genetic code. These results of these experiments confirmed that some amino acids are encoded for by more than one triplet codon, hence the degeneracy of the genetic code. These experiments also established the identity of translational termination codons. An additional important point to come from these early experiments was that the 5' end of the RNA corresponded to the amino terminus of the polypeptide.

This was important since previous labelling experiments had demonstrated that the N-terminus is the beginning of the elongating polypeptide. Therefore, *in vitro* translation experiments established that the RNA is read in the 5' to 3' direction. Crick first postulated that translation of the genetic code would be carried out through mediation of adapter molecules. Each adapter was postulated to carry a specific amino acid and to recognize the corresponding codon. He suggested that the adapters contain RNA because codon recognition could then occur by complementarity to the sequences of the codons in the mRNA.

During the course of *in vitro* protein synthesis and labelling experiments it was shown that the amino acids became transiently bound to a low molecular weight mass fraction of RNA. This fraction of RNAs have been termed transfer RNAs (tRNAs) since they transfer amino acids to the elongating polypeptide. These results indicate that accurate translation requires two equally important recognition steps:

1. The correct choice of amino acid needs to be made for attachment to the correspondingly correct tRNA.
2. Selection of the correct amino acid-charged tRNA by the mRNA. This process is facilitated by the ribosomes.

Summary of Experiments to Determine the Genetic Code:

1. The genetic code is read in a sequential manner starting near the 5' end of the mRNA. This means that translation proceeds along the mRNA in the 5' ——> 3' direction which corresponds to the N-terminal to C-terminal direction of the amino acid sequences within proteins.
2. The code is composed of a triplet of nucleotides.
3. That all 64 possible combinations of the 4 nucleotides code for amino acids, i.e. the code is degenerate since there are only 20 amino acids.

The precise dictionary of the genetic code was determined with the use of *in vitro* translation systems and polyribonucleotides.

The results of these experiments confirmed that some amino acids are encoded by more than one triplet codon, hence the degeneracy of the genetic code. These experiments also established the identity of translational termination codons.

Characteristics of tRNAs

More than 300 different tRNAs have been sequenced, either directly or from their corresponding DNA sequences. tRNAs vary in length from 60–95 nucleotides (18–28 kD). The majority contain 76 nucleotides.

Evidence has shown that the role of tRNAs in translation is to carry activated amino acids to the elongating polypeptide chain. All tRNAs:

1. Exhibit a cloverleaf-like secondary structure.
2. Have a 5'-terminal phosphate.
3. Have a 7 bp stem that includes the 5'-terminal nucleotide and may contain non-Watson-Crick base pairs, e.g. GU. This portion of the tRNA is called the acceptor since the amino acid is carried by the tRNA while attached to the 3'-terminal OH group.
4. Have a D loop and a TΨC loop.

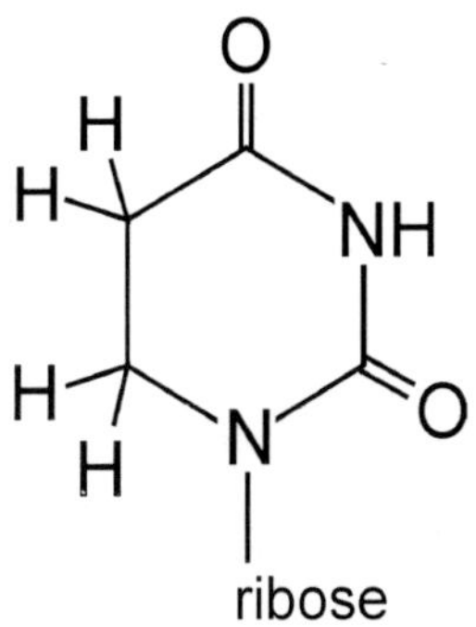

Dihydrouridine (D)

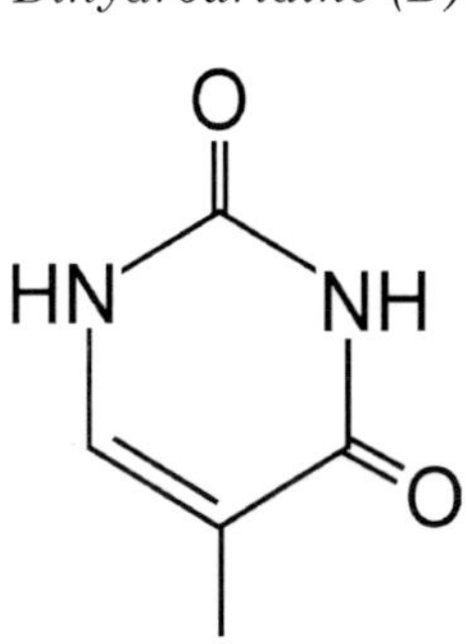

Pseudouridine (Y)

5. Have an anti-codon loop.
6. Terminate at the 3'-end with the sequence 5'–CCA–3'.
7. Contain 13 invariant positions and 8 semi-variant positions.
8. Contain numerous modified nucleotide bases.

Activation of Amino Acids

Activation of amino acids is carried out by a two step process catalysed by aminoacyl-tRNA synthetases. Each tRNA, and the amino acid it carries, are recognized by individual aminoacyl-tRNA synthetases. This means there exists at least 20 different aminoacyl-tRNA synthetases, there are actually at least 21 since the initiator met-tRNA of both prokaryotes and eukaryotes is distinct from non-initiator met-tRNAs. Activation of amino acids requires energy in the form of ATP and occurs in a two step reaction catalysed by the aminoacyl-tRNA synthetases.

First the enzyme attaches the amino acid to the α-phosphate of ATP with the concomitant release of pyrophosphate. This is termed an aminoacyl-adenylate intermediate. In the second step the enzyme catalyzes transfer of the amino acid to either the 2'– or 3'–OH of the ribose portion of the 3'-terminal adenosine residue of the tRNA generating the activated aminoacyl-tRNA.

Although these reaction are freely reversible, the forward reaction is favoured by the coupled hydrolysis of PP_i. Accurate recognition of the correct amino acid as well as the correct tRNA is different for each aminoacyl-tRNA synthetase.

Since the different amino acids have different R groups, the enzyme for each amino acid has a different binding pocket for its specific amino acid. It is not the anticodon that determines the tRNA utilized by the synthetases. Although the exact mechanism is not known for all synthetases, it is likely to be a combination of the presence of specific modified bases and the secondary structure of the tRNA that is correctly recognized by the synthetases.

It is absolutely necessary that the discrimination of correct amino acid and correct tRNA be made by a given synthetase prior to release of the aminoacyl-tRNA from the enzyme. Once the product is released there is no further way to proof-read whether a given tRNA is coupled to its corresponding tRNA.

Erroneous coupling would lead to the wrong amino acid being incorporated into the polypeptide since the discrimination of amino acid during protein synthesis comes from the recognition of the anticodon of a tRNA by the codon of the mRNA and not by recognition of the amino acid. This was demonstrated by reductive desulfuration of cys-$tRNA^{cys}$ with Raney nickel generating ala-$tRNA^{cys}$. Alanine was then incorporated into an elongating polypeptide where cysteine should have been.

The Wobble Hypothesis

The 3 of the possible 64 triplet codons are recognized as translational termination codons. The remaining 61 codons might be considered as being recognized by individual tRNAs. Most cells contain isoaccepting tRNAs, different tRNAs that are specific for the same amino acid, however, many tRNAs bind to two or three codons specifying their cognate amino acids. As an example yeast tRNAphe has the anticodon 5'–GmAA–3' and can recognize the codons 5'–UUC–3' and 5'–UUU–3'. It is, therefore, possible for non-Watson-Crick base pairing to occur at the third codon position, i.e. the 3' nucleotide of the mRNA codon and the 5' nucleotide of the tRNA anticodon. This has phenomenon been termed the wobble hypothesis

Wobble Positions in Anticodon and Codon Interactions

C	A	G	U	I
G	U	C	A	C
		U	G	U
				A

Wobble Positions in Codon and Anticodon Interactions

C	A	G	U
G	U	C	A
I	I	U	G
			I

Diagram showing the various modified nucleotides of tRNAs that are found in the wobble position in the anticodon. The top half shows the wobble nucleotides of the anticodon in blue and the various nucleotides (in red) of the wobble position of the codon that can be found in non-Watson-Crick base-pairs. The lower panel illustrates the opposite showing the wobble nucleotides of the codon in blue and the associated wobble nucleotides of the anticodon in red.

Now that we have charged aminoacyl-tRNAs and the mRNAs to convert nucleotide sequences to amino acid sequences we need to bring the two together accurately and efficiently. This is the job of the ribosomes. Ribosomes are composed of proteins and rRNAs. All living organisms need to synthesis proteins and all cells of an organism need to synthesize proteins, therefore, it is not hard to imagine that ribosomes are a major constituent of all cells of all organisms. The make up of the ribosomes, both rRNA and associated proteins are slightly different between prokaryotes and eukaryotes.

Order of Events in Translation

The ability to begin to identify the roles of the various ribosomal proteins in the processes of ribosome assembly and translation was aided by the discovery that the ribosomal subunits will self assemble in vitro from their constituent parts.

Following assembly of both the small and large subunits onto the mRNA, and given the presence of charged tRNAs, protein synthesis can take place. To reiterate the process of protein synthesis:

1. Synthesis proceeds from the N-terminus to the C-terminus of the protein.
2. The ribosomes "read" the mRNA in the 5' to 3' direction.
3. Active translation occurs on polyribosomes (also termed polysomes). This means that more than one ribosome can be bound to and translate a given mRNA at any one time.
4. Chain elongation occurs by sequential addition of amino acids to the C-terminal end of the ribosome bound polypeptide.

Translation proceeds in an ordered process. First accurate and efficient initiation occurs, then chain elongation and finally accurate and efficient termination must occur. All three of these processes require specific proteins, some of which are ribosome associated and some of which are separate from the ribosome, but may be temporarily associated with it.

Initiation

Initiation of translation in both prokaryotes and eukaryotes requires a specific initiator tRNA, $tRNA^{met}_{i}$, that is used to incorporate the initial methionine residue into all proteins. In *E. coli* a specific version of $tRNA^{met}_{i}$ is required to initiate translation, [$tRNA^{fmet}_{i}$]. The methionine attached to this initiator tRNA is formylated. Formylation requires N^{10}-formy-THF and is carried out after the methionine is attached to the tRNA. The fmet-$tRNA^{fmet}_{i}$ still recognizes the same codon, AUG, as regular $tRNA^{met}$. Although $tRNA^{met}_{i}$ is specific for initiation in eukaryotes it is not a formylated $tRNA^{met}$.

The initiation of translation requires recognition of an AUG codon. In the polycistronic prokaryotic RNAs this AUG codon is located adjacent to a Shine-Delgarno element in the mRNA. The Shine-Delgarno element is recognized by complimentary sequences in the small subunit rRNA (16*S* in *E. coli*). In eukaryotes initiator AUGs are generally, but not always, the first encountered by the ribosome. A specific sequence context, surrounding the initiator AUG, aids ribosomal discrimination. This context is $^{A}/_{G}CC^{A}/_{G}CCAUG^{A}/_{G}$ in most mRNAs.

Specific Steps in Translational Initiation

Initiation of translation requires 4 specific steps:

1. A ribosome must dissociate into its' 40*S* and 60*S* subunits.
2. A ternary complex termed the preinitiation complex is formed consisting of the initiator, GTP, eIF-2 and the 40*S* subunit.

3. The mRNA is bound to the preinitiation complex.
4. The 60*S* subunit associates with the preinitiation complex to form the 80*S* initiation complex.

The initiation factors eIF-1 and eIF-3 bind to the 40*S* ribosomal subunit favouring antiassociation to the 60*S* subunit. The prevention of subunit reassociation allows the preinitiation complex to form.

The first step in the formation of the preinitiation complex is the binding of GTP to eIF-2 to form a binary complex. eIF-2 is composed of three subunits, α, β and γ. The binary complex then binds to the activated initiator tRNA, met-$tRNA^{met}$ forming a ternary complex that then binds to the 40*S* subunit forming the 43S preinitiation complex. The preinitiation complex is stabilized by the earlier association of eIF-3 and eIF-1 to the 40*S* subunit.

The cap structure of eukaryotic mRNAs is bound by specific eIFs prior to association with the preinitiation complex. Cap binding is accomplished by the initiation factor eIF-4F. This factor is actually a complex of 3 proteins; eIF-4E, A and G. The protein eIF-4E is a 24 kDa protein which physically recognizes and binds to the cap structure. eIF-4A is a 46 kDa protein which binds and hydrolyzes ATP and exhibits RNA helicase activity. Unwinding of mRNA secondary structure is necessary to allow access of the ribosomal subunits. eIF-4G aids in binding of the mRNA to the 43S preinitiation complex.

Once the mRNA is properly aligned onto the preinitiation complex and the initiator met-$tRNA^{met}$ is bound to the initiator AUG codon (a process facilitated by eIF-1) the 60*S* subunit associates with the complex. The association of the 60*S* subunit requires the activity of eIF-5 which has first bound to the preinitiation complex.

The energy needed to stimulate the formation of the 80*S* initiation complex comes from the hydrolysis of the GTP bound to eIF-2. The GDP bound form of eIF-2 then binds to eIF-2B which stimulates the exchange of GTP for GDP on eIF-2. When GTP is exchanged eIF-2B dissociates from eIF-2. This is termed the eIF-2 cycle. This cycle is absolutely required in order for eukaryotic translational initiation to occur. The GTP exchange reaction can be affected by phosphorylation of the α-subunit of eIF-2.

At this stage the initiator met-$tRNA^{met}$ is bound to the mRNA within a site of the ribosome termed the P-site, for peptide site. The other site within the ribosome to which incoming charged tRNAs bind is termed the A-site, for amino acid site.

The eIF-2 cycle involves the regeneration of GTP-bound eIF-2 following the hydrolysis of GTP during translational initiation. When the 40*S* preinitiation complex is engaged with the 60*S* ribosome to form the 80*S* initiation complex, the GTP bound to eIF-2 is hydrolysed providing energy for the process. In order for additional rounds of translational initiation to occur, the GDP bound to eIF-2 must be exchanged for GTP. This is the function of eIF-2B which is also called guanine nucleotide exchange factor (GEF).

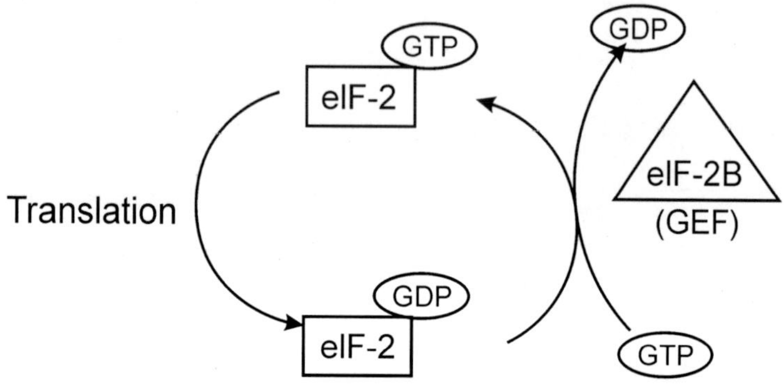

Elongation

The process of elongation, like that of initiation requires specific non-ribosomal proteins. In *E. coli* these are EFs and in eEFs. Elongation of polypeptides occurs in a cyclic manner such that at the end of one complete round of amino acid addition the A site will be empty and ready to accept the incoming aminoacyl-tRNA dictated by the next codon of the mRNA. This means that not only does the incoming amino acid need to be attached to the peptide chain but the ribosome must move down the mRNA to the next codon. Each incoming aminoacyl-tRNA is brought to the ribosome by an eEF-1α-GTP complex. When the correct tRNA is deposited into the A site the GTP is hydrolysed and the eEF-1α-GDP complex dissociates. In order for additional translocation events the GDP must be exchanged for GTP. This is carried out by eEF-1$\beta\gamma$ similarly to the GTP exchange that occurs with eIF-2 catalysed by eIF-2B.

The peptide attached to the tRNA in the P site is transferred to the amino group at the aminoacyl-tRNA in the A site. This reaction is catalysed by peptidyltransferase. This process is termed transpeptidation. The elongated peptide now resides on a tRNA in the A site. The A site needs to be freed in order to accept the next aminoacyl-tRNA. The process of moving the peptidyl-tRNA from the A site to the P site is termed, translocation. Translocation is catalysed by eEF-2 coupled to GTP hydrolysis. In the process of translocation the ribosome is moved along the mRNA such that the next codon of the mRNA resides under the A site. Following translocation eEF-2 is released from the ribosome. The cycle can now begin again. The ability of eEF-2 to carry out translocation is regulated by the state of phosphorylation of the enzyme, when phosphorylated the enzyme is inhibited. Phosphorylation of eEF-2 is catalysed by the enzyme eEF2 kinase (eEF2K). Regulation of eEF2K activity is normally under the control of insulin and Ca^{2+} fluxes. The Ca^{2+}-mediated effects are the result of calmodulin interaction with eEF2K. Activation of eEF2K in skeletal muscle by Ca^{2+} is important to reduce consumption of ATP in the process of protein synthesis during periods of exertion which will lead to release of intracellular Ca^{2+} stores. eEF2K itself is also regulated by phosphorylation and one of the kinases that phosphorylates the enzyme is regulated by mTOR. In addition, the master metabolic regulatory kinase, AMP-activated protein kinase (AMPK) will phosphorylate and activate eEF2K leading to inhibition of eEF-2 activity.

Termination

Like initiation and elongation, translational termination requires specific protein factors identified as releasing factors, RFs in *E. coli* and eRFs in eukaryotes. There are 2 RFs in *E. coli* and one in eukaryotes. The signals for termination are the same in both prokaryotes and eukaryotes. These signals are termination codons present in the mRNA. There are 3 termination codons, UAG, UAA and UGA. In *E. coli* the termination codons UAA and UAG are recognized by RF-1, whereas RF-2 recognizes the termination codons UAA and UGA. The eRF binds to the A site of the ribosome in conjunction with GTP. The binding of eRF to the ribosome stimulates the peptidytransferase activity to transfer the peptidyl group to water instead of an aminoacyl-tRNA. The resulting uncharged tRNA left in the P site is expelled with concomitant hydrolysis of GTP. The inactive ribosome then releases its mRNA and the 80S complex dissociates into the 40*S* and 60*S* subunits ready for another round of translation.

Selenoproteins

Selenium is a trace element and is found as a component of several prokaryotic and eukaryotic enzymes that are involved in redox reactions. The selenium in these selenoproteins is incorporated as a unique amino acid, selenocysteine, during translation. A particularly important eukaryotic selenoenzyme is glutathione peroxidase. This enzyme is required during the oxidation of glutathione by hydrogen peroxide (H_2O_2) and organic hydroperoxides.

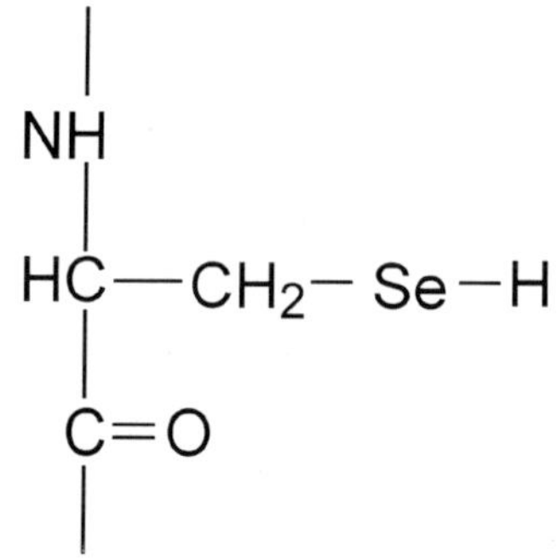

Structure of the Selenocysteine Residue

Incorporation of selenocysteine by the translational machinery occurs via an interesting and unique mechanism. The tRNA for selenocysteine is charged with serine and then enzymatically selenylated to produce the selenocysteinyl-tRNA. The anticodon of selenocysteinyl-tRNA interacts with a stop codon in the mRNA (UGA) instead of a serine codon. The selenocysteinyl-tRNA has a unique structure that is not recognized by the termination machinery and is brought into the ribosome by a dedicated specific elongation factor. An element in the 3' non-translated region (UTR) of selenoprotein mRNAs determines whether UGA is read as a stop codon or as a selenocysteine codon.

Regulation of eIF-4E Activity

The cellular levels of eIF-4E are the lowest of all eukaryotic initiation factors which makes this factor a prime target for regulation. Indeed, at least 3 distinct mechanisms are known to exist that regulate the level and activity of eIF-4E. These include regulation of the level of transcription of the eIF-4E gene, post-translational modification via phosphorylation and inhibition by interaction with binding proteins.

Although the exact mechanisms used to upregulate the transcription of the eIF-4E gene are not yet well understood, it is known that exposure of cells to growth factors as well as activation of T cells leads to increased expression of eIF-4E. The proto-oncogene MYC is believed to play a role in the transcriptional activation of eIF-4E as 2 functional MYC-binding sites have been found in the promoter region of the eIF-4E gene. Of significant note is the finding that cells that are stably over-expressing the MYC gene also have enhanced levels of eIF-4E. Quite strikingly it has been shown that promiscuous elevation in the levels of eIF-4E lead to tumorigenesis placing this translation factor in the category of proto-oncogene.

Numerous extracellular stimuli (e.g. insulin, EGF, angiotensin II and gastrin) that exert a portion of their effects at the level of enhanced translation do so by affecting the state of eIF-4E phosphorylation. However, it should be noted that not all signals that lead to increased eIF-4E phosphorylation lead to increased rates of translation. Changes in eIF-4E phosphorylation correlate well with progression through the cell cycle. In resting (G_0) cells eIF-4E phosphorylation is low, it increases during G_1 and S phase and then declines again in M phase. Phosphorylation of eIF-4E occurs at one major site which is Ser209 (in the human and mouse proteins).

The primary signal transduction pathway leading to eIF-4E phosphorylation is that involving the RAS gene. Many growth factors stimulate activation of RAS in response to binding their cognate receptors. Subsequently, RAS activation leads to the phosphorylation and activation of MAP-interacting kinase-1 (Mnk1) which in turn phosphorylates eIF-4E. Although the exact effect of eIF-4E phosphorylation is not clearly defined, it may be necessary to increase affinity of eIF-4E for the mRNA cap structure and for eIF-4G.

The principal mechanism utilized in the regulation of eIF-4E activity is through its interaction with a family of binding/repressor proteins termed 4EBPs (4E binding proteins) which are widely distributed in numerous vertebrate and invertebrate organisms. In mammalian cells 3 related 4EBPs have been found where 4EBP1 and 4EBP2 are also identified as PHAS-I and PHAS-II (PHAS refers to properties of heat and acid stability).

Binding of 4E-BPs to eIF-4E does not alter the affinity of eIF-4E for the cap structure but prevents the interaction of eIF-4E with eIF-4G which in turn suppresses the formation of the eIF-4F complex. The ability of 4EBPs to interact with eIF-4E is controlled via the phosphorylation of specific Ser and Thr residues in 4EBP. When hypophosphorylated, 4EBPs bind with high efficiency to eIF-4E but lose their binding capacity when phosphorylated. Numerous growth and signal transduction stimulating effectors lead to phosphorylation of 4E-BPs just as these same responses can lead to phosphorylation of eIF-4E.

There are several signal transduction pathways whose activations lead to phosphorylation of 4E-BPs. These include pathways that lead to activation of phosphatidylinositol 3-kinase (PI3K), the Akt Ser/Thr kinase which is also called protein kinase B (PKB) and the FKBP12-rapamycin-associated protein/ mammalian target of rapamycin (FRAP/mTOR) family of proteins. Akt was originally identified as a virally encoded oncogene and there are now at least three members of the PKB/Akt family identified as Akt1, Akt2, and Akt3. The mammalian TOR proteins are homologs of the yeast TOR proteins that were identified in a screen for yeast mutants resistant to rapamycin. Rapamycin is an immunosuppressant used primarily in the prevention of tissue rejection following organ transplantation.

Rapamycin functions within cells by binding the immunophilin FK506-binding protein 12 (FKBP12). Immunophilins are intracellular proteins that binds to immunosuppressive drugs such as FK506 and rapamycin. When rapamycin inhibits the kinase activity of FRAP/mTOR it can no longer phosphorylate 4EB. One of the major effects of insulin is increased protein synthesis and this effect is elicited, in part, via activation of mTOR function.

Targets for mTOR regulation of translational initiation and elongation. AMPK = AMP-activated kinase. TSC1 and TSC2 = Tuberous sclerosis tumour suppressors 1 (hamartin) and 2 (tuberin); Rheb = Ras homolog enriched in brain; PKB/Akt = protein kinase B; 4EBP1 = eIF-4E binding protein; p70S6K = 70kDa ribosomal protein S6 kinase, also called S6K; eEF2K = eukaryotic elongation factor 2 kinase.

Regulation of mTOR activity is effected via several mechanisms. Activation of AMPK results in phosphorylation and activation of the TSC1/TSC2 complex which results in inhibition of mTOR. AMPK can also phosphorylate and inhibit mTOR. Conversely, activation of PKB (as in the case of insulin receptor activation) leads to activation of mTOR either by inhibition of the TSC1/TSC2 complex or by phosphorylation and activation of mTOR directly.

Activation of mTOR leads to phosphorylation of p70S6K and 4EBP1. The net effect of phosphorylation of 4EBP1 is that it is released from eIF-4E allowing eIF-4E to actively bind eIF-4G and recognize the cap structure of mRNAs. Activated p70S6K phosphorylates and inhibits eEF2K. If eEF2K does not phosphorylate eEF2 then translation elongation proceeds uninhibited.

Heme Control of Translation

Regulation of initiation in eukaryotes is effected by phosphorylation of a ser(S) residue in the α-subunit of eIF-2. Phosphorylated eIF-2 in the absence of eIF-2B is just as active an initiator as non-phosphorylated eIF-2. However, when eIF-2 is phosphorylated the GDP-bound complex is stabilized and exchange for GTP is inhibited. The exchange of GDP for GTP is mediated by eIF-2B (also called guanine nucleotide exchange factor, GEF). When eIF-2 is phosphorylated it binds eIF-2B more tightly thus slowing the rate of exchange. It is this inhibited exchange that affects the rate of initiation.

The phosphorylation of eIF-2 is the result of an activity called heme-controlled inhibitor (HCI) which functions as diagrammed below. HCI is generated in the absence of heme, a mitochondrial product. Removal of phosphate is catalysed by a specific eIF-2 phosphatase which is unaffected by heme. The presence of HCI was first seen in *in vitro* translation system derived from lysates of reticulocytes. Reticulocytes synthesize almost exclusively hemoglobin at an extremely high rate. In an intact reticulocyte eIF-2 is protected from phosphorylation by a specific 67 kDa protein.

Interferon Control of Translation

Regulation of translation can also be induced in virally infected cells. It would benefit a virally infected cell to turn off protein synthesis to prevent propagation of the viruses. This is accomplished by the induced synthesis of interferons (IFs). There are 3 classes of IFs. The leukocyte or α-IFs, the fibroblast or β-IFs and the lymphocyte or γ-IFs. IFs are induced by dsRNAs and themselves induce a specific kinase termed RNA-dependent protein kinase (PKR) that phosphorylates eIF-2 thereby shutting off translation in a similar manner to that of heme control of translation. Additionally, IFs induce the synthesis of 2'-5'-oligoadenylate, pppA(2'p5'A)n, that activates a pre-existing ribonuclease, RNase L. RNase L degrades all classes of mRNAs thereby shutting off translation.

Iron Control of Translation

Regulation of the translation of certain mRNAs occurs through the action of specific RNA-binding proteins. Protein of this class have been identified that bind to sequences in either the 5' non-translated region (5'-UTR) or 3'-UTR. Two particularly interesting and important regulatory schemes related to iron metabolism encompass RNA binding proteins that bind to either the 5'-UTR of one mRNA or the 3'-UTR of another. The transferrin receptor is a protein located in the plasma membrane that binds the protein transferrin. Transferrin is the major iron transport protein in the plasma. When iron levels are low the rate of synthesis of the transferrin receptor mRNA increases so that cells can take up more iron. This regulation occurs through the action of an iron response element binding protein (IRBP) that binds to specific iron response elements (IREs) in the 3'-UTR of the transferrin receptor mRNA. These IREs form hair-pin loop structures that are recognized by IRBP. This IRBP is an iron-deficient form of aconitase, the iron-requiring enzyme of the TCA cycle.

When iron levels are low, IRBP is free of iron and can therefore, interact with the IREs in the 3'-UTR of the transferrin receptor mRNA. Transferrin receptor mRNA with IRBP bound is stabilized from degradation. Conversely, when iron levels are high, IRBP binds iron then cannot interact with the IREs in the transferrin receptor mRNA. The effect is an increase in degradation of the transferrin receptor mRNA. A related, but opposite, phenomenon controls the translation of the ferritin mRNA. Ferritin is an iron-binding protein that prevents toxic levels of ionized iron (Fe^{2+}) from building up in cells. The ferritin mRNA has an IRE in its 5'-UTR. As with the transferrin receptor story, when iron levels are high, IRBP cannot bind to the IRE in the 5'-UTR of the ferritin mRNA. This allows the ferritin mRNA to be translated. Conversely, when iron levels are low, the IRBP binds to the IRE in the ferrritin mRNA preventing its translation.

Protein Synthesis Inhibitors

Many of the antibiotics utilized for the treatment of bacterial infections as well as certain toxins function through the inhibition of translation. Inhibition can be effected at all stages of translation from initiation to elongation to termination.

Nucleotide Derivatives in tRNAs

As many as 25% of the nucleotides found in tRNAs are modified post-transcriptionally. At least 80 different modified nucleotides have been identified at >60 different positions in various tRNAs. The two most commonly occurring modified nucleotides are dihydrouridine (abbreviated D) and pseudouridine (abbreviated Ψ) each of which is found in a characteristic loop structure in tRNAs. Dihydrouridine is found in the D-loop (hence the name of the loop) and pseudouridine is found in the TΨC loop (hence the name of the loop.

Synthetic Nucleotide Analogs

Many nucleotide analogues are chemically synthesized and used for their therapeutic potential. The nucleotide analogues can be utilized to inhibit specific enzymatic activities. A large family of analogues are used as anti-tumour agents, for instance, because they interfere with the synthesis of DNA and thereby preferentially kill rapidly dividing cells such as tumour cells. Some of the nucleotide analogues commonly used in chemotherapy are 6-mercaptopurine, 5-fluorouracil, 5-iodo-2'-deoxyuridine and 6-thioguanine. Each of these compounds disrupts the normal replication process by interfering with the formation of correct Watson-Crick base-pairing. Nucleotide analogs also have been targeted for use as antiviral agents. Several analogs are used to interfere with the replication of HIV, such as AZT (azidothymidine) and ddI (dideoxyinosine). Several purine analogs are used to treat gout. The most common is allopurinol, which resembles hypoxanthine. Allopurinol inhibits the activity of xanthine oxidase, an enzyme involved in *de novo* purine biosynthesis. Additionally, several nucleotide analogues are used after organ transplantation in order to suppress the immune system and reduce the likelihood of transplant rejection by the host.

Polynucleotides

Polynucleotides are formed by the condensation of two or more nucleotides. The condensation most commonly occurs between the alcohol of a 5'-phosphate of one nucleotide and the 3'-hydroxyl of a second, with the elimination of H_2O, forming a phosphodiester bond. The formation of phosphodiester bonds in DNA and RNA exhibits directionality. The primary structure of DNA and RNA (the linear arrangement of the nucleotides) proceeds in the 5'—>3' direction. The common representation of the primary structure of DNA or RNA molecules is to write the nucleotide sequences from left to right synonymous with the 5'—>3' direction as shown:

5'-pGpApTpC-3'

Structure of DNA

Utilizing X-ray diffraction data, obtained from crystals of DNA, James Watson and Francis Crick proposed a model for the structure of DNA. This model (subsequently verified by additional data) predicted that DNA would exist as a helix of two complementary antiparallel strands, wound around each other in a rightward direction and stabilized by H-bonding between bases in adjacent strands. In the Watson-Crick model, the bases are in the interior of the helix aligned at a nearly 90 degree angle relative to the axis of the helix. Purine bases form hydrogen bonds with pyrimidines, in the crucial phenomenon of base pairing. Experimental determination has shown that, in any given molecule of DNA, the concentration of adenine (A) is equal to thymine (T) and the concentration of cytidine (C) is equal to guanine (G). This means that A will only base-pair with T, and C with G. According to this pattern, known as Watson-Crick base-pairing, the base-pairs composed of G and C contain three H-bonds, whereas those of A and T contain two H-bonds. This makes G-C base-pairs more stable than A-T base-pairs.

The antiparallel nature of the helix stems from the orientation of the individual strands. From any fixed position in the helix, one strand is oriented in the 5'—>3' direction and the other in the 3'—>5' direction. On its exterior surface, the double helix of DNA contains two deep grooves between the ribose-phosphate chains. These two grooves are of unequal size and termed the major and minor grooves. The difference in their size is due to the asymmetry of the deoxyribose rings and the structurally distinct nature of the upper surface of a base-pair relative to the bottom surface.

The double helix of DNA has been shown to exist in several different forms, depending upon sequence content and ionic conditions of crystal preparation. The B-form of DNA prevails under physiological conditions of low ionic strength and a high degree of hydration. Regions of the helix that are rich in pCpG dinucleotides can exist in a novel left-handed helical conformation termed Z-DNA. This conformation results from a 180 degree change in the orientation of the bases relative to that of the more common A- and B-DNA.

Thermal Properties of DNA

As cells divide it is a necessity that the DNA be copied (replicated), in such a way that each daughter cell acquires the same amount of genetic material. In order for this process to proceed the two strands of the helix must first be separated, in a process termed denaturation. This process can also be carried out in vitro. If a solution of DNA is subjected to high temperature, the H-bonds between bases become unstable and the strands of the helix separate in a process of thermal denaturation.

The base composition of DNA varies widely from molecule to molecule and even within different regions of the same molecule. Regions of the duplex that have predominantly A-T base-pairs will be less thermally stable than those rich in G-C base-pairs. In the process of thermal denaturation, a point is reached at which 50% of the DNA molecule exists as single strands. This point is the melting temperature (T_m), and is characteristic of the base composition of that DNA molecule. The T_m depends upon several factors in addition to the base composition. These include the chemical nature of the solvent and the identities and concentrations of ions in the solution.

When thermally melted DNA is cooled, the complementary strands will again re-form the correct base pairs, in a process is termed annealing or hybridization. The rate of annealing is dependent upon the nucleotide sequence of the two strands of DNA.

Analysis of DNA Structure

Chromatography: Several of the chromatographic techniques available for the characterization of proteins can also be applied to the characterization of DNA. The most commonly used technique is HPLC (high performance liquid chromatography). Affinity chromatographic techniques also can be employed. One common affinity matrix is hydroxyapatite (a form of calcium phosphate), which binds double-stranded DNA with a higher affinity than single-stranded DNA.

Electrophoresis: This procedure can serve the same function with regard to DNA molecules as it does for the analysis of proteins. However, since DNA molecules have much higher molecular weights than proteins, the molecular sieve used in electrophoresis of DNA must be different as well. The material of choice is agarose, a carbohydrate polymer purified from a salt water algae. It is a copolymer of mannose and galactose that when melted and re-cooled forms a gel with pores sizes dependent upon the concentration of agarose. The phosphate backbone of DNA is highly negatively charged, therefore DNA will migrate in an electric field. The size of DNA fragments can then be determined by comparing their migration in the gel to known size standards. Extremely large molecules of DNA (in the range of 30kb–10Mb) are effectively separated in agarose gels using pulsed-field gel electrophoresis (PFGE). This technique employs two or more electrodes, placed orthogonally with respect to the gel, that receive short alternating pulses of current. PFGE allows whole chromosomes and large portions of chromosomes to be analysed.

4

Chemistry of Carbohydrate and Lipids

Carbohydrates or saccharides are the most abundant class of biomolecules. Carbohydrates are used to store energy, though they serve other important functions as well. This is an overview of carbohydrate chemistry, including a look at the types of carbohydrates, their functions, and carbohydrate classification.

What is a Carbohydrate?

Carbohydrates are a common class of simple organic compouds. A carbohydrate is an aldehyde or a ketone that has additional hydroxyl groups. The simplest carbohydrates are called *monosaccharides*, which has the basic structure $(C \cdot H_2O)_n$, where n is three or greater. Monosaccharides link together to form oligosaccharides and polysaccharides. Two monosaccharides link together to form a *disaccharide*.

Functions of Carbohydrates

Carbohydrates serve several biochemical functions:

- Monosaccharides are a fuel for celular metabolism.
- Monosaccharides are used in several biosynthesis reactions.
- Monosaccharides may be converted into space-saving polysaccharides, such as glyocogen and starch. These molecules provide stored energy for plant and animal cells.
- Carbohydrates are used to form structural elements, such as chitin in animals and cellulose in plants.

- Carbohydrates and modified carbohydrates are important for an organism's fertilization, development, blood clotting and immune system function.

Examples of Carbohydrates

- Monosaccharides: glucose, fructose, galactose
- Disaccharides: sucrose, lactose
- Polysaccharides: chitin, cellulose.

Carbohydrate Classification

Three characteristics are used to classify monosaccharides:

- number of carbon atoms in the molecule
- location of the carbonyl group
- the chirality of the carbohydrate.

Aldose - monosaccharide in which the carbonyl group is an aldehyde.

Ketone - monosaccharide in which the carbonyl group is a ketone.

Triose - monosaccharide with 3 carbon atoms.

Tetrose - monosaccharide with 4 carbon atoms.

Pentose - monosaccharide with 5 carbon atoms.

Hexose - monosaccharide with 6 carbon atoms.

Aldohexose - 6-carbon aldehyde (e.g., glucose).

Aldopentose - 5-carbon aldehyde (e.g., ribose).

Ketohexose - 6-carbon hexose (e.g., fructose).

A monosaccharide is D or L depending on the orientation of the asymmetric carbon located furthest from the carbonyl group. In a D sugar, the hydroxyl group is on the right the molecule when written as a Fischer projection. If the hydroxyl group is on the left of the molecule, then it is an L sugar.

Carbohydrates are the most abundant natural organic compounds on Earth. The term "carbohydrate" derives from their general formula of $C_n(H_2O)_n$, first determined in the nineteenth century, and indicates that these compounds are hydrates of carbon. Carbohydrates are more specifically defined as polyhydroxy aldehydes or ketones and the products derived from them. Carbohydrates are synthesized via photosynthesis by plants, algae, and some bacteria. Animals feeding on these organisms then use the energy stored in these compounds.

Energy storage is not the only function of carbohydrates. They have a variety of functions in living organisms, including their contribution to the structure of cell walls and their vital role in communication at the site of cell membranes. Carbohydrates form part of the backbone of RNA and DNA molecules, and they are also found linked to proteins and lipids as glycoproteins and glycolipids. The three basic groups of carbohydrates based on size are: monosaccharides, oligosaccharides, and polysaccharides (saccharide from the Greek *sakcharon,* or "sugar"). The oligosaccharides and polysaccharides are composed of a few and many monosaccharides, respectively. Monosaccharides have two major groups: the aldoses and the ketoses.

Aldoses

The simplest of the aldoses is glyceraldehyde, which is a triose, or three-carbon sugar. Glyceraldehyde has one chiral carbon and therefore two stereoisomers, designated D and L. In nature, only D sugars occur in abundance. Other aldoses can be derived from glyceraldehyde via insertion of additional hydroxy carbons between the carbonyl carbon and the molecule's other carbons. In this way, tetroses, pentoses, and hexoses are formed. Although glyceraldehyde and the tetroses can occur only as simple linear structures, Glucose, an aldohexose, is the most common of the monosaccharides.

In various combinations and permutations, it forms starch, cellulose, sucrose, and lactose (milk sugar), among other things. When metabolized via the glycolytic pathway, it is the major energy source for many living things. Most commonly, glucose forms a ring, its fifth hydroxyl group reacting with the aldehyde carbonyl group to form a hemiacetal. As a result of this reaction, the sugar forms a six-membered ring and the carbonyl carbon becomes chiral. The two new stereoisomers of glucose that revolve on the aldehyde carbon are designated α and β and are considered anomers of one another. The now chiral carbon is called the "anomeric" carbon. These six-membered ring structures are called pyranoses, as they resemble the compound pyran. Thus, in its ring forms, glucose is properly designated α-D-glucopyranose, or β-D-glucopyranose.

These pyranose rings are not flat and can assume several different conformations. In general, the most stable conformation is the "chair" conformation, in which the bulky atomic groups (e.g., the hydroxyl and hydroxymethyl groups) are equatorial or within the plane of the ring. The hydrogens would then be axial or perpendicular to the plane of the ring. Only in β-D-glucopyranose are all of the bulky groups equatorial; thus, β-D-glucopyranose is more stable than α-D-glucopyranose, in which the C-1 hydroxyl group is axial to the ring. The α and β forms of glucose are freely interconvertible in solution. At equilibrium, a solution of glucose contains about two-thirds α-D-glucopyranose, one-third α-D-glucopyranose, and small amounts of the linear and five-membered ring forms of glucose.

Other widely occurring aldoses include mannose, galactose, and ribose. Mannose and galactose are, like glucose, aldohexoses and can form six-membered rings. Mannose is an important part of the complex sugars, or oligosaccharides, that attach to proteins in the formation of glycoproteins. Galactose combines with glucose to form lactose or milk sugar. Structurally, each of these sugars differs from glucose only in the stereo-chemistry that revolves on one carbon: mannose on C-2 and galactose on C-4. Sugars that differ from one another only in respect to the stereo-chemistry at one carbon are considered epimers of each other. Thus mannose is the C-2 epimer of glucose, and galactose the C-4 epimer. Mannose and galactose are not epimers; they differ from each other in respect to the stereochemistry revolving around two carbons. Ribose is an aldopentose. It composes the carbohydrate portion of the ribonucleotides that form a cell's RNA. Ribose, like the aldohexoses, can form a ring. However, ribose forms a five-membered ring, called a "furanose" because of its similarity to the compound furan. Once again, there are two possible forms this ring can have: α-D-ribofuranose and β-D-ribofuranose. RNA contains β-D-ribofuranose.

Ketoses

There are fewer ketoses than there are aldoses because ketoses have one less chiral carbon. The most prevalent of the ketoses are dihydroxyacetone, ribulose, xylulose, and fructose. All four of these sugars are important intermediates in metabolism. Fructose is, along with glucose, part of sucrose or table sugar. Fructose is a ketohexose, and the only one of the four ketoses that can assume a ring structure. Like ribose, fructose forms a five-membered (or furanose) ring and has α and β anomers.

Monosaccharide Derivatives

Monosaccharides undergo incorporation into oligo- and polysaccharides. Individual monosaccharides can also undergo a variety of transformations. One important modification of monosaccharides is the formation of deoxy sugars. The most biologically significant of the deoxy sugars is β-D-2-deoxyribose, in which the C-2 hydroxyl group of ribose has been replaced with hydrogen. This deoxy sugar is the sugar component of DNA.

In the amino sugars, an amino group replaces one or more of the hydroxyl groups. The most common of these sugars are D-glucosamine and D-galactosamine, both of which have an amino group in place of the hydroxyl group on the second carbon. Often, these amino groups are acetylated to give N-acetyl sugars, such as N-acetylglucosamine and N-acetylgalactosamine. These sugars are important components of larger polysaccharides. Other important amino sugars are muramic acid and N-acetylneuraminic acid, which are components of the oligosaccharides of glycoproteins and glycolipids, and of bacterial cell walls. Muramic acid and N-acetylneuraminic acid are glucosamines, which have been linked at either C-3 or C-1 to three-carbon acids. Muramic acid is formed via an ether linkage between the C-3 of glucosamine and the hydroxyl group of lactic acid. N-acetyl-D-neuraminic acid results from the formation of a C–C bond between the C-1 of N-acetyl-D-mannosamine and the C-3 of phosphoenolpyruvate. This and other derivatives of neuraminic acid are collectively called sialic acids and are widely found in bacteria and animals.

Finally, monosaccharides often form ester linkages with phosphate and sulfate ions. In fact it is rare to find free monosaccharide in cells. Glucose is the only unmodified monosaccharide that exists in substantial quantities in living things, in which it exists primarily extracellularly.

Carbohydrate chemistry is a subdiscipline of chemistry primarily concerned with the synthesis, structure, and function of carbohydrate structures. Due to the general structure of carbohydrates, their synthesis is often preoccupied with the selective formation of glycosidic linkages and the selective reaction of hydroxyl groups. As a result, this chemistry relies heavily on the use of protecting groups.

Carbohydrate Synthesis

Carbohydrate synthesis is a sub-field of organic chemistry concerned specifically with the generation of natural and unnatural carbohydrate structures. This can include the synthesis of monosaccharide residues or structures containing more than one monosaccharide, known as oligosaccharides.

Generally speaking, carbohydrates can be classified into two groups, simple sugars and complex carbohydrates. Simple sugars, also called monosaccharides, are carbohydrates which can not be converted into smaller sugars by hydrolysis. When two or more monosaccharide units are connected to one another via a glycoside linkage, complex carbohydrates are formed. Complex carbohydrates, according to the different number of monosaccharide units, can be classed into three groups, disaccharides, oligosaccharides, and polysaccharides. A disaccharide is formed from two monosaccharides. Oligosaccharides can be formed by a small number of monosaccharides linked together. Higher oligosaccharides are called polysaccharides. It is now well known that glycoconjugates play an indispensable role in many biological processes. These biological processes in which carbohydrates are involved are typically associated not to monosaccharides, but to oligosaccharides structures of glycoconjugates. Therefore, the oligosaccharide synthesis becomes more and more important in studying the biological activities.

Oligosaccharide Synthesis

Oligosaccharides have diverse structures. The number of monosaccharides, ring size, the different anomeric stereochemistry, and the existence of the branched-chain sugars all contribute to the amazing complexity of the oligosaccharide structures. The essence of the reducing oligosaccharide synthesis is connecting the anomeric hydroxyl of the glycosyl donors to the alcoholic hydroxyl groups of the glycosyl acceptors. Protection of the hydroxyl groups of the acceptor with the target alcoholic hydroxyl group unprotected can assure the regiochemical control. Additionally, factors such as the different protecting groups, the solvent, and the glycosylation methods can influence the anomeric configurations. This concept is illustrated by an oligosaccharide synthesis in Scheme 1. Oligosaccharide synthesis normally consists of four parts: preparation of the glycosyl donors, preparation of the glycosyl acceptors with a single unprotected hydroxyl group, the coupling of them, and the deprotection proces

Building Blocks

Common donors in oligosaccharide synthesis are glycosyl halides, glycosyl acetates, thioglycosides, trichloroacetimidates, pentenyl glycosides, and glycals. Of all these donors, glycosyl halides are classic donors, which played a historical role in the development of glycosylation reactions. Thioglycoside and trichloroacetimidate donors are used more than others in contemporary glycosylation methods. When it comes to the trichloroacetimidate method, one of the advantages is that there is no need to introduce heavy metal reagents in the activation process. Moreover, using different bases can selectively lead to different anomeric configurations. (Scheme 2) As to the thioglycosides, the greatest strength is that they can offer a temporary protection to the anomeric centre because they can survive after most of the activation processes. Additionally, a variety of activation methods can be employed, such as NIS/AgOTf, NIS/TfOH, IDCP (Iodine dicollidine perchlorate), iodine, and Ph_2SO/Tf_2O. Furthermore, in the preparation of 1, 2-trans glycosidic linkage, using thioglycosides and imidates can promote the rearrangement of the orthoester byproducts, since the reaction mixtures are acidic enough.

Stereoselectivity

The structures of acceptors play a critical role in the rate and stereoselectivity of glycosylations. Generally, the unprotected hydroxyl groups are less reactive when they are between bulky protecting groups. That is the reason why the hydroxyl group at OH-4 in pyranosides is unreactive. Hyperconjugation is involved when OH-4 is anti-periplanar to the ring oxygen, which can also reduce its reactivity. (Scheme 3) Furthermore, acyl protecting groups can reduce the reactivity of the acceptors compared with alkyl protecting groups because of their electron-withdrawing ability. Hydroxyl group at OH-4 of N-acetylglucosamine derivatives is particularly unreactive.

Difficult Linkages

It is more difficult to prepare 1, 2-cis-α glycosidic linkage. Typically, when non-participating groups on O-2 position, 1, 2-cis-α linkage can be achieved either by using the historical important halide ion methods, or by using 2-O-alkylated glycosyl donors, commonly thioglycosides or trichloroacetimidates, in nonpolar solvents. Of all linkages, 1, 2-cis-beta-glycoside is considered to be the most difficult one to synthesize in modern carbohydrate chemistry. Non-participating groups on O-2 position, insoluble silver catalysts, reactive glycosyl donors, and nucleophilic acceptors are basic requires to generate 1, 2-cis-beta-glycosides in a previous method. Recently, several other methods are employed, such as Intramolecular Aglycon Delivery (IAD), beta-D-glycopyranoside to beta-D-mannopyranoside, and Crich beta-mannosylation.

Among these methods, Crich beta-mannosylation is the most powerful one. (Scheme 4) Donors first reacting with triflic anhydride (Tf2O), then using DTBMP (2, 6-di-tert-butyl-4-methylpyridine) as a base, followed by substitution of glycosyl acceptors (R'OH) can assure excellent stereoselectivity.

Glycosidic Bond Formation

Chemical Glycosylation

A chemical glycosylation reaction involves the coupling of a sugar to a glycosyl acceptor forming a glycoside. If the acceptor is another sugar, the product is an oligosaccharide. The reaction involves coupling a glycosyl donor to a glycosyl acceptor via activation utilizing a suitable activator. The reactions often result in a mixture of products due to the creation of a new stereogenic centre at the anomeric position of the glycosyl donor. The formation of a glycosidic linkage allows for the synthesis of complex polysaccharides which may play important roles in biological processes and pathogenesis and therefore having synthetic analogs of these molecules allows for further studies with respect to their biological importance.

Terminology

The glycosylation reaction involves the coupling of a glycosyl donor and a glycosyl acceptor via initiation using an activator under suitable reaction conditions.

- A glycosyl donor is a sugar with a suitable leaving group at the anomeric position. This group, under the reaction conditions, is activated and via the formation of an oxocarbenium is eliminated leaving an electrophilic anomeric carbon.
- A glycosyl acceptor is a sugar with an unprotected nucleophilic hydroxyl group which may attack the carbon of the oxocarbenium ion formed during the reaction and allow for the formation of the glycosidic bond.

An activator is commonly a Lewis acid which enables the leaving group at the anomeric position to leave and results in the formation of the oxocarbenium ion.

Stereochemistry

The formation of a glycosidic linkage results in the formation of a new stereogenic centre and therefore a mixture of products may be expected to result. The linkage formed may either be axial or equatorial (α or β with respect to glucose). To better understand this, the mechanism of a glycosylation reaction must be considered.

Neighbouring Group Participation

The stereochemical outcome of a glycosylation reaction may in certain cases be affected by the type of protecting group employed at position 2 of the glycosyl donour. A participating group, typically one with a carboxyl group present, will predominantly result in the formation of a β-glycoside. Whereas a non-particiapting group, a group usually without a carboxyl group, will often result in an α-glycoside.

Below it can be seen that having an acetyl protecting group at position 2 allows for the formation for an acetoxonium ion intermediate that blocks attack to the bottom face of the ring therefore allowing for the formation of the β-glycoside predominantly.

Protecting Groups

Different protecting groups on either the glycosyl donour or the glycosyl acceptor may affect the reactivity and yield of the glycosylation reaction. Typically, electron-withdrawing groups such as acetyl or benzoyl groups are found to decresase the reactivity of the donour/acceptor and are therefore termed "disarming" groups. Electron-donating groups such as the benzyl group, are found to increase the reactivity of the donour/acceptor and are therefore called "arming" groups.

Current Methods in Glycoside Synthesis

Glycosyl Iodides

Glycosyl iodides were first introduced for use in glycosylation reactions in 1901 by Koenigs and Knorr although were often considered too reactive for synthetic use. Recently several research groups have shown these donours to have unique reactive properties and can differ from other glycosyl chlorides or bromides with respect to reaction time, efficiency, and stereochemistry. Glycosyl iodides may be made under a variety of conditions, one method of note is the reaction of a 1-*O*-acetylpyranoside with TMSI.

Iodide donours may typically be activated under basic conditions to give β-glycosides with good selectivity. The use of tetraalkylammonium iodide salts such as tetrabutylammonium iodide (TBAI) allows for in-situ anomerization of the α-glycosyl halide to the β-glycosyl halide and provides the α-glycoside in good selectivity.

Thioglyosides

Thioglycosides were first reported in 1909 by Fischer and since then have been explored constantly allowing for the development of numerous protocols for their preparation. The advantage of using thioglycosides is their stability under a wide range of reaction conditions allowing for protecting group manipulations. Additionally thioglycosides act as temporary protecting groups at the anomeric position allowing for thioglycosides to be useful as both glycosyl donours as well as glycosyl acceptors. Thioglycosides are usually prepared by reacting per-acetylated sugars with $BF_3 \cdot OEt_2$ and the appropriate thiol.

Trichloroacetimidates

Trichloroacetimidates were first introduced and explored by Schmidt in 1980 and since then have become very popular for glycoside synthesis. The use of trichloroacetimidates provides many advantages including ease of formation, reactivity and stereochemical outcome. *O*-Glycosyl trichloroacetimidates are prepared via the addition of trichloroacetonitrile (Cl_3CCN) under basic conditions to a free anomeric hydroxyl group.

Notable Synthetic Products

Below are a few examples of some notable targets obtained via a series of glycosylation reactions.

Fischer Glycosidation

Fischer glycosidation (or Fischer glycosylation) refers to the formation of a glycoside by the reaction of an aldose or ketose with an alcohol in the presence of an acid catalyst. The reaction is named after the German chemist, Emil Hermann Fischer, winner of the Nobel Prize in chemistry, 1902, who developed this method between 1893 and 1895.

Commonly, the reaction is performed using a solution or suspension of the carbohydrate in the alcohol as the solvent. The carbohydrate is usually completely unprotected. The Fischer glycosidation reaction is an equilibrium process and can lead to a mixture of ring size isomers, and anomers, plus in some cases, small amounts of acyclic forms. With hexoses, short reactions times usually lead to furanose ring forms, and longer reaction times lead to pyranose forms. With long reaction times the most thermodynamically stable product will result which, owing to the anomeric effect, is usually the alpha anomer.

Koenigs–Knorr Reaction

The Koenigs–Knorr reaction in organic chemistry is the substitution reaction of a glycosyl halide with an alcohol to give a glycoside. It is one of the oldest and simplest glycosylation reactions. It is named after Wilhelm Koenigs (1851-1906), student of von Bayer and fellow student with Hermann Emil Fischer, and Edward Knorr, student of Koenigs.

In the above example, the stereochemical outcome is determined by the presence of the neighbouring group at C2 that lends anchimeric assistance, resulting in the formation of a 1,2-trans stereochemical arrangement. Esters (eg acetyl, benzoyl, pivalyl) generally provide good anchimeric assistance, whereas ethers (eg benzyl, methyl etc) do not, leading to mixtures of stereoisomers. Generally, the Koenigs–Knorr reaction refers to the use of glycosyl chlorides, bromides and more recently iodides as glycosyl donors.

The Koenigs–Knorr reaction can be performed with alternative promoters such as various heavy metal salts including mercuric bromide/mercuric oxide, mercuric cyanide and silver triflate. When mercury salts are used, the reaction is normally called the Helferich method. Other glycosidation methods are Fischer glycosidation, use of glycosyl acetates, thioglycosides, glycosyl trichloroacetimidates, glycosyl fluorides or n-pentenyl glycosides as glycosyl donors, or intramolecular aglycon delivery.

Koenigs-Knorr Reaction

The Koenigs–Knorr reaction in organic chemistry is the substitution reaction of a glycosyl halide with an alcohol to give a glycoside. It is one of the oldest and simplest glycosylation reactions. It is named after Wilhelm Koenigs (1851-1906), student of von Bayer and fellow student with Hermann Emil Fischer, and Edward Knorr, student of Koenigs. In the above example, the stereochemical outcome is determined by the presence of the neighbouring group at C2 that lends anchimeric assistance, resulting in the formation of a 1,2-trans stereochemical arrangement. Esters (eg acetyl, benzoyl, pivalyl) generally provide good anchimeric assistance, whereas ethers (eg benzyl, methyl etc) do not, leading to mixtures of stereoisomers.

Generally, the Koenigs–Knorr reaction refers to the use of glycosyl chlorides, bromides and more recently iodides as glycosyl donors. The Koenigs–Knorr reaction can be performed with alternative promoters such as various heavy metal salts including mercuric bromide/mercuric oxide, mercuric cyanide and silver triflate. When mercury salts are used, the reaction is normally called the Helferich method. Other glycosidation methods are Fischer glycosidation, use of glycosyl acetates, thioglycosides, glycosyl trichloroacetimidates, glycosyl fluorides or n-pentenyl glycosides as glycosyl donors, or intramolecular aglycon delivery.

Protecting Groups

Carbohydrate Acetalisation

In carbohydrate chemistry carbohydrate acetalisation is an organic reaction and a very effective means of providing a protecting group. The example below depicts the acetalisation reaction of D-ribose 1. With acetone or 2,2-

dimethoxypropane as the acetalisation reagent the reaction is under thermodynamic reaction control and results in the pentose 2. The latter reagent in itself is an acetal and therefore the reaction is actually a cross-acetalisation.

Oligosaccharides

Reactions of Carbohydrates

Carbohydrates are reactants in many organic reactions. For example:

- Cyanohydrin reaction
- Lobry-de Bruyn-van Ekenstein transformation
- Amadori rearrangement
- Nef reaction
- Wohl degradation
- Tipson-Cohen reaction
- Ferrier rearrangement
- Ferrier II reaction.

Functions of Carbohydrates

Carbohydrates have six major functions within the body:

1. Providing energy and regulation of blood glucose
2. Sparing the use of proteins for energy
3. Breakdown of fatty acids and preventing ketosis
4. Biological recognition processes
5. Flavour and Sweeteners
6. Dietary fiber.

Providing Energy and Regulating Blood Glucose

Glucose is the only sugar used by the body to provide energy for its tissues. Therefore, all digestible polysaccharides, disaccharides, and monosaccharides must eventually be converted into glucose or a metabolite of glucose by various liver enzymes. Because of its significant importance to proper cellular function, blood glucose levels must be kept relatively constant.

Among the enormous metabolic activities the liver performs, it also includes regulating the level of blood glucose. During periods of food consumption, pancreatic beta cells sense the rise in blood glucose and begin to secrete the hormone insulin. Insulin binds to many cells in the body having appropriate receptors for the peptide hormone and causes a general uptake in cellular glucose. In the liver, insulin causes the uptake of glucose as well as the synthesis of glycogen, a glucose storage polymer. In this way, the liver is able to remove excessive levels of blood glucose through the action of insulin.

In contrast, the hormone glucagons is secreted into the bloodstream by pancreatic alpha cells upon sensing falling levels of blood glucose. Upon binding to targeted cells such as skeletal muscle and brain cells, glucagon acts to decrease the amount of glucose in the bloodstream. This hormone inhibits the uptake of glucose by muscle and other cells and promotes the breakdown of glycogen in the liver in order to release glucose into the blood. Glucagon also promotes gluconeogenesis, a process involving the synthesis of glucose from amino acid precursors. Through the effects of both glucagon and insulin, blood glucose can usually be regulated in concentrations between 70 and 115 mg/100 ml of blood.

Other hormones of importance in glucose regulation are epinephrine and cortisol. Both hormones are secreted from the adrenal glands, however, epinephrine mimics the effects of glucagon while cortisol mobilizes glucose during periods of emotional stress or exercise.

Despite the liver's unique ability to maintain homeostatic levels of blood glucose, it only stores enough for a twenty-four hour period of fasting. After twenty four hours, the tissues in the body that preferentially rely on glucose, particularly the brain and skeletal muscle, must seek an alternative energy source. During fasting periods, when the insulin to glucagons ratio is low, adipose tissue begins to release fatty acids into the bloodstream. Fatty acids are long hydrocarbon chains consisting of single carboxylic acid group and are not very soluble in water. Skeletal muscle begins to use fatty acids for energy during resting conditions; however, the brain cannot afford the same luxury. Fatty acids are too long and bulky to cross the blood-brain barrier.

Therefore, proteins from various body tissues are broken down into amino acids and used by the liver to produce glucose for the brain and muscle. This process is known as gluconeogenesis or "the production of new glucose." If fasting is prolonged for more than a day, the body enters a state called ketosis. Ketosis comes from the root word ketones and indicates a carbon atom with two side groups bonded to an oxygen atom. Ketones are produced when there is no longer enough oxaloacetate in the mitochondria of cells to condense with acetyl CoA formed from fatty acids. Oxaloacetate is a four-carbon compound that begins the first reaction of the Krebs Cycle, a cycle containing a series of reactions that produces high-energy species to eventually be used to produce energy for the cell. Since oxaloacetate is formed from pyruvate (a metabolite of glucose), a certain level of carbohydrate is required in order to burn fats. Otherwise, fatty acids cannot be completely broken down and ketones will be produced.

Sparing Protein and Preventing Ketosis

So why are carbohydrates important if the body can use other carbon compounds such as fatty acids and ketones as energy? First of all, maintaining a regular intake of carbohydrates will prevent protein from being used as an energy source. Gluconeogenesis will slow down and amino acids will be freed for the biosyntheses of enzymes, antibodies, receptors and other important proteins. Furthermore, an adequate amount of carbohydrates will prevent the degradation of skeletal muscle and other tissues such as the heart, liver, and kidneys. Most importantly, ketosis will be prevented. Although the brain will adapt to using ketones as a fuel, it preferentially uses carbohydrates and requires a minimum level of glucose circulating in the blood in order to function properly. Before the adaptation process occurs, lower blood glucose levels may cause headaches in some individuals. To prevent these ketotic symptoms, it is recommended that the average person consume at least 50 to 100g of carbohydrates per day.

Although the processes of protein degradation and ketosis can create problems of their own during prolonged fasting, they are adaptive mechanisms during glucose shortages. In summary, the first priority of metabolism during a prolonged fast is to provide enough glucose for the brain and other organs that dependent upon it for energy in order to spare proteins for other cellular functions. The next priority of the body is to shift the use of fuel from glucose to fatty acids and ketone bodies. From then on, ketones become more and more important as a source of fuel while fatty acids and glucose become less important.

Flavour and Sweeteners

A less important function of carbohydrates is to provide sweetness to foods. Receptors located at the tip of the tongue bind to tiny bits of carbohydrates and send what humans perceive as a "sweet" signal to the brain. However, different sugars vary in sweetness. For example, fructose is almost twice as sweet as sucrose and sucrose is approximately 30% sweeter than glucose.

Sweeteners can be classified as either nutritive or alternative. Nutritive sweeteners have all been mentioned before and include sucrose, glucose, fructose, high fructose corn syrup, and lactose. These types of sweeteners not only impart flavour to the food, but can also be metabolized for energy. In contrast, alternative sweeteners provide no food energy and include saccharin, cyclamate, aspartame, and acesulfame. Controversy over saccharin and cyclamate as artificial sweeteners still exists but aspartame and acesulfame are used extensively in many foods in the United States. Aspartame and acesulfame are both hundreds of times sweeter than sucrose but only acesulfame is able to be used in baked goods since it is much more stable than aspartame when heated.

Dietary Fiber

Dietary fibers such as cellulose, hemicellulose, pectin, gum and mucilage are important carbohydrates for several reasons. Soluble dietary fibers like pectin, gum and mucilage pass undigested through the small intestine and are degraded into fatty acids and gases by the large intestine. The fatty acids produced in this way can either be used as a fuel for the large intestine or be absorbed into the bloodstream. Therefore, dietary fiber is essential for proper intestinal health. In general, the consumption of soluble and insoluble fiber makes the elimination of waste much easier. Since dietary fiber is both indigestible and an attractant of water, stools become large and soft. As a result, feces can be expelled with less pressure. However, not enough fiber consumption will change the constitution of the stool and increase the amount of force required during defecation. Excessive pressure during the elimination of waste can force places in the large intestine wall out from between bands of smooth muscle to produce small pouches called diverticula. Hemorrhoids may also result from unnecessary strain during defecation.

The disease of having many diverticula in the large intestine is known as diverticulosis. Although diverticula is often asymptomatic, food particles become trapped in their folds and bacteria begin to metabolize the particles into acids and gases. Eventually, the diverticula may become inflamed, a condition known as diverticulitis. To combat the disease, antibiotics are administered to the patient to destroy the bacteria while the intake of fiber in the diet is decreased until the inflammation has subsided. Once the inflammation has been reduced, a high fiber diet is begun to prevent a relapse.

Besides the prevention of intestinal disease, diets high in fiber have other health benefits. High fiber intake reduces the risk of developing obesity by increasing the bulk of a meal without yielding much energy. An expanded stomach leads to satisfaction despite the fact that the caloric intake has decreased.

Beyond dieters, diabetics can also benefit from consuming a regular amount of dietary fiber. Once in the intestine, it slows the absorption of glucose to prevent a sudden increase in blood glucose levels. A relatively high intake of fiber will also decrease the absorption of cholesterol, a compound that is thought to contribute to atherosclerosis or scarring of the arteries. Serum cholesterol may be further reduced by a reduction in the release of insulin after meals. Since insulin is known to promote cholesterol synthesis in the liver, a reduction in the absorption of glucose after meals through the consumption of fiber can help to control serum cholesterol levels. Furthermore, dietary fiber intake may help prevent colon cancer by diluting potential carcinogens through increased water retention, binding carcinogens to the fiber itself and speeding the passage of food through the intestinal tract so that cancer-causing agents have less time to act.

Biological Recognition Processes

Carbohydrates not only serve nutritional functions, but are also thought to play important roles in cellular recognition processes. For example, many immunoglobulins (antibodies) and peptide hormones contain glycoprotein sequences. These sequences are composed of amino acids linked to carbohydrates. During the course of many hours or days, the carbohydrate polymer linked to the rest of the protein may be cleaved by circulating enzymes or be degraded spontaneously. The liver can recognize differences in length and may internalize the protein in order to begin its own degradation. In this way, carbohydrates may mark the passage of time for proteins.

Carbohydrate-mediated Cell Adhesion

Carbohydrate-mediated cell–cell recognition is emerging as an important component in the repertoire of molecular recognition events that underlie the orderly development and functioning of multicellular organisms. Lectins (carbohydrate-binding proteins) on one cell surface bind to complementary carbohydrates on an apposing cell to initiate physiologically important cell–cell interactions. Genomic analyses indicate <“100 human lectins, and the diversity of lectin protein folds implies that additional families may yet be discovered. Specific carbohydrate–carbohydrate interactions may also mediate cell–cell adhesion. A challenge for glycobiology is to develop robust screening methods to identify and quantify the binding specificities of carbohydrate-mediated cell recognition events. The variety and complexity of naturally occurring glycan structures, and the typical multivalent nature of carbohydrate-mediated adhesion (low monovalent site affinity, high polyvalent affinity) add to the complexity of the challenge.

Microarrays have been applied successfully to the quantification of nucleic acid and protein interactions with immobilized complementary binding partners and recently have been extended to the investigation of protein–carbohydrate interactions. Several methods were published in which natural and/or synthetic saccharides were adsorbed or covalently bound to surfaces in a manner suitable for screening binding antibodies and/or soluble lectins. These methods have been heralded as the birth of glycomics, the first steps to large-scale screening of glycan recognition in biological systems.

Many of the protein–carbohydrate interactions that occur at cell surfaces are intrinsically multivalent. Arrays of lectins (or carbohydrates) on one cell surface interact like Velcro with complementary arrays of carbohydrates on an apposing surface. The best characterized example of the kinetic role of multivalency is binding by the hepatic asialoglycoprotein receptor. On the intact hepatocyte surface, the site affinity of this lectin for lactose is 300 μM, whereas its affinity for a trivalent oligosaccharide with appropriately spaced galactose termini is 7 nM, nearly five orders of magnitude higher. Similarly, carbohydrate–carbohydrate binding may have even lower site affinity, requiring highly multivalent interactions for function (Pincet *et al.*, 2001). The off rate of a monovalent carbohydrate-mediated interaction may be so rapid that solid-phase binding to immobilized glycan arrays, which require a wash step to remove unbound protein, may be difficult to detect. This problem has been circumvented to some extent by using multivalent soluble lectins or by generating multivalency using chimeras or secondary binding proteins.

In nature, multivalency is often generated by lectin expression on cell surfaces, where lectin molecules self-associate or cluster in response to multivalent binding arrays on an apposing surface. Here we report methods that detect specific adhesion of intact cells to covalent carbohydrate microarrays engineered on glass slides. These methods take advantage of the natural multivalency of cell surface carbohydrate binding to extend the applicability of glycan microarrays.

Carbohydrates at the cell surface have been proposed as mediators in cell-cell recognition events involved in embryogenesis, metastasis, and other proliferation processes by calcium-dependent carbohydrate to carbohydrate interactions. They are the most prominently exposed structures on the surface of living cells, and with flexible chains and many binding sites are ideal to serve as the major players in initiating these cellular events.

However, biological relevance of these type interactions is often questioned because of the very low affinity binding of single carbohydrate molecules and that they manifest themselves only through the contact of a large number of molecules tightly arranged in the membrane. Weak interactions are considerably more difficult to study and only a few biologically significant examples of direct carbohydrate-carbohydrate interactions have been reported, e.g. pioneering work showing glycosphingolipid self-interactions through multivalent interaction of Lewis X epitopes. However, there are no reports on the existence of specific proteoglycan self-interactions through carbohydrate-carbohydrate interactions in cellular recognition system, as it has been done with glycosphingolipids. Here, we used sponges, organisms on which the first proteoglycan-mediated cell-cell recognition in the animal kingdom was demonstrated, as a model system to study carbohydrate-mediated cellular recognition.

We show that the interaction between single oligosaccharides from surface proteoglycans is relatively strong and comparable to protein-carbohydrate interactions, highly specific, and dependent on Ca^{2}+-ions. 200 kDa glycans from the core protein of Microciona prolifera cell surface proteoglycans have been previously shown to mediate homotypic Microciona proteoglycan-proteoglycan interactions. Here, 200 kDa glycans from four different sponge species: Microciona prolifera, Halichondria panicea, Suberites fuscus and Cliona celata were purified and investigated for species-specific interactions. Selective recognition of glycans by live cells was studied to confirm the existence of glycan-glycan recognition system in biologically relevant situations.

Mature sponge cells have the ability to reaggregate species-specifically and form homogenous aggregates on a shaker at the right shear forces in the presence of physiological 10 mM Ca^{2+}. Live cells were allowed to aggregate with glycan-coated beads similar in size to small sponge cells in the presence of calcium. They specifically recognized beads coated with their own glycans and did not mix but separated from beads coated with glycans isolated from different species.

The glycan-glycan recognition assay was developed to mimic species-specific cellcell recognition in sponges. 200 kDa glycans immobilized onto beads similar in size to small sponge cells assembled species-specifically in the presence of physiological calcium, at the same shear forces as in cell-cell aggregation. Glycans coated on beads aggregated with glycans from the same species coated on beads, and separated from glycans from other species. The glycan density necessary for specific live cellcell recognition in sponges is 828 molecules/μm^2. In our studies, the glycan density necessary for specific glycan-coated bead was very similar: ~810 molecules/μm^2. Mature live cells demonstrated specific recognition of 200 kDa glycans during selective-binding to glycans coated on surfaces in the presence of calcium.

They strongly adhered to glycans from their own surface proteoglycans coated onto a solid polystyrene phase, while the binding to glycans from different proteoglycans was 3 - 5 times lower. Moreover, homotypic adhesion to glycan-coated plates enhanced sponge cell differentiation and formation of mineral skeleton (spicules). Larval cells, after settlement and spreading of larvae, can fuse species-specifically in nature.

In our studies, live larval cells recognized and adhered specifically to glycans purified from adhesion proteoglycans from their "mother sponge". They showed almost no interaction with glycans from other species. As in cell-glycan adhesion assays, highly species-specific adhesion of 200 kDa glycans to glycan-coated surfaces could be observed in the presence of physiological calcium.

Tested glycans bound strongly to glycans from the same species and showed up to a six fold reduction in binding to glycans from other species. Atomic force microscopy (AFM) was performed to measure for the first time adhesion forces between single glycan molecules obtained from different surface proteoglycans. Measurements revealed equally strong adhesion forces in the range of several hundred piconewtons (pN) between glycan molecules as between proteins and glycans measured in another recognition system. Moreover, statistically significant differences (p value < 0.01) were seen between homotypic (glycans from the same species) and heterotypic (glycans from different species) interactions.

Moreover, the polyvalent character of binding characterized mainly interactions between glycans from the same species. This indicates that not only the higher adhesion force per binding site as such but also the higher amount of multiple interactions between glycans from the same species versus mixture of glycans from different species guaranteed the specificity of the glycan-mediated recognition. These findings confirm for the first time the existence of specific glycan-glycan recognition system between cell surface proteoglycans. We propose that these cell's outermost surface structures serve as important players in initiating the very first contacts between cells through highly species-specific and flexible carbohydratecarbohydrate interactions.

Molecular Recognition and Catalysis

The key to understanding the fundamental processes of catalysis is the transition state (TS): indeed, catalysis is a transition-state molecular recognition event. Practical objectives, such as the design of TS analogues as potential drugs, or the design of synthetic catalysts (including catalytic antibodies), require prior knowledge of the TS structure to be mimicked. Examples, both old and new, of computational modelling studies are discussed, which illustrate this fundamental concept. Free-energy changes along the reaction coordinate for S_N2 methyl transfer catalysed by the enzyme catechol-*O*-methyl transferase are described and compared with those for a model reaction in water, as computed by hybrid quantum-mechanical/molecular-mechanical molecular dynamics simulations. The case is discussed of molecular recognition in a xylanase enzyme that stabilises its sugar substrate in a (normally unfavourable) boat conformation and in which a single-atom mutation affects the free-energy of activation dramatically.

Lipids and Biochemistry

Lipids are a broad group of naturally occurring molecules which includes fats, waxes, sterols, fat-soluble vitamins (such as vitamins A, D, E and K), monoglycerides, diglycerides, phospholipids, and others. The main biological functions of lipids include energy storage, as structural components of cell membranes, and as important signalling molecules.

Lipids may be broadly defined as hydrophobic or amphiphilic small molecules; the amphiphilic nature of some lipids allows them to form structures such as vesicles, liposomes, or membranes in an aqueous environment. Biological lipids originate entirely or in part from two distinct types of biochemical subunits or "building blocks": ketoacyl and isoprene groups. Using this approach, lipids may be divided into eight categories: fatty acyls, glycerolipids, glycerophospholipids, sphingolipids, saccharolipids and polyketides (derived from condensation of ketoacyl subunits); and sterol lipids and prenol lipids (derived from condensation of isoprene subunits).

Although the term *lipid* is sometimes used as a synonym for fats, fats are a subgroup of lipids called triglycerides. Lipids also encompass molecules such as fatty acids and their derivatives (including tri-, di-, and monoglycerides and phospholipids), as well as other sterol-containing metabolites such as cholesterol. Although humans and other mammals use various biosynthetic pathways to both break down and synthesize lipids, some essential lipids cannot be made this way and must be obtained from the diet.

Categories of Lipids

Fatty Acyls

Fatty acyls, a generic term for describing fatty acids, their conjugates and derivatives, are a diverse group of molecules synthesized by chain-elongation of an acetyl-CoA primer with malonyl-CoA or methylmalonyl-CoA groups in a process called fatty acid synthesis.

They are made of a hydrocarbon chain that terminates with a carboxylic acid group; this arrangement confers the molecule with a polar, hydrophilic end, and a nonpolar, hydrophobic end that is insoluble in water. The fatty acid structure is one of the most fundamental categories of biological lipids, and is commonly used as a building block of more structurally complex lipids. The carbon chain, typically between four to 24 carbons long, may be saturated or unsaturated, and may be attached to functional groups containing oxygen, halogens, nitrogen and sulfur.

Where a double bond exists, there is the possibility of either a *cis* or *trans* geometric isomerism, which significantly affects the molecule's molecular configuration. *Cis*-double bonds cause the fatty acid chain to bend, an effect that is more pronounced the more double bonds there are in a chain. This in turn plays an important role in the structure and function of cell membranes. Most naturally occurring fatty acids are of the *cis* configuration, although the *trans* form does exist in some natural and partially hydrogenated fats and oils.

Examples of biologically important fatty acids are the eicosanoids, derived primarily from arachidonic acid and eicosapentaenoic acid, which include prostaglandins, leukotrienes, and thromboxanes.

Other major lipid classes in the fatty acid category are the fatty esters and fatty amides. Fatty esters include important biochemical intermediates such as wax esters, fatty acid thioester coenzyme A derivatives, fatty acid thioester ACP derivatives and fatty acid carnitines. The fatty amides include N-acyl ethanolamines, such as the cannabinoid neurotransmitter anandamide.

Glycerolipids

Glycerolipids are composed mainly of mono-, di- and tri-substituted glycerols, the most well-known being the fatty acid esters of glycerol (triacylglycerols), also known as triglycerides. In these compounds, the three hydroxyl groups of glycerol are each esterified, usually by different fatty acids. Because they function as a food store, these lipids comprise the bulk of storage fat in animal tissues.

The hydrolysis of the ester bonds of triacylglycerols and the release of glycerol and fatty acids from adipose tissue is called fat mobilization. Additional subclasses of glycerolipids are represented by glycosylglycerols, which are characterized by the presence of one or more sugar residues attached to glycerol via a glycosidic linkage. Examples of structures in this category are the digalactosyldiacylglycerols found in plant membranes and seminolipid from mammalian sperm cells.

Glycerophospholipids

Glycerophospholipids, also referred to as phospholipids, are ubiquitous in nature and are key components of the lipid bilayer of cells, as well as being involved in metabolism and cell signalling.

Neural tissue (including the brain) contains relatively high amounts of glycerophospholipids, and alterations in their composition has been implicated in various neurological disorders. Glycerophospholipids may be subdivided into distinct classes, based on the nature of the polar headgroup at the *sn*-3 position of the glycerol backbone in eukaryotes and eubacteria, or the *sn*-1 position in the case of archaebacteria.

Examples of glycerophospholipids found in biological membranes are phosphatidylcholine (also known as PC, GPCho or lecithin), phosphatidylethanolamine (PE or GPEtn) and phosphatidylserine (PS or GPSer).

In addition to serving as a primary component of cellular membranes and binding sites for intra- and intercellular proteins, some glycerophospholipids in eukaryotic cells, such as phosphatidylinositols and phosphatidic acids are either precursors of, or are themselves, membrane-derived second messengers. Typically, one or both of these hydroxyl groups are acylated with long-chain fatty acids, but there are also alkyl-linked and 1Z-alkenyl-linked (plasmalogen) glycerophospholipids, as well as dialkylether variants in archaebacteria.

Sphingolipids

Sphingolipids are a complex family of compounds that share a common structural feature, a sphingoid base backbone that is synthesized *de novo* from the amino acid serine and a long-chain fatty acyl CoA, then converted into ceramides, phosphosphingolipids, glycosphingolipids and other compounds.

The major sphingoid base of mammals is commonly referred to as sphingosine. Ceramides (N-acyl-sphingoid bases) are a major subclass of sphingoid base derivatives with an amide-linked fatty acid. The fatty acids are typically saturated or mono-unsaturated with chain lengths from 16 to 26 carbon atoms. The major phosphosphingolipids of mammals are sphingomyelins (ceramide phosphocholines), whereas insects contain mainly ceramide phosphoethanolamines and fungi have phytoceramide phosphoinositols and mannose-containing headgroups. The glycosphingolipids are a diverse family of molecules composed of one or more sugar residues linked via a glycosidic bond to the sphingoid base. Examples of these are the simple and complex glycosphingolipids such as cerebrosides and gangliosides.

Sterol Lipids

Sterol lipids, such as cholesterol and its derivatives, are an important component of membrane lipids, along with the glycerophospholipids and sphingomyelins. The steroids, all derived from the same fused four-ring core structure, have different biological roles as hormones and signalling molecules. The eighteen-carbon (C18) steroids include the estrogen family whereas the C19 steroids comprise the androgens such as testosterone and androsterone. The C21 subclass includes the progestogens as well as the glucocorticoids and mineralocorticoids. The secosteroids, comprising various forms of vitamin D, are characterized by cleavage of the B ring of the core structure. Other examples of sterols are the bile acids and their conjugates, which in mammals are oxidized derivatives of cholesterol and are synthesized in the liver. The plant equivalents are the phytosterols, such as β-sitosterol, stigmasterol, and brassicasterol; the latter compound is also used as a biomarker for algal growth. The predominant sterol in fungal cell membranes is ergosterol.

Prenol Lipids

Prenol lipids are synthesized from the 5-carbon precursors isopentenyl diphosphate and dimethylallyl diphosphate that are produced mainly via the mevalonic acid (MVA) pathway. The simple isoprenoids (linear alcohols, diphosphates, etc.) are formed by the successive addition of C5 units, and are classified according to number of these terpene units. Structures containing greater than 40 carbons are known as polyterpenes. Carotenoids are important simple isoprenoids that function as antioxidants and as precursors of vitamin A. Another biologically important class of molecules is exemplified by the quinones and hydroquinones, which contain an isoprenoid tail attached to a quinonoid core of non-isoprenoid origin. Vitamin E and vitamin K, as well as the ubiquinones, are examples of this class. Prokaryotes synthesize polyprenols (called bactoprenols) in which the terminal isoprenoid unit attached to oxygen remains unsaturated, whereas in animal polyprenols (dolichols) the terminal isoprenoid is reduced.

Saccharo Lipids

Saccharolipids describe compounds in which fatty acids are linked directly to a sugar backbone, forming structures that are compatible with membrane bilayers. In the saccharolipids, a monosaccharide substitutes for the glycerol backbone present in glycerolipids and glycerophospholipids. The most familiar saccharolipids are the acylated glucosamine precursors of the Lipid A component of the lipopolysaccharides in Gram-negative bacteria. Typical lipid A molecules are disaccharides of glucosamine, which are derivatized with as many as seven fatty-acyl chains. The minimal lipopolysaccharide required for growth in *E. coli* is Kdo_2-Lipid A, a hexa-acylated disaccharide of glucosamine that is glycosylated with two 3-deoxy-D-manno-octulosonic acid (Kdo) residues.

Polyketides

Polyketides are synthesized by polymerization of acetyl and propionyl subunits by classic enzymes as well as iterative and multimodular enzymes that share mechanistic features with the fatty acid synthases. They comprise a large number of secondary metabolites and natural products from animal, plant, bacterial, fungal and marine sources, and have great structural diversity. Many polyketides are cyclic molecules whose backbones are often further modified by glycosylation, methylation, hydroxylation, oxidation, and/or other processes. Many commonly used anti-microbial, anti-parasitic, and anti-cancer agents are polyketides or polyketide derivatives, such as erythromycins, tetracyclines, avermectins, and antitumor epothilones.

Biological Functions

Membranes

Eukaryotic cells are compartmentalized into membrane-bound organelles which carry out different biological functions. The glycerophospholipids are the main structural component of biological membranes, such as the cellular plasma membrane and the intracellular membranes of organelles; in animal cells the plasma membrane physically separates the intracellular components from the extracellular environment.

The glycerophospholipids are amphipathic molecules (containing both hydrophobic and hydrophilic regions) that contain a glycerol core linked to two fatty acid-derived "tails" by ester linkages and to one "head" group by a phosphate ester linkage. While glycerophospholipids are the major component of biological membranes, other non-glyceride lipid components such as sphingomyelin and sterols (mainly cholesterol in animal cell membranes) are also found in biological membranes. In plants and algae, the galactosyldi-acylglycerols, and sulfoquinovosyldiacylglycerol, which lack a phosphate group, are important components of membranes of chloroplasts and related organelles and are the most abundant lipids in photosynthetic tissues, including those of higher plants, algae and certain bacteria.

Bilayers have been found to exhibit high levels of birefringence which can be used to probe the degree of order (or disruption) within the bilayer using techniques such as dual polarisation interferometry.

A biological membrane is a form of lipid bilayer. The formation of lipid bilayers is an energetically preferred process when the glycerophospholipids described above are in an aqueous environment. In an aqueous system, the polar heads of lipids align towards the polar, aqueous environment, while the hydrophobic tails minimize their contact with water and tend to cluster together, forming a vesicle; depending on the concentration of the lipid, this biophysical interaction may result in the formation of micelles, liposomes, or lipid bilayers. Other aggregations are also observed and form part of the polymorphism of amphiphile (lipid) behaviour.

Phase behaviour is an area of study within biophysics and is the subject of current academic research. Micelles and bilayers form in the polar medium by a process known as the hydrophobic effect. When dissolving a lipophilic or amphiphilic substance in a polar environment, the polar molecules (i.e., water in an aqueous solution) become more ordered around the dissolved lipophilic substance, since the polar molecules cannot form hydrogen bonds to the lipophilic areas of the amphiphile. So in an aqueous environment, the water molecules form an ordered "clathrate" cage around the dissolved lipophilic molecule.

Energy Storage

Triacylglycerols, stored in adipose tissue, are a major form of energy storage in animals. The adipocyte, or fat cell, is designed for continuous synthesis and breakdown of triacylglycerols, with breakdown controlled mainly by the activation of hormone-sensitive enzyme lipase. The complete oxidation of fatty acids provides high caloric content, about 9 kcal/g, compared with 4 kcal/g for the breakdown of carbohydrates and proteins. Migratory birds that must fly long distances without eating use stored energy of triacylglycerols to fuel their flights.

Signalling

In recent years, evidence has emerged showing that lipid signalling is a vital part of the cell signalling. Lipid signalling may occur via activation of G protein-coupled or nuclear receptors, and members of several different lipid categories have been identified as signalling molecules and cellular messengers.

These include sphingosine-1-phosphate, a sphingolipid derived from ceramide that is a potent messenger molecule involved in regulating calcium mobilization, cell growth, and apoptosis; diacylglycerol (DAG) and the phosphatidylinositol phosphates (PIPs), involved in calcium-mediated activation of protein kinase C; the prostaglandins, which are one type of fatty-acid derived eicosanoid involved in inflammation and immunity; the steroid hormones such as estrogen, testosterone and cortisol, which modulate a host of functions such as reproduction, metabolism and blood pressure; and the oxysterols such as 25-hydroxy-cholesterol that are liver X receptor agonists.

Other Functions

The "fat-soluble" vitamins (A, D, E and K) – which are isoprene-based lipids – are essential nutrients stored in the liver and fatty tissues, with a diverse range of functions. Acyl-carnitines are involved in the transport and metabolism of fatty acids in and out of mitochondria, where they undergo beta oxidation.

Polyprenols and their phosphorylated derivatives also play important transport roles, in this case the transport of oligosaccharides across membranes. Polyprenol phosphate sugars and polyprenol diphosphate sugars function in extra-cytoplasmic glycosylation reactions, in extracellular polysaccharide biosynthesis (for instance, peptidoglycan polymerization in bacteria), and in eukaryotic protein N-glycosylation. Cardiolipins are a subclass of glycerophospholipids containing four acyl chains and three glycerol groups that are particularly abundant in the inner mitochondrial membrane. They are believed to activate enzymes involved with oxidative phosphorylation.

Metabolism

The major dietary lipids for humans and other animals are animal and plant triglycerides, sterols, and membrane phospholipids. The process of lipid metabolism synthesizes and degrades the lipid stores and produces the structural and functional lipids characteristic of individual tissues.

Biosynthesis

In animals, when there is an oversupply of dietary carbohydrate, the excess carbohydrate is converted to triacylglycerol. This involves the synthesis of fatty acids from acetyl-CoA and the esterification of fatty acids in the production of triacylglycerol, a process called lipogenesis. Fatty acids are made by fatty acid synthases that polymerize and then reduce acetyl-CoA units.

The acyl chains in the fatty acids are extended by a cycle of reactions that add the acetyl group, reduce it to an alcohol, dehydrate it to an alkene group and then reduce it again to an alkane group. The enzymes of fatty acid biosynthesis are divided into two groups, in animals and fungi all these fatty acid synthase reactions are carried out by a single multifunctional protein, while in plant plastids and bacteria separate enzymes perform each step in the pathway. The fatty acids may be subsequently converted to triacylglycerols that are packaged in lipoproteins and secreted from the liver.

The synthesis of unsaturated fatty acids involves a desaturation reaction, whereby a double bond is introduced into the fatty acyl chain. For example, in humans, the desaturation of stearic acid by stearoyl-CoA desaturase-1 produces oleic acid. The doubly unsaturated fatty acid linoleic acid as well as the triply unsaturated α-linolenic acid cannot be synthesized in mammalian tissues, and are therefore essential fatty acids and must be obtained from the diet. Triacylglycerol synthesis takes place in the endoplasmic reticulum by metabolic pathways in which acyl groups in fatty acyl-CoAs are transferred to the hydroxyl groups of glycerol-3-phosphate and diacylglycerol.

Terpenes and isoprenoids, including the carotenoids, are made by the assembly and modification of isoprene units donated from the reactive precursors isopentenyl pyrophosphate and dimethylallyl pyrophosphate. These precursors can be made in different ways. In animals and archaea, the mevalonate pathway produces these compounds from acetyl-CoA, while in plants and bacteria the non-mevalonate pathway uses pyruvate and glyceraldehyde 3-phosphate as substrates. One important reaction that uses these activated isoprene donors is steroid biosynthesis. Here, the isoprene units are joined together to make squalene and then folded up and formed into a set of rings to make lanosterol. Lanosterol can then be converted into other steroids such as cholesterol and ergosterol.

Degradation

Beta oxidation is the metabolic process by which fatty acids are broken down in the mitochondria and/or in peroxisomes to generate acetyl-CoA. For the most part, fatty acids are oxidized by a mechanism that is similar to, but not identical with, a reversal of the process of fatty acid synthesis. That is, two-carbon fragments are removed sequentially from the carboxyl end of the acid after steps of dehydrogenation, hydration, and oxidation to form a beta-keto acid, which is split by thiolysis. The acetyl-CoA is then ultimately converted into ATP, CO_2, and H_2O using the citric acid cycle and the electron transport chain. Hence the Krebs Cycle can start at acetyl-CoA when fat is being broken down for energy if there is little or no glucose available.

The energy yield of the complete oxidation of the fatty acid palmitate is 106 ATP. Unsaturated and odd-chain fatty acids require additional enzymatic steps for degradation.

Nutrition and Health

Most of the lipid found in food is in the form of triacylglycerols, cholesterol and phospholipids. A minimum amount of dietary fat is necessary to facilitate absorption of fat-soluble vitamins (A, D, E and K) and carotenoids. Humans and other mammals have a dietary requirement for certain essential fatty acids, such as linoleic acid (an omega-6 fatty acid) and alpha-linolenic acid (an omega-3 fatty acid) because they cannot be synthesized from simple precursors in the diet. Both of these fatty acids are 18-carbon polyunsaturated fatty acids differing in the number and position of the double bonds. Most vegetable oils are rich in linoleic acid (safflower, sunflower, and corn oils). Alpha-linolenic acid is found in the green leaves of plants, and in selected seeds, nuts and legumes (particularly flax, rapeseed, walnut and soy).

Fish oils are particularly rich in the longer-chain omega-3 fatty acids eicosapentaenoic acid (EPA) and docosahexaenoic acid (DHA). A large number of studies have shown positive health benefits associated with consumption of omega-3 fatty acids on infant development, cancer, cardiovascular diseases, and various mental illnesses, such as depression, attention-deficit hyperactivity disorder, and dementia. In contrast, it is now well-established that consumption of trans fats, such as those present in partially hydrogenated vegetable oils, are a risk factor for cardiovascular disease. A few studies have suggested that total dietary fat intake is linked to an increased risk of obesity and diabetes. However, a number of very large studies, including the Women's Health Initiative Dietary Modification Trial, an eight year study of 49,000 women, the Nurses' Health Study and the Health Professionals Follow-up Study, revealed no such links. None of these studies suggested any connection between percentage of calories from fat and risk of cancer, heart disease or weight gain. The Nutrition Source, a website maintained by the Department of Nutrition at the Harvard School of Public Health, summarizes the current evidence on the impact of dietary fat: "Detailed research—much of it done at Harvard—shows that the total amount of fat in the diet isn't really linked with weight or disease."

Essential Fatty Acid

Essential fatty acids, or EFAs, are fatty acids that humans and other animals must ingest for good health because the body requires them but can't make them from other food components. The term refers to fatty acids required for biological processes, and not those that only act as fuel. There are only two EFAs: alpha-linolenic acid, an omega-3 fatty acid, and linoleic acid, an omega-6 fatty acid. Some researchers consider gamma-linolenic acid (omega-6), lauric acid (saturated fatty acid), and palmitoleic acid (monounsaturated fatty acid) conditionally essential.

They were originally designated as Vitamin F when they were discovered as essential nutrients in 1923. In 1930, work by Burr, Burr and Miller on rats showed that they are better classified with the fats than with the vitamins.

Functions

In the body, essential fatty acids serve multiple functions. In each of these, the balance between dietary ω-3 and ω-6 strongly affects function.

- They are modified to make
  - o the classic eicosanoids (affecting inflammation and many other cellular functions)
  - o the endocannabinoids (affecting mood, behaviour and inflammation)
  - o the lipoxins from ω-6 EFAs and resolvins from ω-3 (in the presence of aspirin, downregulating inflammation.)
  - o the isofurans, neurofurans, isoprostanes, hepoxilins, epoxyeicosatrienoic acids (EETs) and Neuroprotectin D
- They form lipid rafts (affecting cellular signalling)
- They act on DNA (activating or inhibiting transcription factors such as NF-κB, which is linked to pro-inflammatory cytokine production)

Nomenclature and Terminology

Fatty acids are straight chain hydrocarbons possessing a carboxyl (COOH) group at one end. The carbon next to the carboxylate is known as α, the next carbon β, and so forth. Since biological fatty acids can be of different lengths, the last position is labelled as a "ω", the last letter in the Greek alphabet. Since the physiological properties of unsaturated fatty acids largely depend on the position of the first unsaturation relative to the end position and not the carboxylate, the position is signified by (ω minus n). For example, the term ω-3 signifies that the first double bond exists as the third carbon-carbon bond from the terminal CH_3 end (ω) of the carbon chain. The number of carbons and the number of double bonds is also listed. ω-3 18:4 (stearidonic acid) or 18:4 ω-3 or 18:4 n"3 indicates an 18-carbon chain with 4 double bonds, and with the first double bond in the third position from the CH_3 end. Double bonds are *cis* and separated by a single methylene (CH_2) group unless otherwise noted.

Examples

The essential fatty acids start with the short chain polyunsaturated fatty acids (SC-PUFA):

- ω-3 fatty acids:
  - o α-Linolenic acid or ALA (18:3)
- ω-6 fatty acids:
  - o Linoleic acid or LA (18:2)

These two fatty acids cannot be synthesised by humans, as humans lack the desaturase enzymes required for their production. They form the starting point for the creation of longer and more desaturated fatty acids, which are also referred to as long-chain polyunsaturated fatty acids (LC-PUFA):

- ω-3 fatty acids:
  - eicosapentaenoic acid or EPA (20:5)
  - docosahexaenoic acid or DHA (22:6)
- ω-6 fatty acids:
  - gamma-linolenic acid or GLA (18:3)
  - dihomo-gamma-linolenic acid or DGLA (20:3)
  - arachidonic acid or AA (20:4).

ω-9 fatty acids are not essential in humans, because humans generally possess all the enzymes required for their synthesis. Exceptions do occur in older people or people with a liver problem that do not completely produce a sufficient amount, and hence many supplement companies market Omega 3-6-9 blends.

Essential Fatty Acids

Mammals lack the ability to introduce double bonds in fatty acids beyond carbon 9 and 10, hence ω-6 linoleic acid (18:2,9,12), abbreviated LA, and the ω-3 linolenic acid (18:3,9,12,15), abbreviated ALA, are essential for man in the diet. In humans, arachidonic acid (20:4,5,8,11,14) can be synthesized from LA by desaturation and chain elongation (though some carnivores like cats cannot do this, and require arachadonate in the diet). In addition, the human body can make some long-chain ω-3 PUFAs (EPA and DHA) from the ω-3 ALA.

Between 1930 and 1950, arachidonic acid and linolenic acid were termed 'essential' because each was more or less able to meet the growth requirements of rats given fat-free diets. To some extent, any ω-3 and any ω-6 can relieve the worst symptoms of fatty acid deficiency for its class. Particular fatty acids are still needed at critical life stages (e.g. lactation) and in some disease states. In nonscientific writing, common usage is that the term *essential fatty acid* comprises all the ω-3 or -6 fatty acids. Conjugated fatty acids like calendic acid are not normally considered essential. Authoritative sources include the whole families, without qualification. Traditionally speaking the LC-PUFA are not essential. Because the LC-PUFA are sometimes required, they may be considered "conditionally essential", or not essential to healthy adults.

Biologist Ray Peat has claimed there are flaws in the studies showing the need for ω-3 and ω-6 (also called n-3 and n-6) fats. He further claims that EFA deficiencies have sometimes been reversed by adding B vitamins or a fat-free liver extract to the diet. In his view, 'the optional dietary level of the "essential fatty acids" might be close to zero, if other dietary factors were also optimized.' However, these views are contrary to the scientific consensus.

Essential fatty acids should not be confused with essential oils, which are "essential" in the sense of being a concentrated essence.

Food Sources

Almost all the polyunsaturated fat in the human diet is from EFA. Some of the food sources of ω-3 and ω-6 fatty acids are fish and shellfish, flaxseed (linseed), hemp oil, soya oil, canola (rapeseed) oil, chia seeds, pumpkin seeds, sunflower seeds, leafy vegetables, and walnuts. Essential fatty acids play a part in many metabolic processes, and there is evidence to suggest that low levels of essential fatty acids, or the wrong balance of types among the essential fatty acids, may be a factor in a number of illnesses, including osteoporosis.

Plant sources of ω-3 contain neither eicosapentaenoic acid (EPA) nor docosahexaenoic acid (DHA). The human body can (and in case of a purely vegetarian diet often must, unless certain algae or supplements derived from them are consumed) convert α-linolenic acid (ALA) to EPA and subsequently DHA. This however requires more metabolic work, which is thought to be the reason that the absorption of essential fatty acids is much greater from animal rather than plant sources The IUPAC Lipid HandbookPDF (370 KiB) provides a very large and detailed listing of fat contents of animal and vegetable fats, including ω-3 and -6 oils.

The National Institutes of Health's EFA Education group publishes 'Essential Fats in Food Oils.' This lists 40 common oils, more tightly focused on EFAs and sorted by n-6:3 ratio. Stuchlik and Zak, 'Vegetable Lipids as Components of Functional Food'PDF (139 KiB) list notable vegetable sources of EFAs as well as commentary and an overview of the biosynthetic pathways involved. Users can interactively search at Nutrition Data for the richest food sources of particular EFAs or other nutrients. Careful readers will note that these sources are not in excellent agreement. EFA content of vegetable sources varies with cultivation conditions. Animal sources vary widely, both with the animal's feed and that the EFA makeup varies markedly with fats from different body parts.

Lipid Signalling

Lipid signalling, broadly defined, refers to any biological signalling event involving a lipid messenger that binds a protein target, such as a receptor, kinase or phosphatase, which in turn mediate the effects of these lipids on specific cellular responses. Lipid signalling is thought to be qualitatively different from other classical signalling paradigms (such as monoamine neurotransmission) because lipids can freely diffuse through membranes. One consequence of this is that lipid messengers cannot be stored in vesicles prior to release and so are often biosynthesized "on demand" at their intended site of action. As such, many lipid signalling molecules cannot circulate freely in solution but, rather, exist bound to special carrier proteins in serum.

Ceramide

Ceramide (Cer) can be generated by the breakdown of sphingomyelin (SM) by sphingomyelinases (SMases), which are enzymes that hydrolyze the phosphocholine group from the sphingosine backbone. Alternatively, this sphingosine-derived lipid (sphingolipid) can be synthesized from scratch (*de novo*) by the enzymes serine palmitoyl transferase (SPT) and ceramide synthase in organelles such as the endoplasmic reticulum (ER) and possibly, in the mitochondria-associated membranes (MAMs) and the perinuclear membranes. Being located in the metabolic hub, ceramide leads to the formation of other sphingolipids, with the C1 hydroxyl (-OH) group as the major site of modification. A sugar can be attached to ceramide (glycosylation) through the action of the enzymes, glucosyl or galactosyl ceramide synthases. Ceramide can also be broken down by enzymes called ceramidases, leading to the formation of sphingosine, Moreover, a phosphate group can be attached to ceramide (phosphorylation) by the enzyme, ceramide kinase. It is also possible to regenerate sphingomyelin from ceramide by accepting a phosphocholine headgroup from phosphatidylcholine (PC) by the action of an enzyme called sphingomyelin synthase. The latter process results in the formation of diacylglycerol (DAG) from PC.

Ceramide contains two hydrophobic ("water-fearing") chains and a neutral headgroup. Consequently, it has limited solubility in water and is restricted within the organelle where it was formed. Also, because of its hydrophobic nature, ceramide readily flip-flops across membranes as supported by studies in membrane models and membranes from red blood cells (erythrocytes). However, ceramide can possibly interact with other lipids to form bigger regions called microdomains which restrict its flip-flopping abilities. This could have immense effects on the signalling functions of ceramide because it is known that ceramide generated by acidic SMase enzymes in the outer leaflet of an organelle membrane may have different roles compared to ceramide that is formed in the inner leaflet by the action of neutral SMase enzymes. Ceramide mediates many cell-stress responses, including the regulation of programmed cell death (apoptosis) and cell aging (senescence). Numerous research works have focused interest on defining the direct protein targets of action of ceramide. These include enzymes called ceramide-activated Ser-Thr phosphatases (CAPPs), such as protein phosphatase 1 and 2A (PP1 and PP2A), which were found to interact with ceramide in studies done in a controlled environment outside of a living organism (*in vitro*). On the other hand, studies in cells have shown that ceramide-inducing agents such as tumour necrosis factor-alpha α (TNFα) and palmitate induce the ceramide-dependent removal of a phosphate group (dephosphorylation) of the retinoblastoma gene product RB and the enzymes, protein kinases B (AKT protein family) and C α (PKB and PKCα).

Moreover, there is also sufficient evidence which implicates ceramide to the activation of the kinase suppressor of Ras (KSR), PKCæ, and cathepsin D. Interestingly, cathepsin D has been proposed as the main target for ceramide formed in organelles called lysosomes, making lysosomal acidic SMase enzymes one of the key players in the mitochondrial pathway of apoptosis. Ceramide was also shown to activate PKCæ, implicating it to the inhibition of AKT, regulation of the voltage difference between the interior and exterior of the cell (membrane potential) and signalling functions that favour apoptosis. Chemotherapeutic agents such as daunorubicin and etoposide enhance the *de novo* synthesis of ceramide in studies done on mammalian cells. The same results were found for certain inducers of apoptosis particularly stimulators of receptors in a class of lymphocytes (a type of white blood cell) called B-cells. Regulation of the *de novo* synthesis of ceramide by palmitate may have a key role in diabetes and the metabolic syndrome. Experimental evidence shows that there is substantial increase of ceramide levels upon adding palmitate. Ceramide accumulation activates PP2A and the subsequent dephosphorylation and inactivation of AKT, a crucial mediator in metabolic control and insulin signalling. This results in a substantial decrease in insulin responsiveness (i.e. to glucose) and in the death of insulin-producing cells in the pancreas called islets of Langerhans. Inhibition of ceramide synthesis in mice via drug treatments or gene-knockout techniques prevented insulin resistance induced by fatty acids, glucocorticoids or obesity.

An increase in *in vitro* activity of acid SMase has been observed after applying multiple stress stimuli such as ultraviolet (UV) and ionizing radiation, binding of death receptors and chemotherapeutic agents such as platinum, histone deacetylase inhibitors and paclitaxel. In some studies, SMase activation results to its transport to the plasma membrane and the simultaneous formation of ceramide.

Ceramide transfer protein (CERT) transports ceramide from ER to the Golgi for the synthesis of SM. CERT is known to bind phosphatidylinositol phosphates, hinting its potential regulation via phosphorylation, a step of the ceramide metabolism that can be enzymatically regulated by protein kinases and phosphatases, and by inositol lipid metabolic pathways. Up to date, there are at least 26 distinct enzymes with varied subcellular localizations, that act on ceramide as either a substrate or product. Regulation of ceramide levels can therefore be performed by one of these enzymes in distinct organelles by particular mechanisms at various times.

Sphingosine

Sphingosine (Sph) is formed by the action of ceramidase (CDase) enzymes on ceramide in the lysosome. Sph can also be formed in the extracellular (outer leaflet) side of the plasma membrane by the action of neutral CDase enzyme.

Sph then is either recycled back to ceramide or phosphorylated by one of the sphingosine kinase enzymes, SK1 and SK2. The product sphingosine-1-phosphate (S1P) can be dephosphorylated in the ER to regenerate sphingosine by certain S1P phosphatase enzymes within cells, where the salvaged Sph is recycled to ceramide. Sphingosine is a single-chain lipid (usually 18 carbons in length), rendering it to have sufficient solubility in water. This explains its ability to move between membranes and to flip-flop across a membrane. Estimates conducted at physiological pH show that approximately 70% of sphingosine remains in membranes while the remaining 30% is water-soluble. Sph that is formed has sufficient solubility in the liquid found inside cells (cytosol). Thus, Sph may come out of the lysosome and move to the ER without the need for transport via proteins or membrane-enclosed sacs called vesicles. However, its positive charge favors partitioning in lysosomes. It is proposed that the role of SK1 located near or in the lysosome is to 'trap' Sph via phosphorylation.

It is important to note that since sphingosine exerts surfactant activity, it is one of the sphingolipids found at lowest cellular levels. The low levels of Sph and their increase in response to stimulation of cells, primarily by activation of ceramidase by growth-inducing proteins such as platelet-derived growth factor and insulin-like growth factor, is consistent with its function as a second messenger. It was found that immediate hydrolysis of only 3 to 10% of newly-generated ceramide may double the levels of Sph. Treatment of HL60 cells (a type of leukemia cell line) by a plant-derived organic compound called phorbol ester increased Sph levels three-fold, whereby the cells differentiated into white blood cells called macrophages. Treatment of the same cells by exogenous Sph caused apoptosis. A specific protein kinase phosphorylates 14-3-3, otherwise known as sphingosine-dependent protein kinase 1 (SDK1), only in the presence of Sph.

Sph is also known to interact with protein targets such as the protein kinase H homologue (PKH) and the yeast protein kinase (YPK). These targets in turn mediate the effects of Sph and its related sphingoid bases, with known roles in regulating the actin cytoskeleton, endocytosis, the cell cycle and apoptosis. It is important to note however that the second messenger function of Sph is not yet established unambiguously.

Sphingosine-1-Phosphate

Sphingosine-1-phosphate (S1P), like Sph, is composed of a single hydrophobic chain and has sufficient solubility to move between membranes. S1P is formed by phosphorylation of sphingosine by sphingosine kinase (SK). The phosphate group of the product can be detached (dephosphorylated) to regenerate sphingosine via S1P phosphatase enzymes or S1P can be broken down by S1P lyase enzymes to ethanolamine phosphate and hexadecenal.

Similar to Sph, its second messenger function is not yet clear. However, there is substantial evidence that implicates S1P to cell survival, cell migration, and inflammation. Certain growth-inducing proteins such as platelet-derived growth factor (PDGF), insulin-like growth factor (IGF) and vascular endothelial growth factor (VEGF) promote the formation of SK enzymes, leading to increased levels of S1P. Other factors that induce SK include cellular communication molecules called cytokines, such as tumour necrosis factor α (TNFα) and interleukin-1 (IL-1), hypoxia or lack of oxygen supply in cells, oxidized low-density lipoproteins (oxLDL) and several immune complexes.

S1P is probably formed at the inner leaflet of the plasma membrane in response to TNFα and other receptor activity-altering compounds called agonists. S1P, being present in low nanomolar concentrations in the cell, has to interact with high-affinity receptors that are capable of sensing their low levels. So far, the only identified receptors for S1P are the high-affinity G protein-coupled receptors (GPCRs), also known as S1P receptors (S1PRs). S1P is required to reach the extracellular side (outer leaflet) of the plasma membrane to interact with S1PRs and launch typical GPCR signalling pathways. However, the zwitterionic headgroup of S1P makes it unlikely to flip-flop spontaneously.

To overcome this difficulty, the ATP-binding cassette (ABC) transporter C1 (ABCC1) serves as the "exit door" for S1P. On the other hand, the cystic fibrosis transmembrane regulator (CFTR) serves as the means of entry for S1P into the cell. In contrast to its low intracellular concentration, S1P is found in high nanomolar concentrations in serum where it is bound to albumin and lipoproteins. Inside the cell, S1P can induce calcium release independent of the S1PRs—the mechanism of which remains unknown. To date, the intracellular molecular targets for S1P are still unidentified. The SK1-S1P pathway has been extensively studied in relation to cytokine action, with multiple functions connected to effects of TNFα and IL-1 favouring inflammation. Studies show that knockdown of key enzymes such as S1P lyase and S1P phosphatase increased prostaglandin production, parallel to increase of S1P levels.

This strongly suggests that S1P is the mediator of SK1 action and not subsequent compounds. Research done on endothelial and smooth muscle cells is consistent to the hypothesis that S1P has a crucial role in regulating endothelial cell growth, and movement.

Recent work on a sphingosine analogue, FTY270, demonstrates its ability to act as a potent compound that alters the activity of S1P receptors (agonist). FTY270 was further verified in clinical tests to have roles in immune modulation, such as that on multiple sclerosis. This highlights the importance of S1P in the regulation of lymphocyte function and immunity. Most of the studies on S1P are used to further understand diseases such as cancer, arthritis and inflammation, diabetes, immune function and neurodegenerative disorders.

Glucosylceramide

Glucosylceramides (GluCer) are the most widely distributed glycosphingolipids in cells serving as precursors for the formation of over 200 known glycosphingolipids. GluCer is formed by the glycosylation of ceramide in an organelle called Golgi via enzymes called glucosylceramide synthase (GCS) or by the breakdown of complex glycosphingolipids (GSLs) through the action of specific hydrolase enzymes. In turn, certain β-glucosidases hydrolyse these lipids to regenerate ceramide. GluCer appears to be synthesized in the inner leaflet of the Golgi. Studies show that GluCer has to flip to the inside of the Golgi or transfer to the site of GSL synthesis to initiate the synthesis of complex GSLs.

Transferring to the GSL synthesis site is done with the help of a transport protein known as four phosphate adaptor protein 2 (FAPP2) while the flipping to the inside of the Golgi is made possible by the ABC transporter P-glycoprotein, also known as the multi-drug resistance 1 transporter (MDR1). GluCer is implicated in post-Golgi trafficking and drug resistance particularly to chemotherapeutic agents. For instance, a study demonstrated a correlation between cellular drug resistance and modifications in GluCer metabolism.

In addition to their role as building blocks of biological membranes, glycosphingolipids have long attracted attention because of their supposed involvement in cell growth, differentiation, and formation of tumours. The production of GluCer from Cer was found to be important in the growth of neurons or brain cells. On the other hand, pharmacological inhibition of GluCer synthase is being considered a technique to avoid insulin resistance.

Ceramide-1-Phosphate

Ceramide-1-phosphate (C1P) is formed by the action of ceramide kinase (CK) enzymes on Cer. C1P carry ionic charge at neutral pH and contain two hydrophobic chains making it relatively insoluble in aqueous environment. Thus, C1P reside in the organelle where it was formed and is unlikely to spontaneously flip-flop across membrane bilayers.

C1P activate phospholipase A2 and is found, along with CK, to be a mediator of arachidonic acid released in cells in response to a protein called interleukin-1β (IL-1β) and a lipid-soluble molecule that transports calcium ions (Ca) across the bilayer, also known as calcium ionophore.

C1P was also previously reported to encourage cell division (mitogenic) in fibroblasts, block apoptosis by inhibiting acid SMase in white blood cells within tissues (macrophages) and increase intracellular free calcium concentrations in thyroid cells. C1P also has known roles in vesicular trafficking, cell survival, phagocytosis ("cell eating") and macrophage degranulation.

Activators of G-protein Coupled Receptors

Lysophosphatidic Acid (LPA)

LPA is the result of phospholipase A2 action on phosphatidic acid. The SN-1 position can contain either an ester bond or an ether bond, with ether LPA being found at elevated levels in certain cancers. LPA binds the high-affinity G-protein coupled receptors LPA1, LPA2, and LPA3 (also known as EDG2, EDG4, and EDG7, respectively).

Sphingosine-1-phosphate (S1P)

S1P is present at high concentrations in plasma and secreted locally at elevated concentrations at sites of inflammation. It is formed by the regulated phosphorylation of sphingosine. It acts through five dedicated high-affinity G-protein coupled receptors, S1P1 - S1P5. Interestingly, targeted deletion of S1P1 results in lethality in mice and deletion of S1P2 results in seizures and deafness. Additionally, a mere 3- to 5-fold elevation in serum S1P concentrations induces sudden cardiac death by an S1P3-receptor specific mechanism.

Platelet Activating Factor (PAF)

PAF is a potent activator of platelet aggregation, inflammation, and anaphylaxis. It is similar to the ubiquitous membrane phospholipid phosphatidylcholine except that it contains an acetyl-group in the SN-2 position and the SN-1 position contains an ether-linkage. PAF signals through a dedicated G-protein coupled receptor, PAFR and is inactivated by PAF acetylhydrolase.

Endocannabinoids

The endogenous cannabinoids, or endocannabinoids, are endogenous lipids that activate cannabinoid receptors. The first such lipid to be isolated was anandamide which is the arachidonoyl amide of ethanolamine. Anandamide is formed via enzymatic release from N-arachidonoyl phosphatidylethanolamine by enzymes which have not yet been delineated. It activates both the CB1 receptor, found primarily in the central nervous system, and the CB2 receptor which is found primarily in lymphocytes and the periphery. It is found at very low levels (nM) in most tissues and is inactivated by the fatty acid amide hydrolase. Subsequently, another endocannabinoid was isolated, 2-arachidonoylglycerol, which is produced when phospholipase C releases diacylglycerol which is then converted to 2-AG by diacylglycerol lipase. 2-AG can also activate both cannabinoid receptors and is inactivated by monoacylglycerol lipase. It is present at approximately 100-times the concentration of anandamide in most tissues. Elevations in either of these lipids causes analgesia and anti-inflammation but the precise roles played by these two endocannabinoids are still vague and intensive research into their function, metabolism, and regulation is ongoing.

Prostaglandins

Prostaglandins are formed through oxidation of arachidonic acid by cyclooxygenases and other prostaglandin synthases. There are currently nine known G-protein coupled receptors (eicosanoid receptors) that largely mediate prostaglandin physiology (although some prostaglandins activate nuclear receptors).

Retinol Derivatives

Retinaldehyde is a retinol (vitamin A) derivative responsible for vision. It binds rhodopsin, a well-characterized GPCR that binds all-cis retinal in its inactive state. Upon photoisomerization by a photon the cis-retinal is converted to trans-retinal causing activation of rhodopsin which ultimately leads to depolarization of the neuron thereby enabling visual perception.

Activators of Nuclear Receptors

In the field of molecular biology, nuclear receptors are a class of proteins found within cells that are responsible for sensing steroid and thyroid hormones and certain other molecules. In response, these receptors work with other proteins to regulate the expression of specific genes, thereby controlling the development, homeostasis, and metabolism of the organism. Nuclear receptors have the ability to directly bind to DNA and regulate the expression of adjacent genes, hence these receptors are classified as transcription factors. The regulation of gene expression by nuclear receptors happens only when a ligand — a molecule that affects the receptor's behaviour — is present. In more specific terms, ligand binding to a nuclear receptor results in a conformational change in the receptor, which, in turn, activates the receptor, resulting in up-regulation of gene expression. A unique property of nuclear receptors that differentiates them from other classes of receptors is their ability to directly interact with and control the expression of genomic DNA. As a consequence, nuclear receptors play key roles in both embryonic development and adult homeostasis.

Species Distribution

Nuclear receptors are specific to metazoans (animals) and are not found in protists, algae, fungi, or plants. Among animal species, they are not found in sponges, but they are present in cnidarians and all other more advanced animals. There are 270 nuclear receptors in *C. elegans* alone. Humans, mice, and rats have respectively 48, 49, and 47 nuclear receptors each.

Ligands

Ligands that bind to and activate nuclear receptors include lipophilic substances such as endogenous hormones, vitamins A and D, and xenobiotic endocrine disruptors. Because the expression of a large number of genes is regulated by nuclear receptors, ligands that activate these receptors can have profound effects on the organism. Many of these regulated genes are associated with various diseases, which explains why the molecular targets of approximately 13% of U.S. Food and Drug Administration (FDA) approved drugs are nuclear receptors.

A number of nuclear receptors, referred to as orphan receptors, have no known (or at least generally agreed upon) endogenous ligands. Some of these receptors such as FXR, LXR, and PPAR bind a number of metabolic intermediates such as fatty acids, bile acids and/or sterols with relatively low affinity. These receptors hence may function as metabolic sensors. Other nuclear receptors, such as CAR and PXR appear to function as xenobiotic sensors up-regulating the expression of cytochrome P450 enzymes that metabolize these xenobiotics.

Structure

Nuclear receptors are modular in structure and contain the following domains:

- N-terminal regulatory domain: Contains the activation function 1 (AF-1) whose action is independent of the presence of ligand. The transcriptional activation of AF-1 is normally very weak, but it does synergize with AF-2 in the E-domain to produce a more robust upregulation of gene expression. The A-B domain is highly variable in sequence between various nuclear receptors.
- DNA-binding domain (DBD): Highly conserved domain containing two zinc fingers that binds to specific sequences of DNA called hormone response elements (HRE).
- Hinge region: Thought to be a flexible domain that connects the DBD with the LBD. Influences intracellular trafficking and subcellular distribution.
- Ligand binding domain (LBD): Moderately conserved in sequence and highly conserved in structure between the various nuclear receptors. The structure of the LBD is referred to as an alpha helical sandwich fold in which three anti parallel alpha helices (the "sandwich filling") are flanked by two alpha helices on one side and three on the other (the "bread"). The ligand binding cavity is within the interior of the LBD and just below three anti parallel alpha helical sandwich "filling". Along with the DBD, the LBD contributes to the dimerization interface of the receptor and in addition, binds coactivator and corepressor proteins. The LBD also contains the activation function 2 (AF-2) whose action is dependent on the presence of bound ligand.
- C-terminal domain: Variable in sequence between various nuclear receptors.

Mechanism of Action

Mechanism nuclear receptor action. Hormone binding to the NR triggers dissociation of heat shock proteins (HSP), dimerization, and translocation to the nucleus, where the NR binds to a specific sequence of DNA known as a hormone response element (HRE). The nuclear receptor DNA complex in turn recruits other proteins that are responsible for transcription of downstream DNA into mRNA, which is eventually translated into protein, which results in a change in cell function.

Mechanism Nuclear Receptor Action : For the purpose of illustration, the nuclear receptor shown here is the thyroid hormone receptor (TR) heterodimerized to the RXR. In the absence of ligand, the TR is bound to corepressor protein. Ligand binding to TR causes a dissociation of corepressor and recruitment of coactivator protein, which, in turn, recruits additional proteins such as RNA polymerase that are responsible for transcription of downstream DNA into RNA and eventually protein, which results in a change in cell function.

Nuclear receptors (NRs) may be classified into two broad classes according to their mechanism of action and subcellular distribution in the absence of ligand. Small lipophilic substances such as natural hormones diffuse past the cell membrane and bind to nuclear receptors located in the cytosol (type I NR) or nucleus (type II NR) of the cell. This causes a change in the conformation of the receptor, which, depending on the mechanistic class (type I or II), triggers a number of down stream events that eventually results in up or down regulation of gene expression. Accordingly, nuclear receptors may be subdivided into the following four mechanistic classes:

Type I

Ligand binding to type I nuclear receptors in the cytosol results in the dissociation of heat shock proteins, homo-dimerization, translocation (*i.e.*, active transport) from the cytoplasm into the cell nucleus, and binding to specific sequences of DNA known as hormone response elements (HREs). Type I nuclear receptors bind to HREs consisting of two half-sites separated by a variable length of DNA, and the second half-site has a sequence inverted from the first (inverted repeat).

Type I nuclear receptors include members of subfamily 3, such as the androgen receptor, estrogen receptors, glucocorticoid receptor, and progesterone receptor.

The nuclear receptor/DNA complex then recruits other proteins that transcribe DNA downstream from the HRE into messenger RNA and eventually protein, which causes a change in cell function.

Type II

Type II receptors, in contrast to type I, are retained in the nucleus regardless of the ligand binding status and in addition bind as hetero-dimers (usually with RXR) to DNA. In the absence of ligand, type II nuclear receptors are often complexed with corepressor proteins. Ligand binding to the nuclear receptor causes dissociation of corepressor and recruitment of coactivator proteins. Additional proteins including RNA polymerase are then recruited to the NR/DNA complex that transcribe DNA into messenger RNA.

Type II nuclear receptors include principally subfamily 1, for example the retinoic acid receptor, retinoid X receptor and thyroid hormone receptor.

Type III

Type III nuclear receptors (principally NR subfamily 2) are similar to type I receptors in that both classes bind to DNA as homodimers. However, type III nuclear receptors, in contrast to type I, bind to direct repeat instead of inverted repeat HREs.

Type III nuclear receptors are orphan receptors, with their endogenous ligands still unknown.

Type IV

Type IV nuclear receptors bind either as monomers or dimers, but only a single DNA binding domain of the receptor binds to a single half site HRE. Examples of type IV receptors are found in most of the NR subfamilies.

Coregulatory Proteins

Nuclear receptors bound to hormone response elements recruit a significant number of other proteins (referred to as transcription coregulators) that facilitate or inhibit the transcription of the associated target gene into mRNA. The function of these coregulators are varied and include chromatin remodelling (making the target gene either more or less accessible to transcription) or a bridging function to stabilize the binding of other coregulatory proteins.

Coactivators

Binding of agonist ligands to nuclear receptors induces a conformation of the receptor that preferentially binds coactivator proteins. These proteins often have an intrinsic histone acetyltransferase (HAT) activity, which weakens the association of histones to DNA, and therefore promotes gene transcription.

Corepressors

Binding of antagonist ligands to nuclear receptors in contrast induces a conformation of the receptor that preferentially binds corepressor proteins. These proteins, in turn, recruit histone deacetylases (HDACs), which strengthens the association of histones to DNA, and therefore represses gene transcription.

Agonism vs Antagonism

Stuctural basis for the mechanism of nuclear receptor agonist and antagonist action. The structures shown here are of the ligand binding domain (LBD) of the estrogen receptor (green cartoon diagram) complexed with either the agonist diethylstilbestrol (top, PDB 3ERD) or antagonist 4-hydroxytamoxifen (bottom, 3ERT). The ligands are depicted as space filling spheres (white = carbon, red = oxygen). When an agonist is bound to a nuclear receptor, the C terminal alpha helix of the LDB (H12; light blue) is positioned such that a coactivator protein (red) can bind to the surface of the LBD. Antagonists occupy the same ligand binding cavity of the nuclear receptor. However antagonist ligands in addition have a sidechain extension which sterically displaces H12 to occupy roughly the same position in space as coactivators bind. Hence coactivator binding to the LBD is blocked. Depending on the receptor involved, the chemical structure of the ligand and the tissue that is being affected, nuclear receptor ligands may display dramatically diverse effects ranging in a spectrum from agonism to antagonism to inverse agonism.

Agonists

The activity of endogenous ligands (such as the hormones estradiol and testosterone) when bound to their cognate nuclear receptors is normally to upregulate gene expression. This stimulation of gene expression by the ligand is referred to as an agonist response. The agonistic effects of endogenous hormones can also be mimicked by certain synthetic ligands, for example, the glucocorticoid receptor anti-inflammatory drug dexamethasone. Agonist ligands work by inducing a conformation of the receptor which favors coactivator binding.

Antagonists

Other synthetic nuclear receptor ligands have no apparent effect on gene transcription in the absence of endogenous ligand. However they block the effect of agonist through competitive binding to the same binding site in the nuclear receptor. These ligands are referred to as antagonists. An example of antagonistic nuclear receptor drug is mifepristone which binds to the glucocorticoid and progesterone receptors and therefore blocks the activity of the endogenous hormones cortisol and progesterone respectively. Antagonist ligands work by inducing a conformation of the receptor which prevents coactivator and promotes corepressor binding.

Inverse Agonists

Finally, some nuclear receptors promote a low level of gene transcription in the absence of agonists (also referred to as basal or constitutive activity). Synthetic ligands which reduce this basal level of activity in nuclear receptors are known as inverse agonists.

Selective Receptor Modulators

A number of drugs that work through nuclear receptors display an agonist response in some tissues and an antagonistic response in other tissues. This behaviour may have substantial benefits since it may allow retaining the desired beneficial therapeutic effects of a drug while minimizing undesirable side effects. Drugs with this mixed agonist/antagonist profile of action are referred to as selective receptor modulators (SRMs). Examples include Selective Androgen Receptor Modulators (SARMs), Selective Estrogen Receptor Modulators (SERMs) and Selective Progesterone Receptor Modulators (SPRMs). The mechanism of action of SRMs may vary depending on the chemical structure of the ligand and the receptor involved, however it is thought that many SRMs work by promoting a conformation of the receptor that is closely balanced between agonism and antagonism. In tissues where the concentration of coactivator proteins is higher than corepressors, the equilibrium is shifted in the agonist direction. Conversely in tissues where corepressors dominate, the ligand behaves as an antagonist.

Alternative Mechanisms

Transrepression

The most common mechanism of nuclear receptor action involves direct binding of the nuclear receptor to a DNA hormone response element. This mechanism is referred to as transactivation. However some nuclear receptors not only have the ability to directly bind to DNA, but also to other transcription factors. This binding often results in deactivation of the second transcription factor in a process known as transrepression. One example of a nuclear receptor that are able to transrepress is the glucocorticoid receptor (GR). Furthermore certain GR ligands known as Selective Glucocorticoid Receptor Agonists (SEGRAs) are able to activate GR in such a way that GR more strongly transrepresses than transactivates. This selectivity increases the separation between the desired antiinflammatory effects and undesired metabolic side effects of these selective glucocorticoids.

Non-genomic

The classical direct effects of nuclear receptors on gene regulation normally takes hours before a functional effect is seen in cells because of the large number of intermediate steps between nuclear receptor activation and changes in protein expression levels. However it has been observed that some effects from the application of hormones such as estrogen occur within minutes which is inconsistent with the classical mechanism of nuclear receptor action.

While the molecular target for these non-genomic effects of nuclear receptors has not been conclusively demonstrated, it has been hypothesized that there are variants of nuclear receptors which are membrane associated instead of being localized in the cytosol or nucleus. Furthermore these membrane associated receptors function through alternative signal transduction mechanisms not involving gene regulation.

Eicosanoids

In biochemistry, eicosanoids are signalling molecules made by oxidation of twenty-carbon essential fatty acids, (EFAs). They exert complex control over many bodily systems, mainly in inflammation or immunity, and as messengers in the central nervous system. The networks of controls that depend upon eicosanoids are among the most complex in the human body.

Eicosanoids derive from either omega-3 (ω-3) or omega-6 (ω-6) EFAs. The ω-6 eicosanoids are generally pro-inflammatory; ω-3's are much less so. The amounts and balance of these fats in a person's diet will affect the body's eicosanoid-controlled functions, with effects on cardiovascular disease, triglycerides, blood pressure, and arthritis. Anti-inflammatory drugs such as aspirin and other NSAIDs act by downregulating eicosanoid synthesis.

There are four families of eicosanoids—the prostaglandins, prostacyclins, the thromboxanes and the leukotrienes. For each, there are two or three separate series, derived either from an ω-3 or ω-6 EFA. These series' different activities largely explain the health effects of ω-3 and ω-6 fats.

Nomenclature

"Eicosanoid" (*eicosa-*, Greek for "twenty") is the collective term for oxygenated derivatives of three different 20-carbon essential fatty acids:

- Eicosapentaenoic acid (EPA), an ω-3 fatty acid with 5 double bonds;
- Arachidonic acid (AA), an ω-6 fatty acid, with 4 double bonds;
- Dihomo-gamma-linolenic acid (DGLA), an ω-6, with 3 double bonds.

Current usage limits the term to the leukotrienes (LT) and three types of prostanoids—prostaglandins (PG) prostacyclins (PGI), and thromboxanes (TX). This is the definition used in this article. However, several other classes can technically be termed eicosanoid, including the hepoxilins, resolvins, isofurans, isoprostanes, lipoxins, epi-lipoxins, epoxyeicosatrienoic acids (EETs) and endocannabinoids. LTs and prostanoids are sometimes termed 'classic eicosanoids' in contrast to the 'novel', 'eicosanoid-like' or 'nonclassic eicosanoids'.

A particular eicosanoid is denoted by a four-character abbreviation, composed of:

- Its two letter abbreviation (above),
- One A-B-C sequence-letter; and
- A subscript, indicating the number of double bonds.

Examples are:

- The EPA-derived prostanoids have three double bonds, (e.g. PGG_3, PGH_3, PGI_3, TXA_3) while its leukotrienes have five, (LTB_5).
- The AA-derived prostanoids have two double bonds, (e.g. PGG_2, PGH_2, PGI_2, TXA_2) while its leukotrienes have four, (LTB_4).

Furthermore, stereochemistry may differ among the pathways, indicated by Greek letters, e.g. for ($PGF_{2\alpha}$).

Biosynthesis

Two families of enzymes catalyse fatty acid oxygenation to produce the eicosanoids:

- Cyclooxygenase, or COX, generates the prostanoids.
- Lipoxygenase, or LOX, in several forms. 5-lipoxygenase (5-LO) generates the leukotrienes.

Eicosanoids are not stored within cells, but are synthesized as required. They derive from the fatty acids that make up the cell membrane and nuclear membrane. Eicosanoid biosynthesis begins when cell is activated by mechanical trauma, cytokines, growth factors or other stimuli. (The stimulus may even be an eicosanoid from a neighbouring cell; the pathways are complex.)

This triggers the release of a phospholipase at the cell membrane. The phospholipase travels to the nuclear membrane. There, the phospholipase catalyzes ester hydrolysis of phospholipid (by A_2) or diacylglycerol (by phospholipase C). This frees a 20-carbon essential fatty acid. This hydrolysis appears to be the rate-determining step for eicosanoid formation. The fatty acids may be released by any of several phospholipases. Of these, type IV cytosolic phospholipase A_2 ($cPLA_2$) is the key actor, as cells lacking $cPLA_2$ are generally devoid of eicosanoid synthesis. The phospholipase $cPLA_2$ is specific for phospholipids that contain AA, EPA or GPLA at the SN2 position. Interestingly, $cPLA_2$ may also release the lysophospholipid that becomes platelet-activating factor.

Peroxidation and Reactive Oxygen Species

Next, the free fatty acid is oxygenated along any of several pathways. The eicosanoid pathways (*via* lipoxygenase or COX) add molecular oxygen (O_2). Although the fatty acid is symmetric, the resulting eicosanoids are chiral; the oxidation proceeds with high stereospecificity.

The oxidation of lipids is hazardous to cells, particularly when close to the nucleus. There are elaborate mechanisms to prevent unwanted oxidation. COX, the lipoxygenases and the phospholipases are tightly controlled—there are at least eight proteins activated to coordinate generation of leukotrienes. Several of these exist in multiple isoforms.

Oxidation by either COX or lipoxygenase releases reactive oxygen species (ROS) and the initial products in eicosanoid generation are themselves highly reactive peroxides. LTA_4 can form adducts with tissue DNA. Other reactions of lipoxygenases generate cellular damage; murine models implicate 15-lipoxygenase in the pathogenesis of atherosclerosis. The oxidation in eicosanoid generation is compartmentalized; this limits the peroxides' damage. The enzymes which are biosynthetic for eicosanoids (e.g. glutathione-S-transferases, epoxide hydrolases and carrier proteins) belong to families whose functions are largely involved with cellular detoxification. This suggests that eicosanoid signalling may have evolved from the detoxification of ROS. The cell must realize some benefit from generating lipid hydroperoxides close-by its nucleus. PGs and LTs may signal or regulate DNA-transcription there; LTB_4 is ligand for PPARα.

Prostanoid Pathways

Cyclooxygenase (*COX*) catalyzes the conversion of the free essential fatty acids to prostanoids by a two-step process. First, two molecules of O_2 are added as two peroxide linkages, and a 5-member carbon ring is forged near the middle of the fatty acid chain. This forms the short-lived, unstable intermediate Prostaglandin G (PGG). Next, one of the peroxide linkages sheds a single oxygen, forming PGH. All three classes of prostanoids originate from PGH. All have distinctive rings in the centre of the molecule.

They differ in their structures. The PGH compounds (parents to all the rest) have a 5-carbon ring, bridged by two oxygens (a peroxide.) The derived prostaglandins contain a single, unsaturated 5-carbon ring. In prostacyclins, this ring is conjoined to another oxygen-containing ring. In thromboxanes the ring becomes a 6-member ring with one oxygen. The leukotrienes do not have rings.

Leukotriene Pathways

The enzyme 5-lipoxygenase (5-LO) uses 5-lipoxygenase activating protein (FLAP) to convert arachidonic acid into 5-hydroperoxyeicosatetraenoic acid (5-HPETE), which spontaneously reduces to 5-hydroxyeicosatetraenoic acid (5-HETE). The enzyme LTA synthase acts on 5-HPETE to convert it into leukotriene A_4 (LTA_4), which may be converted into LTB_4 by the enzyme leukotriene A4 epoxide hydrolase. Eosinophils, mast cells, and alveolar macrophages use the enzyme leukotriene C4 synthase to conjugate glutathione with LTA_4 to make LTC_4, which is transported outside the cell, where a glutamic acid moiety is removed from it to make LTD_4. The leukotriene LTD_4 is then cleaved by dipeptidases to make LTE_4.

The leukotrienes LTC_4, LTD_4 and LTE_4 all contain cysteine and are collectively known as the cysteinyl leukotrienes.

Function and Pharmacology

Eicosanoids exert complex control over many bodily systems, mainly in inflammation or immunity, and as messengers in the central nervous system. They are found in most living things. In humans, eicosanoids are local hormones that are released by most cells, act on that same cell or nearby cells (i.e., they are autocrine and paracrine mediators), and then are rapidly inactivated.

Eicosanoids have a short half-life, ranging from seconds to minutes. Dietary antioxidants inhibit the generation of some inflammatory eicosanoids, e.g. trans-resveratrol against thromboxane and some leukotrienes. Most eicosanoid receptors are members of the G protein-coupled receptor superfamily.

Receptors: There are specific receptors for all eicosanoids.

Leukotrienes:

- CysLT1 (Cysteinyl leukotriene receptor type 1)
- CysLT2 (Cysteinyl leukotriene receptor type 2)
- BLT1 (Leukotriene B4 receptor).

Prostanoids:

- PGD_2: DP-(PGD_2)
- PGE_2:
    - o EP_1-(PGE_2)
    - o EP_2-(PGE_2)
    - o EP_3-(PGE_2)
    - o EP_4-(PGE_2)
- $PGF_{2\alpha}$: FP-($PGF_{2\alpha}$)
- PGI_2 (prostacyclin): IP-(PGI_2)
- TXA_2 (thromboxane): TP-(TXA_2).

In the inflammatory response, two other groups of dietary essential fatty acids form cascades that parallel and compete with the arachidonic acid cascade. EPA (20:5 ω-3) provides the most important competing cascade. DGLA (20:3 ω-6) provides a third, less prominent cascade. These two parallel cascades soften the inflammatory effects of AA and its products. Low dietary intake of these less-inflammatory essential fatty acids, especially the ω-3s, has been linked to several inflammation-related diseases, and perhaps some mental illnesses. The U.S. National Institutes of Health and the National Library of Medicine state that there is 'A' level evidence that increased dietary ω-3 improves outcomes in hypertriglyceridemia, secondary cardiovascular disease prevention and hypertension.

There is 'B' level evidence ('good scientific evidence') for increased dietary ω-3 in primary prevention of cardiovascular disease, rheumatoid arthritis and protection from ciclosporin toxicity in organ transplant patients. They also note more preliminary evidence showing that dietary ω-3 can ease symptoms in several psychiatric disorders.

Besides the influence on eicosanoids, dietary polyunsaturated fats modulate immune response through three other molecular mechanisms. They (a) alter membrane composition and function, including the composition of lipid rafts; (b) change cytokine biosynthesis and (c) directly activate gene transcription. Of these, the action on eicosanoids is the best explored.

Mechanisms of ω-3 Action

EFA sources: Essential fatty acid production and metabolism to form eicosanoids. At each step, the ω-3 and ω-6 cascades compete for the enzymes. The eicosanoids from AA generally promote inflammation. Those from EPA and from GLA (*via* DGLA) are generally less inflammatory, or inactive, or even anti-inflammatory. Dietary ω-3 and GLA counter the inflammatory effects of AA's eicosanoids in three ways, along the eicosanoid pathways:

- *Displacement*—Dietary ω-3 decreases tissue concentrations of AA, so there is less to form ω-6 eicosanoids.
- *Competitive inhibition*—DGLA and EPA compete with AA for access to the cyclooxygenase and lipoxygenase enzymes. So the presence of DGLA and EPA in tissues lowers the output of AA's eicosanoids.
- *Counteraction*—Some DGLA and EPA derived eicosanoids counteract their AA derived counterparts.

Role in Inflammation

Since antiquity, the cardinal signs of inflammation have been known as: calor (warmth), dolor (pain), tumour (swelling) and rubor (redness). The eicosanoids are involved with each of these signs.

Redness—An insect's sting will trigger the classic inflammatory response. Short acting vasoconstrictors — TXA_2—are released quickly after the injury. The site may momentarily turn pale. Then TXA_2 mediates the release of the vasodilators PGE_2 and LTB_4. The blood vessels engorge and the injury reddens.

Swelling—LTB_4 makes the blood vessels more permeable. Plasma leaks out into the connective tissues, and they swell. The process also loses pro-inflammatory cytokines.

Pain—The cytokines increase COX-2 activity. This elevates levels of PGE_2, sensitizing pain neurons.

Heat—PGE_2 is also a potent pyretic agent. Aspirin and NSAIDS—drugs that block the COX pathways and stop prostanoid synthesis—limit fever or the heat of localized inflammation.

Pharmacy: Eicosanoid, eicosanoid analogs and receptor agonists/ antagonists used as medicines;

Medicine	*Type*	*Medical condition or use*
Alprostadil	PGE_1	Erectile dysfunction, maintaining apatent ductus arteriosus in the fetus
Beraprost	PGI_1 analog	Pulmonary hypertension, avoidingreperfusion injury
Bimatoprost	PG analog	Glaucoma, ocular hypertension
Carboprost	PG analog	Labour induction, abortifacientin early pregnancy
Dinoprostone	PGE_2	Labour induction
Iloprost	PGI_2 analog	Pulmonary arterial hypertension
Latanoprost	PG analog	Glaucoma, ocular hypertension
Misoprostol	PGE_1 analog	Stomach ulcers, labour induction,abortifacient
Montelukast	LT receptorantagonist	Asthma, seasonal allergies
Travoprost	PG analog	Glaucoma, ocular hypertension
Treprostinil	PGI analog	Pulmonary hypertension
U46619	Longer livedTX analog	Research only
Zafirlukast	LT receptorantagonist	Asthma

Action of Prostanoids

Prostanoids mediate local symptoms of inflammation: vasoconstriction or vasodilation, coagulation, pain and fever. Inhibition of cyclooxygenase, specifically the inducible COX-2 isoform, is the hallmark of NSAIDs (non-steroidal anti-inflammatory drugs), such as aspirin. COX-2 is responsible for pain and inflammation, while COX-1 is responsible for platelet clotting actions. Prostanoids activate the $PPAR_\gamma$ members of the steroid/thyroid family of nuclear hormone receptors, directly influencing gene transcription.

Action of Leukotrienes

Leukotrienes play an important role in inflammation. There is a neuroendocrine role for LTC_4 in luteinizing hormone secretion. LTB_4 causes adhesion and chemotaxis of leukocytes and stimulates aggregation, enzyme release, and generation of superoxide in neutrophils. Blocking leukotriene receptors can play a role in the management of inflammatory diseases such as asthma (by the drugs montelukast and zafirlukast), psoriasis, and rheumatoid arthritis.

The slow reacting substance of anaphylaxis comprises the cysteinyl leukotrienes. These have a clear role in pathophysiological conditions such as asthma, allergic rhinitis and other nasal allergies, and have been implicated in atherosclerosis and inflammatory gastrointestinal diseases. They are potent bronchoconstrictors, increase vascular permeability in postcapillary venules, and stimulate mucus secretion. They are released from the lung tissue of asthmatic subjects exposed to specific allergens and play a pathophysiological role in immediate hypersensitivity reactions. Along with PGD, they function in effector cell trafficking, antigen presentation, immune cell activation, matrix deposition, and fibrosis.

5

Nucleic Acid Structure Determination

Structure probing of nucleic acids is the process by which biochemical techniques are used to determine nucleic acid structure. This analysis can be used to define the patterns which can infer the molecular structure, experimental analysis of molecular structure and function, and further understanding on development of smaller molecules for further biological research. Structure probing analysis can be done through many different methods, which include chemical probing, hydroxyl radical probing, SHAPE, nucleotide analog interference mapping (NAIM), and in-line probing.

RNA Sequencing

While methods for each type of probing differ in steps, the probing of secondary structure involves certain steps in order to determine the structure. The first step involved is submitting the structural RNA to the probe of interest and incubating over a certain amount of time to allow the reaction to occur. The RNA is then transcribed using reverse transcriptase PCR, where this results in different lengths of bands due to the modification of the RNA at specific sites, which causes the reverse transcriptase to fall off. These bands are then run on a gel with sequencing data determined from RNA sequencing.

Chemical Probing

RNA chemical probing can involve many different chemicals which serve to modify specific bases at certain sites to show certain locations available for specific modification type.

DMS

Dimethyl sulfate, known as DMS, is a chemical that can be used to modify nucleic acids in order to determine secondary structure. DMS modifies certain bases through methylation. One set of methylation products that is used for RNA is the methylation of N1 adenosine and N3 of cytosine which prevents the natural hydrogen bonds to form between bases modified by DMS. This enables modification sites to be detected by rt-PCR, as modified sites cannot be basepaired, which results in the reverse transcriptase to fall off and different band sizes. DMS can only modify bases that are not basepaired, as single stranded nucleotides have the bases exposed so the chemical can modify.

Detection of these modifications is done by examination of a gel and looking at bands present. One thing of note is that since modification prevents the addition of nucleotides at the modification site each of the bands generated is shifted by one base down on the gel.

The PCR products indicate which adenine and cytosine nucleotides are double stranded, that is protected from DMS modification, or single stranded, that is accessible to DMS modification. This can be used to begin to develop a model of secondary structure for the RNA molecule. Structure probing by DMS allows for detection of secondary structure changes due to binding of RNA molecules along with detecting changes in tertiary structure, but it lacks the ability to determine the origin of these changes, as well as protection of the RNA backbone.

DMS modification can also be used for DNA, for example in footprinting DNA-protein interactions.

CMCT

1-cyclohexyl-(2-morpholinoethyl)carbodiimide metho-p-toluene sulfonate known as CMCT is another chemical that is used in structure probing. CMCT like DMS serves to modify the exposed bases of specific nucleotides, which are uridine, and to a smaller extent guanine. CMCT reacts primarily with N3 of uridine and N1 of guanine modifying two sites responsible for hydrogen bonding on the bases. Modification using CMCT is analogous in detection to DMS as modification prevents basepairing at specific nucleotides which are then detected using rt-PCR and running of the PCR products on an agarose gel. Structure probing of RNA by CMCT indicates the presence of uridine and guanine in single stranded regions by accessibility to modification or the presence of uridine and guanine in double stranded regions by protection from CMCT, which is the absence of a band. CMCT, like DMS can detect secondary and tertiary structure changes, but still has the same weaknesses as the method of modification is the same as DMS.

Kethoxal

The modification by kethoxal prevents single stranded guanine nucleotides to become modified, so that when reverse transcriptase reaches the modified guanine it falls off resulting in a band. After rt-PCR the products will be of different lengths where a band present indicates modification of a guanine base, and the absence of a band indicates that the base was not available for modification. Using this gel in combination with DMS and CMCT a model for structural RNAs can be formed through comparison between the gels which indicate protected and accessible positions in all the bases, which can be used to form a preliminary model.

SHAPE

Selective 22 -hydroxyl acylation analysed by primer extension, or SHAPE, takes advantage of reagents that preferentially modify the backbone of RNA in structurally flexible regions.

Reagents such as N-methylisotoic anhydride (NMIA) and 1-methyl-7-nitroisatoic anhydride (1M7) undergo hydrolysis to form adducts on the 2'-hydroxyl of the RNA backbone. Compared to the chemicals used in other RNA probing techniques, these reagents have the advantage of being largely unbiased to base identity, while remaining very sensitive to conformational dynamics. Nucleotides which are constrained (usually by base-pairing) show less adduct formation than nucleotides which are unpaired. Adduct formation is quantified for each nucleotide in a given RNA by extension of a complementary DNA primer with reverse transcriptase and comparison of the resulting fragments with those from an unmodified control. SHAPE therefore reports on RNA structure at the individual nucleotide level. This data can be used as input to generate highly accurate secondary structure models. SHAPE has been used to analyse diverse RNA structures, including that of an entire HIV-1 genome.

Hydroxyl Radical Probing

Probing with hydroxyl radicals involves an additional step, as hydroxyl radicals are short lived in solution they need to be generated. This can be done using H_2O_2, ascorbic acid, and Fe(II)-EDTA complex which is attached to the backbone through EDTA. The Fe(II) along with ascorbic acid generates hydroxyl radicals which then can react with the nucleic acid molecules. Hydroxyl radical probing is often used in conjunction with chemical probing of nucleic acid molecules that are thought to associate with proteins, this is due to the modification done by hydroxyl radicals. Hydroxyl radicals attack the ribose/ deoxyribose ring and this results in breaking of the phosphate backbone, which is independent of secondary structure, as all backbone is accessible, but is instead resultant of protein or tertiary structure protection.

Probing with hydroxyl radicals shows the protection of structured nucleic acids by the proteins thought to be associated or folding on itself, where cleavage again results in a band formed through gel electrophoresis (after rt-PCR in the case of RNA) that is shorter than the full nucleic acid depending upon where it is cleaved. In this case since the last nucleotide is not modified the band length is indicative of the base that was cleaved. When examining the gel produced by running the gels on a band the areas of various strength of protection where areas of stronger protection for hydroxyl radicals can be said to have tighter association with a protein, or if no protein associates with the nucleic acid it can be caused by the tertiary fold.

In-line Probing

In-line probing does not involve treatment with any type of chemicals or reagents to modify exposed RNA structures. This type of probing assay uses the structure dependent cleavage of RNA, as areas that are single stranded are more flexible and over time will degrade, as RNA structure is not always stable. The process of in-line probing is often used to determine changes in structure due to ligand binding as this can result in different cleavage patterns.

The process of in-line probing involves incubation of structural or functional RNAs over a long period of time, can be several days, but varies in each experiment, and then running the incubated products on a gel to visualize the bands. This experiment is often done using two different conditions: 1) with ligand and 2) in the absence of ligand. Cleavage results in shorter band lengths and is indicative of areas that are not basepaired, as those that are tend to be less sensitive to spontaneous cleavage. In line probing is a functional assay in ligand binding changes of RNA because it can show directly the change in flexibility and binding of regions of DNA in response to a ligand, as well as compare that response to analogous ligands. An in-line probing assay is then commonly used in dynamic studies, specifically when examining riboswitches.

Nucleotide Analog Interference Mapping

Nucleotide analog interference mapping (NAIM) is the process of using nucleotide analogs, molecules that are similar in some ways to nucleotides but lack function, to determine the importance of a functional group at each location of an RNA molecule. The process of NAIM is to insert a single nucleotide analog into a unique site.

This can be done by transcribing a short RNA using T7 RNA polymerase, then synthesizing a short oligonucleotide containing the analog in a specific position, then ligating them together on the DNA template using a ligase. The nucleotide analogs are tagged with a phosphorothioate, the active members of the RNA population are then distinguished from the inactive members, the inactive members then have the phophorothioate tag removed and the analog sites are identified using gel electrophoresis and autoradiography.

This indicates a functionally important nucleotides, as cleavage of the phosphorothioate by iodine results in an RNA that is cleaved at the site of the nucleotide analog insert. By running these truncated RNA molecules on a gel, the nucleotide of interest can be identified against a sequencing experiment Site directed incorporation results indicate positions of importance where when running on a gel, functional RNAs that have the analog incorporated at that position will have a band present, but if the analog results in non-functionality, when the functional RNA molecules are run on a gel there will be no band corresponding to that position on the gel.

This process can be used to evaluate an entire area, where analogs are placed in site specific locations, differing by a single nucleotide, then when functional RNAs are isolated and run on a gel, all areas where bands are produced indicate non-essential nucleotides, but areas where bands are absent from the functional RNA indicate that inserting a nucleotide analog in that position caused the RNA molecule to become non-functional.

Paracrystalline Lattice Models of B-DNA Structures

A paracrystalline lattice, or paracrystal, is a molecular or atomic lattice with significant amounts (e.g., larger than a few percent) of partial disordering of molecular arranegements. Limiting cases of the paracrystal model are nanostructures, such as glasses, liquids, etc., that may possess only local ordering and no global order. Liquid crystals also have paracrystalline rather than crystalline structures.

Highly hydrated B-DNA occurs naturally in living cells in such a paracrystalline state, which is a dynamic one in spite of the relatively rigid DNA double-helix stabilized by parallel hydrogen bonds between the nucleotide base-pairs in the two complementary, helical DNA chains.

For simplicity most DNA molecular models omit both water and ions dynamically bound to B-DNA, and are thus less useful for understanding the dynamic Behaviours of B-DNA *in vivo*. The physical and mathematical analysis of X-ray and spectroscopic data for paracrystalline B-DNA is therefore much more complicated than that of crystalline, A-DNA X-ray diffraction patterns. The paracrystal model is also important for DNA technological applications such as DNA nanotechnology. Novel techniques that combine X-ray diffraction of DNA with X-ray microscopy in hydrated living cells are now also being developed.

Genomic and Biotechnology Applications of DNA Molecular Modelling

There are various uses of DNA molecular modelling in Genomics and Biotechnology research applications, from DNA repair to PCR and DNA nanostructures. Two-dimensional DNA junction arrays have been visualized by Atomic force microscopy.

DNA molecular modelling has various uses in genomics and biotechnology, with research applications ranging from DNA repair to PCR and DNA nanostructures. These include computer molecular models of molecules as varied as RNA polymerase, an E. coli, bacterial DNA primase template suggesting very complex dynamics at the interfaces between the enzymes and the DNA template, and molecular models of the mutagenic, chemical interaction of potent carcinogen molecules with DNA. These are all represented in the gallery below.

Technological application include a DNA biochip and DNA nanostructures designed for DNA computing and other dynamic applications of DNA nanotechnology. The image at right is of self-assembled DNA nanostructures. The DNA "tile" structure in this image consists of four branched junctions oriented at 90° angles. Each tile consists of nine DNA oligonucleotides as shown; such tiles serve as the primary "building block" for the assembly of the DNA nanogrids shown in the AFM micrograph. Quadruplex DNA may be involved in certain cancers. Images of quadruplex DNA are in the gallery below.

Nucleic Acid Structure

The sequence or primary structure of a nucleic acid is the exact specification of its atomic composition and the chemical bonds connecting those atoms. As nucleic acids, e.g. DNA and RNA, are unbranched polymers, this is equivalent to specifying exact sequence of nucleotides that comprise the whole molecule. This sequence is written as a succession of letters representing a real or hypothetical DNA molecule or strand. By convention, the primary structure of a DNA or RNA molecule is reported from the 5' end to the 3' end.

The sequence has capacity to carry information. When used in reference to biological DNA, which carries the information which directs the functions of living beings, the term genetic sequence is often used. Sequences can be read from the biological raw material through DNA sequencing methods. Primary structure is sometimes mistakenly termed *primary sequence*, but there is no such term, as well as no parallel concept of secondary or tertiary sequence.

Nucleotide

Nucleotides are molecules that, when joined together, make up the structural units of RNA and DNA. In addition, nucleotides play central roles in metabolism. In that capacity, they serve as sources of chemical energy (adenosine triphosphate and guanosine triphosphate), participate in cellular signalling (cyclic guanosine monophosphate and cyclic adenosine monophosphate), and are incorporated into important cofactors of enzymatic reactions (coenzyme A, flavin adenine dinucleotide, flavin mononucleotide, and nicotinamide adenine dinucleotide phosphate).

Nucleotide Structure

A nucleotide is composed of a nucleobase (nitrogenous base), a five-carbon sugar (either ribose or 2'-deoxyribose), and one to three phosphate groups. Together, the nucleobase and sugar comprise a nucleoside. The phosphate groups form bonds with either the 2, 3, or 5-carbon of the sugar, with the 5-carbon site most common. Cyclic nucleotides form when the phosphate group is bound to two of the sugar's hydroxyl groups. Ribonucleotides are nucleotides where the sugar is ribose, and deoxyribonucleotides contain the sugar deoxyribose. Nucleotides can contain either a purine or a pyrimidine base.

Nucleic acids are polymeric macromolecules made from nucleotide monomers. In DNA, the purine bases are adenine and guanine, while the pyrimidines are thymine and cytosine. RNA uses uracil in place of thymine. also nucleotide plays an important role in thinking.

Synthesis

Nucleotides can be synthesized by a variety of means both in vitro and in vivo. In vivo, nucleotides can be synthesised de novo or recycled through salvage pathways. Nucleotides undergo breakdown such that useful parts can be reused in synthesis reactions to create new nucleotides. In vitro, protecting groups may be used during laboratory production of nucleotides. A purified nucleoside is protected to create a phosphoramidite, which can then be used to obtain analogues not found in nature and/or to synthesize an oligonucleotide.

Pyrimidine Ribonucleotides

Purine Ribonucleotides

The atoms which are used to build the purine nucleotides come from a variety of sources:

The de novo synthesis of purine nucleotides by which these precursors are incorporated into the purine ring proceeds by a 10-step pathway to the branch-point intermediate IMP, the nucleotide of the base hypoxanthine. AMP and GMP are subsequently synthesized from this intermediate via separate, two-step pathways. Thus, purine moieties are initially formed as part of the ribonucleotides rather than as free bases. Six enzymes take part in IMP synthesis. Three of them are multifunctional:

- GART (reactions 2, 3, and 5)
- PAICS (reactions 6, and 7)
- ATIC (reactions 9, and 10).

Reaction 1: The pathway starts with the formation of PRPP. PRPS1 is the enzyme that activates R5P, which is formed primarily by the pentose phosphate pathway, to PRPP by reacting it with ATP.

The reaction is unusual in that a pyrophosphoryl group is directly transferred from ATP to C1 of R5P and that the product has the α configuration about C1. This reaction is also shared with the pathways for the synthesis of the pyrimidine nucleotides, Trp, and His. As a result of being on (a) such (a) major metabolic crossroad and the use of energy, this reaction is highly regulated.

Reaction 2: In the first reaction unique to purine nucleotide biosynthesis, PPAT catalyses the displacement of PRPP's pyrophosphate group (PP_i) by Gln's amide nitrogen. The reaction occurs with the inversion of configuration about ribose C1, thereby forming β-5-phosphorybosylamine (5-PRA) and establishing the anomeric form of the future nucleotide. This reaction, which is driven to completion by the subsequent hydrolysis of the released PP_i, is the pathway's flux-generating step and is therefore regulated, too.

Length Unit

Nucleotide (abbreviated nt) is a common length unit for single-stranded RNA, similar to how base pair is a length unit for double-stranded DNA.

Abbreviation Codes

The IUPAC has designated the symbols for various nucleotides, these are listed in table below.

IUPAC nucleotide code	*Base*
A	Adenine
C	Cytosine
G	Guanine
T (or U)	Thymine (or Uracil)
R	A or G
Y	C or T (U)
S	G or C
W	A or T (U)
K	G or T (U)
M	A or C
B	C or G or T (U)
D	A or G or T (U)
H	A or C or T (U)
V	A or C or G
N	any base
. or -	gap

Nucleic Acid Notation

The International Union of Pure and Applied Chemistry (IUPAC) first formalized the currently used nucleic acid notation in 1970 . This universally accepted notation uses the Roman characters G, C, A, and T, to represent the four nucleotides commonly found in deoxyribonucleic acids (DNA).

Given the rapidly expanding role for genetic sequencing, synthesis, and analysis in biology, researchers have been compelled to develop alternate notations to further support the analysis and manipulation of genetic data. These notations generally exploit size, shape, symmetry to accomplish these objectives.

Degenerate base symbols in biochemistry are a IUPAC representation for a position on a DNA sequence that can be have multiple possible alternatives. These should not be confused with non-canonical bases because each particular sequence will have in fact one of the regular bases. These are used to encode the consensus sequence of a population of aligned sequences and are used for example in phylogenetic analysis to summarise into one multiple sequences or for BLAST searches, even though IUPAC degenerate symbols are masked (as they are not coded).

IUPAC Notation

Symbol	*Description*	*Bases represented*			
A	adenosine	A			
C	cytidine		C		
G	guanosine			G	
T	thymidine				T
U	uridine				U
W	weak	A			T
S	strong		C	G	
M	amino	A	C		
K	keto			G	T
R	purine	A		G	
Y	pyrimidine		C		T
B	not A (B comes after A)		C	G	T
D	not C (D comes after C)	A		G	T
H	not G (H comes after G)	A	C		T
V	not T (V comes after T (and U))	A	C	G	
N	any base (not a gap)	A	C	G	T

Under the commonly used IUPAC system, nucleic acids are represented by the first letters of their chemical names: [G]uanine, [C]ytosine, [A]denosine, and [T]hymine. This shorthand also includes eleven "ambiguity" characters associated with every possible combination of the four DNA bases. The ambiguity characters were designed to encode positional variations found among families of related genes.

Despite its broad and nearly universal acceptance, the IUPAC system has a number of limitations, which stem from its reliance on the Roman alphabet. The poor legibility of upper-case Roman characters, which are generally used when displaying genetic data, may be chief among these limitations. The value of external projections in distinguishing letters has been well documented. However, these projections are absent from upper case letters, which in some cases are only distinguishable by subtle internal cues. Take for example the upper case C and G used to represent cytosine and guanine. These characters generally comprise half the characters in a genetic sequence but are differentiated by a small internal tick.

Another shortcoming of the IUPAC notation arises from the fact that its eleven ambiguity characters have been selected from the remaining characters of the Roman alphabet. The authors of the notation endeavored to select ambiguity characters with logical mnemonics. For example, S is used to represent the possibility of finding cytosine or guanine at genetic loci, both of which form [S]trong cross-strand binging interactions. Conversely, the weaker interactions of thymine and adenine are represented by a W. However, convenient mnemonics are not as readily available for the other ambiguity characters displayed. This has made ambiguity characters difficult to use and may account for their limited application.

Visually Enhanced Notations

Legibility issues associated with IUPAC-encoded genetic data have led biologists to consider alternate strategies for displaying genetic data. These creative approaches to visualizing DNA sequences have generally relied on the use of spatially distributed symbols and/or visually distinct shapes to encode lengthy nucleic acid sequences. Several of these approaches are summarized below.

Stave Projection

In 1986, Cowin et al. described a novel method for visualizing DNA sequence known as the Stave Projection. Their strategy was to encode nucleotides as circles on series of horizontal bars akin to notes on musical stave. The spacial distribution of the circles made it far easier to distinguish individual bases and compare genetic sequences than IUPAC-encoded data.

Geometric Symbols

Zimmerman et al. took a different approach to visualizing genetic data . Rather than relying on spatially distributed circles to highlight genetic features, they exploited four geometrically diverse symbols found in a standard computer font to distinguish the four bases. The authors developed a simple WordPerfect macro to translate IUPAC characters into the more visually distinct symbols.

DNA Skyline

With the growing availability of font editors, Jarvius and Landegren devised a novel set of genetic symbols, known as the DNA Skyline font, which uses increasingly taller blocks to represent the different DNA bases. While reminiscent of Cowin et al.'s spatially distributed Stave Projection, the DNA Skyline font is easy to download and permits translation to and from the IUPAC notation by simply changing the font in most standard word processing applications.

Functional Ambigraphic Notations

Additional functionality can be found in nucleic acid notations that use ambigrams to mirror structural symmetries found in the DNA double helix. As defined by Douglas Hofstadter, ambigrams are words or symbols that convey the same or different meaning when viewed in a different orientation. It turns out that by assigning ambigraphic characters to complementary bases (i.e. guanine = b, cytosine = q, adenine = n, and thymine = u), it is possible to complement entire DNA sequences by simply rotating the text 180 degrees. An ambigraphic nucleic acid notation also makes it easy to identify genetic palindromes, such as endonuclease restriction sites, as sections of text that can be rotated 180 degrees without changing the sequence.

AmbiScript

The latest in a series of rationally designed nucleic acid notations, AmbiScript combines many of the visual and functional features of its predecessors. As its name implies, AmbiScript is an ambigraphic nucleic acid notation that permits rapid complementation of genetic sequence and identification of biologically significant palindromes. However, notation also uses spatially offset characters to facilitate the visual review and analysis of genetic data. One novel feature that AmbiScript brings to the world of genetic notations is its use of compound symbols to convey the possibility of finding two or more different bases at a given position. This strategy appears to an offer far less cumbersome solution to the use of ambiguity characters first proposed by the IUPAC. As with Jarvius and Landegren's DNA Skyline fonts, AmbiScript fonts are easily downloaded and applied to IUPAC-encoded sequence data.

Genetic Code

The genetic code is the set of rules by which information encoded in genetic material (DNA or mRNA sequences) is translated into proteins (amino acid sequences) by living cells. The code defines a mapping between tri-nucleotide sequences, called codons, and amino acids. With some exceptions, a triplet codon in a nucleic acid sequence specifies a single amino acid. Because the vast majority of genes are encoded with exactly the same code, this particular code is often referred to as the canonical or standard genetic code, or simply *the* genetic code, though in fact there are many variant codes. For example, protein synthesis in human mitochondria relies on a genetic code that differs from the standard genetic code.

Not all genetic information is stored using the genetic code. All organisms' DNA contains regulatory sequences, intergenic segments, and chromosomal structural areas that can contribute greatly to phenotype. Those elements operate under sets of rules that are distinct from the codon-to-amino acid paradigm underlying the genetic code.

Discovery

After the structure of DNA was deciphered by James Watson, Thomas W. Donnellan, Francis Crick, Maurice Wilkins and Rosalind Franklin, serious efforts to understand the nature of the encoding of proteins began. George Gamow postulated that a three-letter code must be employed to encode the 20 standard amino acids used by living cells to encode proteins, because 3 is the smallest integer n such that 4^n is at least 20.

The fact that codons consist of three DNA bases was first demonstrated in the Crick, Brenner et al. experiment. The first elucidation of a codon was done by Marshall Nirenberg and Heinrich J. Matthaei in 1961 at the National Institutes of Health. They used a cell-free system to translate a poly-uracil RNA sequence (i.e., UUUUU...) and discovered that the polypeptide that they had synthesized consisted of only the amino acid phenylalanine. They thereby deduced that the codon UUU specified the amino acid phenylalanine. This was followed by experiments in the laboratory of Severo Ochoa demonstrating that the poly-adenine RNA sequence (AAAAA...) coded for the polypeptide, poly-lysine. and the poly-cytosine RNA sequence (CCCCC...) coded for the polypeptide, poly-proline.

Therefore the codon AAA specified the amino acid lysine, and the codon CCC specified the amino acid proline. Using different copolymers most of the remaining codons were then determined. Extending this work, Nirenberg and Philip Leder revealed the triplet nature of the genetic code and allowed the codons of the standard genetic code to be deciphered. In these experiments various combinations of mRNA were passed through a filter which contained ribosomes, the components of cells that translate RNA into protein. Unique triplets promoted the binding of specific tRNAs to the ribosome. Leder and Nirenberg were able to determine the sequences of 54 out of 64 codons in their experiments. Subsequent work by Har Gobind Khorana identified the rest of the genetic code. Shortly thereafter, Robert W. Holley determined the structure of transfer RNA (tRNA), the adapter molecule that facilitates the process of translating RNA into protein. This work was based upon earlier studies by Severo Ochoa, who received the Nobel prize in 1959 for his work on the enzymology of RNA synthesis. In 1968, Khorana, Holley and Nirenberg received the Nobel Prize in Physiology or Medicine for their work.

Transfer of Information Via the Genetic Code

The genome of an organism is inscribed in DNA, or in the case of some viruses, RNA. The portion of the genome that codes for a protein or an RNA is referred to as a gene. Those genes that code for proteins are composed of tri-nucleotide units called codons, each coding for a single amino acid. Each nucleotide sub-unit consists of a phosphate, deoxyribose sugar and one of the 4 nitrogenous nucleobases. The purine bases adenine (A) and guanine (G) are larger and consist of two aromatic rings. The pyrimidine bases cytosine (C) and thymine (T) are smaller and consist of only one aromatic ring. In the double-helix configuration, two strands of DNA are joined to each other by hydrogen bonds in an arrangement known as base pairing. These bonds almost always form between an adenine base on one strand and a thymine on the other strand and between a cytosine base on one strand and a guanine base on the other. This means that the number of A and T residues will be the same in a given double helix, as will the number of G and C residues. In RNA, thymine (T) is replaced by uracil (U), and the deoxyribose is substituted by ribose.

Each protein-coding gene is transcribed into a template molecule of the related polymer RNA, known as messenger RNA or mRNA. This, in turn, is translated on the ribosome into an amino acid chain or polypeptide. The process of translation requires transfer RNAs specific for individual amino acids with the amino acids covalently attached to them, guanosine triphosphate as an energy source, and a number of translation factors. tRNAs have anticodons complementary to the codons in mRNA and can be "charged" covalently with amino acids at their 3' terminal CCA ends. Individual tRNAs are charged with specific amino acids by enzymes known as aminoacyl tRNA synthetases, which have high specificity for both their cognate amino acids and tRNAs. The high specificity of these enzymes is a major reason why the fidelity of protein translation is maintained.

There are $4^3 = 64$ different codon combinations possible with a triplet codon of three nucleotides; all 64 codons are assigned for either amino acids or stop signals during translation.

If, for example, an RNA sequence, UUUAAACCC is considered and the reading frame starts with the first U (by convention, 5' to 3'), there are three codons, namely, UUU, AAA and CCC, each of which specifies one amino acid. This RNA sequence will be translated into an amino acid sequence, three amino acids long. A comparison may be made with computer science, where the codon is similar to a word, which is the standard "chunk" for handling data (like one amino acid of a protein), and a nucleotide is similar to a bit, in that it is the smallest unit.

These are called forward and reverse codon tables, respectively. For example, the codon AAU represents the amino acid asparagine, and UGU and UGC represent cysteine (standard three-letter designations, Asn and Cys, respectively).

Salient Features

Sequence Reading Frame

A codon is defined by the initial nucleotide from which translation starts. For example, the string GGGAAACCC, if read from the first position, contains the codons GGG, AAA and CCC; and, if read from the second position, it contains the codons GGA and AAC; if read starting from the third position, GAA and ACC. Every sequence can thus be read in three reading frames, each of which will produce a different amino acid sequence (in the given example, Gly-Lys-Pro, Gly-Asn, or Glu-Thr, respectively). With double-stranded DNA there are six possible reading frames, three in the forward orientation on one strand and three reverse on the opposite strand. The actual frame in which a protein sequence is translated is defined by a start codon, usually the first AUG codon in the mRNA sequence.

Start/Stop Codons

Translation starts with a chain initiation codon (start codon). Unlike stop codons, the codon alone is not sufficient to begin the process. Nearby sequences (such as the Shine-Dalgarno sequence in *E. coli*) and initiation factors are also required to start translation. The most common start codon is AUG which is read as methionine or, in bacteria, as formylmethionine. Alternative start codons (depending on the organism), include "GUG" or "UUG", which normally code for valine or leucine, respectively. However, when used as a start codon, these alternative start codons are translated as methionine or formylmethionine.

The three stop codons have been given names: UAG is *amber*, UGA is *opal* (sometimes also called *umber*), and UAA is *ochre*. "Amber" was named by discoverers Richard Epstein and Charles Steinberg after their friend Harris Bernstein, whose last name means "amber" in German. The other two stop codons were named "ochre" and "opal" in order to keep the "colour names" theme. Stop codons are also called "termination" or "nonsense" codons and they signal release of the nascent polypeptide from the ribosome due to binding of release factors in the absence of cognate tRNAs with anticodons complementary to these stop signals.

Effect of Mutations

During the process of DNA replication, errors occasionally occur in the polymerization of the second strand. These errors, called mutations, can have an impact on the phenotype of an organism, especially if they occur within the protein coding sequence of a gene. Error rates are usually very low—1 error in every 10–100 million bases—due to the "proofreading" ability of DNA polymerases.

Missense mutations and nonsense mutations are examples of point mutations, which can cause genetic diseases such as sickle-cell disease and thalassemia respectively. Clinically important missense mutations generally change the properties of the coded amino acid residue between being basic, acidic polar or non-polar, whereas nonsense mutations result in a stop codon.

Mutations that disrupt the reading frame sequence by indels (insertions or deletions) of a non-multiple of 3 nucleotide bases are known as frameshift mutations. These mutations usually result in a completely different translation from the original, and are also very likely to cause a stop codon to be read, which truncates the creation of the protein. These mutations may impair the function of the resulting protein, and are thus rare in *in vivo* protein-coding sequences. One reason inheritance of frameshift mutations is rare is that if the protein being translated is essential for growth under the selective pressures the organism faces, absence of a functional protein may cause death before the organism is viable. Frameshift mutations may result in severe genetic diseases such as Tay-Sachs disease.

Although most mutations that change protein sequences are harmful or neutral, some mutations have a positive effect on an organism. These mutations may enable the mutant organism to withstand particular environmental stresses better than wild-type organisms, or reproduce more quickly. In these cases a mutation will tend to become more common in a population through natural selection. Viruses that use RNA as their genetic material have rapid mutation rates, which can be an advantage since these viruses will evolve constantly and rapidly, and thus evade the defensive responses of e.g. the human immune system. In large populations of asexually reproducing organisms, for example, E. coli, multiple beneficial mutations may co-occur, causing competition among them, this phenomenon is called clonal interference.

Degeneracy of the Genetic Code

The genetic code has redundancy but no ambiguity. For example, although codons GAA and GAG both specify glutamic acid (redundancy), neither of them specifies any other amino acid (no ambiguity). The codons encoding one amino acid may differ in any of their three positions. For example the amino acid glutamic acid is specified by GAA and GAG codons (difference in the third position), the amino acid leucine is specified by UUA, UUG, CUU, CUC, CUA, CUG codons (difference in the first or third position), while the amino acid serine is specified by UCA, UCG, UCC, UCU, AGU, AGC (difference in the first, second or third position). A position of a codon is said to be a fourfold degenerate site if any nucleotide at this position specifies the same amino acid. For example, the third position of the glycine codons (GGA, GGG, GGC, GGU) is a fourfold degenerate site, because all nucleotide substitutions at this site are synonymous; i.e., they do not change the amino acid.

Only the third positions of some codons may be fourfold degenerate. A position of a codon is said to be a twofold degenerate site if only two of four possible nucleotides at this position specify the same amino acid. For example, the third position of the glutamic acid codons (GAA, GAG) is a twofold degenerate site. In twofold degenerate sites, the equivalent nucleotides are always either two purines (A/G) or two pyrimidines (C/U), so only transversional substitutions (purine to pyrimidine or pyrimidine to purine) in twofold degenerate sites are nonsynonymous. A position of a codon is said to be a non-degenerate site if any mutation at this position results in amino acid substitution. There is only one threefold degenerate site where changing to three of the four nucleotides may have no effect on the amino acid (depending on what it is changed to), while changing to the fourth possible nucleotide always results in an amino acid substitution. This is the third position of an isoleucine codon: AUU, AUC, or AUA all encode isoleucine, but AUG encodes methionine. In computation this position is often treated as a twofold degenerate site.

There are three amino acids encoded by six different codons: serine, leucine, and arginine. Only two amino acids are specified by a single codon. One of these is the amino-acid methionine, specified by the codon AUG, which also specifies the start of translation; the other is tryptophan, specified by the codon UGG. The degeneracy of the genetic code is what accounts for the existence of synonymous mutations.

Degeneracy results because there are more codons than encodable amino acids. For example, if there were two bases per codon, then only 16 amino acids could be coded for ($4^2=16$). Because at least 21 codes are required (20 amino acids plus stop), and the next largest number of bases is three, then 4^3 gives 64 possible codons, meaning that some degeneracy must exist. These properties of the genetic code make it more fault-tolerant for point mutations. For example, in theory, fourfold degenerate codons can tolerate any point mutation at the third position, although codon usage bias restricts this in practice in many organisms; twofold degenerate codons can tolerate one out of the three possible point mutations at the third position. Since transition mutations (purine to purine or pyrimidine to pyrimidine mutations) are more likely than transversion (purine to pyrimidine or vice-versa) mutations, the equivalence of purines or that of pyrimidines at twofold degenerate sites adds a further fault-tolerance.

A practical consequence of redundancy is that some errors in the genetic code only cause a silent mutation or an error that would not affect the protein because the hydrophilicity or hydrophobicity is maintained by equivalent substitution of amino acids; for example, a codon of NUN (where N = any nucleotide) tends to code for hydrophobic amino acids. NCN yields amino acid residues that are small in size and moderate in hydropathy; NAN encodes average size hydrophilic residues. These tendencies may result from the shared ancestry of the aminoacyl tRNA synthetases related to these codons.

Even so, single point mutations can still cause dysfunctional proteins. For example, a mutated hemoglobin gene causes sickle-cell disease. In the mutant hemoglobin a hydrophilic glutamate (Glu) is substituted by the hydrophobic valine (Val), that is, GAA or GAG becomes GUA or GUG. The substitution of glutamate by valine reduces the solubility of β-globin which causes hemoglobin to form linear polymers linked by the hydrophobic interaction between the valine groups causing sickle-cell deformation of erythrocytes. Sickle-cell disease is generally not caused by a *de novo* mutation. Rather it is selected for in malarial regions (in a way similar to thalassemia), as heterozygous people have some resistance to the malarial *Plasmodium* parasite (heterozygote advantage).

These variable codes for amino acids are allowed because of modified bases in the first base of the anticodon of the tRNA, and the base-pair formed is called a wobble base pair. The modified bases include inosine and the Non-Watson-Crick U-G basepair.

Variations to the Standard Genetic Code

While slight variations on the standard code had been predicted earlier, none were discovered until 1979, when researchers studying human mitochondrial genes discovered they used an alternative code. Many slight variants have been discovered since, including various alternative mitochondrial codes, as well as small variants such as *Mycoplasma* translating the codon UGA as tryptophan and *Candida* species translating CUG as a serine rather than a leucine. In bacteria and archaea, GUG and UUG are common start codons. However, in rare cases, certain specific proteins may use alternative initiation (start) codons not normally used by that species.

In certain proteins, non-standard amino acids are substituted for standard stop codons, depending upon associated signal sequences in the messenger RNA: UGA can code for selenocysteine and UAG can code for pyrrolysine as discussed in the relevant articles. Selenocysteine is now viewed as the 21st amino acid, and pyrrolysine is viewed as the 22nd.

Notwithstanding these differences, all known codes have strong similarities to each other, and the coding mechanism is the same for all organisms: three-base codons, tRNA, ribosomes, reading the code in the same direction and translating the code three letters at a time into sequences of amino acids.

Expanded Genetic Code

Since 2001, 40 non-natural amino acids have been added into protein by creating a unique codon (recoding) and a corresponding transfer-RNA:aminoacyl – tRNA-synthetase pair to encode it with diverse physicochemical and biological properties in order to be used as a tool to exploring protein structure and function or to create novel or enhanced proteins.

Theories on the Origin of the Genetic Code

Despite the minor variations that exist, the genetic code used by all known forms of life is nearly universal. However, there are a huge number of possible genetic codes.

Phylogenetic analysis of transfer RNA suggests that tRNA molecules evolved before the present set of aminoacyl-tRNA synthetases.

Theoretically the genetic code could be completely random (a "frozen accident"), completely non-random (optimal) or a combination of random and nonrandom. There are sufficient data to refute the first possibility. For a start, a quick view on the table of the genetic code already shows a clustering of amino acid assignments. Furthermore, amino acids that share the same biosynthetic pathway tend to have the same first base in their codons, and amino acids with similar physical properties tend to have similar codons.

There are four themes running through the many theories that seek to explain the evolution of the genetic code (and hence the origin of these patterns):

- *Chemical Principles Govern Specific RNA Interaction with Amino Acids:* Aptamer experiments showed that some amino acids have a selective chemical affinity for the base triplets that code for them. Recent experiments show that of the 8 amino acids tested, 6 show some RNA triplet-amino acid association.

  This has been called the stereochemical code. The stereochemical code could have created an ancient core of assignments. The current complex translation mechanism involving tRNA and associated enzymes may be a later development, and that originally, protein sequences were directly templated on base sequences.

- *Biosynthetic Expansion:* The standard modern genetic code grew from a simpler earlier code through a process of "biosynthetic expansion". Here the idea is that primordial life "discovered" new amino acids (e.g., as by-products of metabolism) and later back-incorporated some of these into the machinery of genetic coding. Although much circumstantial evidence has been found to suggest that fewer different amino acids were used in the past than today, precise and detailed hypotheses about exactly which amino acids entered the code in exactly what order have proved far more controversial.
- Natural selection has led to codon assignments of the genetic code that minimize the effects of mutations. A recent hypothesis suggests that the triplet code was derived from codes that used longer than triplet codons. Longer than triplet decoding has higher degree of codon redundancy and is more error resistant than the triplet decoding. This feature could allow accurate decoding in the absence of highly complex translational machinery such as the ribosome.
- *Information Channels:* Information-theoretic approaches see the genetic code as an error-prone information channel. The inherent noise (i.e. errors) in the channel poses the organism with a fundamental question: how to construct a genetic code that can withstand the impact of noise while accurately and efficiently translating information? These "rate-distortion" models suggest that the genetic code originated as a result of the interplay of the three conflicting evolutionary forces: the needs for diverse amino-acids, for error-tolerance and for minimal cost of resources. The code emerges at a coding transition when the mapping of codons to amino-acids becomes nonrandom. The emergence of the code is governed by the topology defined by the probable errors and is related to the map coloring problem.

Central Dogma of Molecular Biology

The dogma is a framework for understanding the transfer of sequence information between sequential information-carrying biopolymers, in the most common or general case, in living organisms. There are 3 major classes of such biopolymers: DNA and RNA (both nucleic acids), and protein.

There are 3×3 = 9 conceivable direct transfers of information that can occur between these. The dogma classes these into 3 groups of 3: 3 general transfers (believed to occur normally in most cells), 3 special transfers (known to occur, but only under specific conditions in case of some viruses or in a laboratory), and 3 unknown transfers (believed never to occur). The general transfers describe the normal flow of biological information: DNA can be copied to DNA (DNA replication), DNA information can be copied into mRNA, (transcription), and proteins can be synthesized using the information in mRNA as a template (translation).

Biological Sequence Information

Primary Structure

In biochemistry, the Primary Structure of a biological molecule is the exact specification of its atomic composition and the chemical bonds connecting those atoms (including stereochemistry). For a typical unbranched, un-crosslinked biopolymer (such as a molecule of DNA, RNA or typical intracellular protein), the primary structure is equivalent to specifying the sequence of its monomeric subunits, e.g., the nucleotide or peptide sequence.

Primary structure is sometimes mistakenly termed *primary sequence*, but there is no such term, as well as no parallel concept of secondary or tertiary sequence. By convention, the primary structure of a protein is reported starting from the amino-terminal (N) end to the carboxyl-terminal (C) end, while the primary structure of DNA or RNA molecule is reported from the 5' end to the 3' end.

The primary structure of a nucleic acid molecule refers to the exact sequence of nucleotides that comprise the whole molecule. Frequently the primary structure encodes motifs that are of functional importance. Some examples of sequence motifs are: the C/D and H/ACA boxes of snoRNAs, Sm binding site found in spliceosomal RNAs such as U1, U2, U4, U5, U6, U12 and U3, the Shine-Dalgarno sequence, the Kozak consensus sequence and the RNA polymerase III terminator.

Secondary Structure

In biochemistry and structural biology, secondary structure is the general three-dimensional form of *local segments* of biopolymers such as proteins and nucleic acids (DNA/RNA). It does not, however, describe specific atomic positions in three-dimensional space, which are considered to be tertiary structure. Secondary structure is formally defined by the hydrogen bonds of the biopolymer, as observed in an atomic-resolution structure. In proteins, the secondary structure is defined by patterns of hydrogen bonds between backbone amide and carboxyl groups (sidechain-mainchain and sidechain-sidechain hydrogen bonds are irrelevant), where the DSSP definition of a hydrogen bond is used. In nucleic acids, the secondary structure is defined by the hydrogen bonding between the nitrogenous bases.

For proteins, however, the hydrogen bonding is correlated with other structural features, which has given rise to less formal definitions of secondary structure. For example, residues in protein helices generally adopt backbone dihedral angles in a particular region of the Ramachandran plot; thus, a segment of residues with such dihedral angles is often called a "helix", regardless of whether it has the correct hydrogen bonds. Many other less formal definitions have been proposed, often applying concepts from the differential geometry of curves, such as curvature and torsion. Least formally, structural biologists solving a new atomic-resolution structure will sometimes assign its secondary structure "by eye" and record their assignments in the corresponding PDB file.

The secondary structure of a nucleic acid molecule refers to the basepairing interactions within a single molecule or set of interacting molecules. The secondary structure of biological RNA's can often be uniquely decomposed into stems and loops. Frequently these elements, or combinations of them, can be further classified, for example, tetraloops, pseudoknots and stem-loops. There are many secondary structure elements of functional importance to biological RNA's; some famous examples are the Rho-independent terminator stem-loops and the tRNA cloverleaf. There is a minor industry of researchers attempting to determine the secondary structure of RNA molecules. Approaches include both experimental and computational methods.

Tertiary Structure

In biochemistry and molecular biology, the tertiary structure of a protein or any other macromolecule is its three-dimensional structure, as defined by the atomic coordinates. Proteins and nucleic acids are capable of diverse functions ranging from molecular recognition to catalysis. Such functions require a precise three-dimensional tertiary structure. While such structures are diverse and seemingly complex, they are composed of recurring, easily recognizable tertiary structure motifs that serve as molecular building blocks. Tertiary structure is considered to be largely determined by the biomolecule's primary structure, or the sequence of amino acids or nucleotides of which it is composed. Efforts to predict tertiary structure from the primary structure are known generally as structure prediction.

Quaternary Structure

In biochemistry, quaternary structure is the arrangement of multiple folded protein or coiling protein molecules in a multi-subunit complex. For nucleic acids, the term is less common, but can refer to the higher-level organization of DNA in chromatin, including its interactions with histones, or to the interactions between separate RNA units in the ribosome or spliceosome.

Structure Determination

Structure probing is the process by which biochemical techniques are used to determine biomolecular structure. This analysis can be used to define the patterns which can infer the molecular structure, experimental analysis of molecular structure and function, and further understanding on development of smaller molecules for further biological research. Structure probing analysis can be done through many different methods, which include chemical probing, hydroxyl radical probing, nucleotide analog interference mapping (NAIM), and in-line probing.

DNA structures can be determined using either nuclear magnetic resonance spectroscopy or X-ray crystallography. The first published reports of A-DNA X-ray diffraction patterns— and also B-DNA—employed analyses based on Patterson transforms that provided only a limited amount of structural information for oriented fibres of DNA isolated from calf thymus. An alternate analysis was then proposed by Wilkins et al. in 1953 for B-DNA X-ray diffraction/ scattering patterns of hydrated, bacterial oriented DNA fibres and trout sperm heads in terms of squares of Bessel functions. Although the 'B-DNA form' is most common under the conditions found in cells, it is not a well-defined conformation but a family or fuzzy set of DNA-conformations that occur at the high hydration levels present in a wide variety of living cells. Their corresponding X-ray diffraction & scattering patterns are characteristic of molecular paracrystals with a significant degree of disorder (>20%), and concomitantly the structure is not tractable using only the standard analysis.

On the other hand, the standard analysis, involving only Fourier transforms of Bessel functions and DNA molecular models, is still routinely employed for the analysis of A-DNA and Z-DNA X-ray diffraction patterns.

Structure Prediction

Biomolecular structure prediction is the prediction of the three-dimensional structure of a protein from its amino acid sequence, or of a nucleic acid from its base sequence. In other words, it is the prediction of secondary and tertiary structure from its primary structure. Structure prediction is the inverse of biomolecular design. Protein structure prediction is one of the most important goals pursued by bioinformatics and theoretical chemistry. Protein structure prediction is of high importance in medicine (for example, in drug design) and biotechnology (for example, in the design of novel enzymes). Every two years, the performance of current methods is assessed in the CASP experiment.

There has also been a significant amount of bioinformatics research directed at the RNA structure prediction problem. A common problem for researchers working with RNA is to determine the three-dimensional structure of the molecule given just the nucleic acid sequence. However, in the case of RNA much of the final structure is determined by the secondary structure or intra-molecular base-pairing interactions of the molecule. This is shown by the high conservation of base-pairings across diverse species.

Secondary structure of small nucleic acid molecules is largely determined by strong, local interactions such as hydrogen bonds and base stacking. Summing the free energy for such interactions, usually using a nearest-neighbour model, provides an approximation for the stability of given structure. The most straighforward way to find the lowest free energy structure would be to generate all possible structures and calculate the free energy for it, but the number of possible structures for a sequence increases exponentially with the length of the nucleic acid. For longer molecules, the number of possible secondary structures is enormous.

Sequence covariation methods rely on the existence of a data set composed of multiple homologous RNA sequences with related but dissimilar sequences. These methods analyse the covariation of individual base sites in evolution; maintenance at two widely separated sites of a pair of base-pairing nucleotides indicates the presence of a structurally required hydrogen bond between those positions. The general problem of pseudoknot prediction has been shown to be NP-complete.

Design

Biomolecular design can be considered the inverse of structure prediction. In structure prediction, the structure is determined from a known sequence, while in nucleic acid design, a sequence is generated which will form a desired structure.

Other Biomolecules

Other biomolecules, such as polysaccharides and lipids, can also have higher-order structure of biological consequence.

DNA Replication

DNA replication, the basis for biological inheritance, is a fundamental process that occurs in all living organisms that copies their DNA. This process is "replication" in that each strand of the original double-stranded DNA molecule serves as template for the reproduction of the complementary strand. Therefore, following DNA replication, two identical DNA molecules have been produced from a single double-stranded DNA molecule. Cellular proofreading and error toe-checking mechanisms ensure near perfect fidelity for DNA replication.

In a cell, DNA replication begins at specific locations in the genome, called "origins". Unwinding of DNA at the origin, and synthesis of new strands, forms a replication fork. In addition to DNA polymerase, the enzyme that synthesizes the new DNA by adding nucleotides matched to the template strand, a number of other proteins are associated with the fork and assist in the initiation and continuation of DNA synthesis.

DNA replication can also be performed *in vitro* (outside a cell). DNA polymerases, isolated from cells, and artificial DNA primers are used to initiate DNA synthesis at known sequences in a template molecule. The polymerase chain reaction (PCR), a common laboratory technique, employs such artificial synthesis in a cyclic manner to amplify a specific target DNA fragment from a pool of DNA.

DNA Structure

DNA usually exists as a double-stranded structure, with both strands coiled together to form the characteristic double-helix. Each single strand of DNA is a chain of four types of nucleotides: adenine, cytosine, guanine, and thymine. A nucleotide is a mono-, di- or triphosphate deoxyribonucleoside; that is, a deoxyribose sugar is attached to one, two or three phosphates. Chemical interaction of these nucleotides forms phosphodiester linkages, creating the phosphate-deoxribose backbone of the DNA double helix with the bases pointing inward. Nucleotides (bases) are matched between strands through hydrogen bonds to form base pairs. Adenine pairs with thymine and cytosine pairs with guanine.

DNA strands have a directionality, and the different ends of a single strand are called the "3' (three-prime) end" and the "5' (five-prime) end." These terms refer to the carbon atom in deoxyribose to which the next phosphate in the chain attaches. In addition to being complementary, the two strands of DNA are antiparallel: they are orientated in opposite directions. This directionality has consequences in DNA synthesis, because DNA polymerase can only synthesize DNA in one direction by adding nucleotides to the 3' end of a DNA strand. The pairing of bases in DNA through hydrogen bonding means that the information contained within each strand is redundant. The nucleotides on a single strand can be used to reconstruct nucleotides on a newly synthesized partner strand.

DNA Polymerase

A DNA polymerase is an enzyme that catalyses the polymerization of deoxyribonucleotides into a DNA strand. DNA polymerases are best-known for their feedback role in DNA replication, in which the polymerase "reads" an intact DNA strand as a template and uses it to synthesize the new strand. This process copies a piece of DNA. The newly-polymerized molecule is complementary to the template strand and identical to the template's original partner strand. DNA polymerases use magnesium ions as cofactors. Human DNA polymerases are 900-1000 amino acids long.

Function

DNA polymerase can add free nucleotides to only the 3' end of the newly-forming strand. This results in elongation of the new strand in a 5'-3' direction. No known DNA polymerase is able to begin a new chain (*de novo*). DNA polymerase can add a nucleotide onto only a preexisting 3'-OH group, and, therefore, needs a primer at which it can add the first nucleotide.

Primers consist of RNA and/or DNA bases. In DNA replication, the first two bases are always RNA, and are synthesized by another enzyme called primase. An enzyme known as a helicase is required to unwind DNA from a double-strand structure to a single-strand structure to facilitate replication of each strand consistent with the semiconservative model of DNA replication.

Error correction is a property of some, but not all, DNA polymerases. This process corrects mistakes in newly-synthesized DNA. When an incorrect base pair is recognized, DNA polymerase reverses its direction by one base pair of DNA. The 3'-5' exonuclease activity of the enzyme allows the incorrect base pair to be excised (this activity is known as *proofreading*). Following base excision, the polymerase can re-insert the correct base and replication can continue. Various DNA polymerases are extensively used in molecular biology experiments.

Variation Across Species

DNA polymerases have highly-conserved structure, which means that their overall catalytic subunits vary, on a whole, very little from species to species. Conserved structures usually indicate important, irreplaceable functions of the cell, the maintenance of which provides evolutionary advantages.

Some viruses also encode special DNA polymerases, such as Hepatitis B virus DNA polymerase. These may selectively replicate viral DNA through a variety of mechanisms. Retroviruses encode an unusual DNA polymerase called reverse transcriptase, which is an RNA-dependent DNA polymerase (RdDp). It polymerizes DNA from a template of RNA.

DNA Polymerase Families

Based on sequence homology, DNA polymerases can be further subdivided into seven different families: A, B, C, D, X, Y, and RT.

Family A

Polymerases contain both replicative and repair polymerases. Replicative members from this family include the extensively-studied T7 DNA polymerase, as well as the eukaryotic mitochondrial DNA Polymerase γ. Among the repair polymerases are *Escherichia coli* DNA pol I, *Thermus aquaticus* pol I, and *Bacillus stearothermophilus* pol I. These repair polymerases are involved in excision repair and processing of Okazaki fragments generated during lagging strand synthesis.

Family C

Polymerases are the primary bacterial chromosomal replicative enzymes. DNA Polymerase III alpha subunit from *E. coli* is the catalytic subunit and possesses no known nuclease activity. A separate subunit, the epsilon subunit, possesses the 3'-5' exonuclease activity used for editing during chromosomal replication. Recent research has classified Family C polymerases as a subcategory of Family X.

Family D

Polymerases are still not very well characterized. All known examples are found in the Euryarchaeota subdomain of Archaea and are thought to be replicative polymerases.

Family X

Contains the well-known eukaryotic polymerase pol β, as well as other eukaryotic polymerases such as pol σ, pol λ, pol μ, and terminal deoxynucleotidyl transferase (TdT). Pol β is required for short-patch base excision repair, a DNA repair pathway that is essential for repairing abasic sites. Pol λ and Pol μ are involved in non-homologous end-joining, a mechanism for rejoining DNA double-strand breaks. TdT is expressed only in lymphoid tissue, and adds "n nucleotides" to double-strand breaks formed during V(D)J recombination to promote immunological diversity. The yeast *Saccharomyces cerevisiae* has only one Pol X polymerase, Pol IV, which is involved in non-homologous end-joining.

Family Y

Y Polymerases differ from others in having a low fidelity on undamaged templates and in their ability to replicate through damaged DNA. Members of this family are hence called translesion synthesis (TLS) polymerases. Depending on the lesion, TLS polymerases can bypass the damage in an error-free or error-prone fashion, the latter resulting in elevated mutagenesis. Xeroderma pigmentosum variant (XPV) patients for instance have mutations in the gene encoding Pol η (eta), which is error-free for UV-lesions.

In XPV patients, alternative error-prone polymerases, e.g., Pol ζ (zeta) (polymerase ζ is a B Family polymerase a complex of the catalytic subunit REV3L with Rev7, which associates with Rev1), are thought to be involved in mistakes that result in the cancer predisposition of these patients. Other members in humans are Pol ι (iota), Pol κ (kappa), and Rev1 (terminal deoxycytidyl transferase). In *E. coli*, two TLS polymerases, Pol IV (DINB) and Pol V ($UmuD'_2C$), are known.

Family RT

The reverse transcriptase family contains examples from both retroviruses and eukaryotic polymerases. The eukaryotic polymerases are usually restricted to telomerases. These polymerases use an RNA template to synthesize the DNA strand.

Prokaryotic DNA Polymerase

There are 5 known Prokaryotic DNA polymerases:

- Pol I: implicated in DNA repair; has 5'->3' polymerase activity, and both 3'->5' exonuclease activity (proofreading) and 5'->3' exonuclease activity (RNA primer removal).

- Pol II: involved in repairing damaged DNA; has 3'->5' exonuclease activity.
- Pol III: the main polymerase in bacteria (responsible for elongation); has 3'->5' exonuclease activity (proofreading).
- Pol IV: a Y-family DNA polymerase.
- Pol V: a Y-family DNA polymerase; participates in bypassing DNA damage.

Eukaryotic DNA Polymerase

There are at least 15 Eukaryotic DNA polymerase:

- POLA1, POLA2: Pol α (also called RNA primase): forms a complex with a small catalytic (PriS) and a large noncatalytic (PriL) subunit, with the Pri subunits acting as a primase (synthesizing an RNA primer), and then with DNA Pol α elongating that primer with DNA nucleotides. After around 20 nucleotides elongation is taken over by Pol ε (on the leading strand) and α (on the lagging strand).
- POLB: Pol β: Implicated in repairing DNA, in base excision repair and gap-filling synthesis.
- POLG, POLG2: Pol γ: Replicates and repairs mitochondrial DNA and has proofreading 3'->5' exonuclease activity.
- POLD1, POLD2, POLD3, POLD4: Pol δ: Highly processive and has proofreading 3'->5' exonuclease activity. Thought to be the main polymerase involved in lagging strand synthesis, though there is still debate about its role.
- POLE, POLE2, POLE3: Pol ε: Also highly processive and has proofreading 3'->5' exonuclease activity. Highly related to pol δ, and thought to be the main polymerase involved in leading strand synthesis, though there is again still debate about its role.
- POLH, POLI, POLK, : η, ι, κ, and Rev1 are Y-family DNA polymerases and Pol ζ is a B-family DNA polymerase. These polymerases are involved in the bypass of DNA damage.

None of the eukaryotic polymerases can remove primers (5'->3' exonuclease activity); that function is carried out by other enzymes. Only the polymerases that deal with the elongation (γ, δ and ε) have proofreading ability (3'->5' exonuclease).

Origins of Replication

For a cell to divide, it must first replicate its DNA. This process is initiated at particular points within the DNA, known as "origins", which are targeted by proteins that separate the two strands and initiate DNA synthesis. Origins contain DNA sequences recognized by replication initiator proteins (e.g. dnaA in *E coli'* and the Origin Recognition Complex in yeast). These initiator proteins recruit other proteins to separate the two strands and initiate replication forks. Initiator proteins recruit other proteins to separate the DNA strands at the origin, forming a bubble. Origins tend to be "AT-rich" (rich in adenine and thymine bases) to assist this process, because A-T base pairs have two hydrogen bonds (rather than the three formed in a C-G pair)—strands rich in these nucleotides are generally easier to separate due the positive relationship between the number of hydrogen bonds and the difficulty of breaking these bonds. Once strands are separated, RNA primers are created on the template strands. More specifically, the leading strand receives one RNA primer per active origin of replication while the lagging strand receives several; these several fragments of RNA primers found on the lagging strand of DNA are called Okazaki fragments, named after their discoverer.

DNA Polymerase extends the leading strand in one continuous motion and the lagging strand in a discontinuous motion (due to the Okazaki fragments). RNase removes the RNA fragments used to initiate replication by DNA Polymerase, and another DNA Polymerase enters to fill the gaps. When this is complete, a single nick on the leading strand and several nicks on the lagging strand can be found. Ligase works to fill these nicks in, thus completing the newly replicated DNA molecule.

As DNA synthesis continues, the original DNA strands continue to unwind on each side of the bubble, forming 2 replication forks. In bacteria, which have a single origin of replication on their circular chromosome, this process eventually creates a "theta structure" (resembling the Greek letter theta: θ). In contrast, eukaryotes have longer linear chromosomes and initiate replication at multiple origins within these.

Replication Fork

The replication fork is a structure that forms within the nucleus during DNA replication. It is created by helicases, which break the hydrogen bonds holding the two DNA strands together. The resulting structure has two branching "prongs", each one made up of a single strand of DNA. These two strands serve as the template for the leading and lagging strands which will be created as DNA polymerase matches complementary nucleotides to the templates. The templates may be properly referred to as the leading strand template and the lagging strand template.

Replication

When replicating, the original DNA splits in two, forming two "prongs" which resemble a fork (hence the name "replication fork"). DNA has a ladder-like structure; imagine a ladder broken in half vertically, along the steps. Each half of the ladder now requires a new half to match it. Because DNA polymerase can only synthesize a new DNA strand in a 5' to 3' manner, the process of replication goes differently for the two strands comprising the DNA double helix.

Leading Strand

The leading strand template is the template strand of the DNA double helix that is oriented in a 3' to 5' manner. All DNA synthesis occurs 5'-3'. The original DNA strand must be read 3'-5' to produce a 5'-3' nascent strand.

The leading strand is formed along the leading strand template as a polymerase "reads" the template DNA and continuously adds nucleotides to the 3' end of the elongating strand. This polymerase is DNA polymerase III (DNA Pol III) in prokaryotes and presumably Pol ε in eukaryotes.

Lagging Strand

The lagging strand template is the coding strand of the DNA double helix that is oriented in a 5' to 3' manner. The newly made lagging strand still is synthesized 5'-3'. However, since the DNA is oriented in a manner that does not allow continual synthesis, only small sections can be read at a time.

An RNA primer is placed on the DNA strand 3' to the origin of replication. Just as before, DNA Polymerase reads 3'-5' on the original DNA to produce a 5'-3' nascent strand. Polymerase reaches the origin of replication and stops replication until a new RNA primer is placed 3' to the last RNA primer. These fragments of DNA produced on the lagging strand are called Okazaki fragments. The orientation of the original DNA on the lagging strand prevents continual synthesis. As a result, replication of the lagging strand is more complicated than of the leading strand.

On the lagging strand template, primase "reads" the DNA and adds RNA to it in short, separated segments. In eukaryotes, primase is intrinsic to Pol α. DNA polymerase III or Pol δ lengthens the primed segments, forming Okazaki fragments.

Primer removal in eukaryotes is also performed by Pol δ. In prokaryotes, DNA polymerase I "reads" the fragments, removes the RNA using its flap endonuclease domain, and replaces the RNA nucleotides with DNA nucleotides (this is necessary because RNA and DNA use slightly different kinds of nucleotides). DNA ligase joins the fragments together.

Leading Strand

The leading strand is the template strand of the DNA double helix so that the replication fork moves along it in the 3' to 5' direction. This allows the new strand synthesized complementary to it to be synthesized 5' to 3' in the same direction as the movement of the replication fork.

On the leading strand, a polymerase "reads" the DNA and adds nucleotides to it continuously. This polymerase is DNA polymerase III (DNA Pol III) in prokaryotes and presumably Pol ε in eukaryotes.

Lagging Strand

The lagging strand is the strand of the template DNA double helix that is oriented so that the replication fork moves along it in a 5' to 3' manner. Because of its orientation, opposite to the working orientation of DNA polymerase III, which moves on a template in a 3' to 5' manner, replication of the lagging strand is more complicated than that of the leading strand.

On the lagging strand, primase "reads" the DNA and adds RNA to it in short, separated segments. In eukaryotes, primase is intrinsic to Pol α. DNA polymerase III or Pol δ lengthens the primed segments, forming Okazaki fragments. Primer removal in eukaryotes is also performed by Pol δ. In prokaryotes, DNA polymerase I "reads" the fragments, removes the RNA using its flap endonuclease domain (RNA primers are removed by 5'-3' exonuclease activity of polymerase I [weaver, 2005], and replaces the RNA nucleotides with DNA nucleotides (this is necessary because RNA and DNA use slightly different kinds of nucleotides). DNA ligase joins the fragments together.

Dynamics at the Replication Fork

As helicase unwinds DNA at the replication fork, the DNA ahead is forced to rotate. This process results in a build-up of twists in the DNA ahead. This build-up would form a resistance that would eventually halt the progress of the replication fork. DNA topoisomerases are enzymes that solve these physical problems in the coiling of DNA. Topoisomerase I cuts a single backbone on the DNA, enabling the strands to swivel around each other to remove the build-up of twists. Topoisomerase II cuts both backbones, enabling one double-stranded DNA to pass through another, thereby removing knots and entanglements that can form within and between DNA molecules.

Bare single-stranded DNA has a tendency to fold back upon itself and form secondary structures; these structures can interfere with the movement of DNA polymerase. To prevent this, single-strand binding proteins bind to the DNA until a second strand is synthesized, preventing secondary structure formation.

Clamp proteins form a sliding clamp around DNA, helping the DNA polymerase maintain contact with its template and thereby assisting with processivity. The inner face of the clamp enables DNA to be threaded through it. Once the polymerase reaches the end of the template or detects double stranded DNA, the sliding clamp undergoes a conformational change which releases the DNA polymerase. Clamp-loading proteins are used to initially load the clamp, recognizing the junction between template and RNA primers.

Regulation of Replication

Within eukaryotes, DNA replication is controlled within the context of the cell cycle. As the cell grows and divides, it progresses through stages in the cell cycle; DNA replication occurs during the S phase (Synthesis phase). The progress of the eukaryotic cell through the cycle is controlled by cell cycle checkpoints. Progression through checkpoints is controlled through complex interactions between various proteins, including cyclins and cyclin-dependent kinases.

The G1/S checkpoint (or restriction checkpoint) regulates whether eukaryotic cells enter the process of DNA replication and subsequent division. Cells which do not proceed through this checkpoint are quiescent in the "G0" stage and do not replicate their DNA.

Replication of chloroplast and mitochondrial genomes occurs independent of the cell cycle, through the process of D-loop replication.

Bacteria

Most bacteria do not go through a well-defined cell cycle and instead continuously copy their DNA; during rapid growth this can result in multiple rounds of replication occurring concurrently. Within *E coli*, the most well-characterized bacteria, regulation of DNA replication can be achieved through several mechanisms, including: the hemimethylation and sequestering of the origin sequence, the ratio of ATP to ADP, and the levels of protein DnaA. These all control the process of initiator proteins binding to the origin sequences.

Because *E coli* methylates GATC DNA sequences, DNA synthesis results in hemimethylated sequences. This hemimethylated DNA is recognized by a protein (SeqA) which binds and sequesters the origin sequence; in addition, dnaA (required for initiation of replication) binds less well to hemimethylated DNA. As a result, newly replicated origins are prevented from immediately initiating another round of DNA replication. ATP builds up when the cell is in a rich medium, triggering DNA replication once the cell has reached a specific size. ATP competes with ADP to bind to DnaA, and the DnaA-ATP complex is able to initiate replication. A certain number of DnaA proteins are also required for DNA replication — each time the origin is copied the number of binding sites for DnaA doubles, requiring the synthesis of more DnaA to enable another initiation of replication.

Termination of Replication

Because bacteria have circular chromosomes, termination of replication occurs when the two replication forks meet each other on the opposite end of the parental chromosome. *E coli* regulate this process through the use of termination sequences which, when bound by the Tus protein, enable only one direction of replication fork to pass through. As a result, the replication forks are constrained to always meet within the termination region of the chromosome.

Eukaryotes initiate DNA replication at multiple points in the chromosome, so replication forks meet and terminate at many points in the chromosome; these are not known to be regulated in any particular manner. Because eukaryotes have linear chromosomes, DNA replication often fails to synthesize to the very end of the chromosomes (telomeres), resulting in telomere shortening. This is a normal process in somatic cells — cells are only able to divide a certain number of times before the DNA loss prevents further division. (This is known as the Hayflick limit.) Within the germ cell line, which passes DNA to the next generation, telomerase extends the repetitive sequences of the telomere region to prevent degradation. Telomerase can become mistakenly active in somatic cells, sometimes leading to cancer formation.

Leading Strand

The leading strand is the template strand of the DNA double helix so that the replication fork moves along it in the 3' to 5' direction. This allows the new strand synthesized complementary to it to be synthesized 5' to 3' in the same direction as the movement of the replication fork.

On the leading strand, a polymerase "reads" the DNA and adds nucleotides to it continuously. This polymerase is DNA polymerase III (DNA Pol III) in prokaryotes and presumably Pol ε in eukaryotes.

Lagging Strand

The lagging strand is the strand of the template DNA double helix that is oriented so that the replication fork moves along it in a 5' to 3' manner. Because of its orientation, opposite to the working orientation of DNA polymerase III, which moves on a template in a 3' to 5' manner, replication of the lagging strand is more complicated than that of the leading strand.

On the lagging strand, primase "reads" the DNA and adds RNA to it in short, separated segments. In eukaryotes, primase is intrinsic to Pol α. DNA polymerase III or Pol δ lengthens the primed segments, forming Okazaki fragments. Primer removal in eukaryotes is also performed by Pol δ. In prokaryotes, DNA polymerase I "reads" the fragments, removes the RNA using its flap endonuclease domain (RNA primers are removed by 5'-3' exonuclease activity of polymerase I [weaver, 2005], and replaces the RNA nucleotides with DNA nucleotides (this is necessary because RNA and DNA use slightly different kinds of nucleotides). DNA ligase joins the fragments together.

Dynamics at the Replication Fork

As helicase unwinds DNA at the replication fork, the DNA ahead is forced to rotate. This process results in a build-up of twists in the DNA ahead. This build-up would form a resistance that would eventually halt the progress of the replication fork. DNA topoisomerases are enzymes that solve these physical problems in the coiling of DNA. Topoisomerase I cuts a single backbone on the DNA, enabling the strands to swivel around each other to remove the build-up of twists. Topoisomerase II cuts both backbones, enabling one double-stranded DNA to pass through another, thereby removing knots and entanglements that can form within and between DNA molecules.

Bare single-stranded DNA has a tendency to fold back upon itself and form secondary structures; these structures can interfere with the movement of DNA polymerase. To prevent this, single-strand binding proteins bind to the DNA until a second strand is synthesized, preventing secondary structure formation. Clamp proteins form a sliding clamp around DNA, helping the DNA polymerase maintain contact with its template and thereby assisting with processivity. The inner face of the clamp enables DNA to be threaded through it. Once the polymerase reaches the end of the template or detects double stranded DNA, the sliding clamp undergoes a conformational change which releases the DNA polymerase. Clamp-loading proteins are used to initially load the clamp, recognizing the junction between template and RNA primers.

Regulation of Replication

Within eukaryotes, DNA replication is controlled within the context of the cell cycle. As the cell grows and divides, it progresses through stages in the cell cycle; DNA replication occurs during the S phase (Synthesis phase). The progress of the eukaryotic cell through the cycle is controlled by cell cycle checkpoints. Progression through checkpoints is controlled through complex interactions between various proteins, including cyclins and cyclin-dependent kinases. The G1/S checkpoint (or restriction checkpoint) regulates whether eukaryotic cells enter the process of DNA replication and subsequent division. Cells which do not proceed through this checkpoint are quiescent in the "G0" stage and do not replicate their DNA. Replication of chloroplast and mitochondrial genomes occurs independent of the cell cycle, through the process of D-loop replication.

Bacteria

Most bacteria do not go through a well-defined cell cycle and instead continuously copy their DNA; during rapid growth this can result in multiple rounds of replication occurring concurrently. Within *E coli*, the most well-characterized bacteria, regulation of DNA replication can be achieved through several mechanisms, including: the hemimethylation and sequestering of the origin sequence, the ratio of ATP to ADP, and the levels of protein DnaA. These all control the process of initiator proteins binding to the origin sequences.

Because *E coli* methylates GATC DNA sequences, DNA synthesis results in hemimethylated sequences. This hemimethylated DNA is recognized by a protein (SeqA) which binds and sequesters the origin sequence; in addition, dnaA (required for initiation of replication) binds less well to hemimethylated

DNA. As a result, newly replicated origins are prevented from immediately initiating another round of DNA replication.

ATP builds up when the cell is in a rich medium, triggering DNA replication once the cell has reached a specific size. ATP competes with ADP to bind to DnaA, and the DnaA-ATP complex is able to initiate replication. A certain number of DnaA proteins are also required for DNA replication — each time the origin is copied the number of binding sites for DnaA doubles, requiring the synthesis of more DnaA to enable another initiation of replication.

Termination of Replication

Because bacteria have circular chromosomes, termination of replication occurs when the two replication forks meet each other on the opposite end of the parental chromosome. *E coli* regulate this process through the use of termination sequences which, when bound by the Tus protein, enable only one direction of replication fork to pass through. As a result, the replication forks are constrained to always meet within the termination region of the chromosome.

Eukaryotes initiate DNA replication at multiple points in the chromosome, so replication forks meet and terminate at many points in the chromosome; these are not known to be regulated in any particular manner. Because eukaryotes have linear chromosomes, DNA replication often fails to synthesize to the very end of the chromosomes (telomeres), resulting in telomere shortening. This is a normal process in somatic cells — cells are only able to divide a certain number of times before the DNA loss prevents further division. (This is known as the Hayflick limit.) Within the germ cell line, which passes DNA to the next generation, telomerase extends the repetitive sequences of the telomere region to prevent degradation. Telomerase can become mistakenly active in somatic cells, sometimes leading to cancer formation.

DNA-binding Protein

DNA-binding proteins are proteins that are composed of DNA-binding domains and thus have a specific or general affinity for either single or double stranded DNA. Sequence-specific DNA-binding proteins generally interact with the major groove of B-DNA, because it exposes more functional groups that identify a base pair. However there are some known narrow-groove DNA-binding ligands such as Netropsin, Distamycin, Hoechst 33258, Pentamidine and others.

Examples

DNA-binding proteins include transcription factors which modulate the process of transcription, various polymerases, nucleases which cleave DNA molecules, and histones which are involved in chromosome packaging in the cell nucleus. DNA-binding proteins can incorporate such domains as the zinc finger, the helix-turn-helix, and the leucine zipper (among many others) that facilitate binding to nucleic acid.

Non-specific DNA-protein Interactions

Structural proteins that bind DNA are well-understood examples of non-specific DNA-protein interactions. Within chromosomes, DNA is held in complexes with structural proteins. These proteins organize the DNA into a compact structure called chromatin. In eukaryotes this structure involves DNA binding to a complex of small basic proteins called histones, while in prokaryotes multiple types of proteins are involved. The histones form a disk-shaped complex called a nucleosome, which contains two complete turns of double-stranded DNA wrapped around its surface. These non-specific interactions are formed through basic residues in the histones making ionic bonds to the acidic sugar-phosphate backbone of the DNA, and are therefore largely independent of the base sequence.

Chemical modifications of these basic amino acid residues include methylation, phosphorylation and acetylation. These chemical changes alter the strength of the interaction between the DNA and the histones, making the DNA more or less accessible to transcription factors and changing the rate of transcription. Other non-specific DNA-binding proteins in chromatin include the high-mobility group proteins, which bind to bent or distorted DNA. These proteins are important in bending arrays of nucleosomes and arranging them into the larger structures that make up chromosomes.

DNA-binding Proteins that Specifically Bind Single-stranded DNA

A distinct group of DNA-binding proteins are the DNA-binding proteins that specifically bind single-stranded DNA. In humans, replication protein A is the best-understood member of this family and is used in processes where the double helix is separated, including DNA replication, recombination and DNA repair. These binding proteins seem to stabilize single-stranded DNA and protect it from forming stem-loops or being degraded by nucleases.

Binding to Particular DNA Sequences

In contrast, other proteins have evolved to bind to particular DNA sequences. The most intensively studied of these are the various transcription factors, which are proteins that regulate transcription. Each transcription factor binds to one particular set of DNA sequences and activates or inhibits the transcription of genes that have these sequences close to their promoters.

The transcription factors do this in two ways. Firstly, they can bind the RNA polymerase responsible for transcription, either directly or through other mediator proteins; this locates the polymerase at the promoter and allows it to begin transcription. Alternatively, transcription factors can bind enzymes that modify the histones at the promoter; this will change the accessibility of the DNA template to the polymerase.

As these DNA targets can occur throughout an organism's genome, changes in the activity of one type of transcription factor can affect thousands of genes. Consequently, these proteins are often the targets of the signal transduction processes that control responses to environmental changes or cellular differentiation and development. The specificity of these transcription factors' interactions with DNA come from the proteins making multiple contacts to the edges of the DNA bases, allowing them to "read" the DNA sequence. Most of these base-interactions are made in the major groove, where the bases are most accessible. Mathematical descriptions of protein-DNA binding taking into account sequence-specificity, competitive and cooperative binding of proteins of different types are usually performed with the help of the lattice models.

DNA-modifying Enzymes

Nucleases and Ligases

Nucleases are enzymes that cut DNA strands by catalyzing the hydrolysis of the phosphodiester bonds. Nucleases that hydrolyse nucleotides from the ends of DNA strands are called exonucleases, while endonucleases cut within strands. The most frequently used nucleases in molecular biology are the restriction endonucleases, which cut DNA at specific sequences. For instance, the EcoRV enzyme shown to the left recognizes the 6-base sequence 52 - GAT | ATC-32 and makes a cut at the vertical line. In nature, these enzymes protect bacteria against phage infection by digesting the phage DNA when it enters the bacterial cell, acting as part of the restriction modification system. In technology, these sequence-specific nucleases are used in molecular cloning and DNA fingerprinting.

Enzymes called DNA ligases can rejoin cut or broken DNA strands. Ligases are particularly important in lagging strand DNA replication, as they join together the short segments of DNA produced at the replication fork into a complete copy of the DNA template. They are also used in DNA repair and genetic recombination.

Topoisomerases and Helicases

Topoisomerases are enzymes with both nuclease and ligase activity. These proteins change the amount of supercoiling in DNA. Some of these enzymes work by cutting the DNA helix and allowing one section to rotate, thereby reducing its level of supercoiling; the enzyme then seals the DNA break. Other types of these enzymes are capable of cutting one DNA helix and then passing a second strand of DNA through this break, before rejoining the helix. Topoisomerases are required for many processes involving DNA, such as DNA replication and transcription.

Helicases are proteins that are a type of molecular motor. They use the chemical energy in nucleoside triphosphates, predominantly ATP, to break hydrogen bonds between bases and unwind the DNA double helix into single strands. These enzymes are essential for most processes where enzymes need to access the DNA bases.

Polymerases

Polymerases are enzymes that synthesize polynucleotide chains from nucleoside triphosphates. The sequence of their products are copies of existing polynucleotide chains - which are called *templates*. These enzymes function by adding nucleotides onto the 32 hydroxyl group of the previous nucleotide in a DNA strand. As a consequence, all polymerases work in a 52 to 32 direction. In the active site of these enzymes, the incoming nucleoside triphosphate base-pairs to the template: this allows polymerases to accurately synthesize the complementary strand of their template. Polymerases are classified according to the type of template that they use.

In DNA replication, a DNA-dependent DNA polymerase makes a copy of a DNA sequence. Accuracy is vital in this process, so many of these polymerases have a proofreading activity. Here, the polymerase recognizes the occasional mistakes in the synthesis reaction by the lack of base pairing between the mismatched nucleotides. If a mismatch is detected, a 32 to 52 exonuclease activity is activated and the incorrect base removed. In most organisms, DNA polymerases function in a large complex called the replisome that contains multiple accessory subunits, such as the DNA clamp or helicases.

RNA-dependent DNA polymerases are a specialized class of polymerases that copy the sequence of an RNA strand into DNA. They include reverse transcriptase, which is a viral enzyme involved in the infection of cells by retroviruses, and telomerase, which is required for the replication of telomeres. Telomerase is an unusual polymerase because it contains its own RNA template as part of its structure. Transcription is carried out by a DNA-dependent RNA polymerase that copies the sequence of a DNA strand into RNA. To begin transcribing a gene, the RNA polymerase binds to a sequence of DNA called a promoter and separates the DNA strands. It then copies the gene sequence into a messenger RNA transcript until it reaches a region of DNA called the terminator, where it halts and detaches from the DNA. As with human DNA-dependent DNA polymerases, RNA polymerase II, the enzyme that transcribes most of the genes in the human genome, operates as part of a large protein complex with multiple regulatory and accessory subunits.

Genetic Recombination

Genetic recombination is a process by which a molecule of nucleic acid (usually DNA, but can also be RNA) is broken and then joined to a different one. Recombination can occur between similar molecules of DNA, as in homologous recombination, or dissimilar molecules, as in non-homologous end joining. Recombination is a common method of DNA repair in both bacteria and eukaryotes. In eukaryotes, recombination also occurs in meiosis, where it facilitates chromosomal crossover. The crossover process leads to offspring's having different combinations of genes from those of their parents, and can occasionally produce new chimeric alleles. In organisms with an adaptive immune system, a type of genetic recombination called V(D)J recombination helps immune cells rapidly diversify to recognize and adapt to new pathogens. The shuffling of genes brought about by genetic recombination is thought to have many advantages, as it is a major engine of genetic variation and also allows asexually reproducing organisms to avoid Muller's ratchet, in which the genomes of an asexual population accumulate deleterious mutations in an irreversible manner.

In genetic engineering, recombination can also refer to artificial and deliberate recombination of disparate pieces of DNA, often from different organisms, creating what is called recombinant DNA. A prime example of such a use of genetic recombination is gene targeting, which can be used to add, delete or otherwise change an organism's genes. This technique is important to biomedical researchers as it allows them to study the effects of specific genes. Techniques based on genetic recombination are also applied in protein engineering to develop new proteins of biological interest. Genetic recombination is catalysed by many different enzymes, called *recombinases*. RecA, the chief recombinase found in *Escherichia coli*, is responsible for the repair of DNA double strand breaks (DSBs). In yeast and other eukaryotic organisms there are two recombinases required for repairing DSBs. The RAD51 protein is required for mitotic and meiotic recombination, whereas the DMC1 protein is specific to meiotic recombination.

Chromosomal Crossover

Chromosomal crossover (or crossing over) is an exchange of genetic material between homologous chromosomes. It is one of the final phases of genetic recombination, which occurs during prophase I of meiosis (pachytene) in a process called synapsis. Synapsis begins before the synaptonemal complex develops, and is not completed until near the end of prophase I. Crossover usually occurs when matching regions on matching chromosomes break and then reconnect to the other chromosome.

Crossing over was described, in theory, by Thomas Hunt Morgan. He relied on the discovery of the Belgian Professor Frans Alfons Janssens of the University of Leuven who described the phenomenon in 1909 and had called it 'chiasmatypie'. The term *chiasma* is linked if not identical to chromosomal crossover. Morgan immediately saw the great importance of Janssens' cytological interpretation of chiasmata to the experimental results of his research on the heredity of *Drosophila*. The physical basis of crossing over was first demonstrated by Harriet Creighton and Barbara McClintock in 1931.

Chemistry

Meiotic recombination initiates with double-stranded breaks that are introduced into the DNA by the Spo11 protein. One or more exonucleases then digest the 5' ends generated by the double-stranded breaks to produce 3' single-stranded DNA tails. The meiosis-specific recombinase Dmc1 and the general recombinase Rad51 coat the single-stranded DNA to form nucleoprotein filaments. The recombinases catalyse invasion of the opposite chromatid by the single-stranded DNA from one end of the break. Next, the 3' end of the invading DNA primes DNA synthesis, causing displacement of the complementary strand, which subsequently anneals to the single-stranded DNA generated from the other end of the initial double-stranded break. The structure that results is a *cross-strand exchange*, also known as a Holliday junction. The contact between two chromatids that will soon undergo crossing-over is known as a *chiasma*. The Holliday junction is a tetrahedral structure which can be 'pulled' by other recombinases, moving it along the four-stranded structure.

Consequences

In most eukaryotes, a cell carries two copies of each gene, each referred to as an allele. Each parent passes on one allele to each offspring. An individual gamete inherits a complete haploid complement of alleles on chromosomes that are independently selected from each pair of chromatids lined up on the metaphase plate. Without recombination, all alleles for those genes linked together on the same chromosome would be inherited together. Meiotic recombination allows a more independent selection between the two alleles that occupy the positions of single genes, as recombination shuffles the allele content between homologous chromosomes.

Recombination does not have any influence on the statistical probability that another offspring will have the same combination. This theory of "independent assortment" of alleles is fundamental to genetic inheritance. However, there is an exception that requires further discussion.

The frequency of recombination is actually not the same for all gene combinations. This leads to the notion of "genetic distance", which is a measure of recombination frequency averaged over a (suitably large) sample of pedigrees. Loosely speaking, one may say that this is because recombination is greatly influenced by the proximity of one gene to another. If two genes are located close together on a chromosome, the likelihood that a recombination event will separate these two genes is less than if they were farther apart. Genetic linkage describes the tendency of genes to be inherited together as a result of their location on the same chromosome. Linkage disequilibrium describes a situation in which some combinations of genes or genetic markers occur more or less frequently in a population than would be expected from their distances apart. This concept is applied when searching for a gene that may cause a particular disease. This is done by comparing the occurrence of a specific DNA sequence with the appearance of a disease. When a high correlation between the two is found, it is likely that the appropriate gene sequence is really closer.

Problems

Although crossovers typically occur between homologous regions of matching chromosomes, similarities in sequence can result in mismatched alignments. These processes are called unbalanced recombination. Unbalanced recombination is fairly rare compared to normal recombination, but severe problems can arise if a gamete containing unbalanced recombinants becomes part of a zygote. The result can be a local duplication of genes on one chromosome and a deletion of these on the other, a translocation of part of one chromosome onto a different one, or an inversion.

Gene Conversion

Gene conversion is an event in DNA genetic recombination, which occurs at high frequencies during meiotic division but which also occurs in somatic cells. It is a process by which DNA sequence information is transferred from one DNA helix (which remains unchanged) to another DNA helix, whose sequence is altered. It is one of the ways a gene may be mutated. Gene conversion may lead to non-Mendelian inheritance and has often been recorded in fungal crosses.

Mechanism

This conversion of one allele to the other is due to base mismatch repair during recombination: if one of the four strands during meiosis pairs up with one of the four strands of a different chromosome, as can occur if there is sequence homology, mismatch repair can alter the sequence of one of the chromosomes, so that it is identical to the other. Gene conversion can result from the repair of damaged DNA as described by the Double Strand Break Repair Model. Here a break in both strands of DNA is repaired from an intact homologous region of DNA. Resection (degradation) of the DNA strands near the break site leads to stretches of single stranded DNA that can invade the homologous DNA strand. The intact DNA can then function as a template to copy the lost DNA. During this repair process a structure called a double Holliday structure is formed. Depending on how this structure is resolved (taken apart) either cross-over or gene conversion products result.

Effect

Normally, an organism that has inherited different copies of a gene from each of its parents is called heterozygous. This is generically represented as genotype: Aa (i.e. one copy of variant (allele) 'A', and one copy of allele 'a'). When a heterozygote creates gametes by meiosis, the alleles normally split, and end up in a 1:1 ratio in the resulting cells. However, in gene conversion, a ratio other than the expected 1A:1a is observed, in which A and a are the two alleles. Such examples are 3A:1a, 1A:3a, 5A:3a or 3A:5a. In other words there can, for example, be three times as many A alleles as a alleles expressed in the daughter cells, as is the case in 3A:1a.

Importance of Gene Conversion in the Cohesiveness of the Gene Pool

Gene conversion acts to *homogenize* the DNA sequences composing the gene pool of a species. Every gene conversion event takes as its substrate two DNA sequences that are homologous but not identical, because of sequence mismatches, and yields two identical DNA sequences. Gene conversion forms the cohesive force that links DNA sequences within different organisms of a species. Over time, gene conversion events yield a homogenous set of DNA sequences, both for allelic forms of a gene and for multi gene families. Interspersed repeats act to block gene conversion events, thus catalyzing evolution of new genes and species.

Medical Relevance

Gene conversion resulting in mutation of the CYP21A2 gene is a common underlying genetic cause of congenital adrenal hyperplasia.

Somatic gene conversion is one of the mechanisms that can result in familial retinoblastoma, a congenital cancer of the retina. It is theorized that gene conversion may play a role in the development of Huntington's Disease

Non-homologous End Joining

Non-homologous end joining (NHEJ) is a pathway that repairs double-strand breaks in DNA. NHEJ is referred to as "non-homologous" because the break ends are directly ligated without the need for a homologous template, in contrast to homologous recombination, which requires a homologous sequence to guide repair. The term "non-homologous end joining" was coined in 1996 by Moore and Haber.

NHEJ typically utilizes short homologous DNA sequences called microhomologies to guide repair. These microhomologies are often present in single-stranded overhangs on the ends of double-strand breaks. When the overhangs are perfectly compatible, NHEJ usually repairs the break accurately. Imprecise repair leading to loss of nucleotides can also occur, but is much more common when the overhangs are not compatible. Inappropriate NHEJ can lead to translocations and telomere fusion, hallmarks of tumour cells.

NHEJ is evolutionarily conserved throughout all kingdoms of life and is the predominant double-strand break repair pathway in mammalian cells. In budding yeast (*Saccharomyces cerevisiae*), however, homologous recombination dominates when the organism is grown under common laboratory conditions.

When the NHEJ pathway is inactivated, double-strand breaks can be repaired by a more error-prone pathway called microhomology-mediated end joining (MMEJ). In this pathway, end resection reveals short microhomologies on either side of the break, which are then aligned to guide repair. This contrasts with classical NHEJ, which typically uses microhomologies already exposed in single-stranded overhangs on the DSB ends. Repair by MMEJ therefore leads to deletion of the DNA sequence between the microhomologies.

In Bacteria

Many species of bacteria, including *Escherichia coli*, lack an end joining pathway and thus rely completely on homologous recombination to repair double-strand breaks. NHEJ proteins have been identified in a number of bacteria, however, including *Bacillus subtilis*, *Mycobacterium tuberculosis*, and *Mycobacterium smegmatis*. Bacteria utilize a remarkably compact version of NHEJ in which all of the required activities are contained in only two proteins: a Ku homodimer and the multifunctional ligase/polymerase/nuclease LigD.

In mycobacteria, NHEJ is much more error prone than in yeast, with bases often added to and deleted from the ends of double-strand breaks during repair. Many of the bacteria that possess NHEJ proteins spend a significant portion of their life cycle in a stationary haploid phase, in which a template for recombination is not available. NHEJ may have evolved to help these organisms survive DSBs induced during dessication.

Corndog and Omega, two related mycobacteriophages of *Mycobacterium smegmatis*, also encode Ku homologs and exploit the NHEJ pathway to recircularize their genomes during infection. Unlike homologous recombination, which has been studied extensively in bacteria, NHEJ was originally discovered in eukaryotes and was only identified in prokaryotes in the past decade.

In Eukaryotes

In contrast to bacteria, NHEJ in eukaryotes utilizes a number of proteins, which participate in the following steps:

End Binding and Tethering

In yeast, the Mre11-Rad50-Xrs2 (MRX) complex is recruited to DSBs early and is thought to promote bridging of the DNA ends. The corresponding mammalian complex of Mre11-Rad50-Nbs1 (MRN) is also involved in NHEJ, but it may function at multiple steps in the pathway beyond simply holding the ends in proximity. DNA-PKcs is also thought to participate in end bridging during mammalian NHEJ.

Eukaryotic Ku is a heterodimer consisting of Ku70 and Ku80, and forms a complex with DNA-PKcs, which is present in mammals but absent in yeast.

Ku is a basket-shaped molecule that slides onto the DNA end and translocates inward. Ku may function as a docking site for other NHEJ proteins, and is known to interact with the DNA ligase IV complex and XLF.

End Processing

End processing involves removal of damaged or mismatched nucleotides by nucleases and resynthesis by DNA polymerases. This step is not necessary if the ends are already compatible and have 3' hydroxyl and 5' phosphate termini.

Little is known about the function of nucleases in NHEJ. Artemis is required for opening the hairpins that are formed on DNA ends during V(D)J recombination, a specific type of NHEJ, and may also participate in end trimming during general NHEJ. Mre11 has nuclease activity, but it seems to be involved in homologous recombination, not NHEJ. The X family DNA polymerases Pol λ and Pol μ (Pol4 in yeast) fill gaps during NHEJ. Yeast lacking Pol4 are unable to join 3' overhangs that require gap filling, but remain proficient for gap filling at 5' overhangs. This is because the primer terminus used to initiate DNA synthesis is less stable at 3' overhangs, necessitating a specialized NHEJ polymerase.

Ligation

The DNA ligase IV complex, consisting of the catalytic subunit DNA ligase IV and its cofactor XRCC4 (Dnl4 and Lif1 in yeast), performs the ligation step of repair. XLF, also known as Cernunnos, is homologous to yeast Nej1 and is also required for NHEJ. While the precise role of XLF is unknown, it interacts with the XRCC4/DNA ligase IV complex and likely participates in the ligation step. Recent evidence suggests that XLF re-adenylates DNA ligase IV after ligation, recharging the ligase and allowing it to catalyse a second ligation.

Other

In yeast, Sir2 was originally identified as an NHEJ protein, but is now known to be required for NHEJ only because it is required for the transcription of Nej1.

Regulation

The choice between NHEJ and homologous recombination for repair of a double-strand break is regulated at the initial step in recombination, 5' end resection. In this step, the 5' strand of the break is degraded by nucleases to create long 3' single-stranded tails. DSBs that have not been resected can be rejoined by NHEJ, but resection of even a few nucleotides strongly inhibits NHEJ and effectively commits the break to repair by recombination. NHEJ is active throughout the cell cycle, but is most important during G1 when no homologous template for recombination is available. This regulation is accomplished by the cyclin-dependent kinase Cdk1 (Cdc28 in yeast), which is turned off in G1 and expressed in S and G2. Cdk1 phosphorylates the nuclease Sae2, allowing resection to initiate.

V(D)J Recombination

NHEJ plays a critical role in V(D)J recombination, the process by which B-cell and T-cell receptor diversity is generated in the vertebrate immune system. In V(D)J recombination, hairpin-capped double-strand breaks are created by the RAG1/RAG2 nuclease, which cleaves the DNA at recombination signal sequences. These hairpins are then opened by the Artemis nuclease and joined by NHEJ. A specialized DNA polymerase called terminal deoxynucleotidyl transferase (TdT), which is only expressed in lymph tissue, adds nontemplated nucleotides to the ends before the break is joined. This process couples "variable" (V), "diversity" (D), and "joining" (J) regions, which when assembled together create the variable region of a B-cell or T-cell receptor gene. Unlike typical cellular NHEJ, in which accurate repair is the most favourable outcome, error-prone repair in V(D)J recombination is beneficial in that it maximizes diversity in the coding sequence of these genes. Patients with mutations in NHEJ genes are unable to produce functional B cells and T cells and suffer from severe combined immunodeficiency (SCID).

At Telomeres

Telomeres are normally protected by a "cap" that prevents them from being recognized as double-strand breaks. Loss of capping proteins causes telomere shortening and inappropriate joining by NHEJ, producing dicentric chromosomes which are then pulled apart during mitosis. Paradoxically, some NHEJ proteins are involved in telomere capping. For example, Ku localizes to telomeres and its deletion leads to shortened telomeres. Ku is also required for subtelomeric silencing, the process by which genes located near telomeres are turned off.

Consequences of Dysfunction

Several human syndromes are associated with dysfunctional NHEJ. Hypomorphic mutations in LIG4 and XLF cause LIG4 syndrome and XLF-SCID, respectively. These syndromes share many features including cellular radiosensitivity, microcephaly and severe combined immunodeficiency (SCID) due to defective V(D)J recombination. Loss-of-function mutations in Artemis also cause SCID, but these patients do not show the neurological defects associated with LIG4 or XLF mutations. The difference in severity may be explained by the roles of the mutated proteins. Artemis is a nuclease and is thought to be required only for repair of DSBs with damaged ends, whereas DNA Ligase IV and XLF are required for all NHEJ events.

Many NHEJ genes have been knocked out in mice. Deletion of XRCC4 or LIG4 causes embryonic lethality in mice, indicating that NHEJ is essential for viability in mammals. In contrast, mice lacking Ku or DNA-PKcs are viable, probably because low levels of end joining can still occur in the absence of these components. All NHEJ mutant mice show a SCID phenotype, sensitivity to ionizing radiation, and neuronal apoptosis.

Immunoglobulin Class Switching

Immunoglobulin class switching (or isotype switching or isotypic commutation or class switch recombination(CSR)) is a biological mechanism that changes a B cell's production of antibody from one class to another, for example, from an isotype called IgM to an isotype called IgG. During this process, the constant region portion of the antibody heavy chain is changed, but the variable region of the heavy chain stays the same (the terms "constant" and "variable" refer to changes or lack thereof between antibodies that target different epitopes). Since the variable region does not change, class switching does not affect antigen specificity. Instead, the antibody retains affinity for the same antigens, but can interact with different effector molecules.

Mechanism

Class switching occurs after activation of a mature B cell via its membrane-bound antibody molecule (or B cell receptor) to generate the different classes of antibody, all with the same variable domains as the original antibody generated in the immature B cell during the process of V(D)J recombination, but possessing distinct constant domains in their heavy chains.

Naïve mature B cells produce both IgM and IgD, which are the first two heavy chain segments in the immunoglobulin locus. After activation by antigen, these B cells proliferate. If these activated B cells encounter specific signalling molecules via their CD40 and cytokine receptors (both modulated by T helper cells), they undergo antibody class switching to produce IgG, IgA or IgE antibodies. During class switching, the constant region of the immunoglobulin heavy chain changes but the variable regions, and therefore antigenic specificity, stay the same. This allows different daughter cells from the same activated B cell to produce antibodies of different isotypes or subtypes (e.g. IgG1, IgG2 etc.).

Class switching occurs by a mechanism called class switch recombination (CSR) binding. Class switch recombination is a biological mechanism that allows the class of antibody produced by an activated B cell to change during a process known as isotype or class switching. During CSR, portions of the antibody heavy chain locus are removed from the chromosome, and the gene segments surrounding the deleted portion are rejoined to retain a functional antibody gene that produces antibody of a different isotype. Double-stranded breaks are generated in DNA at conserved nucleotide motifs, called switch (S) regions, which are upstream from gene segments that encode the constant regions of antibody heavy chains; these occur adjacent to all heavy chain constant region genes with the exception of the δ-chain. DNA is nicked and broken at two selected S-regions by the activity of a series of enzymes, including Activation-Induced (Cytidine) Deaminase (AID), uracil DNA glycosylase and apyrimidic/apurinic (AP)-endonucleases. The intervening DNA between the S-regions is subsequently deleted from the chromosome, removing unwanted μ or δ heavy chain constant region exons and allowing substitution of a γ, α or ε constant region gene segment. The free ends of the DNA are rejoined by a process called non-homologous end joining (NHEJ) to link the variable domain exon to the desired downstream constant domain exon of the antibody heavy chain. In the absence of non-homologous end joining, free ends of DNA may be rejoined by an alternative pathway biased toward microhomology joins. With the exception of the μ and δ genes, only one antibody class is expressed by a B cell at any point in time.

RNA World Hypothesis

The RNA world hypothesis proposes that a world filled with life based on ribonucleic acid (RNA) predates the current world of life based on deoxyribonucleic acid (DNA) and protein. RNA, which can both store information like DNA and act as an enzyme like proteins, may have supported cellular or pre-cellular life. Some hypotheses as to the origin of life present RNA-based catalysis and information storage as the first step in the evolution of cellular life.

The RNA world is proposed to have evolved into the DNA and protein world of today. DNA, through its greater chemical stability, took over the role of data storage while protein, which is more flexible in catalysis through the great variety of amino acids, became the specialized catalytic molecules. The RNA world hypothesis suggests that RNA in modern cells, in particular rRNA (RNA in the ribosome which catalyses protein production), is an evolutionary remnant of the RNA world.

History

The phrase “RNA World” was first used by Nobel laureate Walter Gilbert in 1986, in a commentary on recent observations of the catalytic properties of various forms of RNA. However, the idea of independent RNA life is older and can be found in Carl Woese’s 1968 book *The Genetic Code*.

In 1963, the molecular biologist Alexander Rich, of the Massachusetts Institute of Technology, had posited much the same idea in an article he contributed to a volume issued in honor of Nobel-laureate physiologist Albert Szent-Gyorgyi.

Properties of RNA

The properties of RNA make the idea of the RNA world hypothesis conceptually possible, although its plausibility as an explanation for the origin of life is debated. RNA is known to form efficient catalysts and its similarity to DNA makes its ability to store information clear.

A slightly different version of the hypothesis is that a different type of nucleic acid, termed *pre-RNA*, was the first one to emerge as a self-reproducing molecule, to be replaced by RNA only later. This claim has been made due to (to date) failed attempts to efficiently synthesize activated RNA nucleotides which have the capability to undergo self-polymerization. Suggestions for 'simpler' nucleic acids include Peptide nucleic acid (PNA), Threose nucleic acid (TNA) or Glycerol nucleic acid (GNA). Despite their structural simplicity and possessing properties as good or greater than RNA, the chemically plausible generation of such monomers or oligomers under potentially prebiotic conditions has yet to be demonstrated.

Ribozyme

A ribozyme (from ribonucleic acid enzyme, also called RNA enzyme or catalytic RNA) is an RNA molecule possessing a well defined tertiary structure that enables it to catalyse a chemical reaction. Many natural ribozymes catalyse either the hydrolysis of one of their own phosphodiester bonds (self-cleaving ribozymes), or the hydrolysis of bonds in other RNAs, but they have also been found to catalyse the aminotransferase activity of the ribosome.

Investigators studying the origin of life have produced ribozymes in the laboratory that are capable of catalyzing their own synthesis under very specific conditions, such as an RNA polymerase ribozyme. Mutagenesis and selection has been performed resulting in isolation of improved variants of the "Round-18" polymerase ribozyme from 2001. "B6.61" is able to add up to 20 nucleotides to a primer template in 24 hours, until it decomposes by hydrolysis of its phosphodiester bonds.

Some ribozymes may play an important role as therapeutic agents, as enzymes which tailor defined RNA sequences, as biosensors, and for applications in functional genomics and gene discovery.

Discovery

Before the discovery of ribozymes, enzymes, which are defined as catalytic proteins, were the only known biological catalysts. In 1967, Carl Woese, Francis Crick, and Leslie Orgel were the first to suggest that RNA could act as a catalyst. This idea was based upon the discovery that RNA can form complex secondary structures. The first ribozymes were discovered in the 1980s by Thomas R. Cech, who was studying RNA splicing in the ciliated protozoan *Tetrahymena thermophila* and Sidney Altman, who was working on the bacterial RNase P complex. These ribozymes were found in the intron of an RNA transcript, which removed itself from the transcript, as well as in the RNA component of the RNase P complex, which is involved in the maturation of pre-tRNAs. In 1989, Thomas R. Cech and Sidney Altman won the Nobel Prize in chemistry for their "discovery of catalytic properties of RNA." The term *ribozyme* was first introduced by Kelly Kruger *et al.* in 1982 in a paper published in *Cell.*

It had been a firmly established belief in biology that catalysis was reserved for proteins. In retrospect, catalytic RNA makes a lot of sense. This is based on the old question regarding the origin of life: Which comes first, enzymes that do the work of the cell or nucleic acids that carry the information required to produce the enzymes? Nucleic acids as catalysts circumvents this problem.

In the 1970s Thomas Cech, at the University of Colorado at Boulder, was studying the excision of introns in a ribosomal RNA gene in *Tetrahymena thermophila.* While trying to purify the enzyme responsible for splicing reaction, he found that intron could be spliced out in the absence of any added cell extract. As much as they tried, Cech and his colleagues could not identify any protein associated with the splicing reaction. After much work, Cech proposed that the intron sequence portion of the RNA could break and reform phosphodiester bonds.

At about the same time, Sidney Altman, who is a Professor at Yale University, was studying the way tRNA molecules are processed in the cell when he and his colleagues isolated an enzyme called RNase-P, which is responsible for conversion of a precursor tRNA into the active tRNA. Much to their surprise, they found that RNase-P contained RNA in addition to protein and that RNA was an essential component of the active enzyme. This was such a foreign idea that they had difficulty publishing their findings. The following year, Altman demonstrated that RNA can act as a catalyst by showing that the RNase-P RNA subunit could catalyse the cleavage of precursor tRNA into active tRNA in the absence of any protein component.

Since Cech's and Altman's discovery, other investigators have discovered other examples of self-cleaving RNA or catalytic RNA molecules. Many ribozymes have either a hairpin – or hammerhead – shaped active centre and a unique secondary structure that allows them to cleave other RNA molecules at specific sequences. It is now possible to make ribozymes that will specifically cleave any RNA molecule. These RNA catalysts may have pharmaceutical applications. For example, a ribozyme has been designed to cleave the RNA of HIV. If such a ribozyme was made by a cell, all incoming virus particles would have their RNA genome cleaved by the ribozyme, which would prevent infection.

Activity

Although most ribozymes are quite rare in the cell, their roles are sometimes essential to life. For example, the functional part of the ribosome, the molecular machine that translates RNA into proteins, is fundamentally a ribozyme, composed of RNA tertiary structural motifs that are often coordinated to metal ions such as Mg as cofactors. There is no requirement for divalent cations in a five-nucleotide RNA that can catalyse *trans*-phenylalanation of a four-nucleotide substrate which has three base complementary sequence with the catalyst. The catalyst and substrate were devised by truncation of the C3 ribozyme.

RNA can also act as a hereditary molecule, which encouraged Walter Gilbert to propose that in the distant past, the cell used RNA as both the genetic material and the structural and catalytic molecule, rather than dividing these functions between DNA and protein as they are today. This hypothesis became known as the "RNA world hypothesis" of the origin of life.

If ribozymes were the first molecular machines used by early life, then today's remaining ribozymes — such as the ribosome machinery — could be considered living fossils of a life based primarily on nucleic acids. A recent test-tube study of prion folding suggests that an RNA may catalyse the pathological protein conformation in the manner of a chaperone enzyme.

RNA in Information Storage

RNA is a very similar molecule to DNA, and only has two chemical differences. The overall structure of RNA and DNA are immensely similar—one strand of DNA and one of RNA can bind to form a double helical structure. This makes the storage of information in RNA possible in a very similar way to the storage of information in DNA.

Comparison of DNA and RNA Structure

The major difference between RNA and DNA is the presence of a hydroxyl group at the 2'-position of the ribose sugar in RNA. This group makes the molecule less stable—in flexible regions of an RNA molecule (i.e., where not constrained in a double helix), it can chemically attack the adjacent phosphodiester bond to cleave the phosphodiester backbone. The hydroxyl group also forces the ribose into the C3'-endo sugar conformation unlike the C2'-endo conformation of the deoxyribose sugar in DNA. This forces a RNA double helix into a slightly different conformation than B-DNA, resembling A-DNA more closely.

RNA also uses a different set of bases than DNA—adenine, guanine, cytosine and uracil, instead of adenine, guanine, cytosine and thymine. Chemically, uracil is similar to thymine, differing only by a methyl group, and its production requires less energy. In terms of base pairing this has no effect, adenine will readily bind uracil or thymine. Uracil is, however, one product of damage to cytosine making RNA particularly susceptible to mutations which replace a GC base pair with a GU (wobble) or AU base pair.

Limitations of Information Storage in RNA

Storing large amounts of information in RNA is not easy. The chemical properties of RNA make large RNA molecules inherently fragile, and they can easily be broken down into their constituent nucleotides through hydrolysis. The aromatic bases also absorb strongly in the ultraviolet region, and would have been susceptible to damage and breakdown by background radiation. These limitations do not make use of RNA as an information storage system impossible, simply energy intensive (to repair or replace damaged RNA molecules) and mutation prone. While this makes it unsuitable for current 'DNA optimised' life, it may have been suitable for primitive life.

Riboswitch

In molecular biology, a riboswitch is a part of an mRNA molecule that can directly bind a small target molecule, and whose binding of the target affects the gene's activity. Thus, an mRNA that contains a riboswitch is directly involved in regulating its own activity, in response to the concentrations of its target molecule. The discovery that modern organisms use RNA to bind small molecules, and discriminate against closely related analogs, significantly expanded the known natural capabilities of RNA beyond its ability to code for proteins or to bind other RNA or protein macromolecules.

The original definition of the term "riboswitch" specified that they directly sense small-molecule metabolite concentrations. Although this definition remains in common use, some biologists have used a broader definition that includes other cis-regulatory RNAs. However, this article will discuss only metabolite-binding riboswitches. Most known riboswitches occur in bacteria, but functional riboswitches of one type (the TPP riboswitch) have been discovered in plants and certain fungi. TPP riboswitches have also been predicted in archaea, but have not been experimentally tested.

History and Discovery of Riboswitches

Before the discovery of riboswitches, the mechanism by which some genes involved in multiple metabolic pathways were regulated remained mysterious. Accumulating evidence increasingly suggested the then-unprecedented idea that the mRNAs involved might bind metabolites directly, to effect their own regulation.

These data included conserved RNA secondary structures often found in the UTRs of the relevant genes and the success of procedures to create artificial small molecule-binding RNAs called aptamers. In 2002, the first comprehensive proofs of multiple classes of riboswitches were published, including protein-free binding assays, and metabolite-binding riboswitches were established as a new mechanism of gene regulation.

Many of the earliest riboswitches to be discovered corresponded to conserved sequence "motifs" (patterns) in 5' UTRs that appeared to correspond to a structured RNA. For example, comparative analysis of upstream regions of several genes expected to be co-regulated led to the description of the S-box (now the SAM-I riboswitch), the THI-box (a region within the TPP riboswitch), the RFN element (now the FMN riboswitch) and the B_{12}-box (a part of the cobalamin riboswitch), and in some cases experimental demonstrations that they were involved in gene regulation via an unknown mechanism.

Bioinformatics has played a role in more recent discoveries, with increasing automation of the basic comparative genomics strategy. Barrick *et al.* (2004) used BLAST to find UTRs homologous to all UTRs in *Bacillus subtilis*. Some of these homologous sets were inspected for conserved structure, resulting in 10 RNA-like motifs. Three of these were later experimentally confirmed as the glmS, glycine and PreQ1-I riboswitches. Subsequent comparative genomics efforts using additional taxa of bacteria and improved computer algorithms have identified further riboswitches that are experimentally confirmed, as well as conserved RNA structures that are hypothesized to function as riboswitches.

Types of Riboswitches

The following is a list of experimentally validated riboswitches, organized by ligand.

- *Cobalamin riboswitch* (also *B_{12}-element*), which binds adenosylcobalamin (the coenzyme form of vitamin B_{12}) to regulate cobalamin biosynthesis and transport of cobalamin and similar metabolites, and other genes.
- *cyclic di-GMP riboswitches* bind the signalling molecule cyclic di-GMP in order to regulate a variety of genes controlled by this second messenger. Two classes of cyclic di-GMP riboswitches are known: cyclic di-GMP-I riboswitches and cyclic di-GMP-II riboswitches. These classes do not appear to be structurally related.
- *FMN riboswitch* (also *RFN-element*) binds flavin mononucleotide (FMN) to regulate riboflavin biosynthesis and transport.
- *glmS riboswitch*, which is a ribozyme that cleaves itself when there is a sufficient concentration of glucosamine-6-phosphate.
- *Glycine riboswitch* binds glycine to regulate glycine metabolism genes, including the use of glycine as an energy source. As of 2010, this riboswitch is the only known natural RNA that exhibits cooperative

binding, which is accomplished by two adjacent aptamer domains in the same mRNA.

- *Lysine riboswitch* (also *L-box*) binds lysine to regulate lysine biosynthesis, catabolism and transport.
- *PreQ1 riboswitches* bind pre-queuosine$_1$, to regulate genes involved in the synthesis or transport of this precursor to queuosine. Two entirely distinct classes of PreQ1 riboswitches are known: PreQ1-I riboswitches and PreQ1-II riboswitches. The binding domain of PreQ1-I riboswitches are unusually small among naturally occurring riboswitches. PreQ1-II riboswitches, which are only found in certain species in the genera *Streptococcus* and *Lactococcus*, have a completely different structure, and are larger.
- *Purine riboswitches* binds purines to regulate purine metabolism and transport. Different forms of the purine riboswitch bind guanine (a form originally known as the *G-box*) or adenine. The specificity for either guanine or adenine depends completely upon Watson-Crick interactions with a single pyrimidine in the riboswitch at position Y74.

  In the guanine riboswitch this residue is always a cytosine (i.e. C74), in the adenine residue it is always a uracil (i.e. U74). Homologous types of purine riboswitches bind deoxyguanosine, but have more significant differences than a single nucleotide mutation.
- *SAH riboswitches* bind S-adenosylhomocysteine to regulate genes involved in recycling this metabolite that is produced when S-adenosylmethionine is used in methylation reactions.
- *SAM riboswitches* bind S-adenosyl methionine (SAM) to regulate methionine and SAM biosynthesis and transport. Three distinct SAM riboswitches are known: *SAM-I* (originally called *S-box*), *SAM-II* and the *S_{MK} box riboswitch*. SAM-I is widespread in bacteria, but SAM-II is found only in alpha-, beta- and a few gamma-proteobacteria.

  The S_{MK} box riboswitch is found only in the order Lactobacillales. These three varieties of riboswitch have no obvious similarities in terms of sequence or structure. A fourth variety, SAM-IV riboswitches, appears to have a similar ligand-binding core to that of SAM-I riboswitches, but in the context of a distinct scaffold.
- *SAM-SAH riboswitches* bind both SAM and SAH with similar affinities. Since they are always found in a position to regulate genes encoding methionine adenosyltransferase, it was proposed that only their binding to SAM is physiologically relevant.

- *Tetrahydrofolate riboswitches* bind tetrahydrofolate to regulate synthesis and transport genes.
- *TPP riboswitches* (also THI-box) binds thiamin pyrophosphate (TPP) to regulate thiamin biosynthesis and transport, as well as transport of similar metabolites. It is the only riboswitch found so far in eukaryotes.

Presumed riboswitches:

- *Moco RNA motif* is presumed to bind molybdenum cofactor, to regulate genes involved in biosynthesis and transport of this coenzyme, as well as enzymes that use it or its derivatives as a cofactor.

Candidate metabolite-binding riboswitches have been identified using bioinformatics, and have moderately complex secondary structures and several highly conserved nucleotide positions, as these features are typical of riboswitches that must specifically bind a small molecule. Riboswitch candidates are also consistently located in the 5' UTRs of protein-coding genes, and these genes are suggestive of metabolite binding, as these are also features of most known riboswitches.

Hypothesized riboswitch candidates highly consistent with the preceding criteria are as follows: crcB RNA Motif, Downstream-peptide motif, glnA RNA motif, manA RNA motif, pfl RNA motif, ydaO/yuaA leader, yjdF RNA motif, ykkC-yxkD leader (and related ykkC-III RNA motif) and the yybP-ykoY leader. The functions of these hypothetical riboswitches remain unknown.

Riboswitches and the RNA World Hypothesis

Riboswitches demonstrate that naturally occurring RNA can bind small molecules specifically, a capability that many previously believed was the domain of proteins or artificially constructed RNAs called aptamers. The existence of riboswitches in all domains of life therefore adds some support to the RNA world hypothesis, which holds that life originally existed using only RNA, and proteins came later; this hypothesis requires that all critical functions performed by proteins could be performed by RNA.

It has been suggested that some riboswitches might represent ancient regulatory systems, or even remnants of RNA-world ribozymes whose bindings domains are conserved.

Riboswitches as Antibiotic Targets

Riboswitches could be a target for novel antibiotics. Indeed, some antibiotics whose mechanism of action was unknown for decades have been shown to operate by targeting riboswitches. For example, when the antibiotic pyrithiamine enters the cell, it is metabolized into pyrithiamine pyrophosphate. Pyrithiamine pyrophosphate has been shown to bind and activate the TPP riboswitch, causing the cell to cease the synthesis and import of TPP. Because pyrithiamine pyrophosphate does not substitute for TPP as a coenzyme, the cell dies.

Engineered Riboswitches

Since riboswitches are an effective method of controlling gene expression in natural organisms, there has been interest in engineering artificial riboswitches.

6

Mechanical Function of Cell

Some cell protuberances has been correlated to the orientation and distribution of microtubules. They are considered to be a framework which determines the shaping of the cell and redistributes its content. The integrity of microtubules is necessary for the individual shape of many cells and the rigidity of elongated processes. These particularly evident in axons and dendrites.

Morphogenesis

Cells's mechanical function is the role that microtubules play in the shaping of the cell during *cell differentiation*. For example, the-elongation of the cells during the induction of the lens placode in the eye is accompanied by the appearance of numerous microtubules. The morphogenetic changes that occur during spermiogenesis provide another interesting example. The enormous elongation that takes place in the nucleus of the spermatid is accompanied by the production of an orderly array of microtubules that are wrapped around the nucleus in a direction perpendicular to the nuclear axis and that have a double helical arrangement.

Cellular Polarity and Motility

The determination of the intrinsic polarity of certain cells is also related to the mechanical function. Treatment of various culture cells with Colcemid results in a change in motion.

The following forms of movement have been observed to persist membrane:

- Ruffling
- Attachment to the surface,
- Endocytosis,
- Extension of microvilli.

Only a saltatory movement of particles was found to be inhibited after the destruction of the microtubules. However, the directional gliding of the cell is replaced by a random movement. Regarding culture cells they are treated in vitro with dibutyryl cyclic AMP become elongated and the number of microtubules increases. All these observations suggests that microtubules are responsible for maintenance of the shape of the cell, the polarity of movement, and the distribution of certain or ganelles.

Circulation and Transport

May also be involved in the *transport* of macromolecules in the cell's interior; to this end they probably form channels in the cytoplasm. A very interesting example is the protozoan *Actinosphaerium* (Heliozoa) which sends out long thin pseudopodia within which cytoplasmic particles migrate back and forth. These pseudopodia contains as many as 500 microtubules disposed in a helical configuration.

When these protozoans are exposed to cold, or high pressure the pseudopodia are withdrawn and the microtubules depolymerize. Another example of an association between microbules and transport of particulate material is the melanocyte, in which the melanin granules move centrifugally and centripetally with different stimuli. The granules have been observed moving between channels created by the microtubules in the cytoplasmic matrix. The erythrophores found in fish scales the pigment granules may move at a speed of 25 to 30 μm per second between the microtubules.

Sensory Transduction

Regularly arranged bundles of microtubules are common in sensory receptors, and a possible function in transduction of different incident energy has been attributed to them.

Microtubular Organelles

Many cells organs can be considered to be derived from special assemblies of microtubules. In fact, cilia, flagella, basal bodies, and centrioles have, as one of their main components, groups of microtubules arranged in a special fashion. For this reason and in spite of the complexity of their structure, these organelles will be considered at this point, along with some of their physiological aspects.

Cillary and Flagellar Motions

Cillary and flagellar all motility is adapted to liquid media and is executed by minute, specially differentiated appendices that vary in size and number. They are called *flagella* if they are few and long, and *cilia,* if short and numerous.

In Protozoa, especially the Infusoria, each cell has hundreds or thousands of minute cilia, and their movement permits a rapid procession of the organism in the liquid medium.

In some special regions of the Infusoria, several cilia fuse and form larger conical appendices, *the cirri*, or membranes observed as *undulating membranes*.

Out of entire class of Protozoa, the Flagellata, is characterized by the presence of flagella. The spermatozoa of metazoans move as isolated cells by means of flagella. On the other hand, epithelial cells that possess vibratile cilia and constitute true ciliated sheets are relatively common.

These may cover large areas of the external surface of the body and determine the motion of the animal. Such is the case with some Platyhelminthes and Nemertea and also with larvae of Echinodermata, Mollusca, and Annelida.

More often, the ciliated epithelial sheets line cavities or internal tubes, such as the air passages of the respiratory system or various parts of the genital tract. In these organs all the cilia move simultaneously in the same direction, and fluid currents are thus produced.

In some cases, the currents serve to eliminate solid particles in suspension (*e.g.*, in the respiratory system). The eggs of amphibians and mammals are driven along the oviduct with the aid of vibratile cilia.

The Cillary Apparatus

Some of the essential components of cillary apparatus are:

- The *cilium*, which is the slender cylindroid process, limited by the plasmalemma (cell membrane), that projects from the free surface of the cell,
- The *basal body*, or granule, the intracellular organelle similar to the centriole from which it originates,
- In some cells fine fibrils—called *ciliary rootlets*—that arise from the basal granule and converge into a conical bundle, the pointed extremity of which ends at one side of the nucleus.

Regarding the basal bodies they are embedded in the ectoplasmic layer beneath the cell surface. In general, they are spaced uniformly and in parallel rows. Cilia and flagella are extremely delicate filaments whose thickness is often at the limit of the resolving power of the light microscope.

Various epithelia have appendices similar in shape to cilia, but immobile; these are called *stereocilia*. Examples are the processes of the epithelial cells of the epididymis. In the macula and crista of the inner ear, there are stereocilia in addition to motile cilia, or *kynocilia*. Stereocilia do not contain microtubules.

The Axoneme

Regarding the term Axoneme the some is applied to the axial basic microtubular structure of cilia and flagella, and is it thought to be the essential motile element. In various flagella *accessory structures* made of fibres, microtubules, crystalline bodies, or mitochondrial derivatives may be found. The axoneme of a cilium or flagellum may range from a few microns to 1 to 2 millimeters in length, but the outside diameter is about 0.21 μm. It is generally surrounded by an *outer ciliary membrane*, which is continuous with the plasma membrane. All the components of the axoneme are embedded within the *ciliary matrix*.

With the fundamental 9 + 2 microtubular pattern and the other essential structural components of the axoneme. A plane perpendicular to the line joining the two central tubules divides the axoneme into a right and a left symmetrical half. It is generally admitted that the plane of the ciliary beat is perpendicular to this plane of symmetry.

The pairs of peripheral microtubules have an ellipsoidal profile, whereas the central ones are circular. The diameter of the peripheral tubules varies from 18 to 25 nm with the central tubules being slightly larger. The two microtubules of each peripheral pair can be clearly distinguished by several morphologic features.

The doublets are skewed at about 10 degrees so that one tubule, designated subfibre A, lies closer to the axis than the other (*i.e.*, subfibre B). The microtubule of subfibre A is smaller, but complete, whereas that of subfibre B is larger and incomlete, since it lacks the wall adjacent to A. In fact, while A has 13 tubulin subunits, B has only 11. Furthermore, subfibre.

A has processes—the so-called *dyein arms* —that are oriented in the same direction in all microtubules. This orientation is clockwise when the axoneme is viewed from base to tip.

Dynein Arms

The arms of subfibre A which are generally called *dynein arms* because they contain dynein, a higher molecular weight ATPase (300,000 to 400,000 daltons). This is a Mg^{2+}-and Ca^{2+} -activated enzyme, which after solubilization can recombine at the same position on the A microtubules.

Two isoenzymatic forms of the ATPase, dynein I and II, have been separated by electrophoresis.[16] Dynein I constitutes the largest proportion of the axonemal ATPase and is present in the arms of subfibre A. As will be mentioned later, the interaction between tubulin and dynein is thought to underlie the basic mechanism of ciliary and flagellar motion.

Nexin Links

The doublets are linked by *interdoublet* or *nexin links.* This name is derived from the fact that another protein called *nexin* has been isolated from these links. This protein has a molecular weight of 150,000 to 160,000 daltons. The function of these links is unknown, but they may serve as structures that maintain the geometric integrity of the axoneme during the sliding motion.

Radial Spokes

The spokes terminate in a dense knob or head, which may have a forklike structure. Certain observations that the spokes are attached perpendicularly to the ciliary axis where it is straight and that they are relatively detached in bent or tilted regions of the axis has led to the hypothesis that they may be active in the conversion of active sliding between the outer doublets into local axial being.

Basal Bodies (Kinetosomes) and Centrioles

After the classic works of Hennegay and Lenhosesk in the year 1897, it has been suggested that basal bodies (or kinetosomes) of cilia and flagella are homologous with the centrioles found in mitotic spindles. In some cells it was discovered that a centriole engaged in mitosis could carry a cilium at the same time. This homology was fully confirmed with the introduction of electron microscopy. Centrioles are cylinders that measure, on the average, 0.2 μm–0.5 μm; at times they may be as long as 2 μm. This cylinder is open on both ends, unless it carries a cilium. In the later case, it is separated from the cilium by a *ciliary plate.*

In freeze-fractured cilia it has been possible to observed the so-called ciliary necklace. This consists of two to six rows of membrane particles present just below the ciliary plate where the central pair of microtubules end.

In the wall of the centriole there are nine groups of microtubules arranged in a circle. Each group is a triplet formed of three tubules (raher than two, as in cilia) that are skewed towards the centre. Since the angle made by the set of tubules with the tangent to the cylinder varies from 70 degrees in the proximal end to 20 to 30 degrees in the distal end, the tubules twist from one end to the other or describe a helical course. The tubules are designated *subfibre A, B,* and *C.* from the centre towards the periphery.

Both subfibres A and B cross the ciliary plate and are continuous with the corresponding subfibres in the axonemene; subfibre C terminates near the *ciliary* or *basal plate.* There are no central microtubules in the centrioles and no central microtubules in the centrioles and no special arms; however, they are linked by connectives. The proximal portion of the centriole has a cartwheel appearance with spokes that radiate from a central hub and connect to subfibre of each triplet. The presence of the cartwheel structure at the proximal end provides the centriole with a structural and functional polarity. The growth of the centriole is from the distal end, and in the case of kinetosomes, it is from this end that the cilium is formed. Further more, the *procentrioles*, which are formed at right angles to the centriole, are located near the proximal end. Both the above given centrioles may give rise to a cilium but in most cases only one does, and the sister centriole remains at a right angle to the kinetosome. In most multiciliated cells there are single basal bodies. It will be shown later that in sensory cells having nonmotile ciliary derivatives and a 9 + 0 configuration, there are two basal bodies. The two centrioles of a pair are always located at a distance from each other of at least 0.8 μm and are perpendicular to each other.

Ciliary Rootlets

In some of the cells, ciliary rootlets originate from the basal body. Most rootlets are striated, having a regular cross-banding with a repeating period of 55 to 70 nm. Five intraperiodic sub-bands have been observed in the cross-striated rootlets. The striated fibres are composed of parallel microfilaments, 3 to 7 nm in diameter, which in turn are formed of globular subunits. These fibres and filaments may serve a structural role such as anchoring the kinetosomes. By analogy with other microfilaments, a contractile role has been postulated for the ciliary rootlets. Furthermore, ATPase has been found associated with the cross-bands of certain rootlets.

Basal Feet and Satellites

Basal feels are dense processes that are arranged perpendicularly to the basal body in a particular direction and originate from two to three of the triplets. These processes impose a structural asymmetry on the basal body that has been related to the direction of the ciliary beat. The basal foot, which is composed of microfilaments that terminate in a dense bar, may be a focal point for the convergence of microtubules. *Satellites* or *pericentriolar bodies*, are electron-dense structures lying near the centriole that probably are nucleating sites for microtubules.

Ciliary Movement

Ciliary movement can be analysed easily by scraping the pharyngeal epithelium of a frog or toad with a spatula and placing the scrapings in a drop of physiologic salt solution between a slide and a coverglass. On the free surface of the epithelial cell, the rapid motion of the vibratile cilia can be seen. If a row of cilia is observed, the contraction is *metachronic* in the plane of the direction of motion; that is, it starts before or after the contraction of the next cilium. In this way true waves of contraction are formed.

Now consider the other part that, in a plane perpendicular to the direction of motion, the contraction is *isochronic*; all the cilia are observed in the same phase of contraction at a given time. The rhythmic contraction of cilia has been interpreted in different ways. A two-step process that involves *intraciliary* excitation followed by *interciliary* conduction has been proposed to explain the metachronic rhythm of the ciliary beat. This mechanism evidently does not depend on the nervous system, since it persists after the epithelium has been separated from the rest of the organism. However, cytoplasmic continuity is indispensable to its maintenance, for if a cut is made in the row of cilia the waves of contraction of the two isolated pieces becomes uncoordinated.

If a piece of epithelium is removed from the pharynx of a frog and implanted with a reversed orientation, the movement is maintained but in the opposite direction. Ciliary contractions are generally rapid (10 to 17 per second in the pharynx of the frog). Analysis of the motion has been facilitated greatly by stroboscopic and ultrarapid microcinematography. Ciliary movement may be pendulous, unciform (hooklike), infundibuliform, or undulant. The first two are carried out in a single plane. In the pendulous movement, typical of the ciliated Protozoa, the cilium is rigid and the motion is carried out by a flexion at its base.

On the other hand, in the unciform movement, the most common type in the Metazoa, the cilium is doubled upon contraction and takes the shape of a hook. In the infundibuliform movement, the cilium or flagellum rotates, describing a conical or funnel-shaped figure. In the undulant motion, characteristic of flagella contraction waves proceed from the site of implantation and pass to the free border. It was shown above that the plane of the beat of a cilium is perpendicular to that passing through the central microtubules. Although the effective stroke is a single plane, the cilium moves away from the original plane during the recovery phase. This movement, which can be beautifully demonstrated by scanning electron microscopy,[23] implies that either the whole cilium rotates at its base or that the central microtubules rotate. Since the last century, ciliary and flagellar motion have been compared to that of muscle. However, the complex structure of the axoneme suggests that this motion may be more complex than the planar sliding of myofilaments in muscle.

The key process appears to be the sliding of the microtubular doublets over each other, with the associated making and breaking of the cross-bridges between adjacent doublets, represented by the dynein arms. The sliding of doublets during the bending of the cilium or flagellum has actually been observed and measured. This displacement of the doublets has been determined because the central microtubules extend to the very tip, whereas subfibres B terminate at a distance from the tip of 1 to 2 μm, and subfibres. A terminate slightly beyond subfibres B. This arrangement has also made it possible to demonstrate that the two subfibres do not move relative to one another within the doublet. In the diagram of the amount of sliding displacement is shown to depend on the distance separating the tubules and on the bend angle. Experiments on ciliary motion has been aimed at demonstrating the involvement of ATP; it was mentioned above that the ATPase activity is associated with the protein *dynein* in the arms of subfibres. A Dynein constitutes about 8 per cent of the total ciliary protein. In glycerin-extracted cilia, flagella, and spermatozoan tails, the addition of ATP produces rhythmic activity that may persist for a few minutes or even hours. Interaction between tubulin and dynein seems to be essential in this type of cell motility. When studying myofilaments in muscle, we shall see that the actin-myosin system is the other process involved in contraction.

Several approaches are current by being used to study the molecular mechanisms in ciliary motion. One of them has been the use of sperm flagella in which the membranes have been removed by detergents. These sperm can be reactivated with ATP. Extraction of dynein arms by KC1 leads to a reduction in the frequency of beating, which can be restored to the original level by the addition of *dynein*. An anti-dynein antiserum inhibits the movement of spermatozoa without membranes, and ATP induces, it. The other observation is that, unlike muscle, cilia and flagella do not require Ca^{2+} for activation. In many cases, however, Ca^{2+} has a regulating effect on the wave form of the beating.

Kartagener's Syndrome

One of the recent approach to the study of ciliary motion is the use of mutants of *Chlamydomonas* in which various deficiencies in axoneme structure have been observed. From the medical point of view, this research is of great importance in the so-called *Kartangener's syndrome,* in which there is infertility with nonmotile sperm chronic bronchitis, and sinusitis. It was been demonstrated that this syndrome is due to a mutation involving the dynein arms. In these patients *cilia and sperm are non-motile because they lack dynein arms, which contain ATPase.*

Photoreceptors

Regarding the studies on the structure of retinal rods and cones have shown that the short, fibrous connection found between the outer and inner segments is of a ciliary nature. Cross sections of this so-called connecting cilium have revealed nine pairs of filaments similar to those found in cilia. In this structure, however, there are no central microtubules.

The outer segment of this cilium is composed of numerous double membrane rod sacs arranged like a stack of coins. The first stage in the development of a rod is a primitive cilium projecting from the bulge of cytoplasm that constitutes the primordium of the inner segment. This cilium contains the nine pairs of filaments and the two basal centrioles.

The apical end is filled with a vesicular material. In the second stage the apical region of the primitive cilium enlarges greatly, owing to the rapid building up of the vesicles and cisternae that constitute the primitive rod sacs. The proximal part of the primitive cilium remains undifferentiated and constitutes the connecting cilium of the adult.

Another structures have also been recognized as ciliary derivatives. For example, the so-called crown cells of the saccus vasculosus found in the third ventricle of fishes are modified cilia with swollen ends that are filled with vesicles. Also, the primitive sensory cells of the pineal eye found in certain lizards have a ciliary structure.

Cilia and Flagella Originate from Basal Bodies

The origin of centrioles and basal bodies, or kinetosomes, is viewed from the standpoint that these are possibly semiautonomous cell organelles and in this way similar to mitochondria and chloroplasts. Isolation of kinetosomes from *Tetrahymena* has verified the presence of RNA and DNA, and this suggests that they are capable of some protein synthesis. More direct evidence of DNA in centrioles was obtained by use of the fluorescent dye acridine orange and by ^{3}H-thymidine incorporation followed by treatment with DNAse. However, most of these studies refer to the pellicle, and the exact localization of DNA has not been determined. Furthermore, these findings do not resolve the question of whether the synthesis of ciliary proteins depends on the DNA-RNA components of the centriole.

Some of the early cytologists realised that the development of cilia and flagella was directly related to the presence of centrioles. Flagella may be experimentally removed and their regeneration followed; thus, in *Chlamydomonas* it was found that the growth rate of a flagellum is about 0.2 μm per minute. During the studies of the origin of centrioles and kinetosomes were hampered by the small size of these structures. Light microscopists thought that centrioles originate by division from preexisting centrioles and that they possessed a genetic continuity comparable to that of chromosomes. The electron microscope has failed to confirm that such a division process takes place.

We find several in directions that centrioles may be generated by two different mechanisms. Centrioles destined to form mitotic spindles of single cilia arise directly from the wall of the preexisting centriole. The *daughter centrioles* appear first as annular structures (*procentrioles*), which lengthen into cylinders. There are three groups of tubules originate from single and double groups that first appear at the base of the procentriole. When they are half grown the daughter centrioles are released into the cytoplasm to complete their maturation. Usually, although not always, daughter centrioles are formed one at a time. Another mechanism in the formation of centrioles is found in those destined to become kinetosomes, as in a ciliated epithelium. The centrioles are assembled progressively from a precursor fibrogranular material located in the apical cytoplasm. The newly formed centrioles become aligned in rows beneath the apical plasma membrane, and each centriole may then produce satellites from the side, a root from its base and a cilium from its apex.

Cilium's development begins with the appearance of a vesicle that becomes attached to the distal end of the centriole. The growing ciliary shaft invaginates the vesicular wall, which forms a temporary ciliary sheath until the permanent one is formed. In many epithelial cells short cilia enclosed within vesicles may be observed.

Under the electron microscopic study of centrioles have revealed that they have a specific tubular structure, that they do not divide, and that they are found in association with other tubular structures. This relationship suggests that the function of the centriole may be to regulate the synthesis and organization of microtubules within the cytoplasm.

Microfilaments

We have studied earlier that the ground cytoplasm or cytosol contains a kind of *cytoskeleton* formed by microtubules and various types of microfilaments. From what was said about the function of cytoplasmic microtubules the reader may have gathered the impression that they play mainly a mechanical function, constituting a kind of *passive portion or framework of the cytoskeleton*.

This concept may not be completely accurate, since these microtubules could store energy by distortion, and during disassembly they could release elastic forces that might transport cytoplasmic structures. Furthermore, we have seen before that the interaction between the tubulin of microtubules and dynein is the main mechanism of motion of cilia and flagella.

Pleasantly the *active* or *motile function of the cytoskeleton* has been attributed to some of the microfilaments that are observed by electron microscopy as thin filaments between 6 and 10 nm thick. Various studies have suggested that such filaments represent contractile systems which may be involved in cytoplasmic streaming, ameboid motion, and other types of cell activities in which a motive force is involved. These studies have recently led to very important conclusions with the demonstration that the contractile proteins of muscle (*i.e.*, *actin* and *myosin*) are also widely found in nonmuscular systems and that they may be chemical components of some of the microfilaments.

Microtrabecular Lattice in Cytosol

High voltage electron microscopy has helped in identifying microfilaments in whole culture cells and in obtaining a three-dimensional view of the so-called *microtrabecular lattice*.

Just beneath the membrane there are bundles of filaments that can be seen to be made of actin and that are in continuity with a lattice of similar filaments that pervades the cytosol. The channels or space found in between this *trabecular lattice* are of the order of 50 to 100 nm and in the living cell may provide for the rapid diffusion of fluid and metabolites through all parts of the cytosol.

The filaments of the lattice are in contact with vesicles of the endoplasmic reticulum, with microtubules and with polysomes, all of which seem to be contained of supported within the lattice. On the other hand, mitochondria appear to be free of the lattice, and this may explain the fact that they move rather freely in the cytosol. This image of the cytoskeleton explains the solgel transformation that occurs in the cyosol. The polymerization and depolymerization of this lattice may also provide for small movements of the contained organelles. The same observations have been made on thin sections of various cells. Networks of microfilaments have been observed in the growth cones of neuroblasts and in the undulating membranes of cultured cells. A thin sheath of microfilaments forming a network is observed below the plasma membrane of fibroblast and glial cells. In nerve endings a three dimensional network of actin filaments has been observed.

Cytochalasin

During the study of microtubules the depolymerizing effect of colchicine has been of paramount importance, in the case of microfilaments the action of the drug *cytochalasin B* has had a decisive influence. Although the mechanism of action is still unknown, cytochalasin B has been found to impair numerous cell activities in which some types of microfilaments are involved. For example, it inhibits smooth muscle contraction, beat of heart cells, migration of cells, cytokinesis, endocytosis, exocytosis, and other processes.

During application to the alga Nilella, cytochalasin B produces rapid cessation of cyclosis, which is reinitiated after removal of the drug. Cytochalasin B disrupts the regular arrangement of microfilaments associated with some of these functions. For example, the contractile ring of microfilaments observed during cell cleavage is altered by this drug. Not all microfilaments, however, are sensitive to the drug. These and other studies have led to the general conclusion that the cytochalasin-B-sensitive microfilaments are the contractile machinery of non-muscle cells.

Actin, Myosin, and other Contractile Protein

The contractile proteins *actin* and *myosin* are present in non-muscle cells has for the first time which is suggested a molecular mechanism for cellular movements. It is now known that all eukaryotic cells contain high concentrations of *actin* and low concentrations of *myosin*, the two *force-generating proteins.*

In addition, there are other proteins, such as *tropomyosin* and *troponin* that, as in muscle, play a regulatory function. Finally, there are *actin-binding* proteins that can cross-link action filaments, and a-*actinin*, a protein found in the Z-line of muscle and which may participate in actin-filament-membrane interactions.

Actin

It comprises a large proportion of the cytoplasmic proteins of many cells; in developing nerve cells, it may constitute up to 20 per cent of these proteins. Actin is present principally in its globular form (G-actin) with a molecular weight of 45,000 and it may quickly polymerize to form the microfilaments of fibrous actin (F-actin). This G to F-actin transition seems to represent the basis of the classical sol-gel transition in the cytoplasm of moving cells.

Cytroplasmic action is very similar in structure to muscle actin and forms identical 6-nm wide microfilaments consisting of a double helical array of globular actin molecules. Actin filaments can be identified under the electron microscope by virtue of the fact that they become *decorated* with myosin (heavy meromyosin) to form arrow-like complexes. *Myosin* has been found in cells having ameboid motion, such as amebae and slime molds, and also in platelets.

Myosin

If we compare with actin, myosin is found in a low concentration and is rather heterogenous in size and composition. In most cases it has a molecular weight of 460,000 daltons, similar to that of muscle myosin, but the two molecules contain different subunits. In *Acanthamoeba* there is a "*minimyosin*" with a MW of 180,000 daltons. In spite of these differences, all myosins bind reversibly to actin filaments and contain a Ca^{2+} activated ATPase. As in muscle, both these functions are localized in the globular heads of the molecule.

Identification of myosin filaments with the electron microscope, at present, is less satisfactory. However, filaments that are 14 to 22 runs thick and 0.7 μm long, which are of probable myosin nature, have been observed.

Contractile and Regulatory Proteins

The microfilaments can be detected by fluorescent antibodies against *actin*, *myosin*, *tropomyosin*, and *a-actinin*. In culture cells actin has a typical pattern of straight and often parallel fibres that are predominantly localized on the lower, contacting side of the cell and that are in close contact with the plasma membrane. This pattern is completely disorganized by the action of cytochalasin B, which does not affect the microtubules.

Staining of culture cells anti-myosin antibodies produces a somewhat similar image to that of actin; however, the majority of the filaments show interruptions or striations. Staining with anti-a-actinin antibodies also reveals a similar pattern, with a periodic arrangement along the filament. These results suggest that *α*-actinin (and also *tropomyosin*) may be involved in the organization of the actin bundles of microfilaments and in the attachment of them to the plasma membrane, especially in thc areas in which there is contact with a solid substrate or with other cells. In these contacting regions actin microfilaments are arranged in parallel, forming bundles that in a living spreading cell may be visible as *stress fibres*. This is very interesting that the microvillus of the intestinc, which contains a bundle of actin filaments that is attached to the plasma membrane by way of a dense material that stains with antibodies against a-actinin. This actin bundle has a polarization similar to that found in amuscle sarcomere except that in this case the Z-line is replaced by the a-actinin at the tip of the microvillus.

Two Recognizable Types of Microfilaments

In addition to microtubules (20 to 30 nm) and actin microfilaments (5 to 7 nm), in a variety of cells there is a third cytoskeletal components composed of filaments of an intermediate size (7 to 12 nm). These have been identified in fibroblasts, muscle cells, and glial cells. Other examples include the *neurofilaments* of nerves, the *tonofilaments* of epithelia, and the *prekeratin filaments* of cells in which keratinization occurs.

Studies with fluorescent antibodies in culture cells have shown the presence of a system of wavy fibres, some of which lie near the cell boundary. This system of fibres is affected neither by cytochalasin, which disorganizes actin filaments, nor by Colcemid, which depolymerizes microtubles. This *intermediate system* of microfilaments appears to be related to the anchorage and maintenance of the epithelioid cell shape in nerve axons the neurofilaments could be involved in axoplasmic transport. Isolation of intermediate filaments is facilitated in cell cultures of baby hamster kidney treated with colchicine. In these cells the 10-nm filaments aggregate to form juxtanuclear caps, which can be isolated. Two proteins of 54,000 and 55,000 daltons were resolved by SDS gel electrophoresis of these intermediary filaments.

The basis of contraction in non-muscular systems is still hypothetical, but the scheme that is evolving is in some ways similar to that of muscle. In non-muscular systems it is based on the interaction of actin and myosin filaments, with the consequent production of a shearing force. The differences between the mechanism here and in muscle are in the more random distribution of theseproteins and the much smaller concentration of myosin. These factors may help to account for the much slower concentration in these primitive contractile systems. The attachment of active filaments to the plasma membrane, how- ever, is favourable to the production of shearing forces acting at the level of the membrane.

Many motile events in non-muscle cells, Like cell spreading and shape changes, locomotion, membrane ruffling, cytokinsis, phagocytosis, and formation of pseudopodia, require a tight coupling between the cell surface and the contractile apparatus, and this could be achieved by a-actinin molecules that bind the actin microfilaments to the plasma membrane

Against muscle cells, in which actin filaments are stable, most non-muscle cells have transitory filaments that may appear or disappear within one minute. Thus the cell must have some mechanism that controls the assembly and disassembly of the actin monomers and the formation of microfilament bundles. Such mechanisms are altered in cells undergoing transformation by oncogenic viruses some of which may show a reduced number of bundles.

We shall now consider two classic types of cell motility in which micro filaments and actinmyosin interactions are involved. This is generally found in the plant cell and another called *ameboid motion* which is found in animal cells and in certain protozoa.

Cytoplasmic Streaming (Cyclosis)

Cytoplasmic streaming, or *cyclosis*, is easily observed in plant cells, in which the cytoplasm is generally reduced to a layer next to the cellulose wall and to fine trabeculae crossing the large central vacuole. Continuous currents can be seen that displace cloroplasts and other cytoplasmic granules.

In some plant cells the protoplasmic current can be initiated by chemicals *(chemodynesis) or* by light *(photodynesis).* Cyclosis is modified by temperature, by The action of ions, or by change in pH. Cyclosis is stopped by mechanical injuries, electric shock, or some anesthetics. Some auxins (plants growth hormones). Increase the rate of cyclosis. In general, all the factors that decrease cell viscosity increase the speed of protoplasmic current and vice versa. Cyclosis decreases progressively in cells subjected to increased hydrostatic pressure at the same time that protoplasm becomes more liquid.

The classic experimental work on cyclosis has utilized the cylindroid cells of *Nitella,* which have a thin protoplasmic layer of about 15 mm surrounding a central vacuole of 0.5 mm by 10 cm. This protoplasmic layer is divided into a cortical region of structured cytoplasm in which are embedded non-motile chloroplasts and a layer of isolated cytoplasm (the endoplasm) where cyclosis takes place. The whole system is surrounded by the plasma membrane, under which is a single layer of microtubules.

Budles of actin filaments of about 0.2 mm are situated just beneath thechloroplasts in the cortical region and in a direction that is parallel to that of cyclosis. (Each bundle is composed of about 100 microfilaments of 5 to 6 nm.) It is assumed that the separation of the ectoplas and endoplasm is labile and the actin filaments may be incorporated into the endoplasm.

It has been proved that plant cytoplasm also contains a myosin-like protein. Thus, the actual motive force may be provided b a mechanism of actin-myosin interaction that takes place in the rgion between the stationary ectoplasm and the moving endoplasm. Furthermore, i has been found that cytoplasmic streaing is stopped when the cortex of the *Nitella* is treated with cytochalasin B.

Ameboid Motion

In ameboid motion the cell changes shape actively, sending forth cytopasmic projections called *pseudopodia,* into which the rotoplasm flows. Although this special form of locomotion can be observed easily in amebae, it also occurs innumerous ther types of cells.One need only to place a drop of blood between a slide and coverglass to see that the leukocytes, at first spheroidal, change their shape, emit pseudopodia, and move about. Intissue cultures, cells move out actively, forming the zone of migration. These changes also occur invivo. For examplein epithelial repair, the cells free themselves and slide along actively towards the depth of the wound. In an inflammatory process, leukocytes wander out of the blood vessels *(dlapedests)* by active ameboid motion and progress towards the focus of infection. Some amebae are predominantly *monopodial* but other may be temporarily or permanently *polypodial.* The shape of pseudopodia varies between a stout, almost cylinadrical *lobopodium* and a fine filamentous or branching *filopodium*. Sometimes these fine processes may be anastomosing (resticulopodia), as in Foraminitera.

As shown by the classic studies of Mast, the protoplasm of the ameba has a clear ectoplasm, which expands consderably towards the end of the pseudopodium. The axial endoplasm is surrounded by a “shear zone” where particles move more freely. At the advancing end is the hyaline cap, and just posterior to it, the “fountain zone,” where the axial endoplasm appears to contract actively and flows below the ectoplasmic tube. At the opposite end is the tail process, also called the *moid,* and near it, the *recruitment zone,* where the endoplasm is recruited from the walls of the ecoplasm in the posterior third of te cell. The rate of progresion varies between 0.5 and 4.6 mm per second among different amebae. In the neutrophilic leukocytes it is approximatly 0.58 mm per second. This rate is modified by temperature and other environmental factors. Insufficient oxygen supply does not stop the movement but does slow it. For this type of locomotion calcium is needed. Serve mechanical injury, electric shock, or ultraviolet radiation causes reraction of pseudopodia.

An important factor in ameboid motion is *adhesion to a solid support.* An ameb that floats freely in the liquid medium can emit pseudopodia but does not progress; only when it adheres to a solid surface doe it commence this type of locomotion. There are substances that at influene the motion by attracting or repelling the cells. This property, d is called *chemotaxi.* It has great importance in defence mechanisms. All that has been said thus far about microfilaments andcell motility aplies to ameboid motion. Actin filaments and myosin aggregates, together forming thicker filaments, have, been found in amebae, and there seems no doubt that the actin-myosin interaction provides the actual motive fore for ameboid motion. However, the views are divided regarding the most likely site of contraction. Whereas some investigtors regard th posterior region of the ectoplasmic tube as more active, others give more importance to the "fountain zone" at the advancing end ofthe ameba.

Types of Plant Cell

Plant cells are eukaryotic cells that differ in several key respects from the cells of other eukarotic organisms. A large central vacuole, a water-filled volume enclosed by a membrane known as the tonoplast maintains the cell's turgor, controls movement of molecules between the cytosol and sap, stores useful material and digests waste proteins and organelles. A cell wall composed of cellulose and hemicellulose, pectin and in many cases lignin, are secreted by the protoplast on the outside of the cell membrane. This contrasts with the cell walls of fungi (which are made of chitin), and of bacteria, which are made of peptidoglycan.

Specialised cell-cell communication pathways known as plasmodesmata, poes in the primary cell wall through which the plasmalemma and endoplasmic reticulum of adjacent cells are continuous. Plastids, notably the chloroplasts which contain chlorophyll and the biochemical systems for light harvesting and photosynthesis, but also amyloplasts specialized for starch storage, elaioplasts specialized for fat storage and chromoplasts specialized for synthesis and storage of pigments.

As in mitochondria, which have a genome encoding 37 genes plastids have their own genomes of about 100-120 unique genes and probably arose as prokaryotic endosymbionts living in the cells of an early eukaryotic ancestor of the land plants and algae. Unlike animal cells, plant cells are stationary. Cell division by construction of a phragmoplast as a template for building a cell plate late in ytokinesis is characteristic of land plants and a few groups of algae, notably the Charophytes and the Order Trentepohliales. The sperm of bryophytes have flagellae similar to those in animals, but higher plants, (including Gymnosperms and flowering plants) lack the flagellae and centrioles that are present in animal cells.

Cell Types

Parenchyma cells are living cells that have diverse functions ranging from storage and support to photosynthesis and phloem loadig (transfer cells). Apart from the xylem and phloem in its vascular bundles, leaves are mainly composed of parenchyma cells. Some parenchyma cells, as in the epidermis, are specialized for light penetration and focusing or regulation of gas exchange, but others are among the least specialized cells in plant tissue, nd may remain totipotent, capabl of dividing to produce new populations of undifferentiated cells, throughot their lives.

Parenchyma cells have thin, permeable primary walls enabling the transport of small molecules between them, and their cytoplasm is responsible for a wide range of biochemical functions such as necta secretion, or the manufacture of secondary products that discourage herbivory. Parenchyma cells which contain many chloroplasts and are concerned primarily with photosynthesis are called chlorenchyma cells. Others, such as the majority of the parenchyma cells in potato tubers andthe seed cotyledons of legumes have a storage function.

Collenchyma cells-collenchyma cells are alive at maturity and have only a primary wall and a secondary wall. These cells mature from meristem derivatives that initially resemble parenchyma, but differences quickly become apparent. Plastids do not develop and the secretory apparatus (ER and Golgi) proliferates to secrete additional primary wall. The wall is most commonly thickest at the corners, where three or more cells come in contact and thinnest where only two cells come in contact, though other arrangements of the wall thickening are possible. Pectin and hemicellulose are the dominant constituents of collenchyma cell walls of dicotyledon angiosperms which may contain as little as 20% of cellulose in Petasites. Collenchyma cells are typically quite elongated, and may divide transversely to give a septate appearance. The role of this cell type is to support the plant in axes still growing in length, and to confer flexibility and tensile strength on tissues. The primary wall lacks lignin that would make it tough and rigid, so this cell type provides what could be called plastic support. Support that can hold a young stem or petiole into the air, but in cells that can be stretched as the cells around them elongate. Stretchable support (without elastic snap-back) is a good way to describe what collenchyma does. Parts of the strings in celery are collenchyma.

Sclerenchyma cells-Sclerenchyma cells (from the Greek skleros, hard) are hard and tough cells with a function in mechanical support. They are of two broad types–sclereids or stone cell and fibres. The cells develop an extensive secondary cell wall that is laid down on the inside of the primary cell wall. The secondary wall is impregnated with lignin, making it hard and impermeable to water. Thus, these cells cannot survive for long as they cannot exchange sufficient material to maintain active metabolism. Sclerenchyma cells are typically dead at functional maturity, and the cytoplasm is missing, leaving an empty central cavity.

Functions for sclereid cells (hard cells that give leaves or fruits a gritty texture) include iscouraging herbivory, by damaging digestive passages in small insect larval stages, and physical protection (a solid tissue of hard sclereid cells form the pit wall in a peach and many other fruit). Functions of fibres include provision of load-bearing support and tensile strength to the leaves and stems of herbaceous plants. Sclerenchyma fibres are not involved in conduction, either of water and nutrients (as in the xylem) or of carbon compounds (as in the phloem), but it is likely that they may have evolved as modifications of xylem and phloem initials in early land plants.

Tissue Types

The major classes of cells differentiate from undifferentiated meristematic cells (analogous to the stem cells of animals) to form the tissue structures of roots, stems, leaves, flowers and reproductive structures. Xylem cells are elongated cells with lignified secondary thickening of the cell walls. Xylem cells are specialised for conduction of water, and first appeared in plants during their transition to land in the Silurian period more than 425 million years ago. The possession of xylem defines the vascular plants or Tracheophytes. Xylem tracheids are pointed, elongated xylem cells, the simplest of which have continuous primary cell walls and lignified secondary wall thickenings in the form of rings, hoops or reticulate networks. More complex tracheids with valve-like perforations called bordered pits characterise the gymnosperms.

The ferns and other pteridophytes and the gymnosperms only have xylem tracheids, while the angiosperms also have xylem vessels. Vessel members are hollow xylem cells aligned end-to-end, without end walls that are assembled into long continuous tubes. The bryophytes lack true xylem cells, but their sporophytes have a water conducting tissue known as the hydrome that is composed of elongated cells of simpler construction. Phloem is a specialised tissue for food conduction in higher plants.The conduction of food is a complex process which is carried in the plant with the help of special cell called phloem cells. These cells conduct inter and intra cellular fluid (food:- proteins and other essential element required by the plant for its metabolism) through the process of osmosis. This phenomena is called as ascent of sap in plants. Phloem consists of two cell types, the sieve tubes and the intimately-associated companion cells.

The sieve tube elements lack nuclei and ribosomes, and their metabolism and functions are regulated by the adjacent nucleate companion cells. Sieve tubes are joined end to end with perforate end-plates between known as sieve plates, which allow transport of photosynthate between the sieve elements. The companion cells, connected to the sieve tubes via plasmodesmata, are responsible for loading the phloem with sugars. The bryophytes lack phloem, but moss sporophytes have a simpler tissue with analogous function known as the leptome. Plant epidermal cells are specialised parenchym cells covering the external surfaces of leaves, stems and roots. The epidermal cells of aerial organs arise from the superficial layer of cells known as the tunica (L1 and

L2 layers) that covers the plant shoot apx, whereas the cortex and vascular tissues arise from innermost layer of the shoot apex known as the corpus (L3 layer). The epidermis of oots originates from the layer of cells immediately beneath the root cap. The epidermis of all aerial organs, but not roots, is covered with a cuticle made of waxes and the polyester cutin. Several cell types may be present in the epidermis. Notable among these are the stomatal guard cells, glandular and clothing hairs or trichomes, and the root hairs of primary roots. In the shoot epidermis of most plants, only the guard cells have chloroplasts. The epidermal cells of the primary shoot are thought to be the only plant cells with the biochemical capacity to synthesize cutin.

Cell Wall of Plant Cells

There is a kind of *exoskeleton* that provides protection and mechanical support for the plant cell. This includes the maintenance of a balance between the osmotic pressure of the intracellular fluid and the tendency of water to penetrate the cell. When plant cells are placed in a solution that has an osmotic pressure similar to that of the intracellular fluid, the cytoplasm remains adherent to the cellulose wall. When the solution of the medium is more concentrated than that of the cell, it loses water, and the cytoplasm retracts from the rigid cell wall. On the other hand, when the solution of the medium is less concentrated than that of the intracellular fluid, the cell swells and eventually bursts.

The formation, from cambial cells, of *xylem vessels* (which become lignified) or of *phloem tissue* (used to transport to the rest of the plant the material that has been photosynthesized in the leaves)—is mainly the result of the special synthesis and assembly of the cell wall. In some cases, the cell wall, can be compared to a piece of plastic reinforced by glass fibres. The wall consists of a *microfibrillar* network lying in a gel-like *matrix* of interlinked molecules. The microfibrils are mostly *cellulose* (the most abundant biological product on earth) consisting of straight polysaccharide chains made of glucose units linked by 1-4 β- bonds. Macrofibrils visible with the light microscope are composed of microfibrils about 25 nm in diameter, each of which is, in turn, composed of about 2000 cellulose chains. About 100 cellulose chains are held together in an elementary fibril. X-ray diffraction reveals a crystalline pattern in cellulose, with a repeating period of 1.03 nm along the fibber axis. This corresponds to a *cellobiose* unit composed of two *β*-glucose molecules.

Extra than some protein, the matrix contains other major polysaccharide fractions are:

- Pectin substances (containing galactose, arabinose, and galacturonic acid), which are soluble in water,
- Hemicelluloses (composed of glucose, xylose, mannose, and glucuronic acid), which are extracted with alkali.

There is only lignin found in maters cell wall which is made of an insoluble aromatic polymer resulting from the polymerization of phenolic alcohols. Certain cell walls may have *cuticular substances (cutin waxes)* and *mineral* deposits in the form of calcium and magnesium carbonates and silicates. In fungi and yeasts the cell wall is composed of *chitin*, a polymer of glucosamine.

Actually, the cell wall determines the *spape* of cells and serves in the classification of plant tissues. (Consideration of specialized plant tissues is beyond the scope of this book. In general, *Primary* and *secondary* walls develop in a special sequence and are differentiated by the disposition of the cellulose microfibrils and by the composition of the matrix.

Regarding the classification of the layers in a matured cell wall depends upon the knowledge of its development during and after cell division. The *cell plate,* which is formed mainly from vesicles of the Golgi complex that become aligned at the equatorial plane, forms the intercellular layer or *middle lamellae* of the more mature cell wall; this layer contains pectin substances. After cell division each daughter cell deposits other layers that constitute the *primary cell wall,* which is composed of pectin, hemicellulose, and a loose network of the cellulose microfibrils, most of which are oriented transverse to the long axis of the cell.

Incase the cell has enlarged to its matter size, then with the development of large vacuolar, the secondary wall appears. When the cell has enlarged to its mature size, usually with the development of large vacuoles the *secondary wall* appears. This consists of material that is added to the inner surface of the primary wall, either a *homogeneous thickening,* as in cells forming sieve tubes *(phloem),* or as *localized thickenings,* as in xylem vessels. In both cases this material consists mainly of cellulose and hemicellulose, with little pectin.

In xylem development a further differentiation consists of the penetration of *lignin* from the outside into the secondary thickenings. This hydrophobic polymer replaces water and finally encrusts all the microfibrils and the matrix. At this stage the is lignified and the cell dies. Regarding the biogenesis of cellulose and other cell wall components two main pathways have been described. One involves the *Golgi complex*, and the other appears to be directly associated with the *plasma membrane* involvement of the Golgi is very evident in some scale-bearing algae. These scales consist of a radial microfibrillar network associated with a spiral one, and an amorphous wall.

The synthesis of polysaccharides in the Golgi of the algae has been followed by the use of periodic acid and silver methenamine. It is evident that the Golgi membranes with their *glucosyl transferase* content are able to polymerize glucan chains into cellulosic microfibrils. The whole process of synthesis, assembly, and release of the scales at the cell surface has been followed under the electron microscope.

The plasma membrane is the most common site of cellulose synthesis in plant cells. However, one must not disregard the essential functions that are probably carried out by the endoplasmic reticulum and the Golgi complex. It is possible that the glycosyl transferases synthesized in the endoplasmic reticulum could be transferred to the Golgi, and from there to the plasma membrane, where they may become active in the synthesis of microfibrils. Using freeze-fracturing, globular complexes situated at the end of growing microfibrils have been observed in the secondary walls of cotton fibres. These globules have been interpreted as enzyme complexes involved in the synthesis of the microfibrils. Such globules, which are mobile, may be guided by cytoplasmic microtubules.

This is the main characteristic of most plant cell is that the presence of bridges of cytoplasmic material that establish a continuity between adjacent cells. These bridges, called *plasmodesmata*, pass through the thickness of the pectocellulose membrane. They have a thin plasma membrane and many contain tubules that establish continuities between the endoplastic reticulum of both cells. The presence of plasmodesmata permits the free circulation of fluid, which is essential to the maintenance of plant cell tonicity, and probably also allows passage of solutes and even of macromolecules. According to these concepts, cell walls do not represent complete partitions between cells, but constitute a vast syncytium supported by a skeleton formed by the pectocellulose membranes.

Regarding the formation of plasmodesmaty, the same is concerned with the formation of the *cell plate*, mentioned earlier, which appears at the equator of dividing cells during telophase. At this time the cell plate is crossed by vesicles and tubules of the endoplasmic reticulum that determine the location of the plasmodesmata.

Cytoplasm

In meristematic cells the membranes of the cytoplasmic vacuolar system are relatively scanty and are masked by the numerous ribosomes that fill the cytoplasmic matrix. Indeed, in undifferentiated cells most of these particles are not attaced to the membranes, but are free in the matrix. Microtubules are preferentially localized below the plasma membrane and are oriented tangentially. A role in wall deposition has been postulated for these structures. Microtubules are found in the regions of cytoplasm that underlie the points where the microfibrils of the secondary wall are being deposited. As mentioned earlier, they may play a role in guiding the enzymes involed in the biosynthesis of the microfibrils, and may thus determine the pattern (spiral or reticular) in which the microfibrils are deposited. Reticulum is plant has rough portions with attached ribosomes and smooth sections in animal cells. It has been postulated that the plasma membrane of the plant cell invaginates and actively takes in fluid (pinocytosis). The same process also occurs in numerous animal cells. The great development of the endomembrane system during cell differentiation os related to the intense hydration of the cytoplasm. This process may give rise to huge vacuoles that are filled with liquid and may be confluent. As a result, the cytoplasm may become compressed in a thin layer against the cellulose membrane and may show cytoplasmic movements, called *cyclosis.*

The endoplasmic reticulum plays a role in the biosynthesis of proteins, lipids, and polysaccharides and in the formation of certain organelles, such as *protein bodies, glyoxysomes,* and vacuoles. We mentioned before that the plasmodesmata are crossed by tubules of smooth endoplasmic reticulum. In cells that differentiate to form sieve tubes, pads of *callose,* a polysaccharide composed of glucose units linked by 1-3 β-bonds, is deposited over the endoplasmic reticulum. Thus, this organelle is also related to the formation of sieve pores.

The seed is the resting stage of the plant in which the cells are in a highly dehydrated state. In addition to the primordia of the root and shoot, the seed is composed of food stores in the form of starch, fat, and proteins.

Seed development involves the following three main stages:

1. A *division phase,* in which the cells of the embryo and the cotyledons are formed.
2. A *phase of cell expansion,* in which proteins and other reserves are synthesized and stored in cotyledons.
3. A *drying phase*, in which there is loss of water with further deposition of the reserves.

Several proteins (*i.e., zein* in corn endosperm, *vicilin* and *legumin* in legume cotyledons) ae localized in special organelles called *protein bodies* or *aleurone grains*. During the expansion phase, the cells have an extensive rough endoplasmic reticulum which is involved in the biosynthesis of stored proteins. These are first accumulated in the central vacuole, and later on smaller protein bodies are formed, possibly within cisternae of the endoplasmic reticulum or the Golgi complex. This process has been followed radioautographically by the use of labeled amino acids. *Zein,* the main protein of corn, is synthesized by mRNA that is specifically associated with ribosomes attached to the endoplasmic reticulum. It has been found that protein bodies also have bound ribosomes that may produce zein.

Seed Germination is germination is accompanied by the metabolism of the protein bodies of cotyledons. An *endopeptidase* synthesized in the seedlings penetrates into the protein bodies, causing hydrolysis of the stored proteins. Enzyme synthesis is preceded by proliferation of the rough endoplasmic reticulum. When treated with the hormone *gibberellic acid,* the aleurone cells of cereal endosperm produce α-amylase and other hydrolytic enzymes, which are released extracellulary and attack the starch reserves. The synthesis of these enzymes is correlated with an extensive proliferation of the rough endoplasmic reticulum in these aleurone cells.

Regarding the germination of the castor bean (*ricinus*) is accompanied by breakdown of a large amount of fat stored in the endosperm. This process is accompanied by the development of *glyoxysomes,* special organelles involved in triglyceride metabolism. Glyoxysomes consist of an amorphous protein matrix surrounded by a limiting membrane. Both the phospholipid components and the enzymes of these organelles originate from the endoplasmic reticulum.

The enzyme of the glyoxysome are used to transform the fat stores of the seed the seed into carbohydrates by way of the *glyoxylate cycle* which is a modification of the Krebs cycle. The overall equation of this cycle is as follows 2 acetyl CoA $\rightarrow$ succinate + $2H^+$. The difference between the glyoxylate cycle and the Krebs cycle is that the former used two auxiliary enzymes, *isocitrate lyase* and *malate synthetase,* and requires two molecules of acetyl CoA, instead of one. The other three enzymes of the cycle (*citrate synthetase, aconitase,* and *malate dehydrogenase*) are the same as those present in the Krebs cycle.

All the enzymes of the glyoxylate cycle have been found in isolated glyoxysomes of the castor bean endosperm. When seeds of the castor bean germinate, considerable change is produced involving the synthesis of glyoxysomic enzymes. If the endosperm is fractionated at different states of germination, it is possible to observe that the marker enzyme malate synthetase hifts from fractions corresponding to the endoplasmic reticulum to others representing the glyoxysomes. These findings support the concept that these organelles are made in the endoplasmic reticulum.

At telophase they aggregate at the periphery of the cell plate and form small vesicles, which fuse to form the plate. As in animal cells, the dictyosomes of some plant cells (*e.g.*, root cap cells of maize) are directly related to secretion. The Golgi cisternae become filled with secretion products, which are then concentrated and discharged.

Dictyosomes and their associated vesicles are numerous in cells involved in the synthesis of mucilage, and in those of the outer root cap of the bean.

The mucilage is produced at the expense of the starch bodies present in plastids. The dictyosomes of plant cells contain specific enzymes, such as thiamine pyrophosphatase and inosinic diphosphatase. Incorporation of labeled glucose is highest during the formation of the cell plate. The role of the Golgi apparatus in cell wall formation was described above.

Plant cell's mitochondria have a structure essentially similar to that of those in animal cells. In meristems, mitochondria have relatively few crests and an abundant matrix. During differentiation this internal structure may vary. In cells engaged in photosynthesis (leaf cells), mitochondria show an increased number of crests; in cells containing starch granules (amyloplasts), mitochondria remain undifferentiated, as in meristems.

One important point is the relationship between mitochondria and chloroplasts. Guillermond studied this problem in leaf meristems and postulated that in early stages there are two types of organelles. One is typically composed of short mitochondria and the other of long filamentous bodies. These elongated organelles increase in thickness and may give rise to vesicles, starch granules, or chloroplasts, whereas the short mitochondria remain unchanged. According to this view, both types of organelles are independent. The introduction of electron microscopy has, in general, confirmed this viewpoint. Although early plastids, also called proplastids, in some ways resemble mitochondria, they are readily distinguishable because of fewer projections of the inner membrane, their large size, and the presence of dense granules in the matrix.

The Chloroplast and other Plastids

Schimper first used the term "plastid" for special cytoplasmic organells present in eukaryotic plant cells in 1883. The most important of all, *the chloroplasts,* are characterized by the presence of pigments such as *chlorophyll* and *cerotenoids* and by their fundamental role in *photosynthesis*.

Extra than chloroplasts, other coloured plastyids may be observed. These are grouped under the name *chromoplasts*. Yellow or orange chromoplasts occur in petals, fruits, and roots of certain higher plants. In general, they have a reduced chlorophyll content and are thus less active photosynthetically. The red colour of ripe tomatoes is the result of chromoplasts that contain the red pigment *lycopene* a member of the carotenoid family. Chromoplasts containing various pigments (*e.g.*, *phyco-erythrin* and *phycocyanin*) are found in algae.

Colourless plastids or leukoplasts are found in embryonic and germ cells. During embryonic development leukoplasts in certain differentiated zones of the root produce starch granules, called *amyloplasts*. They may be seen under the polarization microscope, because of their characteristic birefringence, or they may be distinguished by histochemical reactions for starch. Leukoplasts are also found in meristematic cells and in those regions of the plant not receiving light. Plastids located in the cotyledon and the primordium of the stem are colourless at first but eventually become filled with chlorophyll and acquire the characteristic green colour of chloroplasts. It has already been studied that the protein bodies, other contain starch, fat (*elaioplasts or oleosomes*), or essential oils. It was noted in *Phaseolus vulgaris,* that a single leukoplast may store both starch and protein; *i.e.*, it may qualify as both an amyloplast and a proteinoplast. The starch granules are deposited in the matrix or stroma of the plastid, while the protein, which sometimes shows a crystalline arrangement, is always accumulated within membrane-bound sacs.

It is the greatest biological importance that chloroplasts are the most common of the plastides. By the process of *photosynthesis,* they produce oxygen and most of the chemical energy used on our planet by living organisms. Life is essentially maintained by the chloroplasts; without them there would be no plants or animals, since the latter feed on the foodstuffs produced by plants. From a human point of view we may say that each molecule of oxygen used in respiration and each carbon atom in our bodies has at one time passed through a chloroplast. Chloroplasts are localized mainly in the cells of the leaves of higher plants and in algae.

In the point of view of evolutionary the first living organism were and appeared at least one billion year before photosynthesis. This process appeared first in prokaryotes and then in eukaryotes. With the release of oxygen, a by-product of photosynthesis, the atmosphere favoured aerobic organisms, and anaerobic organisms became a small fraction of all living forms.

In photosynthetic prokaryotes and in the chloropasts of eukaryotes photosynthesis and carried out in species membrances. Photosynthesis is carried out in special membranes in photosynthetic prokaryotes and in the chloroplasts of eukaryotes. These organelles are able to capture light energy and to transform it into chemical energy in the form of ATP, $NADP^+$, and the various foodstuffs that are produced. There is much difference between the shape of the cells within a species and chloroplasts. But these organelles are relatively constant within cells of the same tissue. In leaves of higher plants, each cell contains a large number of spheroid, ovoid, or discoid chloroplasts. Some are club-shaped, having a thin middle zone and bulging ends filled with chlorophyll. Chloroplasts are frequently vesicular, with a colourless centre. The presence of starch granules is detected by the characteristic blue iodine reaction.

Considerably the size of chloroplastes varies. The average diameter in higher plants is 4 to 6 μm. This is constant for a given cell type, but sexual and genetic differences are found. For instance, chloroplasts in polyploid cells are larger than those in the corresponding diploid cells. In general, chloroplasts of plants grown in the shade are larger and contain more chlorophyll than those of plants grown in sunlight. Algae, such as *chlamydomonas,* often possess a single huge chloroplast. In higher plants there are 20 to 40 chloroplasts per cell. It has been calculated that the leaf of *Ricinus communis* contains about 400,000 chloroplasts per square millimeter of surface area. When the number of chloroplasts is insufficient, it is increased by division; when excessive, it is reduced by degeneration. The *distribution* of the chloroplasts within the cell may vary with the amount of light energy Chloroplasts are sometimes distributed homogeneously within the cytoplasm, but are frequently packed near the nucleus, or close to the cell wall.

Observation of living epidermal cells from leaves of Iris..... And other genera has shown that chloroplasts are displaced and deformed by the action of cytoplasmic streaming (cyclosis). In addition to this passive *motility,* active movements of an ameboid or contractile type, which are sometimes related to the degree of illumination, have been observed.

In chloroplasts isolated from spinach, the changes in shape and volume caused by the presence of eight have been obersed. The volume decreases considerably after the chloroplasts are struck by light and photophosphorylation is initiated; this effect is reversible. In the dark contraction of chloroplasts may be induced by addition of ATP. Two proteins having contractile properties, which may account for this phenomenon, have been extracted from isolated chloroplasts. Chloroplasts apparently multiply by division—elongation of the plastid and constriction of the central portion. They have a higher density than the cytoplasm and migrate to the centrifugal pole of the cell when submitted to the action of centrifugal force.

During the study under eight microscope it had already demonstrated that many chloroplasts have a heterogeneous structure made up of small granules called *grana*, which are embedded within the stroma, or matrix. The size of the rana varies from 0.3 to 1.7 μm depending on the species. The smallest ones, within the limit of microscopic visibility, are more numerous. They are flat bodies shaped like platelets or disks, which in a lateral view appear as dense bands perpendicular to the chloroplast surface.

The electron microscope has revealed the true structure of the chloroplast, with its three main components: the envelope, the stroma, and the thylakoids. The envelop is made of a double limiting membrane across which the molecular interchange with the cytosol occurs.

In contrast to the ultrastructure of mitochondria, the inner membrance of mature chloroplasts is not in continuity with the thylakoids (Which could be compared with the mitochondrial crests). In spinach, the outer surface of the envelope os of the order of 500 cm^2 per gm of leaves. Isolated membranes have a yellow colour due to the presence of small amounts of cartenoids. However, these areas lack chlorophyll and cytochromes and contain only 1 to 2 per cent of the protein of the chloroplast. Stroma fill is a kind of gel-fluid phase that surrounds the thylakoids. This component contains about 50 per cent of the chloroplast proteins and most of these are of the soluble type. It has ribosomes and also DNA, both of which are involved in the synthesis of some of the structural proteins of the chloroplast. As will be discussed later, the stroma is where CO_2 fixation occures and where the synthesis of sugars, starch, fatty acids, and some proteins takes place.

The outer surface of the thylakoid is in contact with the stroma, and its inner surface encloses an *intrathylakoid space.* Thylakoids may be *stacked* like a pile of coins, forming the grana or they may be *unstacked (stroma thylakoids), forming a system of anastomosing tubules that are joined to the* grana thylakoids. The number of thylakoids per granum may vary from a few to 50 or more. The thylakoids contain about 50 per cent of the protein and all the components involved in the essential steps of photosynthesis, such as chlorophyll, the cytochromes, plastocyanine, and others.

By the use of the technique of freez fracturing detailed information about the macromolecular organization of the thylakoid membrane has been gained. By the use of freeze-fracturing, the thylakoid membrane appears as a smooth continuum, representing one half of the lipid bilayer, with paticles of various sizes believed to the protein macromolecules or aggregates which are asymmetrically localized within the lipid matrix.

The freezecleaving of the membrane results in a selective segregation of the particles in either the PF or the EF leaflet. Observation of shows that PFu and PFs contain a large number of small particles, while in EFs the particles are larger and less numerous. Very few of these larger particles are found in the EFu face (corresponding to the unstacked intrathylakoid face). A possible interpretation of these observations will be given later on in this discussion of the chloroplast. Proplastids are limited by a double membrane. In the presence of light the inner membrane grows and gives off vesicles that arrange themselves to form larger discs. In the granal regions, piles of closely packed thylakoids are built.

In the mature chloroplasts some compartments of the grana remain connected by intergranal tubules. When plants are grown under low light intensity *(etiolation),* the vesicles formed in the proplastid aggregate, forming one or several *prolamellar bodies.* Sometimes the vesicles form a crystalline pattern consisting of regularly connected tubules. When these plants are re-exposed to light, the vesicles may fuse into layers and again develop into grana.

Molecular Organization of Thylakoid

About the molecular organization of thylakoids, the modern ideas are largely based on the fluid lipid-protein mosaic model of the membrane proposed by Singer and Nicolson in 1972.

This basic concept includes several fundamental notions:

- *Fluidity,* which allows for the free lateral movement of components in the thylakoid membrane;
- *Asymmetry,* in the sense that the crossing of molecules from one half of the bilayer to the other is restricted, permitting the separation of different components;
- *Economy,* since placing the components in a thinmembrane eliminates random movements in the third dimension.

As use have already studied under photosynthesis, in thylakoids there are two *photosystems* (I and II), each consisting of an assembly of many chlorophyll and carotenoid molecules, and a reaction centre where the primary conversion of light into chemical energy takes place. Each assembly of pigments is associated with structural proteins and a system of electron transport to constitute the photosystem. The two photosystems act in series to transfer electrons from water to $NADP^+$. At the same time photophosphorylation is coupled to the electron transport. Here, we will only mention that *photosystem II* (PS II) is involved in the release of O_2 from H_2O, while photosystem I (PS I) plays the major role in the reduction of $NADP^+$.

Lipids represent about 50 per cent of the thylakoid membrane; this includes those directly involved in photosynthesis, such as *chlorophylls* (21 per cent), *carotenoids* (3 per cent). There are also *structural* lipids, which are mainly glycolipids (41 per cent), sulfolipids (4 per cent), and a few phospholipids (10 per cent).

Most of the structural lipids are highly unsaturated. The substance chlorophyll is an asymmetrical molecule having a hydrophilic head made of four pyrrole rings bound to each other and forming a porphyrin. This part of the molecule is similar to some animal pigments, such as hemoglobin and cytochromes. In chlorophyll, however, there is a Mg atom forming a complex with the four rings. In animal pigments the Mg is replaced by Fe. Chlorophyll has a long hydrophobic chain (phytol chain) attached to one of the rings.

Regarding higher plants there are two types of chlorophyll—*a* and *b*. In chlorophyll *b* there is a —CHO group in place of the—CH_3 group, indicated by a circle. Three other types of chlorophylls are found in algae, diatoms, dinoflagellates, and in photosynthetic bacteria. Unlike the cytochromes, the chlorophylls are not bound to proteins. A small amount of pigment absorbing at 700 nm has been found. This pigment is called P700 because it is bleached when it absorbs at that wavelength. The pigments concerned with the group called.

Carotenoids are masked by the green colour of chlorophyll. In autumn, the amount of chlorophyll decreases and the other pigments become apparent. These belong to the *carotenes* and *xanthophylis,* which are both related to vitamin. A Carotenes are characterized chemically by the presence of a short chain of unsaturated hydrocarbon, which makes them completely hydrophobic. Xanthophylls, on the contrary, have several hydroxyl groups.

This is a well known concept that most of the chlorophyll molecules are physically involved with intrinsic proteins, forming chlorophyll-protein complexes. The following complexes have been isolated: the *PS I complex,* containing the reaction centre molecule P700; the *PS II complex,* with the reaction centre molecule P 680; and the *light-harvesting complex.*

Small fragments of membranes rich in PS I and with a high chlorophyll *a*/chlorophyll *b*/ratio have been isolated both from granal and stomal thylakoids; larger fragments rich in PS II and with a low chlorophyll *a*/chlorophyll *b* ratio are derived mainly from granal thylakoids. LHCP (Light Harvesting Chlorophyll Protein) contains 40 to 60 per cent of the total chlorophyll of the mature thylakoid. This complex is photochemically inactive but transfers light energy to PS II and possibly to PS I, and it contains both types of chlorophyll molecules, It has been proposed that the light-harvesting chlorophylls may be located around their specific intrinsic proteins as shown in the model indicated.

The photophosphorylating system comprises an ATPase similar to that of mitochondria, a *coupling factor* CF_1 located on the outer thylakoid surface, and a CF_0 segment across the membrane. Electron microscopic studies show that CF_1 is localized mainly in stromal thylakoids and is excluded from the stacked regions. The CF_1 can diffuse in the lateral plane of the membrane. Also in this system are the electron transport components of both photosystems. In PS I there is evidence of a vectorial disposition, with the acceptor-site ferredoxin and the ferredoxin $NADP^+$ -reductase situated towards the surface of the membrane. The localization of the components of PS II is not well established, however, it is likely that this system also has a vectorial arrangement across the membrane. It is generally accepted that both photosystems I and II have their electron acceptors on the outer or stromal side of the thylakoid membrane and that the electron donors are inside (in the thylakoid space), so that the whole electron transport system is situated across the membrane.

Photosynthesis

In plants photosynthesis is one of the most fundamental biological functions. By means of the chlorophyll contained in the chloroplasts green plants trap the energy of sunlight emitted as photons and transform it into chemical energy. This energy is stored in the chemical bonds that are produced during the synthesis of various foodstuffs. Photosynthesis is somewhat the reverse process. Chloroplasts and mitochondria have many structural and functional similarities, but there are also several differences.

The overall reaction of photosynthesis is:

$$nCO_2 + nH_2O \xrightarrow[\text{chlorophyll}]{\text{light}} (CH_2O)_n + nO_2$$

This shows that, essentially, photosynthesis is the combining of carbon dioxide and water to form various carbohydrates with loss of oxygen.

Each CO_2 molecule from the atmosphere is incorporated into a plant every 200 years, and that all the oxygen in the atmosphere is renewed by plants every 2000 years. Without plants there would be no oxygen in the atmosphere, and life would be almost impossible.

The carbohydrates first formed by photosynthesis are soluble sugars; these can be stored as granules of starch or other polysaccharides inside the chloroplasts or, more usually, inside the amyloplasts. After several steps involving different types of plastids and enzymatic systems, the photosynthesized material is either stored as a reserve product or used as a structural part of the plant (*e.g.*, cellulose). In reaction H_2O is the hydrogen donor in much the same way that, as shown in the following equation, H_2S is the donor in sulfur bacteria:

$$2H_2S + CO_2 \xrightarrow{\text{light}} (CH_2O) + 2S + H_2O$$

Thus, reaction in higher plants can be written as follows:

$$2nH_2O + nCO_2 \xrightarrow{\text{light}} (CH_2O)n + nH_2O + nO_2$$

Reaction shows that water is the H_2 donor and all the O_2 liberated comes from water. Experiments using water labeled with heavy oxygen ($H_2{}^{18}O$) have confirmed this. In this process water participates primarily as a proton and electron donor. Reaction involves a complex series of steps, of which some take place only in the presence of light and the others take place also in darkness—hence the names *light* and *dark reactions.* In the first, light is absorbed and used by chlorophyll; this is the *photochemical (Hill) reaction.*

In the second or dark reaction CO_2 is fixed and reduced by thermochemical mechanisms. Under the study of the photochemical reaction, it is necessary to recall the process of oxidatives phosphorylation in mitochondria. In oxidative phosphorylation the flow of electrons is from $NADH_2$ to O_2, following the path of standard oxidation-reduction potentials (*i.e.*, from —0.6 to +0.81 volts); in photosynthesis, the opposite process takes place. The electrons flow from H2O to $NADPH_2$ (*i.e.*, from +0.81 to —0.6 volts).

Regarding photochemical reaction, the primary reactions of photosynthesis, takes place in tylakoid membrane. When these are illuminated, there is a transfer of electrons from water (E'_0 = 0.81 volts). Transfer against such a negative electrochemical gradient uses the energy provided by the photons of light. Furthermore, the transfer of electrons is accomplished by a chain of *electron carriers* and is coupled with the *phosphorylation* of ADP to ATP. Within the thylakoid membrane, the energy of light is collected by the two *photosystems* PS I and PS II, which act as collector antennae.

Photosystems

There is experimental evidence that the efficiency of photosynthesis is a function of the wavelength of the impinging light.

For example, if chloroplasts are simultaneously excited at 680 and 650 nm, there is considerable stimulation of the photochemical reaction. The existence of two systems of light absorption has been postulated: Photosystem

I is stimulated at the longer wavelength and is not accompanied by the production of O_2, whereas photosystem II is activated at the shorter wavelength and yields O_2.

Both photosystems operate in a sequential and interrelated fashion. When photosystem is excited, electrons are boosted to a higher energy level, thereby reducing ferredoxin. This compound is then reoxidized by the transfer of electrons to $NADP^+$, forming $NADPH_2$. Two photons of light (*i.e.*, two excitons) appear to be needed to produce one molecule of $NADPH_2$. This reducing component, having high energy electrons, can be used in a variety of biochemical reactions—particularly in the synthesis of carbohydrates.

To restore the electrons to photosystem I this process of light absorption must be coordinated with photosystem II. In photosystem II the quanta of light remove electrons from the hydrogen of water and boost them to a higher energy level, thus reducing plastoquinone. From here the electrons are brought back to photosystem I by way of an electron transfer system (*i.e.*, cytochrome b559, cytochrome f, and plastocyanine). From this diagram it is evident that only the excited photosystem II results in the release of O_2 from water.

Electron Transport Systems

Freedoxins plays a role in the transport of electrons from photosystem I to $NADP^+$. Ferredoxin is a protein of about 11,600 daltons and contains iron and sulfur but lacks a porphyrin group. Another component is *ferredoxin-NADP oxireductase,* a flavoprotein that uses $NADP^+$ as an electron acceptor. In the transfer of electrons from photosystem II to I there are other carriers. For example, there is *cytochrome* a hydrophobic protein of 100,000 daltons isolated in non-polar solvents; other carriers include the *cytochromes* b_6 and b_3 (or 559), tightly bound to the membrane structure. Chloroplasts also contain a copper-protein called *plastocyanine* and two quinones—*vitamin* K_1 and *plastoquinone*—the latter similar to ubiquinone in mitochondria.

The sequence in the transfer of electrons from photosystem II to I begins with a non-identified carrier *Q*. This transfer is followed by the sequence: plastoquinone → cytochrome b559 → cytochrome f → plastocyanine. Between the two photosystems the mechanism of interaction is still to be elucidated. Since there is a large pool of plastoquinone molecules, which are lipophilic and very mobile, the link may be not a direct, but a dynamic one.

Photophosphorylation

Regarding mitochondria, in which the electron transport system is coupled with phosphorylation of ADP, in chloroplasts there is a light-induced phosphorylation that is coupled with the electron transfer chain of photosystem II. This process can be uncoupled by 2, 4-dinitrophenol, which also acts in mitochondria.

In mitochondria, the coupling in chloroplasts involves an ATPase having the coupling factor CF_1 on the outer surface of the thylakoid membrane and a CFe hydrophobic segment across the membrane. The mechanisms possibly involved in this coupling are similar to those postulated for mitochondria. Here also, the *chemical-conformational* and the *chemiosmotic* coupling hypotheses have been proposed. The latter, however, is especially favoured in view of the evidence provided by Jagendorf. In his experiments a suspension of spinach chloroplasts with an initial pH of 7 was transferred to a medium with a acid pH 4, they were put in a medium of pH 8, and ADP and Pi were added. Under these conditions there was ATP synthesis driven by the pH difference; this result conforms exactly to the postulate of Mitchell.

In contrast to the oxidative phosphorylation of mitochondria, O_2 is not used in photophosphorylation of chloroplasts. Green plants can produce 30 times as much ATP by photophosphorylation as by oxidative phosphorylation in their own mitochondria. In addition, these plants contain many more chloroplasts than mitochondria. For every one molecule of O_2 liberated eight light quanta are needed. In the overall photosynthetic pathway and two molecules of $NADPH_2$ are formed; the efficiency is close to 25 per cent. Along the electron transport there is the synthesis of three ATP molecules.

The moleculer of ATP and NADPHz produced in the thylakoids provide the chloroplast stroma with the energy needed to fix the CO_2 and to synthesize the carbohydrates. Along with the energy provided by ATP, the reduced $NADPH_2$ can bring about the reduction of atmospheric CO_2 and combine it with the hydrogen to form the various carbohydrates. This process involves many steps, which have been elucidated principally by the use of radioactive CO_2.

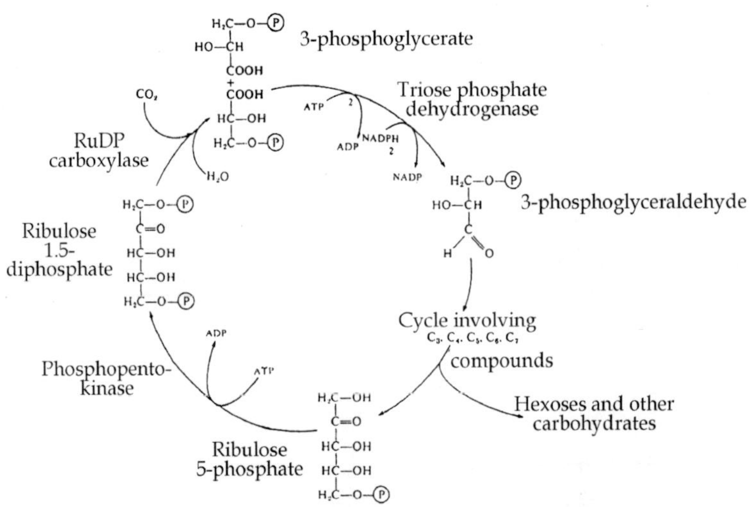

Figure 1: *Diagram of the Calvin or C_3 cycle of photosynthesis in which CO_2 is reduced and carbohydrates are synthesized.*

The reactions involved are so rapid that they appear one second or less after the addition of CO_2. These reactions occur in complete darkness if the plant has been previously exposed to light.

In cells exposed to CO_2 for five seconds, the dominating compound is 3-phosphoglycerate; Two triose phosphate molecules unite to form hexose (fructose) diphosphate, from which glucose phosphate is formed. Then, from glucose phosphate various disaccharides and polysaccharides are formed.

The initial enzyme of the Calvin cycle is *ribulose 1, 5-diphosphate carboxylase* (*i.e.*, *carboxydismutase*), which has a large molecular weight (500,000 daltons), eight large subunits synthesized by the chloroplast, and eight small subunits produced in the cytoplasm. This enzyme, which constitutes half the stromal proteins, plays the central roles of integrating into the pentose (ribulose 1, 5-diphosphate) a molecule of CO_2 from the air and of producing two molecules of *3-phosphoglycerate,* by the use of one molecule of H_2O.

The *3-phosphoglycerate* is then phosphorylated by ATP and forms on activated molecule that is able to accept hydrogens and electrons from NADPH. This molecule is *3-phosphoglyceraldehyde,* which is then reduced to from hexose sugars and more complex carbohydrates. Benson and calvin proposed that in this cycle (C_3) a molecule of ribulose 1, 5-diphosphate is regenerated at each turn. Such regeneration is very complex and involves some twelve enzymatic reactions. To produce a hexose from the CO_2 fixation it is believed that six turns of the cycle are needed.

The classical equation for the synthesis of one hexose is:

$$6CO_2 + 12H_2O \xrightarrow{\text{light}} C_6H_{12}O_6 + 6O_2 + 6H_2O$$

This represents the storage of 686 kcal/mole of hexose. This energy is provided by a total of 12 NADPH and 18 ATP molecules that themselves represent 750 kcal. Thus, the efficiency reached is as high as 90 per cent (686/750 × 100 = 90 per cent).

It is known that the *photosynthetic carbon reduction* (PCR) or *Calvin cycle,* which is present in most higher plants, there is the so-called C_4 *pathway,* which is found in a small number of *angiosperm* species. From the viewpoint of cell biology it is interesting that this C_4 pathway is integrated with the PCR cycle in different cells of the same plant.

The mesophyll cells the CO_2 is assimilated by carboxylation of phosphoenol pyruvate (PEP), which gives rise to C_4 acids, such as malate and aspartate. These are then transferred to the bundle sheath cells, probably by diffusion. Within these cells the CO_2 that is released by decarboxylation enters the PRC cycle, given origin to 3-phosphoglyceric acid, and the C_3 products may go back to the mesophyll cell to enter the C_4 pathway again.

Model of the Chloroplast Membrane

The use of biochemical and biophysical techniques, and especially the freeze-fracturing procedure with electron microscopy, support the concept of a fluid mosaic membrane model for the thylakoid membrane. According to this concept, there is a lipid continuum in which protein molecules are partially or totally embeded. As use already know that by solubilizing the membranes with non-ionic detergents, several integral protein complexes had been isolated and characterized. Three of them the *photosystems I and II* and the *light-harvesting complex,* have already been described. In addition, two more complexes have been separated. One corresponds to a *cytochrome* f-b_6 *complex* which, besides the cytochromes, contains non-heme iron, phospholipids, and carotenoids (but not chlorophyll), and which has a minimum molecular weight of 103,000 daltons.

The other complex is the recently isolated hydrophobic protein component of the chloroplast ATPase (CF_0 or HF_0). This component has six major polypeptides with MW between 11,000 and 42,000 daltons. This HF_0 protein binds the CF_1 (coupling factor of ATPase) to reconstitute the enzyme, but is also capable of proton transfer by itself. It should be stressed here that the CF_1, which is analogous to the F_1 of mitochondria, is a peripheral enzyme with a MW near 350,000 and consisting of five polypeptides.

If we try to establish a structural-functional model of the chloroplast membrane, the data obtained by freeze-fracturing and the existence of the above-mentioned intrinsic protein complexes should be considered. The numbers indicate e that the small PF partides are similarly distributed between the unstocked membranes (PFu) or stromal thylakoids and the stacked membranes (PFs), which correspond to the granal partitions. On the other hand, the large particles of the EF leaflet are more concentrated in the stacked region (EFs). Taking into consideration the surface area of the membranes in both regions, it can be calculated that 80 per cent of the EF particles are in the grana, and only 20 per cent in the stromal thylakoids. Several studies tend to identify the large EF particles with photosystem II and the associated light-harvesting complex (the complete PS II complex.

In fact, the proportion of PS II systems in both stacked and unstacked thylakoids is the same as that of the EF particles.. Furthermore, by analyzing the size of EF particles during the development of chloroplasts it was found that while at the beginning both the PF and EF particles are small, with the grinning process, the EF particles increase in size. These data were interpreted as indicating that the light-harvesting complexes were added to the basic "PS II core". Various integral proton complexes with large particles and a PF leaflet with smaller and more numerous particles. From this model it is evident that while the EF particles can be identified with the complete PS II system, the small PF particles are more difficult to interpret, since they may represent any of the other protein complexes.

Changes in structure of chloroplasts caused by the action of light or darkness have been observed both in vivo and in vitro. In many plant species chloroplasts flatten in the light and become more spherical in the dark. Such changes appear to the mediated by ion fluxes across the membranes of individual thylakoids. In fact, the size of the intrathylakoid space may decrease from 19.5 nm in the dark to 15.2 nm in the light, with a half-time of 1 min. Changes of this type are blocked by drugs that affect ion transport. Similar findings have been observed in vitro in isolated chloroplasts.

When we use freez fracturing it has been demonstrated that during illumination there are changes in the size and density of the particles within the membrane. A more dramatic change was observed in chloroplasts suspended in a solution low in cations. There was unstacking of the membranes coupled with a dispersion of the EF particles, which are normally concentrated in the granal regions. This change is reversible, and upon readdition of cations, stacking is re-established and the EF particles become concentrated in the grana. These observations suggest that the protein complexes have mobility within the fluid membrane.

In the case of chloroplasts the opposite reactions occur; under the action of light H^+ ions are absorbed, and the interior becomes acidified. At the same time K^+ and Mg^{2+} are ejected into the medium. In the dark, H^+ is ejected and K^+ and Mg^{2+} are reabsorbed by the chloroplasts. Thus, the chloroplast membrane has a vectorial organization that is the reverse of the mitochondrial membrane. In physical terms it is understood that dark reactions, being of a themochemical nature, do not need an ordered structural arrangement which are carried out by soluble enzymes. On the contrary, photophosphorylation requires the separation of positive and negative charges into specific pathways of electron flow. As in mitochondria, as per the *chemiosmotic hypothesis* of Mitchell, the proton gradient may serve as a driving force for the formation of ATP at the CF_1 ATPase.

Chloroplasts as Semiautonomous Organelles

Mitochondria one of the chloroplasts, exhibit a certain degree of functional autonomy within the intracellular environment. In fact, they undergo division and may contain some genetic information (cytoplasmic inheritance). Chloroplasts contain their own DNA, special ribosomes, smaller than those present in the cytoplasm, and all the necessary molecular machinery to achieve protein synthesis.

Shimper and Meyer accepted in 1883 that plastids multiply by fission. This is easily observed in unicellular algae that contain only one chloroplast. In the alga *Nitella* a division cycle of 18 hours has been recorded cinematographically. The division process is as orderly as chromosomal division. Plastid reproduction by fission implies a growth process of the daughter plastids. (The way in which plastids become differentiated during development and how they change in fine structure under the action of different external factors such as light have been mentioned earlier).

Inheritance does not follow typical Mendelian segregation, thereby indicating an extrachromosomal type of heredity, which has been shown experimentally. When the unicellular green alga *Euglena* is grown in the dark it contains small, colourless proplastids. If under these conditions the nucleus is irradiated with an ultraviolet microbeam, the proplastids are still capable of developing into chloroplasts.

However, if the cytoplasm is irradiated with the nucleus shielded, a considerable number of colourless colonies (lacking chloroplasts) develop. These results suggest that there is a cytoplasmic DNA in this alga. It has been generally accepted that a characteristic DNA occurs in chloroplasts of algae and higher plants. DNA regions resembling bacterial nucleoids were identified in thin sections under the electron microscope.

The quantity of DNA per chloroplast varies slightly with the species and is in the range of 2 to 5 × 10^{15} gm, or slightly less than in the bacterium *E. coli*. Segments of DNA as long as 150 μm have been separated from chloroplasts. The replication of chloroplast DNA has been followed with ^{3}H-thymidine. The presence of ribosomes within chloroplasts is easily demonstrated by electron microscopy.

These ribosomes are smaller than cytoplasmic ribosomes. Polysomes have also been separated from chloroplasts. The discovery of special ribosomes associated with chloroplasts provided evidence that these organelles contain a specific protein synthesizing system.

In the presence of CO_2 as the sole source of carbon, chloroplasts actively incorporate amino acids into proteins. Protein synthesis is preferentially inhibited by chloramphenicol concentrations that do not affect protein synthesis in the cytoplasm. The involvement of the two types of protein synthesis in the assembly of chloroplasts may be studied by using cycloheximide to inhibit synthesis due to cytoplasmic ribosomes.

Chloroplasts divide, grow, and differentiate; they contain DNA, ribosomal RNA, and messenger RNA, and are able to conduct protein synthesis. It has been suggested that chloroplasts may have resulted from a symbiotic relationship between an autotrophic microorganism, one able to transform energy from light, and a heterotrophic host cell.

However, it is evident that chloroplasts have in their DNA the genetic information to code for only a limited number of proteins and the elements necessary to establish a DNA-RNA-directed protein synthesis (*i.e.*, ribosomes, tRNA, and so forth).

Inchloroplasts protein may be synthesized:

- By an exclusive chloroplastic mechanism,
- By a mechanism involving nuclear genes and chloroplastic ribosomes,
- By nuclear genes and cytoplasmic ribosomes. Although the symbiont hypotheses is attractive, it is evident that synthesis and assembly of the components integrating a chloroplast depend to a great extent on nuclear genes and on cytoplasmically directed protein synthesis.

7

Tissue Culture

Tissue Culture in Cotton

Cotton is an economically important crop that is grown throughout the world. It is grown as a source of fibre, food and feed. Lint, the most economically important product from the cotton plant, provides a source of high quality fibre of the textile industry. Cotton seed is an important source of oil and cotton seed meal is a high protein product used as livestock feed. Although significant progress has been made in the field of cotton improvement with conventional breeding methodology, it has limitations to introduce new alleles.

Genetic engineering offers a directed method of plant breeding that selectively targets one or a few traits for introduction into the crop plant. The first transgenic cotton plants were obtained in 1987. Since then, many laboratories have obtained insect resistant or herbicide resistant transgenic cotton plants. Transgenic cotton has been mainly obtained by Agrobacterium-mediated transformation and particle bombardment. Transgenic plants have also been generated by regeneration from shoot apex tissues in both Gossypium hirsutum and G. barbadense.

More recently a high efficiency, embryogenic calli based method of obtaining transgenic cotton has been standardized in two Chinese cultivars. Another method followed to generate transgenic plants in cotton has been the pollen tube pathway transformation. This method is used to transfer DNA into zygotic embryos via pollen tube pathway. This is a widely used method in China. Nevertheless, one of the major drawbacks of the Agrobacterium-mediated transformation is the recalcitrance of cotton and its difficult-to-regenerate nature. In vitro regeneration in cotton from callus is limited to non-indigenous Coker cultivars or closely related genotypes. Even in the case of Coker cultivars, the low efficiency of somatic embryogenesis, elaborate culture procedures, relatively long time period required for regeneration and high seed to seed variation in response, collectively pose serious technical difficulties and restrict progress in cotton biotechnology. Hence this necessitates development of easy, reliable and efficient transformation protocols for cotton transformation for improvement, particularly in the Indian cultivars which are adapted to local conditions. To tackle the problems pertaining to regeneration in cotton and certain other recalcitrant crops, alternate methods to minimize or eliminate the steps of regeneration are being standardized.

These are called the in planta transformation protocols. Research with Arabidopsis has benefited from the development of high throughput transformation methods that avoid plant tissue culture. In particular, the development of the Agrobacterium tumefaciensmediated vacuum infiltration method has had a major impact on Arabidopsis research. In planta transformation methods have also been standardized for rice, buckwheat, kenaf and mulberry. In all the crops, Agrobacterium is directed towards either the apical meristem or the meristems of axillary buds. One such viable in planta transformation protocol has also been standardized for other crops. The strategy essentially involves in planta inoculation of embryo axes of germinating seeds and allowing them to grow into seedlings ex vitro. These in planta transformation protocols are advantageous over other methods because they do not involve regeneration procedures and therefore the tissue culture-induced somaclonal variations are avoided. This paper presents successful transformation of cotton by following in planta transformation protocol.

Plant Material

Seeds of a breeding line of cotton *viz.*, NC-71 were soaked overnight in distilled water and were surface sterilized first with 1 per cent Bavastin for 10 mins and later with 0.1 per cent HgCl2 for few seconds and washed thoroughly with distilled water after treatment with each sterilant. The seeds were later allowed to germinate on petriplates at 30°C in dark. Two-day old seedlings were taken as explants for Agrobacterium infection.

Bacterial Strain and Vector

Agrobacterium tumefaciens strain LBA4404, harbouring the binary vector pKIWI105, was used for transformation. The vector contains the uid A reporter gene driven by the CaMV 35S promoter and a neomycin phosphotransferase II (npt II) gene driven by the nopaline synthase promoter. The reporter gene of pKIWI105 is a version of uid A that lacks the bacterial ribosome binding site and shows no expression in Agrobacterium but good expression in plant cells.

Transformation and Recovery of Transformants

Agrobacterium strain LBA4404/pKIWI105 was grown overnight at 28°C in LB medium (pH 7.0) containing 50 ìgml-1 kanamycin. The bacterial culture was later resuspended in 100 ml of Winans' AB medium (pH 5.2) and grown for 18 h. For vir gene induction treatments wounded tobacco leaf extract (4 ml in 50 ml Winans' medium) was added separately to the Agrobacterium suspension in Winans' AB medium, 5 h before infection.

The seedlings with just emerging plumule were infected by separating the cotyledons without damaging them such that the meristem is visible, and then pricked at the meristem with a sterile sewing needle and subsequently dunked in the culture of Agrobacterium for 60 mins. Following infection, the seedlings were washed briefly with sterile water and later transferred to autoclaved soilrite moistened with water for germination under aseptic conditions in the growth room in wide mouth capped glass jars of 300 ml capacity, seedlings per jar. After 5 to 6 days, the seedlings were transferred to soilrite in pots and were allowed to grow under growth room conditions for at least 10 days before they were transferred to the greenhouse. The growth chamber was maintained at 26-28°C under a 14 h photoperiod with florescent light of intensity 35 ìmol m-2 s-1.

Expression of α-Glucuronidase

For analysis of the transformants, tissues that were tested and found free of residual Agrobacterium were used. The persistence of Agrobacterium in the putative transformants was largely controlled by a brief agitation (30s) of the co-cultivated embryos with 0.1 per cent sodium hypochlorite followed by thorough washes with distilled water. The method of Jefferson was used to assess histochemical uid A gene expression in the tissues of primary transformants, 4 days post co-cultivation, in the pollen and in the lint. The tissues were incubated overnight at 37°C in a solution containing 0.1M phosphate buffer, pH 7.0, 2 mM X-Gluc, 5 mM each of potassium ferricyanide and ferrocyanide and 0.1 per cent Triton X-100. The tissues were later soaked with 75 per cent ethanol to clear chlorophyll.

For GUS activity in the pollen, flowers with undehisced pollen were selected. After subjecting them to GUS histochemical analysis in an eppendorf tube, the flowers were crushed gently with a glass rod to release the pollen and the pollen was allowed to settle. A Pasteur pipette was used to take in most of the pollen that would have settled down and put on a glass slide. A drop of glycerol was placed on the slide and a cover slip was placed such that the pollen spread evenly on the glass slide. The slides were then observed under the microscope at 40X magnification and photographed. The GUS enzyme activity was quantified by measuring the hydrolysis rate of the fluorescent substrate 4-methyl umbelliferyl-â-D-glucuronide (MUG) as described by

Jefferson. The leaves of T3 generation cotton plants along with the wild type plants were used for the fluorogenic assay. The fresh leaf material (100 mg) was homogenized using a pestle and mortar in 1 ml of GUS extraction buffer (100 mM Potassium phosphate buffer, pH 7.4; 10 mM â-mercaptoethanol; 10 mM EDTA; 0.1 per cent (w/v) triton x-100; 0.2 mM PMSF and 0.1 per cent (w/v) Sodium lauryl sarcosine). The homogenates were centrifuged at 10,000 rpm for 10 min and supernatant was collected and stored on ice until further use. The protein concentration in the supernatant was measured as described by Bradford.

The fluorogenic reaction was carried out in GUS assay buffer (2 mM MUG in 100 ml of GUS extraction buffer). The reaction was set up in 96-well black Nunc microtiter plates. About 2 ìg of total protein was taken in 25 ìl of extraction buffer to which 25 ìl of GUS assay buffer was added and incubated at 37°C for 1 h after sealing the plate with aluminium foil. The reaction was stopped by adding 350 ìl of 0.2 mM Na_2CO_3 (stop buffer).

Fluorescence was measured with excitation and emission filters set at 365 nm and 455 nm using fluoromax-3 with micromax reader, with slit width set at excitation 2 nm and emission 3 nm. The GUS activity was quantified using freshly prepared 4-MU (4-methyl umbelliferone sodium salt) standard of 10nM, 20 nM, 40nM, 80 nM, 120 nM, 160 nM, 200 nM, 240 nM, 280 nM and 320 nM and enzyme blank in the stop buffer (0.2 M Na_2CO_3). The GUS activity was expressed as pmol of 4-MU/mg protein/min.

Molecular Analysis for the Presence and Integration of the Marker Genes

DNA for polymerase chain reaction (PCR) and Southern analysis from leaf tissue was isolated following the procedure of Dellaporta *et al.* Polymerase chain reaction. PCR was performed to amplify a 450 bp uid A gene fragment in the putative transformants.

In order to amplify the uid A gene fragment, PCR was initiated by a hot start at 94°C for 4 min followed by 32 cycles of 1 min at 94°C, 1 min 30 s at 60°C and 1 min at 72°C. Annealing temperature was at 58°C for the amplification of a 750 bp npt II gene. PCR was performed with both the primers in the same reaction mix to check for the co-integration of both the transgenes in the transformants. The products were run on a 1 per cent agarose gel. Southern analysis. In order to analyse the total genomic DNA for transgene integration of npt II gene, 15 ìg of total genomic DNA was digested with the appropriate restriction enzyme. The digested DNA samples were electrophoresed on a 0.8 per cent agarose gel. The separated fragments along with the uncut DNA were transferred onto a nylon membrane. Hybridization was performed at 65°C in Church buffer for 18 h. Membranes were washed for 30 min each in 2X SSC, 0.1 per cent SDS; 0.1X SSC, 0.1 per cent SDS at 65°C. Membranes were wrapped, placed on x-ray film and exposed at -70°C. For probe preparation, uid A gene was amplified using the gene specific primers, the fragment was random prime labeled and used for Southern hybridization.

Plant Transformation and Recovery of Transformants

Optimization of the conditions for transformation was largely based on the protocol standardized earlier for sunflower, groundnut and safflower. The feasibility of the transformation strategy adopted in the study was initially evaluated by the number of cotton embryo axes germinating into normal seedlings following wounding with a needle at the intercotyledonary region, infection with Agrobacterium and transplantation to the greenhouse. The seedlings could withstand infection for a period of 60 min and about 60 per cent of the seedlings survived after infection and developed into plantlets. Out of 50 seedlings that were subjected to Agrobacterium infection, 37 plants that survived the infection process were transferred to the greenhouse. Among them, 25 plants survived, attained maturity, flowered and set seed. However, the growth of these seedlings was slow when compared to that of untransformed controls.

Integration and Expression of the Transgene in T0 Plants.

Infection of the already differentiated embryonic tissue with Agrobacterium may result in the gene integration randomly and the T0 plants will be chimeric. However, the feasibility of gene integration in some tissues by in planta transformation protocol can be ascertained based on the GUS histochemical assay.

The GUS expression in the shoot apex of the primary transformants 5 days after infection whereas endogenous GUS-like activity was not seen in the non-transformed controls. Sections of the GUS-stained tissues revealed the expression of the uid A gene within the cells and not in the apoplastic region. The transgene expression in the cells of the pericycle and not the parenchymatous cells indicating the possibility of stable transformants in the next generation. The section of the wild type tissue did not show any staining. Based on these indicative experiments, the infected seedlings were transferred to pots in the greenhouse.

GUS histochemical analysis of the reproductive structures can provide leads about the possibility of generating stable T1 transformants. Among the large number of plants screened by histochemical GUS assay for expression in the pollen and lint, few plants showed expression in some of the flowers and developing bolls. Two plants, plant no. I and IV out of 25 T0 plants showing consistent GUS expression were chosen for further analysis in the subsequent T1 generation.

Integration and Expression of the Transgene in T1 Plants

Approximately 75 seeds were sown from each of the the two selected primary transformants. From plant I, 26 seeds germinated whereas 59 seeds germinated from plant IV. PCR analysis for uid A gene revealed the presence of a 450 bp amplicon in a total of 13 plants, 3 from I and 10 from IV. Further confirmation of the integration and inheritance of the transgene in the progeny plants was obtained by genomic southern analysis.

The hybridization signal in the uncut DNA of the PCR positive plants which were probed with a 450 bp uid A gene fragment. The variation in the hybridization pattern of the SmaI digested DNA confirms the inheritance and the integration of the uid A gene in the genome of the T1 plants. Three plants, I-E and I-S belonging to plant I of T0 and IV-14 belonging to plant IV of T0 showed the single copy insertions. These progeny plants were phenotypically normal, flowered and set seed in a normal fashion. GUS expression in the pollen of these plants confirmed stability of the transgene.

Stability of the Transgene in T2 Plants

From the selected plants, fifty, forty and sixty seeds were selected randomly from each of the plants and were sown in pots in the greenhouse. PCR was performed using primers for uidA gene in all the 78 plants that germinated *i.e.*, 26 from I-E, 20 from I-S and 32 from IV-14. Amplification of the 450 bp amplified fragment of the uid A gene in all the plants confirm the stable integration of the transgene. Expression of GUS in the pollen supported the PCR analysis.

Analysis of the T3 Generation Plants for the Integration of Transgenes

Following T2 analysis, 13 plants were randomly selected for further analysis into the T3 generation. A total of 130 plants were obtained from these 13 plants. They were subjected to both molecular and expression analysis. PCR with both uid A and npt II primers separately and both the primers in the same reaction confirmed the stability of the gene in all the selected transformants. Among the T3 generation plants, eleven plants (T3) were randomly selected for fluorimetric analysis of GUS activity. Retainment of GUS activity in these ten plants indicated continued inheritance and expression of the gus (uid A) gene.

Transgenic Technology

Recent advances in transgenic technology now make it possible to transfer and express various genes in agriculturally important species like cotton. The rapid development of cotton transformation technology not only provides a valuable method for introducing useful genes into cotton to improve important agronomic traits, but also helps in the study of gene function and regulation.

Although transformation rates have been significantly improved since the first report of success in the transformation of cotton, increase in transformation efficiency is still needed. Transformation techniques that evade tissue culture therefore become important in recalcitrant crops such as cotton. In the present study, a tissue culture-independent in planta transformation protocol was used to develop transformants. Such in planta transformation techniques have also been standardized in other crops like, buckwheat, soybean and rice etc.

In this method, Agrobacterium is targeted to the wounded apical meristem of the differentiated seed embryo. Therefore, Agrobacterium tumefaciens transfers the gene into the genome of diverse cells which are already destined to develop into specific organs and the meristematic cells still to be differentiated. This results in the primary transformants (T0) being chimeric in nature. This is the reason for the analysis of the transgenic plants to be carried out in the T1 generation. Nevertheless, analysis of the T0 generation plants was carried out with an objective to check for chimeras. Uid A gene used in the study that expresses only upon transfer to plant system facilitated the identification of the chimeras.

The first indication of the transformability and chimeric nature in T0 plants was obtained by the GUS histochemical analysis of the seedlings. In the indicative experiment to check the feasibility of the transformation, at least 60 per cent of the T0 seedlings showed GUS expression indicating transgene integration and also chimeric nature of the T0 seedlings. The chimeric plants producing the stable transformants in the T1 generation depends on the type of cells that were transformed in the T0 plants. If the transgene is integrated into undifferentiated meristematic cells which are destined to develop sympodial branches, seeds obtained from these reproductive structures of these branches would produce stable transformants in T1. Therefore, evaluation of the reproductive tissues in T0 is necessary. GUS expression in the lint and pollen of some of the plants clearly suggested that some of the T1 seeds from these plants would produce stable transformants.

In the analysis of T1 generation plants, transgenic nature was ascertained first by PCR and later by Southern analysis. Three plants (I-E, I-S, IV-14) were selected to be the basis for further generations because of the single copy insertion of the transgene as revealed by the Southern analysis. The progeny of these plants in the T2 generation showed the amplification of the transgene in all the randomly selected plants showing transgene inheritance. GUS expression analysis by using the fluorimetric assay in T3 generation plants also supported the transgenic nature as upto 16 fold increase in the GUS activity was seen in the transformants when compared to the controls. Southern analysis of these samples for the release of the gus (uid A) gene reconfirmed the transgenic nature of these selected plants.

Our results indicate that transgenic cotton plants can be obtained by Agrobacterium –based transformation strategy and transgene stability confirmed for three generations. The method therefore is advantageous because it avoids the need for tissue culture. Nevertheless, transformability depends on the susceptibility of the variety to Agrobacterium. A number of factors affect transformability by in planta transformation.

- First, the method depends on the number of T0 seedlings that survive and develop into normal mature plants after infection with Agrobacterium
- Second, chimeras have to be generated suggesting that transformation has occured.
- Third, transgene integration has to occur in the cells that eventually develop into reproductive structures that subsequently produce stable transformants.

With the presumption that 60 per cent of the T0 plants would be chimeric based on an indicative experiment, the surviving 25 T0 plants were maintained in the greenhouse. Nevertheless, analysis of GUS expression in the pollen and lint revealed that 8 per cent of the primary transformants, *i.e.*, 2 plants out of 25 were positive and showed consistent expression and therefore the probability of producing stable transformants. However, the number of stable transformants varied between the T0 plants considered for T1 generation. Plant no. I produced 3 transformants out of 24 plants based on PCR analysis for the uidA gene and 11 transformants from 59 plants belonged to plant no. IV. The transformation efficiency therefore in our study varies between 8.3 per cent and 19 per cent. Similar transformation efficiency based on PCR analysis has also been reported in other crops where in planta transformation methodology has been used.

In summary, this study provides experimental evidences for stable transformation by in planta transformation technique. It opens options to exploit transgene technology in cotton since efficient transformation in diverse genotype is still a constraint.

Tissue Culture of Banana

The propagation of a plant by using a plant part or single cell or group cell in a test tube under very controlled and hygienic conditions is called "Tissue Culture".

Status in India

Banana is a globally important fruit crop with 97.5 million tones of production. In India it supports livelihood of million of people. With total annual production of 16.91 million tones from 490.70 thousand ha., with national average of 33.5 T/ha. Maharashtra ranks first in production with 60 T/ha. Banana contributes 37 per cent to total fruit production in India.

Banana is one of the major and economically important fruit crop of Maharashtra. Banana occupy 20 per cent area among the total area under crop in India. Maharashtra ranks second in area and first in productivity in India. Jalgaon is a major Banana growing district in Maharashtra which occupy 50,000 hectares area under Banana. But most of Banana is grown by planting suckers. The technology development in agriculture is very fast, it results in developing Tissue Culture Technique.

Agro Climate

Banana is basically a tropical crop, grows well in temperature range of 13°C – 38°C with RH regime of 75-85 per cent. In India this crop is being cultivated in climate ranging from humid tropical to dry mild subtropics through selection of appropriate varieties like Grandnaine. Chilling injury occurs at temperatures below 12°C. The normal growth of the banana begins at 18°C, reaches optimum at 27°C, then declines and comes to a halt at 38°C. Higher temperature causes sun scorching. High velocity wind which exceeds 80 km phrs damages the crop.

Soil

Soil for banana should have good drainage, adequate fertility and moisture. Deep, rich loamy soil with pH between 6-7.5 are most preferred for banana cultivation. Ill drained, poorly aerated and nutritionally deficient soils are not suitable for banana. Saline solid, calcareous soil are not suitable for Banana cultivation. Avoided soil of low laying areas, very sandy and heavy black cotton with ill drainage. A soil that is not too acidic and not too alkaline, rich in organic material with high nitrogen content, adequate phosphorus level and plenty of potash are good for banana.

Varieties

In India banana is grown under diverse conditions and production systems. Selection of varieties, therefore is based on a large number of varieties catering to various kinds of needs and situations. However, around 20 cultivars *viz.* Dwarf Cavendish, Robusta, Monthan, Poovan, Nendran, Red banana, Nyali, Safed Velchi, Basarai, Ardhapuri, Rasthali, Karpurvalli, Karthali and Grandnaine etc. Grandnaine is gaining popularity and may soon be the most preferred variety due to its tolerance to biotic stresses and good quality bunches. Bunches have well spaced hands with straight orientation of figures, bigger in size. Fruit develops attractive uniform yellow colour with better self life and quality than other cultivars.

Land Preparation

Prior to planting banana, grow the green manuring crop like daincha, cowpea etc. and burry it in the soil. The land can be ploughed 2-4 times and leveled. Use ratovator or harrow to break the clod and bring the soil to a fine tilt. During soil preparation basal dose of FYM is added and thoroughly mixed into the soil. A pit size of 45cm x 45cm x 45cm is normally required. The pits are to be refilled with topsoil mixed with 10 kg of FYM (well decomposed), 250 gm of Neem cake and 20 gm of conbofuron. Prepared pits are left to solar radiation helps in killing the harmful insects, is effective against soil borne diseases and aids aeration. In saline alkali soil where PH is above 8 Pit mixture is to be modified to incorporate organic matter. Addition of organic matter helps in reducing salinity while addition of purlite improves, porosity and aeration. Alternative to planting in pits is planting in furrows. Depnding on soil strata one can choose appropriate method as well as spacing and depth at which plant is required to be planted.

Planting Material

Sword suckers weighing approximately 500-1000 gm are commonly used as propagating material. Suckers generally may be infected with some pathogens and nematodes. Similarly due to the variation in age and size of sucker the crop is not uniform, harvesting is prolonged and management becomes difficult. Therefore, in-vitro clonal propagation *i.e.* Tissue culture plants are recommended for planting. They are healthy, disease free, uniform and authentic. Properly hardened secondary seedlings are only recommended for planting

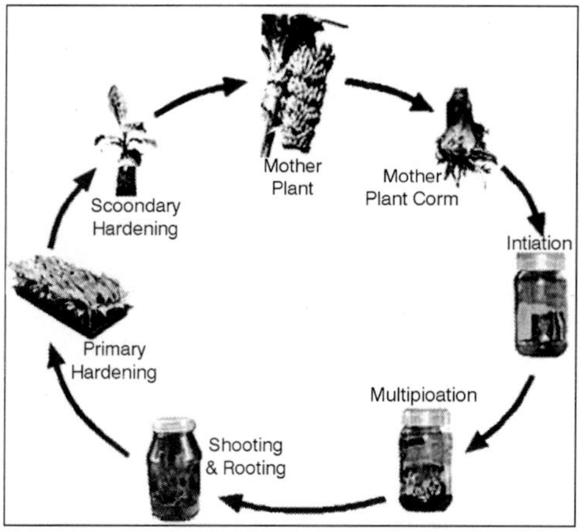

Figure 1: *Clonal Propagation Advantages of Tissue Culture Planting Material*

- True to the type of mother plant under well management.
- Pest and disease free seedlings.
- Uniform growth, increases yield.
- Early maturity of crop-maximum land use is possible in low land holding country like India.
- Round the year planting possible as seedlings are made available throughout the year.

- Two s uccessive ratoons are possible in a short duration which minimizes cost of cultivation.
- No staggered harvesting.
- 95 per cent-98 per cent plants bear bunches.
- New varieties can be introduced and multiplied in a short duration.

Planting Time

Planting of tissue culture Banana can be done throughout the year except when the temperature is too low or too high. Facility of drip irrigation system is important. There are two important seasons in Maharashtra, India; Mrig Baug (Kharif) Month of planting June-July. Kande Baug (Rabi) Month of planting October-November.

Crop Geometry

Traditionally banana growers plant the crop at 1.5 m x 1.5m with high density, however plant growth and yields are poor because of competition for sunlight.

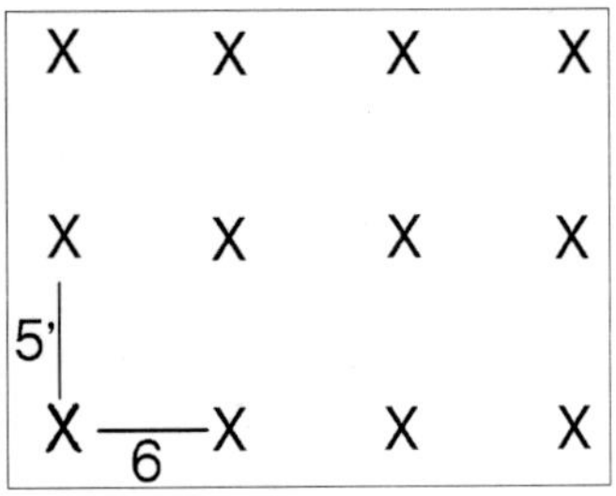

Figure 2: Crop Geometry var. Grand Naine

Various trials are conducted at Jain Irrigation System R&D farm with Grandnaine as cultivar. And then suitable spacing of 1.82 m x 1.52 m is being recommended, it accommodates 1452 plants per acre (3630 plants per hectare) keeping row direction North-South with wide spacing 1.82m between the rows. The region like north India, coastal belt and where humidity is very high and temp falls down upto 5-7°C, the planting distance should not be less than 2.1 m × 1.5 m.

Planting Method

Polybags is separated from the plant without disturbing the root ball of the plant and then plants are planted in the pits keeping the pseudo-stem 2 cm below the ground level. Soil around the plant is gently pressed. Deep planting should be avoided.

Water Management

Banana, a water loving plant, requires a large quantity of water for maximum productivity. But Banana roots are poor withdrawal of water.

Month (Maug Baug)	*Qty. (lpd.)*	*Month (Kande baug)*	*Qty. (lpd.)*
June	06	October	04-06
July	05	November	04
August	06	December	04
September	08	January	06
October	10-12	February	08-10
November	10	March	10-12
December	10	April	16-18
January	10	May	18-20
February	12	June	12
March	16-18	July	12
April	20-22	August	14
May	25-30	September	14-16

Therefore under Indian condition banana production should be supported by an efficient irrigation system like drip irrigation. Water requirement of banana has been worked out to be 2000 mm per Annum. Application of drip irrigation and mulching technology has reported improved water use efficiency. There is saving of 56 per cent of water and increasing yield by 23-32 per cent under drip. Irrigate the plants immediately after planting. Apply sufficient water and maintain field capacity. Excess irrigation will lead to root zone congestion due to removal of air from soil pores, thereby affecting plant establishment and growth. And hence drip method is must for proper water management in Banana.

Fertigation

Banana requires high amount of nutrients, which are often supplied only in part by the soil. Nutrient requirement has been worked out on all India basis is to be 20 kg FYM, 200gm N; 60-70gm P; 300gm K/plant. Banana requires heavy nutrition. Banana crop requires 7-8 Kg N, 0.7- 1.5 Kg P and 17-20 Kg K per metric tonne yield. Banana responds well to application of nutrients. Traditionally farmers use more of urea and less of phosphorous and potash. In order to avoid loss of nutrients from conventional fertilizers *i.e.* loss of N through leaching, volatilization, evaporation and loss of P and K by fixation in the soil, application of water soluble or liquid fertilizers through drip irrigation (fertigation) is encouraged. A 25-30 per cent increase in yield is observed using fertigation. Moreover, it saves labour and time and the distribution of nutrients is uniform.

Schedule of Application

The fertilizer schedule for tissue culture banana variety Grand Naine both in solid and water soluble form is given in the tables below:

Solid Fertilizer Schedule for Grand Naine Banana

Total nutrient requirement		
N - 200 gm/plant	P - 60-70 gm/plant	K - 300 gm/plant
Total quantity of fertilizer required per acre (Spacing 1.8 x 1.5 m; 1452 plants)		
Urea (N)	SSP (P)	MOP (K)
431.0	375.0	500 gm/plant
625.0	545.0	726 kg/acre

Water Soluble Solid Fertilizers

Schedule of water soluble fertilizer application.

Period	Grade	Qty. per 1000 plants (Kg) every 4th day basis	Total Qty. (Kg.)
After planting upto 65 days	Urea	4.13	82.60
	12:61:00	3.00	60.00
	00:00:50	5.00	100.00
65 to 135 days	Urea	6.00	120.00
	12:61:00	2.00	40.00
	00:00:50	5.00	100.00
135 to 165 days	Urea	6.50	65.00
	00:00:50	6.00	60.00
165 to 315 days	Urea	3.00	150.00
	00:00:50	6.00	300.00

Schedule is directive only and may change according to planting season and soil fertility status (soil analysis).

Interculture Operations

The Root system of banana is superficial and easily damaged by cultivation, use of intercrop which is not desirable. However short durational crops (45-60 days) like mung, cowpea, daincha are to be considered as green manuring crops. Crops from cucurbitaceous family should be avoided as these carry viruses.

Weeding

Spraying of Glyphosate (Round up) before planting at the rate of 2 lit/ha is carried out to keep the plantation weed free. One or two manual weedings are necessary.

Micronutrient Foliar Spray

Combined foliar application of $ZnSO_4$ (0.5 per cent), $FcSO_4$ (0.2 per cent), $CuSO_4$ (0.2 per cent) and H_3BO_3 (0.1 per cent) can be adopted to improve morphological, physiological and yield attributes of banana. The micronutrient spray solution is prepared by dissolving the following in 100 lit. of water.

Zinc sulphate	500 gm	}	For every 10 litre of mixture 5-10ml of sticker solution such as Teepol should be added before spraying.
Ferrom sulphate	200 gm		
Copper sulphate	200 gm		
Boric acid	100 gm		

Special Operations

There are operation specific to banana crop which influence productivity and quality.

Desuckering

Removal of unwanted suckers is a critical operation in banana for reducing internal competition with mother plant. Desuckering should be done regularly until shooting.

However in areas where ratoon is also taken for the second crop, a follower is allowed after inflorescence has appeared and this should be managed that planting space is not disturbed. Follower should be opposite to the inflorescence. It should not be far apart from the main plant.

Deflowering

It consists of removal of the withered style and perianth. This is generally not practiced. Therefore, they remain attached to the fruit bunch and then removed after harvesting which is damaging to the fruits. It is therefore suggested that we remove them just after flowering.

Pruning of Leaves

Rubbing leaves damages the fruit, therefore, such leaves should also be pruned during regular check. Older leaves and infected leaves also be pruned as required. Green leaves should not be removed.

Earthing up

Keep the soil loose by harrowing from time to time. Earthing up should be done at 3-4 months after planting *i.e.* raising the soil level around the base of the plant by 10- 12". It is better to prepare a raised bed and keep the drip line on bed 2-3"away from the plant. It also helps to protect plants from wind damage and production losses to some extent.

Removal of Male Buds

Removal of male buds helps fruit development and increases bunch weight. Male buds are removed from the last 1-2 small hands with a clean cut keeping a single finger in the last hand.

Bunch Spray

Spray of monocrotophos (0.2 per cent) after emergence of all hands takes care of the thrips. Thrips attack discolours the fruit skin and makes it unattractive.

Bunch Covering

Covering bunch using dried leaves of the plant is economical and prevents the bunch from direct exposure to sunlight. Bunch cover enhances quality of fruit. But in rainy season this practice should be avoided. Sleeving of bunch is done to protect fruits against dust, spray residue, insect and birds. For this blue plastic sleeves are preferred. This also increases temperature around developing bunch and helps in early maturity.

Dehandling of False Hands of Bunch

In a bunch there are some incomplete hands which are not fit for quality produce. These hands should be removed soon after bloom. This helps in improving the weight of other hands. Sometimes the hand just above the false hand is also removed.

Propping

Due to heavy weight of bunch the plant goes out of balance and the bearing plant may lodge and production and quality are adversely affected. Therefore they should be propped with the help of two bamboos forming a triangle by placing them against the stems on the leaning side. This also helps in uniform development of bunch.

Harvesting

Banana should be harvested at the physiological maturity stage for better post harvest quality. The fruit is climacteric and can reach consumption stage after ripening operation

Maturity Indices

These are established on the basis of fruit shape, angularity, grade or diameter of the median figure of the second hand, starch content and number of days that have elapsed after flowering. Market preferences can also affect the decision for harvesting a slight or full mature fruit.

Removal of Bunch

Bunch should be harvested when figures of second hand from top are 3/4 rounded with the help of sharp sickle 30cm above the first hand. Harvest may be delayed upto 100-110 days after opening of the first hand. Harvested bunch should generally be collected in well padded tray or basket and brought to the collection site. Bunches should be kept out of light after harvest, since this hastens ripening and softening. For local consumption, hands are often left on stalks and sold to retailers. For export, hands are cut into units of 4-16 fingers, graded for both length and girth, and carefully placed in polylined boxes to hold different weight depending on export requirements.

Post Harvest Operations

At collection site injured and over mature fruits are discarded and for local market bunches should be delivered through lorries or wagons. However, for more sophisticated and export market where the quality is predominant, bunches should be dehanded, fruits are cleared in running water or dilute sodium hypochlorite solution to remove the latex and treated with thiobendasole; air dried and graded on the basis of size of fingers as already stated, packed in ventilated CFB boxes of 14.5 kg capacity or as per requirement with polythene lining and pre-cooled at 13-15°C temperature and at 80-90 per cent RH. Such material should than be sent under cool chain at 13°C for marketing.

Yield

The planted crop gets ready for harvest within 11-12 months of planting. First ratoon crop would be ready by 8-10 month from the harvesting of the main crop and second ratoon by 8-9 months after the second crop. Thus over a period to 28-30 months, it is possible to harvest three crops *i.e.* one main crop and two ratoon crops. Under drip irrigation combined with Fertigation yield of Banana as high as 100 T/ha can be obtained with the help of tissue culture technique, even similar yield in the ratoon crops can be achieved if the crop is managed well.

Tissue Culture of Medicinal Plant

Medicinal plants are sources of important therapeutic aid for alleviating human ailments. With increasing realization of the health hazards and toxicity associated with the indiscriminate use of synthetic drugs and antibiotics, interest in the use of plants and plant-based drugs has revived throughout the world. However, a large number of medicinal plants remain to be investigated for their possible pharmacological value.

Most of the pharmaceutical industry is highly dependent on wild populations for the supply of raw materials for extraction of medicinally important compounds. Due to a lack of proper cultivation practices, destruction of plant habitats, and the illegal and indiscriminate collection of plants from these habitats, many medicinal plants are severely threatened. Advanced biotechnological methods of culturing plant cells and tissues should provide new means of conserving and rapidly propagating valuable, rare, and endangered medicinal plants.

Medicinal plants have been the subjects of man's curiosity since time immemorial. Almost every civilization has a history of medicinal plant use. Approximately 80 per cent of the people in the world's developing countries rely on traditional medicine for their primary health care needs, and about 85 per cent of traditional medicine involves the use of plant extracts. Interest in phytomedicine has exploded in the last few years, and about 500 different plant species are used as key ingredients, and many are still being collected from the wild. The resurgence of public interest in plantbased medicine coupled with rapid expansion of pharmaceutical industries have necessitated an increased demand for medicinal plants, leading to over-exploitation that threatens the survival of many rare species. Also, many medicinal plant species are disappearing at an alarming rate due to rapid agricultural and urban development, uncontrolled deforestation and indiscriminate collection. Combinations of in vitro propagation techniques and cryopreservation may help in conservation of biodiversity of locally used medicinal plants. Cryopreservation is a reliable method for long-term storage of the germplasm of endangered species. Several medicinal plant species have been successfully cryopreserved. In vitro cell and tissue culture methodology is envisaged as a mean for germplasm conservation to ensure the survival of endangered plant species, rapid mass propagation for large-scale revegetation, and for genetic manipulation studies.

Plants play a dominant role in the introduction of new therapeutic agents, and also drugs from the higher plants continue to occupy an important niche in modern medicine. Many compounds used in today's medicine have a complex structure, and synthesizing these bioactive compounds chemically at a low price is not easy. With deforestation, medicinal wealth is rapidly lost, such that many valuable plants are threatened with extinction. Pharmaceutical companies depend largely upon materials procured from naturally occurring stands that are being rapidly depleted. Plant tissue culture is an alternative method of propagation and is being used widely for the commercial propagation of a large number of plant species, including many medicinal plants.

The central mountain range of Taiwan is home to many highly valued medicinal herbs. These are being indis-criminately collected in large quantities from the wild to meet the ever-increasing demand for traditional crude drugs. Currently, collection of plants from within a national park is illegal in Taiwan. One of the solutions for protecting rare endemic medicinal herbs is to develop an efficient in vitro propagation method and to encourage commercial cultivation. The National Science Council (NSC) of Taiwan has been promoting research on traditional Chinese medicinal plants since 1988.

The main objectives of this research programme are:

- To collect information about important and rare traditional medicinal herbs,
- To develop simple methods of identifying medicinal herbs,
- To develop methods for mass propagation of medicinal herbs through tissue culture,
- To study active principles and pharmacology for their safer use,
- To promote export of traditional medicinal herbs.

At our research centre, work has been carried out on mass propagation of some important native and traditional Chinese medicinal herbs. This review summarizes the information about tissue culture studies on Chinese medicinal plants and related species reported earlier.

Also, it describes mass propagation of valuable medicinal herbs, Limonium wrightii, Adenophora triphylla, Gentiana davidii var. formosana, Anoectochilus formosanus, Scro-phularia yoshimurae, Pinellia ternata, Bupleurum falcatum, Zingiber zerumbet, Dendrobium linawianum, and Fritillaria hupehensis through shoot morphogenesis, and Angelica sinensis and Corydalis yanhusuo through somatic embryogenesis done at Taiwan Agricultural Research Institute and Chaoyang University of Technology, Taiwan.

Traditional Chinese Medicinal Plants

Plants can be regenerated and mass propagated in vitro either by shoot morphogenesis or somatic embryogenesis. Many important Chinese traditional medicinal herbs have been successfully regenerated in vitro. Each has a particular group of bioactive compounds. Taxol (plaxitaxol), a complex diterpene alkaloid found in the bark of Taxus tree is one of the most promising anticancer agents due to its unique mode of action on microtubular cell system. Latex from the opium poppy, Papaver somniferum, is a commercial source of the analgesics, morphine and codeine. The root of Panax ginseng C.A. Mayer, has been widely used as a tonic and precious medicine since ancient times. The primary bioactive constituents of ginseng are ginsenosides, a group of triterpenoid saponins. Berberine is an isoquinoline alkaloid found in roots of Coptis japonica.

Diosgenin from Dioscorea deltoidea, is a precursor for the chemical synthesis of steroidal drugs and possess tremendous importance to the pharmaceutical industry. Camptothecin, a potent antitumor alkaloid, was isolated from Camptotheca acuminata. Tanshinones are a group of quinoid diterpenoids believed to be active principles of Danshen (Salvia miltiorrhiza), a well-known traditional Chinese medicine. Tanshinone I and cryptotanshinone prevent complications of myocardial ischemia; tanshinone II A has undergone successful clinical trials for the treatment of angina pectoris in China. Podophyllotoxin is an antitumor aryltetralin lignan found in Podophyllum peltatum and Podophyllum hexandrum. It also serves as a starting material for the preparation of its semi-synthetic derivatives, etoposide and teniposide, both widely used in anti-tumor therapy.

We have standardized a protocol for rapid in vitro propagation of this medicinal plant using primary and lateral shoot tip, leaf bases, and inflorescence node explants. The explants induced adventitious shoots on Murashige and Skoog's medium (MS basal medium) supplemented with 8.87 ìM N6-benzyladenine (BA) and 1.07 μM a-naphthaleneacetic acid (NAA) after two months of culture. The adventitious shoots were proliferated by subculturing on MS medium supplemented with BA (2.21-17.75 μM) in combination with

NAA (1.07 μM). Maximum rate of shoot multiplication was observed on MS medium supplemented with 8.87 μM BA and 1.07 μM NAA. The percentage of explants forming shoots and the average number of adventitious shoot buds produced per explant were stimulated by increasing the strength (1/4x, 1/2x, 1x, 2x) of the MS medium. The shoots were rooted on MS basal medium with 4.92 ìM indole-3-butyric acid (IBA) and transferred to soil. The percentage survival of plantlets was 80 per cent. The plants did not show any morphological variation. Results obtained on rapid in vitro propagation for L. wrightii would help in conserving the germplasm and commercial cultivation of this economically important species.

Bibliography

Ashraf, Syed Mahboob: *Handbook of Fruit and Vegetable Products*, Agrobios, Delhi, 2008.

Banerjee, S : *Biochemistry and Biotechnology,* Dominant Pub, Delhi, 2009.

Barrington, E. J. W.: *Biochemistry of Primitive Deuterostomians*, London, Academic Press, 1974.

Bhaskar, Harsh: *Basic Facts on Biochemistry*, Campus Books, Delhi, 2010.

Bhatia, A. L.: *Biochemistry and Endocrinology,* Indus Valley, Delhi, 2002.

Bhatnagar, Vasudev: *Cell Science and Technology*, Campus Books, Delhi, 2009.

Bora, K.K.: *Agros Dictionary of Plant Physiology and Biochemistry*, Agrobios, Delhi, 2001.

Bose, N. Sarath Chandra: *Biochemistry: A Practical Manual,* Pharma Med Press, Delhi, 2010.

Bose, Sarath Chandra: *Biochemistry: A Practical Manual*, Pharma Med Press, Delhi, 2010.

Chanderlekha Goswami: *Biochemistry and Instrumentation*, Manglam Pub, Delhi, 2011.

Chauhan, B.S.: *Principles of Biochemistry and Biophysics,* Laxmi Publications, Delhi, 2008.

Chauhan, Sushil: *Experiments in Biochemistry,* Campus Books, Delhi, 2010.

Collymore L.: *Fruit Production in Barbados*, Port of Spain, Trinidad and Tobago, 1996.

Coultate, T. P.: *Food: The Chemistry of its Components,* London: The Royal Society of Chemistry, 1988.

Daphne C. Elliott: *Biochemistry and Molecular Biology*, Oxford University Press, Delhi, 2005.

Dar, Ghulam Hassan: *Soil Microbiology and Biochemistry,* New India Publishing Agency, Delhi, 2010.

Dutta, Chiranjib: *Fundamentals of Biochemistry: A Practical Approach,* Kanishka, Delhi, 2005.

Elliott, B N: *Biochemistry and Molecular Biology*, Oxford University Press, Delhi, 2005.

Florkin, M.: *Comprehensive Biochemistry: Amsterdam*, Elsevier, 1975.

Goel, A K: *Basic Concept of Organic Chemistry*, Pearl Books, Delhi, 2008.

Golakiya, B A: *DNA: A Bridge Between Biochemistry and Biotechnology,* New India Publishing Agency, Delhi, 2008.

Hemantaranjan, A L: *Developments in Physiology, Biochemistry and Molecular Biology of Plants*, New India Pub, Delhi, 2005.

Jasra, O.P.: *Environmental Biochemistry*, Sarup & Sons, Delhi, 2002.

Khanna, V K: *Objective Genetics, Biotechnology, Biochemistry and Forestry,* I.K. International Publishing House, Delhi, 2008.

Mido, Y.: *An Introduction to Nuclear Chemistry*, Discovery, Delhi, 1995.

Narvekar, Raghunath: *Handbook of Biochemistry,* Adhyayan, Delhi, 2008.

Prakash, M: *Insect Biochemistry,* Discovery Pub, Delhi, 2008.

Prasad, G.P.: *Biochemistry of Proteins*, Discovery, Delhi, 2010.

Purohit, S.S.: *Biochemistry: Fundamentals and Applications,* Agrobios, Delhi, 2009.

Rastogi, Arun: *An Introduction to Practical Biochemistry*, Anmol, Delhi, 2010.

Sangha, S P S: *Basic Techniques in Biochemistry and Molecular Biology,* I K International, Delhi, 2009.

Smriti Ganguly: *Biochemistry of Biomolecules*, Pearl Books, Delhi, 2007.

Srivastava, Seema: *Plant Physiology and Biochemistry,* Campus Books, Delhi, 2009.

Tripathi, Pramila: *Botanical Pesticides in the Management of Post-harvest Fruit Diseases*, Daya, Delhi, 2005.

Whyman, Robin: *Applied Organometallic Chemistry and Catalysis*, Oxford University Press, Delhi, 2005.

Index